GRAPHS OF COMMON FUNCTIONS

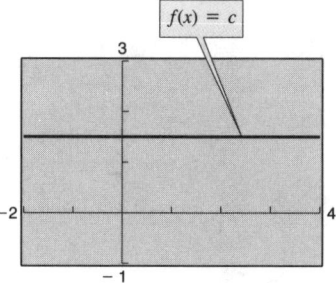

$f(x) = c$

Constant Function

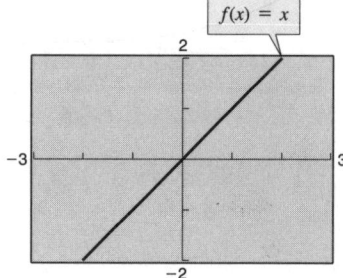

$f(x) = x$

Identity Function

Absolute Value Function

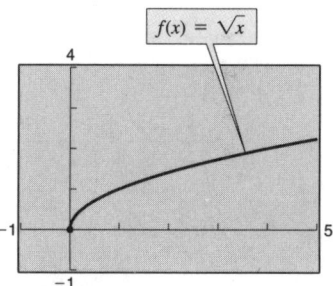

$f(x) = \sqrt{x}$

Square Root Function

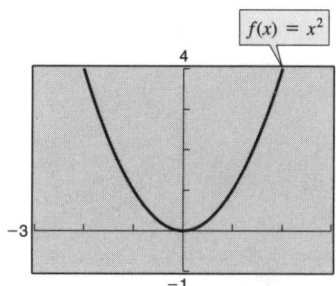

$f(x) = x^2$

Squaring Function

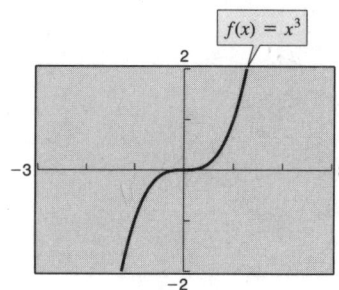

$f(x) = x^3$

Cubing Function

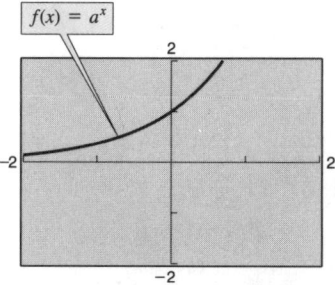

$f(x) = a^x$

Exponential Function ($a > 1$)

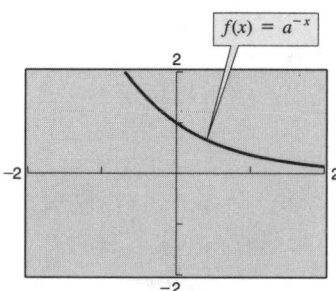

$f(x) = a^{-x}$

Exponential Function ($a > 1$)

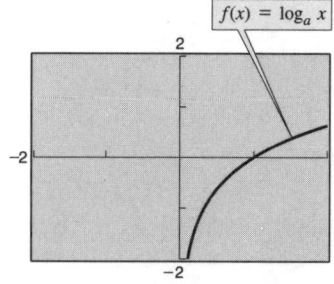

$f(x) = \log_a x$

Logarithmic Function ($a > 1$)

SYMMETRY

x-Axis Symmetry

y-Axis Symmetry

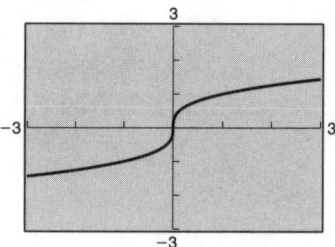

Origin Symmetry

CREATING A VIEWING RECTANGLE

Each of the viewing rectangles shows portions of the graph of $y = 0.1x^4 - x^3 + 2x^2$.
The first viewing rectangle shows the most complete graph of the equation.

PARAMETRIC EQUATIONS OF A GRAPH

Circle

$$x = h + r \cos \theta$$
$$y = k + r \sin \theta$$

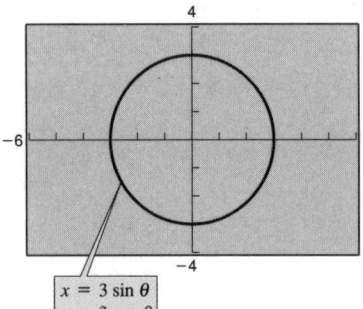

$x = 3 \sin \theta$
$y = 3 \cos \theta$

Ellipse

$$x = h + a \cos \theta$$
$$y = k + b \sin \theta$$

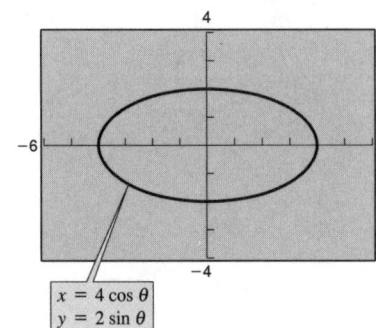

$x = 4 \cos \theta$
$y = 2 \sin \theta$

Hyperbola

$$x = h + a \sec \theta$$
$$y = k + b \tan \theta$$

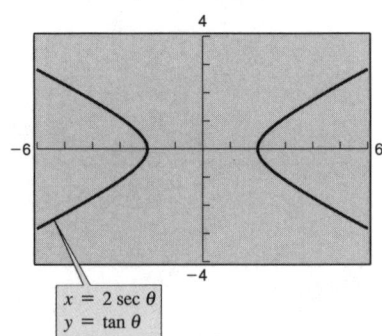

$x = 2 \sec \theta$
$y = \tan \theta$

GRAPHING POLAR EQUATIONS WITH A GRAPHING UTILITY

To graph the polar function $f(\theta)$, write the function in the parametric form $x = f(t) \cos t$ and $y = f(t) \sin t$.

Limaçon

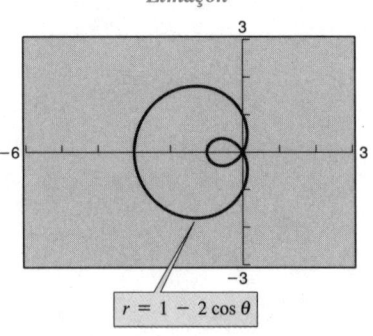

$r = 1 - 2 \cos \theta$

$x = (1 - 2 \cos t) \cos t$

$y = (1 - 2 \cos t) \sin t$

$0 \le t \le 2\pi$

Rose Curve

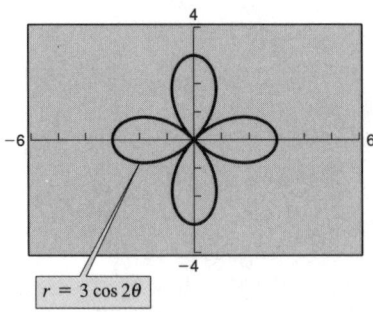

$r = 3 \cos 2\theta$

$x = (3 \cos 2t) \cos t$

$y = (3 \cos 2t) \sin t$

$0 \le t \le 2\pi$

Lemniscate

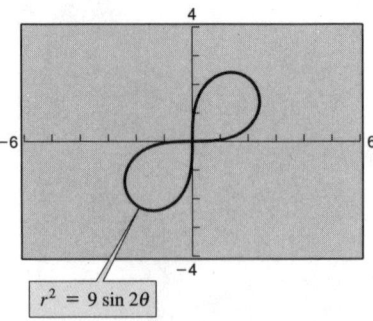

$r^2 = 9 \sin 2\theta$

$x = \left(3\sqrt{\sin(2t)}\right) \cos t$

$y = \left(3\sqrt{\sin(2t)}\right) \sin t$

$0 \le t \le 2\pi$

Precalculus with Limits

Precalculus with Limits
A Graphing Approach

SECOND EDITION

Roland E. Larson

The Pennsylvania State University
The Behrend College

Robert P. Hostetler

The Pennsylvania State University
The Behrend College

Bruce H. Edwards

University of Florida

WITH THE ASSISTANCE OF

David E. Heyd

The Pennsylvania State University
The Behrend College

HOUGHTON MIFFLIN COMPANY   Boston   New York

Sponsoring Editor: Christine B. Hoag
Senior Associate Editor: Maureen Brooks
Managing Editor: Catherine B. Cantin
Assistant Editor: Carolyn Johnson
Supervising Editor: Karen Carter
Associate Project Editor: Rachel D'Angelo Wimberly
Editorial Assistant: Caroline Lipscomb
Production Supervisor: Lisa Merrill
Art Supervisor: Gary Crespo
Marketing Manager: Charles Cavaliere
Marketing Associate: Ros Kane
Marketing Assistant: Kate Burden Thomas

Cover design by Harold Burch Design, NYC

Composition: Meridian Creative Group

Trademark Acknowledgments: TI is a registered trademark of Texas Instruments, Inc. Casio is a registered trademark of Casio, Inc. Sharp is a registered trademark of Sharp Electronics Corp. Hewlett-Packard is a registered trademark.

The *TI-82* and *TI-83* graphing calculator emulators referred to in this text were developed and copyrighted by Meridian Creative Group with the prior written permission of Texas Instruments.

Printed in the U.S.A.

Library of Congress Catalog Card Number: 96-076662

ISBN: 0-669-41758-0

456789–DC–00 99 98 97

Precalculus with Limits: A Graphing Approach, Second Edition, is the premier text for a reform-oriented course. Designed to build a strong foundation in precalculus, the text encourages students to develop a firm grasp of the underlying mathematical concepts while using algebra as a tool for solving real-life problems. The comprehensive text presentation invites discovery and exploration, while the integrated technology and consistent problem-solving strategies help the student develop strong precalculus skills.

Precalculus Reform

The precalculus course has changed over the past few years in response to the growing discussion of reform in mathematics education. Generally speaking, these changes have focused on the following areas: technology, real-life applications, problem-solving, and communicating about mathematics. The Second Edition embodies the spirit of these reform ideals without compromising the mathematical integrity of the course presentation. All text elements from the previous edition were considered for revision and many new examples, exercises, and applications were added.

Technology   Graphing technology is consistently incorporated throughout the Second Edition. The visualization and exploration capabilities of technology encourage the student to participate actively in the learning process, to develop their intuitive understanding of mathematical concepts, and to solve problems using actual data. Thus, students learn how algebra functions as a modeling language for real-life problems. Technology is used as a tool, drawn into the discussion whenever it offers a useful perspective on the topic at hand. For example, the power of graphing technology may be used to guide the students through thought-provoking explorations or to show alternative problem-solving techniques. Where appropriate, situations in which the results obtained through the use of technology may be misleading are also noted.

The Second Edition assumes that the student will use a graphing calculator on a daily basis in the course. Integrated throughout the text at point of use are many opportunities for investigation using technology (e.g., see page 103) and exercises that require the use of a graphing utility (e.g., see page 227). The text also carefully shows how to use graphing technology to best advantage (e.g., see page 217).

Whenever possible, references to graphing technology are generic. In a few cases, however, the text includes programs that will enable the student to investigate particular mathematical concepts (e.g., see page 115). Comparable programs for a wide variety of Texas Instruments, Casio, Sharp, and Hewlett-

Packard graphing calculators—including the most current models—are given in the appendix.

To accommodate a variety of teaching and learning styles, *Precalculus with Limits: A Graphing Approach,* Second Edition, is also available in a multimedia, CD-ROM format. *Interactive Precalculus with Limits: A Graphing Approach; A Self-Guided Study Companion* offers students a variety of additional tutorial assistance, including examples and exercises with detailed solutions; pre-, post-, and self-tests with answers; and *TI-82* and *TI-83* graphing calculator emulators. (See pages xviii–xx for more detailed information.)

Real-Life Applications   To emphasize for students the connection between mathematical concepts and real-world situations, up-to-date, real-life applications are integrated throughout the text. These applications appear as chapter introductions with related exercises (e.g., see pages 237 and 280), examples (e.g., see page 5), exercises (e.g., see page 293), Group Activities (e.g., see page 276), and Chapter Projects (e.g., see page 307).

Students have many opportunities to collect and interpret data, to make conjectures, and to construct mathematical models in the examples, exercises, Group Activities, and Chapter Projects. Students work on modeling problems with experimental and theoretical probabilities (e.g., see page 723), use mathematical models to make predictions or draw conclusions from real data (e.g., see page 149), compare models (e.g., see page 230), and apply curve-fitting techniques to create their own models from data (e.g., see page 144). In the process, the Second Edition gives students many more opportunities to use charts, tables, scatter plots, and graphs to summarize, analyze, and interpret data.

Problem Solving   The primary goal of any mathematics textbook is to encourage students to become competent and confident problem solvers. Many aspects of this revision focused on this goal—including the addition of new features such as Chapter Projects, Explorations, and Group Activities, as well as extensive and careful revision of the examples and exercise sets. Students are asked to use numerical, graphical, and algebraic techniques, and the use of graphing technology as a problem-solving tool is encouraged as appropriate (e.g., see page 155). Throughout, students are encouraged to follow a consistent approach to solving applied problems: Construct a verbal model, label terms, construct an algebraic model, solve the problem using the model, and check the answer in the original statement of the problem.

Like the previous edition, the Second Edition has an abundance of exercises that are designed to develop skills. The text also includes many other types of exercises that offer students the opportunity to refine their problem-solving skills, such as exercises that require interpretations (e.g., see page 251), those having many correct answers (e.g., see page 210), and multipart exercises designed to lead the student through problem-solving strategies (e.g., see page 230).

Communicating about Mathematics Each section in the Second Edition ends with a Group Activity. Designed to be completed in class or as homework assignments, the Group Activities give students the opportunity to work cooperatively as they think, talk, and write about mathematics. Students' understanding is reinforced through interpretation of mathematical concepts and results (e.g., see page 883), problem posing and error analysis (e.g., see page 201), and constructing mathematical models, tables and graphs (e.g., see page 258).

Making connections between algebra and real-world situations also helps students understand the underlying theory. Other connections are emphasized in this text as well, including those to probability (e.g., see Chapter 9), geometry (e.g., see page 847), and statistics (see Chapter 9).

Improved Coverage

As a result of user requests Chapter P, Prerequisites, now begins with an introduction to the Cartesian plane and covers solving equations and inequalities both algebraically and graphically. All or part of this review material may be covered or omitted, offering greater flexibility in designing the course syllabus.

Occurring one chapter earlier are Polynomials and Rational Functions in Chapter 3 and Exponential and Logarithmic Functions in Chapter 4. The chapters covering trigonometry have been expanded to three chapters. Chapter 6 now includes Vectors in the Plane and Vectors and Dot Products previously covered in Chapter 11 of the first edition. Chapter 11 now focuses on the three-dimensional coordinate system, and vectors, lines, and planes in space.

In keeping with the emphasis on real-life applications, sections titled Exploring Data are found throughout the text and include Representing Data Graphically, Linear Models and Scatter Plots, Nonlinear Models, Measures of Central Tendency, and Measures of Dispersion.

Features of the Second Edition

Chapter Opener Each chapter opens with a look at a real-life application. Real data is presented using graphical, numerical, and algebraic techniques.

Theorems, Definitions, and Guidelines All of the important rules, formulas, theorems, guidelines, properties, definitions, and summaries are highlighted for emphasis. Each is also titled for easy reference.

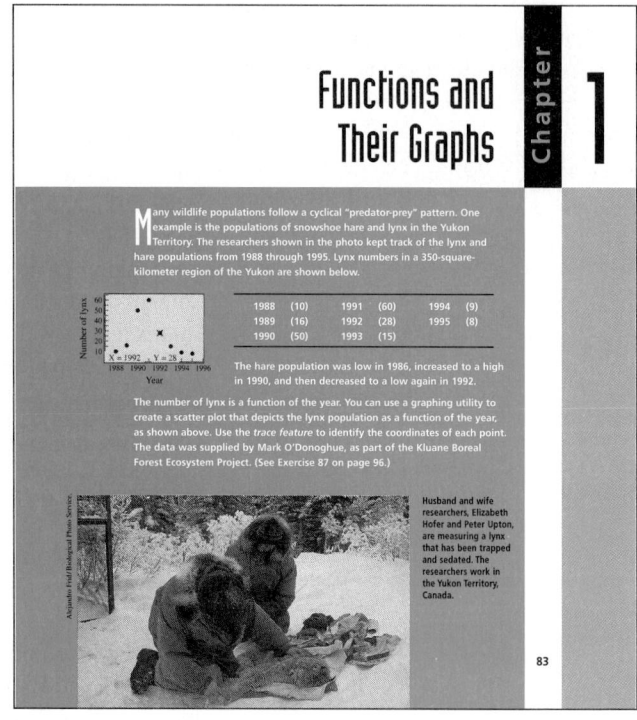

Functions and Their Graphs

Chapter 1

Many wildlife populations follow a cyclical "predator-prey" pattern. One example is the populations of snowshoe hare and lynx in the Yukon Territory. The researchers shown in the photo kept track of the lynx and hare populations from 1988 through 1995. Lynx numbers in a 350-square-kilometer region of the Yukon are shown below.

1988	(10)	1991	(60)	1994	(9)
1989	(16)	1992	(28)	1995	(8)
1990	(50)	1993	(15)		

The hare population was low in 1986, increased to a high in 1990, and then decreased to a low again in 1992.

The number of lynx is a function of the year. You can use a graphing utility to create a scatter plot that depicts the lynx population as a function of the year, as shown above. Use the *trace feature* to identify the coordinates of each point. The data was supplied by Mark O'Donoghue, as part of the Kluane Boreal Forest Ecosystem Project. (See Exercise 87 on page 96.)

Husband and wife researchers, Elizabeth Hofer and Peter Upton, are measuring a lynx that has been trapped and sedated. The researchers work in the Yukon Territory, Canada.

83

---

84 *1 / Functions and Their Graphs*

1.1 Functions

Introduction to Functions / Function Notation /
The Domain of a Function / Applications

Library of Functions

Many functions do not have simple mathematical formulas but are defined by real-life data. Such functions arise when you are using collections of data to model real-life applications. You will see that it is often convenient to *approximate* the data using a mathematical model or formula.

Introduction to Functions

Many everyday phenomena involve pairs of quantities that are related to each other by some rule of correspondence. Here are some examples.

1. The simple interest I earned on $1000 for 1 year is related to the annual interest rate r by the formula $I = 1000r$.
2. The distance d traveled on a bicycle in 2 hours is related to the speed s of the bicycle by the formula $d = 2s$.
3. The area A of a circle is related to its radius r by the formula $A = \pi r^2$.

Not all correspondences between two quantities have simple mathematical formulas. For instance, people commonly match up NFL starting quarterbacks with touchdown passes, and hours of the day with temperature. In each of these cases, however, there is some rule of correspondence that matches each item from one set with exactly one item from a different set. Such a rule of correspondence is called a **function.**

Definition of a Function

A **function** f from a set A to a set B is a rule of correspondence that assigns to each element x in the set A exactly one element y in the set B. The set A is the **domain** (or set of inputs) of the function f, and the set B contains the **range** (or set of outputs).

Figure 1.1

Time of day → Celsius temperature

Set A is the domain.
Inputs : 1, 2, 3, 4, 5, 6

Set B contains the range.
Outputs: 9°, 10°, 12°, 13°, 15°

To help understand this definition, look at the function illustrated in Figure 1.1. This function can be represented by the following ordered pairs.

$$\{(1, 9°), (2, 13°), (3, 15°), (4, 15°), (5, 12°), (6, 10°)\}$$

In each ordered pair, the first coordinate is the input and the second coordinate is the output. In this example, note the following characteristics of a function.

1. Each element in A must be matched with an element of B.
2. Some elements in B may not be matched with any element in A.
3. Two or more elements of A may be matched with the same element of B.

The converse of the third statement is not true. That is, an element of A (the domain) cannot be matched with two different elements of B.

Section Outline Each section begins with a list of the major topics covered in the section. These topics are also the subsection titles and can be used for easy reference and review by students. In addition, an exercise application that uses a skill or illustrates a concept covered in the section is highlighted to emphasize the connection between mathematical concepts and real-life situations.

Library of Functions The concept of the function is introduced in Chapter 1. In the material that follows, the icon appears each time a new type of function is described in detail.

Intuitive Foundation for Calculus   Special emphasis is given to the algebraic skills that are needed in calculus. Many examples in the Second Edition discuss algebraic techniques or graphically show concepts that are used in calculus, providing an intuitive foundation for future work.

Notes   Notes anticipate students' needs by offering additional insights, pointing out common errors, and describing generalizations.

Think About the Proof   Located in the margin adjacent to the corresponding theorem, each Think About the Proof feature offers strategies for proving the theorem. Detailed proofs for all theorems are given in Appendix A.

Technology Technology is integrated throughout the text at point of use as a tool for visualization, investigation, and verification. Instructions for using graphing utilities are given as necessary.

Study Tips Study Tips appear in the margin at point of use and offer students specific suggestions for studying algebra.

Exploration Throughout the text, the Exploration features encourage active participation by students, strengthening their intuition and critical thinking skills by exploring mathematical concepts and discovering mathematical relationships. Using a variety of approaches—including visualization, verification, use of graphing utilities, pattern recognition, and modeling—students are encouraged to develop a conceptual understanding of theoretical topics.

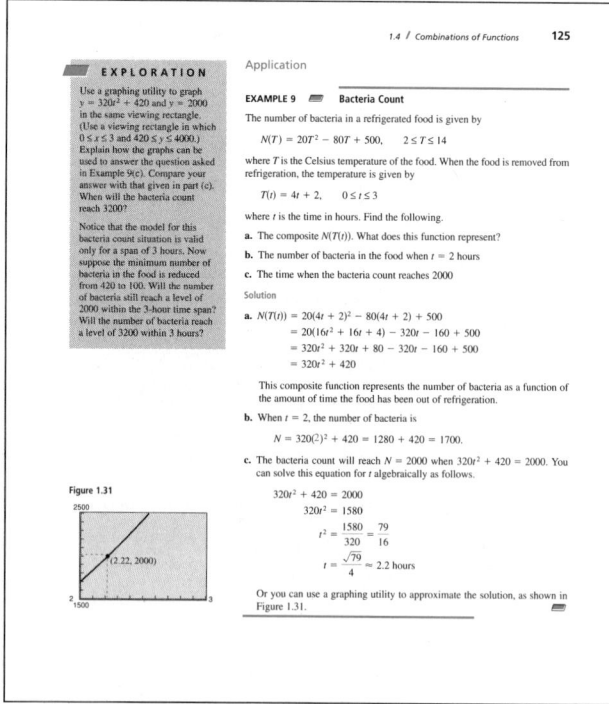

Historical Notes To help students understand that algebra has a past, historical notes featuring mathematicians and their work and mathematical artifacts are included in each chapter.

Graphics Visualization is a critical problem-solving skill. To encourage the development of this ability, the text has more than 3700 figures in examples, exercises, and answers to exercises. Included are graphs of equations and functions, geometric figures, displays of statistical information, scatter plots, and numerous screen outputs from graphing technology. All graphs of equations and functions are computer- or calculator-generated for accuracy, and they are designed to resemble students' actual screen outputs as closely as possible. Graphics are also used to emphasize graphical interpretation, comparison, and estimation.

1.1 / Functions **85**

Note Be sure you see that the *range* of a function is not the same as the use of *range* relating to the viewing rectangle.

The *Interactive* CD-ROM shows every example with its solution; clicking on the *Try It!* button brings up similar problems. Guided Examples and Integrated Examples show step-by-step solutions to additional examples. Integrated Examples are related to several concepts in the section.

Leonhard Euler (1707–1783), a Swiss mathematician, is considered to have been the most prolific and productive mathematician in history. One of his greatest influences on mathematics was his use of symbols, or notation. The function notation $y = f(x)$ was introduced by Euler.

In the following example, you are asked to decide whether different correspondences are functions. To do this, you must decide whether each element in the domain A is matched with exactly one element in the range B. If any element in A is matched with two or more elements in B, the correspondence is not a function.

EXAMPLE 1 ◼ **Testing for Functions**

Let $A = \{a, b, c\}$ and $B = \{1, 2, 3, 4, 5\}$. Which of the following sets of ordered pairs or figures represent functions from set A to set B?

a. $\{(a, 2), (b, 3), (c, 4)\}$ b. $\{(a, 4), (b, 5)\}$

c.   d.

Solution

a. This collection of ordered pairs *does* represent a function from A to B. Each element of A is matched with exactly one element of B.

b. This collection of ordered pairs *does not* represent a function from A to B. Not every element of A is matched with an element of B.

c. This figure *does* represent a function from A to B. It does not matter that each element of A is matched with the same element of B.

d. This figure *does not* represent a function from A to B. The element a in A is matched with *two* elements, 1 and 2, of B. This is also true of the element b. ◼

Representing functions by sets of ordered pairs is common in *discrete mathematics*. In algebra, however, it is more common to represent functions by equations or formulas involving two variables. For instance, the equation

$$y = x^2 \qquad y \text{ is a function of } x.$$

represents the variable y as a function of the variable x. In this equation, x is the **independent variable** and y is the **dependent variable**. The domain of the function is the set of all values taken on by the independent variable x, and the range of the function is the set of all values taken on by the dependent variable y.

---

142 *1 / Functions and Their Graphs*

1.6 Exploring Data: Linear Models and Scatter Plots

Scatter Plots and Correlation / Fitting a Line to Data

Scatter Plots and Correlation

Many real-life situations involve finding relationships between two variables such as the year and the number of people in the labor force. In a typical situation, data is collected and written as a set of ordered pairs. We discussed the graph of such a set, a **scatter plot,** briefly in Section P.1.

Most graphing utilities have built-in statistical programs that can create scatter plots. Use your graphing utility to plot the points given in the table at the right.

EXAMPLE 1 ◼ **Constructing a Scatter Plot**   *Real Life*

The data in the table shows the number of people P (in millions) in the United States who were part of the labor force from 1983 through 1993. In the table, t represents the year, with $t = 3$ corresponding to 1983. Sketch a scatter plot of the data. (Source: U.S. Bureau of Labor Statistics)

t	3	4	5	6	7	8	9	10	11	12	13
P	113	115	117	120	122	123	126	126	127	129	130

Solution

Begin by representing the data with a set of ordered pairs.

(3, 113), (4, 115), (5, 117), (6, 120), (7, 122), (8, 123), (9, 126), (10, 126), (11, 127), (12, 129), (13, 130)

Then plot each point in a coordinate plane, as shown in Figure 1.38. ◼

From the scatter plot in Figure 1.38, it appears that the points describe a relationship that is nearly linear. The relationship is not *exactly* linear because the labor force did not increase by precisely the same amount each year.

A mathematical equation that approximates the relationship between t and P is called a *mathematical model*. When developing a mathematical model, you strive for two (often conflicting) goals—accuracy and simplicity. For the data above, a linear model of the form $P = at + b$ appears to be best. It is simple and relatively accurate.

Figure 1.38

Applications Real-life applications are integrated throughout the text in examples and exercises. These applications offer students constant review of problem-solving skills, and they emphasize the relevance of the mathematics. Many of the applications use recent, real data, and all are titled for easy reference. Photographs with captions in the introduction to the chapter also encourage students to see the link between mathematics and real life.

Examples   Each of the more than 570 text examples was carefully chosen to illustrate a particular mathematical concept, problem-solving approach, or computational technique, and to enhance students' understanding. The examples in the text cover a wide variety of problem types, including theoretical problems, real-life applications (many with real data), and problems requiring the use of graphing technology. Each example is titled for easy reference, and real-life applications are labeled. Many examples include side comments in color that clarify the steps of the solution.

Problem Solving   The text provides ample opportunity for students to hone their problem-solving skills. In both the exercises and the examples in the Second Edition, students are asked to apply verbal, analytical, graphical, and numerical approaches to problem solving. Students are also encouraged to use a graphing utility as a tool for solving problems. Students are taught the following approach to solving applied problems: (1) construct a verbal model; (2) label variable and constant terms; (3) construct an algebraic model; (4) using the model, solve the problem; and (5) check the answer in the original statement of the problem.

CD-ROM   The icon refers to additional features of *Precalculus Functions and Graphs: A Graphing Approach; A Self-Guided Study Companion* that enhance the text presentation, such as exercises, computer animations, examples, tests, and *TI-82* and *TI-83* graphing calculator emulators.

Group Activities   The Group Activities that appear at the ends of sections reinforce students' understanding by studying mathematical concepts in a variety of ways, including talking and writing about mathematics, creating and solving problems, analyzing errors, and developing and using mathematical models. Designed to be completed as group projects in class or as homework assignments, the Group Activities give students opportunities to do interactive learning and to think, talk, and write about mathematics.

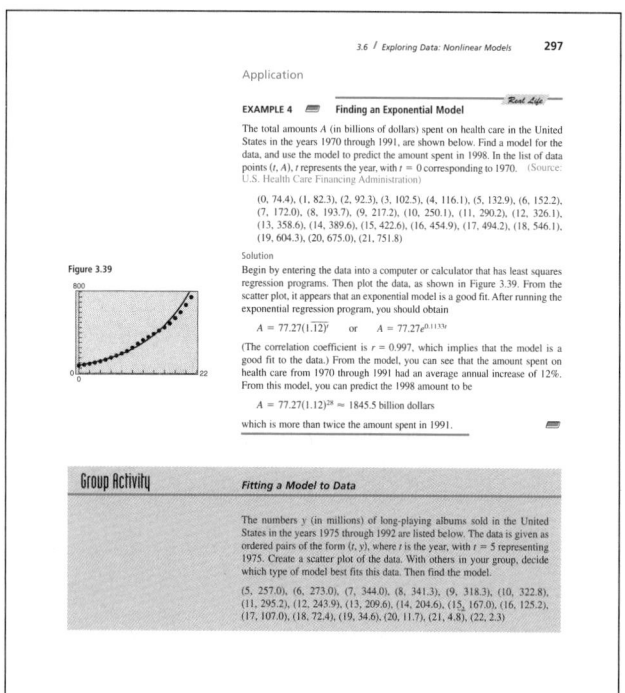

Exercises The exercise sets were completely revised for the Second Edition. More than 5800 exercises with a broad range of conceptual, computational, and applied problems accommodate a variety of teaching and learning styles. Included in the section and review exercise sets are multipart, writing, and more challenging problems with extensive graphics that encourage exploration and discovery, enhance students' skills in mathematical modeling, estimation, and data interpretation and analysis, and encourage the use of graphing technology for conceptual understanding. Applications are labeled for easy reference. The exercise sets are designed to build competence, skill, and understanding; each exercise set is graded in difficulty to allow students to gain confidence as they progress. Detailed solutions to all odd-numbered exercises are given in the *Study and Solutions Guide;* answers to all odd-numbered exercises appear in the back of the text.

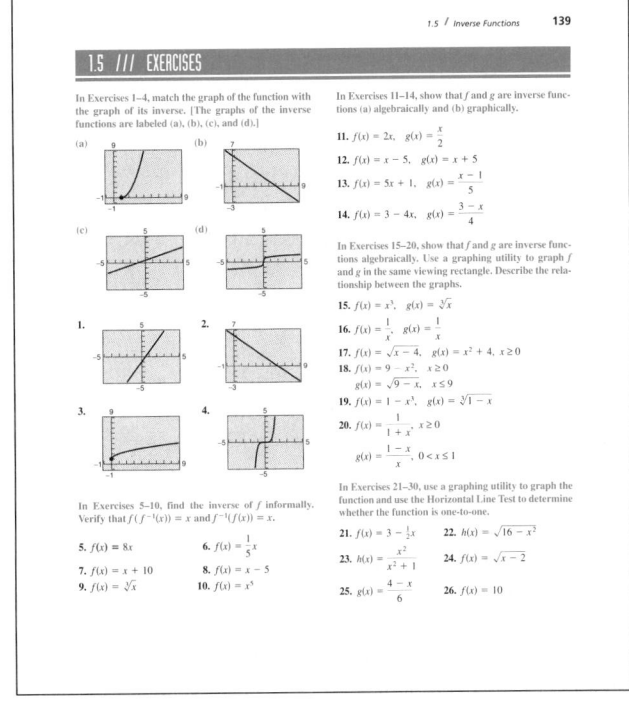

Geometry Geometric formulas and concepts are reviewed throughout the text in examples, Group Activities, and exercises. For reference, common formulas are listed inside the back cover of this text.

Focus on Concepts   Each Focus on Concepts feature is a set of exercises that test students' understanding of the basic concepts covered in the chapter. Answers to all questions are given in the back of the text.

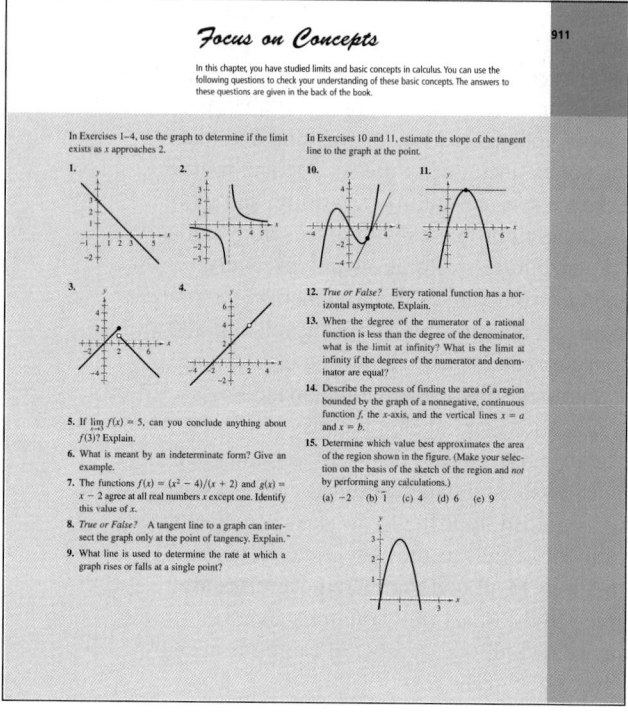

Chapter Projects   Chapter Projects are extended applications that use real data, graphs, and modeling to enhance students' understanding of mathematical concepts. Designed as individual or group projects, they offer additional opportunities to think, discuss, and write about mathematics. Many projects give students the opportunity to collect, analyze, and interpret data.

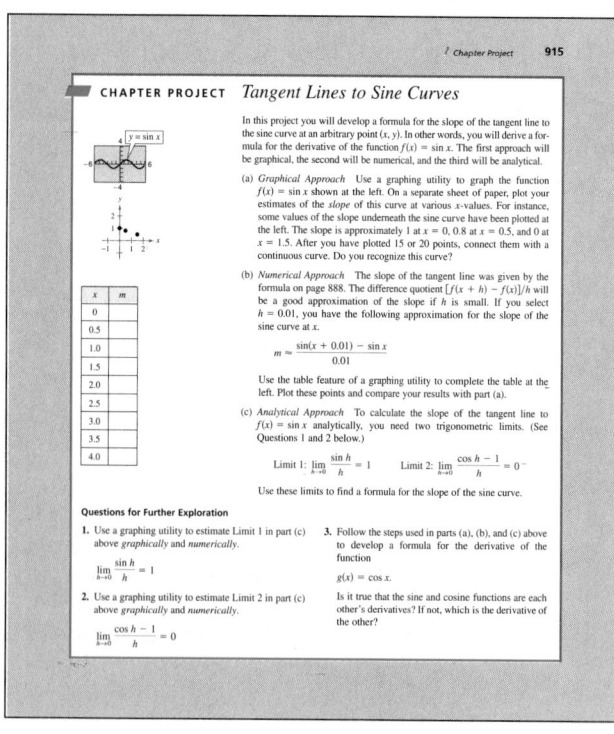

2 /// REVIEW EXERCISES

In Exercises 1–4, use a graphing utility to graph the quadratic function. Identify the vertex and the intercepts.

1. $f(x) = \left(x + \frac{3}{2}\right)^2 + 1$ **2.** $f(x) = (x - 4)^2 - 4$

3. $f(x) = \frac{1}{3}(x^2 + 5x - 4)$

4. $f(x) = 3x^2 - 12x + 11$

In Exercises 5 and 6, find the quadratic function that has the indicated vertex and whose graph passes through the given point.

5. Vertex: $(1, -4)$; Point: $(2, -3)$

6. Vertex: $(2, 3)$; Point: $(-1, 6)$

Graphical Reasoning In Exercises 7 and 8, use a graphing utility to graph each equation in the same viewing rectangle. Describe how each graph differs from the graph of $y = x^2$.

7. (a) $y = 2x^2$ (b) $y = -2x^2$
    (c) $y = x^2 + 2$ (d) $y = (x + 2)^2$

8. (a) $y = x^2 - 4$ (b) $y = 4 - x^2$
    (c) $y = (x - 3)^2$ (d) $y = \frac{1}{2}x^2 - 1$

In Exercises 9–16, find the maximum or minimum value of the quadratic function.

9. $g(x) = x^2 - 2x$ **10.** $f(x) = x^2 + 8x + 10$

11. $f(x) = 6x - x^2$ **12.** $h(x) = 3 + 4x - x^2$

13. $f(t) = -2t^2 + 4t + 1$

14. $h(x) = 4x^2 + 4x + 13$

15. $h(x) = x^2 + 5x - 4$ **16.** $f(x) = 4x^2 + 4x + 5$

17. *Numerical, Graphical, and Analytical Analysis* A rectangle is inscribed in the region bounded by the x-axis, the y-axis, and the graph of $x + 2y - 8 = 0$ (see figure).

(a) Complete six rows of a table like the one below. (The first two rows are shown.)

x	y	Area (xy)
1	$4 - \frac{1}{2}(1)$	$(1)\left[4 - \frac{1}{2}(1)\right] = \frac{7}{2}$
2	$4 - \frac{1}{2}(2)$	$(2)\left[4 - \frac{1}{2}(2)\right] = 6$

(b) Use a graphing utility to generate additional rows of the table in part (a). Use the table to estimate the dimensions that will produce the maximum area.

(c) Write the area A as a function of x. Determine the domain of the function in the context of the problem.

(d) Use a graphing utility to graph the area function. Use the graph to approximate the dimensions that will produce the maximum area.

(e) Write the function in standard form to find analytically the dimensions that will produce the maximum area.

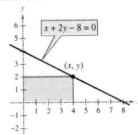

18. *Maximum Profit* Let x be the amount (in hundreds of dollars) a company spends on advertising, and let P be the profit, where

$$P = 230 + 20x - \frac{1}{2}x^2.$$

How much advertising will yield a maximum profit?

In Exercises 19–22, determine the right-hand and left-hand behavior of the graph of the polynomial function.

19. $f(x) = -x^2 + 6x + 9$ **20.** $f(x) = \frac{1}{3}x^3 + 2x$

Review Exercises The Review Exercises at the end of each chapter offer students an opportunity for additional practice. Answers to odd-numbered review exercises are given in the back of the text.

---

4 /// CHAPTER TEST

Take this test as you would take a test in class. After you are done, check your work against the answers given in the back of the book.

 The *Interactive* CD-ROM provides answers to the Chapter Tests and Cumulative Tests. It also offers Chapter Pre-Tests (that test key skills and concepts covered in previous chapters) and Chapter Post-Tests, both of which have randomly generated exercises with diagnostic capabilities.

1. Consider the angle of magnitude $5\pi/4$ radians.
   (a) Sketch the angle in standard position.
   (b) Determine two coterminal angles (one positive and one negative).
   (c) Convert the angle to degree measure.

2. A truck is moving at a rate of 90 kilometers per hour, and the diameter of its wheels is 1 meter. Find the angular speed of the wheels in radians per minute.

3. Find the exact values of the six trigonometric functions of the angle θ shown in the figure.

Figure for 3

4. Given that $\tan \theta = \frac{3}{2}$, find the other five trigonometric functions of θ.

5. Determine the reference angle θ' of the angle $\theta = 290°$ and sketch θ and θ' in standard position.

6. Determine the quadrant in which θ lies if sec $\theta < 0$ and tan $\theta > 0$.

7. Find two values of θ in degrees $(0 \le \theta < 360°)$ if cos $\theta = -\sqrt{3}/2$. (Do not use a calculator.)

8. Use a calculator to approximate two values of θ in radians $(0 \le \theta < 2\pi)$ if csc $\theta = 1.030$. Round the result to two decimal places.

In Exercises 9 and 10, graph the function through two full periods without the aid of a graphing utility.

9. $g(x) = -2 \sin\left(x - \frac{\pi}{4}\right)$ **10.** $f(\alpha) = \frac{1}{2} \tan 2\alpha$

In Exercises 11 and 12, use a graphing utility to graph the function. If the function is periodic, find its period.

11. $y = \sin 2\pi x + 2 \cos \pi x$ **12.** $y = 6e^{-0.12t} \cos(0.25t), \ 0 \le t \le 32$

13. Find a, b, and c for the function $f(x) = a \sin(bx + c)$ so that the graph of f matches the figure.

Figure for 13

14. Find the exact value of $\tan\left(\arccos \frac{2}{3}\right)$ without the aid of a calculator.

15. Graph the function $f(x) = 2 \arcsin\left(\frac{1}{2}x\right)$.

16. A ship leaves port at noon and sails at a speed of 18 knots. Its bearing is N 16° W. If the port is positioned at the origin, determine the coordinates of the position of the ship at 3 P.M.

Chapter Tests Each chapter that is not followed by a Cumulative Test ends with a Chapter Test, an effective tool for student self-assessment.

P–3 /// CUMULATIVE TEST

Take this test as you would take a test in class. After you are done, check your work against the answers given in the back of the book.

The *Interactive* CD-ROM provides answers to the Chapter Tests and Cumulative Tests. It also offers Chapter Pre-Tests (that test key skills and concepts covered in previous chapters) and Chapter Post-Tests, both of which have randomly generated exercises with diagnostic capabilities.

In Exercises 1–6, sketch a graph of the function. Use a graphing utility to verify your result.

1. $h(x) = -(x^2 + 4x)$ **2.** $y = \sqrt{4 - x}$ **3.** $g(s) = \frac{2s}{s - 3}$

4. $g(s) = \frac{2s^2}{s - 3}$ **5.** $f(x) = 6(2^{-x})$ **6.** $g(x) = \log_3 x$

7. Find an equation for the line passing through the points $\left(-\frac{1}{2}, 1\right)$ and $(3, 8)$.

8. Explain why the graph at the right does not represent y as a function of x.

9. Describe how the graph of each function differs from the graph of $y = \sqrt[3]{x}$. It is not necessary to sketch the graphs.
   (a) $r(x) = \frac{1}{2}\sqrt[3]{x}$ (b) $h(x) = \sqrt[3]{x} + 2$ (c) $g(x) = \sqrt[3]{x + 2}$

Figure for 8

$(x - 3)^2 + y^2 = 9$

10. Determine whether the function $h(x) = 5x - 2$ is one-to-one. If so, find its inverse.

In Exercises 11–16, solve (if possible) the equation. Use a graphing utility to verify your result.

11. $2x - 3(x - 4) = 5$ **12.** $\frac{2}{t - 3} + \frac{2}{t - 2} = \frac{10}{t^2 - 5t + 6}$

13. $3y^2 + 6y + 2 = 0$ **14.** $\sqrt{x + 10} = x - 2$

15. $6e^{2x} = 72$ **16.** $\log_2 x + \log_2 5 = 6$

17. Let x be the amount (in hundreds of dollars) that a company spends on advertising, and let P be the profit (in thousands of dollars), where $P = 230 + 20x - \frac{1}{2}x^2$. What amount will yield a maximum profit?

18. Find all the zeros of $f(x) = x^3 + 2x^2 + 4x + 8$.

19. Use a graphing utility to approximate the real zero of the function $g(x) = x^3 + 3x^2 - 6$ to the nearest hundredth.

20. Write $2 \ln x - \frac{1}{2} \ln(x + 5)$ as a logarithm of a single quantity.

21. The numbers of cellular telephone subscribers y (in millions) for the years 1990 through 1993 are given by $(0, 5.3)$, $(1, 7.6)$, $(2, 11.0)$, and $(3, 16.0)$ where x is the time in years, with $x = 0$ corresponding to 1990. Use a graphing utility to fit an exponential model to the data. Create a scatter plot of the data and graph the model in the same viewing rectangle. (Source: Cellular Telecommunications Industry Association)

Cumulative Tests The Cumulative Tests that follow Chapters 3, 6, 10 and 12 help students judge their mastery of previously covered material as well as reinforce the knowledge they have been accumulating throughout the text—preparing them for other exams and for future courses.

Supplements

Precalculus with Limits: A Graphing Approach, Second Edition, by Larson, Hostetler, and Edwards is accompanied by a comprehensive supplements package. Most items are keyed to the text.

Printed Resources

For the student

Study and Solutions Guide by Bruce Edwards, University of Florida, and Dianna L. Zook, Indiana University—Purdue University at Fort Wayne

- Section summaries of key concepts
- Detailed, step-by-step solutions to all odd-numbered exercises
- Key solution steps for Chapter Tests and Cumulative Tests
- Practice tests with solutions
- Study strategies

Graphing Technology Guide

- Keystroke instructions for a wide variety of Texas Instruments, Casio, Sharp, and Hewlett-Packard graphing calculators—including the most current models.
- Examples with step-by-step solutions
- Extensive graphics screen output
- Technology tips

For the instructor

Instructor's Annotated Edition

- Includes the entire student edition of the text, with the student answers section
- Instructor's Answers section: Answers to all even-numbered exercises, and answers to all Explorations, Technology exercises, Group Activities, and Chapter Project exercises
- Annotations at point of use offer specific teaching strategies and suggestions for implementing Group Activities, point out common student errors, and give additional examples, exercises, class activities, and group activities.

Solutions to Even-Numbered Exercises

- Detailed, step-by-step solutions to even-numbered exercises

Test Item File and Instructor's Resource Guide

- Printed test bank with approximately 2000 test items (multiple-choice, open-ended, and writing) coded by level of difficulty
- Technology-required test items coded for easy reference
- Bank of chapter test forms with answer keys

- Two final exam test forms
- Notes to the instructor, including materials for alternative assessment and managing the multicultural and cooperative-learning classrooms

Problem Solving, Modeling, and Data Analysis Labs by Wendy Metzger, Palomar College

- Multipart, guided discovery activities and applications
- Keystroke instructions for Derive and *TI-82*
- Keyed to the text by topic
- Funded in part by NSF (National Science Foundation, Instrumentation and Laboratory Improvement) and California Community College Fund for Instructional Improvement

Media Resources

For the student

Interactive Precalculus with Limits: A Graphing Approach; A Self-Guided Study Companion (See pages xviii–xx for a description, or visit the Houghton Mifflin home page at http://www.hmco.com for a preview.)

- Interactive, multimedia CD-ROM format
- IBM-PC for Windows

Tutor software

- Interactive tutorial software keyed to the text by section
- Diagnostic feedback
- Chapter self-tests
- Guided exercises with step-by-step solutions
- Glossary

Videotapes by Dana Mosely

- Comprehensive, text-specific coverage keyed to the text by section
- Real-life application vignettes introduced where appropriate
- Computer-generated animation
- For media/resource centers
- Additional explanation of concepts, sample problems, and applications
- Instructional graphing calculator videotape also available

For the instructor

Computerized Testing (IBM, Macintosh, Windows)

- New on-line testing
- New grade-management capabilities
- Algorithmic test-generating software provides an unlimited number of tests
- Approximately 2000 test items
- Also available as a printed test bank

Transparency Package

- 70 color transparencies color-coded by topic

Interactive Precalculus with Limits: A Graphing Approach

To accommodate a variety of teaching and learning styles, *Precalculus with Limits: A Graphing Approach; A Self-Guided Study Companion* is also available in a multimedia, CD-ROM format. In this interactive format, the text offers the student additional tutorial assistance with

- Complete solutions to all odd-numbered text exercises.
- Chapter pre-tests, self-tests, and post-tests.
- *TI-82* and *TI-83* emulators.

- Guided examples with step-by-step solutions.
- Editable graphs.
- Animations of mathematical concepts.
- Section and tutorial exercises.
- Glossary of key terms.

These and other pedagogical features of the CD-ROM are illustrated by the screen dumps shown below.

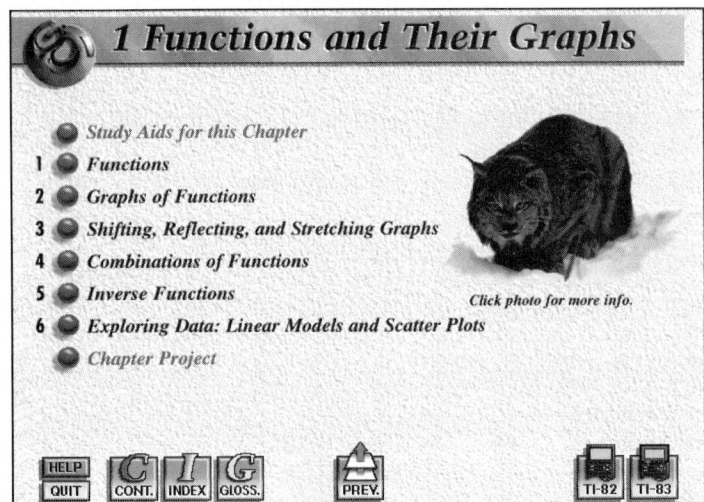

Chapter Topics   Each chapter begins with an outline of the topics to be covered. Using the buttons at the bottom of the screen, the student can quickly move to the appropriate section.

Introductory Chapter Application   Each chapter opens with a real-data application that illustrates the key concepts and techniques to be covered. Clicking on the photo, the student can access additional data and background information that frames the real-world context for a mathematical concept.

Chapter Project   Each chapter is accompanied by a Chapter Project. This offers the student the opportunity to synthesize the algebraic techniques and concepts studied in the chapter. Many projects use real data and emphasize data analysis and mathematical modeling.

Study Aids   Each section offers the student an array of additional study aids, including Chapter Pre-, Post-, and Self-Tests, Review Exercises, and Focus on Concepts. With diagnostics, complete solutions, or answers, these helpful features promote the focused practice needed to master mathematical concepts. Short, informative video segments are also included.

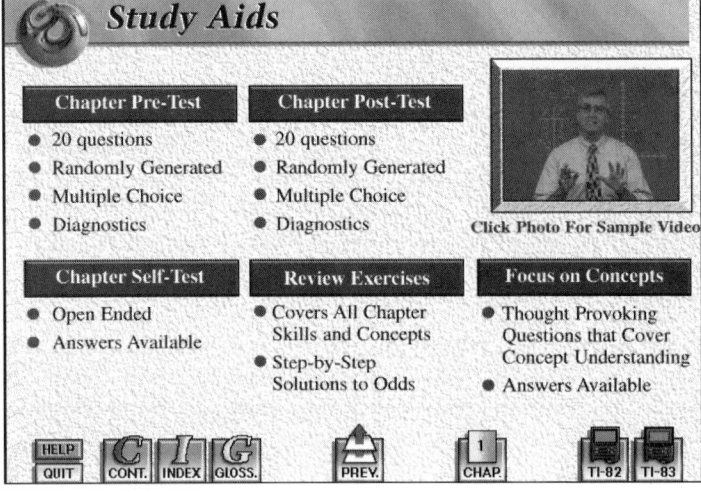

Examples   *Interactive Precalculus with Limits: A Graphing Approach; A Self-Guided Study Companion* illustrates mathematical concepts by featuring all of the Examples found in the text. Guided Examples with step-by-step solutions that appear one line at a time offer additional opportunities for practice and skill development. Group Activities, Exercises, and Integrated Examples help synthesize the concepts in the section.

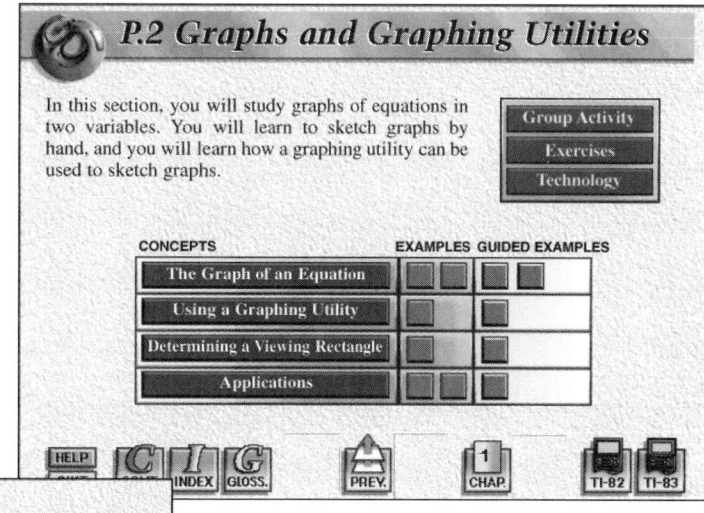

Graphs   Some examples are accompanied by editable graphs for exploration and discovery. Using the keys below the graph, the student can change the function, the graphing window, and the *x*- and *y*-scales, and can trace and zoom.

Try It!   After studying the worked-out example, the student can use the Try It! button to access similar examples—with solutions following on separate screens—to test his or her mastery of mathematical concepts and techniques.

TI-82 and TI-83 Emulators   Accessible on every screen, the *TI-82* and *TI-83* emulators give instant access to graphing utilities as tools for computation and exploration. They are also available for working exercises in the text that require the use of a graphing utility. Instruction on using the emulators is also included at the click of a button.

These emulators were developed and copyrighted by Meridian Creative Group with the prior written permission of Texas Instruments.

In Exercises 1 and 2, complete a table of values. Use the solution points to sketch the graph of the equation.

 $y = -\frac{1}{2}x + 2$ **2.** $y = x^2 - 3x$

In Exercises 3–12, sketch the graph *by hand*.

3. $y - 2x - 3 = 0$ **4.** $3x + 2y + 6 = 0$

5. $x - 5 = 0$ **6.** $y = 8 - |x|$

7. $y = \sqrt{5 - x}$ **8.** $y = \sqrt{x + 2}$

9. $y + 2x^2 = 0$ **10.** $y = x^2 - 4x$

11. $y = \sqrt{25 - x^2}$ **12.** $x^2 + y^2 = 10$

HELP   QUIT   CONT.   INDEX   GLOSS.   PREV.

Section Exercises   Each section is accompanied by a comprehensive set of exercises promoting skills mastery and conceptual understanding. Solutions to all odd-numbered exercises are available for instant feedback.

Tutorial Exercises   Every section has a set of exercises in a multiple-choice format that offer students additional practice. Examples and diagnostics enhance this guided practice.

Chapter Self-Test

Take this test as you would take a test in class. After you are done, check your work with the answers given by selecting the Answer button.

1 2 3 4 5 6 7 8 9 10

1. Use a graphing utility to graph $y = 4 - \frac{3}{4}x$. Check for symmetry and identify x- and y-intercepts.

Answer

HELP   QUIT   CONT.   INDEX   GLOSS.   PREV.   NEXT   CHAP.   TI-82   TI-83

Tests   Every chapter of the interactive text includes tests that are different from those in the textbook: Chapter Pre-Tests (testing key skills and concepts covered in previous chapters) and Chapter Post-Tests test mastery of the material covered in the textbook. The Chapter Self-Tests from the text are also included. Answers to all tests are included.

 The *Interactive* CD-ROM shows every example with its solution; clicking on the *Try It!* button brings up similar problems. Guided Examples and Integrated Examples show step-by-step solutions to additional examples. Integrated Examples are related to several concepts in the section.

Length of
Height of
Length of

Equation: $\dfrac{x}{170.25} =$

$x =$

Thus, the World Trade C

Interactive Precalculus with Limits: A Graphing Approach; A Self-Guided Study Companion supports the mathematical presentation in the text *Precalculus with Limits: A Graphing Approach,* Second Edition, with a variety of tutorial, diagnostic, and demonstration features. Throughout both the student text and the Instructor's Annotated Edition, CD-ROM icons identify these additional functions of the interactive text, as illustrated by the sample text page (at left).

Acknowledgments

We would like to thank the many people who have helped us at various stages of this project to prepare the text and supplements package. Their encouragement, criticisms, and suggestions have been invaluable to us.

Second Edition Reviewers: Daniel D. Anderson, University of Iowa; Joseph A. Bettina, Stevenson High School, Buffalo Grove, IL; Anne E. Brown, Indiana University–South Bend; Elaine N. Daniels, Salve Regina University; Eunice Everett, Seminole Community College; Jeff Frost, Johnson County Community College; Khadiga H. Gamgoum, Northern Virginia Community College; Michele Greenfield, Middlesex County College; Gary Grimes, Mt. Hood Community College; Ruth A. Hartman, Black Hawk College; Zenas Hartvigson, University of Colorado at Denver; Allen Hesse, Rochester Community College; Jean M. Horn, Northern Virginia Community College; Bill Huston, Missouri Western State College; Francine Winston Johnson, Howard Community College; Eleanor Kendrick, San Jose City College; Lynn Krafve, Minnetonka Senior High School, Minnetonka, MN; Peter A. Lappan, Michigan State University; Betty Mayberry, Gallatin High School, Gallatin, TN; Judy McInerney, Sandhills Community College; Richard Morrow, Detroit Country Day School, Beverly Hills, MI; Roger B. Nelsen, Lewis and Clark College; Jon Odell, Richland Community College; Wing M. Park, College of Lake County; Robert Pearce, South Plains College; George W. Shultz, St. Petersburg Junior College; Cathryn U. Stark, Collin County Community College; G. Bryan Stewart, Tarrant County Junior College; Robert Thompson, Lane Community College; Madelyn K. Trainor, San Diego High School, San Diego, CA; Marjorie E. Valentine, San Antonio, TX; Mahbobeh Vezvaei, Kent State University; Georgia Weickel, Schreiber High School, Port Washington, NY; Joel E. Wilson, Eastern Kentucky University; and Karl M. Zilm, Lewis and Clark Community College.

Second Edition Survey Respondents: Marwan A. Abu-Sawwa, Florida Community College at Jacksonville; Barbara C. Armenta, Pima Community College—East; Gladwin E. Bartel, Otero Junior College; Carole A. Bauer, Triton College; Joyce M. Becker, Luther College; Marybeth Beno, South Suburban College; Charles M. Biles, Humboldt State University; Ruthane Bopp, Lake Forest College; Tim Chappell, North Central Missouri College; Michael Davidson, Cabrillo College; Diane L. Doyle, Adirondack Community College; Donna S. Fatheree, University of Louisiana at Lafayette; John R. Formsma, Los Angeles City College; John S. Frohliger, Saint Norbert College; Gary Glaze, Eastern Washington University; Irwin S. Goldfine, Truman College; Elise M. Grabner, Slippery Rock University; Donnie Hallstone, Green River Community College; Lois E. Higbie, Brookdale Community College;

Susan S. Hollar, Kalamazoo Valley Community College; Fran Hopf, Hillsborough Community College; Margaret D. Hovde, Grossmont College; John F. Keating, Massasoit Community College; John G. LaMaster, Indiana University–Purdue University at Fort Wayne; Giles Wilson Maloof, Boise State University, Kenneth Mangels, Concordia State University; Peggy I. Miller, University of Nebraska at Kearney; Gilbert F. Orr, University of Southern Colorado; Jim Paige, Wayne State College; Elise Price, Tarrant County Junior College; Doris Schraeder, McLennan Community College; Linda Schultz, McHenry County College; Fay Sewell, Montgomery County Community College; Patricia G. Shelton, North Carolina A & T State University; Hazel Shows, Hinds Community College; Joseph F. Stokes, Western Kentucky University; Diane Van Nostrand, University of Tulsa; and Raymond D. Wuco, San Joaquin Delta College.

Thanks to all of the people at Houghton Mifflin Company who worked with us in the development and production of the text, especially Chris Hoag, Sponsoring Editor; Cathy Cantin, Managing Editor; Maureen Brooks, Senior Associate Editor; Carolyn Johnson, Assistant Editor; Karen Carter, Supervising Editor; Rachel Wimberly, Associate Project Editor; Gary Crespo, Art Supervisor; Lisa Merrill, Production Supervisor; Ros Kane, Marketing Associate; and Carrie Lipscomb, Editorial Assistant.

We would also like to thank the staff at Larson Texts, Inc. who assisted with proofreading the manuscript, preparing and proofreading the art package, and checking and typesetting the supplements.

On a personal level, we are grateful to our wives, Deanna Gilbert Larson, Eloise Hostetler, and Consuelo Edwards, for their love, patience, and support. Also, special thanks go to R. Scott O'Neil.

If you have suggestions for improving the text, please feel free to write to us. Over the past two decades, we have received many useful comments from both instructors and students, and we value these very much.

Roland E. Larson
Robert P. Hostetler
Bruce H. Edwards

Contents

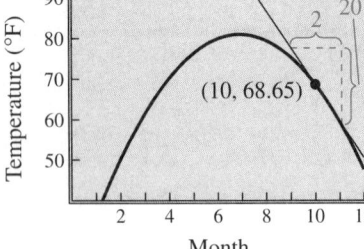

Appendices

Answers

Indices

Prerequisites

The concept for a museum dedicated to the heritage of rock and roll began in 1983 with a group of people from the music industry. This group created the Rock and Roll Hall of Fame Foundation.

The following data shows the numbers of recording artists who were elected to the Rock and Roll Hall of Fame from 1986 through 1995.

1986	(10)	1987	(15)	1988	(5)
1989	(5)	1990	(8)	1991	(7)
1992	(7)	1993	(8)	1994	(8)
1995	(7)				

You can use a graphing utility to represent this data visually as a *line graph*, as shown at the left, which is useful for showing trends in data over periods of time. (See Exercise 89 on page 11.)

The Rock and Roll Hall of Fame and Museum is located in Cleveland, Ohio. The 150,000 square foot building was designed by I. M. Pei. The museum opened in September 1995.

P.1   The Cartesian Plane

The Cartesian Plane / The Distance Formula / The Midpoint Formula /
The Equation of a Circle / Application

Figure P.1

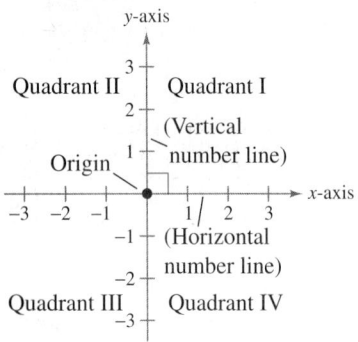

The Cartesian Plane

Just as you can represent real numbers by points on a real number line, you can represent ordered pairs of real numbers by points in a plane called the **rectangular coordinate system,** or the **Cartesian plane,** after the French mathematician René Descartes (1596–1650).

The Cartesian plane is formed by using two real lines intersecting at right angles, as shown in Figure P.1. The horizontal real line is usually called the *x*-**axis,** and the vertical real line is usually called the **y-axis.** The point of intersection of these two axes is the **origin,** and the two axes divide the plane into four parts called **quadrants.**

Each point in the plane corresponds to an **ordered pair** (x, y) of real numbers x and y, called **coordinates** of the point. The *x*-**coordinate** represents the directed distance from the *y*-axis to the point, and the **y-coordinate** represents the directed distance from the *x*-axis to the point, as shown in Figure P.2.

Figure P.2

(x, y)

| Directed distance from *y*-axis | Directed distance from *x*-axis |

Note   The notation (x, y) denotes both a point in the plane and an open interval on the real line. The context will tell you which meaning is intended.

Figure P.3

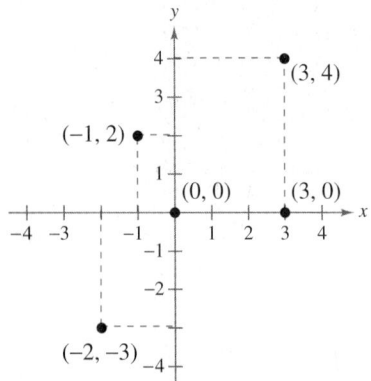

EXAMPLE 1 ▱ **Plotting Points in the Cartesian Plane**

Plot the points $(-1, 2)$, $(3, 4)$, $(0, 0)$, $(3, 0)$, and $(-2, -3)$.

Solution
To plot the point

$$(-1, 2)$$

x-coordinate ⎤ ⎣ *y*-coordinate

imagine a vertical line through -1 on the *x*-axis and a horizontal line through 2 on the *y*-axis. The intersection of these two lines is the point $(-1, 2)$. The other four points can be plotted in a similar way, and are shown in Figure P.3.

▱

Figure P.4

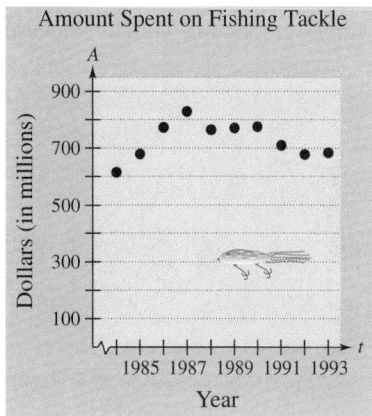

The beauty of a rectangular coordinate system is that it allows you to see relationships between two variables. It would be difficult to overestimate the importance of Descartes's introduction of coordinates to the plane. Today, his ideas are in common use in virtually every scientific and business-related field.

EXAMPLE 2   **Sketching a Scatter Plot**

The amounts A (in millions of dollars) spent on fishing tackle in the United States for the years 1984 to 1993 are given in the table, where t represents the year. Sketch a scatter plot of the data.   (Source: National Sporting Goods Association)

t	1984	1985	1986	1987	1988	1989	1990	1991	1992	1993
A	616	681	773	830	766	769	776	711	678	685

Note   In Example 2, you could have let $t = 1$ represent the year 1984. In that case, the horizontal axis would not have been broken, and the tick marks would have been labeled 1 through 10 (instead of 1984 through 1993).

Solution

To sketch a *scatter plot* of the data given in the table, you simply represent each pair of values by an ordered pair (t, A) and plot the resulting points, as shown in Figure P.4. For instance, the first pair of values is represented by the ordered pair (1984, 616). Note that the break in the t-axis indicates that the numbers between 0 and 1984 have been omitted.

The scatter plot in Example 2 is only one way to represent the data graphically. Two other techniques are shown at the right. The first is a *bar graph* and the second is a *line graph*. All three graphical representations were created with a computer. Try using a graphing utility to graphically represent the data given in Example 2.

Bar graph

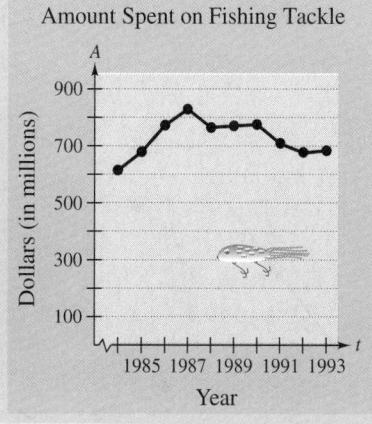

Line graph

The Distance Formula

Recall from the Pythagorean Theorem that, for a right triangle with hypotenuse of length c and sides of lengths a and b, you have

$$a^2 + b^2 = c^2 \qquad \text{Pythagorean Theorem}$$

as shown in Figure P.5. (The converse is also true. That is, if $a^2 + b^2 = c^2$, the triangle is a right triangle.)

Suppose you want to determine the distance d between two points (x_1, y_1) and (x_2, y_2) in the plane. With these two points, a right triangle can be formed, as shown in Figure P.6. The length of the vertical side of the triangle is $|y_2 - y_1|$, and the length of the horizontal side is $|x_2 - x_1|$. By the Pythagorean Theorem, you can write

$$d^2 = |x_2 - x_1|^2 + |y_2 - y_1|^2$$
$$d = \sqrt{|x_2 - x_1|^2 + |y_2 - y_1|^2}$$
$$d = \sqrt{(x_2 - x_1)^2 + (y_2 - y_1)^2}.$$

This result is the **Distance Formula.**

Figure P.5

Figure P.6

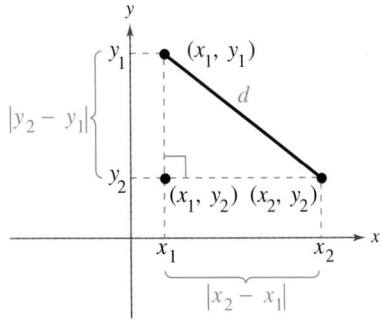

The Distance Formula

The distance d between the points (x_1, y_1) and (x_2, y_2) in the plane is

$$d = \sqrt{(x_2 - x_1)^2 + (y_2 - y_1)^2}.$$

EXAMPLE 3   Finding a Distance

Find the distance between the points $(-2, 1)$ and $(3, 4)$.

Solution
Let $(x_1, y_1) = (-2, 1)$ and $(x_2, y_2) = (3, 4)$. Then apply the Distance Formula as follows.

$$
\begin{aligned}
d &= \sqrt{(x_2 - x_1)^2 + (y_2 - y_1)^2} && \text{Distance Formula} \\
&= \sqrt{[3 - (-2)]^2 + (4 - 1)^2} && \text{Substitute for } x_1, y_1, x_2, \text{ and } y_2. \\
&= \sqrt{(5)^2 + (3)^2} && \text{Simplify.} \\
&= \sqrt{34} && \\
&\approx 5.83 && \text{Use a calculator.}
\end{aligned}
$$

Note in Figure P.7 that a distance of 5.83 looks about right.

Figure P.7

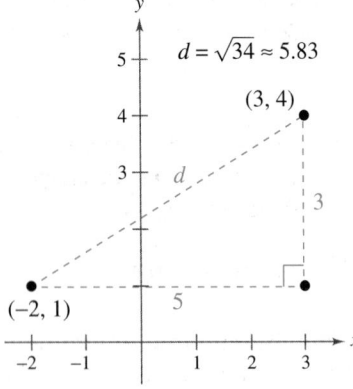

The Midpoint Formula

To find the **midpoint** of the line segment that joins two points in a coordinate plane, you can simply find the average values of the respective coordinates of the two endpoints.

The Midpoint Formula

The midpoint of the segment joining the points (x_1, y_1) and (x_2, y_2) is

$$\text{Midpoint} = \left(\frac{x_1 + x_2}{2}, \frac{y_1 + y_2}{2} \right).$$

 The *Interactive* CD-ROM shows every example with its solution; clicking on the *Try It!* button brings up similar problems. Guided Examples and Integrated Examples show step-by-step solutions to additional examples. Integrated Examples are related to several concepts in the section.

EXAMPLE 4   **Finding a Segment's Midpoint**

Find the midpoint of the line segment joining the points $(-5, -3)$ and $(9, 3)$, as shown in Figure P.8.

Solution

Let $(x_1, y_1) = (-5, -3)$ and $(x_2, y_2) = (9, 3)$.

$$\text{Midpoint} = \left(\frac{x_1 + x_2}{2}, \frac{y_1 + y_2}{2} \right) \qquad \text{Midpoint Formula}$$

$$= \left(\frac{-5 + 9}{2}, \frac{-3 + 3}{2} \right) \qquad \text{Substitute for } x_1, y_1, x_2, \text{ and } y_2.$$

$$= (2, 0) \qquad \text{Simplify.}$$

Figure P.8

Figure P.9

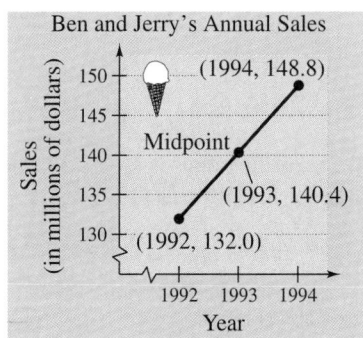

EXAMPLE 5   **Estimating Annual Sales**   *Real Life*

Ben and Jerry's had annual sales of $132.0 million in 1992 and $148.8 million in 1994. Without knowing any additional information, what would you estimate the 1993 sales to have been? (Source: Ben and Jerry's, Inc.)

Solution

One solution to the problem is to assume that sales followed a linear pattern. With this assumption, you can estimate the 1993 sales by finding the midpoint of the segment connecting the points (1992, 132.0) and (1994, 148.8).

$$\text{Midpoint} = \left(\frac{1992 + 1994}{2}, \frac{132.0 + 148.8}{2} \right) = (1993, 140.4)$$

Hence, you would estimate the 1993 sales to have been about $140.4 million, as shown in Figure P.9. (The actual 1993 sales were $140.3 million.)

Figure P.10

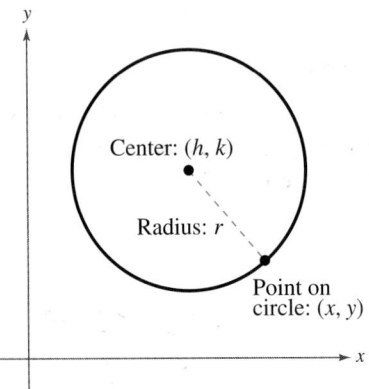

The Equation of a Circle

The Distance Formula provides a convenient way to define circles. A **circle of radius r** with **center** at the point (h, k) is shown in Figure P.10. The point (x, y) is on this circle if and only if its distance from the center (h, k) is r. This means that a **circle** in the plane consists of all points (x, y) that are a given positive distance r from a fixed point (h, k). Using the Distance Formula, you can express this relationship by saying that the point (x, y) lies on the circle if and only if

$$\sqrt{(x - h)^2 + (y - k)^2} = r.$$

By squaring both sides of this equation, you can obtain the **standard form of the equation of a circle.**

Standard Form of the Equation of a Circle

The **standard form of the equation of a circle** is

$$(x - h)^2 + (y - k)^2 = r^2.$$

The point (h, k) is the **center** of the circle, and the positive number r is the **radius** of the circle. The standard form of the equation of a circle whose center is the *origin* is $x^2 + y^2 = r^2$.

EXAMPLE 6 ▱ **Finding an Equation of a Circle**

The point $(3, 4)$ lies on a circle whose center is at $(-1, 2)$, as shown in Figure P.11. Find an equation for the circle.

Solution

The radius r of the circle is the distance between $(-1, 2)$ and $(3, 4)$.

$$r = \sqrt{[3 - (-1)]^2 + (4 - 2)^2}$$
$$= \sqrt{16 + 4}$$
$$= \sqrt{20}$$

Thus, the center of the circle is $(h, k) = (-1, 2)$ and the radius is $r = \sqrt{20}$, and you can write the standard form of the equation of the circle as follows.

$$(x - h)^2 + (y - k)^2 = r^2 \qquad \text{Standard form}$$
$$[x - (-1)]^2 + (y - 2)^2 = \left(\sqrt{20}\right)^2 \qquad \text{Substitute for } h, k, \text{ and } r.$$
$$(x + 1)^2 + (y - 2)^2 = 20 \qquad \text{Equation of circle} \qquad ▱$$

Figure P.11

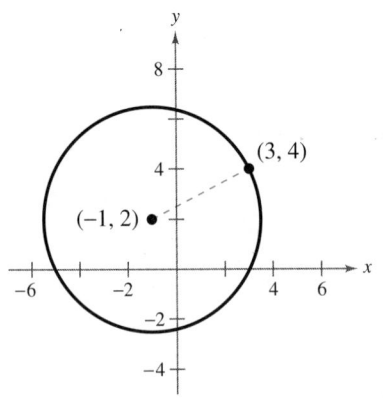

Application

EXAMPLE 7 ▱ **Translating Points in the Plane**

The triangle in Figure P.12(a) has vertices at the points $(-1, 2)$, $(1, -4)$, and $(2, 3)$. Shift the triangle three units to the right and two units up and find the vertices of the shifted triangle, as shown in Figure P.12(b).

Figure P.12

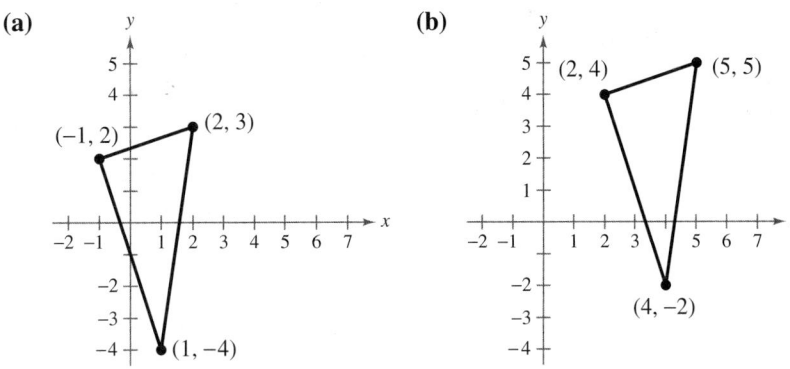

Solution

To shift the vertices three units to the right, add 3 to each x-coordinate. To shift the vertices two units up, add 2 to each y-coordinate.

Original Point	*Translated Point*
$(-1, 2)$	$(-1 + 3, 2 + 2) = (2, 4)$
$(1, -4)$	$(1 + 3, -4 + 2) = (4, -2)$
$(2, 3)$	$(2 + 3, 3 + 2) = (5, 5)$

▱

Group Activity   *Extending the Example*

Example 7 shows how to translate points in a coordinate plane. How are the following transformed points related to the original points?

Original Point	*Transformed Point*
(x, y)	$(-x, y)$
(x, y)	$(x, -y)$
(x, y)	$(-x, -y)$

P.1 /// EXERCISES

In Exercises 1–4, sketch the polygon with the indicated vertices.

1. Triangle: $(-1, 1)$, $(2, -1)$, $(3, 4)$

2. Triangle: $(0, 3)$, $(-1, -2)$, $(4, 8)$

3. Square: $(2, 4)$, $(5, 1)$, $(2, -2)$, $(-1, 1)$

4. Parallelogram: $(5, 2)$, $(7, 0)$, $(1, -2)$, $(-1, 0)$

In Exercises 5–8, approximate the coordinates of the points.

5.

6.

7.

8.

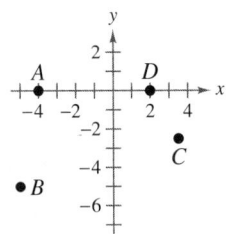

In Exercises 9–12, find the coordinates of the point.

9. The point is located three units to the left of the y-axis and four units above the x-axis.

10. The point is located eight units below the x-axis and four units to the right of the y-axis.

11. The point is located five units below the x-axis and the coordinates of the point are equal.

12. The point is on the x-axis and twelve units to the left of the y-axis.

13. *Think About It*   What is the y-coordinate of any point on the x-axis? What is the x-coordinate of any point on the y-axis?

14. *Think About It*   When plotting points on the rectangular coordinate system, is it true that the scales on the x- and y-axes must be the same? Explain.

In Exercises 15–24, determine the quadrant(s) in which (x, y) is located so that the condition(s) is (are) satisfied.

15. $x > 0$ and $y < 0$

16. $x < 0$ and $y < 0$

17. $x = -4$ and $y > 0$

18. $x > 2$ and $y = 3$

19. $y < -5$

20. $x > 4$

21. $(x, -y)$ is in the second quadrant.

22. $(-x, y)$ is in the fourth quadrant.

23. $xy > 0$

24. $xy < 0$

In Exercises 25 and 26, sketch a scatter plot of the data given in the table.

25. *Normal Temperatures*   The normal temperature y (in degrees Fahrenheit) in Duluth, Minnesota, for each month x, where $x = 1$ represents January, is given in the table.   (Source: NOAA)

x	1	2	3	4	5	6
y	6	12	23	38	50	59

x	7	8	9	10	11	12
y	65	63	54	44	28	14

26. *Wal-Mart*   The number y of Wal-Mart stores for each year x from 1985 through 1994 is given in the table.   (Source: Wal-Mart Annual Report for 1994)

x	1985	1986	1987	1988	1989
y	745	859	980	1114	1259

x	1990	1991	1992	1993	1994
y	1399	1568	1714	1850	1953

In Exercises 27 and 28, the polygon is shifted to a new position in the plane. Find the coordinates of the vertices of the polygon in its *new* position.

27.   **28.**

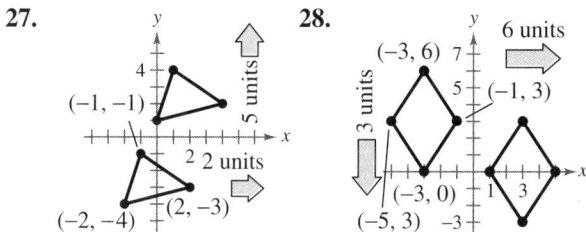

In Exercises 29 and 30, make a table of values for $x = -2, -1, -\frac{1}{2}, 0, \frac{1}{2}, 1$, and 2. Then plot the points on a rectangular coordinate system.

29. $y = 2 - \frac{1}{2}x$ **30.** $y = 2 - \frac{1}{2}x^2$

TV Advertising   In Exercises 31 and 32, refer to the figure.   (Source: Nielson Media Research)

31. Approximate the percent increase in cost of a TV spot from Super Bowl I in 1967 to Super Bowl XXIX in 1995.

32. Estimate the increase in cost of a TV spot (a) from Super Bowl V to Super Bowl XV, and (b) from Super Bowl XV to Super Bowl XXV.

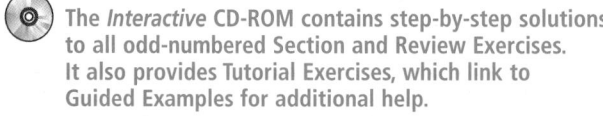

The *Interactive* CD-ROM contains step-by-step solutions to all odd-numbered Section and Review Exercises. It also provides Tutorial Exercises, which link to Guided Examples for additional help.

Milk Prices   In Exercises 33 and 34, refer to the figure, which shows the average price paid to farmers for milk.   (Source: U.S. Department of Agriculture and the National Milk Producers Federation)

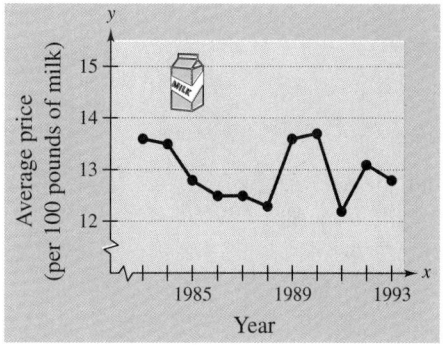

33. Approximate the highest price of milk shown in the graph. When did this occur?

34. Approximate the percent drop in the price of milk from the highest price shown in the graph to the price paid to farmers in January 1993.

Minimum Wage   In Exercises 35 and 36, refer to the figure.   (Source: U.S. Department of Labor)

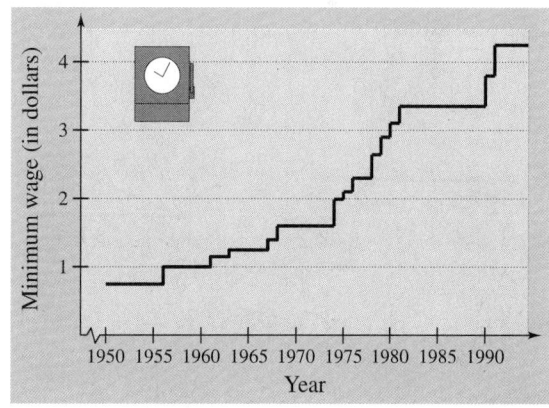

35. During which decade did the minimum wage increase most rapidly?

36. Approximate the percent increase in the minimum wage from 1990 to 1994.

Analyzing Data   In Exercises 37 and 38, refer to the figure, which shows the mathematics entrance test scores x, and the final examination scores y, in an algebra course for a sample of 10 students.

37. Find the entrance exam score of any student with a final exam score in the 80's.

38. Does a higher entrance exam score necessarily imply a higher final exam score? Explain.

In Exercises 39–42, (a) find the length of each side of a right triangle, and (b) show that these lengths satisfy the Pythagorean Theorem.

39.

40.

41.

42.

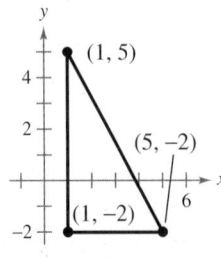

In Exercises 43–46, find the distance between the points. (*Note*: In each case the two points lie on the same horizontal or vertical line.)

43. $(6, -3)$, $(6, 5)$ **44.** $(1, 4)$, $(8, 4)$

45. $(-3, -1)$, $(2, -1)$ **46.** $(-3, -4)$, $(-3, 6)$

In Exercises 47–58, (a) plot the points, (b) find the distance between the points, and (c) find the midpoint of the line segment joining the points.

47. $(1, 1)$, $(9, 7)$ **48.** $(1, 12)$, $(6, 0)$

49. $(-4, 10)$, $(4, -5)$ **50.** $(-7, -4)$, $(2, 8)$

51. $(-1, 2)$, $(5, 4)$ **52.** $(2, 10)$, $(10, 2)$

53. $\left(\frac{1}{2}, 1\right)$, $\left(-\frac{5}{2}, \frac{4}{3}\right)$

54. $\left(-\frac{1}{3}, -\frac{1}{3}\right)$, $\left(-\frac{1}{6}, -\frac{1}{2}\right)$

55. $(6.2, 5.4)$, $(-3.7, 1.8)$

56. $(-16.8, 12.3)$, $(5.6, 4.9)$

57. $(-36, -18)$, $(48, -72)$

58. $(1.451, 3.051)$, $(5.906, 11.360)$

In Exercises 59 and 60, use the **Midpoint Formula** to estimate the sales of a company in 1993, given the sales in 1991 and 1995. Assume the sales followed a linear pattern.

59.

Year	1991	1995
Sales	$520,000	$740,000

60.

Year	1991	1995
Sales	$4,200,000	$5,650,000

In Exercises 61–66, show that the points form the vertices of the polygon.

61. Right triangle: $(4, 0)$, $(2, 1)$, $(-1, -5)$

62. Isosceles triangle: $(1, -3)$, $(3, 2)$, $(-2, 4)$

63. Rhombus: $(0, 0)$, $(1, 2)$, $(2, 1)$, $(3, 3)$

(A rhombus is a parallelogram whose sides are all the same length.)

64. Rhombus: $(4, 0), (0, 6), (-4, 0), (0, -6)$

65. Parallelogram: $(2, 5), (0, 9), (-2, 0), (0, -4)$

66. Parallelogram: $(0, 1), (3, 7), (4, 4), (1, -2)$

In Exercises 67–74, find the standard form of the equation of the specified circle.

67. Center: $(0, 0)$; radius: 3   **68.** Center: $(0, 0)$; radius: 5

69. Center: $(2, -1)$; radius: 4

70. Center: $\left(0, \frac{1}{3}\right)$; radius: $\frac{1}{3}$

71. Center: $(-1, 2)$; solution point: $(0, 0)$

72. Center: $(3, -2)$; solution point: $(-1, 1)$

73. Endpoints of a diameter: $(0, 0), (6, 8)$

74. Endpoints of a diameter: $(-4, -1), (4, 1)$

In Exercises 75–80, find the center and radius, and sketch the circle.

75. $x^2 + y^2 = 4$ **76.** $x^2 + y^2 = 16$

77. $(x - 1)^2 + (y + 3)^2 = 4$

78. $x^2 + (y - 1)^2 = 4$ **79.** $\left(x - \frac{1}{2}\right)^2 + \left(y - \frac{1}{2}\right)^2 = \frac{9}{4}$

80. $(x - 2)^2 + (y + 1)^2 = 2$

81. A line segment has (x_1, y_1) as one endpoint and (x_m, y_m) as its midpoint. Find the other endpoint (x_2, y_2) of the line segment in terms of $x_1, y_1, x_m,$ and y_m.

82. Use the result of Exercise 81 to find the coordinates of the endpoint of a line segment if the coordinates of the other endpoint and midpoint are, respectively,

(a) $(1, -2), (4, -1)$ (b) $(-5, 11), (2, 4)$

83. Use the Midpoint Formula three times to find the three points that divide the line segment joining (x_1, y_1) and (x_2, y_2) into four parts.

84. Use the result in Exercise 83 to find the points that divide the line segment joining the given points into four equal parts.

(a) $(1, -2), (4, -1)$ (b) $(-2, -3), (0, 0)$

85. *Football Pass* In a football game, a quarterback throws a pass from the 15-yard line, 10 yards from the sideline (see figure). The pass is caught on the 40-yard line, 45 yards from the same sideline. How long is the pass?

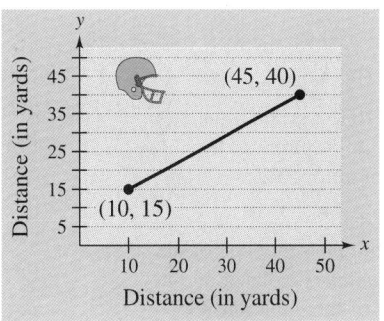

86. *Flying Distance* A plane flies in a straight line to a city that is 100 kilometers east and 150 kilometers north of the point of departure. How far does it fly?

87. *Make a Conjecture* Plot the points $(2, 1), (-3, 5),$ and $(7, -3)$ on a rectangular coordinate system. Then change the sign of the x-coordinate of each point and plot the three new points on the same rectangular coordinate system. What conjecture can you make about the location of a point when the sign of the x-coordinate is changed?

88. Prove that the diagonals of the parallelogram in the figure bisect each other.

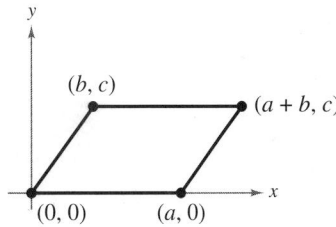

89. *Chapter Opener* Use the graph on page 1.

(a) Describe any trends in the data. From these trends, predict the number of artists elected in 1996.

(b) Why do you think the numbers elected in 1986 and 1987 were greater than in other years?

P.2   Graphs and Graphing Utilities

The Graph of an Equation **/** *Using a Graphing Utility* **/** *Determining a Viewing Rectangle* **/** *Applications*

The Graph of an Equation

News magazines often show graphs comparing the rate of inflation, the federal deficit, wholesale prices, or the unemployment rate to the time of year. Industrial firms and businesses use graphs to report their monthly production and sales statistics. Such graphs provide geometric pictures of the way one quantity changes with respect to another. Frequently, the relationship between two quantities is expressed as an equation. This section introduces the basic procedure for determining the geometric picture associated with an equation.

For an equation in variables x and y, a point (a, b) is a **solution point** if the substitution of $x = a$ and $y = b$ satisfies the equation. Most equations have *infinitely* many solution points. For example, the equation

$$3x + y = 5$$

has solution points $(0, 5)$, $(1, 2)$, $(2, -1)$, $(3, -4)$, and so on. The set of all solution points of an equation is the **graph** of the equation.

Figure P.13

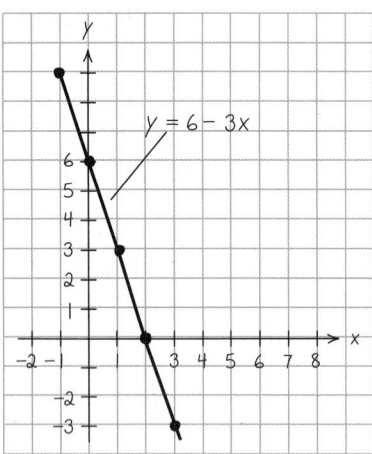

EXAMPLE 1   **Sketching a Graph by Point Plotting**

Use point plotting and graph paper to sketch the graph of

$$3x + y = 6.$$

Solution

In this case you can isolate the variable y to obtain

$$y = 6 - 3x. \qquad \text{Solve equation for } y.$$

Using negative, zero, and positive values for x, you can obtain the following table of values (solution points).

x	-1	0	1	2	3
$y = 6 - 3x$	9	6	3	0	-3

Next, plot these points and connect them, as shown in Figure P.13. It appears that the graph is a straight line. You will study lines extensively in Section P.3.

The points at which a graph touches or crosses an axis are the **intercepts** of the graph. For instance, in Example 1 the point $(0, 6)$ is the y-intercept of the graph because the graph crosses the y-axis at that point. The point $(2, 0)$ is the x-intercept of the graph because the graph crosses the x-axis at that point.

EXAMPLE 2　　Sketching a Graph by Point Plotting

Use point plotting and graph paper to sketch the graph of $y = x^2 - 2$.

Solution
First, make a table of values by choosing several convenient values of x and calculating the corresponding values of y.

x	-2	-1	0	1	2	3
$y = x^2 - 2$	2	-1	-2	-1	2	7

Next, plot the corresponding solution points, as shown in Figure P.14(a). Finally, connect the points with a smooth curve, as shown in Figure P.14(b).

Figure P.14

(a)

(b)

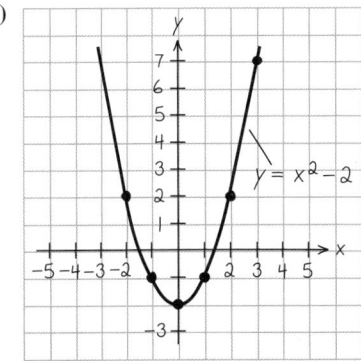

This graph is called a **parabola.** You will study parabolas in Section 2.1.

Note　In this text, you will study two basic ways to create graphs: *by hand* and *using a graphing utility.* For instance, the graphs in Figures P.13 and P.14 were sketched by hand and the graph in Figure P.16 was sketched using a graphing utility.

Using a Graphing Utility

One of the disadvantages of the point-plotting method is that to get a good idea about the shape of a graph you need to plot *many* points. With only a few points, you could badly misrepresent the graph. For instance, consider the equation

$$y = \frac{1}{30} x(x^4 - 10x^2 + 39).$$

Suppose you plotted only five points: $(-3, -3)$, $(-1, -1)$, $(0, 0)$, $(1, 1)$, and $(3, 3)$, as shown in Figure P.15(a). From these five points, you might assume that the graph of the equation is a straight line. That, however, is not correct. By plotting several more points, you can see that the actual graph is not straight at all. See Figure P.15(b).

Figure P.15

(a)   **(b)**

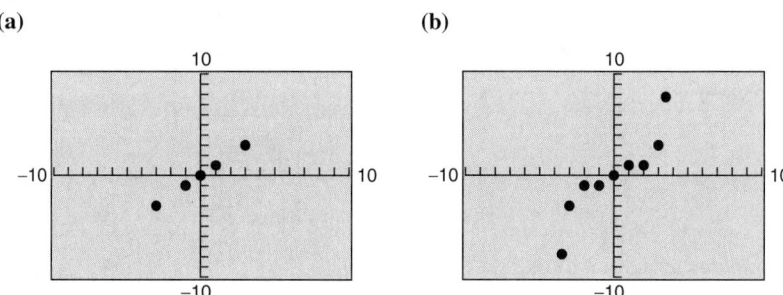

From this, you can see that the point-plotting method leaves you with a dilemma. On the one hand, the method can be very inaccurate if only a few points are plotted. But on the other hand, it is very time-consuming to plot a dozen (or more) points. Technology can help solve this dilemma. Plotting several (even several hundred) points on a rectangular coordinate system is something that a computer or calculator can do easily.

Note   The point-plotting method is the method used by *all* graphing utilities. Each computer or calculator screen is made up of a grid of hundreds or thousands of small areas called **pixels.** Screens that have many pixels per square inch are said to have a higher **resolution** than screens with fewer pixels.

Using a Graphing Utility to Graph an Equation

To graph an equation involving x and y on a graphing utility, use the following procedure.

1. Rewrite the equation so that y is isolated on the left side.
2. Enter the equation into a graphing utility.
3. Determine a **viewing rectangle** that shows all important features of the graph. For some graphing utilities, the standard viewing rectangle ranges between -10 and 10 for both x- and y-values.
4. Activate the graphing utility.

EXAMPLE 3 **Using a Graphing Utility**

Use a graphing utility to graph $2y + x^3 = 4x$.

Solution

To begin, solve the equation for y in terms of x.

$$2y + x^3 = 4x \qquad \text{Original equation}$$
$$2y = -x^3 + 4x \qquad \text{Subtract } x^3 \text{ from both sides.}$$
$$y = -\frac{1}{2}x^3 + 2x \qquad \text{Divide both sides by 2.}$$

Now, by entering this equation into a graphing utility (using a standard viewing rectangle), you can obtain the graph shown in Figure P.16.

Figure P.16

The following table, which can be generated on a *TI-82* or a *TI-83*, shows some points of the graph in Figure P.16.

X	Y₁
−3	7.5
−2	0
−1	−1.5
0	0
1	1.5
2	0
3	−7.5
X = −3	

The *Interactive* CD-ROM offers graphing utility emulators of the *TI-82* and *TI-83*, which can be used with the Examples, Explorations, Technology notes, and Exercises.

EXPLORATION

Use your graphing utility to duplicate the graphs shown below and state the viewing rectangle. Work with a partner. If possible, choose a partner who has the same model of graphing utility as you do.

Graph of $y = 7x - 4$

Graph of $y = -x^2 + 2x - 8$

Determining a Viewing Rectangle

A **viewing rectangle** for a graph is a rectangular portion of the Cartesian plane. A viewing rectangle is determined by six values: the minimum x-value, the maximum x-value, the x-scale, the minimum y-value, the maximum y-value, and the y-scale. The **standard** viewing rectangle for some graphing utilities uses the following values.

```
Xmin = -10
Xmax = 10
Xscl = 1
Ymin = -10
Ymax = 10
Yscl = 1
```

By choosing different viewing rectangles for a graph, it is possible to obtain very different impressions of the graph's shape. For instance, Figure P.17 shows four different viewing rectangles for the graph of

$$y = 0.1x^4 - x^3 + 2x^2.$$

Of these, the view shown in Figure P.17(a) is the most complete because it shows more of the distinguishing portions of the graph.

Figure P.17

(a)

(b)

(c)

(d)

EXAMPLE 4 ▱ **Sketching a Circle with a Graphing Utility**

Use a graphing utility to graph

$$x^2 + y^2 = 9.$$

Solution

The graph of $x^2 + y^2 = 9$ is a circle whose center is the origin and whose radius is 3. (See Section P.1.) To graph the equation, begin by solving the equation for y.

$$x^2 + y^2 = 9 \qquad \text{Original equation}$$
$$y^2 = 9 - x^2 \qquad \text{Solve for } y^2.$$
$$y = \pm\sqrt{9 - x^2} \qquad \text{Solve for } y.$$

The graph of

$$y = \sqrt{9 - x^2} \qquad \text{Upper semicircle}$$

is the upper semicircle. The graph of

$$y = -\sqrt{9 - x^2} \qquad \text{Lower semicircle}$$

is the lower semicircle. Enter *both* equations in your graphing utility and generate the resulting graphs. In Figure P.18(a), note that if you use a standard viewing rectangle, the two graphs do not appear to form a circle. You can overcome this problem by using a **square setting,** in which the horizontal and vertical tick marks have equal spacing, as shown in Figure P.18(b). On many graphing utilities, a square setting can be obtained by using the ratio

$$\frac{Y_{max} - Y_{min}}{X_{max} - X_{min}} = \frac{2}{3}.$$

Note The standard viewing rectangle on many graphing utilities does not give a true geometric perspective. That is, perpendicular lines will not appear to be perpendicular and circles will not appear to be circular. To overcome this, you can use a square setting, as demonstrated in Example 4.

Figure P.18

▱

In applications, it is convenient to use variable names that suggest real-life quantities: *d* for distance, *t* for time, and so on. Most graphing utilities, however, require the variable names to be *x* and *y*.

Applications

The following two applications show how to develop mathematical models to represent real-world situations. You will see that both a graphing utility and algebra can be used to understand and resolve the problems posed.

Once an appropriate viewing rectangle is chosen for a particular graph, the *zoom* and *trace* features of a graphing utility are useful for approximating specific values from the graph.

EXAMPLE 5   **Using the Zoom and Trace Features**

A runner runs at a constant rate of 4.8 miles per hour. The verbal model and algebraic equation relating distance run and elapsed time are as follows.

Verbal Model: Distance = Rate · Time

Equation: $d = 4.8t$

a. Use a graphing utility and an appropriate viewing rectangle to graph the equation $d = 4.8t$. (Represent *d* by *y* and *t* by *x*.)

b. Estimate how far the runner can run in 3.2 hours.

c. Estimate how long it will take to run a 26-mile marathon.

Solution

a. An appropriate viewing rectangle and graph are shown in Figure P.19(a).

b. Figure P.19(b) shows the viewing rectangle after zooming in (near $x = 3.2$) once by a factor of 4. Using the trace feature, you can determine that for $x = 3.2$, the distance is $y \approx 15.36$ miles.

c. Figure P.19(c) shows the viewing rectangle after zooming in (near $y = 26$) twice by a factor of 4. Using the trace feature, you can determine that for $y = 26$, the time is $x \approx 5.42$ hours.

Note   The viewing rectangle on your graphing utility may differ from those shown in parts (b) and (c) of Figure P.19.

Figure P.19

(a)

(b)

(c)

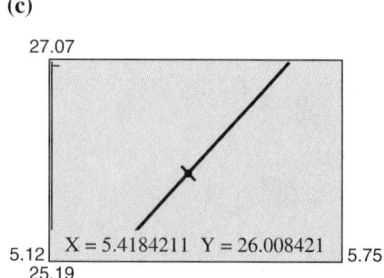

EXAMPLE 6   An Application: Monthly Wages

You receive a monthly salary of $2000 plus a commission of 10% of sales.

a. Find an equation expressing the monthly wages y in terms of the sales x.

b. If sales are $x = 1480$ in August, what are your wages for that month?

c. If you receive $2225 in wages for September, what were your sales for that month?

Figure P.20

(a)

(b)

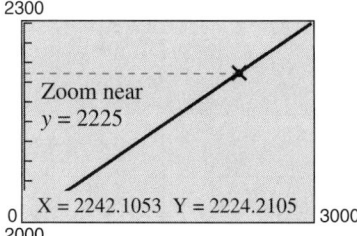

Solution

a. The monthly wages are the sum of the fixed $2000 salary and the 10% commission on sales x.

Verbal Model: Wages = Salary + Commission on Sales

Equation: $y = 2000 + 0.1x$

b. If $x = 1480$, the corresponding wages are

$$y = 2000 + 0.1(1480) = 2000 + 148 = \$2148.$$

You can confirm this result using a graphing utility. Because $x \geq 0$ and the monthly wages are at least $2000, a reasonable viewing rectangle is the one shown in Figure P.20(a). Using the zoom and trace features near $x = 1480$ shows that the wages are about $2148.

c. To answer the third question, you can use the graphing utility to find the value along the x-axis (sales) that corresponds to a y-value of 2225 (wages). Beginning with Figure P.20(b) and using the zoom and trace features, you can estimate x to be $x \approx 2250$. You can verify this answer by observing that

$$\text{Wages} = 2000 + 0.1(2250) = 2000 + 225 = \$2225.$$

Group Activity

Comparison of Wages

Your employer offers you a choice of wage scales: a monthly salary of $3000 plus commission of 7% of sales or a salary of $3400 plus a 5% commission. Discuss how you would choose your option. At what sales level would the options yield the same salary? Be sure to take advantage of your graphing utility.

P.2 /// EXERCISES

In Exercises 1–8, determine whether the points lie on the graph of the equation.

Equation	Points	
1. $y = \sqrt{x} + 4$	(a) $(0, 2)$	(b) $(5, 3)$
2. $y = x^2 - 3x + 2$	(a) $(2, 0)$	(b) $(-2, 8)$
3. $y = 4 - |x - 2|$	(a) $(1, 5)$	(b) $(6, 0)$
4. $y = \frac{1}{3}x^3 - 2x^2$	(a) $\left(2, -\frac{16}{3}\right)$	(b) $(-3, 9)$
5. $2x - y - 3 = 0$	(a) $(1, 2)$	(b) $(1, -1)$
6. $x^2 + y^2 = 20$	(a) $(3, -2)$	(b) $(-4, 2)$
7. $x^2y - x^2 + 4y = 0$	(a) $\left(1, \frac{1}{5}\right)$	(b) $\left(2, \frac{1}{2}\right)$
8. $y = \dfrac{1}{x^2 + 1}$	(a) $(0, 0)$	(b) $(3, 0.1)$

In Exercises 9–16, complete the table. Use the resulting solution points to sketch the graph of the equation. Use a graphing utility to verify the graph.

9. $y = -2x + 3$

x	-1	0	1	$\frac{3}{2}$	2
y					

10. $y = \frac{3}{2}x - 1$

x	-2	0	$\frac{2}{3}$	1	2
y					

11. $y = x^2 - 2x$

x	-1	0	1	2	3
y					

12. $y = 4 - x^2$

x	-2	-1	0	1	2
y					

13. $y = \frac{1}{4}x - 3$

x	-2	-1	0	1	2
y					

14. $y = 3 - |x - 2|$

x	0	1	2	3	4
y					

15. $y = \sqrt{x - 1}$

x	1	2	5	10	17
y					

16. $y = \dfrac{6x}{x^{-2} + 1}$

x	-2	-1	0	1	2
y					

17. *Think About It*   Repeat Exercise 13 for the equation $y = -\frac{1}{4}x - 3$. Use the result to describe any differences in the graphs.

18. *Think About It*   Continue the table in Exercise 16 for x-values of 5, 10, 20, and 40. What is the value of y approaching? Can y be negative for positive values of x? Explain.

In Exercises 19–26, use a graphing utility to graph the equation. Use a standard setting. Approximate any x- or y-intercepts of the graph.

19. $y = x - 5$ **20.** $y = (x + 1)(x - 3)$

21. $y = x^2 + x - 2$ **22.** $y = 9 - x^2$

23. $y = x\sqrt{x + 6}$ **24.** $y = (6 - x)\sqrt{x}$

25. $y = \dfrac{2x}{x - 1}$ **26.** $y = \dfrac{4}{x}$

In Exercises 27–32, match the equation with its graph. [The graphs are labeled (a), (b), (c), (d), (e), and (f).]

(a)

(b)

(c)

(d)

(e)

(f)
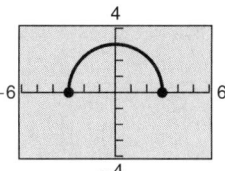

27. $y = 1 - x$

28. $y = x^2 - 2x$

29. $y = \sqrt{9 - x^2}$

30. $y = 2\sqrt{x}$

31. $y = x^3 - x + 1$

32. $y = |x| - 3$

In Exercises 33–46, sketch the graph of the equation.

33. $y = -3x + 2$

34. $y = 2x - 3$

35. $y = 1 - x^2$

36. $y = x^2 - 1$

37. $y = x^2 - 3x$

38. $y = -x^2 - 4x$

39. $y = x^3 + 2$

40. $y = x^3 - 1$

41. $y = \sqrt{x - 3}$

42. $y = \sqrt{1 - x}$

43. $y = |x - 2|$

44. $y = 4 - |x|$

45. $x = y^2 - 1$

46. $x = y^2 - 4$

In Exercises 47–54, use a graphing utility to graph the equation. Use a standard setting. Approximate any intercepts.

47. $y = 3 - \frac{1}{2}x$

48. $y = \frac{2}{3}x - 1$

49. $y = x^2 - 4x + 3$

50. $y = \frac{1}{2}(x + 4)(x - 2)$

51. $y = x(x - 2)^2$

52. $y = \dfrac{4}{x^2 + 1}$

53. $y = \sqrt[3]{x}$

54. $y = \sqrt[3]{x + 1}$

In Exercises 55–58, use a graphing utility to sketch the graph of the equation. Begin by using a standard setting. Then graph the equation a second time using the specified setting. Which setting is better? Explain.

55. $y = \frac{5}{2}x + 5$

56. $y = -3x + 50$

| Xmin = 0 |
| Xmax = 6 |
| Xscl = 1 |
| Ymin = 0 |
| Ymax = 10 |
| Yscl = 1 |

| Xmin = -1 |
| Xmax = 4 |
| Xscl = 1 |
| Ymin = -5 |
| Ymax = 60 |
| Yscl = 5 |

57. $y = -x^2 + 10x - 5$

58. $y = 4(x + 5)\sqrt{4 - x}$

| Xmin = -1 |
| Xmax = 11 |
| Xscl = 1 |
| Ymin = -5 |
| Ymax = 25 |
| Yscl = 2 |

| Xmin = -6 |
| Xmax = 6 |
| Xscl = 1 |
| Ymin = -5 |
| Ymax = 50 |
| Yscl = 4 |

In Exercises 59–62, describe the viewing rectangle.

59. $y = 4x^2 - 25$

60. $y = x^3 - 3x^2 + 4$

61. $y = |x| + |x - 10|$ **62.** $y = 8\sqrt[3]{x - 6}$

In Exercises 63 and 64, use a graphing utility to graph the equation using each of the suggested viewing rectangles. Assume that the equation gives the profit y when x units of a product are sold. Note that a graph can distort the information presented simply by changing the viewing rectangle. Which viewing rectangle would be selected by a person who wishes to argue that profits will increase dramatically with increased sales?

63. $y = 0.25x - 50$

Xmin = -3	Xmin = -3
Xmax = 800	Xmax = 1000
Xscl = 50	Xscl = 100
Ymin = -20	Ymin = -100
Ymax = 100	Ymax = 500
Yscl = 10	Yscl = 40

64. $y = 2.44x - \dfrac{x^2}{20{,}000} - 5000$

Xmin = -5000	Xmin = -5000
Xmax = 22000	Xmax = 22000
Xscl = 5000	Xscl = 5000
Ymin = -20000	Ymin = -5000
Ymax = 60000	Ymax = 24000
Yscl = 10000	Yscl = 5000

In Exercises 65–68, solve for y and use a graphing utility to graph each of the resulting equations in the same viewing rectangle. Adjust the viewing rectangle so that a circle really does appear circular.

65. $x^2 + y^2 = 64$ **66.** $(x - 1)^2 + (y - 2)^2 = 16$

67. $x^2 + y^2 = 49$ **68.** $(x - 3)^2 + (y - 1)^2 = 25$

In Exercises 69–72, explain how to use a graphing utility to verify that $y_1 = y_2$. Identify the rule of algebra that is illustrated.

69. $y_1 = \frac{1}{4}(x^2 - 8)$ **70.** $y_1 = \frac{1}{2}x + (x + 1)$

$y_2 = \frac{1}{4}x^2 - 2$ $y_2 = \frac{3}{2}x + 1$

71. $y_1 = \frac{1}{5}[10(x^2 - 1)]$ **72.** $y_1 = (x - 3) \cdot \dfrac{1}{x - 3}$

$y_2 = 2(x^2 - 1)$ $y_2 = 1$

73. *Depreciation*   A manufacturing plant purchases a new molding machine for $225,000. The depreciated value y after t years is given by

$$y = 225{,}000 - 20{,}000t, \qquad 0 \le t \le 8.$$

(a) Use the constraints of the model to determine an appropriate viewing rectangle.

(b) Use a graphing utility to graph the equation.

74. *Dimensions of a Rectangle*   A rectangle of length x and width w has a perimeter of 12 meters.

(a) Draw a rectangle that gives a visual representation of the problem. Use the specified variables to label the sides of the rectangle.

(b) Show that $w = 6 - x$ is the width of the rectangle and that $A = x(6 - x)$ is its area.

(c) Use a graphing utility to graph the area equation.

(d) From the graph in part (c), estimate the dimensions of the rectangle that yield a maximum area.

75. *Think About It*   Suppose you correctly enter an expression for the variable y on a graphing utility. However, no graph appears on the display when you graph the equation. Give a possible explanation and the steps you could take to remedy the problem. Illustrate your explanation with an example.

Data Analysis In Exercises 76 and 77, (a) sketch a graph comparing the data and the model for the data, (b) use the model to estimate the value of y for the year 1998, and (c) repeat part (b) for the year 2000.

76. *Federal Debt* The table gives the per capita federal debt for the United States for several years. (Source: U.S. Treasury Department)

Year	1950	1960	1970
Per Capita Debt	$1688	$1572	$1807

Year	1980	1990	1994
Per Capita Debt	$3981	$12,848	$15,750

A mathematical model for the per capita debt during this period is

$y = 0.255t^3 - 4.096t^2 + 1570.417$

where y represents the per capita debt and t is the time in years, with $t = 0$ corresponding to 1950.

77. *Life Expectancy* The table gives the life expectancy of a child (at birth) in the United States for selected years from 1920 to 1990. (Source: Department of Health and Human Services)

Year	1920	1930	1940	1950
Life Expectancy	54.1	59.7	62.9	68.2

Year	1960	1970	1980	1990
Life Expectancy	69.7	70.8	73.7	75.4

A mathematical model for the life expectancy during this period is

$$y = \frac{t + 66.93}{0.01t + 1}$$

where y represents the life expectancy and t is the time in years, with $t = 0$ corresponding to 1950.

78. *Think About It* Find a and b if the x-intercepts of the graph of $y = (x - a)(x - b)$ are $(-2, 0)$ and $(5, 0)$.

79. *Dividends Per Share* The dividends per common share of Procter and Gamble Company from 1990 through 1994 can be approximated by the mathematical model

$y = 0.086t + 0.872, \quad 0 \le t \le 4$

where y is the dividend and t is the calendar year, with $t = 0$ corresponding to 1990. Sketch a graph of this equation. (Source: Procter and Gamble Company 1994 Annual Report)

80. *Copper Wire* The resistance y in ohms of 1000 feet of solid copper wire at 77°F can be approximated by the mathematical model

$$y = \frac{10,770}{x^2} - 0.37, \quad 5 \le x \le 100$$

where x is the diameter of the wire in mils (0.001 in.). Use the model to estimate the resistance when $x = 50$. (Source: American Wire Gage)

In Exercises 81–84, use a graphing utility to graph the equation. Move the cursor along the curve to approximate the unknown coordinate of each given solution point accurate to two decimal places. (*Hint:* You may need to use the zoom feature of the graphing utility to obtain the required accuracy.)

81. $y = \sqrt{5 - x}$
(a) $(2, y)$
(b) $(x, 3)$

82. $y = x^3(x - 3)$
(a) $(2.25, y)$
(b) $(x, 20)$

83. $y = x^5 - 5x$
(a) $(-0.5, y)$
(b) $(x, -4)$

84. $y = |x^2 - 6x + 5|$
(a) $(2, y)$
(b) $(x, 1.5)$

P.3   Lines in the Plane

*The Slope of a Line   /   The Point-Slope Form of the Equation of a Line   /
Sketching Graphs of Lines   /   Changing the Viewing Rectangle   /
Parallel and Perpendicular Lines*

Figure P.21

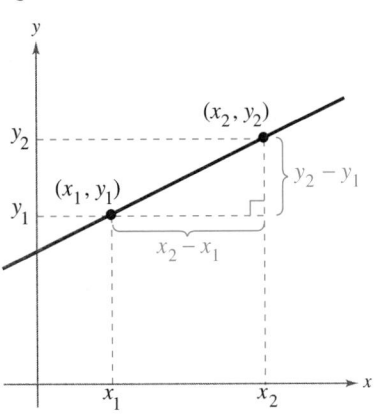

The Slope of a Line

In this section, you will study lines and their equations. The **slope** of a nonvertical line represents the number of units a line rises or falls vertically for each unit of horizontal change from left to right. For instance, consider the two points (x_1, y_1) and (x_2, y_2) on the line shown in Figure P.21. As you move from left to right along this line, a change of $(y_2 - y_1)$ units in the vertical direction corresponds to a change of $(x_2 - x_1)$ units in the horizontal direction. That is,

$$y_2 - y_1 = \text{the change in } y$$

and

$$x_2 - x_1 = \text{the change in } x.$$

The slope of the line is given by the ratio of these two changes.

Definition of the Slope of a Line

The **slope** m of the nonvertical line through (x_1, y_1) and (x_2, y_2) is

$$m = \frac{y_2 - y_1}{x_2 - x_1} = \frac{\text{change in } y}{\text{change in } x}$$

where $x_1 \neq x_2$.

EXPLORATION

Use a graphing utility to compare the slopes of the lines $y = 0.5x$, $y = x$, $y = 2x$, and $y = 4x$. What do you observe about the slopes of the lines? Compare the slopes of the lines $y = -0.5x$, $y = -x$, $y = -2x$, and $y = -4x$. What do you observe about the slopes of these lines? (*Hint:* Use a square setting to guarantee a true geometric perspective.)

When this formula is used, the *order of subtraction* is important. Given two points on a line, you are free to label either one of them as (x_1, y_1) and the other as (x_2, y_2). However, once this has been done, you must form the numerator and denominator using the same order of subtraction.

$$m = \frac{y_2 - y_1}{x_2 - x_1} \qquad m = \frac{y_1 - y_2}{x_1 - x_2} \qquad m = \frac{y_2 - y_1}{x_1 - x_2}$$

Correct   Correct   Incorrect

Note   Throughout this text, the term **line** always means a *straight* line.

EXAMPLE 1 ▰ Finding the Slope of a Line

Find the slope of the line passing through each pair of points.

a. $(-2, 0)$ and $(3, 1)$ **b.** $(-1, 2)$ and $(2, 2)$ **c.** $(0, 4)$ and $(1, -1)$

Solution

Difference in *y*-values

a. $m = \dfrac{\overbrace{y_2 - y_1}}{\underbrace{x_2 - x_1}} = \dfrac{1 - 0}{3 - (-2)} = \dfrac{1}{3 + 2} = \dfrac{1}{5}$

Difference in *x*-values

b. $m = \dfrac{2 - 2}{2 - (-1)} = \dfrac{0}{3} = 0$

Note In Figure P.22, note that the square setting gives the correct "steepness" of the lines.

c. $m = \dfrac{-1 - 4}{1 - 0} = \dfrac{-5}{1} = -5$

The graphs of the three lines are shown in Figure P.22.

Figure P.22

(a)

(b)

(c)

▰

Figure P.23

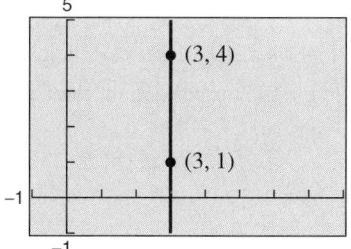

The definition of slope does not apply to vertical lines. For instance, consider the points $(3, 4)$ and $(3, 1)$ on the vertical line shown in Figure P.23. Applying the formula for slope, you obtain $m = (4 - 1)/(3 - 3)$. Because division by zero is undefined, the slope of a vertical line is undefined.

From the slopes of the lines shown in Figures P.22 and P.23, you can make the following generalizations about the slope of a line.

1. A line with positive slope $(m > 0)$ *rises* from left to right.

2. A line with negative slope $(m < 0)$ *falls* from left to right.

3. A line with zero slope $(m = 0)$ is *horizontal*.

4. A line with undefined slope is *vertical*.

The Point-Slope Form of the Equation of a Line

Figure P.24

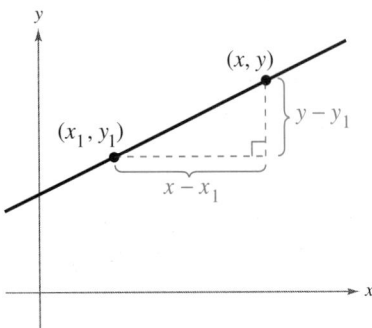

If you know the slope of a line *and* you also know the coordinates of one point on the line, you can find an equation for the line. For instance, in Figure P.24, let (x_1, y_1) be a given point on the line whose slope is m. If (x, y) is any *other* point on a line, it follows that

$$\frac{y - y_1}{x - x_1} = m.$$

This equation in the variables x and y can be rewritten in the **point-slope form** of the equation of a line.

Point-Slope Form of the Equation of a Line

The **point-slope form** of the equation of the line that passes through the point (x_1, y_1) and has a slope of m is

$$y - y_1 = m(x - x_1).$$

EXAMPLE 2   ▱   **The Point-Slope Form of the Equation of a Line**

Find an equation of the line that passes through the point $(1, -2)$ and has a slope of 3.

Figure P.25

Solution

$$y - y_1 = m(x - x_1) \qquad \text{Point-slope form}$$
$$y - (-2) = 3(x - 1) \qquad \text{Substitute for } y_1, m, \text{ and } x_1.$$
$$y + 2 = 3x - 3$$
$$y = 3x - 5 \qquad \text{Equation of line}$$

This line is shown in Figure P.25.   ▱

The point-slope form can be used to find an equation of a nonvertical line passing through two points (x_1, y_1) and (x_2, y_2). First, use the formula for the slope of the line passing through two points.

$$m = \frac{y_2 - y_1}{x_2 - x_1}$$

Then, once you know the slope, use the point-slope form to obtain the equation

$$y - y_1 = \frac{y_2 - y_1}{x_2 - x_1}(x - x_1) = m(x - x_1).$$

This is sometimes called the **two-point form** of the equation of a line.

EXAMPLE 3 ▱ A Linear Model for Sales Prediction

During 1993, L. L. Bean's net sales were $870 million, and in 1994 net sales were $975 million. (Source: L. L. Bean)

a. Write a linear equation giving the net sales y in terms of the year x.

b. Use the equation to estimate the net sales during 1997.

Solution

a. Let $x = 3$ represent 1993. In Figure P.26, let $(3, 870)$ and $(4, 975)$ be two points on the line representing the net sales. The slope of the line passing through these two points is

$$m = \frac{975 - 870}{4 - 3} = 105.$$

Figure P.26

Figure P.27

Linear Extrapolation

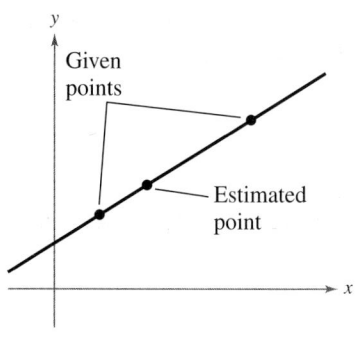

Linear Interpolation

By the point-slope form, the equation of the line is as follows.

$$y - y_1 = m(x - x_1)$$ Point-slope form
$$y - 870 = 105(x - 3)$$ Substitute for y_1, m, and x_1.
$$y = 105x - 315 + 870$$
$$y = 105x + 555$$ Equation of line

b. Using the equation from part (a), estimate the 1997 net sales ($x = 7$) to be $y = 105(7) + 555 = 735 + 555 = \$1290$ million or $1.29 billion. ▱

The approximation method illustrated in Example 3 is **linear extrapolation.** Note in Figure P.27 that for linear extrapolation, the estimated point lies outside of the given points. When the estimated point lies *between* two given points, the procedure is called **linear interpolation.**

Library of Functions

In Chapter 1, you will be introduced to the precise meaning of the term *function*. The simplest type of function is a linear function and has the form

$$y = ax + b.$$

As its name implies, the graph of a linear function is a line that has a slope of a and a y-intercept at $(0, b)$.

Sketching Graphs of Lines

Many problems in coordinate geometry can be classified as follows.

1. Given a graph (or parts of it), find its equation.
2. Given an example, find its graph.

For lines, the first problem is solved easily by using the point-slope form. This formula, however, is not particularly useful for solving the second type of problem. The form that is better suited to graphing linear equations is the **slope-intercept form** $y = mx + b$ of the equation of a line.

EXAMPLE 4   **Determining the Slope and *y*-Intercepts**

a. Graph the lines $y = 2x + 1$, $y = \frac{1}{2}x + 1$, and $y = -2x + 1$ in the same viewing rectangle. What can you observe?

b. Graph the lines $y = 2x + 1$, $y = 2x$, and $y = 2x - 1$ in the same viewing rectangle. What can you observe?

Solution

a. In Figure P.28(a), you can see that all three lines have the same y-intercept $(0, 1)$, but slopes of 2, $\frac{1}{2}$, and -2, respectively.

b. In Figure P.28(b), you can see that all three lines have a slope of 2, but y-intercepts of $(0, 1)$, $(0, 0)$, and $(0, -1)$, respectively.

Figure P.28

(a)

(b)

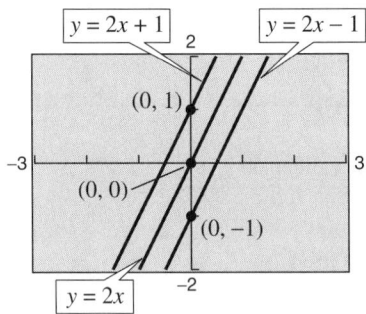

Slope-Intercept Form of the Equation of a Line

The graph of the equation

$$y = mx + b$$

is a line whose slope is m and whose y-intercept is $(0, b)$.

Figure P.29

(a)

(b)

(c)

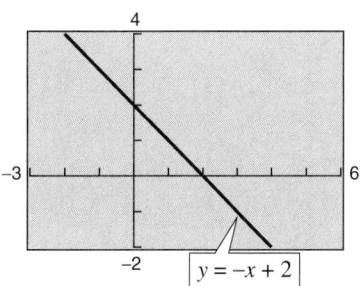

EXAMPLE 5   **Using the Slope-Intercept Form**

Describe the graph of each linear equation.

a. $y = \dfrac{3}{2}x$ **b.** $y = 2$ **c.** $x + y = 2$

Solution

a. Because $b = 0$, the y-intercept is $(0, 0)$. Moreover, because the slope is $m = \frac{3}{2}$, this line *rises* three units for every two units it moves to the right, as shown in Figure P.29(a).

b. By writing the equation $y = 2$ in the form $y = (0)x + 2$, you can see that the y-intercept is $(0, 2)$ and the slope is zero. A zero slope implies that the line is horizontal, as shown in Figure P.29(b).

c. By writing the equation $x + y = 2$ in slope-intercept form, $y = -x + 2$, you can see that the y-intercept is $(0, 2)$. Moreover, because the slope is $m = -1$, this line *falls* one unit for every unit it moves to the right, as shown in Figure P.29(c).

From the slope-intercept form of the equation of a line, you can see that a horizontal line ($m = 0$) has an equation of the form $y = b$. This is consistent with the fact that each point on a horizontal line through $(0, b)$ has a y-coordinate of b.

Similarly, each point on a vertical line through $(a, 0)$ has an x-coordinate of a. Hence, a vertical line has an equation of the form $x = a$. This equation cannot be written in the slope-intercept form, because the slope of a vertical line is undefined. However, *every* line has an equation that can be written in the **general form**

$$Ax + By + C = 0 \qquad \text{General form of the equation of a line}$$

where A and B are not *both* zero.

Summary of Equations of Lines

1. General form: $Ax + By + C = 0$
2. Vertical line: $x = a$
3. Horizontal line: $y = b$
4. Slope-intercept form: $y = mx + b$
5. Point-slope form: $y - y_1 = m(x - x_1)$

Changing the Viewing Rectangle

When a graphing utility is used to sketch a straight line, it is important to realize that the graph of the line may not visually appear to have the slope indicated by its equation. This occurs because of the viewing rectangle used for the graph. For instance, Figure P.30 shows graphs of $y = 2x + 1$ produced on a graphing utility using three different viewing rectangles.

Notice that the slopes in Figure P.30(a) and (b) do not visually appear to be equal to 2. However, if you use the *square* viewing rectangle, as in Figure P.30(c), the slope visually appears to be 2. In general, two graphs of the same equation can appear to be quite different depending on the viewing rectangle selected.

Figure P.30

(a)

(b)

(c)

EXAMPLE 6   Different Viewing Rectangles

The graphs of the two lines

$$y = -x - 1 \qquad \text{and} \qquad y = -10x - 1$$

are shown in Figure P.31. Even though the slopes of these lines are different (-1 and -10, respectively), the graphs seem similar because the viewing rectangles are different.

Figure P.31

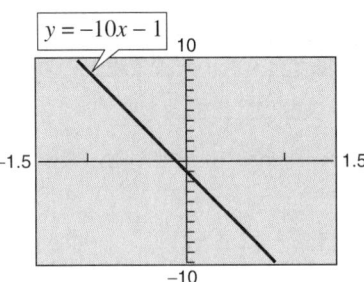

Parallel and Perpendicular Lines

The slope of a line is a convenient tool for determining whether two lines are parallel or perpendicular. Example 4(b) suggests the following property of parallel lines.

Parallel Lines

Two distinct nonvertical lines are **parallel** if and only if their slopes are equal.

EXAMPLE 7 **Equations of Parallel Lines**

Find an equation of the line that passes through the point $(2, -1)$ and is parallel to the line $2x - 3y = 5$, as shown in Figure P.32.

Solution

Begin by finding the slope of the given line.

$$2x - 3y = 5 \qquad \text{Original equation}$$
$$3y = 2x - 5$$
$$y = \frac{2}{3}x - \frac{5}{3} \qquad \text{Slope-intercept form}$$

Therefore, the given line has a slope of $m = \frac{2}{3}$. Because any line parallel to the given line must also have a slope of $\frac{2}{3}$, the required line through $(2, -1)$ has the following equation.

$$y - (-1) = \frac{2}{3}(x - 2) \qquad \text{Point-slope form}$$
$$y = \frac{2}{3}x - \frac{4}{3} - 1$$
$$y = \frac{2}{3}x - \frac{7}{3} \qquad \text{Slope-intercept form}$$

Notice the similarity between the slope-intercept form of the original equation and the slope-intercept form of the parallel equation.

Figure P.32

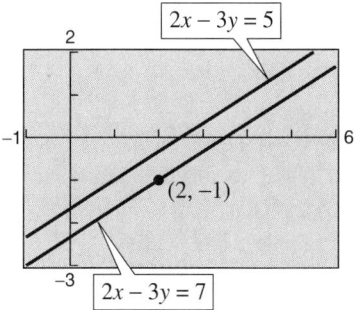

Two nonvertical lines are *perpendicular* if and only if their slopes are negative reciprocals of each other. For instance, the lines $y = 2x$ and $y = -\frac{1}{2}x$ are perpendicular because one has a slope of 2 and the other has a slope of $-\frac{1}{2}$.

Perpendicular Lines

Two nonvertical lines are **perpendicular** if and only if their slopes are negative reciprocals of each other. That is,

$$m_1 = -\frac{1}{m_2}.$$

Figure P.33

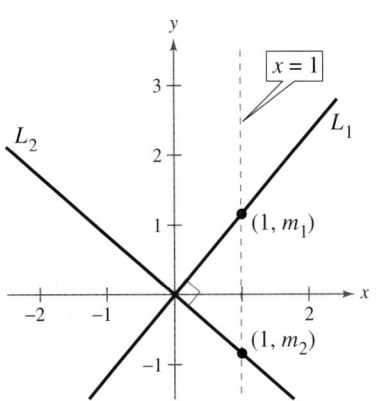

Proof /// The phrase "if and only if" is a way of stating two rules in one. One rule says "If two nonvertical lines are perpendicular, their slopes are negative reciprocals." The other rule, which is the converse, says "If two lines have slopes that are negative reciprocals, the lines must be perpendicular." The proof of the first rule is outlined as follows.

Assume you are given two nonvertical perpendicular lines L_1 and L_2, with slopes of m_1 and m_2. For simplicity's sake, let these two lines intersect at the origin, as shown in Figure P.33. The vertical line $x = 1$ will intersect L_1 and L_2 at the points $(1, m_1)$ and $(1, m_2)$. Because L_1 and L_2 are perpendicular, the triangle formed by these two points and the origin is a right triangle. Using the Pythagorean Theorem, it follows that

$$\left(\sqrt{1 + m_1{}^2}\right)^2 + \left(\sqrt{1 + m_2{}^2}\right)^2 = \left(\sqrt{0^2 + (m_1 - m_2)^2}\right)^2.$$

By simplifying this equation, you can conclude that $m_1 = -1/m_2$.   ///

EXAMPLE 8   Equations of Perpendicular Lines

Find an equation of the line that passes through the point $(2, -1)$ and is perpendicular to the line $2x - 3y = 5$.

Figure P.34

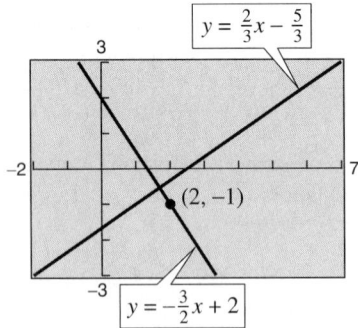

Solution

By writing the given line in the form $y = \frac{2}{3} x - \frac{5}{3}$, you can see that the line has a slope of $\frac{2}{3}$. Hence, any line that is perpendicular to this line must have a slope of $-\frac{3}{2}$ (because $-\frac{3}{2}$ is the negative reciprocal of $\frac{2}{3}$). Therefore, the required line through the point $(2, -1)$ has the following equation.

$$y - (-1) = -\frac{3}{2}(x - 2) \qquad \text{Point-slope form}$$
$$y = -\frac{3}{2}x + 3 - 1$$
$$y = -\frac{3}{2}x + 2 \qquad \text{Slope-intercept form}$$

The graphs of both equations are shown in Figure P.34.

EXAMPLE 9 **Graphs of Perpendicular Lines**

Use a graphing utility to graph the lines given by $y = x + 1$ and $y = -x + 3$. Display *both* graphs in the same viewing rectangle. The lines are supposed to be perpendicular (they have slopes of $m_1 = 1$ and $m_2 = -1$). Do they appear to be perpendicular on the display?

Solution

If the viewing rectangle is nonsquare, as in Figure P.35(a), the two lines will not appear perpendicular. If, however, the viewing rectangle is square, as in Figure P.35(b), the lines will appear perpendicular.

Figure P.35

(a)

(b)

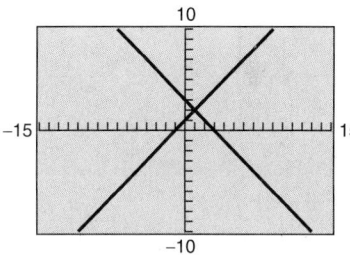

Group Activity

An Application of Slope

In 1985, a college had an enrollment of 5500 students. By 1995, the enrollment had increased to 7000 students.

a. What was the average annual change in enrollment from 1985 to 1995?

b. Use the average annual change in enrollment to estimate the enrollments in 1989, 1993, and 1997.

c. Write the equation of the line that represents the data given in part (b). What is the slope of this line? Interpret the slope in the context of the problem.

d. Discuss the concepts of *slope* and *average rate of change* in a group or in a short paragraph.

P.3 /// EXERCISES

In Exercises 1 and 2, identify the line that has the specified slope.

1. (a) $m = \frac{2}{3}$ (b) m is undefined.   (c) $m = -2$

2. (a) $m = 0$ (b) $m = -\frac{3}{4}$ (c) $m = 1$

Figure for 1   **Figure for 2**

     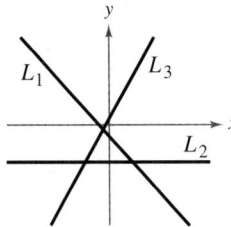

In Exercises 3–8, estimate the slope of the line.

3.   **4.**

5.   **6.**

7.   **8.**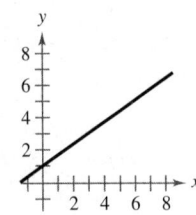

In Exercises 9 and 10, sketch the lines with the given slopes through the given point.

	Point	Slopes			
9.	$(2, 3)$	(a) 0	(b) 1	(c) 2	(d) -3
10.	$(-4, 1)$	(a) 3	(b) -3	(c) $\frac{1}{2}$	(d) undefined

In Exercises 11–16, plot the points and find the slope of the line passing through the points. Use a graphing utility to graph the line segment connecting the two points. (Use the square setting.)

11. $(-3, -2), (1, 6)$ **12.** $(2, 4), (4, -4)$

13. $(-6, -1), (-6, 4)$ **14.** $(0, -10), (-4, 0)$

15. $(1, 2), (-2, -2)$ **16.** $\left(\frac{7}{8}, \frac{3}{4}\right), \left(\frac{5}{4}, -\frac{1}{4}\right)$

In Exercises 17–22, you are given the slope of the line and a point on the line. Find three additional points through which the line passes. (The solution is not unique.)

	Point	Slope
17.	$(2, 1)$	$m = 0$
18.	$(-4, 1)$	m is undefined.
19.	$(5, -6)$	$m = 1$
20.	$(10, -6)$	$m = -1$
21.	$(-8, 1)$	m is undefined.
22.	$(-3, -1)$	$m = 0$

In Exercises 23–26, determine if the lines L_1 and L_2 passing through the pairs of points are parallel, perpendicular, or neither. Use a graphing utility to graph the line segments connecting the pairs of points on the respective lines. (Use a square setting.)

23. L_1: $(0, -1), (5, 9)$
    L_2: $(0, 3), (4, 1)$

24. L_1: $(-2, -1), (1, 5)$
    L_2: $(1, 3), (5, -5)$

25. L_1: $(3, 6), (-6, 0)$
    L_2: $(0, -1), \left(5, \frac{7}{3}\right)$

26. L_1: $(4, 8), (-4, 2)$
    L_2: $(3, -5), \left(-1, \frac{1}{3}\right)$

27. *Essay*   Write a brief paragraph explaining whether or not any pair of points on a line can be used to calculate the slope of the line.

28. *Think About It*   Is it possible for two lines with positive slopes to be perpendicular? Explain.

29. *Rate of Change*   The following are the slopes of lines representing annual sales y in terms of time x in years. Use the slopes to interpret any change in annual sales for a 1-year increase in time.

   (a) $m = 135$ (b) $m = 0$ (c) $m = -40$

30. *Rate of Change*   The following are the slopes of lines representing daily revenues y in terms of time x in days. Use the slopes to interpret any change in daily revenues for a 1-day increase in time.

   (a) $m = 400$ (b) $m = 100$ (c) $m = 0$

31. *Data Analysis*   The data gives the earnings per share of common stock for General Mills for the years 1987 through 1994. Time in years is represented by t, with $t = 0$ corresponding to 1990, and the earnings per share are represented by y.   (Source: General Mills)

$(-3, 1.25), (-2, 1.63), (-1, 2.53), (0, 2.32)$

$(1, 2.87), (2, 2.99), (3, 3.10), (4, 2.95)$

   (a) Use a graphing utility to create a line graph of the data.

   (b) Use the slope to determine the years when earnings decreased most rapidly and increased most rapidly.

32. *Data Analysis*   The data gives the declared dividend per share of common stock for the Procter and Gamble Company for the years 1987 through 1994. Time in years is represented by t, with $t = 0$ corresponding to 1990, and the dividends per share are represented by y.   (Source: Procter and Gamble)

$(-3, 0.675), (-2, 0.688), (-1, 0.750), (0, 0.875)$

$(1, 0.975), (2, 1.025), (3, 1.100), (4, 1.240)$

   (a) Use a graphing utility to create a line graph representing the data.

   (b) Use the slope to determine the year in which earnings increased most rapidly.

In Exercises 33 and 34, use a graphing utility to graph the equation using each of the suggested viewing rectangles. Describe the difference between the two views.

33. $y = 0.5x - 3$

Xmin = -5	Xmin = -2
Xmax = 10	Xmax = 10
Xscl = 1	Xscl = 1
Ymin = -1	Ymin = -4
Ymax = 10	Ymax = 1
Yscl = 1	Yscl = 1

34. $y = -8x + 5$

Xmin = -5	Xmin = -5
Xmax = 5	Xmax = 10
Xscl = 1	Xscl = 1
Ymin = -10	Ymin = -80
Ymax = 10	Ymax = 80
Yscl = 1	Yscl = 20

In Exercises 35–40, find the slope and y-intercept (if possible) of the equation of the line. Sketch a graph of the line by hand. Use a graphing utility to verify your sketch.

35. $5x - y + 3 = 0$ **36.** $2x + 3y - 9 = 0$

37. $5x - 2 = 0$ **38.** $3y + 5 = 0$

39. $7x + 6y - 30 = 0$ **40.** $x - y - 10 = 0$

In Exercises 41–48, find an equation for the line passing through the points. Use a graphing utility to sketch a graph of the line.

41. $(5, -1), (-5, 5)$ **42.** $(4, 3), (-4, -4)$

43. $\left(2, \frac{1}{2}\right), \left(\frac{1}{2}, \frac{5}{4}\right)$ **44.** $(-1, 4), (6, 4)$

45. $(-8, 1), (-8, 7)$ **46.** $(1, 1), \left(6, -\frac{2}{3}\right)$

47. $(1, 0.6), (-2, -0.6)$ **48.** $(-8, 0.6), (2, -2.4)$

49. *Mountain Driving*   When driving down a mountain road, you notice warning signs indicating that it is a "12% grade." This means that the slope of the road is $-\frac{12}{100}$. Approximate the amount of horizontal change in your position if you note from elevation markers that you have descended 2000 feet vertically.

50. *Attic Height*   The "rise to run" ratio that determines the steepness of the roof on a house is 3 to 4. Determine the maximum height in the attic if the house is 32 feet wide (see figure).

32 ft

In Exercises 51–60, find an equation of the line that passes through the given point and has the indicated slope. Sketch a graph of the line by hand. Use a graphing utility to verify your sketch.

Point	Slope
51. $(0, -2)$	$m = 3$
52. $(0, 10)$	$m = -1$
53. $(-3, 6)$	$m = -2$
54. $(0, 0)$	$m = 4$
55. $(4, 0)$	$m = -\frac{1}{3}$
56. $(-2, -5)$	$m = \frac{3}{4}$
57. $(6, -1)$	m is undefined.
58. $(-10, 4)$	$m = 0$
59. $\left(4, \frac{5}{2}\right)$	$m = \frac{4}{3}$
60. $\left(-\frac{1}{2}, \frac{3}{2}\right)$	$m = -3$

Conjecture   In Exercises 61 and 62, use the values of a and b and a graphing utility to graph the equation of the line given by

$$\frac{x}{a} + \frac{y}{b} = 1, \qquad a \neq 0, b \neq 0.$$

Use the graphs to make a conjecture about what a and b represent. Verify your conjecture.

61. $a = 5, \quad b = -3$

62. $a = -6, \quad b = 2$

In Exercises 63 and 64, use the results of Exercises 61 and 62 to write an equation of the line that passes through the points.

63. *x*-intercept: $(2, 0)$
y-intercept: $(0, 3)$

64. *x*-intercept: $\left(-\frac{1}{6}, 0\right)$
y-intercept: $\left(0, -\frac{2}{3}\right)$

In Exercises 65–68, write equations of the lines through the given point (a) parallel to the given line and (b) perpendicular to the given line.

	Point	Line
65.	$(2, 1)$	$4x - 2y = 3$
66.	$\left(\frac{7}{8}, \frac{3}{4}\right)$	$5x + 3y = 0$
67.	$(-1, 0)$	$y = -3$
68.	$(2, 5)$	$x = 4$

Graphical Analysis   In Exercises 69–72, use a graphing utility to graph the three equations in the same viewing rectangle. Adjust the viewing rectangle so that the slope appears visually correct. Identify any lines that are parallel or perpendicular.

69. $L_1: y = 2x$
$L_2: y = -2x$
$L_3: y = \frac{1}{2}x$

70. $L_1: y = \frac{2}{3}x$
$L_2: y = -\frac{3}{2}x$
$L_3: y = \frac{2}{3}x + 2$

71. $L_1: y = -\frac{1}{2}x$
$L_2: y = -\frac{1}{2}x + 3$
$L_3: y = 2x - 4$

72. $L_1: y = x - 8$
$L_2: y = x + 1$
$L_3: y = -x + 3$

Rate of Change   In Exercises 73–76, you are given the dollar value of an item in 1996 *and* the rate at which the value of the item is expected to change during the next 5 years. Use this information to write a linear equation that gives the dollar value V of the item in terms of the year t. (Let $t = 6$ represent 1996.)

	1996 Value	Rate
73.	$2540	$125 increase per year
74.	$156	$4.50 increase per year
75.	$20,400	$2000 decrease per year
76.	$245,000	$5600 decrease per year

Graphical Interpretation In Exercises 77–80, match the description with its graph. Also determine the slope and how it is interpreted in the situation. [The graphs are labeled (a), (b), (c), and (d).]

(a)

(b)

(c)

(d)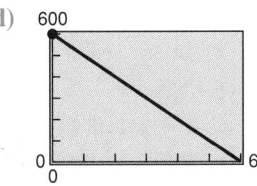

77. A person is paying $10 per week to a friend to repay a $100 loan.

78. An employee is paid $12.50 per hour plus $1.50 for each unit produced per hour.

79. A sales representative receives $20 per day for food plus $0.25 for each mile traveled.

80. A word processor that was purchased for $600 depreciates $100 per year.

In Exercises 81 and 82, find a relationship between x and y such that (x, y) is equidistant from the two points.

81. $(4, -1), (-2, 3)$ **82.** $\left(3, \frac{5}{2}\right), (-7, 1)$

83. *Temperature* Find the equation of the line giving the relationship between the temperature in degrees Celsius C and degrees Fahrenheit F. Remember that water freezes at 0°C (32°F) and boils at 100°C (212°F).

84. *Temperature* Use the result of Exercise 83 to complete the table.

C		$-10°$	$10°$			$177°$
F	$0°$			$68°$	$90°$	

85. *Annual Salary* Suppose that your salary was $28,500 in 1994 and $32,900 in 1996. If your salary follows a linear growth pattern, what will your salary be in 2000?

86. *College Enrollment* A small college had 2546 students in 1994 and 2702 students in 1996. If the enrollment follows a linear growth pattern, how many students will the college have in 2000?

87. *Straight-Line Depreciation* A small business purchases a piece of equipment for $875. After 5 years, the equipment will be outdated and have no value.

(a) Write a linear equation giving the value V of the equipment during the 5 years it will be used.

(b) Use a graphing utility to graph the linear equation representing the depreciation of the equipment, and use the trace key to complete the table.

t	0	1	2	3	4	5
V						

88. *Straight-Line Depreciation* A small business purchases a piece of equipment for $25,000. After 10 years, the equipment will have to be replaced. Its value at that time is expected to be $2000.

(a) Write a linear equation giving the value V of the equipment during the 10 years it will be used.

(b) Use a graphing utility to graph the linear equation representing the depreciation of the equipment and find when $V = \$13,000$.

89. *Sales Price and List Price* A store is offering a 15% discount on all items. Write a linear equation giving the sale price S for an item with a list price L.

90. *Hourly Wages* A manufacturer pays its assembly line workers $11.50 per hour. In addition, workers receive a piecework rate of $0.75 per unit produced. Write a linear equation for the hourly wages W in terms of the number of units x produced per hour.

91. *Contracting Purchase* A contractor purchases a piece of equipment for $36,500. The equipment requires an average expenditure of $5.25 per hour for fuel and maintenance, and the operator is paid $11.50 per hour.

  (a) Write a linear equation giving the total cost C of operating this equipment for t hours. (Include the purchase cost of the equipment.)

  (b) Assuming that customers are charged $27 per hour of machine use, write an equation for the revenue R derived from t hours of use.

  (c) Use the formula $(P = R - C)$ to write an equation for the profit derived from t hours of use.

  (d) *Break-Even Point* Use the result of part (c) to find the number of hours this equipment must be used to yield a profit of 0 dollars.

92. *Perimeter* The length and width of a rectangular garden are 15 meters and 10 meters, respectively. A walkway of width x surrounds the garden.

  (a) Write the outside perimeter y of the walkway in terms of x.

  (b) Use a graphing utility to graph the equation for the perimeter.

  (c) Determine the slope of the graph in part (b). For each additional 1-meter increase in the width of the walkway, determine the increase in its outside perimeter.

93. *Data Analysis* The average annual salaries y of major league baseball players (in thousands of dollars) from 1983 to 1992 are given as ordered pairs (t, y), where t is the time in years, with $t = 0$ corresponding to 1980. The ordered pairs are (3, 289), (4, 329), (5, 371), (6, 413), (7, 412), (8, 439), (9, 497), (10, 598), (11, 851), and (12, 1029). (Source: Major League Baseball Players Association)

  (a) Use the regression capabilities of a graphing utility to find the least squares regression line.

  (b) Use a graphing utility to plot the points and graph the regression line in the same viewing rectangle.

  (c) Use the regression line to predict the average salary in the year 2000.

  (d) Interpret the meaning of the slope of the regression line.

94. *Data Analysis* An instructor gives regular 20-point quizzes and 100-point exams in a mathematics course. Average scores for six students, given as ordered pairs (x, y) where x is the average quiz score and y is the average exam score, are (18, 87), (10, 55), (19, 96), (16, 79), (13, 76), and (15, 82).

  (a) Use the regression capabilities of a graphing utility to find the least squares regression line.

  (b) Use a graphing utility to plot the points and graph the regression line in the same viewing rectangle.

  (c) Use the regression line to predict the average test score for a student whose average quiz score is 17.

  (d) Interpret the meaning of the slope of the regression line.

  (e) If the instructor added 4 points to the average test score of everyone in the class, describe the changes in the positions of the plotted points and the change in the equation of the line.

95. *Real Estate Purchase* A real estate office handles an apartment complex with 50 units. When the rent per unit is $580 per month, all 50 units are occupied. However, when the rent is $625 per month, the average number of occupied units drops to 47. Assume that the relationship between the monthly rent p and the demand x is linear.

  (a) Write the equation of the line giving the demand x in terms of the rent p.

  (b) Use a graphing utility to graph the demand equation and use the trace feature to predict the number of units occupied if the rent is raised to $655.

  (c) Use the demand equation to predict the number of units occupied if the rent is lowered to $595. Verify graphically.

96. *Simple Interest* An inheritance of $12,000 is invested in two different mutual funds. One fund pays $5\frac{1}{2}\%$ simple interest and the other pays 8% simple interest.

  (a) If x dollars is invested in the fund paying $5\frac{1}{2}\%$, how much is invested in the fund paying 8%?

  (b) Write the annual interest y in terms of x.

  (c) Use a graphing utility to graph the function in part (b) over the interval $0 \le x \le 12,000$.

  (d) Explain why the slope of the line in part (c) is negative.

P.4 Solving Equations Algebraically and Graphically

Equations and Solutions of Equations / *Intercepts and Solutions* / *Finding Solutions Graphically* / *Points of Intersection of Two Graphs* / *Polynomial Equations* / *Other Types of Equations*

Equations and Solutions of Equations

An **equation** is a statement that two algebraic expressions are equal. To **solve** an equation in x means to find all values of x for which the equation is true. Such values are **solutions.** For instance, $x = 4$ is a solution of the equation $3x - 5 = 7$, because $3(4) - 5 = 7$ is a true statement.

The solutions of an equation depend upon the kinds of numbers being considered. For instance, in the set of rational numbers, $x^2 = 10$ has no solution because there is no rational number whose square is 10. However, in the set of real numbers the equation has the two solutions $\sqrt{10}$ and $-\sqrt{10}$.

An equation that is true for *every* real number in the domain of the variable is called an **identity.** For example, $x^2 - 9 = (x + 3)(x - 3)$ is an identity because it is a true statement for any real value of x.

An equation that is true for just *some* (or even none) of the real numbers in the domain of the variable is called a **conditional equation.** For example, the equation $x^2 - 9 = 0$ is conditional because $x = 3$ and $x = -3$ are the only values in the domain that satisfy the equation. A **linear equation** in one variable x is an equation that can be written in the standard form $ax + b = 0$, where a and b are real numbers with $a \neq 0$.

An ancient Egyptian papyrus, discovered in 1858, contains one of the earliest examples of mathematical writing in existence. The papyrus itself dates back to around 1650 B.C., but it is actually a copy of writings from two centuries earlier. The algebraic equations on the papyrus were written in words. Diophantus, a Greek who lived around A.D. 250, is often called the Father of Algebra. He was the first to use abbreviated word forms in equations.

EXPLORATION

Use a graphing utility to graph the equation $y = 3x - 6$. Use the result to estimate the x-intercept of the graph. Explain how the x-intercept is related to the solution of the equation $3x - 6 = 0$, as shown in Example 1.

EXAMPLE 1 **Solving a Linear Equation**

Solve $3x - 6 = 0$.

Solution

$3x - 6 = 0$	Original equation
$3x = 6$	Add 6 to both sides.
$x = 2$	Divide both sides by 3.

Check: After solving an equation, you should **check each solution** in the *original* equation.

$3x - 6 = 0$	Original equation
$3(2) - 6 \stackrel{?}{=} 0$	Substitute 2 for x.
$0 = 0$	Solution checks. ✔

To solve an equation involving fractional expressions, find the least common denominator of all terms in the equation and multiply every term by this LCD. This procedure clears the equation of fractions.

EXAMPLE 2 ◢ **Solving an Equation Involving Fractions**

$$\frac{x}{3} + \frac{3x}{4} = 2 \qquad \text{Original equation}$$

$$(12)\frac{x}{3} + (12)\frac{3x}{4} = (12)2 \qquad \text{Multiply by the LCD.}$$

$$4x + 9x = 24 \qquad \text{Reduce and multiply.}$$

$$13x = 24 \qquad \text{Combine like terms.}$$

$$x = \frac{24}{13} \qquad \text{Divide both sides by 13.}$$

The solution is $\frac{24}{13}$. Check this in the original equation. ◢

Study Tip

Students sometimes tell us that a solution looks easy when we work it out in class, but that they don't see where to begin when trying it alone. Keep in mind that no one—not even great mathematicians—can expect to look at every mathematical problem and know immediately where to begin. Many problems involve some trial and error before a solution is found. To make algebra work for you, put in a lot of time, expect to try solution methods that end up not working, and learn from both your successes and your failures.

When multiplying or dividing an equation by a *variable* expression it is possible to introduce an **extraneous** solution—one that does not satisfy the original equation. The next example demonstrates the importance of checking your solution when you have multiplied or divided by a variable expression.

EXAMPLE 3 ◢ **An Equation with an Extraneous Solution**

Solve the equation for x.

$$\frac{1}{x-2} = \frac{3}{x+2} - \frac{6x}{x^2-4}$$

Solution
In this case, the LCD is $x^2 - 4 = (x+2)(x-2)$. Multiplying every term by this LCD and reducing produces the following.

$$\frac{1}{x-2}(x+2)(x-2) = \frac{3}{x+2}(x+2)(x-2) - \frac{6x}{x^2-4}(x+2)(x-2)$$

$$x + 2 = 3(x-2) - 6x, \quad x \neq \pm 2$$

$$x + 2 = 3x - 6 - 6x$$

$$4x = -8$$

$$x = -2$$

A check of $x = -2$ in the original equation shows that it yields a denominator of zero. Thus, $x = -2$ is extraneous, and the equation has *no solution*. ◢

Intercepts and Solutions

In Section P.2, you learned that the **intercepts** of a graph are the points at which the graph intersects the *x*- or *y*-axis.

1. The point $(a, 0)$ is called an ***x*-intercept** of the graph of an equation if it is a solution point of the equation. To find the *x*-intercepts, let *y* be zero and solve the equation for *x*.
2. The point $(0, b)$ is called a ***y*-intercept** of the graph of an equation if it is a solution point of the equation. To find the *y*-intercepts, let *x* be zero and solve the equation for *y*.

It is possible that a particular graph will have no intercepts or several intercepts. For instance, consider the three graphs in Figure P.36.

Figure P.36

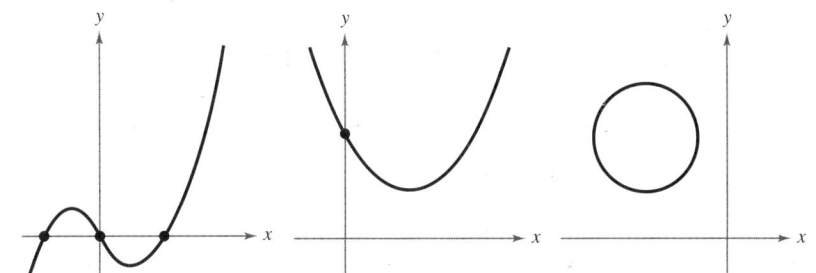

(a) Three *x*-Intercepts
One *y*-Intercept

(b) No *x*-Intercepts
One *y*-Intercept

(c) No Intercepts

EXAMPLE 4 **Finding *x*- and *y*-Intercepts**

Find the *x*- and *y*-intercepts of the graph of $2x + 3y = 5$.

Solution

To find the *x*-intercept, let $y = 0$. This produces

$$2x = 5 \quad \Longrightarrow \quad x = \frac{5}{2}$$

which implies that the graph has one *x*-intercept: $\left(\frac{5}{2}, 0\right)$. To find the *y*-intercept, let $x = 0$. This produces

$$3y = 5 \quad \Longrightarrow \quad y = \frac{5}{3}$$

which implies that the graph has one *y*-intercept: $\left(0, \frac{5}{3}\right)$. See Figure P.37.

Figure P.37

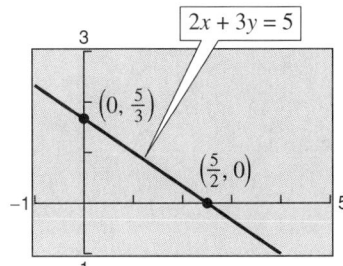

EXPLORATION

Use a graphing utility to graph each function. How many times does each of the graphs intersect the *x*-axis and how many times does each intersect the *y*-axis?

$$y_1 = x^4 - 5x^2 + 4$$
$$y_2 = 3 + x^2$$
$$y_3 = x^2 - 3$$
$$y_4 = 2 \pm \sqrt{4x - x^2 - 3}$$

In general, how many times can a function's graph intersect the *x*-axis? How many times can it intersect the *y*-axis?

Finding Solutions Graphically

Polynomial equations of degree 1 or 2 can be solved in relatively straightforward ways. Polynomial equations of higher degrees can, however, be quite difficult, especially if you rely only on algebraic techniques. For such equations, a graphing utility can be very helpful.

Note In Chapter 2 you will learn techniques for determining the number of solutions of a polynomial equation. For now, you should know that a polynomial equation of degree n cannot have more than n different solutions.

Graphical Approximations of Solutions of an Equation

1. Write the equation in *standard form,* $f(x) = 0$, with the nonzero terms on one side of the equation and zero on the other side.
2. Use a graphing utility to graph the function $y = f(x)$. Be sure the viewing rectangle shows all the relevant features of the graph.
3. Use the zoom and trace features of the graphing utility to approximate each of the x-intercepts of the graph of f. Remember that a graph can have more than one x-intercept, so you may need to change the viewing rectangle a few times.

EXAMPLE 5   **Finding Solutions of an Equation Graphically**

Use a graphing utility to approximate the solutions of $2x^3 - 3x + 2 = 0$.

Solution
Begin by graphing the function $y = 2x^3 - 3x + 2$, as shown in Figure P.38(a). You can see from the graph that there is only one x-intercept. It lies between -1 and -2 and is approximately -1.5. By using the zoom and trace features of a graphing utility you can improve the approximation, as shown in the graph in Figure P.38(b). To three-decimal-place accuracy, the solution is $x \approx -1.476$. Check this approximation on your calculator. You will find that the value of y is $y = 2(-1.476)^3 - 3(-1.476) + 2 \approx -0.003$.

EXPLORATION

Use a graphing utility to graph

$$y = 24x^3 - 36x + 17.$$

Describe a viewing rectangle that allows you to determine the number of real solutions of the equation

$$24x^3 - 36x + 17 = 0.$$

Use the same technique to determine the number of real solutions of

$$97x^3 - 102x^2 - 200x - 63 = 0.$$

Figure P.38

(a)

(b)

Some graphing utilities have built-in programs that will approximate solutions of equations or approximate x-intercepts of graphs. If your graphing utility has such features, try using them to approximate the solutions in Example 6.

Here are some suggestions for using the *zoom-in* feature of a graphing utility.

1. With each successive zoom-in, adjust the x-scale (if necessary) such that the resulting viewing rectangle shows at least the two scale marks between which the solution lies.
2. The accuracy of the approximation will always be such that the error is less than the distance between two scale marks.
3. If you have a *trace* feature on your graphing utility, you can generally add one more decimal place of accuracy without changing the viewing rectangle.

Unless stated otherwise, this book will approximate all real solutions with an error of *at most* 0.01.

EXAMPLE 6   ▱   Rewriting in Standard Form First

Use a graphing utility to approximate the solutions of $x^2 + 3 = 5x$.

Solution

In standard form, this equation is

$$x^2 - 5x + 3 = 0. \qquad \text{Equation in standard form}$$

Thus, you can begin by graphing

$$y = x^2 - 5x + 3 \qquad \text{Function to be graphed}$$

as shown in Figure P.39(a). This graph has two x-intercepts, and by using the zoom and trace features you can approximate the corresponding solutions to be $x \approx 0.70$ and $x \approx 4.30$, as shown in Figures P.39(b) and P.39(c).

Figure P.39

(a)

(b)

(c)

Figure P.40

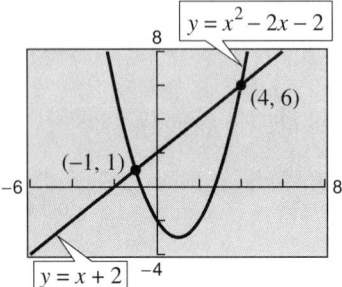

Points of Intersection of Two Graphs

An ordered pair that is a solution of two different equations is called a **point of intersection** of the graphs of the two equations. For instance, in Figure P.40, you can see that the graphs of the following equations have two points of intersection.

$$y = x + 2 \qquad \text{Equation 1}$$
$$y = x^2 - 2x - 2 \qquad \text{Equation 2}$$

The point $(-1, 1)$ is a solution of both equations, and the point $(4, 6)$ is a solution of both equations. To check this algebraically, substitute -1 and 4 into each equation.

Check that $(-1, 1)$ *is a solution.*

Equation 1: $y = -1 + 2 = 1$
Equation 2: $y = (-1)^2 - 2(-1) - 2 = 1$ ✓

Check that $(4, 6)$ *is a solution.*

Equation 1: $y = 4 + 2 = 6$
Equation 2: $y = (4)^2 - 2(4) - 2 = 6$ ✓

The following table generated on a *TI-82* shows some points of the graphs of the equations at the right. Explain how the table can be used to find the points of intersection of the graphs.

X	Y₁	Y₂
-2	0	6
-1	1	1
0	2	-2
1	3	-3
2	4	-2
3	5	1
4	6	6
X = -2		

To find the points of intersection of the graphs of two equations, solve each equation for y (or x) and set the two results equal to each other. The resulting equation will be an equation in one variable, which can be solved using standard procedures, as shown in Example 7.

EXAMPLE 7 ▭ **Finding Points of Intersection**

Find the points of intersection of the graphs of $2x - 3y = -2$ and $4x - y = 6$.

Solution
To begin, solve each equation for y to obtain $y = \frac{2}{3}x + \frac{2}{3}$ and $y = 4x - 6$. Next, set the two expressions for y equal to each other and solve the resulting equation for x, as follows.

$$\frac{2}{3}x + \frac{2}{3} = 4x - 6$$
$$2x + 2 = 12x - 18$$
$$-10x = -20$$
$$x = 2$$

When $x = 2$, the y-value of each of the given equations is 2. Thus, the graphs have one point of intersection, $(2, 2)$, as shown in Figure P.41. ▭

Figure P.41

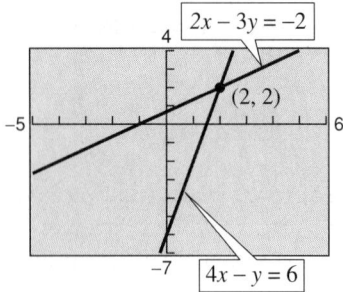

Another way to approximate points of intersection of two graphs is to graph both equations and use the zoom and trace features (or the intersect feature) to find the point or points at which the two graphs intersect.

EXAMPLE 8 **Approximating Points of Intersection**

Approximate the point(s) of intersection of the graphs of each of the following equations.

$$y = x^2 - 3x - 4 \qquad \text{Equation 1 (quadratic function)}$$
$$y = x^3 + 3x^2 - 2x - 1 \qquad \text{Equation 2 (cubic function)}$$

Solution

Begin by using a graphing utility to graph both functions, as shown in Figure P.42. From the display, you can see that the two graphs have only one point of intersection. Then, using the zoom and trace features, approximate the point of intersection to be $(-2.17, 7.25)$. To test the reasonableness of this approximation, you can evaluate both functions at $x = -2.17$.

Quadratic Function:

$$y = (-2.17)^2 - 3(-2.17) - 4$$
$$\approx 7.22$$

Cubic Function:

$$y = (-2.17)^3 + 3(-2.17)^2 - 2(-2.17) - 1$$
$$\approx 7.25$$

Because both functions yield approximately the same y-value, you can conclude that the approximate coordinates of the point of intersection are $x \approx -2.17$ and $y \approx 7.25$.

Figure P.42

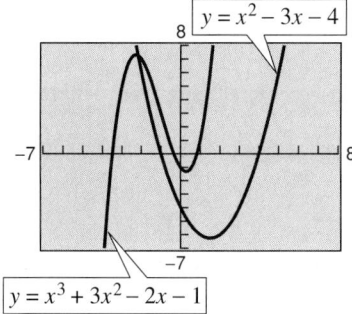

The method shown in Example 8 gives a nice graphical picture of points of intersection of two graphs. However, for actual approximation purposes, it is better to use the procedure described in Example 7. That is, the point of intersection of $y = x^2 - 3x - 4$ and $y = x^3 + 3x^2 - 2x - 1$ coincides with the solution of the equation

$$x^3 + 3x^2 - 2x - 1 = x^2 - 3x - 4 \qquad \text{Equate } y\text{-values.}$$
$$x^3 + 2x^2 + x + 3 = 0. \qquad \text{Write in standard form.}$$

By graphing $y = x^3 + 2x^2 + x + 3$ on a graphing utility and using the zoom and trace features (or the root feature), you can approximate the solution of this equation to be $x \approx -2.17$. The corresponding y-value for *both* of the functions given in Example 8 is $y \approx 7.25$.

Some graphing utilities have built-in programs that approximate the points of intersection of two graphs. If your graphing utility has such a feature, try using it to approximate the result in Example 8.

Polynomial Equations

Polynomial equations can be classified by their degree.

Degree	Name	Example
First degree	Linear equation	$6x + 2 = 4$
Second degree	Quadratic equation	$2x^2 - 5x + 3 = 0$
Third degree	Cubic equation	$x^3 - x = 0$
Fourth degree	Quartic equation	$x^4 - 3x^2 + 2 = 0$
Fifth degree	Quintic equation	$x^5 - 12x^2 + 7x + 4 = 0$

In general, the higher the degree, the more difficult it is to solve the equation—either algebraically or graphically.

You should be familiar with the following four methods for solving quadratic equations *algebraically.*

Solving a Quadratic Equation

Method

Factoring: If $ab = 0$, then $a = 0$ or $b = 0$.

Square Root Principle: If $u^2 = c$, where $c > 0$, then $u = \pm\sqrt{c}$.

Completing the Square: If $x^2 + bx = c$, then

$$x^2 + bx + \left(\frac{b}{2}\right)^2 = c + \left(\frac{b}{2}\right)^2$$

$$\left(x + \frac{b}{2}\right)^2 = c + \frac{b^2}{4}.$$

Quadratic Formula: If $ax^2 + bx + c = 0$, then

$$x = \frac{-b \pm \sqrt{b^2 - 4ac}}{2a}.$$

Example

$$x^2 - x - 6 = 0$$
$$(x - 3)(x + 2) = 0$$
$$x - 3 = 0 \implies x = 3$$
$$x + 2 = 0 \implies x = -2$$

$$(x + 3)^2 = 16$$
$$x + 3 = \pm 4$$
$$x = -3 \pm 4$$
$$x = 1 \quad \text{or} \quad x = -7$$

$$x^2 + 6x = 5$$
$$x^2 + 6x + 3^2 = 5 + 3^2$$
$$(x + 3)^2 = 14$$
$$x + 3 = \pm\sqrt{14}$$
$$x = -3 \pm \sqrt{14}$$

$$2x^2 + 3x - 1 = 0$$
$$x = \frac{-3 \pm \sqrt{3^2 - 4(2)(-1)}}{2(2)}$$
$$= \frac{-3 \pm \sqrt{17}}{4}$$

The methods used to solve quadratic equations can sometimes be extended to polynomial equations of higher degree, as shown in the next two examples.

EXAMPLE 9   ◢   **Solving an Equation of Quadratic Type**

Solve the equation

$$x^4 - 3x^2 + 2 = 0.$$

Solution

$$
\begin{aligned}
x^4 - 3x^2 + 2 &= 0 && \text{Original equation} \\
(x^2)^2 - 3(x^2) + 2 &= 0 && \text{Quadratic form in } x^2 \\
(x^2 - 1)(x^2 - 2) &= 0 && \text{Partially factor.} \\
(x + 1)(x - 1)(x^2 - 2) &= 0 && \text{Factor.}
\end{aligned}
$$

$$x + 1 = 0 \implies x = -1$$
$$x - 1 = 0 \implies x = 1$$
$$x^2 - 2 = 0 \implies x = \pm\sqrt{2}$$

The equation has four solutions: -1, 1, $\sqrt{2}$, and $-\sqrt{2}$. Check these solutions in the original equation. The graph of

$$y = x^4 - 3x^2 + 2$$

as shown in Figure P.43, verifies the solutions graphically.   ◢

Figure P.43

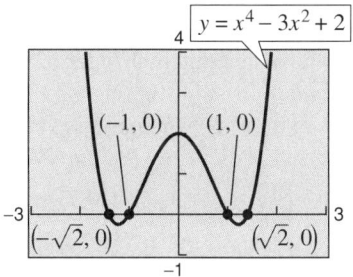

EXAMPLE 10   ◢   **Solving a Polynomial Equation by Factoring**

Solve the equation

$$x^3 - 3x^2 - 3x + 9 = 0.$$

Solution

$$
\begin{aligned}
x^3 - 3x^2 - 3x + 9 &= 0 && \text{Original equation} \\
x^2(x - 3) - 3(x - 3) &= 0 && \text{Group terms.} \\
(x - 3)(x^2 - 3) &= 0 && \text{Factor by grouping.}
\end{aligned}
$$

$$x - 3 = 0 \implies x = 3$$
$$x^2 - 3 = 0 \implies x = \pm\sqrt{3}$$

The equation has three solutions: 3, $\sqrt{3}$, and $-\sqrt{3}$. Check these solutions in the original equation. The graph of

$$y = x^3 - 3x^2 - 3x + 9$$

as shown in Figure P.44, verifies the solutions graphically.   ◢

Figure P.44

Figure P.45

(a)

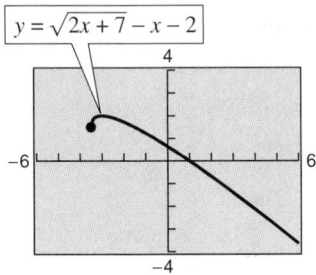

$y = \sqrt{2x + 7} - x - 2$

(b)

$(1, 0)$

X = .99989   Y = 7.0922E–5

Other Types of Equations

An equation involving a radical expression can often be cleared of radicals by raising both sides of the equation to an appropriate power. When using this procedure, remember that it can introduce extraneous solutions—so be sure to check each solution in the original equation.

EXAMPLE 11   ▱   **Solving an Equation Involving a Radical**

Solve the equation $\sqrt{2x + 7} - x = 2$.

Solution
The graph of $y = \sqrt{2x + 7} - x - 2$ is shown in Figure P.45(a). Notice that the domain is $x \geq -\frac{7}{2}$, because the expression under the radical cannot be negative. Furthermore, there seems to be one solution near $x = 1$. Using the zoom and trace features shown in Figure P.45(b), you can verify that $x = 1$ is the only solution. The equation can also be solved algebraically. To do this, first isolate the radical. Then eliminate the square root by squaring both sides of the equation. You will obtain two solutions: -3 and 1. By substituting into the original equation, you can determine that -3 is extraneous whereas 1 is valid. Thus, the equation has only one real solution: $x = 1$.   ▱

EXAMPLE 12   ▱   **Solving an Equation Involving Two Radicals**

Solve the equation

$$\sqrt{2x + 6} - \sqrt{x + 4} = 1.$$

Solution

Figure P.46

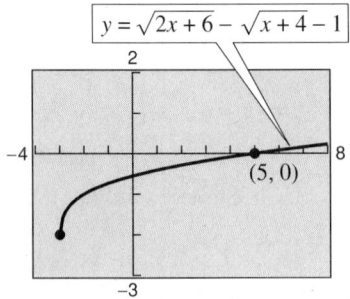

$y = \sqrt{2x + 6} - \sqrt{x + 4} - 1$

$(5, 0)$

$\sqrt{2x + 6} - \sqrt{x + 4} = 1$	Original equation
$\sqrt{2x + 6} = 1 + \sqrt{x + 4}$	Isolate radical.
$2x + 6 = 1 + 2\sqrt{x + 4} + (x + 4)$	Square both sides.
$x + 1 = 2\sqrt{x + 4}$	Isolate radical.
$x^2 + 2x + 1 = 4(x + 4)$	Square both sides.
$x^2 - 2x - 15 = 0$	Standard form
$(x - 5)(x + 3) = 0$	Factor.
$x - 5 = 0 \implies x = 5$	Set 1st factor equal to 0.
$x + 3 = 0 \implies x = -3$	Set 2nd factor equal to 0.

From Figure P.46, you can conclude that $x = 5$ is the only solution.   ▱

As demonstrated in Example 2, you can algebraically solve an equation involving fractions by multiplying both sides of the equation by the least common denominator of each term in the equation. This procedure will "clear the equation of fractions." For instance, in the equation

$$\frac{2}{x^2 + 1} + \frac{1}{x} = \frac{2}{x}$$

you can multiply both sides of the equation by the LCD $x(x^2 + 1)$ to obtain

$$(2x) + (x^2 + 1) = 2(x^2 + 1).$$

Try solving this equation. You should obtain one solution: $x = 1$.

EXPLORATION

Using dot mode, graph the two functions

$$y_1 = \frac{2}{x}$$

$$y_2 = \frac{3}{x - 2} - 1$$

in the same viewing rectangle. How many times do the graphs of the functions intersect each other? What does this tell you about the solution to Example 13?

EXAMPLE 13   **Solving an Equation Involving Fractions**

Solve the equation

$$\frac{2}{x} = \frac{3}{x - 2} - 1.$$

Solution
For this equation, the least common denominator of the three terms is $x(x - 2)$, so you can begin by multiplying each term in the equation by this expression.

$$\frac{2}{x} = \frac{3}{x - 2} - 1$$

$$x(x - 2)\frac{2}{x} = x(x - 2)\frac{3}{x - 2} - x(x - 2)(1)$$

$$2(x - 2) = 3x - x(x - 2), \qquad x \neq 0, 2$$

$$2x - 4 = -x^2 + 5x$$

$$x^2 - 3x - 4 = 0$$

$$(x - 4)(x + 1) = 0$$

$$x - 4 = 0 \quad \Longrightarrow \quad x = 4$$

$$x + 1 = 0 \quad \Longrightarrow \quad x = -1$$

The equation has two solutions: 4 and -1. Check these solutions in the original equation. Use a graphing utility to verify these solutions graphically.

Note Graphs of functions involving variable denominators can be tricky because of the way graphing utilities skip over points at which the denominator is zero. You will study graphs of such functions in Sections 2.6 and 2.7.

EXAMPLE 14 ▱ **Solving an Equation Involving Absolute Value**

Solve $|x^2 - 3x| = -4x + 6$.

Solution

To solve *algebraically* an equation involving an absolute value, you must consider the fact that the expression inside the absolute value sign can be positive or negative. This consideration results in *two* separate equations.

First Equation:

$$x^2 - 3x = -4x + 6 \qquad \text{Use positive expression.}$$
$$x^2 + x - 6 = 0 \qquad \text{Standard form}$$
$$(x + 3)(x - 2) = 0 \qquad \text{Factor.}$$
$$x + 3 = 0 \implies x = -3 \qquad \text{Set 1st factor equal to 0.}$$
$$x - 2 = 0 \implies x = 2 \qquad \text{Set 2nd factor equal to 0.}$$

Second Equation:

$$-(x^2 - 3x) = -4x + 6 \qquad \text{Use negative expression.}$$
$$x^2 - 7x + 6 = 0 \qquad \text{Standard form}$$
$$(x - 1)(x - 6) = 0 \qquad \text{Factor.}$$
$$x - 1 = 0 \implies x = 1 \qquad \text{Set 1st factor equal to 0.}$$
$$x - 6 = 0 \implies x = 6 \qquad \text{Set 2nd factor equal to 0.}$$

Check:

$$|(-3)^2 - 3(-3)| = -4(-3) + 6 \qquad -3 \text{ checks. } ✔$$
$$|2^2 - 3(2)| \neq -4(2) + 6 \qquad 2 \text{ does not check.}$$
$$|1^2 - 3(1)| = -4(1) + 6 \qquad 1 \text{ checks. } ✔$$
$$|6^2 - 3(6)| \neq -4(6) + 6 \qquad 6 \text{ does not check.}$$

The solutions are -3 and 1. This is graphically verified in Figure P.47. ▱

Figure P.47

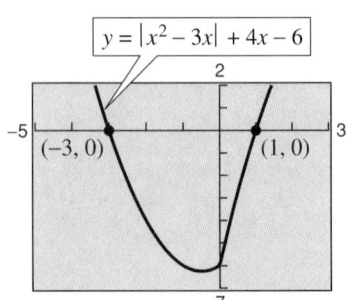

$y = |x^2 - 3x| + 4x - 6$

Group Activity   ***Comparing Strategies***

Partner Activity   With your partner, choose an equation from the exercise set. One of you should solve the equation *graphically* and the other should solve it *algebraically*. Then compare your results. Repeat the process for several equations.

P.4 /// EXERCISES

In Exercises 1–4, determine whether the given values of x are solutions of the equation.

Equation	Values	
1. $5x - 3 = 3x + 5$	(a) $x = 0$	(b) $x = -5$
	(c) $x = 4$	(d) $x = 10$
2. $3 + \dfrac{1}{x + 2} = 4$	(a) $x = -1$	(b) $x = -2$
	(c) $x = 0$	(d) $x = 5$
3. $(x + 5)(x - 3) = 20$	(a) $x = 3$	(b) $x = -2$
	(c) $x = 0$	(d) $x = -7$
4. $\sqrt[3]{x - 8} = 3$	(a) $x = 2$	(b) $x = -5$
	(c) $x = 35$	(d) $x = 8$

In Exercises 5–8, determine whether the equation is an identity or a conditional equation.

5. $2(x - 1) = 2x - 2$ **6.** $3(x + 2) = 5x + 4$

7. $3 + \dfrac{1}{x + 1} = \dfrac{4x}{x + 1}$

8. $x^2 + 2(3x - 2) = x^2 + 6x - 4$

9. *Think About It* What is meant by equivalent equations? Give an example of two equivalent equations.

10. Justify each step of the solution.

$$3(x - 4) + 10 = 7$$
$$3x - 12 + 10 = 7$$
$$3x - 2 = 7$$
$$3x - 2 + 2 = 7 + 2$$
$$3x = 9$$
$$\frac{3x}{3} = \frac{9}{3}$$
$$x = 3$$

11. Solve each equation mentally.

(a) $3x = 15$ (b) $\frac{1}{2}t = 7$

(c) $s + 12 = 18$ (d) $2u - 3 = 25$

12. Solve each equation in two ways. Then explain which way was easier for you.

(a) $3(x - 1) = 4$ (b) $\frac{3}{4}(z - 4) = 6$

In Exercises 13–24, solve the equation and use a graphing utility to verify your solution.

13. $8x - 5 = 3x + 10$ **14.** $7x + 3 = 3x - 13$

15. $2(x + 5) - 7 = 3(x - 2)$

16. $2(13t - 15) + 3(t - 19) = 0$

17. $6[x - (2x + 3)] = 8 - 5x$

18. $3(x + 3) = 5(1 - x) - 1$

19. $\dfrac{5x}{4} + \dfrac{1}{2} = x - \dfrac{1}{2}$ **20.** $\dfrac{x}{5} - \dfrac{x}{2} = 3$

21. $\frac{3}{2}(z + 5) - \frac{1}{4}(z + 24) = 0$

22. $\dfrac{3x}{2} + \dfrac{1}{4}(x - 2) = 10$

23. $0.25x + 0.75(10 - x) = 3$

24. $0.60x + 0.40(100 - x) = 50$

In Exercises 25–40, solve the equation (if possible) and use a graphing utility to verify your solution.

25. $\dfrac{100 - 4u}{3} = \dfrac{5u + 6}{4} + 6$

26. $\dfrac{17 + y}{y} + \dfrac{32 + y}{y} = 100$

27. $\dfrac{5x - 4}{5x + 4} = \dfrac{2}{3}$ **28.** $\dfrac{15}{x} - 4 = \dfrac{6}{x} + 3$

29. $\dfrac{1}{x - 3} + \dfrac{1}{x + 3} = \dfrac{10}{x^2 - 9}$

30. $\dfrac{1}{x - 2} + \dfrac{3}{x + 3} = \dfrac{4}{x^2 + x - 6}$

31. $\dfrac{x}{x + 4} + \dfrac{4}{x + 4} + 2 = 0$

32. $\dfrac{2}{(x - 4)(x - 2)} = \dfrac{1}{x - 4} + \dfrac{2}{x - 2}$

33. $\dfrac{7}{2x + 1} - \dfrac{8x}{2x - 1} = -4$

34. $\dfrac{4}{u-1} + \dfrac{6}{3u+1} = \dfrac{15}{3u+1}$

35. $\dfrac{1}{x} + \dfrac{2}{x-5} = 0$ **36.** $\dfrac{6}{x} - \dfrac{2}{x+3} = \dfrac{3(x+5)}{x(x+3)}$

37. $\dfrac{3}{x(x-3)} + \dfrac{4}{x} = \dfrac{1}{x-3}$

38. $3 = 2 + \dfrac{2}{z+2}$

39. $(x+2)^2 + 5 = (x+3)^2$

40. $(x+1)^2 + 2(x-2) = (x+1)(x-2)$

Statistics Problems In Exercises 41 and 42, suppose you have a uniform beam of length L with a fulcrum x feet from one end (see figure). If objects with weights W_1 and W_2 are placed at opposite ends of the beam, the beam will balance if

$$W_1 x = W_2(L - x).$$

Find x such that the beam will balance.

 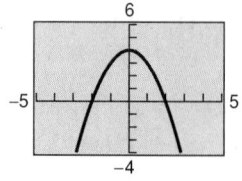

41. Two children weighing 50 pounds and 75 pounds are going to play on a seesaw that is 10 feet long.

42. A person weighing 200 pounds is attempting to move a 550-pound rock with a bar that is 5 feet long.

In Exercises 43–48, find the x- and y-intercepts of the graph of the equation.

43. $y = x - 5$ **44.** $y = 4 - x^2$

45. $y = x\sqrt{x+2}$ **46.** $xy = 4$

 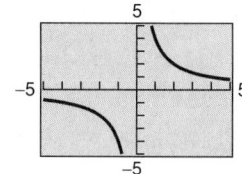

47. $y = |x-2| - 3$ **48.** $x^2 y - x^2 + 4y = 0$

In Exercises 49–54, use a graphing utility to graph the function and verify its zero(s).

Function	Zero(s)
49. $f(x) = 12 - 4x$	$x = 3$
50. $f(x) = 3(x-5) + 9$	$x = 2$
51. $f(x) = x^2 - 2.5x - 6$	$x = -1.5, \ x = 4$
52. $f(x) = x^3 - 9x^2 + 18x$	$x = 0, \ x = 3, \ x = 6$
53. $f(x) = \dfrac{x+2}{3} - \dfrac{x-1}{5} - 1$	$x = 1$
54. $f(x) = x - 3 - \dfrac{10}{x}$	$x = -2, \ x = 5$

Graphical Analysis In Exercises 55–58, use a graphing utility to graph the equation and approximate any x-intercepts. Set $y = 0$ and solve the resulting equation. Compare the results with the x-intercepts of the graph.

55. $y = 2(x-1) - 4$ **56.** $y = \frac{4}{3}x + 2$

57. $y = 20 - (3x - 10)$ **58.** $y = 10 + 2(x - 2)$

In Exercises 59–62, solve the equation algebraically. Then write the equation in the form $f(x) = 0$ and use a graphing utility to verify the algebraic solution.

59. $\dfrac{3x}{2} + \dfrac{1}{4}(x - 2) = 10$

60. $0.60x + 0.40(100 - x) = 50$

61. $3(x + 3) = 5(1 - x) - 1$

62. $(x + 1)^2 + 2(x - 2) = (x + 1)(x - 2)$

In Exercises 63–66, find any points of intersection algebraically.

63. $y = 2 - x$
$y = 2x - 1$

64. $x - y = -4$
$x^2 - y = -2$

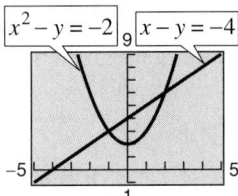

65. $y = \frac{1}{3}x + 2$
$y = \frac{5}{2}x - 11$

66. $y = 4 - x^2$
$y = 2x + 1$

In Exercises 67–74, use a graphing utility to approximate any points of intersection (accurate to three decimal places) of the graphs of the equations.

67. $2x + y = 6$
$-x + y = 0$

68. $3x + y = 2$
$x^3 + y = 0$

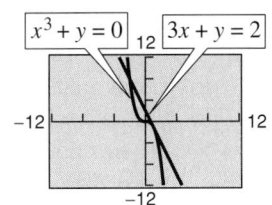

69. $y = 9 - 2x$
$y = x - 3$

70. $y = x^3 - 3$
$y = 5 - 2x$

71. $y = 8$
$y = 3x^2 + 2x$

72. $y = 32$
$y = x^5 - x^2$

73. $y = 2x^2$
$y = x^4 - 2x^2$

74. $y = -x$
$y = 2x - x^2$

In Exercises 75–78, solve the quadratic equation.

75. $x^2 - 2x - 1 = 0$

76. $11x^2 + 33x = 0$

77. $(x + 3)^2 = 81$

78. $x^2 + 3x - \frac{3}{4} = 0$

In Exercises 79–108, find all real solutions of the equation. Use a graphing utility to verify your solutions.

79. $4x^4 - 18x^2 = 0$

80. $20x^3 - 125x = 0$

81. $x^4 - 4x^2 + 3 = 0$

82. $4x^4 - 65x^2 + 16 = 0$

83. $\dfrac{1}{t^2} + \dfrac{8}{t} + 15 = 0$

84. $6\left(\dfrac{s}{s+1}\right)^2 + 5\left(\dfrac{s}{s+1}\right) - 6 = 0$

85. $2x + 9\sqrt{x} - 5 = 0$

86. $3x^{1/3} + 2x^{2/3} = 5$

87. $\sqrt{x - 10} - 4 = 0$

88. $\sqrt{5 - x} - 3 = 0$

89. $\sqrt[3]{2x + 5} + 3 = 0$

90. $\sqrt[3]{3x + 1} - 5 = 0$

91. $\sqrt{x + 1} - 3x = 1$

92. $\sqrt{x + 5} = \sqrt{x - 5}$

93. $\sqrt{x} - \sqrt{x - 5} = 1$

94. $\sqrt{x} + \sqrt{x - 20} = 10$

95. $(x - 5)^{2/3} = 16$

96. $(x^2 - x - 22)^{4/3} = 16$

97. $\dfrac{20 - x}{x} = x$

98. $\dfrac{4}{x} - \dfrac{5}{3} = \dfrac{x}{6}$

99. $\dfrac{1}{x} - \dfrac{1}{x + 1} = 3$

100. $\dfrac{x}{x^2 - 4} + \dfrac{1}{x + 2} = 3$

101. $x = \dfrac{3}{x} + \dfrac{1}{2}$

102. $4x + 1 = \dfrac{3}{x}$

103. $\dfrac{4}{x + 1} - \dfrac{3}{x + 2} = 1$

104. $\dfrac{x + 1}{3} - \dfrac{x + 1}{x + 2} = 0$

105. $|2x - 1| = 5$

106. $|3x + 2| = 7$

107. $|x| = x^2 + x - 3$

108. $|x - 10| = x^2 - 10x$

Graphical Analysis In Exercises 109–120, use a graphing utility to graph the equation. Use the graph to approximate any *x*-intercepts of the graph. Set $y = 0$ and solve the resulting equation. Compare the result with the *x*-intercepts of the graph.

109. $y = x^3 - 2x^2 - 3x$

110. $y = 2x^4 - 15x^3 + 18x^2$

111. $y = x^4 - 10x^2 + 9$

112. $y = x^4 - 29x^2 + 100$

113. $y = \sqrt{11x - 30} - x$

114. $y = 2x - \sqrt{15 - 4x}$

115. $y = \sqrt{7x + 36} - \sqrt{5x + 16} - 2$

116. $y = 3\sqrt{x} - \dfrac{4}{\sqrt{x}} - 4$

117. $y = \dfrac{1}{x} - \dfrac{4}{x - 1} - 1$

118. $y = x + \dfrac{9}{x + 1} - 5$

119. $y = |x + 1| - 2$ **120.** $y = |x - 2| - 3$

In Exercises 121–126, solve for the indicated variable.

121. *Area of a Triangle*

Solve for h: $A = \frac{1}{2}bh$

122. *Investment at Compound Interest*

Solve for P: $A = P\left(1 + \dfrac{r}{n}\right)^{nt}$

123. *Area of a Trapezoid*

Solve for b: $A = \frac{1}{2}(a + b)h$

124. *Geometric Progression*

Solve for r: $S = \dfrac{rL - a}{r - 1}$

125. *Surface Area of a Cone*

Solve for h: $S = \pi r\sqrt{r^2 + h^2}$

126. *Inductance*

Solve for Q: $i = \pm\sqrt{\dfrac{1}{LC}}\sqrt{Q^2 - q}$

127. *Numerical, Graphical, and Analytical Analysis* A rancher has 100 meters of fencing to enclose two adjacent rectangular corrals (see figure).

(a) Write the area of the enclosed region as a function of x.

(b) Use a graphing utility to generate additional rows of the table. Use the table to estimate the dimensions that will produce a maximum area.

x	y	Area $(2xy)$
2	$\frac{92}{3}$	$\frac{368}{3} \approx 123$
4	28	224

(c) Use a graphing utility to graph the area function. Use the graph to estimate the dimensions that will produce a maximum area.

(d) Use the graph to approximate the dimensions such that the enclosed area will be 350 square meters.

(e) Find the required dimensions of part (d) analytically.

$4x + 3y = 100$

128. *Exploration* The graph of the function $f(x) = \frac{1}{5}(x^5 - 20x^3 + 64x) + k$ for $k = 0$ is given in the figure.

(a) Determine the number of zeros of the function when $k = 0$.

(b) Use a graphing utility to graph the function for different values of k. Find a value of k such that the function has one zero. Find k so there are three distinct zeros.

(c) Is there a value of k such that the function has no zero? Explain.

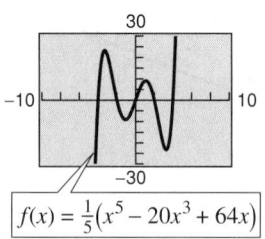

$f(x) = \frac{1}{5}(x^5 - 20x^3 + 64x)$

P.5 Solving Inequalities Algebraically and Graphically

Properties of Inequalities / *Solving a Linear Inequality* /
Inequalities Involving Absolute Value / *Polynomial Inequalities* /
Rational Inequalities / *Applications*

Properties of Inequalities

The inequality symbols $<$, $\leq$, $>$, and $\geq$ are used to compare two numbers and to denote subsets of real numbers. For instance, the simple inequality $x \geq 3$ denotes all real numbers x that are greater than or equal to 3. In this section you will study inequalities that contain more involved statements such as

$$5x - 7 > 3x + 9 \qquad \text{and} \qquad -3 \leq 6x - 1 < 3.$$

As with an equation, you **solve an inequality** in the variable x by finding all values of x for which the inequality is true. These values are **solutions** and are said to **satisfy** the inequality. For instance, the number 9 is a solution of the first inequality listed above because

$$5(9) - 7 > 3(9) + 9.$$

On the other hand, the number 7 is not a solution because

$$5(7) - 7 \not> 3(7) + 9.$$

The set of all real numbers that are solutions of an inequality is the **solution set** of the inequality.

The set of all points on the real number line that represent the solution set is the **graph** of the inequality. Graphs of many types of inequalities consist of intervals on the real number line.

The procedures for solving linear inequalities in one variable are much like those for solving linear equations. To isolate the variable you can make use of the **properties of inequalities.** These properties are similar to the properties of equality, but there are two important exceptions. When both sides of an inequality are multiplied or divided by a negative number, *the direction of the inequality symbol must be reversed.* Here is an example.

$-2 < 5$	Original inequality
$(-3)(-2) > (-3)(5)$	Multiply both sides by -3 and reverse inequality.
$6 > -15$	New inequality

Two inequalities that have the same solution set are **equivalent.** The properties listed at the top of the next page describe operations that can be used to create equivalent inequalities.

Note   Each of the properties at the right is true if the symbol < is replaced by ≤ and > is replaced by ≥. For instance, another form of Property 3 would be as follows.

$$a \leq b \quad \Longrightarrow \quad a + c \leq b + c$$

Properties of Inequalities

Let a, b, c, and d be real numbers.

1. *Transitive Property*

   $a < b$ and $b < c$ $a < c$

2. *Addition of Inequalities*

   $a < b$ and $c < d$ $a + c < b + d$

3. *Addition of a Constant*

   $a < b$ ➡ $a + c < b + c$

4. *Multiplying by a Constant*

   For $c > 0$, $a < b$ ➡ $ac < bc$

   For $c < 0$, $a < b$ ➡ $ac > bc$

Solving a Linear Inequality

The simplest type of inequality to solve is a **linear inequality** in a single variable, such as $2x + 3 > 4$.

EXAMPLE 1   **Solving a Linear Inequality**

Solve the inequality

$$5x - 7 > 3x + 9.$$

Solution

$5x - 7 > 3x + 9$	Original inequality
$5x > 3x + 16$	Add 7 to both sides.
$5x - 3x > 16$	Subtract $3x$ from both sides.
$2x > 16$	Combine like terms.
$x > 8$	Divide both sides by 2.

Thus, the solution set consists of all real numbers that are greater than 8. The interval notation for this solution set is $(8, \infty)$. The number line graph of this solution set is shown in Figure P.48.

Figure P.48   Solution interval: $(8, \infty)$

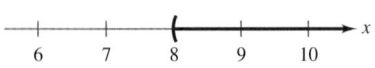

Note   The five inequalities forming the solution steps of Example 1 are all **equivalent** in the sense that each has the same solution set.

Figure P.49   Solution interval: $(-\infty, 2]$

Figure P.50

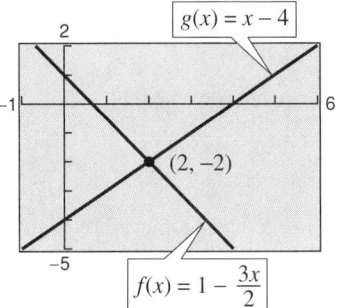

EXAMPLE 2   Solving an Inequality

Solve $1 - \dfrac{3x}{2} \geq x - 4$.

Solution

$1 - \dfrac{3x}{2} \geq x - 4$	Original inequality
$2 - 3x \geq 2x - 8$	Multiply both sides by the LCD.
$-3x \geq 2x - 10$	Subtract 2 from both sides.
$-5x \geq -10$	Subtract 2x from both sides.
$x \leq 2$	Divide both sides by -5 and reverse inequality.

The solution set consists of all real numbers that are less than or equal to 2. The interval notation for this solution set is $(-\infty, 2]$. The number line graph of this solution set is shown in Figure P.49. Graphically, you can solve the given inequality by sketching the graphs of the left and right sides

$$f(x) = 1 - \frac{3x}{2} \quad \text{and} \quad g(x) = x - 4.$$

In Figure P.50, the graphs appear to intersect at the point $(2, -2)$. This is confirmed by the fact that $f(2) = -2 = g(2)$. Moreover, the graph of f lies above the graph of g to the left of their point of intersection, which implies that $f(x) \geq g(x)$ for all $x \leq 2$.

Study Tip

Checking the solution set of an inequality is not as simple as checking the solution of an equation because there are simply too many x-values to substitute into the original inequality. However, you can get an indication of the validity of the solution set by substituting a few convenient values of x. For instance, in Example 2 try substituting $x = 0$ and $x = 4$ into the original inequality.

Sometimes it is possible to write two inequalities as a **double inequality.** For instance, you can write the two inequalities $-4 \leq 5x - 2$ and $5x - 2 < 7$ more simply as $-4 \leq 5x - 2 < 7$. This form allows you to solve the two inequalities together.

EXAMPLE 3   Solving a Double Inequality

$-3 \leq 6x - 1 < 3$	Original inequality
$-2 \leq 6x < 4$	Add 1 to all three parts.
$-\dfrac{1}{3} \leq x < \dfrac{2}{3}$	Divide by 6 and reduce.

The solution set consists of all real numbers that are greater than or equal to $-\frac{1}{3}$ and less than $\frac{2}{3}$. The interval notation for this solution set is $\left[-\frac{1}{3}, \frac{2}{3}\right)$. The number line graph of this solution set is shown in Figure P.51.

Figure P.51   Solution interval: $\left[-\frac{1}{3}, \frac{2}{3}\right)$

Inequalities Involving Absolute Value

Solving an Absolute Value Inequality

Let x be a variable of an algebraic expression and let a be a real number such that $a \geq 0$.

1. The solutions of $|x| < a$ are all values of x that lie between $-a$ and a.

$$|x| < a \qquad \text{if and only if} \qquad -a < x < a.$$

2. The solutions of $|x| > a$ are all values of x that are less than $-a$ or greater than a.

$$|x| > a \qquad \text{if and only if} \qquad x < -a \quad \text{or} \quad x > a.$$

These rules are also valid if $<$ is replaced by $\leq$ and $>$ is replaced by $\geq$.

EXAMPLE 4   **Solving an Absolute Value Inequality**

$|x - 5| < 2$	Original inequality
$-2 < x - 5 < 2$	Equivalent inequalities
$-2 + 5 < x - 5 + 5 < 2 + 5$	Add 5 to all three parts.
$3 < x < 7$	Simplify.

Figure P.52 $|x - 5| < 2$

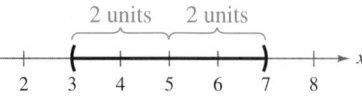

The solution set is all real numbers that are greater than 3 and less than 7. The interval notation for this solution set is $(3, 7)$. The number line graph of this solution set is shown in Figure P.52.

EXAMPLE 5   **Solving an Absolute Value Inequality Graphically**

Use a graphing utility to solve $\left| 2 - \dfrac{x}{3} \right| < 0.01$.

Figure P.53

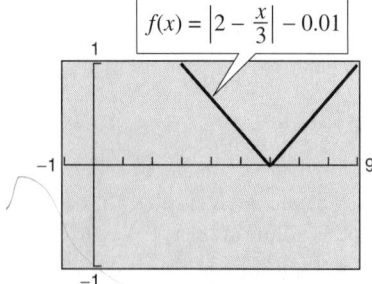

$$f(x) = \left| 2 - \frac{x}{3} \right| - 0.01$$

Solution

You can solve this inequality graphically by graphing the function

$$f(x) = \left| 2 - \frac{x}{3} \right| - 0.01$$

and determining the values of x for which the graph lies *below* the x-axis, as shown in Figure P.53. Using the zoom and trace features, you can determine that the x-intercepts occur at $x \approx 5.97$ and $x \approx 6.03$. Because the graph of f lies below the x-axis between these two values, the solution interval is $(5.97, 6.03)$. Try solving this inequality algebraically to verify the solution interval.

Polynomial Inequalities

To solve a polynomial inequality such as $x^2 - 2x - 3 < 0$, use the fact that a polynomial can change signs only at its zeros (the x-values that make the polynomial equal to zero). Between two consecutive zeros a polynomial must be entirely positive or entirely negative. This means that when the real zeros of a polynomial are put in order, they divide the real number line into intervals in which the polynomial has no sign changes. These zeros are the **critical numbers** of the inequality, and the resulting intervals are the **test intervals** for the inequality. For instance, the polynomial

$$x^2 - 2x - 3 = (x + 1)(x - 3)$$

has two zeros, $x = -1$ and $x = 3$, which divide the real number line into three test intervals: $(-\infty, -1), (-1, 3),$ and $(3, \infty)$. To solve the inequality $x^2 - 2x - 3 < 0$, you only need to test one value from each test interval.

Finding Test Intervals for a Polynomial

To determine the intervals on which the values of a polynomial are entirely negative or entirely positive, use the following steps.

1. Find all real zeros of the polynomial, and arrange the zeros in increasing order (from smallest to largest). The zeros of a polynomial are its **critical numbers.**
2. Use the critical numbers of the polynomial to determine its **test intervals.**
3. Choose one representative x-value in each test interval and evaluate the polynomial at that value. If the value of the polynomial is negative, the polynomial will have negative values for *every* x-value in the interval. If the value of the polynomial is positive, the polynomial will have positive values for *every* x-value in the interval.

$y = 2x^2 + 5x > 12$

Some graphing utilities will produce graphs of inequalities. For instance, on a *TI-82* or *TI-83*, you can graph $2x^2 + 5x > 12$ (see Example 7) by entering

$$Y_1 = 2x^2 + 5x > 12$$

and pressing GRAPH. Using $-10 \leq x \leq 10$ and $-4 \leq y \leq 4$, the graph should look like that at the left. Solve the problem algebraically to verify that the solution is $(-\infty, -4) \cup \left(\frac{3}{2}, \infty\right)$.

EXAMPLE 6   **Investigating Polynomial Behavior**

Determine the intervals on which $x^2 - x - 6$ is entirely negative and those on which it is entirely positive.

Solution

By factoring the quadratic as $x^2 - x - 6 = (x + 2)(x - 3)$, you can see that the critical numbers occur at $x = -2$ and $x = 3$. Therefore, the test intervals for the quadratic are $(-\infty, -2)$, $(-2, 3)$, and $(3, \infty)$. In each test interval, choose a representative x-value and evaluate the polynomial, as shown in the table.

Figure P.54

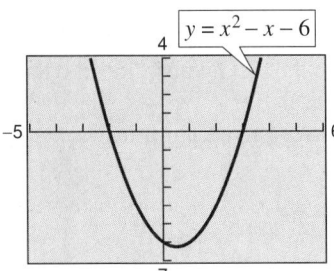

Interval	x-Value	Value of Polynomial	Sign of Polynomial
$(-\infty, -2)$	$x = -3$	$(-3)^2 - (-3) - 6 = 6$	Positive
$(-2, 3)$	$x = 0$	$(0)^2 - (0) - 6 = -6$	Negative
$(3, \infty)$	$x = 4$	$(4)^2 - (4) - 6 = 6$	Positive

The polynomial has positive values for every x in the intervals $(-\infty, -2)$ and $(3, \infty)$ and negative values for every x in the interval $(-2, 3)$. This result is shown graphically in Figure P.54.

Note   To determine the test intervals for a polynomial inequality, the inequality must first be written in standard form with the polynomial on one side and zero on the other side.

EXAMPLE 7   **Solving a Polynomial Inequality**

Solve $2x^2 + 5x > 12$.

Solution

$$2x^2 + 5x > 12 \qquad \text{Original inequality}$$
$$2x^2 + 5x - 12 > 0 \qquad \text{Standard form}$$
$$(x + 4)(2x - 3) > 0 \qquad \text{Factor.}$$

Critical Numbers: $x = -4, x = \frac{3}{2}$

Test Intervals: $(-\infty, -4), \left(-4, \frac{3}{2}\right), \left(\frac{3}{2}, \infty\right)$

Test: Is $(x + 4)(2x - 3) > 0$?

Figure P.55

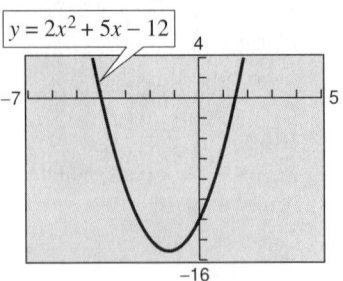

After testing these intervals, you can see that the polynomial $2x^2 + 5x - 12$ is positive in the open intervals $(-\infty, -4)$ and $\left(\frac{3}{2}, \infty\right)$. Therefore, the solution set of the inequality is $(-\infty, -4) \cup \left(\frac{3}{2}, \infty\right)$. You could have solved this example with a graphing utility by graphing the polynomial function $y = 2x^2 + 5x - 12$, and noting where the graph is *above* the x-axis. You can see in Figure P.55 that the polynomial lies above the x-axis for $x < -4$ or $x > \frac{3}{2}$.

EXAMPLE 8 **Unusual Solution Sets**

a. The solution set of $x^2 + 2x + 4 > 0$ consists of the entire set of real numbers, $(-\infty, \infty)$. In other words, the quadratic $x^2 + 2x + 4$ is positive for every real value of x, as indicated in Figure P.56(a). (Note that this quadratic inequality has *no* critical numbers. In such a case, there is only one test interval—the entire real number line.)

b. The solution set of $x^2 + 2x + 1 \le 0$ consists of the single real number -1, because the graph touches the x-axis just at -1, as shown in Figure P.56(b).

c. The solution set of $x^2 + 3x + 5 < 0$ is empty. In other words, the quadratic $x^2 + 3x + 5$ is not less than zero for any value of x, as indicated in Figure P.56(c).

d. The solution set of $x^2 - 4x + 4 > 0$ consists of all real numbers *except* the number 2. In interval notation, this solution set can be written as $(-\infty, 2) \cup (2, \infty)$. The graph of $x^2 - 4x + 4$ lies above the x-axis except at $x = 2$, where it touches it, as indicated in Figure P.56(d).

Study Tip

When solving a polynomial inequality, be sure you have accounted for the particular type of inequality symbol in the inequality. For instance, an inequality containing the < or > symbols should have solution sets consisting of open intervals such as (x_1, x_2), etc. Inequalities containing $\le$ or $\ge$ should have solution sets consisting of closed intervals such as $[x_1, x_2]$, etc.

One of the advantages of technology is that you can solve complicated polynomial inequalities that might be difficult, or even impossible, to factor. For instance, explain how you could use a graphing utility to approximate the solution to the inequality

$$x^3 - 0.26x^2 - 3.1416x + 1.414 < 0.$$

Figure P.56

(a)

(b)

(c)

(d)

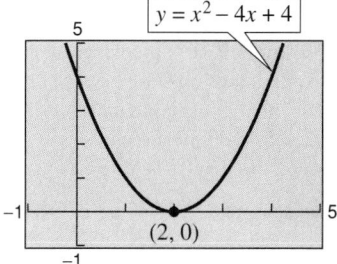

Rational Inequalities

The concepts of critical numbers and test intervals can be extended to inequalities involving rational expressions. To do this, use the fact that the value of a rational expression can change sign only at its *zeros* (the x-values for which its numerator is zero) and its *undefined values* (the x-values for which its denominator is zero). These two types of numbers make up the **critical numbers** of a rational inequality.

Figure P.57

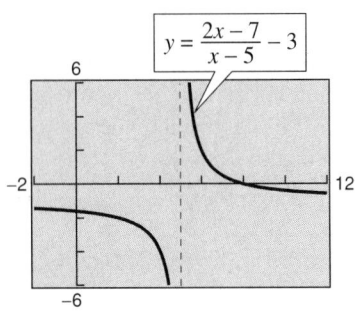

$$y = \frac{2x - 7}{x - 5} - 3$$

EXAMPLE 9   **Solving a Rational Inequality**

Solve $\dfrac{2x - 7}{x - 5} \le 3$.

Solution

In Figure P.57, you can see that the graph of

$$y = \frac{2x - 7}{x - 5} - 3$$

is less than or equal to zero on the intervals $(-\infty, 5)$ and $[8, \infty)$. To solve the inequality algebraically, use the following steps.

$$\frac{2x - 7}{x - 5} \le 3 \qquad\qquad \text{Original inequality}$$

$$\frac{2x - 7}{x - 5} - 3 \le 0 \qquad\qquad \text{Standard form}$$

$$\frac{2x - 7 - 3x + 15}{x - 5} \le 0 \qquad\qquad \text{Add fractions.}$$

$$\frac{-x + 8}{x - 5} \le 0 \qquad\qquad \text{Simplify.}$$

Now, in standard form you can see that the critical numbers are 5 and 8, and you can proceed as follows.

$$\text{Critical Numbers:} \qquad x = 5, x = 8$$

$$\text{Test Intervals:} \qquad (-\infty, 5), (5, 8), (8, \infty)$$

$$\text{Test:} \qquad \text{Is } \frac{-x + 8}{x - 5} \le 0?$$

By testing these intervals, you can determine that the rational expression $(-x + 8)/(x - 5)$ is negative in the open intervals $(-\infty, 5)$ and $(8, \infty)$. Moreover, because $(-x + 8)/(x - 5) = 0$ when $x = 8$, you can conclude that the solution set of the inequality is $(-\infty, 5) \cup [8, \infty)$.

Study Tip

In Example 9, it is incorrect to write the solution set $(-\infty, 5) \cup [8, \infty)$ as a double inequality $8 \le x < 5$. Do you see why?

Applications

EXAMPLE 10   Finding the Domain of an Expression

Find the domain of

$$\sqrt{64 - 4x^2}.$$

Figure P.58

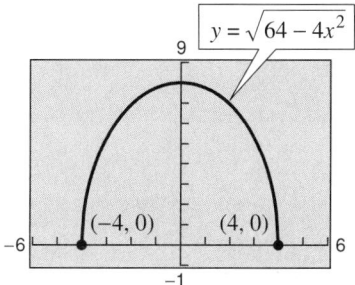

Solution
Remember that the domain of an expression is the set of all x-values for which the expression is defined (has real values). Because $\sqrt{64 - 4x^2}$ is defined only if $64 - 4x^2$ is nonnegative, the domain is given by $64 - 4x^2 \geq 0$.

$$64 - 4x^2 \geq 0 \qquad\qquad \text{Standard form}$$
$$16 - x^2 \geq 0 \qquad\qquad \text{Divide both sides by 4.}$$
$$(4 - x)(4 + x) \geq 0 \qquad\qquad \text{Factor.}$$

The inequality has two critical numbers: -4 and 4. A test shows that $64 - 4x^2 \geq 0$ in the *closed interval* $[-4, 4]$. The graph of $y = \sqrt{64 - 4x^2}$, shown in Figure P.58, confirms that the domain is $[-4, 4]$.

EXAMPLE 11   The Height of a Projectile

Real Life

A projectile is fired straight upward from ground level with an initial velocity of 384 feet per second. During what time period will its height exceed 2000 feet?

Figure P.59

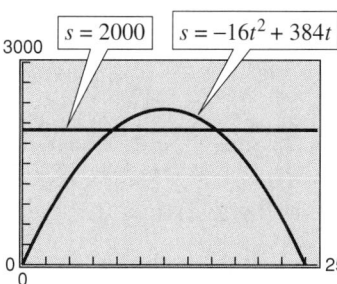

Solution
The position of an object moving vertically is given by $s = -16t^2 + v_0 t + s_0$, where s is the height in feet and t is the time in seconds. In this case, $s_0 = 0$ and $v_0 = 384$. Thus, you need to solve the inequality $-16t^2 + 384t > 2000$. Using a graphing utility, graph $s = -16t^2 + 384t$ and $s = 2000$, as shown in Figure P.59. From the graph, you can determine that $-16t^2 + 384t > 2000$ for t between 7.64 and 16.36. You can verify this result algebraically as follows.

$$-16t^2 + 384t > 2000 \qquad\qquad \text{Original inequality}$$
$$t^2 - 24t < -125 \qquad\qquad \text{Divide by } -16 \text{ and reverse inequality.}$$
$$t^2 - 24t + 125 < 0 \qquad\qquad \text{Standard form}$$

By the Quadratic Formula the critical numbers are 7.64 and 16.36. A test will verify that the height of the projectile will exceed 2000 feet during the time interval 7.64 seconds $< t <$ 16.36 seconds.

Group Activity

Communicating Mathematically

Some people find that it is easier to remember a verbal statement, a numerical example, or a picture than it is to remember a mathematical formula. For instance, you can remember the factoring formula $(u + v)(u - v) = u^2 - v^2$ as "the product of the sum and difference of two terms is the difference of each term squared."

Four different properties of inequalities are listed on page 56. For each property, (a) translate the mathematical statement into a *verbal statement*, (b) compile a list of several numerical examples that demonstrate the property, and (c) construct a number line or series of number lines that graphically illustrates the property.

P.5 /// EXERCISES

In Exercises 1–4, match the inequality with its graph. [The graphs are labeled (a), (b), (c), and (d).]

(a)

(b)

(c)

(d)

1. $x < 3$ 2. $x \geq 5$

3. $-3 < x \leq 4$ 4. $0 \leq x \leq \frac{9}{2}$

In Exercises 5–8, determine whether the given values of x are solutions of the inequality.

Inequality	Values	
5. $5x - 12 > 0$	(a) $x = 3$	(b) $x = -3$
	(c) $x = \frac{5}{2}$	(d) $x = \frac{3}{2}$

Inequality	Values	
6. $-1 < \dfrac{3 - x}{2} \leq 1$	(a) $x = 0$	(b) $x = \sqrt{5}$
	(c) $x = 1$	(d) $x = 5$
7. $|x - 10| \geq 3$	(a) $x = 13$	(b) $x = -1$
	(c) $x = 14$	(d) $x = 9$
8. $|2x - 3| < 15$	(a) $x = -6$	(b) $x = 0$
	(c) $x = 12$	(d) $x = 7$

In Exercises 9–18, solve the inequality and sketch the solution on the real number line. Use a graphing utility to verify your solution graphically.

9. $-10x < 40$ 10. $2x > 3$

11. $4(x + 1) < 2x + 3$ 12. $2x + 7 < 3$

13. $1 < 2x + 3 < 9$

14. $-8 \leq 1 - 3(x - 2) < 13$

15. $-4 < \dfrac{2x - 3}{3} < 4$ 16. $0 \leq \dfrac{x + 3}{2} < 5$

17. $-1 < -\dfrac{x}{3} < 1$ 18. $\dfrac{3}{4} > x + 1 > \dfrac{1}{4}$

Graphical Analysis In Exercises 19–24, use a graphing utility to graph the inequality.

19. $6x > 12$

20. $3x - 1 \le 5$

21. $5 - 2x \ge 1$

22. $3(x + 1) < x + 7$

23. $0 \le 2(x + 4) < 20$

24. $-2 < 3x + 1 < 10$

Graphical Analysis In Exercises 25–28, use a graphing utility to graph the equation. Use the graph to approximate the values of x that satisfy the specified inequalities.

Equation	Inequalities	
25. $y = 2x - 3$	(a) $y \ge 1$	(b) $y \le 0$
26. $y = \frac{2}{3}x + 1$	(a) $y \le 5$	(b) $y \ge 0$
27. $y = -\frac{1}{2}x + 2$	(a) $0 \le y \le 3$	(b) $y \ge 0$
28. $y = -3x + 8$	(a) $-1 \le y \le 3$	(b) $y \le 0$

In Exercises 29 and 30, find the interval(s) on the real number line for which the radicand is nonnegative (greater than or equal to zero).

29. $\sqrt{x - 5}$

30. $\sqrt[4]{6x + 15}$

In Exercises 31–40, solve the inequality and sketch the solution on the real number line.

31. $\left|\dfrac{x}{2}\right| > 3$

32. $|5x| > 10$

33. $|x - 20| \le 4$

34. $|x - 7| < 6$

35. $|x - 20| \ge 4$

36. $|x + 14| + 3 > 17$

37. $\left|\dfrac{x - 3}{2}\right| \ge 5$

38. $|1 - 2x| < 5$

39. $|x - 5| < 0$

40. $3|4 - 5x| \le 9$

Graphical Analysis In Exercises 41 and 42, use a graphing utility to graph the equation. Use the graph to approximate the values of x that satisfy the specified inequalities.

Equation	Inequalities	
41. $y = |x - 3|$	(a) $y \le 2$	(b) $y \ge 4$
42. $y = \left|\frac{1}{2}x + 1\right|$	(a) $y \le 4$	(b) $y \ge 1$

In Exercises 43–48, use absolute value notation to define each interval (or pair of intervals) on the real number line.

43.

44.

45.

46.

47. All real numbers within 10 units of 12

48. All real numbers whose distances from -3 are more than 5

49. *Data Analysis* The college admissions office wants to decide if there is a relationship between IQ scores x and grade-point averages y after the first year. A sample of 12 students yielded the following data.

x	118	131	125	123	133	136
y	2.2	2.4	3.2	2.4	3.5	3.0

x	128	124	116	120	134	131
y	3.0	2.8	2.2	1.8	3.4	3.6

(a) Use a graphing utility to plot the points. What shape does the plot have?

(b) Use the regression capabilities of a graphing utility to find a model appropriate to the shape of the plotted data. Graph the model in the same viewing rectangle as part (a).

(c) Which values of x predict a grade-point average of at least 3.0?

(d) Use the graph to write a statement about the accuracy of the model. If you think the graph indicates that IQ scores are not particularly good predictors of grade-point average, list other factors that may influence college performance.

50. *Teachers' Salaries*   The average salary for elementary and secondary teachers in the United States from 1984 to 1993 is approximated by the model

$$\text{Salary} = 15.812 + 1.472t$$

where the salary is given in thousands of dollars and the time t represents the calendar year, with $t = 4$ corresponding to 1984.   (Source: National Education Association)

(a) Use a graphing utility to graph the model.

(b) Assuming the model is correct, when will the average salary *exceed* $40,000?

In Exercises 51–58, solve the inequality and graph the solution on the real number line. Use a graphing utility to verify your solution graphically.

51. $(x + 2)^2 < 25$

52. $(x + 6)^2 \leq 8$

53. $x^2 + 4x + 4 \geq 9$

54. $x^2 - 6x + 9 < 16$

55. $x^2 + x < 6$

56. $4x^3 - 12x^2 > 0$

57. $x^3 - 4x \geq 0$

58. $x^4(x - 3) \leq 0$

Graphical Analysis   In Exercises 59–62, use a graphing utility to graph the equation. Use the graph to approximate the values of x that satisfy the specified inequalities.

	Equation	*Inequalities*
59.	$y = -x^2 + 2x + 3$	(a) $y \leq 0$ (b) $y \geq 3$
60.	$y = \frac{1}{2}x^2 - 2x + 1$	(a) $y \leq 1$ (b) $y \geq 7$
61.	$y = \frac{1}{8}x^3 - \frac{1}{2}x$	(a) $y \geq 0$ (b) $y \leq 6$
62.	$y = x^3 - x^2 - 16x + 16$	(a) $y \leq 0$ (b) $y \geq 36$

In Exercises 63–66, solve the inequality and graph the solution on the real number line. Use a graphing utility to verify your solution graphically.

63. $\dfrac{1}{x} - x > 0$

64. $\dfrac{1}{x} - 4 < 0$

65. $\dfrac{x + 6}{x + 1} - 2 < 0$

66. $\dfrac{x + 12}{x + 2} - 3 \geq 0$

Graphical Analysis   In Exercises 67–70, use a graphing utility to graph the equation. Use the graph to approximate the values of x that satisfy the specified inequalities.

	Equation	*Inequalities*
67.	$y = \dfrac{3x}{x - 2}$	(a) $y \leq 0$ (b) $y \geq 6$
68.	$y = \dfrac{2(x - 2)}{x + 1}$	(a) $y \leq 0$ (b) $y \geq 8$
69.	$y = \dfrac{2x^2}{x^2 + 4}$	(a) $y \geq 1$ (b) $y \leq 2$
70.	$y = \dfrac{5x}{x^2 + 4}$	(a) $y \geq 1$ (b) $y \leq 0$

In Exercises 71 and 72, find the domain of x in the expression.

71. $\sqrt[4]{4 - x^2}$

72. $\sqrt{x^2 - 4}$

73. *Percent of College Graduates*   The percent P of the American population that graduated from college between 1950 and 1990 is approximated by

$$P = 5.9556 + 1.492t + 0.0056t^2$$

where the time t represents the calendar year, with $t = 0$ corresponding to 1950.   (Source: U.S. Bureau of Census)

(a) Use a graphing utility to graph the model over the indicated years.

(b) According to this model, when will the percent of college graduates exceed 25% of the population? Solve algebraically and verify graphically.

P.6   Exploring Data: Representing Data Graphically

Line Plots / Stem-and-Leaf Plots / Histograms and Frequency Distributions / Line Graphs

Line Plots

Statistics is the branch of mathematics that studies techniques for collecting, organizing, and interpreting data. In this section, you will study several ways to organize data. The first is a **line plot,** which uses a portion of a real number line to order numbers. Line plots are especially useful for ordering small sets of numbers (about 50 or less) by hand.

EXAMPLE 1   Constructing a Line Plot

Use a line plot to organize the following test scores. Which number occurs with the greatest frequency?

93, 70, 76, 67, 86, 93, 82, 78, 83, 86, 64, 78, 76, 66, 83
83, 96, 74, 69, 76, 64, 74, 79, 76, 88, 76, 81, 82, 74, 70

Solution

Begin by scanning the data to find the smallest and largest numbers. For this data, the smallest number is 64 and the largest is 96. Next, draw a portion of a real number line that includes the interval [64, 96].To create the line plot, start with the first number, 93, and enter an × above 93 on the number line. Continue recording ×'s for each number in the list until you obtain the line plot shown in Figure P.60. From the line plot, you can see that 76 had the greatest frequency.

Figure P.60

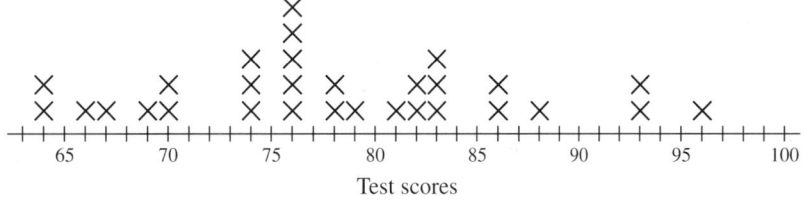

Test scores

Many computer programs and calculators will sort data. Try using a computer or calculator to sort the data in Example 1.

Stem-and-Leaf Plots

Another type of plot that is used to organize sets of numbers is a **stem-and-leaf plot.** A stem-and-leaf plot for the test scores in Example 1 is shown below.

Stems	Leaves
6	4 4 6 7 9
7	0 0 4 4 4 6 6 6 6 8 8 9
8	1 2 2 3 3 3 6 6 8
9	3 3 6

Note that the *leaves* represent the units digits of the numbers and the *stems* represent the tens digits. Stem-and-leaf plots can also be used to compare two sets of data, as shown in the next example.

EXAMPLE 2   **Comparing Two Sets of Data**

Use a stem-and-leaf plot to compare the test scores in Example 1 with the following test scores. Which set of test scores is better?

90, 81, 70, 62, 64, 73, 81, 92, 73, 81, 92, 93, 83, 75, 76
83, 94, 96, 86, 77, 77, 86, 96, 86, 77, 86, 87, 87, 79, 88

Solution
Begin by ordering the second set of scores.

62, 64, 70, 73, 73, 75, 76, 77, 77, 77, 79, 81, 81, 81, 83
83, 86, 86, 86, 86, 87, 87, 88, 90, 92, 92, 93, 94, 96, 96

Now that the data is ordered, you can construct a *double* stem-and-leaf plot by letting the leaves to the right of the stem represent the units digits for the first group of test scores and the leaves to the left of the stem represent the units digits for the second group of test scores.

Leaves (Second Group)	Stems	Leaves (First Group)
4 2	6	4 4 6 7 9
9 7 7 7 6 5 3 3 0	7	0 0 4 4 4 6 6 6 6 8 8 9
8 7 7 6 6 6 6 3 3 1 1 1	8	1 2 2 3 3 3 6 6 8
6 6 4 3 2 2 0	9	3 3 6

From the two sets of leaves, you can see that the second group of test scores is better than the first group.

EXAMPLE 3   ▰   **Using a Stem-and-Leaf Plot**

The table shows the percent of the population of each state (and the District of Columbia) that was 65 or older in 1993. Use a stem-and-leaf plot to organize the data. (Source: U.S. Bureau of Census)

AK 4.4	AL 13.0	AR 15.0	AZ 13.4	CA 10.6	CO 10.0
CT 14.1	D.C. 13.3	DE 12.4	FL 18.6	GA 10.1	HI 11.7
IA 15.5	ID 11.8	IL 12.6	IN 12.7	KS 13.9	KY 12.7
LA 11.3	MA 14.0	MD 11.1	ME 13.7	MI 12.4	MN 12.6
MO 14.2	MS 12.5	MT 13.4	NC 12.5	ND 14.8	NE 14.2
NH 11.9	NJ 13.6	NM 11.0	NV 11.1	NY 13.1	OH 13.3
OK 13.6	OR 13.8	PA 15.8	RI 15.5	SC 11.7	SD 14.7
TN 12.8	TX 10.2	UT 8.9	VA 11.0	VT 12.0	WA 11.6
WI 13.4	WV 15.3	WY 10.9			

Solution

Begin by ordering the numbers, as shown on the left. Next construct the stem-and-leaf plot using the leaves to represent the digits to the right of the decimal points. The stem-and-leaf plot is shown on the right.

4.4, 8.9, 10.0, 10.1, 10.2,
10.6, 10.9, 11.0, 11.0, 11.1,
11.1, 11.3, 11.6, 11.7, 11.7,
11.8, 11.9, 12.0, 12.4, 12.4,
12.5, 12.5, 12.6, 12.6, 12.7,
12.7, 12.8, 13.0, 13.1, 13.3,
13.3, 13.4, 13.4, 13.4, 13.6,
13.6, 13.7, 13.8, 13.9, 14.0,
14.1, 14.2, 14.2, 14.7, 14.8,
15.0, 15.3, 15.5, 15.5, 15.8,
18.6

```
 4. |  4 —— Alaska has the lowest percent.
 5. |
 6. |
 7. |
 8. |  9
 9. |
10. |  0 1 2 6 9
11. |  0 0 1 1 3 6 7 7 8 9
12. |  0 4 4 5 5 6 6 7 7 8
13. |  0 1 3 3 4 4 4 6 6 7 8 9
14. |  0 1 2 2 7 8
15. |  0 3 5 5 8
16. |
17. |       Florida has the highest percent.
18. |  6
```

Histograms and Frequency Distributions

With data such as that given in Example 3, it is useful to group the numbers into intervals and plot the frequency of the data in each interval. For instance, the **frequency distribution** and **histogram** shown in Figure P.61 represent the data given in Example 3.

Figure P.61

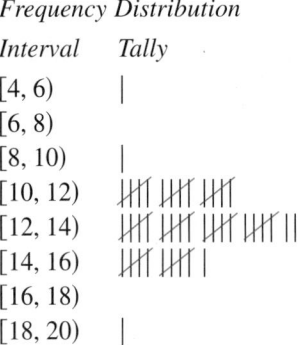

Frequency Distribution

Interval	Tally
[4, 6)	I
[6, 8)	
[8, 10)	I
[10, 12)	‖‖‖ ‖‖‖ ‖‖‖
[12, 14)	‖‖‖ ‖‖‖ ‖‖‖ ‖‖‖ II
[14, 16)	‖‖‖ ‖‖‖ I
[16, 18)	
[18, 20)	I

Histogram

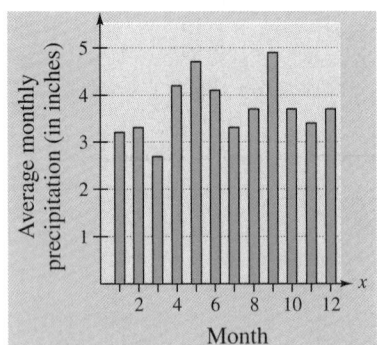

Try using a computer or graphing calculator to create a bar histogram for the data at the right. How does the histogram change when the intervals change?

A histogram has a portion of the real number line as its horizontal axis. A **bar graph** is similar to a histogram, except that the rectangles (bars) can be either horizontal or vertical and the labels of the bars are not necessarily numbers. Another difference between a bar graph and a histogram is that the bars in a bar graph are usually separated by spaces, whereas the bars in a histogram are not separated by spaces.

EXAMPLE 4   **Constructing a Bar Graph**

The data below shows the average monthly precipitation (in inches) in Houston, Texas. Construct a bar graph for this data. What can you conclude? (Source: PC USA)

January	3.2	*February*	3.3	*March*	2.7
April	4.2	*May*	4.7	*June*	4.1
July	3.3	*August*	3.7	*September*	4.9
October	3.7	*November*	3.4	*December*	3.7

Solution

To create a bar graph, begin by drawing a vertical axis to represent the precipitation and a horizontal axis to represent the month. The bar graph is shown in Figure P.62. From the graph, you can see that Houston receives a fairly consistent amount of rain throughout the year—the driest month tends to be March and the wettest month tends to be September.

EXAMPLE 5 ▱ **Constructing a Double Bar Graph**

The table shows the percents of bachelor's degrees awarded to males and females for selected majors in the United States in 1991. Construct a double bar graph for this data.   (Source: U.S. National Center for Education Statistics)

Field of Study	% Female	% Male
Agriculture and Natural Resources	31.5	68.5
Area and Ethnic Studies	80.4	19.6
Education	73.1	26.9
Engineering	8.7	91.3
Home Economics	73.7	26.3
Liberal/General Studies	59.4	40.6
Mathematics	39.0	61.0
Military Sciences	9.4	90.6
Physical Sciences	38.1	61.9
Social Sciences	59.6	40.4

Solution

For this data, a horizontal bar graph seems to be appropriate. Such a graph is shown in Figure P.63.

Figure P.63

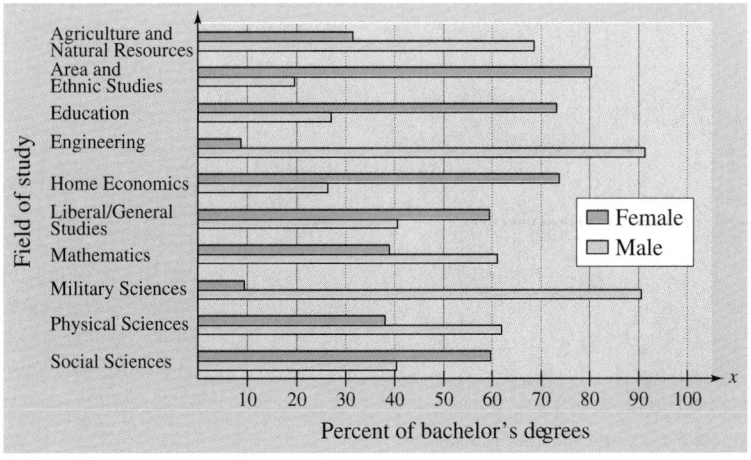

Decade	Number
1851–1860	2598
1861–1870	2315
1871–1880	2812
1881–1890	5247
1891–1900	3688
1901–1910	8795
1911–1920	5736
1921–1930	4107
1931–1940	528
1941–1950	1035
1951–1960	2515
1961–1970	3322
1971–1980	4493
1981–1990	6447

Line Graphs

EXAMPLE 6 ▬ **Constructing a Line Graph**

The table at the left shows the number of immigrants (in thousands) entering the United States in each decade from 1851 to 1990. Construct a line graph of this data. What can you conclude?   (Source: U.S. Bureau of Census)

Solution

Begin by drawing a vertical axis to represent the number of immigrants in thousands. Then label the horizontal axis with decades and plot the points given in the table. Finally, connect the points with line segments, as shown in Figure P.64. From the line graph, you can see that the number of immigrants hit a low point during the depression of the 1930s. Since then the number has steadily increased.

Figure P.64

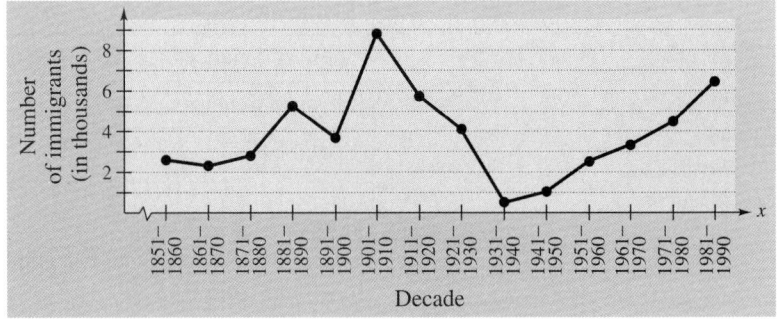

Note A **line graph** is similar to a standard coordinate graph. Line graphs are usually used to show trends over periods of time.

To sketch a line graph using a *TI-82* or *TI-83*, use the following steps.

1. Use the ⎡STAT⎤ key to enter the data into L_1 and L_2.
2. Use the ⎡STAT PLOT⎤ key to select Plot 1 and the line graph icon.
3. Use the ⎡ZOOM⎤ key to select ZoomStat, which will set an appropriate viewing rectangle.

Group Activity   *Organizing Data*

Listed below are the winning times (in seconds) for the 110-meter hurdles in the summer Olympics from 1896 through 1992. With others in your group, decide which type of graph can best represent this data. Sketch the graph and discuss any patterns or trends that you see.

1896 (17.60) 1900 (15.40) 1904 (16.00) 1908 (15.00) 1912 (15.10)
1920 (14.80) 1924 (15.00) 1928 (14.80) 1932 (14.60) 1936 (14.20)
1948 (13.90) 1952 (13.70) 1956 (13.50) 1960 (13.80) 1964 (13.60)
1968 (13.30) 1972 (13.24) 1976 (13.30) 1980 (13.39) 1984 (13.20)
1988 (12.98) 1992 (13.12)

P.6 /// EXERCISES

1. *Gasoline Prices* The line plot shows a sample of prices of unleaded regular gasoline from 25 different cities.

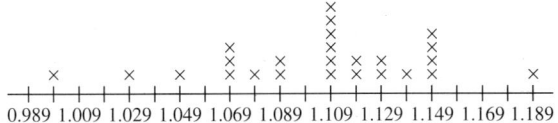

(a) What price occurred with the greatest frequency?

(b) What is the range of prices?

2. *Livestock Weights* The line plot shows the weights (to the nearest hundred pounds) of 30 head of cattle sold by a rancher.

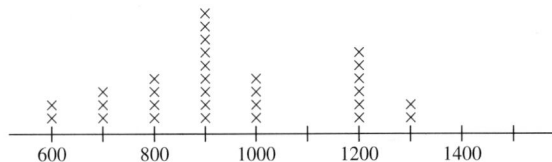

(a) What weight occurred with the greatest frequency?

(b) What is the range of weights?

Quiz and Exam Scores In Exercises 3–8, use the following scores from a math class of 30 students. The scores are for two 25-point quizzes and two 100-point exams.

Quiz #1   20, 15, 14, 20, 16, 19, 10, 21, 24, 15, 15, 14, 15, 21, 19, 15, 20, 18, 18, 22, 18, 16, 18, 19, 21, 19, 16, 20, 14, 12

Quiz #2   22, 22, 23, 22, 21, 24, 22, 19, 21, 23, 23, 25, 24, 22, 22, 23, 23, 23, 23, 22, 24, 23, 22, 24, 21, 24, 16, 21, 16, 14

Exam #1   77, 100, 77, 70, 83, 89, 87, 85, 81, 84, 81, 78, 89, 78, 88, 85, 90, 92, 75, 81, 85, 100, 98, 81, 78, 75, 85, 89, 82, 75

Exam #2   76, 78, 73, 59, 70, 81, 71, 66, 66, 73, 68, 67, 63, 67, 77, 84, 87, 71, 78, 78, 90, 80, 77, 70, 80, 64, 74, 68, 68, 68

3. Construct a line plot for Quiz #1. Which score occurred with the greatest frequency?

4. Construct a line plot for Quiz #2. Which score occurred with the greatest frequency?

5. Construct a line plot for Exam #1. Which score occurred with the greatest frequency?

6. Construct a line plot for Exam #2. Which score occurred with the greatest frequency?

7. Construct a stem-and-leaf plot for Exam #1.

8. Construct a double stem-and-leaf plot to compare the scores for Exam #1 and Exam #2. Which set of scores is higher?

9. *Educational Expenses* The list gives the per capita expenditures for public elementary and secondary education in the 50 states and the District of Columbia in 1993. Use a stem-and-leaf plot to organize the data. (Source: National Education Association)

AK 1800	AL 692	AR 762	AZ 915
CA 919	CO 974	CT 1252	D.C. 1061
DE 974	FL 864	GA 834	HI 932
IA 943	ID 898	IL 865	IN 1029
KS 1024	KY 841	LA 804	MA 913
MD 1003	ME 1098	MI 1155	MN 1124
MO 764	MS 676	MT 1056	NC 818
ND 863	NE 964	NH 916	NJ 1351
NM 919	NV 964	NY 1265	OH 1024
OK 850	OR 1138	PA 1096	RI 903
SC 848	SD 885	TN 663	TX 1049
UT 834	VA 940	VT 1197	WA 1131
WI 1123	WV 1045	WY 1336	

10. *Land Value*   The list gives the average value of land and buildings per acre in the 48 contiguous states in 1993. Use a stem-and-leaf plot to organize the data. (Source: U.S. Department of Agriculture)

AL 863	AR 759	AZ 305	CA 1722
CO 383	CT 4299	DE 2362	FL 2074
GA 964	IA 1245	ID 691	IL 1503
IN 1366	KS 494	KY 1084	LA 945
MA 3662	MD 2521	ME 992	MI 1130
MN 896	MO 715	MS 757	MT 270
NC 1319	ND 388	NE 580	NH 2178
NJ 4536	NM 225	NV 215	NY 1119
OH 1267	OK 512	OR 657	PA 1747
RI 4894	SC 871	SD 370	TN 1049
TX 471	UT 464	VA 1295	VT 1158
WA 782	WI 932	WV 696	WY 149

11. *Cellular-Phone Fraud*   The bar graph gives the industry losses (in millions of dollars) from cellular-phone fraud for the years 1991 through 1994 and the projected loss for 1995. Determine the percent increase in losses from 1991 to 1995. (Source: Cellular Telecommunications Industry Association)

12. *Food Production*   The double bar graph gives the production and exports (in millions of metric tons) of corn, soybeans, and wheat for the year 1992. Determine the percent of each product that is exported. (Source: U.S. Department of Agriculture)

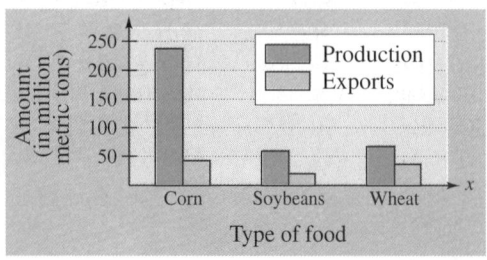

13. *Snowfall*   The list gives the seasonal snowfall (in inches) at Erie, Pennsylvania, starting with the 1960/61 winter and ending with the 1994/95 winter. The amounts are listed in order by year. Organize the data graphically. (Source: National Oceanic and Atmospheric Association)

69.6, 42.5, 75.9, 115.9, 92.9, 84.8, 68.6, 107.9, 79.7, 85.6, 120.0, 92.3, 53.7, 68.6, 66.7, 66.0, 111.5, 142.8, 76.5, 55.2, 89.4, 71.3, 41.2, 110.0, 106.3, 124.9, 68.2, 103.5, 76.5, 114.9, 59.6, 104.8, 108.5, 131.3, 53.7

14. *Fruit Crops*   The list gives farmers' cash receipts (in millions of dollars) from fruit crops in 1992. Construct a bar graph for the data. (Source: U.S. Department of Agriculture)

Apples	1159	Oranges	1616
Cherries	235	Peaches	373
Grapefruit	403	Pears	276
Grapes	1713	Plums and Prunes	245
Lemons	235	Strawberries	685

15. *Travel to the United States*   The places of origin and numbers of travelers (in millions) to the United States in 1992 are as follows: Canada, 18.6; Mexico, 8.3; Europe, 8.3; Latin America, 3.3; Other, 6.3. Construct a horizontal bar graph for this data. (Source: U.S. Travel and Tourism Administration)

16. *Sports Participants*   The list gives the numbers of males and females (in millions) over the age of seven that participated in popular sports activities in the year 1992 in the United States. Construct a double bar graph for the data. (Source: National Sporting Goods Association)

Activity	*Male*	*Female*
Exercise walking	23.2	44.6
Swimming	29.8	33.3
Bicycling	28.0	26.6
Camping	25.7	21.7
Bowling	21.8	20.7
Basketball	20.6	7.6
Running	12.7	9.2
Aerobic exercising	5.1	22.8

17. *Personal Savings*   The line graph shows the percent of disposable personal income saved in the United States in selected years from 1970 to 1993. (Source: U.S. Bureau of Economic Analysis)

   (a) Determine the percent decrease in the rate of saving from 1975 to 1993.

   (b) Is the trend shown in the line graph good for the country? Explain your reasoning.

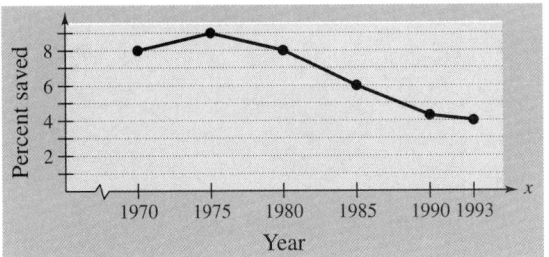

18. *Unemployment Rate* The double line graph shows the percent of the labor force in mining and manufacturing that were unemployed in the years 1987 to 1993. (Source: U.S. Bureau of Labor Statistics)

   (a) Which industry showed the greater variability in unemployment rate?

   (b) During which year was there the greatest difference between the two unemployment rates? During which year was the difference least?

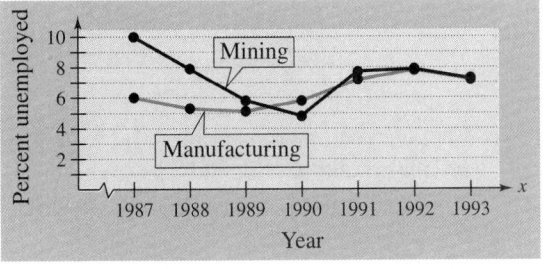

19. *College Attendance*   The table shows the enrollment in a liberal arts college for the years 1988 to 1995. Construct a line graph for the data.

Year	1988	1989	1990	1991
Enrollment	1675	1704	1710	1768

Year	1992	1993	1994	1995
Enrollment	1833	1918	1967	1972

20. *Oil Imports*   The table shows the amount of crude oil imported into the United States (in millions of barrels) for the years 1984 through 1993. Construct a line graph for the data and state what information the graph reveals. (Source: Energy Information Administration)

Year	1984	1985	1986	1987	1988
Imports	1254	1168	1525	1706	1869

Year	1989	1990	1991	1992	1993
Imports	2133	2145	2110	2226	2457

21. *Federal Income*   The list gives the receipts (in billions of dollars) for the federal government of the United States from 1972 to 1993. Construct a line graph for the data. What can you conclude? (Source: U.S. Office of Management and Budget)

Year	Receipts	Year	Receipts
1972	207.3	1983	600.6
1973	230.8	1984	666.5
1974	263.2	1985	734.1
1975	279.1	1986	769.1
1976	298.1	1987	854.1
1977	355.6	1988	909.0
1978	399.6	1989	990.7
1979	463.3	1990	1031.3
1980	517.1	1991	1054.3
1981	599.3	1992	1090.5
1982	617.8	1993	1153.5

Focus on Concepts

In this chapter, you studied several concepts that are required in the study of algebra. You can use the following questions to check your understanding of several of these basic concepts. The answers to these questions are given in the back of the book.

In Exercises 1–4, use the plot of the point (x_0, y_0) in the figure. Match the transformation of the point with the correct plot. [The plots are labeled (a), (b), (c), and (d).]

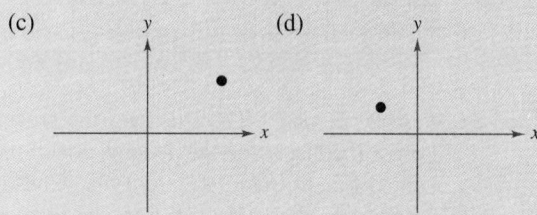

1. $(x_0, -y_0)$

2. $(-2x_0, y_0)$

3. $\left(x_0, \frac{1}{2}y_0\right)$

4. $(-x_0, -y_0)$

5. In your own words, explain the difference between an identity and a conditional equation.

6. Consider the linear equation $ax + b = 0$.

   (a) What is the sign of the solution if $ab > 0$?

   (b) What is the sign of the solution if $ab < 0$?

7. Each graph shows the solution(s) of an equation plotted on the real number line. In each case, determine if the solution(s) is (are) for a linear equation, a quadratic equation, both, or neither. Explain.

   (a) ●—●—● x a b c

   (b) —————●————— x a

   (c) —●——●— x a b

   (d) ●—●—●—● x a b c d

8. To solve the equation $2x^2 + 3x = 15x$, a student divides both sides by x and solves the equation $2x + 3 = 15$. The resulting solution ($x = 6$) satisfies the given equation. Is there an error? Explain.

9. Which graph shows the solution of the inequality $|x - a| \geq 2$? Explain.

   (a) $($————$)$ x $a-2$ a $a+2$

   (b) $]$————$[$ x $a-2$ a $a+2$

   (c) $]$————$[$ x $2-a$ 2 $2+a$

   (d) $[$————$]$ x $2-a$ 2 $2+a$

10. Consider the polynomial $(x - a)(x - b)$ and the real number line (see figure).

    (a) Identify the points on the line at which the polynomial is zero.

    (b) In each of the three subintervals of the line, write the sign of each factor and the sign of the product.

    (c) For which x-values does the polynomial change signs?

P /// REVIEW EXERCISES

Geometry   **In Exercises 1 and 2, plot the points and verify that the points form the polygon.**

1. *Right Triangle:* $(2, 3), (13, 11), (5, 22)$
2. *Parallelogram:* $(1, 2), (8, 3), (9, 6), (2, 5)$

In Exercises 3 and 4, determine the quadrant(s) in which (x, y) is (are) located so that the conditions are satisfied.

3. $x > 0$ and $y = -2$ 4. $xy = 4$

5. (a) Plot the points $(-3, 8)$ and $(1, 5)$, (b) find the distance between the points, and (c) find the midpoint of the line segment joining the points.

6. Complete the table for the equation $y = x^2 - 3x$. Use the resulting solution points to sketch the graph of the equation. Use a graphing utility to verify the graph

   $y = x^2 - 3x$.

x	-1	0	1	2	3
y					

In Exercises 7–14, sketch the graph of the equation *by hand*. Use a graphing utility to verify your graph.

7. $y - 2x - 3 = 0$
8. $x - 5 = 0$
9. $y = 8 - |x|$
10. $y = \sqrt{x + 2}$
11. $y + 2x^2 = 0$
12. $y = x^2 - 4x$
13. $y = \sqrt{25 - x^2}$
14. $x^2 + y^2 = 10$

In Exercises 15–22, use a graphing utility to graph the equation. Approximate any intercepts.

15. $y = \frac{1}{4}(x + 1)^3$
16. $y = 4 - (x - 4)^2$
17. $y = \frac{1}{4}x^4 - 2x^2$
18. $y = \frac{1}{4}x^3 - 3x$
19. $y = x\sqrt{9 - x^2}$
20. $y = x\sqrt{x + 3}$
21. $y = |x - 4| - 4$
22. $y = |x + 2| + |3 - x|$

23. Find the center and radius of the circle given by

    $(x - 3)^2 + (y + 1)^2 = 9$.

    Sketch the graph of the circle.

24. Find the standard form of the equation of a circle if the endpoints of its diameter are $(0, 0)$ and $(4, -6)$.

25. Find a setting on a graphing utility such that the graph of $y = 10x^3 - 21x^2$ agrees with the graph below.

    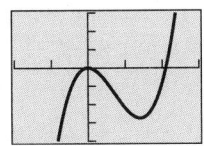

26. *Data Analysis*   The total annual expenditures y for NASA (in billions) for each year from 1990 through 1993 are given in the table. (Source: U.S. Department of Treasury)

x	1990	1991	1992	1993
y	12.4	13.9	14.0	14.3

    A mathematical model for the expenditures for this period is

    $y = 0.58t + 12.78$

    where t is the time in years, with $t = 0$ corresponding to 1990. (a) Use a graphing utility to plot the data and sketch the model for the data, and (b) use the model to estimate the value of y for the years 1998 and 2000.

In Exercises 27–30, plot the points and find the slope of the line passing through the points.

27. $(-4.5, 6), (2.1, 3)$
28. $(-3, 2), (8, 2)$
29. $\left(\frac{3}{2}, 1\right), \left(5, \frac{5}{2}\right)$
30. $(7, -1), (7, 12)$

In Exercises 31–34, use the point on the line and the slope of the line to find three additional points through which the line passes. (The solution is not unique.)

Point	Slope
31. $(2, -1)$	$m = \frac{1}{4}$
32. $(-3, 5)$	$m = -\frac{3}{2}$
33. $(-6, -5)$	$m = -2$
34. $(10, -6)$	m is undefined.

In Exercises 35–40, find an equation of the line that passes through the points.

35. $(0, 0), (0, 10)$ **36.** $(-1, 4), (2, 0)$
37. $(2, 1), (14, 6)$ **38.** $(-2, 2), (3, -10)$
39. $(-1, 0), (6, 2)$ **40.** $(1, 6), (4, 2)$

In Exercises 41–44, find an equation of the line that passes through the given point and has the specified slope. Use a graphing utility to graph the line.

Point	Slope
41. $(0, -5)$	$m = \frac{3}{2}$
42. $(-2, 6)$	$m = 0$
43. $(3, 0)$	$m = -\frac{2}{3}$
44. $(5, 4)$	m is undefined.

In Exercises 45 and 46, write an equation of the line through the point (a) parallel to the given line and (b) perpendicular to the given line. Verify your result with a graphing utility (use a square setting).

Point	Line
45. $(3, -2)$	$5x - 4y = 8$
46. $(-8, 3)$	$2x + 3y = 5$

Rate of Change   In Exercises 47 and 48, you are given the dollar value of a product in 1996 *and* the rate at which the value of the item is expected to change during the next 5 years. Use this information to write a linear equation that gives the dollar value V of the product in terms of the year t. (Let $t = 6$ represent 1996.)

1996 Value	Rate
47. $12,500	$850 increase per year
48. $72.95	$5.15 increase per year

Exploration   In Exercises 49 and 50, find a relationship between x and y such that (x, y) is equidistant from the two points.

49. $(-2, -5), (6, 3)$ **50.** $\left(1, \frac{7}{2}\right), (5, 0)$

In Exercises 51 and 52, determine whether the equation is an identity or a conditional equation.

51. $6 - (x - 2)^2 = 2 + 4x - x^2$
52. $3(x - 2) + 2x = 2(x + 3)$

In Exercises 53–74, solve the equation (if possible) and use a graphing utility to verify your solution.

53. $4(x + 3) - 3 = 2(4 - 3x) - 4$
54. $\frac{1}{2}(x - 3) - 2(x + 1) = 5$
55. $3\left(1 - \frac{1}{5t}\right) = 0$ **56.** $\frac{1}{x - 2} = 3$
57. $6x = 3x^2$ **58.** $15 + x - 2x^2 = 0$
59. $(x + 4)^2 = 18$ **60.** $16x^2 = 25$
61. $x^2 - 12x + 30 = 0$ **62.** $x^2 + 6x - 3 = 0$
63. $5x^4 - 12x^3 = 0$ **64.** $4x^3 - 6x^2 = 0$
65. $\frac{4}{(x - 4)^2} = 1$ **66.** $\frac{1}{(t + 1)^2} = 1$
67. $\sqrt{x + 4} = 3$ **68.** $\sqrt{3x - 2} = 4 - x$
69. $\sqrt{2x + 3} + \sqrt{x - 2} = 2$
70. $5\sqrt{x} - \sqrt{x - 1} = 6$
71. $(x - 1)^{2/3} - 25 = 0$
72. $(x + 2)^{3/4} = 27$
73. $|x - 5| = 10$
74. $|x^2 - 6| = x$

In Exercises 75–80, use a graphing utility to graph the equation. Use the graph to approximate any x-intercepts of the graph. Set $y = 0$ and solve the resulting equation. Compare the result with the x-intercepts of the graph.

75. $y = 4x^3 - 12x^2 + 8x$

76. $y = 12x^3 - 84x^2 + 120x$

77. $y = \dfrac{1}{x} + \dfrac{1}{x + 1} - 2$

78. $y = \dfrac{4}{x - 3} - \dfrac{4}{x} - 1$

79. $y = \sqrt{x^2 + 1} + x - 9$

80. $y = |2x - 3| - 5$

In Exercises 81–84, solve the equation for the indicated variable.

81. Solve for r: $V = \frac{1}{3}\pi r^2 h$

82. Solve for X: $Z = \sqrt{R^2 - X^2}$

83. Solve for p: $L = \dfrac{k}{3\pi r^2 p}$

84. Solve for v: $E = 2kw\left(\dfrac{v}{2}\right)^2$

In Exercises 85–92, solve the inequality. Use a graphing utility to verify your solution.

85. $\frac{1}{2}(3 - x) > \frac{1}{3}(2 - 3x)$

86. $x^2 - 2x \geq 3$

87. $\dfrac{x - 5}{3 - x} < 0$

88. $\dfrac{2}{x + 1} \leq \dfrac{3}{x - 1}$

89. $|x - 2| < 1$

90. $|x| \leq 4$

91. $\left|x - \frac{3}{2}\right| \geq \frac{3}{2}$

92. $|x - 3| > 4$

In Exercises 93–96, use a graphing utility to solve the inequality.

93. $\dfrac{x}{5} - 6 \leq -\dfrac{x}{2} + 6$

94. $2x^2 + x \geq 15$

95. $(x - 4)|x| > 0$

96. $|x(x - 6)| < 5$

97. *Starting Position* A fitness center has two running tracks around a rectangular playing floor. The tracks are 1 meter wide and form semicircles at the narrow ends of the rectangular floor (see figure). Determine the distance between the starting positions if two runners must run the same distance to the finish line in one lap around the track.

98. *Simply Supported Beam* A simply supported beam of length 20 feet supports a uniformly distributed load of 1000 pounds per foot. The bending moment M in foot-pounds x feet from one end of the beam is given by

$M = 500x(20 - x)$.

(a) Use a graphing utility to graph the equation.

(b) Determine any points on the beam where the bending moment is zero.

(c) Use the graph to determine the point on the beam where the bending moment is greatest. What is the bending moment at that point?

(d) Determine the positions on the beam where the bending moment is less than 40,000 foot-pounds.

99. *Weather* The normal daily maximum and minimum temperatures for each month in the city of Chicago are given in the table. Make a double line graph for the data. (Source: NOAA)

Month	Jan.	Feb.	Mar.	Apr.	May	Jun.
Max.	29.0	33.5	45.8	58.6	70.1	79.6
Min.	12.9	17.2	28.5	38.6	47.7	57.5

Month	Jul.	Aug.	Sep.	Oct.	Nov.	Dec.
Max.	83.7	81.8	74.8	63.3	48.4	34.0
Min.	62.6	61.6	53.9	42.2	31.6	19.1

CHAPTER PROJECT *Modeling the Volume of a Box*

Many mathematical results are discovered experimentally by calculating examples and looking for patterns. Prior to the 1950s, this mode of discovery was very time-consuming because the calculations had to be done by hand. The introduction of computer and calculator technology has removed much of this drudgery. In the following project, you are asked to model a real-life situation and solve a problem by looking for patterns in the corresponding data.

Consider a rectangular box with a square base and a surface area of 216 square inches. Let x represent the length (in inches) of each side of the base and let h represent the height (in inches) of the box, as shown at the left. Your goal is to answer the question "Of all rectangular boxes with square bases and surface area of 216 square inches, which has the greatest volume?"

(a) Express the areas of the base, top, and sides in terms of x and h.

(b) Find an expression in terms of x and h for the surface area of the box.

(c) Use the fact that the surface area is 216 square inches to express the variable h in terms of x.

(d) Find an expression for the volume of the box in terms of x alone.

(e) Use the expression in part (d) and a graphing utility to complete the table. Then use the results to decide which box has the greatest volume.

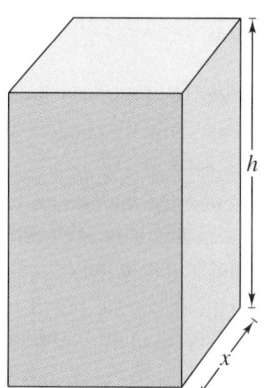

Base, x	Height	Surface Area	Volume
1.0	53.5	216.0	53.5
1.5	35.3	216.0	79.3
2.0	26.0	216.0	104.0
⋮	⋮	⋮	⋮
10.0	0.4	216.0	40.0

Questions for Further Exploration

1. What happens to the height of the box as x gets closer and closer to 0? Of all boxes with square base and a surface area of 216 square inches, is there a tallest? Explain your reasoning.

2. What is the maximum value of x? What happens to the height of the box as x gets closer and closer to this maximum value? Is there a shortest box that has a square base and a surface area of 216 square inches? Explain your reasoning.

3. Complete the table. Does it lend further support to your answer to part (e)? Explain.

x	5.9	5.99	5.999	6.001	6.01	6.1
V						

4. Of all rectangular boxes with surface area of 216 square inches and a base that is x inches by $2x$ inches, which has the maximum volume? Explain your reasoning.

P /// CHAPTER TEST

Take this test as you would take a test in class. After you are done, check your work against the answers given in the back of the book.

The *Interactive* CD-ROM provides answers to the Chapter Tests and Cumulative Tests. It also offers Chapter Pre-Tests (that test key skills and concepts covered in previous chapters) and Chapter Post-Tests, both of which have randomly generated exercises with diagnostic capabilities.

1. Plot the points $(-2, 5)$ and $(6, 0)$. Find the coordinates of the midpoint of the line segment joining the points and the distance between the points.

2. The numbers (in millions) of votes cast for the Democratic candidate for president in 1980, 1984, 1988, and 1992 were 35.5, 37.6, 41.8, and 44.9, respectively. Create a bar graph for the data.

In Exercises 3–8, use a graphing utility to graph the equation. Check for symmetry and identify any *x*- or *y*-intercepts.

3. $y = 4 - \frac{3}{4}|x|$

4. $y = 4 - (x - 2)^2$

5. $y = x - x^3$

6. $y = \sqrt{3 - x}$

7. $2x - 3y = 12$

8. $(x - 3)^2 + y^2 = 9$

9. A line with slope $m = \frac{3}{2}$ passes through the point $(3, -1)$. List three additional points on the line.

10. Find an equation of the line for each of the following.

   (a) Passes through the points $(-4, 0)$ and $(2, 3)$.

   (b) Passes through the point $(0, 4)$ and is perpendicular to the line $5x + 2y = 3$.

In Exercises 11–16, solve (if possible) the equation. Use a graphing utility to verify your solution.

11. $2x - 3(x - 4) = 5$

12. $\dfrac{2}{t - 3} + \dfrac{2}{t - 2} = \dfrac{10}{t^2 - 5t + 6}$

13. $3y^2 + 6y + 2 = 0$

14. $\sqrt{x + 10} = x - 2$

15. $2\sqrt{x} - \sqrt{2x + 1} = 1$

16. $|3x - 1| = 7$

Figure for 19

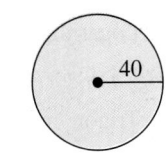

In Exercises 17 and 18, solve the inequality and sketch the solution on the real number line.

17. $-3 \le 2(x + 4) < 14$

18. $\dfrac{2}{x} > \dfrac{5}{x + 6}$

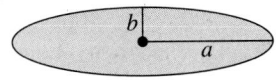

19. The area of the ellipse in the figure is $A = \pi ab$. If a and b satisfy the constraint $a + b = 100$, find a and b if the area of the ellipse equals the area of the circle.

1.1   Functions

Introduction to Functions   /   Function Notation   /
The Domain of a Function   /   Applications

Library of Functions

Many functions do not have simple mathematical formulas but are defined by real-life data. Such functions arise when you are using collections of data to model real-life applications. You will see that it is often convenient to *approximate* the data using a mathematical model or formula.

Introduction to Functions

Many everyday phenomena involve pairs of quantities that are related to each other by some rule of correspondence. Here are some examples.

1. The simple interest I earned on $1000 for 1 year is related to the annual interest rate r by the formula $I = 1000r$.
2. The distance d traveled on a bicycle in 2 hours is related to the speed s of the bicycle by the formula $d = 2s$.
3. The area A of a circle is related to its radius r by the formula $A = \pi r^2$.

Not all correspondences between two quantities have simple mathematical formulas. For instance, people commonly match up NFL starting quarterbacks with touchdown passes, and hours of the day with temperature. In each of these cases, however, there is some rule of correspondence that matches each item from one set with exactly one item from a different set. Such a rule of correspondence is called a **function.**

Definition of a Function

A **function** f from a set A to a set B is a rule of correspondence that assigns to each element x in the set A exactly one element y in the set B. The set A is the **domain** (or set of inputs) of the function f, and the set B contains the **range** (or set of outputs).

To help understand this definition, look at the function illustrated in Figure 1.1. This function can be represented by the following ordered pairs.

$$\{(1, 9°), (2, 13°), (3, 15°), (4, 15°), (5, 12°), (6, 10°)\}$$

In each ordered pair, the first coordinate is the input and the second coordinate is the output. In this example, note the following characteristics of a function.

1. Each element in A must be matched with an element of B.
2. Some elements in B may not be matched with any element in A.
3. Two or more elements of A may be matched with the same element of B.

The converse of the third statement is not true. That is, an element of A (the domain) cannot be matched with two different elements of B.

Figure 1.1

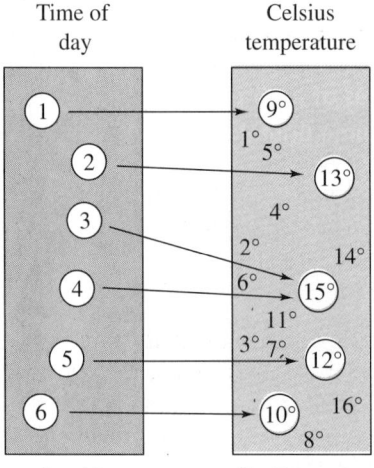

Time of day   Celsius temperature

Set A is the domain.
Inputs : 1, 2, 3, 4, 5, 6

Set B contains the range.
Outputs : 9°, 10°, 12°, 13°, 15°

Note Be sure you see that the *range* of a function is not the same as the use of *range* relating to the viewing rectangle.

The *Interactive* CD-ROM shows every example with its solution; clicking on the *Try It!* button brings up similar problems. Guided Examples and Integrated Examples show step-by-step solutions to additional examples. Integrated Examples are related to several concepts in the section.

In the following example, you are asked to decide whether different correspondences are functions. To do this, you must decide whether each element in the domain A is matched with exactly one element in the range B. If any element in A is matched with two or more elements in B, the correspondence is not a function.

EXAMPLE 1   **Testing for Functions**

Let $A = \{a, b, c\}$ and $B = \{1, 2, 3, 4, 5\}$. Which of the following sets of ordered pairs or figures represent functions from set A to set B?

a. $\{(a, 2), (b, 3), (c, 4)\}$ **b.** $\{(a, 4), (b, 5)\}$

c.   **d.**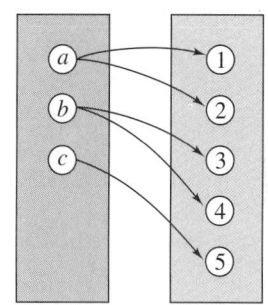

Solution

a. This collection of ordered pairs *does* represent a function from A to B. Each element of A is matched with exactly one element of B.

b. This collection of ordered pairs *does not* represent a function from A to B. Not every element of A is matched with an element of B.

c. This figure *does* represent a function from A to B. It does not matter that each element of A is matched with the same element of B.

d. This figure *does not* represent a function from A to B. The element a in A is matched with *two* elements, 1 and 2, of B. This is also true of the element b.

Leonhard Euler (1707–1783), a Swiss mathematician, is considered to have been the most prolific and productive mathematician in history. One of his greatest influences on mathematics was his use of symbols, or notation. The function notation $y = f(x)$ was introduced by Euler.

Representing functions by sets of ordered pairs is common in *discrete mathematics*. In algebra, however, it is more common to represent functions by equations or formulas involving two variables. For instance, the equation

$$y = x^2 \qquad \text{\small\textit{y} is a function of \textit{x}.}$$

represents the variable y as a function of the variable x. In this equation, x is the **independent variable** and y is the **dependent variable.** The domain of the function is the set of all values taken on by the independent variable x, and the range of the function is the set of all values taken on by the dependent variable y.

The *Interactive* CD-ROM offers graphing utility emulators of the *TI-82* and *TI-83*, which can be used with the Examples, Explorations, Technology notes, and Exercises.

EXPLORATION

Use a graphing utility to graph $x^2 + y = 1$. Then use the graph to write a convincing argument that each x-value has at most one y-value.

Use a graphing utility to graph $-x + y^2 = 1$. (*Hint:* You will need to use two equations.) Then use the graph to find an x-value that corresponds to two y-values. Why does the graph not represent y as a function of x?

EXAMPLE 2   ▭   **Testing for Functions Represented by Equations**

Which of the equations represents y as a function of x?

a. $x^2 + y = 1$ **b.** $-x + y^2 = 1$

Solution
To determine whether y is a function of x, try to solve for y in terms of x.

a. Solving for y yields the following.

$$x^2 + y = 1 \qquad \text{Original equation}$$
$$y = 1 - x^2 \qquad \text{Solve for } y.$$

To each value of x there corresponds exactly one value of y. Thus, y *is* a function of x.

b. Solving for y yields the following.

$$-x + y^2 = 1 \qquad \text{Original equation}$$
$$y^2 = 1 + x \qquad \text{Add } x \text{ to both sides.}$$
$$y = \pm\sqrt{1 + x} \qquad \text{Solve for } y.$$

The $\pm$ indicates that to a given value of x there correspond two values of y. Thus, y *is not* a function of x.   ▭

Function Notation

When an equation is used to represent a function, it is convenient to name the function so that it can be referenced easily. For example, you know that the equation $y = 1 - x^2$ describes y as a function of x. Suppose you give this function the name "f." Then you can use the following **function notation.**

Input	*Output*	*Equation*
x	$f(x)$	$f(x) = 1 - x^2$

The symbol $f(x)$ is read as the **value of f at x** or simply ***f* of *x*.** The symbol $f(x)$ corresponds to the y-value for a given x. Thus, you can write $y = f(x)$. Keep in mind that f is the *name* of the function, whereas $f(x)$ is the *value* of the function at x. For instance, the function given by

$$f(x) = 3 - 2x$$

has *function values* denoted by $f(-1), f(0), f(2)$, and so on. To find these values, substitute the specified input values into the given equation.

For $x = -1$, $f(-1) = 3 - 2(-1) = 3 + 2 = 5.$
For $x = 0$, $f(0) = 3 - 2(0) = 3 - 0 = 3.$
For $x = 2$, $f(2) = 3 - 2(2) = 3 - 4 = -1.$

Library of Functions

The function in Example 4 is a *piecewise-defined* function. This means that the function is defined by two or more equations over a specified domain. In Example 4, you use the top equation for all x-values less than 0, and the bottom equation for all x-values greater than or equal to 0.

Although f is often used as a convenient function name and x is often used as the independent variable, you can use other letters. For instance,

$$f(x) = x^2 - 4x + 7, \quad f(t) = t^2 - 4t + 7, \quad \text{and} \quad g(s) = s^2 - 4s + 7$$

all define the same function. In fact, the role of the independent variable is that of a "placeholder." Consequently, the function could be described by

$$f(\quad) = (\quad)^2 - 4(\quad) + 7.$$

EXAMPLE 3 **Evaluating a Function**

Let $g(x) = -x^2 + 4x + 1$ and find the following.

a. $g(2)$ **b.** $g(t)$ **c.** $g(x + 2)$

Solution

a. Replacing x with 2 in $g(x) = -x^2 + 4x + 1$ yields the following.

$$g(2) = -(2)^2 + 4(2) + 1 = -4 + 8 + 1 = 5$$

b. Replacing x with t yields the following.

$$g(t) = -(t)^2 + 4(t) + 1 = -t^2 + 4t + 1$$

Note   In Example 3, note that $g(x + 2)$ is not equal to $g(x) + g(2)$. In general, $g(u + v) \neq g(u) + g(v)$.

c. Replacing x with $x + 2$ yields the following.

$$
\begin{aligned}
g(x + 2) &= -(x + 2)^2 + 4(x + 2) + 1 \\
&= -(x^2 + 4x + 4) + 4x + 8 + 1 \\
&= -x^2 - 4x - 4 + 4x + 8 + 1 \\
&= -x^2 + 5
\end{aligned}
$$

EXAMPLE 4 **A Piecewise–Defined Function**

Evaluate the function when $x = -1, 0,$ and 1.

$$f(x) = \begin{cases} x^2 + 1, & x < 0 \\ x - 1, & x \geq 0 \end{cases}$$

Most graphing utilities can graph functions that are defined piecewise. For example, on the *TI-82* or *TI-83*, you can obtain the graph of the function in Example 4 as follows.

$$\text{Y}_1 = (\text{X}^2 + 1)(\text{X} < 0) + (\text{X} - 1)(\text{X} \geq 0).$$

Solution

Because $x = -1$ is less than 0, use $f(x) = x^2 + 1$ to obtain

$$f(-1) = (-1)^2 + 1 = 2.$$

For $x = 0$, use $f(x) = x - 1$ to obtain

$$f(0) = (0) - 1 = -1.$$

For $x = 1$, use $f(x) = x - 1$ to obtain $f(1) = (1) - 1 = 0.$

The Domain of a Function

The domain of a function can be described explicitly or it can be *implied* by the expression used to define the function. The **implied domain** is the set of all real numbers for which the expression is defined. For instance, the function given by

$$f(x) = \frac{1}{x^2 - 4}$$

has an implied domain that consists of all real x other than $x = \pm 2$. These two values are excluded from the domain because division by zero is undefined. Another common type of implied domain is that used to avoid even roots of negative numbers. For example, the function given by

$$f(x) = \sqrt{x}$$

is defined only for $x \geq 0$. Hence, its implied domain is the interval $[0, \infty)$. In general, the domain of a function *excludes* values that would cause division by zero *or* result in the even root of a negative number.

EXAMPLE 5   **Finding the Domain of a Function**

Find the domain of each function.

a. f: $\{(-3, 0), (-1, 4), (0, 2), (2, 2), (4, -1)\}$ **b.** $g(x) = \dfrac{1}{x + 5}$

c. Volume of a sphere: $V = \frac{4}{3}\pi r^3$ **d.** $h(x) = \sqrt{4 - x}$

Solution

a. The domain of f consists of all first coordinates in the set of ordered pairs.

Domain $= \{-3, -1, 0, 2, 4\}$.

b. Excluding x-values that yield zero in the denominator, the domain of g is the set of all real numbers $x \neq -5$.

c. Because this function represents the volume of a sphere, the values of the radius r must be positive. Thus, the domain is the set of all real numbers r such that $r > 0$.

d. This function is defined only for x-values for which $4 - x \geq 0$. The domain is all real numbers that are less than or equal to 4.

Library of Functions

The *square root* function $f(x) = \sqrt{x}$ is not defined for $x < 0$. This means that you must be careful when analyzing the domain of complicated functions involving the square root symbol.

Note   In Example 5(c), note that the domain of a function may be implied by the physical context. For instance, from the equation $V = \frac{4}{3}\pi r^3$, you would have no reason to restrict r to positive values, but the physical context implies that a sphere cannot have a negative radius.

Applications

EXAMPLE 6   **The Dimensions of a Container**

Figure 1.2

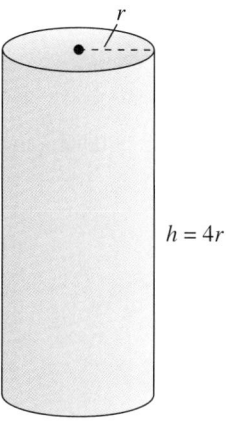

$h = 4r$

You work in the marketing department of a soft-drink company and are experimenting with a new soft-drink can that is slightly narrower and taller than a standard can. For your experimental can, the ratio of the height to the radius is 4, as shown in Figure 1.2.

a. Express the volume of the can as a function of the radius r.

b. Express the volume of the can as a function of the height h.

Solution
The volume of a right circular cylinder is given by the formula

$$V = \pi(\text{radius})^2(\text{height}) = \pi r^2 h.$$

Because the ratio of the height to the radius is 4, you can write $h = 4r$.

a. To write the volume as a function of the radius, use the fact that $h = 4r$.

$$V = \pi r^2 h = \pi r^2(4r) = 4\pi r^3$$

b. To write the volume as a function of the height, use the fact that $r = h/4$.

$$V = \pi \left(\frac{h}{4}\right)^2 h = \frac{\pi h^3}{16}$$

EXAMPLE 7   **The Path of a Baseball**

Figure 1.3

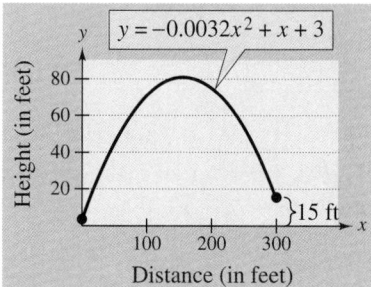

Distance (in feet)

A baseball is hit at a point 3 feet above ground at a velocity of 100 feet per second and an angle of 45°. The path of the baseball is given by the function

$$y = -0.0032x^2 + x + 3$$

where y and x are measured in feet, as shown in Figure 1.3. Will the baseball clear a 10-foot fence located 300 feet from home plate?

Solution
When $x = 300$, the height of the baseball is given by

$$y = -0.0032(300)^2 + 300 + 3 = 15 \text{ feet.}$$

Thus, the ball will clear the fence.

Note In the equation in Example 7, the height of the baseball is a function of the horizontal distance from home plate.

Figure 1.4

Year (0 ↔ 1980)

EXAMPLE 8   **U.S. Travelers Abroad**

The money C (in millions of dollars) spent by U.S. travelers in other countries increased in a linear pattern from 1980 to 1983, as shown in Figure 1.4. Then, in 1984, the money spent took a sharp jump and until 1988 increased in a *different* linear pattern. These two patterns can be approximated by the piecewise-defined function

$$C = \begin{cases} 10,479 + 917.1t, & 0 \le t \le 3 \\ 12,808 + 2350.4t, & 4 \le t \le 8 \end{cases}$$

where $t = 0$ represents 1980. Use this function to approximate the total amount spent by U.S. travelers abroad between 1980 and 1988.   (Source: U.S. Bureau of Economic Analysis)

Solution

From 1980 to 1983, use the formula $C = 10,479 + 917.1t$.

$10,479,$ $11,396,$ $12,313,$ $13,230$

1980   1981   1982   1983

From 1984 to 1988, use the formula $C = 12,808 + 2350.4t$.

$22,210,$ $24,560,$ $26,910,$ $29,261,$ $31,611$

1984   1985   1986   1987   1988

The total of these nine amounts is $181,970, which implies that the total amount spent was approximately $181,970,000,000.

EXAMPLE 9   **From Calculus: Evaluating a Difference Quotient**

For $f(x) = x^2 - 4x + 7$, find $\dfrac{f(x + h) - f(x)}{h}$.

Solution

$$\frac{f(x + h) - f(x)}{h} = \frac{[(x + h)^2 - 4(x + h) + 7] - (x^2 - 4x + 7)}{h}$$

$$= \frac{x^2 + 2xh + h^2 - 4x - 4h + 7 - x^2 + 4x - 7}{h}$$

$$= \frac{2xh + h^2 - 4h}{h}$$

$$= \frac{h(2x + h - 4)}{h}$$

$$= 2x + h - 4, \quad h \ne 0$$

Note   One of the basic definitions in calculus employs the ratio

$$\frac{f(x + h) - f(x)}{h}, \quad h \ne 0$$

which is called a **difference quotient,** as illustrated in Example 9.

Summary of Function Terminology

Function: A **function** is a relationship between two variables such that to each value of the independent variable there corresponds exactly one value of the dependent variable.

Function Notation: $y = f(x)$

f is the **name** of the function.

y is the **dependent variable.**

x is the **independent variable.**

$f(x)$ is the **value of the function at** x.

Domain: The **domain** of a function is the set of all values (inputs) of the independent variable for which the function is defined. If x is in the domain of f, we say that f is **defined** at x. If x is not in the domain of f, we say that f is **undefined** at x.

Range: The **range** of a function is the set of all values (outputs) assumed by the dependent variable (that is, the set of all function values).

Implied Domain: If f is defined by an algebraic expression and the domain is not specified, the **implied domain** consists of all real numbers for which the expression is defined.

Group Activity

Modeling with Piecewise-Defined Functions

x	y
1	5.2
2	5.6
3	6.6
4	8.3
5	11.5
6	15.8
7	12.8
8	10.1
9	8.6
10	6.9
11	4.5
12	2.7

The table at the left shows the monthly revenue y (in thousands of dollars) for one year of a landscaping business, with $x = 1$ representing January.

A mathematical model that represents this data is

$$f(x) = \begin{cases} -1.97x + 26.33 \\ 0.51x^2 - 1.47x + 6.31. \end{cases}$$

For what values of x is each part of the piecewise-defined function defined? How can you tell? Explain your reasoning.

Find $f(5)$ and $f(11)$, and interpret your results in the context of the problem. How do these model values compare with the actual data values?

1.1 /// EXERCISES

In Exercises 1–4, is the relationship a function?

1.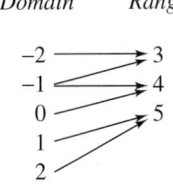
   Domain   *Range*

   $-2 \longrightarrow 5$
   $-1 \quad\quad 6$
   $0 \quad\quad 7$
   $1 \quad\quad 8$
   2

2.   *Domain*   *Range*

   $-2 \quad\quad 3$
   $-1 \quad\quad 4$
   $0 \quad\quad 5$
   1
   2

3.   *Domain*   *Range*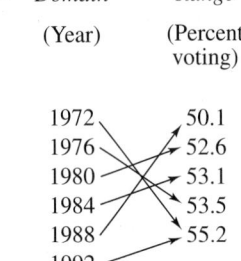

            Cubs
   National
   League   Pirates
            Dodgers

            Orioles
   American
   League   Yankees
            Twins

4.   *Domain*   *Range*

   (Year)   (Percent
              voting)

   1972   50.1
   1976   52.6
   1980   53.1
   1984   53.5
   1988   55.2
   1992

In Exercises 5–8, does the table describe a function? Explain your reasoning.

5.

Input Value	-2	-1	0	1	2
Output Value	-8	-1	0	1	8

6.

Input Value	0	1	2	1	0
Output Value	-4	-2	0	2	4

7.

Input Value	10	7	4	7	10
Output Value	3	6	9	12	15

8.

Input Value	0	3	9	12	15
Output Value	3	3	3	3	3

 The *Interactive* CD-ROM contains step-by-step solutions to all odd-numbered Section and Review Exercises. It also provides Tutorial Exercises, which link to Guided Examples for additional help.

In Exercises 9 and 10, which sets of ordered pairs represent a function from A to B? Explain.

9. $A = \{0, 1, 2, 3\}$ and $B = \{-2, -1, 0, 1, 2\}$
  (a) $\{(0, 1), (1, -2), (2, 0), (3, 2)\}$
  (b) $\{(0, -1), (2, 2), (1, -2), (3, 0), (1, 1)\}$
  (c) $\{(0, 0), (1, 0), (2, 0), (3, 0)\}$

10. $A = \{a, b, c\}$ and $B = \{0, 1, 2, 3\}$
  (a) $\{(a, 1), (c, 2), (c, 3), (b, 3)\}$
  (b) $\{(a, 1), (b, 2), (c, 3)\}$
  (c) $\{(1, a), (0, a), (2, c), (3, b)\}$

Circulation of Newspapers **In Exercises 11 and 12, use the graph, which shows the circulation (in millions) of daily newspapers in the United States.** (Source: Editor & Publisher Company)

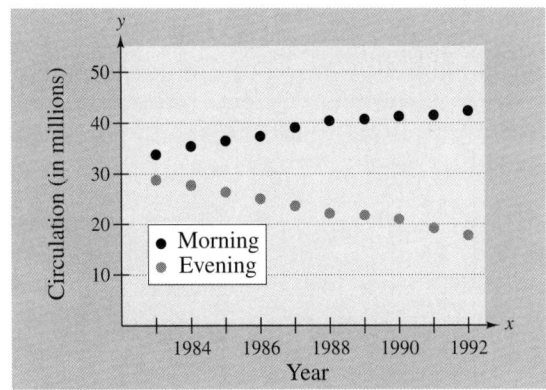

11. Is the circulation of morning newspapers a function of the year? Is the circulation of evening newspapers a function of the year? Explain.

12. Let $f(x)$ represent the circulation of evening newspapers in year x. Find $f(1988)$.

In Exercises 13–22, determine if the equation represents y as a function of x.

13. $x^2 + y^2 = 4$ **14.** $x = y^2$

15. $x^2 + y = 4$ **16.** $x + y^2 = 4$

17. $2x + 3y = 4$

18. $(x - 2)^2 + y^2 = 4$

19. $y^2 = x^2 - 1$

20. $y = \sqrt{x + 5}$

21. $y = |4 - x|$

22. $|y| = 4 - x$

In Exercises 23 and 24, fill in the blanks using the specified function and the given values of the independent variable. Simplify the result.

23. $f(s) = \dfrac{1}{s + 1}$

(a) $f(4) = \dfrac{1}{(\quad) + 1}$

(b) $f(0) = \dfrac{1}{(\quad) + 1}$

(c) $f(4x) = \dfrac{1}{(\quad) + 1}$

(d) $f(x + c) = \dfrac{1}{(\quad) + 1}$

24. $g(x) = x^2 - 2x$

(a) $g(2) = (\quad)^2 - 2(\quad)$

(b) $g(-3) = (\quad)^2 - 2(\quad)$

(c) $g(t + 1) = (\quad)^2 - 2(\quad)$

(d) $g(x + c) = (\quad)^2 - 2(\quad)$

In Exercises 25–36, evaluate the function at the specified values of the independent variable and simplify.

25. $f(x) = 2x - 3$

(a) $f(1)$ (b) $f(-3)$ (c) $f(x - 1)$

26. $g(y) = 7 - 3y$

(a) $g(0)$ (b) $g\left(\frac{7}{3}\right)$ (c) $g(s + 2)$

27. $h(t) = t^2 - 2t$

(a) $h(2)$ (b) $h(1.5)$ (c) $h(x + 2)$

28. $V(r) = \frac{4}{3}\pi r^3$

(a) $V(3)$ (b) $V\left(\frac{3}{2}\right)$ (c) $V(2r)$

29. $f(y) = 3 - \sqrt{y}$

(a) $f(4)$ (b) $f(0.25)$ (c) $f(4x^2)$

30. $f(x) = \sqrt{x + 8} + 2$

(a) $f(-8)$ (b) $f(1)$ (c) $f(x - 8)$

31. $q(x) = \dfrac{1}{x^2 - 9}$

(a) $q(0)$ (b) $q(3)$ (c) $q(y + 3)$

32. $q(t) = \dfrac{2t^2 + 3}{t^2}$

(a) $q(2)$ (b) $q(0)$ (c) $q(-x)$

33. $f(x) = \dfrac{|x|}{x}$

(a) $f(2)$ (b) $f(-2)$ (c) $f(x - 1)$

34. $f(x) = |x| + 4$

(a) $f(2)$ (b) $f(-2)$ (c) $f(x^2)$

35. $f(x) = \begin{cases} 2x + 1, & x < 0 \\ 2x + 2, & x \geq 0 \end{cases}$

(a) $f(-1)$ (b) $f(0)$ (c) $f(2)$

36. $f(x) = \begin{cases} x^2 + 2, & x \leq 1 \\ 2x^2 + 2, & x > 1 \end{cases}$

(a) $f(-2)$ (b) $f(1)$ (c) $f(2)$

In Exercises 37–42, complete the table.

37. $f(x) = x^2 - 3$

x	-2	-1	0	1	2
$f(x)$					

38. $g(x) = \sqrt{x - 3}$

x	3	4	5	6	7
$g(x)$					

39. $h(t) = \frac{1}{2}|t + 3|$

t	-5	-4	-3	-2	-1
$h(t)$					

40. $f(s) = \dfrac{|s - 2|}{s - 2}$

s	0	1	$\frac{3}{2}$	$\frac{5}{2}$	4
$f(s)$					

41. $f(x) = \begin{cases} -\frac{1}{2}x + 4, & x \le 0 \\ (x - 2)^2, & x > 0 \end{cases}$

x	-2	-1	0	1	2
$f(x)$					

42. $h(x) = \begin{cases} 9 - x^2, & x < 3 \\ x - 3, & x \ge 3 \end{cases}$

x	1	2	3	4	5
$h(x)$					

In Exercises 43–46, find all real values of x such that $f(x) = 0$.

43. $f(x) = 15 - 3x$ **44.** $f(x) = \dfrac{3x - 4}{5}$

45. $f(x) = x^2 - 9$ **46.** $f(x) = x^3 - x$

In Exercises 47–50, find the value(s) of x for which $f(x) = g(x)$.

47. $f(x) = x^2$, $g(x) = x + 2$

48. $f(x) = x^2 + 2x + 1$, $g(x) = 3x + 3$

49. $f(x) = \sqrt{3x} + 1$, $g(x) = x + 1$

50. $f(x) = x^4 - 2x^2$, $g(x) = 2x^2$

In Exercises 51–60, find the domain of the function.

51. $f(x) = 5x^2 + 2x - 1$

52. $g(x) = 1 - 2x^2$

53. $h(t) = \dfrac{4}{t}$ **54.** $s(y) = \dfrac{3y}{y + 5}$

55. $g(y) = \sqrt{y - 10}$ **56.** $f(t) = \sqrt[3]{t + 4}$

57. $f(x) = \sqrt[4]{1 - x^2}$ **58.** $h(x) = \dfrac{10}{x^2 - 2x}$

59. $g(x) = \dfrac{1}{x} - \dfrac{3}{x + 2}$ **60.** $f(s) = \dfrac{\sqrt{s - 1}}{s - 4}$

In Exercises 61–64, assume that the domain of f is the set $A = \{-2, -1, 0, 1, 2\}$. Determine the set of ordered pairs representing the function f.

61. $f(x) = x^2$ **62.** $f(x) = \dfrac{2x}{x^2 + 1}$

63. $f(x) = \sqrt{x + 2}$ **64.** $f(x) = |x + 1|$

65. *Essay*   In your own words, explain the meaning of *domain* and *range*.

66. *Think About It*   Describe an advantage of function notation.

Exploration   In Exercises 67–70, select a function from $f(x) = cx$, $g(x) = cx^2$, $h(x) = c\sqrt{|x|}$, or $r(x) = c/x$ and determine the value of the constant c such that the function fits the data given in the table.

67.

x	-4	-1	0	1	4
y	-32	-2	0	-2	-32

68.

x	-4	-1	0	1	4
y	-1	$-\frac{1}{4}$	0	$\frac{1}{4}$	1

69.

x	-4	-1	0	1	4
y	-8	-32	Undef.	32	8

70.

x	-4	-1	0	1	4
y	6	3	0	3	6

In Exercises 71–76, find the difference quotient and simplify your answer.

71. $f(x) = x^2 - x + 1$, $\dfrac{f(2 + h) - f(2)}{h}$, $h \ne 0$

72. $f(x) = 5x - x^2$, $\dfrac{f(5 + h) - f(5)}{h}$, $h \ne 0$

73. $f(x) = x^3$, $\dfrac{f(x + c) - f(x)}{c}$, $c \ne 0$

74. $f(x) = 2x$, $\quad \dfrac{f(x + c) - f(x)}{c}$, $c \neq 0$

75. $g(x) = 3x - 1$, $\quad \dfrac{g(x) - g(3)}{x - 3}$, $x \neq 3$

76. $f(t) = \dfrac{1}{t}$, $\quad \dfrac{f(t) - f(1)}{t - 1}$, $t \neq 0$

77. *Area of a Circle*   Express the area A of a circle as a function of its circumference C.

78. *Area of a Triangle*   Express the area A of an equilateral triangle as a function of the length s of its sides.

79. *Exploration*   An open box of maximum volume is to be made from a square piece of material, 24 centimeters on a side, by cutting equal squares from the corners and turning up the sides (see figure).

(a) Use the table feature of a graphing utility to complete six rows of the table. Use the result to guess the maximum volume.

Height, x	Width	Volume, V
1	$24 - 2(1)$	$1[24 - 2(1)]^2 = 484$
2	$24 - 2(2)$	$2[24 - 2(2)]^2 = 800$

(b) Use a graphing utility to plot the points (x, V). Is V a function of x? If yes, write the volume V as a function of x, and determine its domain.

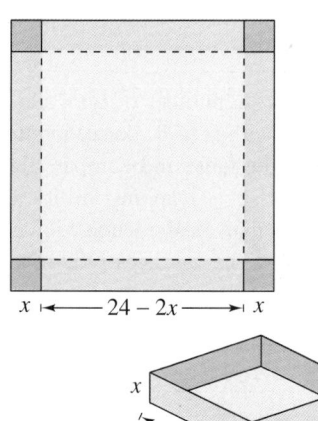

80. *Exploration*   The cost per unit in the production of a certain radio model is $60. The manufacturer charges $90 per unit for orders of 100 or less. To encourage large orders, the manufacturer reduces the charge by $0.15 per radio for each unit ordered in excess of 100.

(a) Use the table feature of a graphing utility to complete six rows of the table. Use the result to estimate the maximum profit.

Units, x	Price, p	Profit, P
102	$90 - 2(0.15)$	$xp - 102(60)$
104	$90 - 4(0.15)$	$xp - 104(60)$

(b) Use a graphing utility to plot the points (x, P). Is P a function of x? If yes, write the profit P as a function of x, and determine its domain.

81. *Area of a Triangle*   A right triangle is formed in the first quadrant by the x and y axes and a line through the point $(2, 1)$ (see figure). Write the area of the triangle as a function of x, and determine the domain of the function.

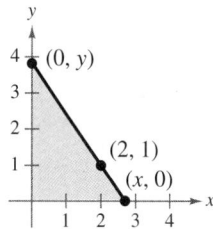

82. *Area of a Rectangle*   A rectangle is bounded by the x-axis and the semicircle $y = \sqrt{36 - x^2}$ (see figure). Write the area of the rectangle as a function of x, and determine the domain of the function.

Figure for 82

Figure for 83

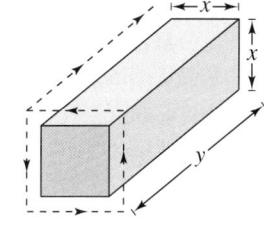

83. *Volume of a Package* A rectangular package to be sent by a postal service can have a maximum combined length and girth (perimeter of a cross section) of 108 inches (see figure). Write the volume of the package as a function of x. What is the domain of the function?

84. *Price of Mobile Homes* The average price p (in thousands of dollars) of a new mobile home in the United States from 1974 to 1993 can be approximated by the piecewise-defined model

$$p(t) = \begin{cases} 19.247 + 1.694t, & -6 \le t \le -1 \\ 19.305 + 0.427t + 0.033t^2, & 0 \le t \le 13 \end{cases}$$

where t = 0 represents 1980 (see figure). Use a graphing utility to graph the model and find the average price of a mobile home in 1978, 1988, and 1993. (Source: U.S. Bureau of Census, Construction Reports)

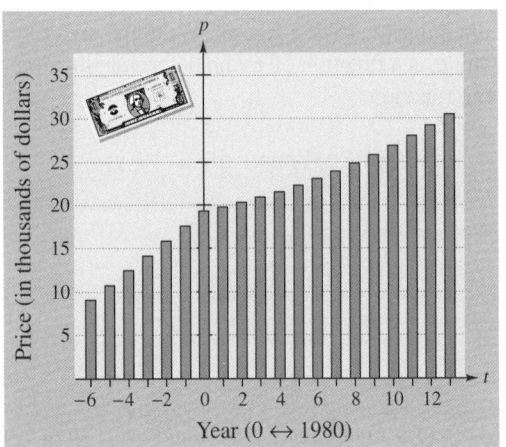

Year (0 ↔ 1980)

85. *Cost, Revenue, and Profit* A company produces a product for which the variable cost is $12.30 per unit and the fixed costs are $98,000. The product sells for $17.98. Let x be the number of units produced and sold.

(a) Write the total cost C as a function of the number of units produced.

(b) Write the revenue R as a function of the number of units sold.

(c) Write the profit P as a function of the number of units sold. (*Note:* P = R − C.)

86. *Charter Bus Fares* For groups of 80 or more people, a charter bus company determines the rate per person according to the formula

$$\text{Rate} = 8 - 0.05(n - 80), \qquad n \ge 80$$

where the rate is given in dollars and n is the number of people.

(a) Express the revenue R for the bus company as a function of n.

(b) Use the function from part (a) to complete the table. What can you conclude?

n	90	100	110	120	130	140	150
R(n)							

(c) Use a graphing utility to graph R and determine the number of people that will produce a maximum revenue. Compare the result with your conclusion from part (b).

87. *Chapter Opener* Use the data on page 83. Let f(t) represent the number of lynx in year t.

(a) Find f(1992).

(b) Find $\dfrac{f(1994) - f(1991)}{1994 - 1991}$

and interpret the result in the context of the problem.

(c) An approximate model for the function is

$$N(t) = \frac{434t + 4387}{45t^2 - 55t + 100}$$

where N is the number of lynx and t is the time in years, with t = 0 corresponding to 1990. Complete the table and compare the result with the data. Use a graphing utility to graph the model and data in the same viewing rectangle. Comment on the validity of the model.

t	1988	1989	1990	1991
N				

t	1992	1993	1994	1995
N				

1.2   Graphs of Functions

The Graph of a Function / *Increasing and Decreasing Functions* / *Relative Minimum and Maximum Values* / *Step Functions* / *Even and Odd Functions*

The Graph of a Function

In Section 1.1 you studied functions from an algebraic point of view. In this section, you will study functions from a geometric perspective. The **graph of a function** f is the collection of ordered pairs $(x, f(x))$ such that x is in the domain of f. As you study this section, remember the following geometrical interpretation of x and $f(x)$.

$$x = \text{the directed distance from the } y\text{-axis}$$
$$f(x) = \text{the directed distance from the } x\text{-axis}$$

Example 1 shows how to use the graph of a function to find the domain and range of the function.

Figure 1.5

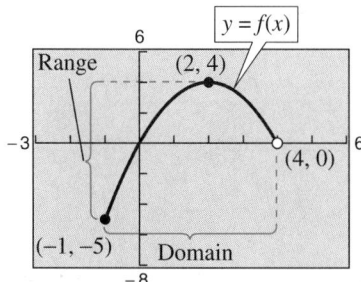

EXAMPLE 1   **Finding the Domain and Range of a Function**

Use the graph of the function f, shown in Figure 1.5.

a. Find the domain of f.

b. Find the function values $f(-1)$ and $f(2)$.

c. Find the range of f.

Solution

a. The closed dot (on the left) indicates that $x = -1$ is in the domain of f, whereas the open dot (on the right) indicates that $x = 4$ is not in the domain. Thus, the domain of f is all x in the interval $[-1, 4)$.

b. Because $(-1, -5)$ is a point on the graph of f, it follows that

$$f(-1) = -5.$$

Similarly, because $(2, 4)$ is a point on the graph of f, it follows that

$$f(2) = 4.$$

c. Because the graph does not extend below $f(-1) = -5$ or above $f(2) = 4$, the range of f is the interval $[-5, 4]$.

By the definition of a function, at most one y-value corresponds to a given x-value. It follows, then, that a vertical line can intersect the graph of a function at most once. This observation provides a convenient visual test for functions.

> ### Vertical Line Test for Functions
>
> A set of points in a coordinate plane is the graph of y as a function of x if and only if no vertical line intersects the graph at more than one point.

EXAMPLE 2   **Vertical Line Test for Functions**

Which of the graphs in Figure 1.6 represent y as a function of x?

Solution

a. This *is not* a graph of y as a function of x because you can find a vertical line that intersects the graph twice.

b. This *is* a graph of y as a function of x because every vertical line intersects the graph at most once.

c. This *is* a graph of y as a function of x. (Note that if a vertical line does not intersect the graph, it simply means that the function is undefined for that particular value of x.)

Figure 1.6

(a)

(b)

(c)

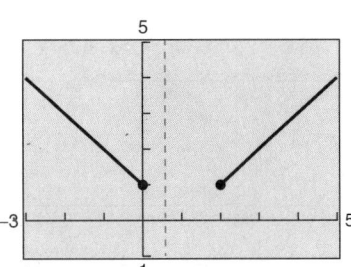

Note   Most graphing utilities are designed to graph functions of x more easily than other types of equations. For instance, the graph shown in Figure 1.6(a) represents the equation $x + y^2 - 4y + 3 = 0$. Try using a graphing utility to duplicate this graph. To sketch the graph, did you need to use *two* equations?

Increasing and Decreasing Functions

The more you know about the graph of a function, the more you know about the function itself. Consider the graph shown in Figure 1.7. Moving from *left to right*, this graph falls from $x = -2$ to $x = 0$, is constant from $x = 0$ to $x = 2$, and rises from $x = 2$ to $x = 4$.

Figure 1.7

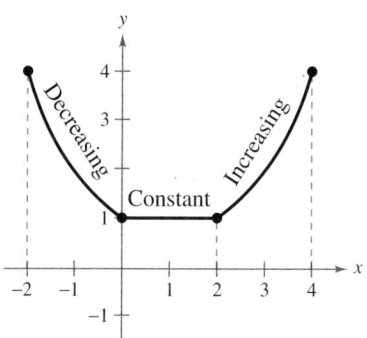

> ### Increasing, Decreasing, and Constant Functions
>
> A function f is **increasing** on an interval if, for any x_1 and x_2 in the interval, $x_1 < x_2$ implies $f(x_1) < f(x_2)$.
>
> A function f is **decreasing** on an interval if, for any x_1 and x_2 in the interval, $x_1 < x_2$ implies $f(x_1) > f(x_2)$.
>
> A function f is **constant** on an interval if, for any x_1 and x_2 in the interval, $f(x_1) = f(x_2)$.

EXAMPLE 3   Increasing and Decreasing Functions

In Figure 1.8, determine the open intervals on which each function is increasing, decreasing, or constant.

Solution

a. Although it might appear that there is an interval in which this function is constant, you can see that if $x_1 < x_2$, then $f(x_1) = x_1^3 < x_2^3 = f(x_2)$. Thus, the function is increasing over the entire real line.

b. This function is increasing on the interval $(-\infty, -1)$, decreasing on the interval $(-1, 1)$, and increasing on the interval $(1, \infty)$.

c. This function is increasing on the interval $(-\infty, 0)$, constant on the interval $(0, 2)$, and decreasing on the interval $(2, \infty)$.

Figure 1.8

(a)

(b)

(c)

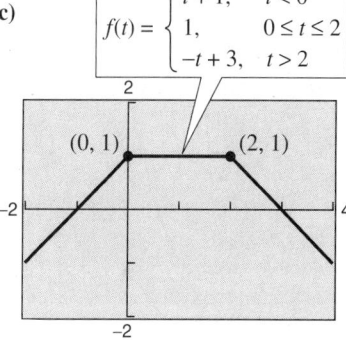

Relative Minimum and Maximum Values

The points at which a function changes its increasing, decreasing, or constant behavior are helpful in determining the relative maximum or relative minimum values of the function.

Definition of Relative Minimum and Relative Maximum

A function value $f(a)$ is called a **relative minimum** of f if there exists an interval (x_1, x_2) that contains a such that

$$x_1 < x < x_2 \quad \text{implies} \quad f(a) \le f(x).$$

A function value $f(a)$ is called a **relative maximum** of f if there exists an interval (x_1, x_2) that contains a such that

$$x_1 < x < x_2 \quad \text{implies} \quad f(a) \ge f(x).$$

Figure 1.9

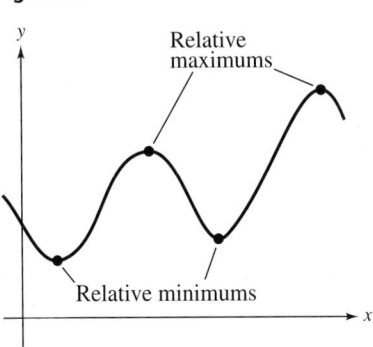

Figure 1.9 shows several different examples of relative minimums and relative maximums. In Section 2.1, you will study a technique for finding the *exact points* at which a second-degree polynomial function has a relative minimum or relative maximum. For the time being, however, you can use a graphing utility to find reasonable approximations of these points.

EXAMPLE 4   ▱   **Approximating a Relative Minimum**

Use a graphing utility to approximate the relative minimum of the function $f(x) = 3x^2 - 4x - 2$.

Solution

Figure 1.10

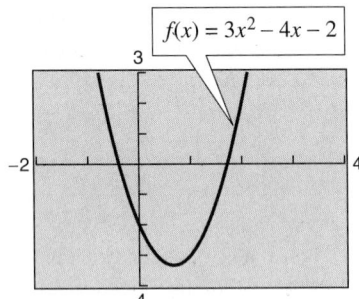

The graph of f is shown in Figure 1.10. By using the zoom and trace features of a graphing utility, you can estimate that the function has a relative minimum at the point

$(0.67, -3.33).$ Relative minimum

Later, in Section 2.1, you will be able to determine that the exact point at which the relative minimum occurs is $\left(\frac{2}{3}, -\frac{10}{3}\right)$.   ▱

Note   When you use a graphing utility to estimate x- and y-values of a relative minimum or relative maximum, the automatic zoom feature will often produce graphs that are nearly flat. To overcome this problem, you can manually change the vertical setting of the viewing rectangle. The graph will vertically stretch if the values of Y_{min} and Y_{max} are closer together.

EXAMPLE 5   **Approximating Relative Minimums and Maximums**

Use a graphing utility to approximate the relative minimum and relative maximum of the function $f(x) = -x^3 + x$.

Solution

A sketch of the graph of f is shown in Figure 1.11. By using the zoom and trace features of the graphing utility, you can estimate that the function has a relative minimum at the point

$(-0.58, -0.38)$ Relative minimum

and a relative maximum at the point

$(0.58, 0.38)$.   Relative maximum

If you go on to take a course in calculus, you will learn a technique for finding the exact points at which this function has a relative minimum and a relative maximum.

Figure 1.11

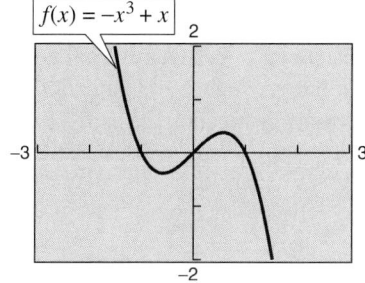

$f(x) = -x^3 + x$

Note   Some graphing utilities have built-in programs that will find minimum or maximum values. For instance, on a *TI-82* or *TI-83*, the minimum program can be accessed by CALC (3:minimum). If your graphing utility has such features, try using them to rework Example 5.

EXAMPLE 6   **The Price of Diamonds**

During the 1980s, the average price of a 1-carat polished diamond decreased and then increased according to the model

$$C = -0.7t^3 + 16.25t^2 - 106t + 388, \qquad 2 \le t \le 10,$$

where C is the average price in dollars (on the Antwerp Index) and t represents the calendar year, with $t = 2$ corresponding to January 1, 1982. According to this model, during which years was the price of diamonds decreasing? During which years was the price of diamonds increasing? Approximate the minimum price of a 1-carat diamond between 1982 and 1990.   (Source: Diamond High Council)

Solution

To solve this problem, sketch an accurate graph of the function, as shown in Figure 1.12. From the graph, you can see that the price of diamonds decreased from 1982 until late 1984. Then, from late 1984 to 1990, the price increased. The minimum price during the 8-year period was approximately $175.

Figure 1.12

$C = -0.7t^3 + 16.25t^2 - 106t + 388$

Average price (in dollars)

Year (2 ↔ 1982)

Step Functions

Figure 1.13

EXAMPLE 7   The Greatest Integer Function

The **greatest integer function** is denoted by $[\![x]\!]$ and is defined by

$$f(x) = [\![x]\!] = \text{the greatest integer less than or equal to } x.$$

The graph of this function is shown in Figure 1.13. Note that the graph of the greatest integer function jumps vertically one unit at each integer and is constant (a horizontal line segment) between each pair of consecutive integers. Because of the jumps in its graph, the greatest integer function is an example of a **step function.** Some values of the greatest integer function are as follows.

$$[\![-1]\!] = -1 \qquad [\![-0.5]\!] = -1$$
$$[\![0]\!] = 0 \qquad [\![0.5]\!] = 0$$
$$[\![1]\!] = 1 \qquad [\![1.5]\!] = 1$$

The range of the greatest integer function is the set of all integers.

Library of Functions

The *greatest integer* function has an infinite number of breaks or steps—one at each integer value in its domain. Could you describe the greatest integer function using a piecewise-defined function? How does the graph of the greatest integer function differ from the graph of a line with zero slope?

EXAMPLE 8   The Cost of a Telephone Call

Suppose the cost of a telephone call between Los Angeles and San Francisco is $0.50 for the first minute and $0.36 for each additional minute. The greatest integer function can be used to create a model for the cost of this call.

$$C = 0.50 + 0.36[\![t]\!], \qquad t > 0,$$

where C is the total cost of the call in dollars and t is the length of the call in minutes. Sketch the graph of this function.

Solution

For calls up to 1 minute, the cost is $0.50. For calls between 1 and 2 minutes, the cost is $0.86, and so on.

Length of Call:	$0 < t < 1$	$1 \leq t < 2$	$2 \leq t < 3$	$3 \leq t < 4$	$4 \leq t < 5$
Cost of Call:	$0.50	$0.86	$1.22	$1.58	$1.94

Using these values, you can sketch the graph shown in Figure 1.14.

Figure 1.14

Note   Most graphing utilities display graphs in *connected mode*, which means that the graph has no breaks. When you are sketching graphs that do have breaks, it is better to use *dot mode*. Graph the greatest integer function [often called Int (x)] in connected and dot modes, and compare the two results.

The *absolute value* function can be expressed as a piecewise-defined function.

$$|x| = \begin{cases} -x, & x < 0 \\ x, & x \geq 0 \end{cases}$$

Use this definition to show that $|-5| = 5$ and $|0| = 0$.

E X P L O R A T I O N

Use a graphing utility to graph the functions $y_1 = x^2$, $y_2 = x^4$, and $y_3 = 1/x^2$ in the standard viewing rectangle. What kind of symmetry do you observe? Then graph and discuss the symmetry of $y_1 = x^3$, $y_2 = x^5$, and $y_3 = 1/x$. Can you predict the symmetry of $y = x^8$ and $y = 1/x^5$ without graphing them?

Even and Odd Functions

 A computer animation of this concept appears in the *Interactive* CD-ROM.

A graph has **symmetry with respect to the y-axis** if, whenever (x, y) is on the graph, so is the point $(-x, y)$. A graph has **symmetry with respect to the origin** if, whenever (x, y) is on the graph, so is the point $(-x, -y)$. A graph has **symmetry with respect to the x-axis** if, whenever (x, y) is on the graph, so is the point $(x, -y)$. These three types of symmetry are illustrated in Figure 1.15.

Figure 1.15

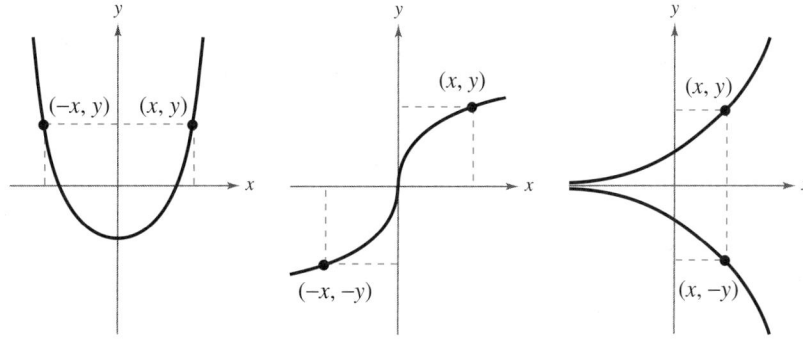

A function whose graph is symmetric with respect to the y-axis is an **even** function. A function whose graph is symmetric with respect to the origin is an **odd** function. The graph of a (nonzero) function cannot be symmetric with respect to the x-axis.

Test for Even and Odd Functions

A function f is **even** if, for each x in the domain of f, $f(-x) = f(x)$.

A function f is **odd** if, for each x in the domain of f, $f(-x) = -f(x)$.

Figure 1.16

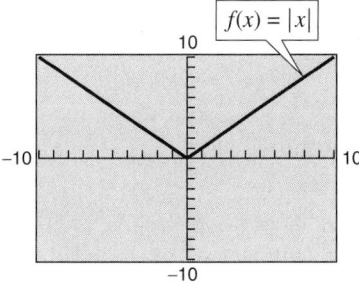

EXAMPLE 9 ▭ **Testing for Evenness and Oddness**

Figure 1.16 shows the $\bigvee$-shape of the **absolute value function,** $f(x) = |x|$. This was produced using the "abs" key and the standard viewing rectangle. Is the function even, odd, or neither?

Solution

Because the graph is symmetric about the y-axis, it is an even function. You can verify this algebraically by observing that

$$f(-x) = |-x|$$
$$= |x|$$
$$= f(x).$$

▭

EXAMPLE 10 ◢ **Even and Odd Functions**

Determine whether each function is even, odd, or neither.

a. $g(x) = x^3 - x$ **b.** $h(x) = x^2 + 1$ **c.** $f(x) = x^3 - 1$

The graphs of the three functions are shown in Figure 1.17.

Solution

a. This function is odd because

$$g(-x) = (-x)^3 - (-x) = -x^3 + x = -(x^3 - x) = -g(x).$$

b. This function is even because

$$h(-x) = (-x)^2 + 1 = x^2 + 1 = h(x).$$

c. Substituting $-x$ for x produces $f(-x) = (-x)^3 - 1 = -x^3 - 1$. Because $f(x) = x^3 - 1$ and $-f(x) = -x^3 + 1$, you can conclude that $f(-x) \neq f(x)$ and $f(-x) \neq -f(x)$. Hence, the function is neither even nor odd.

Figure 1.17

(a)

(b)

(c)

◢

Group Activity   *Increasing and Decreasing Functions*

Describe three different functions that represent quantities between 1980 and 1995. Describe one that decreased during this time, one that increased, and one that was constant. For instance, the value of the dollar decreased, the cost of first-class postage increased, and the land size of the United States remained constant. Present your results graphically.

1.2 /// EXERCISES

In Exercises 1–4, find the domain and range of the function.

1. $f(x) = \sqrt{x^2 - 1}$

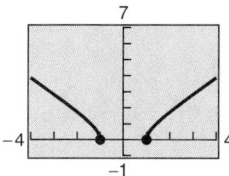

2. $f(x) = \frac{1}{2}|x - 2|$

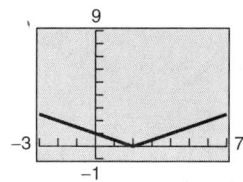

3. $h(x) = \sqrt{16 - x^2}$

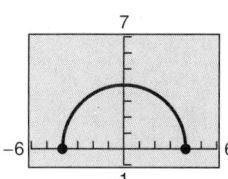

4. $g(x) = \dfrac{|x - 1|}{x - 1}$

In Exercises 5–8, use a graphing utility to graph the function and find its domain and range.

5. $g(x) = 1 - x^2$

6. $f(x) = \sqrt{x - 1}$

7. $f(x) = |x + 3|$

8. $h(t) = \sqrt{4 - t^2}$

In Exercises 9–14, use the Vertical Line Test to determine if y is a function of x. Describe how you can use a graphing utility to produce the given graph.

9. $y = \frac{1}{2}x^2$

10. $y = \frac{1}{4}x^3$

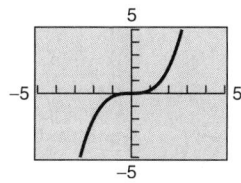

11. $x - y^2 = 1$

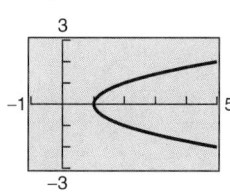

12. $x^2 + y^2 = 25$

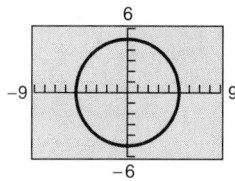

13. $x^2 = 2xy - 1$

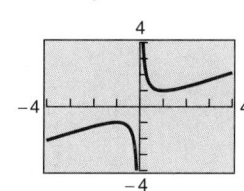

14. $x = |y + 2|$

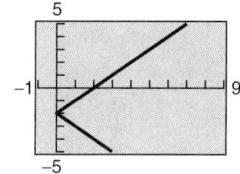

In Exercises 15–18, select the viewing rectangle on a graphing utility that shows the most complete graph of the function.

15. $f(x) = -0.2x^2 + 3x + 32$

Xmin = -2	Xmin = -10	Xmin = 0
Xmax = 20	Xmax = 30	Xmax = 10
Xscl = 1	Xscl = 5	Xscl = 0.5
Ymin = -10	Ymin = -5	Ymin = 0
Ymax = 30	Ymax = 50	Ymax = 200
Yscl = 4	Yscl = 5	Yscl = 25

16. $f(x) = 6[x - (0.1x)^5]$

Xmin = -500	Xmin = -25	Xmin = -20
Xmax = 5000	Xmax = 25	Xmax = 20
Xscl = 50	Xscl = 5	Xscl = 5
Ymin = -500	Ymin = -25	Ymin = -100
Ymax = 500	Ymax = 25	Ymax = 100
Yscl = 50	Yscl = 5	Yscl = 20

17. $f(x) = 4x^3 - x^4$

Xmin = -2	Xmin = -50	Xmin = 0
Xmax = 6	Xmax = 50	Xmax = 2
Xscl = 1	Xscl = 5	Xscl = 0.2
Ymin = -10	Ymin = -50	Ymin = -2
Ymax = 30	Ymax = 50	Ymax = 2
Yscl = 4	Yscl = 5	Yscl = 0.5

18. $f(x) = 10x\sqrt{400 - x^2}$

Xmin = -5	Xmin = -20	Xmin = -25
Xmax = 50	Xmax = 20	Xmax = 25
Xscl = 5	Xscl = 2	Xscl = 5
Ymin = -5000	Ymin = -500	Ymin = -2000
Ymax = 5000	Ymax = 500	Ymax = 2000
Yscl = 500	Yscl = 50	Yscl = 200

In Exercises 19–22, (a) determine the intervals over which the function is increasing, decreasing, or constant, and (b) determine if the function is even, odd, or neither.

19. $f(x) = \frac{3}{2}x$ **20.** $f(x) = x^2 - 4x$

 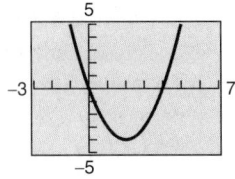

21. $f(x) = x^3 - 3x^2 + 2$ **22.** $f(x) = \sqrt{x^2 - 1}$

 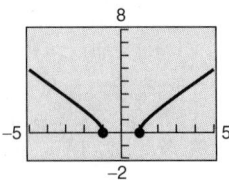

23. *Think About It*   Does the graph in Exercise 11 represent *x* as a function of *y*? Explain.

24. *Think About It*   Does the graph in Exercise 12 represent *x* as a function of *y*? Explain.

In Exercises 25–28, (a) use a graphing utility to graph the function, (b) determine the intervals over which the function is increasing, decreasing, or constant, and (c) determine if the function is even, odd, or neither.

25. $f(x) = 3x^4 - 6x^2$

26. $f(x) = x^{2/3}$

27. $f(x) = x\sqrt{x + 3}$

28. $f(x) = |x + 1| + |x - 1|$

In Exercises 29–34, determine if the function is even, odd, or neither.

29. $f(x) = x^6 - 2x^2 + 3$ **30.** $h(x) = x^3 - 5$

31. $g(x) = x^3 - 5x$ **32.** $f(x) = x\sqrt{1 - x^2}$

33. $f(t) = t^2 + 2t - 3$ **34.** $g(s) = 4s^{2/3}$

Think About It   In Exercises 35 and 36, find the coordinates of a second point on the graph of a function f if the given point is on the graph and the function is (a) even and (b) odd.

35. $\left(-\frac{3}{2}, 4\right)$ **36.** $(4, 9)$

In Exercises 37–46, sketch the graph of the function and determine if the function is even, odd, or neither. Use a graphing utility to verify your sketch.

37. $f(x) = 3$ **38.** $g(x) = x$

39. $f(x) = 5 - 3x$ **40.** $h(x) = x^2 - 4$

41. $f(x) = \sqrt{1 - x}$ **42.** $f(x) = x^{3/2}$

43. $g(t) = \sqrt[3]{t - 1}$ **44.** $f(x) = |x + 2|$

45. $f(x) = \begin{cases} x + 3, & x \le 0 \\ 3, & 0 < x \le 2 \\ 2x - 1, & x > 2 \end{cases}$

46. $f(x) = \begin{cases} 2x + 1, & x \le -1 \\ x^2 - 2, & x > -1 \end{cases}$

In Exercises 47–50, use a graphing utility to graph the piecewise-defined function.

47. $f(x) = \begin{cases} 2x + 3, & x < 0 \\ 3 - x, & x \geq 0 \end{cases}$

48. $f(x) = \begin{cases} \sqrt{4 + x}, & x < 0 \\ \sqrt{4 - x}, & x \geq 0 \end{cases}$

49. $f(x) = \begin{cases} x^2 + 5, & x \leq 1 \\ -x^2 + 4x + 3, & x > 1 \end{cases}$

50. $f(x) = \begin{cases} 1 - (x - 1)^2, & x \leq 2 \\ \sqrt{x - 2}, & x > 2 \end{cases}$

In Exercises 51–56, use a graphing utility to approximate (to two-decimal-place accuracy) any relative minimum or maximum values of the function.

51. $f(x) = x^2 - 6x$

52. $f(x) = (x - 1)^2(x + 2)$

53. $y = 2x^3 + 3x^2 - 12x$

54. $y = x^3 - 6x^2 + 15$

55. $h(x) = (x - 1)\sqrt{x}$

56. $g(x) = x\sqrt{4 - x}$

57. *Maximum Area*   The perimeter of a rectangle (see figure) is 100 meters.

(a) Show that the area of the rectangle is given by $A = x(50 - x)$, where x is its length.

(b) Use a graphing utility to graph the area function.

(c) Use a graphing utility to approximate the maximum area of the rectangle and the dimensions that yield the maximum area.

x

58. *Maximum Profit*   The marketing department of a company estimates that the demand for a product is given by

$$p = 100 - 0.0001x$$

where p is the price per unit and x is the number of units. The cost of producing x units is given by

$$C = 350,000 + 30x$$

and the profit for producing and selling x units is given by

$$P = R - C = xp - C.$$

Use a graphing utility to graph the profit function and estimate the number of units that would produce a maximum profit.

In Exercises 59 and 60, use a graphing utility to graph the function. State the domain and range of the function. Describe the pattern of the graph.

59. $s(x) = 2\left(\frac{1}{4}x - \left[\!\left[\frac{1}{4}x\right]\!\right]\right)$

60. $g(x) = 2\left(\frac{1}{4}x - \left[\!\left[\frac{1}{4}x\right]\!\right]\right)^2$

61. *Cost of a Telephone Call*   The cost of a telephone call between two cities is $0.65 for the first minute and $0.40 for each additional minute.

(a) It is required that a model be created for the cost C of a telephone call between the two cities lasting t minutes. Which of the following is the appropriate model? Explain.

$$C_1(t) = 0.65 + 0.4[\![t - 1]\!]$$
$$C_2(t) = 0.65 - 0.4[\![-(t - 1)]\!]$$

(b) Use a graphing utility to graph the appropriate model and use the zoom and trace features to determine the cost of an 18-minute, 45-second call.

62. *Cost of Overnight Delivery*   Suppose that the cost of sending an overnight package from New York to Atlanta is $9.80 for under one pound and $2.50 for each additional pound. Use the greatest integer function to create a model for the cost C of overnight delivery of a package weighing x pounds where $x > 0$. Sketch the graph of the function.

In Exercises 63–70, graph the function and determine the interval(s) (if any) on the real axis for which $f(x) \geq 0$. Use a graphing utility to verify your results.

63. $f(x) = 4 - x$ **64.** $f(x) = 4x + 2$

65. $f(x) = x^2 - 9$ **66.** $f(x) = x^2 - 4x$

67. $f(x) = 1 - x^4$ **68.** $f(x) = \sqrt{x + 2}$

69. $f(x) = x^2 + 1$ **70.** $f(x) = -\left(1 + |x|\right)$

In Exercises 71–74, write the height h of the rectangle as a function of x.

71.

72.

73.

74.
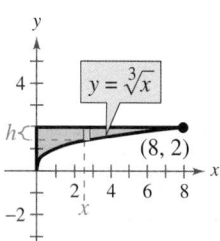

In Exercises 75 and 76, write the length L of the rectangle as a function of y.

75.

76.
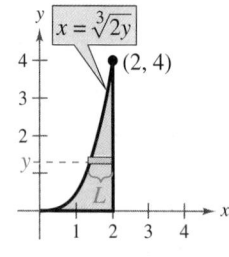

77. *Data Analysis* The amounts y of the merchandise trade balance of the United States in billions of dollars for the years 1986 through 1993 were as follows: 1986(-152.7), 1987(-152.1), 1988(-118.6), 1989 (-109.6), 1990(-101.7), 1991(-65.4), 1992 (-85.4), 1993(-115.8).   (Source: U.S. Bureau of the Census)

(a) Let t be the time in years, with $t = 0$ corresponding to 1990. Use the regression capabilities of a graphing utility to find a cubic model for the data. What is the domain of the model?

(b) Use a graphing utility to graph the data and model.

(c) For which year does the model most accurately estimate the actual data? During which year is it least accurate?

(d) Why would economists be concerned if this model remained valid in the future?

78. *Fluid Flow* The intake pipe of a 100-gallon tank has a flow rate of 10 gallons per minute, and two drain pipes have a flow rate of 5 gallons per minute each. The figure shows the volume V of fluid in the tank as a function of time t. Determine the pipes in which the fluid is flowing in specific subintervals of the 1 hour of time shown on the graph. (There is more than one correct answer.)

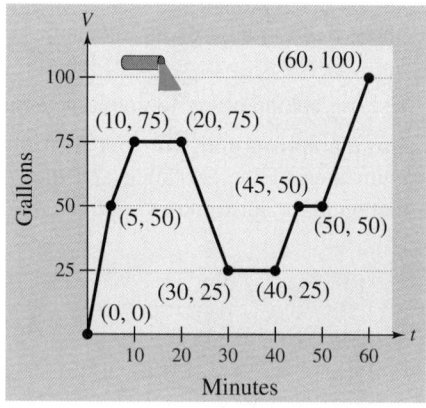

79. Prove that a function of the following form is odd.

$$y = a_{2n+1}x^{2n+1} + a_{2n-1}x^{2n-1} + \cdots + a_3 x^3 + a_1 x$$

80. Prove that a function of the following form is even.

$$y = a_{2n}x^{2n} + a_{2n-2}x^{2n-2} + \cdots + a_2 x^2 + a_0$$

1.3 Shifting, Reflecting, and Stretching Graphs

*Summary of Graphs of Common Functions / Vertical and Horizontal Shifts /
Reflections / Nonrigid Transformations*

Summary of Graphs of Common Functions

One of the goals of this text is to enable you to build your intuition for the basic
shapes of the graphs of different types of functions. For instance, from your
study of lines in Section P.3, you can determine the basic shape of the graph of
the linear function

$$f(x) = ax + b.$$

Specifically, you know that the graph of this function is a line whose slope is a
and whose y-intercept is b.

The six graphs shown in Figure 1.18 represent the most commonly used functions in algebra. Familiarity with the basic characteristics of these simple
graphs will help you analyze the shapes of more complicated graphs. Try using
a graphing utility to verify these graphs.

Figure 1.18

(a) Constant Function

(b) Identity Function

(c) Absolute Value Function

(d) Square Root Function

(e) Squaring Function

(f) Cubing Function

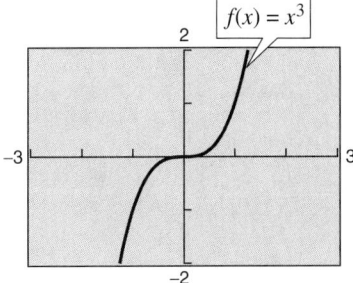

Vertical and Horizontal Shifts

A computer animation of this concept appears in the *Interactive* CD-ROM.

Many functions have graphs that are simple transformations of the common graphs summarized in Figure 1.18. For example, you can obtain the graph of

$$h(x) = x^2 + 2$$

by shifting the graph of $f(x) = x^2$ *up* two units, as shown in Figure 1.19. In function notation, h and f are related as follows.

$$h(x) = x^2 + 2$$
$$= f(x) + 2 \qquad \text{Upward shift of 2}$$

Similarly, you can obtain the graph of

$$g(x) = (x - 2)^2$$

by shifting the graph of $f(x) = x^2$ to the *right* two units, as shown in Figure 1.20. In this case, the functions g and f have the following relationship.

$$g(x) = (x - 2)^2$$
$$= f(x - 2) \qquad \text{Right shift of 2}$$

EXPLORATION

Use a graphing utility to display the graphs of $y = x^2 + c$ where $c = -2, 0, 2,$ and 4. Use the result to describe the effect that c has on the graph.

Use a graphing utility to display the graphs of $y = (x + c)^2$ where $c = -2, 0, 2,$ and 4. Use the result to describe the effect that c has on the graph.

Figure 1.19

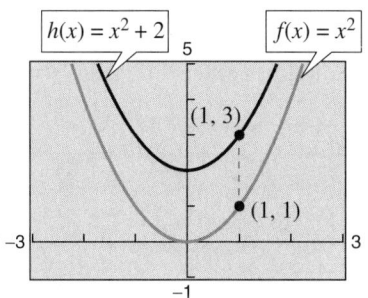

Vertical shift upward: two units

Figure 1.20

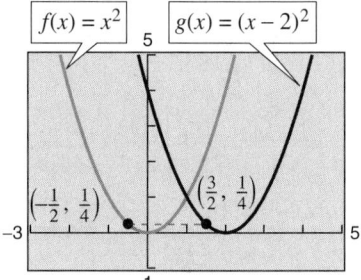

Horizontal shift to the right: two units

The following list summarizes this discussion about horizontal and vertical shifts.

Vertical and Horizontal Shifts

Let c be a positive real number. **Vertical and horizontal shifts** in the graph of $y = f(x)$ are represented as follows.

1. Vertical shift c units **upward:** $h(x) = f(x) + c$
2. Vertical shift c units **downward:** $h(x) = f(x) - c$
3. Horizontal shift c units to the **right:** $h(x) = f(x - c)$
4. Horizontal shift c units to the **left:** $h(x) = f(x + c)$

EXAMPLE 1 ◢ **Shifts in the Graph of a Function**

Compare the graph of each function with the graph of $f(x) = x^3$.

a. $g(x) = x^3 + 1$ **b.** $h(x) = (x - 1)^3$ **c.** $k(x) = (x + 2)^3 + 1$

Solution

a. Begin by graphing $f(x) = x^3$ and $g(x) = x^3 + 1$ in the same viewing rectangle, as shown in Figure 1.21(a). From the graph, you can see that you can obtain the graph of g by shifting the graph of f one unit up.

b. Begin by graphing $f(x) = x^3$ and $h(x) = (x - 1)^3$ in the same viewing rectangle, as shown in Figure 1.21(b). From the graph, you can see that you can obtain the graph of h by shifting the graph of f one unit to the right.

c. Begin by graphing $f(x) = x^3$ and $k(x) = (x + 2)^3 + 1$ in the same viewing rectangle, as shown in Figure 1.21(c). From the graph, you can see that you can obtain the graph of k by shifting the graph of f two units to the left and then one unit up.

Figure 1.21

(a) Vertical shift: one unit up

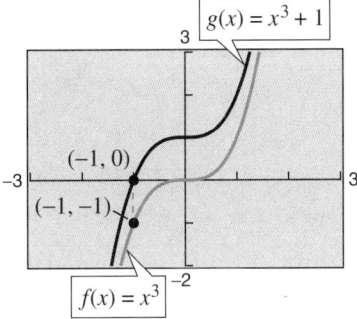

(b) Horizontal shift: one unit right

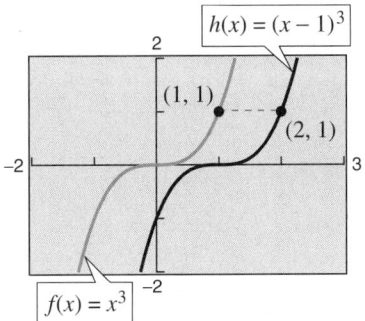

(c) Two units left and one unit up

Figure 1.22

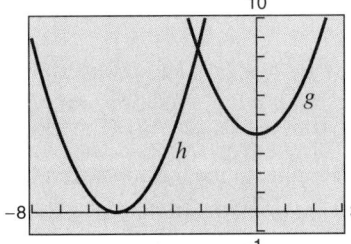

EXAMPLE 2 ◢ **Finding Equations from Graphs**

Each of the graphs shown in Figure 1.22 is a vertical or horizontal shift of the graph of $f(x) = x^2$. Find an equation for each function.

Solution

a. The graph of g is a vertical shift of four units upward of the graph of $f(x) = x^2$. Thus, the equation for g is $g(x) = x^2 + 4$.

b. The graph of h is a horizontal shift of five units to the left of the graph of $f(x) = x^2$. Thus, the equation for h is $h(x) = (x + 5)^2$. ◢

Figure 1.23

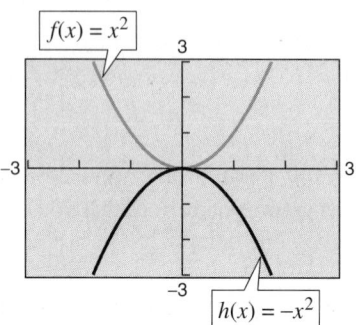

Reflections

The second common type of transformation is called a **reflection.** For instance, if you consider the x-axis to be a mirror, the graph of

$$h(x) = -x^2$$

is the mirror image (or reflection) of the graph of $f(x) = x^2$, as shown in Figure 1.23.

Reflections in the Coordinate Axes

Reflections in the coordinate axes of the graph of $y = f(x)$ are represented as follows.

1. Reflection in the x-axis: $h(x) = -f(x)$
2. Reflection in the y-axis: $h(x) = f(-x)$

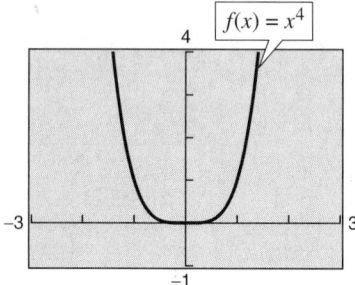

A computer animation of this concept appears in the *Interactive* CD-ROM.

EXAMPLE 3   **Finding Equations from Graphs**

Each of the graphs shown in Figure 1.25 is a transformation of the graph of $f(x) = x^4$ (see Figure 1.24). Find an equation for each function.

Figure 1.24

Figure 1.25

(a)   **(b)**

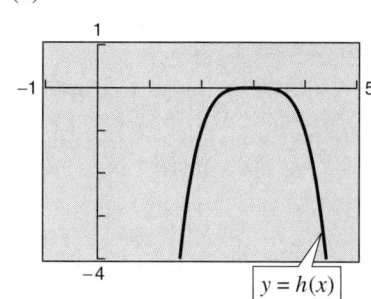

Solution

a. The graph of g is a reflection in the x-axis *followed* by an upward shift of two units of the graph of $f(x) = x^4$. Thus, the equation for g is $g(x) = -x^4 + 2$.

b. The graph of h is a horizontal shift of three units to the right *followed* by a reflection in the x-axis of the graph of $f(x) = x^4$. Thus, the equation for h is $h(x) = -(x - 3)^4$.

EXAMPLE 4 ◢   **Reflections and Shifts**

Compare the graph of each function with the graph of $f(x) = \sqrt{x}$.

a. $g(x) = -\sqrt{x}$ **b.** $h(x) = \sqrt{-x}$ **c.** $k(x) = -\sqrt{x+2}$

Solution

a. Relative to the graph of $f(x) = \sqrt{x}$, the graph of g is a reflection in the x-axis because

$$g(x) = -\sqrt{x} = -f(x).$$

b. The graph of h is a reflection of the graph of $f(x) = \sqrt{x}$ in the y-axis because

$$h(x) = \sqrt{-x} = f(-x).$$

c. From the equation

$$k(x) = -\sqrt{x+2} = -f(x+2)$$

you can conclude that the graph of k is a left shift of two units, followed by a reflection in the x-axis.

The graphs of all three functions are shown in Figure 1.26.

Figure 1.26

(a)

(b)

(c)

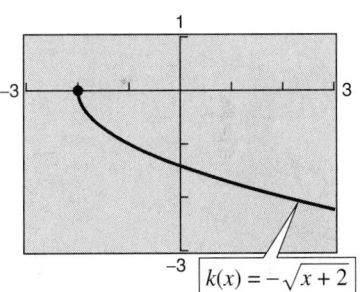

Note   When sketching the graph of a function involving square roots, remember that the domain must be restricted to exclude negative numbers inside the radical. For instance, here are the domains of the functions in Example 4.

Domain of $g(x) = -\sqrt{x}$: $0 \le x$
Domain of $h(x) = \sqrt{-x}$: $0 \ge x$
Domain of $k(x) = -\sqrt{x+2}$: $-2 \le x$

Nonrigid Transformations

Horizontal shifts, vertical shifts, and reflections are called **rigid** transformations because the basic shape of the graph is unchanged. These transformations change only the *position* of the graph in the *xy*-plane. **Nonrigid** transformations are those that cause a *distortion*—a change in the shape of the original graph. For instance, a nonrigid transformation of the graph of $y = f(x)$ is represented by $y = cf(x)$, where the transformation is a **vertical stretch** if $c > 1$ and a **vertical shrink** if $0 < c < 1$.

Figure 1.27

(a)

(b)

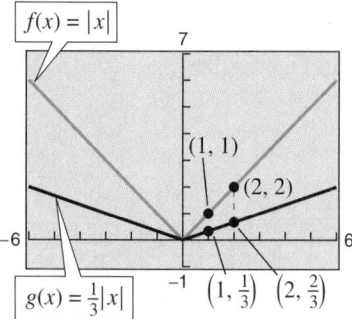

EXAMPLE 5   **Nonrigid Transformations**

Compare the graph of each function with the graph of $f(x) = |x|$.

a. $h(x) = 3|x|$ **b.** $g(x) = \dfrac{1}{3}|x|$

Solution

a. Relative to the graph of $f(x) = |x|$, the graph of

$$h(x) = 3|x| = 3f(x)$$

is a vertical stretch (multiply each *y*-value by 3) of the graph of *f*.

b. Similarly, the equation

$$g(x) = \frac{1}{3}|x| = \frac{1}{3}f(x)$$

indicates that the graph of *g* is a vertical shrink (multiply each *y*-value by $\frac{1}{3}$) of the graph of *f*.

The graphs of all three functions are shown in Figure 1.27.

EXAMPLE 6   **Sequences of Transformations**

Use a graphing utility to graph the two functions

$$g(x) = 5(x^2 - 2) \quad \text{and} \quad h(x) = 5x^2 - 2$$

in the same viewing rectangle. Describe how each function was obtained from $f(x) = x^2$ as a sequence of shifts and stretches.

Solution

Notice that the two graphs in Figure 1.28 are different. The graph of *g* is a downward shift of two units followed by a vertical stretch, whereas *h* is a vertical stretch followed by a downward shift of two units. Hence, the order of applying the transformations is important.

Figure 1.28

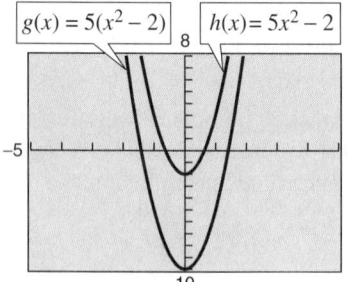

Group Activity   *A Program for Practice*

If you have a programmable calculator, try entering the following program. (These program steps are for a Texas Instruments *TI-82* or *TI-83*. Programs for other graphing calculator models may be found in the appendix.)

```
PROGRAM:PARABOLA              :9 → Xmax
:-6+int (12rand) →H           :1 → Xscl
:-3+int (6rand) →V            :-6 → Ymin
:rand →R                      :6 → Ymax
:If R<.5                      :1 → Yscl
:Then                         :DispGraph
:-1 →R                        :Pause
:Else                         :Disp "Y = R(X+H)²+V"
:1 →R                         :Disp "R = ",R
:End                          :Disp "H = ",H
:"R(X+H)²+V" →Y₁              :Disp "V = ",V
:-9 →Xmin                     :Pause
```

This program will sketch a graph of the function

$$y = R(x + H)^2 + V$$

where $R = \pm 1$, H is an integer between -6 and 6, and V is an integer between -3 and 3. (Each time you run the program, different values of R, H, and V are possible.) From the graph, you should be able to determine the values of R, H, and V. After you have determined the values, press ENTER to see the answer. (To look at the graph again, press GRAPH .)

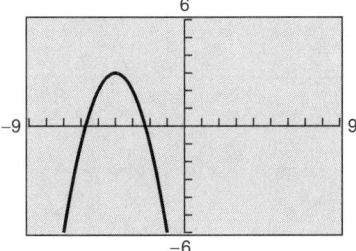

For example, in the graph shown at the left, you can make the following conclusions. Because the graph of $f(x) = x^2$ has been reflected about the *x*-axis, you know that $R = -1$. Because the graph of $f(x) = x^2$ has been shifted four units to the left, you know that $H = 4$. Because the graph of $f(x) = x^2$ has been shifted three units up, you know that $V = 3$. Thus, the equation of the graph shown at the left must be

$$y = -(x + 4)^2 + 3.$$

Try running this program several times. It will give you practice in working with reflections, horizontal shifts, and vertical shifts.

1.3 /// EXERCISES

In Exercises 1–10, sketch the graphs of the three functions *by hand* on the same rectangular coordinate system. Verify your result with a graphing utility.

1. $f(x) = x$

    $g(x) = x - 4$

    $h(x) = 3x$

2. $f(x) = \frac{1}{2}x$

    $g(x) = \frac{1}{2}x + 2$

    $h(x) = \frac{1}{2}(x - 2)$

3. $f(x) = x^2$

    $g(x) = x^2 + 2$

    $h(x) = (x - 2)^2$

4. $f(x) = x^2$

    $g(x) = x^2 - 4$

    $h(x) = (x + 2)^2 + 1$

5. $f(x) = -x^2$

    $g(x) = -x^2 + 1$

    $h(x) = -(x - 2)^2$

6. $f(x) = (x - 2)^2$

    $g(x) = (x - 2)^2 + 2$

    $h(x) = -(x - 2)^2 + 4$

7. $f(x) = x^2$

    $g(x) = \left(\frac{1}{2}x\right)^2$

    $h(x) = (2x)^2$

8. $f(x) = x^2$

    $g(x) = \left(\frac{1}{4}x\right)^2 + 2$

    $h(x) = -\left(\frac{1}{4}x\right)^2$

9. $f(x) = |x|$

    $g(x) = |x| - 1$

    $h(x) = |x - 3|$

10. $f(x) = \sqrt{x}$

    $g(x) = \sqrt{x + 1}$

    $h(x) = \sqrt{x - 2} + 1$

11. Use the graph of *f* to sketch the graphs.

    (a) $y = f(x) + 2$ (b) $y = -f(x)$

    (c) $y = f(x - 2)$ (d) $y = f(x + 3)$

    (e) $y = f(2x)$ (f) $y = f(-x)$

Figure for 11

Figure for 12

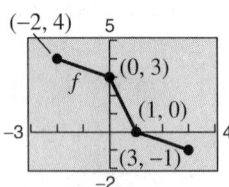

12. Use the graph of *f* to sketch the graphs.

    (a) $y = f(x) - 1$ (b) $y = f(x + 1)$

    (c) $y = f(x - 1)$ (d) $y = -f(x - 2)$

    (e) $y = f(-x)$ (f) $y = \frac{1}{2}f(x)$

In Exercises 13–24, identify the common function and the transformation shown in the graph. Write the formula for the graphed function.

13.

14.

15.

16.

17.

18.

19.

20.

21.

22.

23.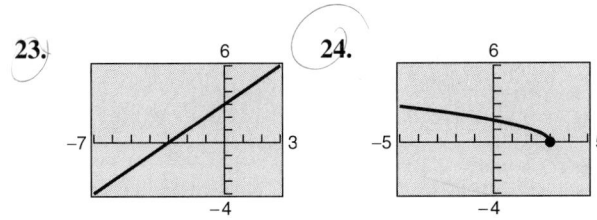

24.

In Exercises 25–30, compare the graph of the function with the graph of $f(x) = \sqrt{x}$.

25. $y = \sqrt{x} + 2$ **26.** $y = -\sqrt{x}$

27. $y = \sqrt{x - 2}$ **28.** $y = \sqrt{x + 3}$

29. $y = \sqrt{2x}$ **30.** $y = \sqrt{-x}$

In Exercises 31–36, compare the graph of the function with the graph of $f(x) = |x|$.

31. $y = |x + 2|$ **32.** $y = |x| - 3$

33. $y = -|x|$ **34.** $y = |-x|$

35. $y = \frac{1}{3}|x|$ **36.** $y = \left|\frac{1}{2}x\right|$

In Exercises 37–40, use a graphing utility to graph the three functions in the same viewing rectangle. Describe the graphs of g and h relative to the graph of f.

37. $f(x) = x^3 - 3x^2$
$g(x) = f(x + 2)$
$h(x) = f(\frac{1}{2}x)$

38. $f(x) = x^3 - 3x^2 + 2$
$g(x) = f(x - 1)$
$h(x) = f(2x)$

39. $f(x) = x^3 - 3x^2$
$g(x) = -\frac{1}{3}f(x)$
$h(x) = f(-x)$

40. $f(x) = x^3 - 3x^2 + 2$
$g(x) = -f(x)$
$h(x) = f(-x)$

In Exercises 41–46, specify the sequence of transformations that will yield the graph of the given function from the graph of the function $f(x) = x^3$.

41. $g(x) = 4 - x^3$ **42.** $g(x) = -(x - 4)^3$

43. $h(x) = \frac{1}{4}(x + 2)^3$ **44.** $h(x) = -2(x - 1)^3 + 3$

45. $p(x) = \left(\frac{1}{3}x\right)^3 + 2$ **46.** $p(x) = [3(x - 2)]^3$

In Exercises 47 and 48, use the graph of $f(x) = x^3 - 3x^2$ (see Exercise 37) to write a formula for the function g shown in the graph.

47.

48.

49. *Profit*   The profit P per week on a certain product is given by the model

$$P(x) = 80 + 20x - 0.5x^2, \qquad 0 \le x \le 20$$

where x is the amount spent on advertising. In this model, x and P are both measured in hundreds of dollars.

(a) Use a graphing utility to graph the profit function.

(b) The business estimates that taxes and operating costs will increase by an average of \$2500 per week during the next year. Rewrite the profit equation to reflect this expected decrease in profits. Identify the type of transformation applied to the graph of the equation.

(c) Rewrite the profit equation so that x measures advertising expenditures in dollars. [Find $P\left(\frac{x}{100}\right)$.] Identify the type of transformation applied to the graph of the profit function.

50. *Automobile Aerodynamics*   The number of horsepower H required to overcome wind drag on a certain automobile is approximated by

$$H(x) = 0.002x^2 + 0.005x - 0.029, \qquad 10 \le x \le 100$$

where x is the speed of the car in miles per hour.

(a) Use a graphing utility to graph the power function.

(b) Rewrite the power function so that x represents the speed in kilometers per hour. [Find $H(1.6x)$.] Identify the type of transformation applied to the graph of the power function.

51. *Exploration*   Use a graphing utility to graph each function. Describe any similarities and differences you observe among the graphs.

(a) $y = x$ (b) $y = x^2$

(c) $y = x^3$ (d) $y = x^4$

(e) $y = x^5$ (f) $y = x^6$

52. *Conjecture*   Use the results of Exercise 51 to make a conjecture about the shapes of the graphs of $y = x^7$ and $y = x^8$. Use a graphing utility to verify your conjecture.

53. Use the results of Exercise 51 to sketch the graph of $y = (x - 3)^3$ by hand. Use a graphing utility to verify your graph.

54. Use the results of Exercise 51 to sketch the graph of $y = (x + 1)^2$ by hand. Use a graphing utility to verify your graph.

Conjecture   In Exercises 55–58, use the results of Exercise 51 to make a conjecture about the shape of the graph of the function. Use a graphing utility to verify your conjecture.

55. $f(x) = x^2(x - 6)^2$

56. $f(x) = x^3(x - 6)^2$

57. $f(x) = x^2(x - 6)^3$

58. $f(x) = x^3(x - 6)^3$

59. *Graphical Reasoning*   An electronically controlled thermostat in a home is programmed to automatically lower the temperature during the night (see figure). The temperature in the house T, in degrees Fahrenheit, is given in terms of t, the time in hours on a 24-hour clock.

(a) Explain why T is a function of t.

(b) Approximate $T(4)$ and $T(15)$.

(c) Suppose the thermostat were reprogrammed to produce a temperature H where $H(t) = T(t - 1)$. How would this change the temperature in the house? Explain.

(d) Suppose the thermostat were reprogrammed to produce a temperature H where $H(t) = T(t) - 1$. How would this change the temperature in the house? Explain.

Figure for 59

60. *Coordinate Axis Scale*   It is necessary to graph the function $f(t)$, which models the specified data for the years 1980 through 1996, with $t = 0$ corresponding to the year 1980. State a possible scale for the vertical axis (e.g., hundreds, thousands, millions, etc.) of the graph and give a reason for your answer.

(a) $f(t)$ represents the average salary of college professors.

(b) $f(t)$ represents the population of the United States.

(c) $f(t)$ represents the percent of the civilian work force that is unemployed.

In Exercises 61–66, use the graph of $f(x)$ (see figure) to sketch the graph of the specified function.

61. $f(x - 4)$

62. $f(x + 2)$

63. $f(x) + 4$

64. $f(x) - 1$

65. $2f(x)$

66. $\frac{1}{2}f(x)$

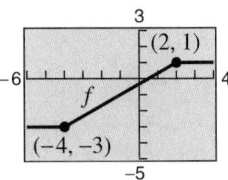

1.4   Combinations of Functions

Arithmetic Combinations of Functions   /   Compositions of Functions   /
Application

Arithmetic Combinations of Functions

Just as two real numbers can be combined by the operations of addition, subtraction, multiplication, and division to form other real numbers, two *functions*
can be combined to create new functions. For example, if

$$f(x) = 2x - 3 \quad \text{and} \quad g(x) = x^2 - 1$$

you can form the sum, difference, product, and quotient of f and g as follows.

$$f(x) + g(x) = (2x - 3) + (x^2 - 1)$$
$$= x^2 + 2x - 4 \qquad \text{Sum}$$

$$f(x) - g(x) = (2x - 3) - (x^2 - 1)$$
$$= -x^2 + 2x - 2 \qquad \text{Difference}$$

$$f(x) \cdot g(x) = (2x - 3)(x^2 - 1)$$
$$= 2x^3 - 3x^2 - 2x + 3 \qquad \text{Product}$$

$$\frac{f(x)}{g(x)} = \frac{2x - 3}{x^2 - 1}, \qquad x \neq \pm 1 \qquad \text{Quotient}$$

The domain of an arithmetic combination of functions f and g consists of all
real numbers that are common to the domains of f and g. In the case of the quotient $f(x)/g(x)$, there is the further restriction that $g(x) \neq 0$.

Sum, Difference, Product, and Quotient of Functions

Let f and g be two functions with overlapping domains. Then, for all x
common to both domains, the **sum, difference, product,** and **quotient** of
f and g are defined as follows.

1. Sum: $(f + g)(x) = f(x) + g(x)$

2. Difference: $(f - g)(x) = f(x) - g(x)$

3. Product: $(fg)(x) = f(x) \cdot g(x)$

4. Quotient: $\left(\dfrac{f}{g}\right)(x) = \dfrac{f(x)}{g(x)}, \quad g(x) \neq 0$

A graphing utility can be used to graph the sum of two functions. For instance, you can illustrate the result of Example 1 by letting $Y_1 = 2X + 1$, $Y_2 = X^2 + 2X - 1$, and $Y_3 = Y_1 + Y_2$. Graph all three functions with your graphing utility set in simultaneous plotting mode. What do you observe? How could you use your graphing utility to illustrate Example 2?

EXAMPLE 1 ▱ **Finding the Sum of Two Functions**

Find $(f + g)(x)$ for the functions

$$f(x) = 2x + 1 \quad \text{and} \quad g(x) = x^2 + 2x - 1.$$

Then evaluate the sum when $x = 2$.

Solution

The sum of the functions f and g is given by

$$\begin{aligned}(f + g)(x) &= f(x) + g(x) \\ &= (2x + 1) + (x^2 + 2x - 1) \\ &= x^2 + 4x.\end{aligned}$$

When $x = 2$, the value of this sum is

$$(f + g)(2) = 2^2 + 4(2) = 12.$$

EXAMPLE 2 ▱ **Finding the Difference of Two Functions**

Find $(f - g)(x)$ for the functions

$$f(x) = 2x + 1 \quad \text{and} \quad g(x) = x^2 + 2x - 1.$$

Then evaluate the difference when $x = 2$.

Solution

The difference of the functions f and g is given by

$$\begin{aligned}(f - g)(x) &= f(x) - g(x) \\ &= (2x + 1) - (x^2 + 2x - 1) \\ &= -x^2 + 2.\end{aligned}$$

When $x = 2$, the value of this difference is

$$(f - g)(2) = -(2)^2 + 2 = -2.$$

In Examples 1 and 2, both f and g have domains that consist of all real numbers. Thus, the domain of both $(f + g)$ and $(f - g)$ is also the set of all real numbers. Remember that any restrictions on the domains of f or g must be taken into account when forming the sum, difference, product, or quotient of f and g. For instance, the domain of $f(x) = 1/x$ is all $x \neq 0$, and the domain of $g(x) = \sqrt{x}$ is $[0, \infty)$. This implies that the domain of $f + g$ is $(0, \infty)$.

You can confirm the results of Example 3 with your graphing utility. Enter the three functions

$$Y_1 = \sqrt{X}$$

$$Y_2 = \sqrt{(4 - X^2)}$$

$$Y_3 = Y_1/Y_2$$

and graph them using a zoom decimal screen (ZDecimal on the *TI-82* or *TI-83*). The zoom decimal screen displays the graph in a new viewing rectangle that sets the change in the *x*-values and the change in the *y*-values to 0.1. Trace along the third curve between X = 0 and X = 2. What happens when you arrive at X = 2? Repeat this experiment for Y3 = Y2/Y1 and observe any changes in the domain.

Figure 1.29

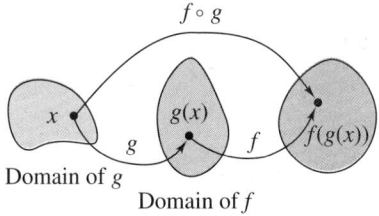

EXAMPLE 3 ▱ Finding the Quotient of Two Functions

Find $(f/g)(x)$ and $(g/f)(x)$ for the functions

$$f(x) = \sqrt{x} \quad \text{and} \quad g(x) = \sqrt{4 - x^2}.$$

Then find the domains of f/g and g/f.

Solution
The quotient of f and g is given by

$$\left(\frac{f}{g}\right)(x) = \frac{f(x)}{g(x)} = \frac{\sqrt{x}}{\sqrt{4 - x^2}},$$

and the quotient of g and f is given by

$$\left(\frac{g}{f}\right)(x) = \frac{g(x)}{f(x)} = \frac{\sqrt{4 - x^2}}{\sqrt{x}}.$$

The domain of f is $[0, \infty)$ and the domain of g is $[-2, 2]$. The intersection of these domains is $[0, 2]$. Thus, the domains for f/g and g/f are as follows.

$$\text{Domain of } \frac{f}{g} : [0, 2) \qquad \text{Domain of } \frac{g}{f} : (0, 2]$$

Can you see why these two domains differ slightly? ▱

Compositions of Functions

Another way of combining two functions is to form the **composition** of one with the other. For instance, if

$$f(x) = x^2 \quad \text{and} \quad g(x) = x + 1,$$

the composition of f with g is

$$f(g(x)) = f(x + 1) = (x + 1)^2.$$

This composition is denoted as $f \circ g$.

Definition of Composition of Two Functions
The **composition** of the function f with g is $$(f \circ g)(x) = f(g(x)).$$ The domain of $f \circ g$ is the set of all x in the domain of g such that $g(x)$ is in the domain of f. (See Figure 1.29.)

EXPLORATION

Let $f(x) = x + 2$ and $g(x) = 4 - x^2$. Are the compositions $f \circ g$ and $g \circ f$ equal? You can use your graphing utility to answer this question by entering and graphing the following functions.

$$Y_1 = (4 - x^2) + 2$$

$$Y_2 = 4 - (x + 2)^2$$

What do you observe? Which function represents $f \circ g$ and which represents $g \circ f$?

EXAMPLE 4   **Forming the Composition of *f* with *g***

Find $(f \circ g)(x)$ for $f(x) = \sqrt{x}$, $x \geq 0$, and $g(x) = x - 1$, $x \geq 1$. If possible, find $(f \circ g)(2)$ and $(f \circ g)(0)$.

Solution

$$\begin{aligned}
(f \circ g)(x) &= f(g(x)) && \text{Definition of } f \circ g \\
&= f(x - 1) && \text{Definition of } g(x) \\
&= \sqrt{x - 1}, \quad x \geq 1 && \text{Definition of } f(x)
\end{aligned}$$

The domain of $f \circ g$ is $[1, \infty)$. Thus,

$$(f \circ g)(2) = \sqrt{2 - 1} = 1$$

is defined, but $(f \circ g)(0)$ is not defined because 0 is not in the domain of $f \circ g$.

The composition of *f* with *g* is generally not the same as the composition of *g* with *f*. This is illustrated in Example 5.

EXAMPLE 5   **Compositions of Functions**

Given $f(x) = x + 2$ and $g(x) = 4 - x^2$, find the following.

a. $(f \circ g)(x)$

b. $(g \circ f)(x)$

Solution

a.
$$\begin{aligned}
(f \circ g)(x) &= f(g(x)) && \text{Definition of } f \circ g \\
&= f(4 - x^2) && \text{Definition of } g(x) \\
&= (4 - x^2) + 2 && \text{Definition of } f(x) \\
&= -x^2 + 6
\end{aligned}$$

b.
$$\begin{aligned}
(g \circ f)(x) &= g(f(x)) && \text{Definition of } g \circ f \\
&= g(x + 2) && \text{Definition of } f(x) \\
&= 4 - (x + 2)^2 && \text{Definition of } g(x) \\
&= 4 - (x^2 + 4x + 4) \\
&= -x^2 - 4x
\end{aligned}$$

Note in this case that $(f \circ g)(x) \neq (g \circ f)(x)$.

Note The following tables help illustrate numerically the composition $f \circ g$ of the functions f and g given in Example 5.

x	0	1	2	3
$g(x)$	4	3	0	−5

$g(x)$	4	3	0	−5
$f(g(x))$	6	5	2	−3

x	0	1	2	3
$f(g(x))$	6	5	2	−3

Note that the first two tables can be combined (or "composed") to produce the values given in the third table.

EXAMPLE 6   ▱   **Finding the Domain of a Composite Function**

Find the composition $(f \circ g)(x)$ for the functions

$$f(x) = x^2 - 9 \quad \text{and} \quad g(x) = \sqrt{9 - x^2}.$$

Then find the domain of $(f \circ g)$.

Solution

The composition of the functions is as follows.

$$
\begin{aligned}
(f \circ g)(x) &= f(g(x)) \\
&= f\left(\sqrt{9 - x^2}\right) \\
&= \left(\sqrt{9 - x^2}\right)^2 - 9 \\
&= 9 - x^2 - 9 \\
&= -x^2
\end{aligned}
$$

From this, it might appear that the domain of the composition is the set of all real numbers. This, however, is not true because the domain of g is $-3 \le x \le 3$. To convince yourself of this, use a graphing utility to graph

$$y = \left(\sqrt{9 - x^2}\right)^2 - 9$$

as shown in Figure 1.30. Notice that the graphing utility does not extend the graph to the left of $x = -3$ or to the right of $x = 3$. Thus, the domain of $f \circ g$ is $-3 \le x \le 3$.   ▱

Figure 1.30

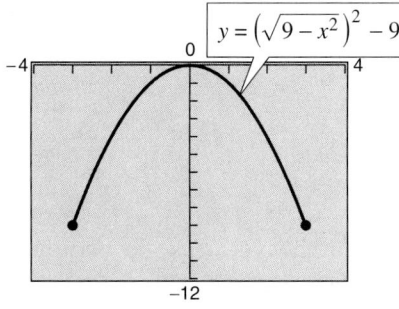

$$y = \left(\sqrt{9 - x^2}\right)^2 - 9$$

EXAMPLE 7   ▱   **A Case in Which $f \circ g = g \circ f$**

Given $f(x) = 2x + 3$ and $g(x) = \frac{1}{2}(x - 3)$, find the following.

a. $(f \circ g)(x)$ **b.** $(g \circ f)(x)$

Note   In Example 7, note that the two composite functions $f \circ g$ and $g \circ f$ are equal, and both represent the identity function. That is,

$$(f \circ g)(x) = x$$
$$(g \circ f)(x) = x.$$

You will study this special case in the next section.

Solution

a.
$$
\begin{aligned}
(f \circ g)(x) &= f(g(x)) \\
&= f\left(\frac{1}{2}(x - 3)\right) \\
&= 2\left[\frac{1}{2}(x - 3)\right] + 3 \\
&= x - 3 + 3 \\
&= x
\end{aligned}
$$

b.
$$
\begin{aligned}
(g \circ f)(x) &= g(f(x)) \\
&= g(2x + 3) \\
&= \frac{1}{2}[(2x + 3) - 3] \\
&= \frac{1}{2}(2x) \\
&= x
\end{aligned}
$$ ▱

In Examples 4, 5, 6, and 7 you formed the composition of two given functions. In calculus, it is also important to be able to identify two functions that make up a given composite function. For instance, the function h given by

$$h(x) = (3x - 5)^3$$

is the composition of f with g, where $f(x) = x^3$ and $g(x) = 3x - 5$. That is,

$$h(x) = (3x - 5)^3 = [g(x)]^3 = f(g(x)).$$

Basically, to "decompose" a composite function, look for an "inner" and an "outer" function. In the function h above, $g(x) = 3x - 5$ is the inner function and $f(x) = x^3$ is the outer function.

EXAMPLE 8 ▰ **Identifying a Composite Function**

Express the function

$$h(x) = \frac{1}{(x - 2)^2}$$

as a composition of two functions.

Solution

One way to write h as a composition of two functions is to take the inner function to be $g(x) = x - 2$ and the outer function to be

$$f(x) = \frac{1}{x^2} = x^{-2}.$$

Then you can write

$$h(x) = \frac{1}{(x - 2)^2} = (x - 2)^{-2} = f(x - 2) = f(g(x)). \qquad ▰$$

EXPLORATION

The function in Example 8 can be decomposed in other ways. For which of the following pairs of functions is $h(x)$ equal to $f(g(x))$?

a. $g(x) = \dfrac{1}{x - 2}$ and $f(x) = x^2$

b. $g(x) = x^2$ and $f(x) = \dfrac{1}{x - 2}$

c. $g(x) = \dfrac{1}{x}$ and $f(x) = (x - 2)^2$

Use a graphing utility to graph $y = 320t^2 + 420$ and $y = 2000$ in the same viewing rectangle. (Use a viewing rectangle in which $0 \le x \le 3$ and $420 \le y \le 4000$.) Explain how the graphs can be used to answer the question asked in Example 9(c). Compare your answer with that given in part (c). When will the bacteria count reach 3200?

Notice that the model for this bacteria count situation is valid only for a span of 3 hours. Now suppose the minimum number of bacteria in the food is reduced from 420 to 100. Will the number of bacteria still reach a level of 2000 within the 3-hour time span? Will the number of bacteria reach a level of 3200 within 3 hours?

Application

EXAMPLE 9   **Bacteria Count**

The number of bacteria in a refrigerated food is given by

$$N(T) = 20T^2 - 80T + 500, \qquad 2 \le T \le 14$$

where T is the Celsius temperature of the food. When the food is removed from refrigeration, the temperature is given by

$$T(t) = 4t + 2, \qquad 0 \le t \le 3$$

where t is the time in hours. Find the following.

a. The composite $N(T(t))$. What does this function represent?

b. The number of bacteria in the food when $t = 2$ hours

c. The time when the bacteria count reaches 2000

Solution

a. $N(T(t)) = 20(4t + 2)^2 - 80(4t + 2) + 500$

$\qquad\qquad = 20(16t^2 + 16t + 4) - 320t - 160 + 500$

$\qquad\qquad = 320t^2 + 320t + 80 - 320t - 160 + 500$

$\qquad\qquad = 320t^2 + 420$

This composite function represents the number of bacteria as a function of the amount of time the food has been out of refrigeration.

b. When $t = 2$, the number of bacteria is

$$N = 320(2)^2 + 420 = 1280 + 420 = 1700.$$

c. The bacteria count will reach $N = 2000$ when $320t^2 + 420 = 2000$. You can solve this equation for t algebraically as follows.

$$320t^2 + 420 = 2000$$

$$320t^2 = 1580$$

$$t^2 = \frac{1580}{320} = \frac{79}{16}$$

$$t = \frac{\sqrt{79}}{4} \approx 2.2 \text{ hours}$$

Or you can use a graphing utility to approximate the solution, as shown in Figure 1.31.

Figure 1.31

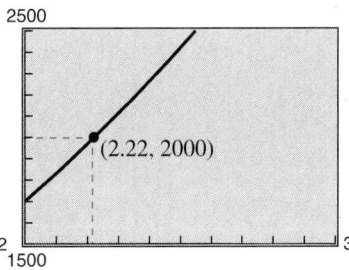

Group Activity   *The Composition of Two Functions*

You are considering buying a new sport-utility vehicle for which the regular price is $20,800. The dealership has advertised a factory rebate of $2000 *and* a 12% discount. Using function notation, the rebate and the discount can be represented as

$$f(x) = x - 2000 \qquad \text{Rebate of \$2000}$$

and

$$g(x) = 0.88x \qquad \text{Discount of 12\%}$$

where x is the price of the vehicle, $f(x)$ is the price after subtracting the rebate, and $g(x)$ is the price after taking the 12% discount. Compare the sale price obtained by subtracting the rebate first and then taking the discount with the sale price obtained by taking the discount first and then subtracting the rebate.

1.4 /// EXERCISES

In Exercises 1–4, use the graphs to graph $h(x) = (f + g)(x)$.

1.

2.

3.

4.
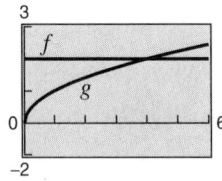

In Exercises 5–12, find the following.

(a) $(f + g)(x)$ (b) $(f - g)(x)$

(c) $(fg)(x)$ (d) $(f/g)(x)$

What is the domain of f/g?

5. $f(x) = x + 1, \quad g(x) = x - 1$

6. $f(x) = 2x - 5, \quad g(x) = 1 - x$

7. $f(x) = x^2, \quad g(x) = 1 - x$

8. $f(x) = 2x - 5, \quad g(x) = 5$

9. $f(x) = x^2 + 5, \quad g(x) = \sqrt{1 - x}$

10. $f(x) = \sqrt{x^2 - 4}, \quad g(x) = \dfrac{x^2}{x^2 + 1}$

11. $f(x) = \dfrac{1}{x}, \quad g(x) = \dfrac{1}{x^2}$

12. $f(x) = \dfrac{x}{x + 1}, \quad g(x) = x^3$

In Exercises 13–24, evaluate the indicated function for $f(x) = x^2 + 1$ and $g(x) = x - 4$.

13. $(f + g)(3)$ **14.** $(f - g)(-2)$

15. $(f - g)(0)$ **16.** $(f + g)(1)$

17. $(f - g)(2t)$ **18.** $(f + g)(t - 1)$

19. $(fg)(4)$ **20.** $(fg)(-6)$

21. $\left(\dfrac{f}{g}\right)(5)$ **22.** $\left(\dfrac{f}{g}\right)(0)$

23. $\left(\dfrac{f}{g}\right)(-1) - g(3)$ **24.** $(2f)(5)$

In Exercises 25–28, graph the functions f, g, and $f + g$ in the same viewing rectangle.

25. $f(x) = \frac{1}{2}x$, $g(x) = x - 1$

26. $f(x) = \frac{1}{3}x$, $g(x) = -x + 4$

27. $f(x) = x^2$, $g(x) = -2x$

28. $f(x) = 4 - x^2$, $g(x) = x$

In Exercises 29 and 30, use a graphing utility to sketch the graphs of f, g, and $f + g$ in the same viewing rectangle. Which function contributes most to the magnitude of the sum when $0 \le x \le 2$? Which function contributes most to the magnitude of the sum when $x > 6$?

29. $f(x) = 3x$, $g(x) = -\dfrac{x^3}{10}$

30. $f(x) = \dfrac{x}{2}$, $g(x) = \sqrt{x}$

31. *Stopping Distance*   A car traveling x miles per hour stops quickly. The distance a car travels during the driver's reaction time is given by $R(x) = \frac{3}{4}x$. The distance traveled while braking is given by $B(x) = \frac{1}{15}x^2$.

(a) Find the stopping-distance function T.

(b) Use a graphing utility to graph the functions R, B, and T in the interval $0 \le x \le 60$.

(c) Which function contributes most to the magnitude of the sum at higher speeds? Explain.

32. *Comparing Sales*   You own two restaurants. From 1990 to 1995, the sales R_1 (in thousands of dollars) for one restaurant can be modeled by

$$R_1 = 480 - 8t - 0.8t^2, \qquad t = 0, 1, 2, 3, 4, 5$$

where $t = 0$ represents 1990. During the same 6-year period, the sales R_2 (in thousands of dollars) for the other restaurant can be modeled by

$$R_2 = 254 + 0.78t, \qquad t = 0, 1, 2, 3, 4, 5.$$

(a) Write a function that represents the total sales for the two restaurants. Use a graphing utility to graph the total sales function.

(b) Use the *stacked bar graph* in the figure, which represents the total sales during the 6-year period, to determine whether the total sales have been increasing or decreasing. Explain.

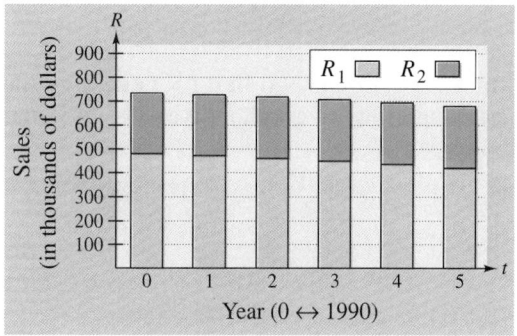

Data Analysis   In Exercises 33 and 34, use the data in the table, which gives the variable costs for operating an automobile in the United States for the years 1985 through 1991. The functions y_1, y_2, and y_3 represent the costs in cents per mile for gas and oil, maintenance, and tires, respectively.   (Source: American Automobile Manufacturers Association)

Year	1985	1986	1987	1988	1989	1990	1991
y_1	6.16	4.48	4.80	5.20	5.20	5.40	6.70
y_2	1.23	1.37	1.60	1.60	1.90	2.10	2.20
y_3	0.65	0.67	0.80	0.80	0.80	0.90	0.90

33. Create a stacked bar graph for the data.

34. Mathematical models for the data are given by

$$y_1 = 0.16t^2 - 2.43t + 13.96$$

$$y_2 = 0.17t + 0.38$$

$$y_3 = 0.04t + 0.44$$

where $t = 5$ represents 1985. Use a graphing utility to graph y_1, y_2, y_3, and $y_1 + y_2 + y_3$ in the same viewing rectangle. Use the model to estimate the total variable cost per mile in 1995.

In Exercises 35–38, find (a) $f \circ g$, (b) $g \circ f$, and (c) $f \circ f$.

35. $f(x) = x^2, \quad g(x) = x - 1$

36. $f(x) = \sqrt[3]{x - 1}, \quad g(x) = x^3 + 1$

37. $f(x) = 3x + 5, \quad g(x) = 5 - x$

38. $f(x) = x^3, \quad g(x) = \dfrac{1}{x}$

In Exercises 39–44, (a) find $f \circ g$ and $g \circ f$. (b) Use a graphing utility to graph $f \circ g$ and $g \circ f$. Determine if $f \circ g = g \circ f$.

39. $f(x) = \sqrt{x + 4}, \quad g(x) = x^2$

40. $f(x) = \sqrt[3]{x + 1}, \quad g(x) = x^3 - 1$

41. $f(x) = \frac{1}{3}x - 3, \quad g(x) = 3x + 1$

42. $f(x) = \sqrt{x}, \quad g(x) = \sqrt{x}$

43. $f(x) = x^{2/3}, \quad g(x) = x^6$

44. $f(x) = |x|, \quad g(x) = x + 6$

In Exercises 45–48, use the graphs of f and g (see figures) to evaluate the functions.

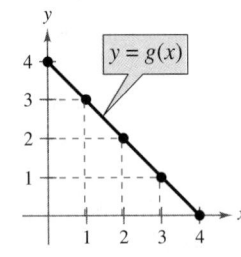

45. (a) $(f + g)(3)$ (b) $(f/g)(2)$

46. (a) $(f - g)(1)$ (b) $(fg)(4)$

47. (a) $(f \circ g)(2)$ (b) $(g \circ f)(2)$

48. (a) $(f \circ g)(1)$ (b) $(g \circ f)(3)$

In Exercises 49–56, find two functions f and g such that $(f \circ g)(x) = h(x)$. (There are many correct answers.)

49. $h(x) = (2x + 1)^2$

50. $h(x) = (1 - x)^3$

51. $h(x) = \sqrt[3]{x^2 - 4}$

52. $h(x) = \sqrt{9 - x}$

53. $h(x) = \dfrac{1}{x + 2}$

54. $h(x) = \dfrac{4}{(5x + 2)^2}$

55. $h(x) = (x + 4)^2 + 2(x + 4)$

56. $h(x) = (x + 3)^{3/2}$

In Exercises 57–60, determine the domains of (a) f, (b) g, and (c) $f \circ g$.

57. $f(x) = \sqrt{x}, \quad g(x) = x^2 + 1$

58. $f(x) = \dfrac{1}{x}, \quad g(x) = x + 3$

59. $f(x) = \dfrac{3}{x^2 - 1}, \quad g(x) = x + 1$

60. $f(x) = 2x + 3, \quad g(x) = \dfrac{x}{2}$

Average Rate of Change In Exercises 61–64, find the difference quotient

$$\frac{f(x + h) - f(x)}{h}$$

and simplify your answer.

61. $f(x) = 3x - 4$

62. $f(x) = 1 - x^2$

63. $f(x) = \dfrac{4}{x}$

64. $f(x) = \sqrt{2x + 1}$

65. *Ripples*   A pebble is dropped into a calm pond, caus-ing ripples in the form of concentric circles (see fig-ure). The radius (in feet) of the outer ripple is given by $r(t) = 0.6t$, where t is the time in seconds after the pebble strikes the water. The area of the circle is given by the function $A(r) = \pi r^2$. Find and interpret $(A \circ r)(t)$.

66. *Area*   A square concrete foundation was prepared as a base for a large cylindrical gasoline tank (see figure).

(a) Express the radius r of the tank as a function of the length x of the sides of the square.

(b) Express the area A of the circular base of the tank as a function of the radius r.

(c) Find and interpret $(A \circ r)(x)$.

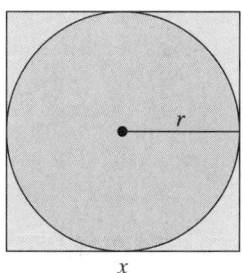

x

67. *Cost*   The weekly cost of producing x units in a manufacturing process is given by the function $C(x) = 60x + 750$. The number of units produced in t hours is given by $x(t) = 50t$.

(a) Find and interpret $(C \circ x)(t)$.

(b) Use a graphing utility to graph the cost as a function of time. Move the cursor along the curve to estimate (to two-decimal-place accuracy) the time that must elapse until the cost increases to $15,000.

68. *Air Traffic Control*   An air traffic controller spots two planes at the same altitude flying toward each other (see figure). Their flight paths form a right angle at point P. One plane is 150 miles from point P and is moving at 450 miles per hour. The other plane is 200 miles from point P and is moving at 450 miles per hour. Write the distance s between the planes as a function of time t.

Distance (in miles)

69. *Think About It*   You are a sales representative for an automobile manufacturer. You are paid an annual salary plus a bonus of 3% of your sales over $500,000. Consider the two functions $f(x) = x - 500,000$ and $g(x) = 0.03x$. If x is greater than $500,000, which of the following represents your bonus? Explain.

(a) $f(g(x))$ (b) $g(f(x))$

70. *Exploration*   The suggested retail price of a new car is p dollars. The dealership advertised a factory rebate of $1200 and an 8% discount.

(a) Write a function R in terms of p, giving the cost of the car after receiving the rebate from the factory.

(b) Write a function S in terms of p, giving the cost of the car after receiving the dealership discount.

(c) Form the composite functions $(R \circ S)(p)$ and $(S \circ R)(p)$, and interpret each.

(d) Find $(R \circ S)(18,400)$ and $(S \circ R)(18,400)$. Which yields the lower cost for the car? Explain.

71. Prove that the product of two odd functions is an even function and the product of two even functions is an even function.

72. *Conjecture*   Use examples to hypothesize whether the product of an odd function and an even function is even or odd. Then prove your hypothesis.

73. Given a function f, prove that $g(x)$ is even and $h(x)$ is odd where

$$g(x) = \tfrac{1}{2}[f(x) + f(-x)] \quad \text{and}$$

$$h(x) = \tfrac{1}{2}[f(x) - f(-x)].$$

74. Use the result of Exercise 73 to prove that any function can be written as a sum of even and odd functions. (*Hint:* Add the two equations in Exercise 73.)

75. Use the result of Exercise 74 to write each function as a sum of even and odd functions.

(a) $f(x) = x^2 - 2x + 1$ (b) $f(x) = \dfrac{1}{x + 1}$

76. *Exploration* Consider the repeated composition of $f(x) = \sqrt{x + k}$ with itself defined by

$$R_n(x) = \underbrace{f(f(f(\cdots f(x) \cdots)))}_{n}.$$

(*Hint:* If you use a *TI-82* or *TI-83*, use the following procedure. Let $Y_1 = \sqrt{x + k}$ and store 0 in x. Using the Y-VARS menu, enter $Y_1 \rightarrow x$ on the home screen. Pressing the ENTER key repeatedly will yield the repeated composition of f.)

(a) Use a graphing utility to complete the table for $k = \tfrac{1}{9}$.

n	1	2	3	4	5
$R_n(0)$					

n	6	7	8	9	10
$R_n(0)$					

(b) Use a graphing utility to complete the table for $k = 9$.

n	1	2	3	4	5
$R_n(0)$					

n	6	7	8	9	10
$R_n(0)$					

(c) Describe what appears to be happening for the repeated composition of the function f.

Data Analysis In Exercises 77 and 78, use the data in the table, which gives the circulations of morning and evening newspapers in the United States for the years 1983 through 1992. The variables y_1 and y_2 represent the circulations in millions of the morning and evening papers, respectively. (Source: Editor & Publisher Co.)

Year	1983	1984	1985	1986	1987
y_1	33.8	35.4	36.4	37.4	39.1
y_2	28.8	27.7	26.4	25.1	23.7

Year	1988	1989	1990	1991	1992
y_1	40.4	40.7	41.3	41.5	42.4
y_2	22.2	21.8	21.0	19.2	17.8

77. Create a stacked bar graph for the data.

78. Use a graphing utility to create a scatter plot of each data set. What type of model would best fit the data? Use the least squares regression capabilities of a graphing utility to find models for y_1 and y_2 (use $t = 3$ to represent 1983). Graph the models for y_1, y_2, and $y_1 - y_2$ in the same viewing rectangle. What does the graph of the difference of the functions indicate about newspaper circulation in general?

Review Solve Exercises 79–82 as a review of the skills and problem-solving techniques you learned in previous sections.

79. Sketch the polygon with vertices $(1, 0)$, $(6, 3)$, $(8, 5)$, and $(3, 2)$.

80. Find the coordinates of a point that is eight units to the left of the y-axis and three units above the x-axis.

81. Determine the quadrant(s) of a point (x, y) where $xy > 0$.

82. Find the distance between the points $(0, 8)$ and $(5, 4)$.

1.5 Inverse Functions

*The Inverse of a Function / The Graph of an Inverse Function /
The Existence of an Inverse Function / Finding Inverse Functions*

Figure 1.32

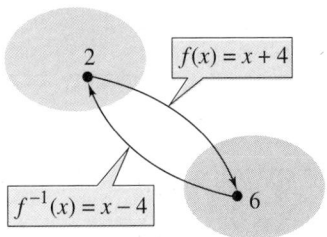

Note A table of values can help
you understand inverse functions. For
instance, the following table shows
several values of the function f in
Example 1.

x	-2	-1	0	1	2
$f(x)$	-8	-4	0	4	8

By interchanging the rows of the table,
you obtain values of the inverse function f^{-1}.

x	-8	-4	0	4	8
$f^{-1}(x)$	-2	-1	0	1	2

In the first table, each output is 4 times
the input, and in the second table each
output is $\frac{1}{4}$ the input.

The Inverse of a Function

Recall from Section 1.1 that a function can be represented by a set of ordered
pairs. For instance, the function $f(x) = x + 4$ from the set $A = \{1, 2, 3, 4\}$ to
the set $B = \{5, 6, 7, 8\}$ can be written as follows.

$$f(x) = x + 4 : \{(1, 5), (2, 6), (3, 7), (4, 8)\}$$

By interchanging the first and second coordinates of each of these ordered
pairs, you can form the **inverse function** of f, which is denoted by f^{-1}. It is a
function from the set B to the set A, and can be written as follows.

$$f^{-1}(x) = x - 4 : \{(5, 1), (6, 2), (7, 3), (8, 4)\}$$

Note that the domain of f is equal to the range of f^{-1}, and vice versa, as
shown in Figure 1.32. Also note that the functions f and f^{-1} have the effect of
"undoing" each other. In other words, when you form the composition of f with
f^{-1} or the composition of f^{-1} with f, you obtain the identity function.

$$f(f^{-1}(x)) = f(x - 4) = (x - 4) + 4 = x$$
$$f^{-1}(f(x)) = f^{-1}(x + 4) = (x + 4) - 4 = x$$

EXAMPLE 1 ▱   Finding Inverse Functions Informally

Find the inverse of $f(x) = 4x$. Then verify that both $f(f^{-1}(x))$ and $f^{-1}(f(x))$ are
equal to the identity function.

Solution

The given function *multiplies* each input by 4. To "undo" this function, you
need to *divide* each input by 4. Thus, the inverse function of $f(x) = 4x$ is

$$f^{-1}(x) = \frac{x}{4}.$$

You can verify that both $f(f^{-1}(x))$ and $f^{-1}(f(x))$ are equal to the identity function as follows.

$$f(f^{-1}(x)) = f\left(\frac{x}{4}\right) = 4\left(\frac{x}{4}\right) = x$$

$$f^{-1}(f(x)) = f^{-1}(4x) = \frac{4x}{4} = x$$

▱

EXAMPLE 2   Finding Inverse Functions Informally

Find the inverse of $f(x) = x - 6$. Then verify that both $f(f^{-1}(x))$ and $f^{-1}(f(x))$ are equal to the identity function.

Solution

The given function *subtracts* 6 from each input. To "undo" this function, you need to *add* 6 to each input. Thus, the inverse function of $f(x) = x - 6$ is

$$f^{-1}(x) = x + 6.$$

You can verify that both $f(f^{-1}(x))$ and $f^{-1}(f(x))$ are equal to the identity function as follows.

$$f(f^{-1}(x)) = f(x + 6) = (x + 6) - 6 = x$$
$$f^{-1}(f(x)) = f^{-1}(x - 6) = (x - 6) + 6 = x$$

The formal definition of the inverse of a function is as follows.

Definition of the Inverse of a Function

Let f and g be two functions such that

$$f(g(x)) = x \qquad \text{for every } x \text{ in the domain of } g$$

and

$$g(f(x)) = x \qquad \text{for every } x \text{ in the domain of } f.$$

Under these conditions, the function g is the **inverse** of the function f. The function g is denoted by f^{-1} (read "f-inverse"). Thus,

$$f(f^{-1}(x)) = x$$

and

$$f^{-1}(f(x)) = x.$$

The domain of f must be equal to the range of f^{-1}, and the range of f must be equal to the domain of f^{-1}.

Note   Don't be confused by the use of -1 to denote the inverse function f^{-1}. In this text, whenever we write f^{-1}, we will *always* be referring to the inverse of the function f and *not* to the reciprocal of $f(x)$.

If the function g is the inverse of the function f, it must also be true that the function f is the inverse of the function g. For this reason, you can say that the functions f and g are *inverses of each other.*

Most graphing utilities can graph $y = x^{1/3}$ in two ways:

$$Y_1 = X \wedge (1/3) \text{ or}$$
$$Y_1 = \sqrt[3]{X}.$$

You may not be able to obtain the complete graph of $y = x^{2/3}$ by entering $Y_1 = X \wedge (2/3)$. If not, you should use

$$Y_1 = (X \wedge (1/3))^2 \text{ or}$$
$$Y_1 = \sqrt[3]{X^2}.$$

EXAMPLE 3   **Verifying Inverse Functions**

Show that the functions are inverses of each other.

$$f(x) = 2x^3 - 1 \quad \text{and} \quad g(x) = \sqrt[3]{\frac{x+1}{2}}$$

Solution

$$f(g(x)) = f\left(\sqrt[3]{\frac{x+1}{2}}\right) = 2\left(\sqrt[3]{\frac{x+1}{2}}\right)^3 - 1$$

$$= 2\left(\frac{x+1}{2}\right) - 1$$

$$= x + 1 - 1$$

$$= x$$

$$g(f(x)) = g(2x^3 - 1) = \sqrt[3]{\frac{(2x^3 - 1) + 1}{2}}$$

$$= \sqrt[3]{\frac{2x^3}{2}}$$

$$= \sqrt[3]{x^3}$$

$$= x$$

EXAMPLE 4   **Verifying Inverse Functions**

Which of the functions is the inverse of $f(x) = \dfrac{5}{x-2}$?

$$g(x) = \frac{x-2}{5} \quad \text{and} \quad h(x) = \frac{5}{x} + 2$$

Solution

By forming the composition of f with g, you have

$$f(g(x)) = f\left(\frac{x-2}{5}\right) = \frac{5}{[(x-2)/5] - 2} = \frac{25}{x-12} \neq x.$$

Because this composition is not equal to the identity function x, it follows that g *is not* the inverse of f. By forming the composition of f with h, you have

$$f(h(x)) = f\left(\frac{5}{x} + 2\right) = \frac{5}{[(5/x) + 2] - 2} = \frac{5}{5/x} = x.$$

Thus, it appears that h *is* the inverse of f. You can confirm this by showing that the composition of h with f is also equal to the identity function.

In Examples 3 and 4, inverse functions were verified algebraically. A graphing utility can also be helpful in checking to see whether one function is the inverse of another function. Use the Graph Reflection Program found in the appendix to verify graphically Example 4.

The Graph of an Inverse Function

The graphs of f and f^{-1} are related to each other in the following way. If the point (a, b) lies on the graph of f, then the point (b, a) lies on the graph of f^{-1} and vice versa. This means that the graph of f^{-1} is a reflection of the graph of f in the line $y = x$, as shown in Figure 1.33.

Figure 1.33

Figure 1.34

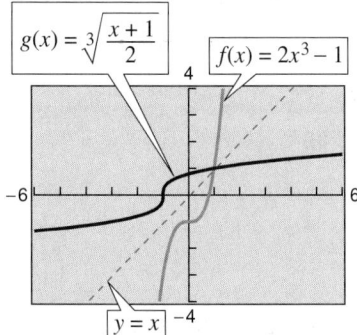

EXAMPLE 5   **Graphical Check for Inverse Functions**

Example 3 shows how to verify *algebraically* that the functions

$$f(x) = 2x^3 - 1 \quad \text{and} \quad g(x) = \sqrt[3]{\frac{x + 1}{2}}$$

are inverses of each other. Verify that f and g are inverses of each other *graphically* by graphing them in the same viewing rectangle and observing that the graph of g is the reflection of the graph of f in the line $y = x$. (Use a square setting.)

Solution

Using a graphing utility, you can obtain the graphs shown in Figure 1.34. From this figure you can see that the graph of g is the reflection of the graph of f in the line $y = x$, which confirms graphically that g is the inverse of f.

---

x	-2	-1	0	1	2
$f(x)$	-17	-3	-1	1	15

x	-17	-3	-1	1	15
$g(x)$	-2	-1	0	1	2

Note   You can verify *numerically* that the functions f and g are inverses by creating two tables of values, as shown at the left. Note that the entries in the two tables are the same except that their rows are interchanged.

The Existence of an Inverse Function

A function need not have an inverse function. For instance, the function

$$f(x) = x^2$$

has no inverse [assuming a domain of $(-\infty, \infty)$]. To have an inverse, a function must be **one-to-one,** which means that no two elements in the domain of f correspond to the same element in the range of f.

> **Definition of a One-to-One Function**
>
> A function f is **one-to-one** if, for a and b in its domain,
>
> $f(a) = f(b)$ implies that $a = b$.

> **Existence of an Inverse Function**
>
> A function f has an inverse function f^{-1} if and only if f is one-to-one.

EXAMPLE 6   **Testing for One-to-One Functions**

Which functions are one-to-one and have inverse functions?

a. $f(x) = x^3 + 1$ **b.** $g(x) = x^2 - x$ **c.** $h(x) = \sqrt{x}$

Solution

a. Let a and b be real numbers with $f(a) = f(b)$.

$$a^3 + 1 = b^3 + 1 \qquad\qquad \text{Set } f(a) = f(b).$$
$$a^3 = b^3$$
$$a = b$$

Therefore, $f(a) = f(b)$ implies that $a = b$. From this, it follows that f is one-to-one and has an inverse function.

b. Because $g(-1) = (-1)^2 - (-1) = 2$ and $g(2) = 2^2 - 2 = 2$, you have two distinct inputs matched with the same output. Thus, g *is not* a one-to-one function and has no inverse function.

c. Let a and b be nonnegative real numbers with $h(a) = h(b)$.

$$\sqrt{a} = \sqrt{b} \qquad\qquad \text{Set } h(a) = h(b).$$
$$a = b$$

Therefore, $h(a) = h(b)$ implies that $a = b$. Thus, h *is* one-to-one and has an inverse function.

The **parametric mode** of a graphing utility allows you to graph the inverse of a function without having to determine the actual equation of the inverse. The key idea is to express both x and y as functions of a third variable t. For instance, you could graph the function $y = 2x^3 - 1$ by setting your graphing utility to parametric mode and graphing the following equations.

$$X_{1T} = T$$
$$Y_{1T} = 2T \wedge 3 - 1.$$

Use the viewing window

$$\text{Tmin} = -4$$
$$\text{Tmax} = 4$$
$$\text{Tstep} = .05$$

with $-6 \le x \le 6$ and $-4 \le y \le 4$.

To graph the inverse of $y = 2x^3 - 1$, all you need to do is reverse the roles of x and y:

$$X_{2T} = 2T \wedge 3 - 1$$
$$Y_{2T} = T.$$

From its graph, it is easy to tell whether a function of x is one-to-one. Simply check to see that every *horizontal* line intersects the graph of the function at most once. For instance, Figure 1.35 shows the graphs of the three functions given in Example 6. On the graph of $g(x) = x^2 - x$, you can find a horizontal line that intersects the graph twice.

Figure 1.35

(a) f is one-to-one.

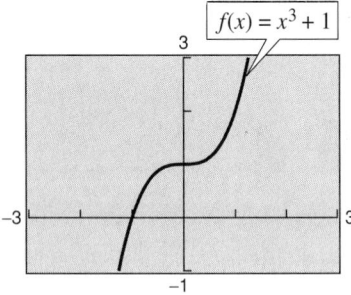

$f(x) = x^3 + 1$

(b) g is not one-to-one.

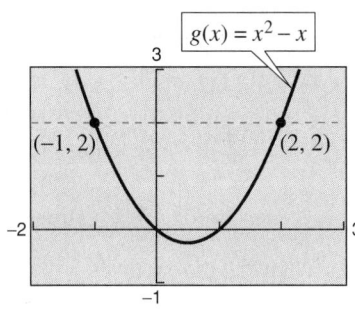

$g(x) = x^2 - x$

$(-1, 2)$ $(2, 2)$

(c) h is one-to-one.

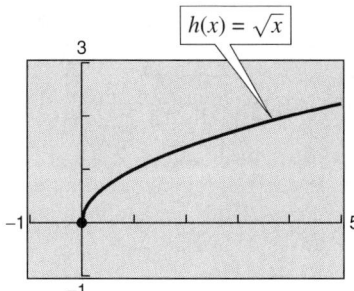

$h(x) = \sqrt{x}$

Two special types of functions that pass the **horizontal line test** are those that are increasing or decreasing on their entire domains.

1. If f is *increasing* on its entire domain, f is one-to-one.
2. If f is *decreasing* on its entire domain, f is one-to-one.

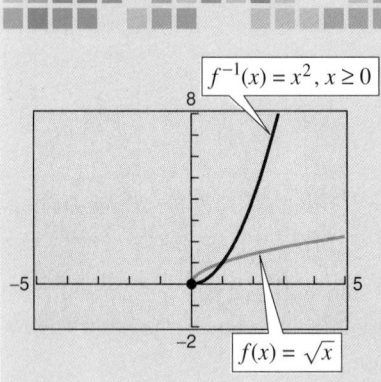

$f^{-1}(x) = x^2, x \geq 0$

$f(x) = \sqrt{x}$

Many graphing utilities, such as the *TI-82* or *TI-83*, have a built-in feature to draw the inverse of a function. To see how this works, consider the function $f(x) = \sqrt{x}$. The inverse of f is $f^{-1}(x) = x^2, x \geq 0$. Enter the function

$$Y_1 = \sqrt{X}$$

and graph it in the standard viewing rectangle. Now go to the DRAW menu and select item 8 (DrawInv), and then select Y_1 from the Y-VARS menu.

DrawInv Y_1

You should obtain the figure on the left, which shows both f and its inverse f^{-1}.

Finding Inverse Functions

For simple functions (such as the ones in Examples 1 and 2) you can find inverse functions by inspection. For instance, the inverse of $f(x) = 8x$ is $f^{-1}(x) = x/8$. For more complicated functions, however, it is best to use the following procedure for finding the inverse of a function.

Note The problem of finding the inverse of a function can be difficult (or even impossible) for two reasons. First, given $y = f(x)$, it may be algebraically difficult to solve for x in terms of y. Second, if f is not one-to-one, then f^{-1} does not exist.

Finding the Inverse of a Function

To find the inverse of f, use the following steps.

1. In the equation for $f(x)$, replace $f(x)$ by y.
2. Interchange the roles of x and y.
3. If the new equation does not represent y as a function of x, the function f does not have an inverse function. If the new equation does represent y as a function of x, solve the new equation for y.
4. Replace y by $f^{-1}(x)$.
5. Verify that f and f^{-1} are inverses of each other by showing that $f(f^{-1}(x)) = x$ and $f^{-1}(f(x)) = x$.

EXAMPLE 7   **Finding the Inverse of a Function**

Find the inverse (if it exists) of

$$f(x) = \frac{5 - 3x}{2}.$$

Solution

From Figure 1.36, you can see that f is one-to-one, and therefore has an inverse.

Figure 1.36

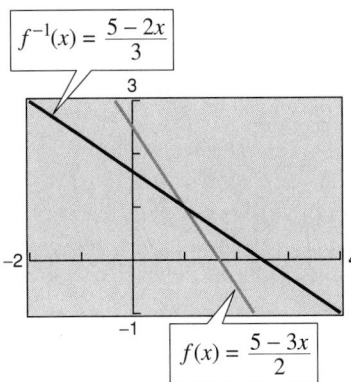

$f^{-1}(x) = \dfrac{5 - 2x}{3}$

$f(x) = \dfrac{5 - 3x}{2}$

$$y = \frac{5 - 3x}{2} \qquad \text{Replace } f(x) \text{ by } y.$$

$$x = \frac{5 - 3y}{2} \qquad \text{Interchange } x \text{ and } y.$$

$$2x = 5 - 3y$$

$$3y = 5 - 2x$$

$$y = \frac{5 - 2x}{3} \qquad \text{Solve for } y.$$

$$f^{-1}(x) = \frac{5 - 2x}{3} \qquad \text{Replace } y \text{ by } f^{-1}(x).$$

The domain and range of both f and f^{-1} consist of all real numbers.

EXAMPLE 8 ▭ **Finding the Inverse of a Function**

Find the inverse of $f(x) = \sqrt{2x - 3}$ and sketch the graphs of f and f^{-1}.

Solution

$$y = \sqrt{2x - 3} \qquad \text{Replace } f(x) \text{ by } y.$$

$$x = \sqrt{2y - 3} \qquad \text{Interchange } x \text{ and } y.$$

$$x^2 = 2y - 3$$

$$2y = x^2 + 3$$

$$y = \frac{x^2 + 3}{2} \qquad \text{Solve for } y.$$

$$f^{-1}(x) = \frac{x^2 + 3}{2}, \quad x \geq 0 \qquad \text{Replace } y \text{ by } f^{-1}(x).$$

The graph of f^{-1} is the reflection of the graph of f in the line $y = x$, as shown in Figure 1.37. Note that the domain of f is the interval $\left[\frac{3}{2}, \infty\right)$ and the range of f is the interval $[0, \infty)$. Moreover, the domain of f^{-1} is the interval $[0, \infty)$ and the range of f^{-1} is the interval $\left[\frac{3}{2}, \infty\right)$. ▭

Figure 1.37

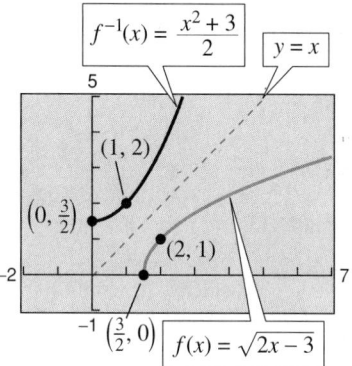

$f^{-1}(x) = \dfrac{x^2 + 3}{2}$ $y = x$

$(1, 2)$

$\left(0, \frac{3}{2}\right)$

$(2, 1)$

$\left(\frac{3}{2}, 0\right)$ $f(x) = \sqrt{2x - 3}$

Group Activity   *The Existence of an Inverse Function*

Write a short paragraph describing why the following functions do or do not have inverse functions. Give a numerical example.

a. Your hourly wage is $7.50 plus $0.90 for each unit x produced per hour. Let $f(x)$ represent your weekly wage for 40 hours of work. Does this function have an inverse?

b. Let x represent the retail price of an item (in dollars), and let $f(x)$ represent the sales tax on the item. Assume that the sales tax is 7% of the retail price *and* that the sales tax is rounded to the nearest cent. Does this function have an inverse? (*Hint:* Can you undo this function? For instance, if you know that the sales tax is $0.14, can you determine *exactly* what the retail price is?)

1.5 /// EXERCISES

In Exercises 1–4, match the graph of the function with the graph of its inverse. [The graphs of the inverse functions are labeled (a), (b), (c), and (d).]

(a)

(b)

(c)

(d)

1.

2.

3.

4.
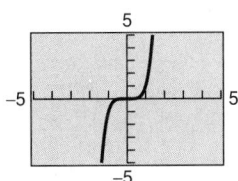

In Exercises 5–10, find the inverse of f informally. Verify that $f(f^{-1}(x)) = x$ and $f^{-1}(f(x)) = x$.

5. $f(x) = 8x$

6. $f(x) = \dfrac{1}{5}x$

7. $f(x) = x + 10$

8. $f(x) = x - 5$

9. $f(x) = \sqrt[3]{x}$

10. $f(x) = x^5$

In Exercises 11–14, show that f and g are inverse functions (a) algebraically and (b) graphically.

11. $f(x) = 2x, \quad g(x) = \dfrac{x}{2}$

12. $f(x) = x - 5, \quad g(x) = x + 5$

13. $f(x) = 5x + 1, \quad g(x) = \dfrac{x - 1}{5}$

14. $f(x) = 3 - 4x, \quad g(x) = \dfrac{3 - x}{4}$

In Exercises 15–20, show that f and g are inverse functions algebraically. Use a graphing utility to graph f and g in the same viewing rectangle. Describe the relationship between the graphs.

15. $f(x) = x^3, \quad g(x) = \sqrt[3]{x}$

16. $f(x) = \dfrac{1}{x}, \quad g(x) = \dfrac{1}{x}$

17. $f(x) = \sqrt{x - 4}, \quad g(x) = x^2 + 4, \ x \geq 0$

18. $f(x) = 9 - x^2, \quad x \geq 0$
$g(x) = \sqrt{9 - x}, \quad x \leq 9$

19. $f(x) = 1 - x^3, \quad g(x) = \sqrt[3]{1 - x}$

20. $f(x) = \dfrac{1}{1 + x}, \quad x \geq 0$
$g(x) = \dfrac{1 - x}{x}, \quad 0 < x \leq 1$

In Exercises 21–30, use a graphing utility to graph the function and use the Horizontal Line Test to determine whether the function is one-to-one.

21. $f(x) = 3 - \frac{1}{2}x$

22. $h(x) = \sqrt{16 - x^2}$

23. $h(x) = \dfrac{x^2}{x^2 + 1}$

24. $f(x) = \sqrt{x - 2}$

25. $g(x) = \dfrac{4 - x}{6}$

26. $f(x) = 10$

27. $h(x) = |x + 4| - |x - 4|$

28. $g(x) = (x + 5)^3$

29. $f(x) = -2x\sqrt{16 - x^2}$

30. $f(x) = \frac{1}{8}(x + 2)^2 - 1$

In Exercises 31–40, find the inverse of the function f. Use a graphing utility to graph both f and f^{-1} in the same viewing rectangle. Describe the relationship between the graphs.

31. $f(x) = 2x - 3$ **32.** $f(x) = 3x$

33. $f(x) = x^5$ **34.** $f(x) = x^3 + 1$

35. $f(x) = \sqrt{x}$ **36.** $f(x) = x^2, \quad x \geq 0$

37. $f(x) = \sqrt{4 - x^2},$ **38.** $f(x) = \dfrac{4}{x}$
$\quad 0 \leq x \leq 2$

39. $f(x) = \sqrt[3]{x - 1}$ **40.** $f(x) = x^{3/5}$

In Exercises 41–56, determine whether the function is one-to-one. If it is, find its inverse.

41. $f(x) = x^4$ **42.** $f(x) = \dfrac{1}{x^2}$

43. $g(x) = \dfrac{x}{8}$ **44.** $f(x) = 3x + 5$

45. $p(x) = -4$ **46.** $f(x) = \dfrac{3x + 4}{5}$

47. $f(x) = (x + 3)^2, \quad x \geq -3$

48. $q(x) = (x - 5)^2$

49. $h(x) = \dfrac{4}{x^2}$

50. $f(x) = |x - 2|, \quad x \leq 2$

51. $f(x) = \sqrt{2x + 3}$ **52.** $f(x) = \sqrt{x - 2}$

53. $g(x) = x^2 - x^4$ **54.** $f(x) = \dfrac{x^2}{x^2 + 1}$

55. $f(x) = 25 - x^2, \quad x \leq 0$

56. $f(x) = ax + b, \quad a \neq 0$

In Exercises 57–60, delete part of the graph of the function so that the part that remains is one-to-one. Find the inverse of the remaining part and give the domain of the inverse. (There is more than one correct answer.)

57. $f(x) = (x - 2)^2$ **58.** $f(x) = 1 - x^4$

 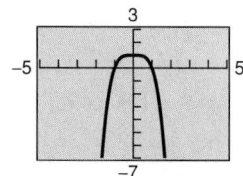

59. $f(x) = |x + 2|$ **60.** $f(x) = |x - 2|$

 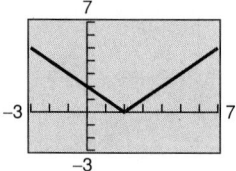

In Exercises 61 and 62, use the graph of the function f to complete the table and sketch the graph of f^{-1}.

61.

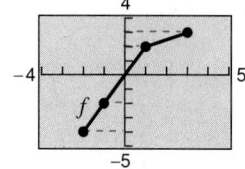

x	$f^{-1}(x)$
-4	
-2	
2	
3	

62.

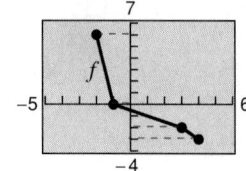

x	$f^{-1}(x)$
-3	
-2	
0	
6	

Graphical Reasoning In Exercises 63–66, (a) use a graphing utility to graph the function, (b) use the DrawInv feature of the graphing utility to draw the inverse of the function, and (c) determine if the graph of the inverse relation is an inverse function (explain).

63. $f(x) = x^3 + x + 1$ **64.** $h(x) = x\sqrt{4 - x^2}$

65. $g(x) = \dfrac{3x^2}{x^2 + 1}$ **66.** $f(x) = \dfrac{4x}{\sqrt{x^2 + 15}}$

True or False? In Exercises 67–70, determine if the statement is true or false. If it is false, give an example to show why.

67. If f is an even function, f^{-1} exists.

68. If the inverse of f exists, the y-intercept of f is an x-intercept of f^{-1}.

69. If $f(x) = x^n$ where n is odd, f^{-1} exists.

70. There exists no function f such that $f = f^{-1}$.

In Exercises 71–76, use the functions $f(x) = \frac{1}{8}x - 3$ and $g(x) = x^3$ to find the indicated value or function.

71. $(f^{-1} \circ g^{-1})(1)$ **72.** $(g^{-1} \circ f^{-1})(-3)$

73. $(f^{-1} \circ f^{-1})(6)$ **74.** $(g^{-1} \circ g^{-1})(-4)$

75. $(f \circ g)^{-1}$ **76.** $g^{-1} \circ f^{-1}$

In Exercises 77–80, use the functions $f(x) = x + 4$ and $g(x) = 2x - 5$ to find the specified functions.

77. $g^{-1} \circ f^{-1}$ **78.** $f^{-1} \circ g^{-1}$

79. $(f \circ g)^{-1}$ **80.** $(g \circ f)^{-1}$

81. Prove that if f and g are one-to-one functions, $(f \circ g)^{-1}(x) = (g^{-1} \circ f^{-1})(x)$.

82. Prove that if f is a one-to-one odd function, f^{-1} is an odd function.

83. *Hourly Wage* Your wage is $8.00 per hour plus $0.75 for each unit produced per hour. Thus, your hourly wage y in terms of the number of units produced is given by $y = 8 + 0.75x$.

(a) Determine the inverse of the function. What does each variable in the inverse function represent?

(b) Use a graphing utility to graph the function and its inverse.

(c) Use the trace key to find the hourly wage if 10 units are produced per hour.

(d) Use the trace key to find the number of units produced when your hourly wage is $22.25.

84. *Cost* Suppose you need 50 pounds of two commodities costing $1.25 and $1.60 per pound, respectively.

(a) Verify that the total cost is $y = 1.25x + 1.60(50 - x)$, where x is the number of pounds of the less expensive commodity.

(b) Find the inverse of the cost function. What does each variable in the inverse function represent?

(c) Use the context of the problem to determine the domain of the inverse function.

(d) Determine the number of pounds of the less expensive commodity purchased if the total cost is $73.

85. *Diesel Engine* The function

$$y = 0.03x^2 + 254.50, \qquad 0 < x < 100$$

approximates the exhaust temperature y of a diesel engine in degrees Fahrenheit, where x is the percent load on the engine.

(a) Determine the inverse of the function. What does each variable in the inverse function represent?

(b) Use a graphing utility to graph the inverse function.

(c) Determine the percent load interval if the exhaust temperature of the engine must not exceed 500° F.

86. *Think About It* The function

$$f(x) = k(2 - x - x^3)$$

is one-to-one and $f^{-1}(3) = -2$. Find k.

1.6   Exploring Data: Linear Models and Scatter Plots

Scatter Plots and Correlation / Fitting a Line to Data

Scatter Plots and Correlation

Many real-life situations involve finding relationships between two variables such as the year and the number of people in the labor force. In a typical situation, data is collected and written as a set of ordered pairs. We discussed the graph of such a set, a **scatter plot,** briefly in Section P.1.

Most graphing utilities have built-in statistical programs that can create scatter plots. Use your graphing utility to plot the points given in the table at the right.

EXAMPLE 1   **Constructing a Scatter Plot**   *Real Life*

The data in the table shows the number of people P (in millions) in the United States who were part of the labor force from 1983 through 1993. In the table, t represents the year, with $t = 3$ corresponding to 1983. Sketch a scatter plot of the data.   (Source: U.S. Bureau of Labor Statistics)

t	3	4	5	6	7	8	9	10	11	12	13
P	113	115	117	120	122	123	126	126	127	129	130

Solution

Begin by representing the data with a set of ordered pairs.

(3, 113), (4, 115), (5, 117), (6, 120), (7, 122), (8, 123),

(9, 126), (10, 126), (11, 127), (12, 129), (13, 130)

Then plot each point in a coordinate plane, as shown in Figure 1.38.

Figure 1.38

From the scatter plot in Figure 1.38, it appears that the points describe a relationship that is nearly linear. The relationship is not *exactly* linear because the labor force did not increase by precisely the same amount each year.

A mathematical equation that approximates the relationship between t and P is called a *mathematical model*. When developing a mathematical model, you strive for two (often conflicting) goals—accuracy and simplicity. For the data above, a linear model of the form $P = at + b$ appears to be best. It is simple and relatively accurate.

Consider a collection of ordered pairs of the form (x, y). If y tends to increase as x increases, the collection is said to have a **positive correlation.** If y tends to decrease as x increases, the collection is said to have a **negative correlation.** Figure 1.39 shows three examples: one with a positive correlation, one with a negative correlation, and one with no (discernible) correlation.

Figure 1.39

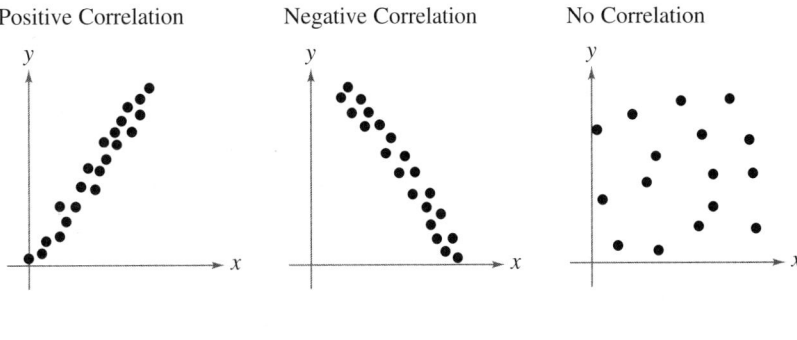

Positive Correlation   Negative Correlation   No Correlation

EXAMPLE 2   *Real Life*   **Interpreting Correlation**

On a Friday, 22 students in a class were asked to keep track of the number of hours they spent studying for a test on Monday and the number of hours they spent watching television. The numbers are shown below. Construct a scatter plot for each set of data. Then determine whether the points are positively correlated, are negatively correlated, or have no discernible correlation. What can you conclude? (The first coordinate is the number of hours and the second coordinate is the score obtained on Monday's test.)

Study Hours: (0, 40), (1, 41), (2, 51), (3, 58), (3, 49), (4, 48), (4, 64), (5, 55), (5, 69), (5, 58), (5, 75), (6, 68), (6, 63), (6, 93), (7, 84), (7, 67), (8, 90), (8, 76), (9, 95), (9, 72), (9, 85), (10, 98)

TV Hours: (0, 98), (1, 85), (2, 72), (2, 90), (3, 67), (3, 93), (3, 95), (4, 68), (4, 84), (5, 76), (7, 75), (7, 58), (9, 63), (9, 69), (11, 55), (12, 58), (14, 64), (16, 48), (17, 51), (18, 41), (19, 49), (20, 40)

Solution

Scatter plots for the two sets of data are shown in Figure 1.40. The scatter plot relating study hours and test scores has a positive correlation. This means that the more a student studied, the higher his or her score tended to be. The scatter plot relating television hours and test scores has a negative correlation. This means that the more time a student spent watching television, the lower his or her score tended to be.

Figure 1.40

Fitting a Line to Data

Finding a linear model to represent the relationship described by a scatter plot is called **fitting a line to data.** You can do this graphically by simply sketching the line that appears to fit the points, finding two points on the line, and then finding the equation of the line that passes through the two points.

Real Life

EXAMPLE 3   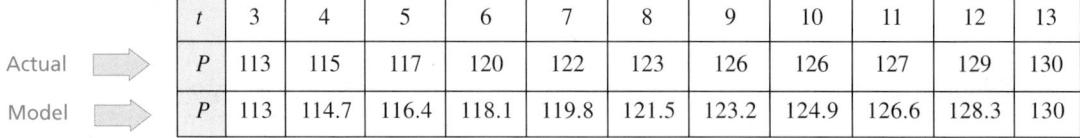   Fitting a Line to Data

Find a linear model that relates the year with the number of people in the United States labor force. (See Example 1.)

t	3	4	5	6	7	8	9	10	11	12	13
P	113	115	117	120	122	123	126	126	127	129	130

Figure 1.41

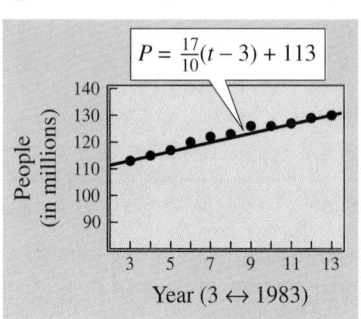

Solution

After plotting the data in the table, draw the line that you think best represents the data, as shown in Figure 1.41. Two points that lie on this line are $(3, 113)$ and $(13, 130)$. Using the point-slope form, you can find the equation of the line to be

$$P = \frac{17}{10}(t - 3) + 113.$$ Linear model

Once you have found a model, you can measure how well the model fits the data by comparing the actual values with the values given by the model, as shown in the following table.

	t	3	4	5	6	7	8	9	10	11	12	13
Actual	P	113	115	117	120	122	123	126	126	127	129	130
Model	P	113	114.7	116.4	118.1	119.8	121.5	123.2	124.9	126.6	128.3	130

The sum of the squares of the differences between the actual values and the model's values is the **sum of the squared differences.** The model that has the least sum is the **least squares regression line** for the data. For the model in Example 3, the sum of the squared differences is 20.85. The least squares regression line for the data is

$$P = 1.7t + 108.9.$$ Best-fitting linear model

Its sum of squared differences is 8.85.

Least Squares Regression Line

The least squares regression line, $y = ax + b$, for the points (x_1, y_1), (x_2, y_2), (x_3, y_3), ..., (x_n, y_n) is given by

$$a = \frac{n\sum\limits_{i=1}^{n}x_iy_i - \sum\limits_{i=1}^{n}x_i \sum\limits_{i=1}^{n}y_i}{n\sum\limits_{i=1}^{n}x_i^2 - \left(\sum\limits_{i=1}^{n}x_i\right)^2} \quad \text{and} \quad b = \frac{1}{n}\left(\sum\limits_{i=1}^{n}y_i - a\sum\limits_{i=1}^{n}x_i\right).$$

EXAMPLE 4 **Finding a Least Squares Regression Line**

Find the least squares regression line for the points $(-3, 0)$, $(-1, 1)$, $(0, 2)$, and $(2, 3)$.

Solution

Begin by constructing a table like that shown below.

x	y	xy	x^2
-3	0	0	9
-1	1	-1	1
0	2	0	0
2	3	6	4
$\sum\limits_{i=1}^{n}x_i = -2$	$\sum\limits_{i=1}^{n}y_i = 6$	$\sum\limits_{i=1}^{n}x_iy_i = 5$	$\sum\limits_{i=1}^{n}x_i^2 = 14$

Applying the formulas for the least squares regression line with $n = 4$ produces

Figure 1.42

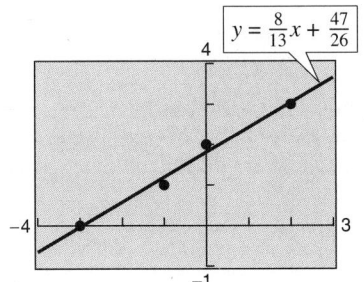

$$y = \tfrac{8}{13}x + \tfrac{47}{26}$$

$$a = \frac{n\sum\limits_{i=1}^{n}x_iy_i - \sum\limits_{i=1}^{n}x_i \sum\limits_{i=1}^{n}y_i}{n\sum\limits_{i=1}^{n}x_i^2 - \left(\sum\limits_{i=1}^{n}x_i\right)^2} = \frac{4(5) - (-2)(6)}{4(14) - (-2)^2} = \frac{32}{52} = \frac{8}{13}$$

and

$$b = \frac{1}{n}\left(\sum\limits_{i=1}^{n}y_i - a\sum\limits_{i=1}^{n}x_i\right) = \frac{1}{4}\left[6 - \frac{8}{13}(-2)\right] = \frac{47}{26}.$$

The least squares regression line is $y = \frac{8}{13}x + \frac{47}{26}$, as shown in Figure 1.42.

EXAMPLE 5 ▱ A Mathematical Model

Real Life

The annual amounts of advertising expenses y (in billions of dollars) in the United States from 1983 to 1992 are given in the table.   (Source: McCann Erickson)

Year	1983	1984	1985	1986	1987	1988	1989	1990	1991	1992
y	75.9	88.1	94.8	102.1	109.8	118.1	125.6	128.6	126.4	131.7

A linear model that approximates this data is

$$y = 6.1703t + 63.8327, \quad 3 \leq t \leq 12$$

where $t = 3$ corresponds to 1983. Plot the actual data *and* the model on the same graph. How closely does the model represent the data?

Solution

The actual data is plotted in Figure 1.43, along with the graph of the linear model. From the figure, it appears that the model is a "good fit" for the actual data. You can see how well the model fits by comparing the actual values of y with the values of y given by the model (these are labeled $y*$ in the table below).

Figure 1.43

$y = 6.1703t + 63.8327$

Advertising expense (in billions of dollars)

Year (3 ↔ 1983)

t	3	4	5	6	7	8	9	10	11	12
y	75.9	88.1	94.8	102.1	109.8	118.1	125.6	128.6	126.4	131.7
$y*$	82.3	88.5	94.7	100.9	107.0	113.2	119.4	125.5	131.7	137.9

▱

Many calculators have "built-in" least squares regression programs. For instance, on the *TI-82* or *TI-83*, you can find a least squares regression line as shown below.

1. Use the stat and edit menus to enter the data in lists L_1 and L_2.
2. For the *TI-82*, use the stat and calc menus to select SetUp. Make sure the XList is set to L_1 and the YList is set to L_2. (These are the default settings on the *TI-83*.)
3. Use the stat and calc menus to select LinReg(ax+b). After running this program, the calculator will display the values of a and b in the model $y = ax + b$. The *TI-82* will also display the correlation coefficient r.

EXAMPLE 6 ▰ **Finding a Least Squares Regression Line**

The following ordered pairs (w, h) represent the shoe sizes w and the heights h (in inches) of 25 men. Use a computer program or a statistical calculator to find the least squares regression line for the data.

(10.0, 70.0)	(10.5, 71.0)	(9.5, 70.0)	(11.0, 72.0)	(12.0, 74.0)
(8.5, 66.0)	(9.0, 68.5)	(13.0, 76.0)	(10.5, 71.5)	(10.5, 70.5)
(10.0, 72.0)	(9.5, 70.0)	(10.0, 71.0)	(10.5, 69.5)	(11.0, 71.5)
(12.0, 73.5)	(12.5, 74.0)	(11.0, 71.5)	(9.0, 67.5)	(10.0, 70.0)
(13.0, 73.5)	(10.5, 72.5)	(10.5, 71.0)	(11.0, 73.0)	(8.5, 68.0)

Solution

A scatter plot for the data is shown in Figure 1.44. Note that the plot does not have 25 points because some of the ordered pairs graph as the same point. After entering the data into a graphing utility, you can obtain

$$a = 1.67 \quad \text{and} \quad b = 53.57.$$

Thus, the least squares regression line for the data is

$$h = 1.67w + 53.57.$$

In Figure 1.44, this line is plotted with the data. Note that the line is a relatively good fit for the data. ▰

Figure 1.44

If you use a graphing calculator or computer program to duplicate the results of Example 6, you will notice that the program also outputs a value of $r \approx 0.918$. This number is the **correlation coefficient** of the data. Correlation coefficients vary between -1 and 1. Basically, the closer $|r|$ is to 1, the better the points can be described by a line. Three examples are shown in Figure 1.45.

Figure 1.45

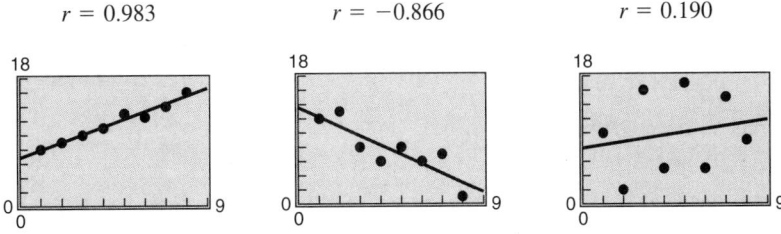

Group Activity

Research Project

Use your school's library or some other reference source to locate data that you think describes a linear relationship. Create a scatter plot of the data, and find the least squares regression line that represents the points. Interpret the slope and y-intercept in the context of the data.

1.6 /// EXERCISES

Correlation   In Exercises 1–4, the scatter plots of sets of data are given. Determine whether there is positive correlation, negative correlation, or very little correlation between the variables.

1.

2.

3.

4.

5. The following ordered pairs give the exposure index x of a carcinogenic substance and the cancer mortality y per 100,000 population. The higher the index level, the higher the level of contamination.

(3.50, 150.1), (3.58, 133.1),
(4.42, 132.9), (2.26, 116.7),
(2.63, 140.7), (4.85, 165.5),
(12.65, 210.7), (7.42, 181.0),
(9.35, 213.4)

(a) Create a scatter plot for the data.

(b) Does the relationship between x and y appear to be approximately linear? Explain.

6. The following ordered pairs give the scores of two consecutive 15-point quizzes for a class of 18 students.

(7, 13), (9, 7), (14, 14),
(15, 15), (10, 15), (9, 7),
(14, 11), (14, 15), (8, 10),
(9, 10), (15, 9), (10, 11),
(11, 14), (7, 14), (11, 10),
(14, 11), (10, 15), (9, 6)

(a) Create a scatter plot for the data.

(b) Does the relationship between consecutive quiz scores appear to be approximately linear? If not, give some possible explanations.

In Exercises 7–10, (a) find the least squares regression line by hand (see Example 4) and use a graphing utility to verify your results, (b) graph the data points and the regression line, and (c) comment on the validity of the model.

7.

8.

9.

10.

11. *Hooke's Law* Hooke's Law states that the force F required to compress or stretch a spring (within its elastic limits) is proportional to the distance d that the spring is compressed or stretched from its original length. That is, $F = kd$, where k is the measure of the stiffness of the spring and is called the *spring constant*. The table gives the elongation d in centimeters of a spring when a force of F kilograms is applied.

F	20	40	60	80	100
d	1.4	2.5	4.0	5.3	6.6

(a) Use the regression capabilities of a graphing utility to find a linear model for the data.

(b) Use a graphing utility to plot the data and graph the model. How well does the model fit the data? Explain your reasoning.

(c) Use the model to estimate the elongation of the spring when a force of 55 kilograms is applied.

12. *Falling Object* In an experiment students measured the speed s (in meters per second) of a falling object t seconds after it was released. The results are given in the table.

t	0	1	2	3	4
s	0	11.0	19.4	29.2	39.4

(a) Use the regression capabilities of a graphing utility to find a linear model for the data.

(b) Use a graphing utility to plot the data and graph the model. How well does the model fit the data? Explain your reasoning.

(c) Use the model to estimate the speed of the object after 2.5 seconds.

13. *Cable TV* The average monthly basic rate R (in dollars) for cable TV for the years 1986 through 1993 in the United States is given in the table. (Source: Paul Kagan Associates, Inc.)

Year	1986	1987	1988	1989
R	11.09	13.27	14.45	15.97

Year	1990	1991	1992	1993
R	17.58	18.61	19.08	19.39

Let t represent the year, with $t = 6$ corresponding to 1986.

(a) Use the regression capabilities of a graphing utility to find a linear model for the data.

(b) Use a graphing utility to plot the data and graph the model.

(c) Interpret the slope of the model in the context of the problem.

(d) Use the model to predict the average monthly basic rate for cable TV for the year 2000.

14. *Price of Homes* The median sales price P (in thousands of dollars) of existing one-family homes for 1986 through 1993 in the United States is given in the table. (Source: National Association of Realtors)

Year	1986	1987	1988	1989
P	80.3	85.6	89.3	93.1

Year	1990	1991	1992	1993
P	95.5	100.3	103.7	106.8

Let t represent the year, with $t = 6$ corresponding to 1986.

(a) Use the regression capabilities of a graphing utility to find a linear model for the data.

(b) Use a graphing utility to plot the data and graph the model.

(c) Interpret the slope of the model in the context of the problem.

(d) Use the model to predict the median price of existing one-family homes for the year 2000.

15. *Advertising and Sales*   The table gives the advertising expenditures x and sales volume y for a company for seven randomly selected months. Both are measured in thousands of dollars.

Month	1	2	3	4	5	6	7
x	2.4	1.6	2.0	2.6	1.4	1.6	2.0
y	202	184	220	240	180	164	186

(a) Use the regression capabilities of a graphing utility to find a linear model for the data.

(b) Use a graphing utility to plot the data and graph the model.

(c) Interpret the slope of the model in the context of the problem.

(d) Use the model to predict sales for advertising expenditures of $1500.

16. *Energy Consumption*   The data gives the per capita energy usage (in thousands of kilograms of coal equivalent) and the per capita gross national product (in thousands of U.S. dollars) for a sample of countries in 1990.   (Source: Statistical Office of the United Nations)

Argentina (1.83, 3.7); Bangladesh (0.01, 0.2); Brazil (0.77, 2.6); Canada (10.46, 21.5); Denmark (4.70, 24.0); Finland (5.93, 26.0); France (3.87, 20.8); Greece (3.05, 6.8); India (0.31, 0.3); Italy (3.86, 19.4); Japan (4.21, 26.2); Mexico (1.75, 3.0); Pakistan (0.28, 0.4); South Korea (2.47, 6.0); Tanzania (0.04, 0.1); United States (10.32, 23.0)

(a) Use the regression capabilities of a graphing utility to find a linear model for the data.

(b) Use a graphing utility to plot the data and graph the model.

(c) Interpret the graph in part (b). Use the graph to identify any countries that appear to differ substantially from most of the others.

17. *Holders of Mortgage Debts*   The table gives the amount of mortgage debt (in billions of dollars) held by savings institutions x and commercial banks y for the years 1988 through 1992 in the United States. (Source: The Federal Reserve Bulletin)

Year	1988	1989	1990	1991	1992
x	925	910	802	705	628
y	674	767	845	876	895

(a) Use the regression capabilities of a graphing utility to find a linear model for the data.

(b) Use a graphing utility to plot the data and graph the model.

(c) Interpret the slope of the model in the context of the problem. What information is given by the sign of the slope?

18. *Voter Registration and Turnout*   The data shows the percent x of the voting-age population that was registered and the percent y that actually voted by state in 1992.   (Source: U.S. Bureau of the Census)

AK (75.1, 68.6)   AL (77.1, 63.6)   AR (66.5, 58.0)
AZ (70.5, 64.7)   CA (57.6, 52.8)   CO (74.7, 68.8)
CT (76.8, 71.7)   D.C. (73.9, 66.1)   DE (70.7, 66.1)
FL (62.7, 55.8)   GA (62.0, 54.1)   HI (60.6, 55.2)
IA (77.9, 70.8)   ID (71.8, 66.9)   IL (72.1, 65.1)
IN (68.0, 63.0)   KS (77.1, 71.9)   KY (64.9, 57.6)
LA (77.0, 68.6)   MA (72.5, 65.8)   MD (71.7, 66.4)
ME (85.4, 74.1)   MI (74.6, 65.9)   MN (86.1, 74.2)
MO (74.2, 66.2)   MS (79.3, 66.7)   MT (77.3, 71.4)
NC (68.7, 60.0)   ND (90.8, 71.1)   NE (73.1, 66.1)
NH (69.9, 64.6)   NJ (68.0, 61.2)   NM (67.1, 62.6)
NV (63.4, 58.1)   NY (62.0, 56.8)   OH (69.6, 64.3)
OK (74.3, 67.5)   OR (74.7, 69.1)   PA (65.1, 59.9)
RI (74.0, 69.0)   SC (67.0, 58.0)   SD (80.1, 70.2)
TN (65.0, 55.6)   TX (64.9, 55.6)   UT (78.8, 72.3)
VA (65.4, 61.2)   VT (76.1, 68.6)   WA (71.5, 66.3)
WI (84.4, 75.3)   WV (64.9, 57.3)   WY (68.1, 64.2)

(a) Use the regression capabilities of a graphing utility to find a linear model for the data.

(b) Use a graphing utility to plot the data and graph the model.

(c) Interpret the graph in part (b). Use the graph to identify any states that appear to differ substantially from most of the others.

(d) Interpret the slope of the model in the context of the problem.

In this chapter, you studied functions and their graphs. You can use the following questions to check your understanding of several of these basic concepts. The answers to these questions are given in the back of the book.

1. Does the relationship shown in the figure represent a function from set A to set B? Explain.

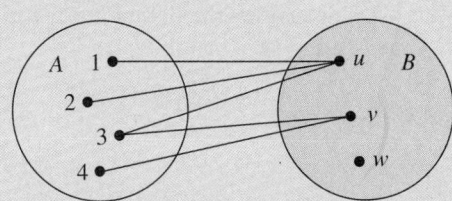

2. Use a graphing utility to select viewing rectangles that would show these graphs.

(a)

(b)

3. If f is an even function, determine if g is even, odd, or neither. Explain.

(a) $g(x) = -f(x)$ (b) $g(x) = f(-x)$

(c) $g(x) = f(x) - 2$ (d) $g(x) = f(x - 2)$

4. Management originally predicted that the profits from the sales of a new product would be approximated by the graph of the function f in the figure. The actual profits are shown by the function g along with a verbal description. Use the concepts of transformations of graphs to write g in terms of f.

(a) The profits were only three-fourths as large as expected.

(b) The profits were consistently $10,000 greater than predicted.

(c) There was a 2-year delay in the introduction of the product. After sales began, profits were as expected.

(d) It consistently required twice the time for profits to reach their predicted levels

5. *True or False?* If the correlation coefficient is close to -1, the regression line cannot be used to describe the data. Explain.

6. *True or False?* The sign of the slope of the regression line and the correlation coefficient are always the same.

1 /// REVIEW EXERCISES

In Exercises 1 and 2, does the table describe a function? Explain your reasoning.

1.

Input Value	10	5	0	5	10
Output Value	-5	-2	1	2	5

2.

Input Value	0	2	4	6	8
Output Value	-2	2	14	34	62

In Exercises 3–6, determine whether the equation represents y as a function of x.

3. $16x - y^4 = 0$ **4.** $2x - y - 3 = 0$

5. $y = \sqrt{1 - x}$ **6.** $|y| = x + 2$

In Exercises 7–12, evaluate the function at the specified values of the independent variable. Simplify your answers.

7. $f(x) = x^2 + 1$

  (a) $f(2)$ (b) $f(t^2)$ (c) $-f(x)$

8. $g(x) = x^{4/3}$

  (a) $g(8)$ (b) $g(t + 1)$ (c) $\dfrac{g(8) - g(1)}{8 - 1}$

9. $h(x) = 6 - 5x^2$

  (a) $h(2)$ (b) $h(x + 3)$ (c) $\dfrac{h(x + t) - h(x)}{t}$

10. $f(t) = \sqrt[4]{t}$

  (a) $f(16)$ (b) $f(t + 5)$ (c) $\dfrac{f(16) - f(0)}{16}$

11. $g(x) = \begin{cases} \frac{1}{2}x + 1, & x \le 2 \\ x - 2, & x > 2 \end{cases}$

  (a) $g(-2)$ (b) $g(2)$ (c) $g(10)$

12. $f(x) = \begin{cases} x^2 + 2, & x < 0 \\ |x - 2|, & x \ge 0 \end{cases}$

  (a) $f(-4)$ (b) $f(0)$ (c) $f(1)$

In Exercises 13–18, determine the domain of the function. Verify your result with a graphing utility.

13. $f(x) = \sqrt{25 - x^2}$ **14.** $f(x) = 3x + 4$

15. $g(s) = \dfrac{5}{3s - 9}$ **16.** $f(x) = \sqrt{x^2 + 8x}$

17. $h(x) = \dfrac{x}{x^2 - x - 6}$ **18.** $h(t) = |t + 1|$

In Exercises 19 and 20, use a graphing utility to select the viewing rectangle that shows the most complete graph of the function.

19. $f(x) = \dfrac{3x}{2(3 - x)}$

Xmin = -4	Xmin = -5	Xmin = 0
Xmax = 4	Xmax = 10	Xmax = 20
Xscl = 1	Xscl = 1	Xscl = 2
Ymin = -3	Ymin = -8	Ymin = 0
Ymax = 3	Ymax = 6	Ymax = 10
Yscl = 1	Yscl = 1	Yscl = 2

20. $f(x) = 4[(0.3x)^3 - 5x]$

Xmin = -200	Xmin = -10	Xmin = -15
Xmax = 200	Xmax = 10	Xmax = 15
Xscl = 50	Xscl = 2	Xscl = 5
Ymin = -500	Ymin = -20	Ymin = -150
Ymax = 500	Ymax = 20	Ymax = 150
Yscl = 50	Yscl = 4	Yscl = 50

In Exercises 21–32, sketch the graph of the function. Use a graphing utility to confirm your graph.

21. $g(x) = \frac{1}{4}x^2$

22. $y = (x - 3)^2$

23. $y = (x - 3)^2 - 4$

24. $y = 9 - (x - 3)^2$

25. $y = \frac{1}{2}x(2 - x)$

26. $f(t) = \sqrt{\dfrac{t}{2}}$

27. $h(x) = 5 - \frac{1}{2}|x|$

28. $s(t) = |t + 1| - 3$

29. $g(x) = \frac{1}{4}x^3, \quad -2 \le x \le 2$

30. $h(x) = x(4 - x), \quad 0 \le x \le 4$

31. $f(x) = \begin{cases} 2 - (x - 1)^2, & x < 1 \\ 2 + (x - 1)^2, & x \ge 1 \end{cases}$

32. $f(x) = \begin{cases} 2x, & x \le 0 \\ x^2 + 1, & x > 0 \end{cases}$

In Exercises 33–36, identify the transformation of the graph of $f(x) = x^4$ and sketch the graph of h.

33. $h(x) = -x^4$

34. $h(x) = x^4 + 1$

35. $h(x) = (x - 2)^4$

36. $h(x) = 2 - x^4$

In Exercises 37–40, determine the transformation of the graph of $y = \sqrt{x}$ shown in the figure and write a formula for the graphed function.

37.

38.

39.

40.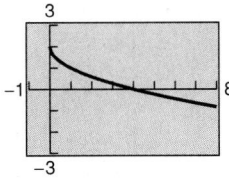

Graphical Analysis In Exercises 41–44, use a graphing utility to graph the function and use the graph to (a) approximate the intervals in which the function is increasing, decreasing, or constant; (b) approximate (to two-decimal-place accuracy) any relative maximum or minimum values of the function; and (c) determine if the function is even, odd, or neither.

41. $g(x) = |x + 2| - |x - 2|$

42. $f(x) = (x^2 - 4)^2$

43. $h(x) = 4x^3 - x^4$

44. $g(x) = \sqrt[3]{x(x + 3)^2}$

45. *Cost and Profit* A company produces a product for which the variable cost is \$5.35 per unit and the fixed costs are \$16,000. The company sells the product for \$8.20 and can sell all that it produces.

(a) Find the total cost as a function of x, the number of units produced.

(b) Find the profit as a function of x.

46. *Exploration* A wire 24 inches long is to be cut into four pieces to form a rectangle whose shortest side has a length of x.

(a) Express the area A of the rectangle as a function of x.

(b) Determine the domain of the function and use a graphing utility to graph the function over that domain.

(c) Use the graph of the function to approximate the maximum area of the rectangle. Make a conjecture about the dimensions of the rectangle.

In Exercises 47–52, (a) find f^{-1}, (b) use a graphing utility to sketch the graphs of f and f^{-1} in the same viewing rectangle, and (c) verify that $f^{-1}(f(x)) = x = f(f^{-1}(x))$.

47. $f(x) = \frac{1}{2}x - 3$

48. $f(x) = 5x - 7$

49. $f(x) = \sqrt{x + 1}$

50. $f(x) = x^3 + 2$

51. $f(x) = x^2 - 5, \quad x \ge 0$

52. $f(x) = \sqrt[3]{x + 1}$

In Exercises 53–56, restrict the domain of the function f to an interval on which the function is increasing and determine f^{-1} over that interval.

53. $f(x) = 2(x - 4)^2$

54. $f(x) = |x - 2|$

55. $f(x) = \sqrt{x^2 - 4}$

56. $f(x) = x^{4/3}$

In Exercises 57–64, let $f(x) = 3 - 2x$, $g(x) = \sqrt{x}$, and $h(x) = 3x^2 + 2$, and find the indicated value.

57. $(f - g)(4)$

58. $(f + h)(5)$

59. $(fh)(1)$

60. $\left(\dfrac{g}{h}\right)(1)$

61. $(h \circ g)(7)$

62. $(g \circ f)(-2)$

63. $g^{-1}(3)$

64. $(h \circ f^{-1})(1)$

65. *Data Analysis*   The total sales of sporting goods in the United States from 1981 through 1992 can be approximated by the model

$$\text{Sales} = 16.8091 + 0.7151t^2 - 0.0446t^3$$

where the sales are measured in billions of dollars and the time t represents the calendar year, with $t = 1$ corresponding to 1981. The actual sales (in billions) are given in the table.

Year	1981	1982	1983	1984	1985	1986
Sales	18.7	18.7	23.1	26.4	27.4	30.6

Year	1987	1988	1989	1990	1991	1992
Sales	33.9	42.1	45.2	44.1	42.8	42.4

(a) Use a graphing utility to plot the data points and graph the model.

(b) Use the model to estimate sales in 1994.

(c) Explain why the model may not be accurate in predicting sales in the future.

66. *Sales*   The sales manager of a company wants to determine if there is a relationship between sales and years of experience for sales personnel that have been with the company 4 years or less. The table gives the years of experience x for eight of the firm's sales personnel and monthly sales y in thousands of dollars.

x	1.5	1.0	0.3	3.0
y	46.7	32.9	19.2	48.4

x	4.0	0.5	2.5	1.8
y	51.2	28.5	53.4	35.5

(a) Use the regression capabilities of a graphing utility to find a linear model for the data.

(b) Use a graphing utility to plot the data and graph the model.

(c) Interpret the slope of the model in the context of the problem.

(d) Use the model to predict monthly sales of a salesperson with 5 years of experience.

67. Order the scatter plots in increasing order by their correlation coefficients.

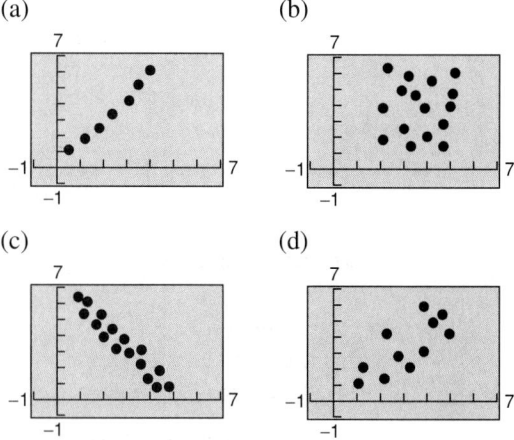

(a)   (b)

(c)   (d)

CHAPTER PROJECT   *Modeling the Area of a Plot*

Many real-life problems can be analyzed from a *graphical,* a *numerical,* and an *analytical* perspective. In this project, you will use all three strategies to determine the maximum size of a rectangular plot that can be enclosed by a fixed amount of fencing.

You have 100 meters of fencing material to enclose a rectangular plot. Your goal is to determine the dimensions of the plot such that you enclose the maximum area possible.

(a) Express the area $A(x)$ of the rectangular plot as a function of the length x of one side, as shown in the figure at the left.

(b) Analyze the problem *numerically* by completing the table.

x	0	5	10	15	20	25	30	35	40	45	50
$A(x)$											

According to this table, what do you think the dimensions of the plot should be to enclose the maximum area? Explain.

(c) Use a graphing utility to graph the area function. What is the domain of the function? Solve the problem *graphically* by using the trace feature of your graphing utility to find the value of x that yields the maximum area.

(d) Solve the problem *analytically* by showing that the area function can be written as $A(x) = 625 - (x - 25)^2$. How does this form of the function allow you to find the dimensions that produce a maximum area?

(e) Discuss the strengths and weaknesses of the three strategies used in parts (b), (c), and (d).

Questions for Further Exploration

1. Suppose you were not restricted to a rectangular plot. Would you be able to use 100 meters of fencing to enclose a greater area? Explain.

2. In the project above, you found the maximum area that can be enclosed in a rectangular plot using 100 meters of fencing. If you doubled the amount of fencing, could you enclose twice as much area? Use numerical, graphical, and analytical approaches and explain your reasoning.

3. Suppose the rectangular plot runs along a building, so that you need to fence only three sides. What dimensions will now yield a maximum area?

1 /// CHAPTER TEST

Take this test as you would take a test in class. After you are done, check your work against the answers given in the back of the book.

 The *Interactive* CD-ROM provides answers to the Chapter Tests and Cumulative Tests. It also offers Chapter Pre-Tests (that test key skills and concepts covered in previous chapters) and Chapter Post-Tests, both of which have randomly generated exercises with diagnostic capabilities.

In Exercises 1–4, use the function $f(x) = 10 - \sqrt{3 - x}$.

1. Evaluate: $f(-6)$ **2.** Simplify: $f(t - 3)$

3. Simplify the difference quotient: $\dfrac{f(x) - f(2)}{x - 2}$

4. Determine the domain of f.

5. The graph of $y^2(4 - x) = x^3$ is shown in the figure. Does the graph represent y as a function of x? Explain.

Figure for 5

Figure for 6

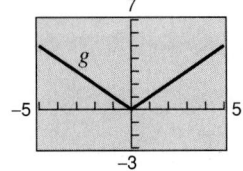

6. The graph of a function g is shown in the figure. Sketch a graph of (a) $\frac{1}{2}g(x - 2)$ and (b) $g\left(\frac{1}{2}x\right) - 1$.

7. A company produces a product for which the variable cost is \$5.60 and the fixed costs are \$24,000. The product sells for \$9.20. Write the total cost C as a function of x. Write the profit P as a function of x.

In Exercises 8 and 9, use a graphing utility to graph the function. Determine if the function is even, odd, or neither. Determine intervals over which the function is increasing, decreasing, and/or constant.

8. $h(x) = \frac{1}{4}x^4 - 2x^2$ **9.** $g(t) = t|t|$

10. Use the functions $f(x) = x^2$ and $g(x) = \sqrt{2 - x}$ to find each specified function and its domain.

(a) $(f - g)(x)$ (b) $\left(\dfrac{f}{g}\right)(x)$ (c) $(f \circ g)(x)$ (d) $g^{-1}(x)$

11. Explain how to determine if a function has an inverse.

12. Explain the relationship between the graph of a function and its inverse.

13. The points $(75, 2.3)$, $(82, 3.2)$, $(90, 3.6)$, and $(65, 2.3)$ give entrance exam scores x and the grade-point averages y after 1 year of college for a sample of four students. Find a least squares regression line for the data and use it to predict the grade-point average of a student with an entrance exam score of 88.

Polynomial and Rational Functions

The Fundamental Theorem of Algebra implies that an *n*th-degree polynomial equation has precisely *n* solutions. This result, however, is true only if repeated and complex solutions are counted. For instance, the equation $x^4 - 2x^3 + 2x^2 - 2x + 1 = 0$ has solutions of 1, 1, *i*, and *−i*.

Imaginary axis

• $2 + 3i$

Real axis

You can verify the real solutions with a graphing utility by graphing the equation $y = x^4 - 2x^3 + 2x^2 - 2x + 1$ and observing that it only touches the *x*-axis at (1, 0).

When first developed, complex numbers were used primarily for theoretical results such as this. Today, however, they have several other uses.

One use of complex numbers is in creating fractals such as that shown in the Zuckerman photograph. To program this fractal with a computer, complex numbers are plotted in the complex plane.

In the complex plane, the point (*a, b*) represents the complex number *a + bi*. For instance, the number $2 + 3i$ is plotted at the left. (See Exercises 58 and 59 on page 211.)

Jim Zuckerman; (inset) Rondi Ballard.

This fractal was created on a computer by photographer Jim Zuckerman. Zuckerman owns and operates a photography and digital imaging company in Northridge, California.

2.1   Quadratic Functions

The Graph of a Quadratic Function / *The Standard Form of a Quadratic Function* / *Applications*

The Graph of a Quadratic Function

In this and the next section, you will study the graphs of polynomial functions.

Definition of Polynomial Function

Let n be a nonnegative integer and let $a_n, a_{n-1}, \ldots, a_2, a_1, a_0$ be real numbers with $a_n \neq 0$. The function given by

$$f(x) = a_n x^n + a_{n-1} x^{n-1} + \cdots + a_2 x^2 + a_1 x + a_0$$

is called a **polynomial function of x with degree n.**

Polynomial functions are classified by degree. For instance, the polynomial function

$$f(x) = a, \qquad a \neq 0 \qquad \text{Constant function}$$

has degree 0 and is called a **constant function.** In Chapter P, you learned that the graph of this type of function is a horizontal line. The polynomial function

$$f(x) = ax + b, \qquad a \neq 0 \qquad \text{Linear function}$$

has degree 1 and is called a **linear function.** In Chapter P, you learned that the graph of the linear function $f(x) = ax + b$ is a line whose slope is a and whose y-intercept is $(0, b)$. In this section you will study second-degree polynomial functions, which are called **quadratic functions.**

Definition of Quadratic Function

Let a, b, and c be real numbers with $a \neq 0$. The function of x given by

$$f(x) = ax^2 + bx + c \qquad \text{Quadratic function}$$

is called a **quadratic function.**

Note   The graph of a quadratic function is a "∪"-shaped curve that is called a **parabola.**

Library of Functions

The graph of a quadratic function is called a parabola. Graph $f(x) = x^2$ and $g(x) = |x|$ in the same viewing rectangle. Zoom in near the origin and compare the shape of the two graphs. Which graph grows faster as x gets larger and larger? Why does the definition of a quadratic function require that $a \neq 0$?

All parabolas are symmetric with respect to a line called the **axis of symmetry,** or simply the **axis** of the parabola. The point where the axis intersects the parabola is the **vertex** of the parabola, as shown in Figure 2.1. If the leading coefficient is positive, the graph of $f(x) = ax^2 + bx + c$ is a parabola that opens upward, and if the leading coefficient is negative, the graph of $f(x) = ax^2 + bx + c$ is a parabola that opens downward.

Figure 2.1

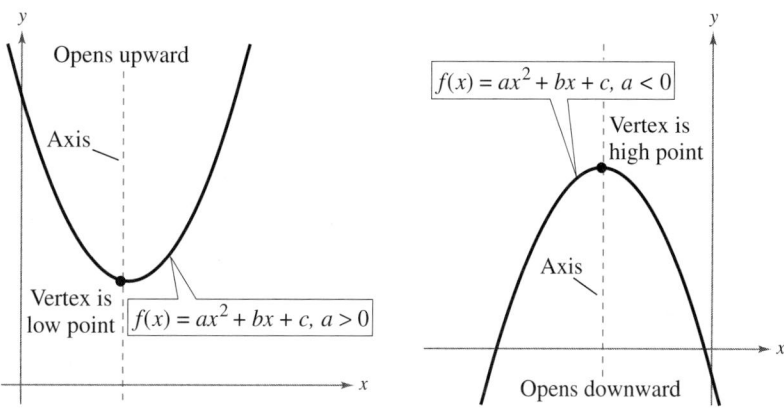

The simplest type of quadratic function is $f(x) = ax^2$. Its graph is a parabola whose vertex is $(0, 0)$. If $a > 0$, the vertex is the *minimum* point on the graph, and if $a < 0$, the vertex is the *maximum* point on the graph, as shown in Figure 2.2.

Figure 2.2

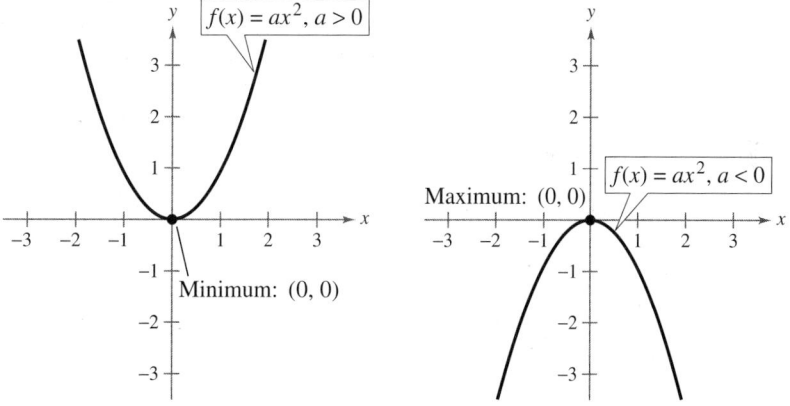

When sketching the graph of $f(x) = ax^2$, it is helpful to use the graph of $y = x^2$ as a reference, as discussed in Section 1.3.

EXAMPLE 1   ◢   **Graphing Simple Quadratic Functions**

Describe how the graph of each function is related to the graph of $y = x^2$.

a. $f(x) = \dfrac{1}{3}x^2$ **b.** $g(x) = 2x^2$

c. $h(x) = -x^2 + 1$ **d.** $k(x) = (x + 2)^2 - 3$

Solution

a. Compared with $y = x^2$, each output of f "shrinks" by a factor of $\frac{1}{3}$. The result is a parabola that opens upward and is broader than the parabola represented by $y = x^2$, as shown in Figure 2.3(a).

b. Compared with $y = x^2$, each output of g "stretches" by a factor of 2, creating a narrower parabola, as shown in Figure 2.3(b).

c. With respect to the graph of $y = x^2$, the negative coefficient in $h(x) = -x^2 + 1$ reflects the graph *downward* and the positive constant term shifts the vertex *up* one unit. The graph of h is shown in Figure 2.3(c).

d. With respect to the graph of $y = x^2$, the graph of $k(x) = (x + 2)^2 - 3$ is obtained by a horizontal shift two units *to the left* and a vertical shift three units *down*, as shown in Figure 2.3(d).

Note In Example 1, note that the coefficient a determines how widely the parabola given by $f(x) = ax^2$ opens. If $|a|$ is small, the parabola opens more widely than if $|a|$ is large.

The *Interactive* CD-ROM shows every example with its solution; clicking on the *Try It!* button brings up similar problems. Guided Examples and Integrated Examples show step-by-step solutions to additional examples. Integrated Examples are related to several concepts in the section.

Figure 2.3

(a)

(b)

(c)

(d)

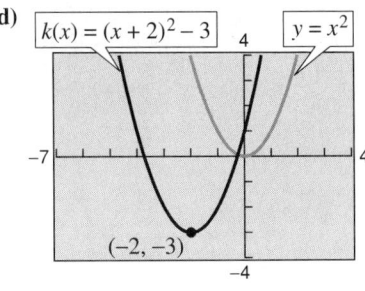

Note Recall from Section 1.3 that the graphs of $y = f(x \pm c)$, $y = f(x) \pm c$, $y = -f(x)$, and $y = f(-x)$ are rigid transformations of the graph of $y = f(x)$.

$y = f(x \pm c)$ Horizontal shift

$y = f(x) \pm c$ Vertical shift

$y = -f(x)$ Reflection in *x*-axis

$y = f(-x)$ Reflection in *y*-axis

◢

The Standard Form of a Quadratic Function

The equation in Example 1(d) is written in the **standard form**

$$f(x) = a(x - h)^2 + k.$$

This form is especially convenient for sketching a parabola because it identifies the vertex of the parabola as (h, k).

> **Standard Form of a Quadratic Function**
>
> The quadratic function
>
> $$f(x) = a(x - h)^2 + k, \qquad a \neq 0$$
>
> is said to be in **standard form.** The graph of f is a parabola whose axis is the vertical line $x = h$ and whose vertex is the point (h, k). If $a > 0$, the parabola opens upward, and if $a < 0$, the parabola opens downward.

EXAMPLE 2   **Writing a Quadratic Function in Standard Form**

Describe the graph of $f(x) = 2x^2 + 8x + 7$.

Solution

Write the quadratic function in standard form by completing the square. Notice that the first step is to factor out any coefficient of x^2 that is different from 1.

$$
\begin{aligned}
f(x) &= 2x^2 + 8x + 7 && \text{Original function} \\
&= 2(x^2 + 4x) + 7 && \text{Factor 2 out of } x\text{-terms.} \\
&= 2(x^2 + 4x + 4 - 4) + 7 && \text{Add and subtract 4 within} \\
&&& \text{parentheses.} \\
&&& \quad (b/2)^2 \\
&= 2(x^2 + 4x + 4) - 2(4) + 7 && \text{Regroup terms.} \\
&= 2(x + 2)^2 - 1 && \text{Standard form}
\end{aligned}
$$

From the standard form, you can see that the graph of f is a parabola that opens upward with vertex $(-2, -1)$, as shown in Figure 2.4.

Figure 2.4

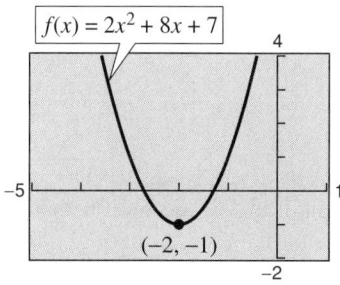

$f(x) = 2x^2 + 8x + 7$

$(-2, -1)$

To find the x-intercepts of the graph of $f(x) = ax^2 + bx + c$, solve the equation $ax^2 + bx + c = 0$. If $ax^2 + bx + c$ does not factor, you can use the Quadratic Formula or a graphing utility to find the x-intercepts. Remember, however, that a parabola may have no x-intercept.

Figure 2.5

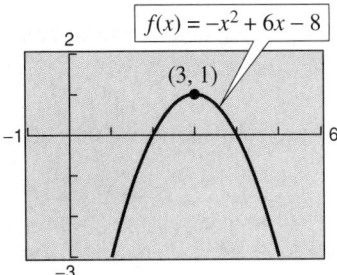

EXAMPLE 3   ▱   **Writing a Quadratic Function in Standard Form**

Describe the graph of $f(x) = -x^2 + 6x - 8$.

Solution

$$
\begin{aligned}
f(x) &= -x^2 + 6x - 8 && \text{Original form} \\
&= -(x^2 - 6x) - 8 && \text{Factor } -1 \text{ out of } x\text{-terms.} \\
&= -(x^2 - 6x + 9 - 9) - 8 && \text{Add and subtract 9 within} \\
& && \text{parentheses.}
\end{aligned}
$$

$$\underset{(b/2)^2}{\underbrace{\qquad\qquad}}$$

$$
\begin{aligned}
&= -(x^2 - 6x + 9) - (-9) - 8 && \text{Regroup terms.} \\
&= -(x - 3)^2 + 1 && \text{Standard form}
\end{aligned}
$$

The graph of f is a parabola that opens downward with vertex at $(3, 1)$, as shown in Figure 2.5.   ▱

Figure 2.6

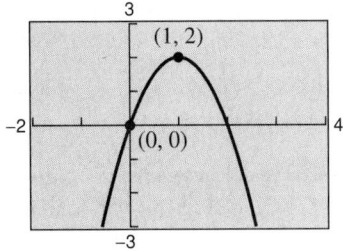

EXAMPLE 4   ▱   **Finding the Equation of a Parabola**

Find an equation for the parabola that has its vertex at $(1, 2)$ and passes through the point $(0, 0)$, as shown in Figure 2.6.

Solution

Because the parabola has a vertex at $(h, k) = (1, 2)$, the equation has the form

$$f(x) = a(x - 1)^2 + 2. \qquad\qquad \text{Standard form}$$

Because the parabola passes through the point $(0, 0)$, it follows that $f(0) = 0$. Thus, you obtain

$$0 = a(0 - 1)^2 + 2,$$

which implies that $a = -2$. The equation is

$$
\begin{aligned}
f(x) &= -2(x - 1)^2 + 2 \\
&= -2x^2 + 4x.
\end{aligned}
$$

Try graphing $f(x) = -2x^2 + 4x$ with a graphing utility to confirm that its vertex is $(1, 2)$ and that it passes through the point $(0, 0)$.   ▱

Note   In Example 4, there are infinitely many different parabolas that have a vertex at $(1, 2)$. Of these, however, the only one that passes through the point $(0, 0)$ is the one given by $f(x) = -2x^2 + 4x$.

Applications

Many applications involve finding the maximum or minimum value of a quadratic function. By writing the quadratic function $f(x) = ax^2 + bx + c$ in standard form,

$$f(x) = a\left(x + \frac{b}{2a}\right)^2 + \left(c - \frac{b^2}{4a}\right)$$

you can see that the vertex occurs at $x = -b/(2a)$, which implies the following.

1. If $a > 0$, f has a *minimum* that occurs at $x = -b/(2a)$.
2. If $a < 0$, f has a *maximum* that occurs at $x = -b/(2a)$.

Real Life

EXAMPLE 5   **The Maximum Height of a Baseball**

A baseball is hit 3 feet above ground at a velocity of 100 feet per second and at an angle of 45 degrees with respect to level ground. The path of the baseball is given by the function

$$f(x) = -0.0032x^2 + x + 3$$

where $f(x)$ is the height of the baseball (in feet) and x is the distance from home plate (in feet). What is the maximum height reached by the baseball? (See Example 7 in Section 1.1.)

Solution
For this quadratic function, you have

$$f(x) = ax^2 + bx + c = -0.0032x^2 + x + 3$$

which implies that $a = -0.0032$ and $b = 1$. Because the function has a maximum when $x = -b/2a$, you can conclude that the baseball reaches its maximum height when

$$x = -\frac{b}{2a} = -\frac{1}{2(-0.0032)} = 156.25 \text{ feet}$$

from home plate. At this distance, the maximum height is

$$f(156.25) = -0.0032(156.25)^2 + 156.25 + 3 = 81.125 \text{ feet.}$$

The path of the baseball is shown in Figure 2.7. You can also solve this problem by using a graphing utility. By using the zoom and trace features, you can determine that the maximum height on the graph occurs when $x \approx 156.25$. Note that you might have to change the y-scale in order to avoid a graph that is "too flat."

Figure 2.7

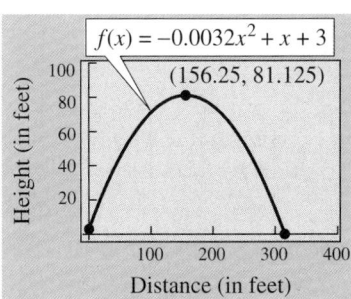

Real Life

EXAMPLE 6 ▱ **Charitable Contributions**

According to a survey conducted by *Independent Sector*, the percent of their income that Americans give to charities is related to their household income. For families with an annual income of $100,000 or less, the percent is approximately given by

$$P = 0.0014x^2 - 0.1529x + 5.855, \qquad 5 \leq x \leq 100$$

where P is the percent of annual income given and x is the annual income in thousands of dollars. According to this model, what income level corresponds to the minimum percent of charitable contributions?

Solution

There are two ways to answer this question. One is to use a graphing utility to graph the quadratic function, as shown in Figure 2.8. From this graph, it appears that the minimum percent corresponds to an income level of about $55,000. By using the zoom and trace features, you can improve the approximation to $x \approx 54,600$. The other way to answer the question is to use the fact that the minimum point of the parabola occurs when $x = -b/2a$. For this function, you have $a = 0.0014$ and $b = -0.1529$. Thus,

$$x = -\frac{b}{2a} = -\frac{-0.1529}{2(0.0014)} \approx 54.6.$$

From this x-value, you can conclude that the minimum percent corresponds to an income level of about $54,600. ▱

Figure 2.8

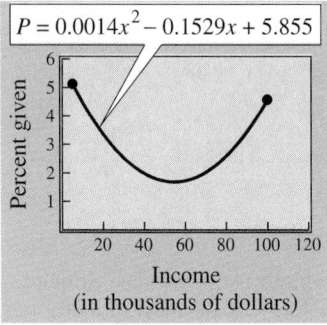

$P = 0.0014x^2 - 0.1529x + 5.855$

Income
(in thousands of dollars)

Group Activity *Finding an Equation for a Curve*

The parabola in the figure at the right has an equation of the form

$$y = ax^2 + bx - 4.$$

Find the equation for this parabola in two different ways, by hand and with technology. Discuss the methods you used. Compare the results of the two methods and compare your results with those of your group members.

2.1 /// EXERCISES

In Exercises 1–8, match the quadratic function with the correct graph. [The graphs are labeled (a), (b), (c), (d), (e), (f), (g), and (h).]

(a)

(b)

(c)

(d)

(e)

(f)

(g)

(h)
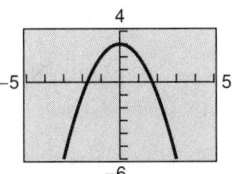

1. $f(x) = (x - 2)^2$
2. $f(x) = (x + 4)^2$
3. $f(x) = x^2 - 2$
4. $f(x) = 3 - x^2$
5. $f(x) = 4 - (x - 2)^2$
6. $f(x) = (x + 1)^2 - 2$
7. $f(x) = x^2 + 3$
8. $f(x) = -(x - 4)^2$

Exploration In Exercises 9–12, use a graphing utility to graph each equation. Describe how each differs from the graph of $y = x^2$.

9. (a) $y = \frac{1}{2}x^2$ (b) $y = -\frac{1}{8}x^2$
    (c) $y = \frac{3}{2}x^2$ (d) $y = -3x^2$

10. (a) $y = x^2 + 1$ (b) $y = x^2 - 1$
    (c) $y = x^2 + 3$ (d) $y = x^2 - 3$

11. (a) $y = (x - 1)^2$ (b) $y = (x + 1)^2$
    (c) $y = (x - 3)^2$ (d) $y = (x + 3)^2$

12. (a) $y = -\frac{1}{2}(x - 2)^2 + 1$ (b) $y = \frac{1}{2}(x - 2)^2 + 1$

In Exercises 13–22, sketch the graph of the function. Identify the vertex, intercepts, and zeros of the function. Use a graphing utility to verify your results.

13. $f(x) = 16 - x^2$
14. $f(x) = \frac{1}{2}x^2 - 4$
15. $h(x) = x^2 - 8x + 16$
16. $g(x) = x^2 + 2x + 1$
17. $f(x) = x^2 - x + \frac{5}{4}$
18. $f(x) = x^2 + 3x + \frac{1}{4}$
19. $f(x) = -x^2 + 2x + 5$
20. $f(x) = -x^2 - 4x + 1$
21. $h(x) = 4x^2 - 4x + 21$
22. $f(x) = 2x^2 - x + 1$

In Exercises 23–26, use a graphing utility to graph the quadratic function. Identify the vertex, intercepts, and zeros of the function. Then check your results algebraically by completing the square.

23. $f(x) = -(x^2 + 2x - 3)$
24. $g(x) = x^2 + 8x + 11$
25. $f(x) = 2x^2 - 16x + 31$
26. $g(x) = \frac{1}{2}(x^2 + 4x - 2)$

The *Interactive* CD-ROM contains step-by-step solutions to all odd-numbered Section and Review Exercises. It also provides Tutorial Exercises, which link to Guided Examples for additional help.

In Exercises 27–30, find an equation for the parabola. Use a graphing utility to graph the equation and verify your result.

27.

28.

29.

30.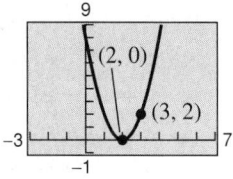

In Exercises 31–36, find the quadratic function that has the indicated vertex and whose graph passes through the given point. Confirm your result with a graphing utility.

31. Vertex: $(-2, 5)$; Point: $(0, 9)$

32. Vertex: $(4, -1)$; Point: $(2, 3)$

33. Vertex: $(3, 4)$; Point: $(1, 2)$

34. Vertex: $(2, 3)$; Point: $(0, 2)$

35. Vertex: $(5, 12)$; Point: $(7, 15)$

36. Vertex: $(-2, -2)$; Point: $(-1, 0)$

Graphical Reasoning   In Exercises 37–40, determine the x-intercepts of the graph visually. How do the x-intercepts correspond to the solutions of the quadratic equation when $y = 0$?

37.

38.

39.

40.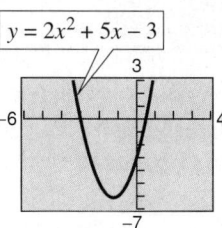

In Exercises 41–44, use a graphing utility to graph the quadratic function. Find the x-intercepts of the graph and compare them with the solutions of the corresponding quadratic equation when $y = 0$.

41. $y = x^2 - 4x$

42. $y = x^2 - 9x + 18$

43. $y = 2x^2 - 7x - 30$

44. $y = -\frac{1}{2}(x^2 - 6x - 7)$

In Exercises 45–48, find two quadratic functions whose graphs—one opening upward and the other downward—have the given x-intercepts. (The answers are not unique.)

45. $(-1, 0), (3, 0)$ **46.** $(4, 0), (8, 0)$

47. $(-3, 0), \left(-\frac{1}{2}, 0\right)$ **48.** $\left(-\frac{5}{2}, 0\right), (2, 0)$

In Exercises 49 and 50, find two positive real numbers that satisfy the requirements.

49. The sum is 110 and the product is a maximum.

50. The sum is S and the product is a maximum.

Maximum Area   In Exercises 51 and 52, consider a rectangle of length x and perimeter P.

(a) Express the area A as a function of x and determine the domain of the function.

(b) Use a graphing utility to graph the area function.

(c) Use the graph to approximate the length and width of the rectangle of maximum area, and verify algebraically.

51. $P = 100$ feet   **52.** $P = 36$ meters

53. *Maximum Revenue* Find the number of units that produces a maximum revenue given by $R = 900x - 0.1x^2$, where R is the total revenue in dollars and x is the number of units sold.

54. *Maximum Profit* Let x be the amount (in hundreds of dollars) a company spends on advertising, and let P be the profit, where $P = 230 + 20x - 0.5x^2$. What expenditure for advertising results in the maximum profit?

55. *Numerical, Graphical, and Analytical Analysis* A rancher has 200 feet of fencing to enclose two adjacent rectangular corrals (see figure). Use the following methods to determine the dimensions that will produce a maximum enclosed area.

(a) Complete six rows of a table such as the one below. (The first two rows are shown.)

x	y	Area
2	$\frac{1}{3}[200 - 4(2)]$	$2xy = 256$
4	$\frac{1}{3}[200 - 4(4)]$	$2xy \approx 491$

(b) Use a graphing utility to generate additional rows of the table in part (a). Use the table to estimate the dimensions that will produce the maximum enclosed area.

(c) Write the area A as a function of x.

(d) Use a graphing utility to graph the area function. Use the graph to approximate the dimensions that will produce the maximum enclosed area.

(e) Write the area function in standard form to find analytically the dimensions that will produce the maximum area.

(f) Compare your results from parts (b), (d), and (e).

56. *Maximum Area* An indoor physical fitness room consists of a rectangular region with a semicircle on each end. The perimeter of the room is to be a 200-meter running track.

(a) Draw a figure that visually represents the problem. Let x and y represent the length and width of the rectangular region.

(b) Determine the radius of the semicircular ends of the track. Determine the distance, in terms of y, around the two semicircular parts of the track.

(c) Use the result of part (b) to write an equation, in terms of x and y, for the distance traveled in one lap around the track. Solve for y.

(d) Use the result of part (c) to write the area A of the rectangular region as a function of x.

(e) Use a graphing utility to graph the area function of part (d). Use the graph to approximate the dimensions that will produce a maximum area of the rectangle.

57. *Trajectory of a Ball* The height y (in feet) of a ball thrown by a child is given by

$$y = -\frac{1}{12}x^2 + 2x + 4$$

where x is the horizontal distance (in feet) from where the ball is thrown (see figure).

(a) Use a graphing utility to graph the path of the ball.

(b) How high is the ball when it leaves the child's hand? (*Note:* Find y when $x = 0$.)

(c) How high is the ball when it is at its maximum height?

(d) How far from the child does the ball strike the ground?

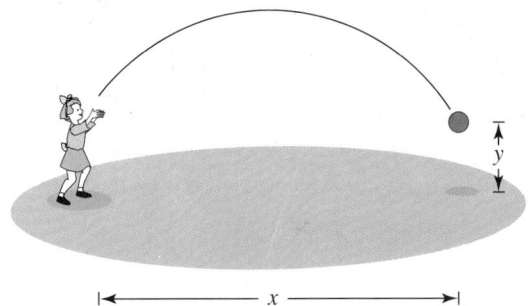

58. *Maximum Height of a Dive*   The path of a diver is given by

$$y = -\frac{4}{9}x^2 + \frac{24}{9}x + 12$$

where y is the height in feet and x is the horizontal distance from the end of the diving board in feet (see figure). What is the maximum height of the dive?

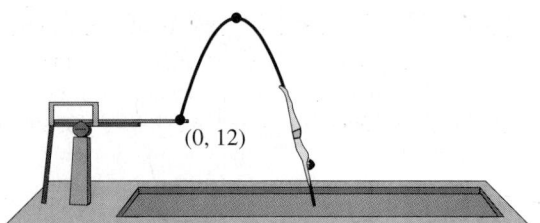

(0, 12)

59. *Forestry*   The number of board feet in a 16-foot log is approximated by the model

$$V = 0.77x^2 - 1.32x - 9.31, \qquad 5 \leq x \leq 40$$

where V is the number of board feet and x is the diameter (in inches) of the log at the small end. (One board foot is a measure of volume equivalent to a board that is 12 inches wide, 12 inches long, and 1 inch thick.)

(a) Use a graphing utility to graph the function.

(b) Estimate the number of board feet in a 16-foot log with a diameter of 16 inches.

(c) Estimate the diameter of a 16-foot log that scaled 500 board feet when the lumber was sold.

60. *Automobile Aerodynamics*   The number of horsepower y required to overcome wind drag on a certain automobile is approximated by

$$y = 0.002s^2 + 0.005s - 0.029, \qquad 0 \leq s \leq 100$$

where s is the speed of the car in miles per hour.

(a) Use a graphing utility to graph the function.

(b) Estimate the maximum speed of the car if the power required to overcome wind drag is not to exceed 10 horsepower.

61. *Graphical Analysis*   From 1950 to 1990, the average annual per capita consumption C of cigarettes by Americans (18 and older) can be modeled by

$$C = 4024.5 + 51.4t - 3.1t^2, \qquad -10 \leq t \leq 30$$

where t is the year, with $t = 0$ corresponding to 1960. (Source: U.S. Center for Disease Control)

(a) Use a graphing utility to graph the model.

(b) Use the graph of the model to approximate the maximum average annual consumption. Beginning in 1966, all cigarette packages were required by law to carry a health warning. Do you think the warning had any effect? Explain.

(c) In 1960, the U.S. population (18 and over) was 116,530,000. Of those, about 48,500,000 were smokers. What was the average annual cigarette consumption *per smoker* in 1960? What was the average daily cigarette consumption *per smoker?*

62. *Data Analysis*   The number y (in millions) of VCRs in use in the United States for the years 1984 through 1993 are given below in the form (t, y). The variable t represents time in years, where $t = 4$ represents 1984. (Source: Television Bureau of Advertising, Inc.)

(4, 9), (5, 18), (6, 31), (7, 43), (8, 51),

(9, 58), (10, 63), (11, 67), (12, 69), (13, 72)

(a) Use the regression capabilities of a graphing utility to fit a quadratic model to the data.

(b) Use a graphing utility to graph the model and the data in the same viewing rectangle.

(c) Do you think the model can be used to predict VCR utilization in the year 2000? Explain.

63. (a) Assume that the function $f(x) = ax^2 + bx + c$ $(a \neq 0)$ has two real zeros. Show that the x-coordinate of the vertex of the graph is the average of the zeros of f. (*Hint:* Use the Quadratic Formula.)

(b) Use a graphing utility to demonstrate the result of part (a) for the function

$$f(x) = \frac{1}{2}(x - 3)^2 - 2.$$

2.2 Polynomial Functions of Higher Degree

Graphs of Polynomial Functions / *The Leading Coefficient Test* /
Zeros of Polynomial Functions / *The Intermediate Value Theorem*

Graphs of Polynomial Functions

You should be able to sketch accurate graphs of polynomial functions of degrees 0, 1, and 2. The graphs of polynomial functions of degree greater than 2 are more difficult to sketch by hand. However, in this section you will learn how to recognize some of the basic features of the graphs of polynomial functions.

The graph of a polynomial function is **continuous.** Essentially, this means that the graph of a polynomial function has no breaks, as shown in Figure 2.9(a). Another feature of the graph of a polynomial function is that it has only smooth, rounded turns, as shown in Figure 2.10(a). It cannot have a sharp, pointed turn such as the one shown in Figure 2.10(b).

EXPLORATION

Use a graphing utility to graph $y = x^n$ with $n = 2, 4$, and 8. (Use $-1.5 \le x \le 1.5$ and $-1 \le y \le 6$.) Compare the graphs. In the interval $(-1, 1)$, which graph is on the bottom? Outside the interval $(-1, 1)$, which graph is on the bottom?

Use a graphing utility to graph $y = x^n$ with $n = 3, 5$, and 7. (Use $-1.5 \le x \le 1.5$ and $-4 \le y \le 4$.) Compare the graphs. In the interval $(-1, 1)$, which graph is on the bottom? Outside the interval $(-1, 1)$, which graph is on the bottom?

The *Interactive* CD-ROM offers graphing utility emulators of the *TI-82* and *TI-83*, which can be used with the Examples, Explorations, Technology notes, and Exercises.

Figure 2.9

(a) Continuous

(b) Discontinuous

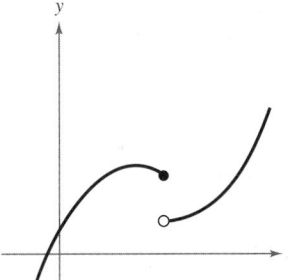

Figure 2.10

(a) Polynomial functions have smooth, rounded graphs.

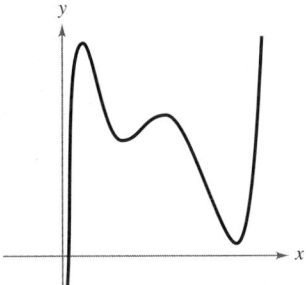

(b) Graphs of polynomial functions cannot have sharp turns.

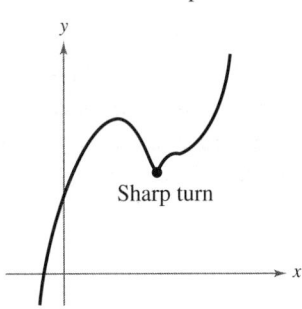

Sharp turn

The polynomial functions that have the simplest graphs are monomials of the form $f(x) = x^n$, where n is an integer greater than zero.

Figure 2.11

If n is even, the graph of $y = x^n$ *touches* axis at x-intercept.

If n is odd, the graph of $y = x^n$ *crosses* axis at x-intercept.

Note In Figure 2.11, you can see that when n is *even* the graph is similar to the graph of $f(x) = x^2$, and when n is *odd* the graph is similar to the graph of $f(x) = x^3$. Moreover, the greater the value of n, the flatter the graph is on the interval $[-1, 1]$.

EXAMPLE 1   **Sketching Transformations of Monomial Functions**

Sketch the graph of each polynomial function.

a. $f(x) = -x^5$ **b.** $g(x) = x^4 + 1$ **c.** $h(x) = (x + 1)^4$

Solution

a. Because the degree of f is odd, the graph is similar to the graph of $y = x^3$. Moreover, the negative coefficient reflects the graph in the x-axis, as shown in Figure 2.12(a).

b. The graph of g is an upward shift, by one unit, of the graph of $y = x^4$, as shown in Figure 2.12(b).

c. The graph of h is a left shift, by one unit, of the graph of $y = x^4$, as shown in Figure 2.12(c).

Figure 2.12

(a)

(b)

(c)

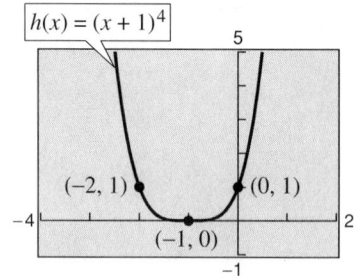

The Leading Coefficient Test

In Example 1, note that all three graphs eventually rise or fall without bound as x moves to the right. Whether the graph of a polynomial eventually rises or falls can be determined by the function's degree (even or odd) and by its leading coefficient, as indicated in the **Leading Coefficient Test.**

Library of Functions

The graphs of polynomials of degree 1 are lines, and those of degree 2 are parabolas. The graphs of polynomials of higher degree are smooth and continuous. The graphs eventually rise or fall without bound as x moves to the right (or left).

Note The dashed portions of the graphs indicate that the test determines *only* the right and left behavior of the graph.

Leading Coefficient Test

As x moves without bound to the left or to the right, the graph of the polynomial function $f(x) = a_n x^n + \cdots + a_1 x + a_0$ eventually rises or falls in the following manner.

1. When n is *odd:*

 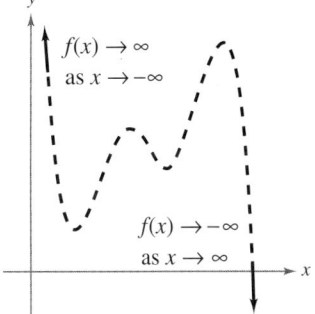

If the leading coefficient is positive ($a_n > 0$), the graph falls to the left and rises to the right.

If the leading coefficient is negative ($a_n < 0$), the graph rises to the left and falls to the right.

2. When n is *even:*

 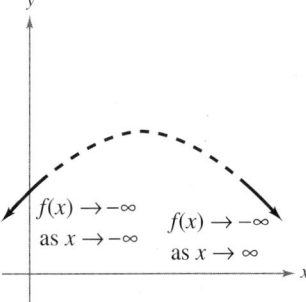

If the leading coefficient is positive ($a_n > 0$), the graph rises to the left and right.

If the leading coefficient is negative ($a_n < 0$), the graph falls to the left and right.

EXAMPLE 2   Applying the Leading Coefficient Test

Use the Leading Coefficient Test to determine the right and left behavior of the graph of each polynomial function.

a. $f(x) = -x^3 + 4x$ **b.** $f(x) = x^4 - 5x^2 + 4$ **c.** $f(x) = x^5 - x$

Solution

a. Because the degree is odd and the leading coefficient is negative, the graph rises to the left and falls to the right, as shown in Figure 2.13(a).

b. Because the degree is even and the leading coefficient is positive, the graph rises to the left and right, as shown in Figure 2.13(b).

c. Because the degree is odd and the leading coefficient is positive, the graph falls to the left and rises to the right, as shown in Figure 2.13(c).

Figure 2.13

(a)

(b)

(c)

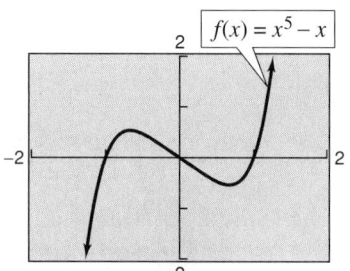

EXPLORATION

For each of the graphs in Figure 2.13, count the number of zeros of the polynomial function and the number of relative extrema, and compare these numbers with the degree of the polynomial. What do you observe?

Zeros of Polynomial Functions

It can be shown that for a polynomial function f of degree n, the following statements are true.

1. The graph of f has at most n real zeros. (This result is discussed in detail in Section 2.5.)

2. The function f has at most $n - 1$ relative **extrema** (local minimums or maximums).

Recall that a **zero** of a function f is a number x for which $f(x) = 0$. Finding the zeros of polynomial functions is one of the most important problems in algebra. You have already seen that there is a strong interplay between graphical and algebraic approaches to this problem. Sometimes you can use information about the graph of a function to help find its zeros. In other cases you can use information about the zeros of a function to find a good viewing rectangle.

Some graphing utilities, such as the *TI-82* and *TI-83*, have two features that analyze the graph of a function: (1) the *root* feature of the *TI-82* or the zero feature of the *TI-83* for finding zeros of a function, and (2) the *minimum* and *maximum* features for finding the relative extrema. If your graphing utility has these features, use them to confirm the results in Examples 3 and 4.

Real Zeros of Polynomial Functions

If f is a polynomial function and a is a real number, the following statements are equivalent.

1. $x = a$ is a *zero* of the function f.
2. $x = a$ is a *solution* of the polynomial equation $f(x) = 0$.
3. $(x - a)$ is a *factor* of the polynomial $f(x)$.
4. $(a, 0)$ is an *x-intercept* of the graph of f.

Finding zeros of polynomial functions is closely related to factoring and finding *x*-intercepts, as demonstrated in Examples 3, 4, and 5.

EXAMPLE 3   Finding Zeros of a Polynomial Function

Find all real zeros of $f(x) = x^3 - x^2 - 2x$.

Solution

$$
\begin{aligned}
f(x) &= x^3 - x^2 - 2x && \text{Original function} \\
&= x(x^2 - x - 2) && \text{Remove common monomial factor.} \\
&= x(x - 2)(x + 1) && \text{Factor completely.}
\end{aligned}
$$

Thus, the real zeros are $x = 0$, $x = 2$, and $x = -1$, and the corresponding *x*-intercepts are $(0, 0)$, $(2, 0)$, and $(-1, 0)$, as shown in Figure 2.14. Note in the figure that the graph has two relative extrema, which is consistent with the fact that a third-degree polynomial can have *at most* two relative extrema.

Figure 2.14

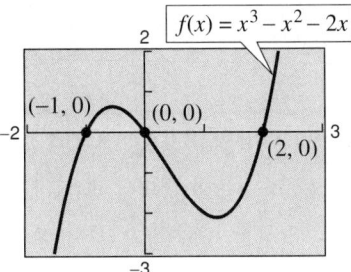

EXAMPLE 4   Analyzing a Polynomial Function

Find all real zeros and relative extrema of $f(x) = -2x^4 + 2x^2$.

Solution

$$
\begin{aligned}
f(x) &= -2x^4 + 2x^2 && \text{Original function} \\
&= -2x^2(x^2 - 1) && \text{Remove common monomial factor.} \\
&= -2x^2(x - 1)(x + 1) && \text{Factor completely.}
\end{aligned}
$$

Thus, the real zeros are $x = 0$, $x = 1$, and $x = -1$, and the corresponding *x*-intercepts are $(0, 0)$, $(1, 0)$, and $(-1, 0)$, as shown in Figure 2.15. Using the minimum and maximum features of a graphing utility, you can determine the three relative extrema to be $(-0.7071, 0.5)$, $(0, 0)$, and $(0.7071, 0.5)$.

Figure 2.15

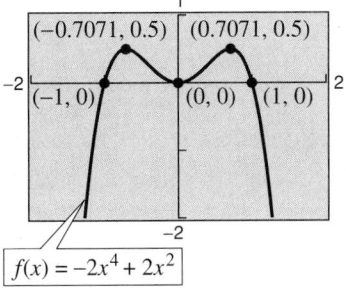

In Example 4, the real zero arising from $-2x^2 = 0$ is a **repeated zero.** In general, a factor of $(x - a)^k$ yields a repeated zero $x = a$ of **multiplicity** k. If k is odd, the graph *crosses* the x-axis at $x = a$. If k is even, the graph *touches* (but does not cross) the x-axis at $x = a$, as shown in Figure 2.15.

EXAMPLE 5   Finding Zeros of a Polynomial Function

Find all real zeros of $f(x) = x^5 - 3x^3 - x^2 - 4x - 1$.

Figure 2.16

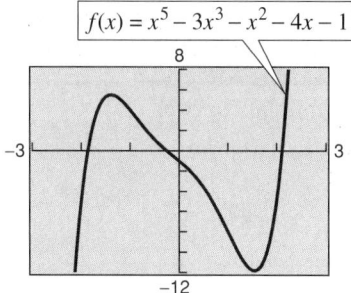

Solution

Use a graphing utility to obtain the graph shown in Figure 2.16. From the graph, you can see that there are three zeros. Using the root feature, you can determine that the zeros are approximately $x = -1.861$, $x = -0.254$, and $x = 2.115$. It should be noted that this fifth-degree polynomial factors as

$$f(x) = x^5 - 3x^3 - x^2 - 4x - 1$$
$$= (x^2 + 1)(x^3 - 4x - 1).$$

The three zeros obtained above are the zeros of the cubic on the right (the quadratic $x^2 + 1$ has two complex zeros and, thus, no *real* zeros).

EXAMPLE 6   Finding a Polynomial Function with Given Zeros

Find polynomial functions with the following zeros. (There are many correct solutions.)

a. $-2, -1, 1, 2$ **b.** $-\dfrac{1}{2}, 3, 3$

EXPLORATION

Use a graphing utility to graph

$$y_1 = x + 2$$
$$y_2 = (x + 2)(x - 1).$$

Predict the shape of the curve $y = (x + 2)(x - 1)(x - 3)$, and verify your answer with a graphing utility.

Solution

a. For each of the given zeros, form a corresponding factor. For instance, the zero given by $x = -2$ corresponds to the factor $(x + 2)$. Thus, you can write the function as

$$f(x) = (x + 2)(x + 1)(x - 1)(x - 2)$$
$$= (x^2 - 4)(x^2 - 1)$$
$$= x^4 - 5x^2 + 4.$$

b. Note that the zero $x = -\frac{1}{2}$ corresponds to either $\left(x + \frac{1}{2}\right)$ or $(2x + 1)$. To avoid fractions, choose the second factor and write

$$f(x) = (2x + 1)(x - 3)^2$$
$$= (2x + 1)(x^2 - 6x + 9)$$
$$= 2x^3 - 11x^2 + 12x + 9.$$

EXAMPLE 7   ▱   **Sketching the Graph of a Polynomial Function**

Sketch the graph of $f(x) = 3x^4 - 4x^3$ by hand.

Solution

Because the leading coefficient is positive and the degree is even, you know that the graph eventually rises to the left and to the right, as shown in Figure 2.17(a). By factoring to obtain

$$f(x) = 3x^4 - 4x^3 = x^3(3x - 4),$$

you can see that the zeros of f are $x = 0$ and $x = \frac{4}{3}$ (both of odd multiplicity). Thus, the x-intercepts occur at $(0, 0)$ and $\left(\frac{4}{3}, 0\right)$. Finally, plot a few additional points, as indicated in the table, and obtain the graph shown in Figure 2.17(b).

x	-1	0.5	1	1.5
$f(x)$	7	-0.3125	-1	1.6875

Figure 2.17

(a)

(b)

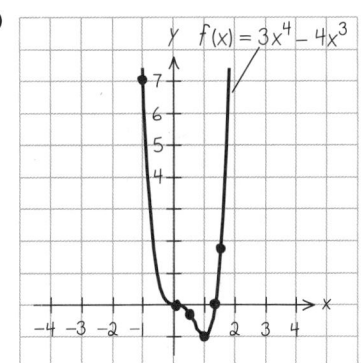

EXPLORATION

Partner Activity Write the equation of a polynomial function that has a degree of 3, 4, or 5. Exchange equations with your partner and sketch, *by hand,* the graph of the equation that your partner wrote. When you are finished, use a graphing utility to check each other's work.

Figure 2.18

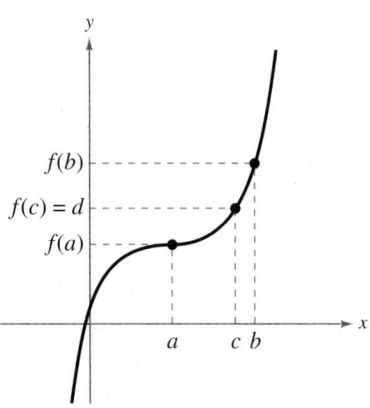

The Intermediate Value Theorem

The **Intermediate Value Theorem** concerns the existence of real zeros of polynomial functions. The theorem states that if $(a, f(a))$ and $(b, f(b))$ are two points on the graph of a polynomial function such that $f(a) \neq f(b)$, then for any number d between $f(a)$ and $f(b)$ there must be a number c between a and b such that $f(c) = d$. (See Figure 2.18.)

Intermediate Value Theorem
Let a and b be real numbers such that $a < b$. If f is a polynomial function such that $f(a) \neq f(b)$, then, in the interval $[a, b]$, f takes on every value between $f(a)$ and $f(b)$.

Figure 2.19

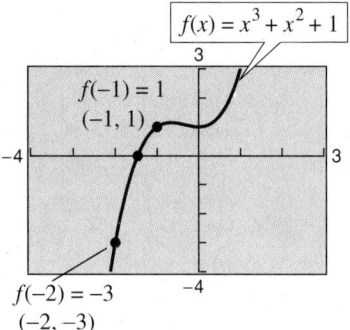

This theorem helps locate the real zeros of a polynomial function in the following way. If you can find a value $x = a$ where a polynomial function is positive, and another $x = b$ where it is negative, you can conclude that the function has at least one real zero between these two values. For example, the function $f(x) = x^3 + x^2 + 1$ is negative when $x = -2$ and positive when $x = -1$. Therefore, it follows from the Intermediate Value Theorem that f must have a real zero somewhere between -2 and -1, as shown in Figure 2.19.

EXAMPLE 8 Approximating a Function's Zeros

Use the Intermediate Value Theorem to find three intervals of length 1 in which the polynomial $f(x) = 12x^3 - 32x^2 + 3x + 5$ is guaranteed to have a zero.

Solution

With a graphing utility, you can determine that there are three real zeros, as shown in Figure 2.20. The following table shows several function values.

Figure 2.20

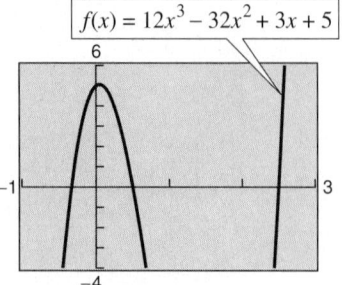

x	-1	0	1	2	3
$f(x)$	-42	5	-12	-21	50

Because $f(-1)$ is negative and $f(0)$ is positive, you can conclude from the Intermediate Value Theorem that the function has a zero between -1 and 0, as indicated in the figure. Similarly, the function has a zero between 0 and 1, and another between 2 and 3. You could obtain more accurate approximations of these zeros by using the root or zero feature of a graphing utility.

Group Activity

The Graphs of Cubic Polynomials

The graphs of cubic polynomials can be categorized according to the four basic shapes below. Match the graph of each function with one of the basic shapes and write a short paragraph describing how you reached your conclusion. Compare your conclusions with those of two classmates and resolve any differences.

1. $f(x) = -x^3$
(a)

2. $f(x) = -x^3 + 4x$
(b)

3. $f(x) = x^3$
(c)

4. $f(x) = x^3 - 4x$
(d)

2.2 /// EXERCISES

In Exercises 1–8, match the polynomial function with its graph. [The graphs are labeled (a) through (h).]

1. $f(x) = -2x + 3$

2. $f(x) = x^2 - 4x$

3. $f(x) = -2x^2 - 5x$

4. $f(x) = 2x^3 - 3x + 1$

5. $f(x) = -\frac{1}{4}x^4 + 3x^2$

6. $f(x) = -\frac{1}{3}x^3 + x^2 - \frac{4}{3}$

7. $f(x) = x^4 + 2x^3$

8. $f(x) = \frac{1}{5}x^5 - 2x^3 + \frac{9}{5}x$

(a)

(b)

(e)

(f)

(c)

(d)

(g)

(h)

In Exercises 9–12, sketch the graphs of $y = x^n$ and the specified transformations.

9. $y = x^3$

    (a) $f(x) = (x - 2)^3$ (b) $f(x) = x^3 - 2$

    (c) $f(x) = -\frac{1}{2}x^3$ (d) $f(x) = (x - 2)^3 - 2$

10. $y = x^5$

    (a) $f(x) = (x + 1)^5$ (b) $f(x) = x^5 + 1$

    (c) $f(x) = 1 - \frac{1}{2}x^5$ (d) $f(x) = -\frac{1}{2}(x + 1)^5$

11. $y = x^4$

    (a) $f(x) = (x + 3)^4$ (b) $f(x) = x^4 - 3$

    (c) $f(x) = 4 - x^4$ (d) $f(x) = \frac{1}{2}(x - 1)^4$

12. $y = x^6$

    (a) $f(x) = -\frac{1}{8}x^6$ (b) $f(x) = x^6 - 4$

    (c) $f(x) = -\frac{1}{4}x^6 + 1$ (d) $f(x) = (x + 2)^6 - 4$

Graphical Analysis   In Exercises 13–16, use a graphing utility to graph the functions f and g in the same viewing rectangle. Zoom out sufficiently far to show that the right-hand and left-hand behavior of f and g appear identical.

13. $f(x) = 3x^3 - 9x + 1$, $g(x) = 3x^3$

14. $f(x) = -\frac{1}{3}(x^3 - 3x + 2)$, $g(x) = -\frac{1}{3}x^3$

15. $f(x) = -(x^4 - 4x^3 + 16x)$, $g(x) = -x^4$

16. $f(x) = 3x^4 - 6x^2$, $g(x) = 3x^4$

In Exercises 17–26, use the Leading Coefficient Test to determine the right-hand and left-hand behavior of the graph of the polynomial function. Verify your result by using a graphing utility to graph the function.

17. $f(x) = 2x^2 - 3x + 1$

18. $f(x) = \frac{1}{3}x^3 + 5x$

19. $g(x) = 5 - \frac{7}{2}x - 3x^2$

20. $f(x) = -2.1x^5 + 4x^3 - 2$

21. $f(x) = 2x^5 - 5x + 7.5$

22. $h(x) = 1 - x^6$

23. $f(x) = 6 - 2x + 4x^2 - 5x^3$

24. $f(x) = \dfrac{3x^4 - 2x + 5}{4}$

25. $h(t) = -\frac{2}{3}(t^2 - 5t + 3)$

26. $f(s) = -\frac{7}{8}(s^3 + 5s^2 - 7s + 1)$

In Exercises 27–34, find all the real zeros of the polynomial function.

27. $f(x) = x^2 - 25$ **28.** $f(x) = 49 - x^2$

29. $h(t) = t^2 - 6t + 9$

30. $f(x) = x^2 + 10x + 25$

31. $f(x) = x^2 + x - 2$ **32.** $f(x) = \frac{1}{2}x^2 + \frac{5}{2}x - \frac{3}{2}$

33. $f(t) = t^3 - 4t^2 + 4t$

34. $f(x) = x^4 - x^3 - 20x^2$

Graphical Analysis   In Exercises 35–46, use a graphing utility to graph the function. Use the graph to approximate any x-intercepts of the graph. Set $y = 0$ and solve the resulting equation. Compare the result with any x-intercepts of the graph.

35. $f(x) = 3x^2 - 12x + 3$

36. $g(x) = 5(x^2 - 2x - 1)$

37. $g(t) = \frac{1}{2}t^4 - \frac{1}{2}$

38. $f(x) = x^5 + x^3 - 6x$

39. $f(x) = 2x^4 - 2x^2 - 40$

40. $g(t) = t^5 - 6t^3 + 9t$

41. $f(x) = 5x^4 + 15x^2 + 10$

42. $f(x) = x^3 - 4x^2 - 25x + 100$

43. $y = 4x^3 - 20x^2 + 25x$

44. $y = 4x^3 + 4x^2 - 7x + 2$

45. $y = x^5 - 5x^3 + 4x$ **46.** $y = \frac{1}{4}x^3(x^2 - 9)$

In Exercises 47–56, find a polynomial function that has the given zeros. (There are many correct answers.)

47. $0, 10$ **48.** $0, -3$

49. $2, -6$ **50.** $-4, 5$

51. $0, -2, -3$ **52.** $0, 2, 5$

53. $4, -3, 3, 0$ **54.** $-2, -1, 0, 1, 2$

55. $1 + \sqrt{3}, 1 - \sqrt{3}$ **56.** $2, 4 + \sqrt{5}, 4 - \sqrt{5}$

In Exercises 57–60, (a) use the Intermediate Value Theorem and a graphing utility to find intervals of length 1 in which the polynomial function is guaranteed to have a zero. (b) Use the root feature of a graphing utility to approximate the zeros of the function.

57. $f(x) = x^3 - 3x^2 + 3$

58. $f(x) = 0.11x^3 - 2.07x^2 + 9.81x - 6.88$

59. $g(x) = 3x^4 + 4x^3 - 3$

60. $h(x) = x^4 - 10x^2 + 2$

In Exercises 61–64, use a graphing utility to graph the function. Describe a viewing rectangle that gives a good view of the basic characteristics of the graph.

61. $f(x) = -\frac{3}{2}$

62. $h(x) = \frac{1}{3}x - 3$

63. $f(t) = \frac{1}{4}(t^2 - 2t + 15)$

64. $g(x) = -x^2 + 10x - 16$

In Exercises 65–72, use a graphing utility to graph the function. Identify any symmetry with respect to the x-axis, y-axis, or origin. Determine the number of x-intercepts of the graph.

65. $f(x) = x^2(x - 4)$

66. $h(x) = \frac{1}{3}x^3(x - 4)^2$

67. $g(t) = -\frac{1}{4}(t - 2)^2(t + 2)^2$

68. $g(x) = \frac{1}{10}(x + 1)^2(x - 3)^3$

69. $f(x) = x^3 - 4x$

70. $f(x) = \frac{1}{4}x^4 - 2x^2$

71. $g(x) = \frac{1}{5}(x + 1)^2(x - 3)(2x - 9)$

72. $h(x) = \frac{1}{5}(x + 2)^2(3x - 5)^2$

73. *Graphical Reasoning* Use a graphing utility to graph the function $f(x) = x^4$. Explain how the graph of g differs (if it does) from the graph of f and confirm your result with a graphing utility. Determine whether g is odd, even, or neither.

(a) $g(x) = f(x) + 2$ (b) $g(x) = f(x + 2)$

(c) $g(x) = f(-x)$ (d) $g(x) = -f(x)$

(e) $g(x) = f\left(\frac{1}{2}x\right)$ (f) $g(x) = \frac{1}{2}f(x)$

(g) $g(x) = f\left(x^{3/4}\right)$ (h) $g(x) = (f \circ f)(x)$

74. *Exploration* Explore the transformations of the form $g(x) = a(x - h)^5 + k$.

(a) Use a graphing utility to graph the functions

$$y_1 = -\frac{1}{3}(x - 2)^5 + 1$$

and

$$y_2 = \frac{3}{5}(x + 2)^5 - 3.$$

Determine whether the graphs are increasing or decreasing. Explain.

(b) Will the graph of g always be increasing or decreasing? If so, is this behavior determined by a, h, or k? Explain.

(c) Use a graphing utility to graph the function

$$H(x) = x^5 - 3x^3 + 2x + 1.$$

Use the graph and the result of part (b) to determine whether H can be written in the form

$$H(x) = a(x - h)^5 + k.$$

Explain your reasoning.

(d) Determine the natural number exponents such that the results of part (b) are true.

75. *Volume of a Box* An open box with locking tabs is to be made from a square piece of material 24 inches on a side. This is done by cutting equal squares from the corners and folding along the dashed lines shown in the figure.

(a) Verify that the volume of the box is given by

$$V(x) = 8x(6 - x)(12 - x).$$

(b) Determine the domain of the function V.

(c) Sketch the graph of the function and estimate the value of x for which $V(x)$ is maximum.

76. *Numerical and Graphical Analysis* An open box is to be made from a square piece of material 36 centimeters on a side by cutting equal squares from the corners and turning up the sides (see figure).

(a) Complete four rows of a table like the one below.

Height	Width	Volume
1	$36 - 2(1)$	$1[36 - 2(1)]^2 = 1156$
2	$36 - 2(2)$	$2[36 - 2(2)]^2 = 2048$

(b) Use a graphing utility to generate additional rows of the table. Use the table to estimate a range of dimensions within which the maximum volume is produced.

(c) Verify that the volume of the box is given by $V(x) = x(36 - 2x)^2$. Determine the domain of the function.

(d) Use a graphing utility to graph V, and use the range of dimensions from part (b) to find the x-value for which $V(x)$ is maximum.

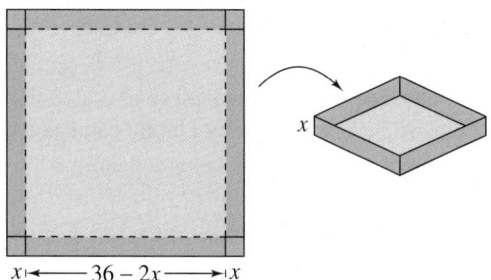

$x \longleftarrow 36 - 2x \longrightarrow x$

77. *Advertising Expenses* The total revenue R (in millions of dollars) for a company is related to its advertising expense by the function

$$R = \frac{1}{100,000}(-x^3 + 600x^2), \qquad 0 \le x \le 400$$

where x is the amount spent on advertising (in tens of thousands of dollars). Use the graph of this function to estimate the point on the graph at which the function is increasing most rapidly. This point is called the **point of diminishing returns** because any expense above this amount will yield less return per dollar invested in advertising. (*Hint:* Use a viewing rectangle in which $-200 \le x \le 600$ and $0 \le y \le 650$.)

78. *Data Analysis* The vertical deflection y of a 4-meter beam is measured in $\frac{1}{2}$-meter intervals x. The measurements are given by the following ordered pairs:

(0, 0), (0.5, 0.06), (1, 0.11), (1.5, 0.15), (2, 0.16)

(2.5, 0.15), (3, 0.11), (3.5, 0.06), (4, 0)

(a) Use the regression capabilities of a graphing utility to fit a quartic equation to the data.

(b) Use a graphing utility to graph the data and the regression equation. How do they compare?

(c) Because $y = 0$ when $x = 0$, what should the constant term in the model be? Does your answer agree with the result of part (a)? Explain.

79. *Data Analysis* The table gives the median values of privately owned U.S. homes for 1981 through 1993. In the table, t is the time in years, with $t = 1$ corresponding to 1981, and y_1 and y_2 are the median prices (in thousands of dollars) in the Northeast and the South, respectively.   (Source: U.S. Department of Housing and Urban Development)

t	1	2	3	4	5
y_1	76.0	78.2	82.2	88.6	103.3
y_2	64.4	66.1	70.9	72.0	75.0

t	6	7	8	9	10
y_1	125.0	140.0	149.0	159.6	159.0
y_2	82.2	88.0	92.0	96.4	99.0

t	11	12	13
y_1	155.9	169.0	162.6
y_2	100.0	105.5	115.0

(a) Use the regression capabilities of a graphing utility to fit a cubic model to the median prices of homes in the Northeast.

(b) Use the regression capabilities of a graphing utility to fit a cubic model to the median prices of homes in the South.

(c) Use the graphs of the models to write a short paragraph about the relationship between the median prices of homes in the two regions.

2.3   Real Zeros of Polynomial Functions

Long Division of Polynomials / Synthetic Division /
The Remainder and Factor Theorems / The Rational Zero Test /
Bounds for Real Zeros of Polynomial Functions

Long Division of Polynomials

Figure 2.21

$f(x) = 6x^3 - 19x^2 + 16x - 4$

Consider the graph of

$$f(x) = 6x^3 - 19x^2 + 16x - 4.$$

Notice that a zero of f occurs at $x = 2$, as shown in Figure 2.21. Because $x = 2$ is a zero of the polynomial function f, you know that $(x - 2)$ is a factor of $f(x)$. This means that there exists a second-degree polynomial $q(x)$ such that $f(x) = (x - 2) \cdot q(x)$. To find $q(x)$, you can use **long division of polynomials.**

EXAMPLE 1   ▱   **Long Division of Polynomials**

Divide $f(x) = 6x^3 - 19x^2 + 16x - 4$ by $x - 2$, and use the result to factor the $f(x)$ completely.

Solution

$$
\begin{array}{r}
\text{Partial quotients} \\
\downarrow \quad \downarrow \quad \downarrow \\
6x^2 - 7x + 2 \\
x - 2 \overline{)\, 6x^3 - 19x^2 + 16x - 4} \\
\underline{6x^3 - 12x^2} \\
-7x^2 + 16x \\
\underline{-7x^2 + 14x} \\
2x - 4 \\
\underline{2x - 4} \\
0
\end{array}
$$

Multiply: $6x^2(x - 2)$.
Subtract.
Multiply: $-7x(x - 2)$.
Subtract.
Multiply: $2(x - 2)$.
Subtract.

You can see that

$$6x^3 - 19x^2 + 16x - 4 = (x - 2)(6x^2 - 7x + 2)$$
$$= (x - 2)(2x - 1)(3x - 2).$$

Note that this factorization agrees with the graph of f (Figure 2.21) in that the three x-intercepts occur at $x = 2$, $x = \frac{1}{2}$, and $x = \frac{2}{3}$.   ▱

In Example 1, $x - 2$ is a factor of the polynomial $6x^3 - 19x^2 + 16x - 4$, and the long division process produces a remainder of zero. Often, long division will produce a nonzero remainder. For instance, if you divide $x^2 + 3x + 5$ by $x + 1$, you obtain the following.

$$
\begin{array}{r}
x + 2 \qquad \text{Quotient}\\
x + 1 \overline{)\, x^2 + 3x + 5} \qquad \text{Dividend}\\
\underline{x^2 + \ x}\\
2x + 5\\
\underline{2x + 2}\\
3 \qquad \text{Remainder}
\end{array}
$$

Divisor $\Rightarrow$ $x + 1 \overline{)\, x^2 + 3x + 5}$

In fractional form, you can write this result as follows.

$$
\underbrace{\frac{x^2 + 3x + 5}{x + 1}}_{\text{Divisor}} = \overbrace{x + 2}^{\text{Quotient}} + \frac{\overset{\text{Remainder}}{3}}{\underset{\text{Divisor}}{x + 1}}
$$

This implies that $x^2 + 3x + 5 = (x + 1)(x + 2) + 3$, which illustrates the following well-known theorem called the **Division Algorithm.**

The Division Algorithm

If $f(x)$ and $d(x)$ are polynomials such that $d(x) \neq 0$, and the degree of $d(x)$ is less than or equal to the degree of $f(x)$, there exist unique polynomials $q(x)$ and $r(x)$ such that

$$
f(x) = d(x)q(x) + r(x)
$$

$\uparrow$ Dividend $\uparrow$ Quotient   Divisor   Remainder

where $r(x) = 0$ *or* the degree of $r(x)$ is less than the degree of $d(x)$. If the remainder $r(x)$ is zero, $d(x)$ **divides evenly** into $f(x)$.

The Division Algorithm can also be written as

$$
\frac{f(x)}{d(x)} = q(x) + \frac{r(x)}{d(x)}.
$$

In the Division Algorithm, the rational expression $f(x)/d(x)$ is **improper** because the degree of $f(x)$ is greater than or equal to the degree of $d(x)$. On the other hand, the rational expression $r(x)/d(x)$ is **proper** because the degree of $r(x)$ is less than the degree of $d(x)$.

EXAMPLE 2   **Long Division of Polynomials**

Divide $x^3 - 1$ by $x - 1$.

Solution

Because there is no x^2-term or x-term in the dividend, you need to line up the subtraction by using zero coefficients (or leaving spaces) for the missing terms.

$$
\begin{array}{r}
x^2 + x + 1 \\
x - 1 \overline{) x^3 + 0x^2 + 0x - 1} \\
\underline{x^3 - x^2} \\
x^2 \\
\underline{x^2 - x} \\
x - 1 \\
\underline{x - 1} \\
0
\end{array}
$$

Thus, $x - 1$ divides evenly into $x^3 - 1$ and you can write

$$
\frac{x^3 - 1}{x - 1} = x^2 + x + 1, \qquad x \neq 1.
$$

Note   You can check the result of a division problem by multiplying. For instance, in Example 2, try checking that $(x - 1)(x^2 + x + 1) = x^3 - 1$.

EXAMPLE 3   **Long Division of Polynomials**

Divide $2x^4 + 4x^3 - 5x^2 + 3x - 2$ by $x^2 + 2x - 3$.

Solution

$$
\begin{array}{r}
2x^2 + 1 \\
x^2 + 2x - 3 \overline{) 2x^4 + 4x^3 - 5x^2 + 3x - 2} \\
\underline{2x^4 + 4x^3 - 6x^2} \\
x^2 + 3x - 2 \\
\underline{x^2 + 2x - 3} \\
x + 1
\end{array}
$$

Note that the first subtraction eliminated two terms from the dividend. When this happens, the quotient skips a term. Thus, you can write

$$
\frac{2x^4 + 4x^3 - 5x^2 + 3x - 2}{x^2 + 2x - 3} = 2x^2 + 1 + \frac{x + 1}{x^2 + 2x - 3}.
$$

Synthetic Division

There is a nice shortcut for long division of polynomials by divisors of the form $x - k$. The shortcut is called **synthetic division.** We summarize the pattern for synthetic division of a cubic polynomial as follows. (The pattern for higher-degree polynomials is similar.)

Synthetic Division (of a Cubic Polynomial)

To divide $ax^3 + bx^2 + cx + d$ by $x - k$, use the following pattern.

Vertical pattern: Add terms.
Diagonal pattern: Multiply by k.

Note Synthetic division works *only* for divisors of the form $x - k$. [Remember that $x + k = x - (-k)$.] You cannot use synthetic division to divide a polynomial by a quadratic such as $x^2 - 3$.

EXAMPLE 4 ▱ **Using Synthetic Division**

Use synthetic division to divide $x^4 - 10x^2 - 2x + 4$ by $x + 3$.

Solution
You should set up the array as follows. Note that a zero is included for each missing term in the dividend.

Divisor: $x + 3$ Dividend: $x^4 - 10x^2 - 2x + 4$

$$
\begin{array}{r|rrrrr}
-3 & 1 & 0 & -10 & -2 & 4 \\
   &   & -3 & 9 & 3 & -3 \\
\hline
   & 1 & -3 & -1 & 1 & \boxed{1} \quad \longleftarrow \text{Remainder: } 1
\end{array}
$$

Quotient: $x^3 - 3x^2 - x + 1$

Thus, you have

$$\frac{x^4 - 10x^2 - 2x + 4}{x + 3} = x^3 - 3x^2 - x + 1 + \frac{1}{x + 3}.$$

The Remainder and Factor Theorems

The remainder obtained in the synthetic division process has an important interpretation, as described in the **Remainder Theorem.**

> ### The Remainder Theorem
>
> If a polynomial $f(x)$ is divided by $x - k$, the remainder is
>
> $$r = f(k).$$

The Remainder Theorem tells you that synthetic division can be used to evaluate a polynomial function. That is, to evaluate a polynomial function $f(x)$ when $x = k$, divide $f(x)$ by $x - k$. The remainder will be $f(k)$, as illustrated in Example 5.

EXAMPLE 5   Using the Remainder Theorem

Use the Remainder Theorem to evaluate the following function at $x = -2$.

$$f(x) = 3x^3 + 8x^2 + 5x - 7$$

Solution

Using synthetic division, you obtain the following.

$$
\begin{array}{r|rrrr}
-2 & 3 & 8 & 5 & -7 \\
   &   & -6 & -4 & -2 \\
\hline
   & 3 & 2 & 1 & -9
\end{array}
$$

Because the remainder is $r = -9$, you can conclude that

$$f(-2) = -9.$$

This means that $(-2, -9)$ is a point on the graph of f. Try checking this by substituting $x = -2$ in the original function.

Another important theorem is the **Factor Theorem,** which is stated below. This theorem states that you can test to see whether a polynomial has $(x - k)$ as a factor by evaluating the polynomial at $x = k$. If the result is 0, $(x - k)$ is a factor.

> ### The Factor Theorem
>
> A polynomial $f(x)$ has a factor $(x - k)$ if and only if $f(k) = 0$.

EXAMPLE 6 ▱ Factoring a Polynomial

Show that $(x - 2)$ and $(x + 3)$ are factors of

$$f(x) = 2x^4 + 7x^3 - 4x^2 - 27x - 18.$$

Then find the remaining factors of $f(x)$.

Solution

Using synthetic division with 2 and -3 *successively,* you obtain the following.

```
2 | 2    7   -4  -27  -18
  |       4   22   36   18          0 remainder
  ---------------------------
    2   11   18    9    0   ⟹   (x − 2) is a factor.
```

```
-3 | 2   11   18    9
   |      -6  -15   -9                0 remainder
   -----------------------
     2    5    3    0   ⟹   (x + 3) is a factor.
```

Because the resulting quadratic factors as

$$2x^2 + 5x + 3 = (2x + 3)(x + 1)$$

the complete factorization of $f(x)$ is

$$f(x) = (x - 2)(x + 3)(2x + 3)(x + 1).$$

Note that this factorization implies that f has four real zeros:

$$2, -3, -\tfrac{3}{2}, \text{ and } -1.$$

This is confirmed by the graph of f, which is shown in Figure 2.22.   ▱

Figure 2.22

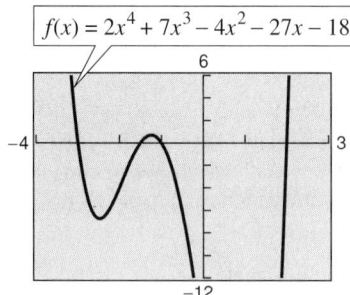

$f(x) = 2x^4 + 7x^3 - 4x^2 - 27x - 18$

In summary, the remainder r, obtained in the synthetic division of $f(x)$ by $x - k$, provides the following information.

1. The remainder r gives the value of f at $x = k$. That is, $r = f(k)$.
2. If $r = 0$, $(x - k)$ is a factor of $f(x)$.
3. If $r = 0$, $(k, 0)$ is an x-intercept of the graph of f.

Note Throughout this text, we have emphasized the importance of developing several problem-solving strategies. In the exercises for this section, try using more than one strategy to solve several of the exercises. For instance, if you find that $x - k$ divides evenly into $f(x)$, try sketching the graph of f. You should find that $(k, 0)$ is an x-intercept of the graph.

The Rational Zero Test

The **Rational Zero Test** relates the possible rational zeros of a polynomial (having integer coefficients) to the leading coefficient and to the constant term of the polynomial.

Note Graph the polynomial $f(x) = x^3 - 53x^2 + 103x - 51$ in the standard viewing rectangle. From the graph alone, you might think that there is only one zero, but the Rational Zero Test says that ±51 might be zeros of the function. If you zoom out several times, you will see a more complete picture of the graph. What are the zeros of f?

The Rational Zero Test

If the polynomial $f(x) = a_n x^n + a_{n-1} x^{n-1} + \cdots + a_2 x^2 + a_1 x + a_0$ has *integer* coefficients, every rational zero of f has the form

$$\text{Rational zero} = \frac{p}{q}$$

where p and q have no common factors other than 1, p is a factor of the constant term a_0, and q is a factor of the leading coefficient a_n.

To use the Rational Zero Test, first list all rational numbers whose numerators are factors of the constant term and whose denominators are factors of the leading coefficient.

$$\text{Possible rational zeros} = \frac{\text{factors of constant term}}{\text{factors of leading coefficient}}$$

Now that you have formed this list of *possible rational zeros,* use a trial-and-error method to determine which, if any, are actual zeros of the polynomial. Note that when the leading coefficient is 1, the possible rational zeros are simply the factors of the constant term. This case is illustrated in Example 7.

EXAMPLE 7 ▬ **Rational Zero Test with Leading Coefficient of 1**

Find the rational zeros of $f(x) = x^3 + x + 1$.

Solution
Because the leading coefficient is 1, the possible rational zeros are simply the factors of the constant term.

Possible Rational Zeros: ±1

By testing these possible zeros, you can see that neither works.

$$f(1) = (1)^3 + 1 + 1 = 3$$
$$f(-1) = (-1)^3 + (-1) + 1 = -1$$

Thus, the polynomial has *no* rational zeros. Note from the graph of f in Figure 2.23 that f does have one real zero (between -1 and 0). However, by the Rational Zero Test, you know that this real zero is *not* a rational number. ▬

Figure 2.23

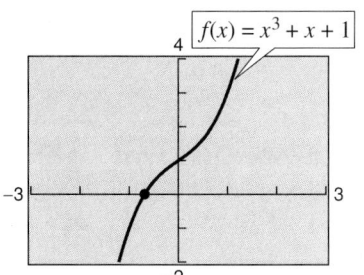

$f(x) = x^3 + x + 1$

If the leading coefficient of a polynomial is not 1, the list of possible rational zeros can increase dramatically. In such cases the search can be shortened in several ways.

1. A programmable calculator can be used to speed up the calculations.
2. A graphing utility can give a good estimate of the locations of the zeros.
3. Synthetic division can be used to test the possible rational zeros.

Finding the first zero is often the most difficult part. After that, the search is simplified by working with the lower-degree polynomial obtained in synthetic division. A graphing utility can help you determine which possible rational zeros to test, as demonstrated in Example 8.

EXAMPLE 8   Using the Rational Zero Test

Find all the real zeros of $f(x) = 10x^3 - 15x^2 - 16x + 12$.

Solution

Because the leading coefficient is 10 and the constant term is 12, there is a long list of possible rational zeros.

Possible Rational Zeros:

$$\frac{\text{Factors of } 12}{\text{Factors of } 10} = \frac{\pm 1, \pm 2, \pm 3, \pm 4, \pm 6, \pm 12}{\pm 1, \pm 2, \pm 5, \pm 10}$$

Figure 2.24

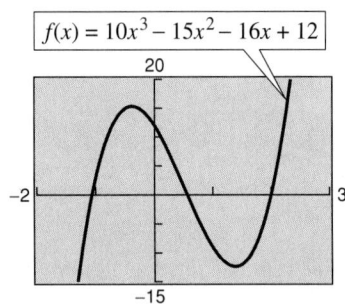
$f(x) = 10x^3 - 15x^2 - 16x + 12$

With so many possibilities (32, in fact), it is worth your time to use a graphing utility to focus on just a few. From Figure 2.24, it looks like three reasonable choices would be $x = -\frac{6}{5}$, $x = \frac{1}{2}$, and $x = 2$. Testing these by synthetic division shows that only $x = 2$ works.

$$
\begin{array}{r|rrrr}
2 & 10 & -15 & -16 & 12 \\
  &   & 20 & 10 & -12 \\
\hline
  & 10 & 5   & -6 & 0
\end{array}
$$

Thus, you have

$$f(x) = (x - 2)(10x^2 + 5x - 6).$$

Using the Quadratic Formula, you find that the two additional zeros are irrational numbers.

$$x = \frac{-5 + \sqrt{265}}{20} \approx 0.5639$$

and

$$x = \frac{-5 - \sqrt{265}}{20} \approx -1.0639$$

Bounds for Real Zeros of Polynomial Functions

The third test for zeros of a polynomial function is related to the sign pattern in the last row of the synthetic division tableau. This test can give you an upper or lower bound of the real zeros of f.

Note A real number b is an **upper bound** for the real zeros of f if no zeros are greater than b. Similarly, b is a **lower bound** if no real zeros of f are less than b.

Upper and Lower Bound Rule

Let $f(x)$ be a polynomial with real coefficients and a positive leading coefficient. Suppose $f(x)$ is divided by $x - c$, using synthetic division.

1. If $c > 0$ and each number in the last row is either positive or zero, c is an *upper bound* for the real zeros of f.
2. If $c < 0$ and the numbers in the last row are alternately positive and negative (zero entries count as positive or negative), c is a *lower bound* for the real zeros of f.

EXPLORATION

Use a graphing utility to graph

$$f(x) = 6x^3 - 4x^2 + 3x - 2.$$

Notice that the graph intersects the x-axis at the point $\left(\frac{2}{3}, 0\right)$. How does this relate to the real zero found in Example 9? Use a graphing utility to graph

$$g(x) = x^4 - 5x^3 + 3x^2 + x.$$

How many times does the graph intersect the x-axis? How many real zeros does g have?

EXAMPLE 9 ▱ **Finding the Zeros of a Polynomial Function**

Find the real zeros of $f(x) = 6x^3 - 4x^2 + 3x - 2$.

Solution
The possible real zeros are as follows.

$$\frac{\text{Factors of } 2}{\text{Factors of } 6} = \frac{\pm 1, \pm 2}{\pm 1, \pm 2, \pm 3, \pm 6}$$

$$= \pm 1, \pm\frac{1}{2}, \pm\frac{1}{3}, \pm\frac{1}{6}, \pm\frac{2}{3}, \pm 2$$

Trying $x = 1$ produces the following.

$$\begin{array}{r|rrrr} 1 & 6 & -4 & 3 & -2 \\ & & 6 & 2 & 5 \\ \hline & 6 & 2 & 5 & 3 \end{array}$$

Thus, $x = 1$ is not a zero, but because the last row has all positive entries, you know that $x = 1$ is an upper bound for the real zeros. Thus, you can restrict the search to zeros less than 1. By trial and error, you can determine that $x = \frac{2}{3}$ is a zero. Thus,

$$f(x) = \left(x - \tfrac{2}{3}\right)\left(6x^2 + 3\right).$$

Because $6x^2 + 3$ has no real zeros, it follows that $x = \frac{2}{3}$ is the only real zero. ▱

2.3 /// EXERCISES

Graphical Analysis In Exercises 1–6, use a graphing utility to graph the two equations in the same viewing rectangle. Use the graphs to verify that the expressions are equivalent. Verify the results algebraically.

1. $y_1 = \dfrac{4x}{x-1}$, $\quad y_2 = 4 + \dfrac{4}{x-1}$

2. $y_1 = \dfrac{3x-5}{x-3}$, $\quad y_2 = 3 + \dfrac{4}{x-3}$

3. $y_1 = \dfrac{x^2}{x+2}$, $\quad y_2 = x - 2 + \dfrac{4}{x+2}$

4. $y_1 = \dfrac{x^4 - 3x^2 - 1}{x^2+5}$, $\quad y_2 = x^2 - 8 + \dfrac{39}{x^2+5}$

5. $y_1 = \dfrac{x^5 - 3x^3}{x^2+1}$, $\quad y_2 = x^3 - 4x + \dfrac{4x}{x^2+1}$

6. $y_1 = \dfrac{x^3 - 2x^2 + 5}{x^2+x+1}$, $\quad y_2 = x - 3 + \dfrac{2(x+4)}{x^2+x+1}$

In Exercises 7–20, divide by long division.

7. Divide $2x^2 + 10x + 12$ by $x + 3$.

8. Divide $5x^2 - 17x - 12$ by $x - 4$.

9. Divide $4x^3 - 7x^2 - 11x + 5$ by $4x + 5$.

10. Divide $6x^3 - 16x^2 + 17x - 6$ by $3x - 2$.

11. Divide $x^4 + 5x^3 + 6x^2 - x - 2$ by $x + 2$.

12. Divide $x^3 + 4x^2 - 3x - 12$ by $x^2 - 3$.

13. Divide $7x + 3$ by $x + 2$.

14. Divide $8x - 5$ by $2x + 1$.

15. $(6x^3 + 10x^2 + x + 8) \div (2x^2 + 1)$

16. $(x^3 - 9) \div (x^2 + 1)$

17. $\dfrac{x^4 + 3x^2 + 1}{x^2 - 2x + 3}$

18. $(x^5 + 7) \div (x^3 - 1)$

19. $\dfrac{2x^3 - 4x^2 - 15x + 5}{(x-1)^2}$

20. $\dfrac{x^4}{(x-1)^3}$

Think About It In Exercises 21 and 22, perform the division by assuming that n is a positive integer.

21. $\dfrac{x^{3n} + 9x^{2n} + 27x^n + 27}{x^n + 3}$

22. $\dfrac{x^{3n} - 3x^{2n} + 5x^n - 6}{x^n - 2}$

In Exercises 23–32, divide by synthetic division.

23. $(4x^3 - 9x + 8x^2 - 18) \div (x + 2)$

24. $(9x^3 - 16x - 18x^2 + 32) \div (x - 2)$

25. $(5x^3 - 6x^2 + 8) \div (x - 4)$

26. $(5x^3 + 6x + 8) \div (x + 2)$

27. $\dfrac{x^3 + 512}{x + 8}$ $\qquad$ **28.** $\dfrac{5x^3}{x + 3}$

29. $\dfrac{-3x^4}{x - 2}$ $\qquad$ **30.** $\dfrac{-3x^4}{x + 2}$

31. $\dfrac{4x^3 + 16x^2 - 23x - 15}{x + \frac{1}{2}}$

32. $\dfrac{3x^3 - 4x^2 + 5}{x - \frac{3}{2}}$

33. *Think About It* What does it mean for a divisor to *divide evenly* into a dividend?

34. *Essay* Explain how you can check polynomial division. Give an example.

In Exercises 35–38, express the function in the form $f(x) = (x - k)q(x) + r$ for the given value of k. Use a graphing utility to demonstrate that $f(k) = r$.

Function	*Value of k*
35. $f(x) = x^3 - x^2 - 14x + 11$	$k = 4$
36. $f(x) = 15x^4 + 10x^3 - 6x^2 + 14$	$k = -\frac{2}{3}$
37. $f(x) = x^3 + 3x^2 - 2x - 14$	$k = \sqrt{2}$
38. $f(x) = 4x^3 - 6x^2 - 12x - 4$	$k = 1 - \sqrt{3}$

In Exercises 39 and 40, use synthetic division to find the required function values. Use a graphing utility to verify your result.

39. $f(x) = 4x^3 - 13x + 10$

(a) $f(1)$ (b) $f(-2)$

(c) $f\left(\frac{1}{2}\right)$ (d) $f(8)$

40. $f(x) = 0.4x^4 - 1.6x^3 + 0.7x^2 - 2$

(a) $f(1)$ (b) $f(-2)$

(c) $f(5)$ (d) $f(-10)$

In Exercises 41–48, use synthetic division to show that x is a solution of the third-degree polynomial equation, and use the result to factor the polynomial completely. List all the real zeros of the function.

Polynomial Equation	Value of x
41. $x^3 - 7x + 6 = 0$	$x = 2$
42. $x^3 - 28x - 48 = 0$	$x = -4$
43. $2x^3 - 15x^2 + 27x - 10 = 0$	$x = \frac{1}{2}$
44. $48x^3 - 80x^2 + 41x - 6 = 0$	$x = \frac{2}{3}$
45. $x^3 + 2x^2 - 3x - 6 = 0$	$x = \sqrt{3}$
46. $x^3 + 2x^2 - 2x - 4 = 0$	$x = \sqrt{2}$
47. $x^3 - 3x^2 + 2 = 0$	$x = 1 + \sqrt{3}$
48. $x^3 - x^2 - 13x - 3 = 0$	$x = 2 - \sqrt{5}$

49. *Data Analysis* The average monthly basic rates R for cable television in the United States for the years 1985 through 1993 are given in the table. The variable t represents the year, with $t = 0$ corresponding to 1990. (Source: *The Cable TV Financial Databook*)

t	-5	-4	-3	-2	-1
R	9.73	10.67	12.18	13.86	15.21

t	0	1	2	3
R	16.78	18.10	19.08	19.39

(a) Use the regression capabilities of a graphing utility to fit a cubic model to the data.

(b) Use a graphing utility to plot the data and graph the model in the same viewing rectangle.

(c) Use a graphing utility and the model to create a table of estimated values of R. Compare the model with the actual data.

(d) Use synthetic division to evaluate the model for the year 1996. Even though the model is relatively accurate for estimating the given data, do you think it is accurate for predicting future cable rates? Explain.

50. *Exploration* Use the form $f(x) = (x - k)q(x) + r$ to create a cubic function that (a) passes through the point $(2, 5)$ and rises to the right, and (b) passes through the point $(-3, 1)$ and falls to the right. (The answers are not unique.)

In Exercises 51–54, use the Rational Zero Test to list all possible rational zeros of f. Verify that the zeros of f shown on the graph are contained in the list.

51. $f(x) = x^3 + 3x^2 - x - 3$

52. $f(x) = x^3 - 4x^2 - 4x + 16$

Figure for 51   **Figure for 52**

    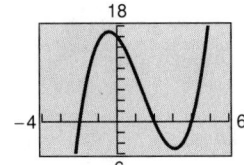

53. $f(x) = 2x^4 - 17x^3 + 35x^2 + 9x - 45$

54. $f(x) = 6x^3 - 71x^2 - 13x + 12$

Figure for 53   **Figure for 54**

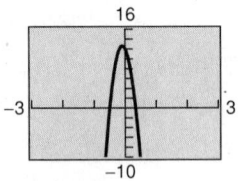

In Exercises 55–64, (a) list the possible rational zeros of f, (b) use a graphing utility to graph f so that some of the possible zeros in part (a) can be disregarded, and (c) determine all the real zeros of f.

55. $f(x) = x^3 + x^2 - 4x - 4$

56. $f(x) = -3x^3 + 20x^2 - 36x + 16$

57. $f(x) = -4x^3 + 15x^2 - 8x - 3$

58. $f(x) = 4x^3 - 12x^2 - x + 15$

59. $f(x) = -2x^4 + 13x^3 - 21x^2 + 2x + 8$

60. $f(x) = 4x^4 - 17x^2 + 4$

61. $f(x) = 32x^3 - 52x^2 + 17x + 3$

62. $f(x) = 6x^3 - x^2 - 13x + 8$

63. $f(x) = 4x^3 + 7x^2 - 11x - 18$

64. $f(x) = 2x^3 + 5x^2 - 21x - 10$

In Exercises 65–70, find all the real solutions of the polynomial equation.

65. $z^4 - z^3 - 2z - 4 = 0$

66. $x^4 - x^3 - 29x^2 - x - 30 = 0$

67. $x^4 - 13x^2 - 12x = 0$

68. $2y^4 + 7y^3 - 26y^2 + 23y - 6 = 0$

69. $2x^4 - 11x^3 - 6x^2 + 64x + 32 = 0$

70. $x^5 - x^4 - 3x^3 + 5x^2 - 2x = 0$

Graphical Analysis In Exercises 71–74, (a) use the root-finding capabilities of a graphing utility to approximate the zeros of the function accurate to three decimal places, and (b) determine one of the exact zeros, use synthetic division to verify your result, and then factor the polynomial completely.

71. $f(x) = x^4 - 3x^2 + 2$

72. $P(t) = t^4 - 7t^2 + 12$

73. $h(x) = x^5 - 7x^4 + 10x^3 + 14x^2 - 24x$

74. $g(x) = 6x^4 - 11x^3 - 51x^2 + 99x - 27$

In Exercises 75–78, use synthetic division to verify the upper bound and/or lower bound of the zeros of f.

75. $f(x) = x^4 - 4x^3 + 15$

Upper bound: $x = 4$;   Lower bound: $x = -1$

76. $f(x) = 2x^3 - 3x^2 - 12x + 8$

Upper bound: $x = 4$;   Lower bound: $x = -3$

77. $f(x) = x^4 - 4x^3 + 16x - 16$

Upper bound: $x = 5$;   Lower bound: $x = -3$

78. $f(x) = 2x^4 - 8x + 3$

Upper bound: $x = 3$;   Lower bound: $x = -4$

In Exercises 79–82, find the rational zeros of the polynomial function.

79. $P(x) = x^4 - \frac{25}{4}x^2 + 9 = \frac{1}{4}(4x^4 - 25x^2 + 36)$

80. $f(x) = x^3 - \frac{3}{2}x^2 - \frac{23}{2}x + 6 = \frac{1}{2}(2x^3 - 3x^2 - 23x + 12)$

81. $f(x) = x^3 - \frac{1}{4}x^2 - x + \frac{1}{4} = \frac{1}{4}(4x^3 - x^2 - 4x + 1)$

82. $f(z) = z^3 + \frac{11}{6}z^2 - \frac{1}{2}z - \frac{1}{3} = \frac{1}{6}(6z^3 + 11z^2 - 3z - 2)$

In Exercises 83–86, match the cubic function with the correct numbers of rational and irrational zeros.

(a) Rational zeros: 0;   Irrational zeros: 1

(b) Rational zeros: 3;   Irrational zeros: 0

(c) Rational zeros: 1;   Irrational zeros: 2

(d) Rational zeros: 1;   Irrational zeros: 0

83. $f(x) = x^3 - 1$ **84.** $f(x) = x^3 - 2$

85. $f(x) = x^3 - x$ **86.** $f(x) = x^3 - 2x$

Think About It In Exercises 87–92, determine (if possible) the zeros of the function g if the function f has zeros at $x = r_1, x = r_2,$ and $x = r_3$.

87. $g(x) = -f(x)$ **88.** $g(x) = 3f(x)$

89. $g(x) = f(x - 5)$ **90.** $g(x) = f(2x)$

91. $g(x) = 3 + f(x)$ **92.** $g(x) = f(-x)$

93. *Dimensions of a Box*   An open box is to be made from a rectangular piece of material 15 centimeters by 9 centimeters by cutting equal squares from the corners and turning up the sides.

(a) Let x represent the length of the sides of the squares. Draw a figure showing the squares removed from the original piece of material and the resulting dimensions of the open box.

(b) Use the figure in part (a) to write the volume V of the box as a function of x. Determine the domain of the function.

(c) Use a graphing utility to graph the function and approximate the dimensions of the box that yield maximum volume.

(d) Find values of x such that $V = 56$. Which of these values is a physical impossibility in the construction of the box? Explain.

94. *Dimensions of a Package*   A rectangular package sent by a shipping service can have a maximum combined length and girth (perimeter of a cross section) of 120 inches (see figure).

(a) Show that the volume of the package is given by
$$V(x) = 4x^2(30 - x).$$

(b) Use a graphing utility to graph the function and approximate the dimensions of the package that yield a maximum volume.

(c) Find values of x such that $V = 13,500$. Which of these values is a physical impossibility in the construction of the package? Explain.

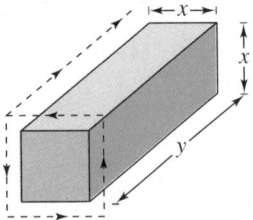

95. *Automobile Emissions*   The number of parts per million of nitric oxide emissions y from a certain car engine is approximated by the model
$$y = -5.05x^3 + 3857x - 38,411.25, \qquad 13 \le x \le 18$$
where x is the air-fuel ratio.

(a) Use a graphing utility to graph the emissions function.

(b) It is observed from the graph that two air-fuel ratios produce 2400 parts per million of nitric oxide, with one being 15. Use the graph to approximate the second air-fuel ratio.

(c) Algebraically approximate the second air-fuel ratio that produces 2400 parts per million of nitric oxide. (*Hint:* Because you know that an air-fuel ratio of 15 produces the specified nitric oxide emission, you can use synthetic division.)

96. *Advertising Costs*   A company that manufactures bicycles estimates that the profit for selling a particular model is given by
$$P = -45x^3 + 2500x^2 - 275,000, \qquad 0 \le x \le 50$$
where P is the profit in dollars and x is the advertising expense in tens of thousands of dollars. According to this model, find the smaller of two advertising amounts that yield a profit of $800,000.

97. *Imports into the United States*   The values of goods (in billions of dollars) imported into the United States for the years 1984 through 1993 are given in the table.   (Source: U.S. General Imports and Imports for Consumption)

Year	1984	1985	1986	1987	1988
Imports	325.7	345.3	370.0	406.2	441.0

Year	1989	1990	1991	1992	1993
Imports	473.4	495.3	487.1	532.7	580.5

(a) A model for this data is given by
$$I = 0.161t^3 + 0.626t^2 + 25.646t + 484.137$$
where I is the annual value of goods imported (in billions of dollars) and t represents the calendar year, with $t = 0$ corresponding to 1990. Use a graphing utility to graph the data and the model in the same viewing rectangle. How do they compare?

(b) According to this model, when did the annual value of imports reach 525 billion dollars?

(c) According to the right-hand behavior of the model, will the value of imports continue to increase? Explain.

2.4   Complex Numbers

*The Imaginary Unit i / Operations with Complex Numbers /
Complex Conjugates and Division / Applications*

The Imaginary Unit *i*

Some quadratic equations have no real solutions. For instance, the quadratic equation

$$x^2 + 1 = 0 \qquad \text{Equation with no real solution}$$

has no real solution because there is no real number x that can be squared to produce -1. To overcome this deficiency, mathematicians created an expanded system of numbers using the **imaginary unit *i*,** defined as

$$i = \sqrt{-1} \qquad \text{Imaginary unit}$$

where $i^2 = -1$. By adding real numbers to real multiples of this imaginary unit, you obtain the set of **complex numbers.** Each complex number can be written in the **standard form, $a + bi$.**

Carl Friedrich Gauss (1777–1855) proved that all the roots of any algebraic equation are "numbers" of the form $a + bi$, where a and b are real numbers and i is the square root of -1. These "numbers" were called complex.

Definition of a Complex Number

For real numbers a and b, the number

$$a + bi$$

is a **complex number.** If $a = 0$ and $b \neq 0$, the complex number bi is an **imaginary number.**

The set of real numbers is a subset of the set of complex numbers, as shown in Figure 2.25. This is true because every real number a can be written as a complex number using $b = 0$. That is, for every real number a, we can write $a = a + 0i$.

Figure 2.25

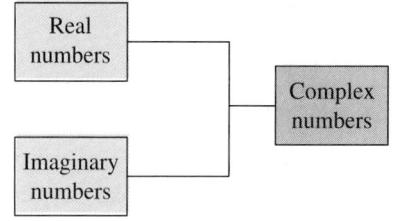

Equality of Complex Numbers

Two complex numbers $a + bi$ and $c + di$, written in standard form, are **equal** to each other

$$a + bi = c + di \qquad \text{Equality of two complex numbers}$$

if and only if $a = c$ and $b = d$.

Operations with Complex Numbers

To add (or subtract) two complex numbers, you add (or subtract) the real and imaginary parts of the numbers separately.

Addition and Subtraction of Complex Numbers
If $a + bi$ and $c + di$ are two complex numbers written in standard form, their sum and difference are defined as follows. *Sum:* $(a + bi) + (c + di) = (a + c) + (b + d)i$ *Difference:* $(a + bi) - (c + di) = (a - c) + (b - d)i$

The **additive identity** in the complex number system is zero (the same as in the real number system). Furthermore, the **additive inverse** of the complex number $a + bi$ is

$$-(a + bi) = -a - bi. \qquad \text{Additive inverse}$$

Thus, you have

$$(a + bi) + (-a - bi) = 0 + 0i = 0.$$

EXAMPLE 1 ▱ **Adding and Subtracting Complex Numbers**

a. $(3 - i) + (2 + 3i) = 3 - i + 2 + 3i$ Remove parentheses.

$\qquad\qquad\qquad\quad = 3 + 2 - i + 3i$ Group like terms.

$\qquad\qquad\qquad\quad = (3 + 2) + (-1 + 3)i$

$\qquad\qquad\qquad\quad = 5 + 2i$ Standard form

b. $2i + (-4 - 2i) = 2i - 4 - 2i$ Remove parentheses.

$\qquad\qquad\qquad = -4 + 2i - 2i$ Group like terms.

$\qquad\qquad\qquad = -4$ Standard form

c. $3 - (-2 + 3i) + (-5 + i) = 3 + 2 - 3i - 5 + i$

$\qquad\qquad\qquad\qquad\qquad = 3 + 2 - 5 - 3i + i$

$\qquad\qquad\qquad\qquad\qquad = 0 - 2i$

$\qquad\qquad\qquad\qquad\qquad = -2i$ ▱

Note In Example 1(b) the sum of two complex numbers can be a real number.

Many of the properties of real numbers are valid for complex numbers as well. Here are some examples.

> *Associative Property of Addition and Multiplication*
> *Commutative Property of Addition and Multiplication*
> *Distributive Property of Multiplication Over Addition*

Notice how these properties are used when two complex numbers are multiplied.

$$(a + bi)(c + di) = a(c + di) + bi(c + di) \qquad \text{Distributive}$$
$$= ac + (ad)i + (bc)i + (bd)i^2 \qquad \text{Distributive}$$
$$= ac + (ad)i + (bc)i + (bd)(-1) \qquad \text{Definition of } i$$
$$= ac - bd + (ad)i + (bc)i \qquad \text{Commutative}$$
$$= (ac - bd) + (ad + bc)i \qquad \text{Associative}$$

Rather than trying to memorize this multiplication rule, we suggest that you simply remember how the distributive property is used to multiply two complex numbers. The procedure is similar to multiplying two polynomials and combining like terms (as in the FOIL Method).

EXAMPLE 2 ▱ **Multiplying Complex Numbers**

a. $(i)(-3i) = -3i^2$ Multiply.

$\qquad\qquad = -3(-1)$ Definition of i

$\qquad\qquad = 3$ Simplify.

b. $(2 - i)(4 + 3i) = 8 + 6i - 4i - 3i^2$ Product of binomials

$\qquad\qquad\qquad = 8 + 6i - 4i - 3(-1)$ Definition of i

$\qquad\qquad\qquad = 8 + 3 + 6i - 4i$ Group like terms.

$\qquad\qquad\qquad = 11 + 2i$ Standard form

c. $(3 + 2i)(3 - 2i) = 9 - 6i + 6i - 4i^2$ Product of binomials

$\qquad\qquad\qquad = 9 - 4(-1)$ Definition of i

$\qquad\qquad\qquad = 9 + 4$ Simplify.

$\qquad\qquad\qquad = 13$ Standard form

d. $(3 + 2i)^2 = 9 + 6i + 6i + 4i^2$ Product of binomials

$\qquad\qquad = 9 + 4(-1) + 12i$ Definition of i

$\qquad\qquad = 9 - 4 + 12i$ Simplify.

$\qquad\qquad = 5 + 12i$ Standard form   ▱

Complex Conjugates and Division

Notice in Example 2(c) that the product of two complex numbers can be a real number. This occurs with pairs of complex numbers of the form $a + bi$ and $a - bi$, called **complex conjugates.**

$$(a + bi)(a - bi) = a^2 - abi + abi - b^2i^2$$
$$= a^2 - b^2(-1)$$
$$= a^2 + b^2$$

To find the quotient of $a + bi$ and $c + di$ where c and d are not both zero, multiply the numerator and denominator by the conjugate of the denominator to obtain

$$\frac{a + bi}{c + di} = \frac{a + bi}{c + di}\left(\frac{c - di}{c - di}\right) = \frac{(ac + bd) + (bc - ad)i}{c^2 + d^2}.$$

EXAMPLE 3   **Dividing Complex Numbers**

$$\frac{1}{1 + i} = \frac{1}{1 + i}\left(\frac{1 - i}{1 - i}\right) \qquad \text{Multiply by conjugate.}$$

$$= \frac{1 - i}{1^2 - i^2} \qquad \text{Expand.}$$

$$= \frac{1 - i}{1 - (-1)} \qquad \text{Definition of } i$$

$$= \frac{1 - i}{2} \qquad \text{Simplify.}$$

$$= \frac{1}{2} - \frac{1}{2}i \qquad \text{Standard form}$$

Some graphing utilities, such as the *TI–92* or *TI-83* from Texas Instruments, can perform operations with complex numbers. For instance, to divide $2 + 3i$ by $4 - 2i$, enter

$$\boxed{(}\;\boxed{2}\;\boxed{+}\;\boxed{3}\;\boxed{i}\;\boxed{)}\;\boxed{\div}$$

$$\boxed{(}\;\boxed{4}\;\boxed{-}\;\boxed{2}\;\boxed{i}\;\boxed{)}\;\boxed{\text{ENTER}}.$$

The display is

$$1/10 + 4/5i.$$

EXAMPLE 4   **Dividing Complex Numbers**

$$\frac{2 + 3i}{4 - 2i} = \frac{2 + 3i}{4 - 2i}\left(\frac{4 + 2i}{4 + 2i}\right) \qquad \text{Multiply by conjugate.}$$

$$= \frac{8 + 4i + 12i + 6i^2}{16 - 4i^2} \qquad \text{Expand.}$$

$$= \frac{8 - 6 + 16i}{16 + 4} \qquad \text{Definition of } i$$

$$= \frac{1}{20}(2 + 16i) \qquad \text{Simplify.}$$

$$= \frac{1}{10} + \frac{4}{5}i \qquad \text{Standard form}$$

Figure 2.26

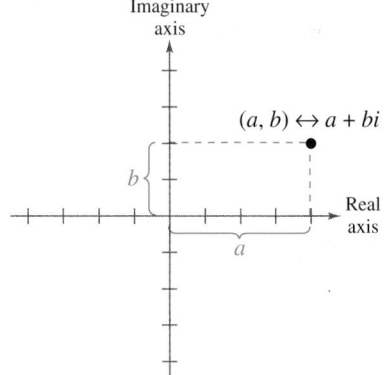

Applications

Most applications involving complex numbers are either theoretical or very technical, and are therefore not appropriate for inclusion in this text. However, to give you some idea of how complex numbers can be used in applications, we give a general description of their use in **fractal geometry.**

To begin, consider a coordinate system called the **complex plane.** Just as every real number corresponds to a point on the real line, every complex number corresponds to a point in the complex plane, as shown in Figure 2.26. In this figure, note that the vertical axis is the **imaginary axis** and the horizontal axis is the **real axis.** The point that corresponds to the complex number $a + bi$ is (a, b).

Figure 2.27

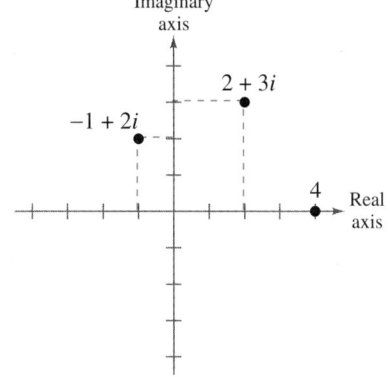

EXAMPLE 5   Plotting Complex Numbers

Plot each complex number in the complex plane.

a. $2 + 3i$ **b.** $-1 + 2i$ **c.** 4

Solution

a. To plot the complex number $2 + 3i$, move (from the origin) two units to the right on the real axis and then three units up, as shown in Figure 2.27. In other words, plotting the complex number $2 + 3i$ in the complex plane is comparable to plotting the point $(2, 3)$ in the Cartesian plane.

b. The complex number $-1 + 2i$ corresponds to the point $(-1, 2)$, as shown in Figure 2.27.

c. The complex number 4 corresponds to the point $(4, 0)$, as shown in Figure 2.27.

In the hands of a person who understands "fractal geometry," the complex plane can become an easel on which stunning pictures, called **fractals,** can be drawn. The most famous such picture is called the **Mandelbrot Set,** named after the Polish-born mathematician Benoit Mandelbrot. To draw the Mandelbrot Set, consider the following sequence of numbers.

$$c, \quad c^2 + c, \quad (c^2 + c)^2 + c, \quad [(c^2 + c)^2 + c]^2 + c, \quad \dots$$

The behavior of this sequence depends on the value of the complex number c. For some values of c this sequence is **bounded,** and for other values it is **unbounded.** If the sequence is bounded, the complex number c is in the Mandelbrot Set, and if the sequence is unbounded, the complex number c is not in the Mandelbrot Set.

EXAMPLE 6 ▱ **Members of the Mandelbrot Set**

a. The complex number -2 is in the Mandelbrot Set because for $c = -2$, the corresponding Mandelbrot sequence is $-2, 2, 2, 2, 2, 2, \ldots$, which is bounded.

b. The complex number i is also in the Mandelbrot Set because for $c = i$, the corresponding Mandelbrot sequence is

$$i, \quad -1 + i, \quad -i, \quad -1 + i, \quad -i, \quad -1 + i, \quad \ldots$$

which is bounded.

c. The complex number $1 + i$ is *not* in the Mandelbrot Set because for $c = 1 + i$, the corresponding Mandelbrot sequence is

$$1 + i, \quad 1 + 3i, \quad -7 + 7i, \quad 1 - 97i, \quad -9407 - 193i,$$
$$88454401 + 3631103i, \quad \ldots$$

which is unbounded.   ▱

Figure 2.29

American Mathematical Society.

With this definition, a picture of the Mandelbrot Set would have only two colors. One color for points that are in the set (the sequence is bounded) and one color for points that are outside the set (the sequence is unbounded). Figure 2.28 shows a black and blue picture of the Mandelbrot Set. The points that are black are in the Mandelbrot Set and the points that are blue are not.

Figure 2.28

American Mathematical Society.

To add more interest to the picture, computer scientists discovered that the points that are not in the Mandelbrot Set can be assigned a variety of colors, depending on "how quickly" their sequences diverge. Figure 2.29 shows three different appendages of the Mandelbrot Set. (The black portions of the picture represent points that are in the Mandelbrot Set.)

Figures 2.30, 2.31, and 2.32 show other types of fractals. From these pictures, you can see why fractals have fascinated people since their discovery (around 1980). The fractals shown were produced on a graphing calculator.

Figure 2.30 The Sierpinski Gasket

Figure 2.31 The Fractal "Dragon"

Figure 2.32 A Fractal Fern

Group Activity

Error Analysis

Suppose you are a math instructor, and one of your students has handed in the following quiz. Find the error(s) in each solution and discuss how to explain each error to your student.

Question

1. Write $\dfrac{5}{3-2i}$ in standard form.

2. Multiply: $(\sqrt{-4} + 3)(i - \sqrt{-3})$.

3. Sketch the graph of $y = -x^2 + 2$.

Answer

$$\dfrac{5}{3-2i} \cdot \dfrac{3+2i}{3+2i} = \dfrac{15+10i}{9-4} = 3+2i$$

$$(\sqrt{-4}+3)(i-\sqrt{-3}) = i\sqrt{-4} - \sqrt{-4}\sqrt{-3} + 3i - 3\sqrt{-3}$$
$$= -2i - \sqrt{12} + 3i - 3i\sqrt{3}$$
$$= (1 - 3\sqrt{3})i - 2\sqrt{3}$$

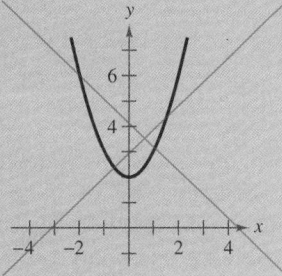

2.4 /// EXERCISES

In Exercises 1–4, solve for a and b.

1. $a + bi = -10 + 6i$ **2.** $a + bi = 13 + 4i$

3. $(a - 1) + (b + 3)i = 5 + 8i$

4. $(a + 6) + 2bi = 6 - 5i$

In Exercises 5–16, write in standard form.

5. $4 + \sqrt{-9}$ **6.** $3 + \sqrt{-16}$

7. $2 - \sqrt{-27}$ **8.** $1 + \sqrt{-8}$

9. $\sqrt{-75}$ **10.** 45

11. $-6i + i^2$ **12.** $-4i^2 + 2i$

13. 8 **14.** $\left(\sqrt{-4}\right)^2 - 5$

15. $\sqrt{-0.09}$ **16.** $\sqrt{-0.0004}$

In Exercises 17–26, perform the addition or subtraction and write the result in standard form.

17. $(5 + i) + (6 - 2i)$ **18.** $(13 - 2i) + (-5 + 6i)$

19. $(8 - i) - (4 - i)$ **20.** $(3 + 2i) - (6 + 13i)$

21. $\left(-2 + \sqrt{-8}\right) + \left(5 - \sqrt{-50}\right)$

22. $\left(8 + \sqrt{-18}\right) - \left(4 + 3\sqrt{2}i\right)$

23. $13i - (14 - 7i)$

24. $22 + (-5 + 8i) + 10i$

25. $-\left(\frac{3}{2} + \frac{5}{2}i\right) + \left(\frac{5}{3} + \frac{11}{3}i\right)$

26. $(1.6 + 3.2i) + (-5.8 + 4.3i)$

In Exercises 27–40, perform the operation and write the result in standard form.

27. $\sqrt{-6} \cdot \sqrt{-2}$ **28.** $\sqrt{-5} \cdot \sqrt{-10}$

29. $\left(\sqrt{-10}\right)^2$ **30.** $\left(\sqrt{-75}\right)^2$

31. $(1 + i)(3 - 2i)$ **32.** $(6 - 2i)(2 - 3i)$

33. $6i(5 - 2i)$ **34.** $-8i(9 + 4i)$

35. $\left(\sqrt{14} + \sqrt{10}i\right)\left(\sqrt{14} - \sqrt{10}i\right)$

36. $\left(3 + \sqrt{-5}\right)\left(7 - \sqrt{-10}\right)$

37. $(4 + 5i)^2$ **38.** $(2 - 3i)^3$

39. $(2 + 3i)^2 + (2 - 3i)^2$

40. $(1 - 2i)^2 - (1 + 2i)^2$

41. *Error Analysis*   Describe the error.

$$\sqrt{-6}\sqrt{-6} = \sqrt{(-6)(-6)} = \sqrt{36} = 6$$

42. *True or False?* There is no complex number that is equal to its conjugate. Explain.

In Exercises 43–50, find the product of the number and its conjugate.

43. $5 + 3i$ **44.** $9 - 12i$

45. $-2 - \sqrt{5}i$ **46.** $-4 + \sqrt{2}i$

47. $20i$ **48.** $\sqrt{-15}$

49. $\sqrt{8}$ **50.** $1 + \sqrt{8}$

In Exercises 51–64, perform the operation and write the result in standard form.

51. $\dfrac{6}{i}$ **52.** $-\dfrac{10}{2i}$

53. $\dfrac{4}{4 - 5i}$ **54.** $\dfrac{3}{1 - i}$

55. $\dfrac{2 + i}{2 - i}$ **56.** $\dfrac{8 - 7i}{1 - 2i}$

57. $\dfrac{6 - 7i}{i}$ **58.** $\dfrac{8 + 20i}{2i}$

59. $\dfrac{1}{(4 - 5i)^2}$ **60.** $\dfrac{(2 - 3i)(5i)}{2 + 3i}$

61. $\dfrac{2}{1 + i} - \dfrac{3}{1 - i}$ **62.** $\dfrac{2i}{2 + i} + \dfrac{5}{2 - i}$

63. $\dfrac{i}{3 - 2i} + \dfrac{2i}{3 + 8i}$ **64.** $\dfrac{1 + i}{i} - \dfrac{3}{4 - i}$

65. Express the first 16 positive integer powers of i as i, $-i$, 1, or -1. Describe any pattern you see.

66. Use your result from Exercise 65 to express each of the following powers of i as i, $-i$, 1, or -1: (a) i^{40} (b) i^{25} (c) i^{50} (d) i^{67}

In Exercises 67–74, simplify the complex number and write it in standard form.

67. $-6i^3 + i^2$

68. $4i^2 - 2i^3$

69. $-5i^5$

70. $(-i)^3$

71. $\left(\sqrt{-75}\right)^3$

72. $\left(\sqrt{-2}\right)^6$

73. $\dfrac{1}{i^3}$

74. $\dfrac{1}{(2i)^3}$

In Exercises 75–78, determine the complex number shown in the complex plane.

75.

76.

77.

78.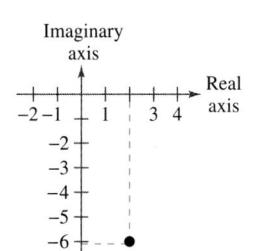

In Exercises 79–82, plot the complex number in the complex plane.

79. $3 - 4i$

80. $2i$

81. -5

82. $-5 + 3i$

Fractals In Exercises 83–88, find the first six terms in the following sequence.

$$c,\ c^2 + c,\ (c^2 + c)^2 + c,\ [(c^2 + c)^2 + c]^2 + c,\ \dots$$

From these terms, do you think the given complex number is in the Mandelbrot Set?

83. $c = 0$

84. $c = 2$

85. $c = \frac{1}{2}i$

86. $c = -i$

87. $c = 1$

88. $c = -1$

89. Cube the complex numbers. What do you notice?

$$2,\quad -1 + \sqrt{3}\,i,\quad -1 - \sqrt{3}\,i$$

90. Raise the numbers to the fourth power.

$$2,\quad -2,\quad 2i,\quad -2i$$

91. Prove that the sum of a complex number $a + bi$ and its conjugate is a real number, and that the difference of a complex number $a + bi$ and its conjugate is an imaginary number.

92. Prove that the product of a complex number $a + bi$ and its conjugate is a real number.

93. Prove that the conjugate of the product of two complex numbers $a_1 + b_1 i$ and $a_2 + b_2 i$ is the product of their conjugates.

94. Prove that the conjugate of the sum of two complex numbers $a_1 + b_1 i$ and $a_2 + b_2 i$ is the sum of their conjugates.

Review Solve Exercises 95–103 as a review of the skills and problem-solving techniques you learned in previous sections.

95. Subtract: $(x^3 - 3x^2) - (6 - 2x - 4x^2)$

96. Add: $(4 + 3x) + (8 - 6x - x^2)$

97. Multiply: $\left(3x - \frac{1}{2}\right)(x + 4)$

98. Expand: $(2x - 5)^2$

99. Expand: $[(x + y) + 3]^2$

100. *Volume of an Oblate Spheroid*

Solve for a: $V = \frac{4}{3}\pi a^2 b$

101. *Newton's Law of Universal Gravitation*

Solve for r: $F = \alpha \dfrac{m_1 m_2}{r^2}$

102. *Mixture Problem* A 5-liter container contains a mixture with a concentration of 50%. How much of this mixture must be withdrawn and replaced by 100% concentrate to bring the mixture up to 60% concentration?

103. *Average Speed* A business executive traveled at an average speed of 100 kilometers per hour on a 200-kilometer trip. Because of heavy traffic, the average speed on the return trip was only 80 kilometers per hour. Find the average speed for the round trip.

Example 2 shows how to use the methods described in Section 2.2 (The Rational Zero Test, synthetic division, and factoring) to find all the zeros of a polynomial function, including complex zeros.

EXAMPLE 2   **Finding the Zeros of a Polynomial Function**

Write $f(x) = x^5 + x^3 + 2x^2 - 12x + 8$ as the product of linear factors, and list all of its zeros.

Solution

The possible rational zeros are $\pm 1, \pm 2, \pm 4$, and ± 8. The graph shown in Figure 2.33 indicates that 1 and -2 are good guesses. Synthetic division produces the following.

Figure 2.33

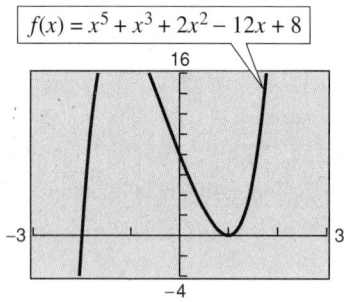

$f(x) = x^5 + x^3 + 2x^2 - 12x + 8$

$$
\begin{array}{r|rrrrrr}
1 & 1 & 0 & 1 & 2 & -12 & 8 \\
  &   & 1 & 1 & 2 & 4 & -8 \\
\hline
  & 1 & 1 & 2 & 4 & -8 & 0
\end{array}
$$

$\Longrightarrow$ 1 is a zero.

$$
\begin{array}{r|rrrrr}
1 & 1 & 1 & 2 & 4 & -8 \\
  &   & 1 & 2 & 4 & 8 \\
\hline
  & 1 & 2 & 4 & 8 & 0
\end{array}
$$

$\Longrightarrow$ 1 is a repeated zero.

$$
\begin{array}{r|rrrr}
-2 & 1 & 2 & 4 & 8 \\
   &   & -2 & 0 & -8 \\
\hline
   & 1 & 0 & 4 & 0
\end{array}
$$

$\Longrightarrow$ -2 is a zero.

Thus, you have

$$f(x) = x^5 + x^3 + 2x^2 - 12x + 8$$
$$= (x - 1)(x - 1)(x + 2)(x^2 + 4).$$

By factoring $x^2 + 4$ as

$$x^2 - (-4) = \left(x - \sqrt{-4}\right)\left(x + \sqrt{-4}\right) = (x - 2i)(x + 2i)$$

you obtain

$$f(x) = (x - 1)(x - 1)(x + 2)(x - 2i)(x + 2i)$$

which gives the following five zeros of f.

$$1, \quad 1, \quad -2, \quad 2i, \quad \text{and} \quad -2i$$

Note from the graph of f shown in Figure 2.33 that the *real* zeros are the only ones that appear as x-intercepts.

Conjugate Pairs

In Example 2, note that the two complex zeros are **conjugates.** That is, they are of the form $a + bi$ and $a - bi$.

Complex Zeros Occur in Conjugate Pairs

Let $f(x)$ be a polynomial function that has *real coefficients*. If $a + bi$, where $b \neq 0$, is a zero of the function, the conjugate $a - bi$ is also a zero of the function.

Note Be sure you see that this result is true only if the polynomial function has *real coefficients*. For instance, the result applies to the function $f(x) = x^2 + 1$, but not to the function $g(x) = x - i$.

EXAMPLE 3   **Finding a Polynomial with Given Zeros**

Find a *fourth-degree* polynomial function, with real coefficients, that has -1, -1, and $3i$ as zeros.

Solution

Because $3i$ is a zero *and* the polynomial is stated to have real coefficients, you know that the conjugate $-3i$ must also be a zero. Thus, from the Linear Factorization Theorem, $f(x)$ can be written as

$$f(x) = a(x + 1)(x + 1)(x - 3i)(x + 3i).$$

For simplicity, let $a = 1$, and obtain

$$f(x) = (x^2 + 2x + 1)(x^2 + 9)$$
$$= x^4 + 2x^3 + 10x^2 + 18x + 9.$$

Factoring a Polynomial

The Linear Factorization Theorem shows that you can write any nth-degree polynomial as the product of n linear factors.

$$f(x) = a(x - c_1)(x - c_2)(x - c_3) \cdots (x - c_n)$$

However, this result includes the possibility that some of the values of c_i are complex. The boxed statement at the top of page 208 implies that even if you do not want to get involved with "complex factors," you can still write $f(x)$ as the product of linear and/or quadratic factors.

Think About the Proof

To prove the theorem at the right, begin by using the Linear Factorization Theorem to write the polynomial function as

$$f(x) = a(x - c_1) \cdots (x - c_n).$$

If each c_i is real, there is nothing to prove. If any of the c_i's are complex, how can you use the theorem on page 207 to conclude that $f(x)$ has a quadratic factor with real coefficients? The details of the proof are given in the appendix.

Factors of a Polynomial

Every polynomial of degree $n > 0$ with real coefficients can be written as the product of linear and quadratic factors with real coefficients, where the quadratic factors have no real zeros.

A quadratic factor with no real zeros is said to be **irreducible over the reals.** Be sure you see that this is not the same as being *irreducible over the rationals.* For example, the quadratic

$$x^2 + 1 = (x - i)(x + i)$$

is irreducible over the reals (and therefore over the rationals). On the other hand, the quadratic

$$x^2 - 2 = (x - \sqrt{2})(x + \sqrt{2})$$

is irreducible over the rationals, but *reducible* over the reals.

EXAMPLE 4   Factoring a Polynomial

Write the polynomial $f(x) = x^4 - x^2 - 20$

a. as the product of factors that are irreducible over the *rationals,*

b. as the product of linear factors and quadratic factors that are irreducible over the *reals,* and

c. in completely factored form.

Solution

a. Begin by factoring the polynomial into the product of two quadratic polynomials.

$$x^4 - x^2 - 20 = (x^2 - 5)(x^2 + 4)$$

Both of these factors are irreducible over the rationals.

b. By factoring over the reals, you have

$$x^4 - x^2 - 20 = (x + \sqrt{5})(x - \sqrt{5})(x^2 + 4)$$

where the quadratic factor is irreducible over the reals.

c. In completely factored form, you have

$$x^4 - x^2 - 20 = (x + \sqrt{5})(x - \sqrt{5})(x - 2i)(x + 2i).$$

Note   In Example 4, notice from the completely factored form that the fourth-degree polynomial has four zeros.

Some graphing utilities can find both real and complex zeros of a function. For instance, to find all the zeros of

$$f(x) = x^4 - 3x^3 + 6x^2 + 2x - 60$$

on a *TI–92*, you can enter the following.

$\boxed{\text{F2}}$ $\boxed{\text{A}}$ 1 (cSolve

Enter $f(x)$.

$\boxed{=}$ 0 $\boxed{,}$ $\boxed{\text{X}}$ $\boxed{)}$ $\boxed{\text{ENTER}}$

The displayed solutions are

$x = 1 + 3i$ or $x = 1 - 3i$

or $x = 3$ or $x = -2$.

EXAMPLE 5 **Finding the Zeros of a Polynomial Function**

Find all the zeros of $f(x) = x^4 - 3x^3 + 6x^2 + 2x - 60$, given that $1 + 3i$ is a zero of f.

Solution

Because complex zeros occur in conjugate pairs, you know that $1 - 3i$ is also a zero of f. This means that both

$$x - (1 + 3i) \quad \text{and} \quad x - (1 - 3i)$$

are factors of $f(x)$. Multiplying these two factors produces

$$[x - (1 + 3i)][x - (1 - 3i)] = [(x - 1) - 3i][(x - 1) + 3i]$$
$$= (x - 1)^2 - 9i^2$$
$$= x^2 - 2x + 10.$$

Using long division, you can divide $x^2 - 2x + 10$ into $f(x)$ to obtain the following.

$$
\require{enclose}
\begin{array}{r}
x^2 - x - 6 \\[-2pt]
x^2 - 2x + 10 \enclose{longdiv}{x^4 - 3x^3 + 6x^2 + 2x - 60} \\
\underline{x^4 - 2x^3 + 10x^2} \\
-x^3 - 4x^2 + 2x \\
\underline{-x^3 + 2x^2 - 10x} \\
-6x^2 + 12x - 60 \\
\underline{-6x^2 + 12x - 60} \\
0
\end{array}
$$

Therefore, you have

$$f(x) = (x^2 - 2x + 10)(x^2 - x - 6) = (x^2 - 2x + 10)(x - 3)(x + 2)$$

and you can conclude that the zeros of f are $1 + 3i$, $1 - 3i$, 3, and -2.

Study Tip

Throughout this chapter, we have basically stated the results and theorems in terms of zeros of polynomial functions. Be sure you see that the same results could have been stated in terms of solutions of polynomial equations. This is true because the zeros of the polynomial function

$$f(x) = a_n x^n + a_{n-1} x^{n-1} + \cdots + a_2 x^2 + a_1 x + a_0$$

are precisely the solutions of the polynomial equation

$$a_n x^n + a_{n-1} x^{n-1} + \cdots + a_2 x^2 + a_1 x + a_0 = 0.$$

Group Activity *Factoring a Polynomial*

Compile a list of all the various techniques for factoring a polynomial that have been covered thus far in the text. Give an example illustrating each technique, and discuss when the use of each technique is appropriate.

2.5 /// EXERCISES

In Exercises 1–4, find all the zeros of the function.

1. $f(x) = x(x - 6)^2$

2. $g(x) = (x - 2)(x + 4)^3$

3. $h(t) = (t - 3)(t - 2)(t - 3i)(t + 3i)$

4. $h(m) = (m - 4)^2(m - 2 + 4i)(m - 2 - 4i)$

Graphical and Analytical Analysis In Exercises 5–8, find all the zeros of the function. Is there a relationship between the number of real zeros and the number of x-intercepts of the graph? Explain.

5. $f(x) = x^3 - 4x^2 + x - 4$

6. $f(x) = x^3 - 4x^2 - 4x + 16$

Figure for 5

Figure for 6

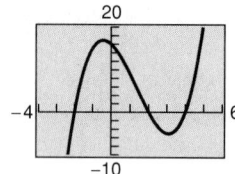

7. $f(x) = x^4 + 4x^2 + 4$

8. $f(x) = x^4 - 3x^2 - 4$

Figure for 7

Figure for 8

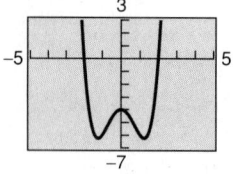

In Exercises 9–28, find all the zeros of the function and write the polynomial as a product of linear factors. Use a graphing utility to graph the function as an aid in finding the zeros and as a check of your results. (If you have a computer algebra system, use it to verify the complex zeros.)

9. $f(x) = x^2 + 25$

10. $f(x) = x^2 - x + 56$

11. $h(x) = x^2 - 4x + 1$

12. $g(x) = x^2 + 10x + 23$

13. $f(x) = x^4 - 81$

14. $f(y) = y^4 - 625$

15. $f(z) = z^2 - 2z + 2$

16. $h(x) = x^3 - 3x^2 + 4x - 2$

17. $f(t) = t^3 - 3t^2 - 15t + 125$

18. $f(x) = x^3 + 11x^2 + 39x + 29$

19. $f(x) = 16x^3 - 20x^2 - 4x + 15$

20. $f(s) = 2s^3 - 5s^2 + 12s - 5$

21. $f(x) = 5x^3 - 9x^2 + 28x + 6$

22. $g(x) = 3x^3 - 4x^2 + 8x + 8$

23. $f(x) = x^4 + 10x^2 + 9$

24. $f(x) = x^4 + 29x^2 + 100$

25. $g(x) = x^4 - 4x^3 + 8x^2 - 16x + 16$

26. $h(x) = x^4 + 6x^3 + 10x^2 + 6x + 9$

27. $f(x) = 2x^4 + 5x^3 + 4x^2 + 5x + 2$

28. $g(x) = x^5 - 8x^4 + 28x^3 - 56x^2 + 64x - 32$

In Exercises 29–36, find a polynomial function with integer coefficients that has the given zeros. (There are many correct answers.)

29. $1, 5i, -5i$

30. $4, 3i, -3i$

31. $2, 4 + i, 4 - i$

32. $6, -5 + 2i, -5 - 2i$

33. $i, -i, 6i, -6i$

34. $2, 2, 2, 4i, -4i$

35. $-5, -5, 1 + \sqrt{3}i$

36. $0, 0, 4, 1 + i$

In Exercises 37–40, write the polynomial (a) as the product of factors that are irreducible over the *rationals*, (b) as the product of linear and quadratic factors that are irreducible over the *reals*, and (c) in completely factored form.

37. $f(x) = x^4 + 6x^2 - 27$

38. $f(x) = x^4 - 2x^3 - 3x^2 + 12x - 18$
(*Hint:* One factor is $x^2 - 6$.)

39. $f(x) = x^4 - 4x^3 + 5x^2 - 2x - 6$
(*Hint:* One factor is $x^2 - 2x - 2$.)

40. $f(x) = x^4 - 3x^3 - x^2 - 12x - 20$
(*Hint:* One factor is $x^2 + 4$.)

In Exercises 41–46, use the given zero to find all the zeros of the function.

Function	Zero
41. $f(x) = 2x^3 + 3x^2 + 50x + 75$	$r = 5i$
42. $f(x) = x^3 + x^2 + 9x + 9$	$r = 3i$
43. $f(x) = 2x^4 - x^3 + 7x^2 - 4x - 4$	$r = 2i$
44. $g(x) = x^3 - 7x^2 - x + 87$	$r = 5 + 2i$
45. $g(x) = 4x^3 + 23x^2 + 34x - 10$	$r = -3 + i$
46. $h(x) = 3x^3 - 4x^2 + 8x + 8$	$r = 1 - \sqrt{3}i$

Graphical Analysis In Exercises 47–50, (a) use the root or zero feature of a graphing utility to approximate the zeros of the function accurate to three decimal places, and (b) determine one of the exact zeros, use synthetic division to verify your result, and find the exact values of the remaining zeros.

47. $f(x) = x^4 + 3x^3 - 5x^2 - 21x + 22$

48. $f(x) = x^3 + 4x^2 + 14x + 20$

49. $h(x) = 8x^3 - 14x^2 + 18x - 9$

50. $f(x) = 25x^3 - 55x^2 - 54x - 18$

51. *Think About It* A zero of the function $f(x) = x^3 + ix^2 + ix - 1$ is $x = -i$.

(a) Show that the conjugate $x = i$ is not a zero of f.

(b) Does part (a) contradict the theorem that complex zeros occur in conjugate pairs? Explain.

52. *Exploration* Use a graphing utility to graph the function $f(x) = x^4 - 4x^2 + k$ for different values of k. Find values of k such that the zeros of f satisfy the specified characteristics. (Some parts do not have unique answers.)

(a) Four real zeros

(b) Two real zeros each of multiplicity 2

(c) Two real zeros and two complex zeros

(d) Four complex zeros

53. *Think About It* Will the answers to Exercise 52 change for the function g?

(a) $g(x) = f(x - 2)$ (b) $g(x) = f(2x)$

54. *Maximum Height* A baseball is thrown upward from ground level with an initial velocity of 48 feet per second, and its height h in feet is given by

$$h = -16t^2 + 48t, \quad 0 \le t \le 3$$

where t is the time in seconds. Suppose you are told that the ball reaches a height of 64 feet. Is this possible? Explain.

55. *Profit* The demand equation for a certain product is given by $p = 140 - 0.0001x$, where p is the unit price (in dollars) of the product and x is the number of units produced and sold. The cost equation for the product is $C = 80x + 150,000$, where C is the total cost (in dollars) and x is the number of units produced. The total profit obtained by producing and selling x units is given by $P = R - C = xp - C$. Suppose you are working in the marketing department that produces this product, and you are asked to determine a price p that would yield a profit of 9 million dollars. Is this possible? Explain.

56. Find a quadratic function f (with integer coefficients) that has $\pm\sqrt{b}i$ as zeros. Assume that b is a positive integer.

57. Find a quadratic function f (with integer coefficients) that has $a \pm bi$ as zeros. Assume that b is a positive integer.

Chapter Opener In Exercises 58 and 59, plot the complex number.

58. $-4 + 2i$ **59.** $\frac{3}{2} + \frac{11}{2}i$

2.6   Rational Functions and Asymptotes

Introduction to Rational Functions / Horizontal and Vertical Asymptotes / Applications

Introduction to Rational Functions

A **rational function** can be written in the form

$$f(x) = \frac{p(x)}{q(x)}$$

where $p(x)$ and $q(x)$ are polynomials and $q(x)$ is not the zero polynomial. In this section we assume that $p(x)$ and $q(x)$ have no common factors.

In general, the *domain* of a rational function of x includes all real numbers except x-values that make the denominator zero. Much of our discussion of rational functions will focus on their graphical behavior near these x-values.

Library of Functions

A rational function $f(x)$ is the quotient of two polynomials, $f(x) = p(x)/q(x)$. A rational function is not defined at points x for which the denominator $q(x) = 0$. Near these points, the graph of the rational function may increase or decrease without bound.

EXAMPLE 1   Finding the Domain of a Rational Function

Find the domain of

$$f(x) = \frac{1}{x}$$

and discuss the behavior of f near any excluded x-values.

Solution
Because the denominator is zero when $x = 0$, the domain of f is all real numbers except $x = 0$. To determine the behavior of f near this excluded value, evaluate $f(x)$ to the left and right of $x = 0$, as indicated in the following two tables.

x	-1	-0.5	-0.1	-0.01	-0.001	$\to 0$
$f(x)$	-1	-2	-10	-100	-1000	$\to -\infty$

x	$0 \leftarrow$	0.001	0.01	0.1	0.5	1
$f(x)$	$\infty \leftarrow$	1000	100	10	2	1

Note that as x approaches 0 *from the left,* $f(x)$ decreases without bound, whereas as x approaches 0 *from the right,* $f(x)$ increases without bound. The graph of f is shown in Figure 2.34.

Figure 2.34

$f(x) = \dfrac{1}{x}$

Figure 2.35

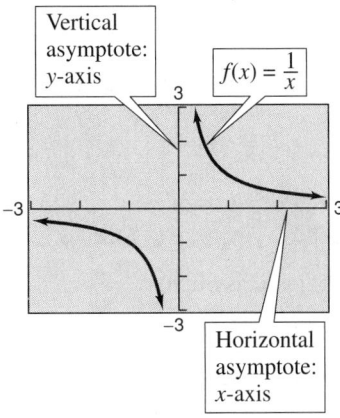

Horizontal and Vertical Asymptotes

In Example 1, the behavior of $f(x) = 1/x$ near $x = 0$ is denoted as follows.

$$f(x) \to -\infty \text{ as } x \to 0^-$$ $$f(x) \to \infty \text{ as } x \to 0^+$$

f(x) decreases without bound as x approaches 0 from the left.

f(x) increases without bound as x approaches 0 from the right.

The line $x = 0$ is a **vertical asymptote** of the graph of f, as shown in Figure 2.35. The graph of f also has a **horizontal asymptote**—the line $y = 0$. This means that the values of $f(x) = 1/x$ approach zero as x increases or decreases without bound.

$$f(x) \to 0 \text{ as } x \to -\infty$$ $$f(x) \to 0 \text{ as } x \to \infty$$

f(x) approaches 0 as x decreases without bound.

f(x) approaches 0 as x increases without bound.

Definition of Vertical and Horizontal Asymptotes

1. The line $x = a$ is a **vertical asymptote** of the graph of f if

$$f(x) \to \infty \quad \text{or} \quad f(x) \to -\infty$$

as $x \to a$, either from the right or from the left.

2. The line $y = b$ is a **horizontal asymptote** of the graph of f if

$$f(x) \to b$$

as $x \to \infty$ or $x \to -\infty$.

Eventually (as $x \to \infty$ or $x \to -\infty$), the distance between the horizontal asymptote and the points on the graph must approach zero. Figure 2.36 shows the horizontal and vertical asymptotes of the graphs of three rational functions.

Figure 2.36

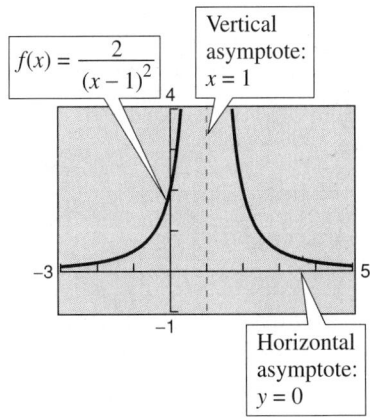

Asymptotes of a Rational Function

Let f be the rational function given by

$$f(x) = \frac{p(x)}{q(x)} = \frac{a_n x^n + a_{n-1} x^{n-1} + \cdots + a_1 x + a_0}{b_m x^m + b_{m-1} x^{m-1} + \cdots + b_1 x + b_0}$$

where $p(x)$ and $q(x)$ have no common factors.

1. The graph of f has vertical asymptotes at the zeros of $q(x)$.
2. The graph of f has at most one horizontal asymptote, as follows.

    a. If $n < m$, the x-axis ($y = 0$) is a horizontal asymptote.

    b. If $n = m$, the line $y = a_n/b_m$ is a horizontal asymptote.

    c. If $n > m$, the graph of f has no horizontal asymptote.

You can apply this theorem by comparing the degrees of the numerator and denominator, as illustrated in Figure 2.37.

a. The graph of

$$f(x) = \frac{2x}{3x^2 + 1} \qquad \text{See Figure 2.37(a).}$$

has the x-axis as a horizontal asymptote.

b. The graph of

$$f(x) = \frac{2x^2}{3x^2 + 1} \qquad \text{See Figure 2.37(b).}$$

has the line $y = \frac{2}{3}$ as a horizontal asymptote.

c. The graph of

$$f(x) = \frac{2x^3}{3x^2 + 1} \qquad \text{See Figure 2.37(c).}$$

has no horizontal asymptote.

EXPLORATION

Use a graphing utility to compare the graphs of

$$y = \frac{3x^3 - 5x^2 + 4x - 5}{2x^2 - 6x + 7}$$

and

$$y = \frac{3x^3}{2x^2}.$$

First, use a viewing rectangle in which $-5 \le x \le 5$ and $-10 \le y \le 10$, then zoom out. Write a convincing argument that the shape of the graph of a rational function eventually behaves like the graph of $y = a_n x^n / b_m x^m$, where $a_n x^n$ is the leading term of the numerator and $b_m x^m$ is the leading term of the denominator.

Figure 2.37

(a)

(b)

(c)

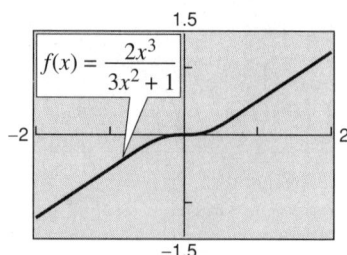

EXAMPLE 2 ◢▬ **A Calculator Experiment**

Find the horizontal asymptote of the graph of $f(x) = \dfrac{3x^3 + 7x^2 + 2}{-4x^3 + 5}$.

Solution

Because the degree of the numerator and denominator are the same, the horizontal asymptote is given by the ratio of the leading coefficients.

$$y = \frac{\text{leading coefficient of numerator}}{\text{leading coefficient of denominator}} = \frac{3}{-4} = -\frac{3}{4}$$

The following tables show how the values of $f(x)$ become closer and closer to $-\frac{3}{4}$ as x becomes increasingly large or small.

The tables at the right can be generated on a *TI-82* or *TI-83* using the Indpnt:Ask feature in the table setup menu.

x Decreases Without Bound

X	Y₁	
−1	0.66667	
−10	−0.5738	
−100	−0.7325	
−1,000	−0.7482	
−10,000	−0.7498	
X =		

x Increases Without Bound

X	Y₁	
1	12	
10	−0.9267	
100	−0.7675	
1,000	−0.7518	
10,000	−0.7502	
X =		

Figure 2.38

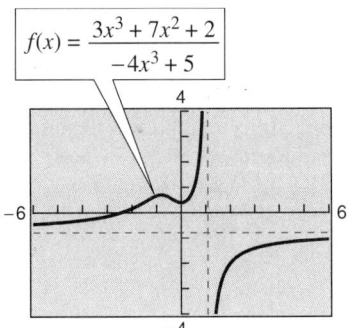

$$f(x) = \frac{3x^3 + 7x^2 + 2}{-4x^3 + 5}$$

The graph is shown in Figure 2.38. Try using a graphing utility to confirm the graph. ▬

EXAMPLE 3 ◢▬ **A Graph with Two Horizontal Asymptotes**

A function that is not rational can have two horizontal asymptotes—one to the left and one to the right. For instance, the graph of

$$f(x) = \frac{x + 10}{|x| + 2}$$

is shown in Figure 2.39. It has the line $y = -1$ as a horizontal asymptote to the left and the line $y = 1$ as a horizontal asymptote to the right. You can confirm this by rewriting the function as follows.

Figure 2.39

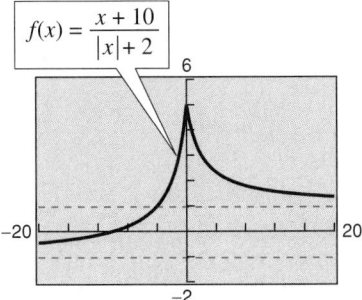

$$f(x) = \frac{x + 10}{|x| + 2}$$

$$f(x) = \begin{cases} \dfrac{x + 10}{-x + 2}, & x \le 0 \\ \dfrac{x + 10}{x + 2}, & x > 0 \end{cases}$$

▬

Applications

There are many examples of asymptotic behavior in real life. For instance, Example 4 shows how a vertical asymptote can be used to analyze the cost of removing pollutants from smokestack emissions.

Real Life

EXAMPLE 4 ▰ Cost-Benefit Model

A utility company burns coal to generate electricity. The cost of removing a certain *percent* of the pollutants from the smokestack emissions is typically not a linear function. That is, if it costs C dollars to remove 25% of the pollutants, it would cost more than $2C$ dollars to remove 50% of the pollutants. As the percent of removed pollutants approaches 100%, the cost tends to become prohibitive. Suppose that the cost C of removing $p\%$ of the smokestack pollutants is given by

$$C = \frac{80,000p}{100 - p}, \qquad 0 \le p < 100.$$

Sketch the graph of this function. Suppose you are a member of a state legislature that is considering a law that would require utility companies to remove 90% of the pollutants from their smokestack emissions. If the current law requires 85% removal, how much additional expense would the new law ask the utility company to spend?

Solution

The graph of this function is shown in Figure 2.40. Note that the graph has a vertical asymptote at $p = 100$. Because the current law requires 85% removal, the current cost to the utility company is

$$C = \frac{80,000(85)}{100 - 85} \qquad \text{Evaluate } C \text{ at } p = 85.$$

$$\approx \$453,333.$$

If the new law increased the percent removal to 90%, the cost to the utility company would be

$$C = \frac{80,000(90)}{100 - 90} \qquad \text{Evaluate } C \text{ at } p = 90.$$

$$= \$720,000.$$

Thus, the new law would require the utility company to spend an additional

$$720,000 - 453,333 = \$266,667.$$

Figure 2.40

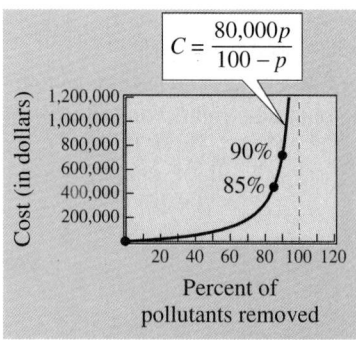

$$C = \frac{80,000p}{100 - p}$$

Cost (in dollars)

1,200,000
1,000,000
800,000
600,000
400,000
200,000

90%
85%

20 40 60 80 100 120

Percent of
pollutants removed

Figure 2.41

The table feature of a graphing utility can be used to find vertical and horizontal asymptotes of rational functions. Use the table feature to find any horizontal or vertical asymptotes of

$$f(x) = \frac{2}{x + 1}.$$

Write a statement explaining how you found the asymptote(s) using the table.

EXAMPLE 5 ▱ **Average Cost of Producing a Product** *Real Life*

A business has a cost function of $C = 0.5x + 5000$, where C is measured in dollars and x is the number of units produced. The *average cost per unit* is

$$\overline{C} = \frac{C}{x} = \frac{0.5x + 5000}{x}.$$

Find the average cost per unit when $x = 1000$, 5000, $10{,}000$, and $100{,}000$. What is the horizontal asymptote for this function, and what does it represent?

Solution

When $x = 1000$, $\overline{C} = \dfrac{0.5(1000) + 5000}{1000} = \$5.50$.

When $x = 5000$, $\overline{C} = \dfrac{0.5(5000) + 5000}{5000} = \$1.50$.

When $x = 10{,}000$, $\overline{C} = \dfrac{0.5(10{,}000) + 5000}{10{,}000} = \$1.00$.

When $x = 100{,}000$, $\overline{C} = \dfrac{0.5(100{,}000) + 5000}{100{,}000} = \$0.55$.

As shown in Figure 2.41, the horizontal asymptote is given by the line $\overline{C} = 0.50$. This line represents the least possible unit cost for the product. ▱

Note that this example points out one of the major problems of a small business. That is, it is difficult to have competitively low prices when the production level is low.

Group Activity

Common Factors in the Numerator and Denominator

When sketching the graph of a rational function, be sure that the rational function has no factor that is common to its numerator and denominator. To see why, consider the function given by

$$f(x) = \frac{x(x - 1)}{x}$$

which has a common factor of x in the numerator and denominator. Sketch the graph of this function. Does it have a vertical asymptote at $x = 0$?

2.6 /// EXERCISES

In Exercises 1–6, (a) complete each table, (b) determine the vertical and horizontal asymptotes of the function, and (c) find the domain of the function.

x	$f(x)$
0.5	
0.9	
0.99	
0.999	

x	$f(x)$
1.5	
1.1	
1.01	
1.001	

x	$f(x)$
5	
10	
100	
1000	

1. $f(x) = \dfrac{1}{x-1}$

2. $f(x) = \dfrac{5x}{x-1}$

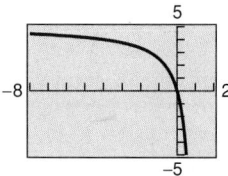

3. $f(x) = \dfrac{3x}{|x-1|}$

4. $f(x) = \dfrac{3}{|x-1|}$

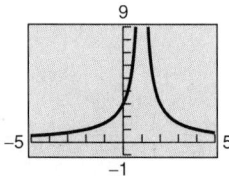

5. $f(x) = \dfrac{3x^2}{x^2-1}$

6. $f(x) = \dfrac{4x}{x^2-1}$

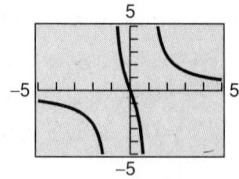

In Exercises 7–12, match the function with its graph. [The graphs are labeled (a), (b), (c), (d), (e), and (f).]

(a)

(b)

(c)

(d)

(e)

(f)

7. $f(x) = \dfrac{2}{x+2}$

8. $f(x) = \dfrac{1}{x-3}$

9. $f(x) = \dfrac{4x+1}{x}$

10. $f(x) = \dfrac{1-x}{x}$

11. $f(x) = \dfrac{x-2}{x-4}$

12. $f(x) = -\dfrac{x+2}{x+4}$

In Exercises 13–20, find the domain of the function and identify any horizontal and vertical asymptotes.

13. $f(x) = \dfrac{1}{x^2}$

14. $f(x) = \dfrac{4}{(x-2)^3}$

15. $f(x) = \dfrac{2 + x}{2 - x}$

16. $f(x) = \dfrac{1 - 5x}{1 + 2x}$

17. $f(x) = \dfrac{x^3}{x^2 - 1}$

18. $f(x) = \dfrac{2x^2}{x + 1}$

19. $f(x) = \dfrac{3x^2 + 1}{x^2 + x + 9}$

20. $f(x) = \dfrac{3x^2 + x - 5}{x^2 + 1}$

Analytical and Numerical Explanation In Exercises 21–24, (a) determine the domain of f and g, (b) find any vertical asymptotes of f, (c) complete the table, and (d) decide what your results suggest.

21. $f(x) = \dfrac{x^2 - 4}{x + 2}$, $g(x) = x - 2$

x	-4	-3	-2.5	-2	-1.5	-1	0
$f(x)$							
$g(x)$							

22. $f(x) = \dfrac{x^2(x - 3)}{x^2 - 3x}$, $g(x) = x$

x	-1	0	1	2	3	3.5	4
$f(x)$							
$g(x)$							

23. $f(x) = \dfrac{x - 3}{x^2 - 3x}$, $g(x) = \dfrac{1}{x}$

x	-1	-0.5	0	0.5	2	3	4
$f(x)$							
$g(x)$							

24. $f(x) = \dfrac{2x - 8}{x^2 - 9x + 20}$, $g(x) = \dfrac{2}{x - 5}$

x	0	1	2	3	4	5	6
$f(x)$							
$g(x)$							

Think About It In Exercises 25–28, write a rational function f having the specified characteristics.

25. Vertical asymptotes: $x = -2, x = 1$

26. Vertical asymptote: None
Horizontal asymptote: $y = 0$

27. Vertical asymptote: None
Horizontal asymptote: $y = 2$

28. Vertical asymptotes: $x = 0, x = \frac{5}{2}$
Horizontal asymptote: $y = -3$

Exploration In Exercises 29–32, (a) determine the value the function f approaches as the magnitude of x increases. Is $f(x)$ greater than or less than this functional value when (b) x is positive and large in magnitude and (c) x is negative and large in magnitude?

29. $f(x) = 4 - \dfrac{1}{x}$

30. $f(x) = 2 + \dfrac{1}{x - 3}$

31. $f(x) = \dfrac{2x - 1}{x - 3}$

32. $f(x) = \dfrac{2x - 1}{x^2 + 1}$

In Exercises 33–36, find the zeros (if any) of the rational function.

33. $g(x) = \dfrac{x^2 - 4}{x + 1}$

34. $h(x) = 4 + \dfrac{5}{x^2 + 2}$

35. $f(x) = 1 - \dfrac{2}{x - 3}$

36. $g(x) = \dfrac{x^3 - 8}{x^2 + 1}$

37. *Cost of Clean Water* The cost in millions of dollars for removing $p\%$ of the industrial and municipal pollutants discharged into a river is given by

$$C = \dfrac{255p}{100 - p}, \qquad 0 \le p < 100.$$

(a) Find the cost of removing 10% of the pollutants.

(b) Find the cost of removing 40% of the pollutants.

(c) Find the cost of removing 75% of the pollutants.

(d) According to this model, would it be possible to remove 100% of the pollutants? Explain.

38. *Recycling Costs*   In a pilot project, a rural township was given recycling bins for separating and storing recyclable products. The cost in dollars for supplying bins to $p\%$ of the population is given by

$$C = \frac{25{,}000p}{100 - p}, \qquad 0 \le p < 100.$$

(a) Find the cost of giving bins to 15% of the population.

(b) Find the cost of giving bins to 50% of the population.

(c) Find the cost of giving bins to 90% of the population.

(d) According to this model, would it be possible to supply bins to 100% of the residents? Explain.

39. *Data Analysis*   Consider a physics laboratory experiment designed to determine an unknown mass. A flexible metal meter stick is clamped to a table with 50 centimeters overhanging the edge (see figure). Known masses M ranging from 200 grams to 2000 grams are attached to the end of the meter stick. For each mass the meter stick is displaced vertically and then allowed to oscillate. The average time t in seconds of one oscillation for each mass is recorded in the table.

M	200	400	600	800	1000
t	0.450	0.597	0.721	0.831	0.906

M	1200	1400	1600	1800	2000
t	1.003	1.088	1.168	1.218	1.338

A model for the data is given by

$$t = \frac{38M + 16{,}965}{10(M + 5000)}.$$

(a) Use a graphing utility to create a table showing the predicted time based on the model for each of the masses shown in the table. What can you conclude?

(b) Use the model to approximate the mass of an object if the average for one oscillation is 1.056 seconds.

Figure for 39

← 50 cm →

M

40. *Data Analysis*   The endpoints of the interval over which distinct vision is possible are called the *near point* and *far point* of the eye (see figure). With increasing age these points normally change. The table gives the approximate near points y in centimeters for various ages x.

x	10	20	30	40	50
y	7	10	14	22	40

(a) Fit a rational model to the data. Take the reciprocals of the near points to generate the points $(x, 1/y)$. Use the regression capabilities of a graphing utility to fit a line to this data. The resulting line has the form

$$\frac{1}{y} = ax + b.$$

Solve for y.

(b) Use a graphing utility to create a table giving the predicted near point based on the model for each of the ages in the given table.

(c) Do you think the model can be used to predict the near point for a person who is 60 years old? Explain.

Object blurry   Object clear   Object blurry

Near point   Far point

41. *Population of Deer* The game commission introduces 100 deer into newly acquired state game lands. The population of the herd is given by

$$N = \frac{20(5 + 3t)}{1 + 0.04t}, \qquad 0 \le t$$

where t is the time in years (see figure).

(a) Find the population when t is 5, 10, and 25.

(b) What is the limiting size of the herd as time increases? Explain your reasoning.

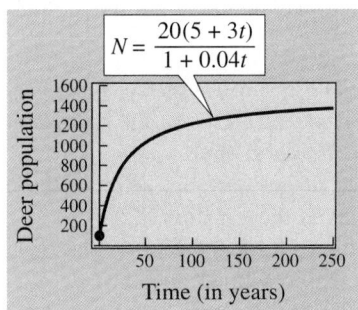

42. *Food Consumption* A biology class performs an experiment comparing the quantity of food consumed by a certain kind of moth with the quantity supplied (see figure). The model for their experimental data is given by

$$y = \frac{1.568x - 0.001}{6.360x + 1}, \qquad 0 < x$$

where x is the quantity (in milligrams) of food supplied and y is the quantity (in milligrams) eaten. At what level of consumption will the moth become satiated?

43. *Human Memory Model* Psychologists have developed mathematical models for predicting performance as a function of the number of trials n of a certain task. Consider the learning curve given by

$$P = \frac{0.5 + 0.9(n - 1)}{1 + 0.9(n - 1)}, \qquad 0 < n$$

where P is the percent of correct responses after n trials.

(a) Complete the table for this model. What does it suggest?

n	1	2	3	4	5	6	7	8	9	10
P										

(b) According to this model, what is the limiting percent of correct responses as n increases? Explain your reasoning.

44. *Human Memory Model* How would the limiting percent of correct responses change if the human memory model in Exercise 43 were changed to

$$P = \frac{0.5 + 0.6(n - 1)}{1 + 0.8(n - 1)}, \qquad 0 < n?$$

Review Solve Exercises 45–48 as a review of the skills and problem-solving techniques you learned in previous sections.

45. $225 - 50x = 0$

46. $t^3 - 50t = 0$

47. $2z^2 - 3z - 35 = 0$

48. $x(10 - x) = 25$

2.7   Graphs of Rational Functions

The Graph of a Rational Function / *Slant Asymptotes* / *Application*

The Graph of a Rational Function

Note Testing for symmetry can be useful, especially for simple rational functions. For example, the graph of $f(x) = 1/x$ is symmetrical with respect to the origin, and the graph of $g(x) = 1/x^2$ is symmetrical with respect to the y-axis.

Analyzing Graphs of Rational Functions

Let $f(x) = p(x)/q(x)$, where $p(x)$ and $q(x)$ are polynomials with no common factors.

1. The y-intercept (if any) is the value of $f(0)$.
2. The x-intercepts (if any) are the zeros of the numerator—that is, the solutions of the equation $p(x) = 0$.
3. The vertical asymptotes (if any) are the zeros of the denominator—that is, the solutions of the equation $q(x) = 0$.
4. The horizontal asymptote (if any) is the value that $f(x)$ approaches as x increases or decreases without bound.
5. Determining the behavior of the graph *between* and *beyond* each x-intercept and vertical asymptote is crucial for describing the complete graph of a rational function.

Graphing utilities have difficulty sketching graphs of rational functions that have vertical asymptotes. Often, the utility will connect parts of the graph that are not supposed to be connected. For instance, the screen at the left shows the graph of

$$f(x) = \frac{1}{x - 2}.$$

Notice that the graph should consist of two *unconnected* portions—one to the left of $x = 2$ and the other to the right of $x = 2$. To eliminate this problem, you can try changing the *mode* of the graphing utility to *dot mode*. The problem with this is that the graph is then represented as a collection of dots rather than as a smooth curve.

EXAMPLE 1 🔳 **Analyzing a Rational Function**

Analyze the function $g(x) = \dfrac{3}{x-2}$.

Solution

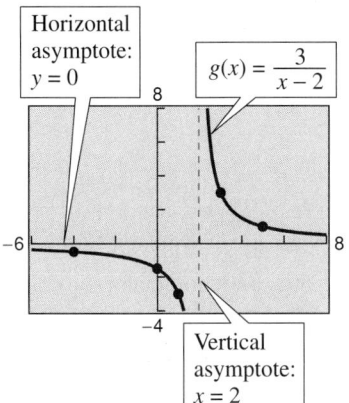

Figure 2.42

Horizontal asymptote: $y = 0$

$g(x) = \dfrac{3}{x-2}$

Vertical asymptote: $x = 2$

y-Intercept:	$\left(0, -\tfrac{3}{2}\right)$, because $g(0) = -\tfrac{3}{2}$.
x-Intercept:	None, because $3 \neq 0$.
Vertical Asymptote:	$x = 2$, zero of denominator
Horizontal Asymptote:	$y = 0$, degree of $p(x)$ < degree of $q(x)$
Additional Points:	

x	-4	1	3	5
$g(x)$	-0.5	-3	3	1

By plotting the intercepts, asymptotes, and a few additional points, you can obtain the graph shown in Figure 2.42. Confirm this with a graphing utility.

🔳

Note The graph of g in Example 1 is a vertical stretch and a right shift of the graph of $f(x) = 1/x$ because

$$g(x) = \frac{3}{x-2} = 3\left(\frac{1}{x-2}\right) = 3f(x-2).$$

EXAMPLE 2 🔳 **Analyzing a Rational Function**

Analyze the function $f(x) = \dfrac{2x-1}{x}$.

Solution

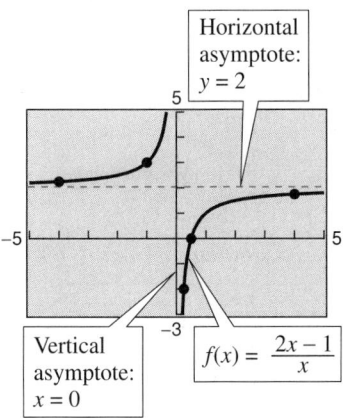

Figure 2.43

Horizontal asymptote: $y = 2$

$f(x) = \dfrac{2x-1}{x}$

Vertical asymptote: $x = 0$

y-Intercept:	None, because $x = 0$ is not in the domain.
x-Intercept:	$\left(\tfrac{1}{2}, 0\right)$, because $2x - 1 = 0$.
Vertical Asymptote:	$x = 0$, zero of denominator
Horizontal Asymptote:	$y = 2$, degree of $p(x)$ = degree of $q(x)$
Additional Points:	

x	-4	-1	$\tfrac{1}{4}$	4
$f(x)$	2.25	3	-2	1.75

By plotting the intercepts, asymptotes, and a few additional points, you can obtain the graph shown in Figure 2.43. Confirm this with a graphing utility.

🔳

EXAMPLE 3 ▱ **Analyzing a Rational Function**

Analyze the function $f(x) = \dfrac{x}{x^2 - x - 2}$.

Solution

By factoring the denominator, you have

$$f(x) = \frac{x}{x^2 - x - 2} = \frac{x}{(x+1)(x-2)}.$$

y-Intercept: (0, 0), because $f(0) = 0$.

x-Intercept: (0, 0)

Vertical Asymptotes: $x = -1$, $x = 2$, zeros of denominator

Horizontal Asymptote: $y = 0$, degree of $p(x) <$ degree of $q(x)$

Additional Points:

x	-3	-0.5	1	3
$f(x)$	-0.3	0.4	-0.5	0.75

The graph is shown in Figure 2.44.

Figure 2.44

Vertical asymptote: $x = -1$

Vertical asymptote: $x = 2$

$f(x) = \dfrac{x}{x^2 - x - 2}$

Horizontal asymptote: $y = 0$

EXAMPLE 4 ▱ **Analyzing a Rational Function**

Analyze the function $f(x) = \dfrac{2(x^2 - 9)}{x^2 - 4}$.

Solution

By factoring the numerator and denominator, you have

$$f(x) = \frac{2(x^2 - 9)}{x^2 - 4} = \frac{2(x-3)(x+3)}{(x-2)(x+2)}.$$

y-Intercept: $\left(0, \frac{9}{2}\right)$, because $f(0) = \frac{9}{2}$.

x-Intercepts: $(-3, 0)$, $(3, 0)$

Vertical Asymptotes: $x = -2$, $x = 2$, zeros of denominator

Horizontal Asymptote: $y = 2$, degree of $p(x) =$ degree of $q(x)$

Symmetry: With respect to y-axis, because $f(-x) = f(x)$.

Additional Points:

x	0.5	2.5	6
$f(x)$	4.67	-2.44	1.69

The graph is shown in Figure 2.45.

Figure 2.45

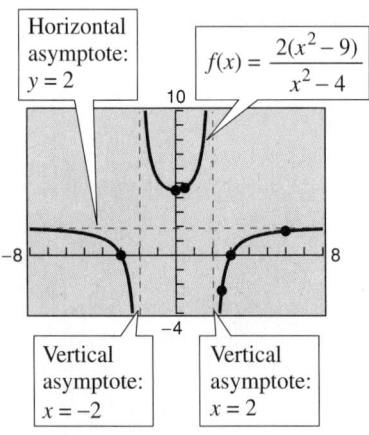

Horizontal asymptote: $y = 2$

$f(x) = \dfrac{2(x^2 - 9)}{x^2 - 4}$

Vertical asymptote: $x = -2$

Vertical asymptote: $x = 2$

Slant Asymptotes

If the degree of the numerator of a rational function is exactly *one more* than the degree of its denominator, the graph of the function has a **slant asymptote**. For example, the graph of

$$f(x) = \frac{x^2 - x}{x + 1}$$

has a slant asymptote, as shown in Figure 2.46. To find the equation of a slant asymptote, use long division. For instance, by dividing $x + 1$ into $x^2 - x$, you have

$$f(x) = \frac{x^2 - x}{x + 1} = \underbrace{x - 2}_{\text{Slant asymptote}} + \frac{2}{x + 1}. \qquad \frac{2}{x + 1} \to 0 \text{ as } x \to \infty$$

Slant asymptote
$(y = x - 2)$

In Figure 2.46, notice that the graph of f approaches the line $y = x - 2$ as x moves to the right or left.

Figure 2.46

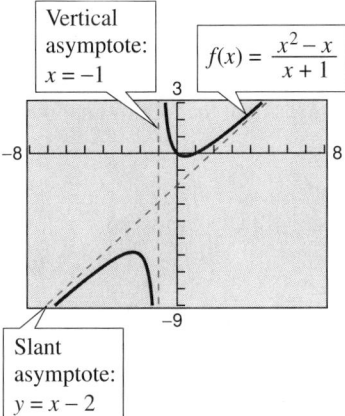

Vertical asymptote: $x = -1$

$f(x) = \dfrac{x^2 - x}{x + 1}$

Slant asymptote: $y = x - 2$

EXAMPLE 5 ▱ **A Rational Function with a Slant Asymptote**

Graph the function $f(x) = \dfrac{x^2 - x - 2}{x - 1}$.

Solution

First write $f(x)$ in two different ways. Factoring the numerator

$$f(x) = \frac{x^2 - x - 2}{x - 1} = \frac{(x - 2)(x + 1)}{x - 1}$$

allows you to recognize the x-intercepts, and long division

$$f(x) = \frac{x^2 - x - 2}{x - 1} = x - \frac{2}{x - 1} \qquad \frac{2}{x - 1} \to 0 \text{ as } x \to \infty$$

allows you to recognize that the line $y = x$ is a slant asymptote of the graph.

y-Intercept:	$(0, 2)$, because $f(0) = 2$.
x-Intercepts:	$(-1, 0), (2, 0)$
Vertical Asymptote:	$x = 1$
Slant Asymptote:	$y = x$

Additional Points:

x	-2	0.5	1.5	3
$f(x)$	-1.33	4.5	-2.5	2

The graph is shown in Figure 2.47.

Figure 2.47

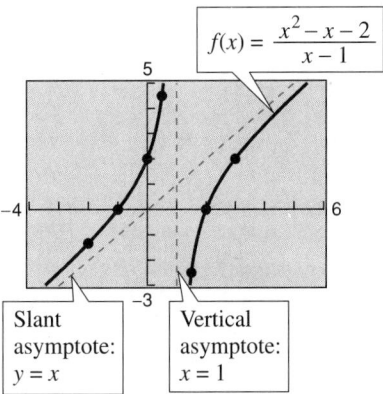

$f(x) = \dfrac{x^2 - x - 2}{x - 1}$

Slant asymptote: $y = x$

Vertical asymptote: $x = 1$

Figure 2.48

Figure 2.49

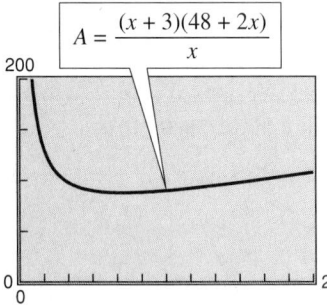

Application

EXAMPLE 3 ◢ **Finding a Minimum Area**

A rectangular page is to contain 48 square inches of print. The margins at the top and bottom of the page are each $1\frac{1}{2}$ inches. The margins on each side are 1 inch. What should the dimensions of the page be so that the minimum amount of paper is used?

Solution

Let A be the area to be minimized. From Figure 2.48, you can write

$$A = (x + 3)(y + 2).$$

The printed area inside the margins is given by $48 = xy$ or $y = 48/x$. To find the minimum area, rewrite the equation for A in terms of just one variable by substituting $48/x$ for y.

$$A = (x + 3)\left(\frac{48}{x} + 2\right) = \frac{(x + 3)(48 + 2x)}{x}, \qquad x > 0$$

The graph of this rational function is shown in Figure 2.49. Because x represents the height of the printed area, you need consider only the portion of the graph for which x is positive. Using the zoom and trace features of a graphing utility, you can approximate the minimum value of A to occur when $x \approx 8.5$ inches. The corresponding value of y is $48/8.5 \approx 5.6$ inches. Thus, the dimensions should be

$$x + 3 \approx 11.5 \text{ inches} \qquad \text{by} \qquad y + 2 \approx 7.6 \text{ inches.}$$

If you go on to take a course in calculus, you will learn a technique for finding the exact value of x that produces a minimum area. In this case, that value is $x = 6\sqrt{2} \approx 8.485$.   ◢

Group Activity   *Asymptotes of Graphs of Rational Functions*

Discuss whether or not it is possible for the graph of a rational function to cross its horizontal asymptote or its slant asymptote. Use the graphs of the following functions to investigate these questions. What can you conclude?

$$f(x) = \frac{x}{x^2 + 1} \qquad \text{and} \qquad g(x) = \frac{x^3}{x^2 + 1}$$

2.7 /// EXERCISES

In Exercises 1–4, use a graphing utility to graph $f(x) = 2/x$ and the function g in the same viewing rectangle. Describe the relationship between the two graphs.

1. $g(x) = f(x) + 1$ **2.** $g(x) = f(x - 1)$

3. $g(x) = -f(x)$ **4.** $g(x) = \frac{1}{2}f(x + 2)$

In Exercises 5–8, use a graphing utility to graph $f(x) = 2/x^2$ and the function g in the same viewing rectangle. Describe the relationship between the two graphs.

5. $g(x) = f(x) - 2$ **6.** $g(x) = -f(x)$

7. $g(x) = f(x - 2)$ **8.** $g(x) = \frac{1}{4}f(x)$

In Exercises 9–12, use a graphing utility to graph $f(x) = 4/x^3$ and the function g in the same viewing rectangle. Describe the relationship between the two graphs.

9. $g(x) = f(x + 2)$ **10.** $g(x) = f(x) + 1$

11. $g(x) = -f(x)$ **12.** $g(x) = \frac{1}{4}f(x)$

In Exercises 13–30, sketch the graph of the rational function. As sketching aids, check for intercepts, symmetry, vertical asymptotes, and horizontal asymptotes. Use a graphing utility to verify your graph.

13. $f(x) = \dfrac{1}{x + 2}$ **14.** $f(x) = \dfrac{1}{x - 3}$

15. $h(x) = \dfrac{-1}{x + 2}$ **16.** $g(x) = \dfrac{1}{3 - x}$

17. $C(x) = \dfrac{5 + 2x}{1 + x}$ **18.** $P(x) = \dfrac{1 - 3x}{1 - x}$

19. $g(x) = \dfrac{1}{x + 2} + 2$ **20.** $f(t) = \dfrac{1 - 2t}{t}$

21. $f(x) = \dfrac{x^2}{x^2 + 9}$ **22.** $f(x) = 2 - \dfrac{3}{x^2}$

23. $h(x) = \dfrac{x^2}{x^2 - 9}$ **24.** $g(x) = \dfrac{x}{x^2 - 9}$

25. $g(s) = \dfrac{s}{s^2 + 1}$ **26.** $f(x) = -\dfrac{1}{(x - 2)^2}$

27. $g(x) = \dfrac{4(x + 1)}{x(x - 4)}$ **28.** $h(x) = \dfrac{2}{x^2(x - 2)}$

29. $f(x) = \dfrac{3x}{x^2 - x - 2}$ **30.** $f(x) = \dfrac{2x}{x^2 + x - 2}$

In Exercises 31–40, use a graphing utility to obtain the graph of the function. Give the domain of the function and identify any vertical or horizontal asymptotes.

31. $f(x) = \dfrac{2 + x}{1 - x}$ **32.** $f(x) = \dfrac{3 - x}{2 - x}$

33. $f(t) = \dfrac{3t + 1}{t}$ **34.** $h(x) = \dfrac{1}{x - 3} + 1$

35. $h(t) = \dfrac{4}{t^2 + 1}$ **36.** $g(x) = -\dfrac{x}{(x - 2)^2}$

37. $f(t) = \dfrac{2t^2}{t^2 - 4}$ **38.** $f(x) = \dfrac{x + 4}{x^2 + x - 6}$

39. $f(x) = \dfrac{20x}{x^2 + 1} - \dfrac{1}{x}$

40. $f(x) = 5\left(\dfrac{1}{x - 4} - \dfrac{1}{x + 2}\right)$

Exploration In Exercises 41 and 42, use a graphing utility to obtain the graph of the function. What do you observe about its asymptotes?

41. $h(x) = \dfrac{6x}{\sqrt{x^2 + 1}}$ **42.** $g(x) = \dfrac{4|x - 2|}{x + 1}$

Exploration In Exercises 43 and 44, use a graphing utility to obtain the graph of the function. What do you observe about its asymptotes?

43. $f(x) = \dfrac{4(x - 1)^2}{x^2 - 4x + 5}$

44. $g(x) = \dfrac{3x^4 - 5x + 3}{x^4 + 1}$

Think About It   In Exercises 45 and 46, use a graphing utility to obtain the graph of the function. Explain why there is no vertical asymptote when a superficial examination of the function may indicate that there should be one.

45. $h(x) = \dfrac{6 - 2x}{3 - x}$ **46.** $g(x) = \dfrac{x^2 + x - 2}{x - 1}$

47. *True or False?*   If the graph of a rational function f has a vertical asymptote at $x = 5$, it is possible to sketch the graph without lifting your pencil from the paper. Explain.

48. *Essay*   Write a paragraph discussing whether every rational function has a vertical asymptote.

In Exercises 49–56, sketch the graph of the rational function. As sketching aids, check for intercepts, symmetry, vertical asymptotes, and slant asymptotes. Use a graphing utility to verify your graph.

49. $f(x) = \dfrac{2x^2 + 1}{x}$ **50.** $f(x) = \dfrac{1 - x^2}{x}$

51. $g(x) = \dfrac{x^2 + 1}{x}$ **52.** $h(x) = \dfrac{x^2}{x - 1}$

53. $f(x) = \dfrac{x^3}{x^2 - 1}$ **54.** $g(x) = \dfrac{x^3}{2x^2 - 8}$

55. $f(x) = \dfrac{x^2 - x + 1}{x - 1}$ **56.** $f(x) = \dfrac{2x^2 - 5x + 5}{x - 2}$

Graphical Reasoning   In Exercises 57–60, (a) use the graph to determine any x-intercepts of the rational function, and (b) set $y = 0$ and solve the resulting equation to confirm your result in part (a).

57. $y = \dfrac{x + 1}{x - 3}$ **58.** $y = \dfrac{2x}{x - 3}$

    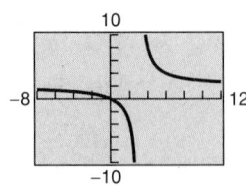

59. $y = \dfrac{1}{x} - x$ **60.** $y = x - 3 + \dfrac{2}{x}$

    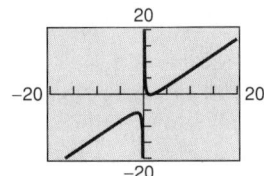

In Exercises 61–64, use a graphing utility to graph the rational function. Give the domain of the function and identify any asymptotes. Then zoom out far enough so that the graph appears as a line. Identify the line.

61. $y = \dfrac{2x^2 + x}{x + 1}$ **62.** $y = \dfrac{x^2 + 5x + 8}{x + 3}$

63. $y = \dfrac{1 + 3x^2 - x^3}{x^2}$ **64.** $y = \dfrac{12 - 2x - x^2}{2(4 + x)}$

Graphical Reasoning   In Exercises 65–68, (a) use a graphing utility to graph the function and determine any x-intercepts, and (b) set $y = 0$ and solve the resulting equation to confirm your result in part (a).

65. $y = \dfrac{1}{x + 5} + \dfrac{4}{x}$ **66.** $y = 20\left(\dfrac{2}{x + 1} - \dfrac{3}{x}\right)$

67. $y = x - \dfrac{6}{x - 1}$ **68.** $y = x - \dfrac{9}{x}$

69. *Concentration of a Mixture*   A 1000-liter tank contains 50 liters of a 25% brine solution. You add x liters of a 75% brine solution to the tank.

(a) Show that the concentration C of the final mixture is given by

$$C = \dfrac{3x + 50}{4(x + 50)}.$$

(b) Determine the domain of the function on the basis of the physical constraints of the problem.

(c) Use a graphing utility to graph the function. As the tank is filled, what happens to the rate at which the concentration of brine increases? What does the concentration of brine appear to approach?

70. *Dimensions of a Rectangle*   A rectangular region of length x and width y has an area of 500 square meters.

(a) Express the width y as a function of x.

(b) Determine the domain of the function on the basis of the physical constraints of the problem.

(c) Sketch a graph of the function and determine the width of the rectangle if $x = 30$ meters.

In Exercises 71 and 72, use a graphing utility to obtain the graph of the function and locate any relative maximum or minimum points on the graph.

71. $f(x) = \dfrac{3(x+1)}{x^2 + x + 1}$ **72.** $C(x) = x + \dfrac{32}{x}$

73. *Page Design*   A page that is x inches wide and y inches high contains 30 square inches of print. The margins at the top and bottom are 2 inches and the margins on each side are 1 inch (see figure).

(a) Show that the total area A of the page is given by

$$A = \dfrac{2x(2x + 11)}{x - 2}.$$

(b) Determine the domain of the function on the basis of the physical constraints of the problem.

(c) Use a graphing utility to graph the area function and approximate the page size such that the minimum amount of paper will be used.

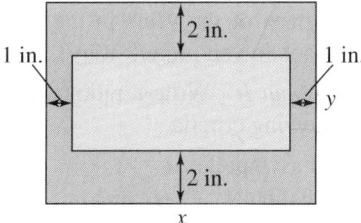

74. *Minimum Area*   A right triangle is formed in the first quadrant by the x-axis, the y-axis, and a line segment through the point $(3, 2)$ (see figure).

(a) Show that an equation of the line segment is

$$y = \dfrac{2(a - x)}{a - 3}, \quad 0 \le x \le a.$$

(b) Show that the area of the triangle is $A = \dfrac{a^2}{a - 3}$.

(c) Use a graphing utility to graph the area function and estimate the value of a that yields a minimum area. Estimate the minimum area.

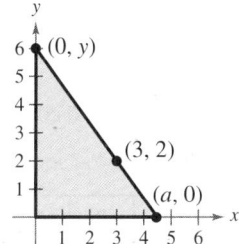

75. *Ordering and Transportation Cost*   The ordering and transportation cost C for the components used in manufacturing a certain product is given by

$$C = 100\left(\dfrac{200}{x^2} + \dfrac{x}{x + 30}\right), \quad 1 \le x$$

where C is measured in thousands of dollars and x is the order size in hundreds. Use a graphing utility to obtain the graph of the cost function and, from the graph, estimate the order size that minimizes cost.

76. *Average Cost*   The cost of producing x units of a product is $C = 0.2x^2 + 10x + 5$, and therefore the average cost per unit is

$$\overline{C} = \dfrac{C}{x} = \dfrac{0.2x^2 + 10x + 5}{x}, \quad 0 < x.$$

Sketch the graph of the average cost function, and estimate the number of units that should be produced to minimize the average cost per unit.

77. *Medicine*   The concentration of a certain chemical in the bloodstream t hours after injection into muscle tissue is given by

$$C = \dfrac{3t^2 + t}{t^3 + 50}, \quad 0 \le t.$$

(a) Determine the horizontal asymptote of the function and interpret its meaning in the context of the problem.

(b) Use a graphing utility to graph the function and approximate the time when the bloodstream concentration is greatest.

2 /// REVIEW EXERCISES

In Exercises 1–4, use a graphing utility to graph the quadratic function. Identify the vertex and the intercepts.

1. $f(x) = \left(x + \frac{3}{2}\right)^2 + 1$ **2.** $f(x) = (x - 4)^2 - 4$

3. $f(x) = \frac{1}{3}(x^2 + 5x - 4)$

4. $f(x) = 3x^2 - 12x + 11$

In Exercises 5 and 6, find the quadratic function that has the indicated vertex and whose graph passes through the given point.

5. Vertex: $(1, -4)$; Point: $(2, -3)$

6. Vertex: $(2, 3)$; Point: $(-1, 6)$

Graphical Reasoning In Exercises 7 and 8, use a graphing utility to graph each equation in the same viewing rectangle. Describe how each graph differs from the graph of $y = x^2$.

7. (a) $y = 2x^2$ (b) $y = -2x^2$

(c) $y = x^2 + 2$ (d) $y = (x + 2)^2$

8. (a) $y = x^2 - 4$ (b) $y = 4 - x^2$

(c) $y = (x - 3)^2$ (d) $y = \frac{1}{2}x^2 - 1$

In Exercises 9–16, find the maximum or minimum value of the quadratic function.

9. $g(x) = x^2 - 2x$ **10.** $f(x) = x^2 + 8x + 10$

11. $f(x) = 6x - x^2$ **12.** $h(x) = 3 + 4x - x^2$

13. $f(t) = -2t^2 + 4t + 1$

14. $h(x) = 4x^2 + 4x + 13$

15. $h(x) = x^2 + 5x - 4$ **16.** $f(x) = 4x^2 + 4x + 5$

17. *Numerical, Graphical, and Analytical Analysis* A rectangle is inscribed in the region bounded by the x-axis, the y-axis, and the graph of $x + 2y - 8 = 0$ (see figure).

(a) Complete six rows of a table like the one below. (The first two rows are shown.)

x	y	Area (xy)
1	$4 - \frac{1}{2}(1)$	$(1)\left[4 - \frac{1}{2}(1)\right] = \frac{7}{2}$
2	$4 - \frac{1}{2}(2)$	$(2)\left[4 - \frac{1}{2}(2)\right] = 6$

(b) Use a graphing utility to generate additional rows of the table in part (a). Use the table to estimate the dimensions that will produce the maximum area.

(c) Write the area A as a function of x. Determine the domain of the function in the context of the problem.

(d) Use a graphing utility to graph the area function. Use the graph to approximate the dimensions that will produce the maximum area.

(e) Write the function in standard form to find analytically the dimensions that will produce the maximum area.

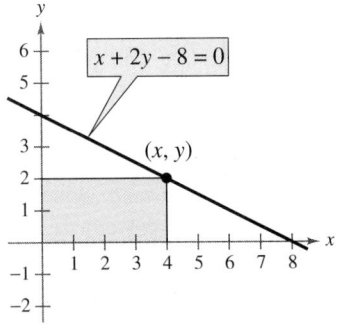

18. *Maximum Profit* Let x be the amount (in hundreds of dollars) a company spends on advertising, and let P be the profit, where

$$P = 230 + 20x - \frac{1}{2}x^2.$$

How much advertising will yield a maximum profit?

In Exercises 19–22, determine the right-hand and left-hand behavior of the graph of the polynomial function.

19. $f(x) = -x^2 + 6x + 9$ **20.** $f(x) = \frac{1}{2}x^3 + 2x$

21. $g(x) = \frac{3}{4}(x^4 + 3x^2 + 2)$

22. $h(x) = -x^5 - 7x^2 + 10x$

Graphical Analysis In Exercises 23 and 24, use a graphing utility to graph the functions f and g in the same viewing rectangle. Zoom out sufficiently far to show that the right-hand and left-hand behavior of f and g appear identical.

23. $f(x) = \frac{1}{2}x^3 - 2x + 1$, $g(x) = \frac{1}{2}x^3$

24. $f(x) = -x^4 + 2x^3$, $g(x) = -x^4$

In Exercises 25–30, use a graphing utility to graph the function.

25. $g(x) = x^4 - x^3 - 2x^2$

26. $h(x) = -2x^3 - x^2 + x$

27. $f(t) = t^3 - 3t$ **28.** $f(x) = -x^3 + 3x - 2$

29. $f(x) = x(x + 3)^2$ **30.** $f(t) = t^4 - 4t^2$

Graphical Analysis In Exercises 31 and 32, use a graphing utility to graph the two equations on the same screen. Use the graphs to verify that the expressions are equivalent. Verify the results analytically.

31. $y_1 = \dfrac{x^2}{x - 2}$, $y_2 = x + 2 + \dfrac{4}{x - 2}$

32. $y_1 = \dfrac{x^4 + 1}{x^2 + 2}$, $y_2 = x^2 - 2 + \dfrac{5}{x^2 + 2}$

In Exercises 33–36, perform the division.

33. $\dfrac{24x^2 - x - 8}{3x - 2}$ **34.** $\dfrac{4x + 7}{3x - 2}$

35. $\dfrac{x^4 - 3x^2 + 2}{x^2 - 1}$ **36.** $\dfrac{3x^4}{x^2 - 1}$

In Exercises 37–40, use synthetic division to perform the division.

37. $\dfrac{0.25x^4 - 4x^3}{x - 2}$ **38.** $\dfrac{2x^3 + 2x^2 - x + 2}{x - (\frac{1}{2})}$

39. $(6x^4 - 4x^3 - 27x^2 + 18x) \div (x - \frac{2}{3})$

40. $(0.1x^3 + 0.3x^2 - 0.5) \div (x - 5)$

In Exercises 41–46, perform the operations and write the result in standard form.

41. $(7 + 5i) + (-4 + 2i)$

42. $\left(\dfrac{\sqrt{2}}{2} - \dfrac{\sqrt{2}}{2}i\right) - \left(\dfrac{\sqrt{2}}{2} + \dfrac{\sqrt{2}}{2}i\right)$

43. $5i(13 - 8i)$

44. $i(6 + i)(3 - 2i)$

45. $\dfrac{6 + i}{i}$

46. $\dfrac{3 + 2i}{5 + i}$

In Exercises 47 and 48, find a polynomial with integer coefficients that has the given zeros.

47. $-1, -1, \frac{1}{3}, -\frac{1}{2}$ **48.** $2, -3, 1 - 2i, 1 + 2i$

In Exercises 49–54, find all the zeros of the function.

49. $f(x) = 4x^3 - 11x^2 + 10x - 3$

50. $f(x) = 10x^3 + 21x^2 - x - 6$

51. $f(x) = 6x^3 - 5x^2 + 24x - 20$

52. $f(x) = x^3 - 1.3x^2 - 1.7x + 0.6$

53. $f(x) = 6x^4 - 25x^3 + 14x^2 + 27x - 18$

54. $f(x) = 5x^4 + 126x^2 + 25$

In Exercises 55–58, use a graphing utility to (a) graph the function, (b) determine the number of real zeros of the function, and (c) approximate the real zeros of the function to the nearest hundredth.

55. $f(x) = x^4 + 2x + 1$

56. $g(x) = x^3 - 3x^2 + 3x + 2$

57. $h(x) = x^3 - 6x^2 + 12x - 10$

58. $f(x) = x^5 + 2x^3 - 3x - 20$

59. *Data Analysis*   Sales in billions of dollars of recreational vehicles in the United States for the years 1982 through 1993 are given in the table.   (Source: National Sporting Goods Association)

Year	1982	1983	1984	1985	1986	1987
Sales	1.7	3.4	4.1	3.5	3.9	4.5

Year	1988	1989	1990	1991	1992	1993
Sales	4.8	4.5	4.1	3.6	4.4	4.8

A model for the data is given by

$$S = -0.8195 + 1.6762t - 0.1347t^2 - 0.0028t^3 + 0.0004t^4$$

where S is the sales in billions of dollars and t is the time in years, with $t = 0$ corresponding to 1980.

(a) Use a graphing utility to plot the data points and graph the model.

(b) The table shows that sales were down from 1989 through 1992. Give a possible explanation. Does the model show the downturn in sales?

(c) Use a graphing utility to approximate the magnitude of the decrease in sales during the slump described in part (b). Was the actual decrease more or less than indicated by the model?

(d) Use the model to estimate sales in 1995.

60. *Age of the Groom*   The average age of the groom in a marriage for a given age of the bride can be approximated by the model

$$y = -0.00428x^2 + 1.442x - 3.136, \quad 20 \le x \le 55$$

where y is the age of the groom and x is the age of the bride. For what age of the bride is the average age of the groom 30?   (Source: U.S. National Center for Health Statistics)

In Exercises 61–68, sketch the graph of the rational function. As sketching aids, check for intercepts, symmetry, vertical asymptotes, horizontal asymptotes, and slant asymptotes.

61. $f(x) = \dfrac{-5}{x^2}$

62. $h(x) = \dfrac{x-3}{x-2}$

63. $P(x) = \dfrac{x^2}{x^2+1}$

64. $f(x) = \dfrac{2x}{x^2+4}$

65. $f(x) = \dfrac{x}{x^2+1}$

66. $h(x) = \dfrac{4}{(x-1)^2}$

67. $f(x) = \dfrac{2x^3}{x^2+1}$

68. $y = \dfrac{2x^2}{x^2-4}$

In Exercises 69–72, use a graphing utility to graph the function. Identify any vertical, horizontal, or slant asymptotes.

69. $s(x) = \dfrac{8x^2}{x^2+4}$

70. $y = \dfrac{5x}{x^2-4}$

71. $g(x) = \dfrac{x^2+1}{x+1}$

72. $y = \dfrac{1}{x+3} + 2$

Think About It   In Exercises 73 and 74, write a rational function f having the specified characteristics.

73. Vertical asymptotes: $x = -3, x = 4$
Horizontal asymptote: $y = 2$

74. Vertical asymptote: $x = 5$
Slant asymptote: $y = 2x$

75. *Numerical and Graphical Analysis*   A right triangle is formed in the first quadrant by the x- and y-axes and a line through the point $(2, 3)$.

(a) Draw a figure that illustrates the problem. Label the known and unknown quantities.

(b) Verify that the area of the triangle is given by

$$A = \dfrac{3x^2}{2(x-2)}, \quad x > 2.$$

(c) Use a graphing utility to generate a table giving the areas for several values of x. Starting the table with $x = 2.5$, increment x in steps of 0.5. Continue until you can approximate the dimensions of the triangle of minimum area.

(d) Use a graphing utility to graph the area function. Use the graph to approximate the dimensions of the triangle of minimum area.

(e) Determine the slant asymptote of the area function. Explain its meaning.

CHAPTER PROJECT   *Graphs of Polynomial and Rational Functions*

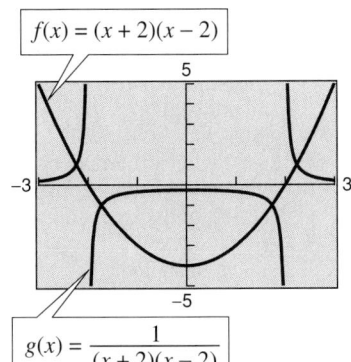

$$f(x) = (x + 2)(x - 2)$$

$$g(x) = \frac{1}{(x + 2)(x - 2)}$$

In this project, you will compare the graph of a polynomial function with the graph of its reciprocal function. Consider the polynomial function

$$f(x) = x^2 - 4 = (x + 2)(x - 2).$$

The reciprocal of this function is the rational function

$$g(x) = \frac{1}{f(x)} = \frac{1}{(x + 2)(x - 2)}.$$

The graphs of f and g are shown at the left.

(a) How are the zeros of f related to the vertical asymptotes of g?

(b) What is the minimum point of the graph of f? How is this point related to the graph of g? In general, how are the relative minimums and relative maximums of a polynomial function f related to the relative minimums and relative maximums of the reciprocal function $g(x) = 1/f(x)$?

(c) The reciprocal function g has a horizontal asymptote at $y = 0$. What does this tell you about the behavior of f for very large and very small values of x?

(d) The graph of a polynomial function P is shown at the left. Use the graph of P to sketch the graph of its reciprocal function $q(x) = 1/P(x)$.

Questions for Further Exploration

1. The polynomial function

$$f(x) = x^2 + 9$$

has no real zeros. What does this indicate about the graph of the reciprocal function $g(x) = 1/f(x)$?

2. Use a graphing utility to graph the function

$$f(x) = (x - 1)(x - 2)(x - 3)$$

and its reciprocal function $g(x) = 1/f(x)$ in the same viewing rectangle. Compare the features of the graphs of f and g.

3. You are given the graphs of two functions at the right. Each function is of the form $g(x) = 1/f(x)$, where $f(x)$ is a polynomial. Use each graph to sketch the graph of the corresponding polynomial function f.

Figure for 3

(a)

(b)

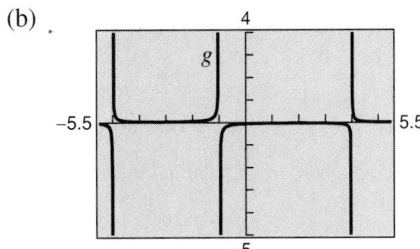

2 /// CHAPTER TEST

Take this test as you would take a test in class. After you are done, check your work against the answers given in the back of the book.

The *Interactive* CD-ROM provides answers to the Chapter Tests and Cumulative Tests. It also offers Chapter Pre-Tests (that test key skills and concepts covered in previous chapters) and Chapter Post-Tests, both of which have randomly generated exercises with diagnostic capabilities.

1. Describe how the graph of g differs from the graph of $f(x) = x^2$.

   (a) $g(x) = 2 - x^2$ (b) $g(x) = \left(x - \frac{3}{2}\right)^2$

2. Identify the vertex and intercepts of the graph of $y = x^2 + 4x + 3$. Sketch a graph of the function.

3. Find an equation of the parabola shown in the figure.

4. The path of a ball is given by $y = -\frac{1}{20}x^2 + 3x + 5$, where y is the height in feet and x is the horizontal distance in feet.

   (a) Find the maximum height of the ball.

   (b) Which constant determines the height at which the ball was thrown? Does changing this constant change the coordinates of the maximum height of the ball? Explain.

5. Divide by long division: $(3x^3 + 4x - 1) \div (x^2 + 1)$

6. Multiply and write the result in standard form: $(10 - 8i)(2 - 3i)$

Figure for 3

(graph showing point $(0, 3)$ and vertex $(3, -6)$)

In Exercises 7 and 8, list all the possible rational zeros of the function. Use a graphing utility to graph the function and find all the rational zeros.

7. $g(t) = 2t^4 - 3t^3 + 16t - 24$ 8. $h(x) = 3x^5 + 2x^4 - 3x - 2$

   den

In Exercises 9 and 10, use the root-finding capabilities of a graphing utility to approximate the real zeros of the function accurate to three decimal places.

9. $f(x) = x^4 - x^3 - 1$ 10. $f(x) = 3x^5 + 2x^4 - 12x - 8$

In Exercises 11 and 12, find a polynomial function with integer coefficients that has the given zeros.

11. $0, 3, 3 + i, 3 - i$ 12. $1 + \sqrt{3}i, 1 - \sqrt{3}i, 2, 2$

In Exercises 13 and 14, use a graphing utility to graph the function.

13. $h(x) = \dfrac{4}{x^2} - 1$ 14. $g(x) = \dfrac{x^2 + 2}{x - 1}$

15. Find a rational function with vertical asymptotes at $x = \pm 3$ and a horizontal asymptote at $y = 4$.

Chapter 3

Exponential and Logarithmic Functions

Automobiles are designed with crumple zones that allow the occupants to move short distances when the automobiles come to abrupt stops. The greater the distance moved, the fewer g's the crash victims experience. (One g is equal to the acceleration due to gravity.) In crash tests with vehicles moving at 90 kilometers per hour, analysts measured the numbers y of g's that were undergone during deceleration by crash dummies that were permitted to move distances of x meters during impact.

x	0.2	0.4	0.6	0.8	1.0
y	158	80	53	40	32

Number of g's

Distance moved (in meters)

You can use a graphing utility to draw a scatter plot of the data and fit an appropriate model to the data. Using natural logarithmic regression capabilities, you can find one model to be $y = 21.37 - 78.58 \ln x$. The graph shows the data points along with the graph of the model. (See Exercise 94 on page 280.)

Brad Trent.

At a General Motors lab, engineer Bonnie Cheung and physicist Stephen Rouhana prepare a dummy for a simulated automobile crash. The laser helps position the dummy.

3.1   Exponential Functions and Their Graphs

Exponential Functions / Graphs of Exponential Functions /
The Natural Base e / Compound Interest / Other Applications

Exponential Functions

So far, this text has dealt mainly with **algebraic functions,** which include polynomial functions and rational functions. In this chapter you will study two types of nonalgebraic functions—*exponential* functions and *logarithmic* functions. These functions are examples of **transcendental functions.**

Note The base $a = 1$ is excluded because it yields

$$f(x) = 1^x = 1.$$

This is a constant function, not an exponential function.

Definition of Exponential Function
The **exponential function** f with base a is denoted by $$f(x) = a^x$$ where $a > 0$, $a \neq 1$, and x is any real number.

You already know how to evaluate a^x for integer and rational values of x. For example, you know that $4^3 = 64$ and $4^{1/2} = 2$. However, to evaluate 4^x for any real number x, you need to interpret forms with *irrational* exponents. For the purposes of this text, it is sufficient to think of

$$a^{\sqrt{2}} \quad (\text{where } \sqrt{2} \approx 1.414214)$$

Library of Functions

The exponential function $f(x) = a^x$ is different from all the functions you have studied so far because the variable x is an *exponent*. The domain, like those of polynomial functions, is the set of all real numbers.

as that value having the successively closer approximations

$$a^{1.4}, a^{1.41}, a^{1.414}, a^{1.4142}, a^{1.41421}, a^{1.414214}, \ldots .$$

Example 1 shows how to use a calculator to evaluate an exponential expression.

EXAMPLE 1   **Evaluating Exponential Expressions**

Use a calculator to evaluate each expression.

a. $2^{-3.1}$ **b.** $2^{-\pi}$

Solution

Number	Graphing Calculator Keystrokes	Display
a. $2^{-3.1}$	2 $\wedge$ (-) 3.1 ENTER	0.1166291
b. $2^{-\pi}$	2 $\wedge$ (-) π ENTER	0.1133147

Graphs of Exponential Functions

The graphs of all exponential functions have similar characteristics, as shown in Examples 2, 3, and 4.

Figure 3.1

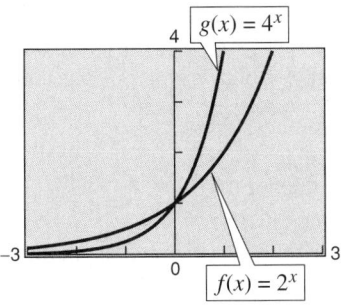

EXAMPLE 2   **Graphs of $y = a^x$**

In the same coordinate plane, sketch the graph of each function.

a. $f(x) = 2^x$

b. $g(x) = 4^x$

Solution

The table below lists some values for each function, and Figure 3.1 shows their graphs. Note that both graphs are increasing. Moreover, the graph of $g(x) = 4^x$ is increasing more rapidly than the graph of $f(x) = 2^x$.

x	-2	-1	0	1	2	3
2^x	$\frac{1}{4}$	$\frac{1}{2}$	1	2	4	8
4^x	$\frac{1}{16}$	$\frac{1}{4}$	1	4	16	64

Figure 3.2

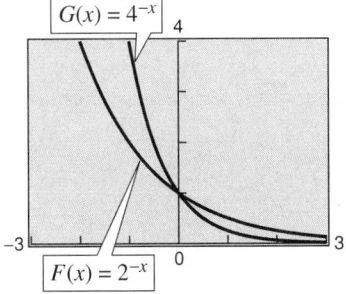

EXAMPLE 3   **Graphs of $y = a^{-x}$**

In the same coordinate plane, sketch the graph of each function.

a. $F(x) = 2^{-x}$

b. $G(x) = 4^{-x}$

Solution

The table below lists some values for each function, and Figure 3.2 shows their graphs. Note that both graphs are decreasing. Moreover, the graph of $G(x) = 4^{-x}$ is decreasing more rapidly than the graph of $F(x) = 2^{-x}$.

x	-3	-2	-1	0	1	2
2^{-x}	8	4	2	1	$\frac{1}{2}$	$\frac{1}{4}$
4^{-x}	64	16	4	1	$\frac{1}{4}$	$\frac{1}{16}$

Note The tables in Examples 2 and 3 were evaluated by hand. You could, of course, use a graphing utility to construct tables with even more values.

> **Study Tip**
>
> Be sure you see that the functions
> $y_1 = \left(\dfrac{1}{2}\right)^x$ and $y_2 = 2^{-x}$ are the
> same. In general, $\left(\dfrac{1}{a}\right)^x = a^{-x}$.

Comparing the functions in Examples 2 and 3, observe that

$$F(x) = 2^{-x} = f(-x) \qquad \text{and} \qquad G(x) = 4^{-x} = g(-x).$$

Consequently, the graph of F is a reflection (in the y-axis) of the graph of f. The graph of G and g have the same relationship. The graphs in Figures 3.1 and 3.2 are typical of the exponential functions a^x and a^{-x}. They have one y-intercept and one horizontal asymptote (the x-axis), and they are continuous. The basic characteristics of these exponential functions are summarized in Figure 3.3.

Graph of $y = a^x$

- Domain: $(-\infty, \infty)$
- Range: $(0, \infty)$
- Intercept: $(0, 1)$
- Increasing
- x-axis is a horizontal asymptote
  $(a^x \to 0$ as $x \to -\infty)$
- Continuous

Graph of $y = a^{-x}$

- Domain: $(-\infty, \infty)$
- Range: $(0, \infty)$
- Intercept: $(0, 1)$
- Decreasing
- x-axis is a horizontal asymptote
  $(a^{-x} \to 0$ as $x \to \infty)$
- Continuous

Figure 3.3

EXPLORATION

Use a graphing utility to graph $y = a^x$ with $a = 3, 5,$ and 7 in the same viewing rectangle. (Use a viewing rectangle in which $-2 \le x \le 1$ and $0 \le y \le 2$.) For instance, the graph of $y = 3^x$ is shown at the left. How do the graphs compare with each other? Which graph is on the top in the interval $(-\infty, 0)$? Which is on the bottom? Which graph is on the top in the interval $(0, \infty)$? Which is on the bottom? Repeat this experiment with the graphs of $y = a^x$ for $a = \frac{1}{3}, \frac{1}{5},$ and $\frac{1}{7}$. (Use a viewing rectangle in which $-1 \le x \le 2$ and $0 \le y \le 2$.) What can you conclude about the relationship between the function's behavior and the value of a?

In the following example, notice how the graph of $y = a^x$ can be used to sketch the graphs of functions of the form

$$f(x) = b \pm a^{x+c}.$$

EXAMPLE 4 Graphs of Exponential Functions

Each of the following graphs is a transformation of the graph of $f(x) = 3^x$, as shown in Figure 3.4.

Note In Figure 3.4, notice that the transformations in parts (a), (c), and (d) keep the *x*-axis as a horizontal asymptote, but the transformation in part (b) yields a new horizontal asymptote of $y = -2$. Also, be sure to note how the *y*-intercept is affected by each transformation.

a. Because $g(x) = 3^{x+1} = f(x + 1)$, the graph of g can be obtained by shifting the graph of f one unit to the left.

b. Because $h(x) = 3^x - 2 = f(x) - 2$, the graph of h can be obtained by shifting the graph of f down two units.

c. Because $k(x) = -3^x = -f(x)$, the graph of k can be obtained by reflecting the graph of f in the *x*-axis.

d. Because $j(x) = 3^{-x} = f(-x)$, the graph of j can be obtained by reflecting the graph of f in the *y*-axis.

Figure 3.4

(a)

(b)

(c)

(d)
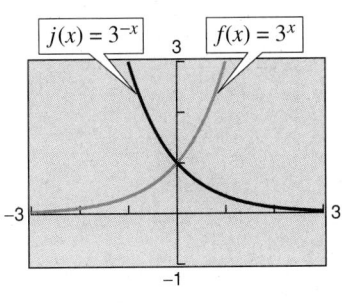

The following table generated on a *TI-82* or a *TI-83* shows some points of the graphs in Figure 3.4(a). The functions $f(x)$ and $g(x)$ are represented by Y1 and Y2, respectively. Explain how you can use the table to describe the transformation.

X	Y1	Y2
-3	0.11111	0.03704
-2	0.33333	0.11111
-1	1	0.33333
0	3	1
1	9	3
2	27	9
3	81	27
X = -3		

EXPLORATION

Use your graphing utility to graph the functions

$$y_1 = 2^x$$
$$y_2 = e^x$$
$$y_3 = 3^x$$

in the same viewing rectangle. From the relative positions of these graphs, make a guess as to the value of the real number e. Then try to find a number a such that the graphs of $y_2 = e^x$ and $y_4 = a^x$ are as close as possible.

The Natural Base e

For many applications, the convenient choice for a base is the irrational number

$$e \approx 2.71828 \ldots .$$

This number is called the **natural base.** The function $f(x) = e^x$ is the **natural exponential function.** Its graph is shown in Figure 3.5. Be sure you see that for the exponential function $f(x) = e^x$, e is the constant $2.71828 \ldots$, whereas x is the variable.

Figure 3.5

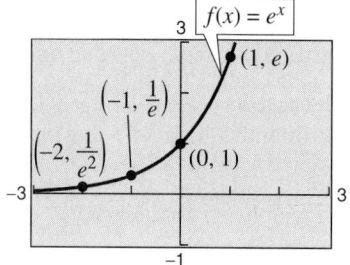

EXAMPLE 5   **Approximation of the Number e**

Evaluate the expression $[1 + (1/x)]^x$ for several large values of x to see that the values approach $e \approx 2.71828$ as x increases without bound.

Solution

Using a calculator, you can complete a table like that shown below. From this table, it seems reasonable to conclude that

$$\left(1 + \frac{1}{x}\right)^x \to e \quad \text{as} \quad x \to \infty.$$

You can further confirm this conclusion by using a graphing utility to graph

$$f(x) = \left(1 + \frac{1}{x}\right)^x$$

and $y = e$ in the same viewing rectangle, as shown in Figure 3.6. Note that as x increases, the graph of f gets closer and closer to the line given by $y = e$.

Figure 3.6

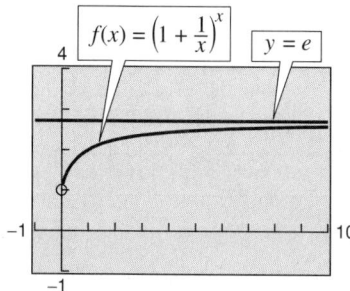

x	10	100	1000	10,000	100,000	1,000,000
$\left(1 + \dfrac{1}{x}\right)^x$	2.59374	2.70481	2.71692	2.71815	2.71827	2.71828

EXPLORATION

Use a graphing utility to graph $y = (1 + x)^{1/x}$. What is the domain of this function? Describe the behavior of the graph near $x = 0$. Is there a y-intercept? How does the behavior of the graph near $x = 0$ relate to the result of Example 5?

The *Interactive* CD-ROM shows every example with its solution; clicking on the *Try It!* button brings up similar problems. Guided Examples and Integrated Examples show step-by-step solutions to additional examples. Integrated Examples are related to several concepts in the section.

EXAMPLE 6   **Evaluating the Natural Exponential Function**

Use a calculator to evaluate each expression.

a. e^{-2} **b.** e^{-1} **c.** e^{1} **d.** e^{2}

Solution

Number	Graphing Calculator Keystrokes	Display
a. e^{-2}	$\boxed{e^x}$ $\boxed{(\text{-})}$ 2 $\boxed{\text{ENTER}}$	0.1353353
b. e^{-1}	$\boxed{e^x}$ $\boxed{(\text{-})}$ 1 $\boxed{\text{ENTER}}$	0.3678794
c. e^{1}	$\boxed{e^x}$ 1 $\boxed{\text{ENTER}}$	2.7182818
d. e^{2}	$\boxed{e^x}$ 2 $\boxed{\text{ENTER}}$	7.3890561

EXAMPLE 7   **Graphing Natural Exponential Functions**

Sketch the graph of each natural exponential function.

a. $f(x) = 2e^{0.24x}$

b. $g(x) = \frac{1}{2}e^{-0.58x}$

Solution

To sketch these two graphs, you can use a calculator to construct a table of values, as shown at the left. After constructing the table, plot the points and connect them with smooth curves, as shown in Figure 3.7. Note that the graph in part (a) is increasing, whereas the graph in part (b) is decreasing. Use a graphing calculator to verify these graphs.

Figure 3.7

x	$f(x)$	$g(x)$
-3	0.974	2.849
-2	1.238	1.595
-1	1.573	0.893
0	2.000	0.500
1	2.543	0.280
2	3.232	0.157
3	4.109	0.088

(a)

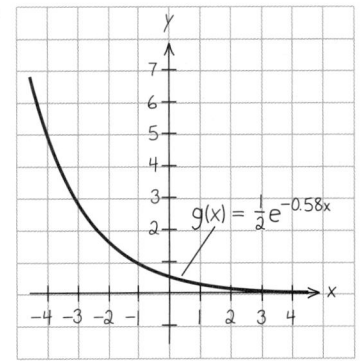
(b)

Compound Interest

One of the most familiar examples of exponential growth is that of an investment earning *continuously compounded interest.* Suppose a principal P is invested at an annual interest rate r, compounded once a year. If the interest is added to the principal at the end of the year, the balance is

$$P_1 = P + Pr$$
$$= P(1 + r).$$

This pattern of multiplying the previous principal by $1 + r$ is then repeated each successive year, as shown in the table.

Time in Years	Balance After Each Compounding
0	$P = P$
1	$P_1 = P(1 + r)$
2	$P_2 = P_1(1 + r) = P(1 + r)(1 + r) = P(1 + r)^2$
3	$P_3 = P_2(1 + r) = P(1 + r)^2(1 + r) = P(1 + r)^3$
$\vdots$	$\vdots$
n	$P_n = P(1 + r)^n$

Study Tip

The interest rate r in the formula for compound interest should be written as a decimal. For example, an interest rate of 7% would be written $r = 0.07$.

To accommodate more frequent (quarterly, monthly, or daily) compounding of interest, let n be the number of compoundings per year and let t be the number of years. Then the rate per compounding is r/n, and the account balance after t years is

$$A = P\left(1 + \frac{r}{n}\right)^{nt}. \qquad \text{Amount with } n \text{ compoundings per year}$$

If you let the number of compoundings n increase without bound, you approach **continuous compounding.** In the formula for n compoundings per year, let $m = n/r$. This produces

$$A = P\left(1 + \frac{r}{n}\right)^{nt} = P\left(1 + \frac{1}{m}\right)^{mrt} = P\left[\left(1 + \frac{1}{m}\right)^m\right]^{rt}.$$

As m increases without bound, you know from Example 5 that $[1 + (1/m)]^m$ approaches e. Hence, for continuous compounding, it follows that

$$P\left[\left(1 + \frac{1}{m}\right)^m\right]^{rt} \rightarrow P[e]^{rt}$$

and you can write $A = Pe^{rt}$. This result is part of the reason that e is the "natural" choice for a base of an exponential function.

Formulas for Compound Interest

After t years, the balance A in an account with principal P and annual interest rate r (expressed as a decimal) is given by the following formulas.

1. For n compoundings per year: $A = P\left(1 + \dfrac{r}{n}\right)^{nt}$

2. For continuous compounding: $A = Pe^{rt}$

Real Life

EXAMPLE 8 ◢ **Finding the Balance for Compound Interest**

A sum of $9000 is invested at an annual interest rate of 8.5%, compounded annually. Find the balance in the account after 3 years.

Solution

In this case, $P = 9000$, $r = 8.5\% = 0.085$, $n = 1$, and $t = 3$. Using the formula

$$A = P\left(1 + \frac{r}{n}\right)^{nt},$$

you have

$$A = 9000(1 + 0.085)^3 = 9000(1.085)^3 \approx \$11{,}495.60.$$

The graph of $f(t) = 9000(1 + 0.085)^t$ is shown in Figure 3.8. ◢

Figure 3.8

$f(t) = 9{,}000(1 + 0.085)^t$

Account balance (in dollars)

20,000
16,000
12,000
8,000
4,000

1 2 3 4 5 6 7 8 9 10

Time (in years)

Real Life

EXAMPLE 9 ◢ **Finding Compound Interest**

A total of $12,000 is invested at an annual interest rate of 9%. Find the balance after 5 years if it is compounded

a. quarterly.   **b.** continuously.

Solution

a. For quarterly compoundings, you have $n = 4$. Thus, in 5 years at 9%, the balance is

$$A = P\left(1 + \frac{r}{n}\right)^{nt} = 12{,}000\left(1 + \frac{0.09}{4}\right)^{4(5)} = \$18{,}726.11.$$

b. For continuous compounding, the balance is

$$A = Pe^{rt} = 12{,}000e^{0.09(5)} = \$18{,}819.75.$$

Note that continuous compounding yields $93.64 more than quarterly compounding. ◢

Other Applications

Figure 3.9

$$y = 10\left(\frac{1}{2}\right)^{t/25}$$

(0, 10)

(25, 5)

(80, 1.088)

Mass (in grams)

Time (in years)

──────── *Real Life* ────────

EXAMPLE 10   ▭   **Radioactive Decay**

Let y represent the mass of a quantity of a radioactive element whose half-life is 25 years. After t years, the mass (in grams) is given by

$$y = 10\left(\frac{1}{2}\right)^{t/25}.$$

a. What is the initial mass (when $t = 0$)?

b. How much of the initial mass is present after 80 years?

Solution

a. When $t = 0$, the mass is

$$y = 10\left(\frac{1}{2}\right)^0 = 10(1) = 10 \text{ grams.}$$

b. When $t = 80$, the mass is

$$y = 10\left(\frac{1}{2}\right)^{80/25} = 10(0.5)^{3.2} \approx 1.088 \text{ grams.}$$

The graph of this function is shown in Figure 3.9.   ▭

──────── *Real Life* ────────

EXAMPLE 11   ▭   **Population Growth**

The approximate number of fruit flies in an experimental population after t hours is given by

$$Q(t) = 20e^{0.03t}, \qquad t \geq 0.$$

a. Find the initial number of fruit flies in the population.

b. How large is the population of fruit flies after 72 hours?

c. Sketch the graph of Q.

Figure 3.10

$$Q(t) = 20e^{0.03t}$$

Number of fruit flies

Time (in hours)

Solution

a. To find the initial population, evaluate $Q(t)$ at $t = 0$.

$$Q(0) = 20e^{0.03(0)} = 20e^0 = 20(1) = 20 \text{ flies}$$

b. After 72 hours, the population size is

$$Q(72) = 20e^{0.03(72)} = 20e^{2.16} \approx 173 \text{ flies.}$$

c. The graph of Q is shown in Figure 3.10.   ▭

Group Activity *Exponential Growth*

Consider the following sequences of numbers.

Sequence 1: 2, 4, 6, 8, 10, 12, . . . , $2n$

Sequence 2: 2, 4, 8, 16, 32, 64, . . . , 2^n

The first sequence, $f(n) = 2n$, is an example of **linear growth.** The second sequence, $f(n) = 2^n$, is an example of **exponential growth.** Which of the following sequences represents linear growth and which represents exponential growth? Can you find a linear function and an exponential function that represent the two sequences? (*Hint:* Look for a *shift*, of one or more units, from a linear or exponential growth pattern.)

a. 2, 5, 8, 11, 14, . . .   **b.** 4, 10, 28, 82, 244, . . .

Later you will see that sequences that represent linear growth are *arithmetic sequences* and sequences that represent exponential growth are *geometric sequences.*

 The *Interactive* CD-ROM contains step-by-step solutions to all odd-numbered Section and Review Exercises. It also provides Tutorial Exercises, which link to Guided Examples for additional help.

3.1 /// EXERCISES

In Exercises 1–10, use a calculator to evaluate the expression. Round your result to three decimal places.

1. $(3.4)^{5.6}$

2. $5000(2^{-1.5})$

3. $(1.005)^{400}$

4. $8^{2\pi}$

5. $5^{-\pi}$

6. $\sqrt[3]{4395}$

7. $100^{\sqrt{2}}$

8. $e^{1/2}$

9. $e^{-3/4}$

10. $e^{3.2}$

Think About It   In Exercises 11–14, use properties of exponents to determine which functions (if any) are the same.

11. $f(x) = 3^{x-2}$
   $g(x) = 3^x - 9$
   $h(x) = \frac{1}{9}(3x)$

12. $f(x) = 4^x + 12$
   $g(x) = 2^{2x+6}$
   $h(x) = 64(4^x)$

13. $f(x) = 16(4^{-x})$
   $g(x) = \left(\frac{1}{4}\right)^{x-2}$
   $h(x) = 16(2^{-2x})$

14. $f(x) = 5^{-x} + 3$
   $g(x) = 5^{3-x}$
   $h(x) = -5^{x-3}$

In Exercises 15–22, graph the exponential function *by hand*. Identify the following features of the graph.

(a) Asymptotes
(b) Intercepts
(c) Increasing or decreasing

15. $g(x) = 5^x$

16. $f(x) = \left(\frac{3}{2}\right)^x$

17. $f(x) = \left(\frac{1}{5}\right)^x = 5^{-x}$

18. $h(x) = \left(\frac{3}{2}\right)^{-x}$

19. $h(x) = 5^{x-2}$

20. $g(x) = \left(\frac{3}{2}\right)^{x+2}$

21. $g(x) = 5^{-x} - 3$

22. $f(x) = \left(\frac{3}{2}\right)^{-x} + 2$

In Exercises 23–30, match the exponential function with its graph. [The graphs are labeled (a), (b), (c), (d), (e), (f), (g), and (h).]

(a)

(b)

(c)

(d)

(e)

(f)

(g)

(h)

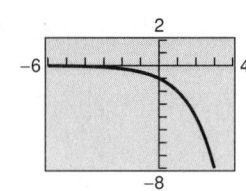

23. $f(x) = 2^x$

24. $f(x) = -2^x$

25. $f(x) = 2^{-x}$

26. $f(x) = -2^{-x}$

27. $f(x) = 2^x - 4$

28. $f(x) = 2^x + 1$

29. $f(x) = -2^{x-2}$

30. $f(x) = 2^{x-2}$

In Exercises 31–40, use a graphing utility to graph the exponential function. Identify any asymptotes of the graph.

31. $y = 2^{-x^2}$

32. $y = 3^{-|x|}$

33. $y = 3^{x-2} + 1$

34. $y = 4^{x+1} - 2$

35. $y = 1.08^{-5x}$

36. $y = 1.08^{5x}$

37. $s(t) = 2e^{0.12t}$

38. $s(t) = 3e^{-0.2t}$

39. $g(x) = 1 + e^{-x}$

40. $h(x) = e^{x-2}$

41. *Exploration* Consider the functions $f(x) = 3^x$ and $g(x) = 4^x$.

(a) Use a graphing utility to complete the table and use the table to estimate the solution of the inequality $4^x < 3^x$.

x	-1	-0.5	0	0.5	1
$f(x)$					
$g(x)$					

(b) Use a graphing utility to graph $f(x)$ and $g(x)$ in the same viewing rectangle. Use the graphs to solve the inequalities.

(i) $4^x < 3^x$ (ii) $4^x > 3^x$

42. *Exploration* Consider the functions $f(x) = \left(\frac{1}{2}\right)^x$ and $g(x) = \left(\frac{1}{4}\right)^x$.

(a) Use a graphing utility to complete the table and use the table to estimate the solution of the inequality $\left(\frac{1}{4}\right)^x < \left(\frac{1}{2}\right)^x$.

x	-1	-0.5	0	0.5	1
$f(x)$					
$g(x)$					

(b) Use a graphing utility to graph $f(x)$ and $g(x)$ in the same viewing rectangle. Use the graphs to solve the inequalities.

(i) $\left(\frac{1}{4}\right)^x < \left(\frac{1}{2}\right)^x$ (ii) $\left(\frac{1}{4}\right)^x > \left(\frac{1}{2}\right)^x$

of your grandchild's birth,
trust fund that pays 8.75%
ontinuously. Determine the
on your grandchild's 25th

rcises 53–56, complete the
ace *A* for *P* dollars invested
mpounded *n* times per year.

	365	Continuous

= 10 years

= 10 years

= 20 years

= 40 years

rcises 57–60, complete the
nt of money *P* that should
roduce a final balance of

	30	40	50

ntinuously

ontinuously

nonthly

ily

demand equation for a
y

$$\overline{}{002x}\Big)$$

is the number of units.

to graph the demand func-
0.

demand of *x* = 500 units.

(a) to approximate the high-
yield a demand of at least

51

62. *Graphical Reasoning*   There are two options for investing $500. The first earns 7% compounded annually and the second earns 7% simple interest. The figure shows the growth of each investment over a 30-year period.

   (a) Identify the two types of investments in the figure. Explain your reasoning.

   (b) Verify your answer in part (a) by finding the equations that model the investment growth and using a graphing utility to graph the models.

63. *Bacteria Growth*   A certain type of bacteria increases according to the model

$$P(t) = 100e^{0.2197t}$$

where t is the time in hours. Find (a) $P(0)$, (b) $P(5)$, and (c) $P(10)$.

64. *Population Growth*   The population of a town increases according to the model

$$P(t) = 2500e^{0.0293t}$$

where t is the time in years, with $t = 0$ corresponding to 1990. Use the model to estimate the population in (a) 2000 and (b) 2010.

65. *Radioactive Decay*   Let Q represent the mass of a quantity of radium 226, whose half-life is 1620 years. The quantity of radium present after t years is given by

$$Q = 25\left(\tfrac{1}{2}\right)^{t/1620}.$$

   (a) Determine the initial quantity (when $t = 0$).

   (b) Determine the quantity present after 1000 years.

   (c) Use a graphing utility to graph the function over the interval $t = 0$ to $t = 5000$.

66. *Data Analysis*   To estimate the amount of defoliation caused by gypsy moths, a forester counts the number x of egg masses on $\frac{1}{40}$ of an acre (circle of radius 18.6 feet) during the fall. The percent of defoliation y the next spring is given in the table. (Source: USDA, Forest Service)

x	0	25	50	75	100
y	12	44	81	96	99

   (a) Use a graphing utility to plot the data points.

   (b) A model for the data is given by

$$y = \frac{300}{3 + 17e^{-0.065x}}.$$

   Use a graphing utility to graph the model in the same viewing rectangle used in part (a). How well does the model fit the data?

   (c) Use a graphing utility to create a table comparing the model with the sample data.

   (d) Estimate the percent of defoliation if 36 egg masses are counted on $\frac{1}{40}$ acre.

   (e) Use the graph to estimate the number of egg masses per $\frac{1}{40}$ acre if it is seen that approximately $\frac{2}{3}$ of the forest is defoliated the next spring.

67. *Data Analysis*   A meteorologist measures the atmospheric pressure P (in kilograms per square meter) at altitude h (in kilometers). The data is shown in the table.

h	0	5	10	15	20
P	10,332	5583	2376	1240	517

   (a) Use a graphing utility to plot the data points.

   (b) A model for the data is given by

$$P = 10,958e^{-0.15h}.$$

   Use a graphing utility to graph the model in the same viewing rectangle used in part (a). How well does the model fit the data?

   (c) Use a graphing utility to create a table comparing the model with the sample data.

   (d) Estimate the pressure at a height of 8 kilometers.

   (e) Estimate the altitude at which the pressure is 2000 kilograms per square meter.

68. *Radioactive Decay* Let Q represent the mass of a quantity of carbon 14, whose half-life is 5730 years. The quantity present after t years is given by

$$Q = 10\left(\tfrac{1}{2}\right)^{t/5730}.$$

(a) Determine the initial quantity (when $t = 0$).

(b) Determine the quantity present after 2000 years.

(c) Sketch the graph of the function over the interval $t = 0$ to $t = 10,000$.

69. *Data Analysis* A cup of water at an initial temperature of 78°C is placed in a room at a constant temperature of 21°C. The temperature of the water is measured every 5 minutes for a period of $\frac{1}{2}$ hour. The results are recorded in the table, where t is the time in minutes and T is the temperature in degrees Celsius.

t	0	5	10	15
T	78.0°	66.0°	57.5°	51.2°

t	20	25	30
T	46.3°	42.5°	39.6°

(a) Use the regression capabilities of a graphing utility to fit a line to the data. Use the graphing utility to plot the data points and the regression line. Does the data appear linear? Explain.

(b) Use the regression capabilities of a graphing utility to fit a parabola to the data. Use the graphing utility to plot the data points and the regression parabola. Does the data appear quadratic? Even though the quadratic model appears to be a "good" fit, explain why it may not be a good model for predicting the temperature of the water when $t = 60$.

(c) The graph of the model should be asymptotic with the temperature of the room. Subtract the room temperature from each of the temperatures in the table. Use a graphing utility to fit an exponential model to the revised data. Add the room temperature to this regression model. Use a graphing utility to plot the original data points and the model.

(d) Explain why the procedure in part (c) was necessary for finding the exponential model.

70. *Inflation* If the annual rate of inflation averages 4% over the next 10 years, the approximate cost C of goods or services during any year in that decade will be given by

$$C(t) = P(1.04)^t$$

where t is the time in years and P is the present cost. If the price of an oil change for your car is presently $23.95, estimate the price 10 years from now.

71. *Depreciation* After t years, the value of a car that costs $20,000 is given by

$$V(t) = 20,000\left(\tfrac{3}{4}\right)^t.$$

Use a graphing utility to graph the function and determine the value of the car 2 years after it was purchased.

72. *True or False?* $e = 271,801/99,990$. Explain.

73. *Think About It* Without using a graphing utility, explain why you know that $2^{\sqrt{2}}$ is greater than 2, but less than 4.

74. *Think About It* Which functions are exponential?

(a) $3x$ (b) $3x^2$

(c) 3^x (d) 2^{-x}

75. *Finding a Pattern* Use a graphing utility to compare the graph of the function $y = e^x$ with the graphs of the following functions.

(a) $y_1 = 1 + \dfrac{x}{1!}$

(b) $y_2 = 1 + \dfrac{x}{1!} + \dfrac{x^2}{2!}$

(c) $y_3 = 1 + \dfrac{x}{1!} + \dfrac{x^2}{2!} + \dfrac{x^3}{3!}$

76. *Finding a Pattern* Identify the pattern of successive polynomials given in Exercise 75. Extend the pattern one more term and compare the graph of the resulting polynomial function with the graph of $y = e^x$. What do you think this pattern implies?

77. Given the exponential function $f(x) = a^x$, show that

(a) $f(u + v) = f(u) \cdot f(v)$ (b) $f(2x) = [f(x)]^2$.

3.2   Logarithmic Functions and Their Graphs

*Logarithmic Functions / Graphs of Logarithmic Functions /
The Natural Logarithmic Function / Application*

Logarithmic Functions

In Section 1.5, you studied the concept of the inverse of a function. There, you learned that if a function has the property that no horizontal line intersects its graph more than once, the function must have an inverse. By looking back at the graphs of the exponential functions introduced in Section 3.1, you will see that every function of the form $f(x) = a^x$ passes the "Horizontal Line Test" and therefore must have an inverse. This inverse function is called the **logarithmic function with base a.**

Note   The equations

$$y = \log_a x \quad \text{and} \quad x = a^y$$

are equivalent. The first equation is in logarithmic form and the second is in exponential form.

Definition of Logarithmic Function

For $x > 0$ and $0 < a \neq 1$,

$$y = \log_a x \quad \text{if and only if} \quad x = a^y.$$

The function given by

$$f(x) = \log_a x$$

is called the **logarithmic function with base a.**

When evaluating logarithms, remember that *a logarithm is an exponent.* This means that $\log_a x$ is the exponent to which a must be raised to obtain x. For instance, $\log_2 8 = 3$ because 2 must be raised to the third power to get 8.

Library of Functions

The logarithmic function is the inverse of the exponential function. Its domain is the set of positive real numbers and its range is the set of all real numbers. Because of the inverse properties of logarithms and exponents, the exponential equation $a^0 = 1$ implies that $\log_a 1 = 0$.

EXAMPLE 1   *Evaluating Logarithms*

a. $\log_2 32 = 5$ because $2^5 = 32$.

b. $\log_3 27 = 3$ because $3^3 = 27$.

c. $\log_4 2 = \frac{1}{2}$ because $4^{1/2} = \sqrt{4} = 2$.

d. $\log_{10} \frac{1}{100} = -2$ because $10^{-2} = \frac{1}{10^2} = \frac{1}{100}$.

e. $\log_3 1 = 0$ because $3^0 = 1$.

f. $\log_2 2 = 1$ because $2^1 = 2$.

The logarithmic function with base 10 is called the **common logarithmic function.** On most calculators, this function is denoted by LOG .

EXAMPLE 2 Evaluating Logarithms on a Calculator

Use a calculator to evaluate each expression.

a. $\log_{10} 10$ **b.** $2\log_{10} 2.5$ **c.** $\log_{10}(-2)$

Solution

Number	Graphing Calculator Keystrokes	Display
a. $\log_{10} 10$	LOG 10 ENTER	1
b. $2\log_{10}2.5$	2 LOG 2.5 ENTER	0.7958800
c. $\log_{10}(-2)$	LOG (-) 2 ENTER	ERROR

Note that the calculator displays an error message when you try to evaluate $\log_{10}(-2)$. The reason for this is that the domain of every logarithmic function is the set of *positive* real numbers.

The following properties follow directly from the definition of the logarithmic function with base a.

> **Properties of Logarithms**
>
> **1.** $\log_a 1 = 0$ because $a^0 = 1$.
> **2.** $\log_a a = 1$ because $a^1 = a$.
> **3.** $\log_a a^x = x$ because $a^x = a^x$.
> **4.** If $\log_a x = \log_a y$, then $x = y$.

EXAMPLE 3 Using Properties of Logarithms

Solve each equation for x.

a. $\log_2 x = \log_2 3$

b. $\log_4 4 = x$

Solution

a. Using Property 4, you can conclude that $x = 3$.

b. Using Property 2, you can conclude that $x = 1$.

Some graphing utilities such as the *TI-85,* do not give an error message for $\log_{10}(-2)$. Instead, the graphing utility will display a complex number. For the purpose of this text, however, we will continue to say that the domain of a logarithmic function is the set of positive *real* numbers.

Graphs of Logarithmic Functions

To sketch the graph of $y = \log_a x$, you can use the fact that the graphs of inverse functions are reflections of each other in the line $y = x$.

Figure 3.11

EXAMPLE 4   Graphs of Exponential and Logarithmic Functions

In the same coordinate plane, sketch the graph of each function.

a. $f(x) = 2^x$ **b.** $g(x) = \log_2 x$

Solution

a. For $f(x) = 2^x$, construct a table of values, as follows.

x	-2	-1	0	1	2	3
2^x	$\frac{1}{4}$	$\frac{1}{2}$	1	2	4	8

By plotting these points and connecting them with a smooth curve, you obtain the graph of $f(x)$ shown in Figure 3.11.

b. Because $g(x) = \log_2 x$ is the inverse of $f(x) = 2^x$, the graph of g is obtained by reflecting the graph of f in the line $y = x$, as shown in Figure 3.11.

Before you can confirm the result of Example 4 with a graphing utility, you need to know how to enter $\log_2 x$. You will learn how to do this using the *Change-of-Base Formula* discussed in Section 3.3.

Figure 3.12

Note Compare Figure 3.12 with that obtained with a graphing utility. Note that the domain and range are $(0, \infty)$ and $(-\infty, \infty)$, respectively.

EXAMPLE 5   Sketching the Graph of a Logarithmic Function

Sketch the graph of the common logarithmic function $f(x) = \log_{10} x$.

Solution

Begin by constructing a table of values. Note that some of the values can be obtained without a calculator, whereas others require a calculator. Next, plot the points and connect them with a smooth curve, as shown in Figure 3.12.

	Without Calculator				With Calculator		
x	$\frac{1}{100}$	$\frac{1}{10}$	1	10	2	5	8
$\log_{10} x$	-2	-1	0	1	0.301	0.699	0.903

The nature of the graph in Figure 3.12 is typical of functions of the form $f(x) = \log_a x, a > 1$. They have one x-intercept and one vertical asymptote. Notice how slowly the graph rises for $x > 1$. In Figure 3.12, you would need to move out to $x = 1000$ before the graph rose to $y = 3$. The basic characteristics of logarithmic graphs are summarized in Figure 3.13.

Note In Figure 3.13, note that the vertical asymptote occurs at $x = 0$, where $\log_a x$ is *undefined*.

Figure 3.13

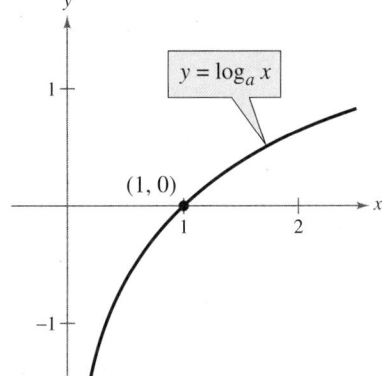

Graph of $y = \log_a x, a > 1$

- Domain: $(0, \infty)$
- Range: $(-\infty, \infty)$
- Intercept: $(1, 0)$
- Increasing
- y-axis is a vertical asymptote
  ($\log_a x \to -\infty$ as $x \to 0^+$)
- Continuous
- Reflection of graph of $y = a^x$
  about the line $y = x$

The *Interactive* CD-ROM offers graphing utility emulators of the *TI-82* and *TI-83*, which can be used with the Examples, Explorations, Technology notes, and Exercises.

EXPLORATION

Use a graphing utility to graph $y = \log_{10} x$ and $y = 8$ in the same viewing rectangle. Do the graphs intersect? If so, find a viewing rectangle that shows the point of intersection. What is the point of intersection?

EXAMPLE 6   **Sketching the Graphs of Logarithmic Functions**

The graph of each of the following functions is similar to the graph of $f(x) = \log_{10} x$, as shown in Figure 3.14.

a. Because $g(x) = \log_{10}(x - 1) = f(x - 1)$, the graph of g can be obtained by shifting the graph of f one unit to the right.

b. Because $h(x) = 2 + \log_{10} x = 2 + f(x)$, the graph of h can be obtained by shifting the graph of f two units up.

Note In Figure 3.14(a), notice how the transformation affects the vertical asymptote.

Figure 3.14

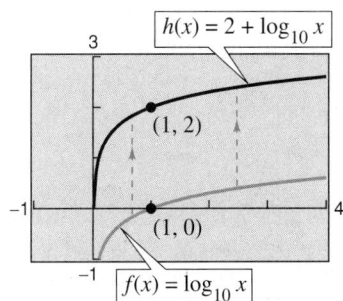

The Natural Logarithmic Function

The four properties of logarithms listed on page 253 are also valid for natural logarithms.

The most widely used base for logarithmic functions is the number e, where

$$e \approx 2.718281828 \ldots .$$

The logarithmic function with base e is the **natural logarithmic function** and is denoted by the special symbol $\ln x$, read as "el en of x."

The Natural Logarithmic Function

The function defined by

$$f(x) = \log_e x = \ln x, \quad x > 0$$

is called the **natural logarithmic function.**

Properties of Natural Logarithms

1. $\ln 1 = 0$ because $e^0 = 1$.
2. $\ln e = 1$ because $e^1 = e$.
3. $\ln e^x = x$ because $e^x = e^x$.
4. If $\ln x = \ln y$, then $x = y$.

Figure 3.15

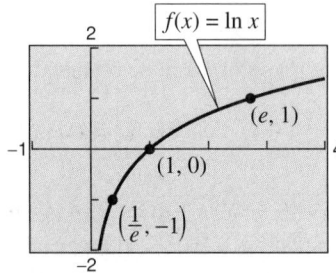

The graph of the natural logarithmic function is shown in Figure 3.15. Try using a graphing utility to confirm this graph. What is the domain of the natural logarithmic function?

EXAMPLE 7   **Using Properties of Natural Logarithms**

a. $\ln \dfrac{1}{e} = \ln e^{-1} = -1$ Property 3

b. $\ln e^2 = 2$ Property 3

c. $\ln e^0 = 0$ Property 1

d. $2 \ln e = 2(1) = 2$ Property 2

On most calculators, the natural logarithm is denoted by $\boxed{\text{LN}}$, as illustrated in Example 8.

EXAMPLE 8 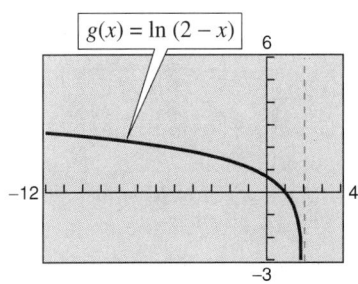 **Evaluating the Natural Logarithmic Function**

Use a calculator to evaluate each expression.

a. $\ln 2$ **b.** $\ln 0.3$ **c.** $\ln e^2$ **d.** $\ln(-1)$

Solution

Number	Graphing Calculator Keystrokes	Display
a. $\ln 2$	$\boxed{\text{LN}}$ 2 $\boxed{\text{ENTER}}$	0.6931472
b. $\ln 0.3$	$\boxed{\text{LN}}$ .3 $\boxed{\text{ENTER}}$	−1.2039728
c. $\ln e^2$	$\boxed{\text{LN}}$ $\boxed{e^x}$ 2 $\boxed{\text{ENTER}}$	2
d. $\ln(-1)$	$\boxed{\text{LN}}$ $\boxed{(-)}$ 1 $\boxed{\text{ENTER}}$	ERROR

In Example 8, be sure you see that $\ln(-1)$ gives an error message on most calculators. This occurs because the domain of $\ln x$ is the set of *positive* real numbers (see Figure 3.15). Hence, $\ln(-1)$ is undefined.

EXAMPLE 9 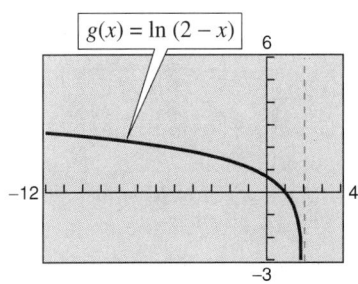 **Finding the Domains of Logarithmic Functions**

Find the domain of each function.

a. $f(x) = \ln(x - 2)$ **b.** $g(x) = \ln(2 - x)$ **c.** $h(x) = \ln x^2$

Solution

a. Because $\ln(x - 2)$ is defined only if $x - 2 > 0$, it follows that the domain of f is $(2, \infty)$.

b. Because $\ln(2 - x)$ is defined only if $2 - x > 0$, it follows that the domain of g is $(-\infty, 2)$. The graph of g is shown in Figure 3.16.

c. Because $\ln x^2$ is defined only if $x^2 > 0$, it follows that the domain of h is all real numbers except $x = 0$.

Note In Example 9, suppose you had been asked to analyze the function $h(x) = \ln |x - 2|$. How would the domain of this function compare with the domains of the functions given in parts (a) and (b) of the example?

Use a graphing utility to graph f and h in Example 9. How can you use the graphs to verify the domains of the functions?

Figure 3.16

$g(x) = \ln (2 - x)$

Application

EXAMPLE 10   Human Memory Model   *Real Life*

Students participating in a psychological experiment attended several lectures on a subject. At the end of the last lecture, and every month for the next year, the students were tested to see how much of the material they remembered. The average scores for the group were given by the *human memory model*

$$f(t) = 75 - 6 \ln(t + 1), \quad 0 \le t \le 12$$

where t is the time in months.

a. What was the average score on the original ($t = 0$) exam?

b. What was the average score at the end of $t = 2$ months?

c. What was the average score at the end of $t = 6$ months?

Solution

a. The original average score was

$$f(0) = 75 - 6 \ln 1 = 75 - 6(0) = 75.$$

b. After 2 months, the average score was

$$f(2) = 75 - 6 \ln 3 \approx 75 - 6(1.0986) \approx 68.4.$$

c. After 6 months, the average score was

$$f(6) = 75 - 6 \ln 7 \approx 75 - 6(1.9459) \approx 63.3.$$

The graph of f in an appropriate viewing rectangle is shown in Figure 3.17.

EXPLORATION

Use a graphing utility to determine how many months it would take for the average score in Example 10 to decrease to 60. Explain your method of solving the problem. Describe another way that you can use a graphing utility to determine the answer.

Figure 3.17

$f(t) = 75 - 6 \ln (t + 1)$

Average score vs. Time (in months)

Group Activity

Transforming Logarithmic Functions

x	$f_1(x)$	$f_2(x)$
−3	Undefined	Undefined
−2.99999	−13.816	−17.816
−2.9999	−9.210	−13.210
−2	0	−4
0	1.099	−2.901
3	1.792	−2.208
6	2.197	−1.803

The table at the left gives selected points for two natural logarithmic functions of the form $f(x) = b \pm \ln(x + c)$. Study the table. What can you infer? Compare the given data with data for $f(x) = \ln x$. Try to find a natural logarithmic function that fits each set of data. Explain the method you used. (*Hint:* You might find it easier to find the value of c first.)

3.2 /// EXERCISES

In Exercises 1–8, write the logarithmic equation in exponential form. For example, the exponential form of $\log_5 25 = 2$ is $5^2 = 25$.

1. $\log_4 64 = 3$

2. $\log_3 81 = 4$

3. $\log_7 \frac{1}{49} = -2$

4. $\log_{10} \frac{1}{1000} = -3$

5. $\log_{32} 4 = \frac{2}{5}$

6. $\log_{16} 8 = \frac{3}{4}$

7. $\ln 1 = 0$

8. $\ln 4 = 1.386\ldots$

In Exercises 9–18, write the exponential equation in logarithmic form.

9. $5^3 = 125$

10. $8^2 = 64$

11. $81^{1/4} = 3$

12. $9^{3/2} = 27$

13. $6^{-2} = \frac{1}{36}$

14. $10^{-3} = 0.001$

15. $e^3 = 20.0855\ldots$

16. $e^0 = 1$

17. $e^x = 4$

18. $u^v = w$

In Exercises 19–30, evaluate the expression without using a calculator.

19. $\log_2 16$

20. $\log_2\left(\frac{1}{8}\right)$

21. $\log_{16} 4$

22. $\log_{27} 9$

23. $\log_7 1$

24. $\log_{10} 1000$

25. $\log_{10} 0.01$

26. $\log_{10} 10$

27. $\ln e^3$

28. $\ln 1$

29. $\log_a a^2$

30. $\log_a \frac{1}{a}$

In Exercises 31–40, use a calculator to evaluate the logarithm. Round to three decimal places.

31. $\log_{10} 345$

32. $\log_{10}\left(\frac{4}{5}\right)$

33. $\log_{10} 145$

34. $\log_{10} 12.5$

35. $\ln 18.42$

36. $\ln\sqrt{42}$

37. $\ln\left(1 + \sqrt{3}\right)$

38. $\ln\left(\sqrt{5} - 2\right)$

39. $\ln 0.32$

40. $\ln 0.75$

In Exercises 41–44, describe the relationship between the graphs of f and g. What is the relationship between the functions f and g?

41. $f(x) = 3^x$
$g(x) = \log_3 x$

42. $f(x) = 5^x$
$g(x) = \log_5 x$

43. $f(x) = e^x$
$g(x) = \ln x$

44. $f(x) = 10^x$
$g(x) = \log_{10} x$

In Exercises 45–50, use the graph of $y = \log_3 x$ to match the given function with its graph. [The graphs are labeled (a), (b), (c), (d), (e), and (f).]

(a)

(b)

(c)

(d)

(e)

(f)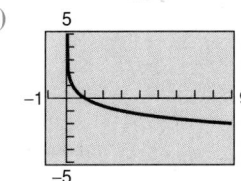

45. $f(x) = \log_3 x + 2$

46. $f(x) = -\log_3 x$

47. $f(x) = -\log_3(x + 2)$

48. $f(x) = \log_3(x - 1)$

49. $f(x) = \log_3(1 - x)$

50. $f(x) = -\log_3(-x)$

In Exercises 51–56, find the domain, vertical asymptote, and x-intercept of the logarithmic function, and sketch its graph by hand.

51. $f(x) = \log_4 x$

52. $g(x) = \log_6 x$

53. $h(x) = \log_4(x - 3)$

54. $f(x) = -\log_6(x + 2)$

55. $y = -\log_3 x + 2$

56. $y = \log_5(x - 1) + 4$

In Exercises 57–62, use a graphing utility to graph the logarithmic function. Determine the domain and identify any vertical asymptote and x-intercept.

57. $y = \log_{10}\left(\dfrac{x}{5}\right)$

58. $y = \log_{10}(-x)$

59. $f(x) = \ln(x - 2)$

60. $h(x) = \ln(x + 1)$

61. $g(x) = \ln(-x)$

62. $f(x) = \ln(3 - x)$

In Exercises 63–66, use a graphing utility to graph the function. What is the domain? Use the graph to determine the intervals in which the function is increasing and decreasing and approximate any relative maximum or minimum values of the function.

63. $f(x) = \dfrac{x}{2} - \ln\dfrac{x}{4}$

64. $g(x) = \dfrac{12 \ln x}{x}$

65. $h(x) = 4x \ln x$

66. $f(x) = \dfrac{x}{\ln x}$

67. *Population Growth*   The population of a town will double in

$$t = \frac{10 \ln 2}{\ln 67 - \ln 50} \text{ years.}$$

Find t.

68. *Human Memory Model*   Students in a mathematics class were given an exam and then tested monthly with an equivalent exam. The average score for the class was given by the human memory model

$$f(t) = 80 - 17 \log_{10}(t + 1), \quad 0 \le t \le 12$$

where t is the time in months.

(a) What was the average score on the original exam ($t = 0$)?

(b) What was the average score after 4 months?

(c) What was the average score after 10 months?

69. *Exploration*   The table of values was obtained by evaluating a function. Determine which of the statements may be true and which must be false.

x	1	2	8
y	0	1	3

(a) y is an exponential function of x.

(b) y is a logarithmic function of x.

(c) x is an exponential function of y.

(d) y is a linear function of x.

70. *Exploration*   Use a graphing utility to compare the graph of the function $y = \ln x$ with the graphs of the following functions.

(a) $y = x - 1$

(b) $y = (x - 1) - \frac{1}{2}(x - 1)^2$

(c) $y = (x - 1) - \frac{1}{2}(x - 1)^2 + \frac{1}{3}(x - 1)^3$

71. *Finding a Pattern*   Identify the pattern of successive polynomials given in Exercise 70. Extend the pattern one more term and compare the graph of the resulting polynomial function with the graph of $y = \ln x$. What do you think the pattern implies?

72. *Graphical Analysis*   Use a graphing utility to graph f and g in the same viewing rectangle and determine which is increasing at the greater rate for "large" values of x. What can you conclude about the rate of growth of the natural logarithmic function?

(a) $f(x) = \ln x$, $\quad g(x) = \sqrt{x}$

(b) $f(x) = \ln x$, $\quad g(x) = \sqrt[4]{x}$

73. *Investment Time*   A principal P, invested at $9\frac{1}{2}\%$ and compounded continuously, increases to an amount K times the original principal after t years, where t is given by

$$t = (\ln K)/0.095.$$

(a) Complete the table and interpret your results.

K	1	2	4	6	8	10	12
t							

(b) Use a graphing utility to graph the function.

74. *Data Analysis*   The table gives the temperatures T (°F) at which water boils at selected pressures p (pounds per square inch). (Source: Standard Handbook of Mechanical Engineers)

p	5	10	14.696 (1atm)
T	162.24°	193.21°	212.00°

p	20	30	40
T	227.96°	250.33°	267.25°

p	60	80	100
T	292.71°	312.03°	327.81°

A model that approximates this data is

$T = 87.97 + 34.96 \ln x + 7.91 \sqrt{x}.$

(a) Use a graphing utility to plot the data points and graph the model in the same viewing rectangle. How well does the model fit the data?

(b) Use the graph to estimate the pressure required for the boiling point of water to exceed 300°F.

75. *Data Analysis*   A meteorologist measures the atmospheric pressure P (in kilograms per square meter) at altitude h (in kilometers). The data is shown below.

h	0	5	10	15	20
P	10,332	5583	2376	1240	517

A model for the data is given by

$P = 10{,}957e^{-0.15h}.$

(a) Use a graphing utility to plot the data points and graph the model in the same viewing rectangle.

(b) Use a graphing utility to plot the points $(h, \ln P)$. Use the regression capabilities of the graphing utility to fit a regression line to the revised data points.

(c) The line in part (b) has the form $\ln P = ah + b$. Write the line in exponential form.

(d) Verify, graphically and analytically, that the result of part (c) is equivalent to the given exponential model for the data.

76. *World Population Growth*   The time t in years for the world population to double if it is increasing at a continuous rate of r is given by

$$t = \frac{\ln 2}{r}.$$

Complete the table. What do your results imply?

r	0.005	0.010	0.015	0.020	0.025	0.030
t						

77. *Tractrix*   A person walking along a dock (the y-axis) drags a boat by a 10-foot rope. The boat travels along a path known as a *tractrix*. The equation of this path is

$$y = 10 \ln \left(\frac{10 + \sqrt{100 - x^2}}{x} \right) - \sqrt{100 - x^2}.$$

(a) Use a graphing utility to obtain a graph of the function. What is the domain of the function?

(b) Identify any asymptotes of the graph.

(c) Determine the position of the person when the x-coordinate of the position of the boat is $x = 2$.

(d) Let $(0, p)$ be the position of the person. Determine p as a function of x, the x-coordinate of the position of the boat.

(e) Use a graphing utility to graph the function p. When does the position of the person change most for a small change in the position of the boat? Explain.

78. *Sound Intensity*   The relationship between the number of decibels β and the intensity of a sound I in watts per square centimeter is given by

$$\beta = 10 \log_{10} \left(\frac{I}{10^{-16}} \right).$$

(a) Determine the number of decibels of a sound with an intensity of 10^{-4} watts per square centimeter.

(b) Determine the number of decibels of a sound with an intensity of 10^{-6} watts per square centimeter.

(c) The intensity of the sound in part (a) is 100 times as great as that in part (b). Is the number of decibels 100 times as great? Explain.

Ventilation Rates   In Exercises 79 and 80, use the model

$$y = 80.4 - 11 \ln x, \qquad 100 \le x \le 1500$$

which approximates the minimum required ventilation rate in terms of the air space per child in a public school classroom. In the model, x is the air space per child in cubic feet and y is the ventilation rate in cubic feet per minute.

79. Use a graphing utility to graph the function and approximate the required ventilation rate if there is 300 cubic feet of air space per child.

80. A classroom is designed for 30 students. The air-conditioning system in the room has the capacity to move 450 cubic feet of air per minute.

(a) Determine the ventilation rate per child, assuming that the room is filled to capacity.

(b) Use the graph of Exercise 79 to estimate the air space required per child.

(c) Determine the minimum number of square feet of floor space required for the room if the ceiling height is 30 feet.

81. *Work*   The work (in foot-pounds) done in compressing an initial volume of 9 cubic feet of air at a pressure of 15 pounds per square inch to a volume of 3 cubic feet is

$$W = 19{,}440(\ln 9 - \ln 3).$$

Find W.

82. (a) Use a graphing utility to complete the table for the function

$$f(x) = \frac{\ln x}{x}.$$

x	1	5	10	10^2	10^4	10^6
$f(x)$						

(b) Use the table to determine what $f(x)$ approaches as x increases without bound.

(c) Use a graphing utility to check the result of part (b).

Monthly Payment   In Exercises 83–86, use the model

$$t = 10.042 \ln\!\left(\frac{x}{x - 1250}\right), \qquad 1250 < x$$

which approximates the length of a home mortgage of $150,000 at 10% in terms of the monthly payment. In the model, t is the length of the mortgage in years and x is the monthly payment in dollars (see figure).

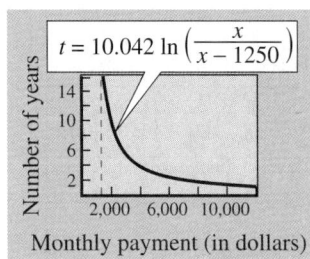

83. Use the model to approximate the length of the mortgage (for $150,000 at 10%) if the monthly payment is $1316.35.

84. Use the model to approximate the length of the mortgage (for $150,000 at 10%) if the monthly payment is $1982.26.

85. Approximate the total amount paid over the term of the mortgage with a monthly payment of $1316.35. What amount of the total is interest costs?

86. Approximate the total amount paid over the term of the mortgage with a monthly payment of $1982.26. What amount of the total is interest costs?

87. *Exploration*   Answer the following for the function $f(x) = \log_{10} x$. Do not use a calculator.

(a) What is the domain of f?

(b) Find f^{-1}.

(c) If x is a real number between 1000 and 10,000, in what interval will $f(x)$ be found?

(d) Determine the interval in which x will be found if $f(x)$ is negative.

(e) If $f(x)$ is increased by one unit, x must have been increased by what factor?

(f) If $f(x_1) = 3n$ and $f(x_2) = n$, what is the ratio of x_1 to x_2?

3.3   Properties of Logarithms

Change of Base / Properties of Logarithms /
Rewriting Logarithmic Expressions / Application

Change of Base

Most calculators have only two types of log keys, one for common logarithms (base 10) and one for natural logarithms (base e). Although common logs and natural logs are the most frequently used, you may occasionally need to evaluate logarithms to other bases. To do this, you can use the following *change-of-base formula.*

Change-of-Base Formula

Let a, b, and x be positive real numbers such that $a \neq 1$ and $b \neq 1$. Then $\log_a x$ is given by

$$\log_a x = \frac{\log_b x}{\log_b a}.$$

One way to look at the change-of-base formula is that logarithms to base a are simply *constant multiples* of logarithms to base b. The constant multiplier is $1/(\log_b a)$.

EXAMPLE 1   🔲   **Changing Bases Using Common Logarithms**

a. $\log_4 30 = \dfrac{\log_{10} 30}{\log_{10} 4} \approx \dfrac{1.47712}{0.60206} \approx 2.4534$

b. $\log_2 14 = \dfrac{\log_{10} 14}{\log_{10} 2} \approx \dfrac{1.14613}{0.30103} \approx 3.8074$ 🔲

EXAMPLE 2   🔲   **Changing Bases Using Natural Logarithms**

a. $\log_4 30 = \dfrac{\ln 30}{\ln 4} \approx \dfrac{3.40120}{1.38629} \approx 2.4534$

b. $\log_2 14 = \dfrac{\ln 14}{\ln 2} \approx \dfrac{2.63906}{0.693147} \approx 3.8074$ 🔲

John Napier, a Scottish mathematician, developed logarithms as a way to simplify some of the tedious calculations of his day. Beginning in 1594, Napier worked about 20 years on the invention of logarithms. Napier was only partially successful in his quest to simplify tedious calculations. Nonetheless, the development of logarithms was a step forward and received immediate recognition.

Note Notice in Examples 1 and 2 that the result is the same whether common logarithms or natural logarithms are used in the change-of-base formula.

Properties of Logarithms

You know from the previous section that the logarithmic function with base a is the *inverse* of the exponential function with base a. Thus, it makes sense that the properties of exponents should have corresponding properties involving logarithms. For instance, the exponential property $a^0 = 1$ has the corresponding logarithmic property $\log_a 1 = 0$.

Note   There is no general property that can be used to rewrite $\log_a(u \pm v)$. Specifically, $\log_a(x + y)$ is not equal to $\log_a x + \log_a y$.

Properties of Logarithms
Let a be a positive number such that $a \neq 1$, and let n be a real number. If u and v are positive real numbers, the following properties are true.

1. $\log_a(uv) = \log_a u + \log_a v$ **1.** $\ln(uv) = \ln u + \ln v$

2. $\log_a \dfrac{u}{v} = \log_a u - \log_a v$ **2.** $\ln \dfrac{u}{v} = \ln u - \ln v$

3. $\log_a u^n = n \log_a u$ **3.** $\ln u^n = n \ln u$

Think About the Proof

To prove Property 1, let $x = \log_a u$ and $y = \log_a v$. The corresponding exponential forms of these two equations are $a^x = u$ and $a^y = v$. Multiplying u and v produces

$$uv = a^x a^y = a^{x+y}.$$

Can you see how to use this equation to complete the proof? The details of this proof are listed in the appendix.

EXAMPLE 3   **Using Properties of Logarithms**

Write the logarithm in terms of ln 2 and ln 3.

a. $\ln 6$ **b.** $\ln \dfrac{2}{27}$

Solution

a. $\ln 6 = \ln(2 \cdot 3)$ Rewrite 6 as 2 · 3.

$\qquad = \ln 2 + \ln 3$ Property 1

b. $\ln \dfrac{2}{27} = \ln 2 - \ln 27$ Property 2

$\qquad = \ln 2 - \ln 3^3$ Rewrite 27 as 3^3.

$\qquad = \ln 2 - 3 \ln 3$ Property 3

EXAMPLE 4   **Using Properties of Logarithms**

Use the properties of logarithms to verify that $-\ln \frac{1}{2} = \ln 2$.

Solution

$$-\ln \tfrac{1}{2} = -\ln(2^{-1}) = -(-1)\ln 2 = \ln 2$$

Try checking this result on your calculator.

Rewriting Logarithmic Expressions

The properties of logarithms are useful for rewriting logarithmic expressions in forms that simplify the operations of algebra. This is true because they convert complicated products, quotients, and exponential forms into simpler sums, differences, and products, respectively.

EXAMPLE 5 ▱ **Rewriting the Logarithm of a Product**

$$\log_{10} 5x^3 y = \log_{10} 5 + \log_{10} x^3 y$$
$$= \log_{10} 5 + \log_{10} x^3 + \log_{10} y$$
$$= \log_{10} 5 + 3 \log_{10} x + \log_{10} y \qquad ▱$$

EXPLORATION

Use a graphing utility to graph the functions

$$y = \ln x - \ln(x - 3)$$

and

$$y = \ln \frac{x}{x - 3}$$

in the same viewing rectangle. Does the graphing utility show the functions with the same domain? If so, should it? Explain your reasoning.

EXAMPLE 6 ▱ **Rewriting the Logarithm of a Quotient**

$$\ln \frac{\sqrt{3x - 5}}{7} = \ln(3x - 5)^{1/2} - \ln 7$$
$$= \frac{1}{2} \ln(3x - 5) - \ln 7 \qquad ▱$$

In Examples 5 and 6, the properties of logarithms were used to *expand* logarithmic expressions. In Examples 7 and 8, this procedure is reversed and the properties of logarithms are used to *condense* logarithmic expressions.

EXAMPLE 7 ▱ **Condensing a Logarithmic Expression**

$$\tfrac{1}{2} \log_{10} x + 3 \log_{10}(x + 1) = \log_{10} x^{1/2} + \log_{10}(x + 1)^3$$
$$= \log_{10}\left[\sqrt{x} \cdot (x + 1)^3 \right] \qquad ▱$$

EXAMPLE 8 ▱ **Condensing a Logarithmic Expression**

$$2 \ln(x + 2) - \ln x = \ln(x + 2)^2 - \ln x$$
$$= \ln \frac{(x + 2)^2}{x} \qquad ▱$$

Application

EXAMPLE 9   **Finding a Mathematical Model**   *Real Life*

The table gives the mean distance from the sun x and the orbital period y of the six planets that are closest to the sun. In the table, the mean distance is given in terms of astronomical units (where the earth's mean distance is defined as 1.0), and the period is given in terms of years. Find an equation that expresses y as a function of x.

Figure 3.18

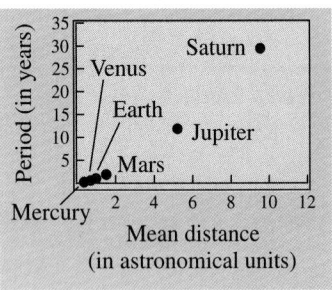

Period (in years) vs. Mean distance (in astronomical units)

Planet	Mercury	Venus	Earth	Mars	Jupiter	Saturn
Period, y	0.241	0.615	1.0	1.881	11.861	29.457
Mean Distance, x	0.387	0.723	1.0	1.523	5.203	9.541

Solution

The points in the table are plotted in Figure 3.18. From this figure it is not clear how to find an equation that relates y and x. To solve this problem, take the natural log of each of the x- and y-values given in the table. This produces the following results.

Planet	Mercury	Venus	Earth	Mars	Jupiter	Saturn
$\ln y$	−1.423	−0.486	0.0	0.632	2.473	3.383
$\ln x$	−0.949	−0.324	0.0	0.421	1.649	2.256

Figure 3.19

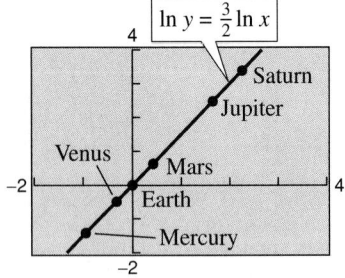

$\ln y = \frac{3}{2} \ln x$

Now, by plotting the points in the second table, you can see that all six of the points appear to lie in a line (see Figure 3.19). You can use a graphical approach or the algebraic approach discussed in Section 1.6 to find that the slope of this line is $\frac{3}{2}$, and you can therefore conclude that $\ln y = \frac{3}{2} \ln x$. [Try to convert this to $y = f(x)$ form.]

Group Activity   *Kepler's Law*

The relationship described in Example 9 was first discovered by Johannes Kepler. Use properties of logarithms to rewrite the relationship so that y is expressed as a function of x. Then verify the values in the first table.

3.3 /// EXERCISES

In Exercises 1 and 2, use a graphing utility to graph the two functions in the same viewing rectangle. What do the graphs suggest? Explain your reasoning.

1. $f(x) = \log_{10} x$

$g(x) = \dfrac{\ln x}{\ln 10}$

2. $f(x) = \ln x$

$g(x) = \dfrac{\log_{10} x}{\log_{10} e}$

In Exercises 3–6, use the change-of-base formula to write the given logarithm as a multiple of a common logarithm. For instance, $\log_2 3 = (1/\log_{10} 2) \log_{10} 3$.

3. $\log_3 5$

4. $\log_4 10$

5. $\log_2 x$

6. $\ln 5$

In Exercises 7–10, use the change-of-base formula to write the given logarithm as a multiple of a natural logarithm. For instance, $\log_2 3 = (1/\ln 2) \ln 3$.

7. $\log_3 5$

8. $\log_4 10$

9. $\log_2 x$

10. $\log_{10} 5$

In Exercises 11–18, evaluate the logarithm using the change-of-base formula. Round your result to three decimal places.

11. $\log_3 7$

12. $\log_7 4$

13. $\log_{1/2} 4$

14. $\log_4 (0.55)$

15. $\log_9 (0.4)$

16. $\log_{20} 125$

17. $\log_{15} 1250$

18. $\log_{1/3} (0.015)$

In Exercises 19–38, use the properties of logarithms to write the expression as a sum, difference, and/or constant multiple of logarithms. (Assume all variables are positive.)

19. $\log_{10} 5x$

20. $\log_{10} 10z$

21. $\log_{10} \dfrac{5}{x}$

22. $\log_{10} \dfrac{y}{2}$

23. $\log_8 x^4$

24. $\log_6 z^{-3}$

25. $\ln \sqrt{z}$

26. $\ln \sqrt[3]{t}$

27. $\ln xyz$

28. $\ln \dfrac{xy}{z}$

29. $\ln \sqrt{a - 1}, \quad a > 1$

30. $\ln \left(\dfrac{x^2 - 1}{x^3} \right), \quad x > 1$

31. $\ln z(z - 1)^2, \quad z > 1$

32. $\ln \sqrt{\dfrac{x^2}{y^3}}$

33. $\ln \sqrt[3]{\dfrac{x}{y}}$

34. $\ln \dfrac{x}{\sqrt{x^2 + 1}}$

35. $\ln \dfrac{x^4 \sqrt{y}}{z^5}$

36. $\ln \sqrt{x^2 (x + 2)}$

37. $\log_b \dfrac{x^2}{y^2 z^3}$

38. $\log_b \dfrac{\sqrt{x y^4}}{z^4}$

Graphical Analysis In Exercises 39 and 40, use a graphing utility to graph the two equations in the same viewing rectangle. What do the graphs suggest? Explain your reasoning.

39. $y_1 = \ln[x^3 (x + 4)], \quad y_2 = 3 \ln x + \ln(x + 4)$

40. $y_1 = \ln\left(\dfrac{\sqrt{x}}{x - 2} \right), \quad y_2 = \tfrac{1}{2} \ln x - \ln(x - 2)$

In Exercises 41–60, write the expression as the logarithm of a single quantity.

41. $\ln x + \ln 2$

42. $\ln y + \ln z$

43. $\log_4 z - \log_4 y$

44. $\log_5 8 - \log_5 t$

45. $2 \log_2 (x + 4)$

46. $-4 \log_6 2x$

47. $\tfrac{1}{3} \log_3 5x$

48. $\tfrac{3}{2} \log_7 (z - 2)$

49. $\ln x - 3 \ln(x + 1)$

50. $2 \ln 8 + 5 \ln z$

51. $\ln(x - 2) - \ln(x + 2)$

52. $3 \ln x + 2 \ln y - 4 \ln z$

53. $\ln x - 2[\ln(x + 2) + \ln(x - 2)]$

54. $4[\ln z + \ln(z + 5)] - 2 \ln(z - 5)$

55. $\frac{1}{3}[2\ln(x+3) + \ln x - \ln(x^2 - 1)]$

56. $2[\ln x - \ln(x+1) - \ln(x-1)]$

57. $\frac{1}{3}[\ln y + 2\ln(y+4)] - \ln(y-1)$

58. $\frac{1}{2}[\ln(x+1) + 2\ln(x-1)] + 3\ln x$

59. $2\ln 3 - \frac{1}{2}\ln(x^2 + 1)$

60. $\frac{3}{2}\ln 5t^6 - \frac{3}{4}\ln t^4$

Graphical Analysis   In Exercises 61 and 62, use a graphing utility to graph the two equations in the same viewing rectangle. What do the graphs suggest? Verify your conclusion algebraically.

61. $y_1 = 2[\ln 8 - \ln(x^2 + 1)], \quad y_2 = \ln\left[\dfrac{64}{(x^2+1)^2}\right]$

62. $y_1 = \ln x + \frac{1}{3}\ln(x+1), \quad y_2 = \ln(x\sqrt[3]{x+1})$

Think About It   In Exercises 63 and 64, use a graphing utility to graph the two equations in the same viewing rectangle. Are the expressions equivalent? Explain.

63. $y_1 = \ln x^2, \quad y_2 = 2\ln x$

64. $y_1 = \frac{1}{4}\ln[x^4(x^2 + 1)], \quad y_2 = \ln x + \frac{1}{4}\ln(x^2 + 1)$

65. *Think About It*   Use a graphing utility to graph

$$f(x) = \ln\frac{x}{2}, \quad g(x) = \frac{\ln x}{\ln 2}, \quad h(x) = \ln x - \ln 2$$

in the same viewing rectangle. Which two functions have identical graphs? Explain why.

66. *Exploration*   Approximate the natural logarithms of as many integers as possible between 1 and 20 given that $\ln 2 \approx 0.6931$, $\ln 3 \approx 1.0986$, and $\ln 5 \approx 1.6094$. (Do not use a calculator.)

In Exercises 67–80, find the exact value of the logarithm without using a calculator. (If this is not possible, state the reason.)

67. $\log_3 9$

68. $\log_6 \sqrt[3]{6}$

69. $\log_4 16^{1.2}$

70. $\log_5 \left(\frac{1}{125}\right)$

71. $\log_3(-9)$

72. $\log_2(-16)$

73. $\log_5 75 - \log_5 3$

74. $\log_4 2 + \log_4 32$

75. $\ln e^2 - \ln e^5$

76. $3\ln e^4$

77. $\log_{10} 0$

78. $\ln 1$

79. $\ln e^{4.5}$

80. $\ln \sqrt[4]{e^3}$

In Exercises 81–88, use the properties of logarithms to simplify the logarithmic expression.

81. $\log_4 8$

82. $\log_5\left(\frac{1}{15}\right)$

83. $\log_7 \sqrt{70}$

84. $\log_2(4^2 \cdot 3^4)$

85. $\log_5\left(\frac{1}{250}\right)$

86. $\log_{10}\left(\frac{9}{300}\right)$

87. $\ln(5e^6)$

88. $\ln \dfrac{6}{e^2}$

89. *Sound Intensity*   The relationship between the number of decibels β and the intensity of a sound I in watts per square centimeter is given by

$$\beta = 10\log_{10}\left(\frac{I}{10^{-16}}\right).$$

Use the properties of logarithms to write the formula in simpler form, and determine the number of decibels of a sound with an intensity of 10^{-10} watts per square centimeter.

90. *Human Memory Model*   Students participating in a psychological experiment attended several lectures. After the last lecture, and every month for the next year, the students were tested to see how much of the material they remembered. The average scores for the group were given by the memory model

$$f(t) = 90 - 15\log_{10}(t + 1), \quad 0 \le t \le 12$$

where t is the time in months.

(a) Use a graphing utility to graph the function over the specified domain.

(b) What was the average score on the original exam $(t = 0)$?

(c) What was the average score after 6 months?

(d) What was the average score after 12 months?

(e) When would the average score have decreased to 75?

91. *Comparing Models*   A cup of water at an initial temperature of 78°C is placed in a room at a constant temperature of 21°C. The temperature of the water is measured every 5 minutes for a period of $\frac{1}{2}$ hour. The results are recorded as ordered pairs of the form (t, T), where t is the time in minutes and T is the temperature in degrees Celsius.

$(0, 78.0°), (5, 66.0°), (10, 57.5°), (15, 51.2°),$
$(20, 46.3°), (25, 42.5°), (30, 39.6°)$

(a) The graph of the model for the data should be asymptotic with the temperature of the room. Subtract the room temperature from each of the temperatures in the ordered pairs. Use a graphing utility to plot the data points (t, T) and $(t, T - 21)$.

(b) Use the regression capabilities of a graphing utility to fit an exponential model to the revised data. This model will be of the form

$$T - 21 = ab^x.$$

Solve for T and graph the model. Compare the result with the plot of the original data.

(c) Take the natural logarithms of the revised temperatures. Use a graphing utility to plot the points $(t, \ln(T - 21))$ and observe that the points appear linear. Use the regression capabilities of a graphing utility to fit a line to this data. The resulting line has the form

$$\ln(T - 21) = at + b.$$

Use the properties of logarithms to solve for T. Verify that the result is equivalent to the model in part (b).

(d) Fit a rational model to the data. Take the reciprocals of the y-coordinates of the revised data to generate the points

$$\left(t, \frac{1}{T - 21} \right).$$

Use a graphing utility to plot these points and observe that they appear linear. Use the regression capabilities of a graphing utility to fit a line

to this data. The resulting line has the form

$$\frac{1}{T - 21} = at + b.$$ Solve for T, and use a graphing utility to graph the rational function and the original data points.

92. *Essay*   Write a short paragraph explaining why the transformations of the data in Exercise 91 were necessary to obtain the models. Why did taking the logarithms of the temperatures lead to a linear scatter plot? Why did taking the reciprocals of the temperatures lead to a linear scatter plot?

In Exercises 93–96, use the change-of-base formula and a graphing utility to graph each function.

93. $f(x) = \log_2 x$ **94.** $f(x) = \log_7 x$

95. $g(x) = \log_3 x^{1/2}$ **96.** $f(t) = \log_5 \dfrac{t}{3}$

True or False?   In Exercises 97–102, determine whether the statement is true or false given that $f(x) = \ln x$. If the statement is false, state why or give an example showing that it is false.

97. $f(0) = 0$

98. $f(ax) = f(a) + f(x), \quad a > 0, x > 0$

99. $f(x - 2) = f(x) - f(2), \quad x > 2$

100. $\sqrt{f(x)} = \frac{1}{2} f(x)$

101. If $f(u) = 2f(v)$, then $v = u^2$.

102. If $f(x) < 0$, then $0 < x < 1$.

103. Prove that $\log_b \dfrac{u}{v} = \log_b u - \log_b v$.

104. Prove that $\log_b u^n = n \log_b u$.

Review   Solve Exercises 105–108 as a review of the skills and problem-solving techniques you learned in previous sections. Simplify the expression.

105. $\dfrac{24xy^{-2}}{16x^{-3}y}$ **106.** $\left(\dfrac{2x^2}{3y} \right)^{-3}$

107. $(18x^3y^4)^{-3}(18x^3y^4)^3$ **108.** $xy(x^{-1} + y^{-1})^{-1}$

3.4 Solving Exponential and Logarithmic Equations

Introduction / Solving Exponential Equations / Solving Logarithmic Equations / Approximating Solutions / Application

Introduction

You can use your graphing utility to verify the inverse properties of logarithmic and exponential functions. For instance, try graphing

$$y_1 = e^{\ln x}$$

$$y_2 = \ln e^x.$$

Which of these graphs shows only part of the line $y = x$? Why?

So far in this chapter, you have studied the definitions, graphs, and properties of exponential and logarithmic functions. In this section, you will study procedures for *solving equations* involving exponential and logarithmic functions. As a simple example, consider the exponential equation $2^x = 32$. One property of exponential functions states that $a^x = a^y$ if and only if $x = y$. You can obtain the solution by rewriting the sample equation in the form $2^x = 2^5$, which implies that $x = 5$.

Although this method works in some cases, it does not work for an equation as simple as $e^x = 7$. In such a case, solution procedures are based on the fact that the exponential and logarithmic functions are inverses of each other. The following comparisons show the **inverse properties** of exponential and logarithmic functions.

Base a	*Base e*
1. $\log_a a^x = x$	$\ln e^x = x$
2. $a^{\log_a x} = x$	$e^{\ln x} = x$

To solve $e^x = 7$, you can take the natural logarithms of both sides to obtain

$e^x = 7$	Original equation
$\ln e^x = \ln 7$	Take logarithms of both sides.
$x = \ln 7$	$\ln e^x = x$ because $e^x = e^x$.

Here are some guidelines for solving exponential and logarithmic equations.

Solving Exponential and Logarithmic Equations

1. *To solve an exponential equation,* first isolate the exponential expression, then take the logarithms of both sides and solve for the variable.

2. *To solve a logarithmic equation,* rewrite the equation in exponential form and solve for the variable.

Solving Exponential Equations

EXAMPLE 1   **Solving an Exponential Equation**

$e^x = 72$	Original equation
$\ln e^x = \ln 72$	Take logarithms of both sides.
$x = \ln 72$	Inverse property of logs and exponents
$x \approx 4.277$	

The solution is $x = \ln 72$. Check this solution in the original equation.

EXAMPLE 2   **Solving an Exponential Equation**

Figure 3.20

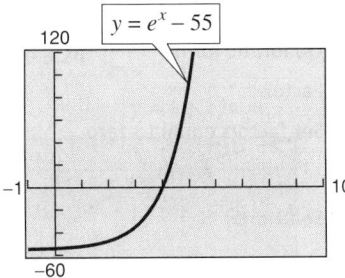

$e^x + 5 = 60$	Original equation
$e^x = 55$	Combine like terms.
$\ln e^x = \ln 55$	Take logarithms of both sides.
$x = \ln 55$	Inverse property of logs and exponents
$x \approx 4.007$	

The solution is $x = \ln 55$. Check this solution in the original equation. The graph of $y = e^x - 55$ is shown in Figure 3.20. Use the root feature of your graphing utility to find the x-intercept and thus confirm the solution.

EXAMPLE 3   **Solving an Exponential Equation**

Figure 3.21

$4e^{2x} = 5$	Original equation
$e^{2x} = \dfrac{5}{4}$	Divide both sides by 4.
$\ln e^{2x} = \ln \dfrac{5}{4}$	Take logarithms of both sides.
$2x = \ln \dfrac{5}{4}$	Inverse property of logs and exponents
$x = \dfrac{1}{2} \ln \dfrac{5}{4}$	Solve for x.
$x \approx 0.112$	

The solution is $x = \frac{1}{2} \ln \frac{5}{4}$. Check this solution in the original equation. The graph of $y = 4e^{2x} - 5$, shown in Figure 3.21, helps to confirm this solution.

Application

Figure 3.28

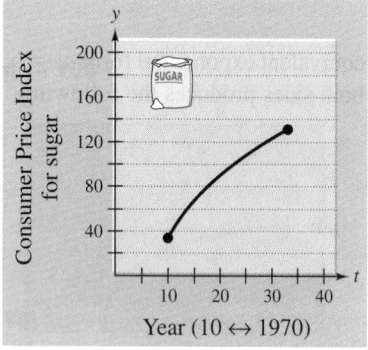

Year (10 ↔ 1970)

EXAMPLE 11 ▱ Consumer Price Index for Sugar

From 1970 to 1993, the Consumer Price Index (CPI) value y for a fixed amount of sugar for the year t can be modeled by the equation $y = -169.8 + 86.8 \ln t$, where $t = 10$ represents 1970 (see Figure 3.28). During which year did the price of sugar reach 4 times its 1970 price of 30.5 on the CPI? (Source: U.S. Bureau of Labor Statistics)

Solution

$-169.8 + 86.8 \ln t = y$	Given model
$-169.8 + 86.8 \ln t = 122$	Let $y = (4)(30.5) = 122$.
$86.8 \ln t = 291.8$	Add 169.8 to both sides.
$\ln t \approx 3.362$	Divide both sides by 86.8.
$e^{\ln t} \approx e^{3.362}$	Exponentiate both sides.
$t \approx 28.8$	Inverse property of exponents and logs

The solution is $t \approx 28.8$ years. Because $t = 10$ represents 1970, it follows that the price of sugar reached four times its 1970 price in 1988. ▱

Group Activity   *Comparing Mathematical Models*

The table below gives the monthly amount of traffic y (in millions of packets) on the NSFNET, the backbone of the Internet, for January of each year x from 1988 through 1995, with $x = 8$ representing 1988. (Source: Network Information Center)

x	8	9	10	11	12	13	14	15
y	85	468	2466	5868	12,153	24,016	45,283	60,399

(a) Create a scatter plot of the data. Find a linear model for the data, and add its graph to your scatter plot. According to this model, when will monthly network traffic reach 150,000 million packets?

(b) Create a new table giving values for $\ln x$ and $\ln y$, and create a scatter plot of the data. Use the method illustrated in Example 9 in Section 3.3 to find a model for the data, and add its graph to your scatter plot. According to this model, when will monthly network traffic reach 150,000 million packets?

(c) Solve the model in part (b) for y, and add its graph to your scatter plot in part (a). Which model best fits the data? Which model will best predict future traffic levels? Explain your reasoning.

3.4 /// EXERCISES

In Exercises 1–6, determine whether the *x*-values are solutions of the equation.

1. $4^{2x-7} = 64$

   (a) $x = 5$

   (b) $x = 2$

2. $2^{3x+1} = 32$

   (a) $x = -1$

   (b) $x = 2$

3. $3e^{x+2} = 75$

   (a) $x = -2 + e^{25}$

   (b) $x = -2 + \ln 25$

   (c) $x \approx 1.2189$

4. $5^{2x+3} = 812$

   (a) $x = -1.5 + \log_5 \sqrt{812}$

   (b) $x \approx 0.5813$

   (c) $x = \frac{1}{2}\left(-3 + \dfrac{\ln 812}{\ln 5}\right)$

5. $\log_4(3x) = 3$

   (a) $x \approx 20.3560$

   (b) $x = -4$

   (c) $x = \frac{64}{3}$

6. $\ln(x - 1) = 3.8$

   (a) $x = 1 + e^{3.8}$

   (b) $x \approx 45.7012$

   (c) $x = 1 + \ln 3.8$

In Exercises 7–10, approximate the point of intersection of the graphs of f and g. Then solve the equation $f(x) = g(x)$ algebraically.

7. $f(x) = 2^x$

   $g(x) = 8$

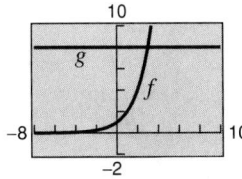

8. $f(x) = 27^x$

   $g(x) = 9$

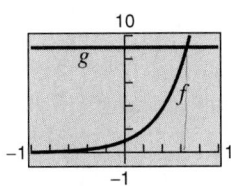

9. $f(x) = \log_3 x$

   $g(x) = 2$

10. $f(x) = \ln (x - 4)$

   $g(x) = 0$

In Exercises 11–20, solve for *x*.

11. $4^x = 16$

12. $3^x = 243$

13. $7^x = \frac{1}{49}$

14. $8^x = 4$

15. $\left(\frac{3}{4}\right)^x = \frac{27}{64}$

16. $3^{x-1} = 27$

17. $\log_4 x = 3$

18. $\log_5 5x = 2$

19. $\log_{10} x = -1$

20. $\ln(2x - 1) = 0$

In Exercises 21–26, simplify the expression.

21. $\ln e^{x^2}$

22. $\ln e^{2x-1}$

23. $e^{\ln(5x+2)}$

24. $-1 + \ln e^{2x}$

25. $e^{\ln x^2}$

26. $-8 + e^{\ln x^3}$

In Exercises 27–34, solve the exponential equation algebraically. Round the result to three decimal places.

27. $e^x = 10$

28. $4e^x = 91$

29. $7 - 2e^x = 5$

30. $-14 + 3e^x = 11$

31. $500e^{-x} = 300$

32. $1000e^{-4x} = 75$

33. $10^x = 42$

34. $10^x = 570$

In Exercises 35–38, let $f(x)$ represent the left side of the equation. Complete the table to obtain a rough estimate of the solution, and then use a graphing utility to graph both sides of the equation to obtain a better estimate of the solution (use three-decimal-place accuracy).

35. $e^{3x} = 12$

x	0.6	0.7	0.8	0.9	1.0
$f(x)$					

36. $e^{2x} = 50$

x	1.6	1.7	1.8	1.9	2.0
$f(x)$					

37. $20(100 - e^{x/2}) = 500$

x	5	6	7	8	9
$f(x)$					

38. $\dfrac{400}{1 + e^{-x}} = 350$

x	0	1	2	3	4
$f(x)$					

In Exercises 39–50, use the root feature of a graphing utility to solve the exponential equation accurate to three decimal places.

39. $e^{2x} - 4e^x - 5 = 0$ **40.** $e^{2x} - 5e^x + 6 = 0$

41. $3^{2x} = 80$ **42.** $6^{5x} = 3000$

43. $5^{-t/2} = 0.20$ **44.** $4^{-3t} = 0.10$

45. $2^{3-x} = 565$ **46.** $\dfrac{3000}{2 + e^{2x}} = 2$

47. $8(10^{3x}) = 12$ **48.** $3(5^{x-1}) = 21$

49. $\left(1 + \dfrac{0.10}{12}\right)^{12t} = 2$ **50.** $\left(1 + \dfrac{0.065}{365}\right)^{365t} = 4$

In Exercises 51–54, use a graphing utility to graph the function and approximate its zero accurate to three decimal places.

51. $g(x) = 6e^{1-x} - 25$ **52.** $f(x) = 3e^{3x/2} - 962$

53. $g(t) = e^{0.09t} - 3$ **54.** $h(t) = e^{0.125t} - 8$

In Exercises 55–60, solve the equation algebraically. Round the result to three decimal places.

55. $\ln x = -3$ **56.** $\ln x = 2$

57. $\ln\sqrt{x+2} = 1$ **58.** $\ln(x+1)^2 = 2$

59. $\log_{10}(x+4) - \log_{10}x = \log_{10}(x+2)$

60. $\log_4 x - \log_4(x-1) = \frac{1}{2}$

In Exercises 61–64, let $f(x)$ represent the left side of the equation. Complete the table to obtain a rough estimate of the solution, and then use a graphing utility to graph both sides of the equation to obtain a better estimate of the solution (use three-decimal-place accuracy).

61. $\ln 2x = 2.4$

x	2	3	4	5	6
$f(x)$					

62. $3 \ln 5x = 10$

x	4	5	6	7	8
$f(x)$					

63. $6 \log_3(0.5x) = 11$

x	12	13	14	15	16
$f(x)$					

64. $5 \log_{10}(x - 2) = 11$

x	150	155	160	165	170
$f(x)$					

In Exercises 65–76, use the root or zero feature of a graphing utility to solve the logarithmic equation accurate to three decimal places.

65. $\log_{10}(z - 3) = 2$ **66.** $\log_{10}x^2 = 6$

67. $2 \ln x = 7$ **68.** $\ln 4x = 1$

69. $\ln x + \ln(x - 2) = 1$

70. $\ln x + \ln(x + 3) = 1$

71. $\log_3 x + \log_3(x^2 - 8) = \log_3 8x$

72. $\log_2 x + \log_2(x + 2) = \log_2(x + 6)$

73. $\ln(x + 5) = \ln(x - 1) - \ln(x + 1)$

74. $\ln(x + 1) - \ln(x - 2) = \ln x^2$

75. $\ln x + \ln(x^2 + 1) = 8$

76. $\log_{10} 8x - \log_{10}(1 + \sqrt{x}) = 2$

In Exercises 77–80, use a graphing utility to approximate the point of intersection of the graphs. Round the result to three decimal places.

77. $y_1 = 7$

$y_2 = 2^x$

78. $y_1 = 500$

$y_2 = 1500e^{-x/2}$

79. $y_1 = 3$

$y_2 = \ln x$

80. $y_1 = 10$

$y_2 = 4 \ln(x - 2)$

Compound Interest In Exercises 81 and 82, find the time required for a $1000 investment to double at interest rate r, compounded continuously.

81. $r = 0.085$ **82.** $r = 0.12$

83. *Think About It* Is the time required for the investments in Exercises 81 and 82 to quadruple twice as long as the time for them to double? Give a reason for your answer and verify your answer algebraically.

84. *Essay* Write a paragraph explaining whether or not the time required for an investment to double is dependent on the size of the investment.

Compound Interest In Exercises 85 and 86, find the time required for a $1000 investment to triple at interest rate r, compounded continuously.

85. $r = 0.085$ **86.** $r = 0.12$

87. *Average Heights* The percents of American males and females between the ages of 18 and 24 who are at least x inches tall are given by

$$m(x) = \frac{100}{1 + e^{-0.6114(x - 69.71)}} \quad \text{Males}$$

$$f(x) = \frac{100}{1 + e^{-0.66607(x - 64.51)}} \quad \text{Females}$$

where m and f are the percents and x is the height in inches. (Source: U.S. National Center for Health Statistics)

(a) Use a graphing utility to graph the two functions in the same viewing rectangle.

(b) Use the graph to determine the horizontal asymptotes of the functions.

(c) What is the median height for each sex?

88. *Human Memory Model* In a group project in learning theory, a mathematical model for the proportion P of correct responses after n trials was found to be

$$P = \frac{0.83}{1 + e^{-0.2n}}.$$

(a) Use a graphing utility to graph the function.

(b) Use the graph to determine the horizontal asymptotes of the function. Interpret the meaning of the upper asymptote in the context of the problem.

(c) After how many trials will 60% of the responses be correct?

89. *Demand Function* The demand equation for a certain product is given by

$$p = 500 - 0.5(e^{0.004x}).$$

Find the demands x for prices of (a) $p = \$350$ and (b) $p = \$300$.

90. *Demand Function* The demand equation for a certain product is given by

$$p = 5000\left(1 - \frac{4}{4 + e^{-0.002x}}\right).$$

Find the demands x for prices of (a) $p = \$600$ and (b) $p = \$400$.

91. *Forest Yield* The yield V (in millions of cubic feet per acre) for a forest at age t years is given by

$$V = 6.7e^{-48.1/t}.$$

(a) Use a graphing utility to graph the function.

(b) Determine the horizontal asymptote of the function. Interpret its meaning in the context of the problem.

(c) Find the time necessary to obtain a yield of 1.3 million cubic feet.

92. *Trees per Acre* The number of trees per acre N of a certain species is approximated by the model

$$N = 68(10^{-0.04x}), \qquad 5 \leq x \leq 40$$

where x is the average diameter of the trees three feet above the ground. Use the model to approximate the average diameter of the trees in a test plot when $N = 21$.

93. *Data Analysis*   An object at a temperature of 160°C is removed from a furnace and placed in a room at 20°C. The temperature *T* of the object is measured each hour *h* and recorded in the table.

h	0	1	2	3	4	5
T	160°	90°	56°	38°	29°	24°

A model for this data is

$$T = 20[1 + 7(2^{-h})].$$

(a) Use a graphing utility to plot the data points and graph the model in the same viewing rectangle.

(b) Use the graph to identify the horizontal asymptote of the model and interpret the asymptote in the context of the problem.

(c) Approximate the time when the temperature of the object is 100°C.

94. *Chapter Opener*   To protect the occupants of automobiles in accidents, the automobiles are designed with crumple zones. Most people will black out under a force of 5 g's (five times the force of gravity) if it is applied for 10 seconds or more, but for very short durations, humans have withstood as much as 40 g's. In crash tests with vehicles moving at 90 kilometers per hour, analysts measured the numbers *y* of g's that were undergone during deceleration by crash dummies that were permitted (by crumple zones and passive restraint systems) to move distances of *x* meters during impact. The data is shown in the table.

x	0.2	0.4	0.6	0.8	1
y	158	80	53	40	32

(a) Use the regression capabilities of a graphing utility to find a natural logarithmic model for the data. A second model for the data is $y = -3.00 + 11.88 \ln x + (36.94/x)$. Graph both models with the data points in the same viewing window. Which model do you think is a better choice? Comment on accuracy and simplicity.

(b) Use both models to estimate the distance traveled during impact if the passenger deceleration must not exceed 30 g's. Comment on the difference in distances traveled given by the two models and the practical implications of choosing a model when designing the crumple zones.

Review   **Solve Exercises 95–100 as a review of the skills and problem-solving techniques you learned in previous sections. Match the equation with its graph, and identify any intercepts. [The graphs are labeled (a), (b), (c), (d), (e), and (f).]**

(a)   (b)

(c)   (d)

(e)   (f)

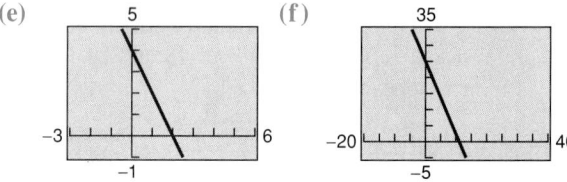

95. $4x - 3y - 9 = 0$

96. $2x + 5y - 10 = 0$

97. $y = 25 - 2.25x$

98. $\dfrac{x}{2} + \dfrac{y}{4} = 1$

99. $y - 3 = 0$

100. $x + 2 = 0$

3.5   Exponential and Logarithmic Models

Introduction / Exponential Growth and Decay / Gaussian Models / Logistics Growth Models / Logarithmic Models

Introduction

The five most common types of mathematical models involving exponential functions and logarithmic functions are as follows.

1. Exponential growth: $y = ae^{bx}, \quad b > 0$
2. Exponential decay: $y = ae^{-bx}, \quad b > 0$
3. Gaussian model: $y = ae^{-(x-b)^2/c}$
4. Logistics growth model: $y = \dfrac{a}{1 + be^{-(x-c)/d}}$
5. Logarithmic models: $y = a + b \ln x, \quad y = a + b \log_{10} x$

Note The compound interest problems studied in Section 3.1 are examples of exponential growth.

The basic shapes of these graphs are shown in Figure 3.29.

Figure 3.29

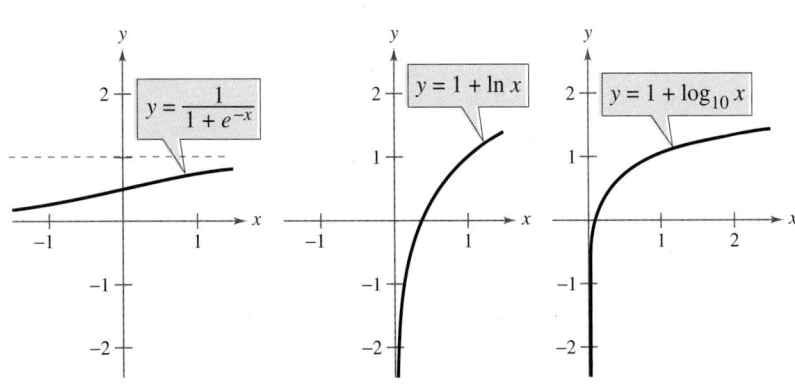

Study Tip

You can often gain quite a bit of insight into a situation modeled by an exponential or logarithmic function by identifying and interpreting the function's asymptotes. Use the graphs in Figure 3.29 to identify the asymptotes of each function.

Exponential Growth and Decay

EXAMPLE 1   ▱   **Population Increase**

Real Life

The world population (in millions) from 1980 through 1992 is shown in the table. The scatter plot of the data is shown in Figure 3.30(a). (Source: Statistical Office of the United Nations)

Figure 3.30

(a)

Year	1980	1985	1986	1987	1988	1989	1990	1991	1992
Population	4453	4850	4936	5024	5112	5202	5294	5384	5478

An exponential growth model that approximates this data is given by

$$P = 4451e^{0.017303t}, \quad 0 \le t \le 12$$

where P is the population (in millions) and $t = 0$ represents 1980. Compare the values given by the model with the estimates given by the United Nations. According to this model, when will the world population reach 6 billion?

(b)

Solution

The following table compares the two sets of population figures. The graph of the model is shown in Figure 3.30(b).

Year	1980	1985	1986	1987	1988	1989	1990	1991	1992
Population	4453	4850	4936	5024	5112	5202	5294	5384	5478
Model	4451	4853	4938	5024	5112	5201	5292	5384	5478

To find when the world population will reach 6 billion, let $P = 6000$ in the model and solve for t.

$4451e^{0.017303t} = P$	Given model
$4451e^{0.017303t} = 6000$	Let $P = 6000$.
$e^{0.017303t} \approx 1.348$	Divide both sides by 4451.
$\ln e^{0.017303t} \approx \ln 1.348$	Take logarithms of both sides.
$0.017303t \approx 0.2986$	Inverse property of logs and exponents
$t \approx 17.26$	Divide both sides by 0.017303.

Note   An exponential model increases (or decreases) by the same percent each year. What is the annual percent increase for the model in Example 1?

According to the model, the world population will reach 6 billion in 1997.

▱

In Example 1, you were given the exponential growth model. But suppose this model had not been given; how could you have found such a model? One technique for doing this is demonstrated in Example 2.

Some graphing utilities have a curve-fitting feature that can be used to find models that represent data. For instance, to find the exponential growth model that approximates the data in the table in Example 1, you can use the following steps using a *TI-82* or a *TI-83*.

1. Press STAT   ENTER .
2. Enter the data in L_1 and L_2.
3. Press STAT   ▷ .
4. Select *ExpReg:* and press
    ENTER   ENTER .

Write the exponential growth model you obtain. How does it compare with the model in Example 1?

EXAMPLE 2   Finding an Exponential Growth Model

Find an exponential growth model whose graph passes through the points $(0, 4453)$ and $(7, 5024)$, as shown in Figure 3.31(a).

Solution

The general form of the model is

$$y = ae^{bx}.$$

From the fact that the graph passes through the point $(0, 4453)$, you know that $y = 4453$ when $x = 0$. By substituting these values into the general form of the model, you have

$$4453 = ae^0$$
$$4453 = a.$$

In a similar way, from the fact that the graph passes through the point $(7, 5024)$, you know that $y = 5024$ when $x = 7$. By substituting these values into the model, you have

$$5024 = 4453e^{7b} \qquad \Longrightarrow \qquad b = \frac{1}{7}\ln\frac{5024}{4453} \approx 0.01724.$$

Thus, the exponential growth model is

$$y = 4453e^{0.01724x}.$$

The graph of the model is shown in Figure 3.31(b).

Figure 3.31

(a)

(b)

Figure 3.32

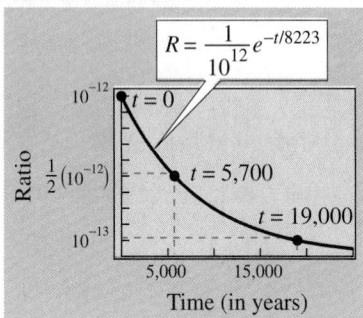

$$R = \frac{1}{10^{12}} e^{-t/8223}$$

In living organic material, the ratio of the content of radioactive carbon (carbon 14) to the content of nonradioactive carbon (carbon 12) is about 1 to 10^{12}. When organic material dies, its carbon 12 content remains fixed, whereas its radioactive carbon 14 begins to decay with a half-life of about 5700 years. To estimate the age of dead organic material, scientists use the following formula, which denotes the ratio of carbon 14 to carbon 12 present at any time t (in years).

$$R = \frac{1}{10^{12}} e^{-t/8223}$$

The graph of R is shown in Figure 3.32. Note that R decreases as t increases.

Real Life

EXAMPLE 3   Carbon Dating

The ratio of carbon 14 to carbon 12 in a newly discovered fossil is

$$R = \frac{1}{10^{13}}.$$

Estimate the age of the fossil.

Solution

In the carbon dating model, substitute the given value of R to obtain the following.

$$\frac{1}{10^{12}} e^{-t/8223} = R \qquad \text{Given model}$$

$$\frac{e^{-t/8223}}{10^{12}} = \frac{1}{10^{13}} \qquad \text{Let } R = \frac{1}{10^{13}}.$$

$$e^{-t/8223} = \frac{1}{10} \qquad \text{Multiply both sides by } 10^{12}.$$

$$\ln e^{-t/8223} = \ln \frac{1}{10} \qquad \text{Take logarithms of both sides.}$$

$$-\frac{t}{8223} \approx -2.3026 \qquad \text{Inverse property of logs and exponents}$$

$$t \approx 18{,}934 \qquad \text{Solve for } t.$$

Thus, to the nearest thousand years, you can estimate the age of the fossil to be 19,000 years. How could you verify this answer with a graphing utility?

Note The carbon dating model in Example 3 assumed that the carbon 14/carbon 12 ratio was one part in 10,000,000,000,000. Suppose an error in measurement occurred and the actual ratio was only one part in 8,000,000,000,000. The fossil age corresponding to the actual ratio would then be approximately 17,000 years. Try checking this result.

Gaussian Models

As mentioned at the beginning of this section, Gaussian models are of the form

$$y = ae^{-(x-b)^2/c}.$$

This type of model is commonly used in probability and statistics to represent populations that are **normally distributed.** For *standard* normal distributions, the model takes the form

$$y = \frac{1}{\sigma\sqrt{2\pi}}e^{-x^2/2\sigma^2}$$

where σ is the standard deviation (σ is the lowercase Greek letter sigma). The graph of a Gaussian model is called a **bell-shaped curve.** Try assigning a value to σ and sketching a normal distribution curve with a graphing utility. Can you see why it is called a bell-shaped curve?

Real Life

EXAMPLE 4   ▭   **SAT Scores**

In 1993, the Scholastic Aptitude Test (SAT) scores for males roughly followed a normal distribution given by

$$y = 0.0026e^{-(x-500)^2/48,000}, \quad 200 \le x \le 800$$

where x is the SAT score for mathematics. Sketch the graph of this function. From the graph, estimate the average SAT score.   (Source: College Board)

Solution

The graph of the function is given in Figure 3.33. From the graph, you can see that the average mathematics score for males in 1988 was 500.

Figure 3.33

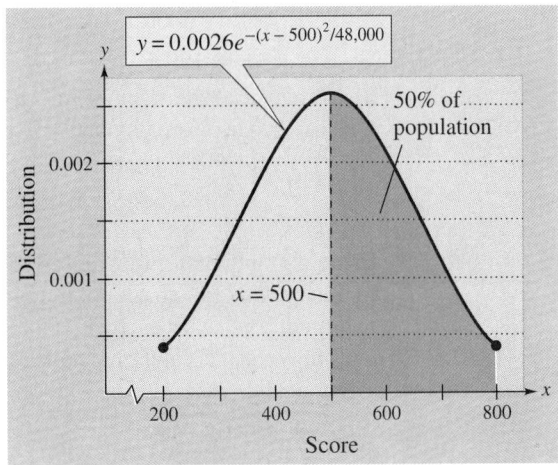

Logistics Growth Models

Some populations initially have rapid growth, followed by a declining rate of growth, as indicated by the graph in Figure 3.34. One model for describing this type of growth pattern is the **logistics curve** given by the function

$$y = \frac{a}{1 + be^{-(x-c)/d}}$$

where y is the population size and x is the time. An example is a bacteria culture allowed to grow initially under ideal conditions, followed by less favorable conditions that inhibit growth. A logistics growth curve is also called a **sigmoidal curve.**

Figure 3.34   Logistics Curve

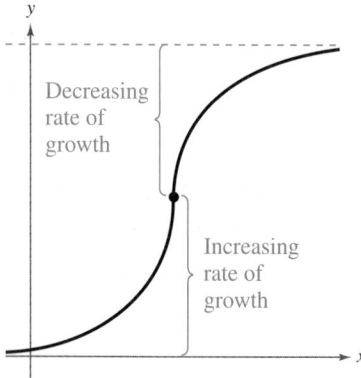

Real Life

EXAMPLE 5   Spread of a Virus

On a college campus of 5000 students, one student returns from vacation with a contagious flu virus. The spread of the virus is modeled by

$$y = \frac{5000}{1 + 4999e^{-0.8t}}, \quad 0 \le t$$

where y is the total number infected after t days. The college will cancel classes when 40% or more of the students are ill.

a. How many students are infected after 5 days?

b. After how many days will the college cancel classes?

Solution

a. After 5 days, the number infected is

$$y = \frac{5000}{1 + 4999e^{-0.8(5)}} = \frac{5000}{1 + 4999e^{-4}} \approx 54.$$

b. In this case, the number infected must be $(0.40)(5000) = 2000$ to cancel classes. Therefore, you can solve for t in the following equation.

$$2000 = \frac{5000}{1 + 4999e^{-0.8t}}$$

Using a graphing utility, graph $y = 2000$ and $y = 5000/(1 + 4999e^{-0.8t})$ and find the point of intersection. In Figure 3.35, you can see that the point of intersection occurs near $t \approx 10.1$. Hence, after 10 days, at least 40% of the students will be infected, and the college will cancel classes. Verify this answer algebraically.

Figure 3.35

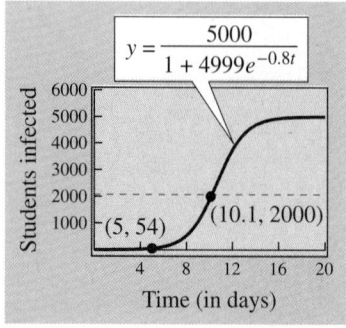

Logarithmic Models

Real Life

EXAMPLE 6   **Magnitudes of Earthquakes**

On the Richter scale, the magnitude R of an earthquake of intensity I is given by

$$R = \log_{10} \frac{I}{I_0}$$

where $I_0 = 1$ is the minimum intensity used for comparison. Find the intensities per unit of area for the following earthquakes. (Intensity is a measure of the wave energy of the earthquake.)

a. Tokyo and Yokohama, Japan, in 1923, $R = 8.3$

b. Kobe, Japan, in 1995, $R = 7.2$

Solution

a. Because $I_0 = 1$ and $R = 8.3$, you have

$$8.3 = \log_{10} I$$
$$I = 10^{8.3} \approx 199{,}526{,}000.$$

b. For $R = 7.2$, you have $7.2 = \log_{10} I$, and $I = 10^{7.2} \approx 15{,}849{,}000$.

Note that an increase of 1.1 units on the Richter scale (from 7.2 to 8.3) represents an intensity change by a factor of

$$\frac{199{,}526{,}000}{15{,}849{,}000} \approx 13.$$

In other words, the earthquake in 1923 had a magnitude about 13 times greater than that of the 1995 quake.

Group Activity   *Identifying Appropriate Models*

Decide for each data set which model from this section would provide the best fit. Discuss your reasoning with others in your group.

Data Set A: (18.4, 1.07), (19, 1.21), (20, 1.45), (22, 1.85), (23.5, 1.99), (25, 1.96), (26.3, 1.80), (27, 1.67), (29.7, 1.04), (31, 0.75)

Data Set B: (1.5, 11.03), (2, 12.47), (3.5, 15.26), (5, 17.05), (7.8, 19.27), (9, 19.99), (10.2, 20.61), (13.6, 22.05), (19.3, 23.80), (27, 25.48)

3.5 /// EXERCISES

In Exercises 1–6, match the function with its graph. [The graphs are labeled (a) through (f).]

(a)

(b)

(c)

(d)

(e)

(f)
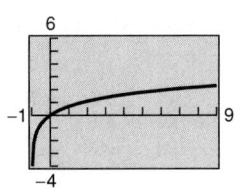

1. $y = 2e^{x/4}$

2. $y = 6e^{-x/4}$

3. $y = \frac{1}{16}(x^2 + 8x + 32)$

4. $y = \frac{12}{x + 4}$

5. $y = \ln(x + 1)$

6. $y = \sqrt{x}$

Compound Interest In Exercises 7–14, complete the table for a savings account in which interest is compounded continuously.

Initial Investment	Annual % Rate	Time to Double	Amount After 10 Years
7. $1000	12%		
8. $20,000	$10\frac{1}{2}\%$		
9. $750		$7\frac{3}{4}$ yr	
10. $10,000		5 yr	

Initial Investment	Annual % Rate	Time to Double	Amount After 10 Years
11. $500			$1292.85
12. $600			$19,205.00
13.	4.5%		$10,000.00
14.	8%		$20,000.00

Compound Interest In Exercises 15 and 16, determine the principal P that must be invested at rate r, compounded monthly, so that $500,000 will be available for retirement in t years.

15. $r = 7\frac{1}{2}\%, t = 20$ **16.** $r = 12\%, t = 40$

Compound Interest In Exercises 17 and 18, determine the time necessary for $1000 to double if it is invested at interest rate r compounded (a) annually, (b) monthly, (c) daily, and (d) continuously.

17. $r = 11\%$ **18.** $r = 10\frac{1}{2}\%$

19. *Compound Interest* Complete the table for the time t necessary for P dollars to triple if interest is compounded continuously at rate r.

r	2%	4%	6%	8%	10%	12%
t						

20. *Modeling Data* Draw a scatter plot of the data in Exercise 19. Use the regression capabilities of a graphing utility to find a model for the data.

21. *Compound Interest* Complete the table for the time t necessary for P dollars to triple if interest is compounded annually at rate r.

r	2%	4%	6%	8%	10%	12%
t						

22. *Modeling Data* Draw a scatter plot of the data in Exercise 21. Use the regression capabilities of a graphing utility to find a model for the data.

23. *Comparing Investments* If $1 is invested in an account over a 10-year period, the amount in the account, where t represents the time in years, is

$$A = 1 + 0.075[\![t]\!] \quad \text{or} \quad A = e^{0.07t}$$

depending on whether the account pays simple interest at $7\frac{1}{2}\%$ or continuous compound interest at 7%. Use a graphing utility to graph each function in the same viewing rectangle. Which grows at a faster rate?

24. *Comparing Investments* If $1 is invested in an account over a 10-year period, the amount in the account, where t represents the time in years, is

$$A = 1 + 0.06[\![t]\!] \quad \text{or} \quad A = \left(1 + \frac{0.055}{365}\right)^{[\![365t]\!]}$$

depending on whether the account pays simple interest at 6% or compound interest at $5\frac{1}{2}\%$ compounded daily. Use a graphing utility to graph each function in the same viewing rectangle. Which grows at a faster rate?

In Exercises 25–28, complete the table for the given radioactive isotope.

Isotope	Half-Life (years)	Initial Quantity	Amount After 1000 Years
25. ^{226}Ra	1620	10g	
26. ^{226}Ra	1620		1.5g
27. ^{14}C	5730	3g	
28. ^{230}Pu	24,360		0.4g

In Exercises 29–32, find the exponential model $y = ae^{bx}$ that fits the points given in the graph or table.

29.

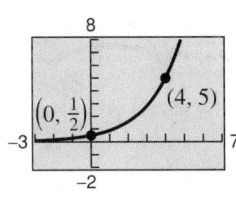

30.

31.

x	y
0	1
3	$\frac{1}{4}$

32.

x	y
0	5
4	1

33. *Population* The population P of a city is given by

$$P = 105{,}300e^{0.015t}$$

where $t = 0$ represents 1990. According to this model, when will the population reach 150,000?

34. *Population* The population P of a city is given by

$$P = 240{,}360e^{0.012t}$$

where $t = 0$ represents 1990. According to this model, when will the population reach 275,000?

35. *Population* The population P of a city is given by

$$P = 2500e^{kt}$$

where $t = 0$ represents 1990. In 1945, the population was 1350. Find the value of k, and use this value to predict the population in the year 2010.

36. *Population* The population P of a city is given by

$$P = 140{,}500e^{kt}$$

where $t = 0$ represents 1990. In 1960, the population was 100,250. Find the value of k, and use this value to predict the population in the year 2000.

Population Exercises 37–40 give the populations (in millions) of four cities in 1990 and the projected populations (in millions) for the year 2000. Find the exponential growth model $y = ae^{bt}$ for each population by letting $t = 0$ correspond to 1990. Use the model to predict the population of the city in 2010. (Source: U.S. Bureau of the Census, International Database)

City	1990	2000
37. Dhaka, Bangladesh	4.22	6.49
38. Houston, Texas	2.30	2.65
39. Detroit, Michigan	3.00	2.74
40. London, United Kingdom	9.17	8.57

41. *Think About It* In Exercises 37 and 38, you can see that the populations of Dhaka and Houston are growing at different rates. What constant in the equation $y = ae^{bt}$ is determined by these different growth rates? Discuss the relationship between the different growth rates and the magnitude of the constant.

42. *Think About It* In Exercises 37 and 39, you can see that one population is increasing while the other is decreasing. What constant in the equation $y = ae^{bt}$ reflects this difference? Explain.

43. *Bacteria Growth* The number of bacteria N in a culture is given by the model

$$N = 100e^{kt}$$

where t is the time in hours. If $N = 300$ when $t = 5$, estimate the time required for the population to double in size. Verify your estimate graphically.

44. *Bacteria Growth* The number of bacteria N in a culture is given by the model

$$N = 250e^{kt}$$

where t is the time in hours. If $N = 280$ when $t = 10$, estimate the time required for the population to double in size. Verify your estimate graphically.

45. *Radioactive Decay* The half-life of radioactive radium (radium 226) is 1620 years. What percent of a present amount of radioactive radium will remain after 100 years?

46. *Radioactive Decay* Carbon 14 dating assumes that the carbon dioxide on earth today has the same radioactive content as it did centuries ago. If this is true, the amount of carbon 14 absorbed by a tree that grew several centuries ago should be the same as the amount of carbon 14 absorbed by a tree growing today. A piece of ancient charcoal contains only 15% as much radioactive carbon as a piece of modern charcoal. How long ago was the tree burned to make the ancient charcoal if the half-life of carbon 14 is 5730 years?

47. *Depreciation* A computer that cost $4600 new has a book value of $3000 after 2 years. Find the value of the computer after 3 years by using the exponential model

$$y = ae^{bt}.$$

48. *Comparing Models* A car that cost $22,000 new has a book value of $13,000 after 2 years.

 (a) Find the straight-line model $V = mt + b$.

 (b) Find the exponential model $V = ae^{kt}$.

 (c) Use a graphing utility to graph the two models in the same viewing rectangle. Which model depreciates faster in the first 2 years?

 (d) Find the book values of the car after 1 year and after 3 years using each model.

 (e) Interpret the slope of the straight-line model.

49. *Sales* The sales S (in thousands of units) of a new product after it has been on the market t years are given by

$$S(t) = 100(1 - e^{kt}).$$

Fifteen thousand units of the new product were sold the first year.

 (a) Complete the model by solving for k.

 (b) Use a graphing utility to graph the model.

 (c) Use the graph to estimate the number of units sold after 5 years.

50. *Sales and Advertising* After discontinuing all advertising for a certain product in 1994, the manufacturer noted that sales began to drop according to the model

$$S = \frac{500,000}{1 + 0.6e^{kt}}$$

where S represents the number of units sold and $t = 0$ represents 1994. In 1996, the company sold 300,000 units.

 (a) Complete the model by solving for k.

 (b) Estimate sales in 1999.

51. *Sales and Advertising* The sales S (in thousands of units) of a product after x hundred dollars is spent on advertising is given by

$$S = 10(1 - e^{kx}).$$

When $500 is spent on advertising, 2500 units are sold.

 (a) Complete the model by solving for k.

 (b) Estimate the number of units that will be sold if advertising expenditures are raised to $700.

52. *Profits* Because of a slump in the economy, a company finds that its annual profits have dropped from $742,000 in 1994 to $632,000 in 1996. If the profit follows an exponential pattern of decline, what is the expected profit for 1997? (Let $t = 0$ represent 1994.)

53. *Learning Curve* The management at a factory has found that the maximum number of units a worker can produce in a day is 30. The learning curve for the number of units N produced per day after a new employee has worked t days is given by

$$N = 30(1 - e^{kt}).$$

After 20 days on the job, a new employee produces 19 units (see figure).

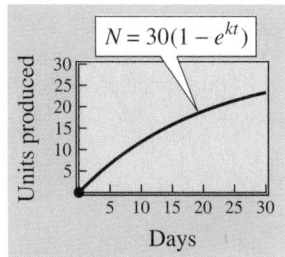

(a) Find the learning curve for this employee (first find the value of k).

(b) How many days should pass before this employee is producing 25 units per day?

(c) Is the employee's production increasing at a linear rate? Explain your reasoning.

54. *Endangered Species* A conservation organization releases 100 animals of an endangered species into a game preserve. The organization believes that the preserve has a carrying capacity of 1000 animals and that the growth of the herd will follow the logistics curve

$$p(t) = \frac{1000}{1 + 9e^{-0.1656t}}$$

where t is measured in months.

(a) Use a graphing utility to graph the function. Use the graph to determine the horizontal asymptotes and interpret the meaning of the larger asymptote in the context of the problem.

(b) Estimate the population after 5 months.

(c) When will the population reach 500?

Earthquake Magnitudes **In Exercises 55 and 56, use the Richter scale (see page 287) for measuring the magnitudes of earthquakes.**

55. Find the magnitude R of an earthquake of intensity I (let $I_0 = 1$).

(a) $I = 80{,}500{,}000$

(b) $I = 48{,}275{,}000$

56. Find the intensity I of an earthquake measuring R on the Richter scale (let $I_0 = 1$).

(a) Colombia in 1906, $R = 8.6$

(b) Los Angeles in 1971, $R = 6.7$

Intensity of Sound **In Exercises 57–60, use the following information. For Exercises 57 and 58, determine the level of sound (in decibels) for the given sound intensity.**

The level of sound β, in decibels, with an intensity I is given by $\beta(I) = 10 \log_{10} \dfrac{I}{I_0}$, where I_0 is an intensity of 10^{-16} watts per square centimeter, corresponding roughly to the faintest sound that can be heard by the human ear.

57. (a) $I = 10^{-14}$ watts per cm² (faint whisper)

(b) $I = 10^{-9}$ watts per cm² (busy street corner)

(c) $I = 10^{-6.5}$ watts per cm² (air hammer)

(d) $I = 10^{-4}$ watts per cm² (threshold of pain)

58. (a) $I = 10^{-13}$ watts per cm² (whisper)

(b) $I = 10^{-7.5}$ watts per cm² (jet 4 miles from takeoff)

(c) $I = 10^{-7}$ watts per cm² (diesel truck at 25 feet)

(d) $I = 10^{-4.5}$ watts per cm² (auto horn at 3 feet)

59. *Noise Level* As a result of the installation of noise suppression materials, the noise level in an auditorium was reduced from 93 to 80 decibels. Find the percent decrease in the intensity level of the noise due to the installation of these materials.

60. *Noise Level* As a result of the installation of a muffler, the level of noise emitted by an engine was reduced from 88 to 72 decibels. Find the percent decrease in the intensity level of the noise due to the installation of the muffler.

Acidity   In Exercises 61–66, use the acidity model given by $pH = -\log_{10}[H^+]$, where acidity (pH) is a measure of the hydrogen ion concentration $[H^+]$ (measured in moles of hydrogen per liter) of a solution.

61. Find the pH if $[H^+] = 2.3 \times 10^{-5}$.

62. Find the pH if $[H^+] = 11.3 \times 10^{-6}$.

63. Compute $[H^+]$ for a solution in which pH = 5.8.

64. Compute $[H^+]$ for a solution in which pH = 3.2.

65. A certain fruit has a pH of 2.5 and an antacid tablet has a pH of 9.5. The hydrogen ion concentration of the fruit is how many times the concentration of the tablet?

66. If the pH of a solution is decreased by one unit, the hydrogen ion concentration is increased by what factor?

67. *Home Mortgage*   A $120,000 home mortgage for 35 years at $9\frac{1}{2}\%$ has a monthly payment of $985.93. Part of the monthly payment goes for the interest charge on the unpaid balance, and the remainder of the payment is used to reduce the principal. The amount that goes for interest is given by

$$u = M - \left(M - \frac{Pr}{12}\right)\left(1 + \frac{r}{12}\right)^{12t}$$

and the amount that goes toward reduction of the principal is given by

$$v = \left(M - \frac{Pr}{12}\right)\left(1 + \frac{r}{12}\right)^{12t}.$$

In these formulas, P is the size of the mortgage, r is the interest rate, M is the monthly payment, and t is the time in years.

(a) Use a graphing utility to graph each function in the same viewing rectangle. (The viewing rectangle should show all 35 years of mortgage payments.)

(b) In the early years of the mortgage, the larger part of the monthly payment goes for what purpose? Approximate the time when the monthly payment is evenly divided between interest and principal reduction.

(c) Repeat parts (a) and (b) for a repayment period of 20 years ($M = \$1118.56$). What can you conclude?

68. *Home Mortgage*   The total interest u paid on a home mortgage of P dollars at interest rate r for t years is given by

$$u = P\left[\frac{rt}{1 - \left(\dfrac{1}{1 + r/12}\right)^{12t}} - 1\right].$$

Consider a $120,000 home mortgage at $9\frac{1}{2}\%$.

(a) Use a graphing utility to graph the total interest function.

(b) Approximate the length of the mortgage when the total interest paid is the same as the size of the mortgage. Is it possible to pay twice as much in interest charges as the size of one's mortgage?

69. *Data Analysis*   The time t (in seconds) required to attain a speed of s miles per hour from a standing start for a 1995 Dodge Avenger is given in the table.   (Source: *Road & Track*, March 1995)

s	30	40	50	60	70	80	90
t	3.4	5.0	7.0	9.3	12.0	15.8	20.0

Two models for this data are

$$t_1 = 40.757 + 0.556s - 15.817 \ln s$$

and

$$t_2 = 1.2259 + 0.0023s^2.$$

(a) Use a graphing utility to fit a linear model t_3 and an exponential model t_4 to the data.

(b) Use a graphing utility to graph the data points and each of the four models.

(c) Use a graphing utility to create a table comparing the given data with the estimates obtained from each model.

(d) Use the results of part (c) to find the sum of the absolute values of the differences between the data and the estimated values given by each model. Based on the four sums, which model do you think best fits the data? Explain.

70. *Comparing Models* A 1990 Chevrolet Beretta with a six-cylinder engine, automatic transmission, and air conditioning had a retail price of $11,500. A local dealership used the following guide for the approximate value of the car for the years 1990 through 1995. (Source: National Automobile Dealer's Association)

Year	1990	1991	1992
Value	$11,500	$9315	$9200

Year	1993	1994	1995
Value	$7935	$7130	$6095

Let *V* represent the value of the automobile in the year *t*, with $t = 0$ corresponding to 1990.

(a) Use the regression capabilities of a graphing utility to find linear and quadratic models for the data. Use the graphing utility to plot the data and graph the models in the same viewing rectangle.

(b) What does the slope represent in the linear model in part (a)?

(c) Do you think the quadratic model is realistic? Explain.

(d) Use the regression capabilities of a graphing utility to fit an exponential model to the data. Use the graphing utility to plot the data and graph the model in the same viewing rectangle. How well does the model fit the data?

(e) Fit a rational model to the data. Take the reciprocal of the depreciated values of the automobile to generate the points $\left(t, \dfrac{1}{V}\right)$. Use the regression capabilities of a graphing utility to fit a line to this data. The resulting line has the form

$$\frac{1}{V} = at + b.$$

Solve for *V* and use the graphing utility to graph the rational function and the original data points in the same viewing rectangle. How well does the model fit the data?

(f) Determine the horizontal asymptotes of the exponential and rational models. Interpret their meanings in the context of the problem.

71. *Essay* Use your school's library or some other reference source to write a paper describing John Napier's work with logarithms.

72. *Essay* Before the development of electronic calculators and graphing utilities, some computations were done on a slide rule. Use your school's library or some other reference source to write a paper describing the use of logarithmic scales on a slide rule.

73. *Estimating the Time of Death* At 8:30 A.M., a coroner was called to the home of a person who had died during the night. In order to estimate the time of death, the coroner took the person's temperature twice. At 9:00 A.M. the temperature was 85.7°, and at 9:30 A.M. the temperature was 82.8°. From these two temperatures the coroner was able to determine that the time elapsed since death and the body temperature were related by the formula

$$t = -2.5 \ln \frac{T - 70}{98.6 - 70}$$

with *t* representing the time in hours elapsed since death and *T* representing the temperature (in degrees Fahrenheit) of the person's body. Assume that the person had a normal body temperature of 98.6° at death and that the room temperature was a constant 70°. (This formula is derived from a general cooling principle called Newton's Law of Cooling.) Use the formula to estimate the time of death of the person.

Review Solve Exercises 74–77 as a review of the skills and problem-solving techniques you learned in previous sections. Divide by synthetic division.

74. $\dfrac{4x^3 + 4x^2 - 39x + 36}{x + 4}$

75. $\dfrac{8x^3 - 36x^2 + 54x - 27}{x - \frac{3}{2}}$

76. $(2x^3 - 8x^2 + 3x - 9) \div (x - 4)$

77. $(x^4 - 3x + 1) \div (x + 5)$

3.6   Exploring Data: Nonlinear Models

Classifying Scatter Plots / *Fitting Nonlinear Models to Data* /
Application

Classifying Scatter Plots

In Section 1.6, you saw how to fit linear models to data. In real life, many relationships between two variables are nonlinear. A scatter plot can be used to give you an idea of which type of model will best fit a set of data.

EXAMPLE 1   Classifying Scatter Plots

Decide whether each set of data could be best modeled by an exponential model, $y = ab^x$, or a logarithmic model, $y = a + b \ln x$.

a. (0.9, 1.9), (1.3, 2.4), (1.3, 2.2), (1.4, 2.4), (1.6, 2.7), (1.8, 3.0), (2.1, 3.4), (2.1, 3.3), (2.5, 4.2), (2.9, 5.1), (3.2, 6.0), (3.3, 6.2), (3.6, 7.3), (4.0, 8.8), (4.2, 9.8), (4.3, 10.2)

b. (0.9, 3.2), (1.3, 4.0), (1.3, 3.8), (1.4, 4.2), (1.6, 4.5), (1.8, 4.8), (2.1, 5.1), (2.1, 5.0), (2.5, 5.5), (2.9, 5.8), (3.2, 6.0), (3.3, 6.2), (3.6, 6.2), (4.0, 6.6), (4.2, 6.7), (4.3, 6.8)

Solution
Begin by entering the data into a graphing utility. You should obtain the scatter plots shown in Figure 3.36.

Figure 3.36

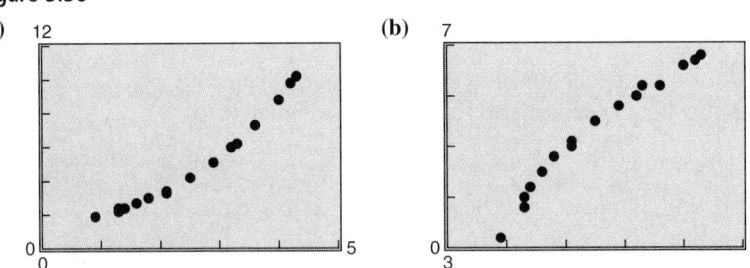

From the scatter plots, it appears that the data in part (a) can be modeled by an exponential function and the data in part (b) can be modeled by a logarithmic function.

Fitting Nonlinear Models to Data

Once you have used a scatter plot to determine the type of model to be fit to a set of data, there are several ways that you can actually find the model. Each method is best used with a computer or calculator, rather than with hand calculations.

EXAMPLE 2   **Fitting a Model to Data**

Fit the following data, from Example 1(a), with a quadratic model, an exponential model, and a power model. Which model do you think fits best?

(0.9, 1.9), (1.3, 2.4), (1.3, 2.2), (1.4, 2.4), (1.6, 2.7), (1.8, 3.0), (2.1, 3.4), (2.1, 3.3), (2.5, 4.2), (2.9, 5.1), (3.2, 6.0), (3.3, 6.2), (3.6, 7.3), (4.0, 8.8), (4.2, 9.8), (4.3, 10.2)

Solution

Begin by entering the data into a calculator or computer that has least squares regression programs. Then run the regression programs for quadratic, exponential, and power models.

Quadratic: $y = ax^2 + bx + c$ $a = 0.589$ $b = -0.685$ $c = 2.188$

Exponential: $y = ab^x$ $a = 1.203$ $b = 1.646$

Power: $y = ax^b$ $a = 1.667$ $b = 1.134$

To decide which model best fits the data, you can compare the y-values given by each model with the actual y-values. The model whose y-values are closest to the actual values is the one that fits best. In this case, the best-fitting model is the exponential model. The graphs of all three models are shown in Figure 3.37.

Note Deciding which model best fits a set of data is a question that is studied in detail in statistics. Basically, the model that fits best is the one whose sum of squared differences is the least. In Example 2, the sums of the squared differences are 0.124 for the quadratic model, 0.057 for the exponential model, and 5.98 for the power model.

Figure 3.37

Quadratic

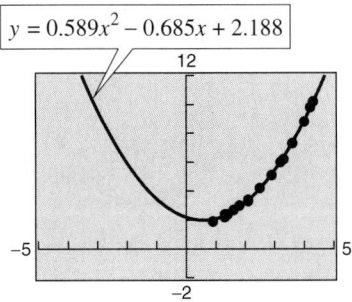

$y = 0.589x^2 - 0.685x + 2.188$

Exponential

$y = 1.203(1.646)^x$

Power

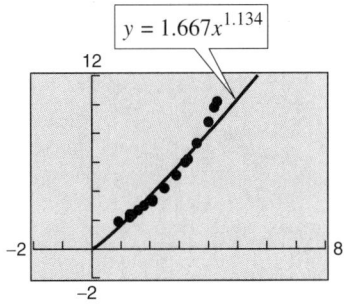

$y = 1.667x^{1.134}$

EXAMPLE 3 ▱ Fitting a Model to Data

Fit the following data, from Example 1(b), with a logarithmic model and a power model. Which model do you think fits best?

(0.9, 3.2), (1.3, 4.0), (1.3, 3.8), (1.4, 4.2), (1.6, 4.5), (1.8, 4.8), (2.1, 5.1), (2.1, 5.0), (2.5, 5.5), (2.9, 5.8), (3.2, 6.0), (3.3, 6.2), (3.6, 6.2), (4.0, 6.6), (4.2, 6.7), (4.3, 6.8)

Solution

Begin by entering the data into a calculator or computer that has least squares regression programs. Then run the regression programs for logarithmic and power models.

Logarithmic: $y = a + b \ln x$ $a = 3.377$ $b = 2.302$

Power: $y = ax^b$ $a = 3.522$ $b = 0.462$

To decide which model best fits the data, you can compare the y-values given by each model with the actual y-values. The model whose y-values are closest to the actual values is the one that fits best. In this case, both models have good fits, but the logarithmic model is slightly better. The graphs of both models are shown in Figure 3.38.

Figure 3.38

Note In Example 3, the sum of the squared differences for the logarithmic model is 0.085 and the sum of the squared differences for the power model is 0.220.

Logarithmic

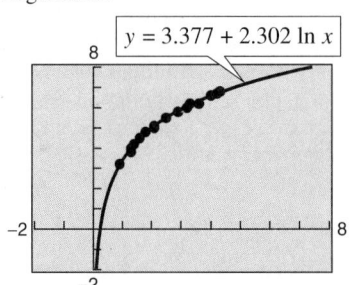

$y = 3.377 + 2.302 \ln x$

Power

$y = 3.522x^{0.462}$

▱

EXPLORATION

Try using a calculator or computer to fit the data in Example 3 with a quadratic model of the form

$y = ax^2 + bx + c.$

Do you think this model fits better than the logarithmic model in Example 3? Explain your reasoning.

Application

EXAMPLE 4 ▰ Finding an Exponential Model

Real Life

The total amounts A (in billions of dollars) spent on health care in the United States in the years 1970 through 1991, are shown below. Find a model for the data, and use the model to predict the amount spent in 1998. In the list of data points (t, A), t represents the year, with $t = 0$ corresponding to 1970. (Source: U.S. Health Care Financing Administration)

(0, 74.4), (1, 82.3), (2, 92.3), (3, 102.5), (4, 116.1), (5, 132.9), (6, 152.2), (7, 172.0), (8, 193.7), (9, 217.2), (10, 250.1), (11, 290.2), (12, 326.1), (13, 358.6), (14, 389.6), (15, 422.6), (16, 454.9), (17, 494.2), (18, 546.1), (19, 604.3), (20, 675.0), (21, 751.8)

Solution

Begin by entering the data into a computer or calculator that has least squares regression programs. Then plot the data, as shown in Figure 3.39. From the scatter plot, it appears that an exponential model is a good fit. After running the exponential regression program, you should obtain

$$A = 77.27(1.12)^t \qquad \text{or} \qquad A = 77.27e^{0.1133t}.$$

(The correlation coefficient is $r = 0.997$, which implies that the model is a good fit to the data.) From the model, you can see that the amount spent on health care from 1970 through 1991 had an average annual increase of 12%. From this model, you can predict the 1998 amount to be

$$A = 77.27(1.12)^{28} \approx 1845.5 \text{ billion dollars}$$

which is more than twice the amount spent in 1991. ▰

Figure 3.39

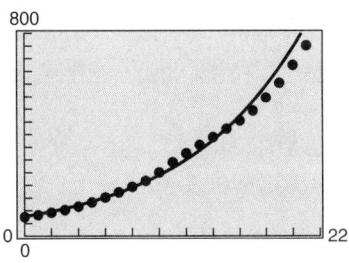

Group Activity

Fitting a Model to Data

The numbers y (in millions) of long-playing albums sold in the United States in the years 1975 through 1992 are listed below. The data is given as ordered pairs of the form (t, y), where t is the year, with $t = 5$ representing 1975. Create a scatter plot of the data. With others in your group, decide which type of model best fits this data. Then find the model.

(5, 257.0), (6, 273.0), (7, 344.0), (8, 341.3), (9, 318.3), (10, 322.8), (11, 295.2), (12, 243.9), (13, 209.6), (14, 204.6), (15, 167.0), (16, 125.2), (17, 107.0), (18, 72.4), (19, 34.6), (20, 11.7), (21, 4.8), (22, 2.3)

3.6 /// EXERCISES

In Exercises 1–8, determine if the scatter plot could best be modeled by a linear model, a quadratic model, an exponential model, a logarithmic model, a Gaussian model, or a logistics model.

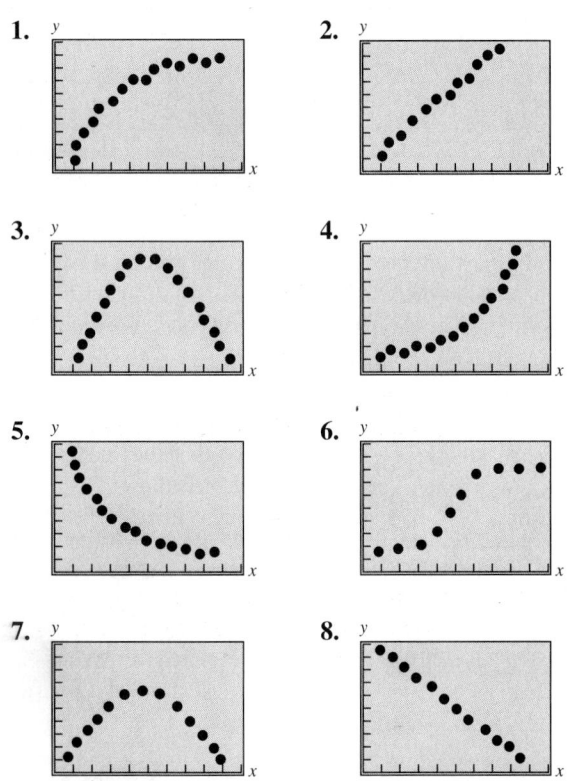

1. **2.**

3. **4.**

5. **6.**

7. **8.**

In Exercises 9–14, sketch a scatter plot of the data. Decide whether the data could best be modeled by a linear, an exponential, or a logarithmic model.

9. (1, 2.0), (1.5, 3.5), (2, 4.0), (4, 5.8), (6, 7.0), (8, 7.8)

10. (1, 5.8), (1.5, 6.0), (2, 6.5), (4, 7.6), (6, 8.9), (8, 10.0)

11. (1, 4.4), (1.5, 4.7), (2, 5.5), (4, 9.9), (6, 18.1), (8, 33.0)

12. (1, 11.0), (1.5, 9.6), (2, 8.2), (4, 4.5), (6, 2.5), (8, 1.4)

13. (1, 7.5), (1.5, 7.0), (2, 6.8), (4, 5.0), (6, 3.5), (8, 2.0)

14. (1, 5.0), (1.5, 6.0), (2, 6.4), (4, 7.8), (6, 8.6), (8, 9.0)

In Exercises 15–18, use a graphing utility to find an exponential model, $y = ab^x$, for the points. Use the graphing utility to obtain a scatter plot and the graph of the model.

15. (0, 4), (1, 5), (2, 6), (3, 8), (4, 12)

16. (0, 6.0), (2, 8.9), (4, 20.0), (6, 34.3), (8, 61.1), (10, 120.5)

17. (0, 10.0), (1, 6.1), (2, 4.2), (3, 3.8), (4, 3.6)

18. (−3, 120.2), (0, 80.5), (3, 64.8), (6, 58.2), (10, 55.0)

In Exercises 19–22, use a graphing utility to find a logarithmic model, $y = a + b \ln x$, for the points. Use the graphing utility to obtain a scatter plot and the graph of the model.

19. (1, 2.0), (2, 3.0), (3, 3.5), (4, 4.0), (5, 4.1), (6, 4.2), (7, 4.5)

20. (1, 8.5), (2, 11.4), (4, 12.8), (6, 13.6), (8, 14.2), (10, 14.6)

21. (1, 10), (2, 6), (3, 6), (4, 5), (5, 3), (6, 2)

22. (3, 14.6), (6, 11.0), (9, 9.0), (12, 7.6), (15, 6.5)

In Exercises 23–26, use a graphing utility to find a power model, $y = ax^b$, for the points. Use the graphing utility to obtain a scatter plot and the graph of the model.

23. (1, 2.0), (2, 3.4), (5, 6.7), (6, 7.3), (10, 12.0)

24. (0.5, 1.0), (2, 12.5), (4, 33.2), (6, 65.7),

     (8, 98.5), (10, 150.0)

25. (1, 10.0), (2, 4.0), (3, 0.7), (4, 0.1)

26. (2, 450), (4, 385), (6, 345), (8, 332), (10, 312)

27. *Breaking Strength*   The breaking strength y (in tons) of a steel cable of diameter d (in inches) is given by the data in the table.

d	0.50	0.75	1.00	1.25	1.50	1.75
y	9.85	21.8	38.3	59.2	84.4	114.0

(a) Use the regression capabilities of a graphing utility to fit an appropriate model to the data.

(b) Use the graphing utility to plot the data and graph the model.

(c) Use the model to predict the breaking strength for a cable of diameter 2 inches.

28. *Foreign Travel*   The numbers of United States citizens y (in millions) who traveled to foreign countries in the years 1984 through 1993 are given in the table, where $t = 4$ represents the year 1984.   (Source: U.S. Bureau of Economic Analysis)

t	4	5	6	7	8
y	34.0	34.7	37.2	39.4	40.7

t	9	10	11	12	13
y	41.1	44.6	41.6	43.9	45.5

(a) Use the regression capabilities of a graphing utility to fit an appropriate model to the data.

(b) Use the graphing utility to plot the data and graph the model.

(c) Use the model to predict the number of foreign travelers in the year 2001.

29. *World Population*   The world population y (in billions) for the years 1983 through 1994 is given in the table, where $x = 3$ corresponds to 1983.   (Source: U.S. Bureau of the Census, International Database)

x	3	4	5	6	7	8
y	4.68	4.77	4.85	4.94	5.02	5.11

x	9	10	11	12	13	14
y	5.20	5.29	5.38	5.48	5.55	5.64

(a) Use the regression capabilities of a graphing utility to fit a linear model to the data.

(b) Use the regression capabilities of a graphing utility to fit an exponential model to the data.

(c) Population growth is often exponential. For the 12 years of data given, is the exponential model better than the linear model? Explain.

(d) Use each model to predict the population in the year 2001.

30. *Atmospheric Pressure*   The atmospheric pressure decreases with increasing altitude. At sea level, the average air pressure is 1.033227 kilograms per square centimeter, and this pressure is called one atmosphere. Variations in weather conditions cause changes in the atmospheric pressure of up to ±5 percent. The table gives the pressures p (in atmospheres) at given altitudes h (in kilometers).

h	0	5	10	15	20	25
p	1	0.55	0.25	0.12	0.06	0.02

(a) Use a graphing utility to attempt to find the logarithmic model $p = a + b \ln h$ for the data. Explain why the result is an error message.

(b) Use a graphing utility to find the logarithmic model $h = a + b \ln p$ for the data.

(c) Use a graphing utility to plot the data and graph the logarithmic model.

(d) Use the model to estimate the altitude at which the pressure is 0.75 atmosphere.

(e) Use the graph to estimate the pressure at an altitude of 13 kilometers.

31. *Lumber* The table gives the total domestic production x and the domestic consumption y of lumber in the United States from 1983 through 1990. The measurements are in millions of board feet. (Source: *Current Industrial Reports*)

x	34.6	37.1	36.4	42.0
y	48.7	52.7	53.5	57.4

x	44.9	44.6	43.6	43.9
y	61.8	58.8	58.8	54.5

(a) Use a graphing utility to find a power model for the data.

(b) Use a graphing utility to plot the data and graph the power model in the same viewing rectangle.

(c) Use the model to estimate consumption for a production level of 50 million board feet.

32. *National Health Expenditures* The table gives the national health expenditures y (in billions of dollars) for the years 1982 through 1991. The years are given by x, where $x = 2$ corresponds to 1982. (Source: U.S. Health Care Financing Administration)

x	2	3	4	5	6
y	326.1	358.6	389.6	422.6	454.9

x	7	8	9	10	11
y	494.2	546.1	604.3	675.0	751.8

(a) Use a graphing utility to find an exponential model for the data. Use the graphing utility to plot the data and graph the exponential model in the same viewing rectangle.

(b) Use a graphing utility to find a logarithmic model for the data. Use the graphing utility to plot the data and graph the logarithmic model in the same viewing rectangle.

(c) Use the graphs plotted in parts (a) and (b) to determine which model is better. If the rate of growth of health care costs could be slowed, which model may be better for the future? Explain.

(d) Use each model to predict health care costs in the year 2001.

33. *Frequency* Students in a physics lab measured the frequency f (in hertz) of a plucked wire with tension T (in newtons). The data is plotted on the graph.

(a) Use a graphing utility to find a power model for the data. Use the graphing utility to plot the data and graph the model in the same viewing rectangle.

(b) Use the model to estimate the frequency at $T = 1.5$.

(c) Based on the limited number of observations, it appears as though the frequency is proportional to what common power of x?

34. *Wind Turbine* Tests on a wind turbine in a wind tunnel measured the power output P (in kilowatts) at various wind speeds S (in miles per hour). The test results are plotted on the graph.

(a) Use a graphing utility to find a power model for the data. Use the graphing utility to plot the data and graph the model in the same viewing rectangle.

(b) Use the model to estimate the frequency at $S = 45$.

(c) Based on the limited number of observations, it appears as though the power output is proportional to what power of x?

35. *Comparing Models*   The amounts y (in billions of dollars) donated to charity (by individuals, foundations, corporations, and charitable bequests) in the years 1983 through 1992 in the United States are given in the table, where $x = 3$ corresponds to 1983. (Source: AAFRC Trust for Philanthropy)

x	3	4	5	6	7
y	63.2	68.6	73.2	83.9	90.3

x	8	9	10	11	12
y	98.4	107.0	111.7	116.8	124.3

(a) Use the regression capabilities of a graphing utility to find the following models for the data.

$$y_1 = ax + b$$
$$y_2 = a + b \ln x$$
$$y_3 = ab^x$$
$$y_4 = ax^b$$

(b) Use the graphing utility to graph the data and each of the models. Use the graphs to select the model that you think best fits the data.

(c) For each of the models y_i $(i = 1, 2, 3, 4)$, complete the following table.

x	y	$y - y_i$	$(y - y_i)^2$
3	63.2		
4	68.6		
5	73.2		
6	83.9		
7	90.3		
8	98.4		
9	107.0		
10	111.7		
11	116.8		
12	124.3		

(d) For each model, find the sum of the entries in the last column of the table in part (c). Use the results to select the best model for the data.

(e) Explain what the sums in part (d) represent.

36. *Comparing Models*   A V8 car engine is coupled to a dynamometer, and the horsepower y is measured at different engine speeds x (in thousands of revolutions per minute). The results are shown in the table.

x	1	2	3	4	5	6
y	40	85	140	200	225	245

(a) Use the regression capabilities of a graphing utility to find the following models for the data.

$$y_1 = ax^3 + bx^2 + cx + d$$
$$y_2 = a + b \ln x$$
$$y_3 = ab^x$$
$$y_4 = ax^b$$

(b) Use the graphing utility to graph the data and each of the models. Use the graphs to select the model that you think best fits the data.

(c) For each of the models y_i $(i = 1, 2, 3, 4)$, complete the following table.

x	y	$y - y_i$	$(y - y_i)^2$
1	40		
2	85		
3	140		
4	200		
5	225		
6	245		

(d) For each model, find the sum of the entries in the last column of the table in part (c). Use the results to select the best model for the data.

(e) Explain what the sums in part (d) represent.

Focus on Concepts

In this chapter, you studied several concepts that are related to exponential and logarithmic functions. You can use the following questions to check your understanding of several of these basic concepts. The answers to these questions are given in the back of the book.

1. *Comparing Graphs* The graphs of $y = e^{kt}$ are shown for $k = a, b, c$, and d. Use the graphs to order a, b, c, and d. Which of the four values are negative? Which are positive?

(a)   (b)

(c)   (d)

2. *True or False?* Rewrite each verbal statement as an equation. Then decide whether the statement is true or false. If it is false, give an example that shows it is false.

   (a) The logarithm of the product of two numbers is equal to the sum of the logarithms of the numbers.

   (b) The logarithm of the sum of two numbers is equal to the product of the logarithms of the numbers.

   (c) The logarithm of the difference of two numbers is equal to the difference of the logarithms of the numbers.

   (d) The logarithm of the quotient of two numbers is equal to the difference of the logarithms of the numbers.

3. *Investing Money* You are investing P dollars at an annual rate r, compounded continuously, for t years. Which of the following would be most advantageous? Explain your reasoning.

   (a) Double the amount you invest.

   (b) Double your interest rate.

   (c) Double the number of years.

4. Identify the model as linear, logarithmic, exponential, logistic, or none of the above. Explain your reasoning.

(a)   (b)

(c)   (d)

(e)   (f)

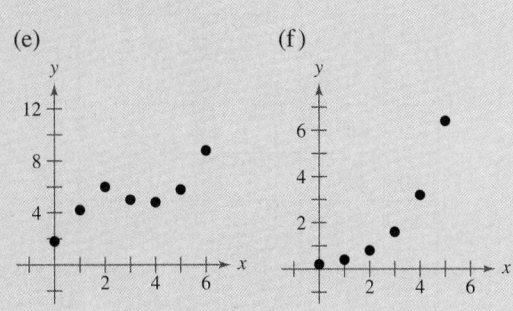

3 /// REVIEW EXERCISES

In Exercises 1–6, match the function with its graph. [The graphs are labeled (a), (b), (c), (d), (e), and (f).]

(a)

(b)

(c)

(d)

(e)

(f)
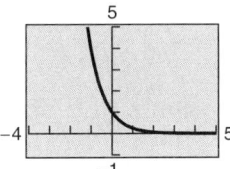

1. $f(x) = 4^x$

2. $f(x) = 4^{-x}$

3. $f(x) = -4^x$

4. $f(x) = 4^x + 1$

5. $f(x) = \log_4 x$

6. $f(x) = \log_4(x - 1)$

In Exercises 7–14, sketch the graph of the function. Use a graphing utility to verify your graph.

7. $f(x) = 6^x$

8. $f(x) = 0.3^x$

9. $g(x) = 6^{-x}$

10. $g(x) = 0.3^{-x}$

11. $h(x) = e^{-x/2}$

12. $h(x) = 2 - e^{-x/2}$

13. $f(x) = e^{x+2}$

14. $s(t) = 4e^{-2/t}, \quad t > 0$

In Exercises 15–18, use a graphing utility to graph the function. Identify any asymptotes.

15. $g(t) = 8 - 0.5e^{-t/4}$

16. $h(x) = 12(1 + e^{-x/2})$

17. $f(x) = \dfrac{10}{1 + 2^{-0.05x}}$

18. $g(x) = 200e^{4/x}$

In Exercises 19 and 20, complete the table to determine the balance A for P dollars invested at rate r for t years and compounded n times per year.

n	1	2	4	12	365	Continuous
A						

19. $P = \$3,500, r = 10.5\%, t = 10$ years

20. $P = \$2,000, r = 12\%, t = 30$ years

In Exercises 21 and 22, complete the table to determine the amount of money P that should be invested at rate r to produce a final balance of \$200,000 in t years.

t	1	10	20	30	40	50
P						

21. $r = 8\%$, compounded continuously

22. $r = 10\%$, compounded monthly

23. *Depreciation*   After t years, the value of a car that originally cost \$14,000 is given by

$$V(t) = 14,000\left(\frac{3}{4}\right)^t.$$

(a) Use a graphing utility to graph the function.

(b) Find the value of the car 2 years after it was purchased.

(c) According to the model, when does the car depreciate most rapidly? Is this realistic? Explain.

24. *Trust Fund*   On the day a child is born, a deposit of $50,000 is made in a trust fund that pays 8.75% interest, compounded continuously. Determine the balance in the account after 35 years.

25. *Drug Decomposition*   A solution of a certain drug contains 500 units per milliliter when prepared. It is analyzed after 40 days and is found to contain 300 units per milliliter. Assuming that the rate of decomposition is proportional to the amount present, the equation giving the amount A after t days is

$$A = 500e^{-0.013t}.$$

Use a graphing utility to graph the model. Approximate A graphically and analytically at t = 60.

26. *Waiting Times*   The average time between incoming calls at a switchboard is 3 minutes. The probability of waiting less than t minutes for the next incoming call is approximated by the model

$$F(t) = 1 - e^{-t/3}.$$

If a call has just come in, find the probability that the next call will come within

(a) $\frac{1}{2}$ minute.   (b) 2 minutes.   (c) 5 minutes.

27. *Fuel Efficiency*   A certain automobile gets 28 miles per gallon of gasoline for speeds up to 50 miles per hour. Over 50 miles per hour, the number of miles per gallon drops at the rate of 12% for each additional 10 miles per hour. If s is the speed and y is the number of miles per gallon, then

$$y = 28e^{0.6 - 0.012s}, \qquad s \geq 50.$$

Use the model to complete the table.

s	50	55	60	65	70
y	28				

28. *Inflation*   If the inflation rate averages 4.5% over the next 10 years, the approximate cost C of goods or services t years from now is given by

$$C(t) = P(1.045)^t$$

where P is the present cost. If the price of a tire is presently $69.95, estimate the price 10 years from now.

In Exercises 29–34, sketch the graph of the function. Use a graphing utility to verify your graph.

29. $g(x) = \log_2 x$

30. $g(x) = \log_5 x$

31. $f(x) = \ln x + 3$

32. $f(x) = \ln(x - 3)$

33. $h(x) = \ln(e^{x-1})$

34. $f(x) = \frac{1}{4} \ln x$

In Exercises 35 and 36, use a graphing utility to graph the function.

35. $y = \log_{10}(x^2 + 1)$

36. $y = \sqrt{x} \ln(x + 1)$

In Exercises 37 and 38, write the equation in logarithmic form.

37. $4^3 = 64$

38. $25^{3/2} = 125$

In Exercises 39–46, evaluate the expression *by hand.*

39. $\log_{10} 1000$

40. $\log_9 3$

41. $\log_3 \frac{1}{9}$

42. $\log_4 \frac{1}{16}$

43. $\ln e^7$

44. $\log_a \frac{1}{a}$

45. $\ln 1$

46. $\ln e^{-3}$

In Exercises 47–50, evaluate the logarithm using the change-of-base formula. Do each problem twice, once with common logarithms and once with natural logarithms. Round the result to three decimal places.

47. $\log_4 9$

48. $\log_{1/2} 5$

49. $\log_{12} 200$

50. $\log_3 0.28$

In Exercises 51–56, write the expression as a sum, difference, and/or multiple of logarithms.

51. $\log_5 5x^2$

52. $\log_7 \frac{\sqrt{x}}{4}$

53. $\log_{10} \frac{5\sqrt{y}}{x^2}$

54. $\ln \left| \frac{x - 1}{x + 1} \right|$

55. $\ln[(x^2 + 1)(x - 1)]$

56. $\ln \sqrt[5]{\frac{4x^2 - 1}{4x^2 + 1}}$

In Exercises 57–62, write the expression as the logarithm of a single quantity.

57. $\log_2 5 + \log_2 x$ **58.** $\log_6 y - 2 \log_6 z$

59. $\frac{1}{2} \ln|2x - 1| - 2 \ln|x + 1|$

60. $5 \ln|x - 2| - \ln|x + 2| - 3 \ln|x|$

61. $\ln 3 + \frac{1}{3} \ln(4 - x^2) - \ln x$

62. $3[\ln x - 2 \ln(x^2 + 1)] + 2 \ln 5$

True or False? In Exercises 63–68, determine whether the statement is true or false.

63. $\log_b b^{2x} = 2x$ **64.** $e^{x-1} = \dfrac{e^x}{e}$

65. The domain of the function $f(x) = \ln x$ is the set of all real numbers.

66. $\ln(x + y) = \ln x + \ln y$

67. $\ln(x + y) = \ln(x \cdot y)$ **68.** $\log\left(\dfrac{10}{x}\right) = 1 - \log x$

In Exercises 69–72, approximate the logarithm using the properties of logarithms and given that $\log_b 2 \approx 0.3562$, $\log_b 3 \approx 0.5646$, and $\log_b 5 \approx 0.8271$.

69. $\log_b 25$ **70.** $\log_b\left(\dfrac{25}{9}\right)$

71. $\log_b \sqrt{3}$ **72.** $\log_b 30$

73. *Climb Rate* The time t, in minutes, for a small plane to climb to an altitude of h feet is given by

$$t = 50 \log_{10} \frac{18{,}000}{18{,}000 - h}$$

where 18,000 feet is the plane's absolute ceiling.

(a) Determine the domain of the function appropriate for the context of the problem.

(b) Use a graphing utility to graph the time function and identify any asymptotes.

(c) As the plane approaches its absolute ceiling, what can be said about the time required to further increase its altitude?

(d) Find the time for the plane to climb to an altitude of 4000 feet.

74. *Snow Removal* The number of miles s of roads cleared of snow is approximated by the model

$$s = 25 - \frac{13 \ln(h/12)}{\ln 3}, \quad 2 \le h \le 15$$

where h is the depth of the snow in inches. Use this model to find s when $h = 10$ inches.

In Exercises 75–80, solve the exponential equation. Round the solution to three decimal places.

75. $e^x = 12$ **76.** $e^{3x} = 25$

77. $3e^{-5x} = 132$ **78.** $14e^{3x+2} = 560$

79. $e^{2x} - 6x + 8 = 0$ **80.** $e^{2x} - 7e^x + 10 = 0$

In Exercises 81–86, solve the logarithmic equation. Round the solution to three decimal places.

81. $\ln 3x = 8.2$ **82.** $2 \ln 4x = 15$

83. $\ln x - \ln 3 = 2$ **84.** $\ln \sqrt{x + 1} = 2$

85. $\log(x - 1) = \log(x - 2) - \log(x + 2)$

86. $\log(1 - x) = -1$

In Exercises 87–90, use a graphing utility to solve the equation. Round the solution to two decimal places.

87. $2^{0.6x} - 3x = 0$

88. $25e^{-0.3x} = 12$

89. $2 \ln(x + 3) + 3x = 8$

90. $6 \log_{10}(x^2 + 1) - x = 0$

In Exercises 91–94, find the exponential function $y = Ae^{bx}$ that passes through the two points.

91. $(0, 2), (4, 3)$ **92.** $\left(0, \frac{1}{2}\right), (5, 5)$

93. $(0, 4), \left(5, \frac{1}{2}\right)$ **94.** $(0, 2), (5, 1)$

95. *Demand Function* The demand equation for a certain product is given by

$$p = 500 - 0.5e^{0.004x}.$$

Find the demands x for prices of (a) $p = \$450$ and (b) $p = \$400$.

96. *Typing Speed* In a typing class, the average number of words per minute typed after t weeks of lessons was found to be

$$N = \frac{157}{1 + 5.4e^{-0.12t}}.$$

Find the time necessary to type (a) 50 words per minute and (b) 75 words per minute.

97. *Compound Interest* A deposit of $750 is made in a savings account for which the interest is compounded continuously. The balance will double in $7\frac{3}{4}$ years.

(a) What is the annual interest rate for this account?

(b) Find the balance in the account after 10 years.

(c) The *effective yield* of a savings plan is the percent increase in the balance after 1 year. Find the effective yield.

98. *Compound Interest* A deposit of $10,000 is made in a savings account for which the interest is compounded continuously. The balance will double in 5 years.

(a) What is the annual interest rate for this account?

(b) Find the balance after 1 year.

(c) The *effective yield* of a savings plan is the percent increase in the balance after 1 year. Find the effective yield.

99. *Sound Intensity* The relationship between the number of decibels β and the intensity of a sound I in watts per centimeter squared is given by

$$\beta = 10 \log_{10}\left(\frac{I}{10^{-16}}\right).$$

Determine the intensity of a sound in watts per centimeter squared if the decibel level is 125.

100. *Earthquake Magnitudes* On the Richter scale, the magnitude R of an earthquake of intensity I is given by

$$R = \log_{10}\frac{I}{I_0}$$

where $I_0 = 1$ is the minimum intensity used for comparison. Find the intensity per unit of area for the following values of R.

(a) $R = 8.4$ (b) $R = 6.85$ (c) $R = 9.1$

101. *Exponential Regression* Use a graphing utility to find an exponential model $y = ab^x$ through the points $(0, 250)$, $(4, 135)$, $(6, 92)$, and $(10, 67)$. Sketch a scatter plot and the graph of the exponential model.

102. *Data Analysis* The data in the table gives the yield y (in milligrams) of a chemical reaction after t minutes.

t	1	2	3	4
y	1.5	7.4	10.2	13.4

t	5	6	7	8
y	15.8	16.3	18.2	18.3

(a) Use a graphing utility to fit the linear model $y = at + b$ to the data.

(b) Use a graphing utility to fit the logarithmic model $y = a + b \ln t$ to the data.

(c) Create a scatter plot for the data, and graph the linear and logarithmic models. Which is a better model for the data? Explain your reasoning.

103. *Sporting Goods Sales* The sales y (in billions of dollars) of sporting goods for the years 1984 through 1993 are given in the table, where $x = 4$ corresponds to 1984. (Source: National Sporting Goods Association)

x	4	5	6	7	8
y	26.4	27.4	30.6	33.9	42.1

x	9	10	11	12	13
y	45.2	44.1	42.9	42.4	44.6

(a) Use a graphing utility to find the power model $y = ax^b$ for the data. Use the graphing utility to plot the data and graph the model.

(b) Use the model to predict sales in the year 2001.

CHAPTER PROJECT *A Graphical Approach to Compound Interest*

In this project, you will use a graphing utility to compare savings plans. For instance, suppose you are depositing $1000 in a savings account and are given the following options.

- 6.2% annual interest rate, compounded annually
- 6.1% annual interest rate, compounded quarterly
- 6.0% annual interest rate, compounded continuously

(a) For each option, write a function that gives the balance as a function of the time t (in years).

(b) Graph all three functions in the same viewing rectangle. Can you find a viewing rectangle that distinguishes among the graphs of the three functions? If so, describe the viewing rectangle.

(c) Find the balances for the three options after 25, 50, 75, and 100 years. Is the option that yields the greatest balance after 25 years the same option that yields the greatest balances after 50, 75, and 100 years? Explain.

(d) The **effective yield** of a savings plan is the percent increase in the balance after 1 year. Find the effective yields for the three options listed above. How can the effective yield be used to decide which option is best?

Questions for Further Exploration

1. You deposit $25,000 in an account to accrue interest for 40 years. The account pays 8% compounded annually. Assume that the income tax on the earned interest is 30%. Which of the following plans produces a larger balance after all income tax is paid?

(a) *Deferred* The income tax on the interest that is earned is paid in one lump sum at the end of 40 years.

(b) *Not Deferred* The income tax on the interest that is earned each year is paid at the end of each year.

2. Which of the following would produce a larger balance? Explain.

(a) 8.02% annual interest rate, compounded monthly

(b) 8.03% annual interest rate, compounded daily

(c) 8% annual interest rate, compounded continuously

3. You deposit $1000 in each of two savings accounts. The interest for the accounts is paid according to the two options described in Question 1. How long would it take for the balance in one of the accounts to exceed the balance in the other account by $100? By $100,000?

4. No income tax is due on the interest earned in some types of investments. You deposit $25,000 into an account. Which of the following plans is better? Explain.

(a) *Tax-Free* The account pays 5%, compounded annually. There is no income tax on the earned interest.

(b) *Tax-Deferred* The account pays 7%, compounded annually. At maturity, the earned interest is taxable at a rate of 40%.

P–3 /// CUMULATIVE TEST

Take this test as you would take a test in class. After you are done, check your work against the answers given in the back of the book.

The *Interactive* CD-ROM provides answers to the Chapter Tests and Cumulative Tests. It also offers Chapter Pre-Tests (that test key skills and concepts covered in previous chapters) and Chapter Post-Tests, both of which have randomly generated exercises with diagnostic capabilities.

In Exercises 1–6, sketch a graph of the function. Use a graphing utility to verify your result.

1. $h(x) = -(x^2 + 4x)$ **2.** $y = \sqrt{4 - x}$ **3.** $g(s) = \dfrac{2s}{s - 3}$

4. $g(s) = \dfrac{2s^2}{s - 3}$ **5.** $f(x) = 6(2^{-x})$ **6.** $g(x) = \log_3 x$

7. Find an equation for the line passing through the points $\left(-\frac{1}{2}, 1\right)$ and $(3, 8)$.

8. Explain why the graph at the right does not represent y as a function of x.

9. Describe how the graph of each function differs from the graph of $y = \sqrt[3]{x}$. It is not necessary to sketch the graphs.

   (a) $r(x) = \frac{1}{2}\sqrt[3]{x}$ (b) $h(x) = \sqrt[3]{x} + 2$ (c) $g(x) = \sqrt[3]{x + 2}$

10. Determine whether the function $h(x) = 5x - 2$ is one-to-one. If so, find its inverse.

Figure for 8

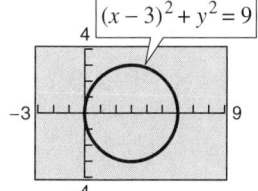

In Exercises 11–16, solve (if possible) the equation. Use a graphing utility to verify your result.

11. $2x - 3(x - 4) = 5$ **12.** $\dfrac{2}{t - 3} + \dfrac{2}{t - 2} = \dfrac{10}{t^2 - 5t + 6}$

13. $3y^2 + 6y + 2 = 0$ **14.** $\sqrt{x + 10} = x - 2$

15. $6e^{2x} = 72$ **16.** $\log_2 x + \log_2 5 = 6$

17. Let x be the amount (in hundreds of dollars) that a company spends on advertising, and let P be the profit (in thousands of dollars), where $P = 230 + 20x - \frac{1}{2}x^2$. What amount will yield a maximum profit?

18. Find all the zeros of $f(x) = x^3 + 2x^2 + 4x + 8$.

19. Use a graphing utility to approximate the real zero of the function $g(x) = x^3 + 3x^2 - 6$ to the nearest hundredth.

20. Write $2 \ln x - \frac{1}{2} \ln(x + 5)$ as a logarithm of a single quantity.

21. The numbers of cellular telephone subscribers y (in millions) for the years 1990 through 1993 are given by $(0, 5.3)$, $(1, 7.6)$, $(2, 11.0)$, and $(3, 16.0)$ where x is the time in years, with $x = 0$ corresponding to 1990. Use a graphing utility to fit an exponential model to the data. Create a scatter plot of the data and graph the model in the same viewing rectangle.   (Source: Cellular Telecommunications Industry Association)

Trigonometric Functions

In 1995, in a huge geological project called *Project Deep Probe,* explosions were set off along a 2100-mile line from northern Canada to the Mexican border. The explosions were recorded by nearly 800 seismographs in Alberta, Montana, and Wyoming.

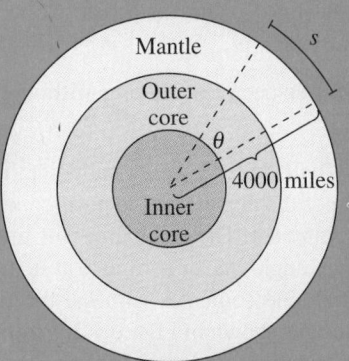

The goal of the project was to use the seismograph readings to obtain a geological profile of the earth's mantle.

The seismographs were located 0.8 mile apart. To find the central angle between two adjacent seismographs, geologists used the formula $s = r\theta$, where θ is measured in radians. Using $s = 0.8$ mile and $r = 4000$ miles, an angle of 0.0002 radians or 0.0115° is obtained. (See Exercises 103 and 104 on page 321.)

© Michael Milstein (Photo and inset).

Project Deep Probe used about 800 portable seismographs from southern Alberta to central Wyoming. Geologists Holger and Reingard Mandler are shown checking a seismograph before burying it in Wyoming.

4.1   Radian and Degree Measure

Angles / *Radian Measure* / *Degree Measure* / *Applications*

Angles

As derived from the Greek language, the word **trigonometry** means "measurement of triangles." Initially, trigonometry dealt with relationships among the sides and angles of triangles and was used in the development of astronomy, navigation, and surveying. With the development of calculus and the physical sciences in the 17th century, a different perspective arose—one that viewed the classic trigonometric relationships as *functions* with the set of real numbers as their domains. Consequently, the applications of trigonometry expanded to include a vast number of physical phenomena involving rotations and vibrations, including sound waves, light rays, planetary orbits, vibrating strings, pendulums, and orbits of atomic particles.

The approach in this text incorporates *both* perspectives, starting with angles and their measure.

An **angle** is determined by rotating a ray (half-line) about its endpoint. The starting position of the ray is the **initial side** of the angle, and the position after rotation is the **terminal side,** as shown in Figure 4.1. The endpoint of the ray is the **vertex** of the angle. This perception of an angle fits a coordinate system in which the origin is the vertex and the initial side coincides with the positive *x*-axis. Such an angle is in **standard position,** as shown in Figure 4.2. **Positive angles** are generated by counterclockwise rotation, and **negative angles** by clockwise rotation, as shown in Figure 4.3. Angles are labeled with Greek letters α (alpha), β (beta), and θ (theta), as well as uppercase letters A, B, and C. In Figure 4.4, note that angles α and β have the same initial and terminal sides. Such angles are **coterminal.**

Figure 4.1

Figure 4.2

Figure 4.3

Figure 4.4

Figure 4.5

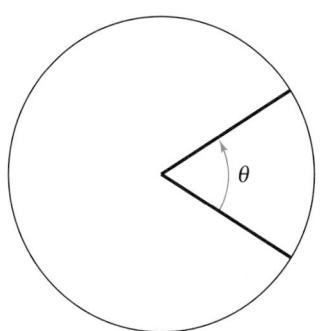

Note One revolution around a circle of radius r corresponds to an angle of 2π radians because

$$\frac{s}{r} = \frac{2\pi r}{r} = 2\pi \text{ radians.}$$

Figure 4.6
(a)

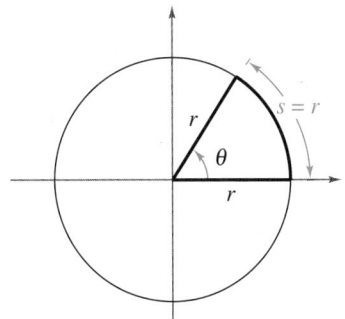

(b)

Radian Measure

The **measure of an angle** is determined by the amount of rotation from the initial side to the terminal side. One way to measure angles is in radians. This type of measure is especially useful in calculus. To define a radian, you can use a **central angle** of a circle, one whose vertex is the center of the circle, as shown in Figure 4.5.

Definition of a Radian

One **radian** is the measure of a central angle θ that intercepts an arc s equal in length to the radius r of the circle. See Figure 4.6(a).

Because the circumference of a circle is $2\pi r$, it follows that a central angle of one full revolution (counterclockwise) corresponds to an arc length of $s = 2\pi r$. Moreover, because $2\pi \approx 6.28$, there are just over six radius lengths in a full circle, as shown in Figure 4.6(b). In general, the radian measure of a central angle θ is obtained by dividing the arc length s by r. That is, $s/r = \theta$, where θ *is measured in radians*. Because the units of measure for s and r are the same, this ratio is unitless—it is simply a real number.

Because the radian measure of an angle of one full revolution is 2π, you can obtain the following.

$$\frac{1}{2}\text{ revolution} = \frac{2\pi}{2} = \pi \text{ radians}$$

$$\frac{1}{4}\text{ revolution} = \frac{2\pi}{4} = \frac{\pi}{2}\text{ radians}$$

$$\frac{1}{6}\text{ revolution} = \frac{2\pi}{6} = \frac{\pi}{3}\text{ radians}$$

These and other common angles are shown in Figure 4.7.

Figure 4.7

Figure 4.8

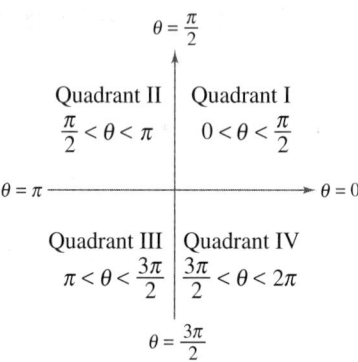

$$\theta = \frac{\pi}{2}$$

Quadrant II	Quadrant I
$\frac{\pi}{2} < \theta < \pi$	$0 < \theta < \frac{\pi}{2}$

$\theta = \pi$ ────────────── $\theta = 0$

Quadrant III	Quadrant IV
$\pi < \theta < \frac{3\pi}{2}$	$\frac{3\pi}{2} < \theta < 2\pi$

$$\theta = \frac{3\pi}{2}$$

Recall that the four quadrants in a coordinate system are numbered I, II, III, and IV. Figure 4.8 shows which angles between 0 and 2π lie in each of the four quadrants. Note that angles between 0 and $\pi/2$ are **acute** and that angles between $\pi/2$ and π are **obtuse.**

Note   The phrase "the terminal side of θ lies in a quadrant" is often abbreviated by simply saying "θ lies in a quadrant." The terminal sides of the "quadrant angles" 0, $\pi/2$, π, and $3\pi/2$ do not lie within quadrants.

Two angles are coterminal if they have the same initial and terminal sides. For instance, the angles 0 and 2π are coterminal, as are the angles $\pi/6$ and $13\pi/6$. You can find an angle that is coterminal to a given angle θ by adding or subtracting 2π (one revolution), as demonstrated in Example 1. A given angle θ has many coterminal angles. For instance, $\theta = \pi/6$ is coterminal with

$$\frac{\pi}{6} + 2n\pi$$

where n is an integer.

EXAMPLE 1   **Sketching and Finding Coterminal Angles**

a. For the positive angle $13\pi/6$, subtract 2π to obtain a coterminal angle

$$\frac{13\pi}{6} - 2\pi = \frac{\pi}{6}.$$ See Figure 4.9(a).

b. For the positive angle $3\pi/4$, subtract 2π to obtain a coterminal angle

$$\frac{3\pi}{4} - 2\pi = -\frac{5\pi}{4}.$$ See Figure 4.9(b).

c. For the negative angle $-2\pi/3$, add 2π to obtain a coterminal angle

$$-\frac{2\pi}{3} + 2\pi = \frac{4\pi}{3}.$$ See Figure 4.9(c).

The *Interactive* CD-ROM shows every example with its solution; clicking on the *Try It!* button brings up similar problems. Guided Examples and Integrated Examples show step-by-step solutions to additional examples. Integrated Examples are related to several concepts in the section.

Figure 4.9

(a)

(b)

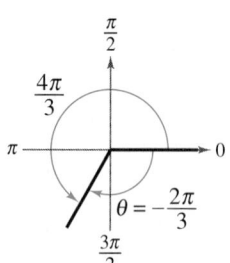

(c)

Two positive angles α and β are **complementary** (complements of each other) if their sum is $\pi/2$. Two positive angles are **supplementary** (supplements of each other) if their sum is π. See Figure 4.10.

EXAMPLE 2 ▱ **Complementary and Supplementary Angles**

If possible, find the complement and the supplement of (a) $2\pi/5$ and (b) $4\pi/5$.

Figure 4.10
(a)

(b)

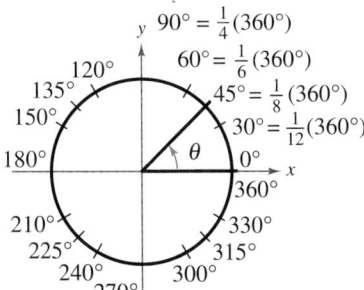

Solution

a. The complement of $2\pi/5$ is
$$\frac{\pi}{2} - \frac{2\pi}{5} = \frac{5\pi}{10} - \frac{4\pi}{10} = \frac{\pi}{10}.$$
The supplement of $2\pi/5$ is
$$\pi - \frac{2\pi}{5} = \frac{3\pi}{5}.$$

b. Because $4\pi/5$ is greater than $\pi/2$, it has no complement. (Remember to use only *positive* angles for complements.) The supplement is
$$\pi - \frac{4\pi}{5} = \frac{\pi}{5}.$$ ▱

Degree Measure

A second way to measure angles is in terms of degrees. A measure of **one degree (1°)** is equivalent to a rotation of $\frac{1}{360}$ of a complete revolution about the vertex. To measure angles, it is convenient to mark degrees on the circumference of a circle, as shown in Figure 4.11. Thus, a full revolution (counterclockwise) corresponds to 360°, a half revolution to 180°, a quarter revolution to 90°, and so on.

Because 2π radians corresponds to one complete revolution, degrees and radians are related by the equations
$$360° = 2\pi \text{ rad}$$
and
$$180° = \pi \text{ rad}.$$

From the latter equation, you obtain
$$1° = \frac{\pi}{180} \text{ rad}$$
and
$$1 \text{ rad} = \left(\frac{180}{\pi}\right)°$$
which lead to the conversion rules at the top of the next page.

Figure 4.11

$90° = \frac{1}{4}(360°)$
$60° = \frac{1}{6}(360°)$
$45° = \frac{1}{8}(360°)$
$30° = \frac{1}{12}(360°)$

Note   When no units of angle measure are specified, *radian measure is implied.* For instance, if you write $\theta = \pi$ or $\theta = 2$, you should mean $\theta = \pi$ radians or $\theta = 2$ radians.

Conversions Between Degrees and Radians

1. To convert degrees to radians, multiply degrees by $\dfrac{\pi \text{ rad}}{180°}$.

2. To convert radians to degrees, multiply radians by $\dfrac{180°}{\pi \text{ rad}}$.

To apply these two conversion rules, use the relationship π rad $= 180°$.

EXAMPLE 3   ▱   **Converting from Degrees to Radians**

a. $135° = (135 \text{ deg})\left(\dfrac{\pi \text{ rad}}{180 \text{ deg}}\right) = \dfrac{3\pi}{4} \text{ rad}$ Multiply by $\pi/180$.

b. $540° = (540 \text{ deg})\left(\dfrac{\pi \text{ rad}}{180 \text{ deg}}\right) = 3\pi \text{ rad}$ Multiply by $\pi/180$.

c. $-270° = (-270 \text{ deg})\left(\dfrac{\pi \text{ rad}}{180 \text{ deg}}\right) = -\dfrac{3\pi}{2} \text{ rad}$ Multiply by $\pi/180$.

▱

Note   If you have a calculator with a "radian-to-degree" conversion key, try using it to verify the result shown in part (c) of Example 4.

EXAMPLE 4   ▱   **Converting from Radians to Degrees**

a. $-\dfrac{\pi}{2} \text{ rad} = \left(-\dfrac{\pi}{2} \text{ rad}\right)\left(\dfrac{180 \text{ deg}}{\pi \text{ rad}}\right) = -90°$ Multiply by $180/\pi$.

b. $\dfrac{9\pi}{2} \text{ rad} = \left(\dfrac{9\pi}{2} \text{ rad}\right)\left(\dfrac{180 \text{ deg}}{\pi \text{ rad}}\right) = 810°$ Multiply by $180/\pi$.

c. $2 \text{ rad} = (2 \text{ rad})\left(\dfrac{180 \text{ deg}}{\pi \text{ rad}}\right) = \dfrac{360}{\pi} \approx 114.59°$ Multiply by $180/\pi$.

▱

$1' = \text{one minute} = \frac{1}{60}(1°)$

$1'' = \text{one second} = \frac{1}{3600}(1°)$.

With calculators it is convenient to use *decimal* degrees to denote fractional parts of degrees. Historically, however, fractional parts of degrees were expressed in *minutes* and *seconds*, using the prime (′) and double prime (″) notations, respectively. Consequently, an angle of 64 degrees, 32 minutes, and 47 seconds was represented by $\theta = 64° \, 32' \, 47''$. Many calculators have special keys for converting angles in degrees, minutes, and seconds (D° M′ S″) into decimal degree form, and vice versa.

Applications

The *radian measure* formula $\theta = s/r$ can be used to measure arc length along a circle. Specifically, for a circle of radius r, a central angle θ intercepts an arc of length s given by

$$s = r\theta \qquad \text{Length of circular arc}$$

where θ is measured in radians.

EXAMPLE 5   ▱   **Finding Arc Length**

A circle has a radius of 4 inches. Find the length of the arc intercepted by a central angle of 240°, as shown in Figure 4.12.

Solution

To use the formula $s = r\theta$, first convert 240° to radian measure.

$$240° = (240 \text{ deg})\left(\frac{\pi \text{ rad}}{180 \text{ deg}}\right) \qquad \text{Convert from degrees to radians.}$$

$$= \frac{4\pi}{3} \text{ radians} \qquad \text{Simplify.}$$

Then, using a radius of $r = 4$ inches, you can find the arc length to be

$$s = r\theta \qquad \text{Length of circular arc}$$

$$= 4\left(\frac{4\pi}{3}\right) \qquad \text{Substitute for } r \text{ and } \theta.$$

$$= \frac{16\pi}{3} \qquad \text{Simplify.}$$

$$\approx 16.76 \text{ inches.} \qquad \text{Use a calculator.}$$

Note that the units for $r\theta$ are determined by the units for r because θ is given in radian measure and therefore has no units.   ▱

Figure 4.12

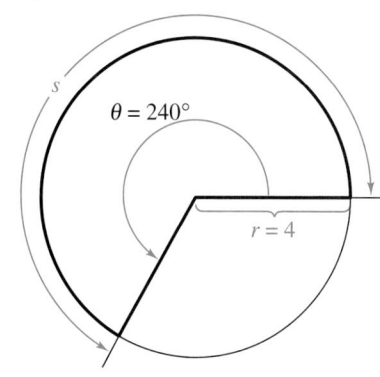

$\theta = 240°$

$r = 4$

s

The formula for the length of a circular arc can be used to analyze the motion of a particle moving at a *constant speed* along a circular path. Consider a circle of radius r. If s is the length of the arc traveled in time t, the **speed** of the particle is

$$\text{Speed} = \frac{\text{distance}}{\text{time}} = \frac{s}{t}.$$

Moreover, if θ is the angle (in radian measure) corresponding to the arc length s, the **angular speed** of the particle is

$$\text{Angular speed} = \frac{\theta}{t}.$$

Real Life

EXAMPLE 6   **Finding the Speed of an Object**

The second hand of a clock is 10.2 centimeters long, as shown in Figure 4.13. Find the speed of the tip of this second hand.

Figure 4.13

10.2 cm

Solution

The time required for the second hand to make one full revolution is

$$t = 60 \text{ seconds} = 1 \text{ minute}.$$

The distance traveled by the tip of the second hand in one revolution is

$$s = 2\pi \,(\text{radius}) = 2\pi(10.2) = 20.4\pi \text{ centimeters}.$$

Therefore, the speed of the tip of the second hand is

$$\text{Speed} = \frac{s}{t} \qquad\qquad \text{Speed = (distance)} \div \text{(time)}$$

$$= \frac{20.4\pi \text{ centimeters}}{60 \text{ seconds}} \qquad\qquad \text{Substitute for } s \text{ and } t.$$

$$\approx 1.068 \text{ centimeters per second.} \quad \text{Simplify.}$$

Real Life

EXAMPLE 7   **Finding Angular Speed and Speed**

A lawn roller that is 30 inches in diameter (see Figure 4.14) makes one revolution every $\frac{5}{6}$ second.

Figure 4.14

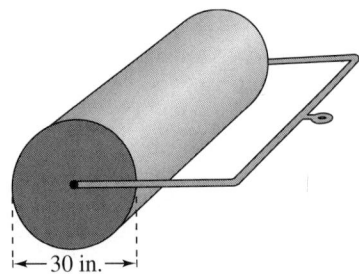

30 in.

a. Find the angular speed of the roller in radians per second.

b. How fast is the roller moving across the lawn?

Solution

a. Because there are 2π radians in one revolution, it follows that the angular speed is

$$\text{Angular speed} = \frac{\theta}{t} = \frac{2\pi \text{ rad}}{5/6 \text{ sec}} = 2.4\pi \text{ radians per second}.$$

b. Because the diameter is 30 inches, $r = 15$ and $s = 2\pi r = 30\pi$ inches. Thus,

$$\text{Speed} = \frac{s}{t} \qquad\qquad \text{Speed = (distance)} \div \text{(time)}$$

$$= \frac{30\pi \text{ inches}}{5/6 \text{ second}} \qquad\qquad \text{Substitute for } s \text{ and } t.$$

$$= 36\pi \text{ inches per second} \qquad \text{Simplify}.$$

$$\approx 113.1 \text{ inches per second}.$$

Group Activity   *An Angle-Drawing Program*

If you have a programmable calculator, try entering the "angle-drawing program." Then use the program to draw several different angles. For instance, the following program is for the *TI-82* or *TI-83*. The figure shows an angle of −750°, as drawn by this program. (Other programs are presented in the appendix.)

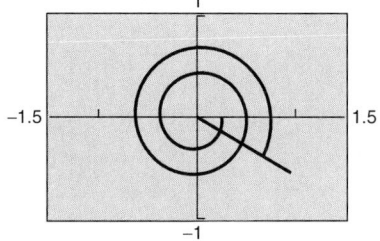

PROGRAM:ANGDRAW	:1 → Ymax
:ClrDraw	:1 → Yscl
:ClrHome	:0 → Tmin
:FnOff	:abs(T) → Tmax
:Radian	:.15 → Tstep
:Param	:cos(T) → A
:Disp "ENTER MODE"	:sin(T) → B
:Disp "0 RADIAN"	:1 → S
:Disp "1 DEGREE"	:If T<0
:Input M	:-1 → S
:Input "ENTER ANGLE",T	:"(.25+.04T)cos(T)" → X1T
:If M=1	:"S(.25+.04T)sin(T)" → Y1T
:πT/180 → T	:DispGraph
:-1.5 → Xmin	:Line(0,0,A,B)
:1.5 → Xmax	:Pause
:1 → Xscl	:Func
:-1 → Ymin	:Stop

To run this program, enter 0 for radian mode or 1 for degree mode. Then enter any angle (the angle can be negative or larger than 360°). After the angle is displayed, press ENTER to clear the display.

The *Interactive* CD-ROM contains step-by-step solutions to all odd-numbered Section and Review Exercises. It also provides Tutorial Exercises, which link to Guided Examples for additional help.

4.1 /// EXERCISES

In Exercises 1–4, estimate the angle to the nearest one-half radian.

1.

2.

3.

4.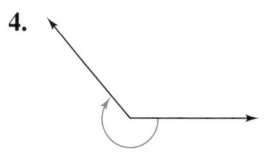

In Exercises 5–10, determine the quadrant in which the angle lies. (The angle measure is given in radians.)

5. (a) $\dfrac{\pi}{5}$ (b) $\dfrac{7\pi}{5}$

6. (a) $\dfrac{5\pi}{4}$ (b) $\dfrac{7\pi}{4}$

7. (a) $-\dfrac{\pi}{12}$ (b) $-\dfrac{11\pi}{9}$

8. (a) -1 (b) -2

9. (a) 3.5 (b) 2.25

10. (a) 5.63 (b) -2.25

In Exercises 11–14, sketch the angle in standard position. Verify your sketch with the angle-drawing program presented on page 317.

11. (a) $\dfrac{5\pi}{4}$ (b) $\dfrac{2\pi}{3}$

12. (a) $-\dfrac{7\pi}{4}$ (b) $-\dfrac{5\pi}{2}$

13. (a) $\dfrac{11\pi}{6}$ (b) 7π

14. (a) 4 (b) -3

In Exercises 15–18, determine two coterminal angles (one positive and one negative) for the given angle. Give your answers in radians.

15. (a)   (b)

16. (a)   (b)

   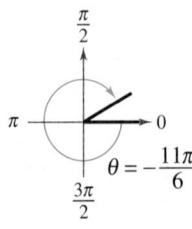

17. (a) $-\dfrac{9\pi}{4}$ (b) $-\dfrac{2\pi}{15}$

18. (a) $\dfrac{8\pi}{9}$ (b) $\dfrac{8\pi}{45}$

In Exercises 19–22, estimate the number of degrees in the angle.

19.   **20.**

21.   **22.**

In Exercises 23–26, determine the quadrant in which the angle lies.

23. (a) $130°$ (b) $285°$

24. (a) $8.3°$ (b) $257°\ 30'$

25. (a) $-132°\ 50'$ (b) $-336°$

26. (a) $-260°$ (b) $-3.4°$

In Exercises 27–30, sketch the angle in standard position. Verify your sketch with the angle-drawing program presented on page 317.

27. (a) $30°$ (b) $150°$

28. (a) $-270°$ (b) $-120°$

29. (a) $405°$ (b) $-480°$

30. (a) $750°$ (b) $-600°$

In Exercises 31–34, determine two coterminal angles (one positive and one negative) for the given angle. Give your answers in degrees.

31. (a)   (b)

32. (a)

$\theta = 120°$

(b) $\theta = -390°$

33. (a) 300°   (b) 740°
34. (a) −420°   (b) 230°

In Exercises 35–38, use the angle-conversion capabilities of a graphing utility to convert the angle measure to decimal degree form.

35. (a) 54° 45′   (b) −128° 30′
36. (a) 245° 10′   (b) 2° 12′
37. (a) 85° 18′ 30″   (b) 330° 25″
38. (a) −135° 36″   (b) −408° 16′ 25″

In Exercises 39–42, use the angle-conversion capabilities of a graphing utility to convert the angle measure to D° M′ S″ form.

39. (a) 240.6°   (b) −145.8°
40. (a) −345.12°   (b) 0.45
41. (a) 2.5   (b) −3.58
42. (a) −0.355   (b) 0.7865

In Exercises 43–46, find (if possible) the complement and supplement of the angle.

43. (a) 18°   (b) 115°
44. (a) 79°   (b) 150°
45. (a) $\dfrac{\pi}{3}$ (b) $\dfrac{3\pi}{4}$
46. (a) 1   (b) 2

In Exercises 47–50, express the angle in radian measure as a multiple of π. (Do not use a calculator.)

47. (a) 30°   (b) 150°
48. (a) 315°   (b) 120°
49. (a) −20°   (b) −240°
50. (a) −270°   (b) 144°

In Exercises 51–54, express the angle in degree measure. (Do not use a calculator.)

51. (a) $\dfrac{3\pi}{2}$ (b) $\dfrac{7\pi}{6}$
52. (a) $-\dfrac{7\pi}{12}$ (b) $\dfrac{\pi}{9}$
53. (a) $\dfrac{7\pi}{3}$ (b) $-\dfrac{11\pi}{30}$
54. (a) $\dfrac{11\pi}{6}$ (b) $\dfrac{34\pi}{15}$

In Exercises 55–62, convert the angle measure from degrees to radians. Round to three decimal places.

55. 115°   **56.** 87.4°
57. −216.35°   **58.** −48.27°
59. 532°   **60.** 0.54°
61. −0.83°   **62.** 345°

In Exercises 63–70, convert the angle measure from radians to degrees. Round to three decimal places.

63. $\dfrac{\pi}{7}$ **64.** $\dfrac{5\pi}{11}$
65. $\dfrac{15\pi}{8}$ **66.** 6.5π
67. −4.2π **68.** 4.8
69. −2   **70.** −0.57

In Exercises 71–74, find the angle in radians.

71.   **72.**

73.   **74.**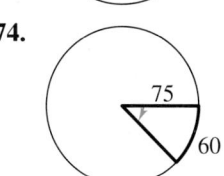

In Exercises 75–78, find the radian measure of the central angle of a circle of radius *r* that intercepts an arc of length *s*.

	Radius, r	*Arc Length, s*
75.	15 inches	4 inches
76.	16 feet	10 feet
77.	14.5 centimeters	25 centimeters
78.	80 kilometers	160 kilometers

In Exercises 79–82, find the length of the arc on a circle of radius *r* intercepted by a central angle *θ*.

	Radius, r	*Central Angle, θ*
79.	15 inches	180°
80.	9 feet	60°
81.	6 meters	2 radians
82.	40 centimeters	$3\pi/4$ radians

Distance Between Cities   In Exercises 83–86, find the distance between the cities. Assume that earth is a sphere of radius 4000 miles and the cities are on the same meridian (one city is due north of the other).

	City	*Latitude*
83.	Dallas	32° 47′ 9″ N
	Omaha	41° 15′ 42″ N
84.	San Francisco	37° 46′ 39″ N
	Seattle	47° 36′ 32″ N
85.	Miami	25° 46′ 37″ N
	Erie	42° 7′ 15″ N
86.	Johannesburg, South Africa	26° 10′ S
	Jerusalem, Israel	31° 47′ N

87. *Difference in Latitudes*   Assuming that earth is a sphere of radius 6378 kilometers, what is the difference in latitude of two cities, one of which is 600 kilometers due north of the other?

88. *Difference in Latitudes*   Assuming that earth is a sphere of radius 6378 kilometers, what is the difference in latitude of two cities, one of which is 800 kilometers due north of the other?

89. *Instrumentation*   The pointer on a voltmeter is 6 centimeters in length (see figure). Find the angle through which the pointer rotates when it moves 2.5 centimeters on the scale.

Figure for 89   **Figure for 90**

90. *Electric Hoist*   An electric hoist is used to lift a piece of equipment (see figure). The diameter of the drum on the hoist is 10 inches, and the equipment must be raised 1 foot. Find the number of degrees through which the drum must rotate.

91. *Angular Speed*   A car is moving at a rate of 50 miles per hour, and the diameter of its wheels is 2.5 feet.

(a) Find the rotational speed of the wheels in revolutions per minute.

(b) Find the angular speed of the wheels in radians per minute.

92. *Angular Speed*   A 2-inch-diameter pulley on an electric motor that runs at 1700 revolutions per minute is connected by a belt to a 4-inch-diameter pulley on a saw arbor.

(a) Find the angular speed (in radians per minute) of each pulley.

(b) Find the speed of the saw in revolutions per minute.

93. *Think About It*   Is a degree or a radian the larger unit of measure? Explain.

94. *Essay*   If the radius of a circle is increasing and the magnitude of a central angle is held constant, how is the length of the intercepted arc changing? Explain your reasoning.

95. *Circular Saw Speed*   The circular blade on a saw has a diameter of 7.5 inches (see figure) and rotates at 2400 revolutions per minute.

(a) Find the angular speed in radians per second.

(b) Find the speed of the saw teeth (in feet per second) as they contact the wood being cut.

|← 7.5 in. →|

96. *Floppy Disk*   The radius of the magnetic disk in a 3.5-inch diskette is 1.68 inches. Find the linear speed of a point on the circumference of the disk if it is rotating at a speed of 360 revolutions per minute.

97. *Speed of a Bicycle*   The radii of the sprocket assemblies and the wheel of a bicycle are 4 inches, 2 inches, and 14 inches, respectively (see figure). If the cyclist is pedaling at a rate of 1 revolution per second, find the speed of the bicycle in (a) feet per second and (b) miles per hour.

14 in.

4 in.

2 in.

98. *Geometry*   Prove that the area of a circular sector of radius r with central angle θ is $A = \frac{1}{2}r^2\theta$, where θ is measured in radians.

Area of a Circular Sector   **In Exercises 99 and 100, use the result of Exercise 98 to find the area of the sector.**

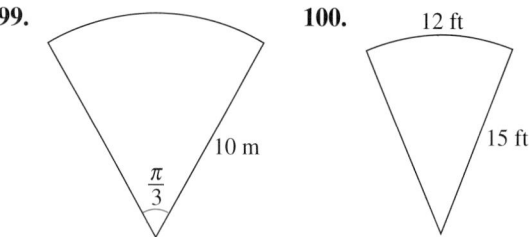

99.

10 m

$\frac{\pi}{3}$

100.

12 ft

15 ft

101. *Think About It*   Write a short paragraph defining, and giving examples of, coterminal angles.

102. *Graphical Reasoning*   The formulas for the area of a circular sector and arc length are $A = \frac{1}{2}r^2\theta$ and $s = r\theta$, respectively. (r is the radius and θ is the angle measured in radians.)

(a) If $\theta = 0.8$, express the area and arc length as functions of r. What is the domain of each function? Use a graphing utility to graph the functions. Use the graphs to determine which function changes more rapidly for changes in r when $r > 1$. Explain.

(b) If $r = 10$ centimeters, express the area and arc length as functions of θ. What is the domain of each function? Use a graphing utility to graph and identify the functions.

Chapter Opener   **In Exercises 103 and 104, use the information given in the chapter opener on page 309.**

103. Suppose the seismographs were 1.2 miles apart. Find the central angle between two adjacent seismographs.

104. Suppose the central angle between two adjacent seismographs was 0.031°. Find the distance between the seismographs.

4.2   Trigonometric Functions: The Unit Circle

The Unit Circle **/** *The Trigonometric Functions* **/** *Domain and Period of Sine and Cosine* **/** *Evaluating Trigonometric Functions with a Calculator*

The Unit Circle

The two historical perspectives of trigonometry incorporate different methods for introducing the trigonometric functions. Our first introduction to these functions is based on the unit circle.

Consider the **unit circle** given by

$$x^2 + y^2 = 1 \qquad \text{Unit circle}$$

as shown in Figure 4.15. Imagine that the real number line is wrapped around this circle, with positive numbers corresponding to a counterclockwise wrapping and negative numbers corresponding to a clockwise wrapping, as shown in Figure 4.16.

Figure 4.15

Figure 4.16

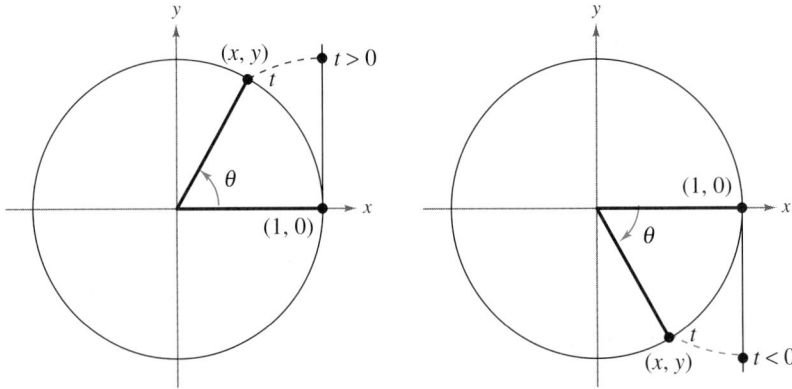

As the real number line is wrapped around the unit circle, each real number t corresponds to a point (x, y) on the circle. For example, the real number 0 corresponds to the point $(1, 0)$. Moreover, because the unit circle has a circumference of 2π, the real number 2π will also correspond to the point $(1, 0)$.

In general, each real number t also corresponds to a central angle θ (in standard position) whose radian measure is t. With this interpretation of t, the arc length formula $s = r\theta$ (with $r = 1$) indicates the real number t is the length of the arc interpreted by the angle θ, given in radians.

The Trigonometric Functions

From the preceding discussion, it follows that the coordinates x and y are two functions of the real variable t. You can use these coordinates to define the six trigonometric functions of t.

sine	cosecant
cosine	secant
tangent	cotangent

These six functions are normally abbreviated as sin, csc, cos, sec, tan, and cot, respectively.

Figure 4.17

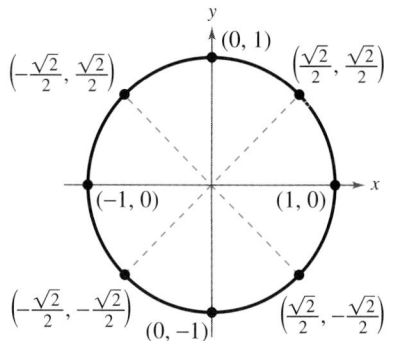

Definition of Trigonometric Functions

Let t be a real number and (x, y) the point on the unit circle corresponding to t.

$$\sin t = y \qquad\qquad \csc t = \frac{1}{y}, \quad y \neq 0$$

$$\cos t = x \qquad\qquad \sec t = \frac{1}{x}, \quad x \neq 0$$

$$\tan t = \frac{y}{x}, \quad x \neq 0 \qquad\qquad \cot t = \frac{x}{y}, \quad y \neq 0$$

Note The functions in the second column are the *reciprocals* of the corresponding functions in the first column.

In the definitions of the trigonometric functions, note that the tangent and secant are not defined when $x = 0$. For instance, because $t = \pi/2$ corresponds to $(x, y) = (0, 1)$, it follows that $\tan(\pi/2)$ and $\sec(\pi/2)$ are *undefined*. Similarly, the cotangent and cosecant are not defined when $y = 0$. For instance, because $t = 0$ corresponds to $(x, y) = (1, 0)$, cot 0 and csc 0 are *undefined*.

In Figure 4.17, the unit circle has been divided into eight equal arcs, corresponding to t-values of

$$0, \frac{\pi}{4}, \frac{\pi}{2}, \frac{3\pi}{4}, \pi, \frac{5\pi}{4}, \frac{3\pi}{2}, \frac{7\pi}{4}, \text{ and } 2\pi.$$

Figure 4.18

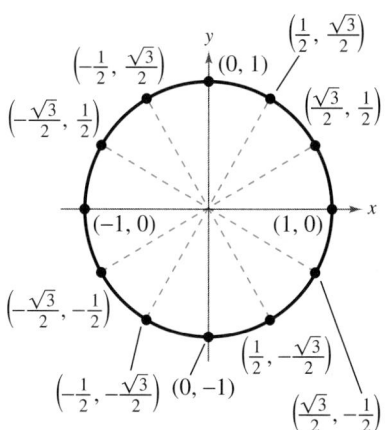

Similarly, in Figure 4.18, the unit circle has been divided into 12 equal arcs, corresponding to t-values of

$$0, \frac{\pi}{6}, \frac{\pi}{3}, \frac{\pi}{2}, \frac{2\pi}{3}, \frac{5\pi}{6}, \pi, \frac{7\pi}{6}, \frac{4\pi}{3}, \frac{3\pi}{2}, \frac{5\pi}{3}, \frac{11\pi}{6}, \text{ and } 2\pi.$$

Using the (x, y) coordinates in Figures 4.17 and 4.18, you can easily evaluate the trigonometric functions for common t-values. This procedure is demonstrated in Examples 1 and 2.

EXAMPLE 1   Evaluating Trigonometric Functions

Evaluate the six trigonometric functions at each real number.

a. $t = \dfrac{\pi}{6}$ **b.** $t = \dfrac{5\pi}{4}$ **c.** $t = 0$ **d.** $t = \pi$

Solution

For each t-value, begin by finding the corresponding point (x, y) on the unit circle. Then use the definition of trigonometric functions listed on page 323.

a. $t = \pi/6$ corresponds to the point $(x, y) = \left(\sqrt{3}/2, 1/2\right)$.

$$\sin \frac{\pi}{6} = y = \frac{1}{2} \qquad\qquad \csc \frac{\pi}{6} = 2$$

$$\cos \frac{\pi}{6} = x = \frac{\sqrt{3}}{2} \qquad\qquad \sec \frac{\pi}{6} = \frac{2}{\sqrt{3}} = \frac{2\sqrt{3}}{3}$$

$$\tan \frac{\pi}{6} = \frac{y}{x} = \frac{1/2}{\sqrt{3}/2} = \frac{1}{\sqrt{3}} \qquad\qquad \cot \frac{\pi}{6} = \sqrt{3}$$

b. $t = 5\pi/4$ corresponds to the point $(x, y) = \left(-\sqrt{2}/2, -\sqrt{2}/2\right)$.

$$\sin \frac{5\pi}{4} = y = -\frac{\sqrt{2}}{2} \qquad\qquad \csc \frac{5\pi}{4} = -\frac{2}{\sqrt{2}} = -\sqrt{2}$$

$$\cos \frac{5\pi}{4} = x = -\frac{\sqrt{2}}{2} \qquad\qquad \sec \frac{5\pi}{4} = -\frac{2}{\sqrt{2}} = -\sqrt{2}$$

$$\tan \frac{5\pi}{4} = \frac{y}{x} = \frac{-\sqrt{2}/2}{-\sqrt{2}/2} = 1 \qquad\qquad \cot \frac{5\pi}{4} = 1$$

c. $t = 0$ corresponds to the point $(x, y) = (1, 0)$.

$$\sin 0 = y = 0 \qquad\qquad\qquad \csc 0 \text{ is undefined.}$$

$$\cos 0 = x = 1 \qquad\qquad\qquad \sec 0 = 1$$

$$\tan 0 = \frac{y}{x} = \frac{0}{1} = 0 \qquad\qquad \cot 0 \text{ is undefined.}$$

d. $t = \pi$ corresponds to the point $(x, y) = (-1, 0)$.

$$\sin \pi = y = 0 \qquad\qquad\qquad \csc \pi \text{ is undefined.}$$

$$\cos \pi = x = -1 \qquad\qquad\qquad \sec \pi = -1$$

$$\tan \pi = \frac{y}{x} = \frac{0}{-1} = 0 \qquad\qquad \cot \pi \text{ is undefined.}$$

EXAMPLE 2   Evaluating Trigonometric Functions

Evaluate the six trigonometric functions at $t = -\dfrac{\pi}{3}$.

Solution

Moving *clockwise* around the unit circle, it follows that $t = -\pi/3$ corresponds to the point $(x, y) = \left(1/2, -\sqrt{3}/2\right)$.

$$\sin\left(-\frac{\pi}{3}\right) = -\frac{\sqrt{3}}{2} \qquad\qquad \csc\left(-\frac{\pi}{3}\right) = -\frac{2}{\sqrt{3}}$$

$$\cos\left(-\frac{\pi}{3}\right) = \frac{1}{2} \qquad\qquad \sec\left(-\frac{\pi}{3}\right) = 2$$

$$\tan\left(-\frac{\pi}{3}\right) = -\sqrt{3} \qquad\qquad \cot\left(-\frac{\pi}{3}\right) = -\frac{1}{\sqrt{3}}$$

Domain and Period of Sine and Cosine

The *domain* of the sine and cosine functions is the set of all real numbers. To determine the *range* of these two functions, consider the unit circle shown in Figure 4.19. Because $r = 1$, it follows that $\sin t = y$ and $\cos t = x$. Moreover, because (x, y) is on the unit circle, you know that $-1 \le y \le 1$ and $-1 \le x \le 1$, and it follows that the values of sine and cosine also range between -1 and 1. That is,

$$-1 \le y \le 1 \qquad\qquad -1 \le x \le 1$$
$$-1 \le \sin t \le 1 \qquad\qquad -1 \le \cos t \le 1$$

Suppose you add 2π to each value of t in the interval $[0, 2\pi]$, thus completing a second revolution around the unit circle, as shown in Figure 4.20. The values of $\sin(t + 2\pi)$ and $\cos(t + 2\pi)$ correspond to those of $\sin t$ and $\cos t$. Similar results can be obtained for repeated revolutions (positive or negative) on the unit circle. This leads to the general result

$$\sin(t + 2\pi n) = \sin t \qquad \text{and} \qquad \cos(t + 2\pi n) = \cos t$$

for any integer n and real number t. Functions that behave in such a repetitive (or cyclic) manner are called **periodic.**

Note In Figure 4.20, note that *positive* multiples of 2π are added to the t-values. You could just as well have added *negative* multiples. For instance, $\pi/4 - 2\pi$ and $\pi/4 - 4\pi$ are also coterminal to $\pi/4$.

Figure 4.19

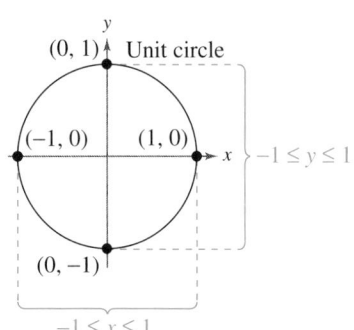

Figure 4.20

$$t = \tfrac{\pi}{2}, \tfrac{\pi}{2} + 2\pi, \tfrac{\pi}{2} + 4\pi, \ldots$$

$$t = \tfrac{3\pi}{4}, \tfrac{3\pi}{4} + 2\pi, \ldots \qquad t = \tfrac{\pi}{4}, \tfrac{\pi}{4} + 2\pi, \ldots$$

$$t = \pi, 3\pi, \ldots$$

$$t = 0, 2\pi, \ldots$$

$$t = \tfrac{5\pi}{4}, \tfrac{5\pi}{4} + 2\pi, \ldots$$

$$t = \tfrac{7\pi}{4}, \tfrac{7\pi}{4} + 2\pi, \tfrac{7\pi}{4} + 4\pi, \ldots$$

The *Interactive* CD-ROM offers graphing utility emulators of the *TI-82* and *TI-83*, which can be used with the Examples, Explorations, Technology notes, and Exercises.

Definition of Periodic Function

A function f is **periodic** if there exists a positive real number c such that

$$f(t + c) = f(t)$$

for all t in the domain of f. The least number c for which f is periodic is the **period** of f.

EXPLORATION

With your graphing utility in radian and parametric modes, enter

$$X_{1T} = \cos T \text{ and } Y_{1T} = \sin T$$

and use the following settings.

Tmin = 0
Tmax = 6.3
Tstep = .1
Xmin = -1.5
Xmax = 1.5
Xscl = 1
Ymin = -1
Ymax = 1
Yscl = 1

1. Graph the entered equations and describe the graph.
2. Use the trace key to move the cursor around the graph. What do the T-values represent? What do the X- and Y-values represent?
3. What are the smallest and greatest values for X and Y?

From this definition it follows that the sine and cosine functions are periodic and have a period of 2π. The other four trigonometric functions are also periodic, and you will study those functions in more detail in Section 4.6.

EXAMPLE 3   Using the Period to Evaluate the Sine and Cosine

a. Because $\dfrac{13\pi}{6} = 2\pi + \dfrac{\pi}{6}$, you have

$$\sin \frac{13\pi}{6} = \sin\left(2\pi + \frac{\pi}{6}\right) = \sin \frac{\pi}{6} = \frac{1}{2}.$$

b. Because $-\dfrac{7\pi}{2} = -4\pi + \dfrac{\pi}{2}$, you have

$$\cos\left(-\frac{7\pi}{2}\right) = \cos\left(-4\pi + \frac{\pi}{2}\right) = \cos \frac{\pi}{2} = 0.$$

Recall from Section 1.2 that a function f is *even* if $f(-t) = f(t)$ and is *odd* if $f(-t) = -f(t)$. Of the six trigonometric functions, two are even and four are odd, as stated in the following theorem. Verification of this theorem, using the unit circle, is left as an exercise.

Even and Odd Trigonometric Functions

The cosine and secant functions are *even*.

$$\cos(-t) = \cos t \qquad \sec(-t) = \sec t$$

The sine, cosecant, tangent, and cotangent functions are *odd*.

$$\sin(-t) = -\sin t \qquad \csc(-t) = -\csc t$$
$$\tan(-t) = -\tan t \qquad \cot(-t) = -\cot t$$

Evaluating Trigonometric Functions with a Calculator

When evaluating a trigonometric function with a calculator, you need to set the calculator to the desired *mode* of measurement (degrees or radians).

Most calculators do not have keys for the cosecant, secant, and cotangent functions. To evaluate these functions, you can use the $\boxed{x^{-1}}$ key with their respective reciprocal functions sine, cosine, and tangent. For example, to evaluate $\csc(\pi/8)$, use the fact that

$$\csc \frac{\pi}{8} = \frac{1}{\sin(\pi/8)}$$

and enter the following keystroke sequence in radian mode.

$\boxed{(}\ \boxed{\text{SIN}}\ \boxed{(}\ \boxed{\pi}\ \boxed{\div}\ \boxed{8}\ \boxed{)}\ \boxed{)}\ \boxed{x^{-1}}\ \boxed{\text{ENTER}}$ Display 2.6131259

EXAMPLE 4 ▱ **Using a Calculator**

Use a calculator to evaluate each expression.

a. $\sin 76.4°$ **b.** $\cot 1.5$

Solution

Function	Mode	Graphing Calculator Keystrokes	Display
a. $\sin 76.4°$	Degree	$\boxed{\text{SIN}}$ 76.4 $\boxed{\text{ENTER}}$	0.9719610
b. $\cot 1.5$	Radian	$\boxed{(}$ $\boxed{\text{TAN}}$ 1.5 $\boxed{)}$ $\boxed{x^{-1}}$ $\boxed{\text{ENTER}}$	0.0709148

▱

Group Activity

Error Analysis

Suppose you are tutoring a student in trigonometry. Your student is asked to evaluate the cosine of 2 radians and, using a calculator, obtains the following.

Keystrokes	*Display*
$\boxed{\text{COS}}$ 2 $\boxed{\text{ENTER}}$	0.999390827

You know that 2 radians lie in the second quadrant. You also know that this implies that the cosine of 2 radians should be negative. What did your student do wrong?

4.2 /// EXERCISES

In Exercises 1–6, find the six trigonometric functions of t that correspond to the point on the unit circle.

1.

$\left(-\frac{3}{5}, \frac{4}{5}\right)$

2.

$\left(\frac{12}{13}, \frac{5}{13}\right)$

3.

$\left(\frac{8}{17}, -\frac{15}{17}\right)$

4.

$\left(-\frac{4}{5}, -\frac{3}{5}\right)$

5.

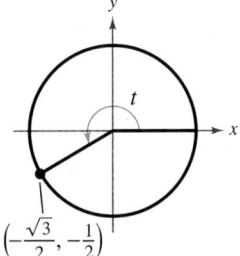

$\left(-\frac{\sqrt{3}}{2}, -\frac{1}{2}\right)$

6. $(-0.8668, 0.4987)$

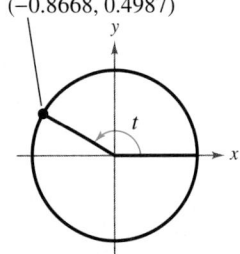

In Exercises 7–14, find the point (x, y) on the unit circle that corresponds to the real number t.

7. $t = \dfrac{\pi}{4}$

8. $t = \dfrac{\pi}{3}$

9. $t = \dfrac{5\pi}{6}$

10. $t = \dfrac{5\pi}{4}$

11. $t = \dfrac{4\pi}{3}$

12. $t = \dfrac{11\pi}{6}$

13. $t = \dfrac{3\pi}{2}$

14. $t = \pi$

In Exercises 15–26, evaluate (if possible) the sine, cosine, and tangent of the real number.

15. $t = \dfrac{\pi}{4}$

16. $t = -\dfrac{\pi}{4}$

17. $t = -\dfrac{\pi}{6}$

18. $t = \dfrac{\pi}{3}$

19. $t = -\dfrac{5\pi}{4}$

20. $t = -\dfrac{5\pi}{6}$

21. $t = \dfrac{11\pi}{6}$

22. $t = \dfrac{2\pi}{3}$

23. $t = \dfrac{4\pi}{3}$

24. $t = \dfrac{7\pi}{4}$

25. $t = -\dfrac{3\pi}{2}$

26. $t = -2\pi$

In Exercises 27–32, evaluate (if possible) the six trigonometric functions of the real number.

27. $t = \dfrac{3\pi}{4}$

28. $t = -\dfrac{2\pi}{3}$

29. $t = \dfrac{\pi}{2}$

30. $t = \dfrac{3\pi}{2}$

31. $t = -\dfrac{4\pi}{3}$

32. $t = -\dfrac{11\pi}{6}$

In Exercises 33–40, evaluate the trigonometric functions using its period as an aid.

33. $\sin 3\pi$

34. $\cos 3\pi$

35. $\cos \dfrac{8\pi}{3}$

36. $\sin \dfrac{9\pi}{4}$

37. $\cos \dfrac{19\pi}{6}$

38. $\sin\left(-\dfrac{13\pi}{6}\right)$

39. $\sin\left(-\dfrac{9\pi}{4}\right)$

40. $\cos\left(-\dfrac{8\pi}{3}\right)$

In Exercises 41–46, use the value of the trigonometric function to evaluate the indicated functions.

41. $\sin t = \frac{1}{3}$

  (a) $\sin(-t)$

  (b) $\csc(-t)$

42. $\sin(-t) = \frac{2}{5}$

  (a) $\sin t$

  (b) $\csc t$

43. $\cos(-t) = -\frac{7}{8}$

  (a) $\cos t$

  (b) $\sec(-t)$

44. $\cos t = -\frac{3}{4}$

  (a) $\cos(-t)$

  (b) $\sec(-t)$

45. $\sin t = \frac{4}{5}$

  (a) $\sin(\pi - t)$

  (b) $\sin(t + \pi)$

46. $\cos t = \frac{4}{5}$

  (a) $\cos(\pi - t)$

  (b) $\cos(t + \pi)$

In Exercises 47–56, use a calculator to evaluate the expression. Round to four decimal places.

47. $\sin \dfrac{\pi}{4}$

48. $\tan \pi$

49. $\cos(-3)$

50. $\cot 1$

51. $\cos(-1.7)$

52. $\csc 2.3$

53. $\csc 0.8$

54. $\sec 1.8$

55. $\sec 22.8$

56. $\sin(-0.9)$

In Exercises 57–60, use the figure and a straightedge.

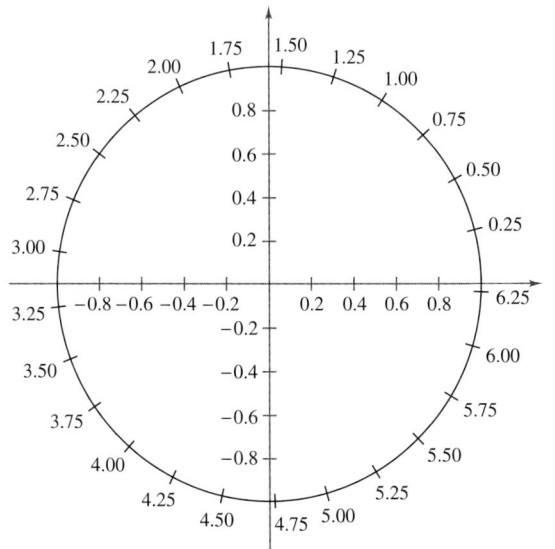

57. Approximate the trigonometric function.

  (a) $\sin 5$ (b) $\cos 2$

58. Approximate the trigonometric function.

  (a) $\sin 0.75$ (b) $\cos 2.5$

59. Approximate t where $0 \le t < 2\pi$.

  (a) $\sin t = 0.25$ (b) $\cos t = -0.25$

60. Approximate t where $0 \le t < 2\pi$.

  (a) $\sin t = -0.75$ (b) $\cos t = 0.75$

In Exercises 61 and 62, verify the statement by choosing specific values of t and showing that the two expressions have different values.

61. $\cos 2t \neq 2\cos t$

62. $\sin(t_1 + t_2) \neq \sin t_1 + \sin t_2$

63. *Exploration* Let (x_1, y_1) and (x_2, y_2) be points on the unit circle corresponding to $t = t_1$ and $t = \pi - t_1$, respectively.

  (a) Identify the symmetry of the points (x_1, y_1) and (x_2, y_2).

  (b) Make a conjecture about any relationship between $\sin t_1$ and $\sin(\pi - t_1)$.

  (c) Make a conjecture about any relationship between $\cos t_1$ and $\cos(\pi - t_1)$.

64. *Exploration* Let (x_1, y_1) and (x_2, y_2) be points on the unit circle corresponding to $t = t_1$ and $t = t_1 + \pi$, respectively.

  (a) Identify the symmetry of the points (x_1, y_1) and (x_2, y_2).

  (b) Make a conjecture about any relationship between $\sin t_1$ and $\sin(t_1 + \pi)$.

  (c) Make a conjecture about any relationship between $\cos t_1$ and $\cos(t_1 + \pi)$.

65. *Harmonic Motion*   The displacement from equilibrium of an oscillating weight suspended by a spring is given by

$$y(t) = \frac{1}{4} \cos 6t$$

where y is the displacement in feet and t is the time in seconds (see figure). Find the displacement when (a) $t = 0$, (b) $t = \frac{1}{4}$, and (c) $t = \frac{1}{2}$.

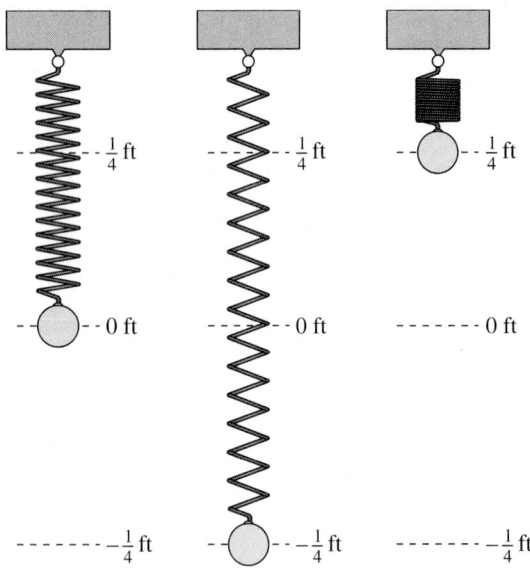

Equilibrium   Maximum negative   Maximum positive
displacement   displacement

Simple Harmonic Motion

66. *Harmonic Motion*   The displacement from equilibrium of an oscillating weight suspended by a spring and subject to the damping effect of friction is given by

$$y(t) = \frac{1}{4} e^{-t} \cos 6t$$

where y is the displacement in feet and t is the time in seconds. Find the displacement when (a) $t = 0$, (b) $t = \frac{1}{4}$, and (c) $t = \frac{1}{2}$.

67. *Electric Circuits*   The initial current and charge in the electrical circuit shown in the figure are zero. When 100 volts is applied to the circuit, the current is given by

$$I = 5e^{-2t} \sin t$$

if the resistance, inductance, and capacitance are 80 ohms, 20 henrys, and 0.01 farads, respectively. Approximate the current $t = 0.7$ seconds after the voltage is applied.

68. Use the unit circle to verify that the cosine and secant functions are even and the sine, cosecant, tangent, and cotangent functions are odd.

69. *Think About It*   Because $f(t) = \sin t$ is an odd function and $g(t) = \cos t$ is an even function, what can be said about the function $h(t) = f(t)g(t)$?

70. *Think About It*   Because $f(t) = \sin t$ and $g(t) = \tan t$ are odd functions, what can be said about the function $h(t) = f(t)g(t)$?

Review   Solve Exercises 71–74 as a review of the skills and problem-solving techniques you learned in previous sections. Find the inverse of the one-to-one function f. Use a graphing utility to graph both f and f^{-1} in the same viewing rectangle.

71. $f(x) = \frac{1}{2}(3x - 2)$

72. $f(x) = \frac{1}{4}x^3 + 1$

73. $f(x) = \sqrt{x^2 - 4}, \quad x \geq 2$

74. $f(x) = \dfrac{2x}{x + 1}, \quad x > -1$

4.3 Right Triangle Trigonometry

The Six Trigonometric Functions / Trigonometric Identities / Evaluating Trigonometric Functions with a Calculator / Applications Involving Right Triangles

The Six Trigonometric Functions

Our second look at the trigonometric functions is from a *right triangle* perspective. Consider a right triangle, one of whose acute angles is labeled θ, as shown in Figure 4.21. The three sides of the triangle are the **hypotenuse,** the **opposite side** (the side opposite the angle θ), and the **adjacent side** (the side adjacent to the angle θ). Using the lengths of these three sides, you can form six ratios that define the six trigonometric functions of the acute angle θ.

In the following definitions, it is important to see that

$$0 < \theta < 90° \qquad \theta \text{ is an acute angle.}$$

and for such angles the value of each trigonometric function is *positive.*

Figure 4.21

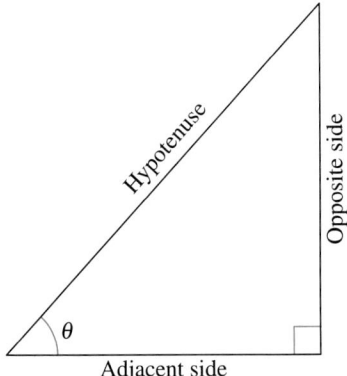

Hypotenuse

Opposite side

θ

Adjacent side

Right Triangle Definitions of Trigonometric Functions

Let θ be an *acute* angle of a right triangle. Then the six trigonometric functions of the angle θ are defined as follows.

$$\sin \theta = \frac{\text{opp}}{\text{hyp}} \qquad \csc \theta = \frac{\text{hyp}}{\text{opp}}$$

$$\cos \theta = \frac{\text{adj}}{\text{hyp}} \qquad \sec \theta = \frac{\text{hyp}}{\text{adj}}$$

$$\tan \theta = \frac{\text{opp}}{\text{adj}} \qquad \cot \theta = \frac{\text{adj}}{\text{opp}}$$

The abbreviations opp, adj, and hyp represent the lengths of the three sides of a right triangle.

$$\text{opp} = \text{the length of the side } opposite \ \theta$$
$$\text{adj} = \text{the length of the side } adjacent \text{ to } \theta$$
$$\text{hyp} = \text{the length of the } hypotenuse$$

The leading Teutonic mathematical astronomer of the 16th century was Georg Joachim Rhaeticus (1514–1576). He was the first to define the trigonometric functions as ratios of the sides of a right triangle.

Note Remember that the functions in the second column are the *reciprocals* of the corresponding functions in the first column.

EXAMPLE 1   ▭   **Evaluating Trigonometric Functions**

Find the values of the six trigonometric functions of θ, as shown in Figure 4.22.

Solution

By the Pythagorean Theorem, $(\text{hyp})^2 = (\text{opp})^2 + (\text{adj})^2$, it follows that

$$\text{hyp} = \sqrt{4^2 + 3^2} = \sqrt{25} = 5.$$

Thus, the six trigonometric functions of θ are

$$\sin \theta = \frac{\text{opp}}{\text{hyp}} = \frac{4}{5} \qquad \csc \theta = \frac{\text{hyp}}{\text{opp}} = \frac{5}{4}$$

$$\cos \theta = \frac{\text{adj}}{\text{hyp}} = \frac{3}{5} \qquad \sec \theta = \frac{\text{hyp}}{\text{adj}} = \frac{5}{3}$$

$$\tan \theta = \frac{\text{opp}}{\text{adj}} = \frac{4}{3} \qquad \cot \theta = \frac{\text{adj}}{\text{opp}} = \frac{3}{4}.$$ ▭

Figure 4.22

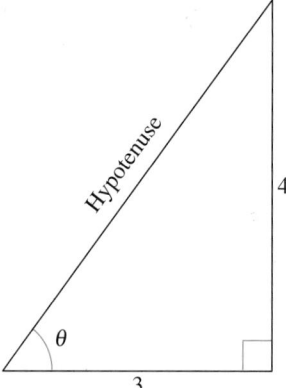

In Example 1, you were given the lengths of the sides of the right triangle, but not the angle θ. A more common problem in trigonometry is finding the trigonometric functions for a *given* acute angle θ. To do this, you can construct a right triangle having θ as one of its angles.

EXAMPLE 2   ▭   **Evaluating Trigonometric Functions of 45°**

Find the values of sin 45°, cos 45°, and tan 45°.

Solution

Construct a right triangle having 45° as one of its acute angles, as shown in Figure 4.23. Choose 1 as the length of the adjacent side. From geometry, you know that the other acute angle is also 45°. Hence, the triangle is isosceles and the length of the opposite side is also 1. Using the Pythagorean Theorem, you find the length of the hypotenuse to be $\sqrt{2}$.

$$\sin 45° = \frac{\text{opp}}{\text{hyp}} = \frac{1}{\sqrt{2}} = \frac{\sqrt{2}}{2}$$

$$\cos 45° = \frac{\text{adj}}{\text{hyp}} = \frac{1}{\sqrt{2}} = \frac{\sqrt{2}}{2}$$

$$\tan 45° = \frac{\text{opp}}{\text{adj}} = \frac{1}{1} = 1$$ ▭

Figure 4.23

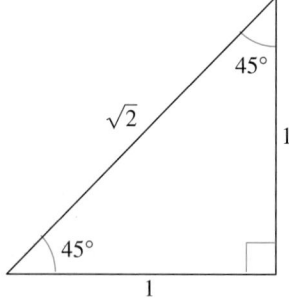

EXAMPLE 3 ▱ Evaluating Trigonometric Functions of 30° and 60°

Use the equilateral triangle shown in Figure 4.24 to find the values of sin 60°, cos 60°, sin 30°, and cos 30°.

Figure 4.24

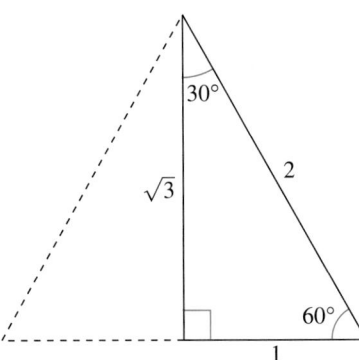

Solution

Try using the Pythagorean Theorem and the equilateral triangle to verify the lengths of the sides given in Figure 4.24. For $\theta = 60°$, you have adj $= 1$, opp $= \sqrt{3}$, and hyp $= 2$. Therefore,

$$\sin 60° = \frac{\text{opp}}{\text{hyp}} = \frac{\sqrt{3}}{2} \qquad \text{and} \qquad \cos 60° = \frac{\text{adj}}{\text{hyp}} = \frac{1}{2}.$$

For $\theta = 30°$, adj $= \sqrt{3}$, opp $= 1$, and hyp $= 2$. Thus,

$$\sin 30° = \frac{\text{opp}}{\text{hyp}} = \frac{1}{2} \qquad \text{and} \qquad \cos 30° = \frac{\text{adj}}{\text{hyp}} = \frac{\sqrt{3}}{2}.$$

Because the angles 30°, 45°, and 60° ($\pi/6$, $\pi/4$, and $\pi/3$) occur frequently in trigonometry, we suggest that you learn to construct the triangles shown in Figures 4.23 and 4.24.

Sines, Cosines, and Tangents of Special Angles

$$\sin 30° = \sin \frac{\pi}{6} = \frac{1}{2} \qquad \cos 30° = \cos \frac{\pi}{6} = \frac{\sqrt{3}}{2} \qquad \tan 30° = \tan \frac{\pi}{6} = \frac{\sqrt{3}}{3}$$

$$\sin 45° = \sin \frac{\pi}{4} = \frac{\sqrt{2}}{2} \qquad \cos 45° = \cos \frac{\pi}{4} = \frac{\sqrt{2}}{2} \qquad \tan 45° = \tan \frac{\pi}{4} = 1$$

$$\sin 60° = \sin \frac{\pi}{3} = \frac{\sqrt{3}}{2} \qquad \cos 60° = \cos \frac{\pi}{3} = \frac{1}{2} \qquad \tan 60° = \tan \frac{\pi}{3} = \sqrt{3}$$

In the box, note that $\sin 30° = \frac{1}{2} = \cos 60°$. This occurs because 30° and 60° are complementary angles, and, in general, it can be shown from the right triangle definitions that *cofunctions of complementary angles are equal.* That is, if θ is an acute angle, the following relationships are true.

$$\sin(90° - \theta) = \cos \theta \qquad\qquad \cos(90° - \theta) = \sin \theta$$

$$\tan(90° - \theta) = \cot \theta \qquad\qquad \cot(90° - \theta) = \tan \theta$$

$$\sec(90° - \theta) = \csc \theta \qquad\qquad \csc(90° - \theta) = \sec \theta$$

Trigonometric Identities

In trigonometry, a great deal of time is spent studying relationships between trigonometric functions (identities).

Fundamental Trigonometric Identities

Reciprocal Identities

$$\sin \theta = \frac{1}{\csc \theta} \qquad \cos \theta = \frac{1}{\sec \theta} \qquad \tan \theta = \frac{1}{\cot \theta}$$

$$\csc \theta = \frac{1}{\sin \theta} \qquad \sec \theta = \frac{1}{\cos \theta} \qquad \cot \theta = \frac{1}{\tan \theta}$$

Quotient Identities

$$\tan \theta = \frac{\sin \theta}{\cos \theta} \qquad \cot \theta = \frac{\cos \theta}{\sin \theta}$$

Pythagorean Identities

$$\sin^2 \theta + \cos^2 \theta = 1 \qquad\qquad 1 + \tan^2 \theta = \sec^2 \theta$$

$$1 + \cot^2 \theta = \csc^2 \theta$$

Note   Note that $\sin^2 \theta$ represents $(\sin \theta)^2$, $\cos^2 \theta$ represents $(\cos \theta)^2$, and so on.

EXAMPLE 4   ▱   Applying Trigonometric Identities

Let θ be an acute angle such that $\sin \theta = 0.6$. Find the values of (a) $\cos \theta$ and (b) $\tan \theta$ using trigonometric identities.

Figure 4.25

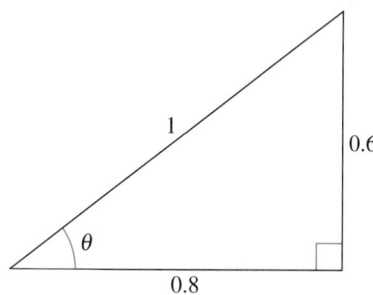

Solution

a. To find the value of $\cos \theta$, use the Pythagorean identity

$$\sin^2 \theta + \cos^2 \theta = 1.$$

Thus, you have

$$(0.6)^2 + \cos^2 \theta = 1$$
$$\cos^2 \theta = 1 - (0.6)^2 = 0.64$$
$$\cos \theta = \sqrt{0.64} = 0.8.$$

b. Now, knowing the sine and cosine of θ, you can find the tangent of θ to be

$$\tan \theta = \frac{\sin \theta}{\cos \theta} = \frac{0.6}{0.8} = 0.75.$$

Try using the definitions of $\cos \theta$ and $\tan \theta$, and the triangle shown in Figure 4.25, to check these results.   ▱

Figure 4.26

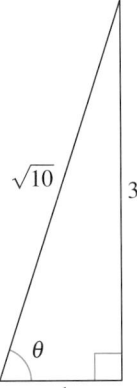

EXAMPLE 5   Applying Trigonometric Identities

Let θ be an acute angle such that $\tan \theta = 3$. Find the values of (a) $\cot \theta$ and (b) $\sec \theta$ using trigonometric identities.

Solution

a. $\cot \theta = \dfrac{1}{\tan \theta}$ Reciprocal identity

  $\cot \theta = \dfrac{1}{3}$

b. $\sec^2 \theta = 1 + \tan^2 \theta$ Pythagorean identity

  $\sec^2 \theta = 1 + 3^2 = 10$

  $\sec \theta = \sqrt{10}$

Try using the definitions of $\cot \theta$ and $\sec \theta$, and the triangle shown in Figure 4.26, to check these results.

Evaluating Trigonometric Functions with a Calculator

To use a calculator to evaluate trigonometric functions of angles measured in degrees, first set the calculator to degree mode and then proceed as demonstrated in Section 4.2. For instance, you can find values of $\cos 28°$ and $\sec 28°$ as follows.

Function	Graphing Calculator Keystrokes	Display
$\cos 28°$	COS 28 ENTER	0.8829476
$\sec 28°$	(COS 28) x^{-1} ENTER	1.1325701

Note Throughout this text, we follow the convention that angles are assumed to be measured in radians unless noted otherwise. For example, sin 1 means the sine of 1 radian and sin 1° means the sine of 1 degree.

EXAMPLE 6   Using a Calculator

Use a calculator to evaluate $\sec(5° \, 40' \, 12'')$.

Solution

Begin by converting to decimal form.

$$5° \, 40' \, 12'' = 5° + \left(\frac{40}{60}\right)° + \left(\frac{12}{3600}\right)° = 5.67°$$

Then use a calculator to evaluate $\sec 5.67°$.

$$\sec(5° \, 40' \, 12'') = \sec 5.67° = \frac{1}{\cos 5.67°} \approx 1.00492$$

Applications Involving Right Triangles

Many applications of trigonometry involve a process called **solving right triangles.** In this type of application, you are usually given one side of a right triangle and one of the acute angles and asked to find one of the other sides, *or* you are given two sides and asked to find one of the acute angles.

Real Life

EXAMPLE 7   ◻   **Solving a Right Triangle**

A surveyor is standing 50 feet from the base of a large tree, as shown in Figure 4.27. The surveyor measures the angle of elevation to the top of the tree as 71.5°. How tall is the tree?

Figure 4.27

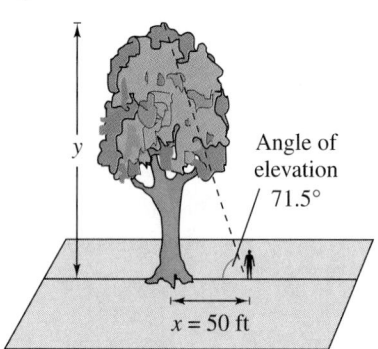

Angle of elevation 71.5°

y

$x = 50$ ft

Solution

From Figure 4.27, you can see that

$$\tan 71.5° = \frac{\text{opp}}{\text{adj}} = \frac{y}{x}$$

where $x = 50$ and y is the height of the tree. Thus, the height of the tree is

$$y = x \tan 71.5° \approx 50(2.98868) \approx 149.4 \text{ feet.}$$ ◻

Real Life

EXAMPLE 8   ◻   **Solving a Right Triangle**

A person is 200 yards from a river. Rather than walking directly to the river, the person walks 400 yards along a straight path to the river's edge. Find the acute angle θ between this path and the river's edge, as illustrated in Figure 4.28.

Figure 4.28

θ

200 yd

400 yd

Solution

From Figure 4.28, you can see that the sine of the angle θ is

$$\sin \theta = \frac{\text{opp}}{\text{hyp}} = \frac{200}{400} = \frac{1}{2}.$$

Now, you recognize that $\theta = 30°$.   ◻

In Example 8, you were able to recognize that the acute angle that satisfies the equation $\sin \theta = \frac{1}{2}$ is $\theta = 30°$. Suppose, however, that you are given the equation $\sin \theta = 0.6$ and asked to find the acute angle θ. Because

$$\sin 30° = \frac{1}{2} = 0.5000 \quad \text{and} \quad \sin 45° = \frac{1}{\sqrt{2}} \approx 0.7071$$

you might guess that θ lies somewhere between $30°$ and $45°$. A more precise value of θ can be found using the *inverse* key on a calculator. To do this, you can use the following keystroke sequence in degree mode.

| SIN⁻¹ | .6 | ENTER | Display 36.8699

Thus, you can conclude that if $\sin \theta = 0.6$, $\theta \approx 36.87°$.

Real Life

EXAMPLE 9   Solving a Right Triangle

A 12-meter flagpole casts a 9-meter shadow, as shown in Figure 4.29. Find θ, the angle of elevation of the sun.

Solution

Figure 4.29 shows that the *opposite* and *adjacent* sides are known. Thus,

$$\tan \theta = \frac{\text{opp}}{\text{adj}} = \frac{12}{9}.$$

With a calculator in degree mode, you use the keystrokes

| TAN⁻¹ | (| 12 | ÷ | 9 |) | ENTER

to obtain $\theta \approx 53.13°$.

Figure 4.29

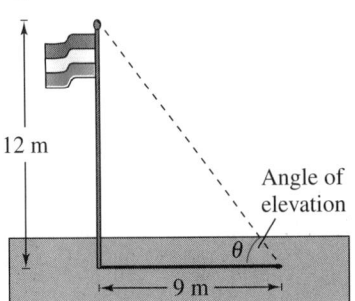

12 m

9 m

Angle of elevation

θ

Group Activity   *Evaluating Trigonometric Functions*

Some functions, such as $f(x) = 5x$, have the property that

$$f(cx) = cf(x).$$

Do any of the six trigonometric functions have this property? Compare the following values and use the results to justify your answer.

a. $\sin 60°$ and $2 \sin 30°$ **b.** $\cos 60°$ and $2 \cos 30°$

c. $\tan 60°$ and $2 \tan 30°$ **d.** $\cot 60°$ and $2 \cot 30°$

e. $\sec 60°$ and $2 \sec 30°$ **f.** $\csc 60°$ and $2 \csc 30°$

4.3 /// EXERCISES

In Exercises 1–4, find the exact values of the six trigonometric functions of the angle θ given in the figure. (Use the Pythagorean Theorem to find the third side of the triangle.)

1.

2.

3.

4.

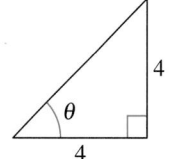

In Exercises 5–8, find the exact values of the six trigonometric functions of the angle θ for each of the triangles. Explain why the function values are the same.

5.

6.

7.

8.

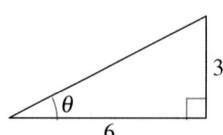

In Exercises 9–16, sketch a right triangle corresponding to the trigonometric function of the acute angle θ. Use the Pythagorean Theorem to determine the third side, and then find the other five trigonometric functions of θ.

9. $\sin \theta = \frac{2}{3}$ **10.** $\cot \theta = 5$

11. $\sec \theta = 2$ **12.** $\cos \theta = \frac{5}{7}$

13. $\tan \theta = 3$ **14.** $\csc \theta = \frac{17}{4}$

15. $\cot \theta = \frac{3}{2}$ **16.** $\sin \theta = \frac{3}{8}$

In Exercises 17–22, use the given function value(s), and trigonometric identities (including the relationship between a trigonometric function and its cofunction of a complementary angle), to find the indicated trigonometric functions.

17. $\sin 60° = \dfrac{\sqrt{3}}{2}, \quad \cos 60° = \dfrac{1}{2}$

    (a) $\tan 60°$ (b) $\sin 30°$
    (c) $\cos 30°$ (d) $\cot 60°$

18. $\sin 30° = \dfrac{1}{2}, \quad \tan 30° = \dfrac{\sqrt{3}}{3}$

    (a) $\csc 30°$ (b) $\cot 60°$
    (c) $\cos 30°$ (d) $\cot 30°$

19. $\csc \theta = 3, \quad \sec \theta = \dfrac{3\sqrt{2}}{4}$

    (a) $\sin \theta$ (b) $\cos \theta$
    (c) $\tan \theta$ (d) $\sec(90° - \theta)$

20. $\sec \theta = 5, \quad \tan \theta = 2\sqrt{6}$

    (a) $\cos \theta$ (b) $\cot \theta$
    (c) $\cot(90° - \theta)$ (d) $\sin \theta$

21. $\cos \alpha = \dfrac{1}{4}$

    (a) $\sec \alpha$ (b) $\sin \alpha$
    (c) $\cot \alpha$ (d) $\sin(90° - \alpha)$

22. $\tan \beta = 5$

    (a) $\cot \beta$ (b) $\cos \beta$
    (c) $\tan(90° - \beta)$ (d) $\csc \beta$

In Exercises 23–32, use trigonometric identities to transform one side of the equation into the other.

23. $\tan \theta \cot \theta = 1$

24. $\cos \theta \sec \theta = 1$

25. $\tan \alpha \cos \alpha = \sin \alpha$

26. $\cot \alpha \sin \alpha = \cos \alpha$

27. $(1 + \cos \theta)(1 - \cos \theta) = \sin^2 \theta$

28. $(1 + \sin \theta)(1 - \sin \theta) = \cos^2 \theta$

29. $(\sec \theta + \tan \theta)(\sec \theta - \tan \theta) = 1$

30. $\sin^2 \theta - \cos^2 \theta = 2 \sin^2 \theta - 1$

31. $\dfrac{\sin \theta}{\cos \theta} + \dfrac{\cos \theta}{\sin \theta} = \csc \theta \sec \theta$

32. $\dfrac{\tan \beta + \cot \beta}{\tan \beta} = \csc^2 \beta$

In Exercises 33–36, evaluate the trigonometric functions by memory or by constructing appropriate triangles for the given special angles.

33. (a) $\cos 60°$ (b) $\tan \dfrac{\pi}{6}$

34. (a) $\csc 30°$ (b) $\sin \dfrac{\pi}{4}$

35. (a) $\cot 45°$ (b) $\cos 45°$

36. (a) $\sin \dfrac{\pi}{3}$ (b) $\csc 45°$

In Exercises 37–46, use a calculator to evaluate each function. Round your answers to four decimal places. (Be sure the calculator is in the correct mode.)

37. (a) $\sin 10°$ (b) $\cos 80°$

38. (a) $\tan 23.5°$ (b) $\cot 66.5°$

39. (a) $\sin 16.35°$ (b) $\csc 16.35°$

40. (a) $\cos 16° \, 18'$ (b) $\sin 73° \, 56'$

41. (a) $\sec 42° \, 12'$ (b) $\csc 48° \, 7'$

42. (a) $\cos 4° \, 50' \, 15''$ (b) $\sec 4° \, 50' \, 15''$

43. (a) $\cot \dfrac{\pi}{16}$ (b) $\tan \dfrac{\pi}{16}$

44. (a) $\sec 0.75$ (b) $\cos 0.75$

45. (a) $\csc 1$ (b) $\tan \dfrac{1}{2}$

46. (a) $\sec\left(\dfrac{\pi}{2} - 1\right)$ (b) $\cot\left(\dfrac{\pi}{2} - \dfrac{1}{2}\right)$

In Exercises 47–52, find the values of θ in degrees $(0° < \theta < 90°)$ and radians $(0 < \theta < \pi/2)$ without the aid of a calculator.

47. (a) $\sin \theta = \dfrac{1}{2}$ (b) $\csc \theta = 2$

48. (a) $\cos \theta = \dfrac{\sqrt{2}}{2}$ (b) $\tan \theta = 1$

49. (a) $\sec \theta = 2$ (b) $\cot \theta = 1$

50. (a) $\tan \theta = \sqrt{3}$ (b) $\cos \theta = \dfrac{1}{2}$

51. (a) $\csc \theta = \dfrac{2\sqrt{3}}{3}$ (b) $\sin \theta = \dfrac{\sqrt{2}}{2}$

52. (a) $\cot \theta = \dfrac{\sqrt{3}}{3}$ (b) $\sec \theta = \sqrt{2}$

In Exercises 53–56, find the values of θ in degrees $(0° < \theta < 90°)$ and radians $(0 < \theta < \pi/2)$ by using a calculator.

53. (a) $\sin \theta = 0.8191$ (b) $\cos \theta = 0.0175$

54. (a) $\cos \theta = 0.9848$ (b) $\cos \theta = 0.8746$

55. (a) $\tan \theta = 1.1920$ (b) $\tan \theta = 0.4663$

56. (a) $\sin \theta = 0.3746$ (b) $\cos \theta = 0.3746$

In Exercises 57–64, solve for x, y, or r, as indicated.

57. Solve for y.

58. Solve for x.

59. Solve for x.

60. Solve for r.

61. Solve for *r*.

62. Solve for *x*.

63. Solve for *y*.

64. Solve for *r*.

65. *Height*   A 6-foot person standing 15 feet from a streetlight casts an 8-foot shadow (see figure). What is the height of the streetlight?

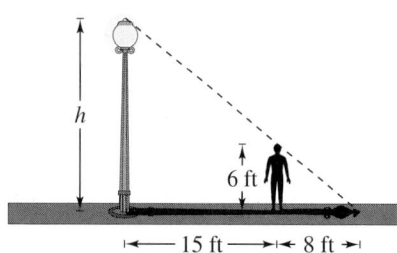

66. *Height*   A 6-foot person walks from the base of a broadcasting tower directly toward the tip of the shadow cast by the tower. When the person was 132 feet from the tower and 3 feet from the tip of the tower's shadow, the person's shadow starts to appear beyond the tower's shadow.

(a) Draw a right triangle that gives a visual representation of the problem. Show the known quantities on the triangle and use a variable to indicate the unknown height of the tower.

(b) Write an equation involving the unknown.

(c) What is the height of the tower?

67. *Length*   A 30-meter line is used to tether a helium-filled balloon. Because of a breeze, the line makes an angle of approximately 75° with the ground.

(a) Draw a right triangle that gives a visual representation of the problem. Show the known quantities on the triangle and use a variable to indicate the unknown height of the balloon.

(b) Use a trigonometric function to write an equation involving the unknown.

(c) What is the height of the balloon?

68. *Width of a River*   A biologist wants to know the width *w* of a river in order to properly set instruments for studying the pollutants in the water. From point *A*, the biologist walks downstream 100 feet and sights to point *C*. From this sighting, it is determined that $\theta = 54°$ (see figure). How wide is the river?

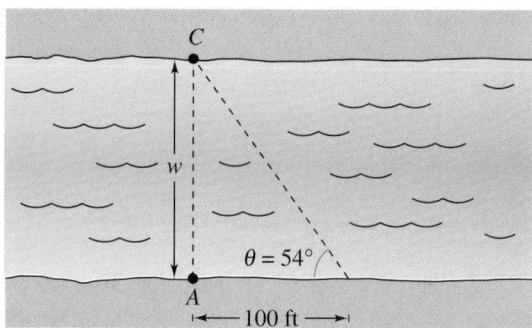

69. *Distance*   From a 60-foot observation tower on the coast, a Coast Guard officer sights a boat in difficulty. The angle of depression of the boat is 3° (see figure). How far is the boat from the shoreline?

(Not drawn to scale)

70. *Angle of Elevation* A ramp 20 feet in length rises to a loading platform that is $3\frac{1}{3}$ feet off the ground.

   (a) Draw a right triangle that gives a visual representation of the problem. Show the known quantities on the triangle and use a variable to indicate the unknown angle of elevation of the ramp.

   (b) Use a trigonometric function to write an equation involving the unknown.

   (c) What is the angle of elevation of the ramp?

71. *Machine Shop Calculations* A steel plate has the form of a quarter of a circle with a radius of 60 centimeters. Two 2-centimeter holes are to be drilled in the plate positioned as shown in the figure. Find the coordinates of the center of each hole.

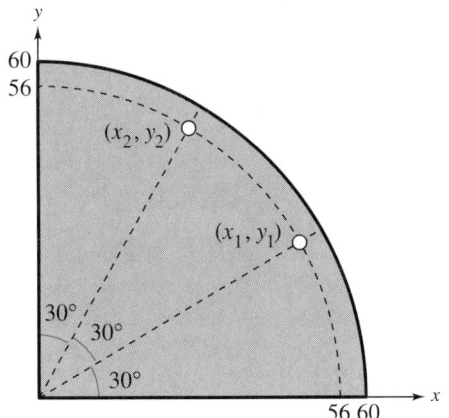

72. *Machine Shop Calculations* A tapered shaft has a diameter of 5 centimeters at the small end and is 15 centimeters long (see figure). If the taper is 3°, find the diameter d of the large end of the shaft.

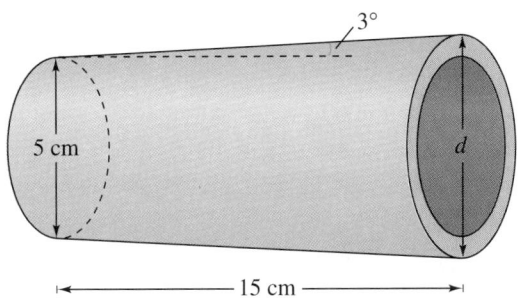

73. *Geometry* Use a compass to sketch a quarter of a circle of radius 10 centimeters. Using a protractor, construct an angle of 20° in standard position (see figure). Drop a perpendicular from the point of intersection of the terminal side of the angle and the arc of the circle. By actual measurement, calculate the coordinates (x, y) of the point of intersection and use these measurements to approximate the six trigonometric functions of a 20° angle.

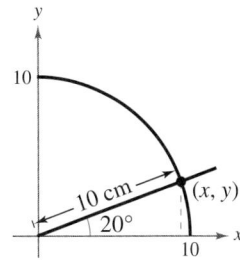

74. *Geometry* Repeat Exercise 73 using a 75° angle.

75. *Exploration*

   (a) Use a graphing utility to complete the table.

θ	0	0.1	0.2	0.3	0.4	0.5
$\sin \theta$						

   (b) Is θ or $\sin \theta$ greater for θ in the interval $(0, 0.5]$?

   (c) As θ approaches 0, how do θ and $\sin \theta$ compare? Explain.

76. *Exploration*

   (a) Use a graphing utility to complete the table.

θ	0	0.3	0.6	0.9	1.2	1.5
$\sin \theta$						
$\cos \theta$						

   (b) Discuss the behavior of the sine function for θ in the interval $[0, 1.5]$.

   (c) Discuss the behavior of the cosine function for θ in the interval $[0, 1.5]$.

   (d) Use the definitions of the sine and cosine functions to explain the results of parts (b) and (c).

77. *Exploration*

(a) Use a graphing utility to complete the table. Round the results to four decimal places.

θ	0°	20°	40°	60°	80°
$\sin \theta$					
$\cos \theta$					
$\tan \theta$					

(b) Classify each of the three trigonometric functions as increasing or decreasing for the table values.

(c) For the values in the table, verify that the tangent function is the quotient of the sine and cosine functions.

78. *Exploration*   Use a graphing utility to complete the table and make a conjecture about the relationship between $\cos \theta$ and $\sin(90° - \theta)$. What are the angles θ and $90° - \theta$ called?

θ	0°	20°	40°	60°	80°
$\cos \theta$					
$\sin(90° - \theta)$					

79. *Numerical Analysis*   The range R of a projectile fired at an angle of θ degrees with the horizontal and with an initial velocity of 50 meters per second is given by

$$R = 530 \sin \theta \cos \theta.$$

(a) Use a graphing utility to complete the table.

θ	20°	30°	40°	50°	60°	70°
R						

(b) Identify a pattern in the range of the projectile versus the angle θ.

(c) Based on the table, make a conjecture about the angle θ that will yield a maximum range.

(d) Test the conjecture by creating a new table with values of θ near the value of θ that you believe yields the maximum range.

80. *Numerical Analysis*   A 3000-pound automobile is negotiating a circular interchange of radius 300 feet at a speed of s miles per hour (see figure). The relationship between the speed and the angle θ (in degrees) at which the roadway should be banked so that no lateral frictional force is exerted on the tires is given by

$$\tan \theta = \frac{0.672s^2}{3000}.$$

(a) Use a graphing utility to complete the table.

s	10	20	30	40	50	60
θ						

(b) The speeds are incremented by 10 miles per hour in the table, but θ does not increase by equal increments. Explain.

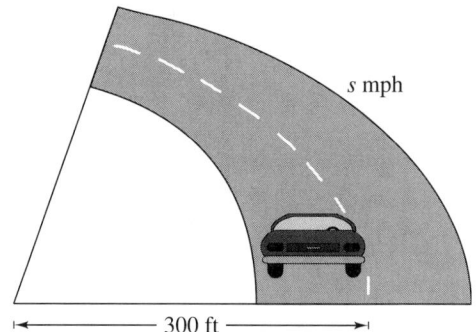

s mph

300 ft

True or False?   **In Exercises 81–86, determine whether the statement is true or false, and give a reason for your answer.**

81. $\sin 60° \csc 60° = 1$

82. $\sec 30° = \csc 60°$

83. $\sin 45° + \cos 45° = 1$

84. $\cot^2 10° - \csc^2 10° = -1$

85. $\dfrac{\sin 60°}{\sin 30°} = \sin 2°$

86. $\tan[(0.8)^2] = \tan^2(0.8)$

4.4   Trigonometric Functions of Any Angle

Introduction **/** *Reference Angles* **/** *Trigonometric Functions of Real Numbers*

Introduction

In Section 4.3, the definitions of trigonometric functions were restricted to *acute* angles. In this section, the definitions are extended to cover *any* angle.

Figure 4.30

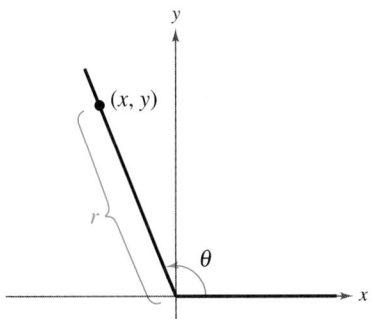

Definitions of Trigonometric Functions of Any Angle

Let θ be an angle in standard position with (x, y) a point on the terminal side of θ and $r = \sqrt{x^2 + y^2} \neq 0$, as shown in Figure 4.30.

$$\sin \theta = \frac{y}{r} \qquad\qquad \cos \theta = \frac{x}{r}$$

$$\tan \theta = \frac{y}{x}, \quad x \neq 0 \qquad\qquad \cot \theta = \frac{x}{y}, \quad y \neq 0$$

$$\sec \theta = \frac{r}{x}, \quad x \neq 0 \qquad\qquad \csc \theta = \frac{r}{y}, \quad y \neq 0$$

Note   If θ is an *acute* angle, the definitions here coincide with those given in the previous section.

Because $r = \sqrt{x^2 + y^2}$ *cannot* be zero, it follows that the sine and cosine functions are defined for any real value of θ. However, if $x = 0$, the tangent and secant of θ are undefined. For example, the tangent of 90° is undefined. Similarly, if $y = 0$, the cotangent and cosecant of θ are undefined.

Figure 4.31

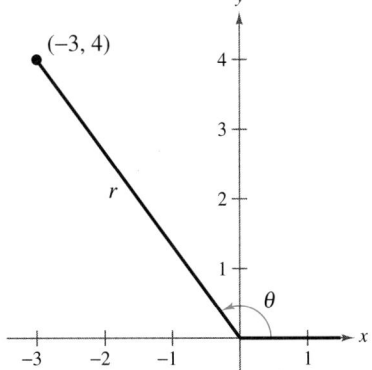

EXAMPLE 1 ▱ **Evaluating Trigonometric Functions**

Let $(-3, 4)$ be a point on the terminal side of θ. Find the sine, cosine, and tangent of θ.

Solution

Referring to Figure 4.31, you see that $x = -3$, $y = 4$, and

$$r = \sqrt{x^2 + y^2} = \sqrt{(-3)^2 + 4^2} = \sqrt{25} = 5.$$

Thus, you have the following.

$$\sin \theta = \frac{y}{r} = \frac{4}{5}, \qquad \cos \theta = \frac{x}{r} = -\frac{3}{5}, \qquad \tan \theta = \frac{y}{x} = -\frac{4}{3} \qquad ▱$$

Figure 4.32

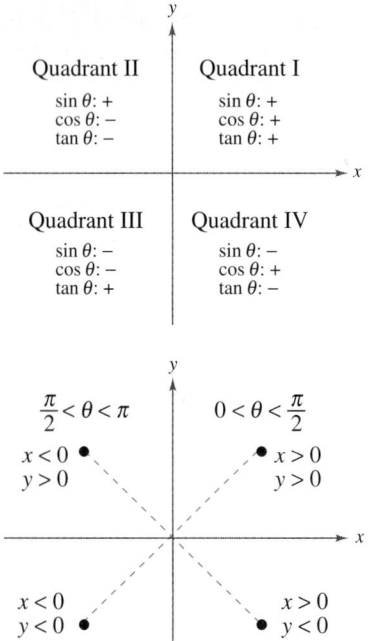

The *signs* of the trigonometric functions in the four quadrants can be determined easily from the definitions of the functions. For instance, because $\cos \theta = x/r$, it follows that $\cos \theta$ is positive wherever $x > 0$, which is in Quadrants I and IV. (Remember, r is always positive.) In a similar manner, you can verify the results shown in Figure 4.32.

EXAMPLE 2   Evaluating Trigonometric Functions

Given $\tan \theta = -\frac{5}{4}$ and $\cos \theta > 0$, find $\sin \theta$ and $\sec \theta$.

Solution

Note that θ lies in Quadrant IV because that is the only quadrant in which the tangent is negative and the cosine is positive. Moreover, using

$$\tan \theta = \frac{y}{x} = -\frac{5}{4}$$

and the fact that y is negative in Quadrant IV, you can let $y = -5$ and $x = 4$. Hence, $r = \sqrt{16 + 25} = \sqrt{41}$, and you have the following.

$$\sin \theta = \frac{y}{r} = \frac{-5}{\sqrt{41}} \approx -0.7809$$

$$\sec \theta = \frac{r}{x} = \frac{\sqrt{41}}{4} \approx 1.6008$$

EXAMPLE 3   Trigonometric Functions of Quadratic Angles

Evaluate the sine function at the four quadrant angles 0, $\dfrac{\pi}{2}$, π, and $\dfrac{3\pi}{2}$.

Solution

Figure 4.33

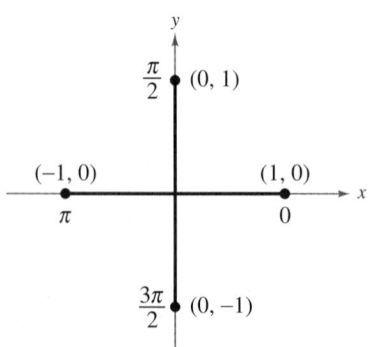

To begin, choose a point on the terminal side of each angle, as shown in Figure 4.33. For each of the four given points, $r = 1$, and you have

$$\sin 0 = \frac{y}{r} = \frac{0}{1} = 0 \qquad\qquad (x, y) = (1, 0)$$

$$\sin \frac{\pi}{2} = \frac{y}{r} = \frac{1}{1} = 1 \qquad\qquad (x, y) = (0, 1)$$

$$\sin \pi = \frac{y}{r} = \frac{0}{1} = 0 \qquad\qquad (x, y) = (-1, 0)$$

$$\sin \frac{3\pi}{2} = \frac{y}{r} = \frac{-1}{1} = -1. \qquad\qquad (x, y) = (0, -1)$$

Try using Figure 4.33 to evaluate some of the other trigonometric functions at the four quadrant angles.

Reference Angles

The values of the trigonometric functions of angles greater than 90° (or less than 0°) can be determined from their values at corresponding acute angles called **reference angles.**

Definition of Reference Angle

Let θ be an angle in standard position. Its **reference angle** is the acute angle θ' formed by the terminal side of θ and the horizontal axis.

Figure 4.34 shows the reference angles for θ in Quadrants II, III, and IV.

Figure 4.34

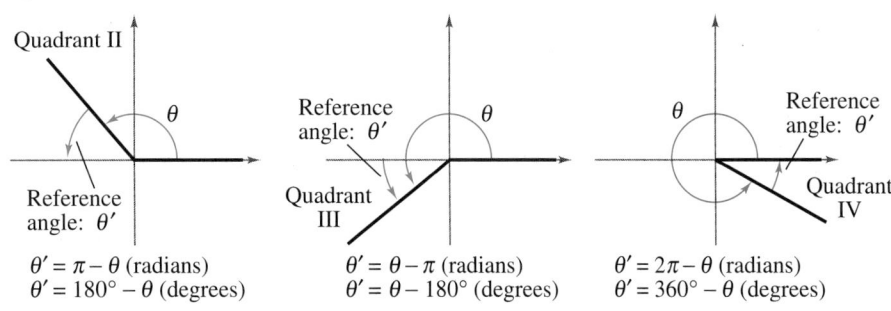

$\theta' = \pi - \theta$ (radians)
$\theta' = 180° - \theta$ (degrees)

$\theta' = \theta - \pi$ (radians)
$\theta' = \theta - 180°$ (degrees)

$\theta' = 2\pi - \theta$ (radians)
$\theta' = 360° - \theta$ (degrees)

Figure 4.35

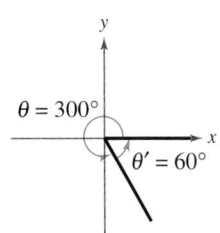

$\theta = 300°$
$\theta' = 60°$

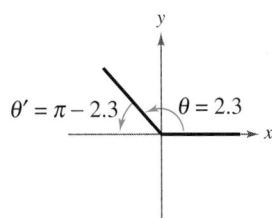

$\theta' = \pi - 2.3$
$\theta = 2.3$

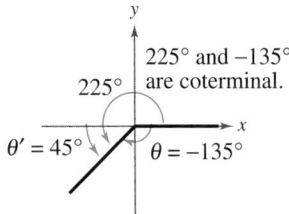

225° and −135° are coterminal.
$225°$
$\theta' = 45°$
$\theta = -135°$

EXAMPLE 4   **Finding Reference Angles**

Find the reference angle θ'.

a. $\theta = 300°$ **b.** $\theta = 2.3$ **c.** $\theta = -135°$

Solution

a. Because 300° lies in Quadrant IV, the angle it makes with the x-axis is

$$\theta' = 360° - 300° = 60°. \qquad \text{Degrees}$$

b. Because 2.3 lies between $\pi/2 \approx 1.5708$ and $\pi \approx 3.1416$, it follows that it is in Quadrant II and its reference angle is

$$\theta' = \pi - 2.3 \approx 0.8416. \qquad \text{Radians}$$

c. First, determine that $-135°$ is coterminal with 225°, which lies in Quadrant III. Hence, the reference angle is

$$\theta' = 225° - 180° = 45°. \qquad \text{Degrees}$$

Figure 4.35 shows each angle θ and its reference angle θ'.

Trigonometric Functions of Real Numbers

Figure 4.36

opp = $|y|$, adj = $|x|$

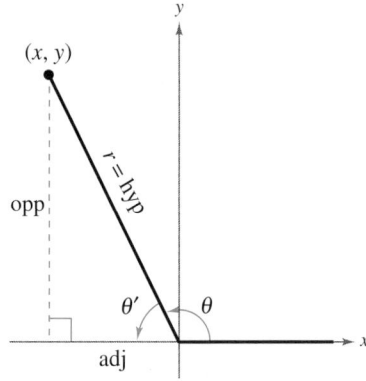

To see how a reference angle is used to evaluate a trigonometric function, consider the point (x, y) on the terminal side of θ, as shown in Figure 4.36. By definition, you know that

$$\sin \theta = \frac{y}{r} \quad \text{and} \quad \tan \theta = \frac{y}{x}.$$

For the right triangle with acute angle θ' and sides of lengths $|x|$ and $|y|$, you have

$$\sin \theta' = \frac{\text{opp}}{\text{hyp}} = \frac{|y|}{r} \quad \text{and} \quad \tan \theta' = \frac{\text{opp}}{\text{adj}} = \frac{|y|}{|x|}.$$

Thus, it follows that $\sin \theta$ and $\sin \theta'$ are equal, *except possibly in sign.* The same is true for $\tan \theta$ and $\tan \theta'$ *and* for the other four trigonometric functions. In all cases, the sign of the function value can be determined by the quadrant in which θ lies.

Evaluating Trigonometric Functions of Any Angle

To find the value of a trigonometric function of any angle θ,

1. determine the function value for the associated reference angle θ';
2. depending on the quadrant in which θ lies, prefix the appropriate sign to the function value.

By using reference angles and the special angles discussed in the previous section, you can greatly extend the scope of *exact* trigonometric values. For instance, knowing the function values of $30°$ means that you know the function values of all angles for which $30°$ is a reference angle. For convenience, the following table gives the exact values of the trigonometric functions of special angles and quadrant angles.

Library of Functions

The trigonometric functions are important for modeling periodic behavior, such as business cycles, planetary orbits, pendulums, and light rays. These functions play a prominent role in calculus.

Trigonometric Values of Common Angles

θ (degrees)	0°	30°	45°	60°	90°	180°	270°
θ (radians)	0	$\dfrac{\pi}{6}$	$\dfrac{\pi}{4}$	$\dfrac{\pi}{3}$	$\dfrac{\pi}{2}$	π	$\dfrac{3\pi}{2}$
$\sin \theta$	0	$\dfrac{1}{2}$	$\dfrac{\sqrt{2}}{2}$	$\dfrac{\sqrt{3}}{2}$	1	0	-1
$\cos \theta$	1	$\dfrac{\sqrt{3}}{2}$	$\dfrac{\sqrt{2}}{2}$	$\dfrac{1}{2}$	0	-1	0
$\tan \theta$	0	$\dfrac{\sqrt{3}}{3}$	1	$\sqrt{3}$	Undef.	0	Undef.

EXAMPLE 5   Trigonometric Functions of Nonacute Angles

Evaluate the following.

a. $\cos \dfrac{4\pi}{3}$ **b.** $\tan(-210°)$ **c.** $\csc \dfrac{11\pi}{4}$

Solution

a. Because $\theta = 4\pi/3$ lies in Quadrant III, the reference angle is $\theta' = (4\pi/3) - \pi = \pi/3$, as shown in Figure 4.37(a). Moreover, the cosine is negative in Quadrant III, so that

$$\cos \frac{4\pi}{3} = (-) \cos \frac{\pi}{3} = -\frac{1}{2}.$$

b. Because $-210° + 360° = 150°$, it follows that $-210°$ is coterminal with the second-quadrant angle $150°$. Therefore, the reference angle is $\theta' = 180° - 150° = 30°$, as shown in Figure 4.37(b). Finally, because the tangent is negative in Quadrant II, you have

$$\tan(-210°) = (-) \tan 30° = -\frac{\sqrt{3}}{3}.$$

c. Because $(11\pi/4) - 2\pi = 3\pi/4$, it follows that $11\pi/4$ is coterminal with the second-quadrant angle $3\pi/4$. Therefore, the reference angle is $\theta' = \pi - (3\pi/4) = \pi/4$, as shown in Figure 4.37(c). Because the cosecant is positive in Quadrant II, you have

$$\csc \frac{11\pi}{4} = (+) \csc \frac{\pi}{4} = \frac{1}{\sin(\pi/4)} = \sqrt{2}.$$

Figure 4.37

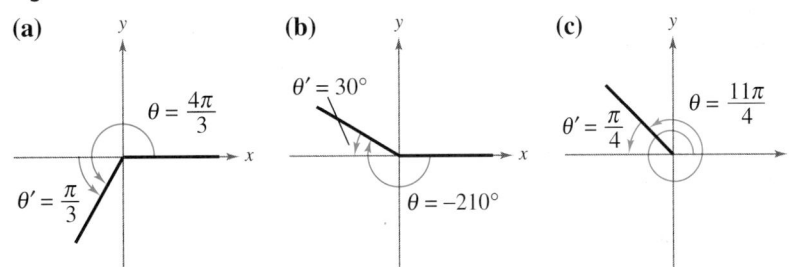

The fundamental trigonometric identities listed in the previous section (for an acute angle θ) are also valid when θ is any angle in the domain of the function.

EXAMPLE 6 ◻ **Using Trigonometric Identities**

Let θ be an angle in Quadrant II such that $\sin \theta = \frac{1}{3}$. Find (a) $\cos \theta$ and (b) $\tan \theta$ by using trigonometric identities.

Solution

a. Using the Pythagorean identity $\sin^2 \theta + \cos^2 \theta = 1$, you obtain

$$\left(\frac{1}{3}\right)^2 + \cos^2 \theta = 1$$

$$\cos^2 \theta = 1 - \frac{1}{9} = \frac{8}{9}.$$

Because $\cos \theta < 0$ in Quadrant II, you can use the negative root to obtain

$$\cos \theta = -\frac{\sqrt{8}}{\sqrt{9}} = -\frac{2\sqrt{2}}{3}.$$

b. Using the trigonometric identity $\tan \theta = \sin \theta / \cos \theta$, you obtain

$$\tan \theta = \frac{1/3}{-2\sqrt{2}/3} = -\frac{1}{2\sqrt{2}} = -\frac{\sqrt{2}}{4}. \qquad \blacksquare$$

EXAMPLE 7 ◻ **Using a Calculator**

a. Use a calculator to evaluate $\cot 410°$ and $\sin(-7)$.

b. Use a calculator to solve $\tan \theta = 4.812$, $0 \le \theta < 2\pi$.

Solution

Function	Mode	Graphing Calculator Keystrokes	Display
a. $\cot 410°$	Degree	(TAN 410) x^{-1} ENTER	0.8390996
$\sin(-7)$	Radian	SIN ((-) 7) ENTER	-0.6569866

b. To solve the equation $\tan \theta = 4.812$, you can use the inverse tangent key, as follows.

Equation	Mode	Graphing Calculator Keystrokes	Display
$\tan \theta = 4.812$	Radian	TAN^{-1} 4.812 ENTER	1.365898912

The angle $\theta \approx 1.366$ lies in Quadrant I. A second value of θ lies in Quadrant III (tangent is positive) and is

$$\theta = \pi + 1.366 \approx 4.507. \qquad \blacksquare$$

Note For your convenience we have included on the inside back cover of this text a summary of basic trigonometry.

At this point, you have completed your introduction to basic trigonometry. You have measured angles in both radians and degrees. You have defined the six trigonometric functions in terms of the unit circle and from a right triangle perspective. In your remaining work with trigonometry you should continue to rely on both perspectives. For instance, in the next two sections on graphing techniques, it helps to think of the trigonometric functions as functions of real numbers. Later, in Section 4.8, you will look at applications involving angles and triangles.

Group Activity   *Patterns in Trigonometric Functions*

Complete the following table. Then identify and discuss any inherent patterns in the trigonometric functions. What can you conclude?

Function	Domain	Range	Even/Odd	Period	Zeros
Sine					
Cosine					
Tangent					
Cosecant					
Secant					
Cotangent					

4.4 /// EXERCISES

In Exercises 1–4, determine the exact values of the six trigonometric functions of the angle θ.

1. (a)   (b)   **2.** (a)   (b)

3. (a)

(b)

4. (a)

(b)

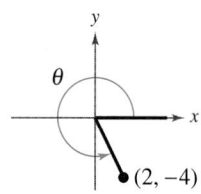

Functional Value	Constraint
15. $\tan \theta = -\frac{15}{8}$	$\sin \theta < 0$
16. $\cos \theta = \frac{8}{17}$	$\tan \theta < 0$
17. $\cot \theta = -3$	$\cos \theta > 0$
18. $\csc \theta = 4$	$\cot \theta < 0$
19. $\sec \theta = -2$	$\sin \theta > 0$
20. $\cot \theta$ is undefined.	$\dfrac{\pi}{2} \le \theta \le \dfrac{3\pi}{2}$
21. $\sin \theta = 0$	$\sec \theta = -1$
22. $\tan \theta$ is undefined.	$\pi \le \theta \le 2\pi$

In Exercises 5–8, the point is on the terminal side of an angle in standard position. Determine the exact values of the six trigonometric functions of the angle.

5. (a) $(7, 24)$ (b) $(7, -24)$

6. (a) $(8, 15)$ (b) $(-9, -40)$

7. (a) $(-4, 10)$ (b) $(3, -5)$

8. (a) $(-5, -2)$ (b) $\left(-\frac{3}{2}, 3\right)$

In Exercises 23–26, find the values (if possible) of the six trigonometric functions of θ if the terminal side of θ lies on the given line in the specified quadrant.

Line	Quadrant
23. $y = -x$	Quadrant II
24. $y = \frac{1}{3}x$	Quadrant III
25. $y = 2x$	Quadrant III
26. $4x + 3y = 0$	Quadrant IV

In Exercises 9–12, determine the quadrant in which θ lies.

9. (a) $\sin \theta < 0$ and $\cos \theta < 0$

     (b) $\sin \theta > 0$ and $\cos \theta < 0$

10. (a) $\sin \theta > 0$ and $\cos \theta > 0$

     (b) $\sin \theta < 0$ and $\cos \theta > 0$

11. (a) $\sin \theta > 0$ and $\tan \theta < 0$

     (b) $\cos \theta > 0$ and $\tan \theta < 0$

12. (a) $\sec \theta > 0$ and $\cot \theta < 0$

     (b) $\csc \theta < 0$ and $\tan \theta > 0$

In Exercises 27–34, find (if possible) the trigonometric function of the quadrant angle.

27. $\cos \pi$ **28.** $\cos \dfrac{3\pi}{2}$

29. $\sec \pi$ **30.** $\sec \dfrac{3\pi}{2}$

31. $\tan \dfrac{\pi}{2}$ **32.** $\tan \pi$

33. $\cot \dfrac{\pi}{2}$ **34.** $\csc \pi$

In Exercises 13–22, find the values (if possible) of the six trigonometric functions of θ using the functional value and constraint.

In Exercises 35–42, find the reference angle θ', and sketch θ and θ' in standard position.

35. (a) $\theta = 203°$ (b) $\theta = 127°$

36. (a) $\theta = 309°$ (b) $\theta = 226°$

37. (a) $\theta = -245°$ (b) $\theta = -72°$

38. (a) $\theta = -145°$ (b) $\theta = -239°$

Functional Value	Constraint
13. $\sin \theta = \frac{3}{5}$	θ lies in Quadrant II.
14. $\cos \theta = -\frac{4}{5}$	θ lies in Quadrant III.

39. (a) $\theta = \dfrac{2\pi}{3}$ (b) $\theta = \dfrac{7\pi}{6}$

40. (a) $\theta = \dfrac{7\pi}{4}$ (b) $\theta = \dfrac{8\pi}{9}$

41. (a) $\theta = 3.5$ (b) $\theta = 5.8$

42. (a) $\theta = \dfrac{11\pi}{3}$ (b) $\theta = -\dfrac{7\pi}{10}$

In Exercises 43–52, evaluate (if possible) the sine, cosine, and tangent of the angles without a calculator.

43. (a) $225°$ (b) $-225°$

44. (a) $300°$ (b) $330°$

45. (a) $750°$ (b) $510°$

46. (a) $-405°$ (b) $-120°$

47. (a) $\dfrac{4\pi}{3}$ (b) $\dfrac{2\pi}{3}$

48. (a) $\dfrac{\pi}{4}$ (b) $\dfrac{5\pi}{4}$

49. (a) $-\dfrac{\pi}{6}$ (b) $\dfrac{5\pi}{6}$

50. (a) $-\dfrac{\pi}{2}$ (b) $\dfrac{\pi}{2}$

51. (a) $\dfrac{11\pi}{4}$ (b) $-\dfrac{13\pi}{6}$

52. (a) $\dfrac{10\pi}{3}$ (b) $\dfrac{17\pi}{3}$

In Exercises 53–62, use a calculator to evaluate the trigonometric function to four decimal places. (Be sure the calculator is set in the correct mode.)

53. (a) $\sin 10°$ (b) $\csc 10°$

54. (a) $\sec 225°$ (b) $\sec 135°$

55. (a) $\cos(-110°)$ (b) $\cos 250°$

56. (a) $\csc 330°$ (b) $\csc 150°$

57. (a) $\tan 240°$ (b) $\cot 210°$

58. (a) $\cot 1.35$ (b) $\tan 1.35$

59. (a) $\tan \dfrac{\pi}{9}$ (b) $\tan \dfrac{10\pi}{9}$

60. (a) $\tan\left(-\dfrac{\pi}{9}\right)$ (b) $\tan\left(-\dfrac{10\pi}{9}\right)$

61. (a) $\sin 0.65$ (b) $\sin(-5.63)$

62. (a) $\sin(-0.65)$ (b) $\sin 5.63$

In Exercises 63–68, find two values of θ that satisfy the equation. Give your answers in degrees ($0° \le \theta < 360°$) and radians ($0 \le \theta < 2\pi$). Do not use a calculator.

63. (a) $\sin \theta = \frac{1}{2}$ (b) $\sin \theta = -\frac{1}{2}$

64. (a) $\cos \theta = \dfrac{\sqrt{2}}{2}$ (b) $\cos \theta = -\dfrac{\sqrt{2}}{2}$

65. (a) $\csc \theta = \dfrac{2\sqrt{3}}{3}$ (b) $\cot \theta = -1$

66. (a) $\sec \theta = 2$ (b) $\sec \theta = -2$

67. (a) $\tan \theta = 1$ (b) $\cot \theta = -\sqrt{3}$

68. (a) $\sin \theta = \dfrac{\sqrt{3}}{2}$ (b) $\sin \theta = -\dfrac{\sqrt{3}}{2}$

In Exercises 69 and 70, use a calculator to approximate two values of θ ($0° \le \theta < 360°$) that satisfy the equation. Round to two decimal places.

69. (a) $\sin \theta = 0.8191$ (b) $\sin \theta = -0.2589$

70. (a) $\cos \theta = 0.8746$ (b) $\cos \theta = -0.2419$

In Exercises 71–74, use a calculator to approximate two values of θ ($0 \le \theta < 2\pi$) that satisfy the equation. Round to three decimal places.

71. (a) $\cos \theta = 0.9848$ (b) $\cos \theta = -0.5890$

72. (a) $\sin \theta = 0.0175$ (b) $\sin \theta = -0.6691$

73. (a) $\tan \theta = 1.192$ (b) $\tan \theta = -8.144$

74. (a) $\cot \theta = 5.671$ (b) $\cot \theta = -1.280$

In Exercises 75–80, find the indicated trigonometric value in the specified quadrant.

Function	Quadrant	Trigonometric Value
75. $\sin \theta = -\frac{3}{5}$	IV	$\cos \theta$
76. $\cot \theta = -3$	II	$\sin \theta$
77. $\tan \theta = \frac{3}{2}$	III	$\sec \theta$
78. $\csc \theta = -2$	IV	$\cot \theta$
79. $\cos \theta = \frac{5}{8}$	I	$\sec \theta$
80. $\sec \theta = -\frac{9}{4}$	III	$\tan \theta$

81. *Average Temperature*   The average daily temperature T (in degrees Fahrenheit) for a certain city is

$$T = 45 - 23 \cos\left[\frac{2\pi}{365}(t - 32)\right]$$

where t is the time in days, with $t = 1$ corresponding to January 1. Find the average daily temperatures on the following days.

(a) January 1

(b) July 4 $(t = 185)$

(c) October 18 $(t = 291)$

82. *Sales*   A company that produces a seasonal product forecasts monthly sales over the next two years to be

$$S = 23.1 + 0.442t + 4.3 \sin\frac{\pi t}{6}$$

where S is measured in thousands of units and t is the time in months, with $t = 1$ representing January 1997. Predict sales for the following months.

(a) February 1997   (b) February 1998

(c) September 1997   (d) September 1998

83. *Distance*   An airplane, flying at an altitude of 6 miles, is on a flight path that passes directly over an observer (see figure). If θ is the angle of elevation from the observer to the plane, find the distance from the observer to the plane when (a) $\theta = 30°$, (b) $\theta = 90°$, and (c) $\theta = 120°$.

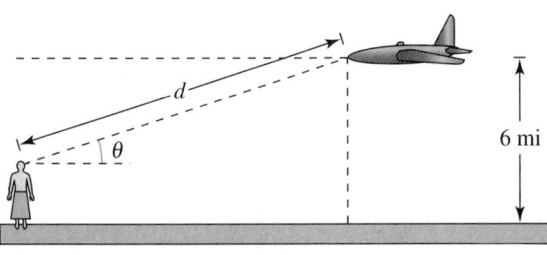

84. *Essay*   Consider an angle in standard position with $r = 12$ centimeters, as shown in the figure. Write a short paragraph describing the change in the magnitudes of x, y, $\sin\theta$, $\cos\theta$, and $\tan\theta$ as θ increases continually from 0° to 90°.

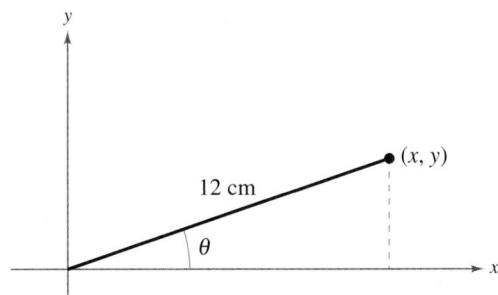

In Exercises 85 and 86, approximate the trigonometric function of the angle θ by selecting a point on the terminal side of the angle and making the necessary measurements.

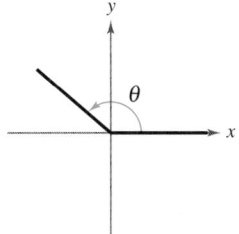

85. $\cos\theta$ **86.** $\tan\theta$

Review   Solve Exercises 87–90 as a review of the skills and problem-solving techniques you learned in previous sections. Graph the function.

87. $y = 2^{x-1}$ **88.** $y = 3^{-x/2}$

89. $y = \ln(x - 1)$ **90.** $y = \ln x^4$

4.5 Graphs of Sine and Cosine Functions

Basic Sine and Cosine Curves / Amplitude and Period of Sine and Cosine Curves / Translations of Sine and Cosine Curves / Mathematical Modeling

Basic Sine and Cosine Curves

In this section you will study techniques for sketching the graphs of the sine and cosine functions. The graph of the sine function is a **sine curve.** In Figure 4.38, the black portion of the graph represents one period of the function and is called **one cycle** of the sine curve. The gray portion of the graph indicates that the basic sine wave repeats indefinitely to the right and left. The graph of the cosine function is shown in Figure 4.39. To produce these graphs with a graphing utility, make sure you have set the mode to radians.

Recall from Section 4.2 that the domain of the sine and cosine functions is the set of all real numbers. Moreover, the range of each function is the interval $[-1, 1]$, and each function has a period of 2π. Do you see how this information is consistent with the basic graphs given in Figures 4.38 and 4.39?

x	$\sin x$	$\cos x$
0	0	1
$\dfrac{\pi}{6}$	$\dfrac{1}{2}$	$\dfrac{\sqrt{3}}{2}$
$\dfrac{\pi}{4}$	$\dfrac{\sqrt{2}}{2}$	$\dfrac{\sqrt{2}}{2}$
$\dfrac{\pi}{3}$	$\dfrac{\sqrt{3}}{2}$	$\dfrac{1}{2}$
$\dfrac{\pi}{2}$	1	0
$\dfrac{3\pi}{4}$	$\dfrac{\sqrt{2}}{2}$	$-\dfrac{\sqrt{2}}{2}$
π	0	-1
$\dfrac{3\pi}{2}$	-1	0
2π	0	1

Figure 4.38

Figure 4.39

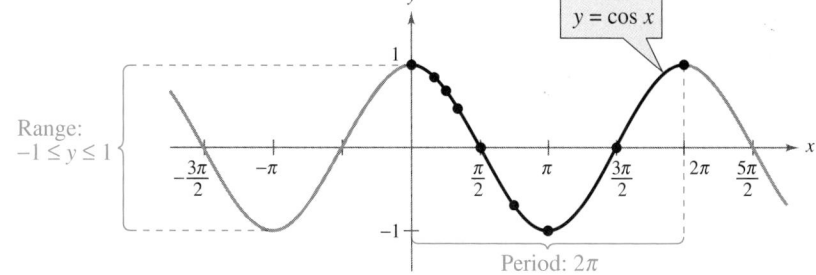

Note Note from Figures 4.38 and 4.39 that the sine graph is symmetric with respect to the *origin,* whereas the cosine graph is symmetric with respect to the *y-axis.* These properties of symmetry follow from the fact that the sine function is odd whereas the cosine function is even.

To sketch the graphs of the basic sine and cosine functions by hand, it helps to note five **key points** in one period of each graph: the *intercepts, maximum points,* and *minimum points.* See Figure 4.40.

Figure 4.40

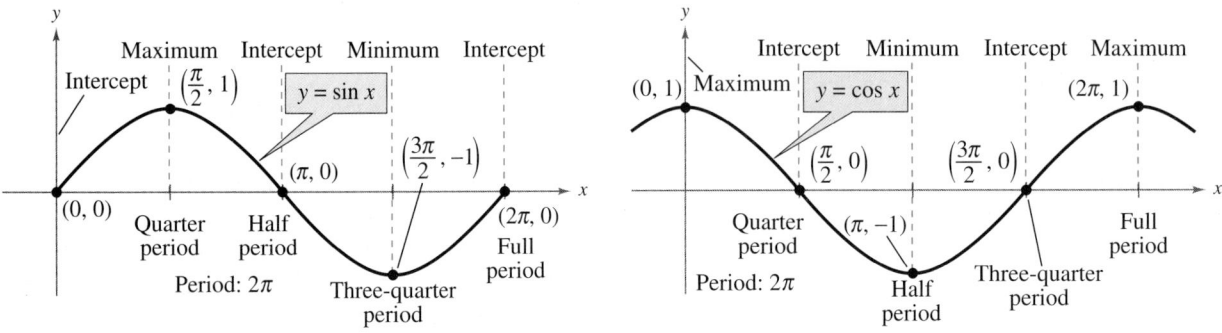

EXAMPLE 1   **Using Key Points to Sketch a Sine Curve**

Sketch the graph of $y = 2 \sin x$ on the interval $[-\pi, 4\pi]$.

Solution

Note that $y = 2 \sin x = 2(\sin x)$ indicates that the y-values for the key points will have twice the magnitude of the graph of $y = \sin x$. Divide the period 2π into four equal parts to get the following key points for $y = 2 \sin x$.

$$(0, 0), \quad \left(\frac{\pi}{2}, 2\right), \quad (\pi, 0), \quad \left(\frac{3\pi}{2}, -2\right), \quad \text{and} \quad (2\pi, 0)$$

By connecting these key points with a smooth curve and extending the curve in both directions over the interval $[-\pi, 4\pi]$, you obtain the graph shown in Figure 4.41. Use a graphing utility to confirm this graph. Be sure to set the graphing utility to radian mode.

Figure 4.41

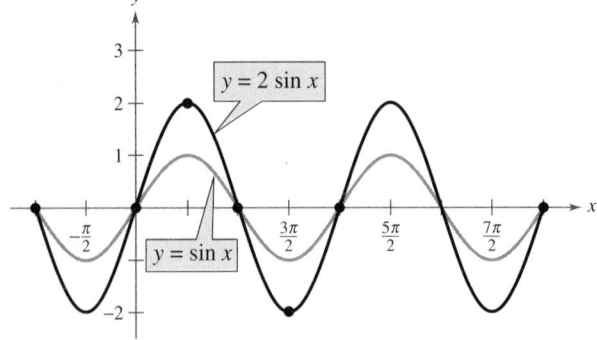

When using a graphing utility to graph trigonometric functions, pay special attention to the viewing rectangle you use. For instance, try graphing $y = [\sin(10x)]/10$ in the standard viewing rectangle in radian mode. What do you observe? Use the zoom feature to find a viewing rectangle that displays a good view of the graph.

Amplitude and Period of Sine and Cosine Curves

In the rest of this section you will study the graphic effect of each of the constants a, b, c, and d in equations of the forms

$$y = d + a \sin(bx - c) \qquad \text{and} \qquad y = d + a \cos(bx - c).$$

A quick review of the transformations studied in Section 1.3 should help in this investigation.

The constant factor a in $y = a \sin x$ acts as a *scaling factor*—a *vertical stretch* or *vertical shrink* of the basic sine curve. If $|a| > 1$, the basic sine curve is stretched, and if $|a| < 1$, the basic sine curve is shrunk. The result is that the graph of $y = a \sin x$ ranges between $-a$ and a instead of between -1 and 1. The absolute value of a is the **amplitude** of the function $y = a \sin x$. The range of the function $y = a \sin x$ is $-a \le y \le a$.

> ### Definition of Amplitude of Sine and Cosine Curves
> The **amplitude** of $y = a \sin x$ and $y = a \cos x$ is the largest value of y and is given by
>
> $$\text{Amplitude} = |a|.$$

EXAMPLE 2 ▱ **Scaling: Vertical Shrinking and Stretching**

On the same coordinate axes, sketch the graphs of

$$y = \frac{1}{2} \cos x \qquad \text{and} \qquad y = 3 \cos x.$$

Solution
Because the amplitude of $y = \frac{1}{2} \cos x$ is $\frac{1}{2}$, the maximum value is $\frac{1}{2}$ and the minimum value is $-\frac{1}{2}$. Divide one cycle, $0 \le x \le 2\pi$, into four equal parts to get the key points

$$\left(0, \frac{1}{2}\right), \quad \left(\frac{\pi}{2}, 0\right), \quad \left(\pi, -\frac{1}{2}\right), \quad \left(\frac{3\pi}{2}, 0\right), \qquad \text{and} \qquad \left(2\pi, \frac{1}{2}\right).$$

A similar analysis shows that the amplitude of $y = 3 \cos x$ is 3, and the key points are

$$(0, 3), \quad \left(\frac{\pi}{2}, 0\right), \quad (\pi, -3), \quad \left(\frac{3\pi}{2}, 0\right), \qquad \text{and} \qquad (2\pi, 3).$$

The graphs of these two functions are shown in Figure 4.42. Use a graphing utility to confirm these graphs. ▱

EXPLORATION

Use a graphing utility to graph $y = a \sin x$, where $a = 0.5, 1, 2,$ and 3. How does the value of a affect the shape of the graph? What happens to the graph if a is negative?

Figure 4.42

Figure 4.43

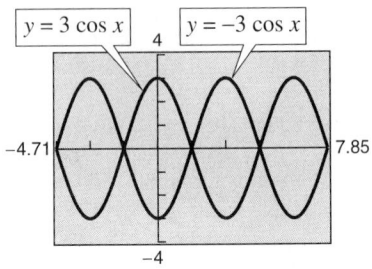

You know from Section 1.3 that the graph of $y = -f(x)$ is a **reflection** in the x-axis of the graph of $y = f(x)$. For instance, the graph of $y = -3 \cos x$ is a reflection of the graph of $y = 3 \cos x$, as shown in Figure 4.43.

Because $y = a \sin x$ completes one cycle from $x = 0$ to $x = 2\pi$, it follows that $y = a \sin bx$ completes one cycle from $x = 0$ to $x = 2\pi/b$.

Period of Sine and Cosine Functions
Let b be a positive real number. The **period** of $y = a \sin bx$ and $y = a \cos bx$ is $2\pi/b$.

Note that if $0 < b < 1$, the period of $y = a \sin bx$ is greater than 2π and represents a *horizontal stretching* of the graph of $y = a \sin x$. Similarly, if $b > 1$, the period of $y = a \sin bx$ is less than 2π and represents a *horizontal shrinking* of the graph of $y = a \sin x$. If b is negative, we use the identities $\sin(-x) = -\sin x$ and $\cos(-x) = \cos x$ to rewrite the function.

EXAMPLE 3   **Scaling: Horizontal Stretching**

Sketch the graph of $y = \sin \dfrac{x}{2}$.

Solution

The amplitude is 1. Moreover, because $b = \frac{1}{2}$, the period is

$$\frac{2\pi}{b} = \frac{2\pi}{\frac{1}{2}} = 4\pi.$$

Now, divide the period-interval $[0, 4\pi]$ into four equal parts with the values π, 2π, and 3π, to obtain the following key points on the graph.

$$(0, 0), \quad (\pi, 1), \quad (2\pi, 0), \quad (3\pi, -1), \quad \text{and} \quad (4\pi, 0)$$

The graph is shown in Figure 4.44. Use a graphing utility to confirm this graph.

Figure 4.44

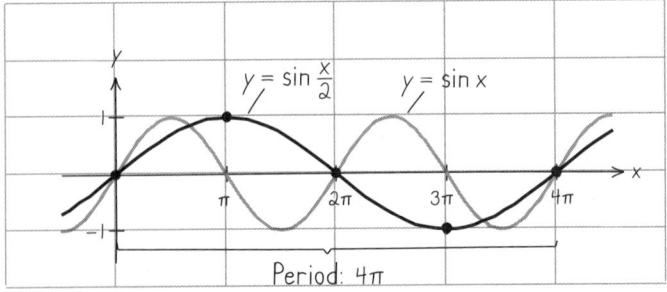

EXPLORATION

Use a graphing utility to graph $y = \sin bx$, where $b = 0.5$, 1, and 2. How does the value of b affect the graph?

Study Tip

In general, to divide a period-interval into four equal parts, successively add "period/4," starting with the left endpoint of the interval. For instance, for the period-interval $[-\pi/6, \pi/2]$ of length $2\pi/3$, you would successively add

$$\frac{2\pi/3}{4} = \frac{\pi}{6}$$

to get $-\pi/6, 0, \pi/6, \pi/3$, and $\pi/2$.

Translations of Sine and Cosine Curves

The constant c in the general equations

$$y = a \sin(bx - c) \qquad \text{and} \qquad y = a \cos(bx - c)$$

creates *horizontal translations* (shifts) of the basic sine and cosine curves. Comparing $y = a \sin bx$ with $y = a \sin(bx - c)$, we find that the graph of $y = a \sin(bx - c)$ completes one cycle from $bx - c = 0$ to $bx - c = 2\pi$. By solving for x, we find the interval for one cycle to be

Left endpoint   Right endpoint

$$\frac{c}{b} \le x \le \frac{c}{b} + \frac{2\pi}{b}.$$

Period

This implies that the period of $y = a \sin(bx - c)$ is $2\pi/b$, and the graph of $y = a \sin bx$ is shifted by an amount c/b. The number c/b is the **phase shift.**

EXPLORATION

Use a graphing utility to graph $y = \sin(x + c)$, where $c = -\pi/4$, 0, and $\pi/4$. Use a viewing rectangle in which $-1.6 \le x \le 6.3$ and $-2 \le y \le 2$. How does the value of c affect the graph?

Graphs of Sine and Cosine Functions

The graphs of $y = a \sin(bx - c)$ and $y = a \cos(bx - c)$ have the following characteristics. (Assume $b > 0$.)

$$\textbf{Amplitude} = |a| \qquad \textbf{Period} = 2\pi/b$$

The left and right endpoints of a one-cycle interval can be determined by solving the equations $bx - c = 0$ and $bx - c = 2\pi$.

EXAMPLE 4   **Horizontal Translation**

Sketch the graph of $y = \dfrac{1}{2} \sin\left(x - \dfrac{\pi}{3}\right)$.

Solution

The amplitude is $\frac{1}{2}$ and the period is 2π. By solving the equations

$$x - \frac{\pi}{3} = 0 \qquad \text{and} \qquad x - \frac{\pi}{3} = 2\pi$$

$$x = \frac{\pi}{3} \qquad\qquad\qquad x = \frac{7\pi}{3}$$

you see that the interval $[\pi/3, 7\pi/3]$ corresponds to one cycle of the graph. Dividing this interval into four equal parts produces the following key points.

$$\left(\frac{\pi}{3}, 0\right), \qquad \left(\frac{5\pi}{6}, \frac{1}{2}\right), \qquad \left(\frac{4\pi}{3}, 0\right), \qquad \left(\frac{11\pi}{6}, -\frac{1}{2}\right), \qquad \text{and} \qquad \left(\frac{7\pi}{3}, 0\right)$$

The graph is shown in Figure 4.45. Check this with a graphing utility.

Figure 4.45

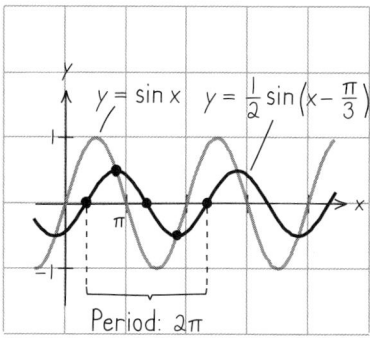

EXAMPLE 5 ▱ **Horizontal Translation**

Use a graphing utility to analyze the graph of

$$y = -3 \cos(2\pi x + 4\pi).$$

Figure 4.46

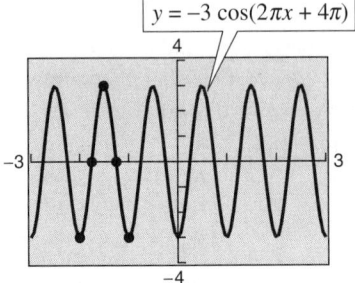

$y = -3 \cos(2\pi x + 4\pi)$

Solution
The amplitude is 3 and the period is $2\pi/2\pi = 1$. By solving the equations

$$2\pi x + 4\pi = 0 \qquad \text{and} \qquad 2\pi x + 4\pi = 2\pi$$
$$2\pi x = -4\pi \qquad\qquad\qquad 2\pi x = -2\pi$$
$$x = -2 \qquad\qquad\qquad\quad x = -1$$

you see that the interval $[-2, -1]$ corresponds to one cycle of the graph. Dividing this interval into four equal parts produces the following key points.

$$(-2, -3), \qquad \left(-\frac{7}{4}, 0\right), \qquad \left(-\frac{3}{2}, 3\right), \qquad \left(-\frac{5}{4}, 0\right), \qquad \text{and} \qquad (-1, -3)$$

The graph is shown in Figure 4.46.   ▱

The final type of transformations is the *vertical translation* caused by the constant d in the equations

$$y = d + a \sin(bx - c) \qquad \text{and} \qquad y = d + a \cos(bx - c).$$

The shift is d units upward for $d > 0$ and downward for $d < 0$. In other words, the graph oscillates about the horizontal line $y = d$ instead of the x-axis.

▱ **EXPLORATION**

Use a graphing utility to graph $y = d + \sin x$, where $d = -2, 1,$ and 3. How does the value of d affect the graph?

EXAMPLE 6 ▱ **Vertical Translation**

Use a graphing utility to analyze the graph of

$$y = 2 + 3 \cos 2x.$$

Figure 4.47

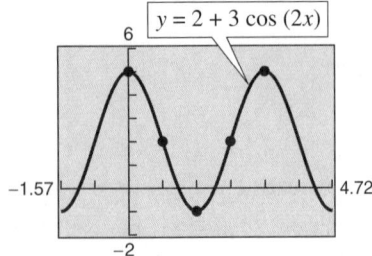

$y = 2 + 3 \cos (2x)$

Solution
The amplitude is 3 and the period is π. The key points over the interval $[0, \pi]$ are

$$(0, 5), \qquad \left(\frac{\pi}{4}, 2\right), \qquad \left(\frac{\pi}{2}, -1\right), \qquad \left(\frac{3\pi}{4}, 2\right), \qquad \text{and} \qquad (\pi, 5).$$

The graph is shown in Figure 4.47. Compared with the graph of $f(x) = 3 \cos 2x$, the graph of $y = 2 + 3 \cos 2x$ is shifted upward two units.   ▱

Mathematical Modeling

Sine and cosine functions can be used to model many real-life situations, including electric currents, musical tones, radio waves, tides, sunrises, and weather patterns.

Real Life

EXAMPLE 7   Finding a Trigonometric Model

Throughout the day, the depth of water at the end of a dock varies with the tides. The table shows the depths (in meters) at various times during the morning.

t (time)	Midnight	2 A.M.	4 A.M.	6 A.M.	8 A.M.	10 A.M.	Noon
y (depth)	2.55	3.80	4.40	3.80	2.55	1.80	2.27

a. Use a trigonometric function to model this data.

b. Find the depths at 9 A.M. and 3 P.M.

c. A boat needs at least 3 meters of water to moor at the dock. During what times in the afternoon can it safely dock?

Solution

Figure 4.48

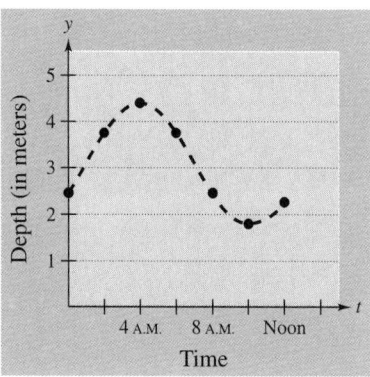

a. Begin by graphing the data, as shown in Figure 4.48. You can use either a sine or cosine model. Suppose you use a cosine model of the form
$$y = a \cos(bt - c) + d.$$
The amplitude is given by
$$a = \tfrac{1}{2}[(\text{high}) - (\text{low})] = \tfrac{1}{2}(4.4 - 1.8) = 1.3.$$
The period is
$$p = 2[(\text{low time}) - (\text{high time})] = 2(10 - 4) = 12$$
which implies that $b = 2\pi/p \approx 0.524$. Because high tide occurs 4 hours after midnight, you can conclude that $c/b = 4$, so $c \approx 2.094$. Moreover, because the average depth is $\tfrac{1}{2}(4.4 + 1.8) = 3.1$, it follows that $d = 3.1$. Thus, you can model the depth with the function
$$y = 1.3 \cos(0.524t - 2.094) + 3.1.$$

Figure 4.49

b. The depths at 9 A.M. and 3 P.M. are as follows:
$$y = 1.3 \cos(0.524 \cdot 9 - 2.094) + 3.1 \approx 1.97 \text{ meters} \quad 9\text{ A.M.}$$
$$y = 1.3 \cos(0.524 \cdot 15 - 2.094) + 3.1 \approx 4.23 \text{ meters}. \quad 3\text{ P.M.}$$

c. Using a graphing utility, you can graph the model with the line $y = 3$, as shown in Figure 4.49. From the graph, it follows that the depth is at least 3 meters between 12:54 P.M. $(t \approx 12.9)$ and 7:06 P.M. $(t \approx 19.1)$.

Figure 4.50

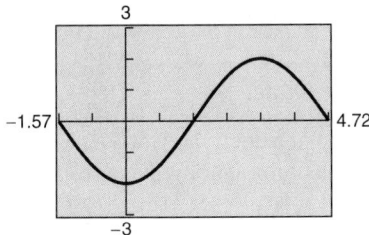

EXAMPLE 8   Finding an Equation for a Graph

Find the amplitude, period, and phase shift for the sine function whose graph is shown in Figure 4.50. Write an equation for this graph.

Solution

The amplitude for this sine curve is 2. The period is 2π, and there is a right phase shift of $\pi/2$. Thus, you can write

$$y = 2 \sin\left(x - \frac{\pi}{2}\right).$$

Try finding a cosine function with this same graph.

Group Activity   *A Sine Show*

Enter the following program in a *TI-82* or *TI-83* graphing calculator. (Program steps for other graphing calculators are given in the appendix.) This program simultaneously draws the unit circle and the corresponding points on the sine curve. After the circle and sine curve are drawn, you can connect the points on the unit circle with their corresponding points on the sine curve by pressing ENTER. Discuss the relationship that is illustrated.

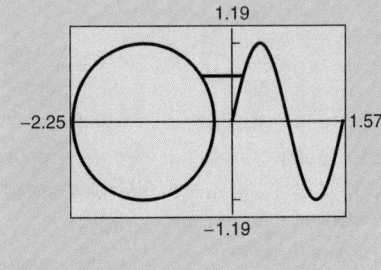

```
PROGRAM:SINESHOW          :"sin(T)"→Y1T
:Radian                   :"T/4"→X2T
:ClrDraw:FnOff            :"sin(T)"→Y2T
:Param:Simul              :DispGraph
:-2.25→Xmin               :For(N,1,12)
:π/2→Xmax                 :N*π/6.5→T
:3→Xscl                   :-1.25+cos(T)→A
:-1.19→Ymin               :sin(T)→B
:1.19→Ymax                :T/4→C
:1→Yscl                   :Line(A,B,C,B)
:0→Tmin                   :Pause
:6.3→Tmax                 :End
:.15→Tstep                :Pause:Func
:"-1.25+cos(T)"→X1T       :Sequential:Disp
```

4.5 /// EXERCISES

In Exercises 1–14, find the period and amplitude. Then describe the viewing rectangle for Exercises 1–6.

1. $y = 3 \sin 2x$

2. $y = 2 \cos 3x$

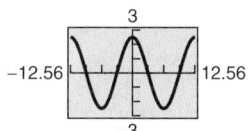

3. $y = \dfrac{5}{2} \cos \dfrac{x}{2}$

4. $y = -3 \sin \dfrac{x}{3}$

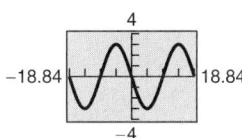

5. $y = \dfrac{2}{3} \sin \pi x$

6. $y = \dfrac{3}{2} \cos \dfrac{\pi x}{2}$

7. $y = -2 \sin x$

8. $y = -\cos \dfrac{2x}{3}$

9. $y = 3 \sin 10x$

10. $y = \dfrac{1}{3} \sin 8x$

11. $y = \dfrac{1}{2} \cos \dfrac{2x}{3}$

12. $y = \dfrac{5}{2} \cos \dfrac{x}{4}$

13. $y = 3 \sin 4\pi x$

14. $y = \dfrac{2}{3} \cos \dfrac{\pi x}{10}$

In Exercises 15–22, describe the relationship between the graphs of f and g.

15. $f(x) = \sin x$
   $g(x) = \sin(x - \pi)$

16. $f(x) = \cos x$
   $g(x) = \cos(x + \pi)$

17. $f(x) = \cos 2x$
   $g(x) = -\cos 2x$

18. $f(x) = \sin 3x$
   $g(x) = \sin(-3x)$

19. $f(x) = \cos x$
   $g(x) = -3 \cos x$

20. $f(x) = \sin x$
   $g(x) = \sin 3x$

21. $f(x) = \sin x$
   $g(x) = 2 + \sin x$

22. $f(x) = \cos 4x$
   $g(x) = -2 + \cos 4x$

In Exercises 23–26, describe the relationship between the graphs of f and g.

23.

24.

25.

26.

27. *Essay* Use a graphing utility to graph the function $y = a \sin x$ for $a = \dfrac{1}{2}, a = \dfrac{3}{2}$, and $a = -3$. Write a paragraph describing the changes in the graph for the specified changes in a.

28. *Essay* Use a graphing utility to graph the function $y = d + \sin x$ for $d = 2, d = 3.5$, and $d = -2$. Write a paragraph describing the changes in the graph for the specified changes in d.

29. *Essay* Use a graphing utility to graph the function $y = \sin bx$ for $b = \dfrac{1}{2}, b = \dfrac{3}{2}$, and $b = 4$. Write a paragraph describing the changes in the graph for the specified changes in b.

30. *Essay*   Use a graphing utility to graph the function $y = \sin(x - c)$ for $c = 1$, $c = 3$, and $c = -2$. Write a paragraph describing the changes in the graph for the specified changes in c.

In Exercises 31–38, sketch the graphs of the two functions in the same coordinate plane. (Include two full periods.)

31. $f(x) = -2 \sin x$
$g(x) = 4 \sin x$

32. $f(x) = \sin x$
$g(x) = \sin \dfrac{x}{3}$

33. $f(x) = \cos x$
$g(x) = 1 + \cos x$

34. $f(x) = 2 \cos 2x$
$g(x) = -\cos 4x$

35. $f(x) = -\dfrac{1}{2} \sin \dfrac{x}{2}$
$g(x) = 3 - \dfrac{1}{2} \sin \dfrac{x}{2}$

36. $f(x) = 4 \sin \pi x$
$g(x) = 4 \sin \pi x - 3$

37. $f(x) = 2 \cos x$
$g(x) = 2 \cos(x + \pi)$

38. $f(x) = -\cos x$
$g(x) = -\cos(x - \pi)$

Conjecture   In Exercises 39–42, use a graphing utility to graph f and g in the same viewing rectangle. (Include two full periods.) Make a conjecture about the functions.

39. $f(x) = \sin x$
$g(x) = \cos\left(x - \dfrac{\pi}{2}\right)$

40. $f(x) = \sin x$
$g(x) = -\cos\left(x + \dfrac{\pi}{2}\right)$

41. $f(x) = \cos x$
$g(x) = -\sin\left(x - \dfrac{\pi}{2}\right)$

42. $f(x) = \cos x$
$g(x) = -\cos(x - \pi)$

In Exercises 43–60, sketch the graph of the function by hand. Use a graphing utility to verify your sketch. (Include two full periods.)

43. $y = -2 \sin 6x$

44. $y = -3 \cos 4x$

45. $y = \cos 2\pi x$

46. $y = \dfrac{3}{2} \sin \dfrac{\pi x}{4}$

47. $y = -\sin \dfrac{2\pi x}{3}$

48. $y = 10 \cos \dfrac{\pi x}{6}$

49. $y = \sin\left(x - \dfrac{\pi}{4}\right)$

50. $y = \dfrac{1}{2} \sin(x - \pi)$

51. $y = 3 \cos(x + \pi)$

52. $y = 4 \cos\left(x + \dfrac{\pi}{4}\right)$

53. $y = \dfrac{1}{10} \cos 60\pi x$

54. $y = -3 + 5 \cos \dfrac{\pi t}{12}$

55. $y = 2 - \sin \dfrac{2\pi x}{3}$

56. $y = 2 \cos x - 3$

57. $y = 3 \cos(x + \pi) - 3$

58. $y = 4 \cos\left(x + \dfrac{\pi}{4}\right) + 4$

59. $y = \dfrac{2}{3} \cos\left(\dfrac{x}{2} - \dfrac{\pi}{4}\right)$

60. $y = -3 \cos(6x + \pi)$

In Exercises 61–68, use a graphing utility to graph the function. (Include two full periods.)

61. $y = -2 \sin(4x + \pi)$

62. $y = -4 \sin\left(\dfrac{2}{3}x - \dfrac{\pi}{3}\right)$

63. $y = \cos\left(2\pi x - \dfrac{\pi}{2}\right) + 1$

64. $y = 3 \cos\left(\dfrac{\pi x}{2} + \dfrac{\pi}{2}\right) - 2$

65. $y = -0.1 \sin\left(\dfrac{\pi x}{10} + \pi\right)$

66. $y = 5 \sin(\pi - 2x) + 10$

67. $y = 5 \cos(\pi - 2x) + 2$

68. $y = \frac{1}{100} \sin 120\pi t$

Graphical Reasoning   In Exercises 69 and 70, find a and d for the function $f(x) = a \cos x + d$ so that the graph of f matches the figure.

69.

70.
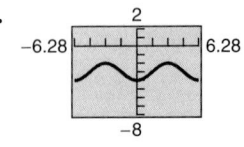

Graphical Reasoning In Exercises 71–74, find a, b, and c for the function $y = a \sin (bx - c)$ so that the graph of f matches the figure.

71.

72.

73.

74.

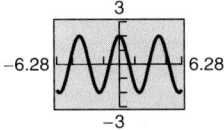

In Exercises 75–78, use a graphing utility to graph y_1 and y_2 for all real numbers x in the interval $[-2\pi, 2\pi]$. Use the graphs to find the real numbers x such that $y_1 = y_2$.

75. $y_1 = \sin x$

$y_2 = -\frac{1}{2}$

76. $y_1 = \cos x$

$y_2 = -1$

77. $y_1 = \cos x$

$y_2 = \dfrac{\sqrt{2}}{2}$

78. $y_1 = \sin x$

$y_2 = \dfrac{\sqrt{3}}{2}$

79. *Exploration* In Section 4.2 it was shown that $f(x) = \cos x$ is an even function and $g(x) = \sin x$ is an odd function. Use a graphing utility to graph h and use the graph to determine if h is even, odd, or neither.

(a) $h(x) = \cos^2 x$

(b) $h(x) = \sin^2 x$

(c) $h(x) = \sin x \cos x$

80. *Conjecture* If f is an even function and g is an odd function, use the results of Exercise 79 to make a conjecture about the following.

(a) $h(x) = [f(x)]^2$

(b) $h(x) = [g(x)]^2$

(c) $h(x) = f(x)g(x)$

81. *Respiratory Cycle* For a person at rest, the velocity v (in liters per second) of air flow during a respiratory cycle is

$$v = 0.85 \sin \frac{\pi t}{3}$$

where t is the time in seconds. (Inhalation occurs when $v > 0$, and exhalation occurs when $v < 0$.)

(a) Use a graphing utility to graph v.

(b) Find the time for one full respiratory cycle.

(c) Find the number of cycles per minute.

82. *Respiratory Cycle* The model in Exercise 81 is for a person at rest. How might the model change for a person who is exercising? Explain.

83. *Exploration* Using calculus, it can be shown that the sine and cosine functions can be approximated by the polynomials

$$\sin x \approx x - \frac{x^3}{3!} + \frac{x^5}{5!} \quad \text{and} \quad \cos x \approx 1 - \frac{x^2}{2!} + \frac{x^4}{4!}$$

where x is in radians.

(a) Use a graphing utility to graph the sine function and its polynomial approximation in the same viewing rectangle.

(b) Use a graphing utility to graph the cosine function and its polynomial approximation in the same viewing rectangle.

(c) Study the patterns in the polynomial approximations of the sine and cosine functions and guess the next term in each. Then repeat parts (a) and (b). How did the accuracy of the approximations change when additional terms were added?

84. *Exploration* Use the polynomial approximations of the sine and cosine functions given in Exercise 83 to approximate the following functional values. Compare the results with those given by a calculator. Is the error in the approximation the same in each case? Explain.

(a) $\sin \dfrac{1}{2}$

(b) $\sin 1$

(c) $\sin \dfrac{\pi}{6}$

(d) $\cos(-0.5)$

(e) $\cos 1$

(f) $\cos \dfrac{\pi}{4}$

Sales   In Exercises 85 and 86, use a graphing utility to graph the sales function over 1 year, where S is the sales in thousands of units and t is the time in months, with $t = 1$ corresponding to January. Determine the months of maximum and minimum sales.

85. $S = 22.3 - 3.4 \cos \dfrac{\pi t}{6}$

86. $S = 74.50 + 43.75 \sin \dfrac{\pi t}{6}$

87. *Fuel Consumption*   The daily consumption C (in gallons) of diesel fuel on a farm is modeled by

$$C = 30.3 + 21.6 \sin\left(\dfrac{2\pi t}{365} + 10.9\right),$$

where t is the time in days, with $t = 1$ corresponding to January 1.

(a) What is the period of the model? Is it what you expected? Explain.

(b) What is the average daily fuel consumption? Which term of the model did you use? Explain.

(c) Use a graphing utility to graph the model. Use the graph to approximate the time of the year when consumption exceeds 40 gallons per day.

88. *Data Analysis*   The motion of an oscillating weight suspended by a spring was measured by a motion detector. The data was collected, and the approximate maximum displacements from equilibrium ($y = 2$) are labeled in the figure. The distance y from the motion detector is measured in centimeters and the time t is measured in seconds.

(a) Is y a function of t? Explain.

(b) Approximate the amplitude and period.

(c) Find a model for the data.

(d) Use a graphing utility to graph the model in part (c). Compare the result with the data in the figure.

89. *Data Analysis*   The table gives the normal daily high temperatures for Honolulu H and Chicago C (in degrees Fahrenheit) for month t, with $t = 1$ corresponding to January.   (Source: NOAA)

t	1	2	3	4	5	6
H	80.1	80.5	81.6	82.8	84.7	86.5
C	29.0	33.5	45.8	58.6	70.1	79.6

t	7	8	9	10	11	12
H	87.5	88.7	88.5	86.9	84.1	81.2
C	83.7	81.8	74.8	63.3	48.4	34.0

(a) A model for Honolulu is given by

$$H(t) = 84.40 + 4.28 \sin\left(\dfrac{\pi t}{6} + 3.86\right).$$

Find a trigonometric model for Chicago.

(b) Use a graphing utility to graph the data points and the model for the temperatures in Honolulu. How well does the model fit?

(c) Use a graphing utility to graph the data points and the model for the temperatures in Chicago. How well does the model fit?

(d) Use the models to estimate the average annual temperature in each city. Which term of the models did you use? Explain.

(e) What are the periods of the two models? Are they what you expected? Explain.

(f) Which city has the greater variability in temperature throughout the year? Which factor of the models determines this variability? Explain.

90. *Graphical Reasoning*   The figure shows the graphs of the functions $f(x) = 1 - \frac{1}{2}x^2$ and $g(x) = \cos x$. Identify the graphs and explain your reasoning.

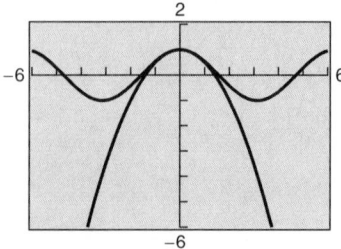

4.6   Graphs of Other Trigonometric Functions

Graph of the Tangent Function / Graph of the Cotangent Function /
Graphs of the Reciprocal Functions / Damped Trigonometric Graphs

Graph of the Tangent Function

Recall from Section 4.2 that the tangent function is odd. That is,

$$\tan(-x) = -\tan x.$$

Consequently, the graph of

$$y = \tan x$$

is symmetric with respect to the origin. You also know from the identity $\tan x = \sin x / \cos x$ that the tangent is undefined when $\cos x = 0$. Two such values are $x = \pm \pi/2 \approx \pm 1.5708$.

x	$-\dfrac{\pi}{2}$	-1.57	-1.5	-1	0	1	1.5	1.57	$\dfrac{\pi}{2}$
$\tan x$	Undef.	-1255.8	-14.1	-1.56	0	1.56	14.1	1255.8	Undef.

tan x approaches $-\infty$ as x approaches $-\pi/2$ from the right

tan x approaches ∞ as x approaches $\pi/2$ from the left

As indicated in the table, $\tan x$ increases without bound as x approaches $\pi/2$ from the left, and decreases without bound as x approaches $-\pi/2$ from the right. Thus, the graph of $y = \tan x$ has *vertical asymptotes* at $x = \pi/2$ and $-\pi/2$, as shown in Figure 4.51. Moreover, because the period of the tangent function is π, vertical asymptotes also occur when $x = \pi/2 + n\pi$, where n is an integer. The domain of the tangent function is the set of all real numbers other than $x = \pi/2 + n\pi$, and the range is the set of all real numbers.

Sketching the graph of a function of the form $y = a\tan(bx - c)$ is similar to sketching the graph of $y = a\sin(bx - c)$ in that you locate key points that identify the intercepts and asymptotes. Two consecutive asymptotes can be found by solving the equations

$$bx - c = -\frac{\pi}{2} \quad \text{and} \quad bx - c = \frac{\pi}{2}.$$

The midpoint between two consecutive asymptotes is an x-intercept of the graph. After plotting the asymptotes and the x-intercept, plot a few additional points between the two asymptotes and sketch one cycle. Finally, sketch one or two additional cycles to the left and right.

Figure 4.51

Period: π
Domain: all $x \neq \frac{\pi}{2} + n\pi$
Range: $(-\infty, \infty)$
Vertical asymptotes: $x = \frac{\pi}{2} + n\pi$

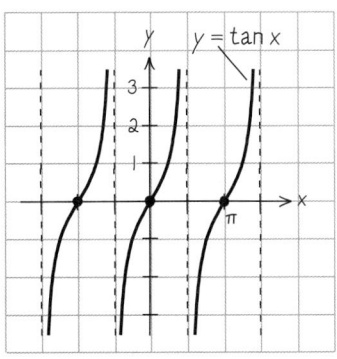

Note The period of the function $y = a\tan(bx - c)$ is the distance between two consecutive asymptotes. The amplitude of a tangent function is not defined.

EXAMPLE 1   **Sketching the Graph of a Tangent Function**

Sketch the graph of $y = \tan \dfrac{x}{2}$.

Solution

By solving the equations

$$\frac{x}{2} = -\frac{\pi}{2} \quad \text{and} \quad \frac{x}{2} = \frac{\pi}{2}$$

$$x = -\pi \qquad\qquad x = \pi$$

you can see that two consecutive asymptotes occur at $x = -\pi$ and $x = \pi$. Between these two asymptotes, plot a few points, including the x-intercept, as shown in the table. Three cycles of the graph are shown in Figure 4.52. Use a graphing utility to confirm this graph.

Figure 4.52

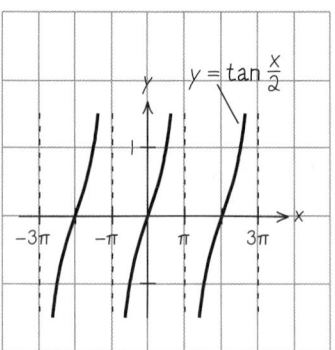

x	$-\pi$	$-\dfrac{\pi}{2}$	0	$\dfrac{\pi}{2}$	π
$\tan \dfrac{x}{2}$	Undef.	-1	0	1	Undef.

EXAMPLE 2   **Sketching the Graph of a Tangent Function**

Sketch the graph of $y = -3 \tan 2x$.

Solution

By solving the equations

$$2x = -\frac{\pi}{2} \quad \text{and} \quad 2x = \frac{\pi}{2}$$

$$x = -\frac{\pi}{4} \qquad\qquad x = \frac{\pi}{4}$$

you can see that two consecutive asymptotes occur at $x = -\pi/4$ and $x = \pi/4$. Between these two asymptotes, plot a few points, including the x-intercept, as shown in the table. Three complete cycles of the graph are shown in Figure 4.53. Use a graphing utility to confirm this graph.

Figure 4.53

x	$-\dfrac{\pi}{4}$	$-\dfrac{\pi}{8}$	0	$\dfrac{\pi}{8}$	$\dfrac{\pi}{4}$
$-3 \tan 2x$	Undef.	3	0	-3	Undef.

By comparing the graphs in Examples 1 and 2, you can see that the graph of $y = a \tan(bx - c)$ is increasing between consecutive vertical asymptotes if $a > 0$, and decreasing between consecutive vertical asymptotes if $a < 0$. In other words, the graph for $a < 0$ is a reflection in the x-axis of the graph for $a > 0$.

Graph of the Cotangent Function

The graph of the cotangent function is similar to the graph of the tangent function. It also has a period of π. However, from the identity

$$y = \cot x = \frac{\cos x}{\sin x}$$

you can see that the cotangent function has vertical asymptotes $x = n\pi$, where n is an integer, because $\sin x$ is zero at these x-values. The graph of the cotangent function is shown in Figure 4.54.

Figure 4.54

Period: π
Domain: all $x \neq n\pi$
Range: $(-\infty, \infty)$
Vertical asymptotes: $x = n\pi$

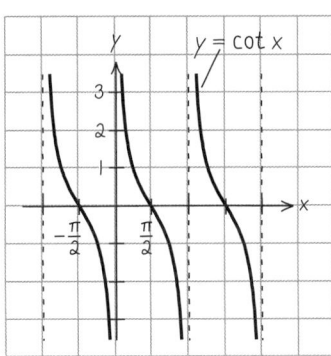

EXAMPLE 3   Sketching the Graph of a Cotangent Function

Sketch the graph of $y = 2 \cot \dfrac{x}{3}$.

Solution
To locate two consecutive vertical asymptotes of the graph, solve the equations $x/3 = 0$ and $x/3 = \pi$, as follows.

$$\frac{x}{3} = 0 \qquad \text{and} \qquad \frac{x}{3} = \pi$$
$$x = 0 \qquad\qquad\qquad x = 3\pi$$

Then, between these two asymptotes, plot a few points, including the x-intercept, as shown in the table. Three cycles of the graph are shown in Figure 4.55. Use a graphing utility to confirm this graph. [Enter the function as $y = 2/\tan(x/3)$.] Note that the period is 3π, the distance between consecutive asymptotes.

Figure 4.55

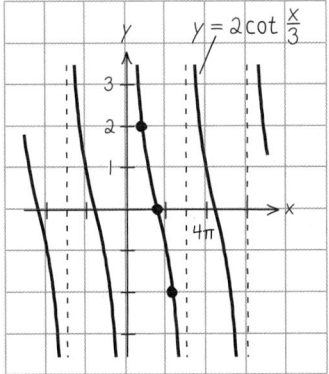

x	0	$\dfrac{3\pi}{4}$	$\dfrac{3\pi}{2}$	$\dfrac{9\pi}{4}$	3π
$2 \cot \dfrac{x}{3}$	Undef.	2	0	-2	Undef.

EXPLORATION

Use a graphing utility to graph the functions $y_1 = \sin x$ and $y_2 = \csc x = 1/\sin x$ in the same viewing rectangle. How are the graphs related? What happens to the graph of the cosecant function as x approaches the zeros of the sine function?

Graphs of the Reciprocal Functions

The graphs of the two remaining trigonometric functions can be obtained from the graphs of the sine and cosine functions using the reciprocal identities

$$\csc x = \frac{1}{\sin x} \quad \text{and} \quad \sec x = \frac{1}{\cos x}.$$

For instance, at a given value of x, the y-coordinate for $\sec x$ is the reciprocal of the y-coordinate for $\cos x$. Of course, when $\cos x = 0$, the reciprocal does not exist. Near such values of x, the behavior of the secant function is similar to that of the tangent function. In other words, the graphs of

$$\tan x = \frac{\sin x}{\cos x} \quad \text{and} \quad \sec x = \frac{1}{\cos x}$$

have vertical asymptotes at $x = \pi/2 + n\pi$, where n is an integer and the cosine is zero at these x-values. Similarly,

$$\cot x = \frac{\cos x}{\sin x} \quad \text{and} \quad \csc x = \frac{1}{\sin x}$$

have vertical asymptotes where $\sin x = 0$—that is, at $x = n\pi$.

To sketch the graph of a secant or cosecant function, we suggest that you first make a sketch of its reciprocal function. For instance, to sketch the graph of $y = \csc x$, first sketch the graph of $y = \sin x$. Then take reciprocals of the y-coordinates to obtain points on the graph of $y = \csc x$. You can use this procedure to obtain the graphs shown in Figure 4.56.

Figure 4.57

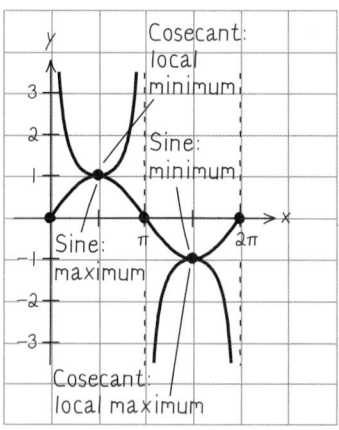

Figure 4.56

Period: 2π
Domain: all $x \neq n\pi$
Range: all y not in $(-1, 1)$
Vertical asymptotes: $x = n\pi$
Symmetry: origin

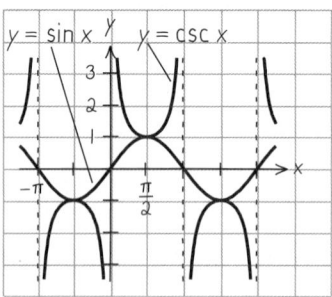

Period: 2π
Domain: all $x \neq \frac{\pi}{2} + n\pi$
Range: all y not in $(-1, 1)$
Vertical asymptotes: $x = \frac{\pi}{2} + n\pi$
Symmetry: y-axis

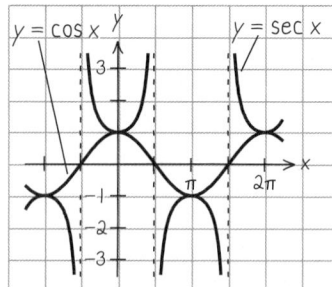

In comparing the graphs of the secant and cosecant functions with those of the sine and cosine functions, note that the "hills" and "valleys" are interchanged. For example, a hill (or maximum point) on the sine curve corresponds to a valley (a local minimum) on the cosecant curve. Similarly, a valley (or minimum point) on the sine curve corresponds to a hill (a local maximum) on the cosecant curve, as shown in Figure 4.57.

EXAMPLE 4   **Comparing Trigonometric Graphs**

Use a graphing utility to compare the graphs of

$$y = 2 \sin\left(x + \frac{\pi}{4}\right) \quad \text{and} \quad y = 2 \csc\left(x + \frac{\pi}{4}\right).$$

Solution
The two graphs are shown in Figure 4.58. Note how the "hills" and "valleys" of the graphs are related. For the function $y = 2 \sin[x + (\pi/4)]$, the amplitude is 2 and the period is 2π. By solving the double inequality

$$0 < x + \frac{\pi}{4} < 2\pi \quad \Longrightarrow \quad -\frac{\pi}{4} < x < \frac{7\pi}{4}$$

you can see that one cycle of the sine function corresponds to the interval from $x = -\pi/4$ to $x = 7\pi/4$. The graph of this sine function is represented by the gray curve in Figure 4.58. Because the sine function is zero at the endpoints of this interval, the corresponding cosecant function

$$y = 2 \csc\left(x + \frac{\pi}{4}\right) = 2\left(\frac{1}{\sin[x + (\pi/4)]}\right)$$

has vertical asymptotes at $x = -\pi/4, 3\pi/4, 7\pi/4$, etc. The graph of the cosecant function is represented by the black curve.

Figure 4.58

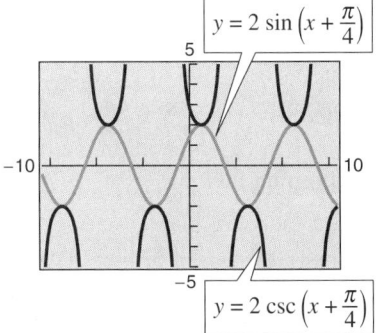

EXAMPLE 5   **Comparing Trigonometric Graphs**

Use a graphing utility to compare the graphs of

$$y = \cos 2x \quad \text{and} \quad y = \sec 2x.$$

Solution
Begin by graphing the two functions, as shown in Figure 4.59. Note that the x-intercepts of $y = \cos 2x$

$$\left(\frac{\pi}{4}, 0\right), \quad \left(\frac{3\pi}{4}, 0\right), \quad \left(\frac{5\pi}{4}, 0\right), \ldots$$

correspond to the vertical asymptotes

$$x = \frac{\pi}{4}, \quad x = \frac{3\pi}{4}, \quad x = \frac{5\pi}{4}, \ldots$$

of the graph of $y = \sec 2x$.

Figure 4.59

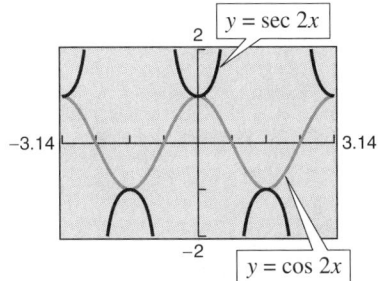

Damped Trigonometric Graphs

A *product* of two functions can be graphed using properties of the individual functions. For instance, consider the function

$$f(x) = x \sin x$$

as the product of the functions $y = x$ and $y = \sin x$. Using properties of absolute value and the fact that $|\sin x| \leq 1$, we have $0 \leq |x| \, |\sin x| \leq |x|$. Consequently,

$$-|x| \leq x \sin x \leq |x|$$

which means that the graph of $f(x) = x \sin x$ lies between the lines $y = -x$ and $y = x$. Furthermore, because

$$f(x) = x \sin x = \pm x \qquad \text{at} \qquad x = \frac{\pi}{2} + n\pi$$

$$f(x) = x \sin x = 0 \qquad \text{at} \qquad x = n\pi$$

the graph of f touches the line $y = -x$ or the line $y = x$ at $x = \pi/2 + n\pi$ and has x-intercepts at $x = n\pi$. A sketch of f is shown in Figure 4.60. In the function $f(x) = x \sin x$, the factor x is called the **damping factor.**

Figure 4.60

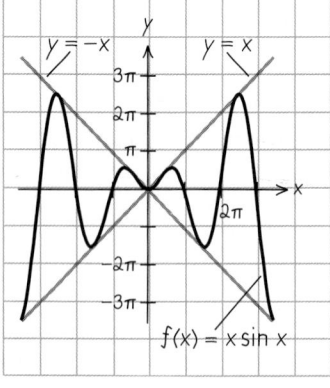

EXAMPLE 6   **Analyzing a Damped Sine Curve**

Analyze the graph of $f(x) = e^{-x} \sin 3x$.

Solution

Consider $f(x)$ as the product of the two functions

$$y = e^{-x} \qquad \text{and} \qquad y = \sin 3x$$

each of which has the set of real numbers as its domain. For any real number x, you know that $e^{-x} \geq 0$ and $|\sin 3x| \leq 1$. Therefore, $|e^{-x}| \, |\sin 3x| \leq e^{-x}$, which means that

$$-e^{-x} \leq e^{-x} \sin 3x \leq e^{-x}.$$

Furthermore, because

$$f(x) = e^{-x} \sin 3x = \pm e^{-x} \quad \text{at} \quad x = \frac{\pi}{6} + \frac{n\pi}{3}$$

and

$$f(x) = e^{-x} \sin 3x = 0 \quad \text{at} \quad x = \frac{n\pi}{3}$$

the graph of f touches the curves $y = -e^{-x}$ and $y = e^{-x}$ at $x = \frac{\pi}{6} + \frac{n\pi}{3}$ and has intercepts at $x = \frac{n\pi}{3}$. The graph is shown in Figure 4.61.

Figure 4.61

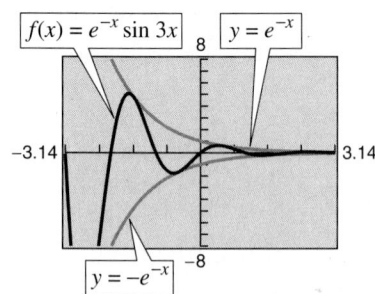

Figure 4.62 summarizes the graphs, domains, ranges, and periods of the six basic trigonometric functions.

Figure 4.62

Domain: all reals
Range: $[-1, 1]$
Period: 2π

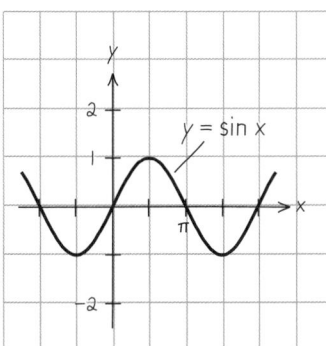

Domain: all reals
Range: $[-1, 1]$
Period: 2π

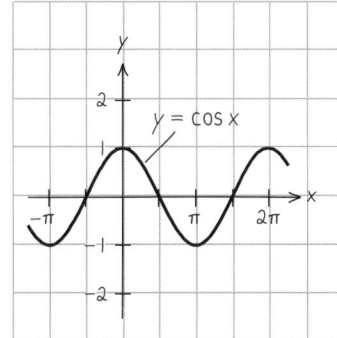

Domain: all $x \neq \frac{\pi}{2} + n\pi$
Range: $(-\infty, \infty)$
Period: π

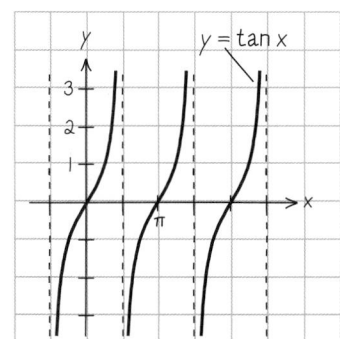

Domain: all $x \neq n\pi$
Range: $(-\infty, -1]$ and $[1, \infty)$
Period: 2π

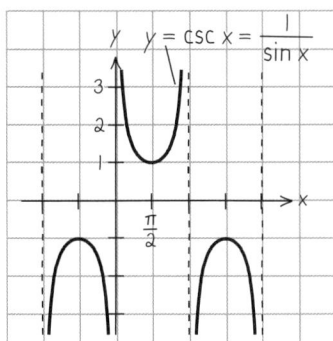

Domain: all $x \neq \frac{\pi}{2} + n\pi$
Range: $(-\infty, -1]$ and $[1, \infty)$
Period: 2π

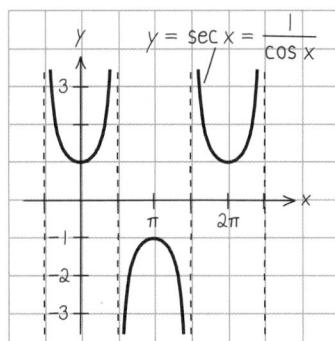

Domain: all $x \neq n\pi$
Range: $(-\infty, \infty)$
Period: π

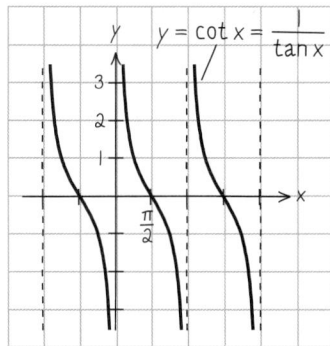

Group Activity   *Combining Trigonometric Functions*

Recall from Section 1.4 that functions can be combined arithmetically. This also applies to trigonometric functions. For each of the functions $h(x) = x + \sin x$ and $h(x) = \cos x - \sin 3x$, (a) identify two simpler functions f and g that comprise the combination, (b) use a table to show how to obtain the numerical values of $h(x)$ from the numerical values of $f(x)$ and $g(x)$, and (c) use graphing utility graphs of f and g to show how h may be formed.

Can you find functions $f(x) = d + a \sin(bx - c)$ and $g(x) = d + a \cos(bx - c)$ such that $f(x) + g(x) = 0$ for all x?

4.6 /// EXERCISES

In Exercises 1–8, match the function with its graph. State the period of the function. [The graphs are labeled (a), (b), (c), (d), (e), (f), (g), and (h).]

(a)

(b)

(c)

(d)

(e)

(f)

(g)

(h)

1. $y = \sec \dfrac{x}{2}$

2. $y = \tan \dfrac{x}{2}$

3. $y = \tan 2x$

4. $y = 2 \csc x$

5. $y = \cot \dfrac{\pi x}{2}$

6. $y = \dfrac{1}{2} \sec \dfrac{\pi x}{2}$

7. $y = -\csc x$

8. $y = -2 \sec 2\pi x$

In Exercises 9–30, sketch the graph of the function. (Include two full periods.) Use a graphing utility to verify your result.

9. $y = \frac{1}{3} \tan x$

10. $y = \frac{1}{4} \tan x$

11. $y = \tan 2x$

12. $y = -3 \tan \pi x$

13. $y = -\frac{1}{2} \sec x$

14. $y = \frac{1}{4} \sec x$

15. $y = -\sec \pi x$

16. $y = 2 \sec 4x$

17. $y = \sec \pi x - 1$

18. $y = -2 \sec 4x + 2$

19. $y = \csc \dfrac{x}{2}$

20. $y = \csc \dfrac{x}{3}$

21. $y = \cot \dfrac{x}{2}$

22. $y = 3 \cot \dfrac{\pi x}{2}$

23. $y = \frac{1}{2} \sec 2x$

24. $y = -\frac{1}{2} \tan x$

25. $y = \tan \dfrac{\pi x}{4}$

26. $y = \sec(x + \pi)$

27. $y = \csc(\pi - x)$

28. $y = \sec(\pi - x)$

29. $y = 2 \cot\left(x + \dfrac{\pi}{2}\right)$

30. $y = \dfrac{1}{4} \csc\left(x + \dfrac{\pi}{4}\right)$

In Exercises 31–40, use a graphing utility to graph the function. (Include two full periods.)

31. $y = \tan \dfrac{x}{3}$

32. $y = -\tan 2x$

33. $y = -2 \sec 4x$

34. $y = \sec \pi x$

35. $y = \tan\left(x - \dfrac{\pi}{4}\right)$

36. $y = -\csc(4x - \pi)$

37. $y = \dfrac{1}{4} \cot\left(x - \dfrac{\pi}{2}\right)$

38. $y = 0.1 \tan\left(\dfrac{\pi x}{4} + \dfrac{\pi}{4}\right)$

39. $y = 2 \sec(2x - \pi)$

40. $y = \dfrac{1}{3} \sec\left(\dfrac{\pi x}{2} + \dfrac{\pi}{2}\right)$

In Exercises 41–44, use a graph to solve the equation in the interval $[-2\pi, 2\pi]$.

41. $\tan x = 1$

42. $\cot x = -\sqrt{3}$

43. $\sec x = -2$

44. $\csc x = \sqrt{2}$

In Exercises 45 and 46, use the graph of the function to determine whether the function is even, odd, or neither.

45. $f(x) = \sec x$ **46.** $f(x) = \tan x$

47. *Essay* Describe the behavior of $f(x) = \tan x$ as x approaches $\pi/2$ from the left and from the right.

48. *Essay* Describe the behavior of $f(x) = \csc x$ as x approaches π from the left and from the right.

49. *Graphical Reasoning* Consider the functions

$$f(x) = 2 \sin x \quad \text{and} \quad g(x) = \tfrac{1}{2} \csc x$$

on the interval $(0, \pi)$.

(a) Use a graphing utility to graph f and g in the same viewing rectangle.

(b) Approximate the interval where $f > g$.

(c) Describe the behavior of each of the functions as x approaches π. How is the behavior of g related to the behavior of f as x approaches π?

50. *Graphical Reasoning* Consider the functions

$$f(x) = \tan \frac{\pi x}{2} \quad \text{and} \quad g(x) = \frac{1}{2} \sec \frac{\pi x}{2}$$

on the interval $(-1, 1)$.

(a) Use a graphing utility to graph f and g in the same viewing rectangle.

(b) Approximate the interval where $f < g$.

(c) Approximate the interval where $2f < 2g$. How does the result compare with that of part (b)? Explain.

In Exercises 51–54, use a graphing utility to graph the two equations in the same viewing rectangle. Use the graphs to lend evidence that the expressions are equivalent. Verify the results analytically.

51. $y_1 = \sin x \csc x$, $y_2 = 1$

52. $y_1 = \sin x \sec x$, $y_2 = \tan x$

53. $y_1 = \dfrac{\cos x}{\sin x}$, $y_2 = \cot x$

54. $y_1 = \sec^2 x - 1$, $y_2 = \tan^2 x$

In Exercises 55–58, match the function with its graph. Describe the behavior of the function as x approaches zero. [The graphs are labeled (a), (b), (c), and (d).]

(a)   (b)

(c)   (d)

55. $f(x) = x \cos x$ **56.** $f(x) = |x \sin x|$

57. $g(x) = |x| \sin x$ **58.** $g(x) = |x| \cos x$

Conjecture In Exercises 59–62, use a graphing utility to graph the functions f and g. Use the graphs to make a conjecture about the relationship between the functions.

59. $f(x) = \sin x + \cos\left(x + \dfrac{\pi}{2}\right), \quad g(x) = 0$

60. $f(x) = \sin x - \cos\left(x + \dfrac{\pi}{2}\right), \quad g(x) = 2 \sin x$

61. $f(x) = \sin^2 x, \quad g(x) = \tfrac{1}{2}(1 - \cos 2x)$

62. $f(x) = \cos^2 \dfrac{\pi x}{2}, \quad g(x) = \dfrac{1}{2}(1 + \cos \pi x)$

In Exercises 63–66, use a graphing utility to graph the function and the damping factor of the function in the same viewing rectangle. Describe the behavior of the function as x increases without bound.

63. $f(x) = 2^{-x/4} \cos \pi x$

64. $f(x) = e^{-x} \cos x$

65. $g(x) = e^{-x^2/2} \sin x$

66. $h(x) = 2^{-x^2/4} \sin x$

67. *Distance*   A plane flying at an altitude of 5 miles over level ground will pass directly over a radar antenna (see figure). Let d be the ground distance from the antenna to the point directly under the plane and let x be the angle of elevation to the plane from the antenna. Write d as a function of x and graph the function over the interval $0 < x < \pi$.

68. *Television Coverage*   A television camera is on a reviewing platform 36 meters from the street on which a parade will be passing from left to right (see figure). Express the distance d from the camera to a particular unit in the parade as a function of the angle x, and graph the function over the interval $-\pi/2 < x < \pi/2$. (Consider x as negative when a unit in the parade approaches from the left.)

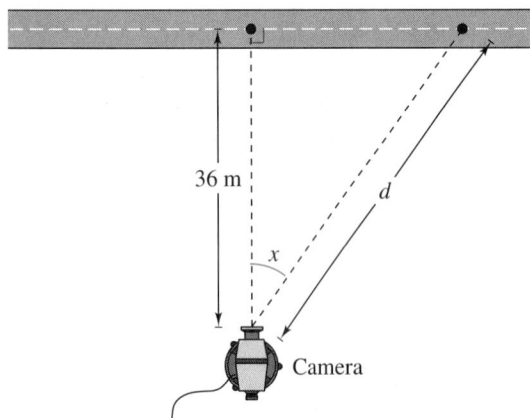

69. *Predatory-Prey Model*   Suppose the population of a certain predator at time t (in months) in a given region

is estimated to be

$$P = 10,000 + 3000 \sin \frac{2\pi t}{24}$$

and the population of its primary food source (its prey) is estimated to be

$$p = 15,000 + 5000 \cos \frac{2\pi t}{24}.$$

Use the graph of the models to explain the oscillations in the size of each population.

70. *Normal Temperatures*   The normal monthly high temperatures in degrees Fahrenheit for Erie, Pennsylvania are approximated by

$$H(t) = 54.33 - 20.38 \cos \frac{\pi t}{6} - 15.69 \sin \frac{\pi t}{6}$$

and the normal monthly low temperatures are approximately

$$L(t) = 39.36 - 15.70 \cos \frac{\pi t}{6} - 14.16 \sin \frac{\pi t}{6}$$

where t is the time in months, with $t = 1$ corresponding to January. (Source: National Oceanic and Atmospheric Association)

(a) Use a graphing utility to graph each function. What is the period of each function?

(b) During what part of the year is the difference between the normal high and low temperatures greatest? When is it smallest?

(c) The sun is the farthest north in the sky around June 21, but the graph shows the warmest temperatures at a later date. Approximate the lag time of the temperatures relative to the position of the sun.

71. *Harmonic Motion*   An object weighing W pounds is suspended from a ceiling by a steel spring (see figure). The weight is pulled downward (positive direction) from its equilibrium position and released. The resulting motion of the weight is described by the function

$$y = \frac{1}{2} e^{-t/4} \cos 4t, \quad t > 0$$

where y is the distance in feet and t is the time in seconds.

(a) Use a graphing utility to graph the function.

(b) Describe the behavior of the displacement function for increasing values of time t.

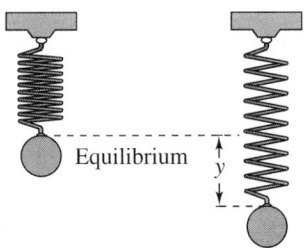

Equilibrium

72. *Exploration*   Consider the function

$$f(x) = x - \cos x.$$

(a) Use a graphing utility to graph the function and verify that there exists a zero between 0 and 1. Use the graph to approximate the zero.

(b) Starting with $x_0 = 1$, generate a sequence

$$x_1, x_2, x_3, \ldots$$

where $x_n = \cos(x_{n-1})$. Verify that the sequence approaches the zero of f.

73. *Approximation*   Using calculus, it can be shown that the tangent function can be approximated by the polynomial

$$\tan x \approx x + \frac{2x^3}{3!} + \frac{16x^5}{5!}$$

where x is in radians. Use a graphing utility to graph the tangent function and its polynomial approximation in the same viewing rectangle. How do the graphs compare?

74. *Approximation*   Using calculus, it can be shown that the secant function can be approximated by the polynomial

$$\sec x \approx 1 + \frac{x^2}{2!} + \frac{5x^4}{4!}$$

where x is in radians. Use a graphing utility to graph the secant function and its polynomial approximation in the same viewing rectangle. How do the graphs compare?

75. *Pattern Recognition*

(a) Use a graphing utility to graph each function.

$$y_1 = \frac{4}{\pi}\left(\sin \pi x + \frac{1}{3} \sin 3 \pi x \right)$$

$$y_2 = \frac{4}{\pi}\left(\sin \pi x + \frac{1}{3} \sin 3 \pi x + \frac{1}{5} \sin 5 \pi x \right)$$

(b) Identify the pattern in part (a) and find a function y_3 that continues the pattern one more term. Use a graphing utility to graph y_3.

(c) The graphs of parts (a) and (b) approximate the periodic function in the figure. Find a function y_4 that is a better approximation.

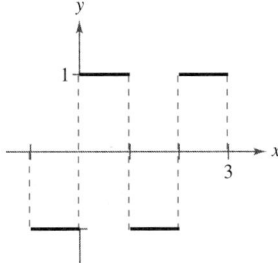

76. *Sales*   The projected monthly sales S (in thousands of units) of a seasonal product is modeled by

$$S = 74 + 3t + 40 \sin \frac{\pi t}{6}$$

where t is the time in months, with $t = 1$ corresponding to January. Graph the sales function over 1 year.

Exploration   **In Exercises 77 and 78, use a graphing utility to graph the function. Describe the behavior of the function as x approaches zero.**

77. $y = \dfrac{6}{x} + \cos x$ **78.** $g(x) = \dfrac{\sin x}{x}$

79. *Data Analysis*   The motion of an oscillating weight suspended by a spring was measured by a motion detector. The data was collected, and the approximate maximum (positive and negative) displacements from equilibrium are shown in the figure. The displacement y is measured in centimeters and the time t is measured in seconds.

(a) Is y a function of t? Explain.

(b) Approximate the frequency of the oscillations.

(c) Fit a model of the form

$$y = ab^t \cos ct.$$

to the data. Use the result of part (b) to approximate c. Use the regression capabilities of a graphing utility to fit an exponential model to the positive maximum displacements of the weight.

(d) Rewrite the model in the form $y = ae^{kt} \cos ct$.

(e) Use a graphing utility to graph the model. Compare the result with the data in the figure.

80. *Essay*   Write a short paragraph describing the specified change in the physical system of Exercise 79.

(a) A spring of less stiffness is used, and so the length of time in each oscillation is greater.

(b) The effect of friction is decreased.

81. *Numerical and Graphical Reasoning*   A crossed belt connects a 10-centimeter pulley on an electric motor with a 20-centimeter pulley on a saw arbor (see figure). The electric motor runs at 1700 revolutions per minute.

(a) Determine the number of revolutions per minute of the saw.

(b) How does crossing the belt affect the saw in relation to the motor?

(c) Let L be the total length of the belt. Write L as a function of ϕ, where ϕ is measured in radians. What is the domain of the function? (*Hint:* Add the lengths of the straight sections of the belt and the length of belt around each pulley.)

(d) Use a graphing utility to complete the table.

ϕ	0.3	0.6	0.9	1.2	1.5
L					

(e) As ϕ increases, do the lengths of the straight sections of the belt change faster or slower than the lengths of the belt around each pulley?

(f) Use a graphing utility to graph the function over the appropriate domain.

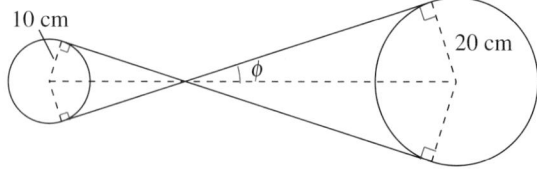

Review   **Solve Exercises 82–85 as a review of the skills and problem-solving techniques you learned in previous sections. Solve the equation. (Round your solution to three decimal places.)**

82. $e^{2x} = 54$ **83.** $\dfrac{300}{1 + e^{-x}} = 100$

84. $\ln(x^2 + 1) = 3.2$ **85.** $\log_8 x + \log_8 (x - 1) = \frac{1}{3}$

4.7 Inverse Trigonometric Functions

Inverse Sine Function / *Other Inverse Trigonometric Functions* / *Compositions of Functions*

Inverse Sine Function

Figure 4.63

Sin x has an inverse on this interval.

Recall from Section 1.5 that, for a function to have an inverse, it must pass the Horizontal Line Test. From Figure 4.63 it is obvious that $y = \sin x$ does not pass the test because different values of x yield the same y-value. However, if you restrict the domain to the interval $-\pi/2 \le x \le \pi/2$ (corresponding to the black portion of the graph in Figure 4.63), the following properties hold.

1. On the interval $[-\pi/2, \pi/2]$, the function $y = \sin x$ is increasing.
2. On the interval $[-\pi/2, \pi/2]$, $y = \sin x$ takes on its full range of values, $-1 \le \sin x \le 1$.
3. On the interval $[-\pi/2, \pi/2]$, $y = \sin x$ passes the Horizontal Line Test.

Thus, on the restricted domain $-\pi/2 \le x \le \pi/2$, $y = \sin x$ has a unique inverse called the **inverse sine function.** It is denoted by

$$y = \arcsin x \quad \text{or} \quad y = \sin^{-1} x.$$

The notation $\sin^{-1} x$ is consistent with the inverse function notation $f^{-1}(x)$. The arcsin x notation (read as "the arcsine of x") comes from the association of a central angle with its intercepted *arc length* on a unit circle. Thus, arcsin x means the angle (or arc) whose sine is x. Both notations, arcsin x and $\sin^{-1} x$, are commonly used in mathematics, so remember that $\sin^{-1} x$ denotes the *inverse* sine function rather than $1/\sin x$. The values of arcsin x lie in the interval

$$-\frac{\pi}{2} \le \arcsin x \le \frac{\pi}{2}.$$

The graph of $y = \arcsin x$ is shown in Example 2.

Library of Functions

The *inverse* trigonometric functions are obtained from the trigonometric functions in much the same way as the logarithmic function was developed from the exponential function. However, unlike the exponential function, the trigonometric functions are not one-to-one, and hence it is necessary to restrict their domains to regions that pass the Horizontal Line Test.

Definition of Inverse Sine Function

The **inverse sine function** is defined by

$$y = \arcsin x \quad \text{if and only if} \quad \sin y = x$$

where $-1 \le x \le 1$ and $-\pi/2 \le y \le \pi/2$. The domain of $y = \arcsin x$ is $[-1, 1]$ and the range is $[-\pi/2, \pi/2]$.

Note When evaluating the inverse sine function, it helps to remember the phrase "the arcsine of x is the angle (or number) whose sine is x."

EXAMPLE 1   **Evaluating the Inverse Sine Function**

If possible, find the exact value.

a. $\arcsin\left(-\dfrac{1}{2}\right)$ **b.** $\sin^{-1}\dfrac{\sqrt{3}}{2}$ **c.** $\sin^{-1} 2$

Solution

a. Because $\sin(-\pi/6) = -\dfrac{1}{2}$ for $-\pi/2 \le y \le \pi/2$, it follows that

$$\arcsin\left(-\frac{1}{2}\right) = -\frac{\pi}{6}.$$

b. Because $\sin(\pi/3) = \sqrt{3}/2$ for $-\pi/2 \le y \le \pi/2$, it follows that

$$\sin^{-1}\frac{\sqrt{3}}{2} = \frac{\pi}{3}.$$

c. It is not possible to evaluate $y = \sin^{-1} x$ when $x = 2$ because there is no angle whose sine is 2. Remember that the domain of the inverse sine function is $[-1, 1]$.

EXAMPLE 2   **Graphing the Arcsine Function**

Sketch a graph of $y = \arcsin x$.

Solution
By definition, the equations

$$y = \arcsin x \quad \text{and} \quad \sin y = x$$

are equivalent for $-\pi/2 \le y \le \pi/2$. Hence, their graphs are the same. From the interval $[-\pi/2, \pi/2]$, you can assign values to y in the second equation to make a table of values.

Figure 4.64

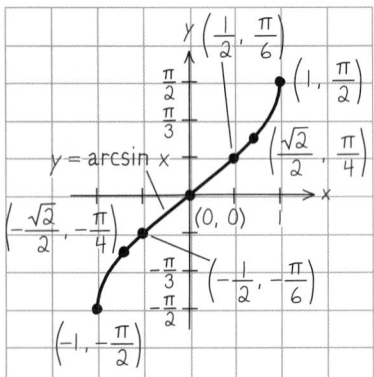

y	$-\dfrac{\pi}{2}$	$-\dfrac{\pi}{4}$	$-\dfrac{\pi}{6}$	0	$\dfrac{\pi}{6}$	$\dfrac{\pi}{4}$	$\dfrac{\pi}{2}$
$x = \sin y$	-1	$-\dfrac{\sqrt{2}}{2}$	$-\dfrac{1}{2}$	0	$\dfrac{1}{2}$	$\dfrac{\sqrt{2}}{2}$	1

The resulting graph for $y = \arcsin x$ is shown in Figure 4.64. Note that it is the reflection (in line $y = x$) of the black portion of Figure 4.63. Use a graphing utility to confirm this graph. Be sure you see that Figure 4.64 shows the *entire* graph of the inverse sine function. Remember that the range of $y = \arcsin x$ is the closed interval $[-\pi/2, \pi/2]$.

Other Inverse Trigonometric Functions

Figure 4.65

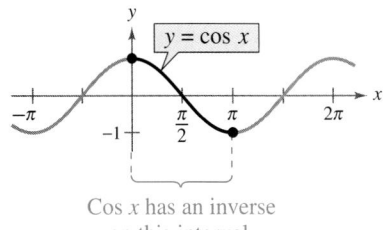

Cos x has an inverse
on this interval.

The cosine function is decreasing on the interval $0 \le x \le \pi$, as shown in Figure 4.65. Consequently, on this interval the cosine function has an inverse function—the **inverse cosine function**—denoted by

$$y = \arccos x \qquad \text{or} \qquad y = \cos^{-1} x.$$

Similarly, you can define an **inverse tangent function** by restricting the domain of $y = \tan x$ to the interval $(-\pi/2, \pi/2)$. The following list summarizes the definitions of the three most common inverse trigonometric functions. The remaining three are discussed in Exercises 77–79. (The graphs, domains, and ranges of *all six* inverse trigonometric functions are summarized in the appendix.)

Definitions of Inverse Trigonometric Functions		
Function	*Domain*	*Range*
$y = \arcsin x$ if and only if $\sin y = x$	$-1 \le x \le 1$	$-\dfrac{\pi}{2} \le y \le \dfrac{\pi}{2}$
$y = \arccos x$ if and only if $\cos y = x$	$-1 \le x \le 1$	$0 \le y \le \pi$
$y = \arctan x$ if and only if $\tan y = x$	$-\infty < x < \infty$	$-\dfrac{\pi}{2} < y < \dfrac{\pi}{2}$

The graphs of these three inverse trigonometric functions are shown in Figure 4.66. A table for $y = \arccos x$ is shown at the left.

X	Y₁
−1	3.1416
−.5	2.0944
−.25	1.8235
0	1.5708
.25	1.3181
.5	1.0472
1	0

X = 1

Figure 4.66

Domain: $[-1, 1]$
Range: $\left[-\dfrac{\pi}{2}, \dfrac{\pi}{2}\right]$

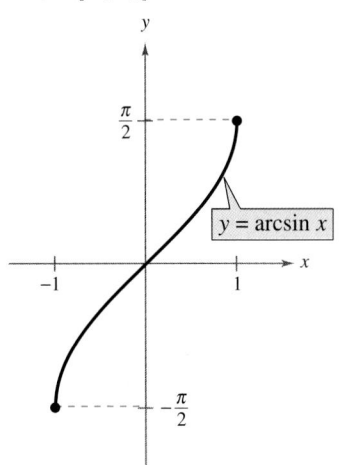

Domain: $[-1, 1]$
Range: $[0, \pi]$

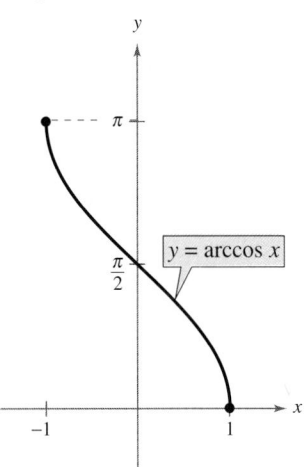

Domain: $(-\infty, \infty)$
Range: $\left(-\dfrac{\pi}{2}, \dfrac{\pi}{2}\right)$

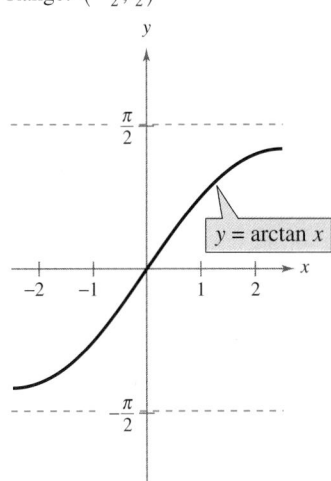

EXAMPLE 3   Evaluating Inverse Trigonometric Functions

Find the exact value.

a. $\arccos \dfrac{\sqrt{2}}{2}$ **b.** $\arccos(-1)$ **c.** $\arctan 0$

Solution

Most graphing utilities do not have function keys for the inverse cosecant function, the inverse secant function, or the inverse cotangent function. To evaluate these functions, you can use the inverse sine, inverse cosine, and inverse tangent functions. For instance, to evaluate arcsec 3.4, enter the expression as $\arccos(1/3.4)$.

a. Because $\cos(\pi/4) = \sqrt{2}/2$, and $\pi/4$ lies in $[0, \pi]$, it follows that

$$\arccos \frac{\sqrt{2}}{2} = \frac{\pi}{4}.$$

b. Because $\cos \pi = -1$, and π lies in $[0, \pi]$, it follows that

$$\arccos(-1) = \pi.$$

c. Because $\tan 0 = 0$, and 0 lies in $(-\pi/2, \pi/2)$, it follows that

$$\arctan 0 = 0.$$

EXAMPLE 4   Calculators and Inverse Trigonometric Functions

Use a calculator to approximate the value (if possible).

a. $\arctan(-8.45)$ **b.** $\arcsin 0.2447$ **c.** $\arccos 2$

Solution

Function	*Mode*	*Graphing Calculator Keystrokes*

a. $\arctan(-8.45)$ Radian   TAN⁻¹ ((-) 8.45) ENTER
From the display, it follows that $\arctan(-8.45) \approx -1.453001$.

b. $\arcsin 0.2447$ Radian   SIN⁻¹ 0.2447 ENTER
From the display, it follows that $\arcsin 0.2447 \approx 0.2472103$.

c. $\arccos 2$ Radian   COS⁻¹ 2 ENTER

In real number mode, the calculator should display an *error* message, because the domain of the inverse cosine function is $[-1, 1]$.

Note In Example 4, if you had set the calculator to degree mode, the display would have been in degrees rather than in radians. This convention is peculiar to calculators. By definition, the values of inverse trigonometric functions are always in *radians*.

Compositions of Functions

Recall from Section 1.5 that inverse functions possess the properties

$$f(f^{-1}(x)) = x \quad \text{and} \quad f^{-1}(f(x)) = x.$$

The inverse trigonometric versions of these properties are given below.

Inverse Properties

If $-1 \le x \le 1$ and $-\pi/2 \le y \le \pi/2$, then

$$\sin(\arcsin x) = x \quad \text{and} \quad \arcsin(\sin y) = y.$$

If $-1 \le x \le 1$ and $0 \le y \le \pi$, then

$$\cos(\arccos x) = x \quad \text{and} \quad \arccos(\cos y) = y.$$

If $-\pi/2 < y < \pi/2$, then

$$\tan(\arctan x) = x \quad \text{and} \quad \arctan(\tan y) = y.$$

Note   Keep in mind that these inverse properties do not apply for arbitrary values of x and y. For instance,

$$\arcsin\left(\sin \frac{3\pi}{2}\right) = \arcsin(-1)$$

$$= -\frac{\pi}{2}$$

$$\ne \frac{3\pi}{2}.$$

In other words, the property

$$\arcsin(\sin y) = y$$

is not valid for values of y outside the interval $[-\pi/2, \pi/2]$.

EXAMPLE 5   ◢   **Using Inverse Properties**

If possible, find the exact value.

a. $\tan[\arctan(-5)]$ **b.** $\arcsin\left(\sin \dfrac{5\pi}{3}\right)$ **c.** $\cos(\cos^{-1} \pi)$

Solution

a. Because -5 lies in the domain of $\arctan x$, the inverse property applies, and you have

$$\tan[\arctan(-5)] = -5.$$

b. In this case, $5\pi/3$ does not lie within the range of the arcsine function, $-\pi/2 \le y \le \pi/2$. However, $5\pi/3$ is coterminal with

$$\frac{5\pi}{3} - 2\pi = -\frac{\pi}{3}$$

which does lie in the range of the arcsine function, and you have

$$\arcsin\left(\sin \frac{5\pi}{3}\right) = \arcsin\left[\sin\left(-\frac{\pi}{3}\right)\right] = -\frac{\pi}{3}.$$

c. The expression $\cos(\cos^{-1} \pi)$ is not defined, because $\cos^{-1} \pi$ is not defined. Remember that the domain of the inverse cosine function is $[-1, 1]$.   ◢

EXPLORATION

Use a graphing utility to graph $y = \arcsin(\sin x)$. What are the domain and range of this function? Explain why $\arcsin(\sin 4)$ does not equal 4.

Now graph $y = \sin(\arcsin x)$ and determine the domain and range. Explain why $\sin(\arcsin 4)$ is not defined.

Example 6 shows how to use right triangles to find exact values of functions of inverse functions. Then, Example 7 shows how to use triangles to convert a trigonometric expression into an algebraic expression. This conversion technique is used frequently in calculus.

Figure 4.67

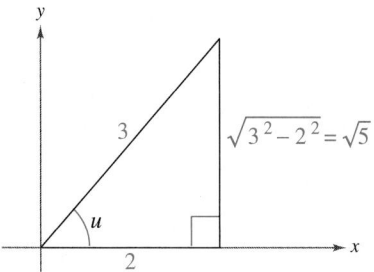

EXAMPLE 6 ◢ **Evaluating Compositions of Functions**

Find the exact value.

a. $\tan\left(\arccos\dfrac{2}{3}\right)$ **b.** $\cos\left[\arcsin\left(-\dfrac{3}{5}\right)\right]$

Solution

a. If you let $u = \arccos\frac{2}{3}$, then $\cos u = \frac{2}{3}$. Because $\cos u$ is positive, u is a *first*-quadrant angle. You can sketch and label angle u as shown in Figure 4.67. Consequently,

$$\tan\left(\arccos\frac{2}{3}\right) = \tan u = \frac{\text{opp}}{\text{adj}} = \frac{\sqrt{5}}{2}.$$

Figure 4.68

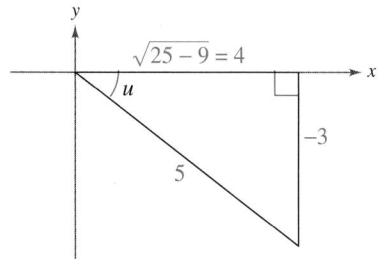

b. If you let $u = \arcsin\left(-\frac{3}{5}\right)$, then $\sin u = -\frac{3}{5}$. Because $\sin u$ is negative, u is a *fourth*-quadrant angle. You can sketch and label angle u as shown in Figure 4.68. Consequently,

$$\cos\left[\arcsin\left(-\frac{3}{5}\right)\right] = \cos u = \frac{\text{adj}}{\text{hyp}} = \frac{4}{5}.$$ ◢

EXAMPLE 7 ◢ **Some Problems from Calculus**

Write each of the following as an algebraic expression in x.

a. $\sin(\arccos 3x), \quad 0 \le x \le \dfrac{1}{3}$ **b.** $\cot(\arccos 3x), \quad 0 \le x \le \dfrac{1}{3}$

Figure 4.69

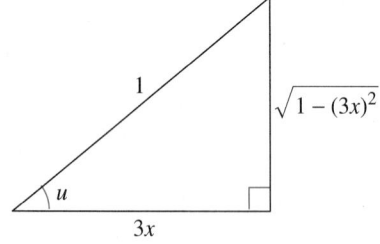

Solution

In you let $u = \arccos 3x$, then $\cos u = 3x$. Because

$$\cos u = \frac{3x}{1} = \frac{\text{adj}}{\text{hyp}}$$

you can sketch a right triangle with acute angle u, as shown in Figure 4.69. From this triangle, you can easily convert each expression to algebraic form.

a. $\sin(\arccos 3x) = \sin u = \dfrac{\text{opp}}{\text{hyp}} = \sqrt{1 - 9x^2}, \quad 0 \le x \le \dfrac{1}{3}$

Note In Example 7, a similar argument can be made for x-values lying in the interval $\left[-\frac{1}{3}, 0\right]$.

b. $\cot(\arccos 3x) = \cot u = \dfrac{\text{adj}}{\text{opp}} = \dfrac{3x}{\sqrt{1 - 9x^2}}, \quad 0 \le x \le \dfrac{1}{3}$ ◢

Group Activity *Inverse Functions*

We have discussed inverse functions for several types of functions. Match the function in the left column with its inverse function in the right column.

1. $f(x) = x$
2. $f(x) = x^2, \quad 0 \le x$
3. $f(x) = x^3$
4. $f(x) = e^x$
5. $f(x) = \ln x$

6. $f(x) = \sin x, \quad -\dfrac{\pi}{2} \le x \le \dfrac{\pi}{2}$

7. $f(x) = \cos x, \quad 0 \le x \le \pi$

8. $f(x) = \tan x, \quad -\dfrac{\pi}{2} < x < \dfrac{\pi}{2}$

a. $f^{-1}(x) = \arcsin x$
b. $f^{-1}(x) = \ln x$
c. $f^{-1}(x) = \sqrt{x}$
d. $f^{-1}(x) = \arctan x$
e. $f^{-1}(x) = \arccos x$
f. $f^{-1}(x) = \sqrt[3]{x}$
g. $f^{-1}(x) = e^x$
h. $f^{-1}(x) = x$

Discuss reasons for your answers. Verify each pair of inverses algebraically, graphically, and numerically.

4.7 /// EXERCISES

1. *Numerical and Graphical Analysis* Consider the function $y = \arcsin x$.

(a) Use a graphing utility to complete the table.

x	-1	-0.8	-0.6	-0.4	-0.2
y					

x	0	0.2	0.4	0.6	0.8	1
y						

(b) Plot the points from the table in part (a) and graph the function. (Do not use a graphing utility.)

(c) Use a graphing utility to graph the inverse sine function and compare the result with your hand drawn graph in part (b).

(d) Determine any intercepts and symmetry of the graph.

2. *Numerical and Graphical Analysis* Consider the function $y = \arccos x$.

(a) Use a graphing utility to complete the table.

x	-1	-0.8	-0.6	-0.4	-0.2
y					

x	0	0.2	0.4	0.6	0.8	1
y						

(b) Plot the points from the table in part (a) and graph the function. (Do not use a graphing utility.)

(c) Use a graphing utility to graph the inverse cosine function and compare the result with your hand drawn graph in part (b).

(d) Determine any intercepts and symmetry of the graph.

3. *Numerical and Graphical Analysis*   Consider the function $y = \arctan x$.

(a) Use a graphing utility to complete the table.

x	-10	-8	-6	-4	-2
y					

x	0	2	4	6	8	10
y						

(b) Plot the points from the table in part (a) and graph the function. (Do not use a graphing utility.)

(c) Use a graphing utility to graph the inverse tangent function and compare the result with your hand drawn graph in part (b).

(d) Determine the horizontal asymptotes of the graph.

4. *True or False?*   Explain your reasoning.

$$\sin\frac{5\pi}{2} = \frac{1}{2} \implies \arcsin\frac{1}{2} = \frac{5\pi}{6}$$

In Exercises 5 and 6, determine the missing coordinates of the points on the graph of the function.

5.

6.

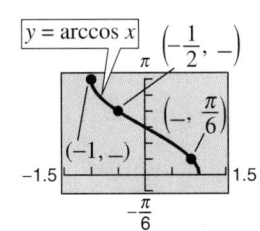

In Exercises 7–14, evaluate the expression without the aid of a calculator.

7. (a) $\arcsin\frac{1}{2}$ (b) $\arcsin 0$

8. (a) $\arccos\frac{1}{2}$ (b) $\arccos 0$

9. (a) $\arctan\dfrac{\sqrt{3}}{3}$ (b) $\arctan(-1)$

10. (a) $\arccos\left(-\dfrac{\sqrt{3}}{2}\right)$ (b) $\arcsin\left(-\dfrac{\sqrt{2}}{2}\right)$

11. (a) $\arctan\left(-\sqrt{3}\right)$ (b) $\arctan\sqrt{3}$

12. (a) $\arccos\left(-\dfrac{1}{2}\right)$ (b) $\arcsin\dfrac{\sqrt{2}}{2}$

13. (a) $\arcsin\dfrac{\sqrt{3}}{2}$ (b) $\arctan\left(-\dfrac{\sqrt{3}}{3}\right)$

14. (a) $\arctan 0$ (b) $\arccos 1$

In Exercises 15–20, use a calculator to approximate the value of the expression. (Round to two decimal places.)

15. (a) $\arccos 0.28$ (b) $\arcsin 0.45$

16. (a) $\arcsin(-0.75)$ (b) $\arccos(-0.7)$

17. (a) $\arctan(-3)$ (b) $\arctan 15$

18. (a) $\arcsin 0.31$ (b) $\arccos 0.26$

19. (a) $\arccos(-0.41)$ (b) $\arcsin(-0.125)$

20. (a) $\arctan 0.92$ (b) $\arctan 2.8$

In Exercises 21 and 22, use a graphing utility to graph f, g, and $y = x$ in the same viewing rectangle to verify geometrically that g is the inverse of f. (Be sure to properly restrict the domain of f.)

21. $f(x) = \tan x$, $g(x) = \arctan x$

22. $f(x) = \sin x$, $g(x) = \arcsin x$

In Exercises 23–26, use an inverse trigonometric function to write θ as a function of x.

23.

24.

25.

26.

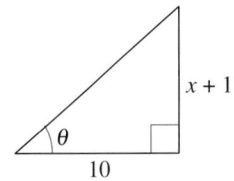

In Exercises 27–32, use the properties of inverse functions to evaluate the expression.

27. $\sin(\arcsin 0.3)$

28. $\tan(\arctan 25)$

29. $\cos[\arccos(-0.1)]$

30. $\sin[\arcsin(-0.2)]$

31. $\arcsin(\sin 3\pi)$

32. $\arccos\left(\cos \dfrac{7\pi}{2}\right)$

In Exercises 33–42, find the exact value of the expression. (*Hint:* Make a sketch of a right triangle.)

33. $\sin\left(\arctan \tfrac{3}{4}\right)$

34. $\sec\left(\arcsin \tfrac{4}{5}\right)$

35. $\cos(\arctan 2)$

36. $\sin\left(\arccos \dfrac{\sqrt{5}}{5}\right)$

37. $\cos\left(\arcsin \tfrac{5}{13}\right)$

38. $\csc\left[\arctan\left(-\tfrac{5}{12}\right)\right]$

39. $\sec\left[\arctan\left(-\tfrac{3}{5}\right)\right]$

40. $\tan\left[\arcsin\left(-\tfrac{3}{4}\right)\right]$

41. $\sin\left[\arccos\left(-\tfrac{2}{3}\right)\right]$

42. $\cot\left(\arctan \tfrac{5}{8}\right)$

In Exercises 43–52, write an algebraic expression that is equivalent to the expression. (*Hint:* Sketch a right triangle, as demonstrated in Example 7.)

43. $\cot(\arctan x)$

44. $\sin(\arctan x)$

45. $\cos(\arcsin 2x)$

46. $\sec(\arctan 3x)$

47. $\sin(\arccos x)$

48. $\sec[\arcsin(x-1)]$

49. $\tan\left(\arccos \dfrac{x}{3}\right)$

50. $\cot\left(\arctan \dfrac{1}{x}\right)$

51. $\csc\left(\arctan \dfrac{x}{\sqrt{2}}\right)$

52. $\cos\left(\arcsin \dfrac{x-h}{r}\right)$

In Exercises 53 and 54, use a graphing utility to graph f and g in the same viewing rectangle to verify that the two are equal. Explain why they are equal. Identify any asymptotes of the graphs.

53. $f(x) = \sin(\arctan 2x)$, $\quad g(x) = \dfrac{2x}{\sqrt{1 + 4x^2}}$

54. $f(x) = \tan\left(\arccos \dfrac{x}{2}\right)$, $\quad g(x) = \dfrac{\sqrt{4 - x^2}}{x}$

In Exercises 55–58, fill in the blanks.

55. $\arctan \dfrac{9}{x} = \arcsin(\rule{1cm}{0.4pt})$, $\quad x \neq 0$

56. $\arcsin \dfrac{\sqrt{36 - x^2}}{6} = \arccos(\rule{1cm}{0.4pt})$, $\quad 0 \leq x \leq 6$

57. $\arccos \dfrac{3}{\sqrt{x^2 - 2x + 10}} = \arcsin(\rule{1cm}{0.4pt})$

58. $\arccos \dfrac{x - 2}{2} = \arctan(\rule{1cm}{0.4pt})$, $\quad |x - 2| \leq 2$

In Exercises 59–66, use a graphing utility to graph the function.

59. $y = 2 \arccos x$

60. $y = \arcsin \dfrac{x}{2}$

61. $f(x) = \arcsin(x - 1)$

62. $g(t) = \arccos(t + 2)$

63. $f(x) = \arctan 2x$

64. $f(x) = \dfrac{\pi}{2} + \arctan x$

65. $h(v) = \tan(\arccos v)$

66. $f(x) = \arccos \dfrac{x}{4}$

In Exercises 67 and 68, write the given function in terms of the sine function by using the identity

$$A \cos \omega t + B \sin \omega t = \sqrt{A^2 + B^2} \sin\left(\omega t + \arctan \dfrac{A}{B}\right).$$

Use a graphing utility to graph both forms of the function. What does the graph imply?

67. $f(t) = 3 \cos 2t + 3 \sin 2t$

68. $f(t) = 4 \cos \pi t + 3 \sin \pi t$

69. *Think About It*   Consider the functions

$f(x) = \sin x$ and $f^{-1}(x) = \arcsin x$.

(a) Use a graphing utility to graph the composite functions $f \circ f^{-1}$ and $f^{-1} \circ f$.

(b) Explain why the graphs in part (a) are not the graph of the line $y = x$. Why do the graphs of $f \circ f^{-1}$ and $f^{-1} \circ f$ differ?

70. *Think About It*   Use a graphing utility to graph the functions $f(x) = \sqrt{x}$ and $g(x) = 6 \arctan x$. For $x > 0$, it appears that $g > f$. Explain why you know that there exists a positive real number a such that $g < f$ for $x > a$. Approximate the number a.

71. *Docking a Boat*   A boat is pulled in by means of a winch located on a dock 10 feet above the deck of the boat (see figure). Let θ be the angle of elevation from the boat to the winch and let s be the length of the rope from the winch to the boat.

(a) Write θ at a function of s.

(b) Find θ when $s = 48$ feet and $s = 24$ feet.

72. *Photography*   A television camera at ground level is filming the lift-off of a space shuttle at a point 750 meters from the launch pad (see figure). Let θ be the angle of elevation to the shuttle and let s be the height of the shuttle.

(a) Write θ as a function of s.

(b) Find θ when $s = 300$ meters and $s = 1200$ meters.

73. *Photography*   A photographer is taking a picture of a 3-foot painting hung in an art gallery. The camera lens is 1 foot below the lower edge of the painting (see figure). The angle β subtended by the camera lens x feet from the painting is given by

$$\beta = \arctan \frac{3x}{x^2 + 4}, \quad x > 0.$$

(a) Use a graphing utility to graph β as a function of x.

(b) Move the cursor along the graph to approximate the distance from the picture when β is maximum.

(c) Identify the asymptote of the graph and discuss its meaning in the context of the problem.

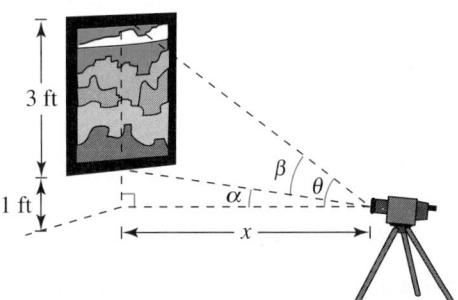

74. *Area*   In calculus, it is shown that the area of the region bounded by the graphs of $y = 0$, $y = 1/(x^2 + 1)$, $x = a$, and $x = b$ is given by

Area $= \arctan b - \arctan a$

(see figure). Find the areas for the following values of a and b.

(a) $a = 0, b = 1$ (b) $a = -1, b = 1$

(c) $a = 0, b = 3$ (d) $a = -1, b = 3$

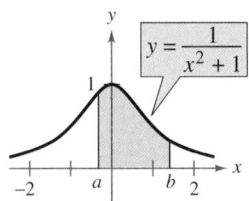

75. *Angle of Elevation* An airplane flies at an altitude of 5 miles toward a point directly over an observer. Consider θ and x as shown in the figure.

(a) Write θ as a function of x.

(b) Find θ when $x = 10$ miles and $x = 3$ miles.

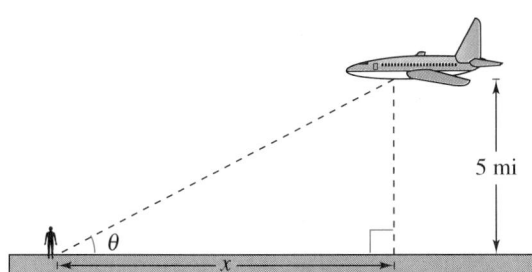

76. *Security Patrol* A security car with its spotlight on is parked 20 meters from a long warehouse. Consider θ and x as shown in the figure.

(a) Write θ as a function of x.

(b) Find θ when $x = 5$ meters and $x = 12$ meters.

77. Define the inverse cotangent function by restricting the domain of the cotangent function to the interval $(0, \pi)$, and sketch its graph.

78. Define the inverse secant function by restricting the domain of the secant function to the intervals $[0, \pi/2)$ and $(\pi/2, \pi]$, and sketch its graph.

79. Define the inverse cosecant function by restricting the domain of the cosecant function to the intervals $[-\pi/2, 0)$ and $(0, \pi/2]$, and sketch its graph.

80. Use the results of Exercises 77–79 to evaluate the following without using a calculator.

(a) $\operatorname{arcsec} \sqrt{2}$ (b) $\operatorname{arcsec} 1$

(c) $\operatorname{arccot}\left(-\sqrt{3}\right)$ (d) $\operatorname{arccsc} 2$

In Exercises 81–86, prove the identity.

81. $\arcsin(-x) = -\arcsin x$

82. $\arctan(-x) = -\arctan x$

83. $\arccos(-x) = \pi - \arccos x$

84. $\arctan x + \arctan \dfrac{1}{x} = \dfrac{\pi}{2}, \quad x > 0$

85. $\arcsin x + \arccos x = \dfrac{\pi}{2}$

86. $\arcsin x = \arctan \dfrac{x}{\sqrt{1 - x^2}}$

Review Solve Exercises 87–90 as a review of the skills and problem-solving techniques you learned in previous sections.

87. *Buy Now or Wait?* A sales representative indicates that if a customer waits another month for a new car that currently costs $23,500, the price will increase by 4%. However, the customer will pay an interest penalty of $725 for the early withdrawal of a certificate of deposit if the car is purchased now. Determine whether the customer should buy now or wait another month.

88. *Insurance Premium* The annual insurance premium for a policyholder is normally $739. However, after having an automobile accident, the policyholder was charged an additional 30%. What is the new annual premium?

89. *Partnership Costs* A group of people agree to share equally in the cost of a $250,000 endowment to a college. If they could find two more people to join the group, each person's share of the cost would decrease by $6250. How many people are presently in the group?

90. *Speed* A boat travels at a speed of 18 miles per hour in still water. It travels 35 miles upstream and then returns to the starting point in a total of 4 hours. Find the speed of the current.

4.8   Applications and Models

Applications Involving Right Triangles / *Trigonometry and Bearings* / *Harmonic Motion*

Note In this section, the three angles of a right triangle are denoted by the letters A, B, and C (where C is the right angle), and the lengths of the sides opposite these angles by the letters a, b, and c (where c is the hypotenuse).

Applications Involving Right Triangles

In keeping with our twofold perspective of trigonometry, this section includes both right triangle applications and applications that emphasize the periodic nature of the trigonometric functions.

Figure 4.70

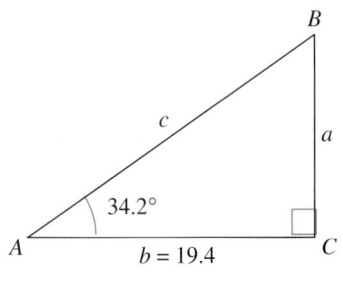

EXAMPLE 1   **Solving a Right Triangle**

Solve the right triangle shown in Figure 4.70.

Solution

Because $C = 90°$, it follows that $A + B = 90°$ and $B = 90° - 34.2° = 55.8°$. To solve for a, use the fact that

$$\tan A = \frac{\text{opp}}{\text{adj}} = \frac{a}{b} \qquad \Longrightarrow \qquad a = b \tan A.$$

Thus, $a = 19.4 \tan 34.2° \approx 13.18$. Similarly, to solve for c, use the fact that

$$\cos A = \frac{\text{adj}}{\text{hyp}} = \frac{b}{c} \qquad \Longrightarrow \qquad c = \frac{b}{\cos A}.$$

Thus, $c = \dfrac{19.4}{\cos 34.2°} \approx 23.46$.

Figure 4.71

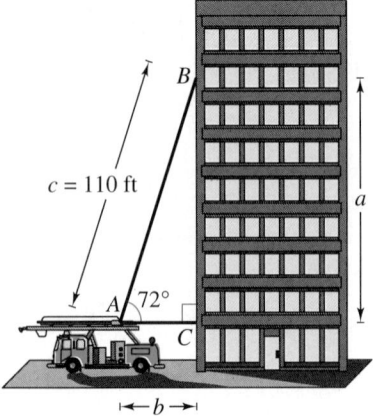

Real Life

EXAMPLE 2   **Finding a Side of a Right Triangle**

A safety regulation states that the maximum angle of elevation for a rescue ladder is 72°. If a fire department's longest ladder is 110 feet, what is the maximum safe rescue height?

Solution

A sketch is shown in Figure 4.71. From the equation $\sin A = a/c$, it follows that

$$a = c \sin A = 110 \sin 72° \approx 104.6.$$

Thus, the maximum safe rescue height is about 104.6 feet above the height of the fire truck.

Figure 4.72

Figure 4.73

In Example 2, the term **angle of elevation** denotes the angle from the horizontal upward to an object. For objects that lie below the horizontal, it is common to use the term **angle of depression,** as shown in Figure 4.72.

Real Life

EXAMPLE 3   **Finding a Side of a Right Triangle**

At a point 200 feet from the base of a building, the angle of elevation to the *bottom* of a smokestack is 35°, and the angle of elevation to the *top* is 53°, as shown in Figure 4.73. Find the height s of the smokestack alone.

Solution
Note from Figure 4.73 that this problem involves two right triangles. In the smaller right triangle, use the fact that $\tan 35° = a/200$ to conclude that the height of the building is

$$a = 200 \tan 35°.$$

Now, in the larger right triangle, use the equation

$$\tan 53° = \frac{a + s}{200}$$

to conclude that $a + s = 200 \tan 53°$. Hence, the height of the smokestack is

$$s = 200 \tan 53° - a$$
$$= 200 \tan 53° - 200 \tan 35°$$
$$\approx 125.4 \text{ feet.}$$

Real Life

EXAMPLE 4   **Finding an Acute Angle of a Right Triangle**

A swimming pool is 20 meters long and 12 meters wide. The bottom of the pool is slanted so that the water depth is 1.3 meters at the shallow end and 4 meters at the deep end, as shown in Figure 4.74. Find the angle of depression of the bottom of the pool.

Figure 4.74

Solution
Using the tangent function, you see that

$$\tan A = \frac{\text{opp}}{\text{adj}} = \frac{2.7}{20} = 0.135.$$

Thus, the angle of depression is given by

$$A = \arctan 0.135$$
$$\approx 0.13419 \text{ radians}$$
$$\approx 7.69°.$$

Figure 4.75

(a)

S 35° E

(b)

N 80° W

(c)

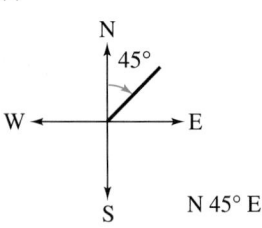

N 45° E

Note   The bearing of S 35° E in Figure 4.75(a) means 35 degrees east of south.

Trigonometry and Bearings

In surveying and navigation, directions are generally given in terms of **bearings.** A bearing measures the acute angle a path or line of sight makes with a fixed north-south line, as shown in Figure 4.75.

Real Life

EXAMPLE 5   **Finding Directions in Terms of Bearings**

A ship leaves port at noon and heads due west at 20 knots (nautical miles per hour). At 2 P.M. the ship changes course to N 54° W, as shown in Figure 4.76. Find the ship's bearing and distance from the port of departure at 3 P.M.

Figure 4.76

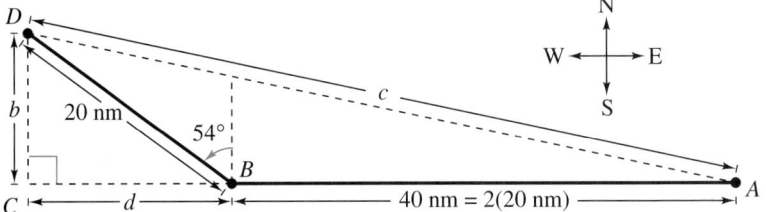

Solution
In triangle *BCD*, you have $B = 90° - 54° = 36°$. The two sides of this triangle can be determined to be

$$b = 20 \sin 36° \quad \text{and} \quad d = 20 \cos 36°.$$

In triangle *ACD*, you find angle *A* as follows.

$$\tan A = \frac{b}{d + 40} = \frac{20 \sin 36°}{20 \cos 36° + 40} \approx 0.2092494$$

$$A \approx \arctan 0.2092494 \approx 0.2062732 \text{ radians} \approx 11.82°$$

The angle with the north-south line is $90° - 11.82° = 78.18°$. Therefore, the bearing of the ship is

N 78.18° W.   Bearing

Finally, from triangle *ACD*, you have $\sin A = b/c$, which yields

$$c = \frac{b}{\sin A} = \frac{20 \sin 36°}{\sin 11.82°}$$

$$\approx 57.4 \text{ nautical miles.}$$ Distance from port

Harmonic Motion

The periodic nature of the trigonometric functions is useful for describing the motion of a point on an object that vibrates, oscillates, rotates, or is moved by wave motion.

For example, consider a ball that is bobbing up and down on the end of a spring, as shown in Figure 4.77. Suppose that 10 centimeters is the maximum distance the ball moves vertically upward or downward from its equilibrium (at-rest) position. Suppose further that the time it takes for the ball to move from its maximum displacement above zero to its maximum displacement below zero and back again is $t = 4$ seconds. Assuming the ideal conditions of perfect elasticity and no friction or air resistance, the ball would continue to move up and down in a uniform and regular manner.

From this spring you can conclude that the period (time for one complete cycle) of the motion is

Period = 4 seconds

and that its amplitude (maximum displacement from equilibrium) is

Amplitude = 10 centimeters.

Motion of this nature can be described by a sine or cosine function, and is called **simple harmonic motion.**

Figure 4.77

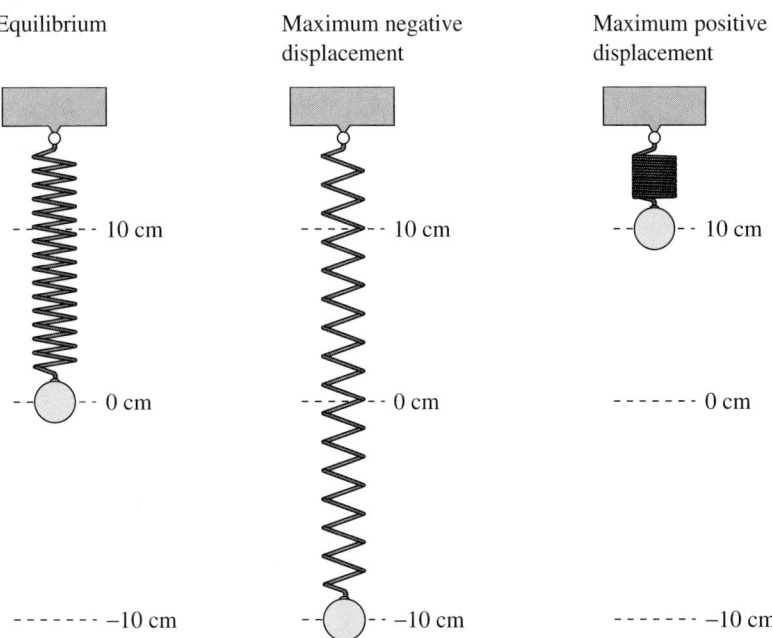

Definition of Simple Harmonic Motion

A point that moves on a coordinate line is said to be in **simple harmonic motion** if its distance d from the origin at time t is given by either

$$d = a \sin \omega t \qquad \text{or} \qquad d = a \cos \omega t$$

where a and ω are real numbers such that $\omega > 0$. The motion has **amplitude** $|a|$, **period** $2\pi/\omega$, and **frequency** $\omega/2\pi$.

Real Life

EXAMPLE 6   **Simple Harmonic Motion**

Write the equation for the simple harmonic motion of the ball illustrated in Figure 4.77, where the period is four seconds. What is the frequency of this motion?

Figure 4.78

Solution

Because the spring is at equilibrium ($d = 0$) when $t = 0$, you use the equation

$$d = a \sin \omega t.$$

Moreover, because the maximum displacement from zero is 10 and the period is 4, you have

$$\text{Amplitude} = |a| = 10$$

$$\text{Period} = \frac{2\pi}{\omega} = 4 \quad \Longrightarrow \quad \omega = \frac{\pi}{2}.$$

Consequently, the equation of motion is

$$d = 10 \sin \frac{\pi}{2} t.$$

Note that the choice of $a = 10$ or $a = -10$ depends on whether the ball initially moves up or down. The frequency is given by

$$\text{Frequency} = \frac{\omega}{2\pi} = \frac{\pi/2}{2\pi} = \frac{1}{4} \text{ cycle per second.}$$

Figure 4.79

One illustration of the relationship between sine waves and harmonic motion is the wave motion that results when a stone is dropped into a calm pool of water. The waves move outward in roughly the shape of sine (or cosine) waves, as shown in Figure 4.78. As an example, suppose you are fishing and your fishing bob is attached so that it does not move horizontally. As the waves move outward from the dropped stone, your fishing bob will move up and down in simple harmonic motion, as shown in Figure 4.79.

EXAMPLE 7 **Simple Harmonic Motion**

Given the equation for simple harmonic motion

$$d = 6 \cos \frac{3\pi}{4} t$$

find (a) the maximum displacement, (b) the frequency, (c) the value of d when $t = 4$, and (d) the least positive value of t for which $d = 0$.

Solution

The given equation has the form $d = a \cos \omega t$, with $a = 6$ and $\omega = 3\pi/4$.

a. The maximum displacement (from the point of equilibrium) is given by the amplitude. Thus, the maximum displacement is 6.

b. Frequency $= \dfrac{\omega}{2\pi} = \dfrac{3\pi/4}{2\pi} = \dfrac{3}{8}$ cycle per unit of time

c. $d = 6 \cos\left[\dfrac{3\pi}{4}(4)\right] = 6 \cos 3\pi = 6(-1) = -6$

d. To find the least positive value of t for which $d = 0$, solve the equation

$$d = 6 \cos \frac{3\pi}{4} t = 0$$

to obtain

$$\frac{3\pi}{4} t = \frac{\pi}{2}, \frac{3\pi}{2}, \frac{5\pi}{2}, \dots \qquad \Longrightarrow \qquad t = \frac{2}{3}, 2, \frac{10}{3}, \dots.$$

Thus, the least positive value of t is $t = \frac{2}{3}$.

Use the root or zero feature of a graphing utility to verify Example 7(d).

Group Activity   *Radio Waves*

Many different physical phenomena can be characterized by wave motion. These include electromagnetic waves such as radio waves, television waves, and microwaves. Radio waves transmit sound in two different ways. For an AM station, the *amplitude* of the wave is modified to carry sound. The letters AM stand for **amplitude modulation.** An FM radio signal has its *frequency* modified in order to carry sound, hence the term **frequency modulation.** Of the two graphs at the left, one shows an AM wave and the other shows an FM wave. Which is which? Explain your reasoning.

(a) **(b)**

4.8 /// EXERCISES

In Exercises 1–10, solve the right triangle shown in the figure. (Round to two decimal places.)

1. $A = 20°$, $b = 10$
2. $B = 54°$, $c = 15$
3. $B = 71°$, $b = 24$
4. $A = 8.4°$, $a = 40.5$
5. $a = 6$, $b = 10$
6. $a = 25$, $c = 35$
7. $b = 16$, $c = 52$
8. $b = 1.32$, $c = 9.45$
9. $A = 12° 15'$, $c = 430.5$
10. $B = 65° 12'$, $a = 14.2$

Figure for 1–10

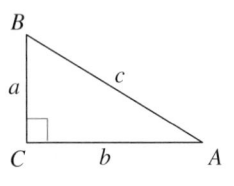

Figure for 11 and 12

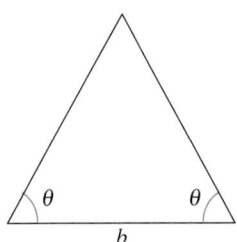

In Exercises 11 and 12, find the altitude of the isosceles triangle shown in the figure. (Round to two decimal places.)

11. $\theta = 52°$, $b = 4$ inches
12. $\theta = 18°$, $b = 10$ meters

13. *Length of a Shadow*   A shadow of length L is created by a 60-foot silo when the sun is $\theta°$ above the horizon.

    (a) Write L as a function of θ.

    (b) Use a graphing utility to complete the table.

θ	10°	20°	30°	40°	50°
L					

    (c) The angle measure increases in equal increments in the table. Does the length of the shadow change in equal increments? Explain.

14. *Length of a Shadow*   A shadow of length L is created by a 600-foot building when the sun is $\theta°$ above the horizon.

    (a) Write L as a function of θ.

    (b) Use a graphing utility to complete the table.

θ	10°	20°	30°	40°	50°
L					

    (c) The angle measure increases in equal increments in the table. Does the length of the shadow change in equal increments? Explain.

15. *Height*   A ladder of length 20 feet leans against the side of a house. The angle of elevation of the ladder is $\theta°$.

    (a) Write the height h of the ladder as a function of θ.

    (b) Use a graphing utility to complete the table.

θ	60°	65°	70°	75°	80°
h					

16. *Height*   The length of a shadow of a tree is 110 feet when the angle of elevation of the sun is $\theta°$.

    (a) Write the height h of the tree as a function of θ.

    (b) Use a graphing utility to complete the table.

θ	10°	15°	20°	25°	30°
h					

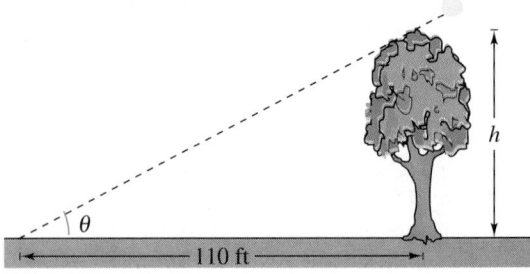

17. *Height* From a point 50 feet in front of a church, the angles of elevation to the base of the steeple and the top of the steeple are 35° and 47° 40′, respectively.

(a) Draw right triangles that give a visual representation of the problem. Label the known quantities and the unknown height of the steeple.

(b) Use a trigonometric function to write an equation involving the unknown.

(c) Find the height of the steeple.

18. *Height* From a point 100 feet in front of a public library, the angles of elevation to the base of the flagpole and the top of the flagpole are 28° and 39° 45′. The flagpole is mounted on the front of the library's roof (see figure). Find the height of the pole.

19. *Depth of a Submarine* The sonar of a navy cruiser detects a submarine that is 4000 feet from the cruiser. The angle between the water level and the submarine is 34° (see figure). How deep is the submarine?

20. *Height of a Kite* A 100-foot line is attached to a kite. When the kite has pulled the line taut, the angle of elevation to the kite is approximately 50°. Approximate the height of the kite.

21. *Angle of Elevation* An amateur radio operator erects a 75-foot vertical tower for an antenna. Find the angle of elevation to the top of the tower at a point at level ground 50 feet from its base.

22. *Angle of Elevation* The height of an outdoor basketball backboard is $12\frac{1}{2}$ feet, and the backboard casts a shadow $17\frac{1}{3}$ feet long.

(a) Draw a right triangle that gives a visual representation of the problem. Label the known and unknown quantities.

(b) Use a trigonometric function to write an equation involving the unknown.

(c) Find the angle of elevation of the sun.

23. *Angle of Depression* A spacecraft is traveling in a circular orbit 150 miles above the surface of the earth (see figure). Find the angle of depression from the spacecraft to the horizon. Assume that the radius of the earth is 4000 miles.

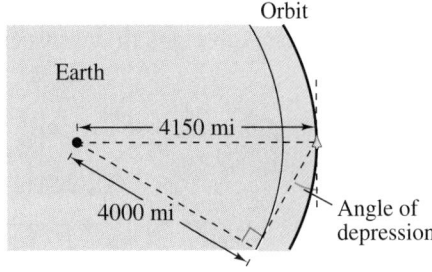

24. *Angle of Depression* Find the angle of depression from the top of a lighthouse 250 feet above water level to the water line of a ship 2 miles offshore.

25. *Airplane Ascent* When an airplane leaves the runway, its angle of climb is 18° and its speed is 275 feet per second. Find the plane's altitude after 1 minute.

26. *Airplane Ascent* How long will it take the plane in Exercise 25 to climb to an altitude of 10,000 feet?

27. *Mountain Descent* A sign on the roadway at the top of a mountain indicates that for the next 4 miles the grade is 10.5° (see figure). Find the change in elevation for a car descending the mountain.

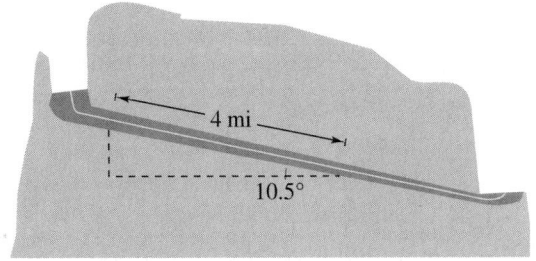

28. *Mountain Descent*   A road sign at the top of a mountain indicates that for the next 4 miles the grade is 12%. Find the angle of the grade and the change in elevation for a car descending the mountain.

29. *Navigation*   An airplane flying at 550 miles per hour has a bearing of N 52° E. After flying 1.5 hours, how far north and how far east has the plane traveled from its point of departure?

30. *Navigation*   A ship leaves port at noon and has a bearing of S 27° W. If the ship sails at 20 knots, how many nautical miles south and how many nautical miles west will the ship have traveled by 6:00 P.M.?

31. *Surveying*   A surveyor wishes to find the distance across a swamp (see figure). The bearing from *A* to *B* is N 32° W. The surveyor walks 50 meters from *A*, and at the point *C* the bearing to *B* is N 68° W. Find (a) the bearing from *A* to *C* and (b) the distance from *A* to *B*.

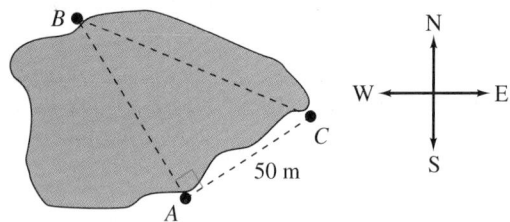

32. *Location of a Fire*   Two fire towers are 30 kilometers apart, tower *A* being due west of tower *B*. A fire is spotted from the towers, and the bearings from *A* and *B* are E 14° N and W 34° N, respectively (see figure). Find the distance *d* of the fire from the line segment *AB*.

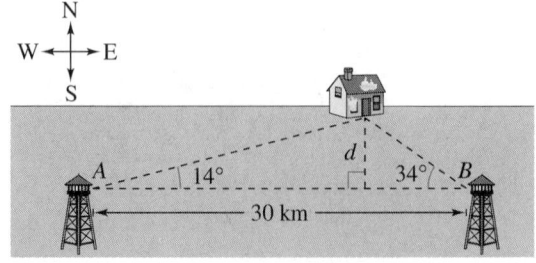

33. *Navigation*   A ship is 45 miles east and 30 miles south of port. If the captain wants to sail directly to port, what bearing should be taken?

34. *Navigation*   A plane is 120 miles north and 85 miles east of an airport. If the pilot wants to fly directly to the airport, what bearing should be taken?

35. *Distance Between Ships*   An observer in a lighthouse 350 feet above sea level observes two ships directly offshore. The angles of depression to the ships are 4° and 6.5° (see figure). How far apart are the ships?

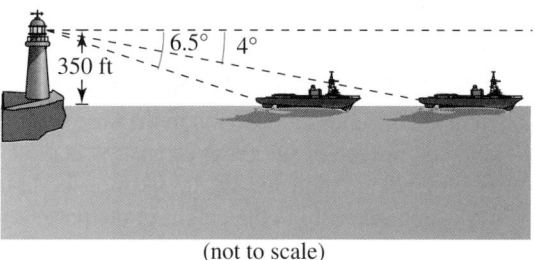

(not to scale)

36. *Distance Between Towns*   A passenger in an airplane flying at an altitude of 10 kilometers sees two towns directly to the left of the plane. The angles of depression to the towns are 28° and 55° (see figure). How far apart are the towns?

37. *Altitude of a Plane*   A plane is observed approaching your home and you assume its speed is 550 miles per hour. If the angle of elevation of the plane is 16° at one time and 57° 1 minute later, approximate the altitude of the plane.

38. *Height of a Mountain*   While traveling across flat land, you notice a mountain directly in front of you. The angle of elevation to the peak is 3.5°. After you drive 13 miles closer to the mountain, the angle of elevation is 9°. Approximate the height of the mountain.

Geometry In Exercises 39 and 40, find the angle α between two nonvertical lines L_1 and L_2. The angle α satisfies the equation

$$\tan \alpha = \left| \frac{m_2 - m_1}{1 + m_2 m_1} \right|$$

where m_1 and m_2 are the slopes of L_1 and L_2, respectively. (Assume $m_1 m_2 \neq -1$.)

39. $3x - 2y = 5$
   $x + y = 1$

40. $2x + y = 8$
   $x - 5y = -4$

41. *Geometry* Determine the angle between the diagonal of a cube and the diagonal of its base, as shown in the figure.

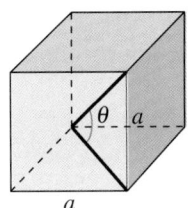

42. *Geometry* Determine the angle between the diagonal of a cube and its edge, as shown in the figure.

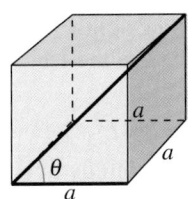

43. *Wrench Size* Express the distance y across the flat sides of a hexagonal nut as a function of r, as shown in the figure.

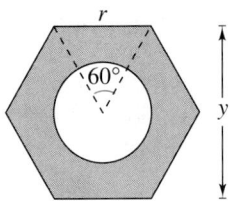

44. *Bolt Circle* The figure shows a circular sheet of diameter 40 centimeters, containing 12 equally spaced bolt holes. Determine the straight-line distance between the centers of the bolt holes.

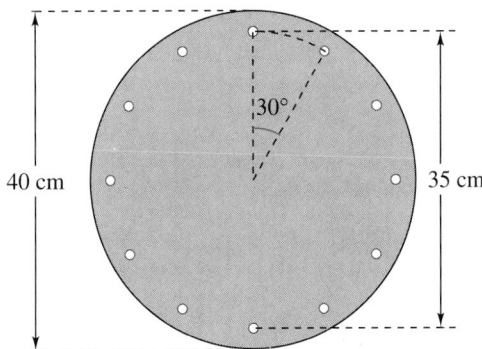

45. *Geometry* A regular pentagon is inscribed in a circle of radius 25 inches. Find the length of the sides of the pentagon.

46. *Geometry* A regular hexagon is inscribed in a circle of radius 25 inches. Find the length of the sides of the hexagon.

Trusses In Exercises 47 and 48, find the lengths of all the unknown members of the truss.

47.

48.

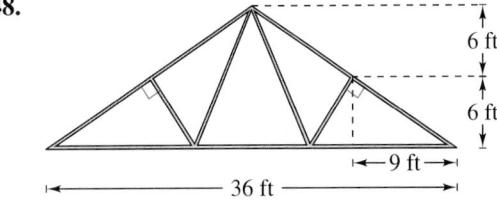

Harmonic Motion   In Exercises 49–52, for the simple harmonic motion described by the trigonometric function, find (a) the maximum displacement, (b) the frequency, and (c) the least positive value of t for which $d = 0$.

49. $d = 4 \cos 8\pi t$ **50.** $d = \frac{1}{2} \cos 20\pi t$

51. $d = \frac{1}{16} \sin 120\pi t$ **52.** $d = \frac{1}{64} \sin 792\pi t$

Harmonic Motion   In Exercises 53–56, find a model for simple harmonic motion satisfying the specified conditions.

	Displacement ($t = 0$)	*Amplitude*	*Period*
53.	0	4 cm	2 sec
54.	0	3 m	6 sec
55.	3 in.	3 in.	1.5 sec
56.	2 ft	2 ft	10 sec

57. *Tuning Fork*   A point on the end of a tuning fork moves in simple harmonic motion described by $d = a \sin \omega t$. Find ω given that the tuning fork for middle C has a frequency of 264 vibrations per second.

58. *Wave Motion*   A buoy oscillates in simple harmonic motion as waves go past. At a given time it is noted that the buoy moves a total of 3.5 feet from its low point to its high point, and that it returns to its high point every 10 seconds (see figure). Write an equation that describes the motion of the buoy if, at $t = 0$, it is at its high point.

59. *Springs*   A weight stretches a spring 1.5 inches. The weight is pushed 3 inches above the point of equilibrium and released. Its motion is modeled by

$$y = \frac{1}{4} \cos 16t, \quad t > 0$$

where y is in feet and t is in seconds.

(a) Use a graphing utility to graph the function.

(b) What is the period of the oscillations?

(c) Determine the first time the weight passes the point of equilibrium ($y = 0$).

60. *Numerical and Graphical Analysis*   A 2-meter-high fence is 3 meters from the side of a grain storage bin. A grain elevator must reach from ground level outside the fence to the storage bin (see figure). The objective is to determine the shortest elevator meeting the constraints.

(a) Complete four rows of the table.

θ	L_1	L_2	$L_1 + L_2$
0.1	$\dfrac{2}{\sin 0.1}$	$\dfrac{3}{\cos 0.1}$	23.0
0.2	$\dfrac{2}{\sin 0.2}$	$\dfrac{3}{\cos 0.2}$	13.1

(b) Use a graphing utility to generate additional rows of the table. Use the table to estimate the minimum length.

(c) Write the length L as a function of θ.

(d) Use a graphing utility to graph the function. Use the graph to estimate the minimum length. How does your estimate compare with that of part (b)?

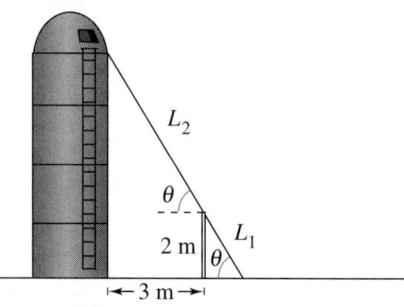

61. *Data Analysis*   The times S of sunset (Greenwich Mean Time) at 40° north latitude on the 15th of each month are: 1(16:59), 2(17:35), 3(18:06), 4(18:38), 5(19:08), 6(19:30), 7(19:28), 8(18:57), 9(18:09), 10(17:21), 11(16:44), 12(16:36). The month is represented by t, with $t = 1$ corresponding to January. A model (where minutes have been converted to the decimal part of an hour) for this data is

$$S(t) = 18.09 + 1.41 \sin\left(\frac{\pi t}{6} + 4.60\right).$$

(a) Use a graphing utility to graph the data points and the model in the same viewing rectangle.

(b) What is the period of the model? Is it what you expected? Explain.

(c) What is the amplitude of the function? What does it represent in the model? Explain.

62. *Numerical and Graphical Analysis*   The cross sections of an irrigation canal are isosceles trapezoids where the length of three of the sides is 8 feet (see figure). The objective is to find the angle θ that maximizes the area of the cross sections.

(a) Complete six rows of the table.

Base 1	Base 2	Altitude	Area
8	$8 + 16 \cos 10°$	$8 \sin 10°$	22.1
8	$8 + 16 \cos 20°$	$8 \sin 20°$	42.5

(b) Use a graphing utility to generate additional rows of the table. Use the table to estimate the maximum cross-sectional area.

(c) Write the area A as a function of θ.

(d) Use a graphing utility to graph the function. Use the graph to estimate the maximum cross-sectional area. How does your estimate compare with that of part (b)?

63. *Data Analysis*   The table gives the average sales S (in millions) of an outerwear manufacturer for each month t, where $t = 1$ represents January.

t	1	2	3	4	5	6
S	13.46	11.15	8.00	4.85	2.54	1.70
t	7	8	9	10	11	12
S	2.54	4.85	8.00	11.15	13.46	14.30

(a) Create a scatter plot of the data.

(b) Find a trigonometric model that fits the data. Graph the model on your scatter plot. How well does the model fit?

(c) What is the period of the model? Do you think it is reasonable given the context? Explain your reasoning.

(d) Interpret the meaning of the model's amplitude in the context of the problem.

64. *Numerical and Graphical Analysis*   Consider the shaded region outside the sector of the circle of radius 10 meters and inside the right triangle (see figure).

(a) Write the area A of the region as a function of θ. Determine the domain of the function.

(b) Use a graphing utility to complete the table.

θ	0	0.3	0.6	0.9	1.2	1.5
A						

(c) Use a graphing utility to graph the function over the appropriate domain.

(d) What does the area approach as θ approaches $\pi/2$?

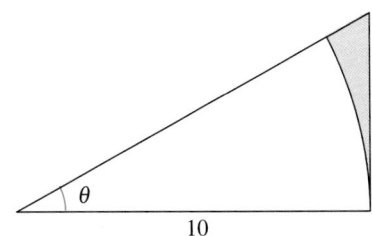

Focus on Concepts

In this chapter, you studied trigonometry. You can use the following questions to check your understanding of several of these basic concepts. The answers to these questions are given in the back of the book.

1. In your own words, explain the meaning of (a) an angle in standard position, (b) a negative angle, (c) a coterminal angle, and (d) an obtuse angle.

2. A fan motor turns at a given angular speed. How does the speed of the tips of the blades change if a fan of greater diameter is installed on the motor? Explain.

3. *True or False?* $y = \sin \theta$ is not a function because $\sin 30° = \sin 150°$. Explain.

4. Describe the behavior of $f(\theta) = \sec \theta$ at the zeros of $g(\theta) = \cos \theta$. Explain.

5. Explain how reference angles are used to find the trigonometric functions of obtuse angles.

In Exercises 6–9, match the function $y = a \sin bx$ with its graph. Base your selection solely on your interpretation of the constants a and b. Explain your reasoning. [The graphs are labeled (a), (b), (c), and (d).]

(a)

(b)

(c)

(d)

6. $y = 3 \sin x$

7. $y = -3 \sin x$

8. $y = 2 \sin \pi x$

9. $y = 2 \sin \dfrac{x}{2}$

10. The function f is periodic with period c. Therefore, $f(t + c) = f(t)$. Are the following equal? Explain.
   (a) $f(t - 2c) \stackrel{?}{=} f(t)$ (b) $f\left(t + \tfrac{1}{2}c\right) \stackrel{?}{=} f\left(\tfrac{1}{2}t\right)$
   (c) $f\left(\tfrac{1}{2}[t + c]\right) \stackrel{?}{=} f\left(\tfrac{1}{2}t\right)$

11. When the sine and cosine functions are graphed, determining the amplitude is part of the analysis. Why is this not true for the other four trigonometric functions?

12. A weight is suspended from a ceiling by a steel spring. The weight is lifted (positive direction) from the equilibrium position and released. The resulting motion of the weight is modeled by

$$y = Ae^{-kt} \cos bt = \tfrac{1}{5}e^{-t/10} \cos 6t, \quad t \geq 0$$

where y is the distance in feet from equilibrium and t is the time in seconds. The graph of the function is given in the figure. In each of the following, describe the change in the system without graphing the resulting function.

   (a) A is changed from $\tfrac{1}{5}$ to $\tfrac{1}{3}$.
   (b) k is changed from $\tfrac{1}{10}$ to $\tfrac{1}{3}$.
   (c) b is changed from 6 to 9.

13. *True or False?*   Explain your reasoning.

$$\tan \frac{3\pi}{4} = -1 \quad\Longrightarrow\quad \arctan(-1) = \frac{3\pi}{4}$$

4 /// REVIEW EXERCISES

In Exercises 1–4, sketch the angle in standard position. List one positive and one negative coterminal angle.

1. $\dfrac{11\pi}{4}$

2. $\dfrac{2\pi}{9}$

3. $-110°$

4. $-405°$

In Exercises 5–8, convert the measure to decimal degree form. Round to two decimal places.

5. $135°\,16'\,45''$

6. $-234°\,50''$

7. $5°\,22'\,53''$

8. $280°\,8'\,50''$

In Exercises 9–12, convert the measure to $D°\,M'\,S''$ form.

9. $135.27°$

10. $25.1°$

11. $-85.15°$

12. $-327.85°$

In Exercises 13–16, convert the measure from radians to degrees. Round to two decimal places.

13. $\dfrac{5\pi}{7}$

14. $-\dfrac{3\pi}{5}$

15. -3.5

16. 1.75

In Exercises 17–20, convert the measure from degrees to radians. Round to four decimal places.

17. $480°$

18. $-16.5°$

19. $-33°\,45'$

20. $84°\,15'$

In Exercises 21–24, find the reference angle for the given angle.

21. $252°$

22. $640°$

23. $-\dfrac{6\pi}{5}$

24. $\dfrac{17\pi}{3}$

In Exercises 25–28, find the unknown in the figure.

25.

26.

27.

28.
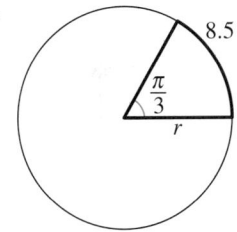

In Exercises 29–32, find the exact values of the six trigonometric functions of the angle θ given in the figure.

29.

30.

31.

32.
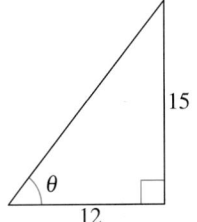

In Exercises 33–38, find the six trigonometric functions of the angle θ (in standard position) whose terminal side passes through the given point.

33. $(12, 16)$ **34.** $(x, 4x), \quad x > 0$

35. $(-7, 2)$ **36.** $(4, -8)$

37. $(-4, -6)$ **38.** $\left(\frac{2}{3}, \frac{5}{2}\right)$

In Exercises 39–42, find the remaining five trigonometric functions of θ satisfying the given conditions.

39. $\sec \theta = \frac{6}{5}, \quad \tan \theta < 0$

40. $\tan \theta = -\frac{12}{5}, \quad \sin \theta > 0$

41. $\sin \theta = \frac{3}{8}, \quad \cos \theta < 0$

42. $\cos \theta = -\frac{2}{5}, \quad \sin \theta > 0$

In Exercises 43–48, evaluate the trigonometric function without using a calculator.

43. $\tan \dfrac{\pi}{3}$ **44.** $\sec \dfrac{\pi}{4}$

45. $\sin \dfrac{5\pi}{3}$ **46.** $\cot\left(-\dfrac{5\pi}{6}\right)$

47. $\cos 495°$ **48.** $\csc 270°$

In Exercises 49–52, use a calculator to evaluate the trigonometric function. Round to two decimal places.

49. $\tan 33°$ **50.** $\csc 105°$

51. $\sec \dfrac{12\pi}{5}$ **52.** $\sin\left(-\dfrac{\pi}{9}\right)$

In Exercises 53–56, find two values of θ in degrees $(0° \le \theta < 360°)$ and in radians $(0 \le \theta < 2\pi)$.

53. $\cos \theta = -\dfrac{\sqrt{2}}{2}$ **54.** $\sec \theta$ is undefined.

55. $\csc \theta = -2$ **56.** $\tan \theta = \dfrac{\sqrt{3}}{3}$

In Exercises 57–60, use a calculator to find two values of θ. Express both values in degrees $(0° \le \theta < 360°)$ and in radians $(0 \le \theta < 2\pi)$.

57. $\sin \theta = 0.8387$ **58.** $\cot \theta = -1.5399$

59. $\sec \theta = -1.0353$ **60.** $\csc \theta = 11.4737$

In Exercises 61–64, find the period and amplitude.

61. $y = 5 \cos \pi x$

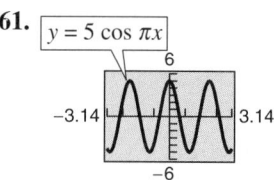

62. $y = -\dfrac{3}{2} \sin \dfrac{x}{2}$

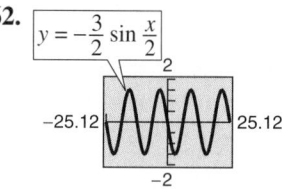

63. $y = -3.4 \sin 2x$

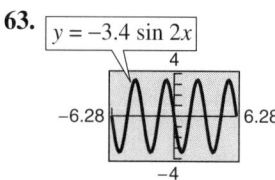

64. $y = 4 \cos \dfrac{\pi x}{2}$

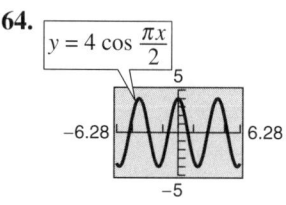

In Exercises 65–78, sketch a graph of the function.

65. $y = 3 \cos 2\pi x$ **66.** $y = -2 \sin \pi x$

67. $f(x) = 5 \sin \dfrac{2x}{5}$ **68.** $f(x) = 8 \cos\left(-\dfrac{x}{4}\right)$

69. $f(x) = -\dfrac{1}{4} \cos \dfrac{\pi x}{4}$ **70.** $f(x) = -\tan \dfrac{\pi x}{4}$

71. $g(t) = \dfrac{5}{2} \sin(t - \pi)$ **72.** $g(t) = 3 \cos(t + \pi)$

73. $h(t) = \tan\left(t - \dfrac{\pi}{4}\right)$ **74.** $h(t) = \sec\left(t - \dfrac{\pi}{4}\right)$

75. $f(t) = \csc\left(3t - \dfrac{\pi}{2}\right)$

76. $f(t) = 3 \csc\left(2t + \dfrac{\pi}{4}\right)$

77. $y = \arcsin \dfrac{x}{2}$

78. $y = 2 \arccos x$

In Exercises 79–88, use a graphing utility to graph the function. If the function is periodic, find the period.

79. $f(x) = \dfrac{x}{4} - \sin x$

80. $y = \dfrac{x}{3} + \cos \pi x$

81. $f(x) = \dfrac{\pi}{2} + \arctan x$

82. $y = 4 - \dfrac{x}{4} + \cos \pi x$

83. $h(\theta) = \theta \sin \pi\theta$

84. $f(\theta) = \cot \dfrac{\pi\theta}{8}$

85. $f(t) = 2.5e^{-t/4} \sin 2\pi t$

86. $f(x) = \arccos(x - \pi)$

87. $g(x) = 3 \sin\left(\dfrac{\pi x}{3} + 1\right)$

88. $E(t) = 110 \cos\left(120\pi t - \dfrac{\pi}{3}\right)$

In Exercises 89–92, use a graphing utility to graph the function. Use the graph to determine if the function is periodic. If the function is periodic, approximate any relative maximum or minimum points through one period.

89. $f(x) = e^{\sin x}$

90. $g(x) = \sin e^x$

91. $g(x) = 2 \sin x \cos^2 x$

92. $h(x) = 4 \sin^2 x \cos^2 x$

In Exercises 93–96, find a, b, and c for $f(x) = a \cos(bx - c)$ so that the graph of f matches the figure.

93.

94.

95.

96.

In Exercises 97 and 98, find a and b for $f(x) = a \tan bx$ so that the graph of f matches the figure.

97.

98.

In Exercises 99–102, write an algebraic expression for the given expression.

99. $\sec[\arcsin(x - 1)]$

100. $\tan\left(\arccos \dfrac{x}{2}\right)$

101. $\sin\left(\arccos \dfrac{x^2}{4 - x^2}\right)$

102. $\csc(\arcsin 10x)$

103. *Altitude of Elevation* Find the altitude of the triangle in the figure.

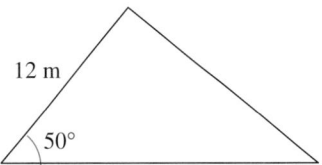

104. *Angle of Elevation* The height of a radio transmission tower is 70 meters, and its casts a shadow of length 30 meters (see figure). Find the angle of elevation of the sun.

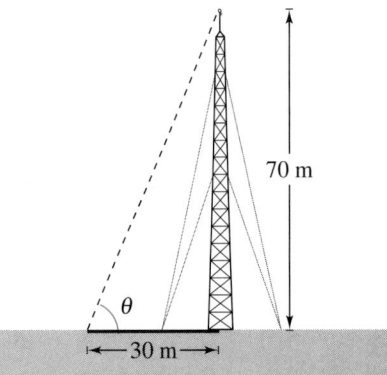

105. *Shuttle Height* An observer 2.5 miles from the launch pad of a space shuttle launch measures the angle of elevation to the base of the shuttle to be 25° soon after lift-off (see figure). How high is the shuttle at that instant? (Assume the shuttle is still moving vertically.)

h

25° Observer

2.5 mi

106. *Distance* From city A to city B, a plane flies 650 miles at a bearing of N 48° E. From city B to city C, the plane flies 810 miles at a bearing of S 65° E. Find the distance from A to C and the bearing from A to C.

107. *Railroad Grade* A train travels 3.5 kilometers on a straight track with a grade of 1° 10′ (see figure). What is the vertical rise of the train in that distance?

3.5 km

1°10′

108. *Distance Between Towns* A passenger in an airplane flying at an altitude of 37,000 feet sees two towns directly to the left of the airplane. The angles of depression to the towns are 32° and 76° (see figure). How far apart are the towns?

32°

76°

37,000 ft

Town 1   Town 2

109. *Exploration* The base of a triangle is also the radius of a circular arc (see figure).

(a) Find the area A of the shaded region as a function of θ for $0 < \theta < \dfrac{\pi}{2}$.

(b) Use a graphing utility to graph the area function over the given domain. Interpret the graph in the context of the problem.

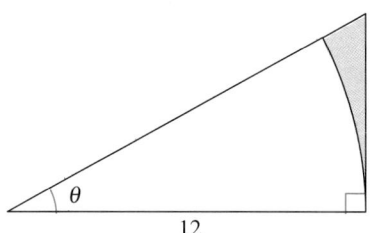

θ

12

110. *Exploration* In calculus it can be shown that the arcsine and arctangent functions can be approximated by the polynomials

$$\arcsin x \approx x + \frac{x^3}{6} + \frac{3x^5}{40} + \frac{5x^7}{112}$$

$$\arctan x \approx x - \frac{x^3}{3} + \frac{x^5}{5} - \frac{x^7}{7} \text{ where } x \text{ is in radians.}$$

(a) Use a graphing utility to graph the arcsine function and its polynomial approximation in the same viewing rectangle. How do the graphs compare?

(b) Use a graphing utility to graph the arctangent function and its polynomial approximation in the same viewing rectangle. How do the graphs compare?

(c) Study the pattern in the polynomial approximation of the arctangent function and guess the next term. Then repeat part (b). How did the accuracy of the approximation change when additional terms were added?

CHAPTER PROJECT   *Fitting a Model to Data*

In this project, you will find and use models relating to the carbon dioxide level of earth's atmosphere.

Since 1958, the Mauna Loa Climate Observatory in Hawaii has been collecting data on the carbon dioxide level of earth's atmosphere. The table shows the average monthly readings for January of each year from 1959 through 1994. The readings measure the carbon dioxide concentration in parts per million.

1959	1960	1961	1962	1963	1964	1965	1966	1967	1968	1969	1970
315.4	316.3	316.7	317.8	318.6	319.4	319.3	320.5	322.2	322.4	323.8	324.9

1971	1972	1973	1974	1975	1976	1977	1978	1979	1980	1981	1982
326.0	326.6	328.4	329.2	330.2	331.6	332.8	334.8	336.1	337.8	339.1	340.6

1983	1984	1985	1986	1987	1988	1989	1990	1991	1992	1993	1994
341.2	343.5	344.8	346.1	347.8	350.3	352.6	353.5	354.6	355.9	356.6	358.3

(a) Enter the data in the table in a graphing utility. (Let $t = 0$ represent 1960.) Draw a scatter plot for the data. Does the data appear to be best modeled with a linear, quadratic, or exponential model?

(b) Find the linear, quadratic, or exponential model that you think best fits the data.

Questions for Further Exploration

1. The data in the table represents the carbon dioxide levels for January of each year. Throughout each year, the level oscillated as follows.

   • In April, the average reading was about 2.5 parts per million higher than the average reading given by the model in part (b) above.

   • In July, the average reading was the same as the average reading given by the model in part (b) above.

   • In October, the average reading was about 2.5 parts per million lower than the average reading given by the model in part (b) above.

   Use a sine function to rewrite the model found in part (b) above so that the model incorporates the described oscillations.

2. Use a graphing utility to sketch the graph of the revised model.

3. Make a careful sketch of the model for 1 year. What physical factors on earth would contribute to the oscillation in the carbon dioxide level during the year?

4. Is the model you found periodic? Explain your reasoning.

5. Use the model to predict the level of carbon dioxide in earth's atmosphere in the following years.
   (a) 2000
   (b) 2010
   (c) 2020

4 /// CHAPTER TEST

Take this test as you would take a test in class. After you are done, check your work against the answers given in the back of the book.

The *Interactive* CD-ROM provides answers to the Chapter Tests and Cumulative Tests. It also offers Chapter Pre-Tests (that test key skills and concepts covered in previous chapters) and Chapter Post-Tests, both of which have randomly generated exercises with diagnostic capabilities.

1. Consider the angle of magnitude $5\pi/4$ radians.

    (a) Sketch the angle in standard position.

    (b) Determine two coterminal angles (one positive and one negative).

    (c) Convert the angle to degree measure.

2. A truck is moving at a rate of 90 kilometers per hour, and the diameter of its wheels is 1 meter. Find the angular speed of the wheels in radians per minute.

3. Find the exact values of the six trigonometric functions of the angle θ shown in the figure.

4. Given that $\tan \theta = \frac{3}{2}$, find the other five trigonometric functions of θ.

5. Determine the reference angle θ' of the angle $\theta = 290°$ and sketch θ and θ' in standard position.

6. Determine the quadrant in which θ lies if $\sec \theta < 0$ and $\tan \theta > 0$.

7. Find two values of θ in degrees $(0 \le \theta < 360°)$ if $\cos \theta = -\sqrt{3}/2$. (Do not use a calculator.)

8. Use a calculator to approximate two values of θ in radians $(0 \le \theta < 2\pi)$ if $\csc \theta = 1.030$. Round the result to two decimal places.

Figure for 3

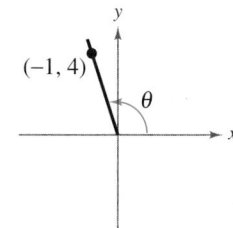

In Exercises 9 and 10, graph the function through two full periods without the aid of a graphing utility.

9. $g(x) = -2 \sin\left(x - \frac{\pi}{4}\right)$ 10. $f(\alpha) = \frac{1}{2} \tan 2\alpha$

In Exercises 11 and 12, use a graphing utility to graph the function. If the function is periodic, find its period.

11. $y = \sin 2\pi x + 2 \cos \pi x$ 12. $y = 6e^{-0.12t} \cos(0.25t), \quad 0 \le t \le 32$

13. Find a, b, and c for the function $f(x) = a \sin(bx + c)$ so that the graph of f matches the figure.

14. Find the exact value of $\tan\left(\arccos \frac{2}{3}\right)$ without the aid of a calculator.

15. Graph the function $f(x) = 2 \arcsin\left(\frac{1}{2}x\right)$.

16. A ship leaves port at noon and sails at a speed of 18 knots. Its bearing is N 16° W. If the port is positioned at the origin, determine the coordinates of the position of the ship at 3 P.M.

Figure for 13

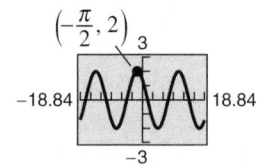

Analytic Trigonometry

The number of hours of daylight that occur at any location on earth depends on two things: the time of year and the latitude of the location.

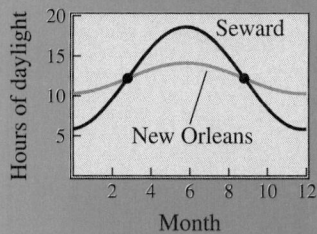

Hours of daylight

Month

Here are models for the numbers of hours of daylight in Seward, Alaska (60 degrees latitude) and New Orleans, Louisiana (30 degrees latitude).

Seward $D = 12.2 - 6.4 \cos [\pi(t + 0.2)/6]$

New Orleans $D = 12.2 - 1.9 \cos [\pi(t + 0.2)/6]$

In these models, D represents the number of hours of daylight and t represents the month, with $t = 0$ representing January 1.

To find the time of year that both cities receive the same amount of daylight, you can use a graphing utility to equate the two models and solve for t. By using the zoom and trace features, or the solve feature, you obtain $t = 2.8$ (spring equinox) and $t = 8.8$ (fall equinox). (See Exercises 93 and 94 on page 415.)

National Center for Atmospheric Research/
National Science Foundation (photo and inset).

Since 1987, Dr. Warren Washington has directed the Climate Global Dynamics Division of NCAR (National Center for Atmospheric Research) in Boulder, Colorado. He helped develop innovative computer models to predict long-term weather pattern.

407

5.1   Using Fundamental Identities

Introduction / Using the Fundamental Identities

Introduction

In Chapter 4, you studied the basic definitions, properties, graphs, and applications of the individual trigonometric functions. In this chapter, you will learn how to use the fundamental identities to

1. evaluate trigonometric functions;
2. simplify trigonometric expressions;
3. develop additional trigonometric identities;
4. solve trigonometric equations.

Fundamental Trigonometric Identities

Reciprocal Identities

$$\sin u = \frac{1}{\csc u} \qquad \cos u = \frac{1}{\sec u} \qquad \tan u = \frac{1}{\cot u}$$

$$\csc u = \frac{1}{\sin u} \qquad \sec u = \frac{1}{\cos u} \qquad \cot u = \frac{1}{\tan u}$$

Quotient Identities

$$\tan u = \frac{\sin u}{\cos u} \qquad \cot u = \frac{\cos u}{\sin u}$$

Pythagorean Identities

$$\sin^2 u + \cos^2 u = 1 \qquad 1 + \tan^2 u = \sec^2 u \qquad 1 + \cot^2 u = \csc^2 u$$

Cofunction Identities

$$\sin\left(\frac{\pi}{2} - u\right) = \cos u \qquad \cos\left(\frac{\pi}{2} - u\right) = \sin u$$

$$\tan\left(\frac{\pi}{2} - u\right) = \cot u \qquad \cot\left(\frac{\pi}{2} - u\right) = \tan u$$

$$\sec\left(\frac{\pi}{2} - u\right) = \csc u \qquad \csc\left(\frac{\pi}{2} - u\right) = \sec u$$

Even/Odd Identities

$$\sin(-u) = -\sin u \qquad \sec(-u) = \sec u \qquad \tan(-u) = -\tan u$$

$$\csc(-u) = -\csc u \qquad \cos(-u) = \cos u \qquad \cot(-u) = -\cot u$$

Note   Pythagorean identities are sometimes used in radical form such as

$$\sin u = \pm\sqrt{1 - \cos^2 u}$$

or

$$\tan u = \pm\sqrt{\sec^2 u - 1}$$

where the sign depends on the choice of u.

Using the Fundamental Identities

One common use of trigonometric identities is to use given values of trigonometric functions to evaluate other trigonometric functions.

The *Interactive* CD-ROM shows every example with its solution; clicking on the *Try It!* button brings up similar problems. Guided Examples and Integrated Examples show step-by-step solutions to additional examples. Integrated Examples are related to several concepts in the section.

You can use a graphing utility to check the result of Example 2. To do this, graph

$$y = \sin x \cos^2 x - \sin x$$

and

$$y = -\sin^3 x$$

in the same viewing rectangle, as shown below. Because Example 2 shows the equivalence algebraically and the two graphs appear to coincide, you can conclude that the expressions are equivalent.

EXAMPLE 1 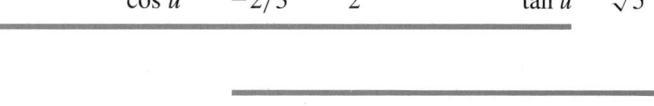 **Using Identities to Evaluate a Function**

Use the given values $\sec u = -\frac{3}{2}$ and $\tan u > 0$ to find the values of all six trigonometric functions.

Solution
Using a reciprocal identity, you have

$$\cos u = \frac{1}{\sec u} = \frac{1}{-3/2} = -\frac{2}{3}.$$

Using a Pythagorean identity, you have

$$\sin^2 u = 1 - \cos^2 u = 1 - \left(-\frac{2}{3}\right)^2 = 1 - \frac{4}{9} = \frac{5}{9}.$$

Because $\sec u < 0$ and $\tan u > 0$, it follows that u lies in Quadrant III. Moreover, because $\sin u$ is negative when u is in Quadrant III, you can choose the negative root and obtain $\sin u = -\sqrt{5}/3$. Now, knowing the values of the sine and cosine, you can find the values of all six trigonometric functions.

$$\sin u = -\frac{\sqrt{5}}{3} \qquad\qquad \csc u = \frac{1}{\sin u} = -\frac{3}{\sqrt{5}}$$

$$\cos u = -\frac{2}{3} \qquad\qquad \sec u = \frac{1}{\cos u} = -\frac{3}{2}$$

$$\tan u = \frac{\sin u}{\cos u} = \frac{-\sqrt{5}/3}{-2/3} = \frac{\sqrt{5}}{2} \qquad \cot u = \frac{1}{\tan u} = \frac{2}{\sqrt{5}}$$

EXAMPLE 2 **Simplifying a Trigonometric Expression**

Simplify $\sin x \cos^2 x - \sin x$.

Solution
First factor out a common monomial factor and then use a fundamental identity.

$$\sin x \cos^2 x - \sin x = \sin x(\cos^2 x - 1) \qquad \text{Monomial factor}$$
$$= -\sin x(1 - \cos^2 x)$$
$$= -\sin x(\sin^2 x) \qquad \text{Pythagorean identity}$$
$$= -\sin^3 x \qquad \text{Multiply.}$$

EXAMPLE 3 Verifying Trigonometric Identities Graphically

Use a graphing utility to determine which of the following is an identity.

a. $\cos 3x \overset{?}{=} 4 \cos^3 x - 3 \cos x$ **b.** $\cos 3x \overset{?}{=} \sin\left(3x - \dfrac{\pi}{2}\right)$

Solution

a. Using a graphing utility, you can see that the graphs of $y = \cos 3x$ and $y = 4 \cos^3 x - 3 \cos x$ appear to coincide, as shown in Figure 5.1(a). Therefore, this appears to be an identity.

b. From the graphs shown in Figure 5.1(b), you can see that this is not an identity.

Figure 5.1

(a)

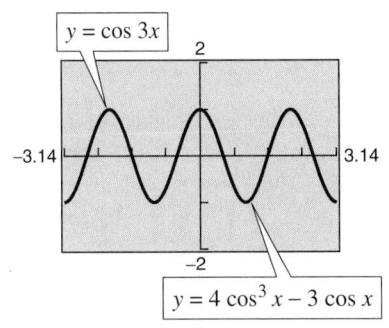

$y = \cos 3x$

$y = 4 \cos^3 x - 3 \cos x$

(b)

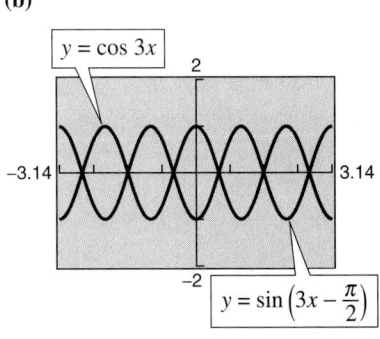

$y = \cos 3x$

$y = \sin\left(3x - \dfrac{\pi}{2}\right)$

The identity in Example 4 can be confirmed using the table feature of a graphing utility

$$y_1 = \frac{\sin x}{1 + \cos x} + \frac{\cos x}{\sin x}$$

and

$$y_2 = \csc x.$$

X	Y₁	Y₂
−1	−1.188	−1.188
−.5	−2.086	−2.086
−.25	−4.042	−4.042
0	ERROR	ERROR
.25	4.042	4.042
.5	2.0858	2.0858
1	1.1884	1.1884

X = −1

EXAMPLE 4 Verifying a Trigonometric Identity Algebraically

Verify the identity $\dfrac{\sin \theta}{1 + \cos \theta} + \dfrac{\cos \theta}{\sin \theta} = \csc \theta.$

Solution

$$\frac{\sin \theta}{1 + \cos \theta} + \frac{\cos \theta}{\sin \theta} = \frac{(\sin \theta)(\sin \theta) + (\cos \theta)(1 + \cos \theta)}{(1 + \cos \theta)(\sin \theta)}$$

$$= \frac{\sin^2 \theta + \cos^2 \theta + \cos \theta}{(1 + \cos \theta)(\sin \theta)} \qquad \text{Multiply.}$$

$$= \frac{1 + \cos \theta}{(1 + \cos \theta)(\sin \theta)} \qquad \text{Pythagorean identity}$$

$$= \frac{1}{\sin \theta} \qquad \text{Cancel common factor.}$$

$$= \csc \theta \qquad \text{Reciprocal identity}$$

EXAMPLE 5 ▱ Factoring Trigonometric Expressions

Factor each expression.

a. $\sec^2\theta - 1$

b. $4\tan^2\theta + \tan\theta - 3$

Solution

a. Here you have the difference of two squares, which factors as
$$\sec^2\theta - 1 = (\sec\theta - 1)(\sec\theta + 1).$$

b. This expression has the polynomial form, $ax^2 + bx + c$, and it factors as
$$4\tan^2\theta + \tan\theta - 3 = (4\tan\theta - 3)(\tan\theta + 1). \quad ▱$$

> **Study Tip**
>
> On occasion, factoring or simplifying can best be done by first rewriting the expression in terms of just *one* trigonometric function or in terms of *sine* or *cosine* *alone*. These strategies are illlustrated in Examples 6 and 7.

EXAMPLE 6 ▱ Factoring a Trigonometric Expression

Factor $\csc^2 x - \cot x - 3$.

Solution

You can use the identity $\csc^2 x = 1 + \cot^2 x$ to rewrite the expression in terms of the cotangent alone.

$$
\begin{aligned}
\csc^2 x - \cot x - 3 &= (1 + \cot^2 x) - \cot x - 3 && \text{Pythagorean identity} \\
&= \cot^2 x - \cot x - 2 && \text{Combine like terms.} \\
&= (\cot x - 2)(\cot x + 1) && \text{Factor.}
\end{aligned}
$$

$$▱$$

EXAMPLE 7 ▱ Simplifying a Trigonometric Expression

Simplify $\sin t + \cot t \cos t$.

Solution

Begin by rewriting the expression in terms of sine and cosine.

$$
\begin{aligned}
\sin t + \cot t \cos t &= \sin t + \left(\frac{\cos t}{\sin t}\right)\cos t && \text{Quotient identity} \\[2mm]
&= \frac{\sin^2 t + \cos^2 t}{\sin t} && \text{Add fractions.} \\[2mm]
&= \frac{1}{\sin t} && \text{Pythagorean identity} \\[2mm]
&= \csc t && \text{Reciprocal identity}
\end{aligned}
$$

$$▱$$

The last two examples in this section involve techniques for rewriting expressions into forms that are useful in calculus.

EXAMPLE 8   Rewriting a Trigonometric Expression

Rewrite $\dfrac{1}{1 + \sin x}$ so that it is *not* in fractional form.

Solution

From the Pythagorean identity $\cos^2 x = 1 - \sin^2 x = (1 - \sin x)(1 + \sin x)$, you can see that by multiplying both the numerator and the denominator by $(1 - \sin x)$ you will produce a monomial denominator.

$$\frac{1}{1 + \sin x} = \frac{1}{1 + \sin x} \cdot \frac{1 - \sin x}{1 - \sin x} \qquad \text{Multiply numerator and denominator by } (1 - \sin x).$$

$$= \frac{1 - \sin x}{1 - \sin^2 x} \qquad \text{Multiply.}$$

$$= \frac{1 - \sin x}{\cos^2 x} \qquad \text{Pythagorean identity}$$

$$= \frac{1}{\cos^2 x} - \frac{\sin x}{\cos^2 x} \qquad \text{Separate fractions.}$$

$$= \frac{1}{\cos^2 x} - \frac{\sin x}{\cos x} \cdot \frac{1}{\cos x} \qquad \text{Separate fractions.}$$

$$= \sec^2 x - \tan x \sec x \qquad \text{Identities}$$

EXAMPLE 9   Trigonometric Substitution

Use the substitution $x = 2 \tan \theta$, $0 < \theta < \pi/2$, to express $\sqrt{4 + x^2}$ as a trigonometric function of θ.

Solution

Begin by letting $x = 2 \tan \theta$. Then you can obtain the following.

$$\sqrt{4 + x^2} = \sqrt{4 + (2 \tan \theta)^2} \qquad \text{Substitute } 2 \tan \theta \text{ for } x.$$

$$= \sqrt{4(1 + \tan^2 \theta)}$$

$$= \sqrt{4 \sec^2 \theta} \qquad \text{Pythagorean identity}$$

$$= 2 \sec \theta \qquad \sec \theta > 0 \text{ for } 0 < \theta < \frac{\pi}{2}$$

Figure 5.2

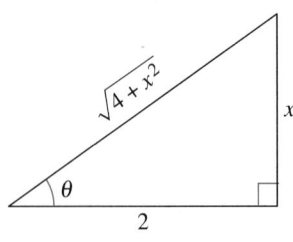

Figure 5.2 shows the right angle illustration of this substitution. For $0 < \theta < \pi/2$, you have opposite $= x$, adjacent $= 2$, and hypotenuse $= \sqrt{4 + x^2}$. Try using these expressions to obtain the result shown above.

Group Activity

Trigonometric Identities

Match each graph with *three* of the trigonometric expressions. Then algebraically verify that the three expressions are equivalent.

1.

2.

3.

a. $1 + \tan^2 x$

d. $(\sin x + 2)(\sin x - 1)$

g. $\dfrac{1}{\cos^2 x}$

b. $\sin^2 x + \sin x - 2$

e. $\sec^2 x$

h. $\dfrac{\sec^2 x - 1}{\sec^2 x}$

c. $1 - \cos^2 x$

f. $-\cos^2 x + \sin x - 1$

i. $\dfrac{1}{\csc^2 x}$

5.1 /// EXERCISES

In Exercises 1–14, use the given values to evaluate (if possible) the other four trigonometric functions.

1. $\sin x = \dfrac{1}{2}, \quad \cos x = \dfrac{\sqrt{3}}{2}$

2. $\tan x = \dfrac{\sqrt{3}}{3}, \quad \cos x = -\dfrac{\sqrt{3}}{2}$

3. $\sec \theta = \sqrt{2}, \quad \sin \theta = -\dfrac{\sqrt{2}}{2}$

4. $\csc \theta = \dfrac{5}{3}, \quad \tan \theta = \dfrac{3}{4}$

5. $\tan x = \dfrac{5}{12}, \quad \sec x = -\dfrac{13}{12}$

6. $\cot \phi = -3, \quad \sin \phi = \dfrac{\sqrt{10}}{10}$

7. $\sec \phi = -1, \quad \sin \phi = 0$

8. $\cos\left(\dfrac{\pi}{2} - x\right) = \dfrac{3}{5}, \quad \cos x = \dfrac{4}{5}$

9. $\sin(-x) = -\dfrac{2}{3}, \quad \tan x = -\dfrac{2\sqrt{5}}{5}$

10. $\csc x = 5, \quad \cos x > 0$

11. $\tan \theta = 2, \quad \sin \theta < 0$

12. $\sec \theta = -3, \quad \tan \theta < 0$

13. $\sin \theta = -1, \quad \cot \theta = 0$

14. $\tan \theta$ is undefined, $\quad \sin \theta > 0$

The *Interactive* CD-ROM contains step-by-step solutions to all odd-numbered Section and Review Exercises. It also provides Tutorial Exercises, which link to Guided Examples for additional help.

In Exercises 15–18, fill in the blanks. (*Note:* $x \to c^+$ indicates that x approaches c from the right, and $x \to c^-$ indicates that x approaches c from the left.)

15. As $x \to \dfrac{\pi^-}{2}$, $\sin x \to$ ▢ and $\csc x \to$ ▢.

16. As $x \to 0^+$, $\cos x \to$ ▢ and $\sec x \to$ ▢.

17. As $x \to \dfrac{\pi^-}{2}$, $\tan x \to$ ▢ and $\cot x \to$ ▢.

18. As $x \to \pi^+$, $\sin x \to$ ▢ and $\csc x \to$ ▢.

In Exercises 19–24, match the trigonometric expression with one of the following.

(a) -1 (b) $\cos x$ (c) $\cot x$

(d) 1 (e) $-\tan x$ (f) $\sin x$

19. $\sec x \cos x$

20. $\cot x \sin x$

21. $\tan^2 x - \sec^2 x$

22. $(1 - \cos^2 x)(\csc x)$

23. $\dfrac{\sin(-x)}{\cos(-x)}$

24. $\dfrac{\sin[(\pi/2) - x]}{\cos[(\pi/2) - x]}$

In Exercises 25–30, match the trigonometric expression with one of the following.

(a) $\csc x$ (b) $\tan x$ (c) $\sin^2 x$

(d) $\sin x \tan x$ (e) $\sec^2 x$ (f) $\sec^2 x + \tan^2 x$

25. $\sin x \sec x$

26. $\cos^2 x(\sec^2 x - 1)$

27. $\sec^4 x - \tan^4 x$

28. $\cot x \sec x$

29. $\dfrac{\sec^2 x - 1}{\sin^2 x}$

30. $\dfrac{\cos^2[(\pi/2) - x]}{\cos x}$

In Exercises 31–44, use the fundamental identities to simplify the expression. Use a graphing utility to verify your result.

31. $\tan \phi \csc \phi$

32. $\sin \phi(\csc \phi - \sin \phi)$

33. $\cos \beta \tan \beta$

34. $\sec^2 x(1 - \sin^2 x)$

35. $\dfrac{\cot x}{\csc x}$

36. $\dfrac{\csc \theta}{\sec \theta}$

37. $\sec \alpha \cdot \dfrac{\sin \alpha}{\tan \alpha}$

38. $\dfrac{1}{\tan^2 x + 1}$

39. $\dfrac{\sin(-x)}{\cos x}$

40. $\dfrac{\tan^2 \theta}{\sec^2 \theta}$

41. $\cos\left(\dfrac{\pi}{2} - x\right)\sec x$

42. $\cot\left(\dfrac{\pi}{2} - x\right)\cos x$

43. $\dfrac{\cos^2 y}{1 - \sin y}$

44. $\cos t(1 + \tan^2 t)$

In Exercises 45–52, factor the expression and use the fundamental identities to simplify. Use a graphing utility to verify your result.

45. $\tan^2 x - \tan^2 x \sin^2 x$

46. $\sec^2 x \tan^2 x + \sec^2 x$

47. $\sin^2 x \sec^2 x - \sin^2 x$

48. $\dfrac{\sec^2 x - 1}{\sec x - 1}$

49. $\tan^4 x + 2\tan^2 x + 1$

50. $1 - 2\cos^2 x + \cos^4 x$

51. $\sin^4 x - \cos^4 x$

52. $\csc^3 x - \csc^2 x - \csc x + 1$

In Exercises 53–56, perform the multiplication and use the fundamental identities to simplify.

53. $(\sin x + \cos x)^2$

54. $(\cot x + \csc x)(\cot x - \csc x)$

55. $(\sec x + 1)(\sec x - 1)$

56. $(3 - 3\sin x)(3 + 3\sin x)$

In Exercises 57–60, perform the addition or subtraction and use the fundamental identities to simplify.

57. $\dfrac{1}{1 + \cos x} + \dfrac{1}{1 - \cos x}$

58. $\dfrac{1}{\sec x + 1} - \dfrac{1}{\sec x - 1}$

59. $\dfrac{\cos x}{1 + \sin x} + \dfrac{1 + \sin x}{\cos x}$

60. $\tan x - \dfrac{\sec^2 x}{\tan x}$

In Exercises 61–64, rewrite the expression so that it is *not* in fractional form.

61. $\dfrac{\sin^2 y}{1 - \cos y}$

62. $\dfrac{5}{\tan x + \sec x}$

63. $\dfrac{3}{\sec x - \tan x}$

64. $\dfrac{\tan^2 x}{\csc x + 1}$

Numerical and Graphical Analysis In Exercises 65–68, use a graphing utility to complete the table and graph the functions. Make a conjecture about y_1 and y_2.

x	0.2	0.4	0.6	0.8	1.0	1.2	1.4
y_1							
y_2							

65. $y_1 = \cos\left(\dfrac{\pi}{2} - x\right), \quad y_2 = \sin x$

66. $y_1 = \cos x + \sin x \tan x, \quad y_2 = \sec x$

67. $y_1 = \dfrac{\cos x}{1 - \sin x}, \quad y_2 = \dfrac{1 + \sin x}{\cos x}$

68. $y_1 = \sec^4 x - \sec^2 x, \quad y_2 = \tan^2 x + \tan^4 x$

In Exercises 69 and 70, use a graphing utility to determine which of the six trigonometric functions is equal to the expression.

69. $\cos x \cot x + \sin x$

70. $\dfrac{1}{2}\left(\dfrac{1 + \sin \theta}{\cos \theta} + \dfrac{\cos \theta}{1 + \sin \theta}\right)$

In Exercises 71–76, use the trigonometric substitution to write the algebraic expression as a trigonometric function of θ, where $0 < \theta < \pi/2$.

71. $\sqrt{25 - x^2}, \quad x = 5 \sin \theta$

72. $\sqrt{16 - 4x^2}, \quad x = 2 \sin \theta$

73. $\sqrt{x^2 - 9}, \quad x = 3 \sec \theta$

74. $\sqrt{x^2 - 4}, \quad x = 2 \sec \theta$

75. $\sqrt{x^2 + 25}, \quad x = 5 \tan \theta$

76. $\sqrt{x^2 + 100}, \quad x = 10 \tan \theta$

In Exercises 77–80, use a graphing utility to solve the equation for θ, where $0 \le \theta < 2\pi$.

77. $\sin \theta = \sqrt{1 - \cos^2 \theta}$

78. $\cos \theta = -\sqrt{1 - \sin^2 \theta}$

79. $\sec \theta = \sqrt{1 + \tan^2 \theta}$

80. $\tan \theta = \sqrt{\sec^2 \theta - 1}$

In Exercises 81 and 82, rewrite the expression as a single logarithm and simplify the result.

81. $\ln|\cos \theta| - \ln|\sin \theta|$

82. $\ln|\cot t| + \ln(1 + \tan^2 t)$

In Exercises 83–86, determine whether or not the equation is an identity, and give a reason for your answer.

83. $\dfrac{\sin k\theta}{\cos k\theta} = \tan \theta, \quad k$ is a constant.

84. $\dfrac{1}{5 \cos \theta} = 5 \sec \theta$

85. $\sin \theta \csc \theta = 1$

86. $\sin \theta \csc \phi = 1$

In Exercises 87–90, use a calculator to demonstrate the identity for the given values of θ.

87. $\csc^2 \theta - \cot^2 \theta = 1$, (a) $\theta = 132°$, (b) $\theta = \dfrac{2\pi}{7}$

88. $\tan^2 \theta + 1 = \sec^2 \theta$, (a) $\theta = 346°$, (b) $\theta = 3.1$

89. $\cos\left(\dfrac{\pi}{2} - \theta\right) = \sin \theta$, (a) $\theta = 80°$, (b) $\theta = 0.8$

90. $\sin(-\theta) = -\sin \theta$, (a) $\theta = 250°$, (b) $\theta = \dfrac{1}{2}$

91. Express each of the other trigonometric functions of θ in terms of $\sin \theta$.

92. Express each of the other trigonometric functions of θ in terms of $\cos \theta$.

93. *Chapter Opener* Which city has the greater variation in the number of daylight hours? Which constant in each model would you use to determine the difference between the greatest and least number of hours of daylight?

94. *Chapter Opener* Determine the period of each model.

5.2   Verifying Trigonometric Identities

Introduction

In this section, you will study techniques for verifying trigonometric identities. In the next section, you will study techniques for solving trigonometric equations. The key to verifying identities *and* solving equations is the ability to use the fundamental identities and the rules of algebra to rewrite trigonometric expressions.

Remember that a *conditional equation* is an equation that is true for only some of the values in its domain. For example, the conditional equation

$$\sin x = 0 \qquad\qquad \text{Conditional equation}$$

is true only for $x = n\pi$, where n is an integer. When you find these values, you are *solving* the equation. On the other hand, an equation that is true for all real values in the domain of the variable is an *identity*. For example, the familiar equation

$$\sin^2 x = 1 - \cos^2 x \qquad\qquad \text{Identity}$$

is true for all real numbers x. Hence, it is an identity.

Although there are similarities, proving that a trigonometric equation is an identity is quite different from solving an equation. There is no well-defined set of rules to follow in verifying trigonometric identities, and the process is best learned by practice.

Guidelines for Verifying Trigonometric Identities

1. **Work with one side of the equation at a time.** It is often better to work with the more complicated side first.
2. Look for opportunities to factor an expression, add fractions, square a binomial, or create a monomial denominator.
3. Look for opportunities to use the fundamental identities. Note which functions are in the final expression you want. Sines and cosines pair up well, as do secants and tangents, and cosecants and cotangents.
4. If the preceding guidelines do not help, try converting all terms to sines and cosines.
5. Do not just sit and stare at the problem. Try something! Even paths that lead to dead ends give you insights.

Verifying Trigonometric Identities

EXAMPLE 1   ▱   **Verifying a Trigonometric Identity**

Verify the identity $\dfrac{\sec^2 \theta - 1}{\sec^2 \theta} = \sin^2 \theta$.

Solution

Because the left side is more complicated, start with it.

Note   Here is another way to verify the identity in Example 1.

$$\frac{\sec^2 \theta - 1}{\sec^2 \theta} = \frac{\sec^2 \theta}{\sec^2 \theta} - \frac{1}{\sec^2 \theta}$$
$$= 1 - \cos^2 \theta$$
$$= \sin^2 \theta$$

$$\frac{\sec^2 \theta - 1}{\sec^2 \theta} = \frac{(\tan^2 \theta + 1) - 1}{\sec^2 \theta} \qquad \text{Pythagorean identity}$$

$$= \frac{\tan^2 \theta}{\sec^2 \theta} \qquad \text{Simplify.}$$

$$= \tan^2 \theta (\cos^2 \theta) \qquad \text{Reciprocal identity}$$

$$= \frac{\sin^2 \theta}{\cos^2 \theta} (\cos^2 \theta) \qquad \text{Quotient identity}$$

$$= \sin^2 \theta \qquad \text{Simplify.} \qquad ▱$$

The *Interactive* CD-ROM offers graphing utility emulators of the *TI-82* and *TI-83*, which can be used with the Examples, Explorations, Technology notes, and Exercises.

As you can see from the note at the upper left, there can be more than one way to verify an identity. Your method may differ from that used by your instructor or fellow students. Here is a good chance to be creative and establish your own style, but try to be as efficient as possible.

EXAMPLE 2   ▱   **Combining Fractions Before Using Identities**

Verify the identity

$$\frac{1}{1 - \sin \alpha} + \frac{1}{1 + \sin \alpha} = 2 \sec^2 \alpha.$$

Solution

Use the table feature of a graphing utility to check the result in Example 2 by entering

$$y_1 = \frac{1}{1 - \sin x} + \frac{1}{1 + \sin x}$$

and

$$y_2 = 2 \sec^2 x.$$

Their tables should be identical.

$$\frac{1}{1 - \sin \alpha} + \frac{1}{1 + \sin \alpha} = \frac{1 + \sin \alpha + 1 - \sin \alpha}{(1 - \sin \alpha)(1 + \sin \alpha)} \qquad \text{Add fractions.}$$

$$= \frac{2}{1 - \sin^2 \alpha} \qquad \text{Simplify.}$$

$$= \frac{2}{\cos^2 \alpha} \qquad \text{Pythagorean identity}$$

$$= 2 \sec^2 \alpha \qquad \text{Reciprocal identity}$$

$$▱$$

EXAMPLE 3   **Verifying a Trigonometric Identity**

Verify the identity

$$(\tan^2 x + 1)(\cos^2 x - 1) = -\tan^2 x.$$

Solution

By applying identities before multiplying, you obtain the following.

$$\begin{aligned}
(\tan^2 x + 1)(\cos^2 x - 1) &= (\sec^2 x)(-\sin^2 x) && \text{Pythagorean identities} \\
&= -\frac{\sin^2 x}{\cos^2 x} && \text{Reciprocal identity} \\
&= -\left(\frac{\sin x}{\cos x}\right)^2 && \text{Rule of exponents} \\
&= -\tan^2 x && \text{Quotient identity}
\end{aligned}$$

Although a graphing utility can be useful in helping to verify an identity, you must use analytical techniques to produce a valid proof. For example, graph the two functions

$$y_1 = \sin 50x$$
$$y_2 = \sin 2x$$

on the ZTrig screen. Although their graphs seem identical, $\sin 50x \neq \sin 2x$.

EXAMPLE 4   **Converting to Sines and Cosines**

Verify the identity

$$\tan x + \cot x = \sec x \csc x.$$

Solution

In this case there appear to be no fractions to add, no products to find, and no opportunity to use one of the Pythagorean identities. Hence, try converting the left side into sines and cosines to see what happens.

$$\begin{aligned}
\tan x + \cot x &= \frac{\sin x}{\cos x} + \frac{\cos x}{\sin x} && \text{Quotient identities} \\
&= \frac{\sin^2 x + \cos^2 x}{\cos x \sin x} && \text{Add fractions.} \\
&= \frac{1}{\cos x \sin x} && \text{Pythagorean identity} \\
&= \frac{1}{\cos x} \cdot \frac{1}{\sin x} && \text{Product of fractions} \\
&= \sec x \csc x && \text{Reciprocal identities}
\end{aligned}$$

Recall from algebra that *rationalizing the denominator* is, on occasion, a powerful simplification technique. A related form of this technique works for simplifying trigonometric expressions as well.

EXAMPLE 5 ◢ **Verifying a Trigonometric Identity**

Verify the identity $\sec y + \tan y = \dfrac{\cos y}{1 - \sin y}$.

Solution

Work with the *right* side. Note that you can create a monomial denominator by multiplying the numerator and denominator by $(1 + \sin y)$.

$$\frac{\cos y}{1 - \sin y} = \frac{\cos y}{1 - \sin y}\left(\frac{1 + \sin y}{1 + \sin y}\right) \qquad \text{Multiply numerator and denominator by } (1 + \sin y).$$

$$= \frac{\cos y + \cos y \sin y}{1 - \sin^2 y} \qquad \text{Multiply.}$$

$$= \frac{\cos y + \cos y \sin y}{\cos^2 y} \qquad \text{Pythagorean identity}$$

$$= \frac{\cos y}{\cos^2 y} + \frac{\cos y \sin y}{\cos^2 y} \qquad \text{Separate fractions.}$$

$$= \frac{1}{\cos y} + \frac{\sin y}{\cos y} \qquad \text{Simplify.}$$

$$= \sec y + \tan y \qquad \text{Identities} \qquad \blacksquare$$

EXAMPLE 6 ◢ **Working with Each Side Separately**

Verify the identity $\dfrac{\cot^2 \theta}{1 + \csc \theta} = \dfrac{1 - \sin \theta}{\sin \theta}$.

Solution

Working with the left side, you have

$$\frac{\cot^2 \theta}{1 + \csc \theta} = \frac{\csc^2 \theta - 1}{1 + \csc \theta} \qquad \text{Pythagorean identity}$$

$$= \frac{(\csc \theta - 1)(\csc \theta + 1)}{1 + \csc \theta} \qquad \text{Factor.}$$

$$= \csc \theta - 1. \qquad \text{Simplify.}$$

Now, simplifying the right side, you have

$$\frac{1 - \sin \theta}{\sin \theta} = \frac{1}{\sin \theta} - \frac{\sin \theta}{\sin \theta} \qquad \text{Separate fractions.}$$

$$= \csc \theta - 1. \qquad \text{Reciprocal identity}$$

The identity is verified because both sides are equal to $\csc \theta - 1$. ◢

In Example 7, powers of trigonometric functions are rewritten as more complicated sums of products of trigonometric functions. This is a common procedure
used in calculus.

EXAMPLE 7   ▱   **Two Examples from Calculus**

Verify each identity.

a. $\tan^4 x = \tan^2 x \sec^2 x - \tan^2 x$

b. $\sin^3 x \cos^4 x = (\cos^4 x - \cos^6 x) \sin x$

Solution

a. $\tan^4 x = (\tan^2 x)(\tan^2 x)$ Separate factors.

         $= \tan^2 x(\sec^2 x - 1)$ Pythagorean identity

         $= \tan^2 x \sec^2 x - \tan^2 x$ Multiply.

b. $\sin^3 x \cos^4 x = \sin^2 x \cos^4 x \sin x$ Separate factors.

         $= (1 - \cos^2 x)\cos^4 x \sin x$ Pythagorean identity

         $= (\cos^4 x - \cos^6 x)\sin x$ Multiply.   ▱

Group Activity   *Error Analysis*

Suppose you are tutoring a student in trigonometry. One of the
homework problems your student encounters asks whether the following statement is an identity.

$$\tan^2 x \sin^2 x \overset{?}{=} \frac{5}{6}\tan^2 x$$

Your student does not attempt to verify the equivalence algebraically,
but mistakenly uses only a graphical approach. Using range settings of
$\text{Xmin} = -3\pi$, $\text{Xmax} = 3\pi$, $\text{Xscl} = \pi/2$, $\text{Ymin} = -20$, $\text{Ymax} = 20$,
and $\text{Yscl} = 1$, your student graphs both sides of the expression on a
graphing utility and concludes that the statement is an identity.

What is wrong with your student's reasoning? Explain.

5.2 /// EXERCISES

In Exercises 1–10, verify the identity.

1. $\sin t \csc t = 1$
2. $\tan y \cot y = 1$

3. $\dfrac{\sec^2 x}{\tan x} = \sec x \csc x$
4. $\cot^2 y (\sec^2 y - 1) = 1$

5. $\cos^2 \beta - \sin^2 \beta = 1 - 2 \sin^2 \beta$

6. $\cos^2 \beta - \sin^2 \beta = 2 \cos^2 \beta - 1$

7. $\tan^2 \theta + 4 = \sec^2 \theta + 3$

8. $2 - \sec^2 z = 1 - \tan^2 z$

9. $\cos x + \sin x \tan x = \sec x$

10. $\dfrac{\cot^3 t}{\csc t} = \cos t (\csc^2 t - 1)$

Numerical, Graphical, and Analytical Analysis   In Exercises 11–18, use a graphing utility to complete the table and graph the functions. Use both as evidence that $y_1 = y_2$. Then verify the identity analytically.

x	0.2	0.4	0.6	0.8	1.0	1.2	1.4
y_1							
y_2							

11. $y_1 = \dfrac{1}{\sec x \tan x}, \quad y_2 = \csc x - \sin x$

12. $y_1 = \dfrac{\sec x - 1}{1 - \cos x}, \quad y_2 = \sec x$

13. $y_1 = \csc x - \sin x, \quad y_2 = \cos x \cot x$

14. $y_1 = \sec x - \cos x, \quad y_2 = \sin x \tan x$

15. $y_1 = \cos x + \sin x \tan x, \quad y_2 = \sec x$

16. $y_1 = \dfrac{\sec x + \tan x}{\sec x - \tan x}, \quad y_2 = (\sec x + \tan x)^2$

17. $y_1 = \dfrac{1}{\tan x} + \dfrac{1}{\cot x}, \quad y_2 = \tan x + \cot x$

18. $y_1 = \dfrac{1}{\sin x} - \dfrac{1}{\csc x}, \quad y_2 = \csc x - \sin x$

In Exercises 19–36, verify the identity.

19. $\sin^{1/2} x \cos x - \sin^{5/2} x \cos x = \cos^3 x \sqrt{\sin x}$

20. $\sec^6 x (\sec x \tan x) - \sec^4 x (\sec x \tan x) = \sec^5 x \tan^3 x$

21. $\cos\left(\dfrac{\pi}{2} - x\right) \csc x = 1$

22. $\dfrac{\cos[(\pi/2) - x]}{\sin[(\pi/2) - x]} = \tan x$

23. $\dfrac{\csc(-x)}{\sec(-x)} = -\cot x$

24. $(1 + \sin y)[1 + \sin(-y)] = \cos^2 y$

25. $\dfrac{\cos(-\theta)}{1 + \sin(-\theta)} = \sec \theta + \tan \theta$

26. $\dfrac{1 + \sec(-\theta)}{\sin(-\theta) + \tan(-\theta)} = -\csc \theta$

27. $\dfrac{\sin x \cos y + \cos x \sin y}{\cos x \cos y - \sin x \sin y} = \dfrac{\tan x + \tan y}{1 - \tan x \tan y}$

28. $\dfrac{\tan x + \tan y}{1 - \tan x \tan y} = \dfrac{\cot x + \cot y}{\cot x \cot y - 1}$

29. $\dfrac{\tan x + \cot y}{\tan x \cot y} = \tan y + \cot x$

30. $\dfrac{\cos x - \cos y}{\sin x + \sin y} + \dfrac{\sin x - \sin y}{\cos x + \cos y} = 0$

31. $\sqrt{\dfrac{1 + \sin \theta}{1 - \sin \theta}} = \dfrac{1 + \sin \theta}{|\cos \theta|}$

32. $\sqrt{\dfrac{1 - \cos \theta}{1 + \cos \theta}} = \dfrac{1 - \cos \theta}{|\sin \theta|}$

33. $\sin^2 x + \sin^2\left(\dfrac{\pi}{2} - x\right) = 1$

34. $\sec^2 y - \cot^2\left(\dfrac{\pi}{2} - y\right) = 1$

35. $\csc x \cos\left(\dfrac{\pi}{2} - x\right) = 1$

36. $\sec^2\left(\dfrac{\pi}{2} - x\right) - 1 = \cot^2 x$

In Exercises 37–48, verify the identity algebraically, and use a graphing utility to confirm it graphically.

37. $2 \sec^2 x - 2 \sec^2 x \sin^2 x - \sin^2 x - \cos^2 x = 1$

38. $\csc x(\csc x - \sin x) + \dfrac{\sin x - \cos x}{\sin x} + \cot x = \csc^2 x$

39. $2 + \cos^2 x - 3 \cos^4 x = \sin^2 x(2 + 3 \cos^2 x)$

40. $4 \tan^4 x + \tan^2 x - 3 = \sec^2 x(4 \tan^2 x - 3)$

41. $\csc^4 x - 2 \csc^2 x + 1 = \cot^4 x$

42. $\sin x(1 - 2 \cos^2 x + \cos^4 x) = \sin^5 x$

43. $\sec^4 \theta - \tan^4 \theta = 1 + 2 \tan^2 \theta$

44. $\csc^4 \theta - \cot^4 \theta = 2 \csc^2 \theta - 1$

45. $\dfrac{\sin \beta}{1 - \cos \beta} = \dfrac{1 + \cos \beta}{\sin \beta}$

46. $\dfrac{\cot \alpha}{\csc \alpha - 1} = \dfrac{\csc \alpha + 1}{\cot \alpha}$

47. $\dfrac{\tan^3 \alpha - 1}{\tan \alpha - 1} = \tan^2 \alpha + \tan \alpha + 1$

48. $\dfrac{\sin^3 \beta + \cos^3 \beta}{\sin \beta + \cos \beta} = 1 - \sin \beta \cos \beta$

Conjecture   In Exercises 49–52, use a graphing utility to graph the trigonometric function. Use the graph to make a conjecture about a simplification of the expression. Verify the resulting identity analytically.

49. $y = \dfrac{1}{\cot x + 1} + \dfrac{1}{\tan x + 1}$

50. $y = \dfrac{\cos x}{1 - \tan x} + \dfrac{\sin x \cos x}{\sin x - \cos x}$

51. $y = \dfrac{1}{\sin x} - \dfrac{\cos^2 x}{\sin x}$

52. $y = \sin t + \dfrac{\cot^2 t}{\csc t}$

In Exercises 53–56, use the properties of logarithms and trigonometric identities to verify the identity.

53. $\ln|\tan \theta| = \ln|\sin \theta| - \ln|\cos \theta|$

54. $\ln|\sec \theta| = -\ln|\cos \theta|$

55. $-\ln(1 + \cos \theta) = \ln(1 - \cos \theta) - 2 \ln|\sin \theta|$

56. $-\ln|\sec \theta + \tan \theta| = \ln|\sec \theta - \tan \theta|$

Think About It   In Exercises 57–60, explain why the equation is *not* an identity and find one value of the variable for which the equation is not true.

57. $\sqrt{\tan^2 x} = \tan x$ **58.** $\sin \theta = \sqrt{1 - \cos^2 \theta}$

59. $\tan \theta = \sqrt{\sec^2 \theta - 1}$

60. $\sqrt{\sin^2 x + \cos^2 x} = \sin x + \cos x$

In Exercises 61–64, use the cofunction identities to evaluate the expression without the aid of a calculator.

61. $\sin^2 25° + \sin^2 65°$

62. $\cos^2 18° + \cos^2 72°$

63. $\cos^2 20° + \cos^2 52° + \cos^2 38° + \cos^2 70°$

64. $\sin^2 12° + \sin^2 40° + \sin^2 50° + \sin^2 78°$

65. Verify that for all integers n

$$\cos\left[\dfrac{(2n + 1)\pi}{2}\right] = 0.$$

66. Verify that for all integers n

$$\sin\left[\dfrac{(12n + 1)\pi}{6}\right] = \dfrac{1}{2}.$$

67. *Friction*   The forces acting on an object weighing W units on an inclined plane positioned at an angle of θ with the horizontal (see figure) are modeled by

$$\mu W \cos \theta = W \sin \theta$$

where μ is the coefficient of friction. Solve the equation for μ and simplify the result.

68. *Rate of Change*   The rate of change of the function $f(x) = \sin x + \csc x$ is given by $\cos x - \csc x \cot x$. Show that the expression for the rate of change can also be given by $-\cos x \cot^2 x$.

5.3 Solving Trigonometric Equations

Introduction / *Equations of Quadratic Type* / *Functions Involving Multiple Angles* / *Using Inverse Functions*

Introduction

To solve a trigonometric equation, use standard algebraic techniques such as collecting like terms and factoring. Your preliminary goal is to isolate the trigonometric function involved in the equation.

EXAMPLE 1 **Solving a Trigonometric Equation**

$2 \sin x - 1 = 0$	Original equation
$2 \sin x = 1$	Add 1 to both sides.
$\sin x = \frac{1}{2}$	Divide both sides by 2.

To solve for x, note that the equation $\sin x = \frac{1}{2}$ has solutions $x = \pi/6$ and $x = 5\pi/6$ in the interval $[0, 2\pi)$. Moreover, because $\sin x$ has a period of 2π, there are infinitely many other solutions, which can be written as

$$x = \frac{\pi}{6} + 2n\pi \qquad \text{and} \qquad x = \frac{5\pi}{6} + 2n\pi \qquad \text{General solution}$$

where n is an integer, as shown in Figure 5.3.

Figure 5.3

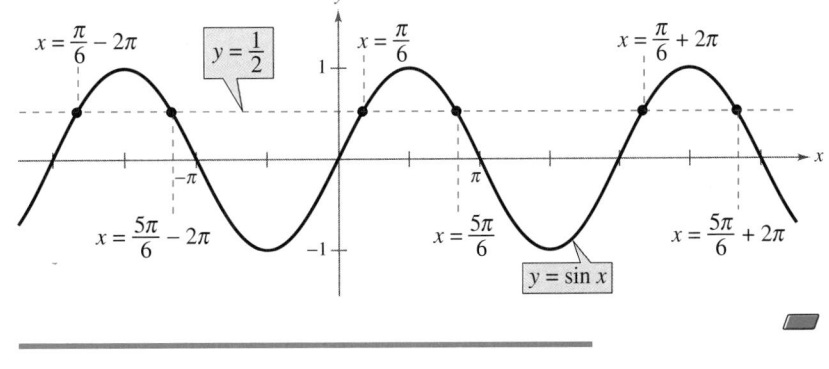

Figure 5.4

$$\sin\left(\frac{5\pi}{6} + 2n\pi\right) = \frac{1}{2} \quad \sin\left(\frac{\pi}{6} + 2n\pi\right) = \frac{1}{2}$$

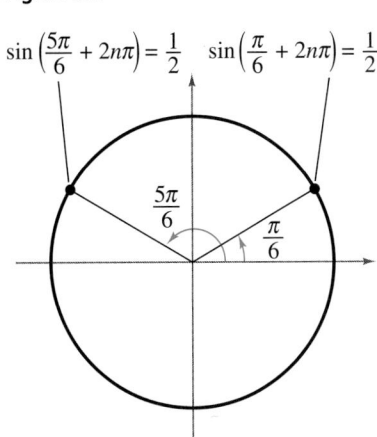

Another way to see that the equation $\sin x = \frac{1}{2}$ has infinitely many solutions is indicated in Figure 5.4. For $0 \le x < 2\pi$, the solutions are $x = \pi/6$ and $x = 5\pi/6$. Any angles that are coterminal with $\pi/6$ or $5\pi/6$ are also solutions of the equation.

EXAMPLE 2 ▱ Collecting Like Terms

Solve $\sin x + \sqrt{2} = -\sin x$.

Solution

Rewrite the equation so that $\sin x$ is isolated on one side of the equation.

$$\sin x + \sqrt{2} = -\sin x \qquad \text{Original equation}$$

$$\sin x + \sin x = -\sqrt{2} \qquad \text{Add } \sin x \text{ and subtract } \sqrt{2} \text{ from both sides.}$$

$$2 \sin x = -\sqrt{2} \qquad \text{Combine like terms.}$$

$$\sin x = -\frac{\sqrt{2}}{2} \qquad \text{Divide both sides by 2.}$$

Because $\sin x$ has a period of 2π, first find all solutions in the interval $[0, 2\pi)$. These are $x = 5\pi/4$ and $x = 7\pi/4$. Finally, add $2n\pi$ to each of these solutions to get the general form

$$x = \frac{5\pi}{4} + 2n\pi \qquad \text{and} \qquad x = \frac{7\pi}{4} + 2n\pi \qquad \text{General solution}$$

where n is an integer. The graph of $y = 2 \sin x + \sqrt{2}$, shown in Figure 5.5, confirms this result.   ▱

Figure 5.5

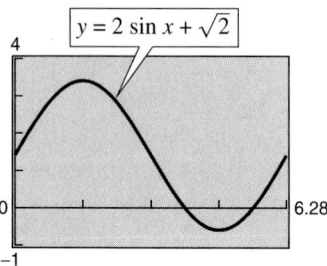

$y = 2 \sin x + \sqrt{2}$

EXAMPLE 3 ▱ Extracting Square Roots

Solve $3 \tan^2 x - 1 = 0$.

Solution

Rewrite the equation so that $\tan x$ is isolated on one side of the equation.

$$3 \tan^2 x - 1 = 0 \qquad \text{Original equation}$$

$$3 \tan^2 x = 1 \qquad \text{Add 1 to both sides.}$$

$$\tan^2 x = \frac{1}{3} \qquad \text{Divide both sides by 3.}$$

$$\tan x = \pm \frac{1}{\sqrt{3}} \qquad \text{Extract square roots.}$$

Because $\tan x$ has a period of π, first find all solutions in the interval $[0, \pi)$. These are $x = \pi/6$ and $x = 5\pi/6$. Finally, add $n\pi$ to each of these solutions to get the general form

$$x = \frac{\pi}{6} + n\pi \qquad \text{and} \qquad x = \frac{5\pi}{6} + n\pi \qquad \text{General solution}$$

where n is an integer. The graph of $y = 3 \tan^2 x - 1$, shown in Figure 5.6, confirms this result.   ▱

Figure 5.6

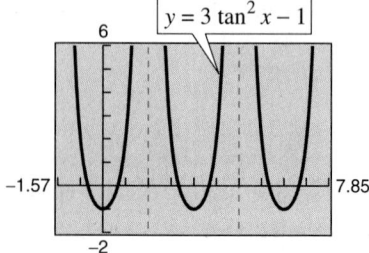

$y = 3 \tan^2 x - 1$

The equations in Examples 1, 2, and 3 involved only one trigonometric function. When two or more functions occur in the same equation, collect all terms on one side and try to separate the functions by factoring or by using appropriate identities. This may produce factors that yield no solutions, as illustrated in Example 4.

EXAMPLE 4 **Factoring**

Solve $\cot x \cos^2 x = 2 \cot x$.

Solution

$$\cot x \cos^2 x = 2 \cot x \qquad \text{Original equation}$$

$$\cot x \cos^2 x - 2 \cot x = 0 \qquad \text{Subtract 2 cot } x \text{ from both sides.}$$

$$\cot x (\cos^2 x - 2) = 0 \qquad \text{Factor.}$$

By setting each of these factors equal to zero, you obtain the following.

$$\cot x = 0 \qquad \text{and} \qquad \cos^2 x - 2 = 0$$

$$x = \frac{\pi}{2} \qquad\qquad\qquad \cos^2 x = 2$$

$$\cos x = \pm\sqrt{2}$$

No solution is obtained from $\cos x = \pm\sqrt{2}$ because $\pm\sqrt{2}$ are outside the range of the cosine function. Therefore, the general form of the solution is obtained by adding multiples of π to $x = \pi/2$, to get

$$x = \frac{\pi}{2} + n\pi \qquad \text{General solution}$$

where n is an integer. The graph of $y = \cot x \cos^2 x - 2 \cot x$, shown in Figure 5.7, confirms this result.

Figure 5.7

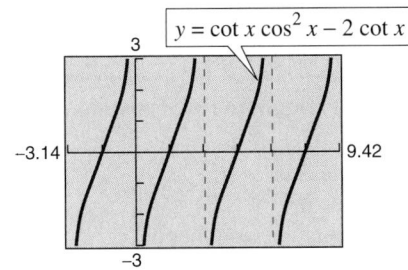

$y = \cot x \cos^2 x - 2 \cot x$

Note In Example 4, don't make the mistake of dividing both sides of the equation by $\cot x$. Doing this would lose the solutions. Can you see why?

Equations of Quadratic Type

Many trigonometric equations are of quadratic type. Here are a couple of examples.

Quadratic in sin x	*Quadratic in sec x*
$2 \sin^2 x - \sin x - 1 = 0$	$\sec^2 x - 3 \sec x - 2 = 0$
$2(\sin x)^2 - (\sin x) - 1 = 0$	$(\sec x)^2 - 3(\sec x) - 2 = 0$

To solve equations of this type, factor the quadratic or, if factoring is not possible, use the Quadratic Formula.

Figure 5.8

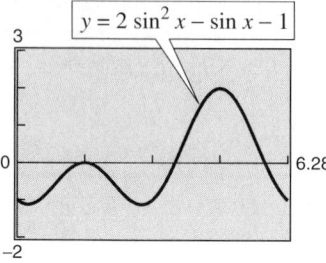

Note In Example 5, the general solution would be

$$x = \frac{7\pi}{6} + 2n\pi$$

$$x = \frac{11\pi}{6} + 2n\pi$$

$$x = \frac{\pi}{2} + 2n\pi$$

where n is an integer.

EXAMPLE 5 ▭ **Factoring an Equation of Quadratic Type**

Find all solutions of $2 \sin^2 x - \sin x - 1 = 0$ in the interval $[0, 2\pi)$.

Solution
The graph of $y = 2 \sin^2 x - \sin x - 1$, shown in Figure 5.8, indicates that there are three solutions in the interval $[0, 2\pi)$. Treating the equation as a quadratic in $\sin x$ and factoring produces the following.

$$2 \sin^2 x - \sin x - 1 = 0 \qquad \text{Original equation}$$
$$(2 \sin x + 1)(\sin x - 1) = 0 \qquad \text{Factor.}$$

Setting each factor equal to zero, you obtain the following solutions.

$$2 \sin x + 1 = 0 \qquad \text{and} \qquad \sin x - 1 = 0$$

$$\sin x = -\frac{1}{2} \qquad\qquad\qquad \sin x = 1$$

$$x = \frac{7\pi}{6}, \frac{11\pi}{6} \qquad\qquad\qquad x = \frac{\pi}{2} \qquad\qquad ▭$$

When working with an equation of quadratic type, be sure that the equation involves a *single* trigonometric function, as shown in the next example.

EXAMPLE 6 ▭ **Rewriting with a Single Trigonometric Function**

Solve $2 \sin^2 x + 3 \cos x - 3 = 0$.

Solution
Begin by rewriting the equation so that it has only cosine functions.

$$2 \sin^2 x + 3\cos x - 3 = 0 \qquad \text{Original equation}$$
$$2(1 - \cos^2 x) + 3 \cos x - 3 = 0 \qquad \text{Pythagorean identity}$$
$$2 \cos^2 x - 3 \cos x + 1 = 0 \qquad \text{Multiply both sides by } -1.$$
$$(2 \cos x - 1)(\cos x - 1) = 0 \qquad \text{Factor.}$$

By setting each factor equal to zero, you can find the solutions in the interval $[0, 2\pi)$ to be $x = 0$, $x = \pi/3$, and $x = 5\pi/3$. The general solution is therefore

$$x = 2n\pi, \qquad x = \frac{\pi}{3} + 2n\pi, \qquad x = \frac{5\pi}{3} + 2n\pi \qquad \text{General solution}$$

Figure 5.9

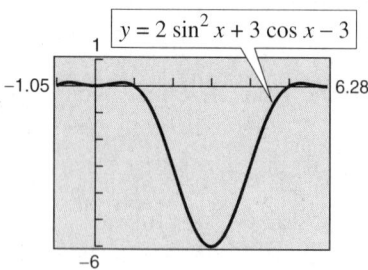

where n is an integer. The graph of $y = 2 \sin^2 x + 3 \cos x - 3$, shown in Figure 5.9, confirms this result. ▭

Sometimes you must square both sides of an equation to obtain a quadratic. This can introduce extraneous solutions.

EXAMPLE 7   **Squaring and Converting to Quadratic Type**

Find all solutions of $\cos x + 1 = \sin x$ in the interval $[0, 2\pi)$.

Solution
It is not clear how to rewrite this equation in terms of a single trigonometric function. See what happens when you square both sides of the equation.

$\cos x + 1 = \sin x$	Original equation
$\cos^2 x + 2\cos x + 1 = \sin^2 x$	Square both sides.
$\cos^2 x + 2\cos x + 1 = 1 - \cos^2 x$	Pythagorean identity
$2\cos^2 x + 2\cos x = 0$	Combine like terms.
$2\cos x(\cos x + 1) = 0$	Factor.

Note In Example 7, the general solution would be

$$x = \frac{\pi}{2} + 2n\pi$$

$$x = \pi + 2n\pi$$

where n is an integer.

Setting each factor equal to zero produces the following.

$$2\cos x = 0 \qquad \text{and} \qquad \cos x + 1 = 0$$
$$\cos x = 0 \qquad\qquad\qquad \cos x = -1$$
$$x = \frac{\pi}{2}, \frac{3\pi}{2} \qquad\qquad\qquad x = \pi$$

Because you squared the original equation, check for extraneous solutions. Of the three possible solutions, $x = 3\pi/2$ is extraneous. (Try checking this.) Thus, in the interval $[0, 2\pi)$, the only two solutions are $x = \pi/2$ and $x = \pi$. The graph of $y = \cos x + 1 - \sin x$, shown in Figure 5.10, confirms this result.

Figure 5.10

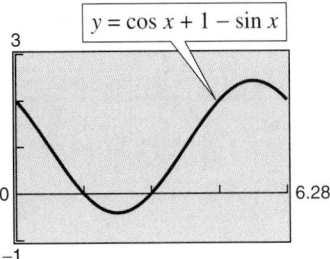

EXPLORATION

Use a graphing utility to confirm the solutions found in Example 7 in two different ways.

1. Graph both sides of the equation and find the x-coordinates of the points at which the graphs intersect.

    Left side: $y = \cos x + 1$ *Right side: $y = \sin x$*

2. Graph the equation $y = \cos x + 1 - \sin x$ and find the x-intercepts of the graph.

Do both methods produce the same x-values? Which method do you prefer? Why?

Functions Involving Multiple Angles

EXAMPLE 8   **Functions of Multiple Angles**

Find all solutions of $2 \cos 3t - 1 = 0$.

Solution

$$2 \cos 3t - 1 = 0 \qquad \text{Original equation}$$

$$2 \cos 3t = 1 \qquad \text{Add 1 to both sides.}$$

$$\cos 3t = \frac{1}{2} \qquad \text{Divide both sides by 2.}$$

In the interval $[0, 2\pi)$, you know that $3t = \pi/3$ and $3t = 5\pi/3$ are the only solutions so that, in general, you have

$$3t = \frac{\pi}{3} + 2n\pi \qquad \text{and} \qquad 3t = \frac{5\pi}{3} + 2n\pi.$$

Dividing this result by 3, you obtain the general solution

$$t = \frac{\pi}{9} + \frac{2n\pi}{3} \qquad \text{and} \qquad t = \frac{5\pi}{9} + \frac{2n\pi}{3} \qquad \text{General solution}$$

where n is an integer. This is confirmed in Figure 5.11.

Figure 5.11

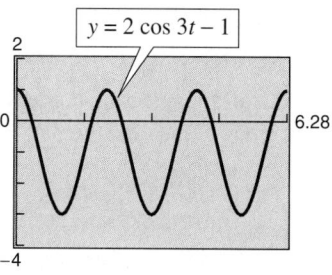

$y = 2 \cos 3t - 1$

EXAMPLE 9   **Functions of Multiple Angles**

Find all solutions of $3 \tan(x/2) + 3 = 0$.

Solution

$$3 \tan \frac{x}{2} + 3 = 0 \qquad \text{Original equation}$$

$$3 \tan \frac{x}{2} = -3 \qquad \text{Subtract 3 from both sides.}$$

$$\tan \frac{x}{2} = -1 \qquad \text{Divide both sides by 3.}$$

In the interval $[0, \pi)$, you know that $x/2 = 3\pi/4$ is the only solution so that, in general, you have

$$\frac{x}{2} = \frac{3\pi}{4} + n\pi.$$

Multiplying this result by 2, you obtain the general solution

$$x = \frac{3\pi}{2} + 2n\pi \qquad \text{General solution}$$

where n is an integer. This is confirmed in Figure 5.12.

Figure 5.12

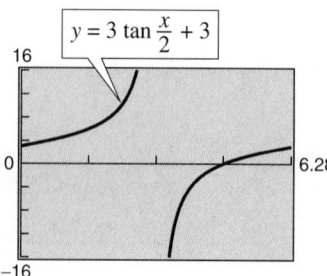

$y = 3 \tan \frac{x}{2} + 3$

Using Inverse Functions

EXAMPLE 10 ◢ **Using Inverse Functions**

Find all solutions of $\sec^2 x - 2 \tan x = 4$.

Solution

$\sec^2 x - 2 \tan x = 4$	Original equation
$1 + \tan^2 x - 2 \tan x - 4 = 0$	Pythagorean identity
$\tan^2 x - 2 \tan x - 3 = 0$	Combine like terms.
$(\tan x - 3)(\tan x + 1) = 0$	Factor.

Setting each factor equal to zero, you obtain two solutions in the interval $(-\pi/2, \pi/2)$. [Recall that the range of the inverse tangent function is $(-\pi/2, \pi/2)$.]

$$\tan x = 3, \qquad \tan x = -1$$

$$x = \arctan 3 \qquad x = -\frac{\pi}{4}$$

Finally, by adding multiples of π, you obtain the general solution

$$x = \arctan 3 + n\pi \quad \text{and} \quad x = -\frac{\pi}{4} + n\pi \qquad \text{General solution}$$

where n is an integer. This is confirmed in Figure 5.13. ◢

Figure 5.13

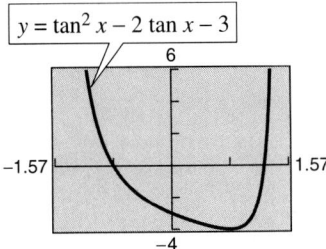

$$y = \tan^2 x - 2 \tan x - 3$$

With some trigonometric equations, there is no reasonable way to find the solutions analytically. In such cases, you can still use a graphing utility to approximate the solutions.

EXAMPLE 11 ◢ **Approximating Solutions**

Approximate the solutions of $x = 2 \sin x$.

Figure 5.14

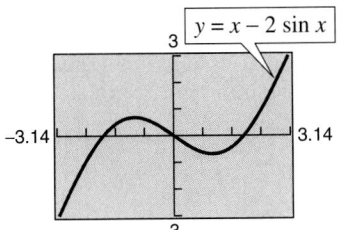

$$y = x - 2 \sin x$$

Solution

The graph of

$$y = x - 2 \sin x$$

is shown in Figure 5.14. From the graph, there appear to be three solutions. You can see that one solution is $x = 0$. Using the root feature of a graphing utility, you can approximate the positive solution as $x \approx 1.8955$. Finally, by symmetry, you can approximate the negative solution as $x \approx -1.8955$. ◢

Figure 5.15

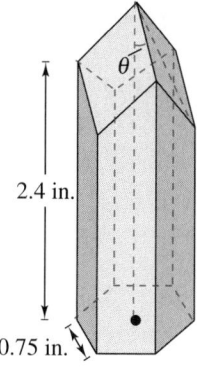

2.4 in.

0.75 in.

EXAMPLE 12   ◼   **Surface Area of a Honeycomb**

The surface area of a honeycomb is given by the equation

$$S = 6hs + \frac{3}{2}s^2\left(\frac{\sqrt{3} - \cos\theta}{\sin\theta}\right), \qquad 0 \le \theta \le 90°$$

where $h = 2.4$ inches, $s = 0.75$ inch, and θ is the angle indicated in Figure 5.15.

a. What value of θ gives a surface area of 12 square inches?

b. What value of θ gives the minimum surface area?

Solution

a. Let $h = 2.4$, $s = 0.75$, and $S = 12$.

$$10.8 + 0.84375\left(\frac{\sqrt{3} - \cos\theta}{\sin\theta}\right) = 12$$

$$0.84375\left(\frac{\sqrt{3} - \cos\theta}{\sin\theta}\right) - 1.2 = 0$$

The graph of

$$y = 0.84375\left(\frac{\sqrt{3} - \cos\theta}{\sin\theta}\right) - 1.2$$

is shown in Figure 5.16. Using the root feature of a graphing utility, you can determine that $\theta \approx 59.9°$ and $\theta \approx 49.9°$.

b. Graph the function

$$S = 10.8 + 0.84375\left(\frac{\sqrt{3} - \cos\theta}{\sin\theta}\right)$$

as shown in Figure 5.17. You can zoom in on the minimum point on the graph, which occurs at $\theta \approx 54.7°$. By using calculus, it can be shown that

$$\theta = \arccos\left(\frac{1}{\sqrt{3}}\right) \approx 54.7356°$$

is the exact minimum value.   ◼

Figure 5.16

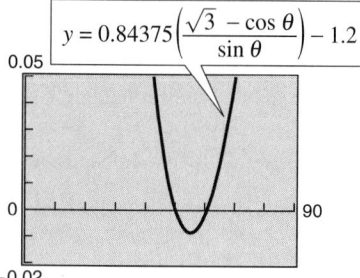

$y = 0.84375\left(\dfrac{\sqrt{3} - \cos\theta}{\sin\theta}\right) - 1.2$

0.05

0

90

−0.02

Figure 5.17

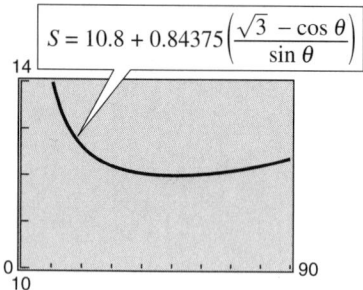

$S = 10.8 + 0.84375\left(\dfrac{\sqrt{3} - \cos\theta}{\sin\theta}\right)$

14

0
10

90

Note   If your graphing utility has a built-in program to find a minimum value of a function, try using the feature to confirm the solution in Example 12(b).

Group Activity   *Equations With No Solutions*

One of the following equations has solutions, and the other two don't. In your group, determine which two equations do not have solutions.

a. $\sin^2 x - 5 \sin x + 6 = 0$

b. $\sin^2 x - 4 \sin x + 6 = 0$

c. $\sin^2 x - 5 \sin x - 6 = 0$

Try to find conditions involving the constants b and c that will guarantee that the equation

$$\sin^2 x + b \sin x + c = 0$$

has at least one solution on some interval of length 2π.

5.3 /// EXERCISES

In Exercises 1–4, find the *x*-intercepts of the graph.

1. $y = \sin \dfrac{\pi x}{2} + 1$

2. $y = \sin \pi x + \cos \pi x$

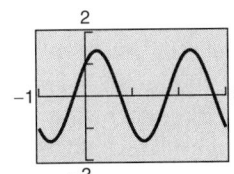

3. $y = \tan^2\left(\dfrac{\pi x}{6}\right) - 3$

4. $y = \sec^4\left(\dfrac{\pi x}{8}\right) - 4$

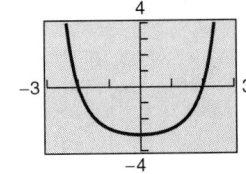

In Exercises 5–10, verify that the *x*-values are solutions of the equation.

5. $2 \cos x - 1 = 0$

(a) $x = \dfrac{\pi}{3}$ (b) $x = \dfrac{5\pi}{3}$

6. $\csc x - 2 = 0$

(a) $x = \dfrac{\pi}{6}$ (b) $x = \dfrac{5\pi}{6}$

7. $3 \tan^2 2x - 1 = 0$

(a) $x = \dfrac{\pi}{12}$ (b) $x = \dfrac{5\pi}{12}$

8. $2 \cos^2 4x - 1 = 0$

(a) $x = \dfrac{\pi}{16}$ (b) $x = \dfrac{3\pi}{16}$

9. $2 \sin^2 x - \sin x - 1 = 0$

(a) $x = \dfrac{\pi}{2}$ (b) $x = \dfrac{7\pi}{6}$

10. $\sec^4 x - 4 \sec^2 x = 0$

(a) $x = \dfrac{2\pi}{3}$ (b) $x = \dfrac{5\pi}{3}$

In Exercises 11–24, solve the equation.

11. $2 \cos x + 1 = 0$ **12.** $2 \sin x - 1 = 0$

13. $\sqrt{3} \csc x - 2 = 0$ **14.** $\tan x + 1 = 0$

15. $3 \sec^2 x - 4 = 0$ **16.** $\csc^2 x - 2 = 0$

17. $2 \sin^2 2x = 1$ **18.** $\tan^2 3x = 3$

19. $4 \sin^2 x - 3 = 0$ **20.** $\sin x(\sin x + 1) = 0$

21. $\sin^2 x = 3 \cos^2 x$ **22.** $\tan 3x(\tan x - 1) = 0$

23. $(3 \tan^2 x - 1)(\tan^2 x - 3) = 0$

24. $\cos 2x(2 \cos x + 1) = 0$

In Exercises 25–40, find all solutions of the equation in the interval $[0, 2\pi)$. Use a graphing utility to confirm your results.

25. $\cos^3 x = \cos x$ **26.** $\tan^2 x - 1 = 0$

27. $3 \tan^3 x = \tan x$ **28.** $2 \sin^2 x = 2 + \cos x$

29. $\sec^2 x - \sec x = 2$ **30.** $\sec x \csc x = 2 \csc x$

31. $2 \sin x + \csc x = 0$ **32.** $\sin 2x = -\dfrac{\sqrt{3}}{2}$

33. $\csc x + \cot x = 1$ **34.** $\tan 3x = 1$

35. $\cos \dfrac{x}{2} = \dfrac{\sqrt{2}}{2}$ **36.** $\sec 4x = 2$

37. $\dfrac{1 + \cos x}{1 - \cos x} = 0$

38. $2 \sin^2 x + 3 \sin x + 1 = 0$

39. $2 \sec^2 x + \tan^2 x - 3 = 0$

40. $\cos x + \sin x \tan x = 2$

In Exercises 41–50, use a graphing utility to approximate the solutions of the equation in the interval $[0, 2\pi)$.

41. $\dfrac{\cos x \cot x}{1 - \sin x} = 3$

42. $2 \cos x - \sin x = 0$

43. $4 \sin^3 x + 2 \sin^2 x - 2 \sin x - 1 = 0$

44. $\dfrac{1 + \sin x}{\cos x} + \dfrac{\cos x}{1 + \sin x} = 4$

45. $2 \sin x - x = 0$ **46.** $x \cos x - 1 = 0$

47. $\sec^2 x + 0.5 \tan x = 1$

48. $\csc^2 x + 0.5 \cot x = 5$

49. $12 \sin^2 x - 13 \sin x + 3 = 0$

50. $3 \tan^2 x + 4 \tan x - 4 = 0$

In Exercises 51–54, consider the function in the interval $[0, 2\pi)$. (a) Use a graphing utility to complete the table. Then make a conjecture about any unit intervals containing a zero of the function. Give a reason for your answer. (b) Use a graphing utility to graph the function and determine any unit intervals containing a zero of the function. Do the results agree with those of part (a)? Explain. (c) Use a graphing utility to approximate the zero of the function.

x	0	1	2	3	4	5	6
$f(x)$							

51. $f(x) = \cot \dfrac{x}{2} - x$ **52.** $f(x) = x \sec x - 1$

53. $f(x) = \sin^2 x + 2 \sin x - 1$

54. $f(x) = 4 \cos^2 x - 4 \cos x - 1$

In Exercises 55 and 56, (a) use a graphing utility to graph the function and approximate the maximum and minimum points on the graph in the interval $[0, 2\pi)$, and (b) solve the trigonometric equation and demonstrate that its solutions are the x-coordinates of the maximum and minimum points of f (calculus is required to find the trigonometric equation).

Function	*Trigonometric Equation*
55. $f(x) = \sin x + \cos x$	$\cos x - \sin x = 0$
56. $f(x) = 2 \sin x + \cos 2x$	$2 \cos x - 4 \sin x \cos x = 0$

Fixed Point **In Exercises 57 and 58, find the smallest positive fixed point of the function f. [A fixed point of a function f is a real number c such that $f(c) = c$.]**

57. $f(x) = \tan \dfrac{\pi x}{4}$ **58.** $f(x) = \cos x$

59. *Graphical Reasoning* Consider the function

$$f(x) = \cos \frac{1}{x}$$

and its graph shown in the figure.

(a) What is the domain of the function?

(b) Identify any symmetry or asymptotes of the graph.

(c) Describe the behavior of the function as $x \to 0$.

(d) How many solutions does the equation

$$\cos \frac{1}{x} = 0$$

have in the interval $[-1, 1]$?

(e) Does the equation $\cos(1/x) = 0$ have a greatest solution? If so, approximate the solution. If not, explain.

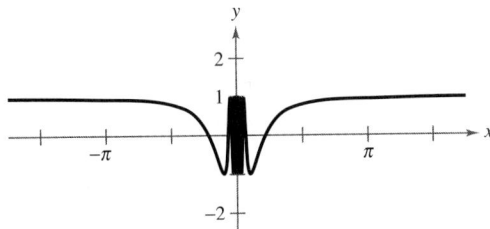

60. *Graphical Reasoning* Consider the function

$$f(x) = \frac{\sin x}{x}$$

and its graph shown in the figure.

(a) What is the domain of the function?

(b) Identify any symmetry or asymptotes of the graph.

(c) Describe the behavior of the function as $x \to 0$.

(d) How many solutions does the equation

$$\frac{\sin x}{x} = 0$$

have in the interval $[-8, 8]$? Find the solutions.

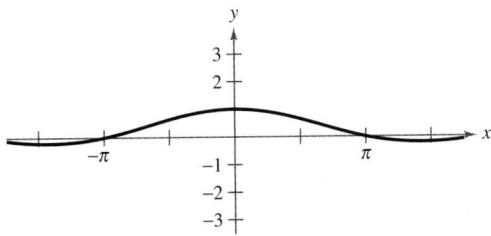

61. *Harmonic Motion* A weight is oscillating on the end of a spring. The position of the weight relative to the point of equilibrium is given by

$$y = \tfrac{1}{12}(\cos 8t - 3 \sin 8t)$$

where y is the displacement in meters and t is the time in seconds (see figure). Find the times when the weight is at the point of equilibrium $(y = 0)$ for $0 \le t \le 1$.

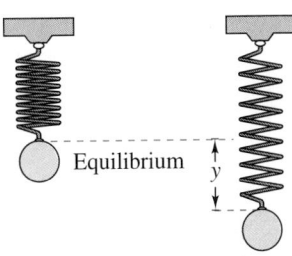

Equilibrium

62. *Sales* The monthly sales (in thousands of units) of a seasonal product are approximated by

$$S = 74.50 + 43.75 \sin \frac{\pi t}{6}$$

where t is the time in months, with $t = 1$ corresponding to January. Determine the months when sales exceed 100,000 units.

63. *Projectile Motion* A batted baseball leaves the bat at an angle of θ with the horizontal and an initial velocity of $v_0 = 100$ feet per second. The ball is caught by an outfielder 300 feet from home plate (see figure). Find θ if the range r of a projectile is given by

$$r = \tfrac{1}{32} v_0^2 \sin 2\theta.$$

$\longleftarrow$ $r = 300$ ft $\longrightarrow$

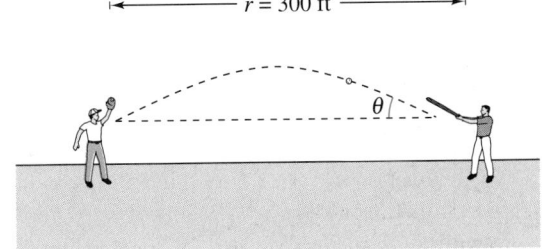

64. *Projectile Motion*   A sharpshooter intends to hit a target at a distance of 1000 yards (see figure) with a gun that has a muzzle velocity of 1200 feet per second. Neglecting air resistance, determine the minimum angle of elevation of the gun if the range is given by

$$r = \frac{1}{32}v_0^2 \sin 2\theta.$$

r = 1000 yd

65. *Area*   The area of a rectangle (see figure) inscribed in one arch of the graph of $y = \cos x$ is given by

$$A = 2x \cos x, \qquad -\frac{\pi}{2} < x < \frac{\pi}{2}.$$

(a) Use a graphing utility to graph the area function, and approximate the area of the largest inscribed rectangle.

(b) Determine the values of x for which $A \geq 1$.

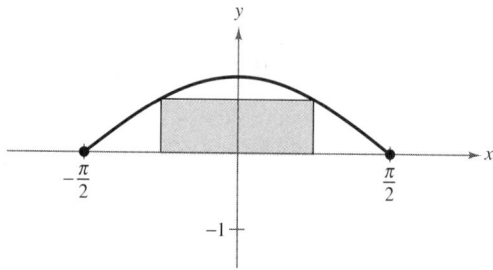

66. *Damped Harmonic Motion*   The displacement from equilibrium of a weight oscillating on the end of a spring is given by

$$y = 1.56e^{-0.22t} \cos 4.9t,$$

where y is the displacement in feet and t is the time in seconds. Use a graphing utility to graph the displacement function for $0 \leq t \leq 10$. Find the time beyond which the displacement does not exceed 1 foot from equilibrium.

67. *Data Analysis*   The table gives the unemployment rate r for the years 1985 through 1994 in the United States. The time t is measured in years, with $t = 0$ corresponding to 1990. (Source: U.S. Bureau of Labor Statistics)

t	-5	-4	-3	-2	-1
r	7.2	7.0	6.2	5.5	5.3

t	0	1	2	3	4
r	5.5	6.7	7.4	6.8	6.1

(a) Create a scatter plot of the data.

(b) Which of the following models best represents the data? Explain your reasoning.

   (i) $r = 1.5 \cos(t + 3.9) + 6.37$

   (ii) $r = 1.03 \sin(0.9t + 0.44) + 6.19$

   (iii) $r = 1.05 \sin[0.95(t + 6.32)] + 6.20$

   (iv) $r = 1.5 \sin[0.5(t + 2.8)] + 6.25$

(c) What term in the model gives the average unemployment rate? What is the rate?

(d) Economists study the lengths of business cycles, such as cycles in the unemployment rate. Based on this short span of time, use the model to give the length of this cycle.

(e) Use the model to estimate the next time the unemployment rate will be 6% or less.

68. *Quadratic Approximation*   Consider the function

$$f(x) = 3 \sin(0.6x - 2).$$

(a) Approximate the zero of the function in the interval $[0, 6]$.

(b) A quadratic approximation agreeing with f at $x = 5$ is given by

$$g(x) = -0.45x^2 + 5.52x - 13.70.$$

Use a graphing utility to graph f and g on the same viewing rectangle. Describe the result.

(c) Use the Quadratic Formula to find the zeros of g. Compare the zero in the interval $[0, 6]$ with the result of part (a).

5.4   Sum and Difference Formulas

Introduction / Using Sum and Difference Formulas

Introduction

In this and the following section, you will study the derivations and uses of several trigonometric identities and formulas.

Sum and Difference Formulas

$$\sin(u + v) = \sin u \cos v + \cos u \sin v \qquad \tan(u + v) = \frac{\tan u + \tan v}{1 - \tan u \tan v}$$

$$\sin(u - v) = \sin u \cos v - \cos u \sin v$$

$$\cos(u + v) = \cos u \cos v - \sin u \sin v$$

$$\cos(u - v) = \cos u \cos v + \sin u \sin v \qquad \tan(u - v) = \frac{\tan u - \tan v}{1 + \tan u \tan v}$$

Figure 5.18

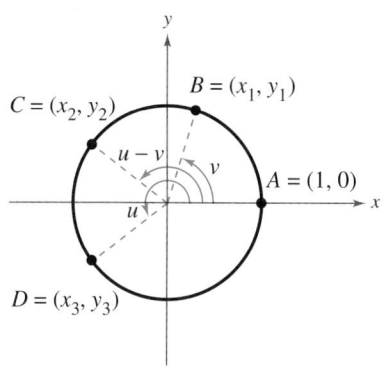

Proof /// Here are proofs for the formulas for $\cos(u \pm v)$. In Figure 5.18, let A be the point $(1, 0)$ and then use u and v to locate the points $B = (x_1, y_1)$, $C = (x_2, y_2)$, and $D = (x_3, y_3)$ on the unit circle. Thus, $x_i^2 + y_i^2 = 1$ for $i = 1, 2, 3$. For convenience, assume that $0 < v < u < 2\pi$. From Figure 5.19, note that arcs AC and BD have the same length. Hence, *line segments AC and BD are also equal in length*, which implies that

$$\sqrt{(x_2 - 1)^2 + (y_2 - 0)^2} = \sqrt{(x_3 - x_1)^2 + (y_3 - y_1)^2}$$

$$x_2^2 - 2x_2 + 1 + y_2^2 = x_3^2 - 2x_1x_3 + x_1^2 + y_3^2 - 2y_1y_3 + y_1^2$$

$$(x_2^2 + y_2^2) + 1 - 2x_2 = (x_3^2 + y_3^2) + (x_1^2 + y_1^2) - 2x_1x_3 - 2y_1y_3$$

$$1 + 1 - 2x_2 = 1 + 1 - 2x_1x_3 - 2y_1y_3$$

$$x_2 = x_3x_1 + y_3y_1.$$

Figure 5.19

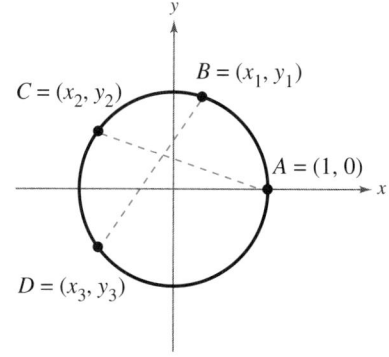

Finally, by substituting the values $x_2 = \cos(u - v)$, $x_3 = \cos u$, $x_1 = \cos v$, $y_3 = \sin u$, and $y_1 = \sin v$, you obtain

$$\cos(u - v) = \cos u \cos v + \sin u \sin v.$$

The formula for $\cos(u + v)$ can be established by considering $u + v = u - (-v)$ and using the formula just derived to obtain

$$\cos(u + v) = \cos[u - (-v)]$$
$$= \cos u \cos(-v) + \sin u \sin(-v)$$
$$= \cos u \cos v - \sin u \sin v. \qquad \qquad ///$$

Note Note that $\sin(u + v) \neq \sin u + \sin v$. Similar statements can be made for $\cos(u + v)$ and $\tan(u + v)$.

Using Sum and Difference Formulas

In the remainder of this section, you will study a variety of uses of sum and difference formulas. For instance, Examples 1 and 2 show how sum and difference formulas can be used to find exact values of trigonometric functions involving sums or differences of special angles.

Note   Try checking the result obtained in Example 1 on your calculator. You will find that $\cos 75° \approx 0.259$.

EXAMPLE 1   Evaluating a Trigonometric Function

Find the exact value of $\cos 75°$.

Solution
To find the *exact* value of $\cos 75°$, use the fact that $75° = 30° + 45°$. Consequently, the formula for $\cos(u + v)$ yields

$$\cos 75° = \cos(30° + 45°)$$
$$= \cos 30° \cos 45° - \sin 30° \sin 45°$$
$$= \frac{\sqrt{3}}{2}\left(\frac{\sqrt{2}}{2}\right) - \frac{1}{2}\left(\frac{\sqrt{2}}{2}\right)$$
$$= \frac{\sqrt{6} - \sqrt{2}}{4}.$$

EXAMPLE 2   Evaluating a Trigonometric Function

Find the exact value of $\sin \dfrac{\pi}{12}$.

EXPLORATION

Use a graphing utility to graph $y = \cos(x + 2)$ and $y = \cos x + \cos 2$ in the same viewing rectangle. What can you conclude about the graphs? Is it true that $\cos(x + 2) = \cos x + \cos 2$?

Use a graphing utility to graph $y = \sin(x + 4)$ and $y = \sin x + \sin 4$ in the same viewing rectangle. What can you conclude about the graphs? Is it true that $\sin(x + 4) = \sin x + \sin 4$?

Solution
Using the fact that

$$\frac{\pi}{12} = \frac{\pi}{3} - \frac{\pi}{4}$$

together with the formula for $\sin(u - v)$, you obtain

$$\sin \frac{\pi}{12} = \sin\left(\frac{\pi}{3} - \frac{\pi}{4}\right)$$
$$= \sin \frac{\pi}{3} \cos \frac{\pi}{4} - \cos \frac{\pi}{3} \sin \frac{\pi}{4}$$
$$= \frac{\sqrt{3}}{2}\left(\frac{\sqrt{2}}{2}\right) - \frac{1}{2}\left(\frac{\sqrt{2}}{2}\right)$$
$$= \frac{\sqrt{6} - \sqrt{2}}{4}.$$

EXAMPLE 3   Evaluating a Trigonometric Expression

Find the exact value of $\sin 42° \cos 12° - \cos 42° \sin 12°$.

Solution

Recognizing that this expression fits the formula for $\sin(u - v)$, you can write

$$\sin 42° \cos 12° - \cos 42° \sin 12° = \sin(42° - 12°)$$

$$= \sin 30° = \frac{1}{2}.$$

Figure 5.20

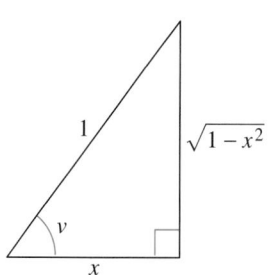

EXAMPLE 4   An Application of a Sum Formula

Evaluate $\cos(\arctan 1 + \arccos x)$.

Solution

This expression fits the formula for $\cos(u + v)$. Angles $u = \arctan 1$ and $v = \arccos x$ are shown in Figure 5.20. Then

$$\cos(u + v) = \cos(\arctan 1)\cos(\arccos x) - \sin(\arctan 1)\sin(\arccos x)$$

$$= \frac{1}{\sqrt{2}} \cdot x - \frac{1}{\sqrt{2}} \cdot \sqrt{1 - x^2}$$

$$= \frac{x - \sqrt{1 - x^2}}{\sqrt{2}}.$$

EXAMPLE 5   Proving a Cofunction Identity

Prove the cofunction identity $\cos\left(\dfrac{\pi}{2} - x\right) = \sin x$.

Solution

Using the formula for $\cos(u - v)$, you have

$$\cos\left(\frac{\pi}{2} - x\right) = \cos\frac{\pi}{2}\cos x + \sin\frac{\pi}{2}\sin x$$

$$= (0)(\cos x) + (1)(\sin x)$$

$$= \sin x.$$

Sum and difference formulas can be used to derive **reduction formulas** involving expressions such as

$$\sin\left(\theta + \frac{n\pi}{2}\right) \quad \text{and} \quad \cos\left(\theta + \frac{n\pi}{2}\right)$$

where n is an integer, as shown in the following example.

EXAMPLE 6   ▱   **Deriving Reduction Formulas**

Simplify each expression.

a. $\cos\left(\theta - \dfrac{3\pi}{2}\right)$ **b.** $\tan(\theta + 3\pi)$

Solution

a. Using the formula for $\cos(u - v)$, you have

$$\cos\left(\theta - \frac{3\pi}{2}\right) = \cos\theta\cos\frac{3\pi}{2} + \sin\theta\sin\frac{3\pi}{2}$$

$$= (\cos\theta)(0) + (\sin\theta)(-1)$$

$$= -\sin\theta.$$

b. Using the formula for $\tan(u + v)$, you have

$$\tan(\theta + 3\pi) = \frac{\tan\theta + \tan 3\pi}{1 - \tan\theta\tan 3\pi}$$

$$= \frac{\tan\theta + 0}{1 - (\tan\theta)(0)}$$

$$= \tan\theta.$$

▱

The next example was taken from calculus. It is used to derive the formula for the derivative of the sine function.

EXAMPLE 7   ▱   **An Application from Calculus**

Verify that

$$\frac{\sin(x + h) - \sin x}{h} = (\cos x)\left(\frac{\sin h}{h}\right) - (\sin x)\left(\frac{1 - \cos h}{h}\right)$$

where $h \neq 0$.

Solution

Using the formula for $\sin(u + v)$, you have

$$\frac{\sin(x + h) - \sin x}{h} = \frac{\sin x\cos h + \cos x\sin h - \sin x}{h}$$

$$= \frac{\cos x\sin h - \sin x(1 - \cos h)}{h}$$

$$= (\cos x)\left(\frac{\sin h}{h}\right) - (\sin x)\left(\frac{1 - \cos h}{h}\right).$$

▱

EXAMPLE 8   Solving a Trigonometric Equation

Find all solutions of

$$\sin\left(x + \frac{\pi}{4}\right) + \sin\left(x - \frac{\pi}{4}\right) = -1$$

in the interval $[0, 2\pi)$.

Figure 5.21

$y = 1 + \sin\left(x + \frac{\pi}{4}\right) + \sin\left(x - \frac{\pi}{4}\right)$

Solution

Using sum and difference formulas, rewrite the given equation as

$$\sin x \cos \frac{\pi}{4} + \cos x \sin \frac{\pi}{4} + \sin x \cos \frac{\pi}{4} - \cos x \sin \frac{\pi}{4} = -1$$

$$2 \sin x \cos \frac{\pi}{4} = -1$$

$$2(\sin x)\left(\frac{\sqrt{2}}{2}\right) = -1$$

$$\sin x = -\frac{1}{\sqrt{2}}$$

$$\sin x = -\frac{\sqrt{2}}{2}.$$

Therefore, the only solutions in the interval $[0, 2\pi)$ are

$$x = \frac{5\pi}{4} \qquad \text{and} \qquad x = \frac{7\pi}{4}.$$

These solutions are checked graphically in Figure 5.21.

Group Activity

The Angle Between Two Lines

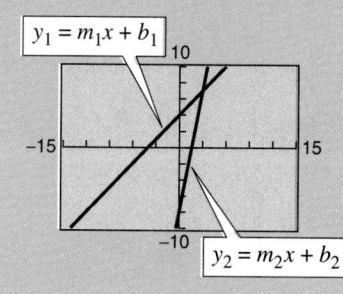

$y_1 = m_1x + b_1$

$y_2 = m_2x + b_2$

The figure at the left shows two lines whose equations are

$$y_1 = m_1x + b_1 \qquad \text{and} \qquad y_2 = m_2x + b_2.$$

Assume that both lines have positive slopes, as shown in the figure. With others in your group, derive a formula for the angle between the two lines. Then use your formula to find the angle between the following two lines.

a. $y = x$ and $y = \sqrt{3}x$

b. $y = x$ and $y = \frac{1}{\sqrt{3}}x$

5.4 /// EXERCISES

In Exercises 1–4, find the exact value of each expression.

1. (a) $\cos\left(\dfrac{\pi}{4} + \dfrac{\pi}{3}\right)$ (b) $\cos\dfrac{\pi}{4} + \cos\dfrac{\pi}{3}$

2. (a) $\sin\left(\dfrac{3\pi}{4} + \dfrac{5\pi}{6}\right)$ (b) $\sin\dfrac{3\pi}{4} + \sin\dfrac{5\pi}{6}$

3. (a) $\sin\left(\dfrac{7\pi}{6} - \dfrac{\pi}{3}\right)$ (b) $\sin\dfrac{7\pi}{6} - \sin\dfrac{\pi}{3}$

4. (a) $\cos\left(\dfrac{2\pi}{3} - \dfrac{\pi}{6}\right)$ (b) $\cos\dfrac{2\pi}{3} + \cos\dfrac{\pi}{6}$

5. *Think About It*   Use the results of Exercises 1–4 to determine if the following are true or false. Explain.

(a) $\sin(u \pm v) = \sin u \pm \sin v$

(b) $\cos(u \pm v) = \cos u \pm \cos v$

6. *True or False?*   It is not possible to find the exact value of sin 75°. If false, find the exact value.

In Exercises 7–16, use the sum and difference identities to find the exact values of the sine, cosine, and tangent of the angle.

7. $75° = 30° + 45°$ **8.** $15° = 45° - 30°$

9. $105° = 60° + 45°$ **10.** $165° = 135° + 30°$

11. $195° = 225° - 30°$ **12.** $255° = 300° - 45°$

13. $\dfrac{11\pi}{12} = \dfrac{3\pi}{4} + \dfrac{\pi}{6}$ **14.** $\dfrac{7\pi}{12} = \dfrac{\pi}{3} + \dfrac{\pi}{4}$

15. $\dfrac{17\pi}{12} = \dfrac{9\pi}{4} - \dfrac{5\pi}{6}$ **16.** $-\dfrac{\pi}{12} = \dfrac{\pi}{6} - \dfrac{\pi}{4}$

In Exercises 17–26, use the sum and difference identities to write the expression as the sine, cosine, or tangent of an angle.

17. $\cos 25° \cos 15° - \sin 25° \sin 15°$

18. $\sin 140° \cos 50° + \cos 140° \sin 50°$

19. $\sin 230° \cos 30° - \cos 230° \sin 30°$

20. $\cos 20° \cos 30° + \sin 20° \sin 30°$

21. $\dfrac{\tan 325° - \tan 86°}{1 + \tan 325° \tan 86°}$

22. $\dfrac{\tan 140° - \tan 60°}{1 + \tan 140° \tan 60°}$

23. $\sin 3 \cos 1.2 - \cos 3 \sin 1.2$

24. $\cos\dfrac{\pi}{7}\cos\dfrac{\pi}{5} - \sin\dfrac{\pi}{7}\sin\dfrac{\pi}{5}$

25. $\dfrac{\tan 2x + \tan x}{1 - \tan 2x \tan x}$

26. $\cos 3x \cos 2y + \sin 3x \sin 2y$

Numerical, Graphical, and Analytical Analysis   In Exercises 27–32, use a graphing utility to complete the table and graph the two functions. Use both as evidence that $y_1 = y_2$. Then verify the identity analytically.

x	0.2	0.4	0.6	0.8	1.0	1.2	1.4
y_1							
y_2							

27. $y_1 = \sin\left(\dfrac{\pi}{2} + x\right)$, $y_2 = \cos x$

28. $y_1 = \sin(3\pi - x)$, $y_2 = \sin x$

29. $y_1 = \sin\left(\dfrac{\pi}{6} + x\right)$, $y_2 = \dfrac{1}{2}(\cos x + \sqrt{3}\sin x)$

30. $y_1 = \cos\left(\dfrac{5\pi}{4} - x\right)$, $y_2 = -\dfrac{\sqrt{2}}{2}(\cos x + \sin x)$

31. $y_1 = \cos(x + \pi)\cos(x - \pi)$, $y_2 = \cos^2 x$

32. $y_1 = \sin(x + \pi)\sin(x - \pi)$, $y_2 = \sin^2 x$

In Exercises 33–36, find the exact value of the trigonometric function given that

$$\sin u = 5/13, \quad \text{where } 0 < u < \pi/2$$
$$\cos v = -3/5, \quad \text{where } \pi/2 < v < \pi.$$

33. $\sin(u + v)$

34. $\cos(v - u)$

35. $\cos(u + v)$

36. $\sin(u - v)$

In Exercises 37–40, find the exact value of the trigonometric function given that

$$\sin u = 7/25, \quad \text{where } \pi/2 < u < \pi$$
$$\cos v = 4/5, \quad \text{where } 3\pi/2 < v < 2\pi.$$

37. $\cos(u + v)$

38. $\sin(u + v)$

39. $\sin(v - u)$

40. $\cos(u - v)$

In Exercises 41–48, verify the identity.

41. $\cos(\pi - \theta) + \sin\left(\dfrac{\pi}{2} + \theta\right) = 0$

42. $\tan\left(\dfrac{\pi}{4} - \theta\right) = \dfrac{1 - \tan \theta}{1 + \tan \theta}$

43. $\sin(x + y) + \sin(x - y) = 2 \sin x \cos y$

44. $\cos(x + y) + \cos(x - y) = 2 \cos x \cos y$

45. $\cos(n\pi + \theta) = (-1)^n \cos \theta, \quad n$ is an integer.

46. $\sin(n\pi + \theta) = (-1)^n \sin \theta, \quad n$ is an integer.

47. $a \sin B\theta + b \cos B\theta = \sqrt{a^2 + b^2} \sin(B\theta + C)$
where $C = \arctan(b/a), a > 0$

48. $a \sin B\theta + b \cos B\theta = \sqrt{a^2 + b^2} \cos(B\theta - C)$
where $C = \arctan(a/b), b > 0$

Conjecture In Exercises 49 and 50, use a graphing utility to graph the trigonometric function. Use the graph to make a conjecture about a simplification of the expression. Verify the resulting identity analytically.

49. $g(x) = \cos(\pi + x)$

50. $h(\theta) = \tan(\pi + \theta)$

In Exercises 51–54, use the formulas given in Exercises 47 and 48 to write the expression in the following forms. Use a graphing utility to verify your results.

(a) $\sqrt{a^2 + b^2} \sin(B\theta + C)$

(b) $\sqrt{a^2 + b^2} \cos(B\theta - C)$

51. $\sin \theta + \cos \theta$

52. $3 \sin 2\theta + 4 \cos 2\theta$

53. $12 \sin 3\theta + 5 \cos 3\theta$

54. $\sin 2\theta - \cos 2\theta$

In Exercises 55 and 56, use the formulas given in Exercises 47 and 48 to write the trigonometric expression in the form $a \sin B\theta + b \cos B\theta$.

55. $2 \sin\left(\theta + \dfrac{\pi}{2}\right)$

56. $5 \cos\left(\theta + \dfrac{3\pi}{4}\right)$

In Exercises 57 and 58, write the trigonometric expression as an algebraic expression.

57. $\sin(\arcsin x + \arccos x)$

58. $\sin(\arctan 2x - \arccos x)$

In Exercises 59–62, find all solutions of the equation in the interval $[0, 2\pi)$. Use a graphing utility to verify your results.

59. $\sin\left(x + \dfrac{\pi}{3}\right) + \sin\left(x - \dfrac{\pi}{3}\right) = 1$

60. $\sin\left(x + \dfrac{\pi}{6}\right) - \sin\left(x - \dfrac{\pi}{6}\right) = \dfrac{1}{2}$

61. $\cos\left(x + \dfrac{\pi}{4}\right) - \cos\left(x - \dfrac{\pi}{4}\right) = 1$

62. $\tan(x + \pi) + 2 \sin(x + \pi) = 0$

In Exercises 63 and 64, use a graphing utility to approximate all solutions of the equation in the interval $[0, 2\pi)$.

63. $\cos\left(x + \dfrac{\pi}{4}\right) + \cos\left(x - \dfrac{\pi}{4}\right) = 1$

64. $\tan(x + \pi) - \cos\left(x + \dfrac{\pi}{2}\right) = 0$

65. *Standing Waves*   The equation of a standing wave is obtained by adding the displacements of two waves traveling in opposite directions (see figure). Assume that each of the waves has amplitude A, period T, and wavelength λ. If the models for these waves are

$$y_1 = A \cos 2\pi \left(\frac{t}{T} - \frac{x}{\lambda} \right)$$

and

$$y_2 = A \cos 2\pi \left(\frac{t}{T} + \frac{x}{\lambda} \right)$$

show that

$$y_1 + y_2 = 2A \cos \frac{2\pi t}{T} \cos \frac{2\pi x}{\lambda}.$$

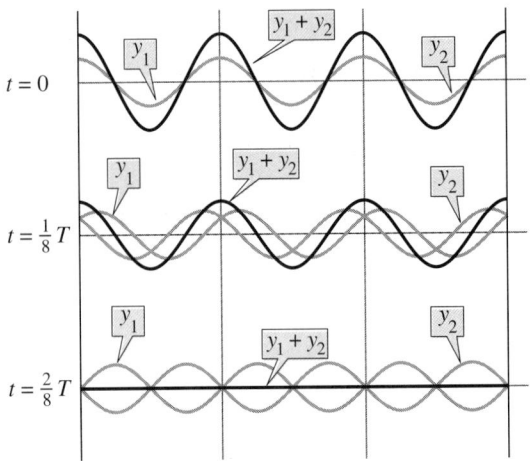

66. *Conjecture*   Three squares of side s are placed side by side (see figure). Make a conjecture about the relationship between the sum $u + v$ and w. Prove your conjecture by using the identity for the tangent of the sum of two angles.

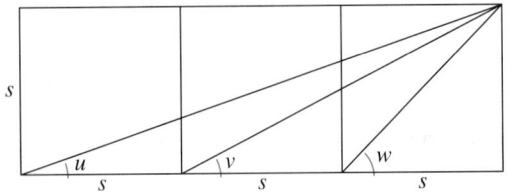

67. *Harmonic Motion*   A weight is attached to a spring suspended vertically from a ceiling. When a driving force is applied to the system, the weight moves vertically from its equilibrium position, and this motion is modeled by

$$y = \tfrac{1}{3} \sin 2t + \tfrac{1}{4} \cos 2t$$

where y is the distance from equilibrium measured in feet and t is the time in seconds.

(a) Use a graphing utility to graph the model.

(b) Write the model in the form

$$y = \sqrt{a^2 + b^2} \sin(Bt + C).$$

(See Exercise 47.) Use a graphing utility to verify your result.

(c) Find the amplitude of the oscillations of the weight.

(d) Find the frequency of the oscillations of the weight.

68. Verify the identity used in calculus.

$$\frac{\cos(x + h) - \cos x}{h} = \frac{\cos x(\cos h - 1)}{h} - \frac{\sin x \sin h}{h}$$

69. *Exploration*   Let $x = \frac{\pi}{6}$ in the identity in Exercise 68 and define the functions f and g as follows.

$$f(h) = \frac{\cos\left(\frac{\pi}{6} + h\right) - \cos \frac{\pi}{6}}{h}$$

$$g(h) = \cos \frac{\pi}{6} \left(\frac{\cos h - 1}{h} \right) - \sin \frac{\pi}{6} \left(\frac{\sin h}{h} \right)$$

(a) What are the domains of these functions?

(b) Use a graphing utility to complete the table.

h	0.01	0.02	0.05	0.1	0.2	0.5
$f(h)$						
$g(h)$						

(c) Use a graphing utility to graph the functions.

(d) Use the table and graph to make a conjecture about the values of the functions as $h \to 0$.

70. Use the sum formulas for sine and cosine to derive the formula

$$\tan(u + v) = \frac{\tan u + \tan v}{1 - \tan u \tan v}.$$

5.5 Multiple-Angle and Product-Sum Formulas

Multiple-Angle Formulas / *Power-Reducing Formulas* /
Half-Angle Formulas / *Product-to-Sum Formulas*

Multiple-Angle Formulas

In this section you will study four other categories of trigonometric identities.

1. The first category involves functions of multiple angles such as $\sin ku$ and $\cos ku$.
2. The second category involves squares of trigonometric functions such as $\sin^2 u$.
3. The third category involves functions of half-angles such as $\sin(u/2)$.
4. The fourth category involves products of trigonometric functions such as $\sin u \cos v$.

The most commonly used multiple-angle formulas are the **double-angle formulas.** They are used often, so you should learn them.

Double-Angle Formulas

$$\sin 2u = 2 \sin u \cos u \qquad \tan 2u = \frac{2 \tan u}{1 - \tan^2 u}$$

$$\cos 2u = \cos^2 u - \sin^2 u$$
$$= 2 \cos^2 u - 1$$
$$= 1 - 2 \sin^2 u$$

Note Note that $\sin 2u \neq 2 \sin u$. Similar statements can be made for $\cos 2u$ and $\tan 2u$.

Proof /// To prove the first formula, let $v = u$ in the formula for $\sin(u + v)$.

$$\sin 2u = \sin(u + u)$$
$$= \sin u \cos u + \cos u \sin u$$
$$= 2 \sin u \cos u$$

To prove the second formula, let $v = u$ in the formula for $\cos(u + v)$.

$$\cos 2u = \cos(u + u)$$
$$= \cos u \cos u - \sin u \sin u$$
$$= \cos^2 u - \sin^2 u$$

The tangent double-angle formula can be proven in a similar way. ///

EXAMPLE 1 ⬭ **Solving a Trigonometric Equation**

Find all solutions of $2 \cos x + \sin 2x = 0$.

Solution

Begin by rewriting the equation so that it involves functions of x (rather than $2x$). Then factor and solve as usual.

$2 \cos x + \sin 2x = 0$	Original equation
$2 \cos x + 2 \sin x \cos x = 0$	Double-angle formula
$2 \cos x(1 + \sin x) = 0$	Factor.
$\cos x = 0, \quad 1 + \sin x = 0$	Set factors equal to zero.
$x = \dfrac{\pi}{2}, \dfrac{3\pi}{2} \qquad x = \dfrac{3\pi}{2}$	Solutions in $[0, 2\pi)$

Therefore, the general solution is

$$x = \frac{\pi}{2} + 2n\pi \quad \text{and} \quad x = \frac{3\pi}{2} + 2n\pi$$

where n is an integer. The graph of $y = 2 \cos x + \sin 2x$, as shown in Figure 5.22, allows you to verify these solutions graphically. ⬭

Figure 5.22

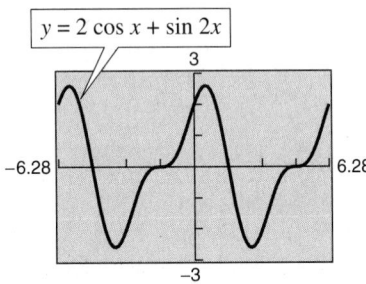

$y = 2 \cos x + \sin 2x$

EXAMPLE 2 ⬭ **Using Double-Angle Formulas in Sketching Graphs**

Analyze the graph of $y = 4 \cos^2 x - 2$ over the interval $[0, 2\pi]$.

Solution

Using a double-angle formula, you can rewrite the given function as

$$
\begin{aligned}
y &= 4 \cos^2 x - 2 \\
&= 2(2 \cos^2 x - 1) \\
&= 2 \cos 2x.
\end{aligned}
$$

Using the techniques discussed in Section 4.5, you can recognize that the graph of this function has an amplitude of 2 and a period of π. The key points in the interval $[0, \pi]$ are as follows.

Maximum	Intercept	Minimum	Intercept	Maximum
$(0, 2)$	$\left(\dfrac{\pi}{4}, 0\right)$	$\left(\dfrac{\pi}{2}, -2\right)$	$\left(\dfrac{3\pi}{4}, 0\right)$	$(\pi, 2)$

Two cycles of the graph are shown in Figure 5.23. ⬭

Figure 5.23

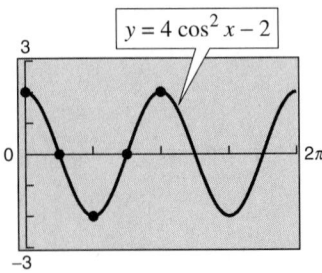

$y = 4 \cos^2 x - 2$

EXAMPLE 3   Evaluating Functions Involving Double Angles

Use the following to find $\sin 2\theta$, $\cos 2\theta$, and $\tan 2\theta$.

$$\cos \theta = \frac{5}{13}, \qquad \frac{3\pi}{2} < \theta < 2\pi$$

Figure 5.24

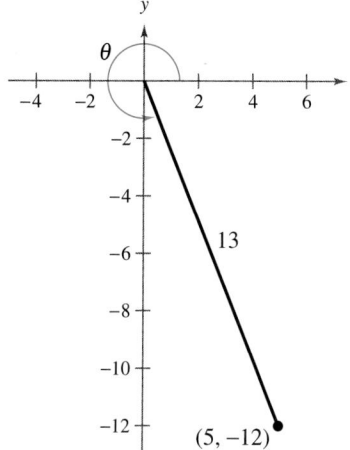

Solution

From Figure 5.24, you can see that $\sin \theta = y/r = -12/13$. Consequently, you can write the following.

$$\sin 2\theta = 2 \sin \theta \cos \theta = 2\left(\frac{-12}{13}\right)\left(\frac{5}{13}\right) = -\frac{120}{169}$$

$$\cos 2\theta = 2 \cos^2 \theta - 1 = 2\left(\frac{25}{169}\right) - 1 = -\frac{119}{169}$$

$$\tan 2\theta = \frac{\sin 2\theta}{\cos 2\theta} = \frac{120}{119}$$

The double-angle formulas are not restricted to angles 2θ and θ. Other *double* combinations, such as 4θ and 2θ or 6θ and 3θ, are also valid. Here are two examples.

$$\sin 4\theta = 2 \sin 2\theta \cos 2\theta \qquad \text{and} \qquad \cos 6\theta = \cos^2 3\theta - \sin^2 3\theta$$

By using double-angle formulas together with the sum formulas derived in the previous section, you can form other multiple-angle formulas.

Study Tip

Notice that you cannot solve
Example 3 by simply evaluating
$\theta = \cos^{-1}(5/13)$, and then calculating $\sin 2\theta$, $\cos 2\theta$, and $\tan 2\theta$.
Do you see why?

EXAMPLE 4   Deriving a Triple-Angle Formula

Express $\sin 3x$ in terms of $\sin x$.

Solution

$$\begin{aligned}
\sin 3x &= \sin(2x + x) \\
&= \sin 2x \cos x + \cos 2x \sin x \\
&= 2 \sin x \cos x \cos x + (1 - 2 \sin^2 x)\sin x \\
&= 2 \sin x \cos^2 x + \sin x - 2 \sin^3 x \\
&= 2 \sin x(1 - \sin^2 x) + \sin x - 2 \sin^3 x \\
&= 2 \sin x - 2 \sin^3 x + \sin x - 2 \sin^3 x \\
&= 3 \sin x - 4 \sin^3 x
\end{aligned}$$

Power-Reducing Formulas

The double-angle formulas can be used to obtain the following **power-reducing formulas.**

Power-Reducing Formulas

$$\sin^2 u = \frac{1 - \cos 2u}{2} \qquad \cos^2 u = \frac{1 + \cos 2u}{2} \qquad \tan^2 u = \frac{1 - \cos 2u}{1 + \cos 2u}$$

Proof /// The first two formulas can be verified by solving for $\sin^2 u$ and $\cos^2 u$, respectively, in the double-angle formulas

$$\cos 2u = 1 - 2 \sin^2 u \qquad \text{and} \qquad \cos 2u = 2 \cos^2 u - 1.$$

The third formula can be verified using the fact that

$$\tan^2 u = \frac{\sin^2 u}{\cos^2 u}. \qquad\qquad ///$$

Example 5 shows a typical power reduction that is used in calculus.

EXAMPLE 5 🔲 **Reducing the Power of a Trigonometric Function**

Rewrite $\sin^4 x$ as a sum of first powers of the cosines of multiple angles.

Solution

Note the repeated use of power-reducing formulas.

$$\sin^4 x = (\sin^2 x)^2 = \left(\frac{1 - \cos 2x}{2} \right)^2$$

$$= \frac{1}{4} (1 - 2 \cos 2x + \cos^2 2x)$$

$$= \frac{1}{4} \left(1 - 2 \cos 2x + \frac{1 + \cos 4x}{2} \right)$$

$$= \frac{1}{4} - \frac{1}{2} \cos 2x + \frac{1}{8} + \frac{1}{8} \cos 4x$$

$$= \frac{3}{8} - \frac{1}{2} \cos 2x + \frac{1}{8} \cos 4x$$

$$= \frac{1}{8} (3 - 4 \cos 2x + \cos 4x) \qquad\qquad 🔲$$

Half-Angle Formulas

You can derive some useful alternative forms of the power-reducing formulas by replacing u with $u/2$. The results are called **half-angle formulas.**

Half-Angle Formulas

$$\sin \frac{u}{2} = \pm \sqrt{\frac{1 - \cos u}{2}}$$

$$\cos \frac{u}{2} = \pm \sqrt{\frac{1 + \cos u}{2}}$$

$$\tan \frac{u}{2} = \frac{1 - \cos u}{\sin u} = \frac{\sin u}{1 + \cos u}$$

The signs of $\sin(u/2)$ and $\cos(u/2)$ depend on the quadrant in which $u/2$ lies.

EXAMPLE 6　　**Using a Half-Angle Formula**

Find the exact value of $\sin 105°$.

Solution

Begin by noting that $105°$ is half of $210°$. Then, using the half-angle formula for $\sin(u/2)$ and the fact that $105°$ lies in Quadrant II, you have

$$\sin 105° = \sqrt{\frac{1 - \cos 210°}{2}}$$

$$= \sqrt{\frac{1 - (-\cos 30°)}{2}}$$

$$= \sqrt{\frac{1 + \left(\sqrt{3}/2\right)}{2}}$$

$$= \frac{\sqrt{2 + \sqrt{3}}}{2}.$$

The positive square root is chosen because $\sin \theta$ is positive in Quadrant II.

Note　Use your calculator to verify the result obtained in Example 6. That is, evaluate $\sin 105°$ and $\left(\sqrt{2 + \sqrt{3}}\right)/2$ and you will see that both values are approximately 0.9659258.

EXAMPLE 7   Solving a Trigonometric Equation

$$2 - \sin^2 x = 2\cos^2\frac{x}{2} \qquad \text{Original equation}$$

$$2 - \sin^2 x = 2\left(\frac{1 + \cos x}{2}\right) \qquad \text{Half-angle formula}$$

$$2 - \sin^2 x = 1 + \cos x \qquad \text{Simplify.}$$

$$2 - (1 - \cos^2 x) = 1 + \cos x \qquad \text{Pythagorean identity}$$

$$\cos^2 x - \cos x = 0 \qquad \text{Simplify.}$$

$$\cos x(\cos x - 1) = 0 \qquad \text{Factor.}$$

By setting the factors $\cos x$ and $(\cos x - 1)$ equal to zero, you find that the solutions in the interval $[0, 2\pi]$ are $x = \pi/2$, $x = 3\pi/2$, and $x = 0$. The graph of $y = 2 - \sin^2 x - 2\cos^2(x/2)$, as shown in Figure 5.25, helps confirm this result.

Figure 5.25

$y = 2 - \sin^2 x - 2\cos^2\left(\frac{x}{2}\right)$

Product-to-Sum Formulas

Each of the following **product-to-sum formulas** is easily verified using the sum and difference formulas discussed in the preceding section.

Product-to-Sum Formulas
$\sin u \sin v = \dfrac{1}{2}[\cos(u - v) - \cos(u + v)]$
$\cos u \cos v = \dfrac{1}{2}[\cos(u - v) + \cos(u + v)]$
$\sin u \cos v = \dfrac{1}{2}[\sin(u + v) + \sin(u - v)]$
$\cos u \sin v = \dfrac{1}{2}[\sin(u + v) - \sin(u - v)]$

EXAMPLE 8   Writing Products as Sums

Rewrite $\cos 5x \sin 4x$ using a product-to-sum formula.

Solution

$$\cos 5x \sin 4x = \frac{1}{2}[\sin(5x + 4x) - \sin(5x - 4x)] = \frac{1}{2}\sin 9x - \frac{1}{2}\sin x$$

Occasionally, it is useful to reverse the procedure and write a sum of trigonometric functions as a product. This can be accomplished with the following **sum-to-product formulas.**

Sum-to-Product Formulas

$$\sin x + \sin y = 2 \sin\left(\frac{x + y}{2}\right) \cos\left(\frac{x - y}{2}\right)$$

$$\sin x - \sin y = 2 \cos\left(\frac{x + y}{2}\right) \sin\left(\frac{x - y}{2}\right)$$

$$\cos x + \cos y = 2 \cos\left(\frac{x + y}{2}\right) \cos\left(\frac{x - y}{2}\right)$$

$$\cos x - \cos y = -2 \sin\left(\frac{x + y}{2}\right) \sin\left(\frac{x - y}{2}\right)$$

Proof /// To prove the first formula, let $x = u + v$ and $y = u - v$. Then substitute $u = (x + y)/2$ and $v = (x - y)/2$ in the product-to-sum formula.

$$\sin u \cos v = \frac{1}{2}[\sin(u + v) + \sin(u - v)]$$

$$\sin\left(\frac{x + y}{2}\right) \cos\left(\frac{x - y}{2}\right) = \frac{1}{2}(\sin x + \sin y)$$

$$2 \sin\left(\frac{x + y}{2}\right) \cos\left(\frac{x - y}{2}\right) = \sin x + \sin y \qquad\qquad ///$$

EXAMPLE 9 ▭ **Using a Sum-to-Product Formula**

Find the exact value of $\cos 195° + \cos 105°$.

Solution

Using the appropriate sum-to-product formula, you obtain

$$\cos 195° + \cos 105° = 2 \cos\left(\frac{195° + 105°}{2}\right) \cos\left(\frac{195° - 105°}{2}\right)$$

$$= 2 \cos 150° \cos 45°$$

$$= 2\left(-\frac{\sqrt{3}}{2}\right)\left(\frac{\sqrt{2}}{2}\right)$$

$$= -\frac{\sqrt{6}}{2}. \qquad\qquad ▭$$

EXAMPLE 10 ▱ **Solving a Trigonometric Equation**

Find all solutions of $\sin 5x + \sin 3x = 0$.

Solution

$$\sin 5x + \sin 3x = 0 \qquad \text{Original equation}$$

$$2 \sin\left(\frac{5x + 3x}{2}\right) \cos\left(\frac{5x - 3x}{2}\right) = 0 \qquad \text{Sum-to-product formula}$$

$$2 \sin 4x \cos x = 0 \qquad \text{Simplify.}$$

Figure 5.26

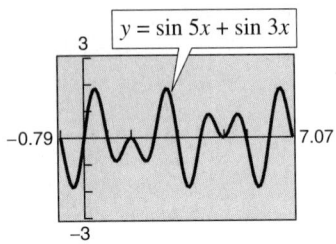

By setting the factor $\sin 4x$ equal to zero, you can find that the solutions in the interval $[0, 2\pi)$ are

$$x = 0, \frac{\pi}{4}, \frac{\pi}{2}, \frac{3\pi}{4}, \pi, \frac{5\pi}{4}, \frac{3\pi}{2}, \frac{7\pi}{4}.$$

Moreover, the equation $\cos x = 0$ yields no additional solutions, and you can conclude that the solutions are of the form

$$x = \frac{n\pi}{4}$$

where n is an integer. These solutions are verified graphically in Figure 5.26. ▱

EXAMPLE 11 ▱ **Verifying a Trigonometric Identity**

Verify the identity $\dfrac{\sin t + \sin 3t}{\cos t + \cos 3t} = \tan 2t$.

Solution

Using appropriate sum-to-product formulas, you have

$$\frac{\sin t + \sin 3t}{\cos t + \cos 3t} = \frac{2 \sin 2t \cos(-t)}{2 \cos 2t \cos(-t)} = \frac{\sin 2t}{\cos 2t} = \tan 2t. \qquad ▱$$

Group Activity   *Deriving an Area Formula*

With others in your group, discuss how you can use a double-angle formula or a half-angle formula to derive a formula for the area of an isosceles triangle. Use a labeled sketch to illustrate your derivation. Then write two examples that show how your formula can be used.

5.5 /// EXERCISES

In Exercises 1–8, use the figure to find the exact value of the trigonometric function.

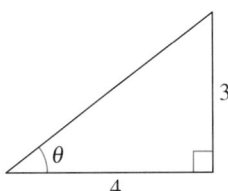

1. $\sin \theta$ **2.** $\tan \theta$

3. $\cos 2\theta$ **4.** $\sin 2\theta$

5. $\tan 2\theta$ **6.** $\sec 2\theta$

7. $\csc 2\theta$ **8.** $\cot 2\theta$

In Exercises 9–18, use a graphing utility to approximate the equation's solutions in the interval $[0, 2\pi)$. If possible, find the exact solutions algebraically.

9. $\sin 2x - \sin x = 0$ **10.** $\sin 2x + \cos x = 0$

11. $4 \sin x \cos x = 1$ **12.** $\sin 2x \sin x = \cos x$

13. $\cos 2x - \cos x = 0$ **14.** $\cos 2x + \sin x = 0$

15. $\tan 2x - \cot x = 0$ **16.** $\tan 2x - 2 \cos x = 0$

17. $\sin 4x = -2 \sin 2x$ **18.** $(\sin 2x + \cos 2x)^2 = 1$

In Exercises 19–22, use a double-angle formula to rewrite the expression. Use a graphing utility to verify that both forms are the same.

19. $6 \sin x \cos x$ **20.** $4 \sin x \cos x + 2$

21. $4 - 8 \sin^2 x$

22. $(\cos x + \sin x)(\cos x - \sin x)$

In Exercises 23–26, find the exact values of $\sin 2u$, $\cos 2u$, and $\tan 2u$ using the double-angle formulas.

23. $\sin u = 3/5, \quad 0 < u < \pi/2$

24. $\cos u = -2/3, \quad \pi/2 < u < \pi$

25. $\tan u = 1/2, \quad \pi < u < 3\pi/2$

26. $\cot u = -4, \quad 3\pi/2 < u < 2\pi$

In Exercises 27–32, rewrite the expression in terms of the first power of the cosine.

27. $\cos^4 x$ **28.** $\sin^4 x$

29. $\sin^2 x \cos^2 x$ **30.** $\cos^6 x$

31. $\sin^2 x \cos^4 x$ **32.** $\sin^4 x \cos^2 x$

In Exercises 33–38, use the figure to find the exact value of the trigonometric function.

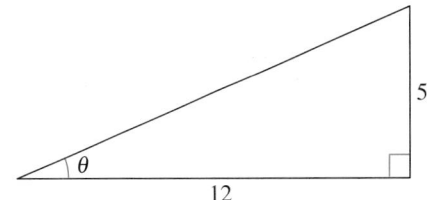

33. $\cos \dfrac{\theta}{2}$ **34.** $\sin \dfrac{\theta}{2}$

35. $\tan \dfrac{\theta}{2}$ **36.** $2 \sin \dfrac{\theta}{2} \cos \dfrac{\theta}{2}$

37. $\csc \dfrac{\theta}{2}$ **38.** $\cot \dfrac{\theta}{2}$

In Exercises 39–44, use the half-angle formulas to determine the exact values of the sine, cosine, and tangent of the angle.

39. $105°$ **40.** $165°$

41. $112° \ 30'$ **42.** $67° \ 30'$

43. $\dfrac{\pi}{8}$ **44.** $\dfrac{\pi}{12}$

In Exercises 45–48, find the exact values of $\sin(u/2)$, $\cos(u/2)$, and $\tan(u/2)$ using the half-angle formulas.

45. $\sin u = 5/13, \quad \pi/2 < u < \pi$

46. $\cos u = 3/5, \quad 0 < u < \pi/2$

47. $\tan u = -5/8, \quad 3\pi/2 < u < 2\pi$

48. $\cot u = 3, \quad \pi < u < 3\pi/2$

In Exercises 49–52, use the half-angle formulas to simplify the expression.

49. $\sqrt{\dfrac{1 - \cos 6x}{2}}$

50. $\sqrt{\dfrac{1 + \cos 4x}{2}}$

51. $-\sqrt{\dfrac{1 - \cos 8x}{1 + \cos 8x}}$

52. $-\sqrt{\dfrac{1 - \cos(x - 1)}{2}}$

In Exercises 53–56, find the exact zeros of the function in the interval $[0, 2\pi)$. Use a graphing utility to graph the function and verify the zeros.

53. $f(x) = \sin \dfrac{x}{2} + \cos x$

54. $h(x) = \sin \dfrac{x}{2} + \cos x - 1$

55. $h(x) = \cos \dfrac{x}{2} - \sin x$

56. $g(x) = \tan \dfrac{x}{2} - \sin x$

In Exercises 57–62, use the product-to-sum formulas to write the product as a sum or difference.

57. $6 \sin \dfrac{\pi}{4} \cos \dfrac{\pi}{4}$

58. $4 \sin \dfrac{\pi}{3} \cos \dfrac{5\pi}{6}$

59. $\sin 5\theta \cos 3\theta$

60. $3 \sin 2\alpha \sin 3\alpha$

61. $5 \cos(-5\beta) \cos 3\beta$

62. $\cos 2\theta \cos 4\theta$

In Exercises 63–72, use the sum-to-product formulas to write the sum or difference as a product.

63. $\sin 60° + \sin 30°$

64. $\cos 120° + \cos 30°$

65. $\cos \dfrac{3\pi}{4} - \cos \dfrac{\pi}{4}$

66. $\sin 5\theta - \sin 3\theta$

67. $\cos 6x + \cos 2x$

68. $\sin x + \sin 5x$

69. $\sin(\alpha + \beta) - \sin(\alpha - \beta)$

70. $\cos\left(\theta + \dfrac{\pi}{2}\right) - \cos\left(\theta - \dfrac{\pi}{2}\right)$

71. $\cos(\phi + 2\pi) + \cos \phi$

72. $\sin\left(x + \dfrac{\pi}{2}\right) + \sin\left(x - \dfrac{\pi}{2}\right)$

In Exercises 73–76, find the exact zeros of the function in the interval $[0, 2\pi)$. Use a graphing utility to graph the function and verify the zeros.

73. $g(x) = \sin 6x + \sin 2x$

74. $h(x) = \cos 2x - \cos 6x$

75. $f(x) = \dfrac{\cos 2x}{\sin 3x - \sin x} - 1$

76. $f(x) = \sin^2 3x - \sin^2 x$

In Exercises 77–80, use the figure to find the exact value of the trigonometric function in two ways.

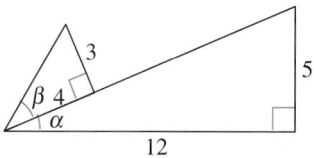

77. $\sin^2 \alpha$

78. $\cos^2 \alpha$

79. $\sin \alpha \cos \beta$

80. $\cos \alpha \sin \beta$

In Exercises 81–94, verify the identity algebraically. Use a graphing utility to confirm the identity graphically.

81. $\csc 2\theta = \dfrac{\csc \theta}{2 \cos \theta}$

82. $\sec 2\theta = \dfrac{\sec^2 \theta}{2 - \sec^2 \theta}$

83. $\cos^2 2\alpha - \sin^2 2\alpha = \cos 4\alpha$

84. $\cos^4 x - \sin^4 x = \cos 2x$

85. $(\sin x + \cos x)^2 = 1 + \sin 2x$

86. $\sin \dfrac{\alpha}{3} \cos \dfrac{\alpha}{3} = \dfrac{1}{2} \sin \dfrac{2\alpha}{3}$

87. $1 + \cos 10y = 2 \cos^2 5y$

88. $\dfrac{\cos 3\beta}{\cos \beta} = 1 - 4 \sin^2 \beta$

89. $\sec \dfrac{u}{2} = \pm \sqrt{\dfrac{2 \tan u}{\tan u + \sin u}}$

90. $\tan \dfrac{u}{2} = \csc u - \cot u$

91. $\cos 3\beta = \cos^3 \beta - 3 \sin^2 \beta \cos \beta$

92. $\sin 4\beta = 4 \sin \beta \cos \beta (1 - 2 \sin^2 \beta)$

93. $\dfrac{\cos 4x - \cos 2x}{2 \sin 3x} = -\sin x$

94. $\dfrac{\cos 3x - \cos x}{\sin 3x - \sin x} = -\tan 2x$

In Exercises 95 and 96, graph the function by using the power-reducing formulas.

95. $f(x) = \sin^2 x$

96. $f(x) = \cos^2 x$

In Exercises 97 and 98, (a) use a graphing utility to graph the function and approximate the maximum and minimum points on the graph in the interval $[0, 2\pi)$, and (b) solve the trigonometric equation and verify that its solutions are the x-coordinates of the maximum and minimum points of f (calculus is required to find the trigonometric equation).

Function	Trigonometric Equation

97. $f(x) = 4 \sin \dfrac{x}{2} + \cos x$ $2 \cos \dfrac{x}{2} - \sin x = 0$

98. $f(x) = \cos 2x - 2 \sin x$ $-2 \cos x (2 \sin x + 1) = 0$

99. *Conjecture*   Consider the function

$$f(x) = 2 \sin x \left[2 \cos^2\left(\dfrac{x}{2}\right) - 1 \right].$$

(a) Use a graphing utility to graph the function.

(b) Make a conjecture about the function that is an identity with f.

(c) Verify your conjecture analytically.

100. *Projectile Motion*   The range of a projectile fired at an angle θ with the horizontal and with an initial velocity of v_0 feet per second is given by

$$r = \tfrac{1}{32} v_0 \sin 2\theta$$

where r is measured in feet. Determine the expression for the range in terms of θ.

101. *Exploration*   Consider the function

$$f(x) = \sin^4 x + \cos^4 x.$$

(a) Use the power-reducing formulas to write the expression in terms of the first power of the cosine.

(b) Determine another way of rewriting the function. Use a graphing utility to rule out incorrectly rewritten functions.

(c) Determine a trigonometric expression such that the sum of the expression and the function is a perfect square trinomial. Rewrite the function as a perfect square trinomial minus the term that you added. Use a graphing utility to rule out incorrectly rewritten functions.

(d) Rewrite the result of part (c) in terms of the sine of a double angle. Use a graphing utility to rule out incorrectly rewritten functions.

(e) In how many ways have you rewritten the trigonometric function? When rewriting a trigonometric expression, your results may not be the same as a friend's results. Does this necessarily mean that one of you is wrong? Explain.

102. *Area*   The length of each of the two equal sides of an isosceles triangle is 10 meters (see figure). The angle between the equal sides is θ.

(a) Express the area of the triangle as a function of $\theta/2$.

(b) Express the area of the triangle as a function of θ and determine the value of θ such that the area is maximum.

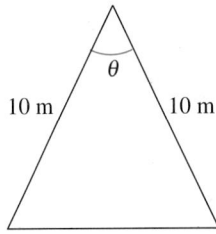

In Exercises 103 and 104, write the trigonometric expression as an algebraic expression.

103. $\sin(2 \arcsin x)$ **104.** $\cos(2 \arccos x)$

Focus on Concepts

In this chapter, you studied the fundamental identities of trigonometry. You can use the following questions to check your understanding of several of these basic concepts. The answers to these questions are given in the back of the book.

1. In your own words, describe the difference between an identity and a conditional equation.

2. Describe the difference between verifying an identity and solving an equation.

3. List the reciprocal identities, quotient identities, and Pythagorean identities by memory.

4. Is $\cos\theta = \sqrt{1 - \sin^2\theta}$ an identity? Explain.

5. *True or False?* Usually there is only one correct set of steps to verify an identity. Explain.

6. By observation, determine which of the following is an identity. Explain.

   (a) $\tan(\theta + \pi) \stackrel{?}{=} \tan\theta$

   (b) $\cos(\theta + \pi) \stackrel{?}{=} \cos\theta$

   (c) $\sec\theta\csc\theta \stackrel{?}{=} 1$

   (d) $\tan\theta\cot\theta \stackrel{?}{=} 1$

   (e) $\sin(\theta - \pi) \stackrel{?}{=} -\sin(\pi - \theta)$

In Exercises 7 and 8, use the graphs of y_1 and y_2 to determine how to change one function to form the identity $y_1 = y_2$.

7. $y_1 = \sec^2\left(\dfrac{\pi}{2} - x\right)$

   $y_2 = \cot^2 x$

8. $y_1 = \dfrac{\cos 3x}{\cos x}$

   $y_2 = (2\sin x)^2$

In Exercises 9 and 10, use the graph to determine the number of points of intersection of the graphs of y_1 and y_2.

9. $y_1 = 2\sin x$

   $y_2 = 3x + 1$

10. $y_1 = 2\sin x$

    $y_2 = \frac{1}{2}x + 1$

In Exercises 11 and 12, use the graph to determine the number of zeros of the function.

11. $y = \sqrt{x+3} + 4\cos x$

12. $y = 2 - \dfrac{1}{2}x^2 + 3\sin\dfrac{\pi x}{2}$

13. Sales of a product are seasonal and can be modeled by the function

$$y = a + bt + c\sin(dt + e),$$

where t is the time in years. What is the value of d?

5 /// REVIEW EXERCISES

In Exercises 1–8, simplify the trigonometric expression.

1. $\dfrac{1}{\cot^2 x + 1}$

2. $\dfrac{\sin 2\alpha}{\cos^2 \alpha - \sin^2 \alpha}$

3. $\dfrac{\sin^2 \alpha - \cos^2 \alpha}{\sin^2 \alpha - \sin \alpha \cos \alpha}$

4. $\dfrac{\sin^3 \beta + \cos^3 \beta}{\sin \beta + \cos \beta}$

5. $\tan^2 \theta(\csc^2 \theta - 1)$

6. $1 - 4 \sin^2 x \cos^2 x$

7. $\dfrac{2 \tan(x + 1)}{1 - \tan^2(x + 1)}$

8. $\sqrt{\dfrac{1 - \cos^2 x}{1 + \cos x}}$

In Exercises 9–26, verify the identity.

9. $\tan x(1 - \sin^2 x) = \frac{1}{2} \sin 2x$

10. $\cos x(\tan^2 x + 1) = \sec x$

11. $\sec^2 x \cot x - \cot x = \tan x$

12. $\sin^3 \theta + \sin \theta \cos^2 \theta = \sin \theta$

13. $\sin^5 x \cos^2 x = (\cos^2 x - 2 \cos^4 x + \cos^6 x)\sin x$

14. $\cos^3 x \sin^2 x = (\sin^2 x - \sin^4 x)\cos x$

15. $\sin 3\theta \sin \theta = \frac{1}{2}(\cos 2\theta - \cos 4\theta)$

16. $\sin 3x \cos 2x = \frac{1}{2}(\sin 5x + \sin x)$

17. $\sqrt{\dfrac{1 - \sin \theta}{1 + \sin \theta}} = \dfrac{1 - \sin\theta}{|\cos\theta|}$

18. $\sqrt{1 - \cos x} = \dfrac{|\sin x|}{\sqrt{1 + \cos x}}$

19. $\cos 3x = 4 \cos^3 x - 3 \cos x$

20. $\cos\left(x + \dfrac{\pi}{2}\right) = -\sin x$

21. $\cot\left(\dfrac{\pi}{2} - x\right) = \tan x$ **22.** $\sin(\pi - x) = \sin x$

23. $\dfrac{\sec x - 1}{\tan x} = \tan \dfrac{x}{2}$ **24.** $\dfrac{2 \cos 3x}{\sin 4x - \sin 2x} = \csc x$

25. $2 \sin y \cos y \sec 2y = \tan 2y$

26. $\dfrac{\sin(\alpha + \beta)}{\cos \alpha \cos \beta} = \tan \alpha + \tan \beta$

In Exercises 27–30, verify the identity algebraically and use a graphing utility to confirm it graphically.

27. $\sin\left(x - \dfrac{3\pi}{2}\right) = \cos x$

28. $\sin 4x = 8 \cos^3 x \sin x - 4 \cos x \sin x$

29. $\tan^2 x = \dfrac{1 - \cos 2x}{1 + \cos 2x}$

30. $\cos^2 5x - \cos^2 x = -\sin 4x \sin 6x$

In Exercises 31–34, find the exact value of the trigonometric function by using the sum, difference, or half-angle formulas.

31. $\sin \dfrac{5\pi}{12} = \sin\left(\dfrac{2\pi}{3} - \dfrac{\pi}{4}\right)$

32. $\cos 285° = \cos(225° + 60°)$

33. $\cos(157° \ 30') = \cos \dfrac{315°}{2}$

34. $\sin \dfrac{3\pi}{8} = \sin\left[\dfrac{1}{2}\left(\dfrac{3\pi}{4}\right)\right]$

In Exercises 35–40, find the exact value of the trigonometric function given that $\sin u = \frac{3}{4}$, $\cos v = -\frac{5}{13}$, and u and v are in Quadrant II.

35. $\sin(u + v)$ **36.** $\tan(u + v)$

37. $\cos(u - v)$ **38.** $\sin 2v$

39. $\cos \dfrac{u}{2}$ **40.** $\tan 2v$

True or False?   In Exercises 41–44, determine if the statement is true or false. If it is false, make the necessary correction.

41. If $\dfrac{\pi}{2} < \theta < \pi$, then $\cos \dfrac{\theta}{2} < 0$.

42. $\sin(x + y) = \sin x + \sin y$

43. $4 \sin(-x) \cos(-x) = -2 \sin 2x$

44. $4 \sin 45° \cos 15° = 1 + \sqrt{3}$

In Exercises 45–50, find all solutions of the equation in the interval $[0, 2\pi)$.

45. $\sin x - \tan x = 0$ **46.** $\csc x - 2 \cot x = 0$

47. $\sin 2x + \sqrt{2} \sin x = 0$

48. $\cos 4x - 7 \cos 2x = 8$

49. $\cos^2 x + \sin x = 1$ **50.** $\sin 4x - \sin 2x = 0$

In Exercises 51–54, use a graphing utility to graph the function and approximate its zeros in the interval $[0, 2\pi)$. If possible, find the exact values of the zeros algebraically.

51. $y = \dfrac{1 + \sin x}{\cos x} + \dfrac{\cos x}{1 + \sin x} - 4$

52. $y = \cos x - \cos \dfrac{x}{2}$

53. $y = \tan^3 x - \tan^2 x + 3 \tan x - 3$

54. $h(s) = \sin s + \sin 3s + \sin 5s$

55. *Think About It*   If a trigonometric equation has an infinite number of solutions, is it true that the equation is an identity? Explain.

56. *Think About It*   Explain how you know from observation that the equation $a \sin x - b = 0$ has no solution if $|a| < |b|$.

In Exercises 57 and 58, write the trigonometric expression as a product.

57. $\cos 3\theta + \cos 2\theta$

58. $\sin\left(x + \dfrac{\pi}{4}\right) - \sin\left(x - \dfrac{\pi}{4}\right)$

In Exercises 59 and 60, write the trigonometric expression as a sum or difference.

59. $\sin 3\alpha \sin 2\alpha$ **60.** $\cos \dfrac{x}{2} \cos \dfrac{x}{4}$

In Exercises 61 and 62, write the trigonometric expression as an algebraic expression.

61. $\cos(2 \arccos 2x)$ **62.** $\sin(2 \arctan x)$

63. *Rate of Change*   The rate of change of the function $f(x) = 2\sqrt{\sin x}$ is given by the expression $\sin^{-1/2} x \cos x$. Show that this expression can also be written as $\cot x \sqrt{\sin x}$.

64. *Projectile Motion*   A baseball leaves the hand of the first baseman at an angle of θ with the horizontal and an initial velocity of $v_0 = 80$ feet per second. The ball is caught by the second baseman 100 feet away. Find θ if the range r of a projectile is given by

$$r = \frac{1}{32} v_0^2 \sin 2\theta.$$

65. *Harmonic Motion*   A weight is attached to a spring suspended vertically from a ceiling. When a driving force is applied to the system, the weight moves vertically from its equilibrium position, and this motion is described by the model

$$y = 1.5 \sin 8t - 0.5 \cos 8t$$

where y is the distance from equilibrium measured in feet and t is the time in seconds.

(a) Write the model in the form
$$y = \sqrt{a^2 + b^2} \sin(Bt + C).$$

(b) Use a graphing utility to obtain a graph of the model.

(c) Find the amplitude of the oscillations of the weight.

(d) Find the frequency of the oscillations of the weight.

66. *Volume*   A trough for feeding cattle is 4 meters long and its cross sections are isosceles triangles with the two equal sides being $\frac{1}{2}$ meter (see figure). The angle between the equal sides is θ.

(a) Express the trough's volume as a function of $\theta/2$.

(b) Express the volume of the trough as a function of θ and determine the value of θ such that the volume is maximum.

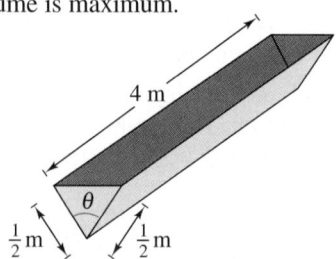

CHAPTER PROJECT   *Projectile Motion*

In this project, you will use parametric equations to model the path of a projectile. For any time t, the horizontal position $x(t)$ and vertical position $y(t)$ of a projectile (ignoring air resistance) launched at ground level is given by the equations

$$x(t) = (v_0 \cos \theta)t$$
$$y(t) = (v_0 \sin \theta)t - 16t^2.$$

In these equations, θ is the angle with the horizontal and v_0 is the initial velocity in feet per second, as indicated in the figure at the left.

Set your graphing utility to parametric and degree modes, and use the viewing window

$$0 \leq t \leq 5$$
$$-20 \leq x \leq 200$$
$$-5 \leq y \leq 20.$$

Let $v_0 = 88$ feet per second and $\theta = 20°$ and graph the parametric equations

$$x(t) = (88 \cos 20°)t$$
$$y(t) = (88 \sin 20°)t - 16t^2.$$

Use the zoom and trace features to (a) find the maximum height attained by the projectile, (b) find the time at which the maximum height occurs, (c) determine the length of time that the projectile is in the air, and (d) determine the range of the projectile.

Questions for Further Exploration

1. Verify analytically the range of the projectile by solving the equation $(88 \sin 20°)t - 16t^2 = 0$ for t and then evaluating $x(t)$ at this t-value.

2. Use a graphing utility to find the maximum height and range of the projectile when $\theta = 30°$ and $v_0 = 132$ feet per second.

3. Let $v_0 = 60$ feet per second and find the maximum range for the angles $\theta = 20°, 30°, 40°, 50°,$ and $60°$. In general, what angle should you use to produce the maximum range?

4. What is the relationship between the time the projectile reaches its maximum height and the time it takes for the projectile to return to the ground? Explain.

5. Eliminate t from the parametric equations

$$x(t) = (v_0 \cos \theta)t \quad \text{and} \quad y(t) = (v_0 \sin \theta)t - 16t^2$$

by solving for t in the first equation and substituting this value into the equation for y. Show in this case that the height of the projectile is given by the equation

$$y = (\tan \theta)x - \frac{16 \sec^2 \theta}{v_0^2}x^2.$$

Use this equation to find the angle θ corresponding to a maximum range of 200 feet and initial velocity of $v_0 = 80$ feet per second.

5 /// CHAPTER TEST

Take this test as you would take a test in class. After you are done, check your work against the answers given in the back of the book.

The *Interactive* CD-ROM provides answers to the Chapter Tests and Cumulative Tests. It also offers Chapter Pre-Tests (that test key skills and concepts covered in previous chapters) and Chapter Post-Tests, both of which have randomly generated exercises with diagnostic capabilities.

1. If $\tan\theta = \frac{3}{2}$ and $\cos\theta < 0$, use the fundamental identities to evaluate the other five trigonometric functions of θ.

2. Use the fundamental identities to simplify $\csc^2\beta(1 - \cos^2\beta)$.

3. Factor and simplify $\dfrac{\sec^4 x - \tan^4 x}{\sec^2 x + \tan^2 x}$.

4. Add and simplify $\dfrac{\cos\theta}{\sin\theta} + \dfrac{\sin\theta}{\cos\theta}$.

5. Determine the values of θ, $0 \le \theta < 2\pi$, for which $\tan\theta = -\sqrt{\sec^2\theta - 1}$ is true.

6. Use a graphing utility to graph the functions $y_1 = \cos x + \sin x \tan x$ and $y_2 = \sec x$. Make a conjecture about y_1 and y_2. Verify the result analytically.

In Exercises 7–12, verify the identity.

7. $\sin\theta\sec\theta = \tan\theta$

8. $\sec^2 x \tan^2 x + \sec^2 x = \sec^4 x$

9. $\dfrac{\csc\alpha + \sec\alpha}{\sin\alpha + \cos\alpha} = \cot\alpha + \tan\alpha$

10. $\cos\left(x + \dfrac{\pi}{2}\right) = -\sin x$

11. $\sin(n\pi + \theta) = (-1)^n \sin\theta,\ \ n$ is an integer

12. $(\sin x + \cos x)^2 - 1 = \sin 2x$

In Exercises 13–16, find all solutions of the equation in the interval $[0, 2\pi)$.

13. $\tan^2 x + \tan x = 0$

14. $\sin 2\alpha - \cos\alpha = 0$

15. $4\cos^2 x - 3 = 0$

16. $\csc^2 x - \csc x - 2 = 0$

17. Use a graphing utility to approximate the solutions of the equation $3\cos x - x = 0$ accurate to three decimal places.

18. Explain why the equation $\cos^2 x + \cos x - 6 = 0$ has no solution.

19. Find the exact value of $\cos 105°$ using the fact that $105° = 135° - 30°$.

20. Use the figure to find the exact values of $\sin 2u$ and $\tan 2u$.

Figure for 20

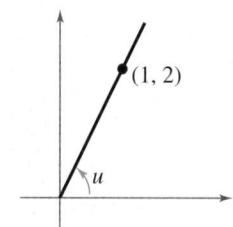

$(1, 2)$

u

Chapter 6

Additional Topics in Trigonometry

Before a bridge, road, tunnel, or other large structure can be designed and built, the land on which it will be built needs to be surveyed. This is one of the jobs of a civil engineer.

To survey a site, civil engineers rely heavily on trigonometry. One of the primary skills a surveyor must master is that of indirect measurement. For instance, the distances, a and c labeled in the diagram below, can be found using the Law of Sines. (See Exercises 27 and 28 on page 514.)

$$\frac{a}{\sin A} = \frac{b}{\sin B}$$

$$\frac{a}{\sin 69°} = \frac{15.6}{\sin 55°}$$

Using a graphing utility in *degree* mode, you find

$$a = \sin 69° \frac{15.6}{\sin 55°} \approx 17.779 \text{ meters.}$$

Julio Esquivel is a civil engineer who specializes in surveying. He often uses aerial photographs such as this one to create topographic maps.

459

6.1   Law of Sines

Introduction / *The Ambiguous Case (SSA)* /
Area of an Oblique Triangle / *Application*

Figure 6.1

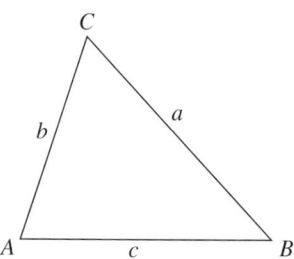

Introduction

In Chapter 4 you looked at techniques for solving right triangles. In this section and the next, you will solve **oblique triangles**—triangles that have no right angles. As standard notation, the angles of a triangle are labeled A, B, and C, and their opposite sides are labeled a, b, and c, as shown in Figure 6.1.

To solve an oblique triangle, you need to know the measure of at least one side and any two other parts of the triangle—two sides, two angles, or one angle and one side. This breaks down into the following four cases.

1. Two angles and any side (AAS or ASA)
2. Two sides and an angle opposite one of them (SSA)
3. Three sides (SSS)
4. Two sides and their included angle (SAS)

The first two cases can be solved using the **Law of Sines,** whereas the last two cases require the **Law of Cosines** (Section 6.2).

Think About the Proof

To prove the Law of Sines, let h be the altitude of either triangle shown in the figure at the right. Then you have

$$\sin A = \frac{h}{b} \text{ or } h = b \sin A$$

and

$$\sin B = \frac{h}{a} \text{ or } h = a \sin B.$$

By equating the two values of h, you can establish part of the Law of Sines. Can you see how to establish the other part? The details of the proof are given in the appendix.

Law of Sines

If ABC is a triangle with sides a, b, and c, then

$$\frac{a}{\sin A} = \frac{b}{\sin B} = \frac{c}{\sin C}.$$

Oblique Triangles

A is acute.

A is obtuse.

Note   The Law of Sines can also be written in the reciprocal form

$$\frac{\sin A}{a} = \frac{\sin B}{b} = \frac{\sin C}{c}.$$

Figure 6.2

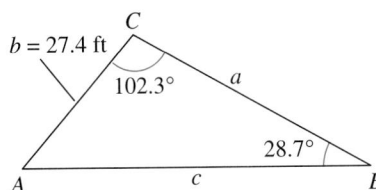

$b = 27.4$ ft

C

$102.3°$

a

$28.7°$

A c B

EXAMPLE 1 🔲   **Given Two Angles and One Side—AAS**

For the triangle in Figure 6.2, $C = 102.3°$, $B = 28.7°$, and $b = 27.4$ feet. Find the remaining angle and sides.

Solution

The third angle of the triangle is

$$A = 180° - B - C = 180° - 28.7° - 102.3° = 49.0°.$$

By the Law of Sines, you have

$$\frac{a}{\sin 49°} = \frac{b}{\sin 28.7°} = \frac{c}{\sin 102.3°}.$$

Using $b = 27.4$ produces

$$a = \frac{27.4}{\sin 28.7°}(\sin 49°) \approx 43.06 \text{ feet}$$

and

$$c = \frac{27.4}{\sin 28.7°}(\sin 102.3°) \approx 55.75 \text{ feet}.$$

Study Tip

When you are solving triangles, a careful sketch is useful as a quick test for the feasibility of an answer. Remember that the longest side lies opposite the largest angle, and the shortest side lies opposite the smallest angle.

Figure 6.3

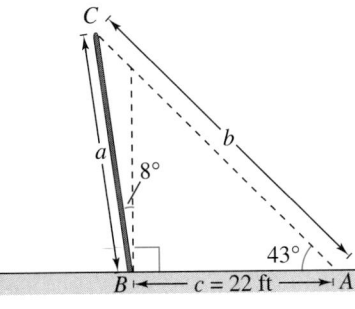

C

a

b

$8°$

$43°$

$B \longleftarrow c = 22$ ft $\longrightarrow A$

Real Life

EXAMPLE 2 🔲   **Given Two Angles and One Side—ASA**

A pole tilts *toward* the sun at an $8°$ angle from the vertical, and it casts a 22-foot shadow. The angle of elevation from the tip of the shadow to the top of the pole is $43°$. How tall is the pole?

Solution

From Figure 6.3, note that $A = 43°$ and $B = 90° + 8° = 98°$. Thus, the third angle is

$$C = 180° - A - B = 180° - 43° - 98° = 39°.$$

By the Law of Sines, you have

$$\frac{a}{\sin 43°} = \frac{c}{\sin 39°}.$$

Because $c = 22$ feet, the length of the pole is

$$a = \frac{22}{\sin 39°}(\sin 43°) \approx 23.84 \text{ feet}.$$

Note For practice, try reworking Example 2 for a pole that tilts *away from* the sun under the same conditions.

The Ambiguous Case (SSA)

In Examples 1 and 2 you saw that two angles and one side determine a unique triangle. However, if two sides and one opposite angle are given, three possible situations can occur: (1) no such triangle exists, (2) one such triangle exists, or (3) two distinct triangles may satisfy the conditions.

The Ambiguous Case (SSA)

Consider a triangle in which you are given a, b, and A. ($h = b \sin A$)

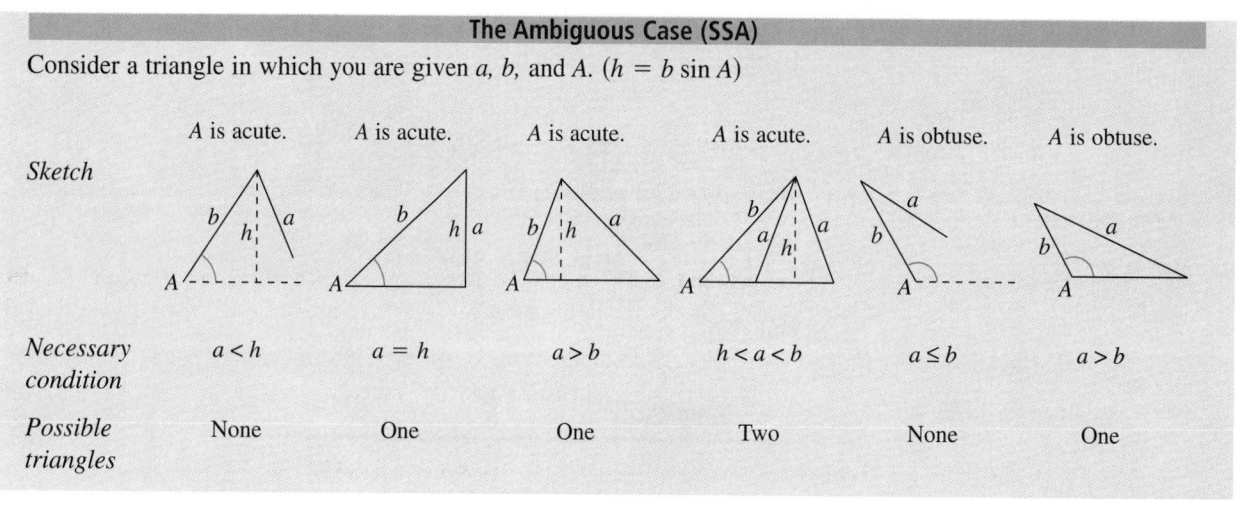

	A is acute.	A is acute.	A is acute.	A is acute.	A is obtuse.	A is obtuse.
Sketch						
Necessary condition	$a < h$	$a = h$	$a > b$	$h < a < b$	$a \leq b$	$a > b$
Possible triangles	None	One	One	Two	None	One

Figure 6.4

One solution: $a > b$

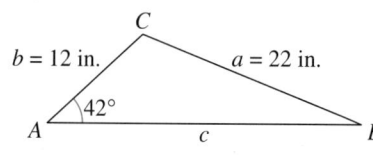

Note   Notice in Example 3 that A is acute and $a > b$, which results in one possible triangle.

EXAMPLE 3   Single-Solution Case—SSA

For the triangle in Figure 6.4, $a = 22$ inches, $b = 12$ inches, and $A = 42°$. Find the remaining side and angles.

Solution
By the Law of Sines, you have

$$\frac{22}{\sin 42°} = \frac{12}{\sin B}$$

$$\sin B = 12\left(\frac{\sin 42°}{22}\right) \approx 0.3649803$$

$$B \approx 21.41°. \qquad \text{B is acute.}$$

Now you can determine that $C \approx 180° - 42° - 21.41° = 116.59°$, and the remaining side is given by

$$\frac{c}{\sin 116.59°} = \frac{22}{\sin 42°}$$

$$c = \sin 116.59°\left(\frac{22}{\sin 42°}\right) \approx 29.40 \text{ inches.}$$

Figure 6.5

No solution: $a < h$

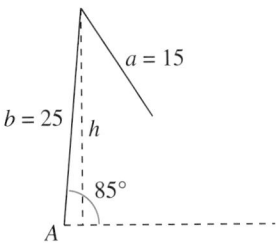

The *Interactive* CD-ROM shows every example with its solution; clicking on the *Try It!* button brings up similar problems. Guided Examples and Integrated Examples show step-by-step solutions to additional examples. Integrated Examples are related to several concepts in the section.

Note Because $h = b \sin A = 31(\sin 20.5°) \approx 10.86$ meters you can conclude that there are two possible triangles (because $h < a < b$).

EXAMPLE 4   **No-Solution Case—SSA**

Show that there is no triangle for which $a = 15$, $b = 25$, and $A = 85°$.

Solution
Begin by making the sketch shown in Figure 6.5. From this figure it appears that no triangle is formed. You can verify this by using the Law of Sines.

$$\frac{a}{\sin A} = \frac{b}{\sin B}$$

$$\frac{15}{\sin 85°} = \frac{25}{\sin B}$$

$$\sin B = 25\left(\frac{\sin 85°}{15}\right) \approx 1.660 > 1$$

This contradicts the fact that $|\sin B| \leq 1$. Hence, no triangle can be formed having sides $a = 15$ and $b = 25$ and an angle of $A = 85°$.

EXAMPLE 5   **Two-Solution Case—SSA**

Find two triangles for which $a = 12$ meters, $b = 31$ meters, and $A = 20.5°$.

Solution
By the Law of Sines, you have

$$\frac{a}{\sin A} = \frac{b}{\sin B}$$

$$\sin B = b\left(\frac{\sin A}{a}\right) = 31\left(\frac{\sin 20.5°}{12}\right) \approx 0.9047.$$

There are two angles $B_1 \approx 64.8°$ and $B_2 \approx 115.2°$ between $0°$ and $180°$ whose sine is 0.9047. For $B_1 \approx 64.8°$, you obtain

$$C \approx 180° - 20.5° - 64.8° = 94.7°$$

$$c = \frac{a}{\sin A}(\sin C) = \frac{12}{\sin 20.5°}(\sin 94.7°) \approx 34.15 \text{ meters.}$$

For $B_2 \approx 115.2°$, you obtain

$$C \approx 180° - 20.5° - 115.2° = 44.3°$$

$$c = \frac{a}{\sin A}(\sin C) = \frac{12}{\sin 20.5°}(\sin 44.3°) \approx 23.93 \text{ meters.}$$

The resulting triangles are shown in Figure 6.6.

Figure 6.6

Area of an Oblique Triangle

The procedure used to prove the Law of Sines leads to a simple formula for the area of an oblique triangle. Referring to Figure 6.7, note that each triangle has a height of

$$h = b \sin A.$$

Consequently, the area of each triangle is given by

$$\text{Area} = \frac{1}{2}(\text{base})(\text{height}) = \frac{1}{2}(c)(b \sin A) = \frac{1}{2}bc \sin A.$$

By similar arguments, you can develop the formulas

$$\text{Area} = \frac{1}{2}ab \sin C = \frac{1}{2}ac \sin B.$$

Figure 6.7

A is acute.   *A* is obtuse.

 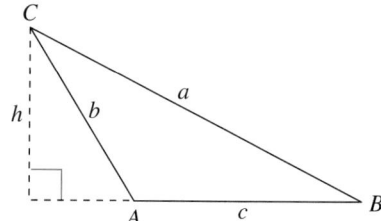

Note Note that if angle *A* is 90°, the formula gives the area for a right triangle as

$$\text{Area} = \frac{1}{2}bc = \frac{1}{2}(\text{base})(\text{height}).$$

Similar results are obtained for angles *C* and *B* equal to 90°.

Area of an Oblique Triangle

The area of any triangle is given by one-half the product of the lengths of two sides times the sine of their included angle. That is,

$$\text{Area} = \frac{1}{2}bc \sin A = \frac{1}{2}ab \sin C = \frac{1}{2}ac \sin B.$$

Real Life

EXAMPLE 6   **Finding the Area of an Oblique Triangle**

Find the area of a triangular lot having two sides of lengths 90 meters and 52 meters and an included angle of 102°.

Solution

Consider $a = 90$ m, $b = 52$ m, and angle $C = 102°$, as shown in Figure 6.8. Then the area of the triangle is

$$\text{Area} = \frac{1}{2}ab \sin C = \frac{1}{2}(90)(52)(\sin 102°) \approx 2289 \text{ square meters.}$$

Figure 6.8

Figure 6.9

Application

Real Life

EXAMPLE 7   An Application of the Law of Sines

The course for a boat race starts at point A and proceeds in the direction S $52°$ W to point B, then in the direction S $40°$ E to point C, and finally back to A, as shown in Figure 6.9. The point C lies 8 kilometers directly south of point A. Approximate the total distance of the race course.

Solution

Because lines BD and AC are parallel, it follows that $\angle BCA \cong \angle DBC$. Consequently, triangle ABC has the measures shown in Figure 6.10. For angle B, you have $B = 180° - 52° - 40° = 88°$. Using the Law of Sines

$$\frac{a}{\sin 52°} = \frac{b}{\sin 88°} = \frac{c}{\sin 40°}$$

you can let $b = 8$ and obtain the following.

$$a = \frac{8}{\sin 88°}(\sin 52°) \approx 6.308 \qquad c = \frac{8}{\sin 88°}(\sin 40°) \approx 5.145$$

The total length of the course is approximately

$$\text{Length} \approx 8 + 6.308 + 5.145 = 19.453 \text{ kilometers.}$$

Figure 6.10

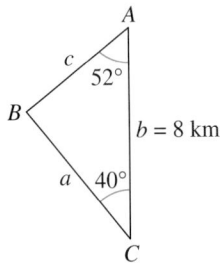

Group Activity   *Using the Law of Sines*

In this section, you have been using the Law of Sines to solve *oblique* triangles. Can the Law of Sines also be used to solve a right triangle? If so, write a short paragraph explaining how to use the Law of Sines to solve the following two triangles. Is there an easier way to solve these triangles?

a. (ASA)

b. (SSA)

The *Interactive* CD-ROM contains step-by-step solutions to all odd-numbered Section and Review Exercises. It also provides Tutorial Exercises, which link to Guided Examples for additional help.

6.1 /// EXERCISES

In Exercises 1–14, use the given information to solve the triangle.

1. $A = 30°$, $a = 20$, $B = 45°$
2. $C = 120°$, $c = 15$, $B = 45°$
3. $A = 10°$, $a = 7.5$, $B = 60°$
4. $C = 135°$, $c = 45$, $B = 10°$
5. $A = 36°$, $a = 8$, $b = 5$
6. $A = 60°$, $a = 9$, $c = 10$
7. $A = 150°$, $C = 20°$, $a = 200$
8. $A = 24.3°$, $C = 54.6°$, $c = 2.68$
9. $A = 83° 20'$, $C = 54.6°$, $c = 18.1$
10. $A = 5° 40'$, $B = 8° 15'$, $b = 4.8$
11. $B = 15° 30'$, $a = 4.5$, $b = 6.8$
12. $C = 85° 20'$, $a = 35$, $c = 50$
13. $A = 110° 15'$, $a = 48$, $b = 16$
14. $B = 2° 45'$, $b = 6.2$, $c = 5.8$

In Exercises 15–18, use the given information to solve the triangle. If two solutions exist, find both.

15. $A = 58°$, $a = 4.5$, $b = 12.8$
16. $A = 58°$, $a = 11.4$, $b = 12.8$
17. $A = 58°$, $a = 4.5$, $b = 5$
18. $A = 110°$, $a = 125$, $b = 100$

In Exercises 19 and 20, find a value for b such that the triangle has (a) one, (b) two, and (c) no solution(s).

19. $A = 36°$, $a = 5$ 20. $A = 60°$, $a = 10$

21. *Height*   A flagpole is located on a slope that makes an angle of $12°$ with the horizontal. The pole casts a 16-meter shadow up the slope when the angle of elevation of the sun is $20°$.

    (a) Draw a triangle that represents the problem.

    (b) Write an equation involving the unknown.

    (c) Find the height of the flagpole.

22. *Height*   Because of prevailing winds, a tree grew so that it was leaning $6°$ from the vertical. At a point 30 meters from the tree, the angle of elevation to the top of the tree is $22° 50'$ (see figure). Find the height h of the tree.

23. *Angle of Elevation*   A 10-meter telephone pole casts a 17-meter shadow directly down a slope when the angle of elevation of the sun is $42°$ (see figure). Find θ, the angle of elevation of the ground.

24. *Flight Path*   A plane flies 500 kilometers with a bearing of N $44°$ W from B to C (see figure). The plane then flies 720 kilometers from C to A. Find the bearing of the flight from C to A.

25. *Bridge Design*   A bridge is to be built across a small lake from B to C (see figure). The bearing from B to C is S 41° W. From a point A, 100 meters from B, the bearings to B and C are S 74° E and S 28° E, respectively. Find the distance from B to C.

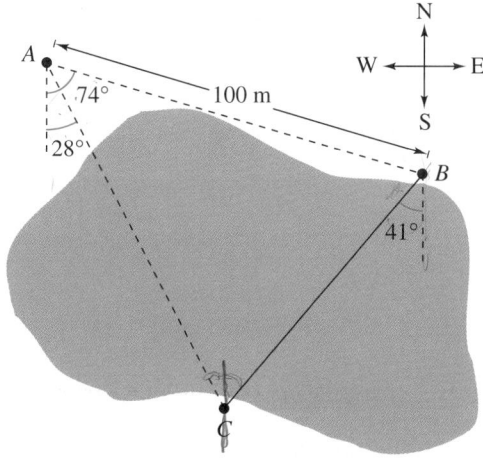

26. *Railroad Track Design*   The circular arc of a railroad curve has a chord of length 3000 feet and a central angle of 40°.

(a) Draw a figure that gives a visual representation of the problem. Show the known quantities on the figure and use variables r and s to represent the radius of the arc and the length of the arc, respectively.

(b) Find the radius r of the circular arc.

(c) Find the length s of the circular arc.

27. *Glide Path*   A pilot has just started on the glide path for landing at an airport where the length of the runway is 9000 feet. The angles of depression from the plane to the ends of the runway are 17.5° and 18.8°.

(a) Draw a figure that gives a visual representation of the problem.

(b) Find the air distance the plane must travel until touching down on the near end of the runway.

(c) Find the ground distance the plane must travel until touching down.

(d) Find the altitude of the plane when the pilot begins the descent.

28. *Altitude*   The angles of elevation to an airplane from two points A and B on level ground are 51° and 68°, respectively. The points A and B are 2.5 miles apart, and the airplane is east of both points in the same vertical plane. Find the altitude of the plane.

29. *Locating a Fire*   Two fire towers A and B are 30 kilometers apart. The bearing from A to B is N 65° E. A fire is spotted by a ranger in each tower, and its bearings from A and B are N 28° E and N 16.5° W, respectively (see figure). Find the distance of the fire from each tower.

30. *Distance*   A boat is sailing due east parallel to the shoreline at a speed of 10 miles per hour. At a given time the bearing to the lighthouse is S 70° E, and 15 minutes later the bearing is S 63° E (see figure). Find the distance from the boat to the shoreline if the lighthouse is at the shoreline.

31. *Distance*   A family is traveling due west on a road that passes a famous landmark. At a given time the bearing to the landmark is N 62° W, and after the family travels 5 miles farther the bearing is N 38° W. What is the closest the family will come to the landmark while on the road?

32. *Engine Design*   The connecting rod in an engine is 6 inches in length and the radius of the crankshaft is $1\frac{1}{2}$ inches (see figure). Let d be the distance the piston is from the top of its stroke for the angle θ.

(a) Use a graphing utility to complete the table.

θ	0°	45°	90°	135°	180°
d					

(b) The spark plug fires at $\theta = 5°$ before top dead center. How far is the piston from the top of its stroke at this time?

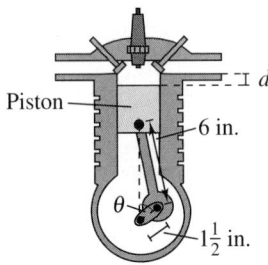

Piston — ⌐⌐ d

—6 in.

θ

$1\frac{1}{2}$ in.

33. *Graphical and Numerical Analysis*   In the figure, α and β are positive angles.

(a) Write α as a function of β.

(b) Use a graphing utility to graph the function. Determine its domain and range.

(c) Write c as a function of β.

(d) Use a graphing utility to graph the function in part (c). Determine its domain and range.

(e) Use a graphing utility to complete the table. What can you infer?

β	0	0.4	0.8	1.2	1.6	2.0	2.4	2.8
α								
c								

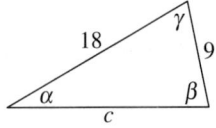

18   γ

9

α β

c

34. *Distance*   The angles of elevation θ and ϕ, to an airplane are being continuously monitored at two observation points A and B, which are 2 miles apart (see figure). Write an equation giving the distance d between the plane and point B in terms of θ and ϕ.

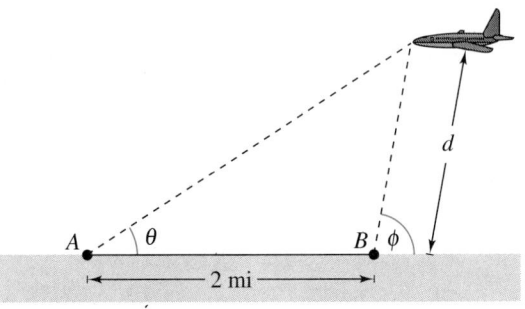

A θ B ϕ

d

|← —— 2 mi —— →|

In Exercises 35–40, find the area of the triangle having the indicated sides and angle.

35. $C = 120°, \quad a = 4, \quad b = 6$

36. $B = 72° \, 30', \quad a = 105, \quad c = 64$

37. $A = 43° \, 45', \quad b = 57, \quad c = 85$

38. $A = 5° \, 15', \quad b = 4.5, \quad c = 22$

39. $B = 130°, \quad a = 62, \quad c = 20$

40. $C = 84° \, 30', \quad a = 16, \quad b = 20$

41. *Graphical Analysis*

(a) Write the area A of the shaded region in the figure as a function of θ.

(b) Use a graphing utility to graph the area function.

(c) Determine the domain of the area function. Explain how the area of the region and the domain of the function would change if the 8-centimeter line segment were decreased in length.

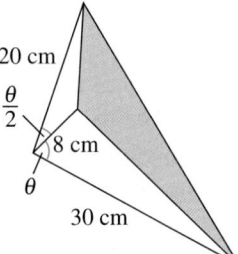

20 cm

$\frac{\theta}{2}$

8 cm

θ

30 cm

6.2   Law of Cosines

Introduction / *Heron's Formula*

Introduction

Two cases remain in the list of conditions needed to solve an oblique triangle—SSS and SAS. The Law of Sines does not work in either of these cases. To see why, consider the three ratios given in the Law of Sines.

$$\frac{a}{\sin A} = \frac{b}{\sin B} = \frac{c}{\sin C}$$

To use the Law of Sines, you must know at least one side and its opposite angle. If you are given three sides (SSS), or two sides and their included angle (SAS), none of the above ratios would be complete. In such cases you can use the **Law of Cosines.**

Law of Cosines	
Standard Form	*Alternative Form*
$a^2 = b^2 + c^2 - 2bc \cos A$	$\cos A = \dfrac{b^2 + c^2 - a^2}{2bc}$
$b^2 = a^2 + c^2 - 2ac \cos B$	$\cos B = \dfrac{a^2 + c^2 - b^2}{2ac}$
$c^2 = a^2 + b^2 - 2ab \cos C$	$\cos C = \dfrac{a^2 + b^2 - c^2}{2ab}$

Figure 6.11

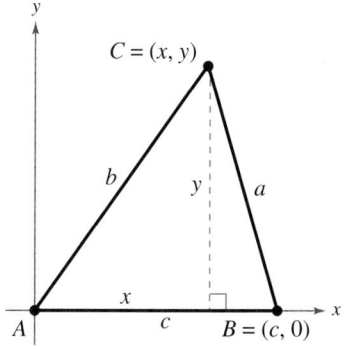

Proof /// Consider a triangle that has three acute angles, as shown in Figure 6.11. In the figure, note that vertex B has coordinates $(c, 0)$. Furthermore, C has coordinates (x, y), where $x = b \cos A$ and $y = b \sin A$. Because a is the distance from vertex C to vertex B, it follows that

$$a = \sqrt{(x - c)^2 + (y - 0)^2}$$
$$a^2 = (b \cos A - c)^2 + (b \sin A)^2$$
$$a^2 = b^2 \cos^2 A - 2bc \cos A + c^2 + b^2 \sin^2 A$$
$$a^2 = b^2(\sin^2 A + \cos^2 A) + c^2 - 2bc \cos A$$
$$a^2 = b^2 + c^2 - 2bc \cos A. \qquad \scriptstyle \sin^2 A + \cos^2 A = 1$$

A similar argument can be used for a triangle having an obtuse angle.   ///

Note that if $A = 90°$ in Figure 6.11, then $\cos A = 0$ and the first form of the Law of Cosines becomes the Pythagorean Theorem.

$$a^2 = b^2 + c^2$$

Thus, the Pythagorean Theorem is actually just a special case of the more general Law of Cosines.

EXAMPLE 1   Three Sides of a Triangle—SSS

Find the three angles of the triangle whose sides have lengths $a = 8$ feet, $b = 19$ feet, and $c = 14$ feet.

Solution

It is a good idea first to find the angle opposite the longest side—side b in this case (see Figure 6.12). Using the Law of Cosines, you find that

$$\cos B = \frac{a^2 + c^2 - b^2}{2ac} = \frac{8^2 + 14^2 - 19^2}{2(8)(14)} \approx -0.45089.$$

Figure 6.12

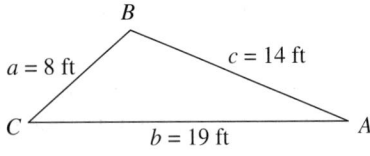

B

$a = 8$ ft

$c = 14$ ft

C

$b = 19$ ft

A

Because $\cos B$ is negative, you know that B is an *obtuse* angle given by $B \approx 116.80°$. At this point you could use the Law of Cosines to find $\cos A$ and $\cos C$. However, knowing that $B \approx 116.80°$, it is simpler to use the Law of Sines to obtain the following.

$$\frac{b}{\sin B} = \frac{a}{\sin A}$$

$$\sin A = a\left(\frac{\sin B}{b}\right) \approx 8\left(\frac{\sin 116.80°}{19}\right) \approx 0.37582$$

Because B is obtuse, you know that A must be acute, because a triangle can have at most one obtuse angle. Thus, $A \approx 22.08°$ and

$$C \approx 180° - 22.08° - 116.80°$$

$$= 41.12°.$$

Do you see why it was wise to find the largest angle *first* in Example 1? Knowing the cosine of an angle, you can determine whether the angle is acute or obtuse. That is,

$$\cos\theta > 0 \quad \text{for} \quad 0° < \theta < 90° \qquad \text{Acute}$$

$$\cos\theta < 0 \quad \text{for} \quad 90° < \theta < 180°. \qquad \text{Obtuse}$$

So, in Example 1, once you found that angle B was obtuse, you subsequently knew that angles A and C were both acute. If the largest angle is acute, the remaining two angles will be acute also.

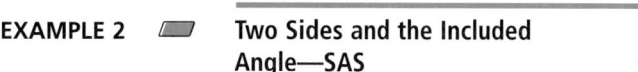

EXAMPLE 2 Two Sides and the Included Angle—SAS

The pitcher's mound on a softball field is 46 feet from home plate and the distance between the bases in 60 feet, as shown in Figure 6.13. How far is the pitcher's mound from first base?

Figure 6.13

The pitcher's mound is not halfway between home plate and second base.

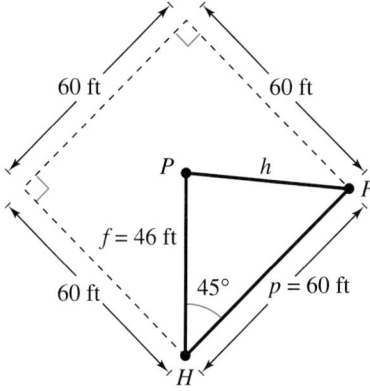

Solution

In triangle *HPF*, $H = 45°$ (line *HP* bisects the right angle at H), $f = 46$, and $p = 60$. Using the Law of Cosines for this SAS case, you have

$$h^2 = f^2 + p^2 - 2fp \cos H$$
$$= 46^2 + 60^2 - 2(46)(60) \cos 45°$$
$$\approx 1812.8.$$

Therefore, the approximate distance from the pitcher's mound to first base is

$$h \approx \sqrt{1812.8} \approx 42.58 \text{ feet.}$$

EXAMPLE 3 Two Sides and the Included Angle—SAS

A ship travels 60 miles due east, then adjusts its course 15° northward, as shown in Figure 6.14. After traveling 80 miles in the new direction, how far is the ship from its point of departure?

Figure 6.14

Solution

You have $c = 60$, $B = 180° - 15° = 165°$, and $a = 80$. Consequently, by the Law of Cosines, it follows that

$$b^2 = a^2 + c^2 - 2ac \cos B$$
$$= 80^2 + 60^2 - 2(80)(60) \cos 165°$$
$$\approx 19,273.$$

Therefore, the distance b is

$$b \approx \sqrt{19,273} \approx 138.8 \text{ miles.}$$

Heron's work provided a scientific foundation for land surveying and engineering. He invented a simple form of the steam engine known as Heron's Fountain.

Heron's Formula

The Law of Cosines can be used to establish the following formula for the area of a triangle. This formula is credited to the Greek mathematician Heron (c. 100 B.C.).

> ### Heron's Area Formula
>
> Given any triangle with sides of lengths a, b, and c, the area of the triangle is
> $$\text{Area} = \sqrt{s(s - a)(s - b)(s - c)}$$
> where $s = (a + b + c)/2$.

Proof /// From the preceding section, you know that

$$\text{Area} = \frac{1}{2} bc \sin A$$

$$= \sqrt{\frac{1}{4} b^2 c^2 \sin^2 A}$$

$$= \sqrt{\frac{1}{4} b^2 c^2 (1 - \cos^2 A)}$$

$$= \sqrt{\left[\frac{1}{2} bc(1 + \cos A)\right]\left[\frac{1}{2} bc(1 - \cos A)\right]}.$$

Using the Law of Cosines, you can show that

$$\frac{1}{2} bc(1 + \cos A) = \frac{a + b + c}{2} \cdot \frac{-a + b + c}{2}$$

and

$$\frac{1}{2} bc(1 - \cos A) = \frac{a - b + c}{2} \cdot \frac{a + b - c}{2}.$$

(See Exercises 41 and 42.) Letting $s = (a + b + c)/2$, these two equations can be rewritten as

$$\frac{1}{2} bc(1 + \cos A) = s(s - a)$$

and

$$\frac{1}{2} bc(1 - \cos A) = (s - b)(s - c).$$

Thus, you can conclude that

$$\text{Area} = \sqrt{s(s - a)(s - b)(s - c)}.$$
///

EXAMPLE 4   **Using Heron's Area Formula**

Find the area of the triangular region having sides of lengths $a = 43$ meters, $b = 53$ meters, and $c = 72$ meters.

Solution
Because

$$s = \frac{1}{2}(a + b + c) = \frac{168}{2} = 84,$$

Heron's Formula yields

$$\text{Area} = \sqrt{s(s - a)(s - b)(s - c)}$$
$$= \sqrt{84(41)(31)(12)}$$
$$\approx 1131.89 \text{ square meters.}$$

Group Activity   *The Area of a Triangle*

You have now discussed three different formulas for the area of a triangle.

Standard Formula: $\text{Area} = \frac{1}{2}bh$
Oblique Triangle: $\text{Area} = \frac{1}{2}bc \sin A = \frac{1}{2}ab \sin C = \frac{1}{2}ac \sin B$
Heron's Formula: $\text{Area} = \sqrt{s(s - a)(s - b)(s - c)}$

Use the most appropriate formula to find the area of each triangle. Show your work and give your reasons for choosing each formula.

a.

b.

c.

d.
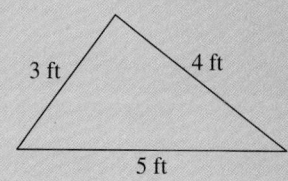

6.2 /// EXERCISES

In Exercises 1–10, use the Law of Cosines to solve the triangle.

1. $a = 6$, $b = 8$, $c = 12$
2. $a = 8$, $b = 3$, $c = 9$
3. $A = 30°$, $b = 15$, $c = 30$
4. $C = 105°$, $a = 10$, $b = 4.5$
5. $a = 9$, $b = 12$, $c = 15$
6. $a = 55$, $b = 25$, $c = 72$
7. $a = 75.4$, $b = 52$, $c = 52$
8. $a = 1.42$, $b = 0.75$, $c = 1.25$
9. $B = 8° 45'$, $a = 25$, $c = 15$
10. $B = 75° 20'$, $a = 6.2$, $c = 9.5$

In Exercises 11–16, solve the parallelogram. (The lengths of the diagonals are given by c and d.)

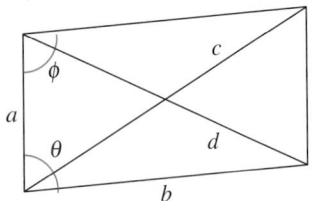

	a	b	c	d	θ	ϕ
11.	4	6			30°	
12.	25	35				120°
13.	10	14	20			
14.	40	60		80		
15.	10		18	12		
16.		25	50	35		

17. *Navigation*   A plane flies 810 miles from A to B with a bearing of N 75° E. Then it flies 648 miles from B to C with a bearing of N 32° E. Draw a figure that gives a visual representation of the problem and find the straight-line distance and bearing from C to A.

18. *Navigation*   A boat race runs along a triangular course marked by buoys A, B, and C. The race starts with the boats headed west for 2500 meters. The other two sides of the course lie to the north of the first side, and their lengths are 1100 meters and 2000 meters. Draw a figure that gives a visual representation of the problem and find the bearings for the last two legs of the race.

19. *Surveying*   To approximate the length of a marsh, a surveyor walks 300 meters from point A to point B, then turns 80° and walks 250 meters to point C (see figure). Approximate the length AC of the marsh.

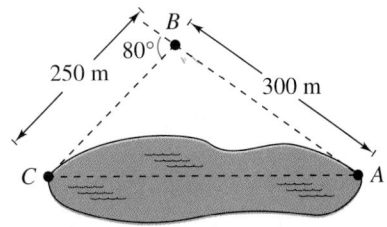

20. *Streetlight Design*   Determine the angle θ in the design of the streetlight shown in the figure.

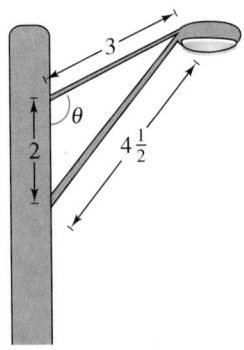

21. *Distance*   Two ships leave a port at 9 A.M. One travels at a bearing of N 53° W at 12 miles per hour and the other travels at a bearing of S 67° W at 16 miles per hour. Approximate how far apart they are at noon that day.

22. *Distance*   A 100-foot vertical tower is to be erected on the side of a hill that makes a 6° angle with the horizontal (see figure). Find the length of each of the two guy wires that will be anchored 75 feet uphill and downhill from the base of the tower.

23. *Navigation*   On a map, Orlando is 178 millimeters due south of Niagara Falls, Denver is 273 millimeters from Orlando, and Denver is 235 millimeters from Niagara Falls (see figure).

(a) Find the bearing of Denver from Orlando.

(b) Find the bearing of Denver from Niagara Falls.

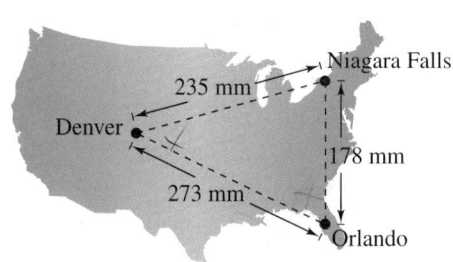

24. *Navigation*   On a map, Minneapolis is 165 millimeters due west of Albany, Phoenix is 216 millimeters from Minneapolis, and Phoenix is 368 millimeters from Albany.

(a) Find the bearing of Minneapolis from Phoenix.

(b) Find the bearing of Albany from Phoenix.

25. *Baseball*   On a baseball diamond with 90-foot sides, the pitcher's mound is 60.5 feet from home plate. How far is it from the pitcher's mound to third base?

26. *Baseball*   The baseball player in center field is playing approximately 330 feet from the television camera that is behind home plate. A batter hits a fly ball that goes to the wall 420 feet from the camera (see figure). Approximate the number of feet that the center fielder has to run to make the catch if the camera turns 8° to follow the play.

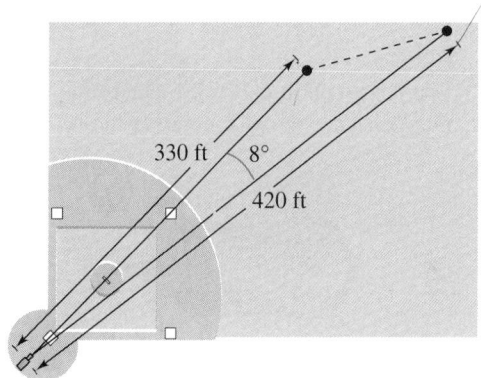

27. *Engineering*   If Q is the midpoint of the line segment $\overline{PR}$ in the truss rafter shown in the figure, find the lengths of the line segments $\overline{PQ}$, $\overline{QS}$, and $\overline{RS}$.

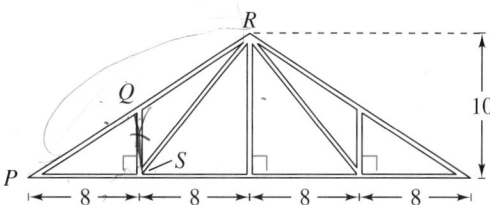

28. *Aircraft Tracking*   To determine the distance between two aircraft, a tracking station continuously determines the distance to each aircraft and the angle A between them. Determine the distance a between the planes when $A = 42°$, $b = 35$ miles, and $c = 20$ miles.

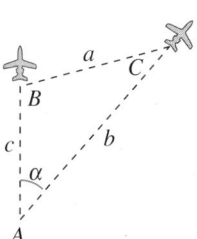

29. *Engine Design* An engine has a 7-inch connecting rod fastened to a crank (see figure).

  (a) Use the Law of Cosines to write an equation giving the relationship between x and θ.

  (b) Write x as a function of θ. (Select the sign that yields positive values of x.)

  (c) Use a graphing utility to graph the function in part (b).

  (d) Use the graph in part (c) to determine the maximum distance the piston moves in one cycle.

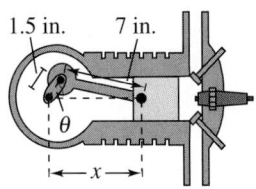

1.5 in.   7 in.

30. *Paper Manufacturing* In a certain process with continuous paper, the paper passes across three rollers of radii 3 inches, 4 inches, and 6 inches (see figure). The centers of the 3-inch and 6-inch rollers are d inches apart, and the length of the arc in contact with the paper on the 4-inch roller is s inches. Complete the following table.

d (inches)	9	10	12	13	14	15	16
θ (degrees)							
s (inches)							

Figure for 30

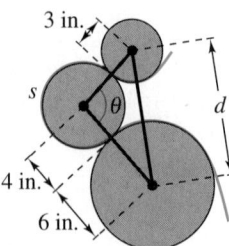

3 in.

s

4 in.

6 in.

d

Figure for 31

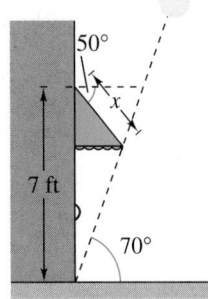

50°

7 ft

70°

31. *Awning Design* A retractable awning lowers at an angle of 50° from the top of a patio door that is 7 feet high (see figure). Find the length x of the awning if no direct sunlight is to enter the door when the angle of elevation of the sun is greater than 70°.

32. *Circumscribed and Inscribed Circles* Let R and r be the radii of the circumscribed and inscribed circles of a triangle ABC, respectively, and let $s = (a + b + c)/2$. Prove the following.

  (a) $2R = \dfrac{a}{\sin A} = \dfrac{b}{\sin B} = \dfrac{c}{\sin C}$

  (b) $r = \sqrt{\dfrac{(s - a)(s - b)(s - c)}{s}}$

Circumscribed and Inscribed Circles **In Exercises 33 and 34, use the results of Exercise 32.**

33. Given a triangle with $a = 25$, $b = 55$, and $c = 72$, find the area of (a) the triangle, (b) the circumscribed circle, and (c) the inscribed circle.

34. Find the length of the largest circular track that can be built on a triangular piece of property whose sides are 200 feet, 250 feet, and 325 feet.

In Exercises 35–40, use Heron's Area Formula to find the area of the triangle.

35. $a = 5$, $b = 7$, $c = 10$

36. $a = 2.5$, $b = 10.2$, $c = 9$

37. $a = 12$, $b = 15$, $c = 9$

38. $a = 75.4$, $b = 52$, $c = 52$

39. $a = 20$, $b = 20$, $c = 10$

40. $a = 4.25$, $b = 1.55$, $c = 3.00$

41. Use the Law of Cosines to prove that

$$\frac{1}{2}bc(1 + \cos A) = \frac{a + b + c}{2} \cdot \frac{-a + b + c}{2}.$$

42. Use the Law of Cosines to prove that

$$\frac{1}{2}bc(1 - \cos A) = \frac{a - b + c}{2} \cdot \frac{a + b - c}{2}.$$

6.3   Vectors in the Plane

Introduction / Component Form of a Vector / Vector Operations /
Unit Vectors / Direction Angles / Applications of Vectors

Introduction

Figure 6.15

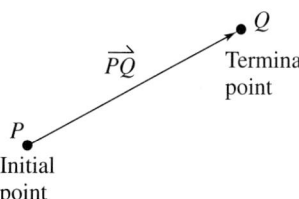

Many quantities in geometry and physics, such as area, time, and temperature, can be represented by a single real number. Other quantities, such as force and velocity, involve both *magnitude* and *direction* and cannot be completely characterized by a single real number. To represent such a quantity, you can use a **directed line segment,** as shown in Figure 6.15. The directed line segment $\overrightarrow{PQ}$ has **initial point** P and **terminal point** Q. Its **length** is denoted by $\|\overrightarrow{PQ}\|$.

Figure 6.16

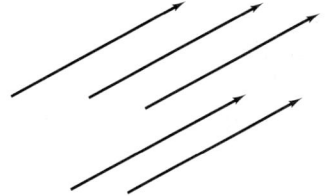

Two directed line segments that have the same length (or magnitude) and direction are called **equivalent.** For example, the directed line segments in Figure 6.16 are all equivalent. The set of all directed line segments that are equivalent to a given directed line segment $\overrightarrow{PQ}$ is a **vector v in the plane,** written $\mathbf{v} = \overrightarrow{PQ}$. Vectors are denoted by lowercase, boldface letters such as $\mathbf{u}$, $\mathbf{v}$, and $\mathbf{w}$.

Be sure you see that a vector in the plane can be represented by many different directed line segments.

EXAMPLE 1   ▰   **Vector Representation by Directed Line Segments**

Let $\mathbf{u}$ be represented by the directed line segment from $P = (0, 0)$ to $Q = (3, 2)$, and let $\mathbf{v}$ be represented by the directed line segment from $R = (1, 2)$ to $S = (4, 4)$, as shown in Figure 6.17. Show that $\mathbf{u} = \mathbf{v}$.

Figure 6.17

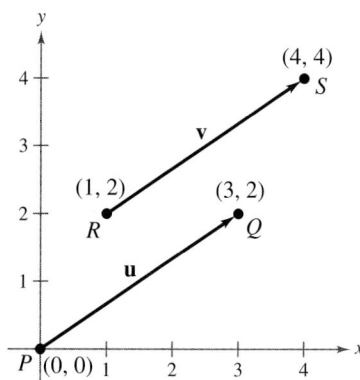

Solution

From the distance formula, it follows that $\overrightarrow{PQ}$ and $\overrightarrow{RS}$ have the *same length.*

$$\|\overrightarrow{PQ}\| = \sqrt{(3 - 0)^2 + (2 - 0)^2} = \sqrt{13}$$
$$\|\overrightarrow{RS}\| = \sqrt{(4 - 1)^2 + (4 - 2)^2} = \sqrt{13}$$

Moreover, both line segments have the *same direction,* because they are both directed toward the upper right on lines having a slope of $\frac{2}{3}$. Thus, $\overrightarrow{PQ}$ and $\overrightarrow{RS}$ have the same length and direction, and it follows that $\mathbf{u} = \mathbf{v}$.   ▰

Component Form of a Vector

The directed line segment whose initial point is the origin is often the most convenient representative of a set of equivalent directed line segments. This representative of the vector **v** is in **standard position.**

A vector whose initial point is at the origin $(0, 0)$ can be uniquely represented by the coordinates of its terminal point (v_1, v_2). This is the **component form of a vector v,** written

$$\mathbf{v} = \langle v_1, v_2 \rangle.$$

The coordinates v_1 and v_2 are the **components** of **v.** If both the initial point and the terminal point lie at the origin, **v** is the **zero vector** and is denoted by $\mathbf{0} = \langle 0, 0 \rangle.$

Note Two vectors $\mathbf{u} = \langle u_1, u_2 \rangle$ and $\mathbf{v} = \langle v_1, v_2 \rangle$ are **equal** if and only if $u_1 = v_1$ and $u_2 = v_2$. For instance, in Example 1, the vector **u** from $P = (0, 0)$ to $Q = (3, 2)$ is

$$\mathbf{u} = \overrightarrow{PQ} = \langle 3 - 0, 2 - 0 \rangle = \langle 3, 2 \rangle$$

and the vector **v** from $R = (1, 2)$ to $S = (4, 4)$ is

$$\mathbf{v} = \overrightarrow{RS} = \langle 4 - 1, 4 - 2 \rangle = \langle 3, 2 \rangle.$$

Component Form of a Vector

The component form of the vector with initial point $P = (p_1, p_2)$ and terminal point $Q = (q_1, q_2)$ is

$$\overrightarrow{PQ} = \langle q_1 - p_1, q_2 - p_2 \rangle = \langle v_1, v_2 \rangle = \mathbf{v}.$$

The **length** (or magnitude) of **v** is given by

$$\|\mathbf{v}\| = \sqrt{(q_1 - p_1)^2 + (q_2 - p_2)^2} = \sqrt{v_1^2 + v_2^2}.$$

If $\|\mathbf{v}\| = 1$, **v** is a **unit vector.** Moreover, $\|\mathbf{v}\| = 0$ if and only if **v** is the zero vector **0.**

EXAMPLE 2   **Finding the Component Form of a Vector**

Find the component form and length of the vector **v** that has initial point $(4, -7)$ and terminal point $(-1, 5)$.

Solution

Let $P = (4, -7) = (p_1, p_2)$ and $Q = (-1, 5) = (q_1, q_2)$. Then, the components of $\mathbf{v} = \langle v_1, v_2 \rangle$ are given by

$$v_1 = q_1 - p_1 = -1 - 4 = -5$$
$$v_2 = q_2 - p_2 = 5 - (-7) = 12.$$

Thus, $\mathbf{v} = \langle -5, 12 \rangle$ and the length of **v** is

$$\|\mathbf{v}\| = \sqrt{(-5)^2 + 12^2} = \sqrt{169} = 13,$$

as shown in Figure 6.18.

Figure 6.18

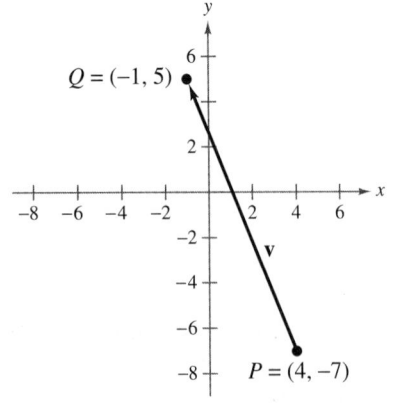

Vector Operations

Figure 6.19

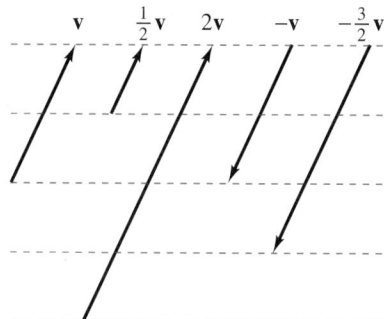

A computer animation of this concept appears in the *Interactive* CD-ROM.

Figure 6.21

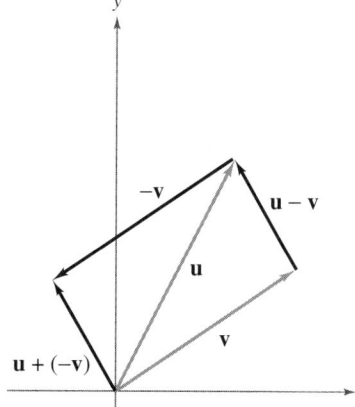

The two basic vector operations are **scalar multiplication** and **vector addition.** Geometrically, the product of a vector **v** and a scalar k is the vector that is $|k|$ times as long as **v**. If k is positive, k**v** has the same direction as **v**, and if k is negative, k**v** has the opposite direction of **v**, as shown in Figure 6.19.

To add two vectors geometrically, position them (without changing length or direction) so that the initial point of one coincides with the terminal point of the other. The sum **u** + **v** is formed by joining the initial point of the second vector **v** with the terminal point of the first vector **u**, as shown in Figure 6.20. This technique is called the **parallelogram law** for vector addition because the vector **u** + **v**, often called the **resultant** of vector addition, is the diagonal of a parallelogram having **u** and **v** as its adjacent sides.

Figure 6.20

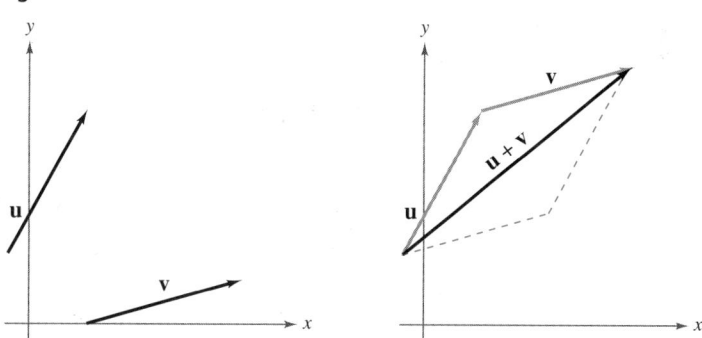

Definition of Vector Addition and Scalar Multiplication

Let $\mathbf{u} = \langle u_1, u_2 \rangle$ and $\mathbf{v} = \langle v_1, v_2 \rangle$ be vectors and let k be a scalar (a real number). Then the **sum** of **u** and **v** is the vector

$$\mathbf{u} + \mathbf{v} = \langle u_1 + v_1, u_2 + v_2 \rangle \qquad \text{Sum}$$

and the **scalar multiple** of k times **u** is the vector

$$k\mathbf{u} = k\langle u_1, u_2 \rangle = \langle ku_1, ku_2 \rangle. \qquad \text{Scalar multiple}$$

The **negative** of $\mathbf{v} = \langle v_1, v_2 \rangle$ is

$$-\mathbf{v} = (-1)\mathbf{v} = \langle -v_1, -v_2 \rangle \qquad \text{Negative}$$

and the **difference** of **u** and **v** is

$$\mathbf{u} - \mathbf{v} = \mathbf{u} + (-\mathbf{v}) = \langle u_1 - v_1, u_2 - v_2 \rangle. \qquad \text{Difference}$$

To represent **u** − **v** graphically, you can use directed line segments with the *same* initial points. The difference **u** − **v** is the vector from the terminal point of **v** to the terminal point of **u**, as shown in Figure 6.21.

The component definitions of vector addition and scalar multiplication are illustrated in Example 3. In this example, notice that each of the vector operations can be interpreted geometrically.

EXAMPLE 3   ◢   **Vector Operations**

Let $\mathbf{v} = \langle -2, 5 \rangle$ and $\mathbf{w} = \langle 3, 4 \rangle$, and find each of the following vectors.

a. $2\mathbf{v}$ **b. w − v**   **c. v + 2w**

Solution

a. Because $\mathbf{v} = \langle -2, 5 \rangle$, you have

$$2\mathbf{v} = \langle 2(-2), 2(5) \rangle$$
$$= \langle -4, 10 \rangle.$$

Note Figure 6.22(b) shows the vector difference $w − v$ as the sum $w + (−v)$.

A sketch of $2\mathbf{v}$ is shown in Figure 6.22(a).

b. The difference of **w** and **v** is given by

$$\mathbf{w} - \mathbf{v} = \langle 3 - (-2), 4 - 5 \rangle$$
$$= \langle 5, -1 \rangle.$$

A sketch of $\mathbf{w} - \mathbf{v}$ is shown in Figure 6.22(b).

c. Because $2\mathbf{w} = \langle 6, 8 \rangle$, it follows that

$$\mathbf{v} + 2\mathbf{w} = \langle -2, 5 \rangle + \langle 6, 8 \rangle$$
$$= \langle -2 + 6, 5 + 8 \rangle$$
$$= \langle 4, 13 \rangle.$$

A sketch of $\mathbf{v} + 2\mathbf{w}$ is shown in Figure 6.22(c).

Figure 6.22

(a)

(b)

(c)

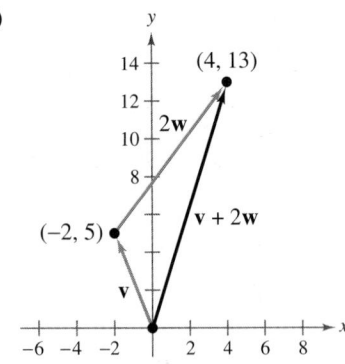

Note Property 9 can be stated as follows: The length of the vector $c\mathbf{v}$ is the absolute value of c times the length of $\mathbf{v}$.

Vector addition and scalar multiplication share many of the properties of ordinary arithmetic.

Properties of Vector Addition and Scalar Multiplication

Let $\mathbf{u}$, $\mathbf{v}$, and $\mathbf{w}$ be vectors and let c and d be scalars. Then the following properties are true.

1. $\mathbf{u} + \mathbf{v} = \mathbf{v} + \mathbf{u}$
2. $(\mathbf{u} + \mathbf{v}) + \mathbf{w} = \mathbf{u} + (\mathbf{v} + \mathbf{w})$
3. $\mathbf{u} + \mathbf{0} = \mathbf{u}$
4. $\mathbf{u} + (-\mathbf{u}) = \mathbf{0}$
5. $c(d\mathbf{u}) = (cd)\mathbf{u}$

6. $(c + d)\mathbf{u} = c\mathbf{u} + d\mathbf{u}$
7. $c(\mathbf{u} + \mathbf{v}) = c\mathbf{u} + c\mathbf{v}$
8. $1(\mathbf{u}) = \mathbf{u}, 0(\mathbf{u}) = \mathbf{0}$
9. $\|c\mathbf{v}\| = |c|\,\|\mathbf{v}\|$

Unit Vectors

In many applications of vectors it is useful to find a unit vector that has the same direction as a given nonzero vector $\mathbf{v}$. To do this, you can divide $\mathbf{v}$ by its length to obtain

$$\mathbf{u} = \text{unit vector} = \frac{\mathbf{v}}{\|\mathbf{v}\|} = \left(\frac{1}{\|\mathbf{v}\|}\right)\mathbf{v}.$$

Note that $\mathbf{u}$ is a scalar multiple of $\mathbf{v}$. The vector $\mathbf{u}$ has length 1 and the same direction as $\mathbf{v}$. The vector $\mathbf{u}$ is called a **unit vector in the direction of v.**

Some of the earliest work with vectors was done by the Irish mathematician William Rowan Hamilton (1805–1865). Hamilton spent many years developing a system of vector-like quantities called quaternions. Although Hamilton was convinced of the benefits of quaternions, the operations he defined did not produce good models for physical phenomena. It wasn't until the latter half of the nineteenth century that the Scottish physicist James Maxwell (1831–1879) restructured Hamilton's quaternions in a form useful for representing physical quantities such as force, velocity, and acceleration.

EXAMPLE 4   **Finding a Unit Vector**

Find a unit vector in the direction of $\mathbf{v} = \langle -2, 5 \rangle$ and verify that the result has length 1.

Solution

The unit vector in the direction of $\mathbf{v}$ is

$$\frac{\mathbf{v}}{\|\mathbf{v}\|} = \frac{\langle -2, 5 \rangle}{\sqrt{(-2)^2 + (5)^2}}$$

$$= \frac{1}{\sqrt{29}}\langle -2, 5 \rangle = \left\langle \frac{-2}{\sqrt{29}}, \frac{5}{\sqrt{29}} \right\rangle.$$

This vector has length 1 because

$$\sqrt{\left(\frac{-2}{\sqrt{29}}\right)^2 + \left(\frac{5}{\sqrt{29}}\right)^2} = \sqrt{\frac{4}{29} + \frac{25}{29}} = \sqrt{\frac{29}{29}} = 1.$$

Figure 6.23

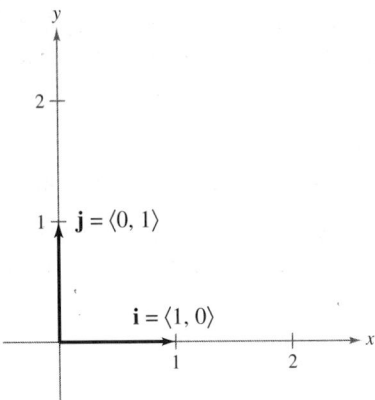

The unit vectors $\langle 1, 0 \rangle$ and $\langle 0, 1 \rangle$ are called the **standard unit vectors** and are denoted by

$$\mathbf{i} = \langle 1, 0 \rangle \quad \text{and} \quad \mathbf{j} = \langle 0, 1 \rangle$$

as shown in Figure 6.23. (Note that the lowercase letter $\mathbf{i}$ is written in boldface to distinguish it from the imaginary number $i = \sqrt{-1}$.) These vectors can be used to represent any vector $\mathbf{v} = \langle v_1, v_2 \rangle$ as follows.

$$\begin{aligned} \mathbf{v} &= \langle v_1, v_2 \rangle \\ &= v_1 \langle 1, 0 \rangle + v_2 \langle 0, 1 \rangle \\ &= v_1 \mathbf{i} + v_2 \mathbf{j} \end{aligned}$$

The scalars v_1 and v_2 are called the **horizontal and vertical components of v,** respectively. The vector sum $v_1\mathbf{i} + v_2\mathbf{j}$ is called a **linear combination** of the vectors $\mathbf{i}$ and $\mathbf{j}$. Any vector in the plane can be expressed as a linear combination of the standard unit vectors $\mathbf{i}$ and $\mathbf{j}$.

Figure 6.24

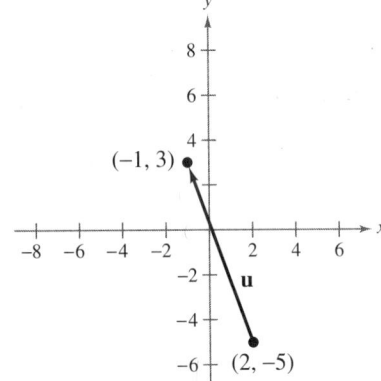

EXAMPLE 5 ▬ **Writing a Linear Combination of Unit Vectors**

Let $\mathbf{u}$ be the vector with initial point $(2, -5)$ and terminal point $(-1, 3)$. Write $\mathbf{u}$ as a linear combination of the standard unit vectors $\mathbf{i}$ and $\mathbf{j}$.

Solution
Begin by writing the component form of the vector $\mathbf{u}$.

$$\begin{aligned} \mathbf{u} &= \langle -1 - 2, 3 + 5 \rangle \\ &= \langle -3, 8 \rangle \\ &= -3\mathbf{i} + 8\mathbf{j} \end{aligned}$$

This result is shown graphically in Figure 6.24. ▬

EXAMPLE 6 ▬ **Vector Operations**

Let $\mathbf{u} = -3\mathbf{i} + 8\mathbf{j}$ and $\mathbf{v} = 2\mathbf{i} - \mathbf{j}$. Find $2\mathbf{u} - 3\mathbf{v}$.

Solution
You could solve this problem by converting $\mathbf{u}$ and $\mathbf{v}$ to component form. This, however, is not necessary. It is just as easy to perform the operations in unit vector form.

$$\begin{aligned} 2\mathbf{u} - 3\mathbf{v} &= 2(-3\mathbf{i} + 8\mathbf{j}) - 3(2\mathbf{i} - \mathbf{j}) \\ &= -6\mathbf{i} + 16\mathbf{j} - 6\mathbf{i} + 3\mathbf{j} \\ &= -12\mathbf{i} + 19\mathbf{j} \end{aligned}$$ ▬

Direction Angles

If **u** is a *unit vector* such that θ is the angle (measured counterclockwise) from the positive x-axis to **u**, the terminal point of **u** lies on the unit circle and you have

$$\mathbf{u} = \langle \cos\theta, \sin\theta \rangle = (\cos\theta)\mathbf{i} + (\sin\theta)\mathbf{j}$$

as shown in Figure 6.25. The angle θ is the **direction angle** of the vector **u**.

Suppose that **u** is a unit vector with direction angle θ. If **v** is any vector that makes an angle θ with the positive x-axis, then it has the same direction as **u** and you can write

$$\mathbf{v} = \|\mathbf{v}\| \langle \cos\theta, \sin\theta \rangle$$
$$= \|\mathbf{v}\|(\cos\theta)\mathbf{i} + \|\mathbf{v}\|(\sin\theta)\mathbf{j}.$$

Because $\mathbf{v} = a\mathbf{i} + b\mathbf{j} = \|\mathbf{v}\|(\cos\theta)\mathbf{i} + \|\mathbf{v}\|(\sin\theta)\mathbf{j}$, it follows that the direction angle θ for **v** is determined from

$$\tan\theta = \frac{\sin\theta}{\cos\theta} = \frac{\|\mathbf{v}\|\sin\theta}{\|\mathbf{v}\|\cos\theta} = \frac{b}{a}.$$

Figure 6.25

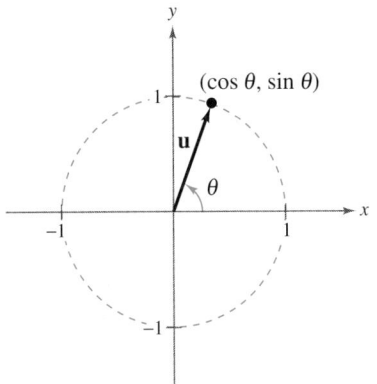

EXAMPLE 7 **Finding Direction Angles of Vectors**

Find the direction angle of each vector.

a. $\mathbf{u} = 3\mathbf{i} + 3\mathbf{j}$ **b.** $\mathbf{v} = 3\mathbf{i} - 4\mathbf{j}$

Solution

a. The direction angle is given by

$$\tan\theta = \frac{b}{a} = \frac{3}{3} = 1.$$

Therefore, $\theta = 45°$, as shown in Figure 6.26.

b. The direction angle is given by

$$\tan\theta = \frac{b}{a} = \frac{-4}{3}.$$

Moreover, because $\mathbf{v} = 3\mathbf{i} - 4\mathbf{j}$ lies in Quadrant IV, θ lies in Quadrant IV and its reference angle is

$$\theta' = \left| \arctan\left(-\frac{4}{3}\right) \right| \approx |-53.13°| = 53.13°.$$

Therefore, it follows that $\theta \approx 360° - 53.13° = 306.87°$, as shown in Figure 6.27.

Figure 6.26

Figure 6.27

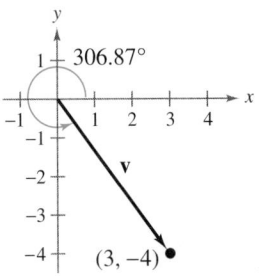

Applications of Vectors

Real Life

EXAMPLE 8   Finding the Component Form of a Vector

Find the component form of the vector that represents the velocity of an airplane descending at a speed of 100 miles per hour at an angle 30° below the horizontal, as shown in Figure 6.28.

Figure 6.28

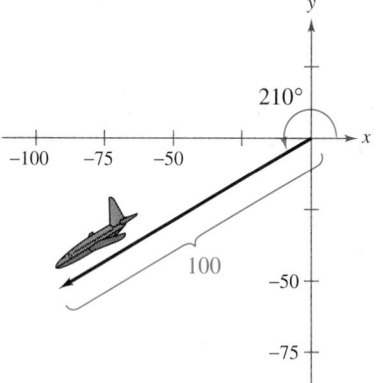

Solution
The velocity vector **v** has a magnitude of 100 and a direction angle of $\theta = 210°$.

$$\mathbf{v} = \|\mathbf{v}\| (\cos \theta)\mathbf{i} + \|\mathbf{v}\| (\sin \theta)\mathbf{j}$$
$$= 100(\cos 210°)\mathbf{i} + 100(\sin 210°)\mathbf{j}$$
$$= 100\left(\frac{-\sqrt{3}}{2}\right)\mathbf{i} + 100\left(\frac{-1}{2}\right)\mathbf{j}$$
$$= -50\sqrt{3}\,\mathbf{i} - 50\mathbf{j}$$
$$= \langle -50\sqrt{3},\ -50 \rangle$$

You should check to see that $\|\mathbf{v}\| = 100$.

Real Life

EXAMPLE 9   An Application

A force of 600 pounds is required to pull a boat and trailer up a ramp inclined at 15° from the horizontal. Find the combined weight of the boat and trailer.

Figure 7.69

Note In Figure 6.29, note that $\overrightarrow{AC}$ is parallel to the ramp.

Solution
Based on Figure 6.29, you can make the following observations.

$\|\overrightarrow{BA}\|$ = force of gravity = combined weight of boat and trailer

$\|\overrightarrow{BC}\|$ = force against ramp

$\|\overrightarrow{AC}\|$ = force required to move boat up ramp = 600 pounds

By construction, triangles *BWD* and *ABC* are similar. Hence, angle *ABC* is 15°. Therefore, in triangle *ABC* you have

$$\sin 15° = \frac{\|\overrightarrow{AC}\|}{\|\overrightarrow{BA}\|} = \frac{600}{\|\overrightarrow{BA}\|}$$

$$\|\overrightarrow{BA}\| = \frac{600}{\sin 15°} \approx 2318.$$

Consequently, the combined weight is approximately 2318 pounds.

Figure 6.30

EXAMPLE 10   **An Application**

An airplane is traveling at a fixed altitude with a negligible wind factor. The airplane is headed N 30° W at a speed of 500 miles per hour, as shown in Figure 6.30. As the airplane reaches a certain point, it encounters a wind with a velocity of 70 miles per hour in the direction N 45° E. What are the resultant speed and direction of the airplane?

Solution

Using Figure 6.30, the velocity of the airplane (alone) is given by

$$\mathbf{v}_1 = 500\langle\cos 120°, \sin 120°\rangle = \langle-250, 250\sqrt{3}\rangle$$

and the velocity of the wind is given by

$$\mathbf{v}_2 = 70\langle\cos 45°, \sin 45°\rangle = \langle35\sqrt{2}, 35\sqrt{2}\rangle.$$

Thus, the velocity of the airplane (in the wind) is given by

$$\mathbf{v} = \mathbf{v}_1 + \mathbf{v}_2 = \langle-250 + 35\sqrt{2}, 250\sqrt{3} + 35\sqrt{2}\rangle \approx \langle-200.5, 482.5\rangle$$

and the speed of the airplane is

$$\|\mathbf{v}\| = \sqrt{(-200.5)^2 + (482.5)^2} \approx 522.5 \text{ miles per hour.}$$

Finally, if θ is the direction angle of the flight path, you have

$$\tan \theta = \frac{482.5}{-200.5} \approx -2.4065$$

which implies that

$$\theta \approx 180° + \arctan(-2.4065) \approx 180° - 67.4° = 112.6°.$$

Group Activity   *Verifying the Associativity of Vector Addition*

On page 481, you learned that vector addition is associative, that is, for vectors **u, v,** and **w,** $(\mathbf{u} + \mathbf{v}) + \mathbf{w} = \mathbf{u} + (\mathbf{v} + \mathbf{w})$. Use graph paper and the information in the table to demonstrate geometrically that the resultant vector is the same regardless of whether you add **u** and **v** or **v** and **w** first.

	u	**v**	**w**
Initial Point	$(-3, 2)$	$(1, 6)$	$(2, -2)$
Terminal Point	$(5, -3)$	$(4, -3)$	$(-4, 3)$

6.3 /// EXERCISES

In Exercises 1–10, find the component form and the magnitude of the vector v.

1.

2.

3.

4.

5.

6.

7.

8.
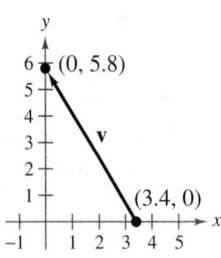

9. Initial point: $(-3, -5)$
Terminal point: $(5, 1)$

10. Initial point: $(-3, 11)$
Terminal point: $(9, 40)$

In Exercises 11–16, use the figure to sketch a graph of the specified vector.

11. $-\mathbf{v}$ **12.** $3\mathbf{v}$
13. $\mathbf{u} + \mathbf{v}$ **14.** $\mathbf{u} + 2\mathbf{v}$
15. $\mathbf{u} - \mathbf{v}$ **16.** $\mathbf{v} - \frac{1}{2}\mathbf{u}$

In Exercises 17–20, find (a) $\mathbf{u} + \mathbf{v}$, (b) $\mathbf{u} - \mathbf{v}$, and (c) $2\mathbf{u} - 3\mathbf{v}$.

17. $\mathbf{u} = \langle 1, 2 \rangle$, $\mathbf{v} = \langle 3, 1 \rangle$
18. $\mathbf{u} = \langle 2, 3 \rangle$, $\mathbf{v} = \langle 4, 0 \rangle$
19. $\mathbf{u} = \mathbf{i} + \mathbf{j}$, $\mathbf{v} = 2\mathbf{i} - 3\mathbf{j}$
20. $\mathbf{u} = 2\mathbf{i} - \mathbf{j}$, $\mathbf{v} = -\mathbf{i} + \mathbf{j}$

In Exercises 21–28, find a unit vector in the direction of the given vector.

21. $\mathbf{u} = \langle 5, 0 \rangle$ **22.** $\mathbf{u} = \langle 0, -3 \rangle$
23. $\mathbf{v} = \langle -2, 2 \rangle$ **24.** $\mathbf{v} = \langle 5, -12 \rangle$
25. $\mathbf{v} = 4\mathbf{i} - 3\mathbf{j}$ **26.** $\mathbf{v} = \mathbf{i} + \mathbf{j}$
27. $\mathbf{w} = 2\mathbf{j}$ **28.** $\mathbf{w} = \mathbf{i} - 2\mathbf{j}$

In Exercises 29–32, find the vector v with the given magnitude and the same direction as u.

Magnitude	Direction
29. $\|\mathbf{v}\| = 5$	$\mathbf{u} = \langle 3, 3 \rangle$
30. $\|\mathbf{v}\| = 3$	$\mathbf{u} = \langle 4, -4 \rangle$
31. $\|\mathbf{v}\| = 7$	$\mathbf{u} = \langle -3, 4 \rangle$
32. $\|\mathbf{v}\| = 10$	$\mathbf{u} = \langle -10, 0 \rangle$

In Exercises 33–38, find the component form of v and sketch the specified vector operations geometrically, where u = 2i − j and w = i + 2j.

33. $v = \frac{3}{2}u$

34. $v = u + w$

35. $v = u + 2w$

36. $v = -u + w$

37. $v = \frac{1}{2}(3u + w)$

38. $v = u - 2w$

In Exercises 39–42, find the magnitude and direction angle of the vector v.

39. $v = 5(\cos 30°i + \sin 30°j)$

40. $v = 8(\cos 135°i + \sin 135°j)$

41. $v = 6i - 6j$

42. $v = -2i + 5j$

In Exercises 43–48, find the component form of v given its magnitude and the angle it makes with the positive x-axis. Sketch v.

Magnitude	Angle
43. $\|v\| = 3$	$\theta = 0°$
44. $\|v\| = 1$	$\theta = 45°$
45. $\|v\| = 3\sqrt{2}$	$\theta = 150°$
46. $\|v\| = 9$	$\theta = 90°$
47. $\|v\| = 2$	v in the direction $i + 3j$
48. $\|v\| = 3$	v in the direction $3i + 4j$

In Exercises 49–52, find the component form of the sum of u and v with direction angles θ_u and θ_v.

Magnitude	Angle
49. $\|u\| = 5,$	$\theta_u = 0°$
$\|v\| = 5,$	$\theta_v = 90°$
50. $\|u\| = 2,$	$\theta_u = 30°$
$\|v\| = 2,$	$\theta_v = 90°$
51. $\|u\| = 20,$	$\theta_u = 45°$
$\|v\| = 50,$	$\theta_v = 180°$
52. $\|u\| = 35,$	$\theta_u = 25°$
$\|v\| = 50,$	$\theta_v = 120°$

In Exercises 53–56, use the Law of Cosines to find the angle α between the given vectors. (Assume $0° \le \alpha \le 180°$.)

53. $v = i + j,\quad w = 2(i - j)$

54. $v = 3i + j,\quad w = 2i - j$

55. $v = i + j,\quad w = 3i - j$

56. $v = i + 2j,\quad w = 2i - j$

In Exercises 57 and 58, find the angle between the forces given the magnitude of their resultant. (*Hint:* Write force one as a vector in the direction of the positive x-axis and force two as a vector at an angle θ with the positive x-axis.)

	Force 1	Force 2	Resultant Force
57.	45 pounds	60 pounds	90 pounds
58.	3000 pounds	1000 pounds	3750 pounds

59. *Think About It* Consider two forces of equal magnitude acting on a point.

(a) If the magnitude of the resultant is the sum of the magnitudes of the two forces, make a conjecture about the angle between the forces.

(b) If the resultant of the forces is **0**, make a conjecture about the angle between the forces.

(c) Can the magnitude of the resultant be greater than the sum of the magnitudes of the two forces? Explain.

60. *Graphical Reasoning* Consider two forces $F_1 = \langle 10, 0 \rangle$ and $F_2 = 5\langle \cos\theta, \sin\theta \rangle$.

(a) Find $\|F_1 + F_2\|$.

(b) Determine the magnitude of the resultant as a function of θ. Use a graphing utility to graph the function for $0 \le \theta < 2\pi$.

(c) Use the graph in part (b) to determine the range of the function. What is its maximum and for what value of θ does it occur? What is its minimum and for what value of θ does it occur?

(d) Explain why the magnitude of the resultant is never 0.

61. *Numerical and Graphical Analysis* Forces with magnitudes of 150 newtons and 220 newtons act on a hook (see figure).

(a) If $\theta = 30°$, find the direction and magnitude of the resultant of these forces.

(b) Express the magnitude M of the resultant and the direction α of the resultant as functions of θ, where $0° \leq \theta \leq 180°$.

(c) Use a graphing utility to complete the table.

θ	0°	30°	60°	90°	120°	150°	180°
M							
α							

(d) Use a graphing utility to graph the two functions.

(e) Explain why one function decreases for increasing θ, whereas the other doesn't.

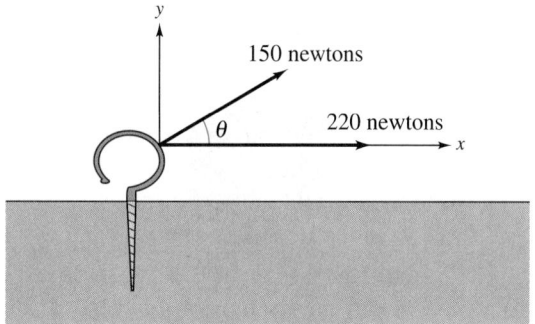

62. *Resultant Force* Forces with magnitudes of 2000 newtons and 900 newtons act on a machine part at angles of 30° and $-45°$, respectively, with the x-axis (see figure). Find the direction and magnitude of the resultant of these forces.

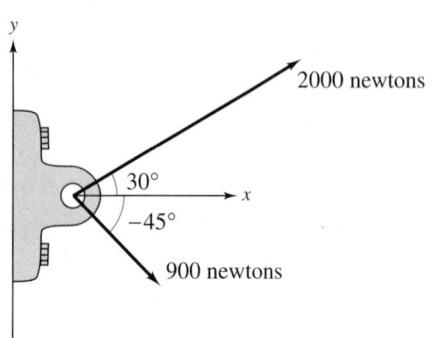

63. *Resultant Force* Three forces with magnitudes of 75 pounds, 100 pounds, and 125 pounds act on an object at angles of 30°, 45°, 120°, respectively, with the positive x-axis. Find the direction and magnitude of the resultant of these forces.

64. *Resultant Force* Three forces with magnitudes of 70 pounds, 40 pounds, and 60 pounds act on an object at angles of $-30°$, 45°, and 135°, respectively, with the positive x-axis. Find the direction and magnitude of the resultant of these forces.

In Exercises 65 and 66, use a graphing utility to graph the vectors and the resultant of the vectors. Find the magnitude and direction of the resultant.

65. **66.**

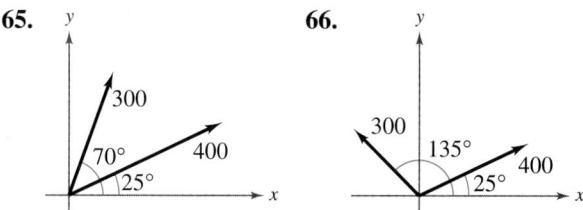

67. *Horizontal and Vertical Components of Velocity* A ball is thrown with an initial velocity of 80 feet per second, at an angle of 40° with the horizontal (see figure). Find the vertical and horizontal components of the velocity.

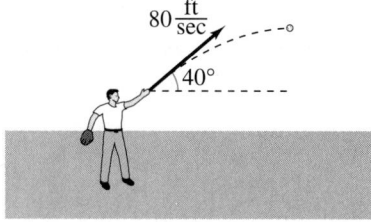

68. *Horizontal and Vertical Components of Velocity* A gun with a muzzle velocity of 1200 feet per second is fired at an angle of 6° with the horizontal. Find the vertical and horizontal components of the velocity.

Cable Tension   In Exercises 69 and 70, use the figure to determine the tension in each cable supporting the given load.

69.

2000 lb

70.

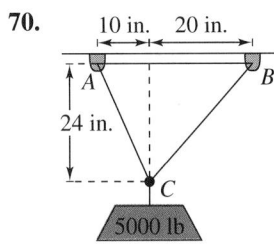

5000 lb

71. *Numerical and Graphical Analysis*   A loaded barge is being towed by two tugboats, and the magnitude of the resultant is 6000 pounds directed along the axis of the barge (see figure). Each tow line makes an angle of θ degrees with the axis of the barge.

(a) Find the tension in the tow lines if $\theta = 20°$.

(b) Write the tension T of each line as a function of θ. Determine the domain of the function.

(c) Use a graphing utility to complete the table.

θ	10°	20°	30°	40°	50°	60°
T						

(d) Use a graphing utility to graph the tension function.

(e) Explain why the tension increases as θ increases.

72. *Numerical and Graphical Analysis*   To carry a 100-pound cylindrical weight, two people lift on the ends of short ropes that are tied to an eyelet on the top center of the cylinder. Each rope makes an angle of θ degrees with the vertical.

(a) Find the tension in the ropes if $\theta = 30°$.

(b) Write the tension T of each rope as a function of θ. Determine the domain of the function.

(c) Use a graphing utility to complete the table.

θ	10°	20°	30°	40°	50°	60°
T						

(d) Use a graphing utility to graph the tension function.

(e) Explain why the tension increases as θ increases.

100 lb

73. *Navigation*   An airplane is flying in the direction S 32° E, with an airspeed of 875 kilometers per hour. Because of the wind, its groundspeed and direction are 800 kilometers per hour and S 40° E, respectively (see figure). Find the direction and speed of the wind.

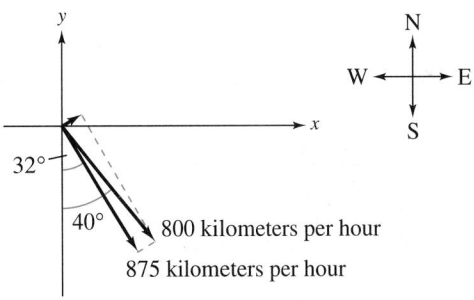

74. *Navigation* An airplane's velocity with respect to the air is 580 miles per hour, and it is heading N 60° W. The wind, at the altitude of the plane, is from the southwest and has a velocity of 60 miles per hour. Draw a figure that gives a visual representation of the problem. What is the true direction of the plane, and what is its speed with respect to the ground?

75. *Work* A heavy implement is pulled 20 feet across a floor, using a force of 85 pounds. Find the work done if the direction of the force is 60° above the horizontal (see figure). (Use the formula for work, $W = FD$, where F is the component of the force in the direction of motion and D is the distance.)

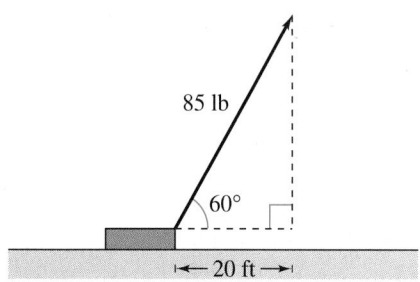

76. *Numerical and Graphical Analysis* A tetherball weighing 1 pound is pulled outward from the pole by a horizontal force **u** until the rope makes an angle of θ degrees with the pole (see figure).

   (a) Determine the resulting tension in the rope and the magnitude of **u** when $\theta = 30°$.

   (b) Write the tension T in the rope and the magnitude of **u** as functions of θ. Determine the domains of the functions.

   (c) Use a graphing utility to complete the table.

θ	0°	10°	20°	30°	40°	50°	60°
T							
$\|\mathbf{u}\|$							

   (d) Use a graphing utility to graph the two functions for $0° \le \theta \le 60°$.

   (e) Compare T and $\|\mathbf{u}\|$ as θ increases.

Figure for 76

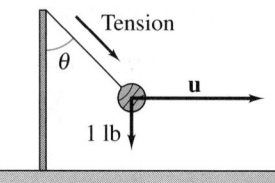

True or False? In Exercises 77–80, decide whether the statement is true or false. If it is false, explain why or give an example that shows it is false.

77. If **u** and **v** have the same magnitude and direction, then **u** = **v**.

78. If **u** is a unit vector in the direction of **v**, then **v** = $\|\mathbf{v}\|$ **u**.

79. If **v** = $a\mathbf{i} + b\mathbf{j} = \mathbf{0}$, then $a = -b$.

80. If **u** = $a\mathbf{i} + b\mathbf{j}$ is a unit vector, then $a^2 + b^2 = 1$.

81. Prove that $(\cos \theta)\mathbf{i} + (\sin \theta)\mathbf{j}$ is a unit vector for any value of θ.

82. *Technology* Write a program for your graphing utility that graphs two vectors and their difference given the vectors in component form.

In Exercises 83 and 84, use the program in Exercise 82 to find the difference of the vectors shown in the figure.

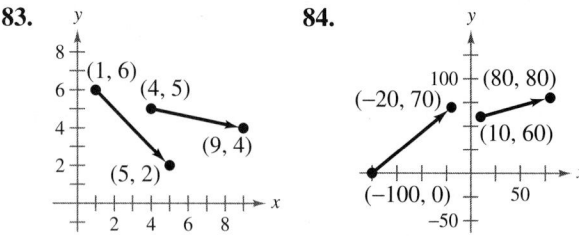

6.4 Vectors and Dot Products

The Dot Product of Two Vectors / *The Angle Between Two Vectors* /
Finding Vector Components / *Work*

The Dot Product of Two Vectors

So far you have studied two vector operations—vector addition and multiplication by a scalar—each of which yields another vector. In this section you will study a third vector operation, the **dot product.** This product yields a scalar, rather than a vector.

Definition of Dot Product

The **dot product** of $\mathbf{u} = \langle u_1, u_2 \rangle$ and $\mathbf{v} = \langle v_1, v_2 \rangle$ is

$$\mathbf{u} \cdot \mathbf{v} = u_1 v_1 + u_2 v_2.$$

Properties of the Dot Product

Let $\mathbf{u}$, $\mathbf{v}$, and $\mathbf{w}$ be vectors in the plane or in space and let c be a scalar.

1. $\mathbf{u} \cdot \mathbf{v} = \mathbf{v} \cdot \mathbf{u}$
2. $\mathbf{0} \cdot \mathbf{v} = 0$
3. $\mathbf{u} \cdot (\mathbf{v} + \mathbf{w}) = \mathbf{u} \cdot \mathbf{v} + \mathbf{u} \cdot \mathbf{w}$
4. $\mathbf{v} \cdot \mathbf{v} = \|\mathbf{v}\|^2$
5. $c(\mathbf{u} \cdot \mathbf{v}) = c\mathbf{u} \cdot \mathbf{v} = \mathbf{u} \cdot c\mathbf{v}$

Think About the Proof

To prove the second, third, and fifth properties of the dot product, consider the component forms of vectors $\mathbf{u}$, $\mathbf{v}$, and $\mathbf{w}$. You are asked to prove these properties in Exercise 60.

Proof /// To prove the first property, let $\mathbf{u} = \langle u_1, u_2 \rangle$ and $\mathbf{v} = \langle v_1, v_2 \rangle$. Then

$$\begin{aligned}
\mathbf{u} \cdot \mathbf{v} &= u_1 v_1 + u_2 v_2 \\
&= v_1 u_1 + v_2 u_2 \\
&= \mathbf{v} \cdot \mathbf{u}.
\end{aligned}$$

For the fourth property, let $\mathbf{v} = \langle v_1, v_2 \rangle$. Then

$$\begin{aligned}
\mathbf{v} \cdot \mathbf{v} &= v_1^2 + v_2^2 \\
&= \left(\sqrt{v_1^2 + v_2^2}\right)^2 \\
&= \|\mathbf{v}\|^2.
\end{aligned}$$

///

Note In Example 1, be sure you see that the dot product of two vectors is a scalar (a real number), not a vector. Moreover, notice that the dot product can be positive, zero, or negative.

A graphing utility can be used to find the angle between two vectors. The following *TI-82/TI-83* program sketches two vectors $\mathbf{u} = \langle a, b \rangle$ and $\mathbf{v} = \langle c, d \rangle$ in standard position and finds the measure of the angle between them. Use the program to verify Example 4. (Before running the program, set an appropriate viewing rectangle.)

```
:VECANGL
:ClrHome
:Disp "ENTER(A,B)"
:Input "ENTER A",A
:Input "ENTER B",B
:ClrHome
:Disp "ENTER(C,D)"
:Input "ENTER C",C
:Input "ENTER D",D
:Line(0,0,A,B)
:Line(0,0,C,D)
:Pause
:AC+BD→E
:√(A²+B²)→U
:√(C²+D²)→V
:cos⁻¹(E/(UV))→θ
:ClrDraw:ClrHome
:Disp "θ=",θ
:Stop
```

EXAMPLE 1   **Finding Dot Products**

Find each dot product.

a. $\langle 4, 5 \rangle \cdot \langle 2, 3 \rangle$

b. $\langle 2, -1 \rangle \cdot \langle 1, 2 \rangle$

c. $\langle 0, 3 \rangle \cdot \langle 4, -2 \rangle$

Solution

a. $\langle 4, 5 \rangle \cdot \langle 2, 3 \rangle = 4(2) + 5(3) = 8 + 15 = 23$

b. $\langle 2, -1 \rangle \cdot \langle 1, 2 \rangle = 2(1) + (-1)(2) = 2 - 2 = 0$

c. $\langle 0, 3 \rangle \cdot \langle 4, -2 \rangle = 0(4) + 3(-2) = 0 - 6 = -6$

EXAMPLE 2   **Using Properties of Dot Products**

Let $\mathbf{u} = \langle -1, 3 \rangle$, $\mathbf{v} = \langle 2, -4 \rangle$, and $\mathbf{w} = \langle 1, -2 \rangle$. Find each dot product.

a. $(\mathbf{u} \cdot \mathbf{v})\mathbf{w}$ **b.** $\mathbf{u} \cdot 2\mathbf{v}$

Solution

Begin by finding the dot product of $\mathbf{u}$ and $\mathbf{v}$.

$$\begin{aligned} \mathbf{u} \cdot \mathbf{v} &= \langle -1, 3 \rangle \cdot \langle 2, -4 \rangle \\ &= (-1)(2) + 3(-4) \\ &= -14 \end{aligned}$$

a. $(\mathbf{u} \cdot \mathbf{v})\mathbf{w} = -14\langle 1, -2 \rangle = \langle -14, 28 \rangle$

b. $\mathbf{u} \cdot 2\mathbf{v} = 2(\mathbf{u} \cdot \mathbf{v}) = 2(-14) = -28$

Notice that the first product is a vector, whereas the second is a scalar. Can you see why?

EXAMPLE 3   **Dot Product and Length**

The dot product of $\mathbf{u}$ with itself is 5. What is the length of $\mathbf{u}$?

Solution

Because $\|\mathbf{u}\|^2 = \mathbf{u} \cdot \mathbf{u} = 5$, it follows that

$$\begin{aligned} \|\mathbf{u}\| &= \sqrt{\mathbf{u} \cdot \mathbf{u}} \\ &= \sqrt{5}. \end{aligned}$$

The Angle Between Two Vectors

The **angle between two nonzero vectors** is the angle θ, $0 \le \theta \le \pi$, between its respective standard position vectors, as shown in Figure 6.31. This angle can be found using the dot product. (Note that the angle between the zero vector and another vector is not defined.)

Angle Between Two Vectors

If θ is the angle between two nonzero vectors $\mathbf{u}$ and $\mathbf{v}$, then

$$\cos \theta = \frac{\mathbf{u} \cdot \mathbf{v}}{\|\mathbf{u}\| \, \|\mathbf{v}\|}.$$

Figure 6.31

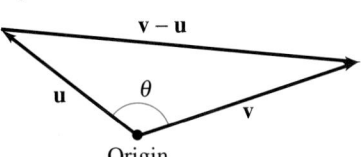

Proof /// Consider the triangle determined by vectors $\mathbf{u}$, $\mathbf{v}$, and $\mathbf{v} - \mathbf{u}$, as shown in Figure 6.31. By the Law of Cosines, you can write

$$\|\mathbf{v} - \mathbf{u}\|^2 = \|\mathbf{u}\|^2 + \|\mathbf{v}\|^2 - 2\|\mathbf{u}\| \, \|\mathbf{v}\| \cos \theta$$
$$(\mathbf{v} - \mathbf{u}) \cdot (\mathbf{v} - \mathbf{u}) = \|\mathbf{u}\|^2 + \|\mathbf{v}\|^2 - 2\|\mathbf{u}\| \, \|\mathbf{v}\| \cos \theta$$
$$(\mathbf{v} - \mathbf{u}) \cdot \mathbf{v} - (\mathbf{v} - \mathbf{u}) \cdot \mathbf{u} = \|\mathbf{u}\|^2 + \|\mathbf{v}\|^2 - 2\|\mathbf{u}\| \, \|\mathbf{v}\| \cos \theta$$
$$\mathbf{v} \cdot \mathbf{v} - \mathbf{u} \cdot \mathbf{v} - \mathbf{v} \cdot \mathbf{u} + \mathbf{u} \cdot \mathbf{u} = \|\mathbf{u}\|^2 + \|\mathbf{v}\|^2 - 2\|\mathbf{u}\| \, \|\mathbf{v}\| \cos \theta$$
$$\|\mathbf{v}\|^2 - 2\mathbf{u} \cdot \mathbf{v} + \|\mathbf{u}\|^2 = \|\mathbf{u}\|^2 + \|\mathbf{v}\|^2 - 2\|\mathbf{u}\| \, \|\mathbf{v}\| \cos \theta$$
$$-2\mathbf{u} \cdot \mathbf{v} = -2\|\mathbf{u}\| \, \|\mathbf{v}\| \cos \theta$$
$$\cos \theta = \frac{\mathbf{u} \cdot \mathbf{v}}{\|\mathbf{u}\| \, \|\mathbf{v}\|}. \qquad ///$$

Figure 6.32

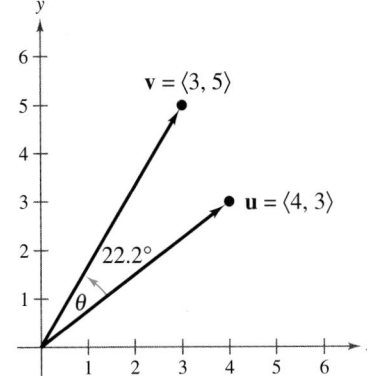

EXAMPLE 4 🔳 **Finding the Angle Between Two Vectors**

Find the angle between $\mathbf{u} = \langle 4, 3 \rangle$ and $\mathbf{v} = \langle 3, 5 \rangle$.

Solution

$$\cos \theta = \frac{\mathbf{u} \cdot \mathbf{v}}{\|\mathbf{u}\| \, \|\mathbf{v}\|} = \frac{\langle 4, 3 \rangle \cdot \langle 3, 5 \rangle}{\|\langle 4, 3 \rangle\| \, \|\langle 3, 5 \rangle\|} = \frac{27}{5\sqrt{34}}$$

This implies that the angle between the two vectors is

$$\theta = \arccos \frac{27}{5\sqrt{34}} \approx 22.2°,$$

as shown in Figure 6.32.

Rewriting the expression for the angle between two vectors in the form

$$\mathbf{u} \cdot \mathbf{v} = \|\mathbf{u}\| \|\mathbf{v}\| \cos \theta \qquad \text{Alternative form of dot product}$$

produces an alternative way to calculate the dot product. From this form, you can see that because $\|\mathbf{u}\|$ and $\|\mathbf{v}\|$ are always positive, $\mathbf{u} \cdot \mathbf{v}$ and $\cos \theta$ will always have the same sign. Figure 6.33 shows the five possible orientations of two vectors.

Figure 6.33

$\theta = \pi$	$\frac{\pi}{2} < \theta < \pi$	$\theta = \frac{\pi}{2}$	$0 < \theta < \frac{\pi}{2}$	$\theta = 0$
$\cos \theta = -1$	$-1 < \cos \theta < 0$	$\cos \theta = 0$	$0 < \cos \theta < 1$	$\cos \theta = 1$
Opposite Direction	Obtuse Angle	90° Angle	Acute Angle	Same Direction

---

Definition of Orthogonal Vectors

The vectors $\mathbf{u}$ and $\mathbf{v}$ are **orthogonal** if $\mathbf{u} \cdot \mathbf{v} = 0$.

The terms "orthogonal" and "perpendicular" mean essentially the same thing—meeting at right angles. By definition, however, the zero vector is orthogonal to every vector $\mathbf{u}$, because $\mathbf{0} \cdot \mathbf{u} = 0$.

Figure 6.34

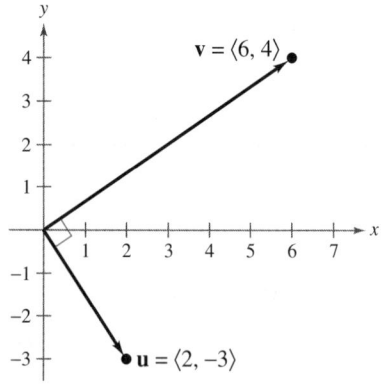

EXAMPLE 5   **Determining Orthogonal Vectors**

Are the vectors $\mathbf{u} = \langle 2, -3 \rangle$ and $\mathbf{v} = \langle 6, 4 \rangle$ orthogonal?

Solution

Begin by finding the dot product of the two vectors.

$$\mathbf{u} \cdot \mathbf{v} = \langle 2, -3 \rangle \cdot \langle 6, 4 \rangle$$
$$= 2(6) + (-3)(4)$$
$$= 0$$

Because the dot product is 0, the two vectors are orthogonal, as shown in Figure 6.34.

Finding Vector Components

Figure 6.35

You have already seen applications in which two vectors are added to produce a resultant vector. Many applications in physics and engineering pose the reverse problem—decomposing a given vector into the sum of two **vector components.**

Consider a boat on an inclined ramp, as shown in Figure 6.35. The force **F** due to gravity pulls the boat *down* the ramp and *against* the ramp. These two orthogonal forces, $\mathbf{w}_1$ and $\mathbf{w}_2$, are vector components of **F**. That is,

$$\mathbf{F} = \mathbf{w}_1 + \mathbf{w}_2. \qquad \text{Vector components of } \mathbf{F}$$

The negative of component $\mathbf{w}_1$ represents the force needed to keep the boat from rolling down the ramp, whereas $\mathbf{w}_2$ represents the force that the tires must withstand against the ramp. A procedure for finding $\mathbf{w}_1$ and $\mathbf{w}_2$ is shown below.

Definition of Vector Components

Let **u** and **v** be nonzero vectors such that

$$\mathbf{u} = \mathbf{w}_1 + \mathbf{w}_2$$

where $\mathbf{w}_1$ and $\mathbf{w}_2$ are orthogonal and $\mathbf{w}_1$ is parallel to **v**, as shown in Figure 6.36. The vectors $\mathbf{w}_1$ and $\mathbf{w}_2$ are called **vector components** of **u**. The vector $\mathbf{w}_1$ is the **projection** of **u** onto **v** and is denoted by

$$\mathbf{w}_1 = \text{proj}_{\mathbf{v}}\mathbf{u}.$$

The vector $\mathbf{w}_2$ is given by $\mathbf{w}_2 = \mathbf{u} - \mathbf{w}_1$.

Figure 6.36

Study Tip

From the definition of vector components, you can see that it is easy to find the component $\mathbf{w}_2$ once you have found the projection of **u** onto **v**. To find the projection, you can use the dot product.

θ is acute.

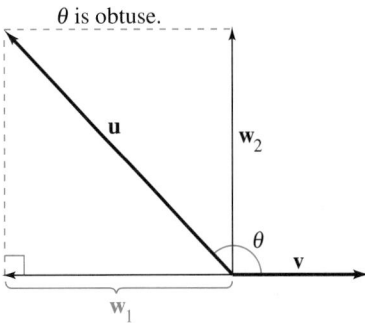

θ is obtuse.

Projection of u onto v

Let **u** and **v** be nonzero vectors. The projection of **u** onto **v** is

$$\text{proj}_{\mathbf{v}}\mathbf{u} = \left(\frac{\mathbf{u} \cdot \mathbf{v}}{\|\mathbf{v}\|^2}\right)\mathbf{v}.$$

Figure 6.37

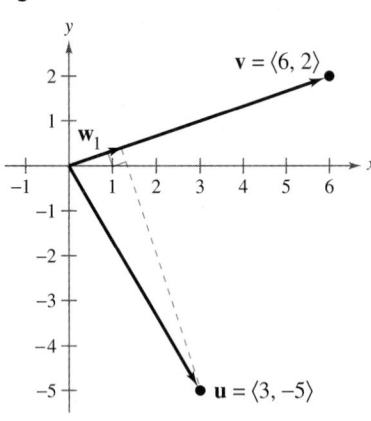

EXAMPLE 6   ▱   Decomposing a Vector into Components

Find the projection of $\mathbf{u} = \langle 3, -5 \rangle$ onto $\mathbf{v} = \langle 6, 2 \rangle$. Then write $\mathbf{u}$ as the sum of two orthogonal vectors, one of which is $\text{proj}_{\mathbf{v}}\mathbf{u}$.

Solution

The projection of $\mathbf{u}$ onto $\mathbf{v}$ is

$$\mathbf{w}_1 = \text{proj}_{\mathbf{v}}\mathbf{u} = \left(\frac{\mathbf{u} \cdot \mathbf{v}}{\|\mathbf{v}\|^2}\right)\mathbf{v} = \left(\frac{8}{40}\right)\langle 6, 2 \rangle = \left\langle \frac{6}{5}, \frac{2}{5} \right\rangle$$

as shown in Figure 6.37. The other component, $\mathbf{w}_2$, is

$$\mathbf{w}_2 = \mathbf{u} - \mathbf{w}_1 = \langle 3, -5 \rangle - \left\langle \frac{6}{5}, \frac{2}{5} \right\rangle = \left\langle \frac{9}{5}, -\frac{27}{5} \right\rangle.$$

Thus, $\mathbf{u} = \mathbf{w}_1 + \mathbf{w}_2 = \left\langle \frac{6}{5}, \frac{2}{5} \right\rangle + \left\langle \frac{9}{5}, -\frac{27}{5} \right\rangle = \langle 3, -5 \rangle.$ ▱

Real Life

EXAMPLE 7   ▱   Finding a Force

A 600-pound boat sits on a ramp inclined at 30°, as shown in Figure 6.38. What force is required to keep the boat from rolling down the ramp?

Solution

Because the force due to gravity is vertical and downward, you can represent the gravitational force by the vector

$$\mathbf{F} = -600\mathbf{j}. \qquad \text{Force due to gravity}$$

Figure 6.38

To find the force required to keep the boat from rolling down the ramp, project $\mathbf{F}$ onto a unit vector $\mathbf{v}$ in the direction of the ramp, as follows.

$$\mathbf{v} = (\cos 30°)\mathbf{i} + (\sin 30°)\mathbf{j} = \frac{\sqrt{3}}{2}\mathbf{i} + \frac{1}{2}\mathbf{j} \qquad \text{Unit vector along ramp}$$

Therefore, the projection of $\mathbf{F}$ onto $\mathbf{v}$ is given by

$$\begin{aligned} \mathbf{w}_1 &= \text{proj}_{\mathbf{v}}\mathbf{F} \\ &= \left(\frac{\mathbf{F} \cdot \mathbf{v}}{\|\mathbf{v}\|^2}\right)\mathbf{v} \\ &= (\mathbf{F} \cdot \mathbf{v})\mathbf{v} = (-600)\left(\frac{1}{2}\right)\mathbf{v} = -300\left(\frac{\sqrt{3}}{2}\mathbf{i} + \frac{1}{2}\mathbf{j}\right). \end{aligned}$$

The magnitude of this force is 300, and therefore a force of 300 pounds is required to keep the boat from rolling down the ramp.   ▱

Figure 6.39

(a)

(b)

Work

The work W done by a constant force **F** acting along the line of motion of an object is given by

$$W = \text{(magnitude of force)(distance)}$$
$$= \|\mathbf{F}\|\,\|\overrightarrow{PQ}\|$$

as shown in Figure 6.39(a). If the constant force **F** is not directed along the line of motion, you can see from Figure 6.39(b) that the work W done by the force is

$$W = \|\text{proj}_{\overrightarrow{PQ}}\,\mathbf{F}\|\,\|\overrightarrow{PQ}\|$$
$$= (\cos\theta)\,\|\mathbf{F}\|\,\|\overrightarrow{PQ}\|$$
$$= \mathbf{F} \cdot PQ.$$

This notion of work is summarized in the following definition.

Definition of Work

The **work** W done by a constant force **F** as its point of application moves along the vector $\overrightarrow{PQ}$ is given by either of the following.

1. $W = \|\text{proj}_{\overrightarrow{PQ}}\,\mathbf{F}\|\,\|\overrightarrow{PQ}\|$ Projection form
2. $W = \mathbf{F} \cdot \overrightarrow{PQ}$ Dot product form

---

Figure 6.40

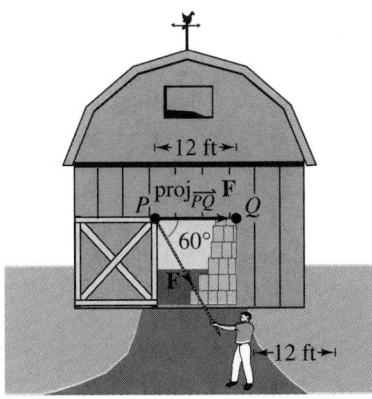

EXAMPLE 8   Finding Work

To close a sliding door, a person pulls on a rope with a constant force of 50 pounds at a constant angle of 60°, as shown in Figure 6.40. Find the work done in moving the door 12 feet to its closed position.

Solution

Using a projection, you can calculate the work as follows.

$$W = \|\text{proj}_{\overrightarrow{PQ}}\,\mathbf{F}\|\,\|\overrightarrow{PQ}\|$$
$$= (\cos 60°)\|\mathbf{F}\|\,\|\overrightarrow{PQ}\|$$
$$= \frac{1}{2}(50)(12)$$
$$= 300 \text{ ft-lb}$$

Thus, the work done is 300 foot-pounds.

Group Activity

The Sign of the Dot Product

On page 494, you were given the alternative form of the dot product of two vectors.

$$\mathbf{u} \cdot \mathbf{v} = \|\mathbf{u}\| \, \|\mathbf{v}\| \cos \theta \qquad \text{Alternative form of dot product}$$

Use this form to determine the sign of the dot product of **u** and **v** for the vectors shown below. Explain your reasoning.

a. b. c.

6.4 /// EXERCISES

In Exercises 1–4, find the dot product of u and v.

1. $\mathbf{u} = \langle 3, 4 \rangle$
   $\mathbf{v} = \langle 2, -3 \rangle$

2. $\mathbf{u} = \langle 5, 12 \rangle$
   $\mathbf{v} = \langle -3, 2 \rangle$

3. $\mathbf{u} = 4\mathbf{i} - 2\mathbf{j}$
   $\mathbf{v} = \mathbf{i} - \mathbf{j}$

4. $\mathbf{u} = 2\mathbf{i} + 5\mathbf{j}$
   $\mathbf{v} = 9\mathbf{i} - 3\mathbf{j}$

In Exercises 5–8, use the vectors $\mathbf{u} = \langle 2, 2 \rangle$ and $\mathbf{v} = \langle -3, 4 \rangle$ to find the indicated quantity. State whether the result is a vector or a scalar.

5. $\mathbf{u} \cdot \mathbf{u}$

6. $\|\mathbf{u}\| - 2$

7. $(\mathbf{u} \cdot \mathbf{v})\mathbf{v}$

8. $\mathbf{u} \cdot 2\mathbf{v}$

In Exercises 9–12, use the dot product to find $\|\mathbf{u}\|$.

9. $\mathbf{u} = \langle -5, 12 \rangle$

10. $\mathbf{u} = \langle 2, -4 \rangle$

11. $\mathbf{u} = 20\mathbf{i} + 25\mathbf{j}$

12. $\mathbf{u} = 6\mathbf{j}$

13. *Revenue* The vector $\mathbf{u} = \langle 1245, 2600 \rangle$ gives the number of units of two products produced by a company. The vector $\mathbf{v} = \langle 12.20, 8.50 \rangle$ gives the price (in dollars) of each unit, respectively. Find the dot product $\mathbf{u} \cdot \mathbf{v}$, and explain what information it gives.

14. *Revenue* Repeat Exercise 13 after increasing the prices by 5%. Identify the vector operation used to increase the prices by 5%.

In Exercises 15–20, find the angle θ between the vectors.

15. $\mathbf{u} = \langle 1, 0 \rangle$
   $\mathbf{v} = \langle 0, -2 \rangle$

16. $\mathbf{u} = \langle 4, 4 \rangle$
   $\mathbf{v} = \langle 2, 0 \rangle$

17. $\mathbf{u} = 3\mathbf{i} + 4\mathbf{j}$
   $\mathbf{v} = -2\mathbf{j}$

18. $\mathbf{u} = 2\mathbf{i} - 3\mathbf{j}$
   $\mathbf{v} = \mathbf{i} - 2\mathbf{j}$

19. $\mathbf{u} = \cos\left(\dfrac{\pi}{3}\right)\mathbf{i} + \sin\left(\dfrac{\pi}{3}\right)\mathbf{j}$

$\mathbf{v} = \cos\left(\dfrac{3\pi}{4}\right)\mathbf{i} + \sin\left(\dfrac{3\pi}{4}\right)\mathbf{j}$

20. $\mathbf{u} = \cos\left(\dfrac{\pi}{4}\right)\mathbf{i} + \sin\left(\dfrac{\pi}{4}\right)\mathbf{j}$

$\mathbf{v} = \cos\left(\dfrac{\pi}{2}\right)\mathbf{i} + \sin\left(\dfrac{\pi}{2}\right)\mathbf{j}$

In Exercises 21–24, use a graphing utility to sketch the vectors and find the degree measure of the angle between the vectors.

21. $\mathbf{u} = 3\mathbf{i} + 4\mathbf{j}$
$\mathbf{v} = -7\mathbf{i} + 5\mathbf{j}$

22. $\mathbf{u} = -6\mathbf{i} - 3\mathbf{j}$
$\mathbf{v} = -8\mathbf{i} + 4\mathbf{j}$

23. $\mathbf{u} = 5\mathbf{i} + 5\mathbf{j}$
$\mathbf{v} = -6\mathbf{i} + 6\mathbf{j}$

24. $\mathbf{u} = 2\mathbf{i} - 3\mathbf{j}$
$\mathbf{v} = 4\mathbf{i} + 3\mathbf{j}$

In Exercises 25 and 26, use vectors to find the interior angles of the triangle with the given vertices.

25. $(1, 2), (3, 4), (2, 5)$

26. $(-3, 0), (2, 2), (0, 6)$

In Exercises 27 and 28, find $\mathbf{u} \cdot \mathbf{v}$, where θ is the angle between $\mathbf{u}$ and $\mathbf{v}$.

27. $\|\mathbf{u}\| = 4, \|\mathbf{v}\| = 10, \ \theta = \dfrac{2\pi}{3}$

28. $\|\mathbf{u}\| = 100, \ \|\mathbf{v}\| = 250, \ \theta = \dfrac{\pi}{6}$

In Exercises 29–34, determine whether $\mathbf{u}$ and $\mathbf{v}$ are orthogonal, parallel, or neither.

29. $\mathbf{u} = \langle -12, 30 \rangle$
$\mathbf{v} = \langle \frac{1}{2}, -\frac{5}{4} \rangle$

30. $\mathbf{u} = \langle 15, 45 \rangle$
$\mathbf{v} = \langle -5, 12 \rangle$

31. $\mathbf{u} = \frac{1}{4}(3\mathbf{i} - \mathbf{j})$
$\mathbf{v} = 5\mathbf{i} + 6\mathbf{j}$

32. $\mathbf{u} = \mathbf{j}$
$\mathbf{v} = \mathbf{i} - 2\mathbf{j}$

33. $\mathbf{u} = 2\mathbf{i} - 2\mathbf{j}$
$\mathbf{v} = -\mathbf{i} - \mathbf{j}$

34. $\mathbf{u} = \langle \cos\theta, \sin\theta \rangle$
$\mathbf{v} = \langle \sin\theta, -\cos\theta \rangle$

In Exercises 35–38, find the projection of u onto v, and the vector component of u orthogonal to v.

35. $\mathbf{u} = \langle 3, 4 \rangle$
$\mathbf{v} = \langle 8, 2 \rangle$

36. $\mathbf{u} = \langle 4, 2 \rangle$
$\mathbf{v} = \langle 1, -2 \rangle$

37. $\mathbf{u} = \langle 0, 3 \rangle$
$\mathbf{v} = \langle 2, 15 \rangle$

38. $\mathbf{u} = \langle -5, -1 \rangle$
$\mathbf{v} = \langle -1, 1 \rangle$

In Exercises 39–42, use the figure to mentally determine the projection of u onto v. (The coordinates of the terminal points of the vectors in standard position are given.) Use the formula to verify your result.

39.

40.

41.

42.

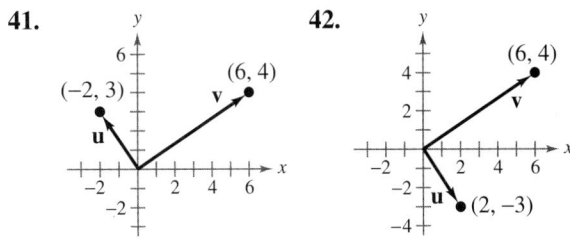

In Exercises 43–46, find two vectors in opposite directions that are orthogonal to the vector u. (The answers are not unique.)

43. $\mathbf{u} = \langle 3, 5 \rangle$

44. $\mathbf{u} = \langle -8, 3 \rangle$

45. $\mathbf{u} = \frac{1}{2}\mathbf{i} - \frac{2}{3}\mathbf{j}$

46. $\mathbf{u} = -\frac{5}{2}\mathbf{i} - 3\mathbf{j}$

47. *Braking Load* A truck with a gross weight of 36,000 pounds is parked on a 10° slope (see figure). Assume the only force to overcome is that due to gravity.

(a) Find the force required to keep the truck from rolling down the hill.

(b) Find the force perpendicular to the hill.

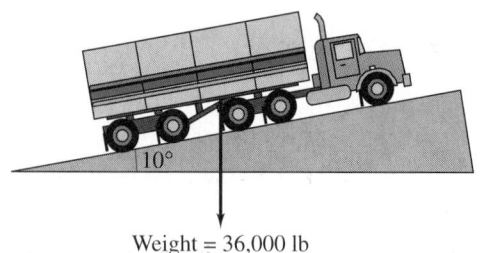

Weight = 36,000 lb

48. *Braking Load* Rework Exercise 47 for a truck that is parked on a 12° slope.

49. *Think About It* What is known about θ, the angle between two nonzero vectors **u** and **v**, if the following are true?

(a) **u** • **v** = 0   (b) **u** • **v** > 0   (c) **u** • **v** < 0

50. *Think About It* What can be said about the vectors **u** and **v** if the following are true?

(a) The projection of **u** onto **v** equals **u**.

(b) The projection of **u** onto **v** equals **0**.

51. *Work* A 25-kilogram (245-newton) bag of sugar is lifted 3 meters. Determine the work done.

52. *Work* Determine the work done by a crane lifting a 2400-pound car 5 feet.

53. *Work* A force of 45 pounds in the direction of 30° above the horizontal is required to slide an implement across a floor (see figure). Find the work done if the implement is dragged 20 feet.

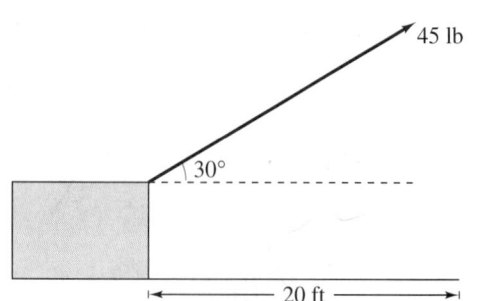

45 lb

30°

20 ft

54. *Work* A tractor pulls a log 800 meters and the tension in the cable connecting the tractor and log is approximately 1600 kilograms (15,691 newtons). Approximate the work done if the direction of the force is 35° above the horizontal.

Work In Exercises 55 and 56, find the work done in moving a particle from P to Q if the magnitude and direction of the force are given by **v**.

55. $P = (0, 0)$, $Q = (4, 7)$, $\mathbf{v} = \langle 1, 4 \rangle$

56. $P = (1, 3)$, $Q = (-3, 5)$, $\mathbf{v} = -2\mathbf{i} + 3\mathbf{j}$

57. For nonzero vectors **u** and **v**, prove that
$$\text{proj}_{\mathbf{v}}\mathbf{u} = \left(\frac{\mathbf{u} \cdot \mathbf{v}}{\|\mathbf{v}\|^2} \right)\mathbf{v}.$$

58. Use vectors to prove that the diagonals of a rhombus are perpendicular.

59. Prove the following.
$$\|\mathbf{u} - \mathbf{v}\|^2 = \|\mathbf{u}\|^2 + \|\mathbf{v}\|^2 - 2\mathbf{u} \cdot \mathbf{v}$$

60. Prove the following properties of the dot product.

(a) $\mathbf{0} \cdot \mathbf{v} = 0$

(b) $\mathbf{u} \cdot (\mathbf{v} + \mathbf{w}) = \mathbf{u} \cdot \mathbf{v} + \mathbf{u} \cdot \mathbf{w}$

(c) $c(\mathbf{u} \cdot \mathbf{v}) = c\mathbf{u} \cdot \mathbf{v} = \mathbf{u} \cdot c\mathbf{v}$

61. Prove that if **u** is orthogonal to **v** and **w**, then **u** is orthogonal to $c\mathbf{v} + d\mathbf{w}$ for any scalars c and d.

Review Solve Exercises 62–65 as a review of the skills and problem-solving techniques you learned in previous sections. Perform the additions or subtractions and, if possible, simplify the results.

62. (a) $\dfrac{y^2}{x} - x$

(b) $\dfrac{\csc^2 x}{\cot x} - \cot x$

63. (a) $1 - \dfrac{1}{x^2}$

(b) $1 - \dfrac{1}{\sec^2 x}$

64. (a) $\dfrac{y}{z} - \dfrac{z}{1+y}$

(b) $\dfrac{\tan x}{\sec x} - \dfrac{\sec x}{1 + \tan x}$

65. (a) $\dfrac{y}{1+z} + \dfrac{1+z}{y}$

(b) $\dfrac{\sin x}{1 + \cos x} + \dfrac{1 + \cos x}{\sin x}$

The Complex Plane / Trigonometric Form of a Complex Number /
Multiplication and Division of Complex Numbers /
Powers of Complex Numbers / Roots of Complex Numbers

The Complex Plane

Recall from Section 2.4 that you can represent a complex number

$$z = a + bi$$

as the point (a, b) in a coordinate plane (the complex plane). The horizontal axis is called the real axis and the vertical axis is called the imaginary axis, as shown in Figure 6.41.

The **absolute value** of the complex number $a + bi$ is defined as the distance between the origin $(0, 0)$ and the point (a, b).

Definition of the Absolute Value of a Complex Number

The **absolute value** of the complex number $z = a + bi$ is given by

$$|a + bi| = \sqrt{a^2 + b^2}.$$

Note If the complex number $a + bi$ is a real number (that is, if $b = 0$), this definition agrees with that given for the absolute value of a real number

$$|a + 0i| = \sqrt{a^2 + 0^2} = |a|.$$

EXAMPLE 1   **Finding the Absolute Value of a Complex Number**

Plot each complex number and find its absolute value.

a. $z = -3i$ **b.** $z = -2 + 5i$

Solution
The points are shown in Figure 6.42.

a. The complex number $z = 0 + (-3)i$ has an absolute value of
$$|z| = \sqrt{0^2 + (-3)^2}$$
$$= 3.$$

b. The complex number $z = -2 + 5i$ has an absolute value of
$$|z| = \sqrt{(-2)^2 + 5^2}$$
$$= \sqrt{29}.$$

Figure 6.41

Figure 6.42

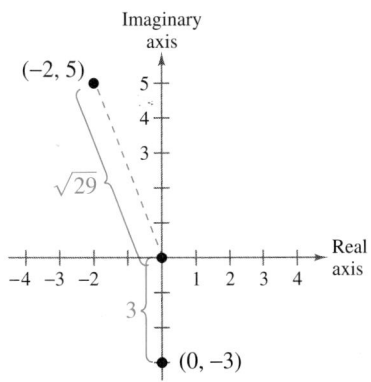

Trigonometric Form of a Complex Number

Figure 6.43

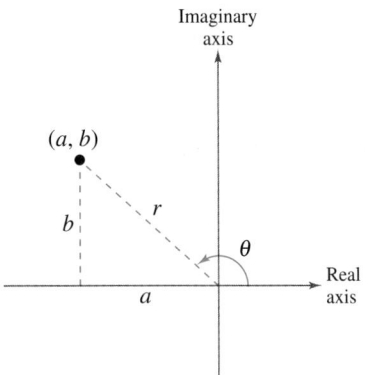

In Section 2.4 you learned how to add, subtract, multiply, and divide complex numbers. To work effectively with *powers* and *roots* of complex numbers, it is helpful to write complex numbers in **trigonometric form.** In Figure 6.43, consider the nonzero complex number $a + bi$. By letting θ be the angle from the positive x-axis (measured counterclockwise) to the line segment connecting the origin and the point (a, b), you can write

$$a = r \cos \theta \qquad \text{and} \qquad b = r \sin \theta$$

where $r = \sqrt{a^2 + b^2}$. Consequently, you have

$$a + bi = (r \cos \theta) + (r \sin \theta)i$$

from which you can obtain the **trigonometric form of a complex number.**

Note The trigonometric form of a complex number is also called the **polar form.** Because there are infinitely many choices for θ, the trigonometric form of a complex number is not unique. Normally, θ is restricted to the interval $0 \le \theta < 2\pi$, although on occasion it is convenient to use $\theta < 0$.

Trigonometric Form of a Complex Number

The **trigonometric form** of the complex number $z = a + bi$ is

$$z = r(\cos \theta + i \sin \theta)$$

where $a = r \cos \theta$, $b = r \sin \theta$, $r = \sqrt{a^2 + b^2}$, and $\tan \theta = b/a$. The number r is the **modulus** of z, and θ is called an **argument** of z.

EXAMPLE 2 ▱ **Writing a Complex Number in Trigonometric Form**

Write the complex number $z = -2 - 2\sqrt{3}i$ in trigonometric form.

Solution

The absolute value of z is

$$r = \left| -2 - 2\sqrt{3}i \right| = \sqrt{(-2)^2 + \left(-2\sqrt{3}\right)^2} = \sqrt{16} = 4$$

and the angle θ is given by

$$\tan \theta = \frac{b}{a} = \frac{-2\sqrt{3}}{-2} = \sqrt{3}.$$

Because $\tan(\pi/3) = \sqrt{3}$ and $z = -2 - 2\sqrt{3}i$ lies in Quadrant III, you choose θ to be $\theta = \pi + \pi/3 = 4\pi/3$. Thus, the trigonometric form is

$$z = r(\cos \theta + i \sin \theta) = 4\left(\cos \frac{4\pi}{3} + i \sin \frac{4\pi}{3} \right).$$

See Figure 6.44.

Figure 6.44

A graphing utility can be used to convert a complex number in polar form to rectangular form, and vice versa. For instance, to illustrate Example 3 on a *TI-82* or *TI-83*, use the following steps.

1. Press $\boxed{\text{ANGLE}}$ and choose P▷Rx(.
2. Enter the values for r and θ as (r, θ). Press $\boxed{\text{ENTER}}$ to obtain the x-coordinate.
3. Press $\boxed{\text{ANGLE}}$ and choose P▷Ry(.
4. Enter the values for r and θ as (r, θ). Press $\boxed{\text{ENTER}}$ to obtain the y-coordinate.

Write the result in rectangular form. Convert the complex number $-1 + \sqrt{3}i$ to trigonometric form using R▷Pr(and R▷Pθ(.

The *Interactive* CD-ROM offers graphing utility emulators of the *TI-82* and *TI-83*, which can be used with the Examples, Explorations, Technology notes, and Exercises.

EXAMPLE 3 ▭ **Writing a Complex Number in Standard Form**

Write the complex number in standard form $a + bi$.

$$z = \sqrt{8}\left[\cos\left(-\frac{\pi}{3}\right) + i\sin\left(-\frac{\pi}{3}\right)\right]$$

Solution

Because $\cos(-\pi/3) = 1/2$ and $\sin(-\pi/3) = -\sqrt{3}/2$, you can write

$$z = \sqrt{8}\left[\cos\left(-\frac{\pi}{3}\right) + i\sin\left(-\frac{\pi}{3}\right)\right]$$

$$= \sqrt{8}\left[\frac{1}{2} - \frac{\sqrt{3}}{2}i\right]$$

$$= 2\sqrt{2}\left[\frac{1}{2} - \frac{\sqrt{3}}{2}i\right]$$

$$= \sqrt{2} - \sqrt{6}i.$$

▭

Multiplication and Division of Complex Numbers

The trigonometric form adapts nicely to multiplication and division of complex numbers. Suppose you are given two complex numbers

$$z_1 = r_1(\cos\theta_1 + i\sin\theta_1) \quad \text{and} \quad z_2 = r_2(\cos\theta_2 + i\sin\theta_2).$$

The product of z_1 and z_2 is

$$z_1 z_2 = r_1 r_2(\cos\theta_1 + i\sin\theta_1)(\cos\theta_2 + i\sin\theta_2)$$

$$= r_1 r_2[(\cos\theta_1\cos\theta_2 - \sin\theta_1\sin\theta_2) + i(\sin\theta_1\cos\theta_2 + \cos\theta_1\sin\theta_2)].$$

Using the sum and difference formulas for cosine and sine, you can rewrite this equation as

$$z_1 z_2 = r_1 r_2[\cos(\theta_1 + \theta_2) + i\sin(\theta_1 + \theta_2)].$$

This establishes the first part of the following rule. The second part is left to you (see Exercise 65).

Product and Quotient of Two Complex Numbers

Let $z_1 = r_1(\cos\theta_1 + i\sin\theta_1)$ and $z_2 = r_2(\cos\theta_2 + i\sin\theta_2)$ be complex numbers.

$$z_1 z_2 = r_1 r_2[\cos(\theta_1 + \theta_2) + i\sin(\theta_1 + \theta_2)] \qquad \text{Product}$$

$$\frac{z_1}{z_2} = \frac{r_1}{r_2}[\cos(\theta_1 - \theta_2) + i\sin(\theta_1 - \theta_2)], \qquad z_2 \neq 0 \qquad \text{Quotient}$$

Note that this rule says that to multiply two complex numbers you multiply moduli and add arguments, whereas to divide two complex numbers you divide moduli and subtract arguments.

EXAMPLE 4   **Multiplying Complex Numbers in Trigonometric Form**

Find the product of the following complex numbers.

$$z_1 = 2\left(\cos\frac{2\pi}{3} + i\sin\frac{2\pi}{3}\right) \qquad z_2 = 8\left(\cos\frac{11\pi}{6} + i\sin\frac{11\pi}{6}\right)$$

Solution

$$z_1 z_2 = 2\left(\cos\frac{2\pi}{3} + i\sin\frac{2\pi}{3}\right) \cdot 8\left(\cos\frac{11\pi}{6} + i\sin\frac{11\pi}{6}\right)$$

$$= 16\left[\cos\left(\frac{2\pi}{3} + \frac{11\pi}{6}\right) + i\sin\left(\frac{2\pi}{3} + \frac{11\pi}{6}\right)\right]$$

$$= 16\left(\cos\frac{5\pi}{2} + i\sin\frac{5\pi}{2}\right)$$

$$= 16\left(\cos\frac{\pi}{2} + i\sin\frac{\pi}{2}\right)$$

$$= 16[0 + i(1)] = 16i$$

Some graphing utilities, such as the *TI-92*, can multiply and divide complex numbers in trigonometric form. If you have access to such a graphing utility, use it to find $z_1 z_2$ and z_1/z_2 in Examples 4 and 5.

Check this result by first converting to the standard forms $z_1 = -1 + \sqrt{3}i$ and $z_2 = 4\sqrt{3} - 4i$ and then multiplying algebraically, as in Section 2.4.

EXAMPLE 5   **Dividing Complex Numbers in Trigonometric Form**

Find z_1/z_2, for the following complex numbers.

$$z_1 = 24(\cos 300° + i\sin 300°) \qquad z_2 = 8(\cos 75° + i\sin 75°)$$

Solution

$$\frac{z_1}{z_2} = \frac{24(\cos 300° + i\sin 300°)}{8(\cos 75° + i\sin 75°)}$$

$$= \frac{24}{8}[\cos(300° - 75°) + i\sin(300° - 75°)]$$

$$= 3(\cos 225° + i\sin 225°)$$

$$= 3\left[\left(-\frac{\sqrt{2}}{2}\right) + i\left(-\frac{\sqrt{2}}{2}\right)\right]$$

$$= -\frac{3\sqrt{2}}{2} - \frac{3\sqrt{2}}{2}i$$

Powers of Complex Numbers

To raise a complex number to a power, consider repeated use of the multiplication rule.

$$z = r(\cos \theta + i \sin \theta)$$
$$z^2 = r(\cos \theta + i \sin \theta)r(\cos \theta + i \sin \theta) = r^2(\cos 2\theta + i \sin 2\theta)$$
$$z^3 = r^2(\cos 2\theta + i \sin 2\theta)r(\cos \theta + i \sin \theta) = r^3(\cos 3\theta + i \sin 3\theta)$$
$$z^4 = r^4(\cos 4\theta + i \sin 4\theta)$$
$$z^5 = r^5(\cos 5\theta + i \sin 5\theta)$$
$$\vdots$$

This pattern leads to the following important theorem, which is named after the French mathematician Abraham DeMoivre (1667–1754).

EXPLORATION

Plot the numbers i, i^2, i^3, i^4, and i^5 in the complex plane. Write each number in trigonometric form and describe what happens to the angle θ as you form higher powers of i^n.

DeMoivre's Theorem

If $z = r(\cos \theta + i \sin \theta)$ is a complex number and n is a positive integer, then

$$z^n = [r(\cos \theta + i \sin \theta)]^n = r^n(\cos n\theta + i \sin n\theta).$$

EXAMPLE 6 ▱ **Finding Powers of a Complex Number**

Use DeMoivre's Theorem to find $\left(-1 + \sqrt{3}i\right)^{12}$.

Solution

First convert to trigonometric form.

$$-1 + \sqrt{3}i = 2\left(\cos \frac{2\pi}{3} + i \sin \frac{2\pi}{3}\right)$$

Then, by DeMoivre's Theorem, you have

$$\left(-1 + \sqrt{3}i\right)^{12} = \left[2\left(\cos \frac{2\pi}{3} + i \sin \frac{2\pi}{3}\right)\right]^{12}$$
$$= 2^{12}\left[\cos\left(12 \cdot \frac{2\pi}{3}\right) + i \sin\left(12 \cdot \frac{2\pi}{3}\right)\right]$$
$$= 4096(\cos 8\pi + i \sin 8\pi)$$
$$= 4096(1 + 0)$$
$$= 4096.$$

Are you surprised to see a real number as the answer? ▱

Roots of Complex Numbers

Recall that a consequence of the Fundamental Theorem of Algebra is that a polynomial equation of degree n has n solutions in the complex number system. Hence, an equation such as $x^6 = 1$ has six solutions, and in this particular case you can find the six solutions by factoring and using the Quadratic Formula.

$$x^6 - 1 = (x^3 - 1)(x^3 + 1)$$
$$= (x - 1)(x^2 + x + 1)(x + 1)(x^2 - x + 1) = 0$$

Consequently, the solutions are

$$x = \pm 1, \qquad x = \frac{-1 \pm \sqrt{3}i}{2}, \qquad \text{and} \qquad x = \frac{1 \pm \sqrt{3}i}{2}.$$

Each of these numbers is a sixth root of 1. In general, the **nth root** of a complex number is defined as follows.

Definition of nth Root of a Complex Number

The complex number $u = a + bi$ is an **nth root** of the complex number z if

$$z = u^n = (a + bi)^n.$$

To find a formula for an nth root of a complex number, let u be an nth root of z, where

$$u = s(\cos \beta + i \sin \beta) \qquad \text{and} \qquad z = r(\cos \theta + i \sin \theta).$$

By DeMoivre's Theorem and the fact that $u^n = z$, you have

$$s^n (\cos n\beta + i \sin n\beta) = r(\cos \theta + i \sin \theta).$$

Taking the absolute value of both sides of this equation, it follows that $s^n = r$. Substituting back into the previous equation and dividing by r, you get

$$\cos n\beta + i \sin n\beta = \cos \theta + i \sin \theta.$$

Thus, it follows that

$$\cos n\beta = \cos \theta \qquad \text{and} \qquad \sin n\beta = \sin \theta.$$

Because both sine and cosine have a period of 2π, these last two equations have solutions if and only if the angles differ by a multiple of 2π. Consequently, there must exist an integer k such that

$$n\beta = \theta + 2\pi k$$
$$\beta = \frac{\theta + 2\pi k}{n}.$$

By substituting this value for β into the trigonometric form of u, you get the result stated in the theorem on the following page.

EXPLORATION

The nth roots of a complex number are useful for solving some polynomial equations. For instance, explain how you can use DeMoivre's Theorem to solve the polynomial equation

$$x^4 + 16 = 0.$$

[*Hint*: Write -16 as

$$16(\cos \pi + i \sin \pi).]$$

Note When k exceeds $n - 1$, the roots begin to repeat. For instance, if $k = n$, the angle

$$\frac{\theta + 2\pi n}{n} = \frac{\theta}{n} + 2\pi$$

is coterminal with θ/n, which is also obtained when $k = 0$.

---

nth Roots of a Complex Number

For a positive integer n, the complex number $z = r(\cos \theta + i \sin \theta)$ has exactly n distinct nth roots given by

$$\sqrt[n]{r}\left(\cos \frac{\theta + 2\pi k}{n} + i \sin \frac{\theta + 2\pi k}{n}\right)$$

where $k = 0, 1, 2, \ldots, n - 1$.

Figure 6.45

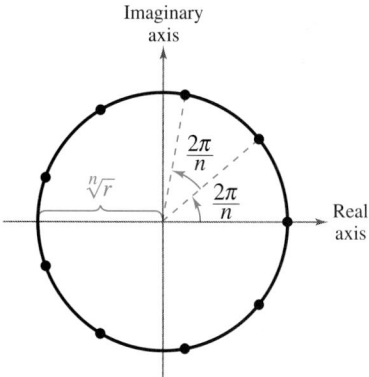

This formula for the nth roots of a complex number z has a nice geometrical interpretation, as shown in Figure 6.45. Note that because the nth roots of z all have the same magnitude $\sqrt[n]{r}$, they all lie on a circle of radius $\sqrt[n]{r}$ with center at the origin. Furthermore, because successive nth roots have arguments that differ by $2\pi/n$, the n roots are equally spaced along the circle.

You have already found the sixth roots of 1 by factoring and by using the Quadratic Formula. Example 7 shows how you can solve the same problem with the formula for nth roots.

EXAMPLE 7 ▬ **Finding the nth Roots of a Real Number**

Find all the sixth roots of 1.

Solution
First write 1 in the trigonometric form $1 = 1(\cos 0 + i \sin 0)$. Then, by the nth root formula, with $n = 6$ and $r = 1$, the roots have the form

$$\sqrt[6]{1}\left(\cos \frac{0 + 2\pi k}{6} + i \sin \frac{0 + 2\pi k}{6}\right)$$

Figure 6.46

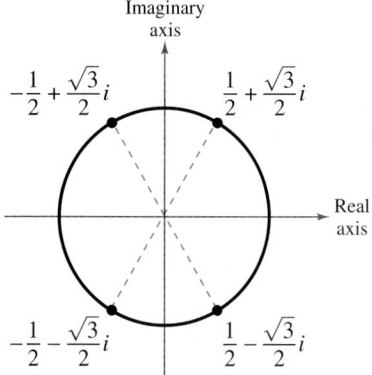

or simply $\cos(\pi k/3) + i \sin(\pi k/3)$. Thus, for $k = 0, 1, 2, 3, 4,$ and 5, the sixth roots are as follows. (See Figure 6.46.)

$$\cos 0 + i \sin 0 = 1$$
$$\cos \frac{\pi}{3} + i \sin \frac{\pi}{3} = \frac{1}{2} + \frac{\sqrt{3}}{2} i$$
$$\cos \frac{2\pi}{3} + i \sin \frac{2\pi}{3} = -\frac{1}{2} + \frac{\sqrt{3}}{2} i$$
$$\cos \pi + i \sin \pi = -1$$
$$\cos \frac{4\pi}{3} + i \sin \frac{4\pi}{3} = -\frac{1}{2} - \frac{\sqrt{3}}{2} i$$
$$\cos \frac{5\pi}{3} + i \sin \frac{5\pi}{3} = \frac{1}{2} - \frac{\sqrt{3}}{2} i$$

In Figure 6.46, notice that the roots obtained in Example 7 all have a magnitude of 1 and are equally spaced around this unit circle. Also notice that the complex roots occur in conjugate pairs, as discussed in Section 2.5. The n distinct nth roots of 1 are called the **nth roots of unity.**

EXAMPLE 8   ▭   **Finding the nth Roots of a Complex Number**

Find the three cube roots of $z = -2 + 2i$.

Solution

Because z lies in Quadrant II, the trigonometric form for z is

$$z = -2 + 2i = \sqrt{8}\,(\cos 135° + i \sin 135°).$$

By the formula for nth roots, the cube roots have the form

$$\sqrt[6]{8}\left(\cos \frac{135° + 360°k}{3} + i \sin \frac{135° + 360°k}{3}\right).$$

Finally, for $k = 0$, 1, and 2, you obtain the roots

$$\sqrt{2}(\cos 45° + i \sin 45°) = 1 + i$$
$$\sqrt{2}(\cos 165° + i \sin 165°) \approx -1.3660 + 0.3660i$$
$$\sqrt{2}(\cos 285° + i \sin 285°) \approx 0.3660 - 1.3660i.$$
▭

Group Activity   *A Famous Mathematical Formula*

The famous formula

$$e^{a+bi} = e^a(\cos b + i \sin b)$$

is called Euler's Formula, after the German mathematician Leonhard Euler (1707–1783). Although the interpretation of this formula is beyond the scope of this text, we decided to include it because it gives rise to one of the most wonderful equations in mathematics.

$$e^{\pi i} + 1 = 0$$

This elegant equation relates the five most famous numbers in mathematics—0, 1, π, e, and i—in a single equation. Show how Euler's Formula can be used to derive this equation.

6.5 /// EXERCISES

In Exercises 1–6, plot the complex number and find its absolute value.

1. $-5i$

2. -5

3. $-4 + 4i$

4. $5 - 12i$

5. $6 - 7i$

6. $-8 + 3i$

In Exercises 7–10, write in trigonometric form.

7.

8.

9.

10.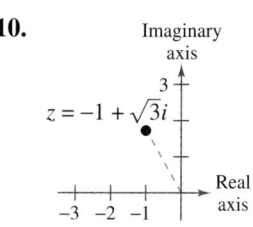

In Exercises 11–26, represent the complex number graphically, and find the trigonometric form of the number.

11. $3 - 3i$

12. $2 + 2i$

13. $\sqrt{3} + i$

14. $-1 + \sqrt{3}i$

15. $-2(1 + \sqrt{3}i)$

16. $\frac{5}{2}(\sqrt{3} - i)$

17. $6i$

18. 4

19. $-7 + 4i$

20. $3 - i$

21. 7

22. $-2i$

23. $1 + 6i$

24. $2\sqrt{2} - i$

25. $-3 - i$

26. $1 + 3i$

In Exercises 27–30, use a graphing utility to represent the complex number in trigonometric form.

27. $5 + 2i$

28. $-3 + i$

29. $3\sqrt{2} - 7i$

30. $-8 - 5\sqrt{3}i$

In Exercises 31–40, represent the complex number graphically, and find the standard form of the number.

31. $2(\cos 150° + i \sin 150°)$

32. $5(\cos 135° + i \sin 135°)$

33. $\frac{3}{2}(\cos 300° + i \sin 300°)$

34. $\frac{3}{4}(\cos 315° + i \sin 315°)$

35. $3.75\left(\cos \dfrac{3\pi}{4} + i \sin \dfrac{3\pi}{4}\right)$

36. $8\left(\cos \dfrac{\pi}{12} + i \sin \dfrac{\pi}{12}\right)$

37. $4\left(\cos \dfrac{3\pi}{2} + i \sin \dfrac{3\pi}{2}\right)$

38. $7(\cos 0 + i \sin 0)$

39. $3[\cos(18° \, 45') + i \sin(18° \, 45')]$

40. $6[\cos(230° \, 30') + i \sin(230° \, 30')]$

In Exercises 41–44, use a graphing utility to represent the complex number in standard form.

41. $5\left(\cos \dfrac{\pi}{9} + i \sin \dfrac{\pi}{9}\right)$

42. $9(\cos 58° + i \sin 58°)$

43. $12\left(\cos \dfrac{3\pi}{5} + i \sin \dfrac{3\pi}{5}\right)$

44. $4(\cos 216.5° + i \sin 216.5°)$

In Exercises 45 and 46, represent the powers z, z^2, z^3, and z^4 graphically. Describe the pattern.

45. $z = \dfrac{\sqrt{2}}{2}(1 + i)$

46. $z = \dfrac{1}{2}(1 + \sqrt{3}i)$

In Exercises 47–58, perform the operation and leave the result in trigonometric form.

47. $\left[3\left(\cos\dfrac{\pi}{3} + i\sin\dfrac{\pi}{3}\right)\right]\left[4\left(\cos\dfrac{\pi}{6} + i\sin\dfrac{\pi}{6}\right)\right]$

48. $\left[\dfrac{3}{2}\left(\cos\dfrac{\pi}{2} + i\sin\dfrac{\pi}{2}\right)\right]\left[6\left(\cos\dfrac{\pi}{4} + i\sin\dfrac{\pi}{4}\right)\right]$

49. $\left[\dfrac{5}{3}(\cos 140° + i\sin 140°)\right]\left[\dfrac{2}{3}(\cos 60° + i\sin 60°)\right]$

50. $\left[\dfrac{1}{2}(\cos 100° + i\sin 100°)\right]\left[\dfrac{4}{5}(\cos 300° + i\sin 300°)\right]$

51. $\left[\dfrac{9}{20}(\cos 310° + i\sin 310°)\right]\left[\dfrac{3}{5}(\cos 200° + i\sin 200°)\right]$

52. $(\cos 5° + i\sin 5°)(\cos 20° + i\sin 20°)$

53. $\dfrac{\cos 40° + i\sin 40°}{\cos 10° + i\sin 10°}$

54. $\dfrac{5(\cos 4.3 + i\sin 4.3)}{4(\cos 2.1 + i\sin 2.1)}$

55. $\dfrac{2(\cos 120° + i\sin 120°)}{4(\cos 40° + i\sin 40°)}$

56. $\dfrac{\cos(5\pi/3) + i\sin(5\pi/3)}{\cos\pi + i\sin\pi}$

57. $\dfrac{12(\cos 52° + i\sin 52°)}{3(\cos 110° + i\sin 110°)}$

58. $\dfrac{9(\cos 20° + i\sin 20°)}{5(\cos 75° + i\sin 75°)}$

In Exercises 59–64, (a) give the trigonometric form of the complex numbers, (b) perform the indicated operation using the trigonometric form, and (c) perform the indicated operation using the standard form, and check your result with that of part (b).

59. $(2 + 2i)(1 - i)$

60. $\left(\sqrt{3} + i\right)(1 + i)$

61. $-2i(1 + i)$

62. $\dfrac{3 + 4i}{1 - \sqrt{3}i}$

63. $\dfrac{5}{2 + 3i}$

64. $\dfrac{4i}{-4 + 2i}$

65. Given two complex numbers $z_1 = r_1(\cos\theta_1 + i\sin\theta_1)$ and $z_2 = r_2(\cos\theta_2 + i\sin\theta_2)$, $z_2 \neq 0$, prove that

$$\dfrac{z_1}{z_2} = \dfrac{r_1}{r_2}[\cos(\theta_1 - \theta_2) + i\sin(\theta_1 - \theta_2)].$$

66. Show that $\bar{z} = r[\cos(-\theta) + i\sin(-\theta)]$ is the complex conjugate of $z = r(\cos\theta + i\sin\theta)$.

67. Use the trigonometric forms of z and $\bar{z}$ in Exercise 66 to find (a) $z\bar{z}$ and (b) $z/\bar{z}$, $\bar{z} \neq 0$.

68. Show that the negative of $z = r(\cos\theta + i\sin\theta)$ is $-z = r[\cos(\theta + \pi) + i\sin(\theta + \pi)]$.

In Exercises 69 and 70, sketch the graph of all complex numbers z satisfying the given condition.

69. $|z| = 2$

70. $\theta = \pi/6$

In Exercises 71–82, use DeMoivre's Theorem to find the indicated power of the complex number. Express the result in standard form.

71. $(1 + i)^5$

72. $(2 + 2i)^6$

73. $(-1 + i)^{10}$

74. $(1 - i)^{12}$

75. $2\left(\sqrt{3} + i\right)^7$

76. $4\left(1 - \sqrt{3}i\right)^3$

77. $[5(\cos 20° + i\sin 20°)]^3$

78. $[3(\cos 150° + i\sin 150°)]^4$

79. $\left(\cos\dfrac{5\pi}{4} + i\sin\dfrac{5\pi}{4}\right)^{10}$

80. $\left[2\left(\cos\dfrac{\pi}{2} + i\sin\dfrac{\pi}{2}\right)\right]^8$

81. $[5(\cos 3.2 + i\sin 3.2)]^4$

82. $(\cos 0 + i\sin 0)^{20}$

In Exercises 83–86, use a graphing utility and DeMoivre's Theorem to find the indicated power of the complex number. Express the result in standard form.

83. $(3 - 2i)^5$

84. $\left(\sqrt{5} - 4i\right)^3$

85. $[3(\cos 15° + i\sin 15°)]^4$

86. $\left[2\left(\cos\dfrac{\pi}{10} + i\sin\dfrac{\pi}{10}\right)\right]^5$

87. Show that $-\dfrac{1}{2}\left(1 + \sqrt{3}i\right)$ is a sixth root of 1.

88. Show that $2^{-1/4}(1 - i)$ is a fourth root of -2.

Graphical Reasoning In Exercises 89 and 90, use the graph of the roots of a complex number. (a) Write each of the roots in trigonometric form. (b) Identify the complex number whose roots are given. (c) Use a graphing utility to verify the results of part (b).

89.

90.

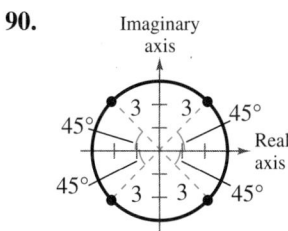

In Exercises 91–102, (a) use the theorem on page 507 to find the indicated roots of the complex number, (b) represent each of the roots graphically, and (c) express each of the roots in standard form.

91. Square roots of $5(\cos 120° + i \sin 120°)$

92. Square roots of $16(\cos 60° + i \sin 60°)$

93. Fourth roots of $16\left(\cos \dfrac{4\pi}{3} + i \sin \dfrac{4\pi}{3}\right)$

94. Fifth roots of $32\left(\cos \dfrac{5\pi}{6} + i \sin \dfrac{5\pi}{6}\right)$

95. Square roots of $-25i$

96. Fourth roots of $625i$

97. Cube roots of $-\dfrac{125}{2}\left(1 + \sqrt{3}i\right)$

98. Cube roots of $-4\sqrt{2}(1 - i)$

99. Cube roots of 8

100. Fourth roots of i

101. Fifth roots of 1

102. Cube roots of 1000

In Exercises 103–106, (a) use the theorem on page 507 and a graphing utility to find the indicated roots of the complex number, (b) represent each of the roots graphically, and (c) express each of the roots in standard form.

103. Cube roots of -125

104. Fourth roots of -4

105. Fifth roots of $128(-1 + i)$

106. Sixth roots of $64i$

In Exercises 107–114, use the theorem on page 507 to find all the solutions of the equation and represent the solutions graphically.

107. $x^4 - i = 0$ **108.** $x^3 + 1 = 0$

109. $x^5 + 243 = 0$ **110.** $x^4 - 81 = 0$

111. $x^3 + 64i = 0$ **112.** $x^6 - 64i = 0$

113. $x^3 - (1 - i) = 0$ **114.** $x^4 + (1 + i) = 0$

Review Solve Exercises 115 and 116 as a review of the skills and problem-solving techniques you learned in previous sections. Find the altitude of the triangle.

115.   **116.**

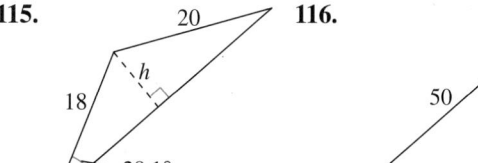

Focus on Concepts

In this chapter, you studied the methods for solving oblique triangles and vectors in the plane. You can use the following questions to check your understanding of several of these basic concepts. The answers to these questions are given in the back of the book.

1. State the Law of Sines from memory.

2. State the Law of Cosines from memory.

3. *True or False?*   The Law of Sines is true if one of the angles in the triangle is a right angle.

4. If one of the angles in the triangle is a right angle, the Law of Cosines simplifies to what famous theorem?

5. *True or False?*   When the Law of Sines is used, the solution is always unique. Explain.

6. What characterizes a vector in the plane?

7. Which vectors in the figure appear to be equivalent?

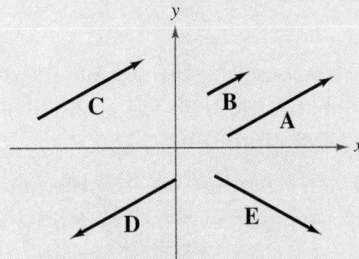

8. The vectors **u** and **v** have the same magnitudes in the two figures. In which figure will the magnitude of the resultant be greater? Give a reason for your answer.

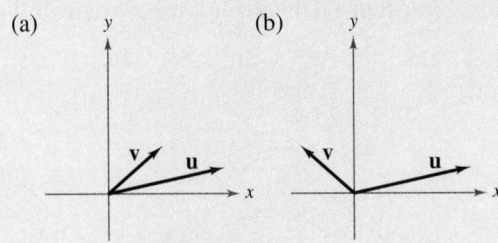

9. Give a geometric description of the scalar multiple $k\mathbf{u}$ of the vector **u**.

10. Give a geometric description of the sum of the vectors **u** and **v**.

11. Which of the two figures shows the difference **u** − **v**? Give a geometric description of the difference and state how you determine its direction.

12. The figure shows complex conjugates z_1 and z_2. Describe $z_1 z_2$ and z_1/z_2.

13. One of the fourth roots of a complex number z is shown in the figure.

   (a) How many roots are not shown?

   (b) Describe the other roots.

6 /// REVIEW EXERCISES

In Exercises 1–16, use the given information to solve the triangle (if possible). If two solutions exist, list both.

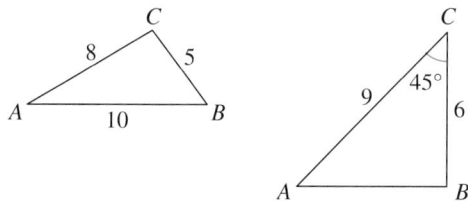

1. $a = 5$, $b = 8$, $c = 10$
2. $a = 6$, $b = 9$, $C = 45°$
3. $A = 12°$, $B = 58°$, $a = 5$
4. $B = 110°$, $C = 30°$, $c = 10.5$
5. $B = 110°$, $a = 4$, $c = 4$
6. $a = 80$, $b = 60$, $c = 100$
7. $A = 75°$, $a = 2.5$, $b = 16.5$
8. $A = 130°$, $a = 50$, $b = 30$
9. $B = 115°$, $a = 7$, $b = 14.5$
10. $C = 50°$, $a = 25$, $c = 22$
11. $A = 15°$, $a = 5$, $b = 10$
12. $B = 150°$, $a = 64$, $b = 10$
13. $B = 150°$, $a = 10$, $c = 20$
14. $a = 2.5$, $b = 15.0$, $c = 4.5$
15. $B = 25°$, $a = 6.2$, $b = 4$
16. $B = 90°$, $a = 5$, $c = 12$

In Exercises 17–20, find the area of the triangle.

17. $a = 4$, $b = 5$, $c = 7$
18. $a = 15$, $b = 8$, $c = 10$
19. $A = 27°$, $b = 5$, $c = 8$
20. $B = 80°$, $a = 4$, $c = 8$

21. *Height*   From a certain distance, the angle of elevation to the top of a building is 17°. At a point 50 meters closer to the building, the angle of elevation is 31°. Approximate the height of the building.

22. *Geometry*   The lengths of the diagonals of a parallelogram are 10 feet and 16 feet. Find the lengths of the sides of the parallelogram if the diagonals intersect at an angle of 28°.

23. *Height of a Tree*   Find the height of a tree that stands on a hillside of slope 28° (from the horizontal) if, from a point 75 feet down the hill, the angle of elevation to the top of the tree is 45° (see figure).

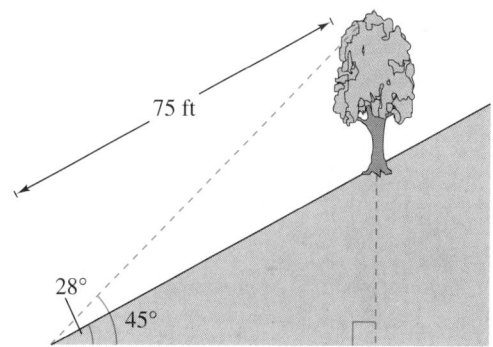

24. *Surveying*   To approximate the length of a marsh, a surveyor walks 425 meters from point *A* to point *B*. Then the surveyor turns 65° and walks 300 meters to point *C*. Approximate the length *AC* of the marsh (see figure).

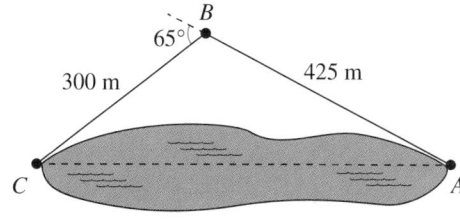

25. *Navigation* Two planes leave an airport at approximately the same time. One is flying at 425 miles per hour at a bearing of N 5° W, and the other is flying at 530 miles per hour at a bearing of N 67° E (see figure). Determine the distance between the planes after flying for 2 hours.

26. *River Width* Determine the width of a river that flows due east, if a surveyor finds that a tree on the opposite bank has a bearing of N 22° 30′ E from a certain point and a bearing of N 15° W from a point 400 feet downstream.

27. *Chapter Opener* Approximate the height of the bridge deck from the water level in the figure on page 531.

28. *Chapter Opener* Approximate *a* in the figure on page 531 if the distance between *A* and *C* is 27 meters.

In Exercises 29–34, find the component form of the vector **v** satisfying the given conditions.

29.

30.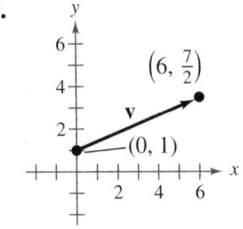

31. Initial point: $(0, 10)$, Terminal point: $(7, 3)$

32. Initial point: $(1, 5)$, Terminal point: $(15, 9)$

33. $\|\mathbf{v}\| = 8$, $\theta = 120°$

34. $\|\mathbf{v}\| = \frac{1}{2}$, $\theta = 225°$

In Exercises 35 and 36, write the vector **v** in the form $\|\mathbf{v}\|(\mathbf{i} \sin \theta + \mathbf{j} \cos \theta)$.

35. $\mathbf{v} = -10\mathbf{i} + 10\mathbf{j}$

36. $\mathbf{v} = 4\mathbf{i} - \mathbf{j}$

In Exercises 37–40, find the component form of the specified vector and sketch its graph given that $\mathbf{u} = 6\mathbf{i} - 5\mathbf{j}$ and $\mathbf{v} = 10\mathbf{i} + 3\mathbf{j}$.

37. $\dfrac{1}{\|\mathbf{u}\|}\mathbf{u}$

38. $3\mathbf{v}$

39. $4\mathbf{u} - 5\mathbf{v}$

40. $\frac{1}{2}\mathbf{v}$

In Exercises 41 and 42, use a graphing utility to graph the vectors and the resultant of the vectors. Find the magnitude and direction of the resultant.

41.

42.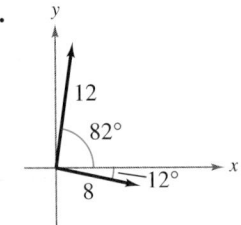

43. *Resultant Force* Find the direction and magnitude of the resultant of the three forces shown in the figure.

Figure for 43   Figure for 45

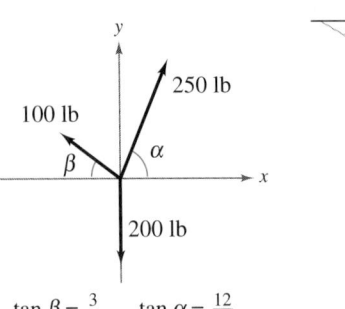

$\tan \beta = \frac{3}{4}$ $\tan \alpha = \frac{12}{5}$

44. *Resultant Force* Forces of 85 pounds and 50 pounds act on a single point. The angle between the forces is 15°. Describe the resultant force.

45. *Rope Tension* A 180-pound weight is supported by two ropes, as shown in the figure. Find the tension in each rope.

46. *Cable Tension* In a manufacturing process, an electric hoist lifts 200-pound ingots (see figure). Find the tension in the supporting cables.

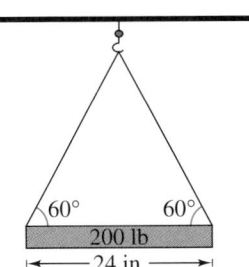

60°   60°
200 lb
|←——— 24 in. ———→|

47. *Braking Force* A 500-pound motorcycle is headed up a hill inclined at 12°. What force is required to keep the motorcycle from rolling back down the hill when stopped at a red light?

48. *Navigation* An airplane has an airspeed of 430 miles per hour at a bearing of S 45° E. If the wind velocity is 35 miles per hour in the direction N 30° E, find the groundspeed and the direction of the plane.

49. *Navigation* An airplane has an airspeed of 724 kilometers per hour at a bearing of N 30° E. If the wind velocity is 32 kilometers per hour from the west, find the groundspeed and the direction of the plane.

50. *Angle Between Forces* Forces of 60 pounds and 100 pounds have a resultant force of 125 pounds. Find the angle between the two forces.

In Exercises 51 and 52, find a unit vector in the direction of $\overrightarrow{PQ}$.

51. $P(7, -4), Q(-3, 2)$

52. $P(0, 3), Q(5, -8)$

In Exercises 53 and 54, decide whether the vectors are orthogonal, parallel, or neither.

53. $\mathbf{u} = \langle 39, -12 \rangle$
$\mathbf{v} = \langle -26, 8 \rangle$

54. $\mathbf{u} = \langle 8, 5 \rangle$
$\mathbf{v} = \langle -2, 4 \rangle$

In Exercises 55–58, find the angle between u and v.

55. $\mathbf{u} = \cos \dfrac{7\pi}{4} \mathbf{i} + \sin \dfrac{7\pi}{4} \mathbf{j}$,

$\mathbf{v} = \cos \dfrac{5\pi}{6} \mathbf{i} + \sin \dfrac{5\pi}{6} \mathbf{j}$

56. $\mathbf{u} = \langle -6, -3 \rangle, \quad \mathbf{v} = \langle 4, 2 \rangle$

57. $\mathbf{u} = \langle 2\sqrt{2}, -4 \rangle, \quad \mathbf{v} = \langle -\sqrt{2}, 1 \rangle$

58. $\mathbf{u} = \langle 3, 1 \rangle, \quad \mathbf{v} = \langle 4, 5 \rangle$

In Exercises 59–62, use a graphing utility to sketch the vectors and find the degree measure of the angle between the vectors.

59. $\mathbf{u} = 4\mathbf{i} + \mathbf{j}$
$\mathbf{v} = \mathbf{i} - 4\mathbf{j}$

60. $\mathbf{u} = 6\mathbf{i} + 2\mathbf{j}$
$\mathbf{v} = -3\mathbf{i} - \mathbf{j}$

61. $\mathbf{u} = 7\mathbf{i} - 5\mathbf{j}$
$\mathbf{v} = 10\mathbf{i} + 3\mathbf{j}$

62. $\mathbf{u} = -5.3\mathbf{i} + 2.8\mathbf{j}$
$\mathbf{v} = -8.1\mathbf{i} - 4\mathbf{j}$

In Exercises 63–66, find $\text{proj}_\mathbf{v}\mathbf{u}$.

63. $\mathbf{u} = \langle -4, 3 \rangle, \quad \mathbf{v} = \langle -8, -2 \rangle$

64. $\mathbf{u} = \langle 5, 6 \rangle, \quad \mathbf{v} = \langle 10, 0 \rangle$

65. $\mathbf{u} = \langle 2, 7 \rangle, \quad \mathbf{v} = \langle 1, -1 \rangle$

66. $\mathbf{u} = \langle -3, 5 \rangle, \quad \mathbf{v} = \langle -5, 2 \rangle$

In Exercises 67–70, find the trigonometric form of the complex number.

67.

68.

69.

70.
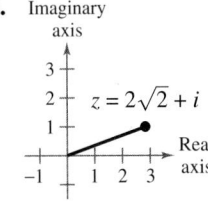

In Exercises 71–74, find the trigonometric form of the complex number.

71. $5 - 5i$

72. $-3\sqrt{3} + 3i$

73. $5 + 12i$

74. -7

In Exercises 75–78, write the complex number in standard form.

75. $100(\cos 240° + i \sin 240°)$

76. $24(\cos 330° + i \sin 330°)$

77. $13(\cos 0 + i \sin 0)$

78. $8\left(\cos \dfrac{5\pi}{6} + i \sin \dfrac{5\pi}{6}\right)$

In Exercises 79 and 80, (a) express the two complex numbers in trigonometric form, and (b) use the trigonometric form to find $z_1 z_2$ and z_1/z_2.

79. $z_1 = 2\sqrt{3} - 2i, \quad z_2 = -10i$

80. $z_1 = -3(1 + i), \quad z_2 = 2(\sqrt{3} + i)$

In Exercises 81–84, use the theorem on page 507 to find the indicated power of the complex number. Express the result in standard form.

81. $\left[5\left(\cos \dfrac{\pi}{12} + i \sin \dfrac{\pi}{12}\right)\right]^4$

82. $\left[2\left(\cos \dfrac{4\pi}{15} + i \sin \dfrac{4\pi}{15}\right)\right]^5$

83. $(2 + 3i)^6$

84. $(1 - i)^8$

Graphical Reasoning In Exercises 85–88, use the graph of the roots of a complex number. (a) Write each of the roots in trigonometric form. (b) Identify the complex number whose roots are given. (c) Use a graphing utility to verify the results of part (b).

85.

86.

87.

88.

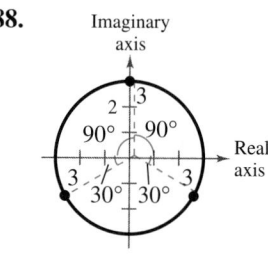

In Exercises 89 and 90, use the theorem on page 507 to find the roots of the complex number.

89. Sixth roots of $-729i$

90. Fourth roots of 256

In Exercises 91–94, find all solutions of the equation and represent the solutions graphically.

91. $x^4 + 81 = 0$

92. $x^5 - 32 = 0$

93. $x^3 + 8i = 0$

94. $(x^3 - 1)(x^2 + 1) = 0$

CHAPTER PROJECT   *Adding Vectors Graphically*

The program below is written for a *TI–82* or *TI–83* graphing calculator. The program sketches two vectors $\mathbf{u} = a\mathbf{i} + b\mathbf{j}$ and $\mathbf{v} = c\mathbf{i} + d\mathbf{j}$ in standard position. Then, using the parallelogram law for vector addition, the program also sketches the vector sum $\mathbf{u} + \mathbf{v}$. *Before* running the program, you should set values that produce an appropriate viewing rectangle.

TI–82 or TI–83 Program

```
PROGRAM:ADDVECT          :Line(0,0,A,B)      :Line(A,B,E,F)
:Input "ENTER A",A       :Line(0,0,C,D)      :Line(C,D,E,F)
:Input "ENTER B",B       :A+C→E              :Pause
:Input "ENTER C",C       :B+D→F              :ClrDraw
:Input "ENTER D",D       :Line(0,0,E,F)      :Stop
```

(a) Use the program listed above to sketch the sum of the vectors $\mathbf{u} = 5\mathbf{i} + 2\mathbf{j}$ and $\mathbf{v} = -4\mathbf{i} + 3\mathbf{j}$. Set your viewing window as indicated at the left. Identify the vectors $\mathbf{u}$, $\mathbf{v}$, and $\mathbf{u} + \mathbf{v}$ in the graph.

(b) An airplane is headed N 60° W at a speed of 400 miles per hour. The airplane encounters wind of velocity 75 miles per hour in the direction N 40° E. Use the program listed above to find the resultant speed and direction of the airplane.

Questions for Further Exploration

In Questions 1–4, use the program listed above (or a comparable program on some other graphing utility) to sketch the sum of the vectors. Use the result to estimate graphically the components of the sum. Then check your result analytically.

1. $\mathbf{u} = 3\mathbf{i} + 4\mathbf{j}$, $\mathbf{v} = -5\mathbf{i} + \mathbf{j}$
2. $\mathbf{u} = 5\mathbf{i} - 4\mathbf{j}$, $\mathbf{v} = 3\mathbf{i} + 2\mathbf{j}$
3. $\mathbf{u} = -4\mathbf{i} + 4\mathbf{j}$, $\mathbf{v} = -2\mathbf{i} - 6\mathbf{j}$
4. $\mathbf{u} = 7\mathbf{i} + 3\mathbf{j}$, $\mathbf{v} = -2\mathbf{i} - 6\mathbf{j}$

5. After encountering the wind, is the airplane in Exercise (b) traveling at a higher speed or a lower speed? Explain.

6. Consider the airplane described in Exercise (b), headed N 60° W at a speed of 400 miles per hour. What wind velocity, in the direction of N 40° E, will produce a resultant direction of N 50° W? Explain how to use the program listed above to obtain the answer *experimentally*. Then explain how to obtain the answer analytically.

7. Consider the airplane described in Exercise (b), headed N 60° W at a speed of 400 miles per hour. What wind direction, at a speed of 75 miles per hour, will produce a resultant direction of N 50° W? Explain how to use the program listed above to obtain the answer *experimentally*. Then explain how to obtain the answer analytically.

4–6 /// CUMULATIVE TEST

Take this test as you would take a test in class. After you are done, check your work against the answers given in the back of the book.

The *Interactive* CD-ROM provides answers to the Chapter Tests and Cumulative Tests. It also offers Chapter Pre-Tests (that test key skills and concepts covered in previous chapters) and Chapter Post-Tests, both of which have randomly generated exercises with diagnostic capabilities.

1. Consider the angle $\theta = -120°$.

    (a) Sketch the angle in standard position.

    (b) Determine a coterminal angle in the interval $[0°, 360°)$.

    (c) Convert the angle to radian measure.

    (d) Find its reference angle θ'.

    (e) Find the exact values of the six trigonometric functions of θ.

2. Convert the angle of magnitude 2.35 radians to degrees. Round the answer to one decimal place.

3. Find $\cos \theta$ if $\tan \theta = -\frac{4}{3}$ and $\sin \theta < 0$.

4. Use a graphing utility to sketch the graphs of (a) $f(x) = 3 - 2 \sin \pi x$ and (b) $g(x) = \frac{1}{2} \tan\left(x - \frac{\pi}{2}\right)$.

5. Find a, b, and c such that the graph of the function $h(x) = a \cos(bx + c)$ matches the graph in the figure.

6. Write an algebraic expression equivalent to $\sin(\arccos 2x)$.

7. Subtract and simplify: $\dfrac{\sin \theta - 1}{\cos \theta} - \dfrac{\cos \theta}{\sin \theta - 1}$.

8. Prove the identities.

    (a) $\cot^2 \alpha(\sec^2 \alpha - 1) = 1$ (b) $\sin(x + y)\sin(x - y) = \sin^2 x - \sin^2 y$

9. Find all solutions of the equations in the interval $[0, 2\pi)$.

    (a) $2\cos^2 \beta - \cos \beta = 0$ (b) $3 \tan \theta - \cot \theta = 0$

10. Use a graphing utility to approximate the remaining side and angles of the triangle shown in the figure. Verify analytically.

    (a) $A = 30°$, $a = 9$, $b = 8$ (b) $A = 30°$, $b = 8$, $c = 10$

11. Find the trigonometric form of the complex number $-2 + 2i$.

12. Find the product of $[4(\cos 30° + i \sin 30°)][6(\cos 120° + i \sin 120°)]$. Write the answer in standard form.

13. Use DeMoivre's Theorem to find the three cube roots of 1.

14. From a point 200 feet from a flagpole, the angles of elevation to the bottom and top of the flag are $16° 45'$ and $18°$, respectively. Approximate the height of the flag to the nearest foot.

15. An airplane's velocity with respect to the air is 500 kilometers per hour, with a bearing of N 30° E. The wind at the altitude of the plane has a velocity of 50 kilometers per hour with a bearing of N 60° E. What is the true direction of the plane, and what is its speed relative to the ground?

Figure for 5

Figure for 10

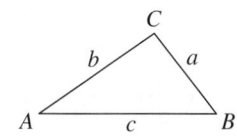

Chapter 7

Systems of Equations and Inequalities

In 1975, pickups, sport utilities, and vans accounted for only 21% of all light vehicles sold in the United States. By 1995, they accounted for almost 40%.

These three types of "trucks" all come in both full-size models and compact models. From 1990 to 1994, the sales (in thousands) of compact pickups P and compact sport utilities S were as follows.

Sales (in thousands)

Year ($0 \leftrightarrow 1990$)

Year	P	S
1990	1102.9	756.3
1991	980.7	807.9
1992	887.8	998.4
1993	954.1	1203.2
1994	1080.8	1359.1

These sales can be modeled by the following quadratic and linear models.

$$P = 47t^2 - 195t + 1109$$
$$S = 160t + 705$$

You can use a graphing utility to graph these two models in the same viewing rectangle. By using the zoom and trace features, you can verify that the curves intersect at approximately (1.4, 928). This indicates that compact sport utilities began outselling compact pickups in 1992. (See Exercises 77 and 78 on page 554.)

In 1994, in the full-size category, pickups outsold sport utilities. But in the compact category, for the third year in a row, sport utilities outsold pickups.

7.1 Solving Systems of Equations

The Method of Substitution / *Graphical Approach to Finding Solutions* / *Applications*

The Method of Substitution

Up to this point in the text, most problems have involved either a function of one variable or a single equation in two variables. However, many problems in science, business, and engineering involve two or more equations in two or more variables. To solve such problems, you need to find solutions of **systems of equations.** Here is an example of a system of two equations in x and y.

$2x + y = 5$	Equation 1
$3x - 2y = 4$	Equation 2

A **solution** of this system is an ordered pair that satisfies each equation in the system. Finding the set of all such solutions is called **solving the system of equations.** For instance, the ordered pair $(2, 1)$ is a solution of this system. To check this, you can substitute 2 for x and 1 for y in *each* equation.

$2x + y = 5$	Equation 1
$2(2) + 1 \stackrel{?}{=} 5$	Substitute 2 for x and 1 for y.
$4 + 1 = 5$	Solution checks in Equation 1. ✓
$3x - 2y = 4$	Equation 2
$3(2) - 2(1) \stackrel{?}{=} 4$	Substitute 2 for x and 1 for y.
$6 - 2 = 4$	Solution checks in Equation 2. ✓

In this chapter you will study three ways to solve equations, beginning with the **method of substitution.**

The Method of Substitution

1. *Solve* one of the equations for one variable in terms of the other.
2. *Substitute* the expression found in Step 1 into the other equation to obtain an equation in one variable.
3. *Solve* the equation obtained in Step 2.
4. *Back-substitute* the solution in Step 3 into the expression obtained in Step 1 to find the value of the other variable.
5. *Check* that the solution satisfies *each* of the original equations.

EXAMPLE 1 Solving a System of Equations

Solve the system of equations.

$$x + y = 4 \qquad \text{Equation 1}$$
$$x - y = 2 \qquad \text{Equation 2}$$

Solution

Begin by solving for y in Equation 1.

$$y = 4 - x \qquad \text{Solve for } y \text{ in Equation 1.}$$

Next, substitute this expression for y into Equation 2 and solve the resulting single-variable equation for x.

$x - y = 2$	Equation 2
$x - (4 - x) = 2$	Substitute $4 - x$ for y.
$x - 4 + x = 2$	Simplify.
$2x = 6$	Combine like terms.
$x = 3$	Solve for x.

Finally, you can solve for y by *back-substituting* $x = 3$ into the equation $y = 4 - x$, to obtain

$y = 4 - x$	Revised Equation 1
$y = 4 - 3$	Substitute 3 for x.
$y = 1.$	Solve for y.

The solution is the ordered pair $(3, 1)$. You can check this as follows.

Check

$x + y = 4$	Equation 1
$3 + 1 \overset{?}{=} 4$	Substitute for x and y. ✓
$4 = 4$	Solution checks in Equation 1.
$x - y = 2$	Equation 2
$3 - 1 \overset{?}{=} 2$	Substitute for x and y. ✓
$2 = 2$	Solution checks in Equation 2.

Note The term *back-substitution* implies that you work *backwards*. First you solve for one of the variables, and then you substitute that value *back* into one of the equations in the system to find the value of the other variable.

EXPLORATION

Use a graphing utility to graph $y = -x + 4$ and $y = x - 2$ in the same viewing rectangle. Use the trace feature to find the coordinates of the point of intersection. Are the coordinates the same as the solution found in Example 1? Explain.

On the *TI-82* or *TI-83*, set the calculator to dot mode and use the following keystrokes.

ZOOM (zoom)(4:ZDecimal)
ENTER

What changes does this make in the coordinates of the point?

Study Tip

Because many steps are required to solve a system of equations, it is very easy to make errors in arithmetic. Thus, we *strongly* suggest that you always *check your solution by substituting it into each equation in the original system.*

Real Life

EXAMPLE 2   **Solving a System by Substitution**

A total of $12,000 is invested in two funds paying 9% and 11% simple interest. The yearly interest is $1180. How much is invested at each rate?

Solution

Verbal Model:

9% fund	$+$	11% fund	$=$	Total investment

9% interest	$+$	11% interest	$=$	Total interest

Labels: Amount in 9% fund $= x$ (dollars)
            Interest for 9% fund $= 0.09x$ (dollars)
            Amount in 11% fund $= y$ (dollars)
            Interest for 11% fund $= 0.11y$ (dollars)
            Total investment $= \$12,000$ (dollars)
            Total interest $= \$1180$ (dollars)

System: $x + \quad y = 12,000$ Equation 1
            $0.09x + 0.11y = 1180$ Equation 2

To begin, it is convenient to multiply both sides of Equation 2 by 100 to obtain $9x + 11y = 118,000$. This eliminates the need to work with decimals.

$9x + 11y = 118,000$ Revised Equation 2

To solve this system, you can solve for x in Equation 1.

$x = 12,000 - y$ Revised Equation 1

Next, substitute this expression for x into Revised Equation 2 and solve the resulting equation for y.

$9x + 11y = 118,000$ Revised Equation 2
$9(12,000 - y) + 11y = 118,000$ Substitute $12,000 - y$ for x.
$108,000 - 9y + 11y = 118,000$ Distributive Property
$2y = 10,000$ Combine like terms.
$y = 5000$ Amount in 11% fund.

Finally, back-substitute the value $y = 5000$ to solve for x.

$x = 12,000 - y$ Revised Equation 1
$x = 12,000 - 5000$ Substitute 5000 for y.
$x = 7000$ Amount in 9% fund

The solution is $(7000, 5000)$. Check this in the original problem.

The *Interactive* CD-ROM offers graphing utility emulators of the *TI-82* and *TI-83*, which can be used with the Examples, Explorations, Technology notes, and Exercises.

One way to check the answers you obtain in this section is to use a graphing utility. For instance, enter the two equations in Example 2

$y_1 = 12,000 - x$

$y_2 = \dfrac{1180 - 0.09x}{0.11}$

and find an appropriate viewing rectangle that shows where the lines intersect. Then use the zoom and trace features to find their point of intersection. Does this point agree with the solution obtained at the right?

The equations in Examples 1 and 2 are linear. Substitution can also be used to solve systems in which one or both of the equations are nonlinear.

EXAMPLE 3   Substitution: Two-Solution Case

Solve the system of equations.

$$x^2 - x - y = 1 \qquad \text{Equation 1}$$
$$-x + y = -1 \qquad \text{Equation 2}$$

Solution

Begin by solving for y in Equation 2 to obtain $y = x - 1$. Next, substitute this expression for y into Equation 1 and solve for x.

$x^2 - x - y = 1$	Equation 1
$x^2 - x - (x - 1) = 1$	Substitute $x - 1$ for y.
$x^2 - 2x + 1 = 1$	Simplify.
$x^2 - 2x = 0$	Standard form
$x(x - 2) = 0$	Factor.
$x = 0, 2$	Solve for x.

Back-substituting these values of x to solve for the corresponding values of y produces the solutions $(0, -1)$ and $(2, 1)$. Check these in the original system.

EXPLORATION

Use a graphing utility to graph the two equations in Example 3

$$y_1 = x^2 - x - 1$$
$$y_2 = x - 1$$

in the same viewing rectangle. How many solutions do you think this system has?

Repeat this experiment for the equations in Example 4. How many solutions does this system have? Explain your reasoning.

EXAMPLE 4   Substitution: No-Solution Case

Solve the system of equations.

$$-x + y = 4 \qquad \text{Equation 1}$$
$$x^2 + y = 3 \qquad \text{Equation 2}$$

Solution

Begin by solving for y in Equation 1 to obtain $y = x + 4$. Next, substitute this expression for y into Equation 2 and solve for x.

$x^2 + y = 3$	Equation 2
$x^2 + (x + 4) = 3$	Substitute $x + 4$ for y.
$x^2 + x + 1 = 0$	Simplify.
$x = \dfrac{-1 \pm \sqrt{1^2 - 4(1)(1)}}{2}$	Quadratic Formula

Because this yields two complex values, the equation $x^2 + x + 1 = 0$ has no *real* solution. Hence, this system has no *real* solution.

The *Interactive* CD-ROM shows every example with its solution; clicking on the *Try It!* button brings up similar problems. Guided Examples and Integrated Examples show step-by-step solutions to additional examples. Integrated Examples are related to several concepts in the section.

Use a graphing utility to graph the equations in Figure 7.1(a) in the same viewing rectangle. To find the point of intersection of the graphs, use the following keystrokes on a *TI-82* or *TI-83*.

$\boxed{\text{CALC}}$ (5:intersect) $\boxed{\text{ENTER}}$

Cursor to the first graph and press $\boxed{\text{ENTER}}$. Then cursor to the second graph and press $\boxed{\text{ENTER}}$

$\boxed{\text{ENTER}}$.

Note Example 5 shows you the value of a graphical approach to solving systems of equations in two variables. Notice what would happen if you tried only the substitution method in Example 5. It would be difficult to solve this equation for x using standard algebraic techniques.

Graphical Approach to Finding Solutions

From Examples 2, 3, and 4, you can see that a system of two equations in two unknowns can have exactly one solution, more than one solution, or no solution. In practice, you can gain insight about the location and number of solutions of a system of equations by graphing each of the equations in the same coordinate plane. The solutions of the system correspond to the **points of intersection** of the graphs. For instance, in Figure 7.1(a), the two equations graph as two lines with a *single point* of intersection. The two equations in Example 3 graph as a parabola and a line with *two points* of intersection, as shown in Figure 7.1(b). The two equations in Example 4 graph as a line and a parabola that happen to have *no points* of intersection, as shown in Figure 7.1(c).

Figure 7.1

(a) One Intersection Point   **(b)** Two Intersection Points   **(c)** No Intersection Points

Figure 7.2

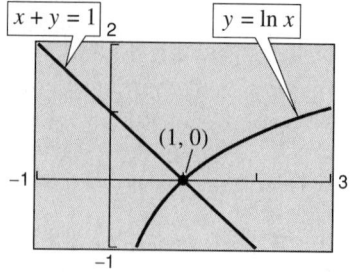

EXAMPLE 5   ◢   **Solving a System of Equations**

Solve the system of equations.

$$y = \ln x \qquad\qquad \text{Equation 1}$$
$$x + y = 1 \qquad\qquad \text{Equation 2}$$

Solution
From the graphs of these equations, shown in Figure 7.2, it is clear that there is only one point of intersection. Also, it appears that $(1, 0)$ is the solution point. You can confirm this by substituting in *both* equations.

Check

$$0 = \ln 1 \qquad\qquad \text{Equation 1 checks.} \checkmark$$
$$1 + 0 = 1 \qquad\qquad \text{Equation 2 checks.} \checkmark$$

Applications

The total cost C of producing x units of a product typically has two components: the initial cost and the cost per unit. When enough units have been sold so that the total revenue R equals the total cost, the sales are said to have reached the **break-even point.** You will find that the break-even point corresponds to the point of intersection of the cost and revenue curves.

Real Life

EXAMPLE 6 **An Application: Break-Even Analysis**

A small business invests $10,000 in equipment to produce a product. Each unit of the product costs $0.65 to produce and is sold for $1.20. How many items must be sold before the business breaks even?

Solution
The total cost of producing x units is

$$
\boxed{\begin{array}{c}\text{Total}\\\text{cost}\end{array}} = \boxed{\begin{array}{c}\text{Cost per}\\\text{unit}\end{array}} \cdot \boxed{\begin{array}{c}\text{Number}\\\text{of units}\end{array}} + \boxed{\begin{array}{c}\text{Initial}\\\text{cost}\end{array}}
$$

$$C = 0.65x + 10,000. \qquad\qquad \text{Equation 1}$$

The revenue obtained by selling x units is

$$
\boxed{\begin{array}{c}\text{Total}\\\text{revenue}\end{array}} = \boxed{\begin{array}{c}\text{Price per}\\\text{unit}\end{array}} \cdot \boxed{\begin{array}{c}\text{Number}\\\text{of units}\end{array}}
$$

$$R = 1.2x. \qquad\qquad \text{Equation 2}$$

Because the break-even point occurs when $R = C$, you have

$$1.2x = 0.65x + 10,000 \qquad\qquad \text{Equate } R \text{ and } C.$$

$$0.55x = 10,000 \qquad\qquad \text{Subtract } 0.65x \text{ from both sides.}$$

$$x = \frac{10,000}{0.55} \qquad\qquad \text{Divide both sides by 0.55.}$$

$$x \approx 18,182 \text{ units.} \qquad\qquad \text{Use a calculator.}$$

Note in Figure 7.3 that sales less than the break-even point correspond to an overall loss, whereas sales greater than the break-even point correspond to a profit. Verify the break-even point using the intersection feature of a graphing utility.

Figure 7.3

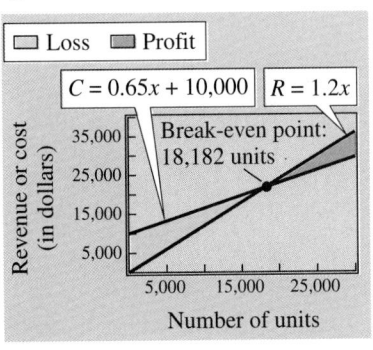

Note Another way to view the solution in Example 6 is to consider the profit function $P = R - C$. The break-even point occurs when the profit is 0, which is the same as saying that $R = C$.

Figure 7.4

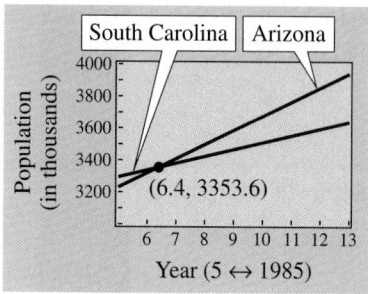

Real Life

EXAMPLE 7   State Populations

From 1985 to 1993, the population of Arizona was increasing at a faster rate than the population of South Carolina. Two models that approximate the populations P are

$$P = 2785.8 + 88.8t \qquad \text{Arizona}$$

$$P = 3079.3 + 42.9t \qquad \text{South Carolina}$$

where $t = 5$ represents 1985 (see Figure 7.4). According to these two models, when would you expect the population of Arizona to have exceeded the population of South Carolina?   (Source: U.S. Bureau of the Census)

Solution

Because the first equation has already been solved for P in terms of t, you can substitute this value into the second equation and solve for t, as follows.

$$2785.8 + 88.8t = 3079.3 + 42.9t$$
$$88.8t - 42.9t = 3079.3 - 2785.8$$
$$45.9t = 293.5$$
$$t \approx 6.4$$

Thus, from the given models, you would expect that the population of Arizona exceeded the population of South Carolina sometime during 1986.

Group Activity   *Points of Intersection*

In this section, you learned that the graphs of two equations can intersect at zero, one, or more points. Use a graphing utility to graph each of the following systems. Decide if the system has no solution, one solution, two solutions, or more than two solutions.

a. $-0.5x^2 + 0.25x + y = -1.5$
$$1.75x + y = -2.625$$

b. $-0.5x^2 + 0.25x + y = -1.5$
$$-1.25x + y = -4.5$$

c. $-0.5x^2 + 0.25x + y = -1.5$
$$3.5x + y = 2.25$$

Create three other systems of equations, one with no solution, one with one solution, and one with two solutions. Trade your systems with a group member. Solve and compare your results.

7.1 /// EXERCISES

In Exercises 1–10, solve the system by the method of substitution. Check your solution graphically.

1. $2x + y = 6$
$-x + y = 0$

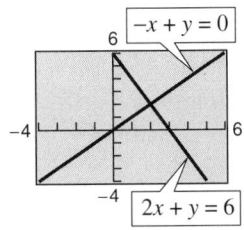

2. $x - y = -4$
$x + 2y = 5$

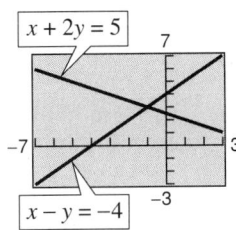

3. $x - y = -4$
$x^2 - y = -2$

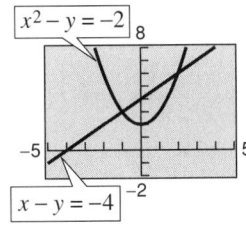

4. $3x + y = 2$
$x^3 + y = 0$

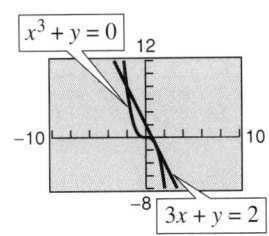

5. $x - 3y = 15$
$x^2 + y^2 = 25$

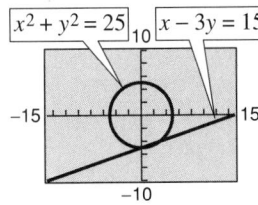

6. $x + y = 0$
$x^3 - 5x - y = 0$

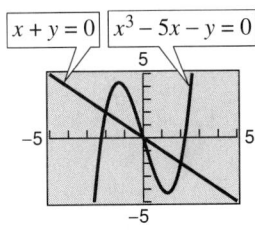

7. $x^2 + y = 0$
$x^2 - 4x - y = 0$

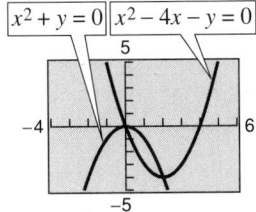

8. $y = -2x^2 + 2$
$y = 2(x^4 - 2x^2 + 1)$

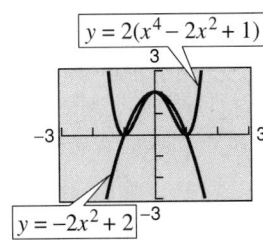

9. $x - 6y = -8$
$x^2 - 4y^3 = 0$

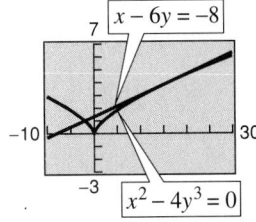

10. $y = x^3 - 3x^2 + 4$
$y = -2x + 4$

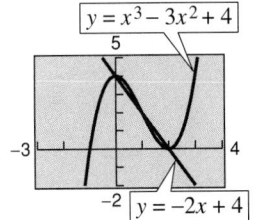

In Exercises 11–22, solve the system by the method of substitution. Use a graphing utility to verify your result.

11. $x - y = 0$
$5x - 3y = 10$

12. $x + 2y = 1$
$5x - 4y = -23$

13. $2x - y + 2 = 0$
$4x + y - 5 = 0$

14. $6x - 3y - 4 = 0$
$x + 2y - 4 = 0$

15. $1.5x + 0.8y = 2.3$
$0.3x - 0.2y = 0.1$

16. $30x - 40y - 33 = 0$
$10x + 20y - 21 = 0$

17. $\frac{1}{5}x + \frac{1}{2}y = 8$
$x + y = 20$

18. $\frac{1}{2}x + \frac{3}{4}y = 10$
$\frac{3}{4}x - y = 4$

19. $2x - y = 4$
$-4x + 2y = -12$

20. $-\frac{2}{3}x + y = -2$
$2x - 3y = 6$

21. $x - y = 0$
$2x + y = 0$

22. $x - 2y = 0$
$3x - y = 0$

In Exercises 23–28, solve the system of equations graphically.

23. $-x + 2y = 2$
$3x + y = 15$

24. $x + y = 0$
$3x - 2y = 10$

25. $x - 3y = -2$
$5x + 3y = 17$

26. $-x + 2y = 1$
$x - y = 2$

27. $x + y = 4$
$x^2 + y^2 - 4x = 0$

28. $x - y + 3 = 0$
$x^2 - 4x + 7 = y$

In Exercises 29–40, use a graphing utility to approximate all points of intersection of the graph of the system of equations.

29. $7x + 8y = 24$
$x - 8y = 8$

30. $x - y = 0$
$5x - 2y = 6$

31. $2x - y + 3 = 0$
$x^2 + y^2 - 4x = 0$

32. $3x - 2y = 0$
$x^2 + y^2 = 4$

33. $x^2 + y^2 = 8$
$y = x^2$

34. $x^2 + y^2 = 25$
$(x - 8)^2 + y^2 = 41$

35. $y = e^x$
$x - y + 1 = 0$

36. $x + 2y = 8$
$y = \log_2 x$

37. $y = \sqrt{x}$
$y = x$

38. $x - y = 3$
$x - y^2 = 1$

39. $x^2 + y^2 = 169$
$x^2 - 8y = 104$

40. $x^2 + y^2 = 4$
$2x^2 - y = 2$

In Exercises 41–52, solve the system graphically or algebraically. Explain why you chose the method you used.

41. $y = 2x$
$y = x^2 + 1$

42. $x + y = 4$
$x^2 + y = 2$

43. $3x - 7y + 6 = 0$
$x^2 - y^2 = 4$

44. $x^2 + y^2 = 25$
$2x + y = 10$

45. $x - 2y = 4$
$x^2 - y = 0$

46. $y = (x + 1)^3$
$y = \sqrt{x - 1}$

47. $y - e^{-x} = 1$
$y - \ln x = 3$

48. $y = x^3 - 2x^2 + x - 1$
$y = -x^2 + 3x - 1$

49. $y = x^4 - 2x^2 + 1$
$y = 1 - x^2$

50. $x^2 + y = 4$
$e^x - y = 0$

51. $xy - 1 = 0$
$2x - 4y + 7 = 0$

52. $x - 2y = 1$
$y = \sqrt{x - 1}$

53. *Think About It* When solving a system of equations by substitution, how do you recognize that the system has no solution?

54. *Essay* Write a brief paragraph describing any advantages of substitution over the graphical method of solving a system of equations.

Break-Even Analysis In Exercises 55–58, use a graphing utility to graph the cost and revenue functions in the same viewing rectangle. Find the sales x necessary to break even ($R = C$) and the corresponding revenue R obtained by selling x units. (Round to the nearest whole unit.)

Cost	Revenue
55. $C = 8650x + 250{,}000$	$R = 9950x$
56. $C = 5.5\sqrt{x} + 10{,}000$	$R = 3.29x$
57. $C = 2.65x + 350{,}000$	$R = 4.15x$
58. $C = 0.08x + 50{,}000$	$R = 0.25x$

59. *Break-Even Point* A small business invests $16,000 to produce an item that will sell for $5.95. Each unit can be produced for $3.45.

(a) Write cost and revenue functions for x units produced and sold.

(b) Use a graphing utility to graph the cost and revenue functions. Use the graph to approximate the number of units that must be sold to break even.

(c) Verify the result of part (b) algebraically.

60. *Break-Even Point* A small business has an initial investment of $5000. The unit cost of the product is $21.60, and the selling price is $34.10. How many units must be sold to break even?

61. *Investment Portfolio* A total of $20,000 is invested in two funds paying 6.5% and 8.5% simple interest. The 6.5% investment has a lower risk. The investor wants a yearly interest income of $1600 from the two investments. What is the most that can be invested at 6.5% to meet this requirement?

 The *Interactive* CD-ROM contains step-by-step solutions to all odd-numbered Section and Review Exercises. It also provides Tutorial Exercises, which link to Guided Examples for additional help.

62. *Investment Portfolio*   A total of $25,000 is invested in two funds paying 6% and 8.5% simple interest. The 6% investment has a lower risk. The investor wants a yearly interest income of $2000 from the two investments.

(a) Write a system of equations in which one equation represents the total amount invested and the other equation represents the $2000 required in interest. Let x and y represent the amounts invested at 6% and 8.5%, respectively.

(b) Use a graphing utility to graph the two equations. As the amount invested at 6% increases, how does the amount invested at 8.5% change and how does the amount of interest change? Explain.

(c) What is the most that can be invested at 6% to meet the requirement of $2000 per year in interest?

63. *Choice of Two Jobs*   You are offered two different jobs selling dental supplies.

- One company offers a straight commission of 6% of sales.
- The other company offers a salary of $250 per week plus 3% of sales.

How much would you have to sell in a week in order to make the straight commission offer better?

64. *Choice of Two Jobs*   You are offered two different jobs selling college textbooks.

- One company offers an annual salary of $20,000 plus a year-end bonus of 1% of your total sales.
- The other company offers an annual salary of $15,000 plus a year-end bonus of 2% of your total sales.

Determine the annual sales that make the second offer better.

65. *Market Equilibrium*   The supply and demand curves for a business dealing with wheat are given by

Supply: $p = 1.45 + 0.00014x^2$

Demand: $p = (2.388 - 0.007x)^2$

where p is the price in dollars per bushel and x is the quantity in bushels per day. Use a graphing utility to graph the supply and demand equations and find the market equilibrium. (The *market equilibrium* is the point of intersection of the graphs for $x > 0$.)

66. *Log Volume*   You are offered two different rules for estimating the number of board feet in a log that is 16 feet long. One is the *Doyle Log Rule* and is modeled by

$$V = (D - 4)^2, \qquad 5 \le D \le 40$$

and the other is the *Scribner Log Rule* and is modeled by

$$V = 0.79D^2 - 2D - 4, \qquad 5 \le D \le 40$$

where D is the diameter of the log and V is its volume in board feet.

(a) Use a graphing utility to graph the log rules in the same viewing rectangle.

(b) For what diameter do the two rules agree?

(c) If you were selling large logs, which rule would you use? Explain your reasoning.

Geometry   In Exercises 67–70, find the dimensions of the rectangle meeting the specified conditions.

Perimeter	Condition
67. 30 meters	The length is 3 meters greater than the width.
68. 280 centimeters	The width is 20 centimeters less than the length.
69. 42 inches	The width is three-fourths the length.
70. 210 feet	The length is one and one-half times the width.

71. *Geometry*   What are the dimensions of a rectangular tract of land if its perimeter is 40 miles and its area is 96 square miles?

72. *Geometry*   What are the dimensions of an isosceles right triangle with a 2-inch hypotenuse and an area of 1 square inch?

73. *Exploration*   Find an equation of a line whose graph intersects the graph of the parabola $y = x^2$ at the following numbers of points. (There is more than one correct answer.)

(a) Two points

(b) One point

(c) No points

74. *Hyperbolic Mirror* In a hyperbolic mirror, light rays directed to one focus will be reflected to the other focus. The mirror in the figure has the equation

$$\frac{x^2}{25} - \frac{y^2}{36} = 1.$$

At which point on the mirror will light from the point $(0, 10)$ reflect to the focus?

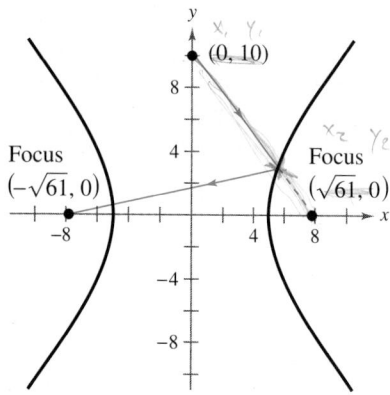

75. *Data Analysis* The table gives the numbers y of millions of short tons of newsprint produced in the years 1990 through 1993 in the United States. (Source: American Paper Institute)

Year	1990	1991	1992	1993
y	6.6	6.8	7.1	7.1

Let t represent the time in years, with $t = 0$ corresponding to 1990.

(a) Use the regression capabilities of a graphing utility to fit a linear model and a quadratic model to the data.

(b) Use the graphing utility to graph the data and the two models in the same viewing rectangle.

(c) Approximate the points of intersection of the graphs of the models.

(d) Use the models to predict newsprint production in 1994. Which model do you think gives the more accurate prediction? Explain.

76. *Conjecture*

(a) Use a graphing utility to graph the system of equations

$$y = b^x$$
$$y = x^b$$

for $b = 2$ and $b = 4$.

(b) For a fixed value of $b > 1$, make a conjecture about the number of points of intersection of the graphs in part (a).

Review Solve Exercises 77–82 as a review of the skills and problem-solving techniques you learned in previous sections. Find the general form of the equation of the line through the two points.

77. $(-2, 7), (5, 5)$

78. $(3.5, 4), (10, 6)$

79. $(6, 3), (10, 3)$

80. $(4, -2), (4, 5)$

81. $\left(\frac{3}{5}, 0\right), (4, 6)$

82. $\left(-\frac{7}{3}, 8\right), \left(\frac{5}{2}, \frac{1}{2}\right)$

7.2 Systems of Linear Equations in Two Variables

The Method of Elimination / *Graphical Interpretation of Solutions* / *Applications*

The Method of Elimination

In Section 7.1, you studied two methods for solving a system of equations: substitution and graphing. Now you will study the **method of elimination.** The key step in this method is to obtain, for one of the variables, coefficients that differ only in sign so that *adding* the equations eliminates the variable.

$$
\begin{aligned}
3x + 5y &= 7 \qquad &\text{Equation 1} \\
-3x - 2y &= -1 \qquad &\text{Equation 2} \\
\hline
3y &= 6 \qquad &\text{Add equations.}
\end{aligned}
$$

Note that by adding the two equations, you eliminate the variable x and obtain a single equation in y. Solving this equation for y produces $y = 2$, which you can then back-substitute into one of the original equations to solve for x.

EXAMPLE 1   **The Method of Elimination**

Solve the system of linear equations.

$$
\begin{aligned}
3x + 2y &= 4 \qquad &\text{Equation 1} \\
5x - 2y &= 8 \qquad &\text{Equation 2}
\end{aligned}
$$

Solution

You can eliminate y by adding the two equations.

$$
\begin{aligned}
3x + 2y &= 4 \qquad &\text{Equation 1} \\
5x - 2y &= 8 \qquad &\text{Equation 2} \\
\hline
8x &= 12 \qquad &\text{Add equations.}
\end{aligned}
$$

Therefore, $x = \frac{3}{2}$. By back-substituting, you can solve for y.

$$
\begin{aligned}
3x + 2y &= 4 \qquad &\text{Equation 1} \\
3\left(\tfrac{3}{2}\right) + 2y &= 4 \qquad &\text{Substitute } \tfrac{3}{2} \text{ for } x. \\
y &= -\tfrac{1}{4} \qquad &\text{Solve for } y.
\end{aligned}
$$

The solution is $\left(\frac{3}{2}, -\frac{1}{4}\right)$. You can check the solution *algebraically* by substituting into the original system, or *graphically,* as shown in Figure 7.5.

Figure 7.5

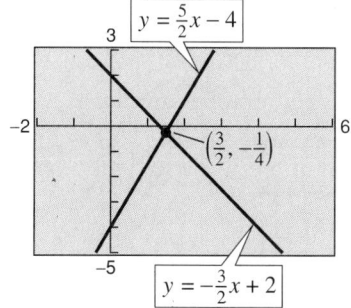

Note   Try using the method of substitution to solve the system given in Example 1. Which method do you think is easier? Many people find that the method of elimination is more efficient.

EXAMPLE 2 ▱ **The Method of Elimination**

Solve the system of linear equations.

$$2x - 3y = -7 \qquad \text{Equation 1}$$
$$3x + y = -5 \qquad \text{Equation 2}$$

Solution

For this system, you can obtain coefficients that differ only in sign by multiplying Equation 2 by 3.

$$2x - 3y = -7 \quad \Longrightarrow \quad 2x - 3y = -7 \qquad \text{Equation 1}$$
$$3x + y = -5 \quad \Longrightarrow \quad \underline{9x + 3y = -15} \qquad \text{Multiply Equation 2 by 3.}$$
$$11x = -22 \qquad \text{Add equations.}$$

Thus, you can see that $x = -2$. By back-substituting this value of x into Equation 1, you can solve for y, as follows.

$$2x - 3y = -7 \qquad \text{Equation 1}$$
$$2(-2) - 3y = -7 \qquad \text{Substitute } -2 \text{ for } x.$$
$$-3y = -3 \qquad \text{Collect like terms.}$$
$$y = 1 \qquad \text{Solve for } y.$$

The solution is $(-2, 1)$. You can check the solution algebraically, as follows, or graphically, as shown in Figure 7.6.

Check

$$2(-2) - 3(1) \overset{?}{=} -7 \qquad \text{Substitute into Equation 1.}$$
$$-4 - 3 = -7 \qquad \text{Equation 1 checks. ✓}$$

$$3(-2) + 1 \overset{?}{=} -5 \qquad \text{Substitute into Equation 2.}$$
$$-6 + 1 = -5 \qquad \text{Equation 2 checks. ✓}$$

▱

Figure 7.6

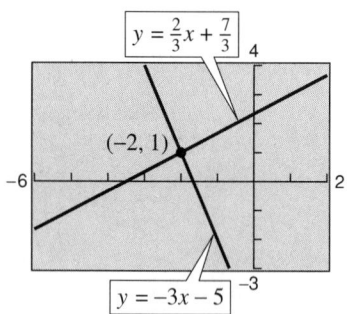

In Example 2, the two systems of linear equations

$$2x - 3y = -7 \qquad \text{and} \qquad 2x - 3y = -7$$
$$3x + y = -5 \qquad\qquad\qquad 9x + 3y = -15$$

are called **equivalent** because they have precisely the same solution set. The operations that can be performed on a system of linear equations to produce an equivalent system are (1) interchanging any two equations, (2) multiplying an equation by a nonzero constant, and (3) adding a multiple of one equation to any other equation in the system.

The Method of Elimination

To use the **method of elimination** to solve a system of two linear equations in x and y, use the following steps.

1. Obtain coefficients for x (or y) that differ only in sign by multiplying all terms of one or both equations by suitably chosen constants.
2. Add the equations to eliminate one variable, and solve the resulting equation.
3. Back-substitute the value obtained in Step 2 into either of the original equations and solve for the other variable.
4. Check your solution in both of the original equations.

EXAMPLE 3 **The Method of Elimination**

Solve the system of linear equations.

$$5x + 3y = 9 \qquad \text{Equation 1}$$
$$2x - 4y = 14 \qquad \text{Equation 2}$$

Solution

You can obtain coefficients that differ only in sign by multiplying Equation 1 by 4 and multiplying Equation 2 by 3.

$$5x + 3y = 9 \quad \Longrightarrow \quad 20x + 12y = 36 \qquad \text{Multiply Equation 1 by 4.}$$
$$\underline{2x - 4y = 14} \quad \Longrightarrow \quad \underline{6x - 12y = 42} \qquad \text{Multiply Equation 2 by 3.}$$
$$26x \qquad\quad = 78 \qquad \text{Add equations.}$$

From this equation, you can see that $x = 3$. By back-substituting this value of x into Equation 2, you can solve for y, as follows.

$$2x - 4y = 14 \qquad \text{Equation 2}$$
$$2(3) - 4y = 14 \qquad \text{Substitute 3 for } x.$$
$$-4y = 8 \qquad \text{Collect like terms.}$$
$$y = -2 \qquad \text{Solve for } y.$$

The solution is $(3, -2)$. You can check the solution algebraically by substituting into the original system, or graphically, as shown in Figure 7.7.

Figure 7.7

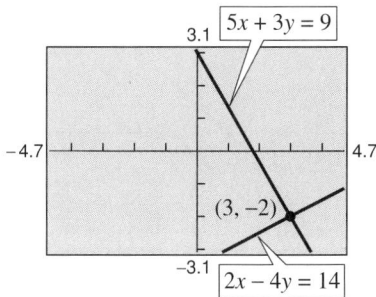

Graphical Interpretation of Solutions

It is possible for a *general* system of equations to have exactly one solution, two or more solutions, or no solution. If a system of *linear* equations has two different solutions, it must have an *infinite* number of solutions. To see why this is true, consider the following graphical interpretations of a system of two linear equations in two variables. (Remember that the graph of a linear equation in two variables is a straight line.)

Graphical Interpretation of Solutions

For a system of two linear equations in two variables, the number of solutions is given by one of the following.

Number of Solutions	*Graphical Interpretation*
1. Exactly one solution	The two lines intersect at one point.
2. Infinitely many solutions	The two lines are identical.
3. No solution	The two lines are parallel.

The graphical interpretations presented above are further illustrated in Figure 7.8. In the second graph, note that the two lines coincide. This case is illustrated in Example 5. In the third graph, note that the two lines are parallel. This case is illustrated in Example 4.

Figure 7.8

Consistent   Consistent   Inconsistent
Two lines that intersect   Two lines that coincide   Two parallel lines
One point of intersection Infinitely many points   No point of intersection

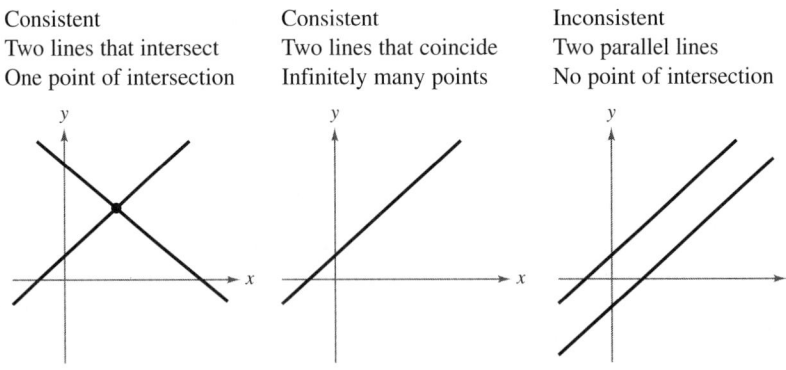

A system of linear equations is called **consistent** if it has at least one solution, and it is called **inconsistent** if it has no solution. In Examples 4 and 5, note how you can use the method of elimination to determine that a system of linear equations has no solution or infinitely many solutions.

EXAMPLE 4 ▱ **The Method of Elimination: No-Solution Case**

Solve the system of linear equations.

$$x - 2y = 3$$ Equation 1
$$-2x + 4y = 1$$ Equation 2

Solution

To obtain coefficients that differ only in sign, multiply Equation 1 by 2.

$$x - 2y = 3$$ ⟹ $$2x - 4y = 6$$ Multiply Equation 1 by 2.
$$-2x + 4y = 1$$ ⟹ $$-2x + 4y = 1$$ Equation 2
$$0 = 7$$ False statement

Because there are no values of x and y for which $0 = 7$, you can conclude that the system is inconsistent and has no solution. The lines corresponding to the two equations in this system are shown in Figure 7.9. Note that the two lines are parallel, and therefore have no point of intersection. ▱

In Example 4, note that the occurrence of a false statement, such as $0 = 7$, indicates that the system has no solution. In the next example, note that the occurrence of a statement that is true for all values of the variables—in this case, $0 = 0$—indicates that the system has infinitely many solutions.

EXAMPLE 5 ▱ **The Method of Elimination: Many-Solutions Case**

Solve the system of linear equations.

$$2x - y = 1$$ Equation 1
$$4x - 2y = 2$$ Equation 2

Solution

To obtain coefficients that differ only in sign, multiply Equation 2 by $-\frac{1}{2}$.

$$2x - y = 1$$ ⟹ $$2x - y = 1$$ Equation 1
$$4x - 2y = 2$$ ⟹ $$-2x + y = -1$$ Multiply Equation 2 by $-\frac{1}{2}$.
$$0 = 0$$ Add equations.

Because the two equations turn out to be equivalent (have the same solution set), you can conclude that the system has infinitely many solutions. The solution set consists of all points (x, y) lying on the line

$$2x - y = 1.$$ See Figure 7.10. ▱

Figure 7.9

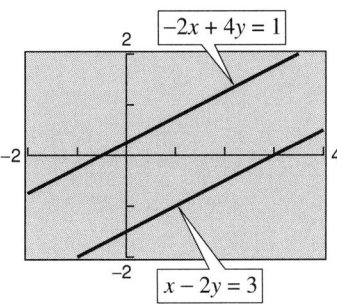

Note In Example 5, notice that you could have reached the conclusion that the equations are equivalent by multiplying Equation 1 by 2 to obtain Equation 2.

Figure 7.10

The general solution of the linear system

$$ax + by = c$$
$$dx + ey = f$$

is $x = (ce - bf)/(ae - db)$ and $y = (af - cd)/(ae - db)$. If $ae - db = 0$, the system does not have a unique solution. A *TI-82* or *TI-83* program for solving such a system is given below. Programs for other graphing calculator models may be found in the appendix. Try using this program to check the solution of the system in Example 6.

```
PROGRAM:SOLVE
:Disp "AX+BY=C"
:Prompt A
:Prompt B
:Prompt C
:Disp "DX+EY=F"
:Prompt D
:Prompt E
:Prompt F
:If AE-DB=0
:Then
:Disp "NO UNIQUE
SOLUTION"
:Else
:(CE-BF)/(AE-DB)→X
:(AF-CD)/(AE-DB)→Y
:Disp X
:Disp Y
:End
```

Example 6 illustrates a strategy for solving a system of linear equations that has decimal coefficients.

EXAMPLE 6   **A Linear System Having Decimal Coefficients**

Solve the system of linear equations.

$$0.02x - 0.05y = -0.38 \qquad \text{Equation 1}$$
$$0.03x + 0.04y = \;\;\;1.04 \qquad \text{Equation 2}$$

Solution

Because the coefficients are two-place decimals, begin by multiplying each equation by 100 to produce a system with integer coefficients.

$$2x - 5y = -38 \qquad \text{Revised Equation 1}$$
$$3x + 4y = 104 \qquad \text{Revised Equation 2}$$

Now, to obtain coefficients that differ only in sign, multiply Revised Equation 1 by 3 and multiply Revised Equation 2 by -2.

$$2x - 5y = -38 \;\;\Longrightarrow\;\; 6x - 15y = -114 \qquad \text{Multiply Revised Equation 1 by 3.}$$
$$3x + 4y = \;104 \;\;\Longrightarrow\;\; -6x - 8y = -208 \qquad \text{Multiply Revised Equation 2 by –2.}$$
$$-23y = -322 \qquad \text{Add equations.}$$

Thus, you can conclude that

$$y = \frac{-322}{-23} = 14.$$

Back-substituting this value into Revised Equation 2 produces the following.

$$3x + 4y = 104 \qquad \text{Revised Equation 2}$$
$$3x + 4(14) = 104 \qquad \text{Substitute 14 for } y.$$
$$3x = 48 \qquad \text{Collect like terms.}$$
$$x = 16 \qquad \text{Solve for } x.$$

The solution is (16, 14). Check this in the original system, as follows.

Check

$$0.02(16) - 0.05(14) \stackrel{?}{=} -0.38 \qquad \text{Substitute into Equation 1.}$$
$$0.32 - 0.70 = -0.38 \qquad \text{Equation 1 checks.} \;\checkmark$$

$$0.03(16) + 0.04(14) \stackrel{?}{=} 1.04 \qquad \text{Substitute into Equation 2.}$$
$$0.48 + 0.56 = 1.04 \qquad \text{Equation 2 checks.} \;\checkmark$$

Applications

At this point, you may be asking the question "How can I tell which application problems can be solved using a system of linear equations?" The answer comes from the following considerations.

1. Does the problem involve more than one unknown quantity?
2. Are there two (or more) equations or conditions to be satisfied?

If one or both of these conditions occur, the appropriate mathematical model for the problem may be a system of linear equations. Example 7 shows how to construct such a model.

Real Life

EXAMPLE 7 **An Application of a Linear System**

An airplane flying into a headwind travels the 2000-mile flying distance between two cities in 4 hours and 24 minutes. On the return flight, the same distance is traveled in 4 hours. Find the airspeed of the plane and the speed of the wind, assuming that both remain constant.

Solution

The two unknown quantities are the speeds of the wind and the plane. If r_1 is the speed of the plane and r_2 is the speed of the wind, then

$$r_1 - r_2 = \text{speed of the plane } against \text{ the wind}$$
$$r_1 + r_2 = \text{speed of the plane } with \text{ the wind}$$

as shown in Figure 7.11. Using the formula

$$\text{Distance} = (\text{rate})(\text{time})$$

for these two speeds, you obtain the following equations.

$$2000 = (r_1 - r_2)\left(4 + \frac{24}{60}\right)$$
$$2000 = (r_1 + r_2)(4)$$

Figure 7.11

These two equations simplify as follows.

$$5000 = 11r_1 - 11r_2 \qquad \text{Equation 1}$$
$$500 = r_1 + r_2 \qquad \text{Equation 2}$$

Original flight

$r_1 - r_2$

By elimination, the solution is

$$r_1 = \frac{5250}{11} \approx 477.27 \text{ miles per hour} \qquad \text{Speed of plane}$$

Return flight

$r_1 + r_2$

$$r_2 = \frac{250}{11} \approx 22.73 \text{ miles per hour.} \qquad \text{Speed of wind}$$

Real Life

EXAMPLE 8   Finding the Point of Equilibrium

The demand and supply functions for a certain type of calculator are given by

$$p = 150 - 0.00001x \qquad \text{Equation 1: demand equation}$$
$$p = 60 + 0.00002x \qquad \text{Equation 2: supply equation}$$

where p is the price in dollars and x is the number of units. Find the point of equilibrium for this market. The point of equilibrium is defined by the price p and number of units x that satisfy both the demand and supply equations.

Figure 7.12

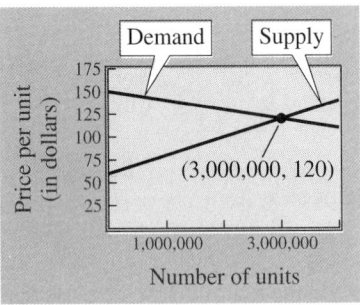

Solution
Begin by substituting the value of p given in Equation 2 into Equation 1.

$$p = 150 - 0.00001x \qquad \text{Equation 1}$$
$$60 + 0.00002x = 150 - 0.00001x \qquad \text{Substitute } 60 + 0.00002x \text{ for } p.$$
$$0.00003x = 90 \qquad \text{Collect like terms.}$$
$$x = 3,000,000 \qquad \text{Solve for } x.$$

Thus, the point of equilibrium occurs when the demand and supply are each 3 million units. (See Figure 7.12.) The price that corresponds to this x-value is obtained by back-substituting $x = 3,000,000$ into either of the original equations. For instance, back-substituting into Equation 1 produces

$$p = 150 - 0.00001(3,000,000) = 150 - 30 = \$120.$$

Try back-substituting $x = 3,000,000$ into Equation 2 to see that you obtain the same price.

Note   In a free market, the demands for many products are related to the prices of the products. As prices decrease, consumer demand increases and the amounts that producers are able or willing to supply decrease.

Group Activity

Creating Consistent and Inconsistent Systems

For each of the following systems, find the value of k such that the system has infinitely many solutions.

a. $4x + 3y = -8$ **b.** $3x - 12y = 9$
     $x + ky = -2$ $x - 4y = k$

Is it possible to find values of k such that the systems have no solutions? Explain why or why not for each system. Is it possible to find values of k such that the systems have unique solutions? Explain why or why not for each system. Summarize the differences between these two systems of equations.

7.2 /// EXERCISES

In Exercises 1–10, solve the system of equations by the method of elimination. Label each graphed line with its equation.

1. $2x + y = 5$
$x - y = 1$

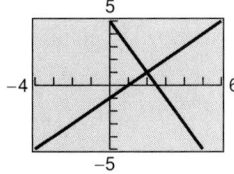

2. $x + 3y = 1$
$-x + 2y = 4$

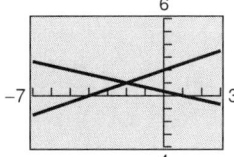

3. $x + y = 0$
$3x + 2y = 1$

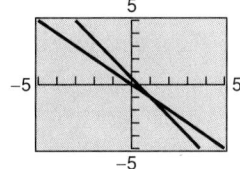

4. $2x - y = 3$
$4x + 3y = 21$

5. $x - y = 2$
$-2x + 2y = 5$

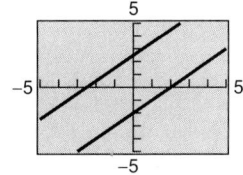

6. $3x + 2y = 3$
$6x + 4y = 14$

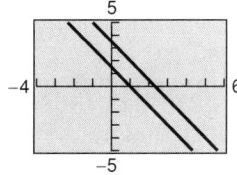

7. $3x - 2y = 5$
$-6x + 4y = -10$

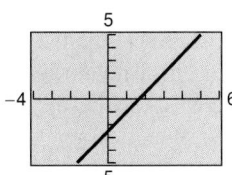

8. $x - 2y = 4$
$6x + 2y = 10$

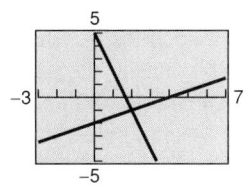

9. $9x + 3y = 1$
$3x - 6y = 5$

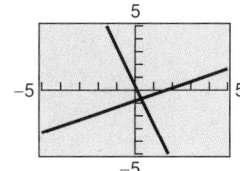

10. $5x + 3y = -18$
$2x - 6y = 1$

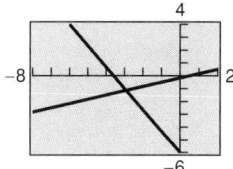

In Exercises 11–20, solve the system by the method of elimination and check any solution algebraically.

11. $x + 2y = 4$
$x - 2y = 1$

12. $3x - 5y = 2$
$2x + 5y = 13$

13. $2x + 3y = 18$
$5x - y = 11$

14. $x + 7y = 12$
$3x - 5y = 10$

15. $3x + 2y = 10$
$2x + 5y = 3$

16. $8r + 16s = 20$
$16r + 50s = 55$

17. $2u + v = 120$
$u + 2v = 120$

18. $5u + 6v = 24$
$3u + 5v = 18$

19. $6r - 5s = 3$
$10r - 12s = 5$

20. $1.8x + 1.2y = 4$
$9x + 6y = 3$

In Exercises 21–30, solve the system by the method of elimination and verify any solution with a graphing utility.

21. $\dfrac{x}{4} + \dfrac{y}{6} = 1$
$x - y = 3$

22. $\frac{2}{3}x + \frac{1}{6}y = \frac{2}{3}$
$4x + y = 4$

23. $\dfrac{x + 3}{4} + \dfrac{y - 1}{3} = 1$
$2x - y = 12$

24. $\dfrac{x - 1}{2} + \dfrac{y + 2}{3} = 4$
$x - 2y = 5$

25. $2.5x - 3y = 1.5$
$10x - 12y = 6$

26. $0.02x - 0.05y = -0.19$
$0.03x + 0.04y = 0.52$

27. $0.05x - 0.03y = 0.21$
$0.07x + 0.02y = 0.16$

28. $0.2x - 0.5y = -27.8$
$0.3x + 0.4y = 68.7$

29. $4b + 3m = 3$
$3b + 11m = 13$

30. $3b + 3m = 7$
$3b + 5m = 3$

In Exercises 31–34, use a graphing utility to graph the lines in the system. Use the graphs to determine if the system is consistent or inconsistent. If the system is consistent, determine the number of solutions.

31. $\frac{1}{5}x - \frac{1}{3}y = 1$
$-3x + 5y = 9$

32. $2x + y = 5$
$x - 2y = -1$

33. $2x - 5y = 0$
$x - y = 3$

34. $4x - 6y = 7$
$2x - 3y = 3.5$

In Exercises 35–38, use a graphing utility to graph the two equations. Use the graphs to approximate the solution of the system.

35. $8x + 9y = 42$
$6x - y = 16$

36. $\frac{3}{2}x - \frac{1}{5}y = 8$
$-2x + 3y = 3$

37. $4y = -8$
$7x - 2y = 25$

38. $0.5x + 2.2y = 9$
$6x + 0.4y = -22$

In Exercises 39–42, use any method to solve the system.

39. $3x - 5y = 7$
$2x + y = 9$

40. $-x + 3y = 17$
$4x + 3y = 7$

41. $y = 2x - 5$
$y = 5x - 11$

42. $7x + 3y = 16$
$y = x + 2$

Exploration In Exercises 43 and 44, find a system of linear equations that has the given solution. (There is more than one correct answer.)

43. $\left(3, \frac{5}{2}\right)$

44. $(8, -2)$

Think About It In Exercises 45 and 46, the graphs of the two equations appear to be parallel. Yet, when the system is solved algebraically, it is found that the system does have a solution. Find the solution and explain why it does not appear on the portion of the graph that is shown.

45. $100y - x = 200$
$99y - x = -198$

46. $21x - 20y = 0$
$13x - 12y = 120$

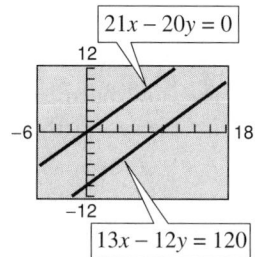

47. *Essay* Briefly explain whether or not it is possible for a consistent system of linear equations to have exactly two solutions.

48. *Think About It* Give examples of (a) a system of linear equations that has no solution and (b) a system that has an infinite number of solutions.

Exploration In Exercises 49 and 50, find the value of k such that the system of equations is inconsistent.

49. $4x - 8y = -3$
$2x + ky = 16$

50. $15x + 3y = 6$
$-10x + ky = 9$

51. *Airplane Speed* An airplane flying into a headwind travels the 1800-mile flying distance between two cities in 3 hours and 36 minutes. On the return flight, the distance is traveled in 3 hours. Find the airspeed of the plane and the speed of the wind, assuming that both remain constant.

52. *Airplane Speed* Two planes start from the same airport and fly in opposite directions. The second plane starts $\frac{1}{2}$ hour after the first plane, but its speed is 80 kilometers per hour faster. Find the groundspeed of each plane if 2 hours after the first plane departs the planes are 3200 kilometers apart.

53. *Acid Mixture* Ten liters of a 30% acid solution is obtained by mixing a 20% solution with a 50% solution.

(a) Write a system of equations in which one equation represents the amount of final mixture required and the other represents the amount of acid in the final mixture. Let x and y represent the amounts of 20% and 50% solutions, respectively.

(b) Use a graphing utility to graph the two equations in part (a). As the amount of the 20% solution increases, how does the amount of the 50% solution change?

(c) How much of each solution is required to obtain the specified concentration in the final mixture?

54. *Fuel Mixture* Five hundred gallons of 89 octane gasoline is obtained by mixing 87 octane gasoline with 92 octane gasoline. How much of each type must be used?

55. *Investment Portfolio* A total of $12,000 is invested in two corporate bonds that pay 10.5% and 12% simple interest. The investor wants an annual interest income of $1350 from the investments. What is the most that can be invested in the 10.5% bond?

56. *Investment Portfolio* A total of $32,000 is invested in two municipal bonds that pay 5.75% and 6.25% simple interest. The investor wants an annual interest income of $1900 from the investments. What is the most that can be invested in the 5.75% bond?

57. *Ticket Sales* Five hundred tickets were sold for a certain performance of a play. The tickets for adults and children sold for $7.50 and $4.00, respectively, and the total receipts for the performance were $3312.50. How many of each kind of ticket were sold?

58. *Shoe Sales* On Saturday night, the manager of a shoe store evaluates the receipts of the previous week's sales. Two hundred and forty pairs of two different styles of tennis shoes were sold. One style sold for $66.95 and the other sold for $84.95. The total receipts were $17,652. The cash register that was supposed to record the number of each type of shoe sold malfunctioned. Can you recover the information? If so, how many shoes of each type were sold?

59. *Driving Distances* In a trip of 300 kilometers, two people drive. One person drives three times as far as the other. Find the distance that each person drives.

60. *Truck Scheduling* A contractor hires two trucking companies to haul 1600 tons of crushed stone for a highway construction project. The contracts state that one company is to haul four times as much as the other. Find the amount hauled by each company.

Fitting a Line to Data In Exercises 61–68, find the least squares regression line $y = ax + b$ for the points

$$(x_1, y_1), (x_2, y_2), \ldots, (x_n, y_n).$$

To find the line, solve the following system for a and b.

$$nb + \left(\sum_{i=1}^{n} x_i\right)a = \sum_{i=1}^{n} y_i$$

$$\left(\sum_{i=1}^{n} x_i\right)b + \left(\sum_{i=1}^{n} x_i^2\right)a = \sum_{i=1}^{n} x_i y_i$$

Then use the linear regression capabilities of a graphing utility to confirm the result.

61. $5b + 10a = 20.2$
$10b + 30a = 50.1$

62. $5b + 10a = 11.7$
$10b + 30a = 25.6$

63. $7b + 21a = 35.1$
$21b + 91a = 114.2$

64. $6b + 15a = 23.6$
$15b + 55a = 48.8$

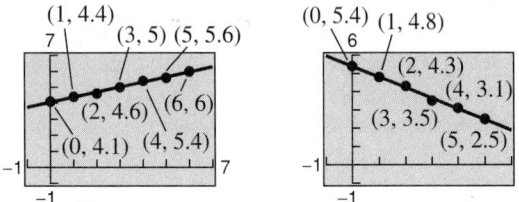

65. $(-2, 0), (0, 1), (2, 3)$

66. $(-3, 0), (-1, 1), (1, 1), (3, 2)$

67. $(0, 4), (1, 3), (1, 1), (2, 0)$

68. $(1, 0), (2, 0), (3, 0), (3, 1), (4, 1), (4, 2), (5, 2), (6, 2)$

69. *Analyzing Data* A farmer used four test plots to determine the relationship between wheat yield in bushels per acre and the amount of fertilizer in hundreds of pounds per acre. The results are given in the table.

Fertilizer, x	1.0	1.5	2.0	2.5
Yield, y	32	41	48	53

Use the technique demonstrated in Exercises 61–68 to find the line that best fits the data. Then use the line to estimate the yield for a fertilizer application of 160 pounds per acre.

70. *Analyzing Data* A store manager wants to know the demand for a certain product as a function of the price. The daily sales for the different prices of the product are given in the table.

Price, x	$1.00	$1.25	$1.50
Demand, y	450	375	330

Use the technique demonstrated in Exercises 61–68 to find the line that best fits the data. Then use the line to predict the demand when the price is $1.40.

Supply and Demand In Exercises 71–74, find the point of equilibrium of the demand and supply equations.

Demand	*Supply*
71. $p = 50 - 0.5x$	$p = 0.125x$
72. $p = 100 - 0.05x$	$p = 25 + 0.1x$
73. $p = 140 - 0.00002x$	$p = 80 + 0.00001x$
74. $p = 400 - 0.0002x$	$p = 225 + 0.0005x$

Advanced Applications In Exercises 75–78, solve the system of equations for u and v. While solving for these variables, consider the transcendental functions as constants. (Systems of this type are found in a course in differential equations.)

75. $u \sin x + v \cos x = 0$
$u \cos x - v \sin x = \sec x$

76. $u \cos 2x + v \sin 2x = 0$
$u(-2 \sin 2x) + v(2 \cos 2x) = \csc 2x$

77. $ue^x + vxe^x = 0$
$ue^x + v(x + 1)e^x = e^x \ln x$

78. $ue^{2x} + vxe^{2x} = 0$
$u(2e^{2x}) + v(2x + 1)e^{2x} = \dfrac{e^{2x}}{x}$

Review Solve Exercises 79–82 as a review of the skills and problem-solving techniques you learned in previous sections. Determine the domain of the function.

79. $f(x) = x^2 - 2x$

80. $g(x) = \sqrt[3]{x - 2}$

81. $h(x) = \sqrt{25 - x^2}$

82. $s(t) = \dfrac{4}{9 - t^2}$

7.3 Multivariable Linear Systems

Row-Echelon Form and Back-Substitution / Gaussian Elimination / Nonsquare Systems / Graphical Interpretation of Three-Variable Systems / Applications

Row-Echelon Form and Back-Substitution

The method of elimination can be applied to a system of linear equations in more than two variables. When elimination is used to solve a system of linear equations, the goal is to rewrite the system in a form to which back-substitution can be applied. To see how this works, consider the following two systems of linear equations.

$$
\begin{aligned}
x - 2y + 3z &= 9 \\
-x + 3y &= -4 \\
2x - 5y + 5z &= 17
\end{aligned}
\qquad
\begin{aligned}
x - 2y + 3z &= 9 \\
y + 3z &= 5 \\
z &= 2
\end{aligned}
$$

The system on the right is said to be in **row-echelon form,** which means that it has a "stair-step" pattern with leading coefficients of 1. After comparing the two systems, it should be clear that it is easier to solve the system on the right.

One of the most influential Chinese mathematics books was the *Chui-chang suan-shu* or *Nine Chapters on the Mathematical Art* (written in approximately 250 B.C.). Chapter Eight of the Nine Chapters contained solutions of systems of linear equations using positive and negative numbers. One such system was

$$
\begin{aligned}
3x + 2y + z &= 39 \\
2x + 3y + z &= 34 \\
x + 2y + 3z &= 26
\end{aligned}
$$

This system was solved using column operations on a matrix. Matrices (plural for matrix) will be discussed in the next chapter.

EXAMPLE 1   Using Back-Substitution

Solve the system of linear equations.

$$
\begin{aligned}
x - 2y + 3z &= 9 && \text{Equation 1} \\
y + 3z &= 5 && \text{Equation 2} \\
z &= 2 && \text{Equation 3}
\end{aligned}
$$

Solution

From Equation 3, you know the value of z. To solve for y, substitute $z = 2$ into Equation 2 to obtain

$$
\begin{aligned}
y + 3(2) &= 5 && \text{Substitute 2 for } z. \\
y &= -1. && \text{Solve for } y.
\end{aligned}
$$

Finally, substitute $y = -1$ and $z = 2$ into Equation 1 to obtain

$$
\begin{aligned}
x - 2(-1) + 3(2) &= 9 && \text{Substitute } -1 \text{ for } y \text{ and } 2 \text{ for } z. \\
x &= 1. && \text{Solve for } x.
\end{aligned}
$$

The solution is $x = 1$, $y = -1$, and $z = 2$, which can be written as the **ordered triple** $(1, -1, 2)$. Check this in the original system of equations.

Gaussian Elimination

Two systems of equations are **equivalent** if they have the same solution set. To solve a system that is not in row-echelon form, first convert it to an *equivalent* system that *is* in row-echelon form by using one or more of the elementary row operations shown at the left. This process is called **Gaussian elimination,** after the German mathematician Carl Friedrich Gauss (1777–1855).

EXAMPLE 2   Using Elimination to Solve a System

Solve the system of linear equations.

$$\begin{cases} x - 2y + 3z = 9 & \text{Equation 1} \\ -x + 3y = -4 & \text{Equation 2} \\ 2x - 5y + 5z = 17 & \text{Equation 3} \end{cases}$$

Solution

There are many ways to begin, but we suggest working from the upper left corner, saving the x in the upper left position and eliminating the other x's from the first column.

$$\begin{aligned} x - 2y + 3z &= 9 \\ y + 3z &= 5 \\ 2x - 5y + 5z &= 17 \end{aligned}$$

> Adding the first equation to the second equation produces a new second equation.

$$\begin{aligned} x - 2y + 3z &= 9 \\ y + 3z &= 5 \\ -y - z &= -1 \end{aligned}$$

> Adding -2 times the first equation to the third equation produces a new third equation.

Now that all but the first x have been eliminated from the first column, go to work on the second column. (You need to eliminate y from the third equation.)

$$\begin{aligned} x - 2y + 3z &= 9 \\ y + 3z &= 5 \\ 2z &= 4 \end{aligned}$$

> Adding the second equation to the third equation produces a new third equation.

Finally, you need a coefficient of 1 for z in the third equation.

$$\begin{aligned} x - 2y + 3z &= 9 \\ y + 3z &= 5 \\ z &= 2 \end{aligned}$$

> Multiplying the third equation by $\frac{1}{2}$ produces a new third equation.

This is the same system that was solved in Example 1, and, as in that example, you can conclude that the solution is $x = 1$, $y = -1$, and $z = 2$. The solution can also be written as the ordered triple $(1, -1, 2)$.

The next example involves an inconsistent system—one that has no solution. The key to recognizing an inconsistent system is that at some stage in the elimination process, you obtain an absurdity such as $0 = -2$.

EXAMPLE 3   An Inconsistent System

Solve the system of linear equations.

$$\begin{aligned} x - 3y + z &= 1 \\ 2x - y - 2z &= 2 \\ x + 2y - 3z &= -1 \end{aligned}$$

Equation 1
Equation 2
Equation 3

Solution

$$\begin{aligned} x - 3y + z &= 1 \\ 5y - 4z &= 0 \\ x + 2y - 3z &= -1 \end{aligned}$$

Adding -2 times the first equation to the second equation produces a new second equation.

$$\begin{aligned} x - 3y + z &= 1 \\ 5y - 4z &= 0 \\ 5y - 4z &= -2 \end{aligned}$$

Adding -1 times the first equation to the third equation produces a new third equation.

$$\begin{aligned} x - 3y + z &= 1 \\ 5y - 4z &= 0 \\ 0 &= -2 \end{aligned}$$

Adding -1 times the second equation to the third equation produces a new third equation.

Because the third "equation" is impossible, you can conclude that this system is inconsistent and therefore has no solution. Moreover, because this system is equivalent to the original system, you can conclude that the original system also has no solution.

As with a system of linear equations in two variables, the number of solutions of a system of linear equations in more than two variables must fall into one of three categories.

The Number of Solutions of a Linear System

For a system of linear equations, exactly one of the following is true.

1. There is exactly one solution.
2. There are infinitely many solutions.
3. There is no solution.

EXAMPLE 4   A System with Infinitely Many Solutions

Solve the system of linear equations.

$$x + y - 3z = -1 \qquad \text{Equation 1}$$
$$y - z = 0 \qquad \text{Equation 2}$$
$$-x + 2y = 1 \qquad \text{Equation 3}$$

Solution

$$x + y - 3z = -1$$
$$y - z = 0$$
$$3y - 3z = 0$$

Adding the first equation to the third equation produces a new third equation.

$$x + y - 3z = -1$$
$$y - z = 0$$
$$0 = 0$$

Adding -3 times the second equation to the third equation produces a new third equation.

This means that Equation 3 depends on Equations 1 and 2 in the sense that it gives us no additional information about the variables. Thus, the original system is equivalent to the system

$$x + y - 3z = -1$$
$$y - z = 0.$$

In the last equation, solve for y in terms of z to obtain $y = z$. Back-substituting for y into the previous equation produces $x = 2z - 1$. Finally, letting $z = a$, the solutions to the given system are all of the form

$$x = 2a - 1, \qquad y = a, \qquad \text{and} \qquad z = a$$

where a is a real number. Thus, every ordered triple of the form

$$(2a - 1, a, a), \qquad a \text{ is a real number}$$

is a solution of the system.

Study Tip

There are an infinite number of solutions to Example 4, but they all are of a specific form. By selecting, for example, a-values of 0, 1, and 3, you can verify that $(-1, 0, 0)$, $(1, 1, 1)$, and $(5, 3, 3)$ are specific solutions. It is incorrect to say simply that the solution to Example 4 is "infinite." You must also specify the form of the solutions.

In Example 4, there are other ways to write the same infinite set of solutions. For instance, the solutions could have been written as

$$\left(b, \tfrac{1}{2}(b + 1), \tfrac{1}{2}(b + 1)\right), \qquad b \text{ is a real number.}$$

Try convincing yourself of this by substituting $a = 0$, $a = 1$, $a = 2$, and $a = 3$ into the solution listed in Example 4. Then substitute $b = -1$, $b = 1$, $b = 3$, and $b = 5$ into the solution listed above. The resulting sets of ordered triples should be the same. Thus, when comparing descriptions of an infinite solution set, keep in mind that there is more than one way to describe the set.

Nonsquare Systems

So far, each system of linear equations has been **square,** which means that the number of equations is equal to the number of variables. In a **nonsquare** system, the number of equations differs from the number of variables. A system of linear equations cannot have a unique solution unless there are at least as many equations as there are variables in the system.

EXAMPLE 5   **A System with Fewer Equations than Variables**

Solve the system of linear equations.

$$x - 2y + z = 2 \qquad \text{Equation 1}$$
$$2x - y - z = 1 \qquad \text{Equation 2}$$

Solution

Begin by rewriting the system in row-echelon form, as follows.

$$\begin{aligned} x - 2y + z &= 2 \\ 3y - 3z &= -3 \end{aligned}$$

> Adding -2 times the first equation to the second equation produces a new second equation.

$$\begin{aligned} x - 2y + z &= 2 \\ y - z &= -1 \end{aligned}$$

> Multiplying the second equation by $\frac{1}{3}$ produces a new second equation.

Solving for y in terms of z, you get $y = z - 1$, and back-substitution into Equation 1 yields

$$x - 2(z - 1) + z = 2$$
$$x - 2z + 2 + z = 2$$
$$x = z.$$

Finally, by letting $z = a$, you have the solution

$$x = a, \qquad y = a - 1, \qquad \text{and} \qquad z = a$$

where a is a real number. Thus, every ordered triple of the form

$$(a, a - 1, a), \qquad a \text{ is a real number}$$

is a solution of the system.

Note   In Example 5, try choosing some values of a to obtain different solutions of the system, such as $(1, 0, 1)$, $(2, 1, 2)$, and $(3, 2, 3)$. Then check each of the solutions in the original system.

Figure 7.13

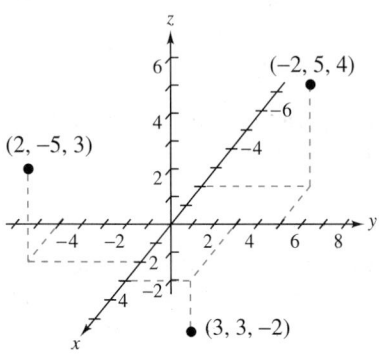

Graphical Interpretation of Three-Variable Systems

Solutions of equations with three variables can be pictured with a **three-dimensional coordinate system.** To construct such a system, begin with the *xy*-coordinate plane in a horizontal position. Then draw the *z*-axis as a vertical line through the origin.

Every ordered triple (x, y, z) corresponds to a point on the three-dimensional coordinate system. For instance, the points corresponding to

$$(-2, 5, 4), \quad (2, -5, 3), \quad \text{and} \quad (3, 3, -2)$$

are shown in Figure 7.13.

The **graph** of an equation in three variables consists of all points (x, y, z) that are solutions of the equation. The graph of a linear equation in three variables is a *plane*. Sketching graphs on a three-dimensional coordinate system is difficult because the sketch itself is only two-dimensional.

Figure 7.14

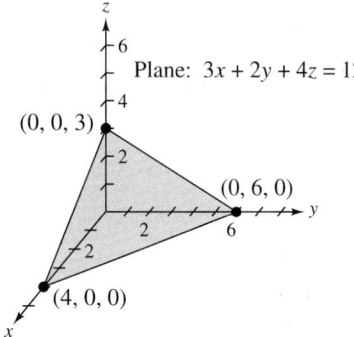

One technique for sketching a plane is to find the three points at which the plane intersects the axes. For instance, the plane given by

$$3x + 2y + 4z = 12$$

intersects the *x*-axis at the point $(4, 0, 0)$, the *y*-axis at the point $(0, 6, 0)$, and the *z*-axis at the point $(0, 0, 3)$.

By plotting these three points, connecting them with line segments, and shading the resulting triangular region, you can picture a portion of the graph, as shown in Figure 7.14.

The graph of a system of three linear equations in three variables consists of *three* planes. When these planes intersect in a single point, the system has exactly one solution. When the three planes have no point in common, the system has no solution. When the three planes contain a common line, the system has infinitely many solutions. (See Figure 7.15.)

Figure 7.15

Exactly One Solution

No Solution

Infinitely Many Solutions

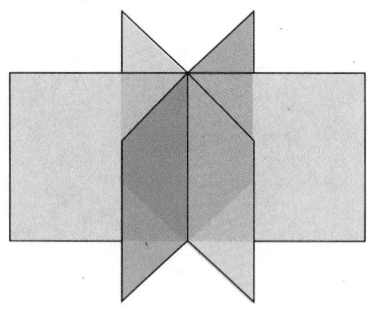

Study Tip

When you use a system of linear equations to solve an application, it is wise to interpret your solution in the context of the problem to see if it makes sense. For instance, in Example 6 the solution results in a position equation of $s = -16t^2 + 48t + 20$, which implies that the object was thrown upward at a velocity of 48 feet per second from a height of 20 feet. The object undergoes a constant downward acceleration of 32 feet per second squared. (Physics will tell you that this is the value of the acceleration due to gravity.)

Applications

EXAMPLE 6   Vertical Motion

Real Life

The height at time t of an object that is moving in a (vertical) line with constant acceleration a is given by the **position equation**

$$s = \tfrac{1}{2}at^2 + v_0 t + s_0.$$

The height s is measured in feet, t is measured in seconds, v_0 is the initial velocity (at $t = 0$) in feet per second, and s_0 is the initial height. Find the values of a, v_0, and s_0, if $s = 52$ at $t = 1$, $s = 52$ at $t = 2$, and $s = 20$ at $t = 3$.

Solution
You can obtain three linear equations in a, v_0, and s_0 as follows.

When $t = 1$: $\tfrac{1}{2}a(1)^2 + v_0(1) + s_0 = 52$ ⟹ $a + 2v_0 + 2s_0 = 104$

When $t = 2$: $\tfrac{1}{2}a(2)^2 + v_0(2) + s_0 = 52$ ⟹ $2a + 2v_0 + s_0 = 52$

When $t = 3$: $\tfrac{1}{2}a(3)^2 + v_0(3) + s_0 = 20$ ⟹ $9a + 6v_0 + 2s_0 = 40$

Solving this system yields $a = -32$, $v_0 = 48$, and $s_0 = 20$.

EXAMPLE 7   Data Analysis: Curve-Fitting

Find a quadratic equation, $y = ax^2 + bx + c$, whose graph passes through the points $(-1, 3)$, $(1, 1)$, and $(2, 6)$.

Solution
Because the graph of $y = ax^2 + bx + c$ passes through the points $(-1, 3)$, $(1, 1)$, and $(2, 6)$, you can write the following.

When $x = -1$, $y = 3$: $a(-1)^2 + b(-1) + c = 3$
When $x = 1$, $y = 1$: $a(1)^2 + b(1) + c = 1$
When $x = 2$, $y = 6$: $a(2)^2 + b(2) + c = 6$

This produces the following system of linear equations.

$a - b + c = 3$ Equation 1
$a + b + c = 1$ Equation 2
$4a + 2b + c = 6$ Equation 3

Figure 7.16

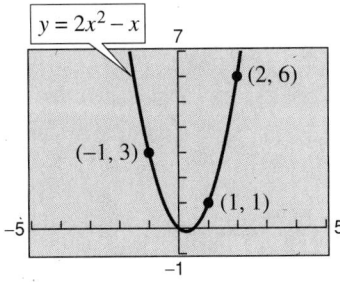

The solution of this system is $a = 2$, $b = -1$, and $c = 0$. Thus, the equation of the parabola is $y = 2x^2 - x$, as shown in Figure 7.16.

Group Activity   *Mathematical Modeling*

x	1	3	5
y	59	49	59

Suppose you work for an outerwear manufacturer, and the marketing department is concerned about sales trends in Georgia. Your manager has asked you to investigate climatic data in hopes of explaining sales patterns and gives you the table at the left, which gives the average monthly temperature y in degrees Fahrenheit for Savannah, Georgia, for the month x, where $x = 1$ corresponds to November.

Construct a scatter plot of the data. Decide which type of mathematical model might be appropriate for the data and use the methods you have learned thus far to fit an appropriate model. Your manager would like to know the average monthly temperatures for December ($x = 2$) and February ($x = 4$). Explain to your manager how you found your model, what it represents, and how it may be used to find the December and February average temperatures. Investigate the usefulness of this model for the rest of the year. Would you recommend using the model to predict average monthly temperatures for the whole year or just part of the year? Explain your reasoning.   (Source: National Climatic Data Center)

7.3 /// EXERCISES

In Exercises 1–6, use back-substitution to solve the system of linear equations.

1.
$$2x - y + 5z = 24$$
$$y + 2z = 6$$
$$z = 4$$

2.
$$4x - 3y - 2z = 21$$
$$6y - 5z = -8$$
$$z = -2$$

3.
$$2x + y - 3z = 10$$
$$y = 2$$
$$y - z = 4$$

4.
$$x = 8$$
$$2x + 3y = 10$$
$$x - y + 2z = 22$$

5.
$$4x - 2y + z = 8$$
$$2z = 4$$
$$-y + z = 4$$

6.
$$5x - 8z = 22$$
$$3y - 5z = 10$$
$$z = -4$$

In Exercises 7 and 8, perform the row operation and write the equivalent system.

7. Add Equation 1 to Equation 2.

$$x - 2y + 3z = 5 \quad \text{Equation 1}$$
$$-x + 3y - 5z = 4 \quad \text{Equation 2}$$
$$2x - 3z = 0 \quad \text{Equation 3}$$

What did this operation accomplish?

8. Add -2 times Equation 1 to Equation 3.

$$x - 2y + 3z = 5 \quad \text{Equation 1}$$
$$-x + 3y - 5z = 4 \quad \text{Equation 2}$$
$$2x - 3z = 0 \quad \text{Equation 3}$$

What did this operation accomplish?

In Exercises 9–34, solve the system of linear equations and check any solution algebraically.

9. $x + y + z = 6$
$2x - y + z = 3$
$3x \quad - z = 0$

10. $x + y + z = 2$
$-x + 3y + 2z = 8$
$4x + y \quad = 4$

11. $2x \quad + 2z = 2$
$5x + 3y \quad = 4$
$3y - 4z = 4$

12. $4x + y - 3z = 11$
$2x - 3y + 2z = 9$
$x + y + z = -3$

13. $6y + 4z = -12$
$3x + 3y \quad = 9$
$2x \quad - 3z = 10$

14. $2x + 4y + z = -4$
$2x - 4y + 6z = 13$
$4x - 2y + z = 6$

15. $3x - 2y + 4z = 1$
$x + y - 2z = 3$
$2x - 3y + 6z = 8$

16. $5x - 3y + 2z = 3$
$2x + 4y - z = 7$
$x - 11y + 4z = 3$

17. $3x + 3y + 5z = 1$
$3x + 5y + 9z = 0$
$5x + 9y + 17z = 0$

18. $2x + y + 3z = 1$
$2x + 6y + 8z = 3$
$6x + 8y + 18z = 5$

19. $x + 2y - 7z = -4$
$2x + y + z = 13$
$3x + 9y - 36z = -33$

20. $2x + y - 3z = 4$
$4x \quad + 2z = 10$
$-2x + 3y - 13z = -8$

21. $3x - 3y + 6z = 6$
$x + 2y - z = 5$
$5x - 8y + 13z = 7$

22. $x \quad + 4z = 13$
$4x - 2y + z = 7$
$2x - 2y - 7z = -19$

23. $x - 2y + 5z = 2$
$4x \quad - z = 0$

24. $x - 3y + 2z = 18$
$5x - 13y + 12z = 80$

25. $2x - 3y + z = -2$
$-4x + 9y \quad = 7$

26. $2x + 3y + 3z = 7$
$4x + 18y + 15z = 44$

27. $x \quad + 3w = 4$
$2y - z - w = 0$
$3y \quad - 2w = 1$
$2x - y + 4z \quad = 5$

28. $x + y + z + w = 6$
$2x + 3y \quad - w = 0$
$-3x + 4y + z + 2w = 4$
$x + 2y - z + w = 0$

29. $x \quad + 4z = 1$
$x + y + 10z = 10$
$2x - y + 2z = -5$

30. $3x - 2y - 6z = -4$
$-3x + 2y + 6z = 1$
$x - y - 5z = -3$

31. $2x + 3y \quad = 0$
$4x + 3y - z = 0$
$8x + 3y + 3z = 0$

32. $4x + 3y + 17z = 0$
$5x + 4y + 22z = 0$
$4x + 2y + 19z = 0$

33. $12x + 5y + z = 0$
$23x + 4y - z = 0$

34. $5x + 5y - z = 0$
$10x + 5y + 2z = 0$
$5x + 15y - 9z = 0$

35. *Think About It* Are the two systems of equations equivalent? Give reasons for your answer.

$x + 3y - z = 6$ $\qquad$ $x + 3y - z = 6$
$2x - y + 2z = 1$ $\qquad$ $-7y + 4z = 1$
$3x + 2y - z = 2$ $\qquad$ $-7y - 4z = -16$

36. *Think About It* When using Gaussian elimination to solve a system of linear equations, how can you recognize that it has no solution? Give an example that illustrates your answer.

Exploration In Exercises 37 and 38, find a system of linear equations that has the given solution. (The answer is not unique.)

37. $(4, -1, 2)$ $\qquad$ **38.** $\left(-\frac{3}{2}, 4, -7\right)$

In Exercises 39–42, find the equation of the parabola

$$y = ax^2 + bx + c$$

that passes through the given points. To verify your result, use a graphing utility to plot the points and graph the parabola.

39. $(0, 0), (2, -2), (4, 0)$

40. $(0, 3), (1, 4), (2, 3)$

41. $(2, 0), (3, -1), (4, 0)$

42. $(1, 3), (2, 2), (3, -3)$

In Exercises 43–46, find the equation of the circle

$$x^2 + y^2 + Dx + Ey + F = 0$$

that passes through the given points. To verify your result, use a graphing utility to plot the points and graph the circle.

43. $(0, 0), (2, 2), (4, 0)$

44. $(0, 0), (0, 6), (3, 3)$

45. $(-3, -1), (2, 4), (-6, 8)$

46. $(0, 0), (0, -2), (3, 0)$

Vertical Motion   In Exercises 47–50, find the position equation $s = \frac{1}{2}at^2 + v_0t + s_0$ for an object at the given heights moving vertically at the specified times.

47. At $t = 1$ second, $s = 128$ feet
   At $t = 2$ seconds, $s = 80$ feet
   At $t = 3$ seconds, $s = 0$ feet

48. At $t = 1$ second, $s = 48$ feet
   At $t = 2$ seconds, $s = 64$ feet
   At $t = 3$ seconds, $s = 48$ feet

49. At $t = 1$ second, $s = 452$ feet
   At $t = 2$ seconds, $s = 260$ feet
   At $t = 3$ seconds, $s = 116$ feet

50. At $t = 1$ second, $s = 132$ feet
   At $t = 2$ seconds, $s = 100$ feet
   At $t = 3$ seconds, $s = 36$ feet

51. *Investments*   An inheritance of $16,000 was divided among three investments yielding a total of $990 in interest per year. The interest rates were 5%, 6%, and 7%. Find the amount in each investment if the 5% and 6% investments were $3000 and $2000 less than the 7% investment, respectively.

52. *Investments*   A total of $1520 a year is received in interest from three investments. The interest rates for the three investments are 5%, 7%, and 8%. The 5% investment is half of the 7% investment, and the 7% investment is $1500 less than the 8% investment. Find the amount in each investment.

53. *Borrowing*   A small corporation borrowed $775,000 to expand its product line. Some of the money was borrowed at 8%, some at 9%, and some at 10%. How much was borrowed at each rate if the annual interest was $67,000 and the amount borrowed at 8% was four times the amount borrowed at 10%?

54. *Borrowing*   A small corporation borrowed $800,000 to expand its product line. Some of the money was borrowed at 8%, some at 9%, and some at 10%. How much was borrowed at each rate if the annual interest was $67,000 and the amount borrowed at 8% was five times the amount borrowed at 10%?

Investment Portfolio   In Exercises 55 and 56, consider an investor with a portfolio totaling $500,000 that is invested in certificates of deposit, municipal bonds, blue-chip stocks, and growth or speculative stocks. How much is put in each type of investment?

55. The certificates of deposit pay 10% annually, and the municipal bonds pay 8% annually. Over a 5-year period, the investor expects the blue-chip stocks to return 12% annually and the growth stocks to return 13% annually. The investor wants a combined annual return of 10% and also wants to have only one-fourth of the portfolio invested in stocks.

56. The certificates of deposit pay 9% annually, and the municipal bonds pay 5% annually. Over a 5-year period, the investor expects the blue-chip stocks to return 12% annually and the growth stocks to return 14% annually. The investor wants a combined annual return of 10% and also wants to have only one-fourth of the portfolio invested in stocks.

57. *Crop Spraying*   A mixture of 12 liters of chemical A, 16 liters of chemical B, and 26 liters of chemical C is required to kill a certain destructive crop insect. Commercial spray X contains 1, 2, and 2 parts, respectively, of these chemicals. Commercial spray Y contains only chemical C. Commercial spray Z contains only chemicals A and B in equal amounts. How much of each type of commercial spray is needed to get the desired mixture?

58. *Chemistry*   A chemist needs 10 liters of a 25% acid solution. The solution is to be mixed from three solutions whose concentrations are 10%, 20%, and 50%. How many liters of each solution should the chemist use to satisfy the following?

(a) Use as little as possible of the 50% solution.

(b) Use as much as possible of the 50% solution.

(c) Use 2 liters of the 50% solution.

59. *Truck Scheduling*   A small company that manufactures products A and B has an order for 15 units of product A and 16 units of product B. The company has trucks of three different sizes that can haul the products in the amounts shown in the table.

Truck	Large	Medium	Small
Product A	6	4	0
Product B	3	4	3

How many trucks of each size are needed to deliver the order? Give *two* possible solutions.

60. *Electrical Networks*   When Kirchhoff's Laws are applied to the electrical network in the figure, the currents I_1, I_2, and I_3 are the solution of the system

$$I_1 - I_2 + I_3 = 0$$
$$3I_1 + 2I_2 \qquad = 7$$
$$\qquad 2I_2 + 4I_3 = 8.$$

Find the currents.

61. *Pulley System*   A system of pulleys is loaded with 128-pound and 32-pound weights (see figure). The tensions t_1 and t_2 in the ropes and the acceleration a of the 32-pound weight are found by solving the system

$$t_1 - 2t_2 \qquad = 0$$
$$t_1 \qquad - 2a = 128$$
$$\qquad t_2 + a = 32$$

where t_1 and t_2 are measured in pounds and a is in feet per second squared. Solve the system.

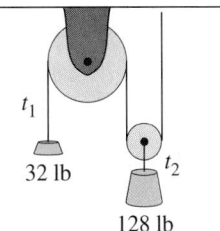

32 lb

128 lb

62. *Pulley System*   If the 32-pound weight is replaced with a 64-pound weight in the pulley system described in Exercise 61, the new pulley system is modeled by the following system of equations.

$$t_1 - 2t_2 \qquad = 0$$
$$t_1 \qquad - 2a = 128$$
$$\qquad t_2 + 2a = 64$$

Solve the system, and use your answer for the acceleration to describe what (if anything) is happening in the system.

Three-Dimensional Graphics   In Exercises 63–66, sketch the plane represented by the linear equation. Then list four points that lie in the plane.

63. $2x + 3y + 4z = 12$

64. $x + y + z = 6$

65. $2x + y + z = 4$

66. $x + 2y + 2z = 6$

Fitting a Parabola In Exercises 67–70, find the least squares regression parabola $y = ax^2 + bx + c$ for the points $(x_1, y_1), (x_2, y_2), \ldots, (x_n, y_n)$. To find the parabola, solve the following system of linear equations for a, b, and c. Then use the least squares regression capabilities of a graphing utility to confirm the result.

$$nc + \left(\sum_{i=1}^{n} x_i\right)b + \left(\sum_{i=1}^{n} x_i^2\right)a = \sum_{i=1}^{n} y_i$$

$$\left(\sum_{i=1}^{n} x_i\right)c + \left(\sum_{i=1}^{n} x_i^2\right)b + \left(\sum_{i=1}^{n} x_i^3\right)a = \sum_{i=1}^{n} x_i y_i$$

$$\left(\sum_{i=1}^{n} x_i^2\right)c + \left(\sum_{i=1}^{n} x_i^3\right)b + \left(\sum_{i=1}^{n} x_i^4\right)a = \sum_{i=1}^{n} x_i^2 y_i$$

67.   **68.**

69.   **70.**

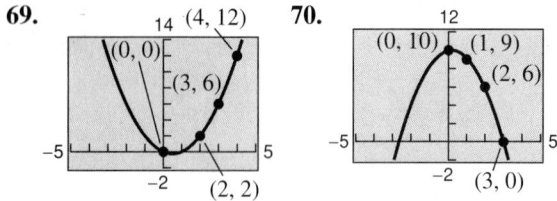

71. *Analyzing Data* In testing of a new braking system on an automobile, the speeds in miles per hour and the stopping distances in feet were recorded as follows.

Speed, x	20	30	40	50	60
Stopping Distance, y	25	55	105	188	300

(a) Find the least squares regression parabola for the data. Then use a graphing utility to graph the parabola and the data on the same set of axes.

(b) Use the model to estimate the stopping distance if the speed is 70 miles per hour.

72. *Analyzing Data* A wildlife management team studied the reproduction rates of deer in five tracts of a wildlife preserve. Each tract contained 5 acres. For each tract, the number of females and the percent of females that had offspring the following year were recorded. The results are given in the table.

Number, x	80	100	120	140	160
Percent, y	80	75	68	55	30

(a) Find the least squares regression parabola for the data.

(b) Use a graphing utility to graph the parabola and the data in the same viewing rectangle.

(c) Use the model to predict the percent of females that would have offspring if $x = 170$.

Advanced Applications In Exercises 73–76, find values of x, y, and λ that satisfy the system. These systems arise in certain optimization problems in calculus; λ is called a Lagrange multiplier.

73.
$$y + \lambda = 0$$
$$x + \lambda = 0$$
$$x + y - 10 = 0$$

74.
$$2x + \lambda = 0$$
$$2y + \lambda = 0$$
$$x + y - 4 = 0$$

75.
$$2x - 2x\lambda = 0$$
$$-2y + \lambda = 0$$
$$y - x^2 = 0$$

76. $2 + 2y + 2\lambda = 0$
$$2x + 1 + \lambda = 0$$
$$2x + y - 100 = 0$$

77. *Chapter Opener* Interpret the slope of the model for sales of compact sport utilities on page 519.

78. *Chapter Opener* If the models on page 519 are assumed to be accurate in forecasting future sales, will there be a time when compact pickup sales again exceed compact utility sales? If so, when?

Review Solve Exercises 79–82 as a review of the skills and problem-solving techniques you learned in previous sections.

79. What is $7\frac{1}{2}\%$ of 85?

80. 225 is what percent of 150?

81. 0.5% of what number is 400?

82. 48% of what number is 132?

7.4   Partial Fractions

Introduction / *Partial Fraction Decomposition*

Introduction

In this section, you will learn to write a rational expression as the sum of two or more simpler rational expressions. For example, the rational expression $(x + 7)/(x^2 - x - 6)$ can be written as the sum of two fractions with linear denominators. That is,

$$\frac{x + 7}{x^2 - x - 6} = \frac{2}{x - 3} + \frac{-1}{x + 2}.$$

Each fraction on the right side of the equation is a **partial fraction,** and together they make up the **partial fraction decomposition** of the left side.

Note One of the most important applications of partial fractions is in calculus. If you go on to take a course in calculus, you will learn how partial fractions can be used in a calculus operation called antidifferentiation.

Decomposition of $N(x)/D(x)$ into Partial Fractions

1. *Divide if improper:* If $N(x)/D(x)$ is an improper fraction, divide the denominator into the numerator to obtain

$$\frac{N(x)}{D(x)} = (\text{polynomial}) + \frac{N_1(x)}{D(x)}$$

and apply Steps 2, 3, and 4 (below) to the proper rational expression $N_1(x)/D(x)$.

2. *Factor denominator:* Completely factor the denominator into factors of the form

$$(px + q)^m \qquad \text{and} \qquad (ax^2 + bx + c)^n$$

where $(ax^2 + bx + c)$ is irreducible.

3. *Linear factors:* For *each* factor of the form $(px + q)^m$, the partial fraction decomposition must include the following sum of m fractions.

$$\frac{A_1}{(px + q)} + \frac{A_2}{(px + q)^2} + \cdots + \frac{A_m}{(px + q)^m}$$

4. *Quadratic factors:* For *each* factor of the form $(ax^2 + bx + c)^n$, the partial fraction decomposition must include the following sum of n fractions.

$$\frac{B_1x + C_1}{ax^2 + bx + c} + \frac{B_2x + C_2}{(ax^2 + bx + c)^2} + \cdots + \frac{B_nx + C_n}{(ax^2 + bx + c)^n}$$

Partial Fraction Decomposition

Algebraic techniques for determining the constants in the numerators of partial fractions are demonstrated in the examples that follow. Note that the techniques vary slightly, depending on the type of factors in the denominator: linear or quadratic, distinct or repeated.

EXAMPLE 1   Distinct Linear Factors

Write the partial fraction decomposition for

$$\frac{x + 7}{x^2 - x - 6}.$$

Solution

Because $x^2 - x - 6 = (x - 3)(x + 2)$, you should include one partial fraction with a constant numerator for each linear factor of the denominator and write

$$\frac{x + 7}{x^2 - x - 6} = \frac{A}{x - 3} + \frac{B}{x + 2}.$$

Multiplying both sides of this equation by the least common denominator, $(x - 3)(x + 2)$, leads to the **basic equation**

$$x + 7 = A(x + 2) + B(x - 3). \qquad \text{Basic equation}$$

Because this equation is true for all x, you can substitute any *convenient* values of x that will help you determine the constants A and B. Values of x that are especially convenient are ones that make the factors $(x + 2)$ and $(x - 3)$ equal to zero. For instance, let $x = -2$. Then

$$-2 + 7 = A(-2 + 2) + B(-2 - 3) \qquad \text{Substitute } -2 \text{ for } x.$$
$$5 = -5B$$
$$-1 = B.$$

To solve for A, let $x = 3$ and obtain

$$3 + 7 = A(3 + 2) + B(3 - 3) \qquad \text{Substitute } 3 \text{ for } x.$$
$$10 = 5A$$
$$2 = A.$$

Therefore, the partial fraction decomposition is

$$\frac{x + 7}{x^2 - x - 6} = \frac{2}{x - 3} - \frac{1}{x + 2}$$

as indicated at the beginning of this section. Check this result by combining the two partial fractions on the right side of the equation.

You can graphically check the decomposition found in Example 1. To do this, graph

$$y_1 = \frac{x + 7}{x^2 - x - 6}$$

and

$$y_2 = \frac{2}{x - 3} - \frac{1}{x + 2}$$

in the same viewing rectangle. Their graphs should be identical, as shown below.

The next example shows how to find the partial fraction decomposition for a rational function whose denominator has a repeated linear factor.

EXAMPLE 2 ◩ **Repeated Linear Factors**

Write the partial fraction decomposition for

$$\frac{5x^2 + 20x + 6}{x^3 + 2x^2 + x}.$$

Solution

Because the denominator factors as

$$x^3 + 2x^2 + x = x(x^2 + 2x + 1) = x(x + 1)^2$$

you should include one partial fraction with a constant numerator for each power of x and $(x + 1)$ and write

$$\frac{5x^2 + 20x + 6}{x(x + 1)^2} = \frac{A}{x} + \frac{B}{x + 1} + \frac{C}{(x + 1)^2}.$$

Multiplying by the LCD, $x(x + 1)^2$, leads to the basic equation

$$
\begin{aligned}
5x^2 + 20x + 6 &= A(x + 1)^2 + Bx(x + 1) + Cx. &&\text{Basic equation}\\
&= Ax^2 + 2Ax + A + Bx^2 + Bx + Cx\\
&= (A + B)x^2 + (2A + B + C)x + A. &&\text{Polynomial form}
\end{aligned}
$$

By equating coefficients of like terms on opposite sides of the equation, you obtain the following system of linear equations.

$$
\begin{aligned}
A + B &= 5\\
2A + B + C &= 20\\
A &= 6
\end{aligned}
$$

Substituting 6 for A in the first equation produces

$$
\begin{aligned}
6 + B &= 5\\
B &= -1.
\end{aligned}
$$

Substituting 6 for A and -1 for B in the second equation produces

$$
\begin{aligned}
2(6) + (-1) + C &= 20\\
C &= 9.
\end{aligned}
$$

Therefore, the partial fraction decomposition is

$$\frac{5x^2 + 20x + 6}{x(x + 1)^2} = \frac{6}{x} - \frac{1}{x + 1} + \frac{9}{(x + 1)^2}.$$

◩

The method of partial fractions was introduced by John Bernoulli (1667–1748), a Swiss mathematician who was instrumental in the early development of calculus. John Bernoulli was a professor at the University of Basel and taught many outstanding students, the most famous of whom was Leonhard Euler.

EXAMPLE 3   Distinct Linear and Quadratic Factors

Write the partial fraction decomposition for

$$\frac{x^2 + 4x + 4}{x^3 - x^2 + 2x - 2}.$$

Solution

Because the denominator factors as

$$x^3 - x^2 + 2x - 2 = (x - 1)(x^2 + 2)$$

you should include one partial fraction with a constant numerator and one partial fraction with a linear numerator and write

$$\frac{x^2 + 4x + 4}{x^3 - x^2 + 2x - 2} = \frac{A}{x - 1} + \frac{Bx + C}{x^2 + 2}.$$

Multiplying by the LCD, $(x - 1)(x^2 + 2)$, yields the basic equation

$$\begin{aligned} x^2 + 4x + 4 &= A(x^2 + 2) + (Bx + C)(x - 1) &&\text{Basic equation} \\ &= Ax^2 + 2A + Bx^2 - Bx + Cx - C \\ &= (A + B)x^2 + (-B + C)x + 2A - C. &&\text{Polynomial form} \end{aligned}$$

By equating coefficients of like terms on opposite sides of the equation, you obtain the following system of linear equations.

$$\begin{aligned} A + B \quad\quad &= 1 \\ -B + C &= 4 \\ 2A \quad\quad - C &= 4 \end{aligned}$$

Using the techniques for solving multivariable linear systems that you learned in Section 7.3, you can solve this system to find that $A = 3$, $B = -2$, and $C = 2$. Therefore, the partial fraction decomposition is

$$\frac{x^2 + 4x + 4}{x^3 - x^2 + 2x - 2} = \frac{3}{x - 1} + \frac{-2x + 2}{x^2 + 2}.$$

Note You can check a partial fraction decomposition algebraically by substituting x-values into each side of the equation. For instance, when $x = 0$, you have

$$\frac{0^2 + 4(0) + 4}{0^3 - 0^2 + 2(0) - 2} \stackrel{?}{=} \frac{3}{0 - 1} + \frac{-2(0) + 2}{0^2 + 2}$$

$$-\frac{4}{2} = -\frac{3}{1} + \frac{2}{2}. \quad \checkmark$$

The next example shows how to find the partial fraction decomposition for a rational function whose denominator has a repeated quadratic factor.

EXAMPLE 4   **Repeated Quadratic Factors**

Write the partial fraction decomposition for

$$\frac{8x^3 + 13x}{(x^2 + 2)^2}.$$

Solution
Include one partial fraction with a linear numerator for each power of $(x^2 + 2)$, and write

$$\frac{8x^3 + 13x}{(x^2 + 2)^2} = \frac{Ax + B}{x^2 + 2} + \frac{Cx + D}{(x^2 + 2)^2}.$$

Multiplying by the LCD, $(x^2 + 2)^2$, yields the basic equation

$$
\begin{aligned}
8x^3 + 13x &= (Ax + B)(x^2 + 2) + Cx + D &&\text{Basic equation}\\
&= Ax^3 + 2Ax + Bx^2 + 2B + Cx + D\\
&= Ax^3 + Bx^2 + (2A + C)x + (2B + D). &&\text{Polynomial form}
\end{aligned}
$$

Equating coefficients of like terms,

$$8x^3 + 0x^2 + 13x + 0 = Ax^3 + Bx^2 + (2A + C)x + (2B + D)$$

produces the following system of linear equations.

$$
\begin{aligned}
A &&&= 8\\
&B &&= 0\\
2A &+ C &&= 13\\
&2B + D &&= 0
\end{aligned}
$$

Substituting 8 for A in the third equation produces

$$C = -3$$

and substituting 0 for B in the fourth equation produces

$$D = 0.$$

Therefore, the partial fraction decomposition is

$$\frac{8x^3 + 13x}{(x^2 + 2)^2} = \frac{8x}{x^2 + 2} + \frac{-3x}{(x^2 + 2)^2}.$$

Group Activity   *Error Analysis*

Suppose you are tutoring a student in algebra. In trying to find a partial fraction decomposition, your student writes the following.

$$\frac{x^2 + 1}{x(x - 1)} = \frac{A}{x} + \frac{B}{x - 1}$$

$$x^2 + 1 = A(x - 1) + Bx \qquad \text{Basic equation}$$

$$x^2 + 1 = (A + B)x - A$$

Your student then forms the following system of linear equations.

$$A + B = 0$$

$$-A = 1$$

Solve the system and check the partial fraction decomposition it yields. Has your student worked the problem correctly? If not, what went wrong?

7.4 /// EXERCISES

In Exercises 1–6, write the form of the partial fraction decomposition of the rational expression. Do not solve for the constants.

1. $\dfrac{7}{x^2 - 14x}$

2. $\dfrac{x - 2}{x^2 + 4x + 3}$

3. $\dfrac{12}{x^3 - 10x^2}$

4. $\dfrac{4x^2 + 3}{(x - 5)^3}$

5. $\dfrac{2x - 3}{x^3 + 10x}$

6. $\dfrac{x - 1}{x(x^2 + 1)^2}$

In Exercises 7–30, write the partial fraction decomposition for the rational expression. Check your result algebraically.

7. $\dfrac{1}{x^2 - 1}$

8. $\dfrac{1}{4x^2 - 9}$

9. $\dfrac{1}{x^2 + x}$

10. $\dfrac{3}{x^2 - 3x}$

11. $\dfrac{1}{2x^2 + x}$

12. $\dfrac{5}{x^2 + x - 6}$

13. $\dfrac{3}{x^2 + x - 2}$

14. $\dfrac{x + 1}{x^2 + 4x + 3}$

15. $\dfrac{x^2 + 12x + 12}{x^3 - 4x}$

16. $\dfrac{x + 2}{x(x - 4)}$

17. $\dfrac{4x^2 + 2x - 1}{x^2(x + 1)}$

18. $\dfrac{2x - 3}{(x - 1)^2}$

19. $\dfrac{3x}{(x - 3)^2}$

20. $\dfrac{6x^2 + 1}{x^2(x - 1)^3}$

21. $\dfrac{x^2 - 1}{x(x^2 + 1)}$

22. $\dfrac{x}{(x - 1)(x^2 + x + 1)}$

23. $\dfrac{x^2}{x^4 - 2x^2 - 8}$

24. $\dfrac{2x^2 + x + 8}{(x^2 + 4)^2}$

25. $\dfrac{x}{16x^4 - 1}$

26. $\dfrac{x^2 - 4x + 7}{(x + 1)(x^2 - 2x + 3)}$

27. $\dfrac{x^2 + 5}{(x + 1)(x^2 - 2x + 3)}$

28. $\dfrac{x + 1}{x^3 + x}$

29. $\dfrac{x^4}{(x - 1)^3}$

30. $\dfrac{x^2 - x}{x^2 + x + 1}$

In Exercises 31–38, write the partial fraction decomposition for the rational expression. Use a graphing utility to check your result graphically.

31. $\dfrac{5 - x}{2x^2 + x - 1}$

32. $\dfrac{3x^2 - 7x - 2}{x^3 - x}$

33. $\dfrac{x - 1}{x^3 + x^2}$

34. $\dfrac{4x^2 - 1}{2x(x + 1)^2}$

35. $\dfrac{x^2 + x + 2}{(x^2 + 2)^2}$

36. $\dfrac{x^3}{(x + 2)^2(x - 2)^2}$

37. $\dfrac{2x^3 - 4x^2 - 15x + 5}{x^2 - 2x - 8}$

38. $\dfrac{x^3 - x + 3}{x^2 + x - 2}$

In Exercises 39–42, write the partial fraction decomposition for the rational expression. Check your result algebraically. Then assign a value to the constant a and check the result graphically.

39. $\dfrac{1}{a^2 - x^2}$

40. $\dfrac{1}{x(x + a)}$

41. $\dfrac{1}{y(a - y)}$

42. $\dfrac{1}{(x + 1)(a - x)}$

Graphical Analysis In Exercises 43–46, write the partial fraction decomposition for the rational function. Identify the graph of the rational function and the graphs of each term of its decomposition. State any relationship between the vertical asymptotes of the rational function and the vertical asymptotes of the terms of the decomposition.

43. $y = \dfrac{x - 12}{x(x - 4)}$

44. $y = \dfrac{2(x + 1)^2}{x(x^2 + 1)}$

45. $y = \dfrac{2(4x - 3)}{x^2 - 9}$

46. $y = \dfrac{2(4x^2 - 15x + 39)}{x^2(x^2 - 10x + 26)}$

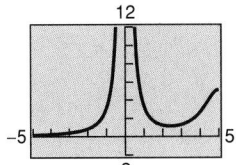

47. *Exhaust Temperatures* The magnitude of the range of exhaust temperatures in degrees Fahrenheit in an experimental diesel engine is approximated by

$$R = \dfrac{2000(4 - 3x)}{(11 - 7x)(7 - 4x)}, \qquad 0 \le x \le 1$$

where x is the relative load.

(a) Write the partial fraction decomposition for the rational function.

(b) The decomposition in part (a) is the difference of two fractions. The absolute values of the terms give the expected maximum and minimum temperatures of the exhaust gases. Use a graphing utility to graph each term.

7.5   Systems of Inequalities

The Graph of an Inequality   /   Systems of Inequalities   /   Applications

The Graph of an Inequality

The following statements are inequalities in two variables.

$$3x - 2y < 6 \qquad \text{and} \qquad 2x^2 + 3y^2 \geq 6$$

An ordered pair (a, b) is a **solution of an inequality** in x and y if the inequality is true when a and b are substituted for x and y, respectively. The **graph** of an inequality is the collection of all solutions of the inequality. To sketch the graph of an inequality, begin by sketching the graph of the *corresponding equation*. The graph of the equation will normally separate the plane into two or more regions. In each such region, one of the following must be true.

1. *All* points in the region are solutions of the inequality.
2. *No* point in the region is a solution of the inequality.

Thus, you can determine whether the points in an entire region satisfy the inequality by simply testing *one* point in the region.

Sketching the Graph of an Inequality in Two Variables

1. Replace the inequality sign with an equal sign, and sketch the graph of the resulting equation. (Use a dashed line for $<$ or $>$ and a solid line for $\leq$ or $\geq$.)
2. Test one point in each of the regions formed by the graph in Step 1. If the point satisfies the inequality, shade the entire region to denote that every point in the region satisfies the inequality.

Figure 7.17

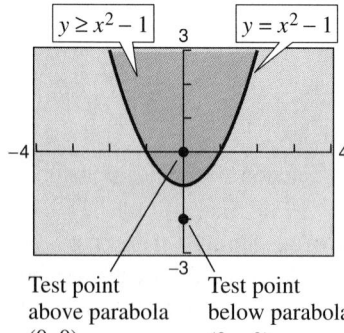

Test point
above parabola
$(0, 0)$

Test point
below parabola
$(0, -2)$

EXAMPLE 1   Sketching the Graph of an Inequality

Sketch the graph of $y \geq x^2 - 1$.

Solution
The graph of the corresponding *equation* $y = x^2 - 1$ is a parabola, as shown in Figure 7.17. By testing the point $(0, 0)$ *above* the parabola and the point $(0, -2)$ *below* the parabola, you can see that the points that satisfy the inequality are those lying above (or on) the parabola.

The inequality given in Example 1 is a nonlinear inequality in two variables. Most of the following examples involve **linear inequalities** such as $ax + by < c$. The graph of a linear inequality is a half-plane lying on one side of the line $ax + by = c$.

EXAMPLE 2   Sketching the Graphs of Linear Inequalities

Sketch the graph of each linear inequality.

a. $x > -2$ **b.** $y \le 3$

Solution

a. The graph of the corresponding equation $x = -2$ is a vertical line. The points that satisfy the inequality $x > -2$ are those lying to the right of this line, as shown in Figure 7.18.

b. The graph of the corresponding equation $y = 3$ is a horizontal line. The points that satisfy the inequality $y \le 3$ are those lying below (or on) this line, as shown in Figure 7.19.

Figure 7.18   **Figure 7.19**

Figure 7.20

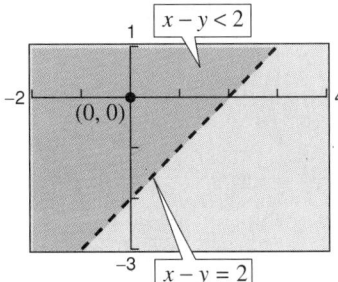

EXAMPLE 3   Sketching the Graph of a Linear Inequality

Sketch the graph of $x - y < 2$.

Solution

The graph of the corresponding equation $x - y = 2$ is a line, as shown in Figure 7.20. Because the origin $(0, 0)$ satisfies the inequality, the graph consists of the half-plane lying above the line. (Try checking a point below the line. Regardless of which point you choose, you will see that it does not satisfy the inequality.)

Systems of Inequalities

Many practical problems in business, science, and engineering involve systems of linear inequalities. A **solution** of a system of inequalities in x and y is a point (x, y) that satisfies each inequality in the system.

To sketch the graph of a system of inequalities in two variables, first sketch the graph of each individual inequality (on the same coordinate system) and then find the region that is *common* to every graph in the system. For systems of *linear* inequalities, it is helpful to find the vertices of the solution region.

EXAMPLE 4   **Solving a System of Inequalities**

Sketch the graph (and label the vertices) of the solution set of the system.

$$x - y < 2$$
$$x > -2$$
$$y \le 3$$

Solution

The graphs of these inequalities are shown in Figures 7.18 to 7.20. The triangular region common to all three graphs can be found by superimposing the graphs on the same coordinate system, as shown in Figure 7.21. To find the vertices of the region, solve the three systems of corresponding equations obtained by taking *pairs* of equations representing the boundaries of the individual regions.

Vertex A: $(-2, -4)$ *Vertex B:* $(5, 3)$ *Vertex C:* $(-2, 3)$

$$x - y = 2 \qquad\qquad x - y = 2 \qquad\qquad x = -2$$
$$x = -2 \qquad\qquad\quad y = 3 \qquad\qquad\qquad y = 3$$

Figure 7.21

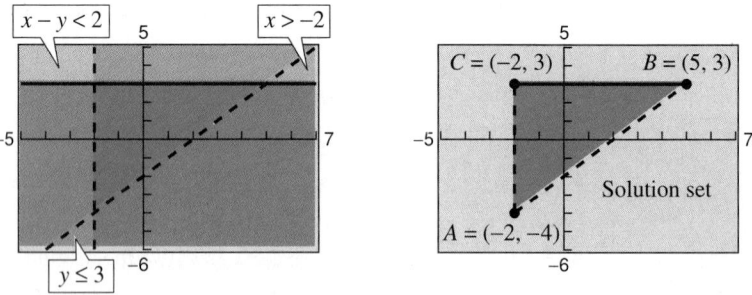

For the triangular region shown in Figure 7.21, each point of intersection of a pair of boundary lines corresponds to a vertex. With more complicated regions, two border lines can sometimes intersect at a point that is not a vertex of the region, as shown in Figure 7.22. To keep track of which points of intersection are actually vertices of the region, we suggest that you sketch the region and refer to your sketch as you find each point of intersection.

Figure 7.22

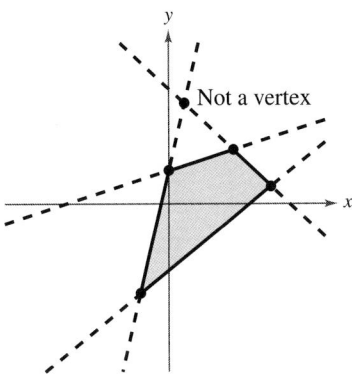

Not a vertex

EXAMPLE 5   **Solving a System of Inequalities**

Sketch the region containing all points that satisfy the system.

$$x^2 - y \leq 1$$
$$-x + y \leq 1$$

Solution

As shown in Figure 7.23, the points that satisfy the inequality $x^2 - y \leq 1$ are the points lying above (or on) the parabola given by

$$y = x^2 - 1. \qquad \text{Parabola}$$

Figure 7.23

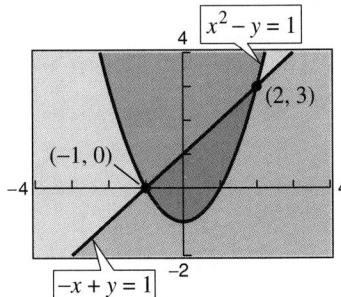

The points that satisfy the inequality $-x + y \leq 1$ are the points lying below (or on) the line given by

$$y = x + 1. \qquad \text{Line}$$

To find the points of intersection of the parabola and the line, solve the system of corresponding equations.

$$x^2 - y = 1$$
$$-x + y = 1$$

Using the method of substitution, you can find the solutions to be $(-1, 0)$ and $(2, 3)$, as shown in Figure 7.23.

When solving a system of inequalities, you should be aware that the system might have no solution. For instance, the system

$$x + y > 3$$
$$x + y < -1$$

has no solution points, because the quantity $(x + y)$ cannot be both less than -1 and greater than 3, as shown in Figure 7.24.

Figure 7.24   No Solution

Figure 7.25   Unbounded Region

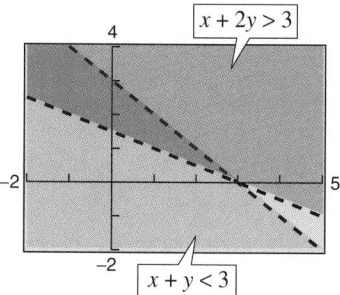

Another possibility is that the solution set of a system of inequalities can be unbounded. For instance, the solution set of

$$x + y < 3$$
$$x + 2y > 3$$

forms an *infinite wedge,* as shown in Figure 7.25.

A graphing utility can be used to graph an inequality. For instance, to graph $y \geq x^2 - 2$ on the *TI-83,* you can use the following steps.

1. Press $\boxed{\text{Y=}}$ and enter $x^2 - 2$ for Y_1.
2. Cursor to the icon on the left of Y_1.
3. Press $\boxed{\text{ENTER}}$ until the ◣ icon appears.
4. $\boxed{\text{GRAPH}}$

The graph is shown at the left. Try using a graphing utility to graph the following inequalities.

a. $y \leq 2x + 2$ **b.** $y \geq \dfrac{1}{2}x^2 - 4$

Figure 7.26

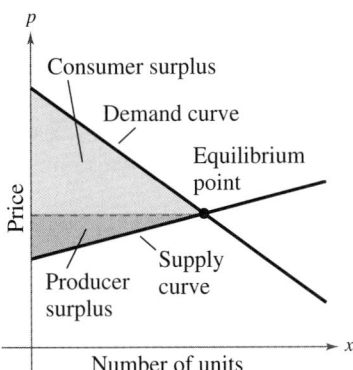

Applications

Example 8 in Section 7.2 discussed the *point of equilibrium* for a demand function and a supply function. The next example discusses two related concepts that economists call **consumer surplus** and **producer surplus.** As shown in Figure 7.26, the consumer surplus is defined as the area of the region that lies *below* the demand curve, *above* the horizontal line passing through the equilibrium point, and to the right of the *p*-axis. Similarly, the producer surplus is defined as the area of the region that lies *above* the supply curve, *below* the horizontal line passing through the equilibrium point, and to the right of the *p*-axis. The consumer surplus is a measure of the amount that consumers would have been willing to pay *above what they actually paid,* whereas the producer surplus is a measure of the amount that producers would have been willing to receive *below what they actually received.*

Real Life

EXAMPLE 6 ▬ **Consumer Surplus and Producer Surplus**

The demand and supply functions for a certain type of calculator are given by

$$p = 150 - 0.00001x \qquad \text{Demand equation}$$
$$p = 60 + 0.00002x \qquad \text{Supply equation}$$

where *p* is the price in dollars and *x* represents the number of units. Find the consumer surplus and producer surplus for these two equations.

Solution

Begin by finding the point of equilibrium by solving the equation

$$60 + 0.00002x = 150 - 0.00001x.$$

In Example 8 in Section 7.2, you saw that the solution is $x = 3{,}000{,}000$, which corresponds to an equilibrium price of $p = \$120$. Thus, the consumer surplus and producer surplus are the areas of the following triangular regions.

Consumer Surplus	*Producer Surplus*
$p \le 150 - 0.00001x$	$p \ge 60 + 0.00002x$
$p \ge 120$	$p \le 120$
$x \ge 0$	$x \ge 0$

In Figure 7.27, you can see that the consumer and producer surpluses are

Figure 7.27

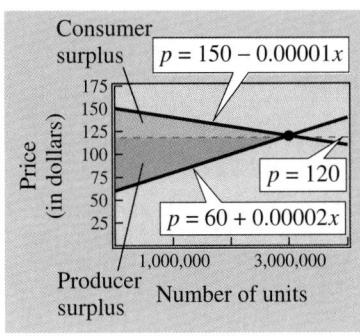

$$\text{Consumer surplus} = \tfrac{1}{2}(\text{base})(\text{height}) = \tfrac{1}{2}(3{,}000{,}000)(30) = \$45{,}000{,}000$$

$$\text{Producer surplus} = \tfrac{1}{2}(\text{base})(\text{height}) = \tfrac{1}{2}(3{,}000{,}000)(60) = \$90{,}000{,}000.$$

▬

Real Life

EXAMPLE 7 ▱ Nutrition

The liquid portion of a diet is to provide at least 300 calories, 36 units of vitamin A, and 90 units of vitamin C daily. A cup of dietary drink X provides 60 calories, 12 units of vitamin A, and 10 units of vitamin C. A cup of dietary drink Y provides 60 calories, 6 units of vitamin A, and 30 units of vitamin C. Set up a system of linear inequalities that describes the minimum daily requirements for calories and vitamins.

Solution

Begin by letting x and y represent the following.

x = number of cups of dietary drink X

y = number of cups of dietary drink Y

To meet the minimum daily requirements, the following inequalities must be satisfied.

For Calories: $60x + 60y \geq 300$

For Vitamin A: $12x + 6y \geq 36$

For Vitamin C: $10x + 30y \geq 90$

$x \geq 0$

$y \geq 0$

The last two inequalities are included because x and y cannot be negative. The graph of this system of inequalities is shown in Figure 7.28. (More is said about this application in Example 7 in Section 7.6.)   ▱

Figure 7.28

Group Activity

Graphing Systems of Inequalities

The Technology feature on page 566 describes how to sketch the graph of a single inequality with a graphing utility. With the user's guide for your graphing utility, find how to graph a *system* of inequalities. Then use the graphing utility to graph the following systems.

a. $y \leq 4 - x^2$

$y \geq 2x - 3$

b. $y \leq 4 - x^2$

$y \geq x^2 - 4$

7.5 /// EXERCISES

In Exercises 1–8, match the inequality with its graph. [The graphs are labeled (a), (b), (c), (d), (e), (f), (g), and (h).]

(a)

(b)

(c)

(d)

(e)

(f)

(g)

(h)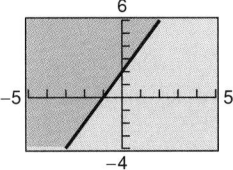

1. $x < 2$

2. $y \geq 3$

3. $2x + 3y \geq 6$

4. $2x - y \leq -2$

5. $x^2 + y^2 < 9$

6. $(x - 2)^2 + (y - 3)^2 > 9$

7. $xy > 1$

8. $y \leq 1 - x^2$

In Exercises 9–18, sketch the graph of the inequality.

9. $x \geq 2$

10. $x \leq 4$

11. $y \geq -1$

12. $y \leq 3$

13. $y < 2 - x$

14. $y > 2x - 4$

15. $2y - x \geq 4$

16. $5x + 3y \geq -15$

17. $y^2 - x < 0$

18. $(x + 1)^2 + y^2 < 9$

In Exercises 19–24, use a graphing utility to graph the inequality. Shade the region representing the solution.

19. $y(1 + x^2) \leq 1$

20. $y < \ln x$

21. $y \geq \frac{2}{3}x - 1$

22. $y \leq 6 - \frac{3}{2}x$

23. $x^2 + 5y - 10 \leq 0$

24. $2x^2 - y - 3 > 0$

In Exercises 25–30, write an inequality for the shaded region shown in the figure.

25.

26.

27.

28.

29.

30.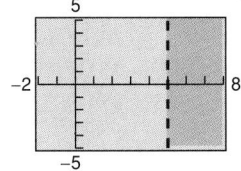

In Exercises 31–44, sketch the graph of the solution of the system of inequalities. Verify your sketch by using a graphing utility to shade the region representing the solution of the system.

31. $x + y \leq 1$
$-x + y \leq 1$
$y \geq 0$

32. $3x + 2y < 6$
$x \qquad > 0$
$y > 0$

33. $x + y \leq 5$
$x \qquad \geq 2$
$y \geq 0$

34. $2x^2 + y \geq 2$
$x \qquad \leq 2$
$y \leq 1$

35. $-3x + 2y < \quad 6$
$x - 4y > -2$
$2x + \; y < \quad 3$

36. $x - 7y > -36$
$5x + 2y > \quad 5$
$6x - 5y > \quad 6$

37. $2x + \; y > 2$
$6x + 3y < 2$

38. $x - 2y < -6$
$5x - 3y > -9$

39. $x \qquad \geq 1$
$x - 2y \leq 3$
$3x + 2y \geq 9$
$x + \; y \leq 6$

40. $x - y^2 > 0$
$x - \; y < 2$

41. $x^2 + y^2 \leq 9$
$x^2 + y^2 \geq 1$

42. $x^2 + y^2 \leq 25$
$4x - 3y \leq \; 0$

43. $x > y^2$
$x < y + 2$

44. $x < 2y - y^2$
$0 < x + \; y$

In Exercises 45–50, use a graphing utility to graph the inequalities. Shade the region representing the solution of the system.

45. $y \leq \sqrt{3x} + 1$
$y \geq x^2 + 1$

46. $y < -x^2 + 2x + 3$
$y > \quad x^2 - 4x + 3$

47. $y < x^3 - 2x + 1$
$y > -2x$
$x \leq 1$

48. $y \geq x^4 - 2x^2 + 1$
$y \leq 1 - x^2$

49. $x^2 y \geq 1$
$0 < x \leq 4$
$y \leq 4$

50. $y \leq e^{-x^2/2}$
$y \geq 0$
$-2 \leq x \leq 2$

In Exercises 51–60, derive a set of inequalities to describe the region.

51.

52.

53.

54.

55.

56.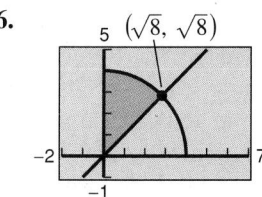

57. Rectangular region with vertices at $(2, 1)$, $(5, 1)$, $(5, 7)$, and $(2, 7)$

58. Parallelogrammatic region with vertices at $(0, 0)$, $(4, 0)$, $(1, 4)$, and $(5, 4)$

59. Triangular region with vertices at $(0, 0)$, $(5, 0)$, and $(2, 3)$

60. Triangular region with vertices at $(-1, 0)$, $(1, 0)$, and $(0, 1)$

In Exercises 61–67, (a) find a system of inequalities that models the problem, and (b) use a graphing utility to shade the region representing the solution of the system.

61. *Investment* A person plans to invest $20,000 in two different interest-bearing accounts. Each account is to contain at least $5000, and one account should have at least twice the amount that is in the other account.

62. *Concert Ticket Sales* Two types of tickets are to be sold for a concert. One type costs $15 per ticket and the other costs $25 per ticket. The promoter of the concert must sell at least 15,000 tickets, including at least 8000 of the $15 tickets and at least 4000 of the $25 tickets, and the gross receipts must total at least $275,000 in order for the concert to be held.

63. *Furniture Production* A furniture company can sell all the tables and chairs it produces. Each table requires 1 hour of assembly and $1\frac{1}{3}$ hours of finishing. Each chair requires $1\frac{1}{2}$ hours of assembly and $1\frac{1}{2}$ hours of finishing. The company's assembly center is available 12 hours per day, and its finishing center is available 15 hours per day.

64. *Computer Inventory* A store sells two models of a certain brand of computer. Because of the demand, it is necessary to stock twice as many units of model A as units of model B. The costs to the store for the two models are $800 and $1200, respectively. The management does not want more than $20,000 in computer inventory at any one time, and it wants at least four model A computers and two model B computers in inventory at all times.

65. *Diet Supplement* A dietitian is asked to design a special diet supplement using two different foods. Each ounce of food X contains 20 units of calcium, 15 units of iron, and 10 units of vitamin B. Each ounce of food Y contains 10 units of calcium, 10 units of iron, and 20 units of vitamin B. The minimum daily requirements in the diet are 280 units of calcium, 160 units of iron, and 180 units of vitamin B.

66. *Diet Supplement* A dietitian is asked to design a special diet supplement using two different foods. Each ounce of food X contains 20 units of calcium, 15 units of iron, and 10 units of vitamin B. Each ounce of food Y contains 10 units of calcium, 10 units of iron, and 20 units of vitamin B. The minimum daily requirements in the diet are 300 units of calcium, 150 units of iron, and 200 units of vitamin B.

67. *Physical Fitness Facility* You plan an indoor running track with an exercise floor inside the track (see figure). The track must be at least 125 meters long, and the exercise floor must have an area of at least 500 square meters.

Figure for 67

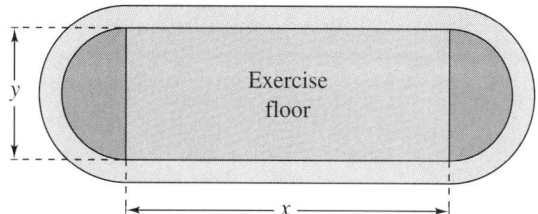

68. *Graphical Reasoning* Two concentric circles have radii of x and y meters, where $y > x$ (see figure). The area between the boundaries of the circles must be at least 10 square meters.

(a) Find an inequality describing the constraints on the circles.

(b) Use a graphing utility to graph the inequality in part (a). Graph the line $y = x$ in the same viewing rectangle.

(c) Identify the graph of the line in relation to the boundary of the inequality. Explain its meaning in the context of the problem.

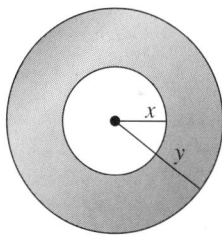

Consumer and Producer Surpluses **In Exercises 69–72, use a graphing utility to shade the regions representing the consumer surplus and producer surplus for the supply and demand equations.**

Demand	*Supply*
69. $p = 50 - 0.5x$	$p = 0.125x$
70. $p = 60 - x$	$p = 10 + \dfrac{7}{3}x$
71. $p = 300 - x$	$p = 100 + x$
72. $p = 140 - 0.00002x$	$p = 80 + 0.00001x$

73. *Think About It* After graphing the boundary of an inequality in x and y, how do you decide on which side of the boundary the solution set of the inequality lies?

7.6   Linear Programming

Linear Programming: A Graphical Approach / *Applications*

Linear Programming: A Graphical Approach

Many applications in business and economics involve a process called **optimization,** in which you are asked to find the minimum or maximum value of a quantity. In this section you will study an optimization strategy called **linear programming.**

A two-dimensional linear programming problem consists of a linear **objective function** and a system of linear inequalities called **constraints.** The objective function gives the quantity that is to be maximized (or minimized), and the constraints determine the set of **feasible solutions.** For example, suppose you are asked to maximize the value of

$$z = ax + by \qquad\qquad \text{Objective function}$$

subject to a set of constraints that determines the region in Figure 7.29. Because every point in the region satisfies each constraint, it is not clear how you should go about finding the point that yields a maximum value of z. Fortunately, it can be shown that if there is an optimal solution, it must occur at one of the vertices. This means that *you can find the maximum value by testing z at each of the vertices.*

Figure 7.29

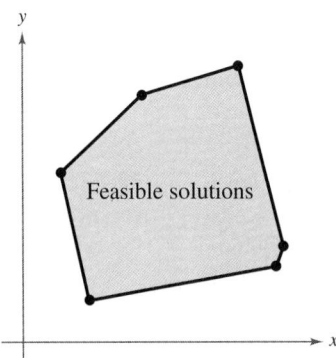

Feasible solutions

Optimal Solution of a Linear Programming Problem
If a linear programming problem has a solution, it must occur at a vertex of the set of feasible solutions. If there is more than one solution, at least one of them must occur at such a vertex. In either case, the value of the objective function is unique.

EXAMPLE 1   Solving a Linear Programming Problem

Find the maximum value of

$$z = 3x + 2y$$ Objective function

subject to the following constraints.

$$\left.\begin{array}{r} x \geq 0 \\ y \geq 0 \\ x + 2y \leq 4 \\ x - y \leq 1 \end{array}\right\}$$ Constraints

Solution

The constraints form the region shown in Figure 7.30. At the four vertices of this region, the objective function has the following values.

At $(0, 0)$: $z = 3(0) + 2(0) = 0$
At $(1, 0)$: $z = 3(1) + 2(0) = 3$
At $(2, 1)$: $z = 3(2) + 2(1) = 8$ Maximum value of z
At $(0, 2)$: $z = 3(0) + 2(2) = 4$

Thus, the maximum value of z is 8, and this occurs when $x = 2$ and $y = 1$.

Figure 7.30

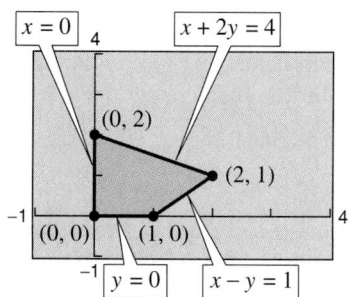

Note   In Example 1, try testing some of the *interior* points in the region. You will see that the corresponding values of z are less than 8. Here are some examples.

At $(1, 1)$: $z = 3(1) + 2(1) = 5$
At $\left(1, \frac{1}{2}\right)$: $z = 3(1) + 2\left(\frac{1}{2}\right) = 4$
At $\left(\frac{1}{2}, \frac{3}{2}\right)$: $z = 3\left(\frac{1}{2}\right) + 2\left(\frac{3}{2}\right) = \frac{9}{2}$

To see why the maximum value of the objective function in Example 1 must occur at a vertex, consider writing the objective function in the form

$$y = -\frac{3}{2}x + \frac{z}{2}$$ Family of lines

where $z/2$ is the y-intercept of the objective function. This equation represents a family of lines, each of slope $-\frac{3}{2}$. Of these infinitely many lines, you want the one that has the largest z-value while still intersecting the region determined by the constraints. In other words, of all the lines with a slope of $-\frac{3}{2}$, you want the one that has the largest y-intercept *and* intersects the given region, as shown in Figure 7.31. It should be clear that such a line will pass through one (or more) of the vertices of the region.

Figure 7.31

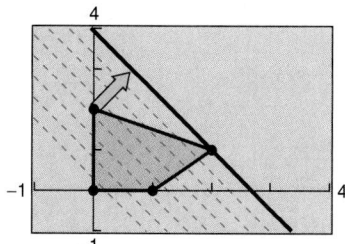

Solving a Linear Programming Problem

To solve a linear programming problem involving two variables by the graphical method, use the following steps.

1. Sketch the region corresponding to the system of constraints. (The points inside or on the boundary of the region are called *feasible solutions*.)
2. Find the vertices of the region.
3. Test the objective function at each of the vertices and select the values of the variables that optimize the objective function. For a bounded region, both a minimum and a maximum value will exist. (For an unbounded region, *if* an optimal solution exists, it will occur at a vertex.)

These guidelines will work whether the objective function is to be maximized or minimized. For instance, the same test used in Example 1 to find the maximum value of z can be used to conclude that the minimum value of z is 0 and that this value occurs at the vertex $(0, 0)$.

EXAMPLE 2   **Solving a Linear Programming Problem**

Find the maximum value of

$$z = 4x + 6y \qquad \text{Objective function}$$

where $x \geq 0$ and $y \geq 0$, subject to the following constraints.

$$\left. \begin{array}{r} -x + y \leq 11 \\ x + y \leq 27 \\ 2x + 5y \leq 90 \end{array} \right\} \quad \text{Constraints}$$

Solution

The region bounded by the constraints is shown in Figure 7.32. By testing the objective function at each vertex, you obtain the following.

At $(0, 0)$: $z = 4(0) + 6(0) = 0$
At $(0, 11)$: $z = 4(0) + 6(11) = 66$
At $(5, 16)$: $z = 4(5) + 6(16) = 116$
At $(15, 12)$: $z = 4(15) + 6(12) = 132$ Maximum value of z
At $(27, 0)$: $z = 4(27) + 6(0) = 108$

Thus, the maximum value of z is 132, and this occurs when $x = 15$ and $y = 12$.

Figure 7.32

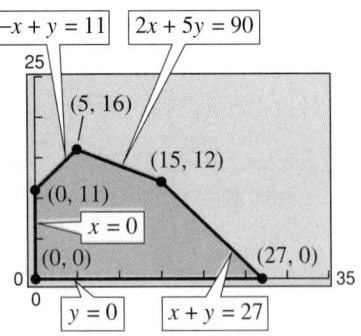

The next example shows that the same basic procedure can be used to solve a problem in which the objective function is to be *minimized*.

EXAMPLE 3 ▱ **Minimizing an Objective Function**

Find the minimum value of

$$z = 5x + 7y \qquad \text{Objective function}$$

where $x \geq 0$ and $y \geq 0$, subject to the following constraints.

$$\left.\begin{array}{r} 2x + 3y \geq 6 \\ 3x - y \leq 15 \\ -x + y \leq 4 \\ 2x + 5y \leq 27 \end{array}\right\} \qquad \text{Constraints}$$

Solution

The region bounded by the constraints is shown in Figure 7.33. By testing the objective function at each vertex, you obtain the following.

At $(0, 2)$: $\quad z = 5(0) + 7(2) = 14 \qquad$ Minimum value of z

At $(0, 4)$: $\quad z = 5(0) + 7(4) = 28$

At $(1, 5)$: $\quad z = 5(1) + 7(5) = 40$

At $(6, 3)$: $\quad z = 5(6) + 7(3) = 51$

At $(5, 0)$: $\quad z = 5(5) + 7(0) = 25$

At $(3, 0)$: $\quad z = 5(3) + 7(0) = 15$

Thus, the minimum value of z is 14, and this occurs when $x = 0$ and $y = 2$. ▱

Figure 7.33

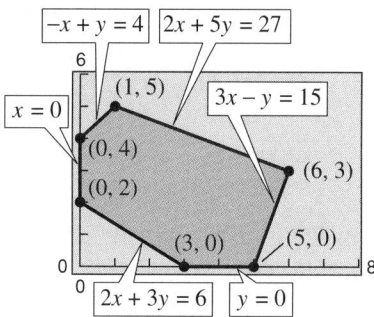

EXAMPLE 4 ▱ **Maximizing an Objective Function**

Find the maximum value of

$$z = 5x + 7y \qquad \text{Objective function}$$

where $x \geq 0$ and $y \geq 0$, subject to the following constraints.

$$\left.\begin{array}{r} 2x + 3y \geq 6 \\ 3x - y \leq 15 \\ -x + y \leq 4 \\ 2x + 5y \leq 27 \end{array}\right\} \qquad \text{Constraints}$$

Solution

This linear programming problem is identical to that given in Example 3 above, except that the objective function is *maximized* instead of minimized. Using the values of z at the vertices shown above, you can conclude that the maximum value of z is 51, and that this value occurs when $x = 6$ and $y = 3$. ▱

Figure 7.34

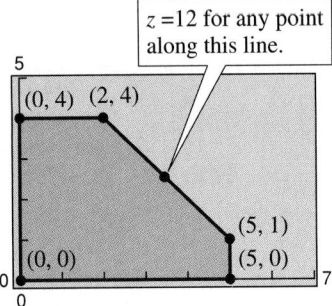

It is possible for the maximum (or minimum) value in a linear programming problem to occur at *two* different vertices. For instance, at the vertices of the region shown in Figure 7.34, the objective function

$$z = 2x + 2y \qquad \text{Objective function}$$

has the following values.

At $(0, 0)$: $z = 2(0) + 2(0) = 0$
At $(0, 4)$: $z = 2(0) + 2(4) = 8$
At $(2, 4)$: $z = 2(2) + 2(4) = 12$ Maximum value of z
At $(5, 1)$: $z = 2(5) + 2(1) = 12$ Maximum value of z
At $(5, 0)$: $z = 2(5) + 2(0) = 10$

In this case, you can conclude that the objective function has a maximum value (of 12) not only at the vertices $(2, 4)$ and $(5, 1)$, but also at *any point on the line segment connecting these two vertices,* as shown in Figure 7.34. Note that the objective function

$$y = -x + \tfrac{1}{2}z$$

has the same slope as the line through the vertices $(2, 4)$ and $(5, 1)$.

Some linear programming problems have no optimal solution. This can occur if the region determined by the constraints is *unbounded.* Example 5 illustrates such a problem.

EXAMPLE 5   **An Unbounded Region**

Find the maximum value of

$$z = 4x + 2y \qquad \text{Objective function}$$

where $x \geq 0$ and $y \geq 0$, subject to the following constraints.

$$\left. \begin{array}{r} x + 2y \geq 4 \\ 3x + y \geq 7 \\ -x + 2y \leq 7 \end{array} \right\} \quad \text{Constraints}$$

Figure 7.35

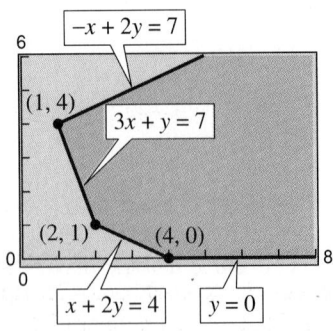

Solution
The region determined by the constraints is shown in Figure 7.35. For this unbounded region, there is no maximum value of z. To see this, note that the point $(x, 0)$ lies in the region for all values of $x \geq 4$. By choosing large values of x, you can obtain values of

$$z = 4(x) + 2(0) = 4x$$

that are as large as you want. Thus, there is no maximum value of z. For this problem, there *is* a minimum value of $z = 10$, which occurs at the vertex $(2, 1)$.

Applications

Example 6 shows how linear programming can be used to find the maximum profit in a business application.

Real Life

EXAMPLE 6 ▱ **Maximum Profit**

A manufacturer wants to maximize the profit for two products. Product I yields a profit of $1.50 per unit, and product II yields a profit of $2.00 per unit. Market tests and available resources have indicated the following constraints.

1. The combined production level should not exceed 1200 units per month.
2. The demand for product II is no more than half the demand for product I.
3. The production level of product I is less than or equal to 600 units plus three times the production level of product II.

Solution

If you let x be the number of units of product I and y be the number of units of product II, the objective function (for the combined profit) is given by

$$P = 1.5x + 2y. \qquad \text{Objective function}$$

The three constraints translate into the following linear inequalities.

1. $x + y \le 1200$ ⟹ $x + y \le 1200$

2. $y \le \frac{1}{2}x$ ⟹ $-x + 2y \le 0$

3. $x \le 3y + 600$ ⟹ $x - 3y \le 600$

Because neither x nor y can be negative, you also have the two additional constraints of $x \ge 0$ and $y \ge 0$. Figure 7.36 shows the region determined by the constraints. To find the maximum profit, test the value of P at the vertices of the region.

At $(0, 0)$: $P = 1.5(0) \quad + 2(0) \quad = \quad 0$
At $(800, 400)$: $P = 1.5(800) + 2(400) = 2000$ Maximum profit
At $(1050, 150)$: $P = 1.5(1050) + 2(150) = 1875$
At $(600, 0)$: $P = 1.5(600) \quad + 2(0) \quad = \quad 900$

Thus, the maximum profit is $2000, and it occurs when the monthly production consists of 800 units of product I and 400 units of product II. ▱

Figure 7.36

Note In Example 6, suppose the manufacturer improved the production of product I so that it yielded a profit of $2.50 per unit. How would this affect the number of units the manufacturer should sell to obtain a maximum profit?

Real Life

EXAMPLE 7 ▬ Minimum Cost

The liquid portion of a diet is to provide at least 300 calories, 36 units of vitamin A, and 90 units of vitamin C daily. A cup of dietary drink X costs $0.12 and provides 60 calories, 12 units of vitamin A, and 10 units of vitamin C. A cup of dietary drink Y costs $0.15 and provides 60 calories, 6 units of vitamin A, and 30 units of vitamin C. How many cups of each drink should be consumed each day to minimize the cost and still meet the daily requirements?

Solution

As in Example 7 on page 568, let x be the number of cups of dietary drink X and let y be the number of cups of dietary drink Y.

Figure 7.37

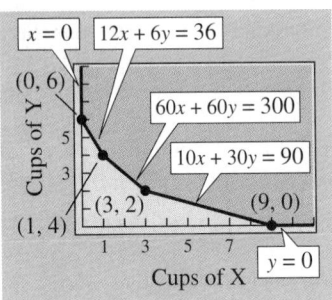

$$
\left.
\begin{array}{ll}
\textit{For Calories:} & 60x + 60y \geq 300 \\
\textit{For Vitamin A:} & 12x + 6y \geq 36 \\
\textit{For Vitamin C:} & 10x + 30y \geq 90 \\
& x \geq 0 \\
& y \geq 0
\end{array}
\right\} \quad \text{Constraints}
$$

The cost C is given by $C = 0.12x + 0.15y$.   Objective function

The graph of the region determined by the constraints is shown in Figure 7.37. To determine the minimum cost, test C at each vertex of the region.

At $(0, 6)$: $C = 0.12(0) + 0.15(6) = 0.90$

At $(1, 4)$: $C = 0.12(1) + 0.15(4) = 0.72$

At $(3, 2)$: $C = 0.12(3) + 0.15(2) = 0.66$ Minimum value of C

At $(9, 0)$: $C = 0.12(9) + 0.15(0) = 1.08$

Thus, the minimum cost is $0.66 per day, and this occurs when three cups of drink X and two cups of drink Y are consumed each day.   ▬

Group Activity

Creating a Linear Programming Problem

Sketch the region determined by the constraints $x \geq 0$, $y \geq 0$, $x + 2y \leq 10$, and $x + y \leq 7$. Find, if possible, an objective function of the form $z = ax + by$ that has a maximum at the indicated vertex of the region.

a. Maximum at $(0, 5)$ **b.** Maximum at $(4, 3)$

c. Maximum at $(7, 0)$ **d.** Maximum at $(0, 0)$

7.6 /// EXERCISES

In Exercises 1–12, find the minimum and maximum values of the objective function, subject to the indicated constraints. (For each exercise, the graph of the region determined by the constraints is provided.)

1. Objective function:

$z = 4x + 5y$

Constraints:

$$x \geq 0$$
$$y \geq 0$$
$$x + y \leq 6$$

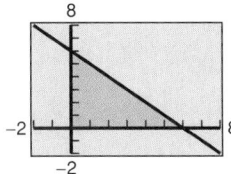

2. Objective function:

$z = 2x + 8y$

Constraints:

$$x \geq 0$$
$$y \geq 0$$
$$2x + y \leq 4$$

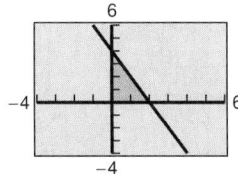

3. Objective function:

$z = 10x + 6y$

Constraints:
(See Exercise 1.)

4. Objective function:

$z = 7x + 3y$

Constraints:
(See Exercise 2.)

5. Objective function:

$z = 3x + 2y$

Constraints:

$$x \geq 0$$
$$y \geq 0$$
$$x + 3y \leq 15$$
$$4x + y \leq 16$$

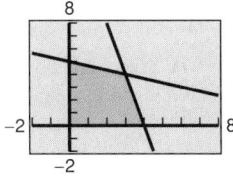

6. Objective function:

$z = 4x + 3y$

Constraints:

$$x \geq 0$$
$$2x + 3y \geq 6$$
$$3x - 2y \leq 9$$
$$x + 5y \leq 20$$

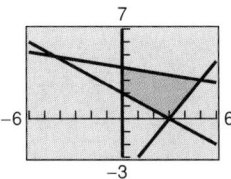

7. Objective function:

$z = 5x + 0.5y$

Constraints:
(See Exercise 5.)

8. Objective function:

$z = x + 6y$

Constraints:
(See Exercise 6.)

9. Objective function:

$z = 10x + 7y$

Constraints:

$$0 \leq x \leq 60$$
$$0 \leq y \leq 45$$
$$5x + 6y \leq 420$$

10. Objective function:

$z = 50x + 35y$

Constraints:

$$x \geq 0$$
$$y \geq 0$$
$$8x + 9y \leq 7200$$
$$8x + 9y \geq 5400$$

11. Objective function:

$z = 25x + 30y$

Constraints:
(See Exercise 9.)

12. Objective function:

$z = 16x + 18y$

Constraints:
(See Exercise 10.)

In Exercises 13–20, sketch the constraint region. Then find the minimum and maximum values of the objective function, subject to the constraints.

13. Objective function:

$z = 6x + 10y$

Constraints:

$$x \geq 0$$
$$y \geq 0$$
$$2x + 5y \leq 10$$

14. Objective function:

$z = 7x + 8y$

Constraints:

$$x \geq 0$$
$$y \geq 0$$
$$x + \tfrac{1}{2}y \leq 4$$

15. Objective function:

$z = 9x + 24y$

Constraints:
(See Exercise 13.)

16. Objective function:

$z = 7x + 2y$

Constraints:
(See Exercise 14.)

17. Objective function:

$z = 4x + 5y$

Constraints:

$x \geq 0$

$y \geq 0$

$x + y \geq 8$

$3x + 5y \geq 30$

18. Objective function:

$z = 4x + 5y$

Constraints:

$x \geq 0$

$y \geq 0$

$2x + 2y \leq 10$

$x + 2y \leq 6$

19. Objective function:

$z = 2x + 7y$

Constraints:
(See Exercise 17.)

20. Objective function:

$z = 2x - y$

Constraints:
(See Exercise 18.)

In Exercises 21–26, use a graphing utility to sketch the region determined by the constraints. Then find the minimum and maximum values of the objective function, subject to the constraints.

21. Objective function:

$z = 4x + y$

Constraints:

$x \geq 0$

$y \geq 0$

$x + 2y \leq 40$

$2x + 3y \geq 72$

22. Objective function:

$z = x$

Constraints:

$x \geq 0$

$y \geq 0$

$2x + 3y \leq 60$

$2x + y \leq 28$

$4x + y \leq 48$

23. Objective function:

$z = x + 4y$

Constraints:
(See Exercise 21.)

24. Objective function:

$z = y$

Constraints:
(See Exercise 22.)

25. Objective function:

$z = 2x + 3y$

Constraints:
(See Exercise 21.)

26. Objective function:

$z = 3x + 2y$

Constraints:
(See Exercise 22.)

Exploration In Exercises 27–30, (a) use a graphing utility to graph the region bounded by the following constraints.

$$3x + y \leq 15$$
$$4x + 3y \leq 30$$
$$x \geq 0$$
$$y \geq 0$$

(b) Graph the objective function for the given maximum value of z in the same viewing rectangle as the graph of the constraints. (c) Use the graph to determine the feasible point or points that yield the maximum. Explain how you arrived at your answer.

Objective Function	Maximum
27. $z = 2x + y$	$z = 12$
28. $z = 5x + y$	$z = 25$
29. $z = x + y$	$z = 10$
30. $z = 3x + y$	$z = 15$

Exploration In Exercises 31–34, (a) use a graphing utility to graph the region bounded by the following constraints.

$$x + 4y \leq 20$$
$$x + y \leq 8$$
$$3x + 2y \leq 21$$
$$x \geq 0$$
$$y \geq 0$$

(b) Graph the objective function for the given maximum value of z in the same viewing rectangle as the graph of the constraints. (c) Use the graph to determine the feasible point or points that yield the maximum. Explain how you arrived at your answer.

Objective Function	Maximum
31. $z = x + 5y$	$z = 25$
32. $z = 2x + 4y$	$z = 24$
33. $z = 4x + 5y$	$z = 36$
34. $z = 4x + y$	$z = 28$

Think About It   In Exercises 35–38, find an objective function that has a maximum or minimum value at the indicated vertex of the constraint region shown below. (There are many correct answers.)

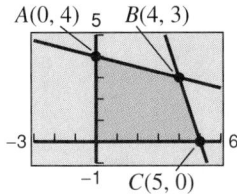

$A(0, 4)$ 5 $B(4, 3)$

-3

6

-1 $C(5, 0)$

35. The maximum occurs at vertex A.

36. The maximum occurs at vertex B.

37. The maximum occurs at vertex C.

38. The minimum occurs at vertex C.

39. *Maximum Profit*   A merchant plans to sell two models of compact disc players at costs of $250 and $400. The $250 model yields a profit of $45, and the $400 model yields a profit of $50. The merchant estimates that the total monthly demand will not exceed 250 units. The merchant does not want to invest more than $70,000 in inventory for these products. Find the number of units of each model that should be stocked in order to maximize profit.

40. *Maximum Profit*   A fruit grower has 150 acres of land available to raise two crops, A and B. It takes 1 day to trim an acre of crop A and 2 days to trim an acre of crop B, and there are 240 days per year available for trimming. It takes 0.3 day to pick an acre of crop A and 0.1 day to pick an acre of crop B, and there are 30 days available for picking. The profits are $140 per acre for crop A and $235 per acre for crop B. Find the number of acres of each fruit that should be planted to maximize profit.

41. *Minimum Cost*   Two gasolines, type A ($1.13 per gallon) and type B ($1.28 per gallon), have octane ratings of 80 and 92, respectively. Determine the blend of minimum cost with an octane rating of at least 90. (*Hint:* Let x be the fraction of each gallon that is type A and let y be the fraction that is type B.)

42. *Maximum Revenue*   An accounting firm has 900 hours of staff time and 100 hours of reviewing time available each week. The firm charges $2000 for an audit and $300 for a tax return. Each audit requires 100 hours of staff time and 10 hours of review time. Each tax return requires 12.5 hours of staff time and 2.5 hours of review time. What numbers of audits and tax returns will yield the maximum revenue?

43. *Maximum Revenue*   The accounting firm in Exercise 42 lowers its charge for an audit to $1000. What numbers of audits and tax returns will yield the maximum revenue?

44. *Maximum Profit*   A manufacturer produces two models of bicycles. The amounts of time (in hours) required for assembling, painting, and packaging the two models are as follows.

	Model A	Model B
Assembling	2	2.5
Painting	4	1
Packaging	1	0.75

The total amounts of time available for assembling, painting, and packaging are 4000, 4800, and 1500 hours, respectively. The profits per unit are $45 (model A) and $50 (model B). How many of each model should be produced to maximize profit?

45. *Maximum Profit*   A manufacturer produces two models of bicycles. The amounts of time (in hours) required for assembling, painting, and packaging the two models are as follows.

	Model A	Model B
Assembling	2.5	3
Painting	2	1
Packaging	0.75	1.25

The total amounts of time available for assembling, painting, and packaging are 4000, 2500, and 1500 hours, respectively. The profits per unit are $50 (model A) and $52 (model B). How many of each model should be produced to maximize profit?

46. *Minimum Cost* A farming cooperative mixes two brands of cattle feed. Brand X costs $25 per bag and contains 2 units of nutritional element A, 2 units of element B, and 2 units of element C. Brand Y costs $20 per bag and contains 1 unit of nutritional element A, 9 units of element B, and 3 units of element C. Find the number of bags of each brand that should be mixed to produce a mixture having a minimum cost per bag. The minimum requirements for nutrients A, B, and C are 12 units, 36 units, and 24 units, respectively.

In Exercises 47–52, the linear programming problem has an unusual characteristic. Sketch a graph of the solution region for the problem and describe the unusual characteristic. The objective function is to be maximized in each case.

47. Objective function:

$z = 2.5x + y$

Constraints:

$x \geq 0$

$y \geq 0$

$3x + 5y \leq 15$

$5x + 2y \leq 10$

48. Objective function:

$z = x + y$

Constraints:

$x \geq 0$

$y \geq 0$

$-x + y \leq 1$

$-x + 2y \leq 4$

49. Objective function:

$z = -x + 2y$

Constraints:

$x \geq 0$

$y \geq 0$

$x \leq 10$

$x + y \leq 7$

50. Objective function:

$z = x + y$

Constraints:

$x \geq 0$

$y \geq 0$

$-x + y \leq 0$

$-3x + y \geq 3$

51. Objective function:

$z = 3x + 4y$

Constraints:

$x \geq 0$

$y \geq 0$

$x + y \leq 1$

$2x + y \leq 4$

52. Objective function:

$z = x + 2y$

Constraints:

$x \geq 0$

$y \geq 0$

$x + 2y \leq 4$

$2x + y \leq 4$

In Exercises 53 and 54, determine values of t such that the objective function has a maximum value at the indicated vertex.

53. Objective function:

$z = 3x + ty$

Constraints:

$x \geq 0$

$y \geq 0$

$x + 3y \leq 15$

$4x + y \leq 16$

(a) $(0, 5)$

(b) $(3, 4)$

54. Objective function:

$z = 3x + ty$

Constraints:

$x \geq 0$

$y \geq 0$

$x + 2y \geq 4$

$x - y \leq 1$

(a) $(2, 1)$

(b) $(0, 2)$

Review Solve Exercises 55–58 as a review of the skills and problem-solving techniques you learned in previous sections. Simplify the compound fraction.

55. $\dfrac{\left(\dfrac{9}{x}\right)}{\left(\dfrac{6}{x} + 2\right)}$

56. $\dfrac{\left(1 + \dfrac{2}{x}\right)}{\left(x - \dfrac{4}{x}\right)}$

57. $\dfrac{\left(\dfrac{4}{x^2 - 9} + \dfrac{2}{x - 2}\right)}{\left(\dfrac{1}{x + 3} + \dfrac{1}{x - 3}\right)}$

58. $\dfrac{\left(\dfrac{1}{x + 1} + \dfrac{1}{2}\right)}{\left(\dfrac{3}{2x^2 + 4x + 2}\right)}$

Focus on Concepts

In this chapter, you studied several concepts that are required for solving systems of equations and inequalities. You can use the following questions to check your understanding of several of these basic concepts. The answers to these questions are given in the back of the book.

1. What is meant by a solution of a system of equations in two variables?

2. When solving a system of equations by substitution, how do you recognize that the system has no solution?

3. When solving a system of equations by elimination, how do you recognize that the system has no solution?

4. Describe any advantages of the algebraic method over the graphical method of solving a system of equations.

5. A system of two equations in two unknowns is solved and has a finite number of solutions. Determine the maximum number of solutions of the system satisfying each of the following conditions.

   (a) Both equations are linear.

   (b) One equation is linear and the other is quadratic.

   (c) Both equations are quadratic.

6. Explain what is meant by an inconsistent system of linear equations.

7. How can you tell graphically that a system of linear equations in two variables has no solution? Give an example.

8. Describe the operations on a system of linear equations that produce equivalent systems of equations.

9. Are the following two systems of equations equivalent? Give reasons for your answer.

$$\begin{aligned} x + 3y - z &= 6 \\ 2x - y + 2z &= 1 \\ 3x + 2y - z &= 2 \end{aligned} \qquad \begin{aligned} x + 3y - z &= 6 \\ -7y + 4z &= 1 \\ -7y - 4z &= -16 \end{aligned}$$

10. One of the following systems is inconsistent and the other has one solution. How can you determine which is which by observation?

$$\begin{aligned} 3x - 5y &= 3 \\ -12x + 20y &= 8 \end{aligned} \qquad \begin{aligned} 3x - 5y &= 3 \\ 9x - 20y &= 6 \end{aligned}$$

In Exercises 11–14, match the system of inequalities with the graph of its solution. [The graphs are labeled (a), (b), (c), and (d).]

(a)   (b)

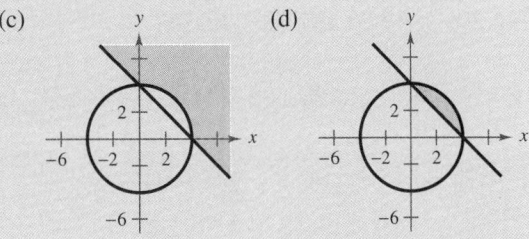

(c)   (d)

11. $x^2 + y^2 \leq 16$ 12. $x^2 + y^2 \leq 16$

    $x + y \geq 4$ $x + y \leq 4$

13. $x^2 + y^2 \geq 16$ 14. $x^2 + y^2 \geq 16$

    $x + y \geq 4$ $x + y \leq 4$

15. The graph of the solution of the inequality $x + 2y < 6$ is given in the figure. Describe how the solution set would change for each of the following.

    (a) $x + 2y \leq 6$ (b) $x + 2y > 6$

7 /// REVIEW EXERCISES

In Exercises 1–6, solve the system by the method of substitution.

1. $x + y = 2$
$\quad x - y = 0$

2. $2x = 3(y - 1)$
$\quad y = x$

3. $x^2 - y^2 = 9$
$\quad x - y = 1$

4. $x^2 + y^2 = 169$
$\quad 3x + 2y = 39$

5. $y = 2x^2$
$\quad y = x^4 - 2x^2$

6. $x = y + 3$
$\quad x = y^2 + 1$

In Exercises 7–10, use a graphing utility to solve the system of equations. If you cannot identify the exact solution, find the solution accurate to two decimal places.

7. $y^2 - 2y + x = 0$
$\quad\quad\quad x + y = 0$

8. $y = 2x^2 - 4x + 1$
$\quad y = x^2 - 4x + 3$

9. $y = 2(6 - x)$
$\quad y = 2^{x-2}$

10. $y = \ln(x - 1) - 3$
$\quad\; y = 4 - \frac{1}{2}x$

In Exercises 11–18, solve the system by elimination.

11. $2x - y = 2$
$\quad 6x + 8y = 39$

12. $40x + 30y = 24$
$\quad\; 20x - 50y = -14$

13. $0.2x + 0.3y = 0.14$
$\quad 0.4x + 0.5y = 0.20$

14. $12x + 42y = -17$
$\quad\; 30x - 18y = 19$

15. $\quad\;\; 3x - 2y = 0$
$\quad 3x + 2(y + 5) = 10$

16. $7x + 12y = 63$
$\quad 2x + 3y = 15$

17. $1.25x - 2y = 3.5$
$\quad\quad 5x - 8y = 14$

18. $1.5x + 2.5y = 8.5$
$\quad\; 6x + 10y = 24$

In Exercises 19 and 20, find a system of linear equations having the given solution. (There is more than one correct answer.)

19. $\left(\frac{4}{3}, 3\right)$

20. $(-6, 8)$

21. *Break-Even Point* You set up a business and make an initial investment of $10,000. The unit cost of the product is $2.85 and the selling price is $4.95. How many units must you sell to break even?

22. *Choice of Two Jobs* You are offered two sales jobs. One company offers an annual salary of $22,500 plus a year-end bonus of 1.5% of your total sales. The other company offers a salary of $20,000 plus a year-end bonus of 2% of total sales. What amount of sales will make the second offer better? Explain.

23. *Flying Speeds* Two planes leave Pittsburgh and Philadelphia at the same time, each going to the other city. One plane flies 25 miles per hour faster than the other. Find the ground speed of each plane if the cities are 275 miles apart and the planes pass one another after 40 minutes of flying time.

24. *Dimensions of a Rectangle* The perimeter of a rectangle is 480 meters and its length is 150% of its width. Find the dimensions of the rectangle.

Supply and Demand In Exercises 25 and 26, find the point of equilibrium.

Demand Function	Supply Function
25. $p = 37 - 0.0002x$	$p = 22 + 0.00001x$
26. $p = 120 - 0.0001x$	$p = 45 + 0.0002x$

In Exercises 27–30, solve the system of equations. Use a graphing utility to verify your solution.

27. $\;\; x + 3y - z = 13$
$\quad 2x \quad\quad - 5z = 23$
$\quad 4x - y - 2z = 14$

28. $\quad\;\; x + 2y + 6z = 4$
$\quad -3x + 2y - z = -4$
$\quad\;\; 4x \quad\quad + 2z = 16$

29. $\;\; x - 2y + z = -6$
$\quad 2x - 3y \quad\quad = -7$
$\quad -x + 3y - 3z = 11$

30. $2x + \quad\quad 6z = -9$
$\quad 3x - 2y + 11z = -16$
$\quad 3x - y + 7z = -11$

Exploration In Exercises 31 and 32, find a system of linear equations having the given solution. (There is more than one correct answer.)

31. $(4, -1, 3)$

32. $\left(5, \frac{3}{2}, 2\right)$

In Exercises 33 and 34, find the equation of the parabola $y = ax^2 + bx + c$ that passes through the given points. Use a graphing utility to verify your result.

33.

34.

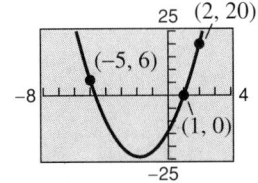

35. *Investments* An inheritance of $20,000 was divided among three investments yielding $1780 in interest per year. The interest rates for the three investments were 7%, 9%, and 11%. Find the amount placed in each investment if the second and third were $3000 and $1000 less than the first, respectively.

36. *Data Analysis* Let x and y represent the median ages at first marriage for women and men, respectively. For 7 selected years since 1970, these median ages are given by the following ordered pairs. (Source: U.S. Center for Health Statistics)

(20.6, 22.5),   (20.8, 22.7),   (21.8, 23.6),
(23.0, 24.8),   (23.3, 25.1),   (23.6, 25.3),
(23.7, 25.5)

(a) Find the least squares regression line $y = ax + b$ for the data by solving the following system of linear equations.

$$7b + \quad 156.8a = \quad 169.5$$
$$156.8b + 3522.78a = 3806.8$$

(b) Use a graphing utility to plot the data and graph the regression line in the same viewing rectangle.

(c) Is the line a good model for the data? Explain.

(d) What information is given by the slope of the regression line? Explain.

In Exercises 37–42, write the partial fraction decomposition for the rational expression.

37. $\dfrac{4 - x}{x^2 + 6x + 8}$

38. $\dfrac{-x}{x^2 + 3x + 2}$

39. $\dfrac{x^2}{x^2 + 2x - 15}$

40. $\dfrac{9}{x^2 - 9}$

41. $\dfrac{x^2 + 2x}{x^3 - x^2 + x - 1}$

42. $\dfrac{3x^3 + 4x}{(x^2 + 1)^2}$

In Exercises 43–50, sketch a graph of the solution set of the system of inequalities. Use a graphing utility to verify your result.

43.
$$x + 2y \le 160$$
$$3x + \quad y \le 180$$
$$x \ge \quad 0$$
$$y \ge \quad 0$$

44.
$$2x + 3y \le 24$$
$$2x + \quad y \le 16$$
$$x \ge \quad 0$$
$$y \ge \quad 0$$

45.
$$3x + 2y \ge 24$$
$$x + 2y \ge 12$$
$$2 \le \quad x \le 15$$
$$y \le 15$$

46.
$$2x + \quad y \ge 16$$
$$x + 3y \ge 18$$
$$0 \le \quad x \le 25$$
$$0 \le \quad y \le 25$$

47.
$$y < x + 1$$
$$y > x^2 - 1$$

48.
$$y \le 6 - 2x - x^2$$
$$y \ge x + 6$$

49.
$$2x - 3y \ge 0$$
$$2x - \quad y \le 8$$
$$y \ge 0$$

50.
$$x^2 + y^2 \le 9$$
$$(x - 3)^2 + y^2 \le 9$$

In Exercises 51 and 52, derive a set of inequalities to describe the region.

51. *Parallelogram:* Vertices at $(1, 5), (3, 1), (6, 10), (8, 6)$

52. *Triangle:* Vertices at $(1, 2), (6, 7), (8, 1)$

In Exercises 53 and 54, find a system of inequalities that models the description. Use a graphing utility to graph and shade the solution of the system.

53. *Fruit Distribution* A Pennsylvania fruit grower has 1500 bushels of apples that are to be divided between markets in Harrisburg and Philadelphia. These two markets need at least 400 bushels and 600 bushels, respectively.

54. *Inventory Costs* A warehouse operator has 24,000 square feet of floor space in which to store two products. Each unit of product I requires 20 square feet of floor space and costs $12 per day to store. Each unit of product II requires 30 square feet of floor space and costs $8 per day to store. The total storage cost per day cannot exceed $12,400.

In Exercises 55 and 56, use a graphing utility to shade the regions representing the consumer surplus and producer surplus for the supply and demand equations.

	Demand	*Supply*
55.	$p = 160 - 0.0001x$	$p = 70 + 0.0002x$
56.	$p = 130 - 0.0002x$	$p = 30 + 0.0003x$

In Exercises 57–60, use a graphing utility to sketch the region determined by the constraints. Then find the minimum or maximum values of the objective function, subject to the constraints.

57. Maximize:

$z = 3x + 4y$

Constraints:

$$x \geq 0$$
$$y \geq 0$$
$$2x + 5y \leq 50$$
$$4x + y \leq 28$$

58. Minimize:

$z = 10x + 7y$

Constraints:

$$x \geq 0$$
$$y \geq 0$$
$$2x + y \geq 100$$
$$x + y \geq 75$$

59. Minimize:

$z = 1.75x + 2.25y$

Constraints:

$$x \geq 0$$
$$y \geq 0$$
$$2x + y \geq 25$$
$$3x + 2y \geq 45$$

60. Maximize:

$z = 50x + 70y$

Constraints:

$$x \geq 0$$
$$y \geq 0$$
$$x + 2y \leq 1500$$
$$5x + 2y \leq 3500$$

61. *Maximum Revenue* A student is working part time as a cosmetologist to pay college expenses. The student may work no more than 24 hours per week. Haircuts cost $17 and require an average of 20 minutes, and permanents cost $60 and require an average

of 1 hour and 10 minutes. What combination of haircuts and/or permanents will yield a maximum revenue?

62. *Maximum Profit* A manufacturer produces products A and B, yielding profits of $18 and $24, respectively. Each product must go through three processes that require the amounts of time per unit shown in the table.

Process	Hours for Product A	Hours for Product B	Hours Available per Day
I	4	2	24
II	1	2	9
III	1	1	8

Find the daily production levels for the two products that will maximize profit.

63. *Minimum Cost* A pet supply company mixes two brands of dry dog food. Brand X costs $15 per bag and contains eight units of nutritional element A, one unit of nutritional element B, and two units of nutritional element C. Brand Y costs $30 per bag and contains two units of nutritional element A, one unit of nutritional element B, and seven units of nutritional element C. Each bag of mixed dog food must contain at least 16 units, 5 units, and 20 units of nutritional elements A, B, and C, respectively. Find the numbers of bags of brands X and Y that should be mixed to produce a mixture meeting the minimum nutritional requirements and having a minimum cost per bag.

64. *Minimum Cost* Two gasolines, type A and type B, have octane ratings of 80 and 92, respectively. Type A costs $1.25 per gallon and type B costs $1.55 per gallon. Determine the blend of minimum cost with an octane rating of at least 88. (*Hint:* Let x be the fraction of each gallon that is type A and let y be the fraction that is type B.)

65. *Exploration* Find k_1 and k_2 such that the following system of equations has an infinite number of solutions.

$$3x - 5y = 8$$
$$2x + k_1 y = k_2$$

CHAPTER PROJECT   *Fitting Models to Data*

In this project, you will find and use models relating to newspaper circulation in the United States.   (Source: Editor and Publisher Company)

(a) The numbers (in millions) of morning, evening, and Sunday newspapers sold each day in the United States in the years 1981 through 1992 are shown in the first table below. Use the linear regression capabilities of a graphing utility to fit least squares regression models to the three sets of data. Then use the models to project the numbers of morning, evening, and Sunday newspapers that will be sold each day in 1998. In the table, $t = 1$ represents 1981.

(b) The numbers of newspaper companies in the United States from 1981 through 1992 are shown in the second table below. Find a linear model for each set of data.

(c) From 1981 through 1992, the circulation of morning newspapers increased. However, because the number of morning newspaper companies also increased, the competition for morning newspaper readers became keener. Did the average circulation per morning newspaper company increase or decrease? Explain.

(d) If you had the opportunity to invest in a company that published only one type of newspaper (morning, evening, or Sunday), which would you choose? Explain your reasoning.

Daily Newspaper Circulation (in millions)

Year, t	1	2	3	4	5	6	7	8	9	10	11	12
Morning	30.6	33.2	33.8	35.4	36.4	37.4	39.1	40.4	40.7	41.3	41.5	42.4
Evening	30.9	29.3	28.8	27.7	26.4	25.1	23.7	22.2	21.8	21.0	19.2	17.8
Sunday	55.2	56.3	56.7	57.5	58.8	58.9	60.1	61.5	62.0	62.6	62.1	62.2

Numbers of Newspaper Companies

Year, t	1	2	3	4	5	6	7	8	9	10	11	12
Morning	408	434	446	458	482	499	511	529	530	559	571	596
Evening	1352	1310	1284	1257	1220	1188	1166	1141	1125	1084	1042	996
Sunday	755	768	772	783	798	802	820	840	847	863	875	891

7 ||| CHAPTER TEST

Take this test as you would take a test in class. After you are done, check your work against the answers given in the back of the book.

In Exercises 1–3, solve the system of equations by the method of substitution.

1. $x - y = 4$
$3x + 2y = 2$

2. $y = x - 1$
$y = (x - 1)^3$

3. $2x - y^2 = 0$
$x - y = 4$

In Exercises 4–6, use a graphing utility to solve the system graphically.

4. $2x - 3y = 0$
$2x + 3y = 12$

5. $y = 9 - x^2$
$y = x + 3$

6. $y = \log_3 x$
$y = -\frac{1}{3}x + 2$

In Exercises 7 and 8, solve the linear system by elimination.

7. $2x + 3y = 17$
$5x - 4y = -15$

8. $x - 2y + 3z = 11$
$2x \quad - z = 3$
$3y + z = -8$

9. Find a system of linear equations that has the solution $\left(\frac{4}{3}, -5\right)$.

10. Find the equation of the parabola $y = ax^2 + bx + c$ passing through the points $(0, 6)$, $(-2, 2)$, and $\left(3, \frac{9}{2}\right)$.

In Exercises 11 and 12, use a graphing utility to graph the inequalities and shade the region representing the solution.

11. $2x + y \leq 4$
$2x - y \geq 0$
$x \geq 0$

12. $y < -x^4 + x^2 + 4$
$y > 4x$

13. Derive a set of inequalities to describe the region in the figure.

14. Find the maximum value of the objective function $z = 20x + 12y$ subject to the constraints $x \geq 0$, $y \geq 0$, $x + 4y \leq 32$, $3x + 2y \leq 36$.

15. A merchant plans to sell two models of compact disc players at costs of $275 and $400. The $275 model yields a profit of $55 and the $400 model yields a profit of $75. The merchant estimates that the total monthly demand will not exceed 300 units. The merchant does not want to invest more than $100,000 in inventory for these products. Find the number of units of each model that should be stocked in order to maximize profit.

Figure for 13

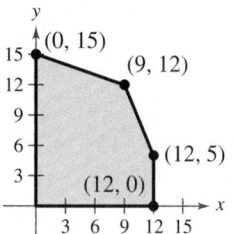

Chapter 8

Matrices and Determinants

The times (in minutes) for the winning men's and women's 1000-meter speed skating events at the winter Olympics are shown below. (In 1994, the winter Olympics occurred only 2 years after the previous winter Olympics.)

Year (0 ↔ 1980)

Year	Men	Women
1976	1.322	1.474
1980	1.253	1.402
1984	1.263	1.360
1988	1.217	1.294
1992	1.248	1.358
1994	1.207	1.312

By using a graphing utility, you can determine that the best-fitting linear models are

$$s = 1.279 - 0.0049t \qquad \text{Men}$$

$$s = 1.411 - 0.0078t \qquad \text{Women}$$

where s is the time (in minutes) and t is the year, with $t = 0$ representing 1980. According to these two models, the women's times are decreasing a little more rapidly than the men's times. (See Exercises 81 and 82 on page 604.)

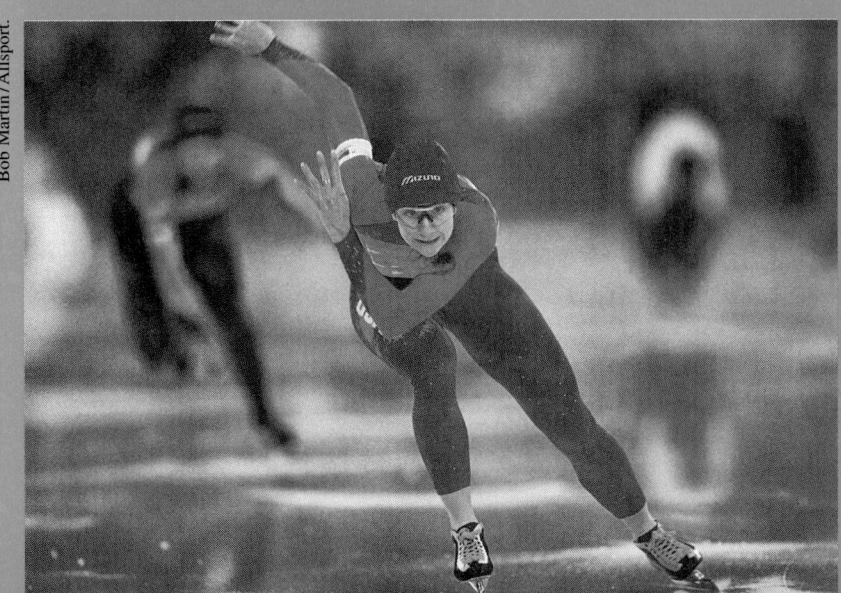

Bob Martin / Allsport.

Bonnie Blair won the 1000-meter women's speed skating event in the 1992 and 1994 winter Olympics. These were the first times this event was ever won by an American.

589

8.1   Matrices and Systems of Equations

Matrices / *Elementary Row Operations* / *Gaussian Elimination with Back-Substitution* / *Gauss-Jordan Elimination*

Matrices

In this section you will study a streamlined technique for solving systems of linear equations. This technique involves the use of a rectangular array of real numbers called a **matrix.**

Note   The plural of matrix is *matrices.*

Definition of Matrix

If m and n are positive integers, an $m \times n$ (read "m by n") **matrix** is a rectangular array

$$\begin{bmatrix} a_{11} & a_{12} & a_{13} & \cdots & a_{1n} \\ a_{21} & a_{22} & a_{23} & \cdots & a_{2n} \\ a_{31} & a_{32} & a_{33} & \cdots & a_{3n} \\ \vdots & \vdots & \vdots & & \vdots \\ a_{m1} & a_{m2} & a_{m3} & \cdots & a_{mn} \end{bmatrix} \Big\} \; m \text{ rows}$$

$$\underbrace{\phantom{a_{11} \quad a_{12} \quad a_{13}}}_{n \text{ columns}}$$

in which each **entry** a_{ij} of the matrix is a real number. An $m \times n$ matrix has m **rows** (horizontal lines) and n **columns** (vertical lines).

The entry in the ith row and jth column is denoted by the *double subscript* notation a_{ij}. A matrix having m rows and n columns is said to be of **order** $m \times n$. If $m = n$, the matrix is **square** of order n. For a square matrix, the entries $a_{11}, a_{22}, a_{33}, \ldots$ are the **main diagonal** entries.

EXAMPLE 1   **Examples of Matrices**

a. *Order:* 1×1

$$[2]$$

Note   A matrix that has only one row is called a **row matrix,** and a matrix that has only one column is called a **column matrix.**

b. *Order:* 1×4

$$\begin{bmatrix} 1 & -3 & 0 & \frac{1}{2} \end{bmatrix}$$

c. *Order:* 2×2

$$\begin{bmatrix} 0 & 0 \\ 0 & 0 \end{bmatrix}$$

d. *Order:* 3×2

$$\begin{bmatrix} 5 & 0 \\ 2 & -2 \\ -7 & 4 \end{bmatrix}$$

A matrix derived from a system of linear equations (each written in standard form with the constant term on the right) is the **augmented matrix** of the system. Moreover, the matrix derived from the coefficients of the system (but not including the constant terms) is the **coefficient matrix** of the system.

$$
\begin{array}{ccc}
\textit{System} & \textit{Augmented Matrix} & \textit{Coefficient Matrix}
\end{array}
$$

$$
\begin{array}{l}
x - 4y + 3z = 5 \\
-x + 3y - z = -3 \\
2x \quad\ - 4z = 6
\end{array}
\quad
\left[\begin{array}{ccc:c}
1 & -4 & 3 & 5 \\
-1 & 3 & -1 & -3 \\
2 & 0 & -4 & 6
\end{array}\right]
\quad
\left[\begin{array}{ccc}
1 & -4 & 3 \\
-1 & 3 & -1 \\
2 & 0 & -4
\end{array}\right]
$$

Note Note the use of 0 for the missing y-variable in the third equation, and also note the fourth column (of constant terms) in the augmented matrix.

When forming either the coefficient matrix or the augmented matrix of a system, you should begin by vertically aligning the variables in the equations and using 0's for the missing variables.

$$
\begin{array}{ccc}
\textit{Given System} & \textit{Line Up Variables.} & \textit{Form Augmented Matrix.}
\end{array}
$$

$$
\begin{array}{l}
x + 3y = 9 \\
-y + 4z = -2 \\
x - 5z = 0
\end{array}
\qquad
\begin{array}{l}
x + \ 3y \quad\ = 9 \\
\quad\ -y + 4z = -2 \\
x \qquad\ - 5z = 0
\end{array}
\qquad
\left[\begin{array}{ccc:c}
1 & 3 & 0 & 9 \\
0 & -1 & 4 & -2 \\
1 & 0 & -5 & 0
\end{array}\right]
$$

Elementary Row Operations

In Section 7.3, you studied three operations that can be used on a system of linear equations to produce an equivalent system.

1. Interchange two equations.
2. Multiply an equation by a nonzero constant.
3. Add a multiple of an equation to another equation.

In matrix terminology, these three operations correspond to **elementary row operations.** An elementary row operation on an augmented matrix of a given system of linear equations produces a new augmented matrix corresponding to a new (but equivalent) system of linear equations. Two matrices are **row-equivalent** if one can be obtained from the other by a sequence of elementary row operations.

Elementary Row Operations

1. Interchange two rows.
2. Multiply a row by a nonzero constant.
3. Add a multiple of a row to another row.

Although elementary row operations are simple to perform, they involve a lot of arithmetic. Because it is easy to make a mistake, we suggest that you get in the habit of noting the elementary row operations performed in each step so that you can go back and check your work.

The *Interactive* CD-ROM shows every example with its solution; clicking on the *Try It!* button brings up similar problems. Guided Examples and Integrated Examples show step-by-step solutions to additional examples. Integrated Examples are related to several concepts in the section.

EXAMPLE 2   **Elementary Row Operations**

a. Interchange the first and second rows.

Original Matrix

$$\begin{bmatrix} 0 & 1 & 3 & 4 \\ -1 & 2 & 0 & 3 \\ 2 & -3 & 4 & 1 \end{bmatrix}$$

New Row-Equivalent Matrix

$$\begin{array}{c} R_2 \\ R_1 \end{array} \begin{bmatrix} -1 & 2 & 0 & 3 \\ 0 & 1 & 3 & 4 \\ 2 & -3 & 4 & 1 \end{bmatrix}$$

b. Multiply the first row by $\frac{1}{2}$.

Original Matrix

$$\begin{bmatrix} 2 & -4 & 6 & -2 \\ 1 & 3 & -3 & 0 \\ 5 & -2 & 1 & 2 \end{bmatrix}$$

New Row-Equivalent Matrix

$$\frac{1}{2}R_1 \to \begin{bmatrix} 1 & -2 & 3 & -1 \\ 1 & 3 & -3 & 0 \\ 5 & -2 & 1 & 2 \end{bmatrix}$$

c. Add -2 times the first row to the third row.

Original Matrix

$$\begin{bmatrix} 1 & 2 & -4 & 3 \\ 0 & 3 & -2 & -1 \\ 2 & 1 & 5 & -2 \end{bmatrix}$$

New Row-Equivalent Matrix

$$-2R_1 + R_3 \to \begin{bmatrix} 1 & 2 & -4 & 3 \\ 0 & 3 & -2 & -1 \\ 0 & -3 & 13 & -8 \end{bmatrix}$$

Note that the elementary row operation is written beside the row that is *changed*.

The *Interactive* CD-ROM offers graphing utility emulators of the *TI-82* and *TI-83*, which can be used with the Examples, Explorations, Technology notes, and Exercises.

Most graphing utilities can perform elementary row operations on matrices. For instance, on a *TI-82* or *TI-83*, you can perform the elementary row operation shown in Example 2(c) as follows.

1. Use the matrix edit feature to enter the matrix as [A].
2. Choose the "* row + (" feature in the matrix math menu.

$$* \text{ row } + (-2, [A], 1, 3) \boxed{\text{ENTER}}$$

The new row-equivalent matrix will be displayed. To do a sequence of row operations, use $\boxed{\text{ANS}}$ in place of [A] in each operation. If you want to save this new matrix, you must do this with separate steps.

In Example 2 of Section 7.3, you used Gaussian elimination with back-substitution to solve a system of linear equations. The next example demonstrates the matrix version of Gaussian elimination. The two methods are essentially the same. The basic difference is that with matrices you do not need to keep writing the variables.

EXAMPLE 3 **Using Elementary Row Operations**

Linear System	*Associated Augmented Matrix*

$$x - 2y + 3z = 9$$
$$-x + 3y \qquad = -4$$
$$2x - 5y + 5z = 17$$

$$\begin{bmatrix} 1 & -2 & 3 & \vdots & 9 \\ -1 & 3 & 0 & \vdots & -4 \\ 2 & -5 & 5 & \vdots & 17 \end{bmatrix}$$

Add the first equation to the second equation.

Add the first row to the second row $(R_1 + R_2)$.

$$x - 2y + 3z = 9$$
$$\quad y + 3z = 5$$
$$2x - 5y + 5z = 17$$

$$R_1 + R_2 \rightarrow \begin{bmatrix} 1 & -2 & 3 & \vdots & 9 \\ 0 & 1 & 3 & \vdots & 5 \\ 2 & -5 & 5 & \vdots & 17 \end{bmatrix}$$

Add -2 times the first equation to the third equation.

Add -2 times the first row to the third row $(-2R_1 + R_3)$.

$$x - 2y + 3z = 9$$
$$\quad y + 3z = 5$$
$$\quad -y - z = -1$$

$$-2R_1 + R_3 \rightarrow \begin{bmatrix} 1 & -2 & 3 & \vdots & 9 \\ 0 & 1 & 3 & \vdots & 5 \\ 0 & -1 & -1 & \vdots & -1 \end{bmatrix}$$

Add the second equation to the third equation.

Add the second row to the third row $(R_2 + R_3)$.

$$x - 2y + 3z = 9$$
$$\quad y + 3z = 5$$
$$\quad 2z = 4$$

$$R_2 + R_3 \rightarrow \begin{bmatrix} 1 & -2 & 3 & \vdots & 9 \\ 0 & 1 & 3 & \vdots & 5 \\ 0 & 0 & 2 & \vdots & 4 \end{bmatrix}$$

Multiply the third equation by $\frac{1}{2}$.

Multiply the third row by $\frac{1}{2}$.

$$x - 2y + 3z = 9$$
$$\quad y + 3z = 5$$
$$\quad z = 2$$

$$\tfrac{1}{2}R_3 \rightarrow \begin{bmatrix} 1 & -2 & 3 & \vdots & 9 \\ 0 & 1 & 3 & \vdots & 5 \\ 0 & 0 & 1 & \vdots & 2 \end{bmatrix}$$

At this point, you can use back-substitution to find that the solution is $x = 1$, $y = -1$, and $z = 2$, as was done in Example 2 of Section 7.3.

Note Remember that you can check a solution by substituting the values of x, y, and z into each equation in the original system.

The last matrix in Example 3 is said to be in **row-echelon form.** The term *echelon* refers to the stair-step pattern formed by the nonzero elements of the matrix. To be in this form, a matrix must have the following properties.

Some graphing utilities, such as the *TI-85, TI-92,* and *HP-48G,* can automatically transform a matrix to row-echelon form and reduced row-echelon form. Read your user's manual to see if your calculator has this capability. If so, use it to verify the results in this section.

Row-Echelon Form and Reduced Row-Echelon Form

A matrix in **row-echelon form** has the following properties.

1. All rows consisting entirely of zeros occur at the bottom of the matrix.

2. For each row that does not consist entirely of zeros, the first nonzero entry is 1 (called a **leading 1**).

3. For two successive (nonzero) rows, the leading 1 in the higher row is farther to the left than the leading 1 in the lower row.

A matrix in *row-echelon form* is in **reduced row-echelon form** if every column that has a leading 1 has zeros in every position above and below its leading 1.

EXAMPLE 4   **Row-Echelon Form**

The following matrices are in row-echelon form.

a. $\begin{bmatrix} 1 & 2 & -1 & 4 \\ 0 & 1 & 0 & 3 \\ 0 & 0 & 1 & -2 \end{bmatrix}$
b. $\begin{bmatrix} 0 & 1 & 0 & 5 \\ 0 & 0 & 1 & 3 \\ 0 & 0 & 0 & 0 \end{bmatrix}$

c. $\begin{bmatrix} 1 & -5 & 2 & -1 & 3 \\ 0 & 0 & 1 & 3 & -2 \\ 0 & 0 & 0 & 1 & 4 \\ 0 & 0 & 0 & 0 & 1 \end{bmatrix}$
d. $\begin{bmatrix} 1 & 0 & 0 & -1 \\ 0 & 1 & 0 & 2 \\ 0 & 0 & 1 & 3 \\ 0 & 0 & 0 & 0 \end{bmatrix}$

The matrices in (b) and (d) also happen to be in *reduced* row-echelon form. The following matrices are not in row-echelon form.

e. $\begin{bmatrix} 1 & 2 & -3 & 4 \\ 0 & 2 & 1 & -1 \\ 0 & 0 & 1 & -3 \end{bmatrix}$
f. $\begin{bmatrix} 1 & 2 & -1 & 2 \\ 0 & 0 & 0 & 0 \\ 0 & 1 & 2 & -4 \end{bmatrix}$

Every matrix has a row-equivalent matrix that is in row-echelon form. For instance, in Example 4, you can change the matrix in part (e) to row-echelon form by multiplying its second row by $\frac{1}{2}$. What elementary row operation could you perform on the matrix in part (f) so that it would be in row-echelon form?

Gaussian Elimination with Back-Substitution

EXAMPLE 5   **Gaussian Elimination with Back-Substitution**

Solve the system.

$$
\begin{aligned}
y + z - 2w &= -3 \\
x + 2y - z &= 2 \\
2x + 4y + z - 3w &= -2 \\
x - 4y - 7z - w &= -19
\end{aligned}
$$

Solution

$$
\begin{array}{c}
R_2 \\
R_1
\end{array}
\left[
\begin{array}{cccc:c}
1 & 2 & -1 & 0 & 2 \\
0 & 1 & 1 & -2 & -3 \\
2 & 4 & 1 & -3 & -2 \\
1 & -4 & -7 & -1 & -19
\end{array}
\right]
$$

First column has leading 1 in upper left corner.

$$
\begin{array}{c}
\\
\\
-2R_1 + R_3 \rightarrow \\
-R_1 + R_4 \rightarrow
\end{array}
\left[
\begin{array}{cccc:c}
1 & 2 & -1 & 0 & 2 \\
0 & 1 & 1 & -2 & -3 \\
0 & 0 & 3 & -3 & -6 \\
0 & -6 & -6 & -1 & -21
\end{array}
\right]
$$

First column has zeros below its leading 1.

$$
\begin{array}{c}
\\
\\
\\
6R_2 + R_4 \rightarrow
\end{array}
\left[
\begin{array}{cccc:c}
1 & 2 & -1 & 0 & 2 \\
0 & 1 & 1 & -2 & -3 \\
0 & 0 & 3 & -3 & -6 \\
0 & 0 & 0 & -13 & -39
\end{array}
\right]
$$

Second column has zeros below its leading 1.

$$
\begin{array}{c}
\\
\\
\tfrac{1}{3}R_3 \rightarrow \\
\\
\end{array}
\left[
\begin{array}{cccc:c}
1 & 2 & -1 & 0 & 2 \\
0 & 1 & 1 & -2 & -3 \\
0 & 0 & 1 & -1 & -2 \\
0 & 0 & 0 & -13 & -39
\end{array}
\right]
$$

Third column has zeros below its leading 1.

$$
\begin{array}{c}
\\
\\
\\
-\tfrac{1}{13}R_4 \rightarrow
\end{array}
\left[
\begin{array}{cccc:c}
1 & 2 & -1 & 0 & 2 \\
0 & 1 & 1 & -2 & -3 \\
0 & 0 & 1 & -1 & -2 \\
0 & 0 & 0 & 1 & 3
\end{array}
\right]
$$

Fourth column has a leading 1.

The matrix is now in row-echelon form, and the corresponding system is

$$
\begin{aligned}
x + 2y - z &= 2 \\
y + z - 2w &= -3 \\
z - w &= -2 \\
w &= 3.
\end{aligned}
$$

Using back-substitution, you can determine that the solution is $x = -1$, $y = 2$, $z = 1$, and $w = 3$. Check this in the original system of equations.   ▇

Gaussian Elimination with Back-Substitution

1. Write the augmented matrix of the system of linear equations.
2. Use elementary row operations to rewrite the augmented matrix in row-echelon form.
3. Write the system of linear equations corresponding to the matrix in row-echelon form, and use back-substitution to find the solution.

When solving a system of linear equations, remember that it is possible for the system to have no solution. If, in the elimination process, you obtain a row with zeros except for the last entry, it is unnecessary to continue the elimination process. You can simply conclude that the system is inconsistent.

EXAMPLE 6 ▱ **A System with No Solution**

Solve the system.

$$\begin{aligned} x - y + 2z &= 4 \\ x \quad\quad + z &= 6 \\ 2x - 3y + 5z &= 4 \\ 3x + 2y - z &= 1 \end{aligned}$$

Solution

$$\begin{bmatrix} 1 & -1 & 2 & \vdots & 4 \\ 1 & 0 & 1 & \vdots & 6 \\ 2 & -3 & 5 & \vdots & 4 \\ 3 & 2 & -1 & \vdots & 1 \end{bmatrix} \begin{matrix} \\ -R_1 + R_2 \rightarrow \\ -2R_1 + R_3 \rightarrow \\ -3R_1 + R_4 \rightarrow \end{matrix} \begin{bmatrix} 1 & -1 & 2 & \vdots & 4 \\ 0 & 1 & -1 & \vdots & 2 \\ 0 & -1 & 1 & \vdots & -4 \\ 0 & 5 & -7 & \vdots & -11 \end{bmatrix}$$

$$\begin{matrix} \\ \\ R_2 + R_3 \rightarrow \\ \\ \end{matrix} \begin{bmatrix} 1 & -1 & 2 & \vdots & 4 \\ 0 & 1 & -1 & \vdots & 2 \\ 0 & 0 & 0 & \vdots & -2 \\ 0 & 5 & -7 & \vdots & -11 \end{bmatrix}$$

Note that the third row of this matrix consists of zeros except for the last entry. This means that the original system of linear equations is *inconsistent*. You can see why this is true by converting back to a system of linear equations.

$$\begin{aligned} x - y + 2z &= 4 \\ y - z &= 2 \\ 0 &= -2 \\ 5y - 7z &= -11 \end{aligned}$$

Because the third equation is not possible, the system has no solution. ▱

Gauss-Jordan Elimination

With Gaussian elimination, elementary row operations are applied to a matrix to obtain a (row-equivalent) row-echelon form. A second method of elimination, called **Gauss-Jordan elimination,** after Carl Friedrich Gauss (1777–1855) and Wilhelm Jordan (1842–1899), continues the reduction process until a *reduced* row-echelon form is obtained. This procedure is demonstrated in the following example.

For a demonstration of a graphical approach to Gauss-Jordan elimination on a 2×3 matrix, see the graphing calculator program for this section in the appendix.

EXAMPLE 7   **Gauss-Jordan Elimination**

Use Gauss-Jordan elimination to solve the system.

$$\begin{aligned} x - 2y + 3z &= 9 \\ -x + 3y &= -4 \\ 2x - 5y + 5z &= 17 \end{aligned}$$

Solution

In Example 3, Gaussian elimination was used to obtain the row-echelon form

$$\begin{bmatrix} 1 & -2 & 3 & \vdots & 9 \\ 0 & 1 & 3 & \vdots & 5 \\ 0 & 0 & 1 & \vdots & 2 \end{bmatrix}.$$

Now, rather than using back-substitution, apply additional elementary row operations until you obtain a matrix in *reduced* row-echelon form. To do this, you must produce zeros above each of the leading 1's, as follows.

$$2R_2 + R_1 \rightarrow \begin{bmatrix} 1 & 0 & 9 & \vdots & 19 \\ 0 & 1 & 3 & \vdots & 5 \\ 0 & 0 & 1 & \vdots & 2 \end{bmatrix}$$

Second column has zeros above its leading 1.

$$\begin{aligned} -9R_3 + R_1 \rightarrow \\ -3R_3 + R_2 \rightarrow \end{aligned} \begin{bmatrix} 1 & 0 & 0 & \vdots & 1 \\ 0 & 1 & 0 & \vdots & -1 \\ 0 & 0 & 1 & \vdots & 2 \end{bmatrix}$$

Third column has zeros above its leading 1.

Now, converting back to a system of linear equations, you have

$$\begin{aligned} x &= 1 \\ y &= -1 \\ z &= 2. \end{aligned}$$

Note Which technique do you prefer: Gaussian elimination or Gauss-Jordan elimination?

The beauty of Gauss-Jordan elimination is that, from the reduced row-echelon form, you can simply read the solution.

The elimination procedures described in this section employ an algorithmic approach that is easily adapted to computer use. However, the procedure makes no effort to avoid fractional coefficients. For instance, if the system given in Example 7 had been listed as

$$\begin{aligned} 2x - 5y + 5z &= 17 \\ x - 2y + 3z &= 9 \\ -x + 3y &= -4 \end{aligned}$$

the procedure would have required multiplication of the first row by $\frac{1}{2}$, which would have introduced fractions in the first row. For hand computations, fractions can sometimes be avoided by judiciously choosing the order in which the elementary row operations are applied.

EXAMPLE 8 ◼   **A System with an Infinite Number of Solutions**

Solve the system.

$$\begin{aligned} 2x + 4y - 2z &= 0 \\ 3x + 5y &= 1 \end{aligned}$$

Solution

$$\begin{bmatrix} 2 & 4 & -2 & \vdots & 0 \\ 3 & 5 & 0 & \vdots & 1 \end{bmatrix} \qquad \frac{1}{2}R_1 \rightarrow \begin{bmatrix} 1 & 2 & -1 & \vdots & 0 \\ 3 & 5 & 0 & \vdots & 1 \end{bmatrix}$$

$$-3R_1 + R_2 \rightarrow \begin{bmatrix} 1 & 2 & -1 & \vdots & 0 \\ 0 & -1 & 3 & \vdots & 1 \end{bmatrix}$$

$$-R_2 \rightarrow \begin{bmatrix} 1 & 2 & -1 & \vdots & 0 \\ 0 & 1 & -3 & \vdots & -1 \end{bmatrix}$$

$$-2R_2 + R_1 \rightarrow \begin{bmatrix} 1 & 0 & 5 & \vdots & 2 \\ 0 & 1 & -3 & \vdots & -1 \end{bmatrix}$$

The corresponding system of equations is。

$$\begin{aligned} x + 5z &= 2 \\ y - 3z &= -1. \end{aligned}$$

Solving for x and y in terms of z, you have $x = -5z + 2$ and $y = 3z - 1$. Then, letting $z = a$, the solution set has the form

$$(-5a + 2, 3a - 1, a)$$

where a is a real number. Try substituting values for a to obtain a few solutions. Then check each solution in the original system of equations. ◼

Note You have seen that the row-echelon form of a given matrix *is not* unique; however, the *reduced* row-echelon form of a given matrix *is* unique. Try applying Gauss-Jordan elimination to the row-echelon matrix given at the right to see that you obtain the same reduced row-echelon form as in Example 7.

It is worth noting that the row-echelon form of a matrix is not unique. That is, two different sequences of elementary row operations may yield different row-echelon forms. For instance, the following sequence of elementary row operations on the matrix in Example 3 produces a slightly different row-echelon form.

$$\begin{bmatrix} 1 & -2 & 3 & \vdots & 9 \\ -1 & 3 & 0 & \vdots & -4 \\ 2 & -5 & 5 & \vdots & 17 \end{bmatrix} \quad \begin{matrix} R_2 \\ R_1 \end{matrix} \begin{bmatrix} -1 & 3 & 0 & \vdots & -4 \\ 1 & -2 & 3 & \vdots & 9 \\ 2 & -5 & 5 & \vdots & 17 \end{bmatrix}$$

$$-R_1 \rightarrow \begin{bmatrix} 1 & -3 & 0 & \vdots & 4 \\ 1 & -2 & 3 & \vdots & 9 \\ 2 & -5 & 5 & \vdots & 17 \end{bmatrix}$$

$$\begin{matrix} -R_1 + R_2 \rightarrow \\ -2R_1 + R_3 \rightarrow \end{matrix} \begin{bmatrix} 1 & -3 & 0 & \vdots & 4 \\ 0 & 1 & 3 & \vdots & 5 \\ 0 & 1 & 5 & \vdots & 9 \end{bmatrix}$$

$$-R_2 + R_3 \rightarrow \begin{bmatrix} 1 & -3 & 0 & \vdots & 4 \\ 0 & 1 & 3 & \vdots & 5 \\ 0 & 0 & 2 & \vdots & 4 \end{bmatrix}$$

$$\tfrac{1}{2}R_3 \rightarrow \begin{bmatrix} 1 & -3 & 0 & \vdots & 4 \\ 0 & 1 & 3 & \vdots & 5 \\ 0 & 0 & 1 & \vdots & 2 \end{bmatrix}$$

The corresponding system of linear equations is

$$\begin{aligned} x - 3y \quad\;\; &= 4 \\ y + 3z &= 5 \\ z &= 2. \end{aligned}$$

Try using back-substitution on this system to see that you obtain the same solution that was obtained in Example 3.

Group Activity   *Error Analysis*

One of your classmates has submitted the following steps for a solution of a system by Gauss-Jordan elimination. Find the error(s) in the solution and discuss how to explain the error(s) to your classmate.

8.1 /// EXERCISES

In Exercises 1–6, determine the order of the matrix.

1. $\begin{bmatrix} 4 & -2 \\ 7 & 0 \\ 0 & 8 \end{bmatrix}$ **2.** $\begin{bmatrix} 5 & -3 & 8 & 7 \end{bmatrix}$

3. $\begin{bmatrix} 2 \\ 36 \\ 3 \end{bmatrix}$ **4.** $\begin{bmatrix} -3 & 7 & 15 & 0 \\ 0 & 0 & 3 & 3 \\ 1 & 1 & 6 & 7 \end{bmatrix}$

5. $\begin{bmatrix} 33 & 45 \\ -9 & 20 \end{bmatrix}$ **6.** $\begin{bmatrix} 4 \end{bmatrix}$

In Exercises 7–10, form the augmented matrix for the system of linear equations.

7. $\begin{aligned} 4x - 3y &= -5 \\ -x + 3y &= 12 \end{aligned}$ **8.** $\begin{aligned} 7x + 4y &= 22 \\ 5x - 9y &= 15 \end{aligned}$

9. $\begin{aligned} x + 10y - 2z &= 2 \\ 5x - 3y + 4z &= 0 \\ 2x + y &= 6 \end{aligned}$ **10.** $\begin{aligned} 7x - 5y + z &= 13 \\ 19x - 8z &= 10 \end{aligned}$

In Exercises 11–14, write the system of linear equations represented by the augmented matrix. (Use variables x, y, z, and w.)

11. $\begin{bmatrix} 1 & 2 & \vdots & 7 \\ 2 & -3 & \vdots & 4 \end{bmatrix}$ **12.** $\begin{bmatrix} 7 & -5 & \vdots & 0 \\ 8 & 3 & \vdots & -2 \end{bmatrix}$

13. $\begin{bmatrix} 2 & 0 & 5 & \vdots & -12 \\ 0 & 1 & -2 & \vdots & 7 \\ 6 & 3 & 0 & \vdots & 2 \end{bmatrix}$

14. $\begin{bmatrix} 9 & 12 & 3 & 0 & \vdots & 0 \\ -2 & 18 & 5 & 2 & \vdots & 10 \\ 1 & 7 & -8 & 0 & \vdots & -4 \end{bmatrix}$

In Exercises 15–18, determine whether the matrix is in row-echelon form. If it is, determine if it is also in reduced row-echelon form.

15. $\begin{bmatrix} 1 & 0 & 0 & 0 \\ 0 & 1 & 1 & 5 \\ 0 & 0 & 0 & 0 \end{bmatrix}$ **16.** $\begin{bmatrix} 1 & 3 & 0 & 0 \\ 0 & 0 & 1 & 8 \\ 0 & 0 & 0 & 0 \end{bmatrix}$

17. $\begin{bmatrix} 2 & 0 & 4 & 0 \\ 0 & -1 & 3 & 6 \\ 0 & 0 & 1 & 5 \end{bmatrix}$

18. $\begin{bmatrix} 1 & 0 & 2 & 1 \\ 0 & 1 & -3 & 10 \\ 0 & 0 & 1 & 0 \end{bmatrix}$

In Exercises 19–22, fill in the blanks using elementary row operations to form a row-equivalent matrix.

19. $\begin{bmatrix} 1 & 4 & 3 \\ 2 & 10 & 5 \end{bmatrix}$ **20.** $\begin{bmatrix} 3 & 6 & 8 \\ 4 & -3 & 6 \end{bmatrix}$

$\begin{bmatrix} 1 & 4 & 3 \\ 0 & & -1 \end{bmatrix}$ $\begin{bmatrix} 1 & & \frac{8}{3} \\ 4 & -3 & 6 \end{bmatrix}$

21. $\begin{bmatrix} 1 & 1 & 4 & -1 \\ 3 & 8 & 10 & 3 \\ -2 & 1 & 12 & 6 \end{bmatrix}$ **22.** $\begin{bmatrix} 2 & 4 & 8 & 3 \\ 1 & -1 & -3 & 2 \\ 2 & 6 & 4 & 9 \end{bmatrix}$

$\begin{bmatrix} 1 & 1 & 4 & -1 \\ 0 & 5 & & \\ 0 & 3 & & \end{bmatrix}$ $\begin{bmatrix} 1 & & & \\ 1 & -1 & -3 & 2 \\ 2 & 6 & 4 & 9 \end{bmatrix}$

$\begin{bmatrix} 1 & 1 & 4 & -1 \\ 0 & 1 & -\frac{2}{5} & \frac{6}{5} \\ 0 & 3 & & \end{bmatrix}$ $\begin{bmatrix} 1 & 2 & 4 & \frac{3}{2} \\ 0 & & -7 & \frac{1}{2} \\ 0 & 2 & & \end{bmatrix}$

23. Perform the *sequence* of row operations on the matrix. What did the operations accomplish?

$\begin{bmatrix} 1 & 2 & 3 \\ 2 & -1 & -4 \\ 3 & 1 & -1 \end{bmatrix}$

(a) Add -2 times Row 1 to Row 2.

(b) Add -3 times Row 1 to Row 3.

(c) Add -1 times Row 2 to Row 3.

(d) Multiply Row 2 by $-\frac{1}{5}$.

(e) Add -2 times Row 2 to Row 1.

The *Interactive* CD-ROM contains step-by-step solutions to all odd-numbered Section and Review Exercises. It also provides Tutorial Exercises, which link to Guided Examples for additional help.

8.1 / *Matrices and Systems of Equations*   **601**

24. Perform the *sequence* of row operations on the matrix. What did the operations accomplish?

$$\begin{bmatrix} 7 & 1 \\ 0 & 2 \\ -3 & 4 \\ 4 & 1 \end{bmatrix}$$

(a) Add Row 3 to Row 4.

(b) Interchange Rows 1 and 4.

(c) Add 3 times Row 1 to Row 3.

(d) Add −7 times Row 1 to Row 4.

(e) Multiply Row 2 by $\frac{1}{2}$.

(f) Add the appropriate multiples of Row 2 to Rows 1, 3, and 4.

In Exercises 25–28, write the matrix in row-echelon form. Remember that the row-echelon form of a matrix is not unique.

25. $\begin{bmatrix} 1 & 1 & 0 & 5 \\ -2 & -1 & 2 & -10 \\ 3 & 6 & 7 & 14 \end{bmatrix}$

26. $\begin{bmatrix} 1 & 2 & -1 & 3 \\ 3 & 7 & -5 & 14 \\ -2 & -1 & -3 & 8 \end{bmatrix}$

27. $\begin{bmatrix} 1 & -1 & -1 & 1 \\ 5 & -4 & 1 & 8 \\ -6 & 8 & 18 & 0 \end{bmatrix}$

28. $\begin{bmatrix} 1 & -3 & 0 & -7 \\ -3 & 10 & 1 & 23 \\ 4 & -10 & 2 & -24 \end{bmatrix}$

In Exercises 29–32, use the matrix capabilities of a graphing utility to write the matrix in *reduced* row-echelon form.

29. $\begin{bmatrix} 3 & 3 & 3 \\ -1 & 0 & -4 \\ 2 & 4 & -2 \end{bmatrix}$ **30.** $\begin{bmatrix} 1 & 3 & 2 \\ 5 & 15 & 9 \\ 2 & 6 & 10 \end{bmatrix}$

31. $\begin{bmatrix} 1 & 2 & 3 & -5 \\ 1 & 2 & 4 & -9 \\ -2 & -4 & -4 & 3 \\ 4 & 8 & 11 & -14 \end{bmatrix}$

32. $\begin{bmatrix} 1 & -3 \\ -1 & 8 \\ 0 & 4 \\ -2 & 10 \end{bmatrix}$

In Exercises 33–36, write the system of linear equations represented by the augmented matrix. Then use back-substitution to find the solution. (Use variables x, y, and z.)

33. $\begin{bmatrix} 1 & -2 & \vdots & 4 \\ 0 & 1 & \vdots & -3 \end{bmatrix}$ **34.** $\begin{bmatrix} 1 & 5 & \vdots & 0 \\ 0 & 1 & \vdots & -1 \end{bmatrix}$

35. $\begin{bmatrix} 1 & -1 & 2 & \vdots & 4 \\ 0 & 1 & -1 & \vdots & 2 \\ 0 & 0 & 1 & \vdots & -2 \end{bmatrix}$

36. $\begin{bmatrix} 1 & 2 & -2 & \vdots & -1 \\ 0 & 1 & 1 & \vdots & 9 \\ 0 & 0 & 1 & \vdots & -3 \end{bmatrix}$

In Exercises 37–40, an augmented matrix that represents a system of linear equations (in variables x, y, and z) has been reduced using Gauss-Jordan elimination. Write the solution represented by the augmented matrix.

37. $\begin{bmatrix} 1 & 0 & \vdots & 7 \\ 0 & 1 & \vdots & -5 \end{bmatrix}$ **38.** $\begin{bmatrix} 1 & 0 & \vdots & -2 \\ 0 & 1 & \vdots & 4 \end{bmatrix}$

39. $\begin{bmatrix} 1 & 0 & 0 & \vdots & -4 \\ 0 & 1 & 0 & \vdots & -8 \\ 0 & 0 & 1 & \vdots & 2 \end{bmatrix}$

40. $\begin{bmatrix} 1 & 0 & 0 & \vdots & 3 \\ 0 & 1 & 0 & \vdots & -1 \\ 0 & 0 & 1 & \vdots & 0 \end{bmatrix}$

In Exercises 41–56, solve the system of equations. Use Gaussian elimination with back-substitution or Gauss-Jordan elimination.

41. $x + 2y = 7$
$2x + y = 8$

42. $2x + 6y = 16$
$2x + 3y = 7$

43. $-3x + 5y = -22$
$3x + 4y = 4$
$4x - 8y = 32$

44. $x + 2y = 0$
$x + y = 6$
$3x - 2y = 8$

45. $8x - 4y = 7$
    $5x + 2y = 1$

46. $2x - y = -0.1$
    $3x + 2y = 1.6$

47. $-x + 2y = 1.5$
    $2x - 4y = 3$

48. $x - 3y = 5$
    $-2x + 6y = -10$

49. $x - 3z = -2$
    $3x + y - 2z = 5$
    $2x + 2y + z = 4$

50. $2x - y + 3z = 24$
    $2y - z = 14$
    $7x - 5y = 6$

51. $x + y - 5z = 3$
    $x - 2z = 1$
    $2x - y - z = 0$

52. $2x + 3z = 3$
    $4x - 3y + 7z = 5$
    $8x - 9y + 15z = 9$

53. $x + 2y + z = 8$
    $3x + 7y + 6z = 26$

54. $4x + 12y - 7z - 20w = 22$
    $3x + 9y - 5z - 28w = 30$

55. $x + 2y = 0$
    $-x - y = 0$

56. $x + 2y = 0$
    $2x + 4y = 0$

In Exercises 57–62, use the matrix capabilities of a graphing utility to reduce the augmented matrix and solve the system of equations.

57. $3x + 3y + 12z = 6$
    $x + y + 4z = 2$
    $2x + 5y + 20z = 10$
    $-x + 2y + 8z = 4$

58. $2x + 10y + 2z = 6$
    $x + 5y + 2z = 6$
    $x + 5y + z = 3$
    $-3x - 15y - 3z = -9$

59. $2x + y - z + 2w = -6$
    $3x + 4y + w = 1$
    $x + 5y + 2z + 6w = -3$
    $5x + 2y - z - w = 3$

60. $x + 2y + 2z + 4w = 11$
    $3x + 6y + 5z + 12w = 30$

61. $x + y + z = 0$
    $2x + 3y + z = 0$
    $3x + 5y + z = 0$

62. $x + 2y + z + 3w = 0$
    $x - y + w = 0$
    $y - z + 2w = 0$

63. *Think About It* The augmented matrix represents a system of linear equations (in variables x, y, and z) that has been reduced using Gauss-Jordan elimination. Write a system of equations with nonzero coefficients that is represented by the reduced matrix. (The answer is not unique.)

$$\begin{bmatrix} 1 & 0 & 3 & \vdots & -2 \\ 0 & 1 & 4 & \vdots & 1 \\ 0 & 0 & 0 & \vdots & 0 \end{bmatrix}$$

64. *Think About It*

(a) Describe the row-echelon form of an augmented matrix that corresponds to a system of linear equations that is inconsistent.

(b) Describe the row-echelon form of an augmented matrix that corresponds to a system of linear equations that has an infinite number of solutions.

65. *Borrowing Money* A small corporation borrowed $1,500,000 to expand its product line. Some of the money was borrowed at 8%, some at 9%, and some at 12%. How much was borrowed at each rate if the annual interest was $133,000 and the amount borrowed at 8% was 4 times the amount borrowed at 12%?

66. *Borrowing Money* A small corporation borrowed $500,000 to expand its product line. Some of the money was borrowed at 9%, some at 10%, and some at 12%. How much was borrowed at each rate if the annual interest was $52,000 and the amount borrowed at 10% was $2\frac{1}{2}$ times the amount borrowed at 9%?

67. *Partial Fractions* Write the partial fraction decomposition for $(4x^2)/[(x + 1)^2(x - 1)]$.

68. *Electrical Network*   The currents in an electrical network are given by the solution of the system

$$I_1 - I_2 + I_3 = 0$$
$$2I_1 + 2I_2 \quad\quad = 7$$
$$\quad\quad 2I_2 + 4I_3 = 8$$

where I_1, I_2, and I_3 are measured in amperes. Solve the system of equations.

In Exercises 69–74, find the specified equation that passes through the points. Use a graphing utility to verify your result.

69. Parabola:

$$y = ax^2 + bx + c$$

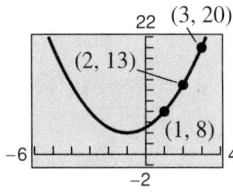

70. Parabola:

$$y = ax^2 + bx + c$$

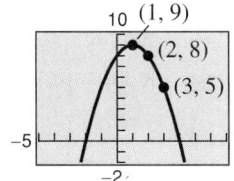

71. Cubic:

$$y = ax^3 + bx^2 + cx + d$$

72. Cubic:

$$y = ax^3 + bx^2 + cx + d$$

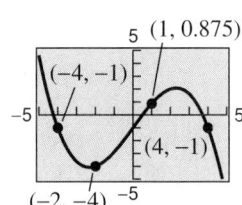

73. Quartic:

$$y = ax^4 + \cdots + dx + e$$

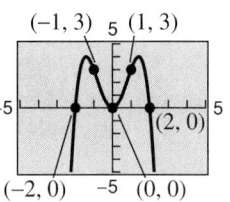

74. Quartic:

$$y = ax^4 + \cdots + dx + e$$

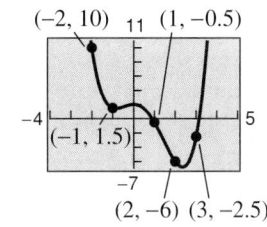

75. *Reading a Graph*   The bar graph gives the value y, in millions of dollars, for new orders of civil jet transport aircraft built by U.S. companies in the years 1990 through 1992.   (Source: Aerospace Industries Association of America)

(a) Find the equation of the parabola that passes through the points. Let $t = 0$ represent 1990.

(b) Use a graphing utility to graph the parabola.

(c) Use the equation in part (a) to estimate y in 1993.

76. *Mathematical Modeling*   After the path of a ball thrown by a baseball player is videotaped, it is analyzed on a television set with a grid covering the screen. The tape is paused three times, and the position of the ball is measured each time. The coordinates are approximately $(0, 5.0)$, $(15, 9.6)$, and $(30, 12.4)$. (The x-coordinate measures the horizontal distance from the player in feet, and the y-coordinate is the height of the ball in feet.)

(a) Find the equation of the parabola $y = ax^2 + bx + c$ that passes through the three points.

(b) Use a graphing utility to graph the parabola. Approximate the maximum height of the ball and the point at which the ball strikes the ground.

(c) Find analytically the maximum height of the ball and the point at which it strikes the ground.

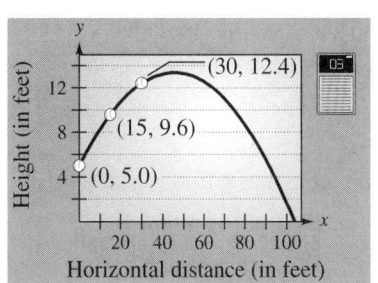

Network Analysis In Exercises 77–80, answer the questions about the specified network. (In a network it is assumed that the total flow into each junction is equal to the total flow out of the junction.)

77. Water is flowing through a network of pipes (in thousands of cubic meters per hour). (See figure.)

    (a) Solve this system for the water flow represented by x_i, $i = 1, 2, 3, 4, 5, 6, 7$.

    (b) Find the network flow pattern when $x_6 = x_7 = 0$.

    (c) Find the network flow pattern when $x_5 = 1000$ and $x_6 = 0$.

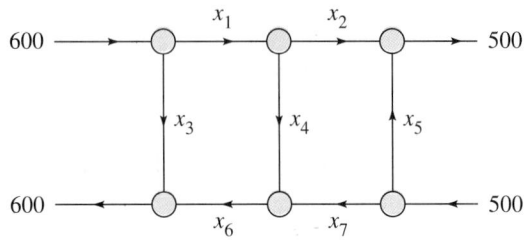

78. The flow of traffic (in vehicles per hour) through a network of streets is shown in the figure.

    (a) Solve this system for the traffic flow represented by x_i, $i = 1, 2, 3, 4, 5$.

    (b) Find the traffic flow when $x_2 = 200$ and $x_3 = 50$.

    (c) Find the traffic flow when $x_2 = 150$ and $x_3 = 0$.

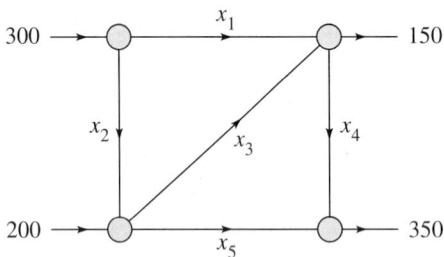

79. The flow of traffic (in vehicles per hour) through a network of streets is shown in the figure.

    (a) Solve this system for the traffic flow represented by x_i, $i = 1, 2, 3, 4$.

    (b) Find the traffic flow when $x_4 = 0$.

    (c) Find the traffic flow when $x_4 = 100$.

Figure for 79

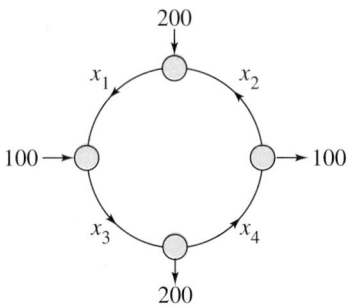

80. The flow of traffic (in vehicles per hour) through a network of streets is shown in the figure.

    (a) Solve this system for the traffic flow represented by x_i, $i = 1, 2, 3, 4, 5$.

    (b) Find the traffic flow when $x_3 = 0$ and $x_5 = 100$.

    (c) Find the traffic flow when $x_3 = x_5 = 100$.

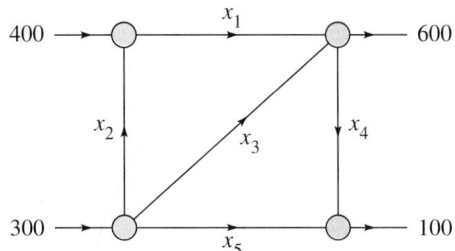

81. *Chapter Opener* Use the models on page 589 to estimate the men's and women's winning times in the 1000-meter speed skating events in the year 2002.

82. *Chapter Opener* If the models on page 589 continue to represent the winning times in the 1000-meter speed skating events, in which winter Olympics will the women's time be less than the men's time?

Review Solve Exercises 83–86 as a review of the skills and problem-solving techniques you learned in previous sections. Graph the function, and check each graph with a graphing utility.

83. $f(x) = 2^{x-1}$

84. $g(x) = 3^{-x/2}$

85. $h(x) = \log_2(x - 1)$

86. $f(x) = 3 + \ln x$

8.2 Operations with Matrices

Equality of Matrices / *Matrix Addition and Scalar Multiplication* /
Matrix Multiplication / *Applications*

Equality of Matrices

In Section 8.1, you used matrices to solve systems of linear equations. Matrices, however, can do much more than this. There is a rich mathematical theory of matrices, and its applications are numerous. This section and the next two introduce some fundamentals of matrix theory. It is standard mathematical convention to represent matrices in any of the following three ways.

1. A matrix can be denoted by an uppercase letter such as A, B, or C.
2. A matrix can be denoted by a representative element enclosed in brackets, such as $[a_{ij}]$, $[b_{ij}]$, or $[c_{ij}]$.
3. A matrix can be denoted by a rectangular array of numbers such as

$$A = [a_{ij}] = \begin{bmatrix} a_{11} & a_{12} & a_{13} & \cdots & a_{1n} \\ a_{21} & a_{22} & a_{23} & \cdots & a_{2n} \\ a_{31} & a_{32} & a_{33} & \cdots & a_{3n} \\ \vdots & \vdots & \vdots & & \vdots \\ a_{m1} & a_{m2} & a_{m3} & \cdots & a_{mn} \end{bmatrix}.$$

Two matrices $A = [a_{ij}]$ and $B = [b_{ij}]$ are **equal** if they have the same order $(m \times n)$ and $a_{ij} = b_{ij}$ for $1 \le i \le m$ and $1 \le j \le n$. In other words, two matrices are equal if their corresponding entries are equal.

A British mathematician, Arthur Cayley, invented matrices around 1858. Cayley was a Cambridge University graduate and a lawyer by profession. His ground-breaking work on matrices was begun as he studied the theory of transformations. Cayley also was instrumental in the development of determinants. Cayley and two American mathematicians, Benjamin Peirce (1809–1880) and his son Charles S. Peirce (1839–1914), are credited with developing "matrix algebra."

EXAMPLE 1 **Equality of Matrices**

Solve for a_{11}, a_{12}, a_{21}, and a_{22} in the following matrix equation.

$$\begin{bmatrix} a_{11} & a_{12} \\ a_{21} & a_{22} \end{bmatrix} = \begin{bmatrix} 2 & -1 \\ -3 & 0 \end{bmatrix}$$

Solution
Because two matrices are equal only if their corresponding entries are equal, you can conclude that

$$a_{11} = 2, \quad a_{12} = -1, \quad a_{21} = -3, \quad \text{and} \quad a_{22} = 0.$$

Matrix Addition and Scalar Multiplication

You can **add** two matrices (of the same order) by adding their corresponding entries.

Definition of Matrix Addition

If $A = [a_{ij}]$ and $B = [b_{ij}]$ are matrices of order $m \times n$, their **sum** is the $m \times n$ matrix given by

$$A + B = [a_{ij} + b_{ij}].$$

The sum of two matrices of different orders is undefined.

Most graphing utilities can perform matrix addition and scalar multiplication. If you have such a graphing utility, duplicate the matrix operations in Examples 2 and 3. Try adding two matrices of different orders such as

$$A = \begin{bmatrix} 1 & 2 \\ 3 & 4 \end{bmatrix} \text{ and}$$

$$B = \begin{bmatrix} 5 \\ 6 \end{bmatrix}.$$

What error message does your utility display?

EXAMPLE 2   Addition of Matrices

a. $\begin{bmatrix} -1 & 2 \\ 0 & 1 \end{bmatrix} + \begin{bmatrix} 1 & 3 \\ -1 & 2 \end{bmatrix} = \begin{bmatrix} -1+1 & 2+3 \\ 0-1 & 1+2 \end{bmatrix} = \begin{bmatrix} 0 & 5 \\ -1 & 3 \end{bmatrix}$

b. $\begin{bmatrix} 0 & 1 & -2 \\ 1 & 2 & 3 \end{bmatrix} + \begin{bmatrix} 0 & 0 & 0 \\ 0 & 0 & 0 \end{bmatrix} = \begin{bmatrix} 0 & 1 & -2 \\ 1 & 2 & 3 \end{bmatrix}$

c. $\begin{bmatrix} 1 \\ -3 \\ -2 \end{bmatrix} + \begin{bmatrix} -1 \\ 3 \\ 2 \end{bmatrix} = \begin{bmatrix} 0 \\ 0 \\ 0 \end{bmatrix}$

d. The sum of

$$A = \begin{bmatrix} 2 & 1 & 0 \\ 4 & 0 & -1 \\ 3 & -2 & 2 \end{bmatrix} \text{ and } B = \begin{bmatrix} 0 & 1 \\ -1 & 3 \\ 2 & 4 \end{bmatrix}$$

is undefined.

In work with matrices, numbers are usually referred to as **scalars.** In this text, scalars will always be real numbers. You can multiply a matrix A by a scalar c by multiplying each entry in A by c.

Definition of Scalar Multiplication

If $A = [a_{ij}]$ is an $m \times n$ matrix and c is a scalar, the **scalar multiple** of A by c is the $m \times n$ matrix given by

$$cA = [ca_{ij}].$$

The symbol $-A$ represents the scalar product $(-1)A$. Moreover, if A and B are of the same order, $A - B$ represents the sum of A and $(-1)B$. That is,

$$A - B = A + (-1)B.$$ Subtraction of matrices

EXAMPLE 3 ◾ **Scalar Multiplication and Matrix Subtraction**

For the following matrices, find (a) $3A$, (b) $-B$, and (c) $3A - B$.

$$A = \begin{bmatrix} 2 & 2 & 4 \\ -3 & 0 & -1 \\ 2 & 1 & 2 \end{bmatrix} \quad \text{and} \quad B = \begin{bmatrix} 2 & 0 & 0 \\ 1 & -4 & 3 \\ -1 & 3 & 2 \end{bmatrix}$$

Solution

a. $3A = 3\begin{bmatrix} 2 & 2 & 4 \\ -3 & 0 & -1 \\ 2 & 1 & 2 \end{bmatrix}$ Scalar multiplication

$$= \begin{bmatrix} 3(2) & 3(2) & 3(4) \\ 3(-3) & 3(0) & 3(-1) \\ 3(2) & 3(1) & 3(2) \end{bmatrix}$$ Multiply each entry by 3.

$$= \begin{bmatrix} 6 & 6 & 12 \\ -9 & 0 & -3 \\ 6 & 3 & 6 \end{bmatrix}$$ Simplify.

b. $-B = (-1)\begin{bmatrix} 2 & 0 & 0 \\ 1 & -4 & 3 \\ -1 & 3 & 2 \end{bmatrix}$ Definition of negation

$$= \begin{bmatrix} -2 & 0 & 0 \\ -1 & 4 & -3 \\ 1 & -3 & -2 \end{bmatrix}$$ Multiply each entry by -1.

c. $3A - B = \begin{bmatrix} 6 & 6 & 12 \\ -9 & 0 & -3 \\ 6 & 3 & 6 \end{bmatrix} - \begin{bmatrix} 2 & 0 & 0 \\ 1 & -4 & 3 \\ -1 & 3 & 2 \end{bmatrix}$ Matrix subtraction

$$= \begin{bmatrix} 4 & 6 & 12 \\ -10 & 4 & -6 \\ 7 & 0 & 4 \end{bmatrix}$$ Subtract corresponding entries.

◾

EXPLORATION

Select two 3×2 matrices A and B. Enter them into your graphing utility and calculate $A + B$ and $B + A$. What do you observe?

Now select a real number c and calculate $c(A + B)$ and $cA + cB$. What do you observe?

It is often convenient to rewrite the scalar multiple cA by factoring c out of every entry in the matrix. For instance, in the following example, the scalar $\frac{1}{2}$ has been factored out of the matrix.

$$\begin{bmatrix} \frac{1}{2} & -\frac{3}{2} \\ \frac{5}{2} & \frac{1}{2} \end{bmatrix} = \frac{1}{2}\begin{bmatrix} 1 & -3 \\ 5 & 1 \end{bmatrix}$$

The properties of matrix addition and scalar multiplication are similar to those of addition and multiplication of real numbers.

Properties of Matrix Addition and Scalar Multiplication

Let A, B, and C be $m \times n$ matrices and let c and d be scalars.

1. $A + B = B + A$	Commutative Property of Matrix Addition
2. $A + (B + C) = (A + B) + C$	Associative Property of Matrix Addition
3. $(cd)A = c(dA)$	Associative Property of Scalar Multiplication
4. $1A = A$	Scalar Identity
5. $c(A + B) = cA + cB$	Distributive Property
6. $(c + d)A = cA + dA$	Distributive Property

Note that the Associative Property of Matrix Addition allows you to write expressions such as $A + B + C$ without ambiguity because the same sum occurs no matter how the matrices are grouped. In other words, you obtain the same sum whether you group $A + B + C$ as $(A + B) + C$ or as $A + (B + C)$. This same reasoning applies to sums of four or more matrices.

EXAMPLE 4   **Addition of More Than Two Matrices**

By adding corresponding entries, you obtain the following sum of four matrices.

$$\begin{bmatrix} 1 \\ 2 \\ -3 \end{bmatrix} + \begin{bmatrix} -1 \\ -1 \\ 2 \end{bmatrix} + \begin{bmatrix} 0 \\ 1 \\ 4 \end{bmatrix} + \begin{bmatrix} 2 \\ -3 \\ -2 \end{bmatrix} = \begin{bmatrix} 2 \\ -1 \\ 1 \end{bmatrix}$$

Most graphing utilities can add and subtract matrices and multiply matrices by scalars. For instance, on a *TI-82* or *TI-83*, you can find the sum of

$$A = \begin{bmatrix} 2 & -3 \\ -1 & 0 \end{bmatrix} \quad \text{and} \quad B = \begin{bmatrix} -1 & 4 \\ 2 & -5 \end{bmatrix}$$

by entering the matrices and then using the following keystrokes.

$[A]$ $\boxed{+}$ $[B]$ $\boxed{\text{ENTER}}$

One important property of addition of real numbers is that the number 0 is the additive identity. That is, $c + 0 = c$ for any real number c. For matrices, a similar property holds. That is, if A is an $m \times n$ matrix and O is the $m \times n$ **zero matrix** consisting entirely of zeros, then $A + O = A$.

In other words, O is the **additive identity** for the set of all $m \times n$ matrices. For example, the following matrices are the additive identities for the set of all 2×3 and 2×2 matrices.

$$O = \begin{bmatrix} 0 & 0 & 0 \\ 0 & 0 & 0 \end{bmatrix} \quad \text{and} \quad O = \begin{bmatrix} 0 & 0 \\ 0 & 0 \end{bmatrix}$$

$$\underbrace{\qquad\qquad}_{\text{Zero } 2 \times 3 \text{ matrix}} \qquad \underbrace{\qquad\qquad}_{\text{Zero } 2 \times 2 \text{ matrix}}$$

Note The algebra of real numbers and the algebra of matrices also have important differences, which will be discussed later.

The algebra of real numbers and the algebra of matrices have many similarities. For example, compare the following solutions.

Real Numbers (Solve for x.)	$m \times n$ *Matrices* (Solve for X.)
$x + a = b$	$X + A = B$
$x + a + (-a) = b + (-a)$	$X + A + (-A) = B + (-A)$
$x + 0 = b - a$	$X + O = B - A$
$x = b - a$	$X = B - A$

EXAMPLE 5 ▱ **Solving a Matrix Equation**

Solve for X in the equation $3X + A = B$, where

$$A = \begin{bmatrix} 1 & -2 \\ 0 & 3 \end{bmatrix} \quad \text{and} \quad B = \begin{bmatrix} -3 & 4 \\ 2 & 1 \end{bmatrix}.$$

Solution

Begin by solving the equation for X to obtain

$$3X = B - A \quad \Longrightarrow \quad X = \frac{1}{3}(B - A).$$

Now, using the matrices A and B, you have

$$X = \frac{1}{3}\left(\begin{bmatrix} -3 & 4 \\ 2 & 1 \end{bmatrix} - \begin{bmatrix} 1 & -2 \\ 0 & 3 \end{bmatrix} \right)$$

$$= \frac{1}{3} \begin{bmatrix} -4 & 6 \\ 2 & -2 \end{bmatrix}$$

$$= \begin{bmatrix} -\frac{4}{3} & 2 \\ \frac{2}{3} & -\frac{2}{3} \end{bmatrix}.$$

8 / *Matrices and Determinants*

Matrix Multiplication

Note The definition of matrix multiplication indicates a *row-by-column* multiplication, where the entry in the ith row and jth column of the product AB is obtained by multiplying the entries in the ith row of A by the corresponding entries in the jth column of B and then adding the results. Example 6 illustrates this process.

The third basic matrix operation is **matrix multiplication.** At first glance, the following definition may seem unusual. You will see later, however, that this definition of the product of two matrices has many practical applications.

Definition of Matrix Multiplication

If $A = [a_{ij}]$ is an $m \times n$ matrix and $B = [b_{ij}]$ is an $n \times p$ matrix, the **product** AB is an $m \times p$ matrix

$$AB = [c_{ij}]$$

where $c_{ij} = a_{i1}b_{1j} + a_{i2}b_{2j} + a_{i3}b_{3j} + \cdots + a_{in}b_{nj}.$

Some graphing utilities, such as the *TI-82* and *TI-83*, are able to add, subtract, and multiply matrices. If you have such a graphing utility, enter the matrices

$$A = \begin{bmatrix} 1 & 2 & 3 \\ 2 & -5 & 1 \end{bmatrix} \quad \text{and}$$

$$B = \begin{bmatrix} -3 & 2 & 1 \\ 4 & -2 & 0 \\ 1 & 2 & 3 \end{bmatrix}$$

and use the following keystrokes to find the product of the matrices.

[A] $\times$ [B] ENTER

You should get:

$$\begin{bmatrix} 8 & 4 & 10 \\ -25 & 16 & 5 \end{bmatrix}.$$

EXAMPLE 6   **Finding the Product of Two Matrices**

Find the product AB where

$$A = \begin{bmatrix} -1 & 3 \\ 4 & -2 \\ 5 & 0 \end{bmatrix} \quad \text{and} \quad B = \begin{bmatrix} -3 & 2 \\ -4 & 1 \end{bmatrix}.$$

Solution

First, note that the product AB is defined because the number of columns of A is equal to the number of rows of B. Moreover, the product AB has order 3×2, and is of the form

$$\begin{bmatrix} -1 & 3 \\ 4 & -2 \\ 5 & 0 \end{bmatrix} \begin{bmatrix} -3 & 2 \\ -4 & 1 \end{bmatrix} = \begin{bmatrix} c_{11} & c_{12} \\ c_{21} & c_{22} \\ c_{31} & c_{32} \end{bmatrix}.$$

To find the entries of the product, multiply each row of A by each column of B, as follows. Use a graphing utility to check this result.

$$AB = \begin{bmatrix} -1 & 3 \\ 4 & -2 \\ 5 & 0 \end{bmatrix} \begin{bmatrix} -3 & 2 \\ -4 & 1 \end{bmatrix}$$

$$= \begin{bmatrix} (-1)(-3) + (3)(-4) & (-1)(2) + (3)(1) \\ (4)(-3) + (-2)(-4) & (4)(2) + (-2)(1) \\ (5)(-3) + (0)(-4) & (5)(2) + (0)(1) \end{bmatrix}$$

$$= \begin{bmatrix} -9 & 1 \\ -4 & 6 \\ -15 & 10 \end{bmatrix}$$

Be sure you understand that for the product of two matrices to be defined, the number of columns of the first matrix must equal the number of rows of the second matrix. That is, the middle two indices must be the same and the outside two indices give the order of the product, as shown in the following diagram.

$$
\begin{matrix}
A & B & = & AB \\
m \times n & n \times p & & m \times p
\end{matrix}
$$

equal
order of AB

EXAMPLE 7   **Matrix Multiplication**

a. $\begin{bmatrix} 1 & 0 & 3 \\ 2 & -1 & -2 \end{bmatrix} \begin{bmatrix} -2 & 4 & 2 \\ 1 & 0 & 0 \\ -1 & 1 & -1 \end{bmatrix} = \begin{bmatrix} -5 & 7 & -1 \\ -3 & 6 & 6 \end{bmatrix}$

        2×3 3×3 2×3

b. $\begin{bmatrix} 3 & 4 \\ -2 & 5 \end{bmatrix} \begin{bmatrix} 1 & 0 \\ 0 & 1 \end{bmatrix} = \begin{bmatrix} 3 & 4 \\ -2 & 5 \end{bmatrix}$

    2×2 2×2 2×2

c. $\begin{bmatrix} 1 & 2 \\ 1 & 1 \end{bmatrix} \begin{bmatrix} -1 & 2 \\ 1 & -1 \end{bmatrix} = \begin{bmatrix} 1 & 0 \\ 0 & 1 \end{bmatrix}$

    2×2 2×2 2×2

d. $\begin{bmatrix} 1 & -2 & -3 \end{bmatrix} \begin{bmatrix} 2 \\ -1 \\ 1 \end{bmatrix} = \begin{bmatrix} 1 \end{bmatrix}$

    1×3 3×1 1×1

e. $\begin{bmatrix} 2 \\ -1 \\ 1 \end{bmatrix} \begin{bmatrix} 1 & -2 & -3 \end{bmatrix} = \begin{bmatrix} 2 & -4 & -6 \\ -1 & 2 & 3 \\ 1 & -2 & -3 \end{bmatrix}$

  3×1 1×3 3×3

f. The product AB for the following matrices is not defined.

$$A = \begin{bmatrix} -2 & 1 \\ 1 & -3 \\ 1 & 4 \end{bmatrix} \quad \text{and} \quad B = \begin{bmatrix} -2 & 3 & 1 & 4 \\ 0 & 1 & -1 & 2 \\ 2 & -1 & 0 & 1 \end{bmatrix}$$

       3×2 3×4

The general pattern for matrix multiplication is as follows. To obtain the entry in the ith row and the jth column of the product AB, use the ith row of A and the jth column of B.

$$\begin{bmatrix} a_{11} & a_{12} & a_{13} & \cdots & a_{1n} \\ a_{21} & a_{22} & a_{23} & \cdots & a_{2n} \\ a_{31} & a_{32} & a_{33} & \cdots & a_{3n} \\ \vdots & \vdots & \vdots & & \vdots \\ a_{i1} & a_{i2} & a_{i3} & \cdots & a_{in} \\ \vdots & \vdots & \vdots & & \vdots \\ a_{m1} & a_{m2} & a_{m3} & \cdots & a_{mn} \end{bmatrix} \begin{bmatrix} b_{11} & b_{12} & \cdots & b_{1j} & \cdots & b_{1p} \\ b_{21} & b_{22} & \cdots & b_{2j} & \cdots & b_{2p} \\ b_{31} & b_{32} & \cdots & b_{3j} & \cdots & b_{3p} \\ \vdots & \vdots & & \vdots & & \vdots \\ b_{n1} & b_{n2} & \cdots & b_{nj} & \cdots & b_{np} \end{bmatrix} = \begin{bmatrix} c_{11} & c_{12} & \cdots & c_{1j} & \cdots & c_{1p} \\ c_{21} & c_{22} & \cdots & c_{2j} & \cdots & c_{2p} \\ \vdots & \vdots & & \vdots & & \vdots \\ c_{i1} & c_{i2} & \cdots & c_{ij} & \cdots & c_{ip} \\ \vdots & \vdots & & \vdots & & \vdots \\ c_{m1} & c_{m2} & \cdots & c_{mj} & \cdots & c_{mp} \end{bmatrix}$$

$$a_{i1}b_{1j} + a_{i2}b_{2j} + a_{i3}b_{3j} + \cdots + a_{in}b_{nj} = c_{ij}$$

Properties of Matrix Multiplication

Let A, B, and C be matrices and let c be a scalar.

1. $A(BC) = (AB)C$ Associative Property of Matrix Multiplication

2. $A(B + C) = AB + AC$ Distributive Property
3. $(A + B)C = AC + BC$ Distributive Property
4. $c(AB) = (cA)B = A(cB)$ Associative Property of Scalar Multiplication

The $n \times n$ matrix that consists of 1's on its main diagonal and 0's elsewhere is called the **identity matrix of order n** and is denoted by

$$I_n = \begin{bmatrix} 1 & 0 & 0 & \cdots & 0 \\ 0 & 1 & 0 & \cdots & 0 \\ 0 & 0 & 1 & \cdots & 0 \\ \vdots & \vdots & \vdots & & \vdots \\ 0 & 0 & 0 & \cdots & 1 \end{bmatrix}.$$ Identity matrix

Note that an identity matrix must be *square*. When the order is understood to be n, you can denote I_n simply by I. If A is an $n \times n$ matrix, the identity matrix has the property that $AI_n = A$ and $I_nA = A$. For example,

$$\begin{bmatrix} 3 & -2 & 5 \\ 1 & 0 & 4 \\ -1 & 2 & -3 \end{bmatrix} \begin{bmatrix} 1 & 0 & 0 \\ 0 & 1 & 0 \\ 0 & 0 & 1 \end{bmatrix} = \begin{bmatrix} 3 & -2 & 5 \\ 1 & 0 & 4 \\ -1 & 2 & -3 \end{bmatrix}$$

and

$$\begin{bmatrix} 1 & 0 & 0 \\ 0 & 1 & 0 \\ 0 & 0 & 1 \end{bmatrix} \begin{bmatrix} 3 & -2 & 5 \\ 1 & 0 & 4 \\ -1 & 2 & -3 \end{bmatrix} = \begin{bmatrix} 3 & -2 & 5 \\ 1 & 0 & 4 \\ -1 & 2 & -3 \end{bmatrix}.$$

Applications

One application of matrix multiplication is representation of a system of linear equations. Note how the system

$$a_{11}x_1 + a_{12}x_2 + a_{13}x_3 = b_1$$
$$a_{21}x_1 + a_{22}x_2 + a_{23}x_3 = b_2$$
$$a_{31}x_1 + a_{32}x_2 + a_{33}x_3 = b_3$$

can be written as the matrix equation $AX = B$, where A is the *coefficient matrix* of the system, and X and B are column matrices.

Note The column matrix B is also called a *constant* matrix. Its entries are the constant terms in the system of equations.

$$\underbrace{\begin{bmatrix} a_{11} & a_{12} & a_{13} \\ a_{21} & a_{22} & a_{23} \\ a_{31} & a_{32} & a_{33} \end{bmatrix}}_{A} \times \underbrace{\begin{bmatrix} x_1 \\ x_2 \\ x_3 \end{bmatrix}}_{X} = \underbrace{\begin{bmatrix} b_1 \\ b_2 \\ b_3 \end{bmatrix}}_{B}$$

EXAMPLE 8 Solving a System of Linear Equations

Solve the matrix equation $AX = B$ for X, where

$$A = \underset{\text{Coefficient matrix}}{\begin{bmatrix} 1 & -2 & 1 \\ 0 & 1 & 2 \\ 2 & 3 & -2 \end{bmatrix}} \quad \text{and} \quad B = \underset{\text{Column matrix}}{\begin{bmatrix} -4 \\ 4 \\ 2 \end{bmatrix}}.$$

Solution
As a system of linear equations, $AX = B$ is as follows.

$$x_1 - 2x_2 + x_3 = -4$$
$$x_2 + 2x_3 = 4$$
$$2x_1 + 3x_2 - 2x_3 = 2$$

Using Gauss-Jordan elimination on the augmented matrix of this system, you obtain the following reduced row-echelon matrix.

$$\begin{bmatrix} 1 & 0 & 0 & \vdots & -1 \\ 0 & 1 & 0 & \vdots & 2 \\ 0 & 0 & 1 & \vdots & 1 \end{bmatrix}$$

Thus, the solution of the system of linear equations is $x_1 = -1$, $x_2 = 2$, and $x_3 = 1$, and the solution of the matrix equation is

$$X = \begin{bmatrix} x_1 \\ x_2 \\ x_3 \end{bmatrix} = \begin{bmatrix} -1 \\ 2 \\ 1 \end{bmatrix}.$$

Use a graphing utility to verify that $AX = B$.

Real Life

EXAMPLE 9 ▭ **Softball Team Expenses**

Two softball teams submit equipment lists to their sponsors.

	Women's Team	Men's Team
Bats	12	15
Balls	45	38
Gloves	15	17

Each bat costs $48, each ball costs $4, and each glove costs $42. Use matrices to find the total cost of equipment for each team.

Solution

The equipment lists and the costs per item can be written in matrix form as

$$E = \begin{bmatrix} 12 & 15 \\ 45 & 38 \\ 15 & 17 \end{bmatrix} \quad \text{and} \quad C = \begin{bmatrix} 48 & 4 & 42 \end{bmatrix}.$$

The total cost of equipment for each team is given by the product

$$CE = \begin{bmatrix} 48 & 4 & 42 \end{bmatrix} \begin{bmatrix} 12 & 15 \\ 45 & 38 \\ 15 & 17 \end{bmatrix} = \begin{bmatrix} 1386 & 1586 \end{bmatrix}.$$

Thus, the total cost of equipment for the women's team is $1386, and the total cost of equipment for the men's team is $1586.   ▭

Group Activity

Problem Posing

Write a matrix multiplication application problem that uses the matrix

$$A = \begin{bmatrix} 20 & 42 & 33 \\ 17 & 30 & 50 \end{bmatrix}.$$

Exchange problems with another student in your class. Form the matrices that represent the problem, and solve the problem. Interpret your solution in the context of the problem. Check with the creator of the problem to see if you are correct. Discuss other ways to represent and/or approach the problem.

8.2 /// EXERCISES

In Exercises 1–4, find x and y.

1. $\begin{bmatrix} x & -2 \\ 7 & y \end{bmatrix} = \begin{bmatrix} -4 & -2 \\ 7 & 22 \end{bmatrix}$

2. $\begin{bmatrix} -5 & x \\ y & 8 \end{bmatrix} = \begin{bmatrix} -5 & 13 \\ 12 & 8 \end{bmatrix}$

3. $\begin{bmatrix} 16 & 4 & 5 & 4 \\ -3 & 13 & 15 & 6 \\ 0 & 2 & 4 & 0 \end{bmatrix} = \begin{bmatrix} 16 & 4 & 2x+1 & 4 \\ -3 & 13 & & 15 & 3x \\ 0 & 2 & 3y-5 & 0 \end{bmatrix}$

4. $\begin{bmatrix} x+2 & 8 & -3 \\ 1 & 2y & 2x \\ 7 & -2 & y+2 \end{bmatrix} = \begin{bmatrix} 2x+6 & 8 & -3 \\ 1 & 18 & -8 \\ 7 & -2 & 11 \end{bmatrix}$

In Exercises 5–10, find (a) $A + B$, (b) $A - B$, (c) $3A$, and (d) $3A - 2B$.

5. $A = \begin{bmatrix} 1 & -1 \\ 2 & -1 \end{bmatrix}$, $B = \begin{bmatrix} 2 & -1 \\ -1 & 8 \end{bmatrix}$

6. $A = \begin{bmatrix} 1 & 2 \\ 2 & 1 \end{bmatrix}$, $B = \begin{bmatrix} -3 & -2 \\ 4 & 2 \end{bmatrix}$

7. $A = \begin{bmatrix} 6 & -1 \\ 2 & 4 \\ -3 & 5 \end{bmatrix}$, $B = \begin{bmatrix} 1 & 4 \\ -1 & 5 \\ 1 & 10 \end{bmatrix}$

8. $A = \begin{bmatrix} 2 & 1 & 1 \\ -1 & -1 & 4 \end{bmatrix}$, $B = \begin{bmatrix} 2 & -3 & 4 \\ -3 & 1 & -2 \end{bmatrix}$

9. $A = \begin{bmatrix} 2 & 2 & -1 & 0 & 1 \\ 1 & 1 & -2 & 0 & -1 \end{bmatrix}$,

$B = \begin{bmatrix} 1 & 1 & -1 & 1 & 0 \\ -3 & 4 & 9 & -6 & -7 \end{bmatrix}$

10. $A = \begin{bmatrix} 3 \\ 2 \\ -1 \end{bmatrix}$, $B = \begin{bmatrix} -4 \\ 6 \\ 2 \end{bmatrix}$

In Exercises 11–14, solve for X given

$A = \begin{bmatrix} -2 & -1 \\ 1 & 0 \\ 3 & -4 \end{bmatrix}$ and $B = \begin{bmatrix} 0 & 3 \\ 2 & 0 \\ -4 & -1 \end{bmatrix}$.

11. $X = 3A - 2B$

12. $2X = 2A - B$

13. $2X + 3A = B$

14. $2A + 4B = -2X$

In Exercises 15–20, find (a) AB, (b) BA, and, if possible, (c) A^2. (*Note:* $A^2 = AA$.)

15. $A = \begin{bmatrix} 1 & 2 \\ 4 & 2 \end{bmatrix}$, $B = \begin{bmatrix} 2 & -1 \\ -1 & 8 \end{bmatrix}$

16. $A = \begin{bmatrix} 2 & -1 \\ 1 & 4 \end{bmatrix}$, $B = \begin{bmatrix} 0 & 0 \\ 3 & -3 \end{bmatrix}$

17. $A = \begin{bmatrix} 3 & -1 \\ 1 & 3 \end{bmatrix}$, $B = \begin{bmatrix} 1 & -3 \\ 3 & 1 \end{bmatrix}$

18. $A = \begin{bmatrix} 1 & -1 \\ 1 & 1 \end{bmatrix}$, $B = \begin{bmatrix} 1 & 3 \\ -3 & 1 \end{bmatrix}$

19. $A = \begin{bmatrix} 1 & -1 & 7 \\ 2 & -1 & 8 \\ 3 & 1 & -1 \end{bmatrix}$, $B = \begin{bmatrix} 1 & 1 & 2 \\ 2 & 1 & 1 \\ 1 & -3 & 2 \end{bmatrix}$

20. $A = \begin{bmatrix} 3 & 2 & 1 \end{bmatrix}$, $B = \begin{bmatrix} 2 \\ 3 \\ 0 \end{bmatrix}$

In Exercises 21–28, find AB, if possible.

21. $A = \begin{bmatrix} 2 & 1 \\ -3 & 4 \\ 1 & 6 \end{bmatrix}$, $B = \begin{bmatrix} 0 & -1 & 0 \\ 4 & 0 & 2 \\ 8 & -1 & 7 \end{bmatrix}$

22. $A = \begin{bmatrix} 0 & -1 & 0 \\ 4 & 0 & 2 \\ 8 & -1 & 7 \end{bmatrix}$, $B = \begin{bmatrix} 2 & 1 \\ -3 & 4 \\ 1 & 6 \end{bmatrix}$

23. $A = \begin{bmatrix} -1 & 3 \\ 4 & -5 \\ 0 & 2 \end{bmatrix}$, $B = \begin{bmatrix} 1 & 2 \\ 0 & 7 \end{bmatrix}$

24. $A = \begin{bmatrix} 1 & 0 & 0 \\ 0 & 4 & 0 \\ 0 & 0 & -2 \end{bmatrix}$, $B = \begin{bmatrix} 3 & 0 & 0 \\ 0 & -1 & 0 \\ 0 & 0 & 5 \end{bmatrix}$

25. $A = \begin{bmatrix} 5 & 0 & 0 \\ 0 & -8 & 0 \\ 0 & 0 & 7 \end{bmatrix}$, $B = \begin{bmatrix} \frac{1}{5} & 0 & 0 \\ 0 & -\frac{1}{8} & 0 \\ 0 & 0 & \frac{1}{2} \end{bmatrix}$

26. $A = \begin{bmatrix} 10 \\ 12 \end{bmatrix}$, $B = \begin{bmatrix} 6 & -2 & 1 & 6 \end{bmatrix}$

27. $A = \begin{bmatrix} 0 & 0 & 5 \\ 0 & 0 & -3 \\ 0 & 0 & 4 \end{bmatrix}$, $B = \begin{bmatrix} 6 & -11 & 4 \\ 8 & 16 & 4 \\ 0 & 0 & 0 \end{bmatrix}$

28. $A = \begin{bmatrix} 1 & 0 & 3 & -2 \\ 6 & 13 & 8 & -17 \end{bmatrix}$, $B = \begin{bmatrix} 1 & 6 \\ 4 & 2 \end{bmatrix}$

In Exercises 29–34, use the matrix capabilities of a graphing utility to find *AB*.

29. $A = \begin{bmatrix} 5 & 6 & -3 \\ -2 & 5 & 1 \\ 10 & -5 & 5 \end{bmatrix}$, $B = \begin{bmatrix} 1 & -1 & 2 \\ 8 & 1 & 4 \\ 4 & -2 & 9 \end{bmatrix}$

30. $A = \begin{bmatrix} 11 & -12 & 4 \\ 14 & 10 & 12 \\ 6 & -2 & 9 \end{bmatrix}$, $B = \begin{bmatrix} 12 & 10 \\ -5 & 12 \\ 15 & 16 \end{bmatrix}$

31. $A = \begin{bmatrix} -3 & 8 & -6 & 8 \\ -12 & 15 & 9 & 6 \\ 5 & -1 & 1 & 5 \end{bmatrix}$,

$B = \begin{bmatrix} 3 & 1 & 6 \\ 24 & 15 & 14 \\ 16 & 10 & 21 \\ 8 & -4 & 10 \end{bmatrix}$

32. $A = \begin{bmatrix} -2 & 4 & 8 \\ 21 & 5 & 6 \\ 13 & 2 & 6 \end{bmatrix}$,

$B = \begin{bmatrix} 2 & 0 \\ -7 & 15 \\ 32 & 14 \\ 0.5 & 1.6 \end{bmatrix}$

33. $A = \begin{bmatrix} 9 & 10 & -38 & 18 \\ 100 & -50 & 250 & 75 \end{bmatrix}$,

$B = \begin{bmatrix} 52 & -85 & 27 & 45 \\ 40 & -35 & 60 & 82 \end{bmatrix}$

34. $A = \begin{bmatrix} 15 & -18 \\ -4 & 12 \\ -8 & 22 \end{bmatrix}$,

$B = \begin{bmatrix} -7 & 22 & 1 \\ 8 & 16 & 24 \end{bmatrix}$

In Exercises 35–38, find matrices *A*, *X*, and *B* such that the system of linear equations can be written as the matrix equation $AX = B$. Solve the system of equations. Use a graphing utility to check your result.

35. $\begin{aligned} -x + y &= 4 \\ -2x + y &= 0 \end{aligned}$

36. $\begin{aligned} x - 2y + 3z &= 9 \\ -x + 3y - z &= -6 \\ 2x - 5y + 5z &= 17 \end{aligned}$

37. $\begin{aligned} 2x + 3y &= 5 \\ x + 4y &= 10 \end{aligned}$

38. $\begin{aligned} x + y - 3z &= -1 \\ -x + 2y &= 1 \\ -y + z &= 0 \end{aligned}$

In Exercises 39–42, use the matrix capabilities of a graphing utility to find

$$f(A) = a_0 I_n + a_1 A + a_2 A^2 + \cdots + a_n A^n.$$

39. $f(x) = x^2 - 5x + 2$, $A = \begin{bmatrix} 2 & 0 \\ 4 & 5 \end{bmatrix}$

40. $f(x) = x^2 - 7x + 6$, $A = \begin{bmatrix} 5 & 4 \\ 1 & 2 \end{bmatrix}$

41. $f(x) = x^3 - 10x^2 + 31x - 30$, $A = \begin{bmatrix} 3 & 1 & 4 \\ 0 & 2 & 6 \\ 0 & 0 & 5 \end{bmatrix}$

42. $f(x) = x^2 - 10x + 24$, $A = \begin{bmatrix} 8 & -4 \\ 2 & 2 \end{bmatrix}$

43. *Think About It* If *a*, *b*, and *c* are real numbers such that $c \neq 0$ and $ac = bc$, then $a = b$. However, if *A*, *B*, and *C* are nonzero matrices such that $AC = BC$, then *A* is *not necessarily* equal to *B*. Illustrate this using the following matrices.

$A = \begin{bmatrix} 0 & 1 \\ 0 & 1 \end{bmatrix}$, $B = \begin{bmatrix} 1 & 0 \\ 1 & 0 \end{bmatrix}$, $C = \begin{bmatrix} 2 & 3 \\ 2 & 3 \end{bmatrix}$

44. *Think About It* If *a* and *b* are real numbers such that $ab = 0$, then $a = 0$ or $b = 0$. However, if *A* and *B* are matrices such that $AB = 0$, it is *not necessarily* true that $A = 0$ or $B = 0$. Illustrate this using the following matrices.

$A = \begin{bmatrix} 3 & 3 \\ 4 & 4 \end{bmatrix}$, $B = \begin{bmatrix} 1 & -1 \\ -1 & 1 \end{bmatrix}$

Think About It In Exercises 45–54, use matrices A and B each of order 2×3, C of order 3×2, and D of order 2×2. Determine whether the matrices are of proper order to perform the operation(s). If so, give the order of the answer.

45. $A + 2C$

46. $B - 3C$

47. AB

48. BC

49. $BC - D$

50. $CB - D$

51. $(CA)D$

52. $(BC)D$

53. $D(A - 3B)$

54. $(BC - D)A$

55. *Factory Production* A certain corporation has three factories, each of which manufactures two products. The number of units of product i produced at factory j in one day is represented by a_{ij} in the matrix

$$A = \begin{bmatrix} 60 & 40 & 20 \\ 30 & 90 & 60 \end{bmatrix}.$$

Find the production levels if production is increased by 20%. (*Hint:* Because an increase of 20% corresponds to 100% + 20%, multiply the given matrix by 1.2.)

56. *Factory Production* A certain corporation has four factories, each of which manufactures two products. The number of units of product i produced at factory j in one day is represented by a_{ij} in the matrix

$$A = \begin{bmatrix} 100 & 90 & 70 & 30 \\ 40 & 20 & 60 & 60 \end{bmatrix}.$$

Find the production levels if production is increased by 10%.

57. *Crop Production* A fruit grower raises two crops, which are shipped to three outlets. The number of units of crop i that are shipped to outlet j is represented by a_{ij} in the matrix

$$A = \begin{bmatrix} 100 & 75 & 75 \\ 125 & 150 & 100 \end{bmatrix}.$$

The profit per unit is represented by the matrix

$$B = [\$3.75 \quad \$7.00].$$

Find the product BA, and state what each entry of the product represents.

58. *Revenue* A manufacturer produces three models of a product, which are shipped to two warehouses. The number of units of model i that are shipped to warehouse j is represented by a_{ij} in the matrix

$$A = \begin{bmatrix} 5,000 & 4,000 \\ 6,000 & 10,000 \\ 8,000 & 5,000 \end{bmatrix}.$$

The price per unit is represented by the matrix

$$B = [\$20.50 \quad \$26.50 \quad \$29.50].$$

Compute BA and interpret the result.

Exploration In Exercises 59 and 60, let $i = \sqrt{-1}$.

59. Consider the matrix

$$A = \begin{bmatrix} i & 0 \\ 0 & i \end{bmatrix}.$$

Find A^2, A^3, and A^4. Identify any similarities with i^2, i^3, and i^4.

60. Find and identify A^2 for the matrix

$$A = \begin{bmatrix} 0 & -i \\ i & 0 \end{bmatrix}.$$

61. *Inventory Levels* A company sells five models of computers through three retail outlets. The inventories are given by S.

Model

$$\begin{array}{c} \quad\;\; A \;\; B \;\; C \;\; D \;\; E \\ S = \begin{bmatrix} 3 & 2 & 2 & 3 & 0 \\ 0 & 2 & 3 & 4 & 3 \\ 4 & 2 & 1 & 3 & 2 \end{bmatrix} \begin{array}{l} 1 \\ 2 \\ 3 \end{array} \end{array} \text{Outlet}$$

The wholesale and retail prices are given by T.

Price

$$\begin{array}{c} \text{Wholesale} \;\; \text{Retail} \\ T = \begin{bmatrix} \$840 & \$1100 \\ \$1200 & \$1350 \\ \$1450 & \$1650 \\ \$2650 & \$3000 \\ \$3050 & \$3200 \end{bmatrix} \begin{array}{l} A \\ B \\ C \\ D \\ E \end{array} \end{array} \text{Model}$$

Compute ST and interpret the result.

62. *Labor/Wage Requirements*   A company that manufactures boats has the following labor-hour and wage requirements.

Labor per Boat

Department

$$S = \begin{bmatrix} 1.0 \text{ hr} & 0.5 \text{ hr} & 0.2 \text{ hr} \\ 1.6 \text{ hr} & 1.0 \text{ hr} & 0.2 \text{ hr} \\ 2.5 \text{ hr} & 2.0 \text{ hr} & 0.4 \text{ hr} \end{bmatrix} \begin{matrix} \text{Small} \\ \text{Medium} \\ \text{Large} \end{matrix} \text{ Boat size}$$

Cutting Assembly Packaging

Wages per Hour

Plant

A   B

$$T = \begin{bmatrix} \$12 & \$10 \\ \$9 & \$8 \\ \$6 & \$5 \end{bmatrix} \begin{matrix} \text{Cutting} \\ \text{Assembly} \\ \text{Packaging} \end{matrix} \text{ Department}$$

Compute ST and interpret the result.

63. *Voting Preference*   The matrix

From

R   D   I

$$P = \begin{bmatrix} 0.6 & 0.1 & 0.1 \\ 0.2 & 0.7 & 0.1 \\ 0.2 & 0.2 & 0.8 \end{bmatrix} \begin{matrix} \text{R} \\ \text{D} \\ \text{I} \end{matrix} \text{ To}$$

is called a stochastic matrix. Each entry $p_{ij}(i \neq j)$ represents the proportion of the voting population that changes from party i to party j, and p_{ii} represents the proportion that remains loyal to the party from one election to the next. Compute and interpret P^2.

64. *Voting Preference*   Use a graphing utility to find P^3, P^4, P^5, P^6, P^7, and P^8 for the matrix given in Exercise 63. Can you detect a pattern as P is raised to higher powers?

65. *Exploration*   Let A and B be unequal diagonal matrices of the same order. (A *diagonal matrix* is a square matrix in which each entry not on the main diagonal is zero.) Determine the products AB for several pairs of such matrices. Make a conjecture about a quick rule for such products.

66. *Exploration*   Consider matrices of the form

$$A = \begin{bmatrix} 0 & a_{12} & a_{13} & a_{14} & \cdots & a_{1n} \\ 0 & 0 & a_{23} & a_{24} & \cdots & a_{2n} \\ 0 & 0 & 0 & a_{34} & \cdots & a_{3n} \\ \vdots & \vdots & \vdots & \vdots & \cdots & \vdots \\ 0 & 0 & 0 & 0 & \cdots & a_{(n-1)n} \\ 0 & 0 & 0 & 0 & \cdots & 0 \end{bmatrix}.$$

(a) Write a 2×2 matrix and a 3×3 matrix in the form of A.

(b) Use a graphing utility to raise each of the matrices to higher powers. Describe the result.

(c) Use the result of part (b) to make a conjecture about powers of A if A is a 4×4 matrix. Use a graphing utility to test your conjecture.

(d) Use the results of parts (b) and (c) to make a conjecture about powers of A if A is an $n \times n$ matrix.

Review   **Solve Exercises 67–70 as a review of the skills and problem-solving techniques you learned in previous sections. Simplify the logarithmic expression.**

67. $\log_4 \sqrt{32}$

68. $\log_5(5^3 \cdot 4)$

69. $\log_2 \dfrac{12}{5}$

70. $\ln \dfrac{100}{e^2}$

8.3   The Inverse of a Square Matrix

The Inverse of a Matrix / *Finding Inverse Matrices* /
The Inverse of a 2 × 2 Matrix / *Systems of Linear Equations*

The Inverse of a Matrix

This section further develops the algebra of matrices. To begin, consider the real number equation $ax = b$. To solve this equation for x, multiply both sides of the equation by a^{-1} (provided that $a \neq 0$).

$$ax = b$$
$$(a^{-1}a)x = a^{-1}b$$
$$(1)x = a^{-1}b$$
$$x = a^{-1}b$$

The number a^{-1} is called the *multiplicative inverse of a* because $a^{-1}a = 1$. The definition of the multiplicative inverse of a matrix is similar.

Note The symbol A^{-1} is read "*A* inverse."

Definition of the Inverse of a Square Matrix

Let A be an $n \times n$ matrix. If there exists a matrix A^{-1} such that

$$AA^{-1} = I_n = A^{-1}A$$

A^{-1} is called the **inverse** of A.

Note Recall that it is not always true that $AB = BA$, even if both products are defined. However, if A and B are both square matrices and $AB = I_n$, it can be shown that $BA = I_n$. Hence, in Example 1, you need only to check that $AB = I_2$.

EXAMPLE 1   The Inverse of a Matrix

Show that B is the inverse of A, where

$$A = \begin{bmatrix} -1 & 2 \\ -1 & 1 \end{bmatrix} \quad \text{and} \quad B = \begin{bmatrix} 1 & -2 \\ 1 & -1 \end{bmatrix}.$$

Solution

To show that B is the inverse of A, show that $AB = I = BA$, as follows.

$$AB = \begin{bmatrix} -1 & 2 \\ -1 & 1 \end{bmatrix}\begin{bmatrix} 1 & -2 \\ 1 & -1 \end{bmatrix} = \begin{bmatrix} -1+2 & 2-2 \\ -1+1 & 2-1 \end{bmatrix} = \begin{bmatrix} 1 & 0 \\ 0 & 1 \end{bmatrix}$$

$$BA = \begin{bmatrix} 1 & -2 \\ 1 & -1 \end{bmatrix}\begin{bmatrix} -1 & 2 \\ -1 & 1 \end{bmatrix} = \begin{bmatrix} -1+2 & 2-2 \\ -1+1 & 2-1 \end{bmatrix} = \begin{bmatrix} 1 & 0 \\ 0 & 1 \end{bmatrix}$$

If a matrix A has an inverse, A is called **invertible** (or **nonsingular**); otherwise, A is called **singular.** A nonsquare matrix cannot have an inverse. To see this, note that if A is of order $m \times n$ and B is of order $n \times m$ (where $m \neq n$), the products AB and BA are of different orders and therefore cannot be equal to each other. Not all square matrices possess inverses (see the matrix at the bottom of page 622). If, however, a matrix does have an inverse, that inverse is unique. The following example shows how to use systems of equations to find the inverse of a matrix.

Most graphing utilities have the capability of finding the inverse of a square matrix. For instance, to find the inverse of the matrix

$$A = \begin{bmatrix} 2 & -3 & 1 \\ -1 & 2 & -1 \\ -2 & 0 & 1 \end{bmatrix}$$

on a *TI-82* or *TI-83*, enter the matrix. Then use the following keystrokes

After you find A^{-1}, store it as $[\text{B}]$ and use the graphing utility to find $[\text{A}] \times [\text{B}]$ and $[\text{B}] \times [\text{A}]$. What can you conclude?

EXAMPLE 2   **Finding the Inverse of a Matrix**

Find the inverse of

$$A = \begin{bmatrix} 1 & 4 \\ -1 & -3 \end{bmatrix}.$$

Solution

To find the inverse of A, try to solve the matrix equation $AX = I$ for X.

$$\underset{A}{\begin{bmatrix} 1 & 4 \\ -1 & -3 \end{bmatrix}} \underset{X}{\begin{bmatrix} x_{11} & x_{12} \\ x_{21} & x_{22} \end{bmatrix}} = \underset{I}{\begin{bmatrix} 1 & 0 \\ 0 & 1 \end{bmatrix}}$$

$$\begin{bmatrix} x_{11} + 4x_{21} & x_{12} + 4x_{22} \\ -x_{11} - 3x_{21} & -x_{12} - 3x_{22} \end{bmatrix} = \begin{bmatrix} 1 & 0 \\ 0 & 1 \end{bmatrix}$$

Equating corresponding entries, you obtain the following two systems of linear equations.

$$x_{11} + 4x_{21} = 1 \qquad\qquad x_{12} + 4x_{22} = 0$$
$$-x_{11} - 3x_{21} = 0 \qquad\qquad -x_{12} - 3x_{22} = 1$$

From the first system you can determine that $x_{11} = -3$ and $x_{21} = 1$, and from the second system you can determine that $x_{12} = -4$ and $x_{22} = 1$. Therefore, the inverse of A is

$$X = A^{-1} = \begin{bmatrix} -3 & -4 \\ 1 & 1 \end{bmatrix}.$$

You can use matrix multiplication to check this result.

Check

$$AA^{-1} = \begin{bmatrix} 1 & 4 \\ -1 & -3 \end{bmatrix} \begin{bmatrix} -3 & -4 \\ 1 & 1 \end{bmatrix} = \begin{bmatrix} 1 & 0 \\ 0 & 1 \end{bmatrix} \checkmark$$

$$A^{-1}A = \begin{bmatrix} -3 & -4 \\ 1 & 1 \end{bmatrix} \begin{bmatrix} 1 & 4 \\ -1 & -3 \end{bmatrix} = \begin{bmatrix} 1 & 0 \\ 0 & 1 \end{bmatrix} \checkmark$$

Finding Inverse Matrices

In Example 2, note that the two systems of linear equations have the *same coefficient matrix A*. Rather than solve the two systems represented by

$$\begin{bmatrix} 1 & 4 & \vdots & 1 \\ -1 & -3 & \vdots & 0 \end{bmatrix} \quad \text{and} \quad \begin{bmatrix} 1 & 4 & \vdots & 0 \\ -1 & -3 & \vdots & 1 \end{bmatrix}$$

separately, you can solve them *simultaneously* by **adjoining** the identity matrix to the coefficient matrix to obtain

$$\begin{matrix} A & & & I \\ \begin{bmatrix} 1 & 4 & \vdots & 1 & 0 \\ -1 & -3 & \vdots & 0 & 1 \end{bmatrix} \end{matrix}.$$

Then, applying Gauss-Jordan elimination to this matrix, you can solve *both* systems with a single elimination process, as follows.

$$\begin{bmatrix} 1 & 4 & \vdots & 1 & 0 \\ -1 & -3 & \vdots & 0 & 1 \end{bmatrix}$$

$$R_1 + R_2 \rightarrow \begin{bmatrix} 1 & 4 & \vdots & 1 & 0 \\ 0 & 1 & \vdots & 1 & 1 \end{bmatrix}$$

$$-4R_2 + R_1 \rightarrow \begin{bmatrix} 1 & 0 & \vdots & -3 & -4 \\ 0 & 1 & \vdots & 1 & 1 \end{bmatrix}$$

Thus, from the "doubly augmented" matrix $[A \ \vdots \ I]$, you obtained the matrix $[I \ \vdots \ A^{-1}]$.

$$\begin{matrix} A & & I & & & I & & A^{-1} \\ \begin{bmatrix} 1 & 4 & \vdots & 1 & 0 \\ -1 & -3 & \vdots & 0 & 1 \end{bmatrix} & \implies & \begin{bmatrix} 1 & 0 & \vdots & -3 & -4 \\ 0 & 1 & \vdots & 1 & 1 \end{bmatrix} \end{matrix}$$

This procedure (or algorithm) works for any square matrix that has an inverse.

Finding an Inverse Matrix

Let A be a square matrix of order n.

1. Write the $n \times 2n$ matrix that consists of the given matrix A on the left and the $n \times n$ identity matrix I on the right to obtain $[A \ \vdots \ I]$. Note that we separate the matrices A and I by a dotted line. We call this process **adjoining** the matrices A and I.
2. If possible, row reduce A to I using elementary row operations on the *entire* matrix $[A \ \vdots \ I]$. The result will be the matrix $[I \ \vdots \ A^{-1}]$. If this is not possible, A is not invertible.
3. Check your work by multiplying to see that $AA^{-1} = I = A^{-1}A$.

EXAMPLE 3   **Finding the Inverse of a Matrix**

Find the inverse of $A = \begin{bmatrix} 1 & -1 & 0 \\ 1 & 0 & -1 \\ 6 & -2 & -3 \end{bmatrix}$.

Solution

Begin by adjoining the identity matrix to A to form the matrix

$$[A \;\vdots\; I] = \begin{bmatrix} 1 & -1 & 0 & \vdots & 1 & 0 & 0 \\ 1 & 0 & -1 & \vdots & 0 & 1 & 0 \\ 6 & -2 & -3 & \vdots & 0 & 0 & 1 \end{bmatrix}.$$

Using elementary row operations to obtain the form $[I \;\vdots\; A^{-1}]$ results in

$$\begin{bmatrix} 1 & 0 & 0 & \vdots & -2 & -3 & 1 \\ 0 & 1 & 0 & \vdots & -3 & -3 & 1 \\ 0 & 0 & 1 & \vdots & -2 & -4 & 1 \end{bmatrix}.$$

Therefore, the matrix A is invertible and its inverse is

$$A^{-1} = \begin{bmatrix} -2 & -3 & 1 \\ -3 & -3 & 1 \\ -2 & -4 & 1 \end{bmatrix}.$$

Try using a graphing utility to confirm this result by multiplying A by A^{-1} to obtain I.

The process shown in Example 3 applies to any $n \times n$ matrix A. If A has an inverse, this process will find it. If A does not have an inverse, the process will tell us so. For instance, the following matrix has no inverse.

$$A = \begin{bmatrix} 1 & 2 & 0 \\ 3 & -1 & 2 \\ -2 & 3 & -2 \end{bmatrix}$$

To confirm that matrix A above has no inverse, begin by adjoining the identity matrix to A to form

$$[A \;\vdots\; I] = \begin{bmatrix} 1 & 2 & 0 & \vdots & 1 & 0 & 0 \\ 3 & -1 & 2 & \vdots & 0 & 1 & 0 \\ -2 & 3 & -2 & \vdots & 0 & 0 & 1 \end{bmatrix}.$$

Then use elementary row operations to obtain

$$\begin{bmatrix} 1 & 2 & 0 & \vdots & 1 & 0 & 0 \\ 0 & -7 & 2 & \vdots & -3 & 1 & 0 \\ 0 & 0 & 0 & \vdots & -1 & 1 & 1 \end{bmatrix}.$$

At this point in the elimination process you can see that it is impossible to obtain the identity matrix I on the left. Therefore, A is not invertible.

Verify the computations in Example 3 with your graphing utility. Enter the 3×6 matrix $[A \;\vdots\; I]$ and row reduce it to the matrix $[I \;\vdots\; A^{-1}]$, as follows.

$$\begin{bmatrix} 1 & -1 & 0 & 1 & 0 & 0 \\ 1 & 0 & -1 & 0 & 1 & 0 \\ 6 & -2 & -3 & 0 & 0 & 1 \end{bmatrix}$$

$$\downarrow$$

$$\begin{bmatrix} 1 & 0 & 0 & -2 & -3 & 1 \\ 0 & 1 & 0 & -3 & -3 & 1 \\ 0 & 0 & 1 & -2 & -4 & 1 \end{bmatrix}$$

What happens if you try this method to find the inverse of

$$A = \begin{bmatrix} 1 & 2 & 0 \\ 3 & -1 & 2 \\ -2 & 3 & -2 \end{bmatrix}?$$

The Inverse of a 2 × 2 Matrix

Using Gauss-Jordan elimination to find the inverse of a matrix works well (even as a computer technique) for matrices of order 3 × 3 or greater. For 2 × 2 matrices, however, many people prefer to use a formula for the inverse rather than Gauss-Jordan elimination. This simple formula, which works *only* for 2 × 2 matrices, is explained as follows. If A is a 2 × 2 matrix given by

$$A = \begin{bmatrix} a & b \\ c & d \end{bmatrix}$$

then A is invertible if and only if $ad - bc \neq 0$. If $ad - bc \neq 0$, the inverse is given by

$$A^{-1} = \frac{1}{ad - bc} \begin{bmatrix} d & -b \\ -c & a \end{bmatrix}.$$

Try verifying this inverse by multiplication.

Note   The denominator $ad - bc$ is called the **determinant** of the 2 × 2 matrix A. You will study determinants in the next section.

EXPLORATION

Use a graphing utility with matrix operations to find the inverse of the matrix

$$A = \begin{bmatrix} 1 & -3 \\ -2 & 6 \end{bmatrix}.$$

What message appears on the screen? Why does the graphing utility display this message?

EXAMPLE 4   ▱   **Finding the Inverse of a 2 × 2 Matrix**

If possible, find the inverse of the matrix.

a. $A = \begin{bmatrix} 3 & -1 \\ -2 & 2 \end{bmatrix}$

b. $B = \begin{bmatrix} 3 & -1 \\ -6 & 2 \end{bmatrix}$

Solution

a. For the matrix A, apply the formula for the inverse of a 2 × 2 matrix to obtain

$$ad - bc = (3)(2) - (-1)(-2) = 4.$$

Because this quantity is not zero, the inverse is formed by interchanging the entries on the main diagonal, changing the signs of the other two entries, and multiplying by the scalar $\frac{1}{4}$, as follows.

$$A^{-1} = \frac{1}{4} \begin{bmatrix} 2 & 1 \\ 2 & 3 \end{bmatrix} = \begin{bmatrix} \frac{1}{2} & \frac{1}{4} \\ \frac{1}{2} & \frac{3}{4} \end{bmatrix}$$

b. For the matrix B, you have

$$ad - bc = (3)(2) - (-1)(-6) = 0$$

which means that B is not invertible.   ▱

Systems of Linear Equations

The formula $X = A^{-1}B$ is used on most graphing utilities to solve linear systems that have invertible coefficient matrices. That is, you enter the $n \times n$ coefficient matrix [A] and the $n \times 1$ column matrix [B]. The solution X is given by $[A]^{-1}[B]$.

You know that a system of linear equations can have exactly one solution, infinitely many solutions, or no solution. If the coefficient matrix A of a *square* system (a system that has the same number of equations as variables) is invertible, the system has a unique solution, which is given as follows.

A System of Equations with a Unique Solution

If A is an invertible matrix, the system of linear equations represented by $AX = B$ has a unique solution given by

$$X = A^{-1}B.$$

Note Use Gauss-Jordan elimination or a graphing utility to verify A^{-1} for the system of equations in Example 5.

EXAMPLE 5   Solving a System of Equations Using an Inverse

Use an inverse matrix to solve the system.

$$\begin{aligned}
2x + 3y + z &= -1 \\
3x + 3y + z &= 1 \\
2x + 4y + z &= -2
\end{aligned}$$

Solution

$$X = A^{-1}B = \begin{bmatrix} -1 & 1 & 0 \\ -1 & 0 & 1 \\ 6 & -2 & -3 \end{bmatrix} \begin{bmatrix} -1 \\ 1 \\ -2 \end{bmatrix} = \begin{bmatrix} 2 \\ -1 \\ -2 \end{bmatrix}$$

Thus, the solution is $x = 2$, $y = -1$, and $z = -2$.

Group Activity

Finding an Inverse Matrix

Use a graphing utility to decide which of the following matrices is (are) invertible.

a. $A = \begin{bmatrix} -3 & 2 \\ 7 & 4 \end{bmatrix}$ **b.** $B = \begin{bmatrix} 1 & -4 & 2 \\ 2 & -9 & 5 \\ 1 & -5 & 4 \end{bmatrix}$

c. $C = \begin{bmatrix} -4 & 6 & -4 \\ 2 & 4 & 0 \\ 6 & -2 & 4 \end{bmatrix}$

8.3 /// EXERCISES

In Exercises 1–8, show that B is the inverse of A.

1. $A = \begin{bmatrix} 2 & 1 \\ 5 & 3 \end{bmatrix}$, $B = \begin{bmatrix} 3 & -1 \\ -5 & 2 \end{bmatrix}$

2. $A = \begin{bmatrix} 1 & -1 \\ -1 & 2 \end{bmatrix}$, $B = \begin{bmatrix} 2 & 1 \\ 1 & 1 \end{bmatrix}$

3. $A = \begin{bmatrix} 1 & 2 \\ 3 & 4 \end{bmatrix}$, $B = \begin{bmatrix} -2 & 1 \\ \frac{3}{2} & -\frac{1}{2} \end{bmatrix}$

4. $A = \begin{bmatrix} 1 & -1 \\ 2 & 3 \end{bmatrix}$, $B = \begin{bmatrix} \frac{3}{5} & \frac{1}{5} \\ -\frac{2}{5} & \frac{1}{5} \end{bmatrix}$

5. $A = \begin{bmatrix} -2 & 2 & 3 \\ 1 & -1 & 0 \\ 0 & 1 & 4 \end{bmatrix}$, $B = \frac{1}{3}\begin{bmatrix} -4 & -5 & 3 \\ -4 & -8 & 3 \\ 1 & 2 & 0 \end{bmatrix}$

6. $A = \begin{bmatrix} 2 & -17 & 11 \\ -1 & 11 & -7 \\ 0 & 3 & -2 \end{bmatrix}$, $B = \begin{bmatrix} 1 & 1 & 2 \\ 2 & 4 & -3 \\ 3 & 6 & -5 \end{bmatrix}$

7. $A = \begin{bmatrix} 2 & 0 & 1 & 1 \\ 3 & 0 & 0 & 1 \\ -1 & 1 & -2 & 1 \\ 4 & -1 & 1 & 0 \end{bmatrix}$,

$B = \begin{bmatrix} -1 & 2 & -1 & -1 \\ -4 & 9 & -5 & -6 \\ 0 & 1 & -1 & -1 \\ 3 & -5 & 3 & 3 \end{bmatrix}$

8. $A = \begin{bmatrix} -1 & 1 & 0 & -1 \\ 1 & -1 & 1 & 0 \\ -1 & 1 & 2 & 0 \\ 0 & -1 & 1 & 1 \end{bmatrix}$,

$B = \frac{1}{3}\begin{bmatrix} -3 & 1 & 1 & -3 \\ -3 & -1 & 2 & -3 \\ 0 & 1 & 1 & 0 \\ -3 & -2 & 1 & 0 \end{bmatrix}$

In Exercises 9–24, find the inverse of the matrix (if it exists).

9. $\begin{bmatrix} 2 & 0 \\ 0 & 3 \end{bmatrix}$

10. $\begin{bmatrix} 1 & 2 \\ 3 & 7 \end{bmatrix}$

11. $\begin{bmatrix} 1 & -2 \\ 2 & -3 \end{bmatrix}$

12. $\begin{bmatrix} -7 & 33 \\ 4 & -19 \end{bmatrix}$

13. $\begin{bmatrix} -1 & 1 \\ -2 & 1 \end{bmatrix}$

14. $\begin{bmatrix} 11 & 1 \\ -1 & 0 \end{bmatrix}$

15. $\begin{bmatrix} 2 & 4 \\ 4 & 8 \end{bmatrix}$

16. $\begin{bmatrix} 2 & 3 \\ 1 & 4 \end{bmatrix}$

17. $\begin{bmatrix} 2 & 7 & 1 \\ -3 & -9 & 2 \end{bmatrix}$

18. $\begin{bmatrix} -2 & 5 \\ 6 & -15 \\ 0 & 1 \end{bmatrix}$

19. $\begin{bmatrix} 1 & 1 & 1 \\ 3 & 5 & 4 \\ 3 & 6 & 5 \end{bmatrix}$

20. $\begin{bmatrix} 1 & 2 & 2 \\ 3 & 7 & 9 \\ -1 & -4 & -7 \end{bmatrix}$

21. $\begin{bmatrix} 1 & 0 & 0 \\ 3 & 4 & 0 \\ 2 & 5 & 5 \end{bmatrix}$

22. $\begin{bmatrix} 1 & 0 & 0 \\ 3 & 0 & 0 \\ 2 & 5 & 5 \end{bmatrix}$

23. $\begin{bmatrix} -8 & 0 & 0 & 0 \\ 0 & 1 & 0 & 0 \\ 0 & 0 & 4 & 0 \\ 0 & 0 & 0 & -5 \end{bmatrix}$

24. $\begin{bmatrix} 1 & 3 & -2 & 0 \\ 0 & 2 & 4 & 6 \\ 0 & 0 & -2 & 1 \\ 0 & 0 & 0 & 5 \end{bmatrix}$

In Exercises 25–34, use the matrix capabilities of a graphing utility to find the inverse of the matrix (if it exists).

25. $\begin{bmatrix} 1 & 2 & -1 \\ 3 & 7 & -10 \\ -5 & -7 & -15 \end{bmatrix}$

26. $\begin{bmatrix} 10 & 5 & -7 \\ -5 & 1 & 4 \\ 3 & 2 & -2 \end{bmatrix}$

27. $\begin{bmatrix} 1 & 1 & 2 \\ 3 & 1 & 0 \\ -2 & 0 & 3 \end{bmatrix}$

28. $\begin{bmatrix} 3 & 2 & 2 \\ 2 & 2 & 2 \\ -4 & 4 & 3 \end{bmatrix}$

29. $\begin{bmatrix} 0.1 & 0.2 & 0.3 \\ -0.3 & 0.2 & 0.2 \\ 0.5 & 0.4 & 0.4 \end{bmatrix}$

30. $\begin{bmatrix} 2 & 0 & 0 \\ 0 & 3 & 0 \\ 0 & 0 & 5 \end{bmatrix}$

31. $\begin{bmatrix} 1 & 0 & 3 & 0 \\ 0 & 2 & 0 & 4 \\ 1 & 0 & 3 & 0 \\ 0 & 2 & 0 & 4 \end{bmatrix}$

32. $\begin{bmatrix} -1 & 0 & 1 & 0 \\ 0 & 2 & 0 & -1 \\ 2 & 0 & -1 & 0 \\ 0 & -1 & 0 & 1 \end{bmatrix}$

33. $\begin{bmatrix} 1 & -2 & -1 & -2 \\ 3 & -5 & -2 & -3 \\ 2 & -5 & -2 & -5 \\ -1 & 4 & 4 & 11 \end{bmatrix}$

34. $\begin{bmatrix} 4 & 8 & -7 & 14 \\ 2 & 5 & -4 & 6 \\ 0 & 2 & 1 & -7 \\ 3 & 6 & -5 & 10 \end{bmatrix}$

35. If A is a 2×2 matrix given by

$$A = \begin{bmatrix} a & b \\ c & d \end{bmatrix}$$

then A is invertible if and only if $ad - bc \neq 0$. If $ad - bc \neq 0$, verify that the inverse is given by

$$A^{-1} = \frac{1}{ad - bc} \begin{bmatrix} d & -b \\ -c & a \end{bmatrix}.$$

36. Use the result of Exercise 35 to find the inverse of each matrix.

(a) $\begin{bmatrix} 5 & -2 \\ 2 & 3 \end{bmatrix}$

(b) $\begin{bmatrix} 7 & 12 \\ -8 & -5 \end{bmatrix}$

In Exercises 37–40, use an inverse matrix to solve the system of linear equations. (Use the inverse matrix found in Exercise 11.)

37. $x - 2y = 5$
$2x - 3y = 10$

38. $x - 2y = 0$
$2x - 3y = 3$

39. $x - 2y = 4$
$2x - 3y = 2$

40. $x - 2y = 1$
$2x - 3y = -2$

In Exercises 41 and 42, use an inverse matrix to solve the system of linear equations. (Use the inverse matrix found in Exercise 19.)

41. $x + y + z = 0$
$3x + 5y + 4z = 5$
$3x + 6y + 5z = 2$

42. $x + y + z = -1$
$3x + 5y + 4z = 2$
$3x + 6y + 5z = 0$

In Exercises 43 and 44, use an inverse matrix and the matrix capabilities of a graphing utility to solve the system of linear equations. (Use the inverse matrix found in Exercise 33.)

43. $x_1 - 2x_2 - x_3 - 2x_4 = 0$
$3x_1 - 5x_2 - 2x_3 - 3x_4 = 1$
$2x_1 - 5x_2 - 2x_3 - 5x_4 = -1$
$-x_1 + 4x_2 + 4x_3 + 11x_4 = 2$

44. $x_1 - 2x_2 - x_3 - 2x_4 = 1$
$3x_1 - 5x_2 - 2x_3 - 3x_4 = -2$
$2x_1 - 5x_2 - 2x_3 - 5x_4 = 0$
$-x_1 + 4x_2 + 4x_3 + 11x_4 = -3$

In Exercises 45–52, use an inverse matrix to solve (if possible) the system of linear equations.

45. $3x + 4y = -2$
$5x + 3y = 4$

46. $18x + 12y = 13$
$30x + 24y = 23$

47. $-0.4x + 0.8y = 1.6$
$2x - 4y = 5$

48. $13x - 6y = 17$
$26x - 12y = 8$

49. $3x + 6y = 6$
$6x + 14y = 11$

50. $3x + 2y = 1$
$2x + 10y = 6$

51. $4x - y + z = -5$
$2x + 2y + 3z = 10$
$5x - 2y + 6z = 1$

52. $4x - 2y + 3z = -2$
$2x + 2y + 5z = 16$
$8x - 5y - 2z = 4$

In Exercises 53–56, use the matrix capabilities of a graphing utility to solve (if possible) the system of linear equations.

53. $5x - 3y + 2z = 2$
$\quad 2x + 2y - 3z = 3$
$\quad -x + 7y - 8z = 4$

54. $2x + 3y + 5z = 4$
$\quad 3x + 5y + 9z = 7$
$\quad 5x + 9y + 17z = 13$

55. $\quad 7x - 3y \quad + 2w = \quad 41$
$\quad -2x + y \quad - w = -13$
$\quad 4x \quad + z - 2w = \quad 12$
$\quad -x + y \quad - w = \quad -8$

56. $2x + 5y \quad + w = \quad 11$
$\quad x + 4y + 2z - 2w = -7$
$\quad 2x - 2y + 5z + w = \quad 3$
$\quad x \quad - 3w = -1$

Bond Investments In Exercises 57–60, consider a person who invests in AAA-rated bonds, A-rated bonds, and B-rated bonds. The average yields are 6.5% on AAA-bonds, 7% on A-bonds, and 9% on B-bonds. The person invests twice as much in B-bonds as in A-bonds. Let x, y, and z represent the amounts invested in AAA-, A-, and B-bonds, respectively.

$$x + \quad y + \quad z = \text{(total investment)}$$
$$0.065x + 0.07y + 0.09z = \text{(annual return)}$$
$$2y - \quad z = 0$$

Use the inverse of the coefficient matrix of this system to find the amount invested in each type of bond.

57. Total investment = $25,000
Annual return = $1900

58. Total investment = $45,000
Annual return = $3750

59. Total investment = $12,000
Annual return = $835

60. Total investment = $500,000
Annual return = $38,000

61. *Essay* Write a brief paragraph explaining the advantage of using an inverse matrix to solve the systems of linear equations in Exercises 37–44.

62. *True or False?* Multiplication of an invertible matrix and its inverse is commutative. Give an example to demonstrate your answer.

Circuit Analysis In Exercises 63 and 64, consider the circuit in the figure. The currents I_1, I_2, and I_3, in amperes, are given by the solution of the system of linear equations

$$2I_1 \quad + 4I_3 = E_1$$
$$I_2 + 4I_3 = E_2$$
$$I_1 + I_2 - I_3 = 0$$

where E_1 and E_2 are voltages. Use the inverse of the coefficient matrix of this system to find the unknown currents for the given voltages.

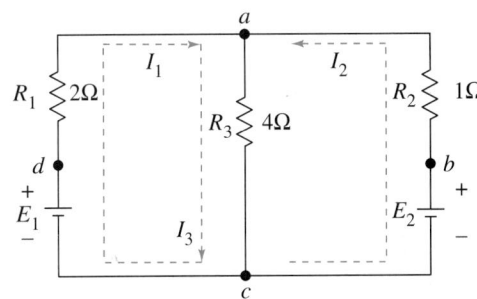

63. $E_1 = 14$ V, $E_2 = 28$ V
64. $E_1 = 10$ V, $E_2 = 10$ V

65. *Exploration* Consider the matrices of the form

$$A = \begin{bmatrix} a_{11} & 0 & 0 & 0 & \cdots & 0 \\ 0 & a_{22} & 0 & 0 & \cdots & 0 \\ 0 & 0 & a_{33} & 0 & \cdots & 0 \\ \vdots & \vdots & \vdots & \vdots & \cdots & \vdots \\ 0 & 0 & 0 & 0 & \cdots & a_{nn} \end{bmatrix}.$$

(a) Write a 2×2 matrix and a 3×3 matrix in the form of A. Find the inverse of each.

(b) Use the result of part (a) to make a conjecture about the inverse of a matrix of the form of A.

8.4   The Determinant of a Square Matrix

The Determinant of a 2 × 2 Matrix

Every *square* matrix can be associated with a real number called its
determinant. Determinants have many uses, and several will be discussed in
this and the next section. Historically, the use of determinants arose from
special number patterns that occur when systems of linear equations are solved.
For instance, the system

$$a_1 x + b_1 y = c_1$$
$$a_2 x + b_2 y = c_2$$

has a solution given by

$$x = \frac{c_1 b_2 - c_2 b_1}{a_1 b_2 - a_2 b_1} \quad \text{and} \quad y = \frac{a_1 c_2 - a_2 c_1}{a_1 b_2 - a_2 b_1}$$

provided that $a_1 b_2 - a_2 b_1 \neq 0$. Note that the denominator of each fraction is
the same. This denominator is called the **determinant** of the coefficient matrix
of the system.

Coefficient Matrix   *Determinant*

$$A = \begin{bmatrix} a_1 & b_1 \\ a_2 & b_2 \end{bmatrix} \qquad \det(A) = a_1 b_2 - a_2 b_1$$

The determinant of the matrix A can also be denoted by vertical bars on both
sides of the matrix, as indicated in the following definition.

Definition of the Determinant of a 2 × 2 Matrix

The **determinant** of the matrix

$$A = \begin{bmatrix} a_1 & b_1 \\ a_2 & b_2 \end{bmatrix}$$

is given by

$$\det(A) = |A| = \begin{vmatrix} a_1 & b_1 \\ a_2 & b_2 \end{vmatrix} = a_1 b_2 - a_2 b_1.$$

Note   In this text, $\det(A)$ and $|A|$ are used interchangeably to represent the
determinant of A. Although vertical bars are also used to denote the absolute
value of a real number, the context will show which use is intended.

A convenient method for remembering the formula for the determinant of a 2×2 matrix is shown in the following diagram.

$$\det(A) = \begin{vmatrix} a_1 & b_1 \\ a_2 & b_2 \end{vmatrix} = a_1 b_2 - a_2 b_1$$

Note that the determinant is given by the difference of the products of the two diagonals of the matrix.

EXAMPLE 1   The Determinant of a 2 $\times$ 2 Matrix

Find the determinant of each matrix.

a. $A = \begin{bmatrix} 2 & -3 \\ 1 & 2 \end{bmatrix}$ **b.** $B = \begin{bmatrix} 2 & 1 \\ 4 & 2 \end{bmatrix}$ **c.** $C = \begin{bmatrix} 0 & \frac{3}{2} \\ 2 & 4 \end{bmatrix}$

Solution

Note Notice in Example 1 that the determinant of a matrix can be positive, zero, or negative.

a. $\det(A) = \begin{vmatrix} 2 & -3 \\ 1 & 2 \end{vmatrix} = 2(2) - 1(-3) = 4 + 3 = 7$

b. $\det(B) = \begin{vmatrix} 2 & 1 \\ 4 & 2 \end{vmatrix} = 2(2) - 4(1) = 4 - 4 = 0$

c. $\det(C) = \begin{vmatrix} 0 & \frac{3}{2} \\ 2 & 4 \end{vmatrix} = 0(4) - 2\left(\frac{3}{2}\right) = 0 - 3 = -3$

The determinant of a matrix of order 1×1 is defined simply as the entry of the matrix. For instance, if $A = [-2]$, $\det(A) = -2$.

Most graphing utilities can evaluate the determinant of a matrix. For instance, on a *TI-82* or *TI-83,* you can evaluate the determinant of

$$A = \begin{bmatrix} 2 & -3 \\ 1 & 2 \end{bmatrix}$$

by entering the matrix as $[A]$ and then choosing the "det" feature in the matrix math menu.

det $[A]$ | ENTER |

The result should be 7, as in Example 1(a). Try evaluating determinants of other matrices. What happens when you try to evaluate the determinant of a nonsquare matrix?

Minors and Cofactors

To define the determinant of a square matrix of order 3×3 or higher, it is convenient to introduce the concepts of **minors** and **cofactors.**

> ### Minors and Cofactors of a Square Matrix
>
> If A is a square matrix, the **minor** M_{ij} of the entry a_{ij} is the determinant of the matrix obtained by deleting the ith row and jth column of A. The **cofactor** C_{ij} of the entry a_{ij} is given by
>
> $$C_{ij} = (-1)^{i+j} M_{ij}.$$

Sign Pattern for Cofactors

$$\begin{bmatrix} + & - & + \\ - & + & - \\ + & - & + \end{bmatrix}$$

3×3 matrix

$$\begin{bmatrix} + & - & + & - \\ - & + & - & + \\ + & - & + & - \\ - & + & - & + \end{bmatrix}$$

4×4 matrix

$$\begin{bmatrix} + & - & + & - & + & \cdots \\ - & + & - & + & - & \cdots \\ + & - & + & - & + & \cdots \\ - & + & - & + & - & \cdots \\ + & - & + & - & + & \cdots \\ \vdots & \vdots & \vdots & \vdots & \vdots & \end{bmatrix}$$

$n \times n$ matrix

Note In the sign pattern for cofactors above, notice that *odd* positions (where $i + j$ is odd) have negative signs and *even* positions (where $i + j$ is even) have positive signs.

EXAMPLE 2   ▱   **Finding the Minors and Cofactors of a Matrix**

Find all the minors and cofactors of

$$A = \begin{bmatrix} 0 & 2 & 1 \\ 3 & -1 & 2 \\ 4 & 0 & 1 \end{bmatrix}.$$

Solution

To find the minor M_{11}, delete the first row and first column of A and evaluate the determinant of the resulting matrix.

$$\begin{bmatrix} 0 & 2 & 1 \\ 3 & -1 & 2 \\ 4 & 0 & 1 \end{bmatrix}, \quad M_{11} = \begin{vmatrix} -1 & 2 \\ 0 & 1 \end{vmatrix} = -1(1) - 0(2) = -1$$

Similarly, to find M_{12}, delete the first row and second column.

$$\begin{bmatrix} 0 & 2 & 1 \\ 3 & -1 & 2 \\ 4 & 0 & 1 \end{bmatrix}, \quad M_{12} = \begin{vmatrix} 3 & 2 \\ 4 & 1 \end{vmatrix} = 3(1) - 4(2) = -5$$

Continuing this pattern, you obtain the following minors.

$$\begin{array}{lll} M_{11} = -1 & M_{12} = -5 & M_{13} = 4 \\ M_{21} = 2 & M_{22} = -4 & M_{23} = -8 \\ M_{31} = 5 & M_{32} = -3 & M_{33} = -6 \end{array}$$

Now, to find the cofactors, combine these minors with the checkerboard pattern of signs shown at the left (for a 3×3 matrix) to obtain the following.

$$\begin{array}{lll} C_{11} = -1 & C_{12} = 5 & C_{13} = 4 \\ C_{21} = -2 & C_{22} = -4 & C_{23} = 8 \\ C_{31} = 5 & C_{32} = 3 & C_{33} = -6 \end{array}$$

▱

The Determinant of a Square Matrix

The following definition is called **inductive** because it uses determinants of matrices of order $n - 1$ to define the determinant of a matrix of order n.

Note   Try checking that for a 2×2 matrix this definition yields

$$|A| = a_{11}a_{22} - a_{21}a_{12}$$

as previously defined.

Determinant of a Square Matrix

If A is a square matrix (of order 2×2 or greater), the determinant of A is the sum of the entries in any row (or column) of A multiplied by their respective cofactors. For instance, expanding along the first row yields

$$|A| = a_{11}C_{11} + a_{12}C_{12} + \cdots + a_{1n}C_{1n}.$$

Applying this definition to find a determinant is called **expanding by cofactors.**

EXAMPLE 3   **The Determinant of a Matrix of Order 3×3**

Find the determinant of

$$A = \begin{bmatrix} 0 & 2 & 1 \\ 3 & -1 & 2 \\ 4 & 0 & 1 \end{bmatrix}.$$

Solution
Note that this is the same matrix that was given in Example 2. There you found the cofactors of the entries in the first row to be

$$C_{11} = -1, \quad C_{12} = 5, \quad \text{and} \quad C_{13} = 4.$$

Therefore, by the definition of the determinant of a square matrix, you have

$$\begin{aligned} |A| &= a_{11}C_{11} + a_{12}C_{12} + a_{13}C_{13} \qquad \text{First-row expansion} \\ &= 0(-1) + 2(5) + 1(4) \\ &= 14. \end{aligned}$$

In Example 3, the determinant was found by expanding by the cofactors in the first row. You could have used any row or column. For instance, you could have expanded along the second row to obtain

$$\begin{aligned} |A| &= a_{21}C_{21} + a_{22}C_{22} + a_{23}C_{23} \qquad \text{Second-row expansion} \\ &= 3(-2) + (-1)(-4) + 2(8) \\ &= 14. \end{aligned}$$

When expanding by cofactors, you do not need to find cofactors of zero entries, because zero times its cofactor is zero.

$$a_{ij}C_{ij} = (0)C_{ij} = 0$$

Thus, the row (or column) containing the most zeros is usually the best choice for expansion by cofactors. This is demonstrated in the next example.

EXAMPLE 4   **The Determinant of a Matrix of Order 4 × 4**

Find the determinant of

$$A = \begin{bmatrix} 1 & -2 & 3 & 0 \\ -1 & 1 & 0 & 2 \\ 0 & 2 & 0 & 3 \\ 3 & 4 & 0 & 2 \end{bmatrix}.$$

Solution

After inspecting this matrix, you can see that three of the entries in the third column are zeros. Thus, you can eliminate some of the work in the expansion by using the third column.

$$|A| = 3(C_{13}) + 0(C_{23}) + 0(C_{33}) + 0(C_{43})$$

Because C_{23}, C_{33}, and C_{43} have zero coefficients, you need only find the cofactor C_{13}. To do this, delete the first row and third column of A and evaluate the determinant of the resulting matrix.

$$C_{13} = (-1)^{1+3} \begin{vmatrix} -1 & 1 & 2 \\ 0 & 2 & 3 \\ 3 & 4 & 2 \end{vmatrix} \qquad \text{Delete 1st row and 3rd column.}$$

$$= \begin{vmatrix} -1 & 1 & 2 \\ 0 & 2 & 3 \\ 3 & 4 & 2 \end{vmatrix} \qquad \text{Simplify.}$$

Expanding by cofactors in the second row yields the following.

$$C_{13} = 0(-1)^3 \begin{vmatrix} 1 & 2 \\ 4 & 2 \end{vmatrix} + 2(-1)^4 \begin{vmatrix} -1 & 2 \\ 3 & 2 \end{vmatrix} + 3(-1)^5 \begin{vmatrix} -1 & 1 \\ 3 & 4 \end{vmatrix}$$

$$= 0 + 2(1)(-8) + 3(-1)(-7)$$

$$= 5$$

> **Study Tip**
>
> Although most graphing utilities can calculate the determinant of a square matrix, it is also important to know how to calculate them *by hand*.

Thus, you obtain

$$|A| = 3C_{13} = 3(5) = 15.$$

Note   Try using a graphing utility to confirm the result of Example 4.

Triangular Matrices

Evaluating determinants of matrices of order 4 or higher can be tedious. There is, however, an important exception: the determinant of a **triangular** matrix. A square matrix is **upper triangular** if it has all zero entries below its main diagonal and **lower triangular** if it has all zero entries above its main diagonal. A matrix that is both upper and lower triangular is called **diagonal.** That is, a diagonal matrix is one in which all entries above and below the main diagonal are zero.

Upper Triangular Matrix

$$\begin{bmatrix} a_{11} & a_{12} & a_{13} & \cdots & a_{1n} \\ 0 & a_{22} & a_{23} & \cdots & a_{2n} \\ 0 & 0 & a_{33} & \cdots & a_{3n} \\ \vdots & \vdots & \vdots & & \vdots \\ 0 & 0 & 0 & \cdots & a_{nm} \end{bmatrix}$$

Lower Triangular Matrix

$$\begin{bmatrix} a_{11} & 0 & 0 & \cdots & 0 \\ a_{21} & a_{22} & 0 & \cdots & 0 \\ a_{31} & a_{32} & a_{33} & \cdots & 0 \\ \vdots & \vdots & \vdots & & \vdots \\ a_{n1} & a_{n2} & a_{n3} & \cdots & a_{nm} \end{bmatrix}$$

To find the determinant of a triangular matrix of any order, simply form the product of the entries on the main diagonal.

EXAMPLE 5 **The Determinant of a Triangular Matrix**

a.
$$\begin{vmatrix} 2 & 0 & 0 & 0 \\ 4 & -2 & 0 & 0 \\ -5 & 6 & 1 & 0 \\ 1 & 5 & 3 & 3 \end{vmatrix} = (2)(-2)(1)(3) = -12$$

b.
$$\begin{vmatrix} -1 & 0 & 0 & 0 & 0 \\ 0 & 3 & 0 & 0 & 0 \\ 0 & 0 & 2 & 0 & 0 \\ 0 & 0 & 0 & 4 & 0 \\ 0 & 0 & 0 & 0 & -2 \end{vmatrix} = (-1)(3)(2)(4)(-2) = 48$$

Group Activity

The Determinant of a Triangular Matrix

Write an argument that explains why the determinant of a 3×3 triangular matrix is the product of its main-diagonal entries.

$$\begin{bmatrix} a_{11} & a_{12} & a_{13} \\ 0 & a_{22} & a_{23} \\ 0 & 0 & a_{33} \end{bmatrix} = a_{11}a_{22}a_{33}$$

8.4 /// EXERCISES

In Exercises 1–16, find the determinant of the matrix.

1. $[5]$

2. $[-8]$

3. $\begin{bmatrix} 2 & 1 \\ 3 & 4 \end{bmatrix}$

4. $\begin{bmatrix} -3 & 1 \\ 5 & 2 \end{bmatrix}$

5. $\begin{bmatrix} 5 & 2 \\ -6 & 3 \end{bmatrix}$

6. $\begin{bmatrix} 2 & -2 \\ 4 & 3 \end{bmatrix}$

7. $\begin{bmatrix} -7 & 6 \\ \frac{1}{2} & 3 \end{bmatrix}$

8. $\begin{bmatrix} 4 & -3 \\ 0 & 0 \end{bmatrix}$

9. $\begin{bmatrix} 2 & 6 \\ 0 & 3 \end{bmatrix}$

10. $\begin{bmatrix} 2 & -3 \\ -6 & 9 \end{bmatrix}$

11. $\begin{bmatrix} 2 & -1 & 0 \\ 4 & 2 & 1 \\ 4 & 2 & 1 \end{bmatrix}$

12. $\begin{bmatrix} -2 & 2 & 3 \\ 1 & -1 & 0 \\ 0 & 1 & 4 \end{bmatrix}$

13. $\begin{bmatrix} 6 & 3 & -7 \\ 0 & 0 & 0 \\ 4 & -6 & 3 \end{bmatrix}$

14. $\begin{bmatrix} 1 & 1 & 2 \\ 3 & 1 & 0 \\ -2 & 0 & 3 \end{bmatrix}$

15. $\begin{bmatrix} -1 & 2 & -5 \\ 0 & 3 & 4 \\ 0 & 0 & 3 \end{bmatrix}$

16. $\begin{bmatrix} 1 & 0 & 0 \\ -4 & -1 & 0 \\ 5 & 1 & 5 \end{bmatrix}$

In Exercises 17–20, use the matrix capabilities of a graphing utility to find the determinant of the matrix.

17. $\begin{bmatrix} 0.3 & 0.2 & 0.2 \\ 0.2 & 0.2 & 0.2 \\ -0.4 & 0.4 & 0.3 \end{bmatrix}$

18. $\begin{bmatrix} 0.1 & 0.2 & 0.3 \\ -0.3 & 0.2 & 0.2 \\ 0.5 & 0.4 & 0.4 \end{bmatrix}$

19. $\begin{bmatrix} 1 & 4 & -2 \\ 3 & 6 & -6 \\ -2 & 1 & 4 \end{bmatrix}$

20. $\begin{bmatrix} 2 & 3 & 1 \\ 0 & 5 & -2 \\ 0 & 0 & -2 \end{bmatrix}$

In Exercises 21–24, find all (a) minors and (b) cofactors of the matrix.

21. $\begin{bmatrix} 3 & 4 \\ 2 & -5 \end{bmatrix}$

22. $\begin{bmatrix} 11 & 0 \\ -3 & 2 \end{bmatrix}$

23. $\begin{bmatrix} 3 & -2 & 8 \\ 3 & 2 & -6 \\ -1 & 3 & 6 \end{bmatrix}$

24. $\begin{bmatrix} -2 & 9 & 4 \\ 7 & -6 & 0 \\ 6 & 7 & -6 \end{bmatrix}$

In Exercises 25–30, find the determinant of the matrix by the method of expansion by cofactors. Expand using the indicated row or column.

25. $\begin{bmatrix} -3 & 2 & 1 \\ 4 & 5 & 6 \\ 2 & -3 & 1 \end{bmatrix}$ (a) Row 1   (b) Column 2

26. $\begin{bmatrix} -3 & 4 & 2 \\ 6 & 3 & 1 \\ 4 & -7 & -8 \end{bmatrix}$ (a) Row 2   (b) Column 3

27. $\begin{bmatrix} 5 & 0 & -3 \\ 0 & 12 & 4 \\ 1 & 6 & 3 \end{bmatrix}$ (a) Row 2   (b) Column 2

28. $\begin{bmatrix} 10 & -5 & 5 \\ 30 & 0 & 10 \\ 0 & 10 & 1 \end{bmatrix}$ (a) Row 3   (b) Column 1

29. $\begin{bmatrix} 6 & 0 & -3 & 5 \\ 4 & 13 & 6 & -8 \\ -1 & 0 & 7 & 4 \\ 8 & 6 & 0 & 2 \end{bmatrix}$ (a) Row 2   (b) Column 2

30. $\begin{bmatrix} 10 & 8 & 3 & -7 \\ 4 & 0 & 5 & -6 \\ 0 & 3 & 2 & 7 \\ 1 & 0 & -3 & 2 \end{bmatrix}$ (a) Row 3   (b) Column 1

In Exercises 31–40, find the determinant of the matrix. Expand by cofactors on the row or column that appears to make the computations easiest.

31. $\begin{bmatrix} 1 & 4 & -2 \\ 3 & 2 & 0 \\ -1 & 4 & 3 \end{bmatrix}$

32. $\begin{bmatrix} 2 & -1 & 3 \\ 1 & 4 & 4 \\ 1 & 0 & 2 \end{bmatrix}$

33. $\begin{bmatrix} 2 & 4 & 6 \\ 0 & 3 & 1 \\ 0 & 0 & -5 \end{bmatrix}$

34. $\begin{bmatrix} -3 & 0 & 0 \\ 7 & 11 & 0 \\ 1 & 2 & 2 \end{bmatrix}$

35. $\begin{bmatrix} 2 & 6 & 6 & 2 \\ 2 & 7 & 3 & 6 \\ 1 & 5 & 0 & 1 \\ 3 & 7 & 0 & 7 \end{bmatrix}$

36. $\begin{bmatrix} 3 & 6 & -5 & 4 \\ -2 & 0 & 6 & 0 \\ 1 & 1 & 2 & 2 \\ 0 & 3 & -1 & -1 \end{bmatrix}$

37. $\begin{bmatrix} 5 & 3 & 0 & 6 \\ 4 & 6 & 4 & 12 \\ 0 & 2 & -3 & 4 \\ 0 & 1 & -2 & 2 \end{bmatrix}$

38. $\begin{bmatrix} 1 & 4 & 3 & 2 \\ -5 & 6 & 2 & 1 \\ 0 & 0 & 0 & 0 \\ 3 & -2 & 1 & 5 \end{bmatrix}$

39. $\begin{bmatrix} 3 & 2 & 4 & -1 & 5 \\ -2 & 0 & 1 & 3 & 2 \\ 1 & 0 & 0 & 4 & 0 \\ 6 & 0 & 2 & -1 & 0 \\ 3 & 0 & 5 & 1 & 0 \end{bmatrix}$

40. $\begin{bmatrix} 5 & 2 & 0 & 0 & -2 \\ 0 & 1 & 4 & 3 & 2 \\ 0 & 0 & 2 & 6 & 3 \\ 0 & 0 & 3 & 4 & 1 \\ 0 & 0 & 0 & 0 & 2 \end{bmatrix}$

In Exercises 41–48, use the matrix capabilities of a graphing utility to evaluate the determinant.

41. $\begin{vmatrix} 3 & 8 & -7 \\ 0 & -5 & 4 \\ 8 & 1 & 6 \end{vmatrix}$

42. $\begin{vmatrix} 5 & -8 & 0 \\ 9 & 7 & 4 \\ -8 & 7 & 1 \end{vmatrix}$

43. $\begin{vmatrix} 7 & 0 & -14 \\ -2 & 5 & 4 \\ -6 & 2 & 12 \end{vmatrix}$

44. $\begin{vmatrix} 3 & 0 & 0 \\ -2 & 5 & 0 \\ 12 & 5 & 7 \end{vmatrix}$

45. $\begin{vmatrix} 1 & -1 & 8 & 4 \\ 2 & 6 & 0 & -4 \\ 2 & 0 & 2 & 6 \\ 0 & 2 & 8 & 0 \end{vmatrix}$

46. $\begin{vmatrix} 0 & -3 & 8 & 2 \\ 8 & 1 & -1 & 6 \\ -4 & 6 & 0 & 9 \\ -7 & 0 & 0 & 14 \end{vmatrix}$

47. $\begin{vmatrix} 3 & -2 & 4 & 3 & 1 \\ -1 & 0 & 2 & 1 & 0 \\ 5 & -1 & 0 & 3 & 2 \\ 4 & 7 & -8 & 0 & 0 \\ 1 & 2 & 3 & 0 & 2 \end{vmatrix}$

48. $\begin{vmatrix} -2 & 0 & 0 & 0 & 0 \\ 0 & 3 & 0 & 0 & 0 \\ 0 & 0 & -1 & 0 & 0 \\ 0 & 0 & 0 & 2 & 0 \\ 0 & 0 & 0 & 0 & -4 \end{vmatrix}$

In Exercises 49–52, evaluate the determinants to verify the equation.

49. $\begin{vmatrix} w & x \\ y & z \end{vmatrix} = -\begin{vmatrix} y & z \\ w & x \end{vmatrix}$

50. $\begin{vmatrix} w & cx \\ y & cz \end{vmatrix} = c\begin{vmatrix} w & x \\ y & z \end{vmatrix}$

51. $\begin{vmatrix} w & x \\ y & z \end{vmatrix} = \begin{vmatrix} w & x + cw \\ y & z + cy \end{vmatrix}$

52. $\begin{vmatrix} w & x \\ cw & cx \end{vmatrix} = 0$

In Exercises 53 and 54, evaluate the determinant to verify the equation.

53. $\begin{vmatrix} 1 & x & x^2 \\ 1 & y & y^2 \\ 1 & z & z^2 \end{vmatrix} = (y - x)(z - x)(z - y)$

54. $\begin{vmatrix} a + b & a & a \\ a & a + b & a \\ a & a & a + b \end{vmatrix} = b^2(3a + b)$

In Exercises 55 and 56, solve for x.

55. $\begin{vmatrix} x - 1 & 2 \\ 3 & x - 2 \end{vmatrix} = 0$

56. $\begin{vmatrix} x - 2 & -1 \\ -3 & x \end{vmatrix} = 0$

In Exercises 57–62, evaluate the determinant, where the entries are functions. Determinants of this type occur in calculus.

57. $\begin{vmatrix} 4u & -1 \\ -1 & 2v \end{vmatrix}$

58. $\begin{vmatrix} 3x^2 & -3y^2 \\ 1 & 1 \end{vmatrix}$

59. $\begin{vmatrix} e^{2x} & e^{3x} \\ 2e^{2x} & 3e^{3x} \end{vmatrix}$

60. $\begin{vmatrix} e^{-x} & xe^{-x} \\ -e^{-x} & (1 - x)e^{-x} \end{vmatrix}$

61. $\begin{vmatrix} x & \ln x \\ 1 & 1/x \end{vmatrix}$

62. $\begin{vmatrix} x & x \ln x \\ 1 & 1 + \ln x \end{vmatrix}$

636 8 / *Matrices and Determinants*

In Exercises 63–66, find (a) $|A|$, (b) $|B|$, (c) AB, and (d) $|AB|$.

63. $A = \begin{bmatrix} -1 & 0 \\ 0 & 3 \end{bmatrix}$, $B = \begin{bmatrix} 2 & 0 \\ 0 & -1 \end{bmatrix}$

64. $A = \begin{bmatrix} -2 & 1 \\ 4 & -2 \end{bmatrix}$, $B = \begin{bmatrix} 1 & 2 \\ 0 & -1 \end{bmatrix}$

65. $A = \begin{bmatrix} -1 & 2 & 1 \\ 1 & 0 & 1 \\ 0 & 1 & 0 \end{bmatrix}$, $B = \begin{bmatrix} -1 & 0 & 0 \\ 0 & 2 & 0 \\ 0 & 0 & 3 \end{bmatrix}$

66. $A = \begin{bmatrix} 2 & 0 & 1 \\ 1 & -1 & 2 \\ 3 & 1 & 0 \end{bmatrix}$, $B = \begin{bmatrix} 2 & -1 & 4 \\ 0 & 1 & 3 \\ 3 & -2 & 1 \end{bmatrix}$

67. *Exploration* Find square matrices A and B to demonstrate that

$$|A + B| \neq |A| + |B|.$$

68. *Exploration* Consider square matrices in which the entries are consecutive integers. An example of such a matrix is

$$\begin{bmatrix} 4 & 5 & 6 \\ 7 & 8 & 9 \\ 10 & 11 & 12 \end{bmatrix}.$$

(a) Use a graphing utility to evaluate four determinants of this type. Make a conjecture based on the results.

(b) Verify your conjecture.

69. *Essay* Write a brief paragraph explaining the difference between a square matrix and its determinant.

70. *Think About It* If A is a matrix of order 3×3 such that $|A| = 5$, is it possible to find $|2A|$? Explain.

In Exercises 71–73, a property of determinants is given. State how the property has been applied to the given determinants and use a graphing utility to verify the results.

71. If A and B are square matrices and B is obtained from A by interchanging two rows of A or interchanging two columns of A, then $|B| = -|A|$.

(a) $\begin{vmatrix} 1 & 3 & 4 \\ -7 & 2 & -5 \\ 6 & 1 & 2 \end{vmatrix} = -\begin{vmatrix} 1 & 4 & 3 \\ -7 & -5 & 2 \\ 6 & 2 & 1 \end{vmatrix}$

(b) $\begin{vmatrix} 1 & 3 & 4 \\ -2 & 2 & 0 \\ 1 & 6 & 2 \end{vmatrix} = -\begin{vmatrix} 1 & 6 & 2 \\ -2 & 2 & 0 \\ 1 & 3 & 4 \end{vmatrix}$

72. If A and B are square matrices and B is obtained from A by adding a multiple of a row of A to another row of A or by adding a multiple of a column of A to another column of A, then $|B| = |A|$.

(a) $\begin{vmatrix} 1 & -3 \\ 5 & 2 \end{vmatrix} = \begin{vmatrix} 1 & -3 \\ 0 & 17 \end{vmatrix}$

(b) $\begin{vmatrix} 5 & 4 & 2 \\ 2 & -3 & 4 \\ 7 & 6 & 3 \end{vmatrix} = \begin{vmatrix} 1 & 10 & -6 \\ 2 & -3 & 4 \\ 7 & 6 & 3 \end{vmatrix}$

73. If A and B are square matrices and B is obtained from A by multiplying a row of A by a nonzero constant c or multiplying a column of A by a nonzero constant c, then $|B| = c|A|$.

(a) $\begin{vmatrix} 5 & 10 & 15 \\ 2 & -3 & 4 \\ 2 & -7 & 1 \end{vmatrix} = 5\begin{vmatrix} 1 & 2 & 3 \\ 2 & -3 & 4 \\ 2 & -7 & 1 \end{vmatrix}$

(b) $\begin{vmatrix} 1 & 8 & -3 \\ 3 & -12 & 6 \\ 7 & 4 & 9 \end{vmatrix} = 12\begin{vmatrix} 1 & 2 & -1 \\ 3 & -3 & 2 \\ 7 & 1 & 3 \end{vmatrix}$

8.5 Applications of Matrices and Determinants

Area of a Triangle / *Lines in the Plane* / *Cramer's Rule* / *Cryptography*

Area of a Triangle

In this section, you will study some additional applications of matrices and determinants. The first involves a formula for finding the area of a triangle whose vertices are given by three points on a rectangular coordinate system.

Area of a Triangle

The area of a triangle with vertices (x_1, y_1), (x_2, y_2), and (x_3, y_3) is given by

$$\text{Area} = \pm\frac{1}{2}\begin{vmatrix} x_1 & y_1 & 1 \\ x_2 & y_2 & 1 \\ x_3 & y_3 & 1 \end{vmatrix}$$

where the symbol $(\pm)$ indicates that the appropriate sign should be chosen to yield a positive area.

EXAMPLE 1   **Finding the Area of a Triangle**

Find the area of a triangle whose vertices are $(1, 0)$, $(2, 2)$, and $(4, 3)$, as shown in Figure 8.1.

Solution

Let $(x_1, y_1) = (1, 0)$, $(x_2, y_2) = (2, 2)$, and $(x_3, y_3) = (4, 3)$. Then, to find the area of a triangle, evaluate the determinant

$$\begin{vmatrix} x_1 & y_1 & 1 \\ x_2 & y_2 & 1 \\ x_3 & y_3 & 1 \end{vmatrix} = \begin{vmatrix} 1 & 0 & 1 \\ 2 & 2 & 1 \\ 4 & 3 & 1 \end{vmatrix}$$

$$= 1(-1)^2\begin{vmatrix} 2 & 1 \\ 3 & 1 \end{vmatrix} + 0(-1)^3\begin{vmatrix} 2 & 1 \\ 4 & 1 \end{vmatrix} + 1(-1)^4\begin{vmatrix} 2 & 2 \\ 4 & 3 \end{vmatrix}$$

$$= 1(-1) + 0 + 1(-2)$$

$$= -3.$$

Using this value, you can conclude that the area of the triangle is

$$\text{Area} = -\frac{1}{2}\begin{vmatrix} 1 & 0 & 1 \\ 2 & 2 & 1 \\ 4 & 3 & 1 \end{vmatrix} = -\frac{1}{2}(-3) = \frac{3}{2}.$$

Figure 8.1

Figure 8.2

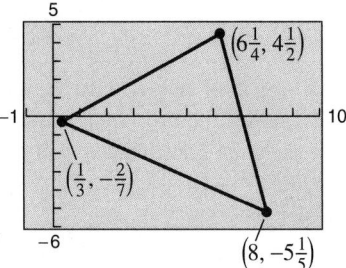

EXAMPLE 2   Finding the Area of a Triangle

Find the area of the triangle whose vertices are $\left(\frac{1}{3}, -\frac{2}{7}\right)$, $\left(6\frac{1}{4}, 4\frac{1}{2}\right)$, and $\left(8, -5\frac{1}{5}\right)$, as shown in Figure 8.2.

Solution

Let $(x_1, y_1) = \left(\frac{1}{3}, -\frac{2}{7}\right)$, $(x_2, y_2) = \left(6\frac{1}{4}, 4\frac{1}{2}\right)$, and $(x_3, y_3) = \left(8, -5\frac{1}{5}\right)$. Then, to find the area of the triangle, evaluate the determinant

$$\begin{vmatrix} x_1 & y_1 & 1 \\ x_2 & y_2 & 1 \\ x_3 & y_3 & 1 \end{vmatrix} = \begin{vmatrix} \frac{1}{3} & -\frac{2}{7} & 1 \\ 6\frac{1}{4} & 4\frac{1}{2} & 1 \\ 8 & -5\frac{1}{5} & 1 \end{vmatrix}.$$

Using the matrix capabilities of a graphing utility, you find the value of the determinant to be $-65.7\overline{6}$.

Now you can use this value to conclude that the area of the triangle is

$$\text{Area} = -\frac{1}{2}(-65.7\overline{6}) = 32.88\overline{3}.$$

Figure 8.3

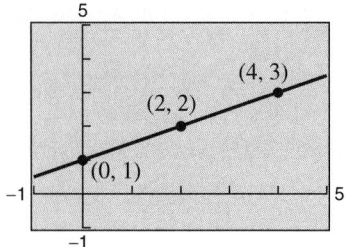

Lines in the Plane

Suppose the three points in Example 1 had been on the same line. What would have happened had the area formula been applied to three such points? The answer is that the determinant would have been zero. Consider, for instance, the three collinear points $(0, 1)$, $(2, 2)$, and $(4, 3)$, as shown in Figure 8.3. The area of the "triangle" that has these three points as vertices is

$$\frac{1}{2}\begin{vmatrix} 0 & 1 & 1 \\ 2 & 2 & 1 \\ 4 & 3 & 1 \end{vmatrix} = \frac{1}{2}\left[0(-1)^2\begin{vmatrix} 2 & 1 \\ 3 & 1 \end{vmatrix} + 1(-1)^3\begin{vmatrix} 2 & 1 \\ 4 & 1 \end{vmatrix} + 1(-1)^4\begin{vmatrix} 2 & 2 \\ 4 & 3 \end{vmatrix}\right]$$

$$= \frac{1}{2}[0 + (-1)(-2) + 1(-2)] = 0.$$

This result is generalized as follows.

Test for Collinear Points

Three points (x_1, y_1), (x_2, y_2), and (x_3, y_3) are collinear (lie on the same line) if and only if

$$\begin{vmatrix} x_1 & y_1 & 1 \\ x_2 & y_2 & 1 \\ x_3 & y_3 & 1 \end{vmatrix} = 0.$$

Figure 8.4

EXAMPLE 3 ▭ Testing for Collinear Points

Determine whether the points $(-2, -2)$, $(1, 1)$, and $(7, 5)$ lie on the same line. (See Figure 8.4.)

Solution

Letting $(x_1, y_1) = (-2, -2)$, $(x_2, y_2) = (1, 1)$, and $(x_3, y_3) = (7, 5)$, you have

$$\begin{vmatrix} x_1 & y_1 & 1 \\ x_2 & y_2 & 1 \\ x_3 & y_3 & 1 \end{vmatrix} = \begin{vmatrix} -2 & -2 & 1 \\ 1 & 1 & 1 \\ 7 & 5 & 1 \end{vmatrix}$$

$$= -2(-1)^2 \begin{vmatrix} 1 & 1 \\ 5 & 1 \end{vmatrix} + (-2)(-1)^3 \begin{vmatrix} 1 & 1 \\ 7 & 1 \end{vmatrix} + 1(-1)^4 \begin{vmatrix} 1 & 1 \\ 7 & 5 \end{vmatrix}$$

$$= -2(-4) + 2(-6) + 1(-2)$$

$$= -6.$$

Because the value of this determinant *is not* zero, you can conclude that the three points do not lie on the same line. ▭

The test for collinear points can be adapted to another use. That is, if you are given two points on a rectangular coordinate system, you can find an equation of the line passing through the two points, as follows.

Two-Point Form of the Equation of a Line

An equation of the line passing through the distinct points (x_1, y_1) and (x_2, y_2) is given by

$$\begin{vmatrix} x & y & 1 \\ x_1 & y_1 & 1 \\ x_2 & y_2 & 1 \end{vmatrix} = 0.$$

Note that this method of finding the equation of a line works for all lines, including horizontal and vertical lines. For instance, the equation of the vertical line through $(2, 0)$ and $(2, 2)$ is

$$\begin{vmatrix} x & y & 1 \\ 2 & 0 & 1 \\ 2 & 2 & 1 \end{vmatrix} = 0$$

$$-2x + 4 = 0$$

$$x = 2.$$

You can use the following steps on a *TI-82* or *TI-83* graphing calculator to check whether three points are collinear.

1. Plot the points by entering Pt-On(x, y) for each point. [Pt-On(can be found in the DRAW POINTS menu.]
2. Draw a line from the two farthest points by entering Line (x_1, y_1, x_2, y_2).

Use the steps above to check the results of Example 3. Explain how the graph shows that the points are not collinear. Why is Step 2 important in determining if points are collinear?

Figure 8.5

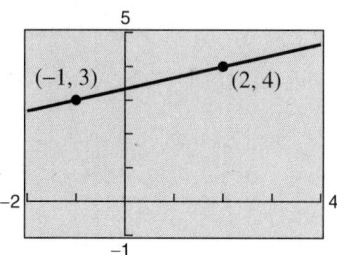

EXAMPLE 4 ▰ Finding an Equation of a Line

Find an equation of the line passing through the two points $(2, 4)$ and $(-1, 3)$, as shown in Figure 8.5.

Solution

Applying the determinant formula for the equation of a line produces

$$\begin{vmatrix} x & y & 1 \\ 2 & 4 & 1 \\ -1 & 3 & 1 \end{vmatrix} = 0.$$

To evaluate this determinant, you can expand by cofactors along the first row to obtain the following.

$$x(-1)^2 \begin{vmatrix} 4 & 1 \\ 3 & 1 \end{vmatrix} + y(-1)^3 \begin{vmatrix} 2 & 1 \\ -1 & 1 \end{vmatrix} + 1(-1)^4 \begin{vmatrix} 2 & 4 \\ -1 & 3 \end{vmatrix} = x - 3y + 10$$
$$= 0$$

Therefore, an equation of the line is

$$x - 3y + 10 = 0.$$

▰

Note There are a variety of ways to check that the equation of the line in Example 4 is correct. You can check it algebraically using the techniques you learned in Section P.3, or you can check it graphically by plotting the points and graphing the line in the same viewing rectangle.

Cramer's Rule

So far, you have studied three methods for solving a system of linear equations: substitution, elimination (with equations), and elimination (with matrices). You will now study one more method, **Cramer's Rule,** named after Gabriel Cramer (1704–1752). This rule uses determinants to write the solution of a system of linear equations. To see how Cramer's Rule works, take another look at the solution described at the beginning of Section 8.4. There, it was pointed out that the system

$$a_1 x + b_1 y = c_1$$
$$a_2 x + b_2 y = c_2$$

has a solution given by

$$x = \frac{c_1 b_2 - c_2 b_1}{a_1 b_2 - a_2 b_1} \quad \text{and} \quad y = \frac{a_1 c_2 - a_2 c_1}{a_1 b_2 - a_2 b_1}$$

provided that $a_1 b_2 - a_2 b_1 \neq 0$. Each numerator and denominator in this solution can be expressed as a determinant, as follows.

$$x = \frac{c_1 b_2 - c_2 b_1}{a_1 b_2 - a_2 b_1} = \frac{\begin{vmatrix} c_1 & b_1 \\ c_2 & b_2 \end{vmatrix}}{\begin{vmatrix} a_1 & b_1 \\ a_2 & b_2 \end{vmatrix}}, \qquad y = \frac{a_1 c_2 - a_2 c_1}{a_1 b_2 - a_2 b_1} = \frac{\begin{vmatrix} a_1 & c_1 \\ a_2 & c_2 \end{vmatrix}}{\begin{vmatrix} a_1 & b_1 \\ a_2 & b_2 \end{vmatrix}}$$

Relative to the original system, the denominator for x and y is simply the determinant of the *coefficient* matrix of the system. This determinant is denoted by D. The numerators for x and y are denoted by D_x and D_y, respectively. They are formed by using the column of constants as replacements for the coefficients of x and y, as follows.

Coefficient *Matrix*	D	D_x	D_y
$\begin{bmatrix} a_1 & b_1 \\ a_2 & b_2 \end{bmatrix}$	$\begin{vmatrix} a_1 & b_1 \\ a_2 & b_2 \end{vmatrix}$	$\begin{vmatrix} c_1 & b_1 \\ c_2 & b_2 \end{vmatrix}$	$\begin{vmatrix} a_1 & c_1 \\ a_2 & c_2 \end{vmatrix}$

EXAMPLE 5   ▱   **Using Cramer's Rule for a 2 × 2 System**

Use Cramer's Rule to solve the following system of linear equations.

$$4x - 2y = 10$$
$$3x - 5y = 11$$

Solution

To begin, find the determinant of the coefficient matrix.

$$D = \begin{vmatrix} 4 & -2 \\ 3 & -5 \end{vmatrix} = -20 - (-6) = -14$$

Because this determinant is not zero, you can apply Cramer's Rule to find the solution, as follows.

$$x = \frac{D_x}{D} = \frac{\begin{vmatrix} 10 & -2 \\ 11 & -5 \end{vmatrix}}{-14} = \frac{(-50) - (-22)}{-14} = \frac{-28}{-14} = 2$$

$$y = \frac{D_y}{D} = \frac{\begin{vmatrix} 4 & 10 \\ 3 & 11 \end{vmatrix}}{-14} = \frac{44 - 30}{-14} = \frac{14}{-14} = -1$$

Therefore, the solution is $x = 2$ and $y = -1$. Check this in the original system.

▱

Cramer's Rule generalizes easily to systems of n equations in n variables. The value of each variable is given as the quotient of two determinants. The denominator is the determinant of the coefficient matrix, and the numerator is the determinant of the matrix formed by replacing the column corresponding to the variable (being solved for) with the column representing the constants. For instance, the solution for x_3 in the system

$$a_{11}x_1 + a_{12}x_2 + a_{13}x_3 = b_1$$
$$a_{21}x_1 + a_{22}x_2 + a_{23}x_3 = b_2$$
$$a_{31}x_1 + a_{32}x_2 + a_{33}x_3 = b_3$$

is given by

$$x_3 = \frac{|A_3|}{|A|} = \frac{\begin{vmatrix} a_{11} & a_{12} & b_1 \\ a_{21} & a_{22} & b_2 \\ a_{31} & a_{32} & b_3 \end{vmatrix}}{\begin{vmatrix} a_{11} & a_{12} & a_{13} \\ a_{21} & a_{22} & a_{23} \\ a_{31} & a_{32} & a_{33} \end{vmatrix}}.$$

Cramer's Rule

If a system of n linear equations in n variables has a coefficient matrix A with a *nonzero* determinant $|A|$, the solution is given by

$$x_1 = \frac{|A_1|}{|A|}, \quad x_2 = \frac{|A_2|}{|A|}, \quad \cdots, \quad x_n = \frac{|A_n|}{|A|}$$

where the ith column of A_i is the column of constants in the system of equations. If the coefficient matrix is zero, the system has either no solution or infinitely many solutions.

EXAMPLE 6   Using Cramer's Rule for a 3 × 3 System

Use Cramer's Rule, if possible, to solve the following system of linear equations.

$$-x \qquad + z = 4$$
$$2x - y + z = -3$$
$$y - 3z = 1$$

Solution

Using the matrix capabilities of a graphing utility to evaluate the determinant of the coefficient matrix A, you find that Cramer's Rule cannot be applied because $|A| = 0$.

Cryptography

A **cryptogram** is a message written according to a secret code. (The Greek word *kryptos* means "hidden.") Matrix multiplication can be used to **encode** and **decode** messages. To begin, you need to assign a number to each letter in the alphabet (with 0 assigned to a blank space), as follows.

0 = _	9 = I	18 = R
1 = A	10 = J	19 = S
2 = B	11 = K	20 = T
3 = C	12 = L	21 = U
4 = D	13 = M	22 = V
5 = E	14 = N	23 = W
6 = F	15 = O	24 = X
7 = G	16 = P	25 = Y
8 = H	17 = Q	26 = Z

Then the message is converted to numbers and partitioned into **uncoded row matrices,** each having n entries, as demonstrated in Example 7.

EXAMPLE 7 ▱ **Forming Uncoded Row Matrices**

Write the uncoded row matrices of order 1×3 for the message

MEET ME MONDAY.

Solution

Partitioning the message (including blank spaces, but ignoring punctuation) into groups of three produces the following uncoded row matrices.

$$\begin{bmatrix} 13 & 5 & 5 \end{bmatrix} \quad \begin{bmatrix} 20 & 0 & 13 \end{bmatrix} \quad \begin{bmatrix} 5 & 0 & 13 \end{bmatrix} \quad \begin{bmatrix} 15 & 14 & 4 \end{bmatrix} \quad \begin{bmatrix} 1 & 25 & 0 \end{bmatrix}$$
$$ M \quad E \quad E \quad\quad T \quad\quad M \quad E \quad\quad M \quad O \quad N \quad D \quad A \quad Y$$

Note that a blank space is used to fill out the last uncoded row matrix. ▱

To **encode** a message, choose an $n \times n$ invertible matrix A and multiply the uncoded row matrices by A (on the right) to obtain **coded row matrices.** Here is an example.

Uncoded Matrix   Encoding Matrix A   Coded Matrix

$$\begin{bmatrix} 13 & 5 & 5 \end{bmatrix} \begin{bmatrix} 1 & -2 & 2 \\ -1 & 1 & 3 \\ 1 & -1 & -4 \end{bmatrix} = \begin{bmatrix} 13 & -26 & 21 \end{bmatrix}$$

This technique is further illustrated in Example 8.

EXAMPLE 8   Encoding a Message

Use the following matrix to encode the message MEET ME MONDAY.

$$A = \begin{bmatrix} 1 & -2 & 2 \\ -1 & 1 & 3 \\ 1 & -1 & -4 \end{bmatrix}$$

Solution

The coded row matrices are obtained by multiplying each of the uncoded row matrices found in Example 7 by the matrix A, as follows.

Uncoded Matrix   Encoding Matrix A   Coded Matrix

$$[13 \quad 5 \quad 5] \begin{bmatrix} 1 & -2 & 2 \\ -1 & 1 & 3 \\ 1 & -1 & -4 \end{bmatrix} = [13 \quad -26 \quad 21]$$

$$[20 \quad 0 \quad 13] \begin{bmatrix} 1 & -2 & 2 \\ -1 & 1 & 3 \\ 1 & -1 & -4 \end{bmatrix} = [33 \quad -53 \quad -12]$$

$$[5 \quad 0 \quad 13] \begin{bmatrix} 1 & -2 & 2 \\ -1 & 1 & 3 \\ 1 & -1 & -4 \end{bmatrix} = [18 \quad -23 \quad -42]$$

$$[15 \quad 14 \quad 4] \begin{bmatrix} 1 & -2 & 2 \\ -1 & 1 & 3 \\ 1 & -1 & -4 \end{bmatrix} = [5 \quad -20 \quad 56]$$

$$[1 \quad 25 \quad 0] \begin{bmatrix} 1 & -2 & 2 \\ -1 & 1 & 3 \\ 1 & -1 & -4 \end{bmatrix} = [-24 \quad 23 \quad 77]$$

Thus, the sequence of coded row matrices is

$$[13 \; -26 \; 21][33 \; -53 \; -12][18 \; -23 \; -42][5 \; -20 \; 56][-24 \; 23 \; 77].$$

Finally, removing the matrix notation produces the following cryptogram.

$$13 \; -26 \; 21 \quad 33 \; -53 \; -12 \quad 18 \; -23 \; -42 \quad 5 \; -20 \; 56 \quad -24 \; 23 \; 77$$

An efficient method for encoding the message at the right with your graphing utility is to enter A as a 3×3 matrix. Let B be the 5×3 matrix whose rows are the uncoded row matrices,

$$B = \begin{bmatrix} 13 & 5 & 5 \\ 20 & 0 & 13 \\ 5 & 0 & 13 \\ 15 & 14 & 4 \\ 1 & 25 & 0 \end{bmatrix}.$$

The product BA gives the coded row matrices.

For those who do not know the matrix A, decoding the cryptogram found in Example 8 is difficult. But for an authorized receiver who knows the matrix A, decoding is simple. The receiver need only multiply the coded row matrices by A^{-1} (on the right) to retrieve the uncoded row matrices. Here is an example.

$$\underbrace{[13 \; -26 \; 21]}_{\text{Coded}} A^{-1} = \underbrace{[13 \quad 5 \quad 5]}_{\text{Uncoded}}$$

EXAMPLE 9   ▱   **Decoding a Message**

Use the inverse of the matrix $A = \begin{bmatrix} 1 & -2 & 2 \\ -1 & 1 & 3 \\ 1 & -1 & -4 \end{bmatrix}$ to decode the cryptogram

13 −26 21 33 −53 −12 18 −23 −42 5 −20 56 −24 23 77.

Solution

Partition the message into groups of three to form the coded row matrices. Then, multiply each coded row matrix by A^{-1} (on the right).

Coded Matrix   *Decoding Matrix A^{-1}*   *Decoded Matrix*

$$[13 \quad -26 \quad 21] \begin{bmatrix} -1 & -10 & -8 \\ -1 & -6 & -5 \\ 0 & -1 & -1 \end{bmatrix} = [13 \quad 5 \quad 5]$$

$$[33 \quad -53 \quad -12] \begin{bmatrix} -1 & -10 & -8 \\ -1 & -6 & -5 \\ 0 & -1 & -1 \end{bmatrix} = [20 \quad 0 \quad 13]$$

$$[18 \quad -23 \quad -42] \begin{bmatrix} -1 & -10 & -8 \\ -1 & -6 & -5 \\ 0 & -1 & -1 \end{bmatrix} = [5 \quad 0 \quad 13]$$

$$[5 \quad -20 \quad 56] \begin{bmatrix} -1 & -10 & -8 \\ -1 & -6 & -5 \\ 0 & -1 & -1 \end{bmatrix} = [15 \quad 14 \quad 4]$$

$$[-24 \quad 23 \quad 77] \begin{bmatrix} -1 & -10 & -8 \\ -1 & -6 & -5 \\ 0 & -1 & -1 \end{bmatrix} = [1 \quad 25 \quad 0]$$

Thus, the message is as follows.

[13 5 5] [20 0 13] [5 0 13] [15 14 4] [1 25 0]

M E E T   M E   M O N D A Y   ▱

Group Activity   *Cryptography*

Create your own numeric code for the alphabet (such as on page 643), and use it to convert a message of your own into numbers. Create an invertible $n \times n$ matrix A to encode your message. Exchange your numeric code, encoded message, and matrix A with another group. Find the necessary decoding matrix and decode the message you received.

8.5 /// EXERCISES

In Exercises 1–10, use a determinant to find the area of the triangle with the given vertices.

1.

2.

3. $(0, 0), (1, 5), (3, 1)$ **4.** $(0, 0), (4, 5), (5, -2)$

5. $\left(0, \frac{1}{2}\right), \left(\frac{5}{2}, 0\right), (4, 3)$ **6.** $(0, 4), (2, 3), (5, 0)$

7. $(4, 5), (6, 1), (7, 9)$

8. $(0, -2), (-1, 4), (3, 5)$

9. $(-3, 5), (2, 6), (3, -5)$

10. $(-2, 4), (1, 5), (3, -2)$

In Exercises 11 and 12, find a value of x such that the triangle has an area of 4.

11. $(-5, 1), (0, 2), (-2, x)$

12. $(-4, 2), (-3, 5), (-1, x)$

In Exercises 13–16, use Cramer's Rule to solve (if possible) the system of equations.

13. $3x + 4y = -2$
  $5x + 3y = 4$

14. $-0.4x + 0.8y = 1.6$
  $0.2x + 0.3y = 2.2$

15. $4x - y + z = -5$
  $2x + 2y + 3z = 10$
  $5x - 2y + 6z = 1$

16. $4x - 2y + 3z = -2$
  $2x + 2y + 5z = 16$
  $8x - 5y - 2z = 4$

In Exercises 17 and 18, use a graphing utility and Cramer's Rule to solve (if possible) the system of equations.

17. $3x + 3y + 5z = 1$
  $3x + 5y + 9z = 2$
  $5x + 9y + 17z = 4$

18. $2x + 3y + 5z = 4$
  $3x + 5y + 9z = 7$
  $5x + 9y + 17z = 13$

19. *Area of a Region*   A large region of forest has been infected with gypsy moths. The region is roughly triangular, as shown in the figure. From the northernmost vertex A of the region, the distances to the other vertices are 25 miles south and 10 miles east (for vertex B), and 20 miles south and 28 miles east (for vertex C). Use a graphing utility to approximate the number of square miles in this region.

20. *Area of a Region*   You own a triangular tract of land, as shown in the figure. To estimate the number of square feet in the tract, you start at one vertex, walk 65 feet east and 50 feet north to the second vertex, and then walk 85 feet west and 30 feet north to the third vertex. Use a graphing utility to determine how many square feet there are in the tract of land.

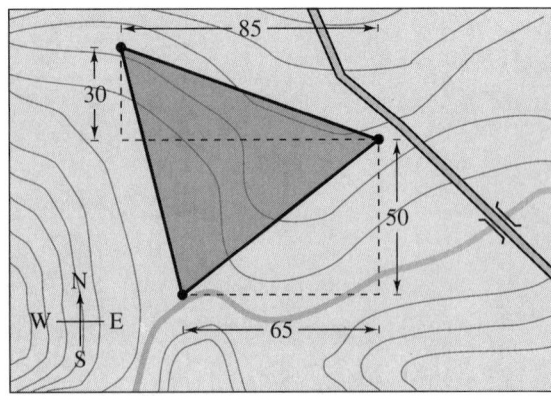

In Exercises 21–26, use the determinant feature of a graphing utility to decide if the points are collinear.

21. $(3, -1), (0, -3), (12, 5)$

22. $(-3, -5), (6, 1), (10, 2)$

23. $\left(2, -\frac{1}{2}\right), (-4, 4), (6, -3)$

24. $(0, 1), (4, -2), (-8, 7)$

25. $(0, 2), (1, 2.4), (-1, 1.6)$

26. $(2, 3), (3, 3.5), (-1, 2)$

In Exercises 27–32, use a determinant to find an equation of the line through the points.

27. $(0, 0), (5, 3)$ **28.** $(0, 0), (-2, 2)$

29. $(-4, 3), (2, 1)$ **30.** $(10, 7), (-2, -7)$

31. $\left(-\frac{1}{2}, 3\right), \left(\frac{5}{2}, 1\right)$ **32.** $\left(\frac{2}{3}, 4\right), (6, 12)$

In Exercises 33 and 34, find x such that the points are collinear.

33. $(2, -5), (4, x), (5, -2)$

34. $(-6, 2), (-5, x), (-3, 5)$

In Exercises 35 and 36, find the uncoded 1×3 row matrices for the message. Then encode the message using the matrix.

Message	Matrix
35. TROUBLE IN RIVER CITY	$\begin{bmatrix} 1 & -1 & 0 \\ 1 & 0 & -1 \\ -6 & 2 & 3 \end{bmatrix}$
36. PLEASE SEND MONEY	$\begin{bmatrix} 4 & 2 & 1 \\ -3 & -3 & -1 \\ 3 & 2 & 1 \end{bmatrix}$

In Exercises 37–40, write a cryptogram for the message using the matrix

$$A = \begin{bmatrix} 1 & 2 & 2 \\ 3 & 7 & 9 \\ -1 & -4 & -7 \end{bmatrix}.$$

37. LANDING SUCCESSFUL

38. BEAM ME UP SCOTTY

39. HAPPY BIRTHDAY

40. OPERATION OVERLORD

In Exercises 41 and 42, use A^{-1} to decode the cryptogram.

41. $A = \begin{bmatrix} 1 & 2 \\ 3 & 5 \end{bmatrix}$

11, 21, 64, 112, 25, 50, 29, 53, 23, 46, 40, 75, 55, 92

42. $A = \begin{bmatrix} 1 & -1 & 0 \\ 1 & 0 & -1 \\ -6 & 2 & 3 \end{bmatrix}$

9, -1, -9, 38, -19, -19, 28, -9, -19, -80, 25, 41, -64, 21, 31, 9, -5, -4

In Exercises 43 and 44, decode the cryptogram by using the inverse of the matrix

$$A = \begin{bmatrix} 1 & 2 & 2 \\ 3 & 7 & 9 \\ -1 & -4 & -7 \end{bmatrix}.$$

43. 20, 17, -15, -12, -56, -104, 1, -25, -65, 62, 143, 181

44. 13, -9, -59, 61, 112, 106, -17, -73, -131, 11, 24, 29, 65, 144, 172

45. The following cryptogram was encoded with a 2×2 matrix.

8, 21, -15, -10, -13, -13, 5, 10, 5, 25, 5, 19, -1, 6, 20, 40, -18, -18, 1, 16

The last word of the message is _RON. What is the message?

46. The following cryptogram was encoded with a 2×2 matrix.

5, 2, 25, 11, -2, -7, -15, -15, 32, 14, -8, -13, 38, 19, -19, -19, 37, 16

The last word of the message is _SUE. What is the message?

Focus on Concepts

In this chapter, you studied several concepts that are required in the study of matrices and determinants and their applications. You can use the following questions to check your understanding of several of these basic concepts. The answers to these questions are given in the back of the book.

1. Describe the three elementary row operations that can be performed on an augmented matrix.

2. What is the relationship between the three elementary row operations on an augmented matrix and the operations that lead to equivalent systems of equations?

3. In your own words, describe the difference between a matrix in row-echelon form and a matrix in reduced row-echelon form.

In Exercises 4–7, the row-echelon form of an augmented matrix that corresponds to a system of linear equations is given. Use the matrix to determine whether the system is consistent or inconsistent, and if it is consistent, determine the number of solutions.

4. $\begin{bmatrix} 1 & 2 & 3 & \vdots & 9 \\ 0 & 1 & -2 & \vdots & 2 \\ 0 & 0 & 0 & \vdots & 0 \end{bmatrix}$

5. $\begin{bmatrix} 1 & 2 & 3 & \vdots & 9 \\ 0 & 1 & -2 & \vdots & 2 \\ 0 & 0 & 0 & \vdots & 8 \end{bmatrix}$

6. $\begin{bmatrix} 1 & 2 & 3 & \vdots & 9 \\ 0 & 1 & -2 & \vdots & 2 \\ 0 & 0 & 1 & \vdots & -3 \end{bmatrix}$

7. $\begin{bmatrix} 1 & 2 & 3 & 10 & 6 & \vdots & 0 \\ 0 & 1 & -5 & -2 & 0 & \vdots & 5 \\ 0 & 0 & 1 & 12 & 0 & \vdots & -2 \\ 0 & 0 & 0 & 1 & 1 & \vdots & 0 \end{bmatrix}$

In Exercises 8–10, determine if the matrix operations (a) $A + 3B$ and (b) AB can be performed. If not, state why.

8. $A = \begin{bmatrix} 2 & -2 \\ 3 & 5 \end{bmatrix}$, $B = \begin{bmatrix} -3 & 10 \\ 12 & 8 \end{bmatrix}$

9. $A = \begin{bmatrix} 5 & 4 \\ -7 & 2 \\ 11 & 2 \end{bmatrix}$, $B = \begin{bmatrix} 4 & 12 \\ 20 & 40 \\ 15 & 30 \end{bmatrix}$

10. $A = \begin{bmatrix} 5 & 4 \\ -7 & 2 \\ 11 & 2 \end{bmatrix}$, $B = \begin{bmatrix} 4 & 12 \\ 20 & 40 \end{bmatrix}$

11. Under what conditions does a matrix have an inverse?

12. Explain the difference between a square matrix and its determinant.

13. Is it possible to find the determinant of a 4×5 matrix? Explain.

14. What is meant by the cofactor of an entry of a matrix? How is it used to find the determinant of the matrix?

15. Three people were asked to solve a system of equations using an augmented matrix. Each person reduced the matrix to row-echelon form. The reduced matrices were

$\begin{bmatrix} 1 & 2 & \vdots & 3 \\ 0 & 1 & \vdots & 1 \end{bmatrix}$

$\begin{bmatrix} 1 & 0 & \vdots & 1 \\ 0 & 1 & \vdots & 1 \end{bmatrix}$

and

$\begin{bmatrix} 1 & 2 & \vdots & 3 \\ 0 & 0 & \vdots & 0 \end{bmatrix}.$

Could all three be right? Explain.

8 /// REVIEW EXERCISES

In Exercises 1 and 2, form the augmented matrix for the system of linear equations.

1. $3x - 10y = 15$
$5x + 4y = 22$

2. $8x - 7y + 4z = 12$
$3x - 5y + 2z = 20$
$5x + 3y - 3z = 26$

In Exercises 3 and 4, write the system of linear equations represented by the augmented matrix. (Use variables x, y, z, and w.)

3. $\begin{bmatrix} 5 & 1 & 7 & \vdots & -9 \\ 4 & 2 & 0 & \vdots & 10 \\ 9 & 4 & 2 & \vdots & 3 \end{bmatrix}$

4. $\begin{bmatrix} 13 & 16 & 7 & 3 & \vdots & 2 \\ 1 & 21 & 8 & 5 & \vdots & 12 \\ 4 & 10 & -4 & 3 & \vdots & -1 \end{bmatrix}$

In Exercises 5 and 6, write the matrix in *reduced* row-echelon form.

5. $\begin{bmatrix} 0 & 1 & 1 \\ 1 & 2 & 3 \\ 2 & 2 & 2 \end{bmatrix}$

6. $\begin{bmatrix} 1 & 1 & 1 & 0 \\ 1 & 1 & 0 & 1 \\ 1 & 0 & 1 & 1 \\ 0 & 1 & 1 & 1 \end{bmatrix}$

In Exercises 7–10, use the matrix capabilities of a graphing utility to write the matrix in *reduced* row-echelon form.

7. $\begin{bmatrix} 3 & -2 & 1 & 0 \\ 4 & -3 & 0 & 1 \end{bmatrix}$

8. $\begin{bmatrix} 1 & 1 & 2 & 1 & 0 & 0 \\ -1 & 0 & 3 & 0 & 1 & 0 \\ 1 & 2 & 8 & 0 & 0 & 1 \end{bmatrix}$

9. $\begin{bmatrix} 1 & 3 & 4 \\ 0 & 1 & 1 \\ 2 & 4 & 6 \end{bmatrix}$

10. $\begin{bmatrix} 4 & 8 & 16 \\ 3 & -1 & 2 \\ -2 & 10 & 12 \end{bmatrix}$

In Exercises 11–22, use matrices and elementary row operations to solve (if possible) the system of equations.

11. $5x + 4y = 2$
$-x + y = -22$

12. $2x - 5y = 2$
$3x - 7y = 1$

13. $2x + y = 0.3$
$3x - y = -1.3$

14. $0.2x - 0.1y = 0.07$
$0.4x - 0.5y = -0.01$

15. $2x + y + 2z = 4$
$2x + 2y = 5$
$2x - y + 6z = 2$

16. $2x + 3y + z = 10$
$2x - 3y - 3z = 22$
$4x - 2y + 3z = -2$

17. $4x + 4y + 4z = 5$
$4x - 2y - 8z = 1$
$5x + 3y + 8z = 6$

18. $2x + 3y + 3z = 3$
$6x + 6y + 12z = 13$
$12x + 9y - z = 2$

19. $-x + y + 2z = 1$
$2x + 3y + z = -2$
$5x + 4y + 2z = 4$

20. $3x + 21y - 29z = -1$
$2x + 15y - 21z = 0$

21. $x + 2y + 6z = 1$
$2x + 5y + 15z = 4$
$3x + y + 3z = -6$

22. $x + 2y + w = 3$
$-3y + 3z = 0$
$4x + 4y + z + 2w = 0$
$2x + z = 3$

23. *Think About It* Describe the row-echelon form of an augmented matrix that corresponds to a system of linear equations that has a unique solution.

24. *Partial Fractions*   Write the partial fraction decomposition for the rational expression

$$\frac{x+9}{(x+1)(x+2)^2} = \frac{A}{x+1} + \frac{B}{x+2} + \frac{C}{(x+2)^2}.$$

In Exercises 25–28, use the matrix capabilities of a graphing utility to reduce the augmented matrix and solve the system of equations.

25. $x - 3y = -2$
   $x + y = 2$

26. $-x + 3y = 5$
   $4x - y = 2$

27. $x + 2y - z = 7$
   $-y - z = 4$
   $4x - z = 16$

28. $3x + 6z = 0$
   $-2x + y = 5$
   $y + 2z = 3$

In Exercises 29–36, perform the matrix operations. If it is not possible, explain why.

29. $\begin{bmatrix} 2 & 1 & 0 \\ 0 & 5 & -4 \end{bmatrix} - 3\begin{bmatrix} 5 & 3 & -6 \\ 0 & -2 & 5 \end{bmatrix}$

30. $-2\begin{bmatrix} 1 & 2 \\ 5 & -4 \\ 6 & 0 \end{bmatrix} + 8\begin{bmatrix} 7 & 1 \\ 1 & 2 \\ 1 & 4 \end{bmatrix}$

31. $\begin{bmatrix} 1 & 2 \\ 5 & -4 \\ 6 & 0 \end{bmatrix}\begin{bmatrix} 6 & -2 & 8 \\ 4 & 0 & 0 \end{bmatrix}$

32. $\begin{bmatrix} 1 & 5 & 6 \\ 2 & -4 & 0 \end{bmatrix}\begin{bmatrix} 6 & -2 & 8 \\ 4 & 0 & 0 \end{bmatrix}$

33. $\begin{bmatrix} 1 & 5 & 6 \\ 2 & -4 & 0 \end{bmatrix}\begin{bmatrix} 6 & 4 \\ -2 & 0 \\ 8 & 0 \end{bmatrix}$

34. $\begin{bmatrix} 4 \\ 6 \end{bmatrix}\begin{bmatrix} 6 & -2 \end{bmatrix}$

35. $\begin{bmatrix} 1 & 3 & 2 \\ 0 & 2 & -4 \\ 0 & 0 & 3 \end{bmatrix}\begin{bmatrix} 4 & -3 & 2 \\ 0 & 3 & -1 \\ 0 & 0 & 2 \end{bmatrix}$

36. $\begin{bmatrix} 2 & 1 \\ 6 & 0 \end{bmatrix}\left(\begin{bmatrix} 4 & 2 \\ -3 & 1 \end{bmatrix} + \begin{bmatrix} -2 & 4 \\ 0 & 4 \end{bmatrix}\right)$

In Exercises 37–40, use a graphing utility to perform the matrix operations.

37. $3\begin{bmatrix} 8 & -2 & 5 \\ 1 & 3 & -1 \end{bmatrix} + 6\begin{bmatrix} 4 & -2 & -3 \\ 2 & 7 & 6 \end{bmatrix}$

38. $-5\begin{bmatrix} 2 & 0 \\ 7 & -2 \\ 8 & 2 \end{bmatrix} + 4\begin{bmatrix} 4 & -2 \\ 6 & 11 \\ -1 & 3 \end{bmatrix}$

39. $\begin{bmatrix} 4 & 1 \\ 11 & -7 \\ 12 & 3 \end{bmatrix}\begin{bmatrix} 3 & -5 & 6 \\ 2 & -2 & -2 \end{bmatrix}$

40. $\begin{bmatrix} -2 & 3 & 10 \\ 4 & -2 & 2 \end{bmatrix}\begin{bmatrix} 1 & 1 \\ -5 & 2 \\ 3 & 2 \end{bmatrix}$

In Exercises 41–44, solve for X given

$$A = \begin{bmatrix} -4 & 0 \\ 1 & -5 \\ -3 & 2 \end{bmatrix} \quad \text{and} \quad B = \begin{bmatrix} 1 & 2 \\ -2 & 1 \\ 4 & 4 \end{bmatrix}.$$

41. $X = 3A - 2B$

42. $6X = 4A + 3B$

43. $3X + 2A = B$

44. $2A - 5B = 3X$

45. Write the system of linear equations represented by the matrix equation

$$\begin{bmatrix} 5 & 4 \\ -1 & 1 \end{bmatrix}\begin{bmatrix} x \\ y \end{bmatrix} = \begin{bmatrix} 2 \\ -22 \end{bmatrix}.$$

46. Write the matrix equation $AX = B$ for the following system of linear equations.

$$2x + 3y + z = 10$$
$$2x - 3y - 3z = 22$$
$$4x - 2y + 3z = -2$$

In Exercises 47–50, use a graphing utility to find the inverse of the matrix (if it exists).

47. $\begin{bmatrix} 2 & 6 \\ 3 & -6 \end{bmatrix}$

48. $\begin{bmatrix} 3 & -10 \\ 4 & 2 \end{bmatrix}$

49. $\begin{bmatrix} 2 & 0 & 3 \\ -1 & 1 & 1 \\ 2 & -2 & 1 \end{bmatrix}$

50. $\begin{bmatrix} 1 & 4 & 6 \\ 2 & -3 & 1 \\ -1 & 18 & 16 \end{bmatrix}$

In Exercises 51–58, evaluate the determinant. Use a graphing utility to verify your result.

51. $\begin{vmatrix} 50 & -30 \\ 10 & 5 \end{vmatrix}$ **52.** $\begin{vmatrix} 8 & 5 \\ 2 & -4 \end{vmatrix}$

53. $\begin{vmatrix} 10 & 8 \\ -6 & -4 \end{vmatrix}$ **54.** $\begin{vmatrix} x & x^2 \\ 1 & 2x \end{vmatrix}$

55. $\begin{vmatrix} 1 & 0 & -2 \\ 0 & 1 & 0 \\ -2 & 0 & 1 \end{vmatrix}$ **56.** $\begin{vmatrix} 0 & 3 & 1 \\ 5 & -2 & 1 \\ 1 & 6 & 1 \end{vmatrix}$

57. $\begin{vmatrix} 3 & 0 & -4 & 0 \\ 0 & 8 & 1 & 2 \\ 6 & 1 & 8 & 2 \\ 0 & 3 & -4 & 1 \end{vmatrix}$ **58.** $\begin{vmatrix} -5 & 6 & 0 & 0 \\ 0 & 1 & -1 & 2 \\ -3 & 4 & -5 & 1 \\ 1 & 6 & 0 & 3 \end{vmatrix}$

In Exercises 59–66, use a graphing utility to solve (if possible) the system of linear equations using the inverse of the coefficient matrix.

59. $x + 2y = -1$
$3x + 4y = -5$

60. $x + 3y = 23$
$-6x + 2y = -18$

61. $-3x - 3y - 4z = 2$
$y + z = -1$
$4x + 3y + 4z = -1$

62. $x - 3y - 2z = 8$
$-2x + 7y + 3z = -19$
$x - y - 3z = 3$

63. $x + 3y + 2z = 2$
$-2x - 5y - z = 10$
$2x + 4y = -12$

64. $2x + 4y = -12$
$3x + 4y - 2z = -14$
$-x + y + 2z = -6$

65. $-x + y + z = 6$
$4x - 3y + z = 20$
$2x - y + 3z = 8$

66. $2x + 3y - 4z = 1$
$x - y + 2z = -4$
$3x + 7y - 10z = 0$

In Exercises 67–70, use Cramer's Rule to solve (if possible) the system of equations.

67. $x + 2y = 5$
$-x + y = 1$

68. $2x - y = -10$
$3x + 2y = -1$

69. $20x + 8y = 11$
$12x - 24y = 21$

70. $13x - 6y = 17$
$26x - 12y = 8$

In Exercises 71–74, use a graphing utility and Cramer's Rule to solve (if possible) the system of equations.

71. $3x + 6y = 5$
$6x + 14y = 11$

72. $-0.4x + 0.8y = 1.6$
$0.2x + 0.3y = 2.2$

73. $5x - 3y + 2z = 2$
$2x + 2y - 3z = 3$
$x - 7y + 8z = -4$

74. $14x - 21y - 7z = 10$
$-4x + 2y - 2z = 4$
$56x - 21y + 7z = 5$

75. *Mixture Problem* A florist wants to arrange a dozen flowers consisting of two varieties: carnations and roses. Carnations cost $0.75 each and roses cost $1.50 each. How many of each should the florist use so that the arrangement will cost $12.00?

76. *Mixture Problem* One hundred liters of a 60% acid solution is obtained by mixing a 75% solution with a 50% solution. How many liters of each must be used to obtain the desired mixture?

77. *Fitting a Parabola to Three Points* Find an equation of the parabola $y = ax^2 + bx + c$ that passes through the points $(-1, 2)$, $(0, 3)$, and $(1, 6)$.

78. *Break-Even Point* A small business invests $25,000 in equipment to produce a product. Each unit of the product costs $3.75 to produce and is sold for $5.25. How many items must be sold before the business breaks even?

79. *Data Analysis*   The median prices y (in thousands of dollars) of one-family houses sold in the United States in the years 1981 through 1993 are shown in the figure. The least squares regression line $y = a + bt$ for this data is found by solving the system

$$13a + 91b = 1107$$
$$91a + 819b = 8404.7$$

where $t = 1$ represents 1981.   (Source: National Association of Realtors)

(a) Use a graphing utility to solve this system.

(b) Use a graphing utility to graph the regression line.

(c) Interpret the meaning of the slope of the regression line in the context of the problem.

(d) Use the regression line to estimate the median price of homes in 1995.

Year $(1 \leftrightarrow 1981)$

80. Solve the equation $\begin{vmatrix} 2 - \lambda & 5 \\ 3 & -8 - \lambda \end{vmatrix} = 0.$

In Exercises 81–84, use a determinant to find the area of the triangle with the given vertices.

81. $(1, 0), (5, 0), (5, 8)$

82. $(-4, 0), (4, 0), (0, 6)$

83. $(1, 2), (4, -5), (3, 2)$

84. $\left(\frac{3}{2}, 1\right), \left(4, -\frac{1}{2}\right), (4, 2)$

In Exercises 85–88, use a determinant to find an equation of the line through the given points.

85. $(-4, 0), (4, 4)$ **86.** $(2, 5), (6, -1)$

87. $\left(-\frac{5}{2}, 3\right), \left(\frac{7}{2}, 1\right)$ **88.** $(-0.8, 0.2), (0.7, 3.2)$

89. Verify that

$$
\begin{vmatrix}
a_{11} & a_{12} & a_{13} \\
a_{21} & a_{22} & a_{23} \\
a_{31} + c_1 & a_{32} + c_2 & a_{33} + c_3
\end{vmatrix}
$$

$$
=
\begin{vmatrix}
a_{11} & a_{12} & a_{13} \\
a_{21} & a_{22} & a_{23} \\
a_{31} & a_{32} & a_{33}
\end{vmatrix}
+
\begin{vmatrix}
a_{11} & a_{12} & a_{13} \\
a_{21} & a_{22} & a_{23} \\
c_1 & c_2 & c_3
\end{vmatrix}.
$$

90. *Circuit Analysis*   Consider the circuit in the figure. The currents I_1, I_2, and I_3 in amperes are given by the solution of the system of linear equations. Use the inverse of the coefficient matrix of this system to find the unknown currents.

$$I_1 + I_2 + I_3 = 0$$
$$4I_1 - 10I_2 = 12$$
$$10I_2 - 2I_3 = -6$$

91. *Think About It*   If A is a 3×3 matrix and $|A| = 2$, what is the value of $|4A|$? Give the reason for your answer.

CHAPTER PROJECT   *Row Operations and Graphing*

In this project, you will investigate the graphical interpretation of elementary row operations.

(a) Solve the following systems by hand using Gauss-Jordan elimination.

$$2x - 4y = 9 \qquad 6x + 2y = -19$$
$$x + 5y = 15 \qquad 3x - y = -5$$

(b) Enter the row operations program listed in the appendix into a graphing calculator. This program demonstrates how elementary row operations used in Gauss-Jordan elimination may be depicted graphically. For each system in part (a), run the program using a 2×3 matrix that corresponds to the system of equations. Compare the results of the program with those you obtained in part (a).

Graph of the system
$$2x - 4y = 9$$
$$x + 5y = 15$$

(c) During the running of the program, a row of the matrix is multiplied by a constant. What effect does this operation have on the graph of the corresponding linear equation?

(d) During the running of the program, a multiple of the first row of the matrix is added to the second row to obtain a 0 below the leading 1. What effect does this operation have on the graph of the corresponding linear equation?

(e) Each time the 2×3 matrix is transformed, the graph of the corresponding linear equations is displayed. What do you notice about the point of intersection each time?

Questions for Further Exploration

1. Is finding a point of intersection using the program more or less accurate than finding the point of intersection using the zoom and trace features? Explain your reasoning and give an example.

2. Run the program to find the solution to the following linear system.

$$2x = -15$$
$$3x + 5y = 3$$

Why is only one line drawn in all but the last screen? Verify the program's solution by hand.

3. Run the program using the following linear system.

$$2y = -3$$
$$-2x + y = 3$$

Describe what happens and why.

4. A system of equations with three variables has a corresponding 3×4 augmented matrix. Write a program that will transform a 3×4 matrix into reduced row-echelon form. At the end of the program display the final matrix.

8 /// CHAPTER TEST

Take this test as you would take a test in class. After you are done, check your
work against the answers given in the back of the book.

The *Interactive* CD-ROM
provides answers to the
Chapter Tests and Cumulative
Tests. It also offers Chapter
Pre-Tests (that test key skills
and concepts covered in previous chapters) and Chapter
Post-Tests, both of which have
randomly generated exercises
with diagnostic capabilities.

In Exercises 1 and 2, write the matrix in reduced row-echelon form. Use a graphing
utility to verify your result.

1. $\begin{bmatrix} 1 & -1 & 5 \\ 6 & 2 & 3 \\ 5 & 3 & -3 \end{bmatrix}$ 2. $\begin{bmatrix} 1 & 0 & -1 & 2 \\ -1 & 1 & 1 & -3 \\ 1 & 1 & -1 & 1 \\ 3 & 2 & -3 & 4 \end{bmatrix}$

3. Use the matrix capabilities of a graphing utility to reduce the augmented matrix and
   solve the system of equations.

   $$4x + 3y - 2z = 14$$
   $$-x - y + 2z = -5$$
   $$3x + y - 4z = 8$$

4. Find the equation of the parabola $y = ax^2 + bx + c$ that passes through
   the points in the figure. Use a graphing utility to verify your result.

5. Find (a) $A - B$, (b) $3A$, and (c) $3A - 2B$.

   $$A = \begin{bmatrix} 5 & 4 & 4 \\ -4 & -4 & 0 \end{bmatrix}, \quad B = \begin{bmatrix} 4 & -1 & 6 \\ -4 & 0 & -3 \end{bmatrix}$$

6. Find AB, if possible.

   $$A = \begin{bmatrix} 2 & -2 & 6 \\ 3 & -1 & 7 \\ 2 & 0 & -2 \end{bmatrix}, \quad B = \begin{bmatrix} 4 & 4 \\ 3 & 2 \\ 1 & -2 \end{bmatrix}$$

7. Find A^{-1} for $A = \begin{bmatrix} -6 & 4 \\ 10 & -5 \end{bmatrix}$.

8. Use the result of Exercise 7 to solve the system.

   $$-6x + 4y = 10$$
   $$10x - 5y = 20$$

9. Evaluate the determinant of the matrix

   $$\begin{bmatrix} 4 & 0 & 3 \\ 1 & -8 & 2 \\ 3 & 2 & 2 \end{bmatrix}.$$

10. Use a determinant to find the area of the triangle in the figure.

Figure for 4

Figure for 10

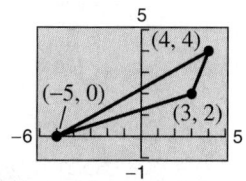

Sequences, Probability, and Statistics

Burney Le Boeuf, shown in the inset photo, is a marine biologist at the University of California at Santa Cruz. He has spent many years studying elephant seals.

Of all mammals, reptiles, and birds, elephant seals are the champions of deep sea dives. One bull elephant seal was measured at a depth of nearly 1 mile! "Our theories were blown clear out of the water," says Le Boeuf. (Source: Smithsonian Magazine, September, 1995, "Elephant Seals, the Champion Divers of the Deep.")

The frequency distribution on the left shows the recorded depths of over 10,000 dives by female elephant seals. From this data, you can make predictions about the depth of a particular dive. For instance, because 54.6% of all dives were in depths greater than 400 meters, you can conclude that the probability that a particular dive will exceed 400 meters is 0.546. (See Exercises 59 and 60 on page 728)

Frans Lanting—Minden Pictures.

Researchers from the University of California at Santa Cruz are trying to block a tranquilized bull elephant seal from entering the water. They will then attach instruments to the bull to measure the depths and durations of its dives.

9.1   Sequences and Summation Notation

Sequences   / Factorial Notation   / Summation Notation   / Application

Sequences

In mathematics, the word *sequence* is used in much the same way as in ordinary English. Saying that a collection is listed *in sequence* means that it is ordered so that it has a first member, a second member, a third member, and so on.

Mathematically, you can think of a sequence as a *function* whose domain is the set of positive integers. Rather than using function notation, however, sequences are usually written using subscript notation, as indicated in the following definition.

Note   On occasion, it is convenient to begin subscripting a sequence with 0 instead of 1 so that the terms of the sequence become

$$a_0, a_1, a_2, a_3, \ldots .$$

Definition of Sequence

An **infinite sequence** is a function whose domain is the set of positive integers. The function values

$$a_1, a_2, a_3, a_4, \ldots , a_n, \ldots$$

are the **terms** of the sequence. If the domain of the function consists of the first n positive integers only, the sequence is a **finite sequence.**

To graph a sequence using a graphing utility, set the utility to "seq" and "dot" modes and enter the sequence into Un. The graph of the sequence in Example 1(a) is shown below.

Graph the sequence in Example 1(b) and use the trace key to identify its terms.

EXAMPLE 1   **Finding the Terms of a Sequence**

a. The first four terms of the sequence given by $a_n = 3n - 2$ are

$a_1 = 3(1) - 2 = 1$	1st term
$a_2 = 3(2) - 2 = 4$	2nd term
$a_3 = 3(3) - 2 = 7$	3rd term
$a_4 = 3(4) - 2 = 10.$	4th term

b. The first four terms of the sequence given by $a_n = 3 + (-1)^n$ are

$a_1 = 3 + (-1)^1 = 3 - 1 = 2$	1st term
$a_2 = 3 + (-1)^2 = 3 + 1 = 4$	2nd term
$a_3 = 3 + (-1)^3 = 3 - 1 = 2$	3rd term
$a_4 = 3 + (-1)^4 = 3 + 1 = 4.$	4th term

EXAMPLE 2 ▭ **Finding the Terms of a Sequence**

The first four terms of the sequence given by $a_n = \dfrac{(-1)^n}{2n - 1}$ are

$$a_1 = \frac{(-1)^1}{2(1) - 1} = \frac{-1}{2 - 1} = -1 \qquad \text{1st term}$$

$$a_2 = \frac{(-1)^2}{2(2) - 1} = \frac{1}{4 - 1} = \frac{1}{3} \qquad \text{2nd term}$$

$$a_3 = \frac{(-1)^3}{2(3) - 1} = \frac{-1}{6 - 1} = -\frac{1}{5} \qquad \text{3rd term}$$

$$a_4 = \frac{(-1)^4}{2(4) - 1} = \frac{1}{8 - 1} = \frac{1}{7}. \qquad \text{4th term}$$

Simply listing the first few terms is not sufficient to define a unique sequence—the *n*th term *must be given*. To see this, consider the following sequences, both of which have the same first three terms.

$$\frac{1}{2}, \frac{1}{4}, \frac{1}{8}, \frac{1}{16}, \ \cdots \ \cdots \ \frac{1}{2^n}, \ \cdots$$

$$\frac{1}{2}, \frac{1}{4}, \frac{1}{8}, \frac{1}{15}, \ \cdots \ \cdots \ \frac{6}{(n + 1)(n^2 - n + 6)}, \ \cdots$$

EXAMPLE 3 ▭ **Finding the *n*th Term of a Sequence**

Write an expression for the apparent *n*th term (a_n) of each sequence.

a. 1, 3, 5, 7, . . .   **b.** 2, 5, 10, 17, . . .

Solution

a.   *n:* 1 2 3 4 . . . *n*

    Terms: 1 3 5 7 . . . a_n

    Apparent Pattern: Each term is 1 less than twice *n*, which implies that

$$a_n = 2n - 1.$$

b.   *n:* 1 2 3 4 . . . *n*

    Terms: 2 5 10 17 . . . a_n

    Apparent Pattern: Each term is 1 more than the square of *n*, which implies that

$$a_n = n^2 + 1.$$

Note Some sequences are defined **recursively.** To define a sequence recursively, you need to be given one or more of the first few terms. All other terms of the sequence are then defined using previous terms. A well-known example is the Fibonacci Sequence shown in Example 4.

The *Interactive* CD-ROM shows every example with its solution; clicking on the *Try It!* button brings up similar problems. Guided Examples and Integrated Examples show step-by-step solutions to additional examples. Integrated Examples are related to several concepts in the section.

EXAMPLE 4   A Sequence That Is Defined Recursively

The Fibonacci Sequence is defined recursively, as follows.

$$a_0 = 1, a_1 = 1, a_k = a_{k-2} + a_{k-1}, \qquad \text{where } k \geq 2$$

Write the first six terms of this sequence.

Solution

$a_0 = 1$	0th term is given.
$a_1 = 1$	1st term is given.
$a_2 = a_0 + a_1 = 1 + 1 = 2$	Use recursive formula.
$a_3 = a_1 + a_2 = 1 + 2 = 3$	Use recursive formula.
$a_4 = a_2 + a_3 = 2 + 3 = 5$	Use recursive formula.
$a_5 = a_3 + a_4 = 3 + 5 = 8$	Use recursive formula.

Factorial Notation

Some very important sequences in mathematics involve terms that are defined with special types of products called **factorials.**

Definition of Factorial

If n is a positive integer, n **factorial** is defined by

$$n! = 1 \cdot 2 \cdot 3 \cdot 4 \cdots (n - 1) \cdot n.$$

As a special case, zero factorial is defined as $0! = 1$.

Here are some values of $n!$ for the first several nonnegative integers. Notice that $0! = 1$ by definition.

$$0! = 1$$
$$1! = 1$$
$$2! = 1 \cdot 2 = 2$$
$$3! = 1 \cdot 2 \cdot 3 = 6$$
$$4! = 1 \cdot 2 \cdot 3 \cdot 4 = 24$$
$$5! = 1 \cdot 2 \cdot 3 \cdot 4 \cdot 5 = 120$$

The value of n does not have to be very large before the value of $n!$ becomes huge. For instance, $10! = 3,628,800$.

Factorials follow the same conventions for order of operations as do exponents. For instance,

$$2n! = 2(n!) = 2(1 \cdot 2 \cdot 3 \cdot 4 \cdots n)$$

whereas $(2n)! = 1 \cdot 2 \cdot 3 \cdot 4 \cdots 2n$.

EXAMPLE 5   **Finding the Terms of a Sequence Involving Factorials**

List the first five terms of the sequence given by $a_n = 2^n / n!$. Begin with $n = 0$.

Solution

$$a_0 = \frac{2^0}{0!} = \frac{1}{1} = 1 \qquad\qquad \text{0th term}$$

$$a_1 = \frac{2^1}{1!} = \frac{2}{1} = 2 \qquad\qquad \text{1st term}$$

$$a_2 = \frac{2^2}{2!} = \frac{4}{2} = 2 \qquad\qquad \text{2nd term}$$

$$a_3 = \frac{2^3}{3!} = \frac{8}{6} = \frac{4}{3} \qquad\qquad \text{3rd term}$$

$$a_4 = \frac{2^4}{4!} = \frac{16}{24} = \frac{2}{3} \qquad\qquad \text{4th term}$$

Most graphing utilities have the capability to compute $n!$. Use your utility to compare $3 \cdot 5!$ and $(3 \cdot 5)!$. How large a value of $n!$ will your graphing utility allow you to compute?

When working with fractions involving factorials, you will often find that the fractions can be reduced.

EXAMPLE 6   **Evaluating Factorial Expressions**

Evaluate each factorial expression.

a. $\dfrac{8!}{2! \cdot 6!}$ **b.** $\dfrac{2! \cdot 6!}{3! \cdot 5!}$ **c.** $\dfrac{n!}{(n-1)!}$

Solution

a. $\dfrac{8!}{2! \cdot 6!} = \dfrac{1 \cdot 2 \cdot 3 \cdot 4 \cdot 5 \cdot 6 \cdot 7 \cdot 8}{1 \cdot 2 \cdot 1 \cdot 2 \cdot 3 \cdot 4 \cdot 5 \cdot 6} = \dfrac{7 \cdot 8}{2} = 28$

b. $\dfrac{2! \cdot 6!}{3! \cdot 5!} = \dfrac{1 \cdot 2 \cdot 1 \cdot 2 \cdot 3 \cdot 4 \cdot 5 \cdot 6}{1 \cdot 2 \cdot 3 \cdot 1 \cdot 2 \cdot 3 \cdot 4 \cdot 5} = \dfrac{6}{3} = 2$

c. $\dfrac{n!}{(n-1)!} = \dfrac{1 \cdot 2 \cdot 3 \cdots (n-1) \cdot n}{1 \cdot 2 \cdot 3 \cdots (n-1)} = n$

Summation Notation

There is a convenient notation for the sum of the terms of a finite sequence. It is called **summation notation** or **sigma notation** because it involves the use of the uppercase Greek letter sigma, written as Σ.

Definition of Summation Notation

The sum of the first n terms of a sequence is represented by

$$\sum_{i=1}^{n} a_i = a_1 + a_2 + a_3 + a_4 + \cdots + a_n$$

where i is called the **index of summation,** n is the **upper limit of summation,** and 1 is the **lower limit of summation.**

EXAMPLE 7 ▱ **Sigma Notation for Sums**

a. $\displaystyle\sum_{i=1}^{5} 3i = 3(1) + 3(2) + 3(3) + 3(4) + 3(5)$

$\qquad\qquad = 3(1 + 2 + 3 + 4 + 5)$

$\qquad\qquad = 3(15) = 45$

b. $\displaystyle\sum_{k=3}^{6} (1 + k^2) = (1 + 3^2) + (1 + 4^2) + (1 + 5^2) + (1 + 6^2)$

$\qquad\qquad\qquad = 10 + 17 + 26 + 37 = 90$

c. $\displaystyle\sum_{i=0}^{8} \frac{1}{i!} = \frac{1}{0!} + \frac{1}{1!} + \frac{1}{2!} + \frac{1}{3!} + \frac{1}{4!} + \frac{1}{5!} + \frac{1}{6!} + \frac{1}{7!} + \frac{1}{8!}$

$\qquad\qquad = 1 + 1 + \frac{1}{2} + \frac{1}{6} + \frac{1}{24} + \frac{1}{120} + \frac{1}{720} + \frac{1}{5040} + \frac{1}{40,320}$

$\qquad\qquad \approx 2.71828$

For this summation, note that the sum is very close to the irrational number $e \approx 2.718281828$. It can be shown that as more terms of the sequence whose nth term is $1/n!$ are added, the sum becomes closer and closer to e. ▱

Note In Example 7, note that the lower limit of a summation does not have to be 1. Also note that the index of summation does not have to be the letter i. For instance, in part (b), the letter k is the index of summation.

Think About the Proof

Use the Distributive Property for real numbers to prove Property 1 of the Properties of Sums. The details of the proof are given in the appendix.

Properties of Sums

1. $\displaystyle\sum_{i=1}^{n} ca_i = c\sum_{i=1}^{n} a_i$, c is any constant

2. $\displaystyle\sum_{i=1}^{n} (a_i + b_i) = \sum_{i=1}^{n} a_i + \sum_{i=1}^{n} b_i$

3. $\displaystyle\sum_{i=1}^{n} (a_i - b_i) = \sum_{i=1}^{n} a_i - \sum_{i=1}^{n} b_i$

Variations in the upper and lower limits of summation can produce quite different-looking summation notations for *the same sum*. For example, consider the following two sums.

$$\sum_{i=1}^{5} 3(2^i) = 3\sum_{i=1}^{5} 2^i = 3(2^1 + 2^2 + 2^3 + 2^4 + 2^5)$$

$$\sum_{i=0}^{4} 3(2^{i+1}) = 3\sum_{i=0}^{4} 2^{i+1} = 3(2^1 + 2^2 + 2^3 + 2^4 + 2^5)$$

Notice that both sums have identical terms.

The sum of the terms of any *finite* sequence must be a finite number. An important discovery in mathematics was that for some special types of *infinite* sequences, the sum of *all* the terms is a finite number. For instance, it can be shown that the sum of all of the terms of the sequence whose nth term is $1/2^n$ (with n beginning at 0) is

$$\sum_{n=0}^{\infty} \frac{1}{2^n} = 1 + \frac{1}{2} + \frac{1}{4} + \frac{1}{8} + \frac{1}{16} + \frac{1}{32} + \cdots = 2.$$

See Figure 9.1.

Figure 9.1   The total area is 2.

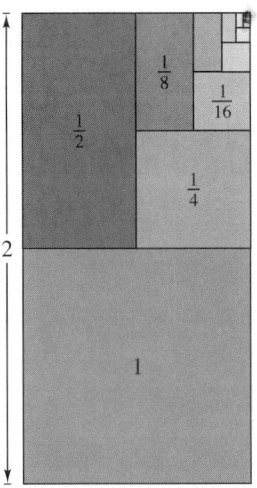

Note   The summation of the terms of a sequence is a **series.** The summation

$$\sum_{i=1}^{\infty} a_i = a_1 + a_2 + a_3 + \cdots$$

is an **infinite series.** Infinite series have important uses in calculus.

EXAMPLE 8   ▱   **Finding the Sum of an Infinite Sequence**

$$\sum_{n=1}^{\infty} \frac{3}{10^n} = \frac{3}{10^1} + \frac{3}{10^2} + \frac{3}{10^3} + \frac{3}{10^4} + \frac{3}{10^5} + \cdots$$

$$= 0.3 + 0.03 + 0.003 + 0.0003 + 0.00003 + \cdots$$

$$= 0.33333\cdots$$

$$= \frac{1}{3}$$

Application

Sequences have many applications in business and science. One is illustrated in Example 9.

Real Life

EXAMPLE 9   ▱   **Population of the United States**

From 1950 to 1993, the resident population of the United States can be approximated by the model

$$a_n = \sqrt{23{,}107 + 874.7n + 2.74n^2}, \qquad n = 0, 1, \ldots, 43$$

where a_n is the population in millions and n represents the calendar year, with $n = 0$ corresponding to 1950. Find the last five terms of this finite sequence. (Source: U.S. Bureau of Census)

Solution

The last five terms of this finite sequence are as follows.

$$a_{39} = \sqrt{23{,}107 + 874.7(39) + 2.74(39)^2} \approx 247.8 \qquad \text{1989 population}$$

$$a_{40} = \sqrt{23{,}107 + 874.7(40) + 2.74(40)^2} \approx 250.0 \qquad \text{1990 population}$$

$$a_{41} = \sqrt{23{,}107 + 874.7(41) + 2.74(41)^2} \approx 252.1 \qquad \text{1991 population}$$

$$a_{42} = \sqrt{23{,}107 + 874.7(42) + 2.74(42)^2} \approx 254.3 \qquad \text{1992 population}$$

$$a_{43} = \sqrt{23{,}107 + 874.7(43) + 2.74(43)^2} \approx 256.5 \qquad \text{1993 population}$$

The bar graph in Figure 9.2 graphically represents the populations given by this sequence for the entire 44-year period from 1950 to 1993.

Figure 9.2

Group Activity

A Summation Program

A graphing calculator can be programmed to calculate the sum of the first *n* terms of a sequence. For instance, the program below lists the program steps for the *TI-82* and *TI-83*. See the appendix for programs for other graphing calculator models.

PROGRAM:SUM	Title of program
:Prompt M	M is the lower limit of summation.
:Prompt N	N is the upper limit of summation.
:0 → S	Initialize S for use as stored sum.
:For (X, M, N)	Loop from lower limit to upper limit.
:S + Y₁ → S	Evaluate function stored in Y₁ and add to sum.
:Disp S	Display partial sum.
:End	End For loop.

To use this program to find the sum of a sequence, store the *n*th term of the sequence as Y₁ and run the program. For instance, to find the sum

$$\sum_{i=1}^{10} 2^i$$

you should press $\boxed{\text{Y=}}$, enter 2^X, and then use the keystroke sequence

$\boxed{\text{PRGM}}$, cursor to program SUM, $\boxed{\text{ENTER}}$ $\boxed{\text{ENTER}}$

1 $\boxed{\text{ENTER}}$ 10 $\boxed{\text{ENTER}}$.

The result should be 2046. (Note that the calculator displays 10 numbers—the last number displayed is the final sum.) To pause the program after each partial sum, insert :Pause after :Disp S. Then press $\boxed{\text{ENTER}}$ to display the next partial sum. Try using this program to find the following sums.

n	Uₙ
1	2
2	6
3	14
4	30
6	126
8	510
10	2046
n = 10	

a. $\sum_{i=1}^{15} (i - 2)^2$ **b.** $\sum_{i=1}^{20} 3^{(i-1)}$ **c.** $\sum_{i=1}^{10} \sqrt{e^i}$

Another way to calculate the sum of a sequence using the *TI-82* or *TI-83* is to combine the **sum** and **seq(** commands under the LIST menu. The partial sums can be displayed in a table using seq mode by choosing the "ask" feature in the TblSet menu, as shown at the left. Try using this method to find the sums in (a), (b), and (c).

The *Interactive* CD-ROM contains step-by-step solutions to all odd-numbered Section and Review Exercises. It also provides Tutorial Exercises, which link to Guided Examples for additional help.

9.1 /// EXERCISES

In Exercises 1–22, write the first five terms of the sequence. (Assume n begins with 1.)

1. $a_n = 2n + 1$ **2.** $a_n = 4n - 3$

3. $a_n = 2^n$ **4.** $a_n = \left(\frac{1}{2}\right)^n$

5. $a_n = (-2)^n$ **6.** $a_n = \left(-\frac{1}{2}\right)^n$

7. $a_n = \dfrac{n + 1}{n}$ **8.** $a_n = \dfrac{n}{n + 1}$

9. $a_n = \dfrac{6n}{3n^2 - 1}$ **10.** $a_n = \dfrac{3n^2 - n + 4}{2n^2 + 1}$

11. $a_n = \dfrac{1 + (-1)^n}{n}$ **12.** $a_n = 1 + (-1)^n$

13. $a_n = 3 - \dfrac{1}{2^n}$ **14.** $a_n = \dfrac{3^n}{4^n}$

15. $a_n = \dfrac{1}{n^{3/2}}$ **16.** $a_n = \dfrac{10}{n^{2/3}}$

17. $a_n = \dfrac{3^n}{n!}$ **18.** $a_n = \dfrac{n!}{n}$

19. $a_n = \dfrac{(-1)^n}{n^2}$ **20.** $a_n = (-1)^n\left(\dfrac{n}{n + 1}\right)$

21. $a_n = \frac{2}{3}$ **22.** $a_n = n(n - 1)(n - 2)$

In Exercises 23 and 24, find the indicated term of the sequence.

23. $a_n = (-1)^n(3n - 2)$ **24.** $a_n = \dfrac{2^n}{n!}$

$a_{25} = $ ▨ $a_{10} = $ ▨

In Exercises 25–28, write the first five terms of the sequence defined recursively.

25. $a_1 = 28, \quad a_{k+1} = a_k - 4$

26. $a_1 = 15, \quad a_{k+1} = a_k + 3$

27. $a_1 = 3, \quad a_{k+1} = 2(a_k - 1)$

28. $a_1 = 32, \quad a_{k+1} = \frac{1}{2}a_k$

In Exercises 29–34, use a graphing utility to graph the first 10 terms of the sequence.

29. $a_n = \dfrac{2}{3}n$ **30.** $a_n = 2 - \dfrac{4}{n}$

31. $a_n = 16(-0.5)^{n-1}$ **32.** $a_n = 8(0.75)^{n-1}$

33. $a_n = \dfrac{2n}{n + 1}$ **34.** $a_n = \dfrac{3n^2}{n^2 + 1}$

In Exercises 35–38, match the sequence with its graph. [The graphs are labeled (a), (b), (c), and (d).]

(a)

(b)

(c)

(d)

35. $a_n = \dfrac{8}{n + 1}$ **36.** $a_n = \dfrac{8n}{n + 1}$

37. $a_n = 4(0.5)^{n-1}$ **38.** $a_n = \dfrac{4^n}{n!}$

In Exercises 39–46, simplify the ratio of factorials.

39. $\dfrac{4!}{6!}$ **40.** $\dfrac{4!}{7!}$

41. $\dfrac{10!}{8!}$ **42.** $\dfrac{25!}{23!}$

43. $\dfrac{(n + 1)!}{n!}$ **44.** $\dfrac{(n + 2)!}{n!}$

45. $\dfrac{(2n - 1)!}{(2n + 1)!}$ **46.** $\dfrac{(2n + 2)!}{(2n)!}$

In Exercises 47–60, write an expression for the *most apparent* nth term of the sequence. (Assume *n* begins with 1.)

47. 1, 4, 7, 10, 13, . . .

48. 3, 7, 11, 15, 19, . . .

49. 0, 3, 8, 15, 24, . . .

50. $1, \frac{1}{4}, \frac{1}{9}, \frac{1}{16}, \frac{1}{25}, \ldots$

51. $\frac{2}{3}, \frac{3}{4}, \frac{4}{5}, \frac{5}{6}, \frac{6}{7}, \ldots$

52. $\frac{2}{1}, \frac{3}{3}, \frac{4}{5}, \frac{5}{7}, \frac{6}{9}, \ldots$

53. $\frac{1}{2}, \frac{-1}{4}, \frac{1}{8}, \frac{-1}{16}, \ldots$

54. $\frac{1}{3}, \frac{2}{9}, \frac{4}{27}, \frac{8}{81}, \ldots$

55. $1 + \frac{1}{1}, 1 + \frac{1}{2}, 1 + \frac{1}{3}, 1 + \frac{1}{4}, 1 + \frac{1}{5}, \ldots$

56. $1 + \frac{1}{2}, 1 + \frac{3}{4}, 1 + \frac{7}{8}, 1 + \frac{15}{16}, 1 + \frac{31}{32}, \ldots$

57. $1, \frac{1}{2}, \frac{1}{6}, \frac{1}{24}, \frac{1}{120}, \ldots$

58. 2, −4, 6, −8, 10, . . .

59. 1, −1, 1, −1, 1, . . .

60. $1, 2, \frac{2^2}{2}, \frac{2^3}{6}, \frac{2^4}{24}, \frac{2^5}{120}, \ldots$

In Exercises 61–64, write the first five terms of the sequence defined recursively. Use the pattern to write the *n*th term of the sequence as a function of *n*. (Assume *n* begins with 1.)

61. $a_1 = 6, \quad a_{k+1} = a_k + 2$

62. $a_1 = 25, \quad a_{k+1} = a_k - 5$

63. $a_1 = 81, \quad a_{k+1} = \frac{1}{3}a_k$

64. $a_1 = 14, \quad a_{k+1} = -2a_k$

In Exercises 65–76, find the sum.

65. $\displaystyle\sum_{i=1}^{5}(2i + 1)$

66. $\displaystyle\sum_{i=1}^{6}(3i - 1)$

67. $\displaystyle\sum_{k=1}^{4} 10$

68. $\displaystyle\sum_{k=1}^{5} 6$

69. $\displaystyle\sum_{i=0}^{4} i^2$

70. $\displaystyle\sum_{i=0}^{5} 3i^2$

71. $\displaystyle\sum_{k=0}^{3} \frac{1}{k^2 + 1}$

72. $\displaystyle\sum_{j=3}^{5} \frac{1}{j}$

73. $\displaystyle\sum_{i=1}^{4}[(i - 1)^2 + (i + 1)^3]$

74. $\displaystyle\sum_{k=2}^{5}(k + 1)(k - 3)$

75. $\displaystyle\sum_{i=1}^{4} 2^i$

76. $\displaystyle\sum_{j=0}^{4}(-2)^j$

In Exercises 77–80, use a calculator to find the sum.

77. $\displaystyle\sum_{j=1}^{6}(24 - 3j)$

78. $\displaystyle\sum_{j=1}^{10} \frac{3}{j + 1}$

79. $\displaystyle\sum_{k=0}^{4} \frac{(-1)^k}{k + 1}$

80. $\displaystyle\sum_{k=0}^{4} \frac{(-1)^k}{k!}$

In Exercises 81–90, use sigma notation to write the sum.

81. $\dfrac{1}{3(1)} + \dfrac{1}{3(2)} + \dfrac{1}{3(3)} + \cdots + \dfrac{1}{3(9)}$

82. $\dfrac{5}{1 + 1} + \dfrac{5}{1 + 2} + \dfrac{5}{1 + 3} + \cdots + \dfrac{5}{1 + 15}$

83. $\left[2\left(\frac{1}{8}\right) + 3\right] + \left[2\left(\frac{2}{8}\right) + 3\right] + \cdots + \left[2\left(\frac{8}{8}\right) + 3\right]$

84. $\left[1 - \left(\frac{1}{6}\right)^2\right] + \left[1 - \left(\frac{2}{6}\right)^2\right] + \cdots + \left[1 - \left(\frac{6}{6}\right)^2\right]$

85. $3 - 9 + 27 - 81 + 243 - 729$

86. $1 - \dfrac{1}{2} + \dfrac{1}{4} - \dfrac{1}{8} + \cdots - \dfrac{1}{128}$

87. $\dfrac{1}{1^2} - \dfrac{1}{2^2} + \dfrac{1}{3^2} - \dfrac{1}{4^2} + \cdots - \dfrac{1}{20^2}$

88. $\dfrac{1}{1 \cdot 3} + \dfrac{1}{2 \cdot 4} + \dfrac{1}{3 \cdot 5} + \cdots + \dfrac{1}{10 \cdot 12}$

89. $\dfrac{1}{4} + \dfrac{3}{8} + \dfrac{7}{16} + \dfrac{15}{32} + \dfrac{31}{64}$

90. $\dfrac{1}{2} + \dfrac{2}{4} + \dfrac{6}{8} + \dfrac{24}{16} + \dfrac{120}{32} + \dfrac{720}{64}$

91. *Compound Interest*   A deposit of $5000 is made in an account that earns 8% interest compounded quarterly. The balance in the account after n quarters is given by

$$A_n = 5000\left(1 + \frac{0.08}{4}\right)^n, \qquad n = 1, 2, 3, \ldots .$$

   (a) Compute the first eight terms of this sequence.

   (b) Find the balance in this account after 10 years by computing the 40th term of the sequence.

92. *Compound Interest*   A deposit of $100 is made *each month* in an account that earns 12% interest compounded monthly. The balance in the account after n months is given by

$$A_n = 100(101)[(1.01)^n - 1], \qquad n = 1, 2, 3, \ldots .$$

   (a) Compute the first six terms of this sequence.

   (b) Find the balance in this account after 5 years by computing the 60th term of the sequence.

   (c) Find the balance in this account after 20 years by computing the 240th term of the sequence.

93. *Per Capita Hospital Care*   The average per capita annual cost for hospital care from 1981 to 1991 is approximated by the model

$$a_n = 510.13 + 16.37n + 3.23n^2, \quad n = 1, \ldots, 11$$

where a_n is the per capita cost in dollars and n is the year, with $n = 1$ corresponding to 1981. Find the terms of this finite sequence and use a graphing utility to construct a bar graph that represents the sequence. (Source: U.S. Health Care Financing Administration)

94. *Federal Debt*   It took more than 200 years for the United States to accumulate a $1 trillion debt. Then it took just 8 years to get to a $3 trillion debt. The federal debt during the decade of the 1980s is approximated by the model

$$a_n = 0.1\sqrt{82 + 9n^2}, \qquad n = 0, \ldots, 10$$

where a_n is the debt in trillions and n is the year, with $n = 0$ corresponding to 1980. Find the terms of this finite sequence and use a graphing utility to construct a bar graph that represents the sequence. (Source: Treasury Department)

95. *Corporate Income*   The net incomes a_n (in millions of dollars) of Wal-Mart for the years 1985 through 1994 are shown in the figure. These incomes can be approximated by the model

$$a_n = 129.9 + 0.9n^3, \qquad n = 5, \ldots, 14$$

where $n = 5$ represents 1985. Use this model to approximate the total net income from 1985 through 1994. Compare this sum with the result of adding the incomes shown in the figure. (Source: Wal-Mart)

96. *Corporate Dividends*   From 1985 through 1994, the dividends a_n declared per share of common stock of Procter & Gamble Company can be approximated by the model

$$a_n = 1.39 + 0.18n - 1.02 \ln n, \quad n = 5, \ldots, 14$$

where $n = 5$ represents 1985. Use this model to approximate the total dividends per share of common stock from 1985 through 1994. (Source: Procter & Gamble Company)

Fibonacci Sequence   In Exercises 97 and 98, use the Fibonacci Sequence. (See Example 4.)

97. Write the first 12 terms of the Fibonacci sequence a_n and the first 10 terms of the sequence given by

$$b_n = \frac{a_{n+1}}{a_n}, \qquad n > 1.$$

98. Using the definition for b_n in Exercise 97, show that b_n can be defined recursively by

$$b_n = 1 + \frac{1}{b_{n-1}}.$$

9.2   Arithmetic Sequences

Arithmetic Sequences   / The Sum of an Arithmetic Sequence   / Applications

Arithmetic Sequences

A sequence whose consecutive terms have a common difference is called an **arithmetic sequence.**

> **Definition of Arithmetic Sequence**
>
> A sequence is **arithmetic** if the differences between consecutive terms are the same. Thus, the sequence
>
> $$a_1, a_2, a_3, a_4, \ldots, a_n, \ldots$$
>
> is arithmetic if there is a number d such that
>
> $$a_2 - a_1 = d, \quad a_3 - a_2 = d, \quad a_4 - a_3 = d$$
>
> and so on. The number d is the **common difference** of the arithmetic sequence.

EXAMPLE 1   **Examples of Arithmetic Sequences**

a. The sequence whose nth term is $4n + 3$ is arithmetic. For this sequence, the common difference between consecutive terms is 4.

$$7, 11, 15, 19, \ldots, 4n + 3, \ldots$$

$11 - 7 = 4$

b. The sequence whose nth term is $7 - 5n$ is arithmetic. For this sequence, the common difference between consecutive terms is -5.

$$2, -3, -8, -13, \ldots, 7 - 5n, \ldots$$

$-3 - 2 = -5$

c. The sequence whose nth term is $\frac{1}{4}(n + 3)$ is arithmetic. For this sequence, the common difference between consecutive terms is $\frac{1}{4}$.

$$1, \frac{5}{4}, \frac{3}{2}, \frac{7}{4}, \ldots, \frac{n + 3}{4}, \ldots$$

$\frac{5}{4} - 1 = \frac{1}{4}$

In Example 1, notice that each of the arithmetic sequences has an *n*th term that is of the form $dn + c$, where the common difference of the sequence is *d*. This result is summarized as follows.

The *n*th Term of an Arithmetic Sequence

The *n*th term of an arithmetic sequence has the form

$$a_n = dn + c$$

where *d* is the common difference between consecutive terms of the sequence and $c = a_1 - d$.

EXAMPLE 2   **Finding the *n*th Term of an Arithmetic Sequence**

Find a formula for the *n*th term of the arithmetic sequence whose common difference is 3 and whose first term is 2.

Solution

Because the sequence is arithmetic, you know that the formula for the *n*th term is of the form $a_n = dn + c$. Moreover, because the common difference is $d = 3$, the formula must have the form

$$a_n = 3n + c.$$

Because $a_1 = 2$, it follows that

$$c = a_1 - d = 2 - 3 = -1.$$

Thus, the formula for the *n*th term is

$$a_n = 3n - 1.$$

The sequence therefore has the following form.

$$2, 5, 8, 11, 14, \ldots, 3n - 1, \ldots$$

Another way to find a formula for the *n*th term of the sequence in Example 2 is to begin by writing the terms of the sequence.

a_1	a_2	a_3	a_4	a_5	a_6	a_7	. . .
2	2 + 3	5 + 3	8 + 3	11 + 3	14 + 3	17 + 3	. . .
2	5	8	11	14	17	20	. . .

From these terms, you can reason that the *n*th term is of the form

$$a_n = dn + c = 3n - 1.$$

EXAMPLE 3 ▱ **Finding the *n*th Term of an Arithmetic Sequence**

The fourth term of an arithmetic sequence is 20, and the 13th term is 65. Write the first several terms of this sequence.

Solution

The fourth and 13th terms of the sequence are related by

$$a_{13} = a_4 + 9d.$$

Using $a_4 = 20$ and $a_{13} = 65$, you can conclude that $d = 5$, which implies that the sequence is as follows.

a_1	a_2	a_3	a_4	a_5	a_6	a_7	a_8	a_9	a_{10}	a_{11}
5,	10,	15,	20,	25,	30,	35,	40,	45,	50,	55, . . . ▱

> **Study Tip**
>
> If you substitute $a_1 - d$ for c in the formula $a_n = dn + c$, the *n*th term of an arithmetic sequence has the alternative recursive formula
>
> $$a_n = a_1 + (n - 1)d.$$
>
> Use this formula to solve Example 4 and Example 9.

If you know the *n*th term of an arithmetic sequence *and* you know the common difference of the sequence, you can find the $(n + 1)$th term by using the *recursive formula*

$$a_{n+1} = a_n + d.$$

With this formula, you can find any term of an arithmetic sequence, *provided* that you know the previous term. For instance, if you know the first term, you can find the second term. Then, knowing the second term, you can find the third term, and so on.

EXAMPLE 4 ▱ **Using a Recursive Formula**

Find the ninth term of the arithmetic sequence that begins with 2 and 9.

Solution

For this sequence, the common difference is $d = 9 - 2 = 7$. There are two ways to find the ninth term. One way is simply to write out the first nine terms (by repeatedly adding 7).

$$2, 9, 16, 23, 30, 37, 44, 51, 58$$

Another way to find the ninth term is first to find a formula for the *n*th term. Because the first term is 2, it follows that

$$c = a_1 - d = 2 - 7 = -5.$$

Therefore, a formula for the *n*th term is $a_n = 7n - 5$, which implies that the ninth term is $a_9 = 7(9) - 5 = 58$. ▱

The Sum of an Arithmetic Sequence

There is a simple formula for the *sum* of a finite arithmetic sequence.

> **The Sum of a Finite Arithmetic Sequence**
>
> The sum of a finite arithmetic sequence with n terms is given by
>
> $$S_n = \frac{n}{2}(a_1 + a_n).$$

Note Be sure you see that this formula works only for *arithmetic* sequences.

EXAMPLE 5 **Finding the Sum of an Arithmetic Sequence**

Find the sum: $1 + 3 + 5 + 7 + 9 + 11 + 13 + 15 + 17 + 19$.

Solution
To begin, notice that the sequence is arithmetic (with a common difference of 2). Moreover, the sequence has 10 terms. Thus, the sum of the sequence is

$$S_n = 1 + 3 + 5 + 7 + 9 + 11 + 13 + 15 + 17 + 19$$
$$= \frac{n}{2}(a_1 + a_n)$$
$$= \frac{10}{2}(1 + 19) \qquad n = 10$$
$$= 5(20)$$
$$= 100.$$

Think About the Proof

To prove the formula for the sum of a finite arithmetic sequence, consider writing the sum in two different ways. One way is to write the sum as

$$S_n = a_1 + (a_1 + d)$$
$$+ (a_1 + 2d) + \cdots$$
$$+ [a_1 + (n - 1)d].$$

In the second way, you repeatedly subtract d from the nth term to obtain

$$S_n = a_n + (a_n - d)$$
$$+ (a_n - 2d) + \cdots$$
$$+ [a_n - (n - 1)d].$$

Can you discover a way to combine these two versions of S_n to obtain the following formula?

$$S_n = \frac{n}{2}(a_1 + a_n)$$

The details of this proof are shown in the appendix.

EXAMPLE 6 **Finding the Sum of an Arithmetic Sequence**

Find the sum of the integers from 1 to 100.

Solution
The integers from 1 to 100 form an arithmetic sequence that has 100 terms. Thus, you can use the formula for the sum of an arithmetic sequence, as follows.

$$S_n = 1 + 2 + 3 + 4 + 5 + 6 + \cdots + 99 + 100$$
$$= \frac{n}{2}(a_1 + a_n)$$
$$= \frac{100}{2}(1 + 100) \qquad n = 100$$
$$= 50(101)$$
$$= 5050$$

EXAMPLE 7   **Finding the Sum of an Arithmetic Sequence**

Find the sum of the first 150 terms of the arithmetic sequence

$$5, 16, 27, 38, 49, \ldots .$$

Solution
For this arithmetic sequence, you have $a_1 = 5$ and $d = 16 - 5 = 11$. Thus, $c = a_1 - d = 5 - 11 = -6$, and the nth term is

$$a_n = 11n - 6.$$

Therefore, $a_{150} = 11(150) - 6 = 1644$, and the sum of the first 150 terms is

Note   The sum of the first n terms of an infinite sequence is called the **nth partial sum.**

$$S_n = \frac{n}{2}(a_1 + a_n)$$

$$= \frac{150}{2}(5 + 1644)$$

$$= 75(1649)$$

$$= 123{,}675.$$

Applications

Real Life

EXAMPLE 8   **Seating Capacity**

An auditorium has 20 rows of seats. There are 20 seats in the first row, 21 seats in the second row, 22 seats in the third row, and so on. (See Figure 9.3.) How many seats are there in all 20 rows?

Solution
The numbers of seats in the 20 rows form an arithmetic sequence in which the common difference is $d = 1$. Because $c = a_1 - d = 20 - 1 = 19$, you can determine that the formula for the nth term of the sequence is $a_n = n + 19$. Therefore, the 20th term in the sequence is $a_{20} = 20 + 19 = 39$, and the total number of seats is

Figure 9.3

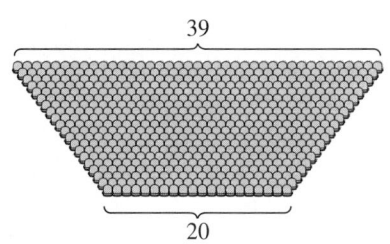
39

20

$$S_n = 20 + 21 + 22 + \cdots + 39$$

$$= \frac{n}{2}(a_1 + a_{20})$$

$$= \frac{20}{2}(20 + 39)$$

$$= 10(59)$$

$$= 590.$$

Real Life

EXAMPLE 9 ▬ **Total Sales**

A small business sells $10,000 worth of products during its first year. The owner of the business has set a goal of increasing annual sales by $7500 each year for 9 years. Assuming that this goal is met, find the total sales during the first 10 years this business is in operation.

Solution

The annual sales form an arithmetic sequence in which $a_1 = 10,000$ and $d = 7500$. Thus, $c = a_1 - d = 10,000 - 7500 = 2500$, and the nth term of the sequence is

$$a_n = 7500n + 2500.$$

This implies that the 10th term of the sequence is $a_{10} = 77,500$. The sum of the first 10 terms of the sequence is

$$S_n = \frac{n}{2}(a_1 + a_{10})$$

$$= \frac{10}{2}(10,000 + 77,500)$$

$$= 5(87,500)$$

$$= 437,500.$$

Thus, the total sales for the first 10 years is $437,500.   ▬

Group Activity

Numerical Relationships

Decide whether it is possible to fill in the blanks in each of the following such that the resulting sequence is arithmetic. If so, find a recursive formula for the sequence.

a. -7,   ,   ,   ,   , 11

b. 17,   ,   ,   ,   ,   ,   , 71

c. $2, 6$,   , 162

d. $4, 7.5$,   ,   ,   ,   ,   ,   , 39

e. $8, 12$,   ,   , 60.75

9.2 /// EXERCISES

In Exercises 1–10, determine whether the sequence is arithmetic. If it is, find the common difference.

1. $10, 8, 6, 4, 2, \ldots$

2. $4, 7, 10, 13, 16, \ldots$

3. $1, 2, 4, 8, 16, \ldots$

4. $3, \frac{5}{2}, 2, \frac{3}{2}, 1, \ldots$

5. $\frac{9}{4}, 2, \frac{7}{4}, \frac{3}{2}, \frac{5}{4}, \ldots$

6. $\frac{1}{3}, \frac{2}{3}, \frac{4}{3}, \frac{8}{3}, \frac{16}{3}, \ldots$

7. $-12, -8, -4, 0, 4, \ldots$

8. $\ln 1, \ln 2, \ln 3, \ln 4, \ln 5, \ldots$

9. $5.3, 5.7, 6.1, 6.5, 6.9, \ldots$

10. $1^2, 2^2, 3^2, 4^2, 5^2, \ldots$

In Exercises 11–18, write the first five terms of the sequence. Determine whether the sequence is arithmetic, and if it is, find the common difference.

11. $a_n = 5 + 3n$

12. $a_n = (2^n)n$

13. $a_n = \dfrac{1}{n+1}$

14. $a_n = 1 + (n-1)4$

15. $a_n = 100 - 3n$

16. $a_n = 2^{n-1}$

17. $a_n = 3 + \dfrac{(-1)^n 2}{n}$

18. $a_n = (-1)^n$

In Exercises 19–24, write the first five terms of the arithmetic sequence. Find the common difference and write the nth term of the sequence as a function of n.

19. $a_1 = 15, \quad a_{k+1} = a_k + 4$

20. $a_1 = 2, \quad a_{k+1} = a_k + \frac{2}{3}$

21. $a_1 = 200, \quad a_{k+1} = a_k - 10$

22. $a_1 = 72, \quad a_{k+1} = a_k - 6$

23. $a_1 = \frac{3}{2}, \quad a_{k+1} = a_k - \frac{1}{4}$

24. $a_1 = 0.375, \quad a_{k+1} = a_k + 0.25$

In Exercises 25–32, write the first five terms of the arithmetic sequence.

25. $a_1 = 5, d = 6$

26. $a_1 = 5, d = -\frac{3}{4}$

27. $a_1 = -2.6, d = -0.4$

28. $a_1 = 16.5, d = 0.25$

29. $a_1 = 2, a_{12} = 46$

30. $a_4 = 16, a_{10} = 46$

31. $a_8 = 26, a_{12} = 42$

32. $a_3 = 19, a_{15} = -1.7$

In Exercises 33–44, find a formula for a_n for the arithmetic sequence.

33. $a_1 = 1, d = 3$

34. $a_1 = 15, d = 4$

35. $a_1 = 100, d = -8$

36. $a_1 = 0, d = -\frac{2}{3}$

37. $a_1 = x, d = 2x$

38. $a_1 = -y, d = 5y$

39. $4, \frac{3}{2}, -1, -\frac{7}{2}, \ldots$

40. $10, 5, 0, -5, -10, \ldots$

41. $a_1 = 5, a_4 = 15$

42. $a_1 = -4, a_5 = 16$

43. $a_3 = 94, a_6 = 85$

44. $a_5 = 190, a_{10} = 115$

In Exercises 45–48, match the sequence with its graph. [The graphs are labeled (a), (b), (c), and (d).]

(a)

(b)

(c)

(d)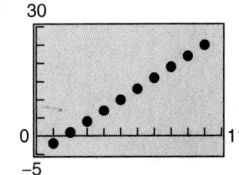

45. $a_n = -\frac{2}{3}n + 6$

46. $a_n = 3n - 5$

47. $a_n = 2 + \frac{3}{4}n$

48. $a_n = 25 - 3n$

In Exercises 49–52, use a graphing utility to graph the first 10 terms of the sequence.

49. $a_n = 15 - \frac{3}{2}n$ **50.** $a_n = -5 + 2n$

51. $a_n = 0.2n + 3$ **52.** $a_n = -0.3n + 8$

53. *Think About It* Describe the geometric pattern of the graph of an arithmetic sequence. Explain.

54. *Think About It* Explain how to use the first two terms of an arithmetic sequence to find the nth term.

In Exercises 55–62, find the nth partial sum of the arithmetic sequence.

55. $8, 20, 32, 44, \ldots, \quad n = 10$

56. $2, 8, 14, 20, \ldots, \quad n = 25$

57. $-6, -2, 2, 6, \ldots, \quad n = 50$

58. $0.5, 0.9, 1.3, 1.7, \ldots, \quad n = 10$

59. $40, 37, 34, 31, \ldots, \quad n = 10$

60. $1.50, 1.45, 1.40, 1.35, \ldots, \quad n = 20$

61. $a_1 = 100, \quad a_{25} = 220, \quad n = 25$

62. $a_1 = 15, \quad a_{100} = 307, \quad n = 100$

In Exercises 63–70, find the sum.

63. $\displaystyle\sum_{n=1}^{50} n$

64. $\displaystyle\sum_{n=1}^{100} 2n$

65. $\displaystyle\sum_{n=1}^{100} 5n$

66. $\displaystyle\sum_{n=51}^{100} 7n$

67. $\displaystyle\sum_{n=11}^{30} n - \sum_{n=1}^{10} n$

68. $\displaystyle\sum_{n=51}^{100} n - \sum_{n=1}^{50} n$

69. $\displaystyle\sum_{n=1}^{500} (n + 3)$

70. $\displaystyle\sum_{n=1}^{250} (1000 - n)$

In Exercises 71–76, use a calculator to find the sum.

71. $\displaystyle\sum_{n=1}^{20} (2n + 5)$

72. $\displaystyle\sum_{n=1}^{100} \frac{n + 4}{2}$

73. $\displaystyle\sum_{n=0}^{50} (1000 - 5n)$

74. $\displaystyle\sum_{n=0}^{100} \frac{8 - 3n}{16}$

75. $\displaystyle\sum_{i=1}^{60} \left(250 - \frac{8}{3}i\right)$

76. $\displaystyle\sum_{j=1}^{200} (4.5 + 0.025j)$

77. Find the sum of the first 100 positive odd integers.

78. Find the sum of the integers from -10 to 50.

Job Offer In Exercises 79 and 80, consider a job offer with the given starting salary and guaranteed salary increase for the first 5 years of employment.

(a) Determine the person's salary during the sixth year of employment.

(b) Determine the person's total compensation from the company through 6 full years of employment.

	Starting Salary	Annual Raise
79.	$32,500	$1500
80.	$36,800	$1750

81. *Seating Capacity* Determine the seating capacity of an auditorium with 30 rows of seats if there are 20 seats in the first row, 24 seats in the second row, 28 seats in the third row, and so on.

82. *Seating Capacity* Determine the seating capacity of an auditorium with 36 rows of seats if there are 15 seats in the first row, 18 seats in the second row, 21 seats in the third row, and so on.

83. *Brick Pattern* A brick patio has the approximate shape of a trapezoid (see figure). The patio has 18 rows of bricks. The first row has 14 bricks and the 18th row has 31 bricks. How many bricks are in the patio?

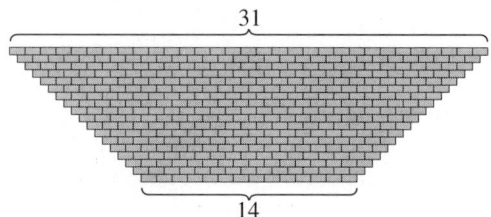

31

14

84. *Number of Logs* Logs are stacked in a pile as shown in the figure. The top row has 15 logs and the bottom row has 21 logs. How many logs are in the stack?

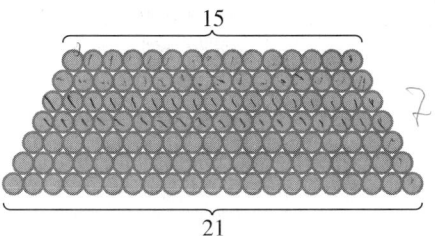

85. *Auditorium Seating* Each row in a small auditorium has two more seats than the preceding row (see figure). Find the seating capacity of the auditorium if the front row seats 25 people and there are 15 rows of seats.

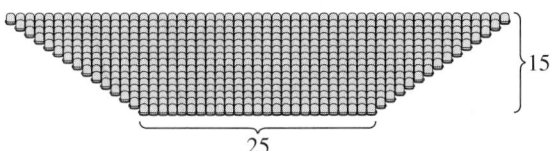

86. *Baling Hay* In the first two trips around a field baling hay, a farmer makes 93 bales and 89 bales, respectively (see figure). Because each trip is shorter than the preceding trip, the farmer estimates that the same pattern will continue. Estimate the total number of bales made if there are another six trips around the field.

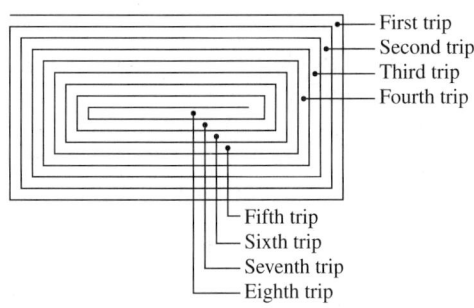

87. *Grandfather Clock* Each hour, a grandfather clock strikes the number of times corresponding to the hour of the day. How many times does the clock strike in a day?

88. *Falling Object* An object (with negligible air resistance) is dropped from an airplane. During the first second of fall, the object falls 4.9 meters; during the second second, it falls 14.7 meters; during the third second, it falls 24.5 meters; during the fourth second, it falls 34.3 meters. If this arithmetic pattern continues, how many meters will the object have fallen after 10 seconds?

89. *Pattern Recognition*

(a) Compute the following five sums of positive odd integers.

$1 + 3 = $

$1 + 3 + 5 = $

$1 + 3 + 5 + 7 = $

$1 + 3 + 5 + 7 + 9 = $

$1 + 3 + 5 + 7 + 9 + 11 = $

(b) Use the sums in part (a) to make a conjecture about the sums of positive odd integers. Check your conjecture for the sum

$1 + 3 + 5 + 7 + 9 + 11 + 13 = $.

(c) Verify your conjecture analytically.

90. *Think About It* In each of the following, an operation is performed on each term of an arithmetic sequence. Determine if the resulting sequence is arithmetic and, if so, state the common difference.

(a) A constant C is added to each term.

(b) Each term is multiplied by a nonzero constant C.

(c) Each term is squared.

91. *Think About It* The sum of the first 20 terms of an arithmetic sequence with a common difference of 3 is 650. Find the first term.

92. *Think About It* The sum of the first n terms of an arithmetic sequence with first term a_1 and common difference d is S_n. Determine the sum if the first term is increased by 5. Explain.

9.3	Geometric Sequences

Geometric Sequences / *The Sum of a Geometric Sequence* / *Application*

EXPLORATION

Consider the following three sequences.

$$1, 3, 9, 27, 81, 243, \ldots, \frac{3^n}{3}, \ldots$$

$$-2, 4, -8, 16, -32, 64, \ldots,$$
$$(-2)^n, \ldots$$

$$\frac{1}{2}, \frac{1}{4}, \frac{1}{8}, \frac{1}{16}, \ldots, \left(\frac{1}{2}\right)^n, \ldots$$

What relationship do you observe between successive terms of these sequences? Sequences with this property are called **geometric sequences.**

Geometric Sequences

In Section 9.2, you learned that a sequence whose consecutive terms have a common *difference* is an arithmetic sequence. In this section, you will study another important type of sequence called a **geometric sequence.** Consecutive terms of a geometric sequence have a common *ratio.*

Definition of Geometric Sequence

A sequence is **geometric** if the ratios of consecutive terms are the same.

$$\frac{a_2}{a_1} = r, \quad \frac{a_3}{a_2} = r, \quad \frac{a_4}{a_3} = r, \ldots, \qquad r \neq 0$$

The number r is the **common ratio** of the sequence.

EXAMPLE 1   ▱   **Examples of Geometric Sequences**

a. The sequence whose nth term is 2^n is geometric. For this sequence, the common ratio between consecutive terms is 2.

$$2, 4, 8, 16, \ldots, 2^n, \ldots$$
$$\frac{4}{2} = 2$$

b. The sequence whose nth term is $4(3^n)$ is geometric. For this sequence, the common ratio between consecutive terms is 3.

$$12, 36, 108, 324, \ldots, 4(3^n), \ldots$$
$$\frac{36}{12} = 3$$

c. The sequence whose nth term is $\left(-\frac{1}{3}\right)^n$ is geometric. For this sequence, the common ratio between consecutive terms is $-\frac{1}{3}$.

$$-\frac{1}{3}, \frac{1}{9}, -\frac{1}{27}, \frac{1}{81}, \ldots, \left(-\frac{1}{3}\right)^n, \ldots$$
$$\frac{1/9}{-1/3} = -\frac{1}{3}$$

In Example 1, notice that each of the geometric sequences has an nth term that is of the form ar^n, where the common ratio of the sequence is r.

The nth Term of a Geometric Sequence

The nth term of a geometric sequence has the form

$$a_n = a_1 r^{n-1}$$

where r is the common ratio of consecutive terms of the sequence. Thus, every geometric sequence can be written in the following form.

$$a_1, \ a_2, \ a_3, \quad a_4, \quad a_5, \ \ldots \ldots, \ a_n, \ \ldots \ldots$$

$$a_1, \ a_1 r, \ a_1 r^2, \ a_1 r^3, \ a_1 r^4, \ \ldots, \ a_1 r^{n-1}, \ \ldots$$

Note If you know the nth term of a geometric sequence, you can find the $(n + 1)$th term by multiplying by r. That is, $a_{n+1} = ra_n$.

You can generate the geometric sequence in Example 2 with a graphing utility using the following steps.

3 | ENTER |

2 | × | ANS |

Now press | ENTER | repeatedly to generate the terms of the sequence.

EXAMPLE 2   Finding the Terms of a Geometric Sequence

Write the first five terms of the geometric sequence whose first term is $a_1 = 3$ and whose common ratio is $r = 2$.

Solution
Starting with 3, repeatedly multiply by 2 to obtain the following.

$$a_1 = 3 \qquad\qquad \text{1st term}$$
$$a_2 = 3(2^1) = 6 \qquad \text{2nd term}$$
$$a_3 = 3(2^2) = 12 \qquad \text{3rd term}$$
$$a_4 = 3(2^3) = 24 \qquad \text{4th term}$$
$$a_5 = 3(2^4) = 48 \qquad \text{5th term}$$

EXAMPLE 3   Finding a Term of a Geometric Sequence

Find the 15th term of the geometric sequence whose first term is 20 and whose common ratio is 1.05.

Solution

$$a_{15} = a_1 r^{n-1} \qquad\qquad \text{Formula for geometric sequence}$$
$$= 20(1.05)^{15-1} \qquad \text{Substitute for } a_1, r, \text{ and } n.$$
$$\approx 39.599 \qquad\qquad \text{Use a calculator.}$$

EXAMPLE 4 ⬛ **Finding a Term of a Geometric Sequence**

Find the 12th term of the geometric sequence

$$5, 15, 45, \ldots .$$

Solution

The common ratio of this sequence is $r = 15/5 = 3$. Because the first term is $a_1 = 5$, you can determine the 12th term ($n = 12$) to be

$$
\begin{aligned}
a_{12} &= a_1 r^{n-1} &&\text{Formula for geometric sequence}\\
&= 5(3)^{12-1} &&\text{Substitute for } a_1, r, \text{ and } n.\\
&= 5(177{,}147) &&\text{Use a calculator.}\\
&= 885{,}735. &&\text{Simplify.}
\end{aligned}
$$

⬛

If you know any two terms of a geometric sequence, you can use that information to find a formula for the nth term of the sequence.

Study Tip

Remember that r is the common ratio of consecutive terms of a geometric sequence. So, in Example 5,

$$
\begin{aligned}
a_{10} &= a_1 r^9\\
&= a_1 \cdot r \cdot r \cdot r \cdot r^6\\
&= a_1 \cdot \frac{a_2}{a_1} \cdot \frac{a_3}{a_2} \cdot \frac{a_4}{a_3} \cdot r^6\\
&= a_4 r^6.
\end{aligned}
$$

EXAMPLE 5 ⬛ **Finding a Term of a Geometric Sequence**

The fourth term of a geometric sequence is 125, and the 10th term is 125/64. Find the 14th term. (Assume that the terms of the sequence are positive.)

Solution

The 10th term is related to the fourth term by the equation

$$a_{10} = a_4 r^6. \qquad \text{Multiply 4th term by } r^{10-4}.$$

Because $a_{10} = 125/64$ and $a_4 = 125$, you can solve for r as follows.

$$\frac{125}{64} = 125 r^6$$

$$\frac{1}{64} = r^6$$

$$\frac{1}{2} = r$$

You can obtain the 14th term by multiplying the 10th term by r^4.

$$
\begin{aligned}
a_{14} &= a_{10} r^4\\
&= \frac{125}{64}\left(\frac{1}{2}\right)^4 = \frac{125}{1024}
\end{aligned}
$$

⬛

The Sum of a Geometric Sequence

The formula for the sum of a *finite* geometric sequence is as follows.

Think About the Proof

To prove the formula for the sum of a finite geometric sequence, consider the following two equations.

$$S_n = a_1 + a_1 r + a_1 r^2$$
$$+ \cdots + a_1 r^{n-2}$$
$$+ a_1 r^{n-1}$$
$$rS_n = a_1 r + a_1 r^2 + a_1 r^3$$
$$+ \cdots + a_1 r^{n-1} + a_1 r^n$$

Can you discover a way to simplify the difference of these two equations to obtain the formula for the sum of a geometric sequence? The details of the proof are given in the appendix.

The Sum of a Finite Geometric Sequence

The sum of the geometric sequence

$$a_1, \ a_1 r, \ a_1 r^2, \ a_1 r^3, \ a_1 r^4, \ldots, a_1 r^{n-1}$$

with common ratio $r \neq 1$ is given by

$$S_n = a_1 \left(\frac{1 - r^n}{1 - r} \right).$$

EXAMPLE 6   **Finding the Sum of a Finite Geometric Sequence**

Find the sum $\displaystyle\sum_{n=1}^{12} 4(0.3)^n$.

Solution

By writing out a few terms, you have

$$\sum_{n=1}^{12} 4(0.3)^n = 4(0.3) + 4(0.3)^2 + 4(0.3)^3 + \cdots + 4(0.3)^{12}.$$

Now, because $a_1 = 4(0.3)$, $r = 0.3$, and $n = 12$, you can apply the formula for the sum of a finite geometric sequence to obtain

$$\sum_{n=1}^{12} 4(0.3)^n = a_1 \left(\frac{1 - r^n}{1 - r} \right)$$

$$= 4(0.3) \left[\frac{1 - (0.3)^{12}}{1 - 0.3} \right]$$

$$\approx 1.714.$$

Using a *TI-82* or *TI-83* graphing calculator, you can calculate the sum of the sequence in Example 6 as

sum(seq(4(.3)∧N, N, 1, 12, 1)) = 1.7142848.

Calculate the sum beginning at $n = 0$. You should obtain a sum of 5.7142848.

When using the formula for the sum of a geometric sequence, be careful to check that the index begins at $i = 1$. If the index begins at $i = 0$, you must adjust the formula for the nth partial sum. For instance, if the index in Example 6 had begun with $n = 0$, the sum would have been

$$\sum_{n=0}^{12} 4(0.3)^n = 4 + \sum_{n=1}^{12} 4(0.3)^n \approx 4 + 1.714 = 5.714.$$

Study Tip

Notice that the formula for the sum of an infinite geometric sequence requires that $|r| < 1$. What happens if $r = 1$ or $r = -1$? Give examples of infinite geometric sequences for which $|r| > 1$ and convince yourself that they do not have finite sums.

The formula for the sum of a *finite* geometric sequence can, depending on the value of r, be extended to produce a formula for the sum of an *infinite* geometric sequence. Specifically, if the common ratio r has the property that $|r| < 1$, it can be shown that r^n becomes arbitrarily close to zero as n increases without bound. Consequently,

$$a_1\left(\frac{1 - r^n}{1 - r}\right) \longrightarrow a_1\left(\frac{1 - 0}{1 - r}\right) \quad \text{as} \quad n \longrightarrow \infty.$$

This result is summarized as follows.

The Sum of an Infinite Geometric Sequence

If $|r| < 1$, the infinite geometric sequence

$$a_1, a_1r, a_1r^2, a_1r^3, \ldots, a_1r^{n-1}, \ldots$$

has the sum

$$S = \frac{a_1}{1 - r}.$$

Use a graphing utility to create a table showing the partial sums of the sequence in Example 7(a) by entering

$$Un = \text{sum}(\text{seq}(4(.6)\wedge(N-1), N, 1, n, 1)).$$

How can you determine the upper limit of the sequence from the table?

EXAMPLE 7   Finding the Sum of an Infinite Geometric Sequence

Find the sum of each infinite geometric sequence.

a. $\displaystyle\sum_{n=1}^{\infty} 4(0.6)^{n-1}$ **b.** $\displaystyle\sum_{n=1}^{\infty} 3(0.1)^{n-1}$

Solution

a. $\displaystyle\sum_{n=1}^{\infty} 4(0.6)^{n-1} = 4 + 4(0.6) + 4(0.6)^2 + 4(0.6)^3 + \cdots + 4(0.6)^{n-1} + \cdots$

$$= \frac{4}{1 - (0.6)} \qquad \frac{a_1}{1 - r}$$

$$= 10$$

b. $\displaystyle\sum_{n=1}^{\infty} 3(0.1)^{n-1} = 3 + 3(0.1) + 3(0.1)^2 + 3(0.1)^3 + \cdots + 3(0.1)^{n-1} + \cdots$

$$= \frac{3}{1 - (0.1)} \qquad \frac{a_1}{1 - r}$$

$$= \frac{10}{3}$$

$$\approx 3.33$$

Application

Note The type of investment account in Example 8 is an *annuity*.

EXAMPLE 8   🗆   **Compound Interest**

Real Life

A deposit of $50 is made on the first day of each month in a savings account that pays 6% compounded monthly. What is the balance at the end of 2 years?

Solution

The first deposit will gain interest for 24 months, and its balance will be

$$A_{24} = 50\left(1 + \frac{0.06}{12}\right)^{24} = 50(1.005)^{24}.$$

The second deposit will gain interest for 23 months, and its balance will be

$$A_{23} = 50\left(1 + \frac{0.06}{12}\right)^{23} = 50(1.005)^{23}.$$

The last deposit will gain interest for only 1 month, and its balance will be

$$A_1 = 50\left(1 + \frac{0.06}{12}\right)^{1} = 50(1.005).$$

The total balance in the account will be the sum of the balances of the 24 deposits. Using the formula for the sum of a finite geometric sequence, with $A_1 = 50(1.005)$ and $r = 1.005$, you have

$$S_n = 50(1.005)\left[\frac{1 - (1.005)^{24}}{1 - 1.005}\right] = \$1277.96.$$

🗆

Group Activity   *An Experiment*

You will need a piece of string or yarn, a pair of scissors, and a tape measure. Measure out any length of string at least 5 feet long. Double over the string and cut it in half. Take one of the resulting halves, double it over, and cut it in half. Continue this process until you are no longer able to cut a length of string in half. How many cuts were you able to make? Construct a sequence of the resulting string lengths after each cut, starting with the original length of the string. Find a formula for the *n*th term of this sequence. How many cuts could you theoretically make? Discuss why you were not able to make that many cuts.

9.3 /// EXERCISES

In Exercises 1–10, determine whether the sequence is geometric. If it is, find the common ratio.

1. 5, 15, 45, 135, . . .   **2.** 3, 12, 48, 192, . . .

3. 3, 12, 21, 30, . . .   **4.** 1, −2, 4, −8, . . .

5. $1, -\frac{1}{2}, \frac{1}{4}, -\frac{1}{8}, \ldots$ **6.** 5, 1, 0.2, 0.04, . . .

7. $\frac{1}{2}, \frac{2}{3}, \frac{3}{4}, \frac{4}{5}, \ldots$ **8.** $9, -6, 4, -\frac{8}{3}, \ldots$

9. $1, \frac{1}{2}, \frac{1}{3}, \frac{1}{4}, \ldots$ **10.** $\frac{1}{5}, \frac{2}{7}, \frac{3}{9}, \frac{4}{11}, \ldots$

In Exercises 11–20, write the first five terms of the geometric sequence.

11. $a_1 = 2, r = 3$ **12.** $a_1 = 6, r = 2$

13. $a_1 = 1, r = \frac{1}{2}$ **14.** $a_1 = 1, r = \frac{1}{3}$

15. $a_1 = 5, r = -\frac{1}{10}$ **16.** $a_1 = 6, r = -\frac{1}{4}$

17. $a_1 = 1, r = e$ **18.** $a_1 = 2, r = \sqrt{3}$

19. $a_1 = 3, r = \dfrac{x}{2}$ **20.** $a_1 = 5, r = 2x$

In Exercises 21–26, write the first five terms of the geometric sequence. Determine the common ratio and write the nth term of the sequence as a function of n.

21. $a_1 = 64, \quad a_{k+1} = \frac{1}{2}a_k$

22. $a_1 = 81, \quad a_{k+1} = \frac{1}{3}a_k$

23. $a_1 = 4, \quad a_{k+1} = 3a_k$

24. $a_1 = 5, \quad a_{k+1} = -2a_k$

25. $a_1 = 6, \quad a_{k+1} = -\frac{3}{2}a_k$

26. $a_1 = 36, \quad a_{k+1} = -\frac{2}{3}a_k$

In Exercises 27–38, find the nth term of the geometric sequence.

27. $a_1 = 4, r = \frac{1}{2}, n = 10$

28. $a_1 = 5, r = \frac{3}{2}, n = 8$

29. $a_1 = 6, r = -\frac{1}{3}, n = 12$

30. $a_1 = 8, r = \sqrt{5}, n = 9$

31. $a_1 = 100, r = e^x, n = 9$

32. $a_1 = 1, r = -\dfrac{x}{3}, n = 7$

33. $a_1 = 500, r = 1.02, n = 40$

34. $a_1 = 1000, r = 1.005, n = 60$

35. $a_1 = 16, a_4 = \frac{27}{4}, n = 3$

36. $a_2 = 3, a_5 = \frac{3}{64}, n = 1$

37. $a_2 = -18, a_5 = \frac{2}{3}, n = 6$

38. $a_3 = \frac{16}{3}, a_5 = \frac{64}{27}, n = 7$

In Exercises 39–42, match the sequence with its graph. [The graphs are labeled (a), (b), (c), and (d).]

(a)

(b)

(c)

(d)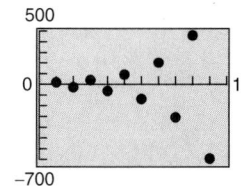

39. $a_n = 18\left(\frac{2}{3}\right)^{n-1}$ **40.** $a_n = 18\left(-\frac{2}{3}\right)^{n-1}$

41. $a_n = 18\left(\frac{3}{2}\right)^{n-1}$ **42.** $a_n = 18\left(-\frac{3}{2}\right)^{n-1}$

In Exercises 43–46, use a graphing utility to graph the first 10 terms of the sequence.

43. $a_n = 12(-0.75)^{n-1}$ **44.** $a_n = 12(-0.4)^{n-1}$

45. $a_n = 2(1.3)^{n-1}$ **46.** $a_n = 2(-1.4)^{n-1}$

47. *Essay* Write a brief paragraph explaining why the terms of a geometric sequence decrease in magnitude when $-1 < r < 1$.

48. *Essay* Write a brief paragraph explaining how to use the first two terms of a geometric sequence to find the nth term.

49. *Compound Interest* A principal of $1000 is invested at 10% interest. Find the amount after 10 years if the interest is compounded (a) annually, (b) semiannually, (c) quarterly, (d) monthly, and (e) daily.

50. *Compound Interest* A principal of $2500 is invested at 12% interest. Find the amount after 20 years if the interest is compounded (a) annually, (b) semiannually, (c) quarterly, (d) monthly, and (e) daily.

51. *Depreciation* A company buys a machine for $135,000 and it depreciates at a rate of 30% per year. (In other words, at the end of each year the depreciated value is 70% of what it was at the beginning of the year.) Find the depreciated value of the machine after 5 full years.

52. *Population Growth* A city of 250,000 people is growing at a rate of 1.3% per year. Estimate the population of the city 30 years from now.

In Exercises 53 and 54, find the first four terms of the sequence of partial sums of the geometric sequence. In a sequence of partial sums, the term S_n is the sum of the first n terms of the sequence. For instance, S_2 is the sum of the first two terms.

53. $8, -4, 2, -1, \frac{1}{2}, \ldots$

54. $8, 12, 18, 27, \frac{81}{2}, \ldots$

In Exercises 55–64, find the sum.

55. $\sum_{n=1}^{9} 2^{n-1}$

56. $\sum_{n=1}^{9} (-2)^{n-1}$

57. $\sum_{i=1}^{7} 64\left(-\frac{1}{2}\right)^{i-1}$

58. $\sum_{i=1}^{6} 32\left(\frac{1}{4}\right)^{i-1}$

59. $\sum_{n=0}^{20} 3\left(\frac{3}{2}\right)^{n}$

60. $\sum_{n=0}^{15} 2\left(\frac{4}{3}\right)^{n}$

61. $\sum_{i=1}^{10} 8\left(-\frac{1}{4}\right)^{i-1}$

62. $\sum_{i=1}^{10} 5\left(-\frac{1}{3}\right)^{i-1}$

63. $\sum_{n=0}^{5} 300(1.06)^{n}$

64. $\sum_{n=0}^{6} 500(1.04)^{n}$

In Exercises 65 and 66, use summation notation to express the sum.

65. $5 + 15 + 45 + \cdots + 3645$

66. $2 - \frac{1}{2} + \frac{1}{8} - \cdots + \frac{1}{2048}$

67. *Annuities* A deposit of $100 is made at the beginning of each month in an account that pays 10% interest, compounded monthly. The balance A in the account at the end of 5 years is given by

$$A = 100\left(1 + \frac{0.10}{12}\right)^{1} + \cdots + 100\left(1 + \frac{0.10}{12}\right)^{60}.$$

Find A.

68. *Annuities* A deposit of $50 is made at the beginning of each month in an account that pays 12% interest, compounded monthly. The balance A in the account at the end of 5 years is given by

$$A = 50\left(1 + \frac{0.12}{12}\right)^{1} + \cdots + 50\left(1 + \frac{0.12}{12}\right)^{60}.$$

Find A.

69. *Annuities* A deposit of P dollars is made at the beginning of each month in an account earning an annual interest rate r, compounded monthly. The balance A after t years is

$$A = P\left(1 + \frac{r}{12}\right) + P\left(1 + \frac{r}{12}\right)^{2} + \cdots + P\left(1 + \frac{r}{12}\right)^{12t}.$$

Show that the balance is given by

$$A = P\left[\left(1 + \frac{r}{12}\right)^{12t} - 1\right]\left(1 + \frac{12}{r}\right).$$

70. *Annuities*   A deposit of P dollars is made at the beginning of each month in an account earning an annual interest rate r, compounded continuously. The balance A after t years is

$$A = Pe^{r/12} + Pe^{2r/12} + \cdots + Pe^{12tr/12}.$$

Show that the balance is given by

$$A = \frac{Pe^{r/12}(e^{rt} - 1)}{e^{r/12} - 1}.$$

Annuities   In Exercises 71–74, consider making monthly deposits of P dollars in a savings account earning an annual interest rate r. Use the results of Exercises 69 and 70 to find the balance A after t years if the interest is compounded (a) monthly and (b) continuously.

71. $P = \$50$, $r = 7\%$, $t = 20$ years
72. $P = \$75$, $r = 9\%$, $t = 25$ years
73. $P = \$100$, $r = 10\%$, $t = 40$ years
74. $P = \$20$, $r = 6\%$, $t = 50$ years

75. *Annuities*   Consider an initial deposit of P dollars in an account earning an annual interest rate r, compounded monthly. At the end of each month, a withdrawal of W dollars will occur and the account will be depleted in t years. The amount of the initial deposit required is given by

$$P = W\left(1 + \frac{r}{12}\right)^{-1} + W\left(1 + \frac{r}{12}\right)^{-2} + \cdots$$
$$+ W\left(1 + \frac{r}{12}\right)^{-12t}.$$

Show that the initial deposit is given by

$$P = W\left(\frac{12}{r}\right)\left[1 - \left(1 + \frac{r}{12}\right)^{-12t}\right].$$

76. *Annuities*   Determine the amount required in an individual retirement account for an individual who retires at age 65 and wants an income of $2000 from the account each month for 20 years. Use the result of Exercise 75, and assume that the account earns 9% compounded monthly.

77. *Geometry*   The sides of a square are 16 inches in length. A new square is formed by connecting the midpoints of the sides of the original square, and two of the triangles are shaded (see figure). If this process is repeated five more times, determine the total area of the shaded region.

   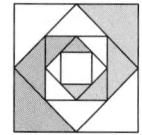

78. *Corporate Revenue*   The annual revenues a_n (in billions of dollars) for the Coca-Cola Company for 1985 through 1994 can be approximated by the model

$$a_n = 3.49e^{0.108n}, \quad n = 5, 6, 7, \ldots, 14$$

where $n = 5$ represents 1985. Use this model and the formula for the sum of a geometric sequence to approximate the total revenue earned during this 10-year period.   (Source: The Coca-Cola Company)

79. *Think About It*   Suppose you go to work for a company that pays $0.01 the first day, $0.02 the second day, $0.04 the third day, and so on. If the daily wage keeps doubling, what will your total income be for working (a) 29 days? (b) 30 days? (c) 31 days?

80. *Salary*   A company is offering a job with a salary of $30,000 for the first year. Suppose that during the next 39 years, there is a 5% raise each year. Determine the total compensation over the 40-year period.

In Exercises 81–88, find the sum.

81. $\displaystyle\sum_{n=0}^{\infty} \left(\tfrac{1}{2}\right)^n$

82. $\displaystyle\sum_{n=0}^{\infty} 2\left(\tfrac{2}{3}\right)^n$

83. $\displaystyle\sum_{n=0}^{\infty} \left(-\tfrac{1}{2}\right)^n$

84. $\displaystyle\sum_{n=0}^{\infty} 2\left(-\tfrac{2}{3}\right)^n$

85. $\displaystyle\sum_{n=0}^{\infty} 4\left(\tfrac{1}{4}\right)^n$

86. $\displaystyle\sum_{n=0}^{\infty} \left(\tfrac{1}{10}\right)^n$

87. $8 + 6 + \frac{9}{2} + \frac{27}{8} + \cdots$

88. $3 - 1 + \frac{1}{3} - \frac{1}{9} + \cdots$

In Exercises 89–92, find the rational number representation of the repeating decimal.

89. $0.\overline{36}$

90. $0.\overline{297}$

91. $0.3\overline{18}$

92. $1.3\overline{8}$

Graphical Reasoning In Exercises 93 and 94, use a graphing utility to graph the function. Identify the horizontal asymptote of the graph and determine its relationship to the sum.

93. $f(x) = 6\left[\dfrac{1 - (0.5)^x}{1 - (0.5)}\right]$, $\displaystyle\sum_{n=0}^{\infty} 6\left(\dfrac{1}{2}\right)^n$

94. $f(x) = 2\left[\dfrac{1 - (0.8)^x}{1 - (0.8)}\right]$, $\displaystyle\sum_{n=0}^{\infty} 2\left(\dfrac{4}{5}\right)^n$

95. *Distance*   A ball is dropped from a height of 16 feet. Each time it drops h feet, it rebounds $0.81h$ feet.

(a) Find the total distance traveled by the ball.

(b) The ball takes the following time for each fall.

$s_1 = -16t^2 + 16,$ $s_1 = 0$ if $t = 1$

$s_2 = -16t^2 + 16(0.81),$ $s_2 = 0$ if $t = 0.9$

$s_3 = -16t^2 + 16(0.81)^2,$ $s_3 = 0$ if $t = (0.9)^2$

$s_4 = -16t^2 + 16(0.81)^3,$ $s_4 = 0$ if $t = (0.9)^3$

$\vdots$ $\vdots$

$s_n = -16t^2 + 16(0.81)^{n-1},$ $s_n = 0$ if $t = (0.9)^{n-1}$

Beginning with s_2, the ball takes the same amount of time to bounce up as it does to fall, and thus the total time elapsed before it comes to rest is

$$t = 1 + 2\sum_{n=1}^{\infty} (0.9)^n.$$

Find this total.

Review   Solve Exercises 96–99 as a review of the skills and problem-solving techniques you learned in previous sections.

96. The ratio of cement to sand in a 90-pound bag of dry mix is 1 to 4. Find the number of pounds of sand in the bag.

97. A truck traveled at an average speed of 50 miles per hour on a 200-mile trip. On the return trip, the average speed was 42 miles per hour. Find the average speed for the round trip.

98. Find two consecutive positive even integers whose product is 624.

99. Suppose your friend can mow a lawn in 4 hours and you can mow it in 6 hours. How long will it take both of you to mow the lawn?

9.4	Mathematical Induction

Introduction / *Sums of Powers of Integers* / *Pattern Recognition* /
Finite Differences

Introduction

In this section you will study a form of mathematical proof called **mathematical induction.** It is important that you clearly see the logical need for it, so let's take a closer look at a problem discussed on page 670.

$$S_1 = 1 = 1^2$$
$$S_2 = 1 + 3 = 2^2$$
$$S_3 = 1 + 3 + 5 = 3^2$$
$$S_4 = 1 + 3 + 5 + 7 = 4^2$$
$$S_5 = 1 + 3 + 5 + 7 + 9 = 5^2$$

Judging from the pattern formed by these first five sums, it appears that the sum of the first n integers is

$$S_n = 1 + 3 + 5 + 7 + 9 + \cdots + (2n - 1) = n^2.$$

Although this particular formula *is* valid, it is important for you to see that recognizing a pattern and then simply *jumping to the conclusion* that the pattern must be true for all values of n is *not* a logically valid method of proof. There are many examples in which a pattern appears to be developing for small values of n but then fails at some point. One of the most famous cases of this was the conjecture by the French mathematician Pierre de Fermat (1601–1665), who speculated that all numbers of the form

$$F_n = 2^{2^n} + 1, \quad n = 0, 1, 2, \ldots$$

are prime. For $n = 0, 1, 2, 3$, and 4, the conjecture is true.

$$F_0 = 3, F_1 = 5, F_2 = 17, F_3 = 257, F_4 = 65{,}537$$

The size of the next Fermat number ($F_5 = 4{,}294{,}967{,}297$) is so great that it was difficult for Fermat to determine whether or not it was prime. However, another well-known mathematician, Leonhard Euler (1707–1783), later found a factorization

$$F_5 = 4{,}294{,}967{,}297 = 641(6{,}700{,}417)$$

which proved that F_5 is not prime, and therefore Fermat's conjecture was false.

Just because a rule, pattern, or formula seems to work for several values of n, you cannot simply decide that it is valid for *all* values of n without going through a *legitimate proof.*

The Principle of Mathematical Induction

Let P_n be a statement involving the positive integer n. If

1. P_1 is true, and
2. the truth of P_k implies the truth of P_{k+1}, for every positive k,

then P_n must be true for all positive integers n.

Note It is important to recognize that both parts of the Principle of Mathematical Induction are necessary.

To apply the Principle of Mathematical Induction, you need to be able to determine the statement P_{k+1} for a given statement P_k.

EXAMPLE 1   **A Preliminary Example**

Find P_{k+1} for the following.

a. $P_k : S_k = \dfrac{k^2(k+1)^2}{4}$

b. $P_k : S_k = 1 + 5 + 9 + \cdots + [4(k-1) - 3] + (4k - 3)$

c. $P_k : 3^k \geq 2k + 1$

Solution

a. $P_{k+1} : S_{k+1} = \dfrac{(k+1)^2(k+1+1)^2}{4}$ Replace k by $k + 1$.

$= \dfrac{(k+1)^2(k+2)^2}{4}.$ Simplify.

b. $P_{k+1} : S_{k+1} = 1 + 5 + 9 + \cdots + \{4[(k+1) - 1] - 3\} + [4(k+1) - 3]$
$= 1 + 5 + 9 + \cdots + (4k - 3) + (4k + 1).$

c. $P_{k+1} : 3^{k+1} \geq 2(k+1) + 1$
$3^{k+1} \geq 2k + 3.$

Figure 9.4

A well-known illustration used to explain why the Principle of Mathematical Induction works is the unending line of dominoes represented by Figure 9.4. If the line actually contains infinitely many dominoes, it is clear that you could not knock down the entire line by knocking down only *one domino* at a time. However, suppose it were true that each domino would knock down the next one as it fell. Then you could knock them all down simply by pushing the first one and starting a chain reaction. Mathematical induction works in the same way. If the truth of P_k implies the truth of P_{k+1} and if P_1 is true, the chain reaction proceeds as follows: P_1 implies P_2, P_2 implies P_3, P_3 implies P_4, and so on.

EXAMPLE 2   Using Mathematical Induction

Use mathematical induction to prove the following formula.

$$S_n = 1 + 3 + 5 + 7 + \cdots + (2n - 1)$$
$$= n^2$$

Solution

Mathematical induction consists of two distinct parts. First, you must show that the formula is true when $n = 1$.

1. When $n = 1$, the formula is valid, because

$$S_1 = 1 = 1^2.$$

The second part of mathematical induction has two steps. The first step is to assume that the formula is valid for *some* integer k. The second step is to use this assumption to prove that the formula is valid for the next integer, $k + 1$.

2. Assuming that the formula

$$S_k = 1 + 3 + 5 + 7 + \cdots + (2k - 1)$$
$$= k^2$$

is true, you must show that the formula $S_{k+1} = (k + 1)^2$ is true.

$$
\begin{aligned}
S_{k+1} &= 1 + 3 + 5 + 7 + \cdots + (2k - 1) + [2(k + 1) - 1] \\
&= [1 + 3 + 5 + 7 + \cdots + (2k - 1)] + (2k + 2 - 1) \\
&= S_k + (2k + 1) && \text{Group terms to form } S_k. \\
&= k^2 + 2k + 1 && \text{Replace } S_k \text{ by } k^2. \\
&= (k + 1)^2
\end{aligned}
$$

Combining the results of parts (1) and (2), you can conclude by mathematical induction that the formula is valid for *all* positive integer values of n.

Note When using mathematical induction to prove a *summation* formula (such as the one in Example 2), it is helpful to think of S_{k+1} as $S_{k+1} = S_k + a_{k+1}$, where a_{k+1} is the $(k + 1)$ term of the original sum.

It occasionally happens that a statement involving natural numbers is not true for the first $k - 1$ positive integers but is true for all values of $n \geq k$. In these instances, you use a slight variation of the Principle of Mathematical Induction in which you verify P_k rather than P_1. This variation is called the **extended principle of mathematical induction.** To see the validity of this, note from Figure 9.4 that all but the first $k - 1$ dominoes can be knocked down by knocking over the kth domino. This suggests that you can prove a statement P_n to be true for $n \geq k$ by showing that P_k is true and that P_k implies P_{k+1}. In Exercises 35–38 in this section, you are asked to apply this extension of mathematical induction.

EXAMPLE 3 ▱ **Using Mathematical Induction**

Use mathematical induction to prove the following formula.

$$S_n = 1^2 + 2^2 + 3^2 + 4^2 + \cdots + n^2$$
$$= \frac{n(n+1)(2n+1)}{6}$$

Solution

1. When $n = 1$, the formula is valid, because

$$S_1 = 1^2 = \frac{1(2)(3)}{6}.$$

2. Assuming that

$$S_k = 1^2 + 2^2 + 3^2 + 4^2 + \cdots + k^2$$
$$= \frac{k(k+1)(2k+1)}{6}$$

you must show that

$$S_{k+1} = \frac{(k+1)(k+2)(2k+3)}{6}.$$

To do this, write the following.

$$S_{k+1} = S_k + a_{k+1}$$
$$= (1^2 + 2^2 + 3^2 + 4^2 + \cdots + k^2) + (k+1)^2$$
$$= \frac{k(k+1)(2k+1)}{6} + (k+1)^2$$
$$= \frac{k(k+1)(2k+1) + 6(k+1)^2}{6}$$
$$= \frac{(k+1)[k(2k+1) + 6(k+1)]}{6}$$
$$= \frac{(k+1)(2k^2 + 7k + 6)}{6}$$
$$= \frac{(k+1)(k+2)(2k+3)}{6}$$

Note When proving a formula with mathematical induction, the only statement that you *need* to verify is P_1. As a check, however, it is a good idea to try verifying some of the other statements. For instance, in Example 3, try verifying S_2 and S_3.

Combining the results of parts (1) and (2), you can conclude by mathematical induction that the formula is valid for *all* $n \geq 1$. ▱

Sums of Powers of Integers

The formula in Example 3 is one of a collection of useful summation formulas. We summarize this and other formulas dealing with the sums of various powers of the first n positive integers, as follows.

Sums of Powers of Integers

1. $1 + 2 + 3 + 4 + \cdots + n = \dfrac{n(n+1)}{2}$

2. $1^2 + 2^2 + 3^2 + 4^2 + \cdots + n^2 = \dfrac{n(n+1)(2n+1)}{6}$

3. $1^3 + 2^3 + 3^3 + 4^3 + \cdots + n^3 = \dfrac{n^2(n+1)^2}{4}$

4. $1^4 + 2^4 + 3^4 + 4^4 + \cdots + n^4 = \dfrac{n(n+1)(2n+1)(3n^2+3n-1)}{30}$

5. $1^5 + 2^5 + 3^5 + 4^5 + \cdots + n^5 = \dfrac{n^2(n+1)^2(2n^2+2n-1)}{12}$

Note   Each of these formulas for sums can be proven by mathematical induction. (See Exercises 11–13, 15, 16.)

EXAMPLE 4   ▱   **Finding a Sum of Powers of Integers**

Find $\displaystyle\sum_{n=1}^{7} n^3 = 1^3 + 2^3 + 3^3 + 4^3 + 5^3 + 6^3 + 7^3.$

Solution
Using the formula for the sum of the cubes of the first n positive integers, you obtain the following.

$$\sum_{n=1}^{7} n^3 = 1^3 + 2^3 + 3^3 + 4^3 + 5^3 + 6^3 + 7^3$$

$$= \frac{7^2(7+1)^2}{4}$$

$$= \frac{49(64)}{4}$$

$$= 784$$

Check this sum by adding the numbers 1, 8, 27, 64, 125, 216, and 343.   ▱

EXAMPLE 5   **Proving an Inequality by Mathematical Induction**

Prove that $n < 2^n$ for all positive integers n.

Solution

1. For $n = 1$, the formula is true, because

$$1 < 2^1.$$

2. Assuming that

$$k < 2^k$$

you need to show that $k + 1 < 2^{k+1}$. For $n = k$, you have

$$2^{k+1} = 2(2^k) > 2(k) = 2k. \qquad \text{By assumption}$$

Because $2k = k + k > k + 1$ for all $k > 1$, it follows that

$$2^{k+1} > 2k > k + 1$$

or

$$k + 1 < 2^{k+1}.$$

Therefore, $n < 2^n$ for all integers $n \geq 1$.

Pattern Recognition

Note   Some common patterns to look for when you must find the nth term of a sequence are:

Linear	$an \pm b$
Quadratic	$n^2 \pm b$
Cubic	$n^3 \pm b$
Exponential	$2^n \pm b,\ 3^n \pm b$
Factorial	$n!,\ (2n)!,\ (2n-1)!$

Although choosing a formula on the basis of a few observations *does not* guarantee the validity of a formula, pattern recognition *is* important. Once you have a pattern or formula that you think works, you can try using mathematical induction to prove your formula.

Finding a Formula for the nth Term of a Sequence

To find a formula for the nth term of a sequence, consider the following guidelines.

1. Calculate the first several terms of the sequence. It is often a good idea to write the terms in both simplified and factored forms.
2. Try to find a recognizable pattern for the terms and write a formula for the nth term of the sequence. This is your *hypothesis* or *conjecture*. You might try computing one or two more terms in the sequence to test your hypothesis.
3. Use mathematical induction to prove your hypothesis.

EXAMPLE 6 ◢ **Finding a Formula for a Finite Sum**

Find a formula for the following finite sum.

$$\frac{1}{1 \cdot 2} + \frac{1}{2 \cdot 3} + \frac{1}{3 \cdot 4} + \frac{1}{4 \cdot 5} + \cdots + \frac{1}{n(n + 1)}$$

Solution

Begin by writing out the first few sums.

$$S_1 = \frac{1}{1 \cdot 2} = \frac{1}{2} = \frac{1}{1 + 1}$$

$$S_2 = \frac{1}{1 \cdot 2} + \frac{1}{2 \cdot 3} = \frac{4}{6} = \frac{2}{3} = \frac{2}{2 + 1}$$

$$S_3 = \frac{1}{1 \cdot 2} + \frac{1}{2 \cdot 3} + \frac{1}{3 \cdot 4} = \frac{9}{12} = \frac{3}{4} = \frac{3}{3 + 1}$$

$$S_4 = \frac{1}{1 \cdot 2} + \frac{1}{2 \cdot 3} + \frac{1}{3 \cdot 4} + \frac{1}{4 \cdot 5} = \frac{48}{60} = \frac{4}{5} = \frac{4}{4 + 1}$$

From this sequence, it appears that the formula for the kth sum is

$$S_k = \frac{1}{1 \cdot 2} + \frac{1}{2 \cdot 3} + \frac{1}{3 \cdot 4} + \frac{1}{4 \cdot 5} + \cdots + \frac{1}{k(k + 1)}$$

$$= \frac{k}{k + 1}.$$

To prove the validity of this hypothesis, use mathematical induction, as follows. Note that you have already verified the formula for $n = 1$, so you can begin by assuming that the formula is valid for $n = k$ and trying to show that it is valid for $n = k + 1$.

$$S_{k+1} = \left[\frac{1}{1 \cdot 2} + \frac{1}{2 \cdot 3} + \frac{1}{3 \cdot 4} + \frac{1}{4 \cdot 5} + \cdots + \frac{1}{k(k + 1)} \right] + \frac{1}{(k + 1)(k + 2)}$$

$$= \frac{k}{k + 1} + \frac{1}{(k + 1)(k + 2)}$$

$$= \frac{k(k + 2) + 1}{(k + 1)(k + 2)}$$

$$= \frac{k^2 + 2k + 1}{(k + 1)(k + 2)}$$

$$= \frac{(k + 1)^2}{(k + 1)(k + 2)}$$

$$= \frac{k + 1}{k + 2}$$

Thus, the hypothesis is valid. ◢

> *Study Tip*
>
> To show that the formula in Example 6 is valid for $n = k + 1$, you have to verify that
>
> $$S_{k+1} = \frac{k + 1}{(k + 1) + 1} = \frac{k + 1}{k + 2}.$$

Finite Differences

The **first differences** of a sequence are found by subtracting consecutive terms. The **second differences** are found by subtracting consecutive first differences. The first and second differences of the sequence 3, 5, 8, 12, 17, 23, . . . are as follows.

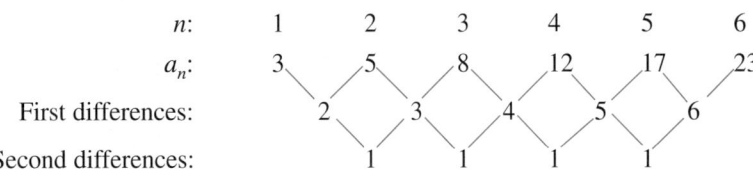

For this sequence, the second differences are all the same. When this happens, the sequence has a perfect quadratic model. If the first differences are all the same, the sequence has a linear model—that is, it is arithmetic.

EXAMPLE 7 ▰ **Finding a Quadratic Model**

Find the quadratic model for the sequence

$$3, 5, 8, 12, 17, 23, ...$$

Solution

You know the model has the form

$$a_n = an^2 + bn + c.$$

By substituting 1, 2, and 3 for n, you can obtain a system of three linear equations in three variables.

$$a_1 = a(1)^2 + b(1) + c = 3 \qquad \text{Substitute 1 for } n.$$
$$a_2 = a(2)^2 + b(2) + c = 5 \qquad \text{Substitute 2 for } n.$$
$$a_3 = a(3)^2 + b(3) + c = 8 \qquad \text{Substitute 3 for } n.$$

You now have a system of three equations in a, b, and c.

$$a + b + c = 3 \qquad \text{Equation 1}$$
$$4a + 2b + c = 5 \qquad \text{Equation 2}$$
$$9a + 3b + c = 8 \qquad \text{Equation 3}$$

Using the techniques discussed in Chapter 7, you can find the solution to be $a = \frac{1}{2}$, $b = \frac{1}{2}$, and $c = 2$. Thus, the quadratic model is

$$a_n = \frac{1}{2}n^2 + \frac{1}{2}n + 2.$$

Try checking the values of a_1, a_2, and a_3. ▰

Group Activity

The Sum of the Angles of a Regular Polygon

A *regular n-sided polygon* is a polygon that has *n* equal sides and *n* equal angles. For instance, an equilateral triangle is a regular three-sided polygon. Each angle of an equilateral triangle measures $60°$, and the sum of all three angles is $180°$. Similarly, the sum of the four angles of a regular four-sided polygon (a square) is $360°$. The figure below shows the sums of the angles of the first eight regular polygons.

(a) Equilateral Triangle $(180°)$ **(b)** Square $(360°)$ **(c)** Regular Pentagon $(540°)$ **(d)** Regular Hexagon $(720°)$

(e) Regular Heptagon $(900°)$ **(f)** Regular Octagon $(1080°)$ **(g)** Regular Nonagon $(1260°)$ **(h)** Regular Decagon $(1440°)$

a. Use the data given in the figure to complete the first row of the table. Then complete the second and third rows of the table by finding the first and second differences.

n	3	4	5	6	7	8	9	10
a_n	180	360	?	?	?	?	?	?
First Differences		?	?	?	?	?	?	?
Second Differences			?	?	?	?	?	?

b. From the result of the table, does the data fit a linear model, a quadratic model, or neither? Explain your reasoning.

c. Find a model that fits the data. Then use the result to find a model that gives the number of degrees in each angle of a regular *n*-sided polygon.

9.4 /// EXERCISES

In Exercises 1–4, find P_{k+1} for the given P_k

1. $P_k = \dfrac{5}{k(k+1)}$

2. $P_k = \dfrac{1}{(k+1)(k+3)}$

3. $P_k = \dfrac{k^2(k+1)^2}{4}$

4. $P_k = \dfrac{k}{2}(3k-1)$

In Exercises 5–18, use mathematical induction to prove the formula for every positive integer n.

5. $2 + 4 + 6 + 8 + \cdots + 2n = n(n+1)$

6. $3 + 7 + 11 + 15 + \cdots + (4n-1) = n(2n+1)$

7. $2 + 7 + 12 + 17 + \cdots + (5n-3) = \dfrac{n}{2}(5n-1)$

8. $1 + 4 + 7 + 10 + \cdots + (3n-2) = \dfrac{n}{2}(3n-1)$

9. $1 + 2 + 2^2 + 2^3 + \cdots + 2^{n-1} = 2^n - 1$

10. $2(1 + 3 + 3^2 + 3^3 + \cdots + 3^{n-1}) = 3^n - 1$

11. $1 + 2 + 3 + 4 + \cdots + n = \dfrac{n(n+1)}{2}$

12. $1^2 + 2^2 + 3^2 + 4^2 + \cdots + n^2 = \dfrac{n(n+1)(2n+1)}{6}$

13. $1^3 + 2^3 + 3^3 + 4^3 + \cdots + n^3 = \dfrac{n^2(n+1)^2}{4}$

14. $\left(1 + \dfrac{1}{1}\right)\left(1 + \dfrac{1}{2}\right)\left(1 + \dfrac{1}{3}\right) \cdots \left(1 + \dfrac{1}{n}\right) = n + 1$

15. $\displaystyle\sum_{i=1}^{5} i^5 = \dfrac{n^2(n+1)^2(2n^2 + 2n - 1)}{12}$

16. $\displaystyle\sum_{i=1}^{n} i^4 = \dfrac{n(n+1)(2n+1)(3n^2 + 3n - 1)}{30}$

17. $\displaystyle\sum_{i=1}^{n} i(i+1) = \dfrac{n(n+1)(n+2)}{3}$

18. $\displaystyle\sum_{i=1}^{n} \dfrac{1}{(2i-1)(2i+1)} = \dfrac{n}{2n+1}$

In Exercises 19–28, find the sum using the formulas for the sums of powers of integers.

19. $\displaystyle\sum_{n=1}^{20} n$

20. $\displaystyle\sum_{n=1}^{50} n$

21. $\displaystyle\sum_{n=1}^{6} n^2$

22. $\displaystyle\sum_{n=1}^{10} n^3$

23. $\displaystyle\sum_{n=1}^{5} n^4$

24. $\displaystyle\sum_{n=1}^{8} n^5$

25. $\displaystyle\sum_{n=1}^{6} (n^2 - n)$

26. $\displaystyle\sum_{n=1}^{10} (n^3 - n^2)$

27. $\displaystyle\sum_{i=1}^{6} (6i - 8i^3)$

28. $\displaystyle\sum_{j=1}^{4} \left(2 + \tfrac{5}{2}j - \tfrac{3}{2}j^2\right)$

In Exercises 29–34, find a formula for the sum of the first n terms of the sequence.

29. $1, 5, 9, 13, \ldots$

30. $25, 22, 19, 16, \ldots$

31. $1, \dfrac{9}{10}, \dfrac{81}{100}, \dfrac{729}{1000}, \ldots$

32. $3, -\dfrac{9}{2}, \dfrac{27}{4}, -\dfrac{81}{8}, \ldots$

33. $\dfrac{1}{4}, \dfrac{1}{12}, \dfrac{1}{24}, \dfrac{1}{40}, \ldots, \dfrac{1}{2n(n-1)}, \ldots$

34. $\dfrac{1}{2 \cdot 3}, \dfrac{1}{3 \cdot 4}, \dfrac{1}{4 \cdot 5}, \dfrac{1}{5 \cdot 6}, \ldots, \dfrac{1}{(n+1)(n+2)}, \ldots$

In Exercises 35–38, prove the inequality for the indicated integer values of n.

35. $n! > 2^n, \quad n \ge 4$

36. $\left(\tfrac{4}{3}\right)^n > n, \quad n \ge 7$

37. $\dfrac{1}{\sqrt{1}} + \dfrac{1}{\sqrt{2}} + \dfrac{1}{\sqrt{3}} + \cdots + \dfrac{1}{\sqrt{n}} > \sqrt{n}, \quad n \ge 2$

38. $\left(\dfrac{x}{y}\right)^{n+1} < \left(\dfrac{x}{y}\right)^{n}, \quad n \ge 1 \text{ and } 0 < x < y$

In Exercises 39–48, use mathematical induction to prove the given property for all positive integers n.

39. $(ab)^n = a^n b^n$

40. $\left(\dfrac{a}{b}\right)^n = \dfrac{a^n}{b^n}$

41. If $x_1 \neq 0$, $x_2 \neq 0$, . . . , $x_n \neq 0$, then

$$(x_1 x_2 x_3 \cdot \cdot \cdot x_n)^{-1} = x_1^{-1} x_2^{-1} x_3^{-1} \cdot \cdot \cdot x_n^{-1}.$$

42. If $x_1 > 0$, $x_2 > 0$, . . . , $x_n > 0$, then

$$\ln(x_1 x_2 x_3 \cdot \cdot \cdot x_n) = \ln x_1 + \ln x_2 + \ln x_3$$
$$+ \cdot \cdot \cdot + \ln x_n.$$

43. Generalized Distributive Law:

$$x(y_1 + y_2 + \cdot \cdot \cdot + y_n) = xy_1 + xy_2 + \cdot \cdot \cdot + xy_n$$

44. $(a + bi)^n$ and $(a - bi)^n$ are complex conjugates for all $n \geq 1$.

45. *Trigonometry* $\sin(x + n\pi) = (-1)^n \sin x$

46. *Trigonometry* $\tan(x + n\pi) = \tan x$

47. A factor of $(n^3 + 3n^2 + 2n)$ is 3.

48. A factor of $(2^{2n-1} + 3^{2n-1})$ is 5.

49. *Essay*   In your own words, explain what is meant by a proof by mathematical induction.

50. *Think About It*   What conclusion can be drawn from the given information about the sequence of statements P_n?

(a) P_3 is true and P_k implies P_{k+1}.

(b) $P_1, P_2, P_3, \ldots, P_{50}$ are all true.

(c) P_1, P_2, and P_3 are all true, but the truth of P_k does not imply that P_{k+1} is true.

(d) P_2 is true and P_{2k} implies P_{2k+2}.

In Exercises 51–54, write the first five terms of the sequence.

51. $a_0 = 1$

$a_n = a_{n-1} + 2$

53. $a_0 = 4$

$a_1 = 2$

$a_n = a_{n-1} - a_{n-2}$

52. $a_0 = 10$

$a_n = 4a_{n-1}$

54. $a_0 = 0$

$a_1 = 2$

$a_n = a_{n-1} + 2a_{n-2}$

In Exercises 55–64, write the first five terms of the sequence where $a_1 = f(1)$. Then calculate the first and second differences of the sequence. Does the sequence have a linear model, a quadratic model, or neither?

55. $f(1) = 0$

$a_n = a_{n-1} + 3$

57. $f(1) = 3$

$a_n = a_{n-1} - n$

59. $a_0 = 0$

$a_n = a_{n-1} + n$

61. $f(1) = 2$

$a_n = a_{n-1} + 2$

63. $a_0 = 1$

$a_n = a_{n-1} + n^2$

56. $f(1) = 2$

$a_n = n - a_{n-1}$

58. $f(2) = -3$

$a_n = -2a_{n-1}$

60. $a_0 = 2$

$a_n = (a_{n-1})^2$

62. $f(1) = 0$

$a_n = a_{n-1} + 2n$

64. $a_0 = 0$

$a_n = a_{n-1} - 1$

In Exercises 65–68, find a quadratic model for the sequence with the indicated terms.

65. $a_0 = 3$, $a_1 = 3$, $a_4 = 15$

66. $a_0 = 7$, $a_1 = 6$, $a_3 = 10$

67. $a_0 = -3$, $a_2 = 1$, $a_4 = 9$

68. $a_0 = 3$, $a_2 = 0$, $a_6 = 36$

Review   Solve Exercises 69–72 as a review of the skills and problem-solving techniques you learned in previous sections. Solve the system of equations.

69. $y = x^2$

$-3x + 2y = 2$

70. $x - y^3 = 0$

$x - 2y^2 = 0$

71. $x - y \quad\quad = -1$

$x + 2y - 2z = \quad 3$

$3x - y + 2z = \quad 3$

72. $2x + \quad y - 2z = \quad 1$

$x \quad\quad - z = \quad 1$

$3x + 3y + \quad z = 12$

9.5   The Binomial Theorem

Binomial Coefficients / Pascal's Triangle / Binomial Expansions

Binomial Coefficients

Recall that a **binomial** is a polynomial that has two terms. In this section, you will study a formula that provides a quick method of raising a binomial to a power. To begin, let's look at the expansion of $(x + y)^n$ for several values of n.

$$(x + y)^0 = 1$$
$$(x + y)^1 = x + y$$
$$(x + y)^2 = x^2 + 2xy + y^2$$
$$(x + y)^3 = x^3 + 3x^2y + 3xy^2 + y^3$$
$$(x + y)^4 = x^4 + 4x^3y + 6x^2y^2 + 4xy^3 + y^4$$
$$(x + y)^5 = x^5 + 5x^4y + 10x^3y^2 + 10x^2y^3 + 5xy^4 + y^5$$

There are several observations you can make about these expansions.

1. In each expansion, there are $n + 1$ terms.
2. In each expansion, x and y have symmetric roles. The powers of x decrease by 1 in successive terms, whereas the powers of y increase by 1.
3. The sum of the powers of each term is n. For instance, in the expansion of $(x + y)^5$, the sum of the powers of each term is 5.

$$4 + 1 = 5 \quad 3 + 2 = 5$$
$$(x + y)^5 = x^5 + 5x^4y^1 + 10x^3y^2 + 10x^2y^3 + 5x^1y^4 + y^5$$

4. The coefficients increase and then decrease in a symmetric pattern.

The coefficients of a binomial expansion are called **binomial coefficients.** To find them, you can use the following theorem.

Think About the Proof

Use mathematical induction to prove the Binomial Theorem. The details of the proof are given in the appendix.

The Binomial Theorem

In the expansion of $(x + y)^n$

$$(x + y)^n = x^n + nx^{n-1}y + \cdots +{}_n C_r\, x^{n-r}y^r + \cdots + nxy^{n-1} + y^n$$

the coefficient of $x^{n-r}y^r$ is given by

$${}_n C_r = \frac{n!}{(n-r)!\,r!}.$$

Note The symbol $\binom{n}{r}$ is often used in place of ${}_n C_r$ to denote binomial coefficients.

EXAMPLE 1 ▱ **Finding Binomial Coefficients**

Find the binomial coefficients.

a. $_8C_2$ **b.** $_{10}C_3$ **c.** $_7C_0$ **d.** $_8C_8$

Solution

a. $_8C_2 = \dfrac{8!}{6! \cdot 2!} = \dfrac{(8 \cdot 7) \cdot \cancel{6!}}{\cancel{6!} \cdot 2!} = \dfrac{8 \cdot 7}{2 \cdot 1} = 28$

b. $_{10}C_3 = \dfrac{10!}{7! \cdot 3!} = \dfrac{(10 \cdot 9 \cdot 8) \cdot \cancel{7!}}{\cancel{7!} \cdot 3!} = \dfrac{10 \cdot 9 \cdot 8}{3 \cdot 2 \cdot 1} = 120$

c. $_7C_0 = \dfrac{\cancel{7!}}{\cancel{7!} \cdot 0!} = 1$

d. $_8C_8 = \dfrac{\cancel{8!}}{0! \cdot \cancel{8!}} = 1$ ▱

Note   When $r \neq 0$ and $r \neq n$, as in parts (a) and (b) above, there is a simple pattern for evaluating binomial coefficients.

$$_8C_2 = \frac{\overbrace{8 \cdot 7}^{2 \text{ factors}}}{\underbrace{2 \cdot 1}_{2 \text{ factorial}}} \quad \text{and} \quad _{10}C_3 = \frac{\overbrace{10 \cdot 9 \cdot 8}^{3 \text{ factors}}}{\underbrace{3 \cdot 2 \cdot 1}_{3 \text{ factorial}}}$$

EXAMPLE 2 ▱ **Finding Binomial Coefficients**

Find the binomial coefficients.

a. $_7C_3$ **b.** $_7C_4$ **c.** $_{12}C_1$ **d.** $_{12}C_{11}$

Solution

a. $_7C_3 = \dfrac{7 \cdot \cancel{6} \cdot 5}{\cancel{3} \cdot \cancel{2} \cdot 1} = 35$

b. $_7C_4 = \dfrac{7 \cdot \cancel{6} \cdot 5 \cdot \cancel{4}}{\cancel{4} \cdot \cancel{3} \cdot \cancel{2} \cdot 1} = 35$

c. $_{12}C_1 = \dfrac{12}{1} = 12$

d. $_{12}C_{11} = \dfrac{12!}{1! \cdot 11!} = \dfrac{(12) \cdot \cancel{11!}}{1! \cdot \cancel{11!}} = \dfrac{12}{1} = 12$ ▱

Note   It is not a coincidence that the results in parts (a) and (b) of Example 2 are the same and that the results in parts (c) and (d) are the same. In general, it is true that

$$_nC_r = {}_nC_{n-r}.$$

This shows the symmetric property of binomial coefficients that was identified earlier.

Pascal's Triangle

There is a convenient way to remember a pattern for binomial coefficients. By arranging the coefficients in a triangular pattern, you obtain the following array, which is called **Pascal's Triangle.** This triangle is named after the famous French mathematician Blaise Pascal (1623–1662).

$$
\begin{array}{ccccccccccccccc}
&&&&&&&1\\
&&&&&&1&&1\\
&&&&&1&&2&&1\\
&&&&1&&3&&3&&1\\
&&&1&&4&&6&&4&&1\\
&&1&&5&&10&&10&&5&&1\\
&1&&6&&15&&20&&15&&6&&1\\
1&&7&&21&&35&&35&&21&&7&&1
\end{array}
$$

Note The top row in Pascal's Triangle is called the *zero row* because it corresponds to the binomial expansion

$$(x + y)^0 = 1.$$

Similarly, the next row is called the *first row* because it corresponds to the binomial expansion

$$(x + y)^1 = 1(x) + 1(y).$$

In general, the *nth row* in Pascal's Triangle gives the coefficients of $(x + y)^n$.

The first and last number in each row of Pascal's Triangle is 1. Every other number in each row is formed by adding the two numbers immediately above the number. Pascal noticed that numbers in this triangle are precisely the same numbers that are the coefficients of binomial expansions, as follows.

$$
\begin{aligned}
(x + y)^0 &= 1\\
(x + y)^1 &= 1x + 1y\\
(x + y)^2 &= 1x^2 + 2xy + 1y^2\\
(x + y)^3 &= 1x^3 + 3x^2y + 3xy^2 + 1y^3\\
(x + y)^4 &= 1x^4 + 4x^3y + 6x^2y^2 + 4xy^3 + 1y^4\\
(x + y)^5 &= 1x^5 + 5x^4y + 10x^3y^2 + 10x^2y^3 + 5xy^4 + 1y^5\\
(x + y)^6 &= 1x^6 + 6x^5y + 15x^4y^2 + 20x^3y^3 + 15x^2y^4 + 6xy^5 + 1y^6\\
(x + y)^7 &= 1x^7 + 7x^6y + 21x^5y^2 + 35x^4y^3 + 35x^3y^4 + 21x^2y^5 + 7xy^6 + 1y^7
\end{aligned}
$$

EXAMPLE 3   **Using Pascal's Triangle**

Use the seventh row of Pascal's Triangle to find the binomial coefficients.

$$_8C_0, \; _8C_1, \; _8C_2, \; _8C_3, \; _8C_4, \; _8C_5, \; _8C_6, \; _8C_7, \; _8C_8$$

Solution

$$
\begin{array}{ccccccccccccccccc}
1 && 7 && 21 && 35 && 35 && 21 && 7 && 1\\
1 && 8 && 28 && 56 && 70 && 56 && 28 && 8 && 1
\end{array}
$$

$$_8C_0 \quad _8C_1 \quad _8C_2 \quad _8C_3 \quad _8C_4 \quad _8C_5 \quad _8C_6 \quad _8C_7 \quad _8C_8$$

"Pascal's" Triangle and forms of the Binomial Theorem were known in Eastern cultures prior to the Western "discovery" of the theorem. A Chinese text *Precious Mirror* contains a triangle of binomial expansions through the eighth power.

Binomial Expansions

As mentioned at the beginning of this section, when you write out the coefficients for a binomial that is raised to a power, you are **expanding a binomial.** The formulas for binomial coefficients give you an easy way to expand binomials, as demonstrated in the next three examples.

EXAMPLE 4 ▱ **Expanding a Binomial**

Write the expansion for the expression

$(x + 1)^3$.

Solution

The binomial coefficients from the third row of Pascal's Triangle are

1, 3, 3, 1.

Therefore, the expansion is as follows.

$$(x + 1)^3 = (1)x^3 + (3)x^2(1) + (3)x(1^2) + (1)(1^3)$$
$$= x^3 + 3x^2 + 3x + 1$$

▱

To expand binomials representing *differences*, rather than sums, you alternate signs. Here are two examples.

$$(x - 1)^3 = x^3 - 3x^2 + 3x - 1$$
$$(x - 1)^4 = x^4 - 4x^3 + 6x^2 - 4x + 1$$

EXAMPLE 5 ▱ **Expanding a Binomial**

Write the expansion for the expression

$(x - 3)^4$.

Solution

The binomial coefficients from the fourth row of Pascal's Triangle are

1, 4, 6, 4, 1.

Therefore, the expansion is as follows.

$$(x - 3)^4 = (1)x^4 - (4)x^3(3) + (6)x^2(3^2) - (4)x(3^3) + (1)(3^4)$$
$$= x^4 - 12x^3 + 54x^2 - 108x + 81$$

▱

EXAMPLE 6 🔲 **Expanding a Binomial**

Write the expansion for $(x - 2y)^4$.

Solution

Use the fourth row of Pascal's Triangle, as follows.

$$(x - 2y)^4 = (1)x^4 - (4)x^3(2y) + (6)x^2(2y)^2 - (4)x(2y)^3 + (1)(2y)^4$$
$$= x^4 - 8x^3y + 24x^2y^2 - 32xy^3 + 16y^4$$ 🔲

EXAMPLE 7 🔲 **Finding a Term in a Binomial Expansion**

Find the sixth term of $(a + 2b)^8$.

Solution

From the Binomial Theorem, you can see that the $r + 1$ term is $_nC_r\, x^{n-r}\, y^r$. So in this case, $6 = r + 1$ means that $r = 5$. Because $n = 8$, $x = a$, and $y = 2b$, the sixth term in the binomial expansion is

$$_8C_5 a^{8-5}(2b)^5 = 56 \cdot a^3 \cdot (2b)^5$$
$$= 56(2^5)a^3b^5$$
$$= 1792a^3b^5$$ 🔲

Group Activity   *Error Analysis*

Suppose you are a math instructor and receive the following solutions from one of your students on a quiz. Find the error(s) in each solution, and discuss ways that your student could avoid the error(s) in the future.

a. Find the second term in the expansion of

$$(2x - 3y)^5.$$
$$\overline{5(2x)^4(3y)^2 = 720x^4y^2}$$

b. Find the fourth term in the expansion of

$$\left(\tfrac{1}{2}x + 7y\right)^6.$$
$$\overline{_6C_4\left(\tfrac{1}{2}x\right)^2(7y)^4 = 9003.75x^2y^4}$$

9.5 /// EXERCISES

In Exercises 1–10, evaluate $_nC_r$.

1. $_5C_3$

2. $_8C_6$

3. $_{12}C_0$

4. $_{20}C_{20}$

5. $_{20}C_{15}$

6. $_{12}C_5$

7. $_{100}C_{98}$

8. $_{10}C_4$

9. $_{100}C_2$

10. $_{10}C_6$

11. *Essay* In your own words, explain how to form the rows of Pascal's Triangle.

12. Form the first nine rows of Pascal's Triangle.

In Exercises 13–16, evaluate using Pascal's Triangle.

13. $_7C_4$

14. $_6C_3$

15. $_8C_5$

16. $_8C_7$

In Exercises 17–36, use the Binomial Theorem to expand and simplify the expression.

17. $(x + 1)^4$

18. $(x + 1)^6$

19. $(a + 2)^3$

20. $(a + 3)^4$

21. $(y - 2)^4$

22. $(y - 2)^5$

23. $(x + y)^5$

24. $(x + y)^6$

25. $(r + 3s)^6$

26. $(x + 2y)^4$

27. $(x - y)^5$

28. $(2x - y)^5$

29. $(1 - 2x)^3$

30. $(5 - 3y)^3$

31. $(x^2 + 5)^4$

32. $(x^2 + y^2)^6$

33. $\left(\dfrac{1}{x} + y\right)^5$

34. $\left(\dfrac{1}{x} + 2y\right)^6$

35. $2(x - 3)^4 + 5(x - 3)^2$

36. $3(x + 1)^5 - 4(x + 1)^3$

In Exercises 37–40, expand the binomial using Pascal's Triangle to determine the coefficients.

37. $(2t - s)^5$

38. $(x + 2y)^5$

39. $(3 - 2z)^4$

40. $(3y + 2)^5$

In Exercises 41–48, find the coefficient a of the given term in the expansion of the binomial.

Binomial	Term
41. $(x + 3)^{12}$	ax^5
42. $(x^2 + 3)^{12}$	ax^8
43. $(x - 2y)^{10}$	ax^8y^2
44. $(4x - y)^{10}$	ax^2y^8
45. $(3x - 2y)^9$	ax^4y^5
46. $(2x - 3y)^8$	ax^6y^2
47. $(x^2 + y)^{10}$	ax^8y^6
48. $(z^2 - 1)^{12}$	az^6

49. *Think About It* How many terms are in the expansion of $(x + y)^n$?

50. *Think About It* How do the expansions of $(x + y)^n$ and $(x - y)^n$ differ?

In Exercises 51–54, use the Binomial Theorem to expand and simplify the expression.

51. $\left(\sqrt{x} + 3\right)^4$

52. $\left(2\sqrt{t} - 1\right)^3$

53. $(x^{2/3} - y^{1/3})^3$

54. $(u^{3/5} + 2)^5$

In Exercises 55–58, expand the binomial in the difference quotient and simplify.

$$\frac{f(x + h) - f(x)}{h}$$

55. $f(x) = x^3$

56. $f(x) = x^4$

57. $f(x) = \sqrt{x}$

58. $f(x) = \dfrac{1}{x}$

In Exercises 59–64, use the Binomial Theorem to expand the complex number. Simplify your result.

59. $(1 + i)^4$

60. $(2 - i)^5$

61. $(2 - 3i)^6$

62. $\left(5 + \sqrt{-9}\right)^3$

63. $\left(-\dfrac{1}{2} + \dfrac{\sqrt{3}}{2}i\right)^3$

64. $\left(5 - \sqrt{3}i\right)^4$

Probability In Exercises 65–68, consider n independent trials of an experiment where each trial has two possible outcomes called success and failure. The probability of a success on each trial is p and the probability of a failure is $q = 1 - p$. In this context the term $_nC_k\, p^k q^{n-k}$ in the expansion of $(p + q)^n$ gives the probability of k successes in the n trials of the experiment.

65. A fair coin is tossed seven times. To find the probability of obtaining 4 heads, evaluate the term

$$_7C_4\left(\tfrac{1}{2}\right)^4\left(\tfrac{1}{2}\right)^3$$

in the expansion $\left(\tfrac{1}{2} + \tfrac{1}{2}\right)^7$.

66. The probability of a baseball player getting a hit on any given time at bat is $\tfrac{1}{4}$. To find the probability that the player gets 3 hits during the next 10 times at bat, evaluate the term

$$_{10}C_3\left(\tfrac{1}{4}\right)^3\left(\tfrac{3}{4}\right)^7$$

in the expansion $\left(\tfrac{1}{4} + \tfrac{3}{4}\right)^{10}$.

67. The probability of a sales representative making a sale with any one customer is $\tfrac{1}{3}$. The sales representative makes 8 contacts a day. To find the probability of making 4 sales, evaluate the term

$$_8C_4\left(\tfrac{1}{3}\right)^4\left(\tfrac{2}{3}\right)^4$$

in the expansion $\left(\tfrac{1}{3} + \tfrac{2}{3}\right)^8$.

68. To find the probability that the sales representative in Exercise 67 makes 4 sales if the probability of a sale with any one customer is $\tfrac{1}{2}$, evaluate the term

$$_8C_4\left(\tfrac{1}{2}\right)^4\left(\tfrac{1}{2}\right)^4$$

in the expansion $\left(\tfrac{1}{2} + \tfrac{1}{2}\right)^8$.

Approximation In Exercises 69–72, use the Binomial Theorem to approximate the given quantity accurate to three decimal places. For example, in Exercise 69, use the expansion

$$(1.02)^8 = (1 + 0.02)^8 = 1 + 8(0.02) + 28(0.02)^2 + \ldots$$

69. $(1.02)^8$

70. $(2.005)^{10}$

71. $(2.99)^{12}$

72. $(1.98)^9$

Graphical Reasoning In Exercises 73–76, use a graphing utility to obtain the graph of f and g in the same viewing rectangle. What is the relationship between the two graphs? Use the Binomial Theorem to write the polynomial function g in standard form.

73. $f(x) = x^3 - 4x, \quad g(x) = f(x + 4)$

74. $f(x) = -x^4 + 4x^2 - 1, \quad g(x) = f(x - 3)$

75. $f(x) = -x^2 + 3x + 2, \quad g(x) = f(x - 2)$

76. $f(x) = 2x^2 - 4x + 1, \quad g(x) = f(x + 3)$

In Exercises 77–80, prove the given property for all integers r and n where $0 \le r \le n$.

77. $_nC_r = {_nC_{n-r}}$

78. $_nC_0 - {_nC_1} + {_nC_2} - \cdots \pm {_nC_n} = 0$

79. $_{n+1}C_r = {_nC_r} + {_nC_{r-1}}$

80. The sum of the numbers in the nth row of Pascal's Triangle is 2^n.

Exploration In Exercises 81 and 82, use a graphing utility to evaluate and determine which two are equal.

81. (a) $_{12}C_5$

    (b) $\left(_6C_5\right)^2$

    (c) $_{11}C_5 + {_{11}C_4}$

    (d) $_6C_5 + {_6C_5}$

82. (a) $_{25}C_6$

    (b) $2\left(_{25}C_2 + {_{25}C_4}\right)$

    (c) $\displaystyle\sum_{k=0}^{5}\left[\left(_{10}C_k\right)\left(_8C_{5-k}\right)\right]$

    (d) $_{18}C_5$

Graphical Reasoning   In Exercises 83 and 84, use a graphing utility to obtain the graphs of the functions in the given order and in the same viewing rectangle. Compare the graphs. Which two functions have identical graphs and why?

83. (a) $f(x) = (1 - x)^3$

(b) $g(x) = 1 - 3x$

(c) $h(x) = 1 - 3x + 3x^2$

(d) $p(x) = 1 - 3x + 3x^2 - x^3$

84. (a) $f(x) = \left(1 - \frac{1}{2}x\right)^4$

(b) $g(x) = 1 - 2x + \frac{3}{2}x^2$

(c) $h(x) = 1 - 2x + \frac{3}{2}x^2 - \frac{1}{2}x^3$

(d) $p(x) = 1 - 2x + \frac{3}{2}x^2 - \frac{1}{2}x^3 + \frac{1}{16}x^4$

85. *Life Insurance*   The average amount of life insurance per household $f(t)$ (in thousands of dollars) from 1970 through 1992 can be approximated by the model

$$f(t) = 0.1506t^2 + 0.7361t + 21.1374, \qquad 0 \le t \le 22$$

where $t = 0$ represents 1970 (see figure). You want to adjust this model so that $t = 0$ corresponds to 1980 rather than 1970. To do this, you shift the graph of f 10 units *to the left* and obtain

$$g(t) = f(t + 10).$$

(Source: American Council of Life Insurance)

(a) Write $g(t)$ in standard form.

(b) Use a graphing utility to graph f and g in the same viewing rectangle.

86. *Health Maintenance Organizations*   The number of people $f(t)$ (in millions) enrolled in health maintenance organizations in the United States from 1976 through 1992 can be approximated by the model

$$f(t) = 0.1043t^2 + 0.7100t + 4.6852, \qquad 0 \le t \le 16$$

where $t = 0$ represents 1976 (see figure). You want to adjust this model so that $t = 0$ corresponds to 1980 rather than 1976. To do this, you shift the graph of f four units *to the left* and obtain

$$g(t) = f(t + 4).$$

(Source: Group Health Insurance Association of America)

(a) Write $g(t)$ in standard form.

(b) Use a graphing utility to graph f and g in the same viewing rectangle.

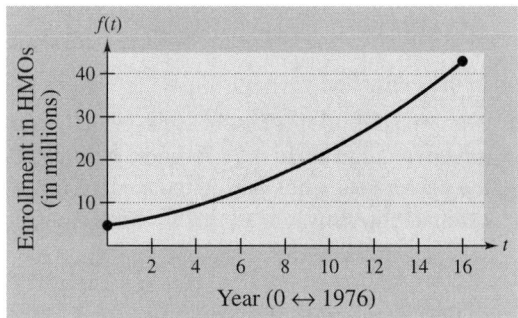

Review   Solve Exercises 87–90 as a review of the skills and problem-solving techniques you learned in previous sections. Describe the relationship between the graphs of f and g.

87. $g(x) = f(x) + 8$

88. $g(x) = f(x - 3)$

89. $g(x) = f(-x)$

90. $g(x) = -f(x)$

9.6   Counting Principles

Simple Counting Problems / Counting Principles / Permutations / Combinations

Simple Counting Problems

This and the next section of this chapter present a brief introduction to some of the basic counting principles and their application to probability. In the next section, you will see that much of probability has to do with counting the number of ways an event can occur.

Real Life

EXAMPLE 1   **Selecting Pairs of Numbers at Random**

Eight pieces of paper are numbered from 1 to 8 and placed in a box. One piece of paper is drawn from the box, its number is written down, and the piece of paper is replaced in the box. Then, a piece of paper is again drawn from the box, and its number is written down. Finally, the two numbers are added together. How many different ways can a total of 12 be obtained?

Solution

To solve this problem, count the different ways that a total of 12 can be obtained using two numbers from 1 to 8.

First number	4	5	6	7	8
Second number	8	7	6	5	4

From this list, you can see that a total of 12 can occur in five different ways.

Real Life

EXAMPLE 2   **Selecting Pairs of Numbers at Random**

Note   The difference between the counting problems in Examples 1 and 2 can be distinguished by saying that the random selection in Example 1 occurs **with replacement,** whereas the random selection in Example 2 occurs **without replacement,** which eliminates the possibility of choosing two 6's.

Eight pieces of paper are numbered from 1 to 8 and placed in a box. Two pieces of paper are drawn from the box, and the numbers on the paper are written down and totaled. How many different ways can a total of 12 be obtained?

Solution

To solve this problem, count the different ways that a total of 12 can be obtained *using two different numbers* from 1 to 8.

First number	4	5	7	8
Second number	8	7	5	4

Thus, a total of 12 can be obtained in four different ways.

Counting Principles

Examples 1 and 2 describe simple counting problems in which you can *list* each possible way that an event can occur. When it is possible, this is always the best way to solve a counting problem. However, some events can occur in so many different ways that it is not feasible to write out the entire list. In such cases, you must rely on formulas and counting principles. The most important of these is the **Fundamental Counting Principle.**

Note The Fundamental Counting Principle can be extended to three or more events. For instance, the number of ways that three events E_1, E_2, and E_3 can occur is $m_1 \cdot m_2 \cdot m_3$.

> ### Fundamental Counting Principle
>
> Let E_1 and E_2 be two events. The first event E_1 can occur in m_1 different ways. After E_1 has occurred, E_2 can occur in m_2 different ways. The number of ways that the two events can occur is $m_1 \cdot m_2$.

Real Life

EXAMPLE 3   **Using the Fundamental Counting Principle**

How many different pairs of letters from the English alphabet are possible?

Solution
This experiment has two events. The first event is the choice of the first letter, and the second event is the choice of the second letter. Because the English alphabet contains 26 letters, it follows that the number of letter pairs is $26 \cdot 26 = 676$.

Real Life

EXAMPLE 4   **Using the Fundamental Counting Principle**

Telephone numbers in the United States have 10 digits. The first three are the *area code* and the next seven are the *local telephone number.* How many different telephone numbers are possible within each area code? (Note that a local telephone number cannot begin with 0 or 1.)

Solution
Because the first digit cannot be 0 or 1, there are only eight choices for the first digit. For each of the other six digits, there are 10 choices.

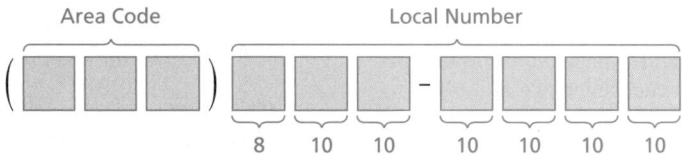

Thus, the number of local telephone numbers that are possible within each area code is $8 \cdot 10 \cdot 10 \cdot 10 \cdot 10 \cdot 10 \cdot 10 = 8,000,000$.

Permutations

One important application of the Fundamental Counting Principle is in determining the number of ways that n elements can be arranged (in order). An ordering of n elements is called a **permutation** of the elements.

Definition of Permutation

A **permutation** of n different elements is an ordering of the elements such that one element is first, one is second, one is third, and so on.

EXAMPLE 5   **Finding the Number of Permutations of *n* Elements**

How many permutations are possible for the letters A, B, C, D, E, and F?

Solution

Consider the following reasoning.

First position:	Any of the *six* letters.
Second position:	Any of the remaining *five* letters.
Third position:	Any of the remaining *four* letters.
Fourth position:	Any of the remaining *three* letters.
Fifth position:	Either of the remaining *two* letters.
Sixth position:	The *one* remaining letter.

Thus, the number of choices for the six positions are as follows.

Permutations of six letters

6	5	4	3	2	1

The total number of permutations of the six letters is $6! = 720$.

Number of Permutations of *n* Elements

The number of permutations of n elements is given by

$$n \cdot (n - 1) \cdots 4 \cdot 3 \cdot 2 \cdot 1 = n!.$$

In other words, there are $n!$ different ways that n elements can be ordered.

Occasionally, you are interested in ordering a *subset* of a collection of elements rather than the entire collection. For example, you might want to choose (and order) *r* elements out of a collection of *n* elements. Such an ordering is called a **permutation of *n* elements taken *r* at a time.**

Real Life

EXAMPLE 6   **Counting Horse Race Finishes**

Eight horses are running in a race. In how many different ways can these horses come in first, second, and third? (Assume that there are no ties.)

Solution

Here are the different possibilities.

Win (first position):   *Eight* choices
Place (second position):   *Seven* choices
Show (third position):   *Six* choices

Using the Fundamental Counting Principle, multiply these three numbers together to obtain the following.

Different orders of horses

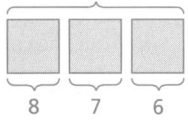

8   7   6

Thus, there are $8 \cdot 7 \cdot 6 = 336$ different orders.

Permutations of *n* Elements Taken *r* at a Time

The number of permutations of *n* elements taken *r* at a time is

$$_nP_r = \frac{n!}{(n-r)!} = n(n-1)(n-2)\cdots(n-r+1).$$

Using this formula, you can rework Example 6 to find that the number of permutations of eight horses taken three at a time is

$$_8P_3 = \frac{8!}{5!}$$

$$= \frac{8 \cdot 7 \cdot 6 \cdot 5!}{5!}$$

$$= 336$$

which is the same answer obtained in Example 6.

Remember that for permutations, order is important. Thus, if you are looking at the possible permutations of the letters A, B, C, and D taken three at a time, the permutations (A, B, D) and (B, A, D) would be different because the *order* of the elements is different.

Suppose, however, that you are asked to find the possible permutations of the letters A, A, B, and C. The total number of permutations of the four letters would be $_4P_4 = 4!$. However, not all of these arrangements would be *distinguishable* because there are two A's in the list. To find the number of distinguishable permutations, you can use the following formula.

Distinguishable Permutations

Suppose a set of n objects has n_1 of one kind of object, n_2 of a second kind, n_3 of a third kind, and so on, with $n = n_1 + n_2 + n_3 + \cdots + n_k$. Then the number of **distinguishable permutations** of the n objects is

$$\frac{n!}{n_1! \cdot n_2! \cdot n_3! \cdots n_k!}.$$

EXAMPLE 7 ▱ **Distinguishable Permutations**

In how many distinguishable ways can the letters in BANANA be written?

Solution
This word has six letters, of which three are A's, two are N's, and one is a B. Thus, the number of distinguishable ways the letters can be written is

$$\frac{6!}{3! \cdot 2! \cdot 1!} = \frac{6 \cdot 5 \cdot 4 \cdot 3!}{3! \cdot 2!} = 60.$$

The 60 different "words" are as follows.

AAABNN	AAANBN	AAANNB	AABANN	AABNAN	AABNNA
AANABN	AANANB	AANBAN	AANBNA	AANNAB	AANNBA
ABAANN	ABANAN	ABANNA	ABNAAN	ABNANA	ABNNAA
ANAABN	ANAANB	ANABAN	ANABNA	ANANAB	ANANBA
ANBAAN	ANBANA	ANBNAA	ANNAAB	ANNABA	ANNBAA
BAAANN	BAANAN	BAANNA	BANAAN	BANANA	BANNAA
BNAAAN	BNAANA	BNANAA	BNNAAA	NAAABN	NAAANB
NAABAN	NAABNA	NAANAB	NAANBA	NABAAN	NABANA
NABNAA	NANAAB	NANABA	NANBAA	NBAAAN	NBAANA
NBANAA	NBNAAA	NNAAAB	NNAABA	NNABAA	NNBAAA

▱

Combinations

When one counts the number of possible permutations of a set of elements, order is important. As a final topic in this section, we look at a method of selecting subsets of a larger set in which order *is not* important. Such subsets are called **combinations of *n* elements taken *r* at a time.** For instance, the combinations

$$\{A, B, C\} \quad \text{and} \quad \{B, A, C\}$$

are equivalent because both sets contain the same three elements, and the order in which the elements are listed is not important. Hence, you would count only one of the two sets. A common example of how a combination occurs is a card game in which the player is free to reorder the cards after they have been dealt.

EXAMPLE 8   ◢   **Combinations of *n* Elements Taken *r* at a Time**

In how many different ways can three letters be chosen from the letters A, B, C, D, and E? (The order of the three letters is not important.)

Solution
The following subsets represent the different combinations of three letters that can be chosen from five letters.

$$\{A, B, C\} \qquad \{A, B, D\}$$
$$\{A, B, E\} \qquad \{A, C, D\}$$
$$\{A, C, E\} \qquad \{A, D, E\}$$
$$\{B, C, D\} \qquad \{B, C, E\}$$
$$\{B, D, E\} \qquad \{C, D, E\}$$

From this list, you can conclude that there are 10 different ways that three letters can be chosen from five letters.   ◢

Most graphing utilities have keys that will evaluate the formulas for the number of permutations or combinations of *n* elements taken *r* at a time. For instance, on a *TI-82* or *TI-83*, you can evaluate $_8C_5$ as follows.

8 [MATH] (PRB) (3 : nCr) 5 [ENTER]

The display should be 56. You can evaluate $_8P_5$ (the number of permutations of eight elements taken five at a time) in a similar way.

Combinations of n Elements Taken r at a Time

The number of combinations of n elements taken r at a time is

$$_nC_r = \frac{n!}{(n-r)!r!}.$$

Note that the formula for $_nC_r$ is the same one given for binomial coefficients. To see how this formula is used, let's solve the counting problem in Example 8. In that problem, you are asked to find the number of combinations of five elements taken three at a time. Thus, $n = 5$, $r = 3$, and the number of combinations is

$$_5C_3 = \frac{5!}{2!3!} = \frac{5 \cdot \overset{2}{\cancel{4}} \cdot \cancel{3!}}{\cancel{2} \cdot 1 \cdot \cancel{3!}} = 10$$

which is the same answer obtained in Example 8.

Real Life

EXAMPLE 9 ▱ **Counting Card Hands**

A standard poker hand consists of five cards dealt from a deck of 52. How many different poker hands are possible? (After the cards are dealt, the player may reorder them, and therefore order is not important.)

Solution

You can find the number of different poker hands by using the formula for the number of combinations of 52 elements taken five at a time, as follows.

$$_{52}C_5 = \frac{52!}{47!5!} = \frac{52 \cdot 51 \cdot 50 \cdot 49 \cdot 48 \cdot 47!}{5 \cdot 4 \cdot 3 \cdot 2 \cdot 1 \cdot 47!} = 2{,}598{,}960 \qquad ▱$$

Group Activity

Problem Posing

According to NASA, each space shuttle astronaut consumes an average of 3000 calories per day. An evening meal normally consists of a main dish, a vegetable dish, and two different desserts. The space shuttle food list contains 10 items classified as main dishes, eight vegetable dishes, and 13 desserts. How many different evening meal menus are possible? Create two other problems that could be asked about the evening meal menus, and solve them.   (Source: NASA)

9.6 /// EXERCISES

Random Selection In Exercises 1–8, determine the number of ways a computer can randomly generate one or more such integers from 1 through 12.

1. An odd integer

2. An even integer

3. A prime integer

4. An integer that is greater than 9

5. An integer that is divisible by 4

6. An integer that is divisible by 7

7. Two integers whose sum is 8

8. Two *distinct* integers whose sum is 8

9. *Entertainment Systems* A customer can choose one of two amplifiers, one of four disc players, and one of six speaker models for an entertainment system. Determine the number of possible system configurations.

10. *Computer Systems* A customer in a computer store can choose one of three monitors, one of two keyboards, and one of four computers. If all the choices are compatible, determine the number of possible system configurations.

11. *Job Applicants* A college needs two additional faculty members: a chemist and a statistician. In how many ways can these positions be filled if there are three applicants for the chemistry position and four applicants for the statistics position?

12. *Course Schedule* A college student is preparing a course schedule for the next semester. The student may select one of two mathematics courses, one of three science courses, and one of five courses from the social sciences and humanities. How many schedules are possible?

13. *True-False Exam* In how many ways can a six-question true-false exam be answered? (Assume that no questions are omitted.)

14. *True-False Exam* In how many ways can a 10-question true-false exam be answered? (Assume that no questions are omitted.)

15. *Toboggan Ride* Four people are lining up for a ride on a toboggan, but only two of the four are willing to take the first position. With that constraint, in how many ways can the four people be seated on the toboggan?

16. *Aircraft Boarding* Ten people are boarding an aircraft. Four have tickets for first class and board before those in the economy class. In how many ways can the 10 people board the aircraft?

17. *License Plate Numbers* In a certain state the automobile license plates consist of two letters followed by a four-digit number. How many distinct license plate numbers can be formed?

18. *License Plate Numbers* In a certain state the automobile license plates consist of two letters followed by a four-digit number. To avoid confusion between "O" and "zero" and "I" and "one," the letters "O" and "I" are not used. How many distinct license plate numbers can be formed?

19. *Three-Digit Numbers* How many three-digit numbers can be formed under the following conditions?

    (a) The leading digit cannot be zero.

    (b) The leading digit cannot be zero and no repetition of digits is allowed.

    (c) The leading digit cannot be zero and the number must be a multiple of 5.

    (d) The number is at least 400.

20. *Four-Digit Numbers* How many four-digit numbers can be formed under the following conditions?

    (a) The leading digit cannot be zero.

    (b) The leading digit cannot be zero and no repetition of digits is allowed.

    (c) The leading digit cannot be zero and the number must be less than 5000.

    (d) The leading digit cannot be zero and the number must be even.

21. *Combination Lock* A combination lock will open when the right choice of three numbers (from 1 to 40, inclusive) is selected. How many different lock combinations are possible?

22. *Combination Lock* A combination lock will open when the right choice of three numbers (from 1 to 50, inclusive) is selected. How many different lock combinations are possible?

23. *Concert Seats* Three couples have reserved seats in a given row for a concert. In how many different ways can they be seated if

(a) there are no seating restrictions?

(b) the two members of each couple wish to sit together?

24. *Single File* In how many orders can three girls and two boys walk through a doorway single file if

(a) there are no restrictions?

(b) the girls walk through before the boys?

In Exercises 25–30, evaluate $_nP_r$.

25. $_4P_4$

26. $_5P_5$

27. $_8P_3$

28. $_{20}P_2$

29. $_5P_4$

30. $_7P_4$

In Exercises 31 and 32, solve for n.

31. $14 \cdot {}_nP_3 = {}_{n+2}P_4$

32. $_nP_5 = 18 \cdot {}_{n-2}P_4$

In Exercises 33–38, evaluate using a calculator.

33. $_{20}P_5$

34. $_{100}P_5$

35. $_{100}P_3$

36. $_{10}P_8$

37. $_{20}C_5$

38. $_{10}C_7$

39. *Think About It* Can your calculator evaluate $_{100}P_{80}$? If not, explain why.

40. *Essay* Explain in words the meaning of $_nP_r$.

41. Write all permutations of the letters A, B, C, and D.

42. Write all the permutations of the letters A, B, C, and D if the letters B and C must remain between the letters A and D.

43. *Posing for a Photograph* In how many ways can five children line up in a row?

44. *Riding in a Car* In how many ways can six people sit in a six-passenger car?

45. *Choosing Officers* From a pool of 12 candidates, the offices of president, vice-president, secretary, and treasurer will be filled. In how many different ways can the offices be filled?

46. *Assembly Line Production* Four processes are involved in assembling a certain product, and they can be performed in any order. The management wants to test each order to determine which is the least time-consuming. How many different orders will have to be tested?

In Exercises 47–50, find the number of distinguishable permutations of the group of letters.

47. A, A, G, E, E, E, M

48. B, B, B, T, T, T, T, T

49. A, L, G, E, B, R, A

50. M, I, S, S, I, S, S, I, P, P, I

51. Write all the possible selections of two letters that can be formed from the letters A, B, C, D, E, and F. (The order of the two letters is not important.)

52. Write all the possible selections of three letters that can be formed from the letters A, B, C, D, E, and F. (The order of the three letters is not important.)

53. *Forming an Experimental Group* In order to conduct a certain experiment, four students are randomly selected from a class of 20. How many different groups of four students are possible?

54. *Test Questions* You can answer any 10 questions from a total of 12 questions on an exam. In how many different ways can you select the questions?

55. *Lottery Choices* There are 40 numbers in a particular state lottery. In how many ways can a player select six of the numbers?

56. *Lottery Choices* There are 50 numbers in a particular state lottery. In how many ways can a player select six of the numbers?

57. *Number of Subsets*   How many subsets of four elements can be formed from a set of 100 elements?

58. *Number of Subsets*   How many subsets of five elements can be formed from a set of 80 elements?

59. *Geometry*   Three points that are not on a line determine three lines. How many lines are determined by seven points, no three of which are on a line?

60. *Defective Units*   A shipment of 12 microwave ovens contains three defective units. In how many ways can a vending company purchase four of these units and receive (a) all good units, (b) two good units, and (c) at least two good units?

61. *Job Applicants*   An employer interviews eight people for four openings in the company. Three of the eight people are women. If all eight are qualified, in how many ways can the employer fill the four positions if (a) the selection is random and (b) exactly two women are selected?

62. *Poker Hand*   Five cards are selected from an ordinary deck of 52 playing cards. In how many ways càn you get a full house? (A full house consists of three of one kind and two of another. For example, A-A-A-5-5 and K-K-K-10-10 are full houses.)

63. *Forming a Committee*   Four people are to be selected at random from a group of four couples. In how many ways can this be done, given the following conditions?

   (a) There are no restrictions.

   (b) The group must have at least one couple.

   (c) Each couple must be represented in the group.

64. *Interpersonal Relationships*   The complexity of the interpersonal relationships increases dramatically as the size of a group increases. Determine the number of different two-person relationships in a group of people of size (a) 3, (b) 8, (c) 12, and (d) 20.

In Exercises 65–68, find the number of diagonals of the polygon. (A line segment connecting any two non-adjacent vertices is called a *diagonal* of the polygon.)

65. Pentagon   **66.** Hexagon

67. Octagon   **68.** Decagon

 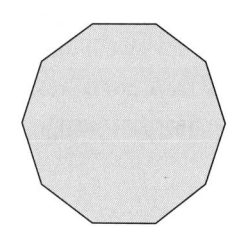

In Exercises 69–72, prove the identity.

69. $_nP_{n-1} = {_nP_n}$ **70.** $_nC_n = {_nC_0}$

71. $_nC_{n-1} = {_nC_1}$ **72.** $_nC_r = \dfrac{_nP_r}{r!}$

Review   Solve Exercises 73–76 as a review of the skills and problem-solving techniques you learned in previous sections. Solve the equation. (Round approximate answers to two decimal places.)

73. $\sqrt{x-3} = x-6$ **74.** $\dfrac{4}{t} + \dfrac{3}{2t} = 1$

75. $\log_2(x-3) = 5$ **76.** $e^{x/3} = 16$

9.7   Probability

*The Probability of an Event   / Mutually Exclusive Events /
Independent Events   / The Complement of an Event*

The Probability of an Event

Any happening whose result is uncertain is called an **experiment.** The possible results of the experiment are **outcomes,** the set of all possible outcomes of the experiment is the **sample space** of the experiment, and any subcollection of a sample space is an **event.**

For instance, when a six-sided die is tossed, the sample space can be represented by the numbers from 1 through 6. For this experiment, each of the outcomes is *equally likely.*

To describe sample spaces in such a way that each outcome is equally likely, you must sometimes distinguish between various outcomes in ways that appear artificial. Example 1 illustrates such a situation.

Blaise Pascal (1623–1662) was a French mathematician and scientist. He, along with Pierre de Fermat, laid the foundation for the mathematical theory of probability.

EXAMPLE 1   **Finding the Sample Space**   *Real Life*

Find the sample space for each of the following.

a. One coin is tossed.   **b.** Two coins are tossed.   **c.** Three coins are tossed.

Solution

a. Because the coin will land either heads up (denoted by H) or tails up (denoted by T), the sample space is $S = \{H, T\}$.

b. Because either coin can land heads up or tails up, the possible outcomes are as follows.

　　HH = heads up on both coins

　　HT = heads up on first coin and tails up on second coin

　　TH = tails up on first coin and heads up on second coin

　　TT = tails up on both coins

Thus, the sample space is $S = \{HH, HT, TH, TT\}$. Note that this list distinguishes between the two cases HT and TH, even though these two outcomes appear to be similar.

c. Following the notation of part (b), the sample space is
$$S = \{HHH, HHT, HTH, HTT, THH, THT, TTH, TTT\}.$$

To calculate the probability of an event, count the number of outcomes in the event and in the sample space. The *number of outcomes* in event E is denoted by $n(E)$ and the number of outcomes in the sample space S is denoted by $n(S)$. The probability that event E will occur is given by $n(E)/n(S)$.

The Probability of an Event

If an event E has $n(E)$ equally likely outcomes and its sample space S has $n(S)$ equally likely outcomes, the **probability** of event E is

$$P(E) = \frac{n(E)}{n(S)}.$$

Because the number of outcomes in an event must be less than or equal to the number of outcomes in the sample space, the probability of an event must be a number from 0 to 1, inclusive. That is,

$$0 \le P(E) \le 1.$$

If $P(E) = 0$, event E *cannot occur*, and E is called an **impossible event.** If $P(E) = 1$, event E *must occur*, and E is called a **certain event.**

Real Life

EXAMPLE 2   **Finding the Probability of an Event**

a. Two coins are tossed. What is the probability that both land heads up?

b. A card is drawn from a standard deck of playing cards. What is the probability that it is an ace?

Solution

a. Following the procedure in Example 1(b), let

$$E = \{HH\}$$

and

$$S = \{HH, HT, TH, TT\}.$$

The probability of getting two heads is

$$P(E) = \frac{n(E)}{n(S)} = \frac{1}{4}.$$

b. Because there are 52 cards in a standard deck of playing cards and there are four aces (one in each suit), the probability of drawing an ace is

$$P(E) = \frac{n(E)}{n(S)} = \frac{4}{52} = \frac{1}{13}.$$

Figure 9.5

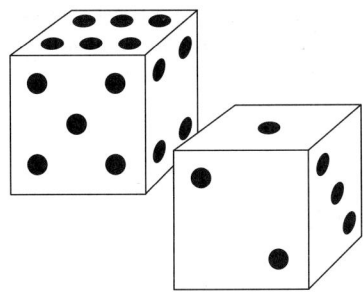

EXAMPLE 3 ▱ **Finding the Probability of an Event**

Two six-sided dice are tossed. What is the probability that the total of the two dice is 7? (See Figure 9.5.)

Solution

Because there are six possible outcomes on each die, you can use the Fundamental Counting Principle to conclude that there are 6 • 6 or 36 different outcomes when two dice are tossed. To find the probability of rolling a total of 7, you must first count the number of ways this can occur.

First Die	1	2	3	4	5	6
Second Die	6	5	4	3	2	1

Thus, a total of 7 can be rolled in six ways, which means that the probability of rolling a 7 is

$$P(E) = \frac{n(E)}{n(S)} = \frac{6}{36} = \frac{1}{6}.$$

You could have written out each sample space in Examples 2 and 3 and simply counted the outcomes in the desired events. For larger sample spaces, however, you should use the counting principles discussed in Section 9.6.

EXAMPLE 4 ▱ **Finding the Probability of an Event**

Twelve-sided dice, as shown in Figure 9.6, can be constructed (in the shape of regular dodecahedrons) so that each of the numbers from 1 to 6 appears twice on each die. Prove that these dice can be used in any game requiring ordinary six-sided dice without changing the probabilities of different outcomes.

Figure 9.6

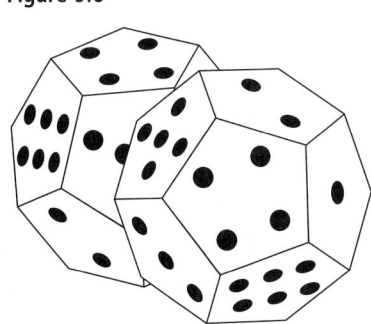

Solution

For an ordinary six-sided die, each of the numbers 1, 2, 3, 4, 5, and 6 occurs only once, so the probability of any particular number coming up is

$$P(E) = \frac{n(E)}{n(S)} = \frac{1}{6}.$$

For one of the 12-sided dice, each number occurs twice, so the probability of any particular number coming up is

$$P(E) = \frac{n(E)}{n(S)} = \frac{2}{12} = \frac{1}{6}.$$

Real Life

EXAMPLE 5 ▭ The Probability of Winning a Lottery

In a state lottery, a player chooses six different numbers from 1 to 40. If these six numbers match the six numbers drawn by the lottery commission, the player wins (or shares) the top prize. What is the probability of winning the top prize?

Solution

To find the number of elements in the sample space, use the formula for the number of combinations of 40 elements taken six at a time.

$$n(S) = {}_{40}C_6 = \frac{40 \cdot 39 \cdot 38 \cdot 37 \cdot 36 \cdot 35}{6 \cdot 5 \cdot 4 \cdot 3 \cdot 2 \cdot 1} = 3{,}838{,}380$$

If a person buys only one ticket, the probability of winning is

$$P(E) = \frac{n(E)}{n(S)} = \frac{1}{3{,}838{,}380}.$$

▭

Real Life

EXAMPLE 6 ▭ Random Selection

The numbers of colleges and universities in the United States in 1992 are shown in Figure 9.7. One institution is selected at random. What is the probability that the institution is in one of the three southern regions?   (Source: U.S. National Center for Education Statistics)

Solution

From the figure, the total number of colleges and universities is 3628. Because there are $600 + 272 + 289 = 1161$ colleges and universities in the three southern regions, the probability that the institution is in one of these regions is

$$P(E) = \frac{n(E)}{n(S)} = \frac{1161}{3628} \approx 0.320.$$

Figure 9.7

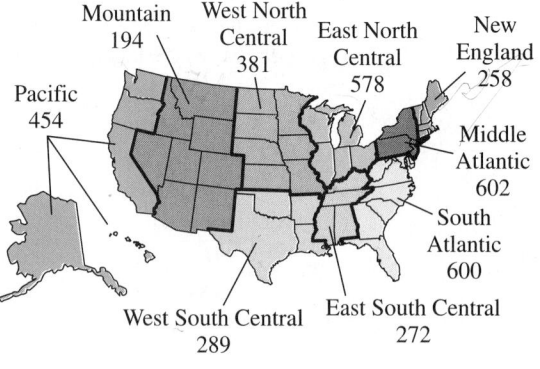

Mountain 194
West North Central 381
East North Central 578
New England 258
Pacific 454
Middle Atlantic 602
South Atlantic 600
West South Central 289
East South Central 272

▭

Mutually Exclusive Events

Two events A and B (from the same sample space) are **mutually exclusive** if A and B have no outcomes in common. In the terminology of sets, the **intersection of A and B** is the empty set and

$$P(A \cap B) = 0.$$

For instance, if two dice are tossed, the event A of rolling a total of 6 and the event B of rolling a total of 9 are mutually exclusive. To find the probability that one or the other of two mutually exclusive events will occur, you can *add* their individual probabilities.

Probability of the Union of Two Events

If A and B are events in the same sample space, the probability of A *or* B occurring is given by

$$P(A \cup B) = P(A) + P(B) - P(A \cap B).$$

If A and B are mutually exclusive, then

$$P(A \cup B) = P(A) + P(B).$$

Real Life

EXAMPLE 7   ▱   **The Probability of a Union**

One card is selected from a standard deck of 52 playing cards. What is the probability that the card is either a heart or a face card?

Solution

Because the deck has 13 hearts, the probability of selecting a heart (event A) is $P(A) = \frac{13}{52}$. Similarly, because the deck has 12 face cards, the probability of selecting a face card (event B) is $P(B) = \frac{12}{52}$. Because three of the cards are hearts and face cards (see Figure 9.8), it follows that $P(A \cap B) = \frac{3}{52}$. Finally, applying the formula for the probability of the union of two events, you can conclude that the probability of selecting a heart or a face card is

$$P(A \cup B) = P(A) + P(B) - P(A \cap B)$$

$$= \frac{13}{52} + \frac{12}{52} - \frac{3}{52}$$

$$= \frac{22}{52}$$

$$\approx 0.423.$$

▱

Figure 9.8

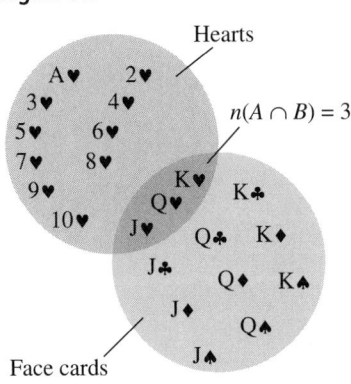

Hearts

$n(A \cap B) = 3$

Face cards

Real Life

EXAMPLE 8 ▭ Probability of Mutually Exclusive Events

The personnel department of a company has compiled data on employees' numbers of years of service. The results are shown in the table.

Years of Service	Number of Employees
0–4	157
5–9	89
10–14	74
15–19	63
20–24	42
25–29	38
30–34	37
35–39	21
40–44	8

If an employee is chosen at random, what is the probability that the employee has had nine or fewer years of service?

Solution

To begin, add the number of employees and find that the total is 529. Next, let event A represent choosing an employee with 0 to 4 years of service and let event B represent choosing an employee with 5 to 9 years of service. Then

$$P(A) = \frac{157}{529} \quad \text{and} \quad P(B) = \frac{89}{529}.$$

Because A and B have no outcomes in common, you can conclude that these two events are mutually exclusive and that

$$P(A \cup B) = P(A) + P(B) = \frac{157}{529} + \frac{89}{529}$$

$$= \frac{246}{529}$$

$$\approx 0.465.$$

Thus, the probability of choosing an employee who has nine or fewer years of service is about 0.465. ▭

Independent Events

Two events are **independent** if the occurrence of one has no effect on the occurrence of the other. To find the probability that two independent events will occur, *multiply* the probabilities of each. For instance, rolling a total of 12 with two six-sided dice has no effect on the outcome of future rolls of the dice.

> **Probability of Independent Events**
>
> If A and B are independent events, the probability that both A and B will occur is
>
> $$P(A \text{ and } B) = P(A) \cdot P(B).$$

Real Life

EXAMPLE 9   **Probability of Independent Events**

A random number generator on a computer selects three integers from 1 to 20. What is the probability that all three numbers are less than or equal to 5?

Solution
The probability of selecting a number from 1 to 5 is

$$P(A) = \frac{5}{20} = \frac{1}{4}.$$

Thus, the probability that all three numbers are less than or equal to 5 is

$$P(A) \cdot P(A) \cdot P(A) = \left(\frac{1}{4}\right)\left(\frac{1}{4}\right)\left(\frac{1}{4}\right)$$

$$= \frac{1}{64}.$$

Real Life

EXAMPLE 10   **Probability of Independent Events**

In 1992, 56% of the population of the United States were 30 years old or older. Suppose that in a survey, 10 people were chosen at random from the population. What is the probability that all 10 were 30 years old or older?   (Source: U.S. Bureau of the Census)

Solution
Let A represent choosing a person who was 30 years old or older. Because the probability of choosing a person who was 30 years old or older was 0.56, you can conclude that the probability that all 10 people were 30 years old or older is

$$[P(A)]^{10} = (0.56)^{10} \approx 0.0030.$$

9.7 /// EXERCISES

In Exercises 1–6, determine the sample space for the given experiment.

1. A coin and a six-sided die are tossed.

2. A six-sided die is tossed twice and the sum of the points is recorded.

3. A taste tester has to rank three varieties of yogurt, A, B, and C, according to preference.

4. Two marbles are selected from a sack containing two red marbles, two blue marbles, and one black marble. The color of each marble is recorded.

5. Two county supervisors are selected from five supervisors, A, B, C, D, and E, to study a recycling plan.

6. A sales representative makes presentations about a product in three homes per day. In each home there may be a sale (denote by S) or there may be no sale (denote by F).

Heads or Tails   In Exercises 7–10, find the probability in the experiment of tossing a coin three times. Use the sample space $S = \{HHH, HHT, HTH, HTT, THH, THT, TTH, TTT\}$.

7. The probability of getting exactly one tail

8. The probability of getting a head on the first toss

9. The probability of getting at least one head

10. The probability of getting at least two heads

Drawing a Card   In Exercises 11–14, find the probability in the experiment of selecting one card from a standard deck of 52 playing cards.

11. The card is a face card.

12. The card is not a face card.

13. The card is a red face card.

14. The card is a 6 or lower. (Aces are low.)

Tossing a Die   In Exercises 15–20, find the probability in the experiment of tossing a six-sided die twice.

15. The sum is 4.

16. The sum is at least 7.

17. The sum is less than 11.

18. The sum is 2, 3, or 12.

19. The sum is odd and no more than 7.

20. The sum is odd or prime.

Drawing Marbles   In Exercises 21–24, find the probability in the experiment of drawing two marbles (without replacement) from a bag containing one green, two yellow, and three red marbles.

21. Both marbles are red.

22. Both marbles are yellow.

23. Neither marble is yellow.

24. The marbles are of different colors.

In Exercises 25 and 26, you are given the probability that an event *will* happen. Find the probability that the event *will not* happen.

25. $P(E) = 0.7$

26. $P(E) = 0.36$

In Exercises 27 and 28, you are given the probability that an event *will not* happen. Find the probability that the event *will* happen.

27. $P(E') = 0.15$

28. $P(E') = 0.84$

29. *Graphical Reasoning* At the end of 1994 there were approximately 2.5 million minimum-wage workers in the United States. The figure gives the age profile of these workers. (Source: U.S. Bureau of Labor Statistics)

  (a) Estimate the number of minimum-wage workers in the age category 16–19.

  (b) A person is selected at random from the population of minimum-wage workers. What is the probability that the person is in the 25–34 age group?

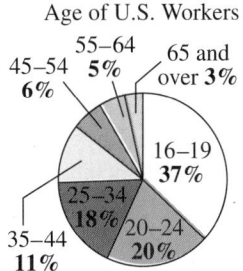

Age of U.S. Workers

30. *Graphical Reasoning* In 1991 there were approximately 101 million workers in the civilian labor force in the United States. The figure gives the educational attainments of these workers. (Source: U.S. Bureau of Labor Statistics)

  (a) Estimate the number of workers whose highest educational attainment was a high school education.

  (b) A person is selected at random from the civilian work force. What is the probability that the person has 1 to 3 years of college education?

  (c) A person is selected at random from the civilian work force. What is the probability that the person has more than a high school education?

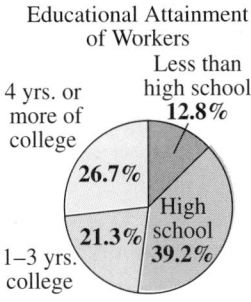

Educational Attainment
of Workers

31. *Data Analysis* A study of the effectiveness of a flu vaccine was conducted with a sample of 500 people. Some in the study were given no vaccine, some were given one injection, and some were given two injections. The results of the study are given below.

	No Vaccine	One Injection	Two Injections	Total
Flu	7	2	13	22
No flu	149	52	277	478
Total	156	54	290	500

A person is selected at random from the sample. Find the specified probability.

  (a) The person had two injections.

  (b) The person did not get the flu.

  (c) The person got the flu and had one injection.

32. *Data Analysis* A sample of 100 college students were interviewed to determine their political party affiliations and whether they favored a balanced-budget amendment to the constitution. The results of the study are given in the table.

	Favor	Not favor	Unsure	Total
Democrat	23	25	7	55
Republican	32	9	4	45
Total	55	34	11	100

A person is selected at random from the sample. Find the probability that the described person is selected.

  (a) A person who doesn't favor the amendment

  (b) A Republican

  (c) A Democrat who favors the amendment

33. *Alumni Association* A college sends a survey to selected members of the class of 1995. Of the 1254 people who graduated that year, 672 are women, and of those, 124 went on to graduate school. Of the 582 male graduates, 198 went on to graduate school. If an alumni member is selected at random, what is the probability that the person is (a) female, (b) male, and (c) female and did not attend graduate school?

34. *Post–High School Education*   In a high school graduating class of 72 students, 28 are on the honor roll. Of these, 18 are going on to college, and of the other 44 students, 12 are going on to college. If a student is selected at random from the class, what is the probability that the person chosen is (a) going to college, (b) not going to college, and (c) on the honor roll, but not going to college?

35. *Winning an Election*   Taylor, Moore, and Jenkins are candidates for public office. It is estimated that Moore and Jenkins have about the same probability of winning, and Taylor is believed to be twice as likely to win as either of the others. Find the probability of each candidate winning the election.

36. *Winning an Election*   Three people have been nominated for president of a class. From a poll, it is estimated that the first has a 37% chance of winning and the second has a 44% chance of winning. What is the probability that the third candidate will win?

In Exercises 37–48, the sample spaces are large and you should use the counting principles discussed in Section 9.6.

37. *Preparing for a Test*   A class is given a list of 20 study problems from which 10 will be part of an upcoming exam. If a given student knows how to solve 15 of the problems, find the probability that the student will be able to answer (a) all 10 questions on the exam, (b) exactly eight questions on the exam, and (c) at least nine questions on the exam.

38. *Preparing for a Test*   A class is given a list of eight study problems from which five will be part of an upcoming exam. If a given student knows how to solve six of the problems, find the probability that the student will be able to answer (a) all five questions on the exam, (b) exactly four questions on the exam, and (c) at least four questions on the exam.

39. *Letter Mix-Up*   Four letters and envelopes are addressed to four different people. If the letters are randomly inserted into the envelopes, what is the probability that (a) exactly one is inserted in the correct envelope and (b) at least one is inserted in the correct envelope?

40. *Payroll Mix-Up*   Five paychecks and envelopes are addressed to five different people. If the paychecks are randomly inserted into the envelopes, what is the probability that (a) exactly one is inserted in the correct envelope and (b) at least one is inserted in the correct envelope?

41. *Game Show*   On a game show you are given five digits to arrange in the proper order to form the price of a car. If you are correct, you win the car. What is the probability of winning, given the following conditions?

   (a) You guess the position of each digit.

   (b) You know the first digit and guess the others.

42. *Game Show*   On a game show you are given four digits to arrange in the proper order to form the price of a car. If you are correct, you win the car. What is the probability of winning, given the following conditions?

   (a) You guess the position of each digit.

   (b) You know the first digit and guess the others.

43. *Drawing Cards from a Deck*   Two cards are selected at random from an ordinary deck of 52 playing cards. Find the probability that two aces are selected, given the following conditions.

   (a) The cards are drawn in sequence, with the first card being replaced and the deck reshuffled prior to the second drawing.

   (b) The two cards are drawn consecutively, without replacement.

44. *Poker Hand*   Five cards are drawn from an ordinary deck of 52 playing cards. What is the probability of getting a full house?

45. *Defective Units*   A shipment of 12 microwave ovens contains three defective units. A vending company has ordered four of these units, and because all are packaged identically, the selection will be random.

   (a) What is the probability that all four units are good?

   (b) What is the probability that exactly two units are good?

   (c) What is the probability that at least two units are good?

46. *Defective Units*   A shipment of 20 compact disc players contains four defective units. A retail outlet has ordered five of these units.

(a) What is the probability that all five units are good?

(b) What is the probability that exactly four units are good?

(c) What is the probability that at least one unit is good?

47. *Random Number Generator*   Two integers (from 1 through 30) are chosen by a random number generator. What is the probability that (a) the numbers are both even, (b) one number is even and one is odd, (c) both numbers are less than 10, and (d) the same number is chosen twice?

48. *Random Number Generator*   Two integers (from 1 through 40) are chosen by a random number generator. What is the probability that (a) the numbers are both even, (b) one number is even and one is odd, (c) both numbers are less than 30, and (d) the same number is chosen twice?

49. *Backup System*   A space vehicle has an independent backup system for one of its communication networks. The probability that either system will function satisfactorily for the duration of a flight is 0.985. What is the probability that during a given flight (a) both systems function satisfactorily, (b) at least one system functions satisfactorily, and (c) both systems fail?

50. *Backup Vehicle*   A fire company keeps two rescue vehicles to serve the community. Because of the demand on the company's time and the chance of mechanical failure, the probability that a specific vehicle is available when needed is 90%. If the availability of one vehicle is *independent* of the other, find the probability that (a) both vehicles are available at a given time, (b) neither vehicle is available at a given time, and (c) at least one vehicle is available at a given time.

51. *Making a Sale*   A sales representative makes sales at approximately one-fourth of all calls. If, on a given day, the representative contacts five potential clients, what is the probability that a sale will be made with (a) all five contacts, (b) none of the contacts, and (c) at least one contact?

52. *A Boy or a Girl?*   Assume that the probability of the birth of a child of a particular sex is 50%. In a family with four children, what is the probability that (a) all the children are boys, (b) all the children are the same sex, and (c) there is at least one boy?

53. *Is That Cash or Charge?*   Suppose that the methods used by shoppers to pay for merchandise are as shown in the pie graph. If two shoppers are chosen at random, what is the probability that both shoppers paid for their purchases only in cash?

How Shoppers Pay
for Merchandise

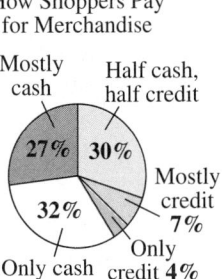

54. *Flexible Work Hours*   In a survey, people were asked if they would prefer to work flexible hours—even if it meant slower career advancement—so they could spend more time with their families. The results of the survey are shown in the figure below. Suppose that three people from the survey were chosen at random. What is the probability that all three people would prefer flexible work hours?

Flexible Work Hours

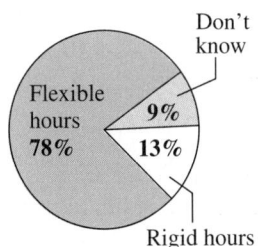

55. *Geometry*   You and a friend agree to meet at your favorite fast-food restaurant between 5:00 and 6:00 P.M. The one who arrives first will wait 15 minutes for the other, after which the first person will leave (see figure). What is the probability that the two of you will actually meet, assuming that your arrival times are random within the hour?

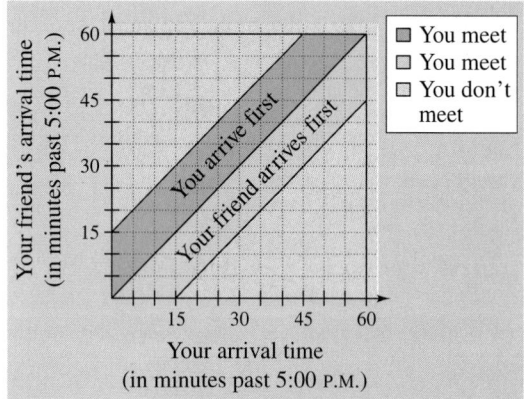

56. *Estimating* π A coin of diameter d is dropped onto a paper that contains a grid of squares d units on a side (see figure).

(a) Find the probability that the coin covers a vertex of one of the squares on the grid.

(b) Perform the experiment 100 times and use the results to approximate π.

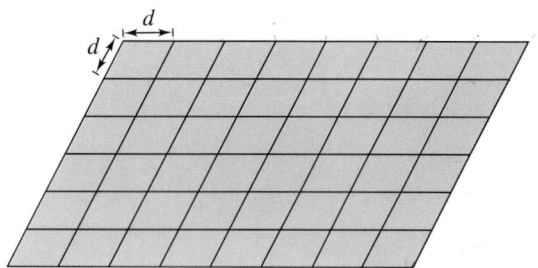

57. *Pattern Recognition and Exploration*   Consider a group of n people.

(a) Explain why the following gives the probability that the n people have distinct birthdays.

$$n = 2: \quad \frac{365}{365} \cdot \frac{364}{365} = \frac{365 \cdot 364}{365^2}$$

$$n = 3: \quad \frac{365}{365} \cdot \frac{364}{365} \cdot \frac{363}{365} = \frac{365 \cdot 364 \cdot 363}{365^3}$$

(b) Use the pattern in part (a) to write an expression for the probability that four people ($n = 4$) have distinct birthdays.

(c) Let P_n be the probability that the n people have distinct birthdays. Verify that this probability can be obtained recursively by

$$P_1 = 1 \quad \text{and} \quad P_n = \frac{365 - (n - 1)}{365} P_{n-1}.$$

(d) Explain why $Q_n = 1 - P_n$ gives the probability that at least two people in a group of n people have the same birthday.

(e) Use the results of parts (c) and (d) to complete the table.

n	10	15	20	23	30	40	50
P_n							
Q_n							

(f) How many people must be in a group before the probability of at least two of them having the same birthday is greater than $\frac{1}{2}$? Explain.

58. *Think About It*   The weather forecast indicates that the probability of rain is 40%. Explain what this means?

59. *Chapter Opener*   If the probability that a dive of an elephant seal will exceed 400 meters is 0.546, determine the probability that a dive will be 400 meters or less.

60. *Chapter Opener*   Use the histogram on page 655 to estimate the probability that a dive will be between 300 and 500 meters.

9.8 Exploring Data: Measures of Central Tendency

Mean, Median, and Mode / *Working with Organized Data* / *Choosing a Measure of Central Tendency*

Mean, Median, and Mode

In many real-life situations, it is helpful to describe data by a single number that is most representative of an entire collection of numbers. Such a number is called a **measure of central tendency.** The most commonly used measures are as follows.

1. The **mean,** or **average,** of n numbers is the sum of the numbers divided by n.
2. The **median** of n numbers is the middle number when the numbers are written in order. If n is even, the median is the average of the two middle numbers.
3. The **mode** of n numbers is the number that occurs most frequently. If two numbers tie for most frequent occurrence, the collection has two modes and is called **bimodal.**

EXAMPLE 1   **Finding Measures of Central Tendency**

Find the mean, median, and mode of the following numbers.

5, 8, 3, 10, 8, 7, 9, 7, 5, 6, 7, 8, 4, 7, 6,
6, 5, 9, 4, 6, 7, 4, 8, 7, 9, 5, 4, 6, 7, 8

Solution

The collection has 30 numbers. Thus, the mean is

$$\text{Mean} = \frac{5 + 8 + 3 + \cdots + 7 + 8}{30} = \frac{195}{30} = 6.5.$$

To find the median, order the numbers as follows.

3, 4, 4, 4, 4, 5, 5, 5, 5, 6, 6, 6, 6, 6, 7,
7, 7, 7, 7, 7, 8, 8, 8, 8, 8, 9, 9, 9, 10

Because the two middle numbers are 7's, the median is 7. To find the mode of the numbers, construct a frequency distribution, as shown in Figure 9.9. From the distribution, you can see that the mode is 7.

Figure 9.9

Frequency Distribution

Number	Tally
3	|
4	||||
5	||||
6	卌
7	卌 ||
8	卌
9	|||
10	|

In Example 1, all three measures of central tendency are about the same. If this always happened, there would be no need for different types of measures. The next example shows how different the values of the three measures can be.

Real Life

EXAMPLE 2   **Comparing Measures of Central Tendency**

You are interviewing for a job. The person who is interviewing you tells you that the average income of the 25 employees is $60,849. The actual annual incomes of the 25 employees are shown below. What are the mean, median, and mode of the incomes? Was the person telling you the truth?

$17,305,	$478,320,	$45,678,	$18,980,	$17,408,
$25,676,	$28,906,	$12,500,	$24,540,	$33,450,
$12,500,	$33,855,	$37,450,	$20,432,	$28,956,
$34,983,	$36,540,	$250,921,	$36,853,	$16,430,
$32,654,	$98,213,	$48,980,	$94,024,	$35,671

Solution

The mean of the incomes is

$$\text{Mean} = \frac{17{,}305 + 478{,}320 + 45{,}678 + 18{,}980 + \cdots + 35{,}671}{25}$$

$$= \frac{1{,}521{,}225}{25}$$

$$= \$60{,}849.$$

To find the median, order the incomes as follows.

$12,500,	$12,500,	$16,430,	$17,305,	$17,408,
$18,980,	$20,432,	$24,540,	$25,676,	$28,906,
$28,956,	$32,654,	$33,450,	$33,855,	$34,983,
$35,671,	$36,540,	$36,853,	$37,450,	$45,678,
$48,980,	$94,024,	$98,213,	$250,921,	$478,320

From this list, you can see that the median (the middle number) is $33,450. From the same list, you can see that $12,500 is the only income that occurs more than once. Thus, the mode is $12,500. Technically, the person was telling the truth because the average is (generally) defined to be the mean. However, of the three measures of central tendency

Mean: $60,849

Median: $33,450

Mode: $12,500

it seems clear that the median is the most representative. The mean is inflated by the two highest salaries.

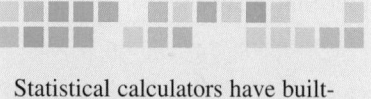

Statistical calculators have built-in programs for calculating the mean of a collection of numbers. Try using your calculator to find the mean shown in Example 2. Then use your calculator to sort the incomes shown in Example 2.

Working with Organized Data

In Examples 1 and 2, you were asked to find the mean, median, and mode of "raw data." Suppose, however, that the data were already organized in a frequency distribution or histogram. To find (or approximate) the measures of central tendency in such cases, follow the procedure demonstrated in the next two examples.

Real Life

EXAMPLE 3   **Working with Frequency Distributions**

A pair of six-sided dice is tossed 500 times.* The frequency of the totals is shown in the following frequency distribution. What are the mean, median, and mode of the totals?

Total	2	3	4	5	6	7	8	9	10	11	12
Frequency	17	22	46	50	63	87	64	65	47	23	16

Solution
The mean of the numbers is

$$\text{Mean} = \frac{17(2) + 22(3) + 46(4) + \cdots + 16(12)}{500} = \frac{3533}{500} = 7.066.$$

Both the median and mode of the totals are 7. A histogram of the data is shown in Figure 9.10. This type of histogram is called *bell-shaped*. For such distributions, the mean, median, and mode are approximately the same—they all lie near the center of the histogram.

Most computers and calculators that have statistical programs are capable of working with grouped data. Try entering the data given in Example 3 into your calculator. Then use the calculator to find the mean of the data.

Figure 9.10

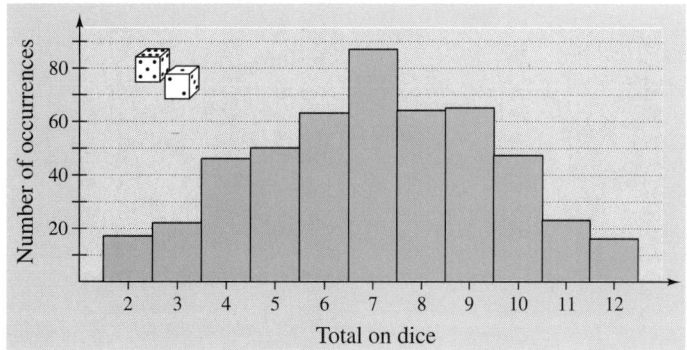

Total on dice

---

*The totals given in the table were simulated by a computer with a random number generator. Such simulations are called **Monte Carlo simulations.**

Real Life

EXAMPLE 4   Working with Grouped Data

Figure 9.11 shows the heights of 78 women who were chosen at random from a first-year college class. Approximate the mean height of the women.

Figure 9.11

Height (in inches)

Solution

Notice that the histogram groups the heights into 1-inch intervals. Because you are not given the exact heights, you cannot find the exact mean. You can, however, approximate the mean by using the midpoint of each interval.

Height Interval	Midpoint, h	Frequency, f	hf
58–59	58.5	1	(58.5)(1) = 58.5
59–60	59.5	3	(59.5)(3) = 178.5
60–61	60.5	4	(60.5)(4) = 242.0
61–62	61.5	7	(61.5)(7) = 430.5
62–63	62.5	7	(62.5)(7) = 437.5
63–64	63.5	8	(63.5)(8) = 508.0
64–65	64.5	10	(64.5)(10) = 645.0
65–66	65.5	11	(65.5)(11) = 720.5
66–67	66.5	7	(66.5)(7) = 465.5
67–68	67.5	8	(67.5)(8) = 540.0
68–69	68.5	5	(68.5)(5) = 342.5
69–70	69.5	2	(69.5)(2) = 139.0
70–71	70.5	3	(70.5)(3) = 211.5
71–72	71.5	0	(71.5)(0) = 0.0
72–73	72.5	1	(72.5)(1) = 72.5
73–74	73.5	1	(73.5)(1) = 73.5
			Total 5065.0

From the above table, you can approximate the mean height to be

$$\text{Mean} = \frac{5065}{78} \approx 64.9.$$

Choosing a Measure of Central Tendency

Which of the three measures of central tendency is the most representative? The answer is that it depends on the distribution of the data *and* the way in which you plan to use the data. For instance, in Example 2, the mean salary of $60,849 does not seem very representative to a potential employee. To a city income tax collector who wants to estimate 1% of the total income of the 25 employees, however, the mean is precisely the right measure.

EXAMPLE 5 ▱ **Choosing a Measure of Central Tendency**

Which measure of central tendency is the most representative of the data given in the frequency distribution?

a. *Number*	*Tally*	**b.** *Number*	*Tally*	**c.** *Number*	*Tally*
1	7	1	9	1	6
2	20	2	8	2	1
3	15	3	7	3	2
4	11	4	6	4	3
5	8	5	5	5	5
6	3	6	6	6	5
7	2	7	7	7	4
8	0	8	8	8	3
9	15	9	9	9	0

Solution

a. For the data, the mean is 4.23, the median is 3, and the mode is 2. Of these, the mode is probably the most representative.

b. For the data, the mean and median are each 5 and the modes are 1 and 9 (the distribution is bimodal). Of these, the mean or median is the most representative.

c. For the data, the mean is 4.59, the median is 5, and the mode is 1. Of these, the mean or median is the most representative. ▱

Group Activity

Comparing Measures of Central Tendency

Find a set of numbers that has each property. If it is not possible, explain.

a. Mean = 6,   Median = 4,   Mode = 4

b. Mean = 6,   Median = 6,   Mode = 4

c. Mean = 6,   Median = 4,   Mode = 6

9.8 /// EXERCISES

In Exercises 1–6, find the mean, median, and mode of the set of measurements.

1. 5, 12, 7, 14, 8, 9, 7
2. 30, 37, 32, 39, 33, 34, 32
3. 5, 12, 7, 24, 8, 9, 7
4. 20, 37, 32, 39, 33, 34, 32
5. 5, 12, 7, 14, 9, 7
6. 30, 37, 32, 39, 34, 32

7. *Think About It*   Compare the answers for Exercises 1 and 3. Which of the measures of central tendency is sensitive to extreme measurements? Explain.

8. *Exploration*
   (a) Add 6 to each measurement in Exercise 1 and calculate the mean, median, and mode of the revised measurements. How are the measures of central tendency changed?
   (b) If a constant k is added to each measurement in a set of data, how will the measures of central tendency change? Use properties of summation to prove the conjecture for the mean.

In Exercises 9–12, use a graphing utility to find the mean, median, and mode of the set of measurements.

9. 31, 31, 32, 25, 23, 22, 26, 26, 23, 23
10. 10, 6, 13, 6, 14, 10, 11, 15, 5, 17
11. 39, 31, 26, 37, 32, 25, 41, 21, 23, 29, 30
12. 51.0, 62.9, 49.9, 41.9, 27.7, 49.0, 70.5, 55.0, 41.1, 35.2, 48.1

13. *Electric Bills*   Determine the mean and median of the following monthly electric bills.

January	$77.92	February	$69.84
March	$62.00	April	$62.50
May	$67.99	June	$75.35
July	$91.76	August	$84.98
September	$97.82	October	$93.98
November	$75.35	December	$67.00

14. *Car Rental*   A car rental company kept the following record of the number of miles driven in one of their cars. Determine the mean, median, and mode of the data.

Monday	410	Tuesday	260
Wednesday	320	Thursday	320
Friday	460	Saturday	150
Sunday	385		

15. *Six-Child Families*   A study was done on families having six children. The table gives the number of families in the study having the specified number of girls.
    (a) How many families were in the study?
    (b) Determine the mean, median, and mode of the data.

Number of Girls	0	1	2	3	4	5	6
Frequency	1	24	45	54	50	19	7

16. *Baseball*   A baseball fan examined the records of a favorite baseball player's performance during the last 50 games. The number of games in which the player had 0, 1, 2, 3, and 4 hits are recorded in the table.

Number of Hits	0	1	2	3	4
Frequency	14	26	7	2	1

   (a) Determine the average number of hits per game.
   (b) Determine the player's batting average if the player batted 200 times during the 50 games.

17. *Shoe Sales*   A salesperson sold eight pairs of a certain style of men's shoes. The sizes of the eight pairs were $10\frac{1}{2}$, 8, 12, $10\frac{1}{2}$, 10, $9\frac{1}{2}$, 11 and $10\frac{1}{2}$. Which measure of central tendency best describes the typical shoe size for the data?

18. *Hourly Wage*   The median hourly wage at a particular company is $14.18. If there are 254 employees who earn hourly wages, is it true that at least 127 employees earn at least $14.18 per hour?

19. *Tire Wear* A tire company tested a new tire design on 100 cars. The useful tread lives (in thousands of miles) for the cars are given in the table.

Miles	26–28	28–30	30–32	32–34
Frequency	1	7	5	11

Miles	34–36	36–38	38–40	40–42
Frequency	19	25	12	11

Miles	42–44	44–46	46–48
Frequency	5	3	1

(a) Create a histogram for the data.

(b) Approximate the average tread life of the tires.

20. *Health* The frequency distribution gives the number of deaths (in thousands) attributed to heart disease and accidents by age group in the United States in 1990. (Source: U.S. National Center for Health Statistics)

Age	15–24	25–34	35–44
Heart Disease	0.9	3.3	11.8
Accidents	16.2	16.0	11.7

Age	45–54	55–64	65–74
Heart Disease	30.2	77.5	161.4
Accidents	7.4	7.2	8.4

(a) Create a histogram for the heart disease data and approximate the mean age of death by heart disease.

(b) Create a histogram for the accident data and approximate the mean age of death by accidents.

21. *Battery Life* The lifetimes (in months) of a sample of 25 batteries are given below.

37	39	20	38	62
19	71	52	35	27
52	16	26	10	25
16	14	56	15	69
27	19	62	32	14

(a) Use a graphing utility to create a histogram for the data. Use the following window.

```
Xmin = 5
Xmax = 80
Xscl = 10
Ymin = -2
Ymax = 8
Yscl = 1
```

(b) Repeat part (a) with Xscl = 5. Describe the change in the histogram.

(c) Repeat part (a) with Xmin = 10. Describe the change in the histogram.

(d) Use a graphing utility to sort the data in ascending order. Determine the median of the data.

(e) Use a graphing utility to find the mean of the data.

22. *Weight* The weights (in pounds) of a sample of 30 bales of hay are given below.

49	57	50	58	44
46	51	46	59	37
43	57	41	50	38
52	45	49	51	47
52	54	56	43	62
51	47	49	56	57

(a) Use a graphing utility to create a histogram for the data. Use the following window.

```
Xmin = 35
Xmax = 70
Xscl = 3
Ymin = -2
Ymax = 8
Yscl = 1
```

(b) Repeat part (a) with Xscl = 4. Describe the change in the histogram.

(c) Repeat part (a) with Xmin = 36. Describe the change in the histogram.

(d) Use a graphing utility to sort the data in ascending order. Determine the median of the data.

(e) Use a graphing utility to find the mean of the data.

9.9 Exploring Data: Measures of Dispersion

Variance and Standard Deviation / *Computing Standard Deviation* / *Box-and-Whisker Plots*

Variance and Standard Deviation

In Section 9.8, you saw that very different sets of numbers can have the same mean. In this section, you will study two **measures of dispersion,** which give you an idea of how much the numbers in the set differ from the mean of the set. These two measures are called the *variance* of the set and the *standard deviation* of the set.

> **Definitions of Variance and Standard Deviation**
>
> Consider a set of numbers, $(x_1, x_2, \ldots, x_n)$, with a mean of $\bar{x}$. The **variance** of the set is
>
> $$v = \frac{(x_1 - \bar{x})^2 + (x_2 - \bar{x})^2 + \cdots + (x_n - \bar{x})^2}{n}$$
>
> and the **standard deviation** of the set is
>
> $$\sigma = \sqrt{v}$$
>
> where σ is the lowercase Greek letter *sigma*.

The standard deviation of a set is a measure of how much a typical number in the set differs from the mean. The greater the standard deviation, the more the numbers in the set *vary* from the mean. For instance, each of the following sets has a mean of 5.

$$\{5, 5, 5, 5\}, \quad \{4, 4, 6, 6\}, \quad \text{and} \quad \{3, 3, 7, 7\}$$

The standard deviations of the sets are 0, 1, and 2.

$$\sigma_1 = \sqrt{\frac{(5 - 5)^2 + (5 - 5)^2 + (5 - 5)^2 + (5 - 5)^2}{4}} = 0$$

$$\sigma_2 = \sqrt{\frac{(4 - 5)^2 + (4 - 5)^2 + (6 - 5)^2 + (6 - 5)^2}{4}} = 1$$

$$\sigma_3 = \sqrt{\frac{(3 - 5)^2 + (3 - 5)^2 + (7 - 5)^2 + (7 - 5)^2}{4}} = 2$$

EXAMPLE 1 ▱ **Estimations of Standard Deviation**

Consider the three sets of data represented by the bar graphs in Figure 9.12. Which set has the smallest standard deviation? Which has the largest?

Figure 9.12

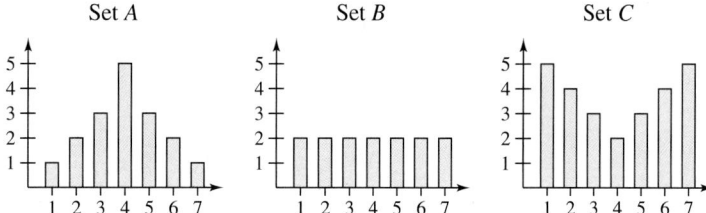

Solution

Of the three sets, the numbers in set A are grouped most closely to the center and the numbers in set C are the most dispersed. Thus, set A has the smallest standard deviation and set C has the largest standard deviation. ▱

EXAMPLE 2 ▱ **Finding Standard Deviation**

Find the standard deviation of each set shown in Example 1.

Solution

Because of the symmetry of each bar graph, you can conclude that each has a mean of $\bar{x} = 4$. The standard deviation of set A is

$$\sigma = \sqrt{\frac{(-3)^2 + 2(-2)^2 + 3(-1)^2 + 5(0)^2 + 3(1)^2 + 2(2)^2 + (3)^2}{17}}$$

$$\approx 1.53.$$

The standard deviation of set B is

$$\sigma = \sqrt{\frac{2(-3)^2 + 2(-2)^2 + 2(-1)^2 + 2(0)^2 + 2(1)^2 + 2(2)^2 + 2(3)^2}{14}}$$

$$= 2.$$

The standard deviation of set C is

$$\sigma = \sqrt{\frac{5(-3)^2 + 4(-2)^2 + 3(-1)^2 + 2(0)^2 + 3(1)^2 + 4(2)^2 + 5(3)^2}{26}}$$

$$\approx 2.22.$$

These values confirm the results of Example 1. That is, set A has the smallest standard deviation and set C has the largest. ▱

Computing Standard Deviation

Most computations in statistics are performed with calculators and computers that are programmed with formulas for the mean, variance, and standard deviation. The following two examples, however, show how to organize computations that are done by hand.

EXAMPLE 3   Using a Table

Find the standard deviation of the following set of numbers.

5, 6, 6, 7, 7, 8, 8, 8, 9, 10

Solution

Begin by finding the mean of the set.

$$\bar{x} = \frac{5 + 6 + 6 + 7 + 7 + 8 + 8 + 8 + 9 + 10}{10} = \frac{74}{10} = 7.4$$

To find the variance, you can use the following tabular format.

If you have access to a computer or calculator with a standard deviation program, try using it to obtain the result given in Example 3. If you do this, the program will probably output two versions of the standard deviation. In one, the sum of the squared differences is divided by n, and in the other it is divided by $n - 1$. In this text, we always divide by n.

x	$x - \bar{x}$	$(x - \bar{x})^2$
5	−2.4	5.76
6	−1.4	1.96
6	−1.4	1.96
7	−0.4	0.16
7	−0.4	0.16
8	0.6	0.36
8	0.6	0.36
8	0.6	0.36
9	1.6	2.56
10	2.6	6.76
		20.4

⇐ Sum of squared differences

To find the variance, divide the sum of the squared differences by 10 (the number of entries in the set).

$$v = \frac{20.4}{10} = 2.04$$

Finally, the standard deviation is the square root of the variance.

$$\sigma = \sqrt{2.04} \approx 1.43$$

Because the mean of a set is often a rounded decimal, the technique shown in Example 3 can be cumbersome *and* it can involve round-off error. The following alternative formula provides a more efficient way to compute the standard deviation.

Alternative Formula for Standard Deviation

The standard deviation of $\{x_1, x_2, \ldots, x_n\}$ is

$$\sigma = \sqrt{\frac{x_1^2 + x_2^2 + \cdots + x_n^2}{n} - \bar{x}^2}.$$

Because of messy computations, this formula is difficult to verify. Conceptually, however, the process is straightforward. It consists of showing that the expressions

$$\sqrt{\frac{(x_1 - \bar{x})^2 + (x_2 - \bar{x})^2 + \cdots + (x_n - \bar{x})^2}{n}}$$

and

$$\sqrt{\frac{x_1^2 + x_2^2 + \cdots + x_n^2}{n} - \bar{x}^2}$$

are equivalent. Try verifying this equivalence for the set $\{x_1, x_2, x_3\}$ with $\bar{x} = (x_1 + x_2 + x_3)/3$.

EXAMPLE 4 **Using the Alternative Formula**

Use the alternative formula for standard deviation to find the standard deviation of the set given in Example 3.

5, 6, 6, 7, 7, 8, 8, 8, 9, 10

Solution

From Example 3, you know the mean is 7.4. Thus, the standard deviation is

$$\sigma = \sqrt{\frac{5^2 + 2(6^2) + 2(7^2) + 3(8^2) + 9^2 + 10^2}{10} - (7.4)^2}$$

$$= \sqrt{\frac{568}{10} - 54.76}$$

$$= \sqrt{2.04}$$

$$\approx 1.43.$$

Note that this result agrees with that obtained in Example 3.

A well-known theorem in statistics, called *Chebychev's Theorem*, states that at least

$$1 - \frac{1}{k^2}$$

of the numbers in a distribution must lie within k standard deviations of the mean. Thus, at least 75% of the numbers in a collection must lie within two standard deviations of the mean, and at least 88.9% of the numbers must lie within three standard deviations of the mean. For most distributions, these percents are low. For instance, in all three distributions shown in Example 1, 100% of the numbers lie within two standard deviations of the mean.

Real Life

EXAMPLE 5   Describing a Distribution

The following table shows the number of dentists (per 100,000 people) in each state. Find the mean and standard deviation of the numbers. What percent of the numbers lie within two standard deviations of the mean?   (Source: American Dental Association)

AK	66	AL	40	AR	39	AZ	51	CA	62
CO	69	CT	80	D.C.	94	DE	44	FL	50
GA	46	HI	80	IA	55	ID	53	IL	61
IN	47	KS	51	KY	53	LA	45	MA	74
MD	68	ME	47	MI	62	MN	67	MO	53
MS	37	MT	62	NC	42	ND	47	NE	63
NH	59	NJ	77	NM	45	NV	49	NY	73
OH	55	OK	47	OR	70	PA	61	RI	56
SC	41	SD	49	TN	53	TX	47	UT	66
VA	54	VT	57	WA	68	WI	65	WV	43
WY	52								

Figure 9.13

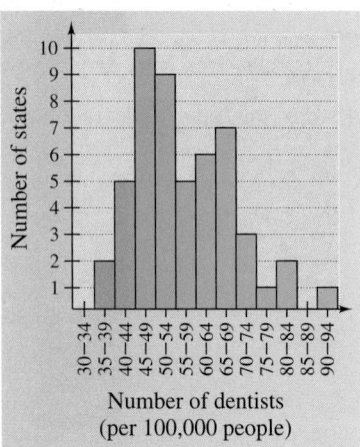

Number of dentists
(per 100,000 people)

Solution

Begin by entering the numbers into a computer or calculator that has a standard deviation program. After running the program, you should obtain

$$\bar{x} \approx 56.76 \quad \text{and} \quad \sigma \approx 12.14.$$

The interval that contains all numbers that lie within two standard deviations of the mean is

$$[56.76 - 2(12.14), 56.76 + 2(12.14)] \quad \text{or} \quad [32.48, 81.04].$$

From the histogram in Figure 9.13, you can see that all but one of the numbers (98%) lie in the interval [35, 84]. (The one exception is the number of dentists per 100,000 people in Washington, DC.)

Box-and-Whisker Plots

Standard deviation is the measure of dispersion that is associated with the mean. **Quartiles** measure dispersion associated with the median.

Definition of Quartiles

Consider an ordered set of numbers whose median is m. The **lower quartile** is the median of the numbers that occur before m. The **upper quartile** is the median of the numbers that occur after m.

EXAMPLE 6   **Finding Quartiles of a Set**

Find the lower and upper quartiles for the following set.

34, 14, 24, 16, 12, 18, 20, 24, 16, 26, 13, 27

Solution

Begin by ordering the set.

12, 13, 14, 16, 16, 18, 20, 24, 24, 26, 27, 34

1st 25%   2nd 25%   3rd 25%   4th 25%

The median of the entire set is 19. The median of the six numbers that are less than 19 is 15. Thus, the lower quartile is 15. The median of the six numbers that are greater than 19 is 25. Thus, the upper quartile is 25.

Quartiles are represented graphically by a **box-and-whisker plot,** as shown in Figure 9.14. In the plot, notice that five numbers are listed: the smallest number, the lower quartile, the median, the upper quartile, and the largest number. Also notice that the numbers are spaced proportionally, as though they were on a real number line.

Figure 9.14

| 12 | 15 | 19 | 25 | 34 |

The next example shows how to find quartiles when the number of elements in a set is not divisible by 4.

EXAMPLE 7 ▰ Sketching Box-and-Whisker Plots

Sketch a box-and-whisker plot for each of the sets shown below.

a. 27, 28, 30, 42, 45, 50, 50, 61, 62, 64, 66

b. 82, 82, 83, 85, 87, 89, 90, 94, 95, 95, 96, 98, 99

c. 11, 13, 13, 15, 17, 18, 20, 24, 24, 27

Solution

a. This set has 11 numbers. The median is 50 (the sixth number). The lower quartile is 30 (the median of the first five numbers). The upper quartile is 62 (the median of the last five numbers).

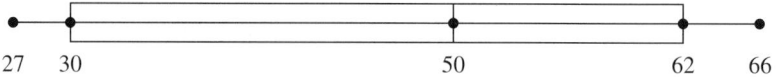

b. This set has 13 numbers. The median is 90 (the seventh number). The lower quartile is 84 (the median of the first six numbers). The upper quartile is 95.5 (the median of the last six numbers).

c. This set has 10 numbers. The median is 17.5 (the average of the fifth and sixth numbers). The lower quartile is 13 (the median of the first five numbers). The upper quartile is 24 (the median of the last five numbers).

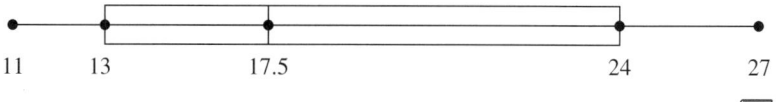

Group Activity

Estimating Standard Deviation

Graphically estimate the standard deviation of the data represented by the bar graph. Then calculate the standard deviation. Compare your calculated results with your estimates.

a.

b.

9.9 /// EXERCISES

In Exercises 1–6, find the mean, variance, and standard deviation of the data without the aid of a graphing utility.

1. 4, 10, 8, 2

2. 3, 15, 6, 9, 2

3. 0, 1, 1, 2, 2, 2, 3, 3, 4

4. 2, 2, 2, 2, 2, 2

5. 1, 2, 3, 4, 5, 6, 7

6. 1, 1, 1, 5, 5, 5

In Exercises 7–10, use the alternative formula for standard deviation to find the standard deviation of the data.

7. 2, 4, 6, 6, 13, 5

8. 10, 25, 50, 26, 15, 33, 29, 4

9. 18, 17, 4, 10, 8, 14, 2

10. 6, 13, 5, 11, 13, 5, 7, 3, 12, 4

In Exercises 11–16, use a graphing utility to find the mean, variance, and standard deviation of the data.

11. 49, 62, 40, 29, 32, 70

12. 246, 336, 473, 167, 219, 359

13. 1.5, 0.4, 2.1, 0.7, 0.8

14. 6.0, 9.1, 4.4, 8.7, 10.4

15. 8.1, 6.9, 3.7, 4.2, 6.1

16. 9.0, 7.5, 3.3, 7.4, 6.0

In Exercises 17 and 18, line plots of sets of data are given. Determine the mean and standard deviation of each set.

17. (a)

(b)

(c)

(d)

18. (a)

(b)

(c)

(d)

19. *Think About It* An instructor adds five points to each student's exam score. Will this change the mean or standard deviation of the exam scores? Explain.

20. *Think About It* Without calculating the standard deviation, explain why the set {4, 4, 20, 20} has a standard deviation of 8.

21. *Think About It*   Consider the four sets of data represented by the histograms. Order the sets from the smallest to the largest standard deviation.

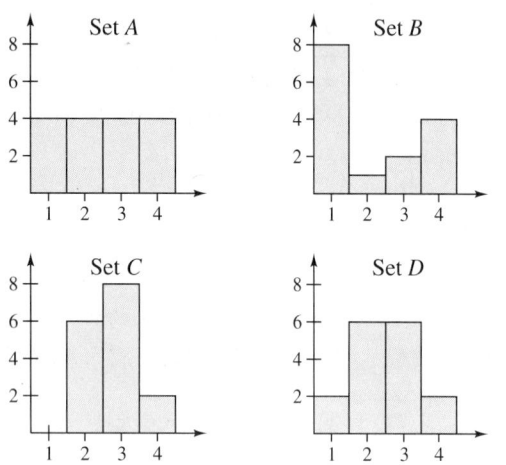

22. *Think About It*   The histograms represent the test scores of two classes of a college course in mathematics. Which histogram has the smaller standard deviation?

23. *Test Scores*   The scores of a mathematics exam given to 600 science and engineering students at a college had a mean and a standard deviation of 235 and 28, respectively.

    (a) Use Chebychev's Theorem to determine the intervals containing at least $\frac{3}{4}$ and at least $\frac{8}{9}$ of the scores.

    (b) How would the intervals in part (a) change if the standard deviation were 16?

    (a) How would the intervals in part (a) change if the mean were 225?

24. *Heights*   The heights (in inches) of the players on a high school football team had a mean and a standard deviation of 71.1 inches and 3.3 inches, respectively. Use Chebychev's Theorem to determine the intervals containing at least $\frac{3}{4}$ and at least $\frac{8}{9}$ of the heights of the players.

25. *Grocery Shopping*   Over the past year a family went grocery shopping every two weeks. The costs of the groceries were as follows.

$210.55	$178.60	$197.22	$181.23	$172.45
$201.74	$211.38	$199.71	$193.30	$202.01
$190.64	$180.13	$180.95	$192.76	$219.82
$220.70	$192.02	$194.40	$195.06	$202.36
$185.12	$198.46	$194.62	$204.28	$188.60
$242.15				

    (a) Use a graphing utility to find the mean and standard deviation of the data.

    (b) What percent of the data lie within two standard deviations of the mean?

26. *Precipitation*   The following data represents the annual precipitation (in inches) at Erie, Pennsylvania for the years 1960 through 1995.   (Source: National Oceanic and Atmospheric Administration)

27.41	36.50	36.90	28.11	36.47
38.41	37.74	37.78	34.33	36.58
41.50	34.06	43.55	38.04	41.83
43.03	43.85	61.70	35.04	55.31
47.04	41.97	41.56	46.25	37.79
45.87	47.30	44.86	38.87	41.88
51.96	31.71	44.09	38.10	40.50
34.39				

    (a) Use a graphing utility to find the mean and standard deviation of the data.

    (b) What percent of the data lie within two standard deviations of the mean?

27. *Household Sizes* The following data gives the average number of people per household in 1993 for each of the 50 states. (Source: U.S. Bureau of the Census)

AK	2.81	AL	2.62	AR	2.58	AZ	2.63
CA	2.82	CO	2.52	CT	2.58	DE	2.61
FL	2.49	GA	2.66	HI	2.99	IA	2.53
ID	2.75	IL	2.66	IN	2.59	KS	2.55
KY	2.60	LA	2.77	MA	2.57	MD	2.66
ME	2.55	MI	2.65	MN	2.58	MO	2.55
MS	2.75	MT	2.55	NC	2.55	ND	2.54
NE	2.56	NH	2.61	NJ	2.71	NM	2.75
NV	2.54	NY	2.64	OH	2.59	OK	2.56
OR	2.54	PA	2.57	RI	2.56	SC	2.67
SD	2.62	TN	2.57	TX	2.75	UT	3.15
VA	2.60	VT	2.56	WA	2.56	WI	2.62
WV	2.55	WY	2.65					

(a) Use a graphing utility to find the mean and standard deviation of the data.

(b) Identify the states where the number of people per household differed from the mean by more than two standard deviations.

28. *Weight* The weights (in pounds) of a sample of 30 bales of hay are given below.

49	57	50	58	44
46	51	46	59	37
43	57	41	50	38
52	45	49	51	47
52	54	56	43	62
51	47	49	56	57

(a) Use a graphing utility to find the mean and standard deviation of the data.

(b) How many bales have weights that differ from the mean by more than two standard deviations?

In Exercises 29–32, sketch a box-and-whisker plot for the data without the aid of a graphing utility.

29. 23, 15, 14, 23, 13, 14, 13, 20, 12

30. 11, 10, 11, 14, 17, 16, 14, 11, 8, 14, 20

31. 46, 48, 48, 50, 52, 47, 51, 47, 49, 53

32. 25, 20, 22, 28, 24, 28, 25, 19, 27, 29, 28, 21

In Exercises 33–36, use a graphing utility to create a box-and-whisker plot for the data.

33. 19, 12, 14, 9, 14, 15, 17, 13, 19, 11, 10, 19

34. 9, 5, 5, 5, 6, 5, 4, 12, 7, 10, 7, 11, 8, 9, 9

35. 20.1, 43.4, 34.9, 23.9, 33.5, 24.1, 22.5, 42.4, 25.7, 17.4, 23.8, 33.3, 17.3, 36.4, 21.8

36. 78.4, 76.3, 107.5, 78.5, 93.2, 90.3, 77.8, 37.1, 97.1, 75.5, 58.8, 65.6

37. *Product Lifetime* A company has redesigned a product in an attempt to increase the lifetime of the product. The two sets of data give the lifetimes (in months) of samples of 20 units before and 20 units after the design change. Create a box-and-whisker plot for each set of data, and then comment on the differences between the plots.

Original Design

15.1	78.3	56.3	68.9	30.6
27.2	12.5	42.7	72.7	20.2
53.0	13.5	11.0	18.4	85.2
10.8	38.3	85.1	10.0	12.6

New Design

55.8	71.5	25.6	19.0	23.1
37.2	60.0	35.3	18.9	80.5
46.7	31.1	67.9	23.5	99.5
54.0	23.2	45.5	24.8	87.8

Focus on Concepts

In this chapter, you studied several concepts that are required in the study of sequences, counting principles, and probability. You can use the following questions to check your understanding of several of these basic concepts. The answers to these questions are given in the back of the book.

1. An infinite sequence is a function. What is the domain of the function?

2. How do the two sequences differ?

   (a) $a_n = \dfrac{(-1)^n}{n}$ (b) $a_n = \dfrac{(-1)^{n+1}}{n}$

In Exercises 3–6, decide whether the statement is true. Explain your reasoning.

3. $\dfrac{(n+2)!}{n!} = (n+2)(n+1)$

4. $\displaystyle\sum_{i=1}^{5}(i^3 + 2i) = \sum_{i=1}^{5}i^3 + \sum_{i=1}^{5}2i$

5. $\displaystyle\sum_{k=1}^{8}3k = 3\sum_{k=1}^{8}k$

6. $\displaystyle\sum_{j=1}^{6}2^j = \sum_{j=3}^{8}2^{j-2}$

7. In your own words, explain what makes a sequence (a) arithmetic and (b) geometric.

8. The graphs of two sequences are shown below. Identify each sequence as arithmetic or geometric. Explain your reasoning.

   (a)   (b)

9. Explain what a recursion formula is.

10. Explain why the terms of a geometric sequence decrease when $0 < r < 1$.

In Exercises 11–14, match the sequence or sum of a sequence with its graph without doing any calculations. Explain your reasoning. [The graphs are labeled (a), (b), (c), and (d).]

11. $a_n = 4\left(\tfrac{1}{2}\right)^{n-1}$ 12. $a_n = 4\left(-\tfrac{1}{2}\right)^{n-1}$

13. $a_n = \displaystyle\sum_{k=1}^{n}4\left(\tfrac{1}{2}\right)^{k-1}$ 14. $a_n = \displaystyle\sum_{k=1}^{n}4\left(-\tfrac{1}{2}\right)^{k-1}$

15. How do the expansions of $(x+y)^n$ and $(x-y)^n$ differ?

16. What is the relationship between $_nC_r$ and $_nC_{n-r}$?

17. Without calculating the numbers, determine which of the following is greater. Explain.

    (a) The combination of 10 elements taken 6 at a time

    (b) The permutation of 10 elements taken 6 at a time

18. The weather forecast indicates that the probability of rain is 60%. Explain what this means.

19. Which measure of central tendency is affected by an extreme measure in the data list?

20. What is implied about a set of measurements if the standard deviation of the set is 0?

9 /// REVIEW EXERCISES

In Exercises 1–4, write the first five terms of the sequence. (Assume n begins with 1.)

1. $a_n = 2 + \dfrac{6}{n}$

2. $a_n = \dfrac{5n}{2n - 1}$

3. $a_n = \dfrac{72}{n!}$

4. $a_n = n(n - 1)$

In Exercises 5–8, use a graphing utility to graph the first 10 terms of the sequence.

5. $a_n = \dfrac{3}{2}n$

6. $a_n = 4(0.4)^{n-1}$

7. $a_n = \dfrac{3n}{n + 2}$

8. $a_n = 5 - \dfrac{3}{n}$

In Exercises 9–12, use sigma notation to write the sum.

9. $\dfrac{1}{2(1)} + \dfrac{1}{2(2)} + \dfrac{1}{2(3)} + \cdots + \dfrac{1}{2(20)}$

10. $2(1^2) + 2(2^2) + 2(3^2) + \cdots + 2(9^2)$

11. $\dfrac{1}{2} + \dfrac{2}{3} + \dfrac{3}{4} + \cdots + \dfrac{9}{10}$

12. $1 - \dfrac{1}{3} + \dfrac{1}{9} - \dfrac{1}{27} + \cdots$

In Exercises 13–20, find the sum.

13. $\displaystyle\sum_{i=1}^{6} 5$

14. $\displaystyle\sum_{k=2}^{5} 4k$

15. $\displaystyle\sum_{j=1}^{4} \dfrac{6}{j^2}$

16. $\displaystyle\sum_{i=1}^{8} \dfrac{i}{i + 1}$

17. $\displaystyle\sum_{k=1}^{10} 2k^3$

18. $\displaystyle\sum_{j=0}^{4} (j^2 + 1)$

19. $\displaystyle\sum_{n=0}^{10} (n^2 + 3)$

20. $\displaystyle\sum_{n=1}^{100} \left(\dfrac{1}{n} - \dfrac{1}{n + 1} \right)$

In Exercises 21–24, write the first five terms of the arithmetic sequence.

21. $a_1 = 3, \ d = 4$

22. $a_1 = 8, \ d = -2$

23. $a_4 = 10, \ a_{10} = 28$

24. $a_2 = 14, \ a_6 = 22$

In Exercises 25–28, write the first five terms of the arithmetic sequence defined recursively. Determine the common difference and write the nth term of the sequence as a function of n.

25. $a_1 = 35, \qquad a_{k+1} = a_k - 3$

26. $a_1 = 15, \qquad a_{k+1} = a_k + \dfrac{5}{2}$

27. $a_1 = 9, \qquad a_{k+1} = a_k + 7$

28. $a_1 = 100, \qquad a_{k+1} = a_k - 5$

In Exercises 29 and 30, write an expression for the nth term of the arithmetic sequence and find the sum of the first 20 terms of the sequence.

29. $a_1 = 100, \ d = -3$

30. $a_1 = 10, \ a_3 = 28$

In Exercises 31–34, find the sum.

31. $\displaystyle\sum_{j=1}^{10} (2j - 3)$

32. $\displaystyle\sum_{j=1}^{8} (20 - 3j)$

33. $\displaystyle\sum_{k=1}^{11} \left(\dfrac{2}{3}k + 4 \right)$

34. $\displaystyle\sum_{k=1}^{25} \left(\dfrac{3k + 1}{4} \right)$

35. Find the sum of the first 100 positive multiples of 5.

36. Find the sum of the integers from 20 to 80 (inclusive).

37. *Job Offer*   The starting salary for a job is $34,000 with a guaranteed salary increase of $2250 per year for the first four years of employment. Determine (a) the salary during the fifth year and (b) the total compensation through five full years of employment.

38. *Baling Hay*   In the first two trips baling hay around a large field, a farmer obtains 123 bales and 112 bales, respectively. Because each round gets shorter, the farmer estimates that the same pattern will continue. Estimate the total number of bales made if there are another six trips around the field.

In Exercises 39–42, write the first five terms of the geometric sequence.

39. $a_1 = 4, \; r = -\frac{1}{4}$

40. $a_1 = 2, \; r = 2$

41. $a_1 = 9, \; a_3 = 4$

42. $a_1 = 2, \; a_3 = 12$

In Exercises 43–46, write the first five terms of the geometric sequence defined recursively. Determine the common ratio and write the nth term of the sequence as a function of n.

43. $a_1 = 120, \quad a_{k+1} = \frac{1}{3}a_k$

44. $a_1 = 200, \quad a_{k+1} = 0.1a_k$

45. $a_1 = 25, \quad a_{k+1} = -\frac{3}{5}a_k$

46. $a_1 = 18, \quad a_{k+1} = \frac{5}{3}a_k$

In Exercises 47 and 48, write an expression for the nth term of the geometric sequence and find the sum of the first 20 terms of the sequence.

47. $a_1 = 16, \; a_2 = -8$ **48.** $a_1 = 100, \; r = 1.05$

In Exercises 49–54, find the sum.

49. $\displaystyle\sum_{i=1}^{7} 2^{i-1}$ **50.** $\displaystyle\sum_{i=1}^{5} 3^{i-1}$

51. $\displaystyle\sum_{i=1}^{\infty} \left(\frac{7}{8}\right)^{i-1}$ **52.** $\displaystyle\sum_{i=1}^{\infty} \left(\frac{1}{3}\right)^{i-1}$

53. $\displaystyle\sum_{k=1}^{\infty} 4\left(\frac{2}{3}\right)^{k-1}$ **54.** $\displaystyle\sum_{k=1}^{\infty} 1.3\left(\frac{1}{10}\right)^{k-1}$

In Exercises 55 and 56, use a graphing utility to find the sum.

55. $\displaystyle\sum_{i=1}^{10} 10\left(\frac{3}{5}\right)^{i-1}$ **56.** $\displaystyle\sum_{i=1}^{25} 100(1.06)^{i-1}$

57. *Depreciation*   A company buys a machine for \$120,000. During the next 5 years it depreciates at a rate of 30% per year. (That is, at the end of each year the depreciated value is 70% of what it was at the beginning of the year.)

(a) Find the formula for the nth term of a geometric sequence that gives the value of the machine after t full years after it was purchased.

(b) Find the depreciated value of the machine at the end of 5 full years.

58. *Total Compensation*   A job pays a salary of \$32,000 the first year. During the next 39 years, suppose there is a 5.5% raise each year. What would the total salary be over the 40-year period?

59. *Compound Interest*   A deposit of \$200 is made at the beginning of each month for 2 years in an account that pays 6%, compounded monthly. What is the balance in the account at the end of 2 years?

60. *Compound Interest*   A deposit of \$100 is made at the beginning of each month for 10 years in an account that pays 6.5%, compounded monthly. What is the balance in the account at the end of 10 years?

In Exercises 61–64, use mathematical induction to prove the formula for every positive integer n.

61. $1 + 4 + \cdots + (3n - 2) = \dfrac{n}{2}(3n - 1)$

62. $1 + \dfrac{3}{2} + 2 + \dfrac{5}{2} + \cdots + \dfrac{1}{2}(n + 1) = \dfrac{n}{4}(n + 3)$

63. $\displaystyle\sum_{i=0}^{n-1} ar^i = \dfrac{a(1 - r^n)}{1 - r}$

64. $\displaystyle\sum_{k=0}^{n-1} (a + kd) = \dfrac{n}{2}[2a + (n - 1)d]$

In Exercises 65–68, evaluate $_nC_r$ or $_nP_r$. Use the $\boxed{_nC_r}$ or $\boxed{_nP_r}$ feature of a graphing utility to verify your answer.

65. $_6C_4$

66. $_{10}C_7$

67. $_8P_5$

68. $_{12}P_3$

In Exercises 69–74, use the Binomial Theorem to expand the binomial. Simplify your answer. (Remember that $i = \sqrt{-1}$.)

69. $\left(\dfrac{x}{2} + y\right)^4$

70. $(a - 3b)^5$

71. $\left(\dfrac{2}{x} - 3x\right)^6$

72. $(3x + y^2)^7$

73. $(5 + 2i)^4$

74. $(4 - 5i)^3$

75. *Amateur Radio*   A Novice Amateur Radio license consists of two letters, one digit, and then three more letters. How many different licenses can be issued if no restrictions are placed on the letters or digits?

76. *Morse Code*   In Morse code, all characters are transmitted using a sequence of dits and dahs. How many different characters can be formed by a sequence of three dits and dahs? (These can be repeated. For example, dit-dit-dit represents the letter *s*.)

In Exercises 77 and 78, use a tree diagram to determine the possible arrangements needed to answer the question.

77. *Itineraries*   A college theater company is planning a tour with performances in Phoenix, Dallas, and New Orleans. If there are no restrictions on the order of the performances, how many ways can the tour be arranged?

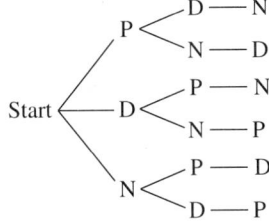

To determine the possible arrangements, count the paths. A path moves from the start to the end of the last branch. In this case, there are six paths, and so there are six possible ways to arrange the itineray of the tour. The tree diagram verifies visually what you know by using counting principles. That is, the number of possible arrangements is $3 \cdot 2 \cdot 1 = 6$. Use a tree diagram to show the possible arrangements for the college theater company if the city of Houston is added to their tour.

78. *Gender*   Use a tree diagram to show the possible arrangements for the genders of three children within a family.

79. *Matching Socks*   A man has five pairs of socks (no two pairs are the same color). If he randomly selects two socks from a drawer, what is the probability that he gets a matched pair?

80. *Bookshelf Order*   A child carries a five-volume set of books to a bookshelf. The child is not able to read, and hence cannot distinguish one volume from another. What is the probability that the books are shelved in the correct order?

81. *Roll of the Dice*   Are the chances of rolling a 3 with one die the same as rolling a total of 6 with two dice? If not, which has the higher probability?

82. *Roll of the Dice*   A six-sided die is rolled six times. What is the probability that each side appears exactly once?

83. *Tossing a Coin*   Find the probability of obtaining at least one tail when a coin is tossed five times.

84. *Parental Independence*   Suppose in a survey senior citizens were asked if they would live with their children when they reached the point of not being able to live alone. The results are shown in the figure. Suppose three senior citizens who could not live alone are randomly selected. What is the probability that all three are not living with their children?

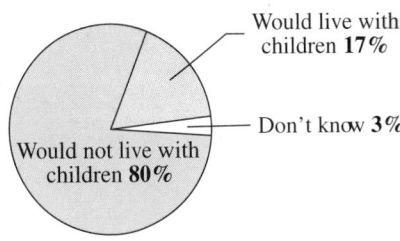

85. *Card Game*   Five cards are drawn from an ordinary deck of 52 playing cards. Find the probability of getting exactly two pairs. (For example, the hand could be A-A-5-5-Q or 4-4-7-7-K.)

86. *Data Analysis*   A sample of college students, faculty, and administration were asked whether they favored a proposed increase in the annual activity fee to enhance student life on campus. The results of the study are given in the table.

	Students	Faculty	Admin.	Total
Favor	237	37	18	292
Oppose	163	38	7	208
Total	400	75	25	500

A person is selected at random from the sample. Find the specified probability.

(a) The person is not in favor of the proposal.

(b) The person is a student.

(c) The person is a faculty member and is in favor of the proposal.

In Exercises 87 and 88, line plots of sets of data are given. Determine the mean and standard deviation of each set.

87. (a)   (b)

(c)   (d)

88. (a)   (b)

(c)   (d)

In Exercises 89–92, use a graphing utility to find the mean, median, and standard deviation of the data.

89. 14, 12, 6, 4, 11, 6, 18, 14, 6, 16

90. 5, 18, 8, 11, 12, 10, 18, 20, 15

91. 104.0, 143.9, 167.1, 182.4, 194.3, 206.0

92. 228.1, 298.6, 309.1, 335.2, 356.6, 393.8

Automotive Batteries   In Exercises 93 and 94, use the following data, which represent the lifetimes (in months) of a sample of 45 automotive batteries.

35	48	38	41	42	40	48	50	55
52	55	52	47	37	57	46	42	53
44	59	62	56	56	47	47	56	56
48	63	51	52	56	36	59	53	64
62	46	44	51	46	58	54	46	56

93. Use a graphing utility to determine the mean, median, and standard deviation of the data. How many of the batteries had a lifetime more than two standard deviations from the mean?

94. Organize the data graphically using a box-and-whisker plot.

95. *Log Length*   The lengths (in feet) of 30 logs are listed below.

25	26	26	27	27	28
28	28	29	29	30	30
30	30	30	31	31	31
31	32	32	32	32	32
33	34	34	35	35	36

(a) Find the mean, median, and mode of the data.

(b) Find the standard deviation of the data.

(c) What percent of the logs have lengths that differ from the mean by more than one standard deviation?

CHAPTER PROJECT   *Exploring Difference Quotients*

In this project, you will explore difference quotients and see how they can be used to find average rates.

Suppose you are driving from Atlanta to Miami. The trip is about 700 miles and takes you 12 hours. Your average speed is

$$\text{Average speed} = \frac{\text{distance}}{\text{time}} = \frac{700}{12} \approx 58.3 \text{ miles per hour.}$$

This concept can be generalized as follows. Let f be a function defined on the interval $[a, b]$. The **average rate of change of f** from a to b is given by

$$\text{Average rate of change} = \frac{f(b) - f(a)}{b - a}.$$

This expression is called a **difference quotient.**

(a) Calculate the average rate of change of $f(x) = 3x + 4$ on the interval $[2, 6]$. Select any other interval and show that you obtain the same average rate of change.

(b) Calculate the average rates of change of $f(x) = x^2$ on the intervals $[1, 3]$ and $[4, 6]$. Are they equal? Explain.

(c) During a 1-hour trip, your average speed is 50 miles per hour. Discuss the relationship between this speed and the speeds shown on your speedometer.

Questions for Further Exploration

1. Let $f(x) = x^2 - 2$. Calculate the average rates of change of f on the intervals $[1, 3]$, $[1, 2]$, $[1, 1.5]$, and $[1, 1.1]$.

(a) Find an expression for the average rate of change of f on the interval $[1, 1 + h]$, where h is any real number.

(b) Create a table that shows the average rate of change of f for several values of h. Choose values of h that get closer and closer to 0. What value does the average rate of change of f approach as h approaches 0?

(c) The answer to part (b) is the slope of the tangent line to the graph of f at the point $(1, f(1))$. Graph this line and f in the same viewing rectangle. Describe the behavior of the graphs near the point $(1, f(1))$.

2. The data in the table gives the costs of first-class postage in the United States for selected years from 1968 to 1995, where $t = 0$ corresponds to 1900. (Source: U.S. Postal Service)

t	68	71	74	75	78
Postage	$0.06	$0.08	$0.10	$0.13	$0.15

t	81	85	88	91	95
Postage	$0.20	$0.22	$0.25	$0.29	$0.32

(a) Find the average rate of change of the cost of postage between each two adjacent time intervals.

(b) When was the average rate of change largest? smallest?

(c) Are there intervals over which the average rate of change was zero? What does this mean?

9 /// CHAPTER TEST

Take this test as you would take a test in class. After you are done, check your work against the answers given in the back of the book.

 The *Interactive* CD-ROM provides answers to the Chapter Tests and Cumulative Tests. It also offers Chapter Pre-Tests (that test key skills and concepts covered in previous chapters) and Chapter Post-Tests, both of which have randomly generated exercises with diagnostic capabilities.

In Exercises 1 and 2, write the first five terms of the sequence.

1. $a_n = \left(-\frac{2}{3}\right)^{n-1}$ (Begin with $n = 1$.)

2. $a_1 = 12$, $a_{k+1} = a_k + 4$

In Exercises 3 and 4, find a formula for the nth term of the sequence.

3. Arithmetic: $a_1 = 5000$, $d = -100$

4. Geometric: $a_1 = 4$, $a_{k+1} = \frac{1}{2}a_k$

5. Use sigma notation to write: $\dfrac{2}{3(1) + 1} + \dfrac{2}{3(2) + 1} + \cdots + \dfrac{2}{3(12) + 1}$.

6. Find the sum of the first 50 positive integers that are multiples of 3.

7. Find the common ratio: $2, -3, \frac{9}{2}, -\frac{27}{4}, \frac{81}{8}, \ldots$.

8. Find the sum of the first eight terms of the geometric sequence in which $a_1 = 3$ and $a_2 = 6$.

9. Find the balance in an increasing annuity when a principal of $50 is deposited at the beginning of each month for 25 years. Assume the amount in the fund is compounded monthly at 8%.

10. Evaluate: $_{20}C_3$.

11. Find the coefficient of the term x^3y^5 in the expansion of $(x + y)^8$.

12. How many distinct license plates can be issued with one letter followed by a three-digit number?

13. Four students are randomly selected from a class of 25 to answer questions from a reading assignment. In how many ways can the four be selected?

Figure for 15

14. Suppose that two spark plugs require replacement in a four-cylinder engine. If the mechanic randomly removes two plugs, find the probability that they are the two defective plugs.

15. Find the mean and standard deviation of the data represented by the histogram in the figure.

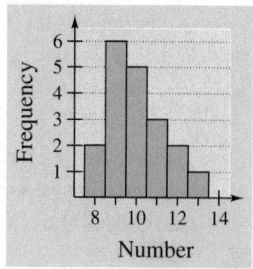

16. Sketch a box-and-whisker plot of the following data.
    46, 30, 28, 46, 26, 28, 26, 40, 24

Topics in
Analytic Geometry

On July 16, 1994, the comet Shoemaker-Levy 9 collided with Jupiter, the largest planet in our solar system. The dark spots in the inset photo of Jupiter show the results of the collision. The impact was filmed from the space probe Galileo, on its way to Jupiter.

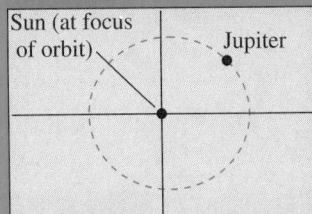

Sun (at focus of orbit)

Jupiter

All the planets in our solar system travel in elliptical paths about the sun. Comets, on the other hand, can have varying paths. Some are elliptical, some are parabolic, and some are hyperbolic.

Since the solar system's formation, thousands of comets have collided with the planets and their moons. For evidence of this, all you need to do is look at the moon's surface through a telescope. (See Exercises 51 and 52 on page 771.)

Alan Levenson; (inset) Courtesy of the
Space Telescope Science Institute/NASA.

Astronomers David Levy, Carolyn Shoemaker, and Eugene Shoemaker (from left to right) are shown with the 18-inch Schmidt telescope they used to discover the comet that is now named after them.

753

10.1   Introduction to Conics: Parabolas

Conics / Parabolas / Applications

Conics

Conic sections were discovered during the classical Greek period, 600 to 300 B.C. The early Greeks were concerned largely with the geometric properties of conics. It was not until the early 17th century that the broad applicability of conics became apparent, and they then played a prominent role in the early development of calculus.

Each **conic section** (or simply **conic**) is the intersection of a plane and a double-napped cone. Notice in Figure 10.1(a) that in the formation of the four basic conics, the intersecting plane does not pass through the vertex of the cone. When the plane does pass through the vertex, the resulting figure is a **degenerate conic,** as shown in Figure 10.1(b).

Figure 10.1

(a) Circle   Ellipse   Parabola   Hyperbola   **(b)** Point   Line   Two intersecting lines

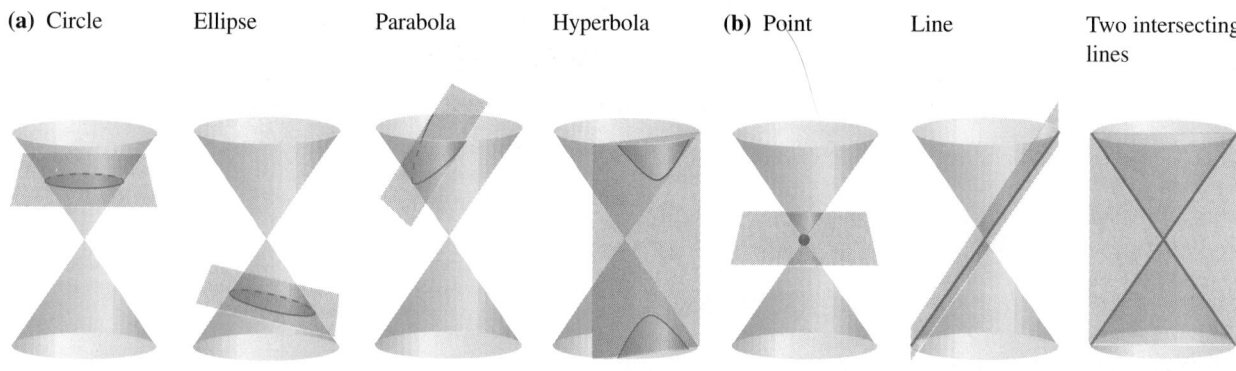

There are several ways to approach the study of conics. You could begin by defining conics in terms of the intersections of planes and cones, as the Greeks did, or you could define them algebraically, in terms of the general second-degree equation

$$Ax^2 + Bxy + Cy^2 + Dx + Ey + F = 0.$$

Note   The *Bxy* term in the general second-degree equation indicates that the axes of the conic are not parallel to the *x*- or *y*-axis. Equations with the *Bxy* term will be studied in Section 10.4.

However, you will study a third approach, in which each of the conics is defined as a **locus** (collection) of points satisfying a geometric property. For example, the definition of a circle as the collection of all points (x, y) that are equidistant from a fixed point (h, k) leads to the standard equation of a circle

$$(x - h)^2 + (y - k)^2 = r^2.$$ Equation of circle

Figure 10.2

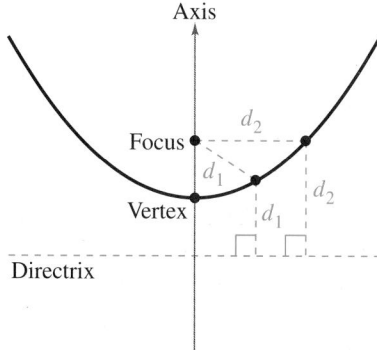

Parabolas

Definition of a Parabola

A **parabola** is the set of all points (x, y) that are equidistant from a fixed line (**directrix**) and a fixed point (**focus**) not on the line.

The midpoint between the focus and the directrix is called the **vertex,** and the line passing through the focus and the vertex is called the **axis** of the parabola. Note in Figure 10.2 that a parabola is symmetric with respect to its axis. Using the definition of a parabola, you can derive the following **standard form** of the equation of a parabola whose directrix is parallel to the x-axis or to the y-axis.

Standard Equation of a Parabola

The **standard form** of the equation of a parabola with vertex at (h, k) is as follows.

$$(x - h)^2 = 4p(y - k), \ p \neq 0 \quad \text{Vertical axis; directrix: } y = k - p$$

$$(y - k)^2 = 4p(x - h), \ p \neq 0 \quad \text{Horizontal axis; directrix: } x = h - p$$

The focus lies on the axis p units (*directed distance*) from the vertex. If the vertex is at the origin $(0, 0)$, the equation takes one of the following forms.

$$x^2 = 4py \qquad \text{Vertical axis}$$

$$y^2 = 4px \qquad \text{Horizontal axis}$$

See Figure 10.3.

Figure 10.3

(a) Vertical axis: $p > 0$ **(b)** Vertical axis: $p < 0$ **(c)** Horizontal axis: $p > 0$ **(d)** Horizontal axis: $p < 0$

$$(x - h)^2 = 4p(y - k) \qquad\qquad\qquad (y - k)^2 = 4p(x - h)$$

The *Interactive* CD-ROM shows every example with its solution; clicking on the *Try It!* button brings up similar problems. Guided Examples and Integrated Examples show step-by-step solutions to additional examples. Integrated Examples are related to several concepts in the section.

Figure 10.4

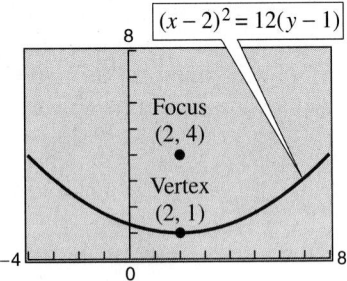

EXAMPLE 1 ▱ **Finding the Standard Equation of a Parabola**

Find the standard form of the equation of the parabola with vertex $(2, 1)$ and focus $(2, 4)$.

Solution

Because the axis of the parabola is vertical, consider the equation

$$(x - h)^2 = 4p(y - k)$$

where $h = 2$, $k = 1$, and $p = 4 - 1 = 3$. Thus, the standard form is

$$(x - 2)^2 = 12(y - 1).$$

The graph of this parabola is shown in Figure 10.4.   ▱

Note   By expanding the standard equation in Example 1, you can obtain the more common quadratic form $y = \frac{1}{12}(x^2 - 4x + 16)$. Try using a graphing utility to confirm the graph shown in Figure 10.4.

EXAMPLE 2 ▱ **Finding the Focus of a Parabola**

Find the focus of the parabola given by

$$y = -\frac{1}{2}x^2 - x + \frac{1}{2}.$$

Solution

To find the focus, convert to standard form by completing the square.

$$y = -\frac{1}{2}x^2 - x + \frac{1}{2}$$ Original equation

$$-2y = x^2 + 2x - 1$$ Multiply both sides by –2.

$$1 - 2y = x^2 + 2x$$ Group terms.

$$2 - 2y = x^2 + 2x + 1$$ Complete the square.

$$-2(y - 1) = (x + 1)^2$$ Standard form

Figure 10.5

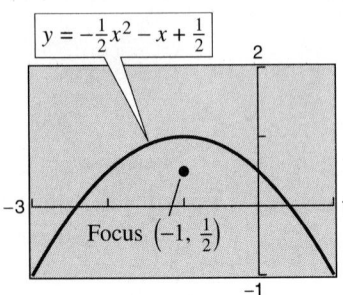

Comparing this equation with $(x - h)^2 = 4p(y - k)$, you can conclude that $h = -1$, $k = 1$, and $p = -\frac{1}{2}$. Because p is negative, the parabola opens downward, as shown in Figure 10.5. Therefore, the focus of the parabola is

$$(h, k + p) = \left(-1, \frac{1}{2}\right).$$ Focus   ▱

Figure 10.6

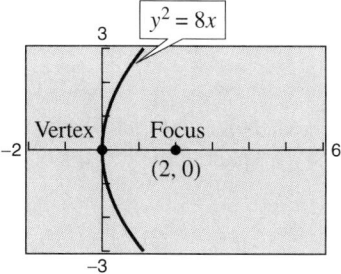

EXAMPLE 3   **Vertex at the Origin**

Find the standard equation of the parabola with vertex at the origin and focus $(2, 0)$.

Solution
The axis of the parabola is horizontal, passing through $(0, 0)$ and $(2, 0)$, as shown in Figure 10.6. Thus, the standard form is

$$y^2 = 4px$$

where $h = k = 0$ and $p = 2$. Therefore, the equation is

$$y^2 = 8x.$$

Note Try using a graphing utility to confirm the equation found in Example 3. To do this, it helps to split the equation into two parts: $y = \sqrt{8x}$ (upper part) and $y = -\sqrt{8x}$ (lower part).

Applications

A line segment that passes through the focus of a parabola and has endpoints on the parabola is called a **focal chord.** The specific focal chord perpendicular to the axis of the parabola is called the **latus rectum.**

Parabolas occur in a wide variety of applications. For instance, a parabolic reflector can be formed by revolving a parabola around its axis. The resulting surface has the property that all incoming rays parallel to the axis are reflected through the focus of the parabola; this is the principle behind the construction of the parabolic mirrors used in reflecting telescopes. Conversely, the light rays emanating from the focus of a parabolic reflector used in a flashlight are all parallel to one another, as shown in Figure 10.7.

A line is **tangent** to a parabola at a point on the parabola if the line intersects, but does not cross, the parabola at the point. Tangent lines to parabolas have special properties related to the use of parabolas in constructing reflective surfaces.

Figure 10.7

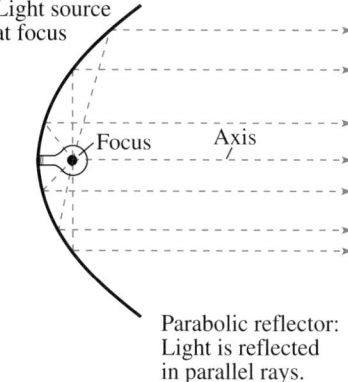

Parabolic reflector: Light is reflected in parallel rays.

Figure 10.8

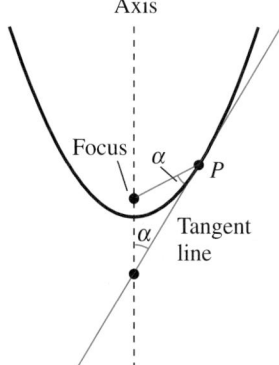

Reflective Property of a Parabola

The tangent line to a parabola at a point P makes equal angles with the following two lines (see Figure 10.8).

1. The line passing through P and the focus.
2. The axis of the parabola.

EXAMPLE 4   ◢   **Finding the Tangent Line at a Point on a Parabola**

Find the equation of the tangent line to the parabola given by $y = x^2$ at the point $(1, 1)$.

Solution

For this parabola, $p = \frac{1}{4}$ and the focus is $\left(0, \frac{1}{4}\right)$, as shown in Figure 10.9. You can find the y-intercept $(0, b)$ of the tangent line by equating the lengths of the two sides of the isosceles triangle

$$d_1 = \frac{1}{4} - b$$

Figure 10.9

and

$$d_2 = \sqrt{(1 - 0)^2 + \left[1 - \left(\frac{1}{4}\right)\right]^2} = \frac{5}{4}$$

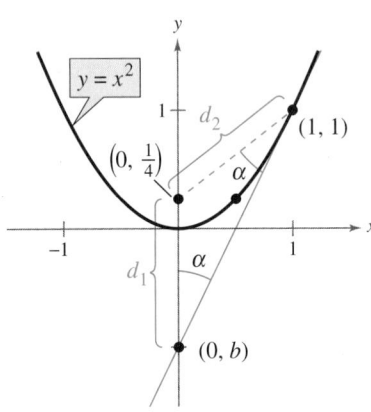

shown in Figure 10.9. Setting $d_1 = d_2$ produces

$$\frac{1}{4} - b = \frac{5}{4}$$

$$b = -1.$$

Thus, the slope of the tangent line is

$$m = \frac{1 - (-1)}{1 - 0} = 2$$

and its slope-intercept equation is

$$y = 2x - 1. \qquad\qquad\qquad ◢$$

Note   Try using a graphing utility to confirm the result of Example 4. By graphing

$$y = x^2 \qquad \text{and} \qquad y = 2x - 1$$

in the same viewing rectangle, you should be able to see that the line touches the parabola at the point $(1, 1)$.

Group Activity   *Television Antenna Dishes*

Cross sections of television antenna dishes are parabolic in shape. Write a paragraph describing why these dishes are parabolic. Include a graphical representation of your description.

10.1 /// EXERCISES

In Exercises 1–6, match the equation with its graph. [The graphs are labeled (a), (b), (c), (d), (e), and (f).]

(a)

(b)

(c)

(d)

(e)

(f)

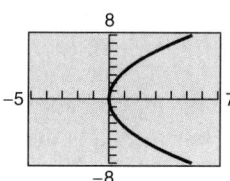

1. $y^2 = -4x$ — Sideways   **2.** $x^2 = 2y$

3. $x^2 = -8y$ **4.** $y^2 = 12x$

5. $(y - 1)^2 = 4(x - 3)$ **6.** $(x + 3)^2 = -2(y - 1)$

In Exercises 7–20, find the vertex, focus, and directrix of the parabola and sketch its graph.

7. $y = \frac{1}{2}x^2$ **8.** $y = 2x^2$

9. $y^2 = -6x$ **10.** $y^2 = 3x$

11. $x^2 + 8y = 0$ **12.** $x + y^2 = 0$

13. $(x - 1)^2 + 8(y + 2) = 0$

14. $(x + 3) + (y - 2)^2 = 0$

15. $\left(y + \frac{1}{2}\right)^2 = 2(x - 5)$

16. $\left(x + \frac{1}{2}\right)^2 = 4(y - 3)$

17. $y = \frac{1}{4}(x^2 - 2x + 5)$

18. $4x - y^2 - 2y - 33 = 0$

19. $y^2 + 6y + 8x + 25 = 0$

20. $y^2 - 4y - 4x = 0$

In Exercises 21–24, find the vertex, focus, and directrix of the parabola. Use a graphing utility to graph the parabola.

21. $y = -\frac{1}{6}(x^2 + 4x - 2)$

22. $x^2 - 2x + 8y + 9 = 0$

23. $y^2 + x + y = 0$

24. $y^2 - 4x - 4 = 0$

In Exercises 25 and 26, the equations of a parabola and a tangent line to the parabola are given. Use a graphing utility to graph both in the same viewing rectangle. Determine the coordinates of the point of tangency.

Parabola	*Tangent Line*
25. $y^2 - 8x = 0$	$x - y + 2 = 0$
26. $x^2 + 12y = 0$	$x + y - 3 = 0$

In Exercises 27 and 28, change the equation so that its graph matches the given graph.

27. $y^2 = -6x$ **28.** $y^2 = 9x$

The *Interactive* CD-ROM contains step-by-step solutions to all odd-numbered Section and Review Exercises. It also provides Tutorial Exercises, which link to Guided Examples for additional help.

29. *Exploration* Consider the parabola $x^2 = 4py$.

(a) Use a graphing utility to graph the parabola for $p = 1, p = 2, p = 3,$ and $p = 4$. Describe the effect on the graph when p increases.

(b) Locate the focus for each parabola in part (a).

(c) For each parabola in part (a), find the length of the chord passing through the focus and parallel to the directrix. How can the length of this chord be determined directly from the standard form of the equation of the parabola?

(d) Explain how the result of part (c) can be used as a sketching aid when graphing parabolas.

30. *True or False?* Determine whether the following statement is true or false. If it is false, explain why. "It is possible for a parabola to intersect its directrix."

In Exercises 31–42, find an equation of the parabola with its vertex at the origin.

31.

32.

33. Focus: $\left(0, -\frac{3}{2}\right)$

34. Focus: $(2, 0)$

35. Focus: $(-2, 0)$

36. Focus: $(0, -2)$

37. Directrix: $y = -1$

38. Directrix: $x = 3$

39. Directrix: $y = 2$

40. Directrix: $x = -2$

41. Horizontal axis and passes through the point $(4, 6)$

42. Vertical axis and passes through the point $(-2, -2)$

In Exercises 43–52, find an equation of the parabola.

43.

44.

45.

46.

47. Vertex: $(3, 2)$; Focus: $(1, 2)$

48. Vertex: $(-1, 2)$; Focus: $(-1, 0)$

49. Vertex: $(0, 4)$; Directrix: $y = 2$

50. Vertex: $(-2, 1)$; Directrix: $x = 1$

51. Focus: $(2, 2)$; Directrix: $x = -2$

52. Focus: $(0, 0)$; Directrix: $y = 4$

In Exercises 53 and 54, change the equation so its graph matches the description.

53. $(y - 3)^2 = 6(x + 1)$; upper half of parabola

54. $(y + 1)^2 = 2(x - 2)$; lower half of parabola

55. *Satellite Antenna* The receiver in a parabolic television dish antenna is 3.5 feet from the vertex and is located at the focus (see figure). Find an equation of a cross section of the reflector. (Assume that the dish is directed upward and the vertex is at the origin.)

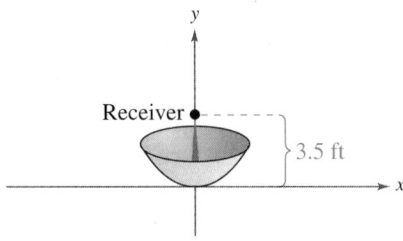

56. *Suspension Bridge* Each cable of a suspension bridge is suspended (in the shape of a parabola) between two towers that are 120 meters apart, and the top of each tower is 20 meters above the roadway. The cables touch the roadway midway between the towers.

(a) Create a sketch for solving the problem. Draw a rectangular coordinate system on the bridge with the center of the bridge at the origin. Identify the coordinates of the known points.

(b) Find an equation for the parabolic shape of each cable.

(c) Complete the table by finding the height of the suspension cables above the roadway at a distance of x meters from the center of the bridge.

x	0	20	40	60
y				

57. *Beam Deflection* A simply supported beam is 16 meters long and has a load at the center (see figure). The deflection of the beam at its center is 3 centimeters. Assume the shape of the deflected beam is parabolic.

(a) Find an equation of the parabola. (Assume that the origin is at the center of the beam.)

(b) How far from the center of the beam is the deflection 1 centimeter?

3 cm

16 m

(not to scale)

58. *Revenue* The revenue R generated by the sale of x units is given by

$$R = 375x - \frac{3}{2}x^2.$$

Use a graphing utility to graph the function and approximate the number of sales that will maximize revenue.

59. *Highway Design* Highway engineers design a parabolic curve for an entrance ramp from a straight street to an interstate highway (see figure). Find an equation of the parabola.

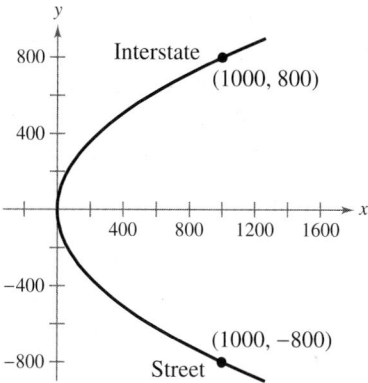

Interstate

(1000, 800)

400   800   1200   1600

(1000, −800)

Street

60. *Satellite Orbit* An earth satellite in a 100-mile-high circular orbit around the earth has a velocity of approximately 17,500 miles per hour. If this velocity is multiplied by $\sqrt{2}$, the satellite will have the minimum velocity necessary to escape the earth's gravity and it will follow a parabolic path with the center of the earth as the focus (see figure).

(a) Find the escape velocity of the satellite.

(b) Find an equation of its path (assume the radius of the earth is 4000 miles).

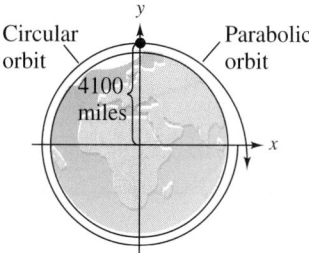

Circular orbit

Parabolic orbit

4100 miles

61. *Projectile Motion* A bomber is flying at an altitude of 30,000 feet and a speed of 540 miles per hour (792 feet per second). When should a bomb be dropped so that it will hit the target if the path of the bomb is modeled by

$$y = 30,000 - \frac{x^2}{39,204}?$$

62. *Path of a Projectile* The path of a softball is given by the equation

$$y = -0.08x^2 + x + 4.$$

The coordinates x and y are measured in feet, with $x = 0$ corresponding to the position where the ball was thrown.

  (a) Use a graphing utility to graph the trajectory of the softball.

  (b) Move the cursor along the path to approximate the highest point and the range of the trajectory.

63. *Exploration* Let (x_1, y_1) be the coordinates of a point on the parabola $x^2 = 4py$. The equation of the line tangent to the parabola at the point is

$$y - y_1 = \frac{x_1}{2p}(x - x_1).$$

What is the slope of the tangent line?

64. *Area* The area of the shaded region in the figure is given by

$$A = \frac{8}{3}p^{1/2}b^{3/2}.$$

  (a) Find the area if $p = 2$ and $b = 4$.

  (b) Give a geometric explanation of why the area approaches 0 as p approaches 0.

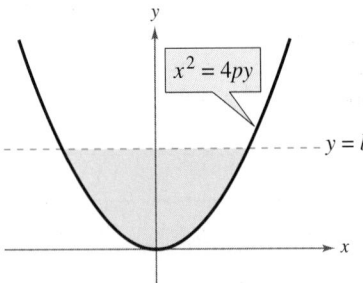

In Exercises 65–68, find an equation of the tangent line to the parabola at the given point and find the x-intercept of the line.

65. $x^2 = 2y$, $(4, 8)$ **67.** $y = -2x^2$, $(-1, -2)$

66. $x^2 = 2y$, $\left(-3, \frac{9}{2}\right)$ **68.** $y = -2x^2$, $(3, -18)$

Projectile Motion In Exercises 69 and 70, consider the path of a projectile projected horizontally with a velocity of v feet per second at a height of s feet, where the model for the path is

$$y = -\frac{16}{v^2}x^2 + s.$$

In this model, air resistance is disregarded and y is the height (in feet) of the projectile t seconds after its release.

69. A ball is thrown from the top of a 75-foot tower with a velocity of 32 feet per second.

  (a) Find the equation of the parabolic path.

  (b) How far does the ball travel horizontally before striking the ground?

70. A bomber flying due east at 550 miles per hour at an altitude of 42,000 feet releases a bomb. Determine the distance the bomb travels horizontally before striking the ground.

Review Solve Exercises 71–74 as a review of the skills and problem-solving techniques you learned in previous sections.

71. Find a polynomial function with integer coefficients that has the zeros 3, $2 + i$, and $2 - i$.

72. Find all the zeros of

$$f(x) = 2x^3 - 3x^2 + 50x - 75$$

if one of the zeros is $x = \frac{3}{2}$.

73. Find all the zeros of the function

$$g(x) = 6x^4 + 7x^3 - 29x^2 - 28x + 20$$

if two of the zeros are $x = \pm2$.

74. Use a graphing utility to graph the function

$$h(x) = 2x^4 + x^3 - 19x^2 - 9x + 9.$$

Use the graph to approximate the zeros of h.

10.2   Ellipses

Introduction

The second type of conic is called an **ellipse.** It is defined as follows.

Definition of an Ellipse

An **ellipse** is the set of all points (x, y) the sum of whose distances from two distinct fixed points (**foci**) is constant. (See Figure 10.10.)

Note   The line through the foci intersects the ellipse at two points (**vertices**). The chord joining the vertices is the **major axis,** and its midpoint is the **center** of the ellipse. The chord perpendicular to the major axis at the center is the **minor axis** of the ellipse.

Figure 10.10 $d_1 + d_2$ is a constant.

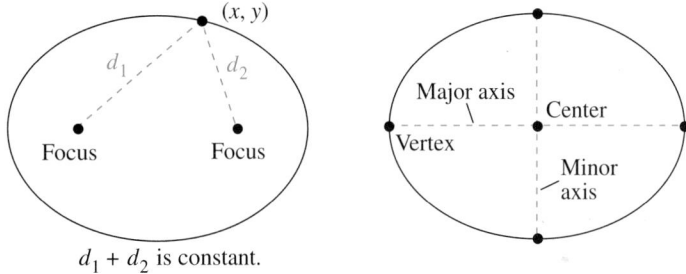

$d_1 + d_2$ is constant.

To derive the standard form of the equation of an ellipse, consider the ellipse in Figure 10.11 with the following points: center, (h, k); vertices, $(h \pm a, k)$; foci, $(h \pm c, k)$. The sum of the distances from any point on the ellipse to the two foci is constant. Using a vertex point, this constant sum is

$$(a + c) + (a - c) = 2a \qquad \text{Length of major axis}$$

or simply the length of the major axis. Now, if you let (x, y) be *any* point on the ellipse, the sum of the distances between (x, y) and the two foci must also be $2a$. That is,

$$\sqrt{[x - (h - c)]^2 + (y - k)^2} + \sqrt{[x - (h + c)]^2 + (y - k)^2} = 2a.$$

Finally, using Figure 10.11 and $b^2 = a^2 - c^2$, you obtain the following equation of the ellipse.

$$b^2(x - h)^2 + a^2(y - k)^2 = a^2b^2$$
$$\frac{(x - h)^2}{a^2} + \frac{(y - k)^2}{b^2} = 1$$

Figure 10.11

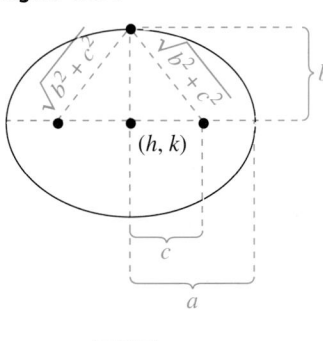

$$2\sqrt{b^2 + c^2} = 2a$$
$$b^2 + c^2 = a^2$$

In the development on page 763, you would obtain a similar equation if you began with a vertical major axis. Both results are summarized as follows.

Standard Equation of an Ellipse

The standard form of the equation of an ellipse, with center (h, k) and major and minor axes of lengths $2a$ and $2b$, where $0 < b < a$, is

$$\frac{(x - h)^2}{a^2} + \frac{(y - k)^2}{b^2} = 1 \qquad \text{Major axis is horizontal.}$$

$$\frac{(x - h)^2}{b^2} + \frac{(y - k)^2}{a^2} = 1. \qquad \text{Major axis is vertical.}$$

The foci lie on the major axis, c units from the center, with $c^2 = a^2 - b^2$. If the center is at the origin $(0, 0)$, the equation takes one of the following forms.

$$\frac{x^2}{a^2} + \frac{y^2}{b^2} = 1 \qquad \text{Major axis is horizontal.}$$

$$\frac{x^2}{b^2} + \frac{y^2}{a^2} = 1 \qquad \text{Major axis is vertical.}$$

Figure 10.12 shows both the vertical and horizontal orientations for an ellipse.

Figure 10.13

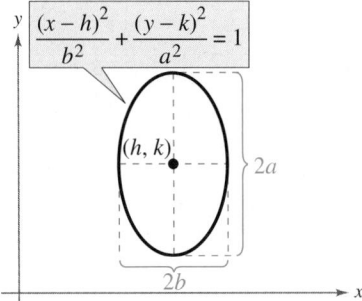

Figure 10.12

A computer animation of this concept appears in the *Interactive* CD-ROM.

You can visualize the definition of an ellipse by imagining two thumbtacks placed at the foci, as shown in Figure 10.13. If the ends of a fixed length of string are fastened to the thumbtacks and the string is drawn taut with a pencil, the path traced by the pencil will be an ellipse.

EXAMPLE 1   Finding the Standard Equation of an Ellipse

Find the standard form of the equation of the ellipse having foci at $(0, 1)$ and $(4, 1)$ and a major axis of length 6, as shown in Figure 10.14.

Solution

Because the foci occur at $(0, 1)$ and $(4, 1)$, the center of the ellipse is $(2, 1)$. This implies that the distance from the center to one of the foci is $c = 2$, and because $2a = 6$ you know that $a = 3$. Now, from $c^2 = a^2 - b^2$, you have

$$b = \sqrt{a^2 - c^2}$$
$$= \sqrt{9 - 4}$$
$$= \sqrt{5}.$$

Because the major axis is horizontal, the standard equation is

$$\frac{(x - 2)^2}{9} + \frac{(y - 1)^2}{5} = 1.$$

Figure 10.14

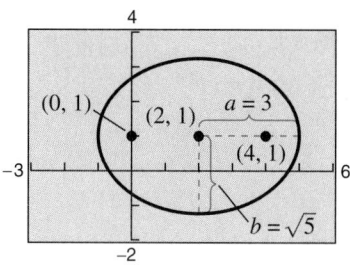

Note In Example 1, note the use of the equation $c^2 = a^2 - b^2$. Don't confuse this equation with the Pythagorean Theorem—there is a sign difference.

EXAMPLE 2   Writing an Equation in Standard Form

Sketch the graph of the ellipse whose equation is

$$x^2 + 4y^2 + 6x - 8y + 9 = 0.$$

Solution

Begin by writing the given equation in standard form. In the fourth step, note that 9 and 4 are added to *both* sides of the equation.

$$x^2 + 4y^2 + 6x - 8y + 9 = 0 \quad \text{Original equation}$$
$$\left(x^2 + 6x + \quad\right) + \left(4y^2 - 8y + \quad\right) = -9 \quad \text{Group terms.}$$
$$\left(x^2 + 6x + \quad\right) + 4\left(y^2 - 2y + \quad\right) = -9 \quad \text{Factor 4 out of } y\text{-terms.}$$
$$(x^2 + 6x + 9) + 4(y^2 - 2y + 1) = -9 + 9 + 4(1)$$
$$(x + 3)^2 + 4(y - 1)^2 = 4 \quad \text{Completed square form}$$
$$\frac{(x + 3)^2}{4} + \frac{(y - 1)^2}{1} = 1 \quad \text{Standard form}$$

Now you see that the center is at $(h, k) = (-3, 1)$. Because the denominator of the x-term is $a^2 = 2^2$, you can locate the endpoints of the major axis two units to the right and left of the center. Similarly, because the denominator of the y-term is $b^2 = 1^2$, you can locate the endpoints of the minor axis one unit up and down from the center. The graph of this ellipse is shown in Figure 10.15.

Figure 10.15

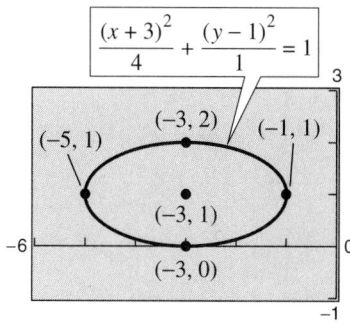

You can use a graphing utility to graph an ellipse by graphing the upper and lower portions in the same viewing rectangle. For instance, to graph the ellipse in Example 2, first solve for y to get

$$y_1 = 1 + \sqrt{1 - \frac{(x+3)^2}{4}} \qquad \text{Top of ellipse}$$

and

$$y_2 = 1 - \sqrt{1 - \frac{(x+3)^2}{4}}. \qquad \text{Bottom of ellipse}$$

Use a viewing rectangle in which $-6 \le x \le 0$ and $-1 \le y \le 3$. You should obtain the graph shown in Figure 10.15.

EXPLORATION

Use the parametric mode of a graphing utility to graph the ellipse given by

$$x = -3 + 2\cos t$$
$$y = 1 + \sin t$$

for $0 \le t \le 6.3$. How does the graph compare with Figure 10.15?

 The *Interactive* CD-ROM offers graphing utility emulators of the *TI-82* and *TI-83*, which can be used with the Examples, Explorations, Technology notes, and Exercises.

Figure 10.16

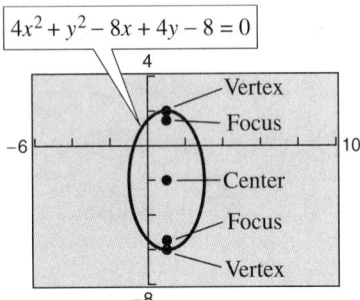

EXAMPLE 3   Analyzing an Ellipse

Find the center, vertices, and foci of the ellipse given by

$$4x^2 + y^2 - 8x + 4y - 8 = 0.$$

Solution

By completing the square, you can write the given equation in standard form.

$$4x^2 + y^2 - 8x + 4y - 8 = 0$$
$$\left(4x^2 - 8x + \boxed{}\right) + \left(y^2 + 4y + \boxed{}\right) = 8$$
$$4\left(x^2 - 2x + \boxed{}\right) + \left(y^2 + 4y + \boxed{}\right) = 8$$
$$4(x^2 - 2x + 1) + (y^2 + 4y + 4) = 8 + 4(1) + 4$$
$$4(x - 1)^2 + (y + 2)^2 = 16$$
$$\frac{(x - 1)^2}{4} + \frac{(y + 2)^2}{16} = 1$$

Thus, the major axis is vertical, where $h = 1, k = -2, a = 4, b = 2$, and

$$c = \sqrt{16 - 4} = 2\sqrt{3}.$$

Therefore, you have the following.

     Center: $(1, -2)$ Vertices: $(1, -6)$ Foci: $\left(1, -2 - 2\sqrt{3}\right)$

                                                $(1, 2)$ $\left(1, -2 + 2\sqrt{3}\right)$

The graph of the ellipse is shown in Figure 10.16.

Applications

Ellipses have many practical and aesthetic uses. For instance, machine gears, supporting arches, and acoustical designs often involve elliptical shapes. The orbits of satellites and planets are also ellipses. Example 4 investigates the elliptical orbit of the moon about the earth.

Figure 10.17

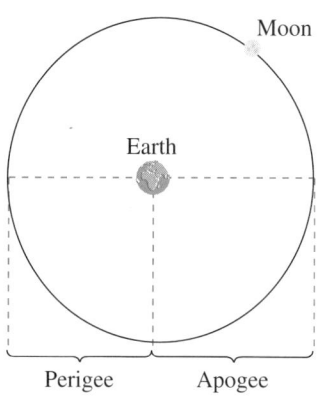

Real Life

EXAMPLE 4   **An Application Involving an Elliptical Orbit**

The moon travels about the earth in an elliptical orbit with the earth at one focus, as shown in Figure 10.17. The major and minor axes of the orbit have lengths of 768,806 kilometers and 767,746 kilometers, respectively. Find the greatest and least distances (the apogee and perigee) from the earth's center to the moon's center.

Solution

Because $2a = 768,806$ and $2b = 767,746$, you have $a = 384,403$ and $b = 383,873$, which implies that

$$\begin{aligned} c &= \sqrt{a^2 - b^2} \\ &= \sqrt{384,403^2 - 383,873^2} \\ &\approx 20,179. \end{aligned}$$

Therefore, the greatest distance between the center of the earth and the center of the moon is

$$a + c \approx 404,582 \text{ km}$$

and the least distance is

$$a - c \approx 364,224 \text{ km}.$$

Eccentricity

One of the reasons it was difficult for early astronomers to detect that the orbits of the planets are ellipses is that the foci of the planetary orbits are relatively close to their centers, thus making the orbits nearly circular. To measure the ovalness of an ellipse, you can use the concept of **eccentricity.**

Definition of Eccentricity

The **eccentricity** e of an ellipse is given by the ratio

$$e = \frac{c}{a}.$$

To see how this ratio is used to describe the shape of an ellipse, note that because the foci of an ellipse are located along the major axis between the vertices and the center, it follows that

$$0 < c < a.$$

For an ellipse that is nearly circular, the foci are close to the center and the ratio c/a is small [Figure 10.18a]. On the other hand, for an elongated ellipse, the foci are close to the vertices and the ratio c/a is close to 1 [Figure 10.18(b)].

Figure 10.18

(a) **(b)**

Note Note that $0 < e < 1$ for every ellipse.

The orbit of the moon has an eccentricity of $e = 0.0549$, and the eccentricities of the nine planetary orbits are as follows.

Mercury:	$e = 0.2056$	Saturn:	$e = 0.0543$
Venus:	$e = 0.0068$	Uranus:	$e = 0.0460$
Earth:	$e = 0.0167$	Neptune:	$e = 0.0082$
Mars:	$e = 0.0934$	Pluto:	$e = 0.2481$
Jupiter:	$e = 0.0484$		

Group Activity

Graphing Ellipses

Write an equation of an ellipse and graph it on graph paper. Do not write the equation on your graph. Exchange graphs with another student. Use the graph you receive to reconstruct the equation of the ellipse it represents. Find the eccentricity of the ellipse. Compare and discuss your results.

10.2 /// EXERCISES

In Exercises 1–6, match the equation with its graph. [The graphs are labeled (a), (b), (c), (d), (e), and (f).]

(a)

(b)

(c)

(d)

(e)

(f)

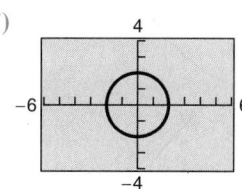

1. $\dfrac{x^2}{4} + \dfrac{y^2}{9} = 1$

2. $\dfrac{x^2}{9} + \dfrac{y^2}{4} = 1$

3. $\dfrac{x^2}{4} + \dfrac{y^2}{25} = 1$

4. $\dfrac{y^2}{4} + \dfrac{x^2}{4} = 1$

5. $\dfrac{(x-2)^2}{16} + (y+1)^2 = 1$

6. $\dfrac{(x+2)^2}{4} + \dfrac{(y+2)^2}{16} = 1$

In Exercises 7–18, find the center, vertices, foci, and eccentricity of the ellipse, and sketch its graph.

7. $\dfrac{x^2}{25} + \dfrac{y^2}{16} = 1$

8. $\dfrac{x^2}{144} + \dfrac{y^2}{169} = 1$

9. $\dfrac{x^2}{16} + \dfrac{y^2}{25} = 1$

10. $\dfrac{x^2}{169} + \dfrac{y^2}{144} = 1$

11. $\dfrac{x^2}{9} + \dfrac{y^2}{5} = 1$

12. $\dfrac{x^2}{28} + \dfrac{y^2}{64} = 1$

13. $\dfrac{(x-1)^2}{9} + \dfrac{(y-5)^2}{25} = 1$

14. $(x+2)^2 + \dfrac{(y+4)^2}{1/4} = 1$

15. $9x^2 + 4y^2 + 36x - 24y + 36 = 0$

16. $9x^2 + 4y^2 - 36x + 8y + 31 = 0$

17. $16x^2 + 25y^2 - 32x + 50y + 16 = 0$

18. $9x^2 + 25y^2 - 36x - 50y + 61 = 0$

In Exercises 19–22, use a graphing utility to graph the ellipse. Find the center, foci, and vertices. (Recall that it may be necessary to solve the equation for y and obtain two functions.)

19. $5x^2 + 3y^2 = 15$

20. $x^2 + 4y^2 = 4$

21. $12x^2 + 20y^2 - 12x + 40y - 37 = 0$

22. $36x^2 + 9y^2 + 48x - 36y + 43 = 0$

In Exercises 23 and 24, change the equation so that its graph matches the description.

23. $\dfrac{(x-3)^2}{9} + \dfrac{y^2}{4} = 1$; right half of ellipse

24. $\dfrac{(x+1)^2}{16} + \dfrac{(y-2)^2}{25} = 1$; bottom half of ellipse

In Exercises 25–32, find an equation of the ellipse with its center at the origin.

25. Vertices: $(\pm 5, 0)$; Foci: $(\pm 2, 0)$

26. Vertices: $(0, \pm 8)$; Foci: $(0, \pm 4)$

27.

28.

29. Foci: $(\pm 5, 0)$; Major axis of length 12

30. Foci: $(\pm 2, 0)$; Major axis of length 8

31. Vertices: $(0, \pm 5)$; Passes through the point $(4, 2)$

32. Major axis is vertical;
Passes through the points $(0, 4)$ and $(2, 0)$

In Exercises 33–44, find an equation for the ellipse.

33.

34.

35.

36.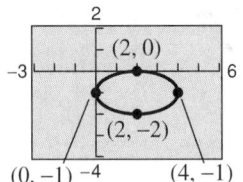

37. Vertices: $(0, 2)$, $(4, 2)$; Minor axis of length 2

38. Foci: $(0, 0)$, $(4, 0)$; Major axis of length 8

39. Foci: $(0, 0)$, $(0, 8)$; Major axis of length 16

40. Center: $(2, -1)$; Vertex: $(2, \frac{1}{2})$;
Minor axis of length 2

41. Vertices: $(3, 1)$, $(3, 9)$; Minor axis of length 6

42. Center: $(3, 2)$; $a = 3c$; Foci: $(1, 2)$, $(5, 2)$

43. Center: $(0, 4)$; $a = 2c$; Vertices: $(-4, 4)$, $(4, 4)$

44. Vertices: $(5, 0)$, $(5, 12)$;
Endpoints of the minor axis: $(0, 6)$, $(10, 6)$

45. *Think About It*   At the beginning of this section it was noted that an ellipse can be drawn using two thumbtacks, a string of fixed length (greater than the distance between the two tacks), and a pencil (see Figure 10.13). If the ends of the string are fastened at the tacks and the string is drawn taut with a pencil, the path traced by the pencil is an ellipse.

(a) What is the length of the string in terms of a?

(b) Explain why the path is an ellipse.

46. *Fireplace Arch*   A fireplace arch is to be constructed in the shape of a semiellipse. The opening is to have a height of 2 feet at the center and a width of 6 feet along the base (see figure). The contractor draws the outline of the ellipse by the method described in Exercise 45. Give the required positions of the tacks and the length of the string.

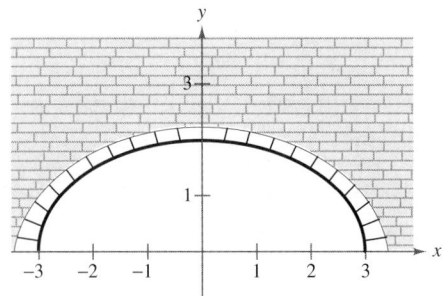

47. *Exploration*   The area A of the ellipse

$$\frac{x^2}{a^2} + \frac{y^2}{b^2} = 1$$

is $A = \pi ab$. For a particular application, $a + b = 20$.

(a) Write the area of the ellipse as a function of a.

(b) Find the equation of an ellipse with an area of 264 square centimeters.

(c) Complete the table and make a conjecture about the shape of the ellipse with a maximum area.

a	8	9	10	11	12	13
A						

(d) Use a graphing utility to graph the area function, and use the graph to make a conjecture about the shape of the ellipse that yields a maximum area.

48. *Geometry* The area of the ellipse in the figure is twice the area of the circle. What is the length of the major axis? (*Hint: A* $= \pi ab$.)

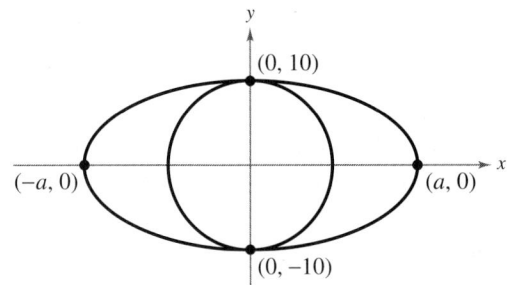

49. Find an equation of the ellipse with vertices $(\pm 5, 0)$ and eccentricity $e = \frac{3}{5}$.

50. Find an equation of the ellipse with vertices $(0, \pm 8)$ and eccentricity $e = \frac{1}{2}$.

51. *Chapter Opener* Halley's comet has an elliptical orbit with the sun at one focus. The eccentricity of the orbit is approximately 0.97. The length of the major axis of the orbit is about 36.23 astronomical units. (An astronomical unit is about 93 million miles.) Find an equation for the orbit. Place the center of the orbit at the origin and place the major axis on the x-axis.

52. *Chapter Opener* The comet Encke has an elliptical orbit with the sun at one focus. Encke ranges from 0.34 to 4.08 astronomical units from the sun. Find an equation of the orbit. Place the center of the orbit at the origin and place the major axis on the x-axis.

53. *Exploration*

(a) Show that the equation of an ellipse can be written as
$$\frac{(x-h)^2}{a^2} + \frac{(y-k)^2}{a^2(1-e^2)} = 1.$$

(b) Use a graphing utility to graph the ellipse
$$\frac{(x-2)^2}{4} + \frac{(y-3)^2}{4(1-e^2)} = 1$$
for $e = 0.95$, $e = 0.75$, $e = 0.5$, $e = 0.25$, and $e = 0$.

(c) Use the results of part (b) to make a conjecture about the change in the shape of the ellipse as e approaches 0.

54. *Geometry* A line segment through a focus with endpoints on the ellipse and perpendicular to the major axis is called a **latus rectum** of the ellipse. Therefore, an ellipse has two latera recta. Knowing the length of the latera recta is helpful in sketching an ellipse because it yields other points on the curve (see figure). Show that the length of each latus rectum is $2b^2/a$.

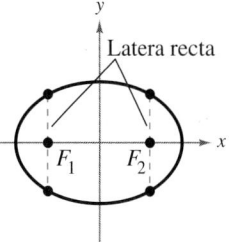

In Exercises 55–58, sketch the graph of the ellipse, making use of the latera recta (see Exercise 54).

55. $\dfrac{x^2}{4} + \dfrac{y^2}{1} = 1$

56. $\dfrac{x^2}{9} + \dfrac{y^2}{16} = 1$

57. $9x^2 + 4y^2 = 36$

58. $5x^2 + 3y^2 = 15$

59. *True or False?* The graph of $(x^2/4) + y^4 = 1$ is an ellipse. Explain.

60. *True or False?* The graph of $(x^4/4) + y^4 = 1$ is an ellipse. Explain.

10.3   Hyperbolas

Introduction / Asymptotes of a Hyperbola / Applications

Introduction

The definition of a hyperbola parallels that of an ellipse. The difference is that for an ellipse the *sum* of the distances between the foci and a point on the ellipse is fixed, whereas for a hyperbola the *difference* of these distances is fixed.

> ### Definition of a Hyperbola
>
> A **hyperbola** is the set of all points (x, y) the difference of whose distances from two distinct fixed points **(foci)** is constant. (See Figure 10.19.)

Figure 10.19

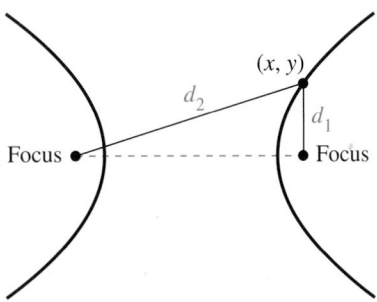

$d_2 - d_1$ is constant.

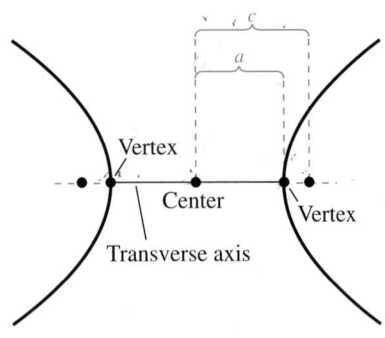

Every hyperbola has two disconnected **branches.** The line through the two foci intersects a hyperbola at its two **vertices.** The line segment connecting the vertices is called the **transverse axis,** and the midpoint of the transverse axis is called the **center** of the hyperbola. The development of the standard form of the equation of a hyperbola is similar to that of an ellipse.

> ### Standard Equation of a Hyperbola
>
> The standard form of the equation of a hyperbola with center at (h, k) is
>
> $$\frac{(x - h)^2}{a^2} - \frac{(y - k)^2}{b^2} = 1 \qquad \text{Transverse axis is horizontal.}$$
>
> $$\frac{(y - k)^2}{a^2} - \frac{(x - h)^2}{b^2} = 1. \qquad \text{Transverse axis is vertical.}$$
>
> The vertices are a units from the center, and the foci are c units from the center. Moreover, $c^2 = a^2 + b^2$. If the center of the hyperbola is at the origin $(0, 0)$, the equation takes one of the following forms.
>
> $$\frac{x^2}{a^2} - \frac{y^2}{b^2} = 1 \qquad \text{Transverse axis is horizontal.}$$
>
> $$\frac{y^2}{a^2} - \frac{x^2}{b^2} = 1 \qquad \text{Transverse axis is vertical.}$$

Note   Note that a, b, and c are related differently for hyperbolas than for ellipses.

Figure 10.20 shows both the horizontal and vertical orientations for a hyperbola.

Figure 10.20

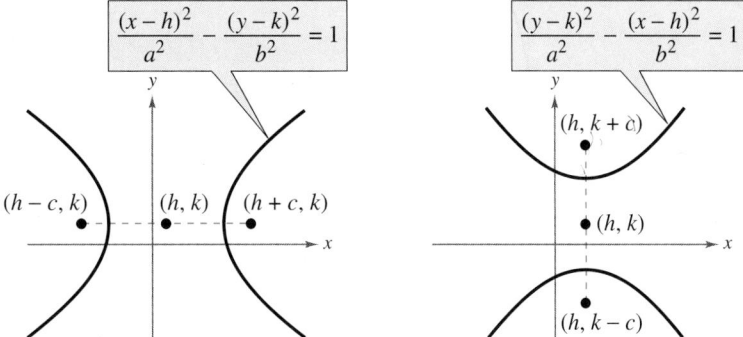

EXAMPLE 1 ▱   **Finding the Standard Equation of a Hyperbola**

Find the standard form of the equation of the hyperbola with foci at $(-1, 2)$ and $(5, 2)$ and vertices at $(0, 2)$ and $(4, 2)$.

Solution

By the Midpoint Formula, the center of the hyperbola occurs at the point $(2, 2)$. Furthermore, $c = 3$ and $a = 2$, and it follows that

$$b^2 = c^2 - a^2 = 3^2 - 2^2 = 9 - 4 = 5.$$

Thus, the equation of the hyperbola is

$$\frac{(x-2)^2}{4} - \frac{(y-2)^2}{5} = 1.$$

Figure 10.21 shows the hyperbola.   ▱

Figure 10.21

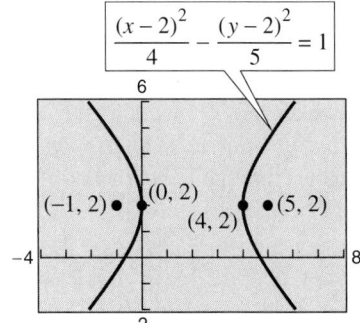

  E X P L O R A T I O N

Most graphing utilities have a parametric mode. Try using parametric mode to graph the hyperbola given by $x = 2 + 2 \sec t$ and $y = 2 + \sqrt{5} \tan t$. How does the result compare with the graph given in Figure 10.21? (Let the parameter vary from Tmin = 0 to Tmax = 6.28, with Tstep = 0.13.)

Figure 10.22

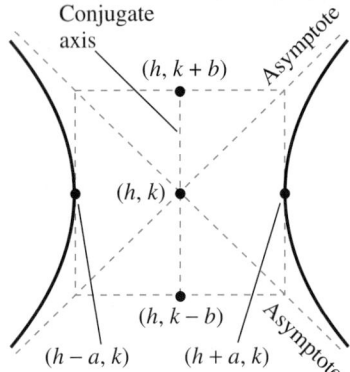

Asymptotes of a Hyperbola

Each hyperbola has two **asymptotes** that intersect at the center of the hyperbola, as shown in Figure 10.22. The asymptotes pass through the vertices of a rectangle of dimensions $2a$ by $2b$, with its center at (h, k).

Asymptotes of a Hyperbola	
$y = k \pm \dfrac{b}{a}(x - h)$	Asymptotes for horizontal transverse axis
$y = k \pm \dfrac{a}{b}(x - h)$	Asymptotes for vertical transverse axis

EXAMPLE 2   **Using Asymptotes to Sketch a Hyperbola**

Sketch the hyperbola whose equation is $4x^2 - y^2 = 16$.

Solution

$$4x^2 - y^2 = 16 \qquad \text{Original equation}$$

$$\frac{4x^2}{16} - \frac{y^2}{16} = \frac{16}{16} \qquad \text{Divide both sides by 16.}$$

$$\frac{x^2}{4} - \frac{y^2}{16} = 1 \qquad \text{Standard form}$$

From this, you can conclude that the transverse axis is horizontal and the vertices occur at $(-2, 0)$ and $(2, 0)$. Moreover, the ends of the conjugate axis occur at $(0, -4)$ and $(0, 4)$, and you are able to sketch the rectangle shown in Figure 10.23(a). Finally, by drawing the asymptotes through the corners of this rectangle, complete the sketch, as shown in Figure 10.23(b).

Note   The **conjugate axis** of a hyperbola is defined as follows.

Horizontal Transverse Axis: The conjugate axis is the line segment of length $2b$ joining $(h, k + b)$ and $(h, k - b)$.

Vertical Transverse Axis: The conjugate axis is the line segment of length $2b$ joining $(h + b, k)$ and $(h - b, k)$.

Figure 10.23

(a)

(b)

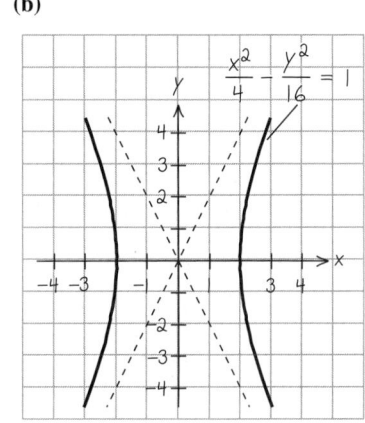

EXAMPLE 3 ▭ **Finding the Asymptotes of a Hyperbola**

Sketch the hyperbola given by $4x^2 - 3y^2 + 8x + 16 = 0$ and find the equations of its asymptotes.

Solution

$$4x^2 - 3y^2 + 8x + 16 = 0 \qquad \text{Original equation}$$

$$4(x^2 + 2x) - 3y^2 = -16$$

$$4(x^2 + 2x + 1) - 3y^2 = -16 + 4$$

$$4(x + 1)^2 - 3y^2 = -12 \qquad \text{Complete the square.}$$

$$\frac{y^2}{4} - \frac{(x + 1)^2}{3} = 1 \qquad \text{Standard form}$$

Figure 10.24

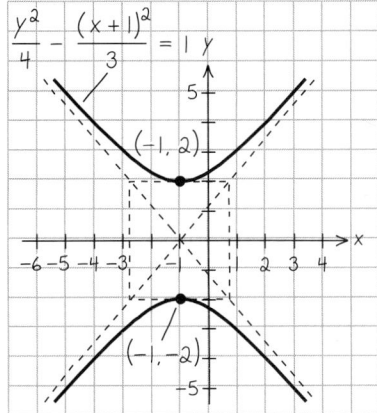

From this equation you can conclude that the hyperbola is centered at $(-1, 0)$, has vertices at $(-1, 2)$ and $(-1, -2)$, and the ends of the conjugate axis occur at $\left(-1 - \sqrt{3}, 0\right)$ and $\left(-1 + \sqrt{3}, 0\right)$. To sketch the hyperbola, draw a rectangle through these four points. The asymptotes are the lines passing through the corners of the rectangle, as shown in Figure 10.24. Finally, using $a = 2$ and $b = \sqrt{3}$, you can conclude that the equations of the asymptotes are

$$y = \frac{2}{\sqrt{3}}(x + 1) \qquad \text{and} \qquad y = -\frac{2}{\sqrt{3}}(x + 1). \qquad ▭$$

Note If the constant term F in the equation in Example 3 had been $F = 4$ instead of 16, you would have obtained the following degenerate case.

$$\text{Two Intersecting Lines:} \quad \frac{y^2}{4} - \frac{(x + 1)^2}{3} = 0$$

You can use a graphing utility to graph a hyperbola by graphing the upper and lower portions in the same viewing rectangle. For instance, to graph the hyperbola in Example 3, first solve for y to get

$$y_1 = 2\sqrt{1 + \frac{(x + 1)^2}{3}} \qquad \text{and} \qquad y_2 = -2\sqrt{1 + \frac{(x + 1)^2}{3}}.$$

Use a viewing rectangle in which $-8 \leq x \leq 6$ and $-6 \leq y \leq 6$. You should obtain the graph shown at the left.

Figure 10.25

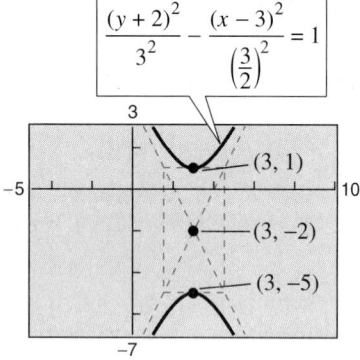

$$\frac{(y+2)^2}{3^2} - \frac{(x-3)^2}{\left(\frac{3}{2}\right)^2} = 1$$

EXAMPLE 4 ▭ **Using Asymptotes to Find the Standard Equation**

Find the standard form of the equation of the hyperbola having vertices at $(3, -5)$ and $(3, 1)$ and with asymptotes $y = 2x - 8$ and $y = -2x + 4$, as shown in Figure 10.25.

Solution

By the Midpoint Formula, the center of the hyperbola is at $(3, -2)$. Furthermore, the hyperbola has a vertical transverse axis with $a = 3$. From the given equations, you can determine the slopes of the asymptotes to be

$$m_1 = 2 = \frac{a}{b} \quad \text{and} \quad m_2 = -2 = -\frac{a}{b}$$

and because $a = 3$, you can conclude that $b = \frac{3}{2}$. Thus, the standard equation is

$$\frac{(y+2)^2}{9} - \frac{(x-3)^2}{9/4} = 1. \qquad \qquad ▭$$

As with ellipses, the **eccentricity** of a hyperbola is

$$e = \frac{c}{a} \qquad \qquad \text{Eccentricity}$$

and because $c > a$ it follows that $e > 1$. If the eccentricity is large, the branches of the hyperbola are nearly flat. If the eccentricity is close to 1, the branches of the hyperbola are more pointed, as shown in Figure 10.26.

Figure 10.26

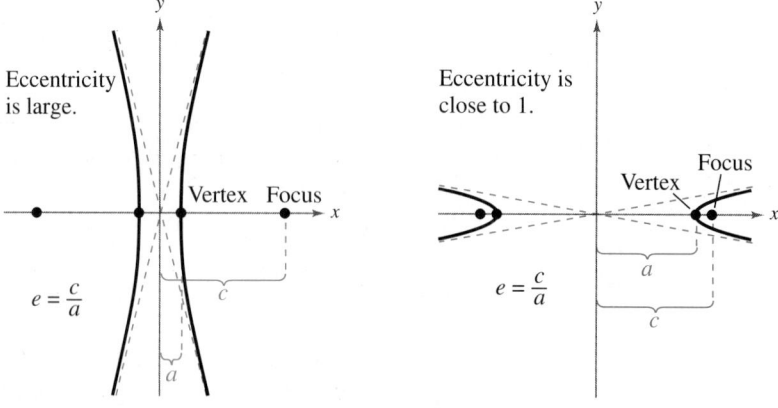

Applications

The following application was developed during World War II. It shows how the properties of hyperbolas can be used in radar and other detection systems.

Real Life

EXAMPLE 5 ▭ An Application Involving Hyperbolas

Two microphones, 1 mile apart, record an explosion. Microphone A received the sound 2 seconds before microphone B. Where was the explosion?

Solution

Assuming sound travels at 1100 feet per second, you know that the explosion took place 2200 feet further from B than from A, as shown in Figure 10.27. The locus of all points that are 2200 feet closer to A than to B is one branch of the hyperbola $(x^2/a^2) - (y^2/b^2) = 1$, where

$$c = \frac{5280}{2} = 2640 \quad \text{and} \quad a = \frac{2200}{2} = 1100.$$

Thus, $b^2 = c^2 - a^2 = 5{,}759{,}600$, and you can conclude that the explosion occurred somewhere on the right branch of the hyperbola given by

$$\frac{x^2}{1{,}210{,}000} - \frac{y^2}{5{,}759{,}600} = 1.$$

▭

Figure 10.27

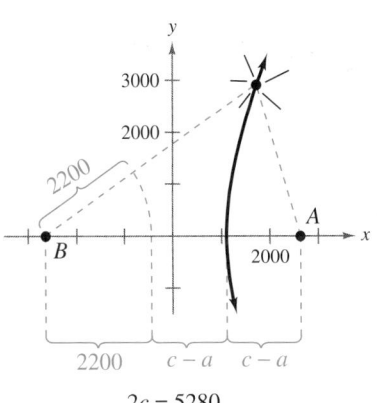

$$2c = 5280$$
$$2200 + 2(c - a) = 5280$$

Another interesting application of conic sections involves the orbits of comets in our solar system. Of the 610 comets identified prior to 1970, 245 have elliptical orbits, 295 have parabolic orbits, and 70 have hyperbolic orbits. The center of the sun is a focus of each of these orbits, and each orbit has a vertex at the point where the comet is closest to the sun, as shown in Figure 10.28. Undoubtedly, there have been many comets with parabolic or hyperbolic orbits that have not been identified. You get to see such comets only *once*. Comets with elliptical orbits, such as Halley's comet, are the only ones that remain in our solar system.

If p is the distance between the vertex and the focus in meters, and v is the velocity of the comet at the vertex in meters per second, the type of orbit is determined as follows.

Figure 10.28

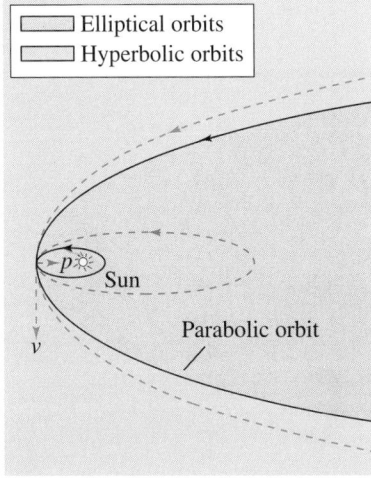

1. Ellipse: $v < \sqrt{2GM/p}$
2. Parabola: $v = \sqrt{2GM/p}$
3. Hyperbola: $v > \sqrt{2GM/p}$

In each of these equations, $M \approx 1.991 \times 10^{30}$ kilograms (the mass of the sun) and $G \approx 6.67 \times 10^{-11}$ cubic meters per gram-second squared.

Note The test at the right is valid *if* the graph is a conic. The test does not apply to equations such as $x^2 + y^2 = -1$, which is not a conic.

Classifying a Conic from Its General Equation
The graph of $Ax^2 + Cy^2 + Dx + Ey + F = 0$ is one of the following.

1. Circle: $A = C$
2. Parabola: $AC = 0$ $A = 0$ or $C = 0$, but not both.
3. Ellipse: $AC > 0$ A and C have like signs.
4. Hyperbola: $AC < 0$ A and C have unlike signs.

EXAMPLE 6   **Classifying Conics from General Equations**

Classify each graph.

a. $4x^2 - 9x + y - 5 = 0$

b. $4x^2 - y^2 + 8x - 6y + 4 = 0$

c. $2x^2 + 4y^2 - 4x + 12y = 0$

Solution

a. For the equation $4x^2 - 9x + y - 5 = 0$, you have
$$AC = 4(0) = 0. \qquad \text{Parabola}$$
Thus, the graph is a parabola.

b. For the equation $4x^2 - y^2 + 8x - 6y + 4 = 0$, you have
$$AC = 4(-1) < 0. \qquad \text{Hyperbola}$$
Thus, the graph is a hyperbola.

c. For the equation $2x^2 + 4y^2 - 4x + 12y = 0$, you have
$$AC = 2(4) > 0. \qquad \text{Ellipse}$$
Thus, the graph is an ellipse.

The first woman to be credited with detecting a new comet was the English astronomer Caroline Herschel (1750–1848). During her long life, Caroline Herschel discovered a total of eight new comets.

Group Activity

Sketching Conics

Sketch each of the conics described in Example 6. With others in your group, discuss procedures that allow you to sketch the conic efficiently.

10.3 /// EXERCISES

In Exercises 1–4, match the equation with its graph. [The graphs are labeled (a), (b), (c), and (d).]

(a)

(b)

(c)

(d)
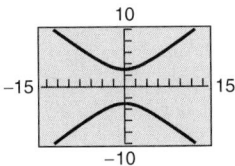

1. $\dfrac{x^2}{16} - \dfrac{y^2}{4} = 1$

2. $\dfrac{y^2}{16} - \dfrac{x^2}{4} = 1$

3. $\dfrac{y^2}{9} - \dfrac{x^2}{16} = 1$

4. $\dfrac{y^2}{16} - \dfrac{x^2}{9} = 1$

In Exercises 5–18, find the center, vertices, foci, and asymptotes of the hyperbola, and sketch its graph.

5. $x^2 - y^2 = 1$

6. $\dfrac{x^2}{9} - \dfrac{y^2}{16} = 1$

7. $\dfrac{y^2}{1} - \dfrac{x^2}{4} = 1$

8. $\dfrac{y^2}{9} - \dfrac{x^2}{1} = 1$

9. $\dfrac{y^2}{25} - \dfrac{x^2}{144} = 1$

10. $\dfrac{x^2}{36} - \dfrac{y^2}{4} = 1$

11. $\dfrac{(x-1)^2}{4} - \dfrac{(y+2)^2}{1} = 1$

12. $\dfrac{(x+1)^2}{144} - \dfrac{(y-4)^2}{25} = 1$

13. $(y+6)^2 - (x-2)^2 = 1$

14. $\dfrac{(y-1)^2}{1/4} - \dfrac{(x+3)^2}{1/9} = 1$

15. $9x^2 - y^2 - 36x - 6y + 18 = 0$

16. $x^2 - 9y^2 + 36y - 72 = 0$

17. $x^2 - 9y^2 + 2x - 54y - 80 = 0$

18. $16y^2 - x^2 + 2x + 64y + 63 = 0$

In Exercises 19–22, find the center, vertices, and foci of the hyperbola and use a graphing utility to graph the hyperbola and its asymptotes.

19. $2x^2 - 3y^2 = 6$

20. $3y^2 - 5x^2 = 15$

21. $9y^2 - x^2 + 2x + 54y + 62 = 0$

22. $9x^2 - y^2 + 54x + 10y + 55 = 0$

In Exercises 23–28, find an equation of the hyperbola with its center at the origin.

23. Vertices: $(0, \pm 2)$; Foci: $(0, \pm 4)$

24. Vertices: $(\pm 3, 0)$; Foci: $(\pm 5, 0)$

25. Vertices: $(\pm 1, 0)$; Asymptotes: $y = \pm 3x$

26. Vertices: $(0, \pm 3)$; Asymptotes: $y = \pm 3x$

27. Foci: $(0, \pm 8)$; Asymptotes: $y = \pm 4x$

28. Foci: $(\pm 10, 0)$; Asymptotes: $y = \pm \frac{3}{4}x$

In Exercises 29–36, find an equation for the hyperbola.

29. Vertices: $(2, 0), (6, 0)$; Foci: $(0, 0), (8, 0)$

30. Vertices: $(2, 3), (2, -3)$; Foci: $(2, 5), (2, -5)$

31. Vertices: $(4, 1), (4, 9)$; Foci: $(4, 0), (4, 10)$

32. Vertices: $(-2, 1), (2, 1)$; Foci: $(-3, 1), (3, 1)$

33. Vertices: $(2, 3), (2, -3)$;
      Passes through the point $(0, 5)$

34. Vertices: $(-2, 1), (2, 1)$;
      Passes through the point $(4, 3)$

35. Vertices: $(0, 2)$, $(6, 2)$;
Asymptotes: $y = \frac{2}{3}x$, $y = 4 - \frac{2}{3}x$

36. Vertices: $(3, 0)$, $(3, 4)$;
Asymptotes: $y = \frac{2}{3}x$, $y = 4 - \frac{2}{3}x$

In Exercises 37 and 38, determine what part of the graph of the hyperbola

$$\frac{(x - 3)^2}{4} - \frac{(y - 1)^2}{9} = 1$$

is given by the equation.

37. $x = 3 - \frac{2}{3}\sqrt{9 + (y - 1)^2}$

38. $y = 1 + \frac{3}{2}\sqrt{(x - 3)^2 - 4}$

39. *LORAN* Long distance radio navigation for aircraft and ships uses synchronized pulses transmitted by widely separated transmitting stations. These pulses travel at the speed of light (186,000 miles per second). The difference in the times of arrival of these pulses at an aircraft or ship is constant on a hyperbola having the transmitting stations as foci. Assume that two stations, 300 miles apart, are positioned on the rectangular coordinate system at points with coordinates $(-150, 0)$ and $(150, 0)$, and that a ship is traveling on a path with coordinates $(x, 75)$ (see figure). Find the x-coordinate of the position of the ship if the time difference between the pulses from the transmitting stations is 1000 microseconds (0.001 second).

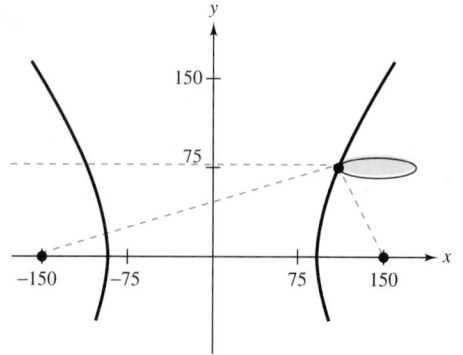

40. *Sound Location* Three listening stations located at $(4400, 0)$, $(4400, 1100)$, and $(-4400, 0)$ monitor an explosion. If the last two stations detect the explosion 1 second and 5 seconds after the first, respectively, determine the coordinates of the explosion. (Assume that the coordinate system is measured in feet and that sound travels at 1100 feet per second.)

41. *Hyperbolic Mirror* A hyperbolic mirror (used in some telescopes) has the property that a light ray directed at a focus will be reflected to the other focus (see figure). The focus of a hyperbolic mirror has coordinates $(24, 0)$. Find the vertex of the mirror if its mount has coordinates $(24, 24)$.

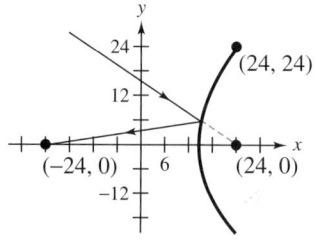

42. Consider a hyperbola centered at the origin with a horizontal transverse axis. Use the definition of a hyperbola to derive its standard form.

In Exercises 43–50, classify the graph of the equation as a circle, a parabola, an ellipse, or a hyperbola.

43. $x^2 + y^2 - 6x + 4y + 9 = 0$

44. $x^2 + 4y^2 - 6x + 16y + 21 = 0$

45. $4x^2 - y^2 - 4x - 3 = 0$

46. $y^2 - 4y - 4x = 0$

47. $4x^2 + 3y^2 + 8x - 24y + 51 = 0$

48. $4y^2 - 2x^2 - 4y - 8x - 15 = 0$

49. $25x^2 - 10x - 200y - 119 = 0$

50. $4x^2 + 4y^2 - 16y + 15 = 0$

10.4 Rotation and Systems of Quadratic Equations

Rotation / Invariants Under Rotation / Systems of Quadratic Equations

Rotation

In the previous section you learned that the equation of a conic with axes parallel to the coordinate axes has a standard form that can be written in the general form

$$Ax^2 + Cy^2 + Dx + Ey + F = 0. \qquad \text{Horizontal or vertical axes}$$

Figure 10.29

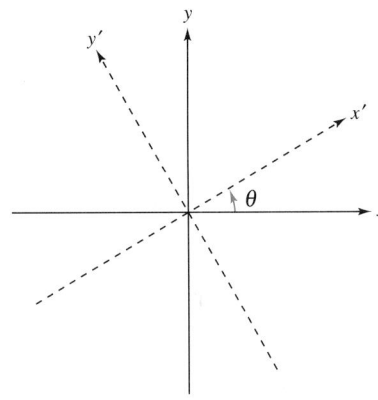

In this section you will study the equations of conics whose axes are rotated so that they are not parallel to either the *x*-axis or the *y*-axis. The general equation for such conics contains an *xy*-term.

$$Ax^2 + Bxy + Cy^2 + Dx + Ey + F = 0 \qquad \text{Equation in xy-plane}$$

To eliminate this *xy*-term, you can use a procedure called **rotation of axes.** The objective is to rotate the *x*- and *y*-axes until they are parallel to the axes of the conic. The rotated axes are denoted as the *x′*-axis and the *y′*-axis, as shown in Figure 10.29. After the rotation, the equation of the conic in the new *x′y′*-plane will have the form

$$A'(x')^2 + C'(y')^2 + D'x' + E'y' + F' = 0. \qquad \text{Equation in } x'y'\text{-plane}$$

Because this equation has no *xy*-term, you can obtain a standard form by completing the square. The following theorem identifies how much to rotate the axes to eliminate the *xy*-term and also the equations for determining the new coefficients A', C', D', E', and F'.

Rotation of Axes to Eliminate an *xy*-Term

The general second-degree equation $Ax^2 + Bxy + Cy^2 + Dx + Ey + F = 0$ can be rewritten as

$$A'(x')^2 + C'(y')^2 + D'x' + E'y' + F' = 0$$

by rotating the coordinate axes through an angle θ, where

$$\cot 2\theta = \frac{A - C}{B}.$$

The coefficients of the new equation are obtained by making the substitutions

$$x = x' \cos \theta - y' \sin \theta \qquad \text{and} \qquad y = x' \sin \theta + y' \cos \theta.$$

Figure 10.30

Rotated: $x' = r\cos(\alpha - \theta)$
$y' = r\sin(\alpha - \theta)$

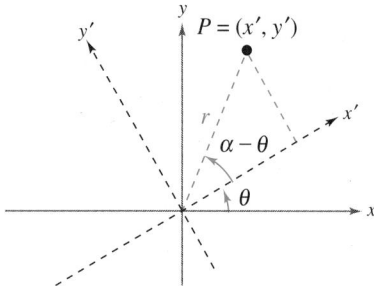

Original: $x = r\cos\alpha$
$y = r\sin\alpha$

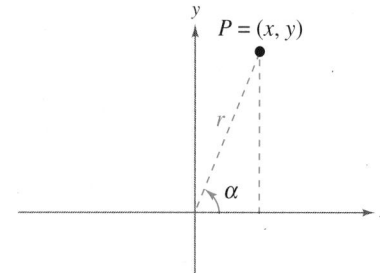

Proof /// You need to discover how the coordinates in the xy-system are related to the coordinates in the $x'y'$-system. To do this, choose a point $P = (x, y)$ in the original system and attempt to find its coordinates (x', y') in the rotated system. In either system, the distance r between the point P and the origin is the same. Thus, the equations for x, y, x', and y' are those given in Figure 10.30. Using the formulas for the sine and cosine of the difference of two angles, you have the following.

$$x' = r\cos(\alpha - \theta)$$
$$= r(\cos\alpha\cos\theta + \sin\alpha\sin\theta)$$
$$= r\cos\alpha\cos\theta + r\sin\alpha\sin\theta$$
$$= x\cos\theta + y\sin\theta$$
$$y' = r\sin(\alpha - \theta)$$
$$= r(\sin\alpha\cos\theta - \cos\alpha\sin\theta)$$
$$= r\sin\alpha\cos\theta - r\cos\alpha\sin\theta$$
$$= y\cos\theta - x\sin\theta$$

Solving this system for x and y yields

$$x = x'\cos\theta - y'\sin\theta$$
$$y = x'\sin\theta + y'\cos\theta.$$

Finally, by substituting these values for x and y into the original equation and collecting terms, you obtain

$$A' = A\cos^2\theta + B\cos\theta\sin\theta + C\sin^2\theta$$
$$C' = A\sin^2\theta - B\cos\theta\sin\theta + C\cos^2\theta$$
$$D' = D\cos\theta + E\sin\theta$$
$$E' = -D\sin\theta + E\cos\theta$$
$$F' = F.$$

To eliminate the $x'y'$-term, you must select θ so that $B' = 0$.

$$B' = 2(C - A)\sin\theta\cos\theta + B(\cos^2\theta - \sin^2\theta)$$
$$= (C - A)\sin 2\theta + B\cos 2\theta$$
$$= B(\sin 2\theta)\left(\frac{C - A}{B} + \cot 2\theta\right)$$
$$= 0, \quad \sin 2\theta \neq 0$$

If $B = 0$, no rotation is necessary because the xy-term is not present in the original equation. If $B \neq 0$, the only way to make $B' = 0$ is to let

$$\cot 2\theta = \frac{A - C}{B}, \quad B \neq 0.$$

Thus, you have established the desired results. ///

EXAMPLE 1   Rotation of Axes for a Hyperbola

Write the equation $xy - 1 = 0$ in standard form.

Solution

Because $A = 0$, $B = 1$, and $C = 0$, you have

$$\cot 2\theta = \frac{A - C}{B} = 0 \quad \Longrightarrow \quad 2\theta = \frac{\pi}{2} \quad \Longrightarrow \quad \theta = \frac{\pi}{4}$$

which implies that

$$x = x' \cos \frac{\pi}{4} - y' \sin \frac{\pi}{4}$$

$$= x'\left(\frac{\sqrt{2}}{2}\right) - y'\left(\frac{\sqrt{2}}{2}\right)$$

$$= \frac{x' - y'}{\sqrt{2}}$$

and

$$y = x' \sin \frac{\pi}{4} + y' \cos \frac{\pi}{4}$$

$$= x'\left(\frac{\sqrt{2}}{2}\right) + y'\left(\frac{\sqrt{2}}{2}\right)$$

$$= \frac{x' + y'}{\sqrt{2}}.$$

The equation in the $x'y'$-system is obtained by substituting these expressions into the equation $xy - 1 = 0$.

$$\left(\frac{x' - y'}{\sqrt{2}}\right)\left(\frac{x' + y'}{\sqrt{2}}\right) - 1 = 0$$

$$\frac{(x')^2 - (y')^2}{2} - 1 = 0$$

$$\frac{(x')^2}{(\sqrt{2})^2} - \frac{(y')^2}{(\sqrt{2})^2} = 1 \qquad \text{Standard form}$$

In the $x'y'$-system, this is a hyperbola centered at the origin with vertices at $(\pm\sqrt{2}, 0)$, as shown in Figure 10.31. To find the coordinates of the vertices in the xy-system, substitute the coordinates $(\pm\sqrt{2}, 0)$ into the equations

$$x = \frac{x' - y'}{\sqrt{2}} \qquad \text{and} \qquad y = \frac{x' + y'}{\sqrt{2}}.$$

This substitution yields the vertices $(1, 1)$ and $(-1, -1)$ in the xy-system. Note also that the asymptotes of the hyperbola have equations $y' = \pm x'$, which correspond to the original x- and y-axes.

Figure 10.31

Vertices:
In $x'y'$-system: $(\sqrt{2}, 0), (-\sqrt{2}, 0)$
In xy-system: $(1, 1), (-1, -1)$

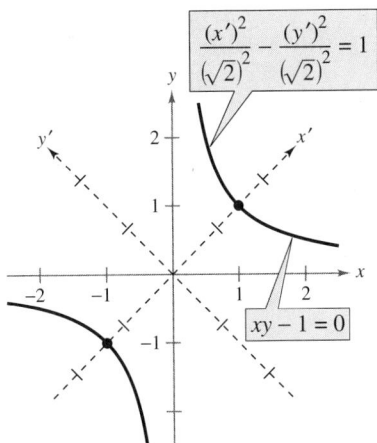

$$\frac{(x')^2}{(\sqrt{2})^2} - \frac{(y')^2}{(\sqrt{2})^2} = 1$$

$$xy - 1 = 0$$

EXAMPLE 2 　　　 **Rotation of Axes for an Ellipse**

Sketch the graph of $7x^2 - 6\sqrt{3}xy + 13y^2 - 16 = 0$.

Solution

Because $A = 7$, $B = -6\sqrt{3}$, and $C = 13$, you have

$$\cot 2\theta = \frac{A - C}{B} = \frac{7 - 13}{-6\sqrt{3}} = \frac{1}{\sqrt{3}}$$

which implies that $\theta = \pi/6$. The equation in the $x'y'$-system is obtained by making the substitutions

$$x = x' \cos \frac{\pi}{6} - y' \sin \frac{\pi}{6}$$

$$= x'\left(\frac{\sqrt{3}}{2}\right) - y'\left(\frac{1}{2}\right)$$

$$= \frac{\sqrt{3}x' - y'}{2}$$

and

$$y = x' \sin \frac{\pi}{6} + y' \cos \frac{\pi}{6}$$

$$= x'\left(\frac{1}{2}\right) + y'\left(\frac{\sqrt{3}}{2}\right)$$

$$= \frac{x' + \sqrt{3}y'}{2}$$

into the original equation. Thus, you have

$$7x^2 - 6\sqrt{3}xy + 13y^2 - 16 = 0$$

$$7\left(\frac{\sqrt{3}x' - y'}{2}\right)^2 - 6\sqrt{3}\left(\frac{\sqrt{3}x' - y'}{2}\right)\left(\frac{x' + \sqrt{3}y'}{2}\right) + 13\left(\frac{x' + \sqrt{3}y'}{2}\right)^2 - 16 = 0$$

which simplifies to

$$4(x')^2 + 16(y')^2 - 16 = 0$$

$$4(x')^2 + 16(y')^2 = 16$$

$$\frac{(x')^2}{4} + \frac{(y')^2}{1} = 1. \qquad \text{Standard form}$$

This is the equation of an ellipse centered at the origin with vertices $(\pm 2, 0)$ in the $x'y'$-system, as shown in Figure 10.32. 　　■

Figure 10.32

Vertices:
In $x'y'$-system: $(\pm 2, 0), (0, \pm 1)$
In xy-system: $\left(\pm\sqrt{3}, \pm 1\right), \left(\pm\frac{1}{2}, \pm\frac{\sqrt{3}}{2}\right)$

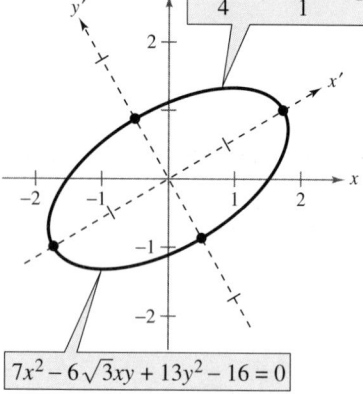

$$\frac{(x')^2}{4} + \frac{(y')^2}{1} = 1$$

$$7x^2 - 6\sqrt{3}xy + 13y^2 - 16 = 0$$

EXAMPLE 3 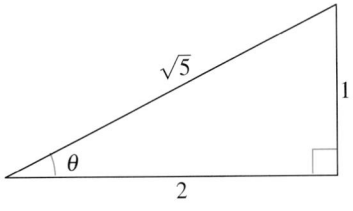 **Rotation of Axes for a Parabola**

Sketch the graph of $x^2 - 4xy + 4y^2 + 5\sqrt{5}y + 1 = 0$.

Figure 10.33

Solution

Because $A = 1$, $B = -4$, and $C = 4$, you have

$$\cot 2\theta = \frac{A - C}{B} = \frac{1 - 4}{-4} = \frac{3}{4}.$$

Using the identity $\cot 2\theta = (\cot^2 \theta - 1)/(2 \cot \theta)$ produces

$$\cot 2\theta = \frac{3}{4} = \frac{\cot^2 \theta - 1}{2 \cot \theta}$$

from which you obtain the equation

$$4 \cot^2 \theta - 4 = 6 \cot \theta$$
$$4 \cot^2 \theta - 6 \cot \theta - 4 = 0$$
$$(2 \cot \theta - 4)(2 \cot \theta + 1) = 0.$$

Considering $0 < \theta < \pi/2$, you have $2 \cot \theta = 4$. Thus,

$$\cot \theta = 2 \quad \Longrightarrow \quad \theta \approx 26.6°.$$

From the triangle in Figure 10.33, you obtain $\sin \theta = 1/\sqrt{5}$ and $\cos \theta = 2/\sqrt{5}$. Consequently, you use the substitutions

$$x = x' \cos \theta - y' \sin \theta = x'\left(\frac{2}{\sqrt{5}}\right) - y'\left(\frac{1}{\sqrt{5}}\right) = \frac{2x' - y'}{\sqrt{5}}$$

$$y = x' \sin \theta + y' \cos \theta = x'\left(\frac{1}{\sqrt{5}}\right) + y'\left(\frac{2}{\sqrt{5}}\right) = \frac{x' + 2y'}{\sqrt{5}}.$$

Figure 10.34

Vertex:

In $x'y'$-system: $\left(\frac{4}{5}, -1\right)$

In xy-system: $\left(\frac{13}{5\sqrt{5}}, -\frac{6}{5\sqrt{5}}\right)$

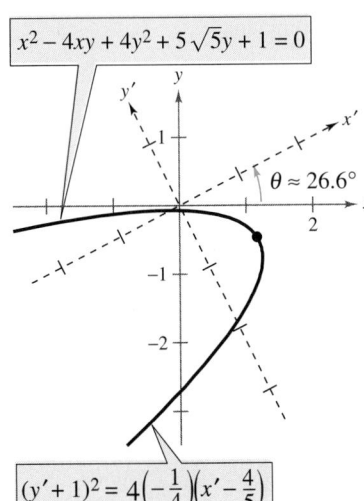

$x^2 - 4xy + 4y^2 + 5\sqrt{5}y + 1 = 0$

$\theta \approx 26.6°$

$(y' + 1)^2 = 4\left(-\frac{1}{4}\right)\left(x' - \frac{4}{5}\right)$

Substituting these expressions into the original equation, you have

$$x^2 - 4xy + 4y^2 + 5\sqrt{5}y + 1 = 0$$

$$\left(\frac{2x' - y'}{\sqrt{5}}\right) - 4\left(\frac{2x' - y'}{\sqrt{5}}\right)\left(\frac{x' + 2y'}{\sqrt{5}}\right) + 4\left(\frac{x' + 2y'}{\sqrt{5}}\right)^2 + 5\sqrt{5}\left(\frac{x' + 2y'}{\sqrt{5}}\right) + 1 = 0$$

which simplifies as follows.

$$5(y')^2 + 5x' + 10y' + 1 = 0$$
$$5(y' + 1)^2 = -5x' + 4 \qquad \text{Complete the square.}$$
$$(y' + 1)^2 = (-1)\left(x' - \frac{4}{5}\right) \qquad \text{Standard form}$$

The graph of this equation is a parabola with vertex at $\left(\frac{4}{5}, -1\right)$. Its axis is parallel to the x'-axis in the $x'y'$-system, as shown in Figure 10.34.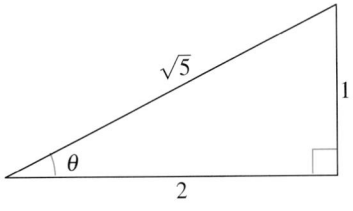

Invariants Under Rotation

In the rotation of axes theorem listed at the beginning of this section, note that the constant term is the same in both equations—that is, $F' = F$. Such quantities are **invariant under rotation.** The next theorem lists some other rotation invariants.

Rotation Invariants

The rotation of the coordinate axes through an angle θ that transforms the equation $Ax^2 + Bxy + Cy^2 + Dx + Ey + F = 0$ into the form

$$A'(x')^2 + C'(y')^2 + D'x' + E'y' + F' = 0$$

has the following rotation invariants.

1. $F = F'$
2. $A + C = A' + C'$
3. $B^2 - 4AC = (B')^2 - 4A'C'$

You can use the results of this theorem to classify the graph of a second-degree equation *with* an xy-term in much the same way you do for a second-degree equation *without* an xy-term. Note that because $B' = 0$, the invariant $B^2 - 4AC$ reduces to

$$B^2 - 4AC = -4A'C'. \qquad \text{Discriminant}$$

This quantity is called the **discriminant** of the equation

$$Ax^2 + Bxy + Cy^2 + Dx + Ey + F = 0.$$

Now, from the classification procedure given in Section 10.3, you know that the sign of $A'C'$ determines the type of graph for the equation

$$A'(x')^2 + C'(y')^2 + D'x' + E'y' + F' = 0.$$

Consequently, the sign of $B^2 - 4AC$ will determine the type of graph for the original equation, as given in the following classification.

Classification of Conics by the Discriminant

The graph of the equation $Ax^2 + Bxy + Cy^2 + Dx + Ey + F = 0$ is, except in degenerate cases, determined by its discriminant as follows.

1. Ellipse or circle: $B^2 - 4AC < 0$
2. Parabola: $B^2 - 4AC = 0$
3. Hyperbola: $B^2 - 4AC > 0$

EXAMPLE 4 ▭ **Using the Discriminant**

Classify the graph of each of the following equations.

a. $4xy - 9 = 0$

b. $2x^2 - 3xy + 2y^2 - 2x = 0$

c. $x^2 - 6xy + 9y^2 - 2y + 1 = 0$

d. $3x^2 + 8xy + 4y^2 - 7 = 0$

Solution

a. Because $B^2 - 4AC = 16 - 0 > 0$, the graph is a hyperbola.

b. Because $B^2 - 4AC = 9 - 16 < 0$, the graph is a circle or an ellipse.

c. Because $B^2 - 4AC = 36 - 36 = 0$, the graph is a parabola.

d. Because $B^2 - 4AC = 64 - 48 > 0$, the graph is a hyperbola.   ▭

Systems of Quadratic Equations

To find the points of intersection of two conics, you can use elimination or substitution, as demonstrated in Examples 5 and 6.

EXAMPLE 5 ▭ **Solving a Quadratic System by Elimination**

Solve the system of quadratic equations.

$$x^2 + y^2 - 16x + 39 = 0 \qquad \text{Equation 1}$$
$$x^2 - y^2 - 9 = 0 \qquad \text{Equation 2}$$

Figure 10.35

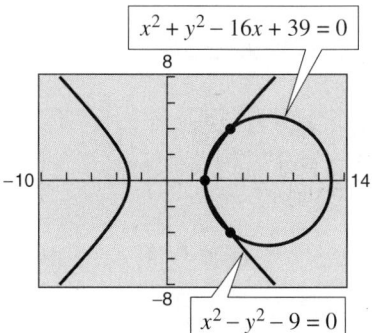

$x^2 + y^2 - 16x + 39 = 0$

$x^2 - y^2 - 9 = 0$

Solution

You can eliminate the y^2-term by adding the two equations. The resulting equation can then be solved for x.

$$2x^2 - 16x + 30 = 0$$
$$2(x - 3)(x - 5) = 0$$

There are two real solutions. $x = 3$ and $x = 5$. The corresponding y-values are $y = 0$ and $y = \pm 4$. Thus the graphs have three points of intersection:

$$(3, 0), \quad (5, 4), \quad \text{and} \quad (5, -4)$$

as shown in Figure 10.35.   ▭

EXAMPLE 6 ▱ **Solving a Quadratic System by Substitution**

Solve the system of quadratic equations.

$$x^2 + 4y^2 - 4x - 8y + 4 = 0 \qquad \text{Equation 1}$$
$$x^2 + 4y - 4 = 0 \qquad \text{Equation 2}$$

Solution

Because Equation 2 has no y^2-term, solve the equation for y to obtain

$$y = 1 - \frac{1}{4}x^2.$$

Next, substitute this expression for y into Equation 1 and solve for x.

$$x^2 + 4y^2 - 4x - 8y + 4 = 0$$

$$x^2 + 4\left(1 - \frac{1}{4}x^2\right)^2 - 4x - 8\left(1 - \frac{1}{4}x^2\right) + 4 = 0$$

$$x^2 + 4 - 2x^2 + \frac{1}{4}x^4 - 4x - 8 + 2x^2 + 4 = 0$$

$$\frac{1}{4}x^4 + x^2 - 4x = 0$$

$$x^4 + 4x^2 - 16x = 0$$

$$x(x - 2)(x^2 + 2x + 8) = 0$$

Figure 10.36

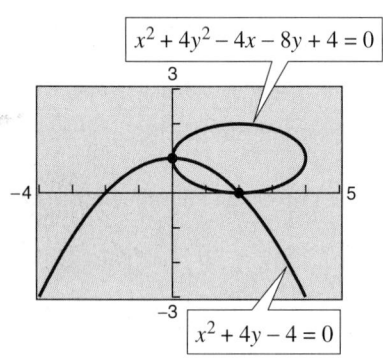

$x^2 + 4y^2 - 4x - 8y + 4 = 0$

$x^2 + 4y - 4 = 0$

In factored form, you can see that the equation has two real solutions: $x = 0$ and $x = 2$. The corresponding values of y are $y = 1$ and $y = 0$. This implies that the solutions of the system of equations are

$$(0, 1) \quad \text{and} \quad (2, 0)$$

as shown in Figure 10.36. ▱

Group Activity

Classifying a Graph as a Hyperbola

In Section 2.6, you studied the graphs of rational functions such as

$$f(x) = \frac{1}{x} \quad \text{and} \quad g(x) = \frac{1}{x^3}.$$

Use a graphing utility to graph each function. Are the graphs hyperbolas? Explain how you can decide.

10.4 /// EXERCISES

In Exercises 1–4, the $x'y'$-coordinate system has been rotated θ degrees from the xy-coordinate system. The coordinates of a point on the xy-coordinate system are given. Find the coordinates of the point on the rotated coordinate system.

1. $\theta = 90°, (0, 3)$ **2.** $\theta = 45°, (3, 3)$
3. $\theta = 30°, (1, 4)$ **4.** $\theta = 60°, (3, 1)$

In Exercises 5–16, rotate the axes to eliminate the xy-term. Sketch the graph of the resulting equation, showing both sets of axes.

5. $xy + 1 = 0$ **6.** $xy - 4 = 0$
7. $x^2 - 10xy + y^2 + 1 = 0$
8. $xy + x - 2y + 3 = 0$
9. $xy - 2y - 4x = 0$
10. $13x^2 + 6\sqrt{3}xy + 7y^2 - 16 = 0$
11. $5x^2 - 2xy + 5y^2 - 12 = 0$
12. $2x^2 - 3xy - 2y^2 + 10 = 0$
13. $3x^2 - 2\sqrt{3}xy + y^2 + 2x + 2\sqrt{3}y = 0$
14. $16x^2 - 24xy + 9y^2 - 60x - 80y + 100 = 0$
15. $9x^2 + 24xy + 16y^2 + 90x - 130y = 0$
16. $9x^2 + 24xy + 16y^2 + 80x - 60y = 0$

In Exercises 17–22, use a graphing utility to graph the conic. Determine the angle θ through which the axes are rotated. Explain how you used the graphing utility to obtain the graph.

17. $x^2 + xy + y^2 = 10$
18. $x^2 - 4xy + 2y^2 = 6$
19. $17x^2 + 32xy - 7y^2 = 75$
20. $40x^2 + 36xy + 25y^2 = 52$
21. $32x^2 + 50xy + 7y^2 = 52$
22. $4x^2 - 12xy + 9y^2 + (4\sqrt{13} - 12)x$
$- (6\sqrt{13} + 8)y = 91$

In Exercises 23–28, match the graph with its equation. [The graphs are labeled (a), (b), (c), (d), (e), and (f).]

(a)

(b)

(c)

(d)

(e)

(f)
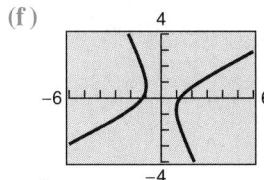

23. $xy + 3 = 0$ **24.** $x^2 + 2xy + y^2 = 0$
25. $-2x^2 + 3xy + 2y^2 + 3 = 0$
26. $x^2 - xy + 3y^2 - 5 = 0$
27. $3x^2 + 2xy + y^2 - 10 = 0$
28. $x^2 - 4xy + 4y^2 + 10x - 30 = 0$

In Exercises 29–36, use the discriminant to determine whether the graph of the equation is a parabola, an ellipse, or a hyperbola.

29. $16x^2 - 24xy + 9y^2 - 30x - 40y = 0$
30. $x^2 - 4xy - 2y^2 - 6 = 0$
31. $13x^2 - 8xy + 7y^2 - 45 = 0$
32. $2x^2 + 4xy + 5y^2 + 3x - 4y - 20 = 0$

33. $x^2 - 6xy - 5y^2 + 4x - 22 = 0$

34. $36x^2 - 60xy + 25y^2 + 9y = 0$

35. $x^2 + 4xy + 4y^2 - 5x - y - 3 = 0$

36. $x^2 + xy + 4y^2 + x + y - 4 = 0$

In Exercises 37–40, sketch (if possible) the graph of the degenerate conic.

37. $y^2 - 4x^2 = 0$

38. $x^2 + y^2 - 2x + 6y + 10 = 0$

39. $x^2 + 2xy + y^2 - 1 = 0$

40. $x^2 - 10xy + y^2 = 0$

In Exercises 41–48, use a graphing utility to graph the equations and find any points of intersection of the graphs by the method of elimination.

41. $-x^2 + y^2 + 4x - 6y + 4 = 0$
$x^2 + y^2 - 4x - 6y + 12 = 0$

42. $-x^2 - y^2 - 8x + 20y - 7 = 0$
$x^2 + 9y^2 + 8x + 4y + 7 = 0$

43. $-4x^2 - y^2 - 32x + 24y - 64 = 0$
$4x^2 + y^2 + 56x - 24y + 304 = 0$

44. $x^2 - 4y^2 - 20x - 64y - 172 = 0$
$16x^2 + 4y^2 - 320x + 64y + 1600 = 0$

45. $x^2 - y^2 - 12x + 12y - 36 = 0$
$x^2 + y^2 - 12x - 12y + 36 = 0$

46. $x^2 + 4y^2 - 2x - 8y + 1 = 0$
$-x^2 + 2x - 4y - 1 = 0$

47. $-16x^2 - y^2 + 24y - 80 = 0$
$16x^2 + 25y^2 - 400 = 0$

48. $16x^2 - y^2 + 16y - 128 = 0$
$y^2 - 48x - 16y - 32 = 0$

In Exercises 49–54, use a graphing utility to graph the equations and find any points of intersection of the graphs by the method of substitution.

49. $x^2 + y^2 - 25 = 0$
$9x - 4y^2 = 0$

50. $4x^2 + 9y^2 - 36y = 0$
$x^2 + 9y - 27 = 0$

51. $x^2 + 2y^2 - 4x + 6y - 5 = 0$
$x + y + 5 = 0$

52. $x^2 + 2y^2 - 4x + 6y - 2 = 0$
$x^2 - 4x - y + 4 = 0$

53. $xy + x - 2y + 3 = 0$
$x^2 + 4y^2 - 9 = 0$

54. $5x^2 - 2xy + 5y^2 - 12 = 0$
$x + y - 1 = 0$

55. Show that the equation $x^2 + y^2 = r^2$ is invariant under rotation of axes.

56. Find the lengths of the major and minor axes of the ellipse in Exercise 10.

Review   Solve Exercises 57–60 as a review of the skills and problem-solving techniques you learned in previous sections. Graph the rational function.

57. $g(x) = \dfrac{2}{2 - x}$ **58.** $f(x) = \dfrac{2x}{2 - x}$

59. $h(t) = \dfrac{t^2}{2 - t}$ **60.** $g(s) = \dfrac{2}{4 - s^2}$

10.5   Parametric Equations

Plane Curves **/** *Sketching a Plane Curve* **/** *Eliminating the Parameter* **/** *Finding Parametric Equations for a Graph*

Plane Curves

Up to this point, you have been representing a graph by a single equation involving the *two* variables x and y. In this section, you will study situations in which it is useful to introduce a *third* variable to represent a curve in the plane.

To see the usefulness of this procedure, consider the path followed by an object that is propelled into the air at an angle of 45°. If the initial velocity of the object is 48 feet per second, it can be shown that it follows the parabolic path given by

$$y = -\frac{x^2}{72} + x, \qquad \text{Rectangular equation}$$

as shown in Figure 10.37. However, this equation does not tell the whole story. Although it does tell us *where* the object has been, it doesn't tell us *when* the object was at a given point (x, y) on the path. To determine this time, you can introduce a third variable t, which is called a **parameter.** It is possible to write both x and y as functions of t to obtain the **parametric equations**

$$x = 24\sqrt{2}t \qquad \text{Parametric equation for } x$$

and

$$y = -16t^2 + 24\sqrt{2}t. \qquad \text{Parametric equation for } y$$

From this set of equations you can determine that at time $t = 0$, the object is at the point $(0, 0)$. Similarly, at time $t = 1$, the object is at the point $\left(24\sqrt{2}, 24\sqrt{2} - 16\right)$, and so on.

For this particular motion problem, x and y are continuous functions of t, and the resulting path is a **plane curve.** (Recall that a *continuous function* is one whose graph can be traced without lifting the pencil from the paper.)

Figure 10.37

Curvilinear motion:
two variables for position
one variable for time

Rectangular equation:

$$y = -\frac{x^2}{72} + x$$

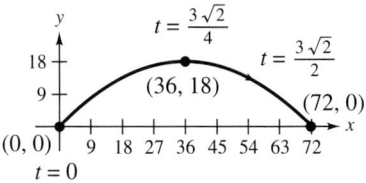

Parametric equations:

$x = 24\sqrt{2}t$

$y = -16t^2 + 24\sqrt{2}t$

Definition of a Plane Curve

If f and g are continuous functions of t on an interval I, the set of ordered pairs $(f(t), g(t))$ is a **plane curve** C. The equations

$$x = f(t) \qquad \text{and} \qquad y = g(t)$$

are **parametric equations** for C, and t is the **parameter.**

Sketching a Plane Curve

One way to sketch a curve represented by a pair of parametric equations is to plot points in the *xy* plane. Each set of coordinates (*x*, *y*) is determined from a value chosen for the parameter *t*. By plotting the resulting points in order of *increasing* values of *t*, you trace the curve in a specific direction. This is called the **orientation** of the curve.

EXAMPLE 1 ▱ Sketching a Curve

Sketch the curve given by the parametric equations

$$x = t^2 - 4 \quad \text{and} \quad y = \frac{t}{2}, \quad -2 \le t \le 3.$$

Describe the orientation of the curve.

Solution
Using values of *t* in the given interval, the parametric equations yield the points (*x*, *y*), as shown in the table.

t	-2	-1	0	1	2	3
x	0	-3	-4	-3	0	5
y	-1	$-\frac{1}{2}$	0	$\frac{1}{2}$	1	$\frac{3}{2}$

By plotting these points in order of increasing *t*, you obtain the curve shown in Figure 10.38. The arrows on the curve indicate its orientation as *t* increases from −2 to 3. Thus, if a particle were moving on this curve, it would start at $(0, -1)$ and move along the curve to the point $\left(5, \frac{3}{2}\right)$. ▱

Figure 10.38

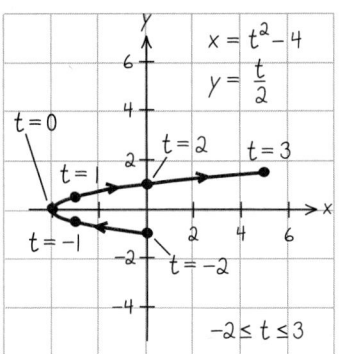

The graph shown in Figure 10.38 does not define *y* as a function of *x*. This points out one benefit of parametric equations—they can be used to represent graphs that are more general than graphs of functions.

Two different sets of parametric equations can have the same graph. For example, the set of parametric equations

$$x = 4t^2 - 4 \quad \text{and} \quad y = t, \quad -1 \le t \le \frac{3}{2}$$

has the same graph as the set given in Example 1. However, by comparing the values of *t* in Figures 10.38 and 10.39, you can see that this second graph is traced out more *rapidly* (considering *t* as time) than the first graph. Thus, in applications, different parametric representations can be used to represent various *speeds* at which objects travel along a given path.

Figure 10.39

Another way to display a curve represented by a pair of parametric equations is to use a graphing utility.

EXAMPLE 2   Using a Graphing Utility in Parametric Mode

Use a graphing utility to graph the curves represented by the parametric equations. For which curve is y a function of x? (Use $-4 \le t \le 4$.)

a. $x = t^2$ $\quad$ **b.** $x = t$ $\quad$ **c.** $x = t^2$
$\quad\;\; y = t^3$ $\qquad\quad\;\; y = t^3$ $\qquad\quad\;\; y = t$

Solution

Begin by setting the graphing utility to parametric mode. When choosing a viewing rectangle, you must set not only minimum and maximum values of x and y, but also minimum and maximum values of t.

a. Enter the parametric equations for x and y.

$\qquad X_{1T} = T^2, \qquad Y_{1T} = T^3$

The curve is shown in Figure 10.40(a). From the graph, you can see that y *is not* a function of x.

b. Enter the parametric equations for x and y.

$\qquad X_{1T} = T, \qquad Y_{1T} = T^3$

The curve is shown in Figure 10.40(b). From the graph, you can see that y *is* a function of x.

c. Enter the parametric equations for x and y.

$\qquad X_{1T} = T^2, \qquad Y_{1T} = T$

The curve is shown in Figure 10.40(c). From the graph, you can see that y *is not* a function of x.

Figure 10.40

(a)

(b)

(c)

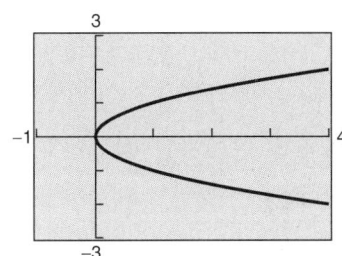

Eliminating the Parameter

Many curves that are represented by sets of parametric equations have graphs that can also be represented by rectangular equations (in x and y). The process of finding the rectangular equation is called **eliminating the parameter.**

Parametric equations	Solve for t in one equation	Substitute into second equation	Rectangular equation

$$x = t^2 - 4 \qquad t = 2y \qquad x = (2y)^2 - 4 \qquad x = 4y^2 - 4$$
$$y = \tfrac{1}{2}t$$

After eliminating the parameter, you can recognize that the curve is a parabola with a horizontal axis and vertex at $(-4, 0)$.

Converting equations from parametric to rectangular form can change the ranges of x and y. In such cases, you should restrict x and y in the rectangular equation such that its graph matches the graph of the parametric equations.

Figure 10.41

(a)

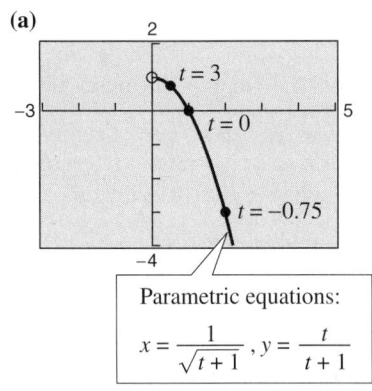

Parametric equations:
$$x = \frac{1}{\sqrt{t+1}}, \, y = \frac{t}{t+1}$$

(b)

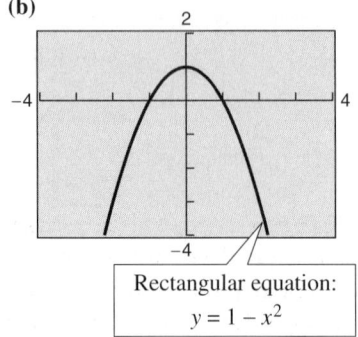

Rectangular equation:
$$y = 1 - x^2$$

EXAMPLE 3 **Eliminating the Parameter**

Identify the curve represented by the equations

$$x = \frac{1}{\sqrt{t+1}} \qquad \text{and} \qquad y = \frac{t}{t+1}.$$

Solution

Solving for t in the equation for x produces

$$x = \frac{1}{\sqrt{t+1}} \qquad \text{or} \qquad \frac{1}{x^2} = t + 1$$

which implies that $t = (1/x^2) - 1$. Substituting into the equation for y, you obtain

$$y = \frac{t}{t+1}$$
$$= \frac{(1/x^2) - 1}{(1/x^2) - 1 + 1}$$
$$= 1 - x^2.$$

From the rectangular equation, you can recognize the curve to be a parabola that opens downward and has its vertex at $(0, 1)$. The rectangular equation is defined for all values of x. From the parametric equation for x, however, you can see that the curve is defined only when $-1 < t$. Thus, you should restrict the domain of x to positive values, as shown in Figure 10.41(a).

It is not necessary for the parameter in a set of parametric equations to represent time. The next example uses an *angle* as the parameter.

EXAMPLE 4   **Eliminating the Parameter**

Sketch the curve represented by

$$x = 3 \cos \theta \quad \text{and} \quad y = 4 \sin \theta, \quad 0 \le \theta \le 2\pi$$

by eliminating the parameter.

Solution

Begin by solving for $\cos \theta$ and $\sin \theta$ in the given equations.

$$\cos \theta = \frac{x}{3} \quad \text{and} \quad \sin \theta = \frac{y}{4} \qquad \text{Solve for } \cos \theta \text{ and } \sin \theta.$$

Make use of the identity $\sin^2\theta + \cos^2\theta = 1$ to form an equation involving only x and y.

$$\cos^2 \theta + \sin^2 \theta = 1 \qquad \text{Pythagorean identity}$$

$$\cos^2 \theta + \sin^2 \theta = \left(\frac{x}{3}\right)^2 + \left(\frac{y}{4}\right)^2 = 1 \qquad \text{Substitute.}$$

$$\frac{x^2}{9} + \frac{y^2}{16} = 1 \qquad \text{Rectangular equation}$$

From this rectangular equation, you can see that the graph is an ellipse centered at $(0, 0)$, with vertices at $(0, 4)$ and $(0, -4)$, and minor axis of length $2b = 6$, as shown in Figure 10.42. Note that the elliptic curve is traced out *counterclockwise* as θ varies from 0 to 2π.

Figure 10.42

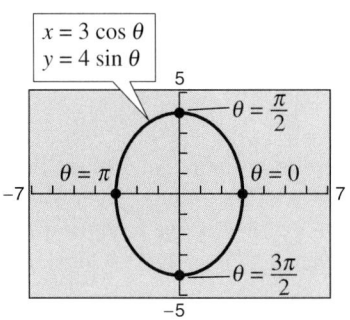

In Examples 3 and 4 it is important to realize that eliminating the parameter is primarily an *aid to curve sketching*. If the parametric equations represent the path of a moving object, the graph alone is not sufficient to describe the object's motion. You still need the parametric equations to tell you the *position, direction,* and *speed* at a given time.

Finding Parametric Equations for a Graph

You have been studying techniques for sketching the graph represented by a set of parametric equations. Now, consider the reverse problem. That is, how can you find a set of parametric equations for a given graph or a given physical description? From the discussion following Example 1, you know that such a representation is not unique. This is further demonstrated in Example 5.

Figure 10.43

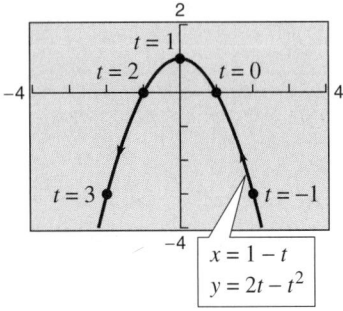

$x = 1 - t$
$y = 2t - t^2$

EXAMPLE 5 ▭ **Finding Parametric Equations for a Given Graph**

Find a set of parametric equations to represent the graph of $y = 1 - x^2$, using the following parameters.

a. $t = x$ **b.** $t = 1 - x$

Solution

a. Letting $t = x$, you obtain the parametric equations

$$x = t \quad \text{and} \quad y = 1 - x^2 = 1 - t^2.$$

b. Letting $t = 1 - x$, you obtain the parametric equations

$$x = 1 - t \quad \text{and} \quad y = 1 - (1 - t)^2 = 2t - t^2.$$

In Figure 10.43, note how the curve for part (b) is oriented by the increasing values of t. For part (a), the curve would have the opposite orientation. ▭

EXAMPLE 6 ▭ **Parametric Equations for a Cycloid**

Describe the **cycloid** traced out by a point P on the circumference of a circle of radius a as the circle rolls along a straight line in a plane.

Solution

As the parameter, let θ be the measure of the circle's rotation, and let the point $P = (x, y)$ begin at the origin. When $\theta = 0$, P is at the origin; when $\theta = \pi$, P is at a maximum point $(\pi a, 2a)$; and when $\theta = 2\pi$, P is back on the x-axis at $(2\pi a, 0)$. From Figure 10.44, you can see that $\angle APC = 180° - \theta$. Hence, you have

$$\sin \theta = \sin(180° - \theta) = \sin(\angle APC) = \frac{AC}{a} = \frac{BD}{a}$$

$$\cos \theta = -\cos(180° - \theta) = -\cos(\angle APC) = \frac{AP}{-a}$$

which implies that $AP = -a \cos \theta$ and $BD = a \sin \theta$. Because the circle rolls along the x-axis, you know that $OD = \overset{\frown}{PD} = a\theta$. Furthermore, because $BA = DC = a$, you have

$$x = OD - BD = a\theta - a \sin \theta$$
$$y = BA + AP = a - a \cos \theta.$$

Therefore, the parametric equations are

$$x = a(\theta - \sin \theta) \quad \text{and} \quad y = a(1 - \cos \theta). \quad ▭$$

A computer animation of this concept appears in the *Interactive* CD-ROM.

Figure 10.44

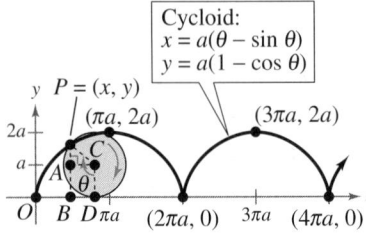

Cycloid:
$x = a(\theta - \sin \theta)$
$y = a(1 - \cos \theta)$

Group Activity

Changing the Orientation of a Curve

The **orientation** of a curve refers to the direction in which the curve is traced as the values of the parameter increase. For instance, as t increases, the ellipse given by

$$x = \cos t \quad \text{and} \quad y = 2 \sin t$$

is traced out *counterclockwise*. Find a parametric representation for which the ellipse is traced out *clockwise*.

10.5 /// EXERCISES

1. Consider the parametric equations $x = \sqrt{t}$ and $y = 1 - t$.

(a) Use a graphing utility to complete the table.

t	0	1	2	3	4
x					
y					

(b) Plot the points (x, y) generated in part (a), and sketch a graph of the parametric equations.

(c) Find the rectangular equation by eliminating the parameter. Use a graphing utility to sketch its graph. How do the graphs differ?

2. Consider the parametric equations $x = 4 \cos^2 \theta$ and $y = 2 \sin \theta$.

(a) Use a graphing utility to complete the table.

θ	$-\dfrac{\pi}{2}$	$-\dfrac{\pi}{4}$	0	$\dfrac{\pi}{4}$	$\dfrac{\pi}{2}$
x					
y					

(b) Plot the points (x, y) generated in part (a), and sketch a graph of the parametric equations.

(c) Find the rectangular equation by eliminating the parameter. Use a graphing utility to sketch its graph. How do the graphs differ?

In Exercises 3–16, sketch the curve represented by the parametric equations (indicate the direction of the curve). Use a graphing utility to confirm your result. Then eliminate the parameter and write the corresponding rectangular equation whose graph represents the curve.

3. $x = t$

  $y = -2t$

4. $x = t$

  $y = \frac{1}{2}t$

5. $x = 3t - 1$

  $y = 2t + 1$

6. $x = 3 - 2t$

  $y = 2 + 3t$

7. $x = \frac{1}{4}t$

  $y = t^2$

8. $x = t$

  $y = t^3$

9. $x = t + 1$

  $y = t^2$

10. $x = \sqrt{t}$

  $y = 1 - t$

11. $x = t^3$

  $y = \dfrac{1}{2}t^2$

12. $x = t - 1$

  $y = \dfrac{t}{t - 1}$

13. $x = 2t$

  $y = |t - 2|$

14. $x = |t - 1|$

  $y = t + 2$

15. $x = 3 \cos \theta$

  $y = 3 \sin \theta$

16. $x = \cos \theta$

  $y = 3 \sin \theta$

In Exercises 17–28, use a graphing utility to sketch the curve represented by the parametric equations (indicate the direction of the curve). Then eliminate the parameter and write the corresponding rectangular equation whose graph represents the curve.

17. $x = 4 \sin 2\theta$
$y = 2 \cos 2\theta$

18. $x = \cos \theta$
$y = 2 \sin 2\theta$

19. $x = 4 + 2 \cos \theta$
$y = -1 + \sin \theta$

20. $x = 4 + 2 \cos \theta$
$y = -1 + 2 \sin \theta$

21. $x = 4 + 2 \cos \theta$
$y = -1 + \sin \theta$

22. $x = \sec \theta$
$y = \tan \theta$

23. $x = 4 \sec \theta$
$y = 3 \tan \theta$

24. $x = \sec \theta$
$y = \cos \theta$

25. $x = e^{-t}$
$y = e^{3t}$

26. $x = e^{2t}$
$y = e^{t}$

27. $x = t^3$
$y = 3 \ln t$

28. $x = \ln 2t$
$y = t^2$

In Exercises 29 and 30, determine how the plane curves differ from each other.

29. (a) $x = t$
$y = 2t + 1$

    (b) $x = \cos \theta$
$y = 2 \cos \theta + 1$

    (c) $x = e^{-t}$
$y = 2e^{-t} + 1$

    (d) $x = e^{t}$
$y = 2e^{t} + 1$

30. (a) $x = t$
$y = t$

    (b) $x = t^2$
$y = t^2$

    (c) $x = -t$
$y = -t$

    (d) $x = t^3$
$y = t^3$

In Exercises 31–34, eliminate the parameter and obtain the standard form of the rectangular equation.

31. Line through (x_1, y_1) and (x_2, y_2):
$x = x_1 + t(x_2 - x_1)$
$y = y_1 + t(y_2 - y_1)$

32. Circle: $x = h + r \cos \theta$
$y = k + r \sin \theta$

33. Ellipse: $x = h + a \cos \theta$
$y = k + b \sin \theta$

34. Hyperbola: $x = h + a \sec \theta$
$y = k + b \tan \theta$

In Exercises 35–40, use the results of Exercises 31–34 to find a set of parametric equations for the line or conic.

35. Line: Passes through $(0, 0)$ and $(5, -2)$
36. Line: Passes through $(1, 4)$ and $(5, -2)$
37. Circle: Center: $(2, 1)$; Radius: 4
38. Circle: Center: $(-3, 1)$; Radius: 3
39. Ellipse: Vertices: $(\pm 5, 0)$; Foci: $(\pm 4, 0)$
40. Hyperbola: Vertices: $(0, \pm 1)$; Foci: $(0, \pm 2)$

In Exercises 41 and 42, find two different sets of parametric equations for the given rectangular equation.

41. $y = 3x - 2$ **42.** $y = x^2$

In Exercises 43–48, use a graphing utility to obtain a graph of the curve represented by the parametric equations.

43. Cycloid: $x = 2(\theta - \sin \theta)$
$y = 2(1 - \cos \theta)$

44. Prolate cycloid: $x = 2\theta - 4 \sin \theta$
$y = 2 - 4 \cos \theta$

45. Hypocycloid: $x = 3 \cos^3 \theta$
$y = 3 \sin^3 \theta$

46. Curtate cycloid: $x = 2\theta - \sin \theta$
$y = 2 - \cos \theta$

47. Witch of Agnesi: $x = 2 \cot \theta$
$y = 2 \sin^2 \theta$

48. Folium of Descartes: $x = \dfrac{3t}{1 + t^3}$

$y = \dfrac{3t^2}{1 + t^3}$

In Exercises 49–52, match the parametric equations with the correct graph. [The graphs are labeled (a), (b), (c), and (d).]

(a)

(b)

(c)

(d)

49. Lissajous curve: $x = 2 \cos \theta$, $y = \sin 2\theta$

50. Evolute of ellipse: $x = 2 \cos^3 \theta$, $y = 4 \sin^3 \theta$

51. Involute of circle: $x = \frac{1}{2}(\cos \theta + \theta \sin \theta)$

$y = \frac{1}{2}(\sin \theta - \theta \cos \theta)$

52. Serpentine curve: $x = \frac{1}{2} \cot \theta$, $y = 4 \sin \theta \cos \theta$

Projectile Motion   **In Exercises 53–56, consider a projectile launched at a height h feet above the ground at an angle θ with the horizontal. If the initial velocity is v_0 feet per second, the path of the projectile is modeled by the parametric equations**

$$x = (v_0 \cos \theta)t \quad \text{and} \quad y = h + (v_0 \sin \theta)t - 16t^2.$$

53. *Baseball*   The center-field fence in a ballpark is 10 feet high and 400 feet from home plate. The baseball is hit 3 feet above the ground. It leaves the bat at an angle of θ degrees with the horizontal at a speed of 100 miles per hour (see figure).

   (a) Write a set of parametric equations for the path of the baseball.

   (b) Use a graphing utility to sketch the path of the baseball for $\theta = 15°$. Is the hit a home run?

   (c) Use a graphing utility to sketch the path of the baseball for $\theta = 23°$. Is the hit a home run?

   (d) Find the minimum angle required for the hit to be a home run.

Figure for 53

54. *Football*   The quarterback of a football team releases a pass at a height of 7 feet above the playing field, and the football is caught by a receiver at a height of 4 feet, 30 yards directly downfield. The pass is released at an angle of 35° with the horizontal.

   (a) Write a set of parametric equations for the path of the football.

   (b) Find the speed of the football when it is released.

   (c) Use a graphing utility to graph the path of the football and approximate its maximum height.

   (d) Find the time the receiver has to position himself after the quarterback releases the football.

55. *Projectile Motion*   Eliminate the parameter t from the position function for the motion of a projectile to show that the rectangular equation is

$$y = -\frac{16 \sec^2 \theta}{v_0^2}x^2 + (\tan \theta)x + h.$$

56. *Path of a Projectile*   The path of a projectile is given by the rectangular equation $y = 5 + x - 0.005x^2$.

   (a) Use the result of Exercise 55 to find h, v_0, and θ. Find the parametric equations of the path.

   (b) Use a graphing utility to graph the rectangular equation for the path of the projectile. Confirm your answer in part (a) by sketching the curve represented by the parametric equations.

   (c) Use a graphing utility to approximate the maximum height of the projectile and its range.

57. *True or False?*   The graph of the parametric equations $x = t^2$ and $y = t^2$ is the line $y = x$. Explain.

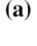

10.6   Polar Coordinates

Introduction / *Coordinate Conversion* / *Equation Conversion*

Introduction

Figure 10.45

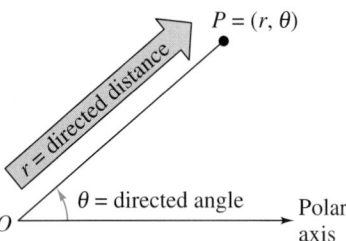

So far, you have been representing graphs of equations as collections of points (x, y) on the rectangular coordinate system, where x and y represent the directed distances from the coordinate axes to the point (x, y). In this section, you will study a second system called the **polar coordinate system.**

To form the polar coordinate system in the plane, fix a point O, called the **pole** (or **origin**), and construct from O an initial ray called the **polar axis,** as shown in Figure 10.45. Then each point P in the plane can be assigned **polar coordinates** (r, θ) as follows.

1. $r = $ *directed distance* from O to P
2. $\theta = $ *directed angle*, counterclockwise from polar axis to segment $\overline{OP}$

EXAMPLE 1   **Plotting Points in the Polar Coordinate System**

a. The point $(r, \theta) = (2, \pi/3)$ lies two units from the pole on the terminal side of the angle $\theta = \pi/3$, as shown in Figure 10.46(a).

b. The point $(r, \theta) = (3, -\pi/6)$ lies three units from the pole on the terminal side of the angle $\theta = -\pi/6$, as shown in Figure 10.46(b).

c. The point $(r, \theta) = (3, 11\pi/6)$ coincides with the point $(3, -\pi/6)$, as shown in Figure 10.46(c).

Figure 10.46

(a)

(b)

(c)

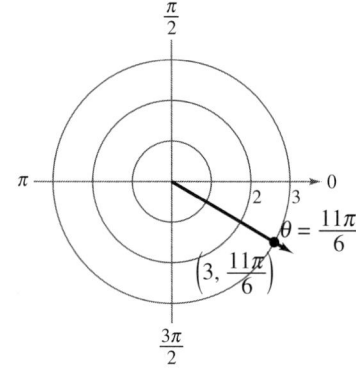

In rectangular coordinates, each point (x, y) has a unique representation. This is not true for polar coordinates. For instance, the coordinates (r, θ) and $(r, 2\pi + \theta)$ represent the same point, as illustrated in Example 1. Another way to obtain multiple representations of a point is to use negative values for r. Because r is a *directed distance,* the coordinates (r, θ) and $(-r, \theta + \pi)$ represent the same point. In general, the point (r, θ) can be represented as

$$(r, \theta) = (r, \theta \pm 2n\pi)$$

or

$$(r, \theta) = (-r, \theta \pm (2n + 1)\pi)$$

where n is any integer. Moreover, the pole is represented by $(0, \theta)$, where θ is any angle.

EXAMPLE 2   Multiple Representation of Points

Plot the point $(3, -3\pi/4)$ and find three additional polar representations of this point, using $-2\pi < \theta < 2\pi$.

Figure 10.47

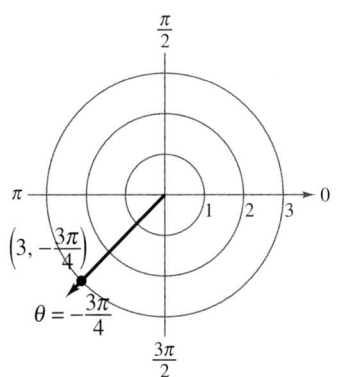

$$\left(3, -\frac{3\pi}{4}\right) = \left(3, \frac{5\pi}{4}\right) = \left(-3, -\frac{7\pi}{4}\right) = \left(-3, \frac{\pi}{4}\right) = \dots$$

Solution

The point is shown in Figure 10.47. Three other representations are as follows.

$$\left(3, -\frac{3\pi}{4} + 2\pi\right) = \left(3, \frac{5\pi}{4}\right) \qquad \text{Add } 2\pi \text{ to } \theta.$$

$$\left(-3, -\frac{3\pi}{4} - \pi\right) = \left(-3, -\frac{7\pi}{4}\right) \qquad \text{Replace } r \text{ by } -r; \text{ subtract } \pi \text{ from } \theta.$$

$$\left(-3, -\frac{3\pi}{4} + \pi\right) = \left(-3, \frac{\pi}{4}\right) \qquad \text{Replace } r \text{ by } -r; \text{ add } \pi \text{ to } \theta.$$

EXPLORATION

Set your graphing utility to polar mode. Then graph the equation $r = 3$. (Use a setting of $0 \le \theta \le 2\pi$, $-6 \le x \le 6$, and $-4 \le y \le 4$.) You should obtain a circle of radius 3.

a. Use the trace feature to cursor around the circle. Can you locate the point $(3, 5\pi/4)$?

b. Can you locate other names for the point $(3, 5\pi/4)$? If so, explain how you did it.

Figure 10.48

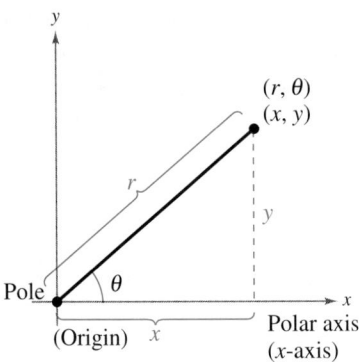

y

(r, θ)
(x, y)

r

y

Pole   θ   x
(Origin)   x   Polar axis
                  (x-axis)

Coordinate Conversion

To establish the relationship between polar and rectangular coordinates, let the polar axis coincide with the positive x-axis and the pole with the origin, as shown in Figure 10.48. Because (x, y) lies on a circle of radius r, it follows that $r^2 = x^2 + y^2$. Moreover, for $r > 0$, the definitions of the trigonometric functions imply that

$$\tan\theta = \frac{y}{x}, \quad \cos\theta = \frac{x}{r}, \quad \text{and} \quad \sin\theta = \frac{y}{r}.$$

You can show that the same relationships hold for $r < 0$.

Coordinate Conversion

The polar coordinates (r, θ) are related to the rectangular coordinates (x, y) as follows.

$$x = r\cos\theta \quad \text{and} \quad \tan\theta = \frac{y}{x}$$

$$y = r\sin\theta \qquad\qquad r^2 = x^2 + y^2$$

EXAMPLE 3   Polar-to-Rectangular Conversion

Convert the points (a) $(2, \pi)$ and (b) $\left(\sqrt{3}, \pi/6\right)$ to rectangular coordinates. (See Figure 10.49.)

Solution

a. For the point $(r, \theta) = (2, \pi)$, you have

$$x = r\cos\theta = 2\cos\pi = -2$$

and

$$y = r\sin\theta = 2\sin\pi = 0.$$

The rectangular coordinates are $(x, y) = (-2, 0)$.

Figure 10.49

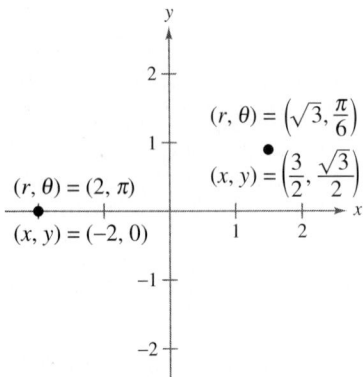

y

2

1

$(r, \theta) = \left(\sqrt{3}, \frac{\pi}{6}\right)$

$(x, y) = \left(\frac{3}{2}, \frac{\sqrt{3}}{2}\right)$

$(r, \theta) = (2, \pi)$

$(x, y) = (-2, 0)$ 1   2   x

−1

−2

b. For the point $(r, \theta) = \left(\sqrt{3}, \pi/6\right)$, you have

$$x = \sqrt{3}\cos\frac{\pi}{6} = \sqrt{3}\left(\frac{\sqrt{3}}{2}\right) = \frac{3}{2}$$

and

$$y = \sqrt{3}\sin\frac{\pi}{6} = \sqrt{3}\left(\frac{1}{2}\right) = \frac{\sqrt{3}}{2}.$$

The rectangular coordinates are $(x, y) = \left(3/2, \sqrt{3}/2\right)$.

Figure 10.50

(a)

(b)

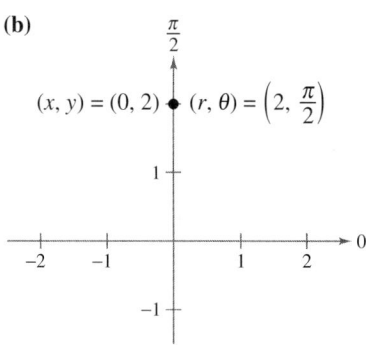

EXAMPLE 4   ▱   **Rectangular-to-Polar Conversion**

Convert the points (a) $(-1, 1)$ and (b) $(0, 2)$ to polar coordinates.

Solution

a. For the second-quadrant point $(x, y) = (-1, 1)$, you have

$$\tan \theta = \frac{y}{x}$$

$$\tan \theta = -1$$

$$\theta = \frac{3\pi}{4}.$$

Because θ lies in the same quadrant as (x, y), use positive r.

$$r = \sqrt{x^2 + y^2}$$
$$= \sqrt{(-1)^2 + (1)^2}$$
$$= \sqrt{2}$$

Thus, *one* set of polar coordinates is $(r, \theta) = \left(\sqrt{2}, 3\pi/4\right)$, as shown in Figure 10.50(a).

b. Because the point $(x, y) = (0, 2)$ lies on the positive y-axis, choose

$$\theta = \pi/2 \quad \text{and} \quad r = 2.$$

This implies that one set of polar coordinates is $(r, \theta) = (2, \pi/2)$, as shown in Figure 10.50(b).   ▱

Equation Conversion

By comparing Examples 3 and 4, you see that point conversion from the polar to the rectangular system is straightforward, whereas point conversion from the rectangular to the polar system is more involved. For equations, the opposite is true. To convert a rectangular equation to polar form, you simply replace x by $r \cos \theta$ and y by $r \sin \theta$. For instance, the rectangular equation $y = x^2$ can be written in polar form as follows.

$$y = x^2 \qquad \text{Rectangular equation}$$
$$r \sin \theta = (r \cos \theta)^2 \qquad \text{Polar equation}$$

On the other hand, converting a polar equation to rectangular form requires considerable ingenuity.

Example 5 demonstrates several polar-to-rectangular conversions that enable you to sketch the graphs of some polar equations.

Figure 10.51

(a)

(b)

(c)

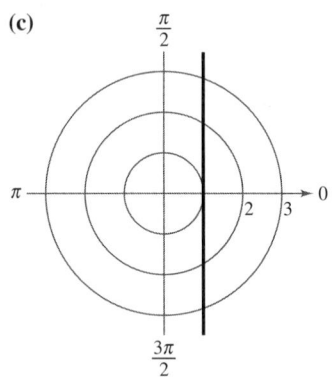

EXAMPLE 5 ▱ **Converting Polar Equations to Rectangular Form**

Describe the graph of each polar equation and find the corresponding rectangular equation.

a. $r = 2$ **b.** $\theta = \dfrac{\pi}{3}$ **c.** $r = \sec\theta$

Solution

a. The graph of the polar equation $r = 2$ consists of all points that are two units from the pole. In other words, this graph is a circle centered at the origin with a radius of 2, as shown in Figure 10.51(a). You can confirm this by converting to rectangular form, using the relationship $r^2 = x^2 + y^2$.

$$r = 2 \quad \Longrightarrow \quad r^2 = 2^2 \quad \Longrightarrow \quad x^2 + y^2 = 2^2$$

Polar equation   Rectangular equation

b. The graph of the polar equation $\theta = \pi/3$ consists of all points on the line that makes an angle of $\pi/3$ with the positive x-axis, as shown in Figure 10.51(b). To convert to rectangular form, you make use of the relationship $\tan\theta = y/x$.

$$\theta = \frac{\pi}{3} \quad \Longrightarrow \quad \tan\theta = \sqrt{3} \quad \Longrightarrow \quad y = \sqrt{3}x$$

Polar equation   Rectangular equation

c. The graph of the polar equation $r = \sec\theta$ is not evident by simple inspection, so you convert to rectangular form by using the relationship $r\cos\theta = x$.

$$r = \sec\theta \quad \Longrightarrow \quad r\cos\theta = 1 \quad \Longrightarrow \quad x = 1$$

Polar equation   Rectangular equation

Now you see that the graph is a vertical line, as shown in Figure 10.51(c).

▱

Curve sketching by converting to rectangular form is not always convenient. The next section demonstrates how to use a graphing utility to analyze the graph of a polar equation.

Group Activity   *Using a Graphing Utility*

Use a graphing utility to confirm the graphs shown in Figure 10.51.

10.6 /// EXERCISES

In Exercises 1–4, a point in polar coordinates is given. Find the corresponding rectangular coordinates for the point.

1. $\left(4, \dfrac{\pi}{2}\right)$

2. $\left(4, \dfrac{3\pi}{2}\right)$

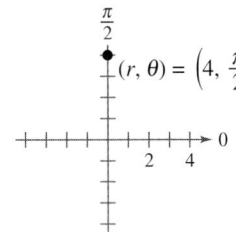

3. $\left(-1, \dfrac{5\pi}{4}\right)$

4. $(0, -\pi)$

 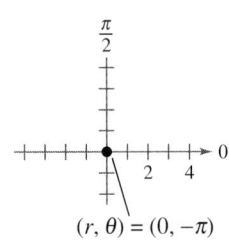

In Exercises 5–10, plot the point given in polar coordinates and find the corresponding rectangular coordinates for the point.

5. $(4, -\pi/3)$

6. $(-1, -3\pi/4)$

7. $(0, -7\pi/6)$

8. $(32, 5\pi/2)$

9. $\left(\sqrt{2}, 2.36\right)$

10. $(-3, -1.57)$

In Exercises 11–14, use a graphing utility to find the rectangular coordinates for the point given in polar coordinates.

11. $(2, 3\pi/4)$

12. $(-2, 7\pi/6)$

13. $(-4.5, 1.3)$

14. $(8.25, 3.5)$

In Exercises 15–24, the rectangular coordinates of a point are given. Plot the point and find *two* sets of polar coordinates for the point for $0 \le \theta < 2\pi$.

15. $(1, 1)$

16. $(0, -5)$

17. $(-6, 0)$

18. $(-3, -3)$

19. $(-3, 4)$

20. $(3, -1)$

21. $\left(-\sqrt{3}, -\sqrt{3}\right)$

22. $(2, 0)$

23. $(4, 6)$

24. $(5, 12)$

In Exercises 25–30, use a graphing utility to find one set of polar coordinates for the point given in rectangular coordinates.

25. $(3, -2)$

26. $(-4, 1)$

27. $\left(\sqrt{3}, 2\right)$

28. $\left(3\sqrt{2}, 3\sqrt{2}\right)$

29. $\left(\dfrac{5}{2}, \dfrac{4}{3}\right)$

30. $(0, -5)$

True or False?   In Exercises 31 and 32, determine whether the statement is true or false. If it is false, explain why or give an example that shows it is false.

31. If (r_1, θ_1) and (r_2, θ_2) represent the same point in the polar coordinate system, then $|r_1| = |r_2|$.

32. If (r, θ_1) and (r, θ_2) represent the same point in the polar coordinate system, then $\theta_1 = \theta_2 + 2\pi n$, for some integer n.

In Exercises 33–46, convert the rectangular equation to polar form.

33. $x^2 + y^2 = 9$

34. $x^2 + y^2 = a^2$

35. $x^2 + y^2 - 2ax = 0$

36. $x^2 + y^2 - 2ay = 0$

37. $y = 4$

38. $y = b$

39. $x = 10$

40. $x = a$

41. $3x - y + 2 = 0$

42. $4x + 7y - 2 = 0$

43. $xy = 4$ **44.** $y = x$

45. $(x^2 + y^2)^2 - 9(x^2 - y^2) = 0$

46. $y^2 - 8x - 16 = 0$

In Exercises 47–56, convert the polar equation to rectangular form.

47. $r = 4 \sin \theta$ **48.** $r = 4 \cos \theta$

49. $\theta = \dfrac{\pi}{6}$ **50.** $r = 4$

51. $r = 2 \csc \theta$ **52.** $r^2 = \sin 2\theta$

53. $r = 2 \sin 3\theta$ **54.** $r = \dfrac{1}{1 - \cos \theta}$

55. $r = \dfrac{6}{2 - 3 \sin \theta}$

56. $r = \dfrac{6}{2 \cos \theta - 3 \sin \theta}$

In Exercises 57–62, convert the polar equation to rectangular form and sketch its graph.

57. $r = 3$ **58.** $r = 8$

59. $\theta = \dfrac{\pi}{4}$ **60.** $\theta = \dfrac{5\pi}{6}$

61. $r = 3 \sec \theta$ **62.** $r = 2 \csc \theta$

63. Convert the polar equation

$$r = 2(h \cos \theta + k \sin \theta)$$

to rectangular form and verify that it is the equation of a circle. Find the radius and the rectangular coordinates of the center of the circle.

64. Convert the polar equation $r = \cos \theta + 3 \sin \theta$ to rectangular form and identify the graph.

65. *Think About It*

(a) Show that the distance between the points (r_1, θ_1) and (r_2, θ_2) is given by

$$\sqrt{r_1{}^2 + r_2{}^2 - 2r_1 r_2 \cos(\theta_1 - \theta_2)}.$$

(b) Describe the position of the points relative to each other if $\theta_1 = \theta_2$. Simplify the distance formula for this case. Is the simplification what you expected? Explain.

(c) Simplify the distance formula if $\theta_1 - \theta_2 = 90°$. Is the simplification what you expected? Explain.

(d) Choose two points on the polar coordinate system and find the distance between them. Then choose different polar representations of the same two points and apply the distance formula again. Discuss the result.

66. *Exploration*

(a) Set the window format of your graphing utility to rectangular coordinates and locate the cursor at any position off the coordinate axes. Move the cursor horizontally and observe any changes in the displayed coordinates of the points. Explain the changes. Now repeat the process moving the cursor vertically.

(b) Set the window format of your graphing utility to polar coordinates and locate the cursor at any position off the coordinate axes. Move the cursor horizontally and observe any changes in the displayed coordinates of the points. Explain the changes. Now repeat the process moving the cursor vertically.

(c) Explain why the results of parts (a) and (b) are not the same.

Review   Solve Exercises 67–72 as a review of the skills and problem-solving techniques you learned in previous sections. Use determinants to solve the system of equations.

67. $5x - 7y = -11$ **68.** $3x + 5y = 10$
      $-3x + y = -3$ $4x - 2y = -5$

69. $3a - 2b + c = 0$ **70.** $5u + 7v + 9w = 15$
     $2a + b - 3c = 0$ $u - 2v - 3w = 7$
     $a - 3b + 9c = 8$ $8u - 2v + w = 0$

71. $x + y + z - 3w = -8$
     $3x - y - 2z + w = 7$
    $-x + y - z + 2w = -2$
        $2y + w = -6$

72. $2y + 5z + 6w = 32$
   $2x + 4y - 5z - w = -7$
   $3x - 6y + z + 5w = 6$
   $4x - 2y - z = -12$

10.7   Graphs of Polar Equations

Introduction / *Symmetry* / *Zeros and Maximum r-Values* /
Special Polar Graphs

Introduction

Figure 10.52

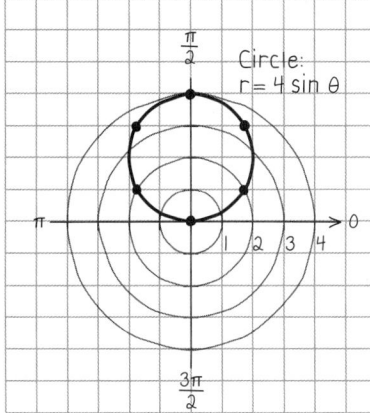

In previous chapters you spent a lot of time learning how to sketch graphs in rectangular coordinates. You began with the basic point-plotting method. Then you used sketching aids such as a graphing utility, symmetry, intercepts, asymptotes, periods, and shifts to further investigate the nature of the graph. This section approaches curve sketching in the polar coordinate system similarly.

EXAMPLE 1 **Graphing a Polar Equation by Point Plotting**

Sketch the graph of the polar equation $r = 4 \sin \theta$.

Solution

The sine function is periodic, so you can get a full range of r-values by considering values of θ in the interval $0 \le \theta \le 2\pi$, as shown in the table.

θ	0	$\dfrac{\pi}{6}$	$\dfrac{\pi}{3}$	$\dfrac{\pi}{2}$	$\dfrac{2\pi}{3}$	$\dfrac{5\pi}{6}$	π	$\dfrac{7\pi}{6}$	$\dfrac{3\pi}{2}$	$\dfrac{11\pi}{6}$	2π
r	0	2	$2\sqrt{3}$	4	$2\sqrt{3}$	2	0	-2	-4	-2	0

If you plot these points as shown in Figure 10.52, it appears that the graph is a circle of radius 2 whose center is at the point $(x, y) = (0, 2)$. ▱

The table feature of the *TI-82* and *TI-83* is very useful in constructing tables of values for polar equations. Set your graphing utility to polar mode and enter the polar equation in Example 1. You can verify the table of values in Example 1 by using TblMin = 0 (or TblStart = 0) and ΔTbl = $\pi/6$ from the TblSet menu.

▱   EXPLORATION

You can confirm the graph found in Example 1 in three ways.

1. *Convert to Rectangular Form*   Multiply both sides of the polar equation by r and convert the result to rectangular form.
2. *Use a Polar Coordinate Mode*   Set your graphing utility to polar mode and graph the polar equation. (Use $0 \le \theta \le 2\pi$, $-6 \le x \le 6$, and $-4 \le y \le 4$.)
3. *Use a Parametric Mode*   Set your graphing utility to parametric mode and graph $x = (4 \sin t) \cos t$ and $y = (4 \sin t) \sin t$.

Most graphing utilities have a polar-coordinate graphing mode. If yours doesn't, you can use the following parametric conversion to graph a polar equation.

Polar Equations in Parametric Form

The graph of the polar equation $r = f(\theta)$ can be written in parametric form, using t as a parameter, as follows.

$$x = f(t) \cos t \qquad \text{and} \qquad y = f(t) \sin t.$$

Symmetry

In Figure 10.52, note that as θ increases from 0 to 2π the graph is traced out twice. Moreover, note that the graph is *symmetric with respect to the line* $\theta = \pi/2$. Had you known about this symmetry and retracing ahead of time, you could have used fewer points.

Symmetry with respect to the line $\theta = \pi/2$ is one of three important types of symmetry to consider in polar curve sketching. (See Figure 10.53.)

Figure 10.53

Symmetry with Respect
to the Line $\theta = \frac{\pi}{2}$

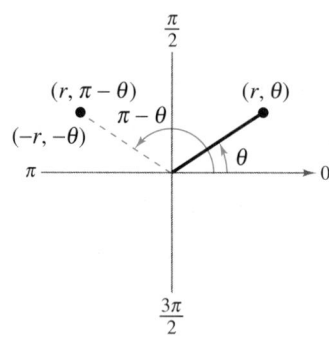

Symmetry with Respect
to the Polar Axis

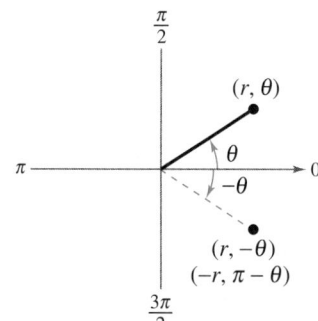

Symmetry with Respect
to the Pole

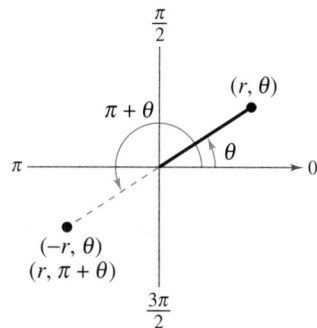

Tests for Symmetry in Polar Coordinates

The graph of a polar equation is symmetric with respect to the following if the given substitution yields an equivalent equation.

1. **The line $\theta = \pi/2$:** Replace (r, θ) by $(r, \pi - \theta)$ or $(-r, -\theta)$.

2. **The polar axis:** Replace (r, θ) by $(r, -\theta)$ or $(-r, \pi - \theta)$.

3. **The pole:** Replace (r, θ) by $(r, \pi + \theta)$ or $(-r, \theta)$.

Figure 10.54

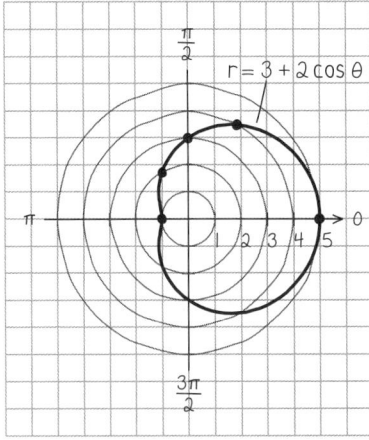

r = 3 + 2 cos θ

Note The graph in Figure 10.54 is called a **limaçon.**

EXAMPLE 2 ▱ **Using Symmetry to Sketch a Polar Graph**

Use symmetry to sketch the graph of $r = 3 + 2 \cos \theta$.

Solution
Replacing (r, θ) by $(r, -\theta)$ produces

$$r = 3 + 2 \cos(-\theta)$$

$$= 3 + 2 \cos \theta.$$

Thus, you can conclude that the curve is symmetric with respect to the polar axis. Plotting the points in the table and using polar axis symmetry, you obtain the graph shown in Figure 10.54.

θ	0	$\pi/3$	$\pi/2$	$2\pi/3$	π
r	5	4	3	2	1

Use a graphing utility to confirm this graph. ▱

The three tests for symmetry in polar coordinates on page 808 are sufficient to guarantee symmetry, but they are not necessary. For instance, Figure 10.55 shows the graph of

$$r = \theta + 2\pi. \qquad \text{Spiral of Archimedes}$$

From the figure, you can see that the graph is symmetric with respect to the line $\theta = \pi/2$. Yet the tests on page 808 fail to indicate symmetry because neither of the following replacements yields an equivalent equation.

Figure 10.55

Spiral of Archimedes:
$r = \theta + 2\pi, \ -4\pi \le \theta \le 0$

Original Equation	*Replacement*	*New Equation*
$r = \theta + 2\pi$	(r, θ) by $(-r, -\theta)$	$-r = -\theta + 2\pi$
$r = \theta + 2\pi$	(r, θ) by $(r, \pi - \theta)$	$r = -\theta + 3\pi$

The equations discussed in Examples 1 and 2 are of the form

$$r = 4 \sin \theta = f(\sin \theta) \quad \text{and} \quad r = 3 + 2 \cos \theta = g(\cos \theta).$$

The graph of the first equation is symmetric with respect to the line $\theta = \pi/2$, and the graph of the second equation is symmetric with respect to the polar axis. This observation can be generalized to yield the following *quick test for symmetry.*

1. The graph of $r = f(\sin \theta)$ is symmetric with respect to the line $\theta = \pi/2$.
2. The graph of $r = g(\cos \theta)$ is symmetric with respect to the polar axis.

Zeros and Maximum *r*-Values

Two additional aids to sketching graphs of polar equations involve knowing the θ-values for which $|r|$ is maximum and knowing the θ-values for which $r = 0$. For instance, in Example 1, the maximum value of $|r|$ for $r = 4 \sin \theta$ is $|r| = 4$, and this occurs when $\theta = \pi/2$, as shown in Figure 10.52. Moreover, $r = 0$ when $\theta = 0$.

Figure 10.56

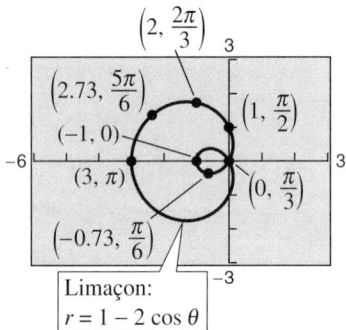

$\left(2, \frac{2\pi}{3}\right)$

$\left(2.73, \frac{5\pi}{6}\right)$

$(-1, 0)$

$\left(1, \frac{\pi}{2}\right)$

$(3, \pi)$

$\left(0, \frac{\pi}{3}\right)$

$\left(-0.73, \frac{\pi}{6}\right)$

Limaçon:
$r = 1 - 2 \cos \theta$

EXAMPLE 3   Finding Maximum *r*-Values of a Polar Graph

Find the maximum value of r for the graph of $r = 1 - 2 \cos \theta$.

Solution

Because the polar equation is of the form

$$r = 1 - 2 \cos \theta = g(\cos \theta)$$

you know the graph is symmetric with respect to the polar axis. You can confirm this by graphing the polar equation, as shown in Figure 10.56. (In the graph, θ varies from 0 to 2π.) To find the maximum r-value for the graph, use your graphing utility's trace feature. When you do this, you should find that the graph has a maximum r-value of 3. This value of r occurs when $\theta = \pi$. In the graph, note that the point $(3, \pi)$ is farthest from the pole.

Note   Note how the negative r-values determine the *inner loop* of the graph in Figure 10.56. This type of graph is a limaçon.

EXPLORATION

The graph of the polar equation

$$r = e^{\cos \theta} - 2 \cos 4\theta + \sin^5 \frac{\theta}{12}$$

is called *the butterfly curve*, as shown at the left.

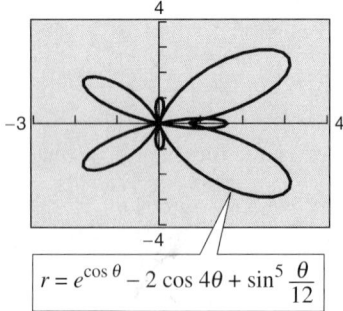

$r = e^{\cos \theta} - 2 \cos 4\theta + \sin^5 \frac{\theta}{12}$

a. The graph at the left was produced using $0 \le \theta \le 2\pi$. Does this show the entire graph? Explain your reasoning.

b. As θ increases from 0 to 2π, how many loops are produced on the graph? Justify your answer.

c. Use the trace feature of your graphing calculator to approximate the maximum r-value of the graph. Does this value change if you use $0 \le \theta \le 4\pi$ instead of $0 \le \theta \le 2\pi$? Explain.

EXAMPLE 4 ▰ Analyzing a Polar Graph

Analyze the graph of $r = 2 \cos 3\theta$.

Solution

Symmetry	With respect to the polar axis
Maximum value of $\|r\|$	$\|r\| = 2$ when $3\theta = 0, \pi, 2\pi, 3\pi$
	or $\theta = 0, \pi/3, 2\pi/3, \pi$
Zeros of r	$r = 0$ when $3\theta = \pi/2, 3\pi/2, 5\pi/2$
	or $\theta = \pi/6, \pi/2, 5\pi/6$

Note The graph shown in Figure 10.57 is called a **rose curve,** and each of the loops on the graph is called a *petal* of the rose curve.

θ	0	$\pi/12$	$\pi/6$	$\pi/4$	$\pi/3$	$5\pi/12$	$\pi/2$
r	2	$\sqrt{2}$	0	$-\sqrt{2}$	-2	$-\sqrt{2}$	0

By plotting these points and using the specified symmetry, zeros, and maximum values, you can obtain the graph shown in Figure 10.57. Note how the entire curve is generated as θ increases from 0 to π.

A computer animation of this concept appears in the *Interactive* CD-ROM.

Figure 10.57

$0 \le \theta \le \dfrac{\pi}{6}$

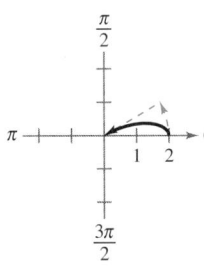

$0 \le \theta \le \dfrac{\pi}{3}$

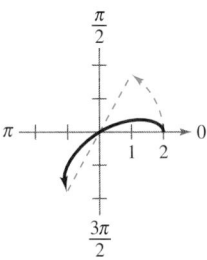

$0 \le \theta \le \dfrac{\pi}{2}$

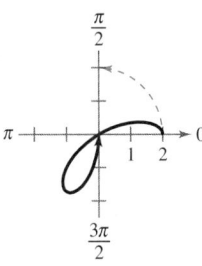

$0 \le \theta \le \dfrac{2\pi}{3}$

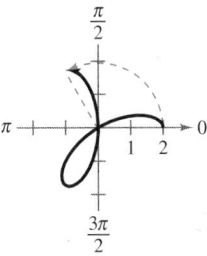

$0 \le \theta \le \dfrac{5\pi}{6}$

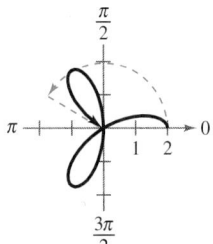

$0 \le \theta \le \pi$

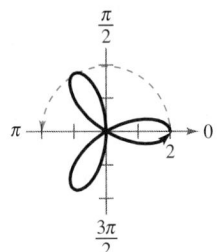

EXPLORATION

Notice that the rose curve in Example 4 has three petals. How many petals does the rose curve $r = 2 \cos 4\theta$ have? Experiment with other rose curves and determine the number of petals for the curves $r = 2 \cos n\theta$ and $r = 2 \sin n\theta$, where n is a positive integer.

Special Polar Graphs

Several important types of graphs have equations that are simpler in polar form than in rectangular form. For example, the circle $r = 4 \sin \theta$ in Example 1 has the more complicated rectangular equation $x^2 + (y - 2)^2 = 4$. The following list gives several other types of graphs that have simple polar equations.

Limaçons
$$r = a \pm b \cos \theta$$
$$r = a \pm b \sin \theta$$
$$(0 < a, \, 0 < b)$$

$\dfrac{a}{b} < 1$	$\dfrac{a}{b} = 1$	$1 < \dfrac{a}{b} < 2$	$\dfrac{a}{b} \geq 2$
Limaçon with inner loop	Cardioid (heart-shaped)	Dimpled limaçon	Convex limaçon

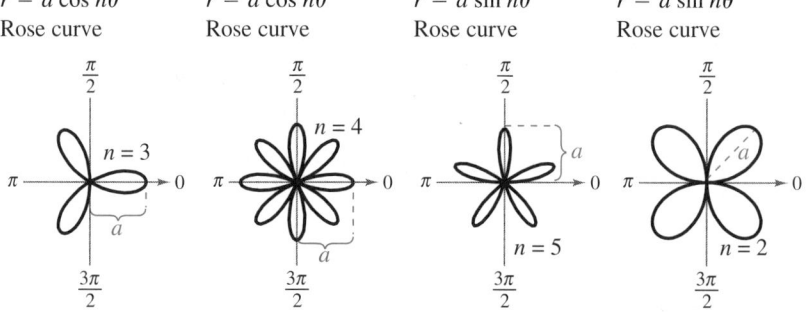

Rose Curves
n petals if n is odd
$2n$ petals if n is even
$(n \geq 2)$

$r = a \cos n\theta$	$r = a \cos n\theta$	$r = a \sin n\theta$	$r = a \sin n\theta$
Rose curve	Rose curve	Rose curve	Rose curve

Circles and Lemniscates

$r = a \cos \theta$	$r = a \sin \theta$	$r^2 = a^2 \sin 2\theta$	$r^2 = a^2 \cos 2\theta$
Circle	Circle	Lemniscate	Lemniscate

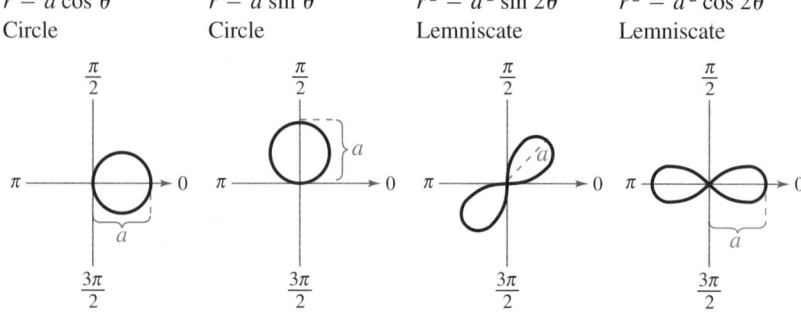

EXAMPLE 5 📖 **Analyzing a Rose Curve**

Analyze the graph of $r = 3 \cos 2\theta$.

Solution
Begin with an analysis of the basic features of the graph.

Type of curve	Rose curve with $2n = 4$ petals
Symmetry	With respect to polar axis and the line $\theta = \pi/2$
Maximum value of $\lvert r \rvert$	$\lvert r \rvert = 3$ when $\theta = 0, \pi/2, \pi, 3\pi/2$
Zeros of r	$r = 0$ when $\theta = \pi/4, 3\pi/4$

Using a graphing utility (with $0 \le \theta \le 2\pi$), you can obtain the graph shown in Figure 10.58. ▱

Figure 10.58

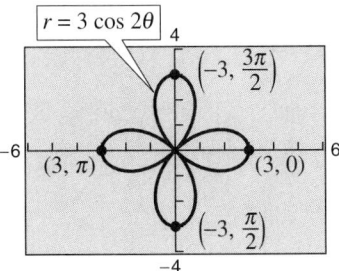

EXAMPLE 6 📖 **Analyzing a Lemniscate**

Analyze the graph $r^2 = 9 \sin 2\theta$.

Solution
Begin with an analysis of the basic features of the graph.

Type of curve	Lemniscate
Symmetry	With respect to pole
Maximum value of $\lvert r \rvert$	$\lvert r \rvert = 3$ when $\theta = \pi/4$
Zeros of r	$r = 0$ when $\theta = 0, \pi/2$

Using a graphing utility (with $r = \sqrt{9 \sin 2\theta}$ and $0 \le \theta \le 2\pi$), you can obtain the graph shown in Figure 10.59. ▱

Figure 10.59

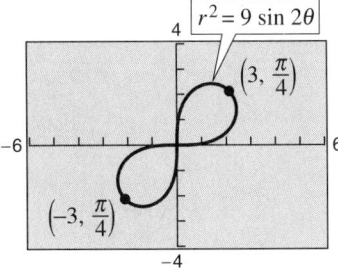

Group Activity   *Analyzing Polar Graphs*

Choose one of the equations below and graph it with your graphing utility. Show the graph to your partner and ask him or her to match the graph with its equation. Then switch roles and continue until you have graphed all six equations.

a. $r = 3 \cos \theta$ **b.** $r = 3 \sin 2\theta$ **c.** $r = 3(2 - \cos \theta)$

d. $r = 3(1 - 2 \cos \theta)$ **e.** $r = 3\pi/2$ **f.** $r^2 = 16 \cos 2\theta$

10.7 /// EXERCISES

In Exercises 1–6, identify the type of polar graph.

1.

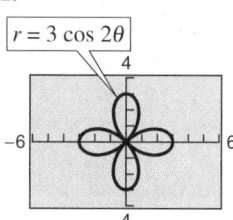

$r = 3 \cos 2\theta$

2.

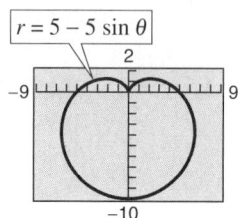

$r = 5 - 5 \sin \theta$

3.

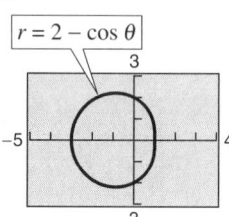

$r = 2 - \cos \theta$

4.

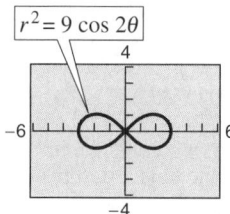

$r^2 = 9 \cos 2\theta$

5.

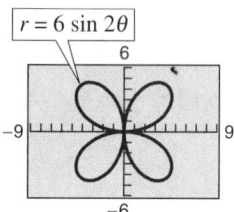

$r = 6 \sin 2\theta$

6.

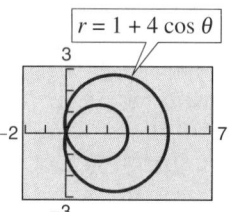

$r = 1 + 4 \cos \theta$

15. $r = 4 \cos 3\theta$

$r = 4 \cos 3\theta$

16. $r = 5 \sin 2\theta$

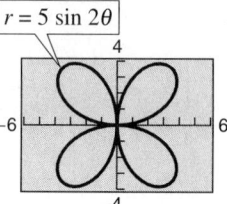

$r = 5 \sin 2\theta$

In Exercises 17–36, sketch the graph of the polar equation.

17. $r = 5$ **18.** $r = 2$

19. $r = \dfrac{\pi}{6}$ **20.** $r = -\dfrac{\pi}{4}$

21. $r = 3 \sin\theta$ **22.** $r = 3 \cos\theta$

23. $r = 3(1 - \cos \theta)$ **24.** $r = 2(1 - \sin\theta)$

25. $r = 4(1 + \sin\theta)$ **26.** $r = 1 + \cos\theta$

27. $r = 3 - 2 \cos\theta$ **28.** $r = 5 - 4 \sin\theta$

29. $r = 2 + \sin\theta$ **30.** $r = 4 + 3 \cos\theta$

31. $r = 2 \cos 3\theta$ **32.** $r = -\sin 5\theta$

33. $r = 3 \sin 2\theta$ **34.** $r = 3 \cos 2\theta$

35. $r = \dfrac{\theta}{2}$ **36.** $r = \theta$

In Exercises 7–12, test for symmetry with respect to $\theta = \pi/2$, the polar axis, and the pole.

7. $r = 10 + 6 \cos \theta$ **8.** $r = 16 \cos 3\theta$

9. $r = \dfrac{2}{1 + \sin \theta}$ **10.** $r = 6 \sin \theta$

11. $r = 4 \sec \theta \csc \theta$ **12.** $r^2 = 25 \sin 2\theta$

In Exercises 13–16, find the maximum value of $|r|$ and any zeros of r.

13. $r = 10(1 - \sin \theta)$

14. $r = 6 + 12 \cos \theta$

In Exercises 37–54, use a graphing utility to graph the polar equation.

37. $r = 6 \cos \theta$ **38.** $r = \dfrac{\theta}{4}$

39. $r = 3(2 - \sin\theta)$ **40.** $r = \cos 2\theta$

41. $r = 2 + 4 \sin\theta$ **42.** $r = 1 - 2 \cos\theta$

43. $r = 3 - 4 \cos\theta$ **44.** $r = 2(1 - 2 \sin\theta)$

45. $r = \dfrac{3}{\sin\theta - 2 \cos\theta}$ **46.** $r = \dfrac{6}{2 \sin\theta - 3 \cos\theta}$

47. $r^2 = 4 \cos 2\theta$ **48.** $r^2 = 4 \sin\theta$

49. $r = 4 \sin\theta \cos^2\theta$ **50.** $r = 2 \cos(3\theta - 2)$

51. $r = 2 \sec\theta$ **52.** $r = 3 \csc\theta$

53. $r = 2 \csc\theta + 5$ **54.** $r = 2 - \sec\theta$

In Exercises 55–62, use a graphing utility to graph the polar equation. Find an interval for θ for which the graph is traced *only once*.

55. $r = 3 - 4\cos\theta$ **56.** $r = 2(1 - 2\sin\theta)$

57. $r = 2 + \sin\theta$ **58.** $r = 4 + 3\cos\theta$

59. $r = 2\cos\left(\dfrac{3\theta}{2}\right)$ **60.** $r = 3\sin\left(\dfrac{5\theta}{2}\right)$

61. $r^2 = 4\sin 2\theta$ **62.** $r^2 = \dfrac{1}{\theta}$

In Exercises 63–66, use a graphing utility to graph the polar equation and show that the indicated line is an asymptote of the graph.

	Name of Graph	*Polar Equation*	*Asymptote*
63.	Conchoid	$r = 2 - \sec\theta$	$x = -1$
64.	Conchoid	$r = 2 + \csc\theta$	$y = 1$
65.	Hyperbolic spiral	$r = \dfrac{2}{\theta}$	$y = 2$
66.	Strophoid	$r = 2\cos 2\theta \sec\theta$	$x = -2$

67. *Exploration* Sketch the graph of $r = 4\sin\theta$ over each of the following intervals. Describe the part of the graph obtained in each case.

(a) $0 \le \theta \le \dfrac{\pi}{2}$ (b) $\dfrac{\pi}{2} \le \theta \le \pi$

(c) $-\dfrac{\pi}{2} \le \theta \le \dfrac{\pi}{2}$ (d) $\dfrac{\pi}{4} \le \theta \le \dfrac{3\pi}{4}$

68. *Graphical Reasoning* Use a graphing utility to graph the polar equation

$r = 6[1 + \cos(\theta - \phi)]$

for (a) $\phi = 0$, (b) $\phi = \pi/4$, and (c) $\phi = \pi/2$. Use the graphs to describe the effect of the angle ϕ. Write the equation as a function of $\sin\theta$ for part (c).

69. The graph of $r = f(\theta)$ is rotated about the pole through an angle ϕ. Show that the equation of the rotated graph is $r = f(\theta - \phi)$.

70. Consider the graph of $r = f(\sin\theta)$.

(a) Show that if the graph is rotated counterclockwise $\pi/2$ radians about the pole, the equation of the rotated graph is $r = f(-\cos\theta)$.

(b) Show that if the graph is rotated counterclockwise π radians about the pole, the equation of the rotated graph is $r = f(-\sin\theta)$.

(c) Show that if the graph is rotated counterclockwise $3\pi/2$ radians about the pole, the equation of the rotated graph is $r = f(\cos\theta)$.

In Exercises 71–74, use Exercise 69's and 70's results.

71. Write an equation for the limaçon $r = 2 - \sin\theta$ after it has been rotated by the given amount.

(a) $\dfrac{\pi}{4}$ (b) $\dfrac{\pi}{2}$ (c) π (d) $\dfrac{3\pi}{2}$

72. Write an equation for the rose curve $r = 2\sin 2\theta$ after it has been rotated by the given amount.

(a) $\dfrac{\pi}{6}$ (b) $\dfrac{\pi}{2}$ (c) $\dfrac{2\pi}{3}$ (d) π

73. Sketch the graph of each equation.

(a) $r = 1 - \sin\theta$ (b) $r = 1 - \sin\left(\theta - \dfrac{\pi}{4}\right)$

74. Use a graphing utility to graph each equation.

(a) $r = 3\sec\theta$ (b) $r = 3\sec\left(\theta - \dfrac{\pi}{4}\right)$

(c) $r = 3\sec\left(\theta + \dfrac{\pi}{3}\right)$ (d) $r = 3\sec\left(\theta - \dfrac{\pi}{2}\right)$

75. *Exploration* Use a graphing utility to graph the polar equation $r = 2 + k\cos\theta$ for $k = 0$, $k = 1$, $k = 2$, and $k = 3$. Identify each graph.

76. *Exploration* Consider the polar equation $r = 3\sin k\theta$.

(a) Use a graphing utility to graph the equation for $k = 1.5$. Find the interval for θ for which the graph is traced only once.

(b) Use a graphing utility to graph the equation for $k = 2.5$. Find the interval for θ for which the graph is traced only once.

(c) Is it possible to find an interval for θ for which the graph is traced only once for any rational number k? Explain.

10.8   Polar Equations of Conics

Alternative Definition of Conics / Polar Equations of Conics /
Application

Alternative Definition of Conics

In Sections 10.2 and 10.3, you learned that the rectangular equations of ellipses and hyperbolas take simple forms when the origin lies at the *center.* As it happens, there are many important applications of conics in which it is more convenient to use one of the *foci* as the origin for the coordinate system. For example, the sun lies at one focus of the earth's orbit. Similarly the light source of a parabolic reflector lies at its focus. In this section you will learn that polar equations of conics take simple forms if one of the foci lies at the pole.

To begin, consider the following alternative definition of a conic that uses the concept of eccentricity.

Alternative Definition of a Conic

The locus of a point in the plane that moves so that its distance from a fixed point (focus) is in constant ratio to its distance from a fixed line (directrix) is a **conic.** The constant ratio is the **eccentricity** of the conic and is denoted by *e*. Moreover, the conic is an **ellipse** if $e < 1$, a **parabola** if $e = 1$, and a **hyperbola** if $e > 1$.

> ### EXPLORATION
>
> Use a graphing utility to graph the following polar curves in the same viewing rectangle, $-4 \le x \le 4$, $-3 \le y \le 3$, and $0 \le \theta \le 2\pi$.
>
> $$r_1 = \frac{0.5}{1 + 0.5 \cos \theta}$$
>
> $$r_2 = \frac{1}{1 + \cos \theta}$$
>
> $$r_3 = \frac{2}{1 + 2 \cos \theta}$$
>
> What do you observe? Where do you think the foci are for each conic? What can you say about the eccentricity of each conic?

In Figure 10.60, note that for each type of conic, the pole corresponds to the fixed point (focus) given in the definition.

Figure 10.60

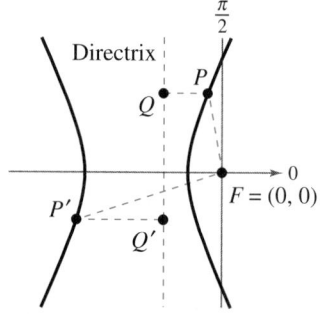

Polar Equations of Conics

The benefit of locating a focus at the pole can be seen in the proof of the following theorem.

Polar Equations of Conics

The graph of a polar equation of the form

1. $r = \dfrac{ep}{1 \pm e \cos \theta}$

2. $r = \dfrac{ep}{1 \pm e \sin \theta}$

is a conic, where $e > 0$ is the eccentricity and $|p|$ is the distance between the focus (pole) and the directrix.

Proof /// A proof for $r = ep/(1 + e \cos \theta)$ with $p > 0$ is listed here. The proofs of the other cases are similar. In Figure 10.61, consider a vertical directrix, p units to the right of the focus $F = (0, 0)$. If $P = (r, \theta)$ is a point on the graph of

$$r = \frac{ep}{1 + e \cos \theta}$$

the distance between P and the directrix is

$$
\begin{aligned}
PQ &= |p - x| \\
&= |p - r \cos \theta| \\
&= \left| p - \left(\frac{ep}{1 + e \cos\theta} \right) \cos \theta \right| \\
&= \left| p \left(1 - \frac{e \cos\theta}{1 + e \cos \theta} \right) \right| \\
&= \left| \frac{p}{1 + e \cos \theta} \right| \\
&= \left| \frac{r}{e} \right|.
\end{aligned}
$$

Moreover, because the distance between P and the pole is simply $PF = |r|$, the ratio of PF to PQ is

$$
\begin{aligned}
\frac{PF}{PQ} &= \frac{|r|}{|r/e|} \\
&= |e| \\
&= e
\end{aligned}
$$

and by definition, the graph of the equation must be a conic. ///

Figure 10.61

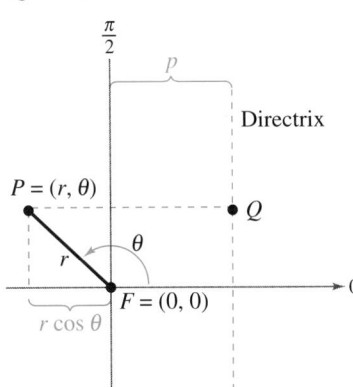

By completing the proofs of the other three cases, you can see that the equations

$$r = \frac{ep}{1 \pm e \cos \theta} \qquad \text{Vertical directrix}$$

correspond to conics with vertical directrices and the equations

$$r = \frac{ep}{1 \pm e \sin \theta} \qquad \text{Horizontal directrix}$$

correspond to conics with horizontal directrices. Moreover, the converse is also true—that is, any conic with a focus at the pole and having a horizontal or vertical directrix can be represented by one of the given equations.

EXAMPLE 1   Determining a Conic from Its Equation

Sketch the graph of the conic given by

$$r = \frac{15}{3 - 2 \cos \theta}.$$

Solution
To determine the type of conic, rewrite the equation as

$$r = \frac{15}{3 - 2 \cos \theta} = \frac{5}{1 - (2/3) \cos \theta}. \qquad \begin{array}{l}\text{Divide numerator and} \\ \text{denominator by 3.}\end{array}$$

From this form you can conclude that the graph is an ellipse with $e = \frac{2}{3}$. The graph is shown in Figure 10.62.

Figure 10.62

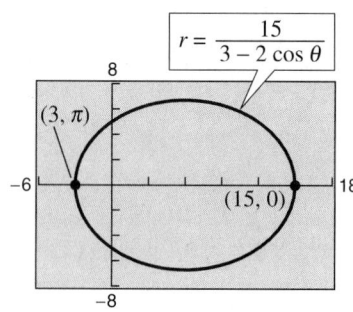

$$r = \frac{15}{3 - 2 \cos \theta}$$

For the ellipse in Figure 10.62, the major axis is horizontal and the vertices lie at $(r, \theta) = (15, 0)$ and $(r, \theta) = (3, \pi)$. Thus, the length of the *major* axis is $2a = 18$. To find the length of the *minor* axis, you can use the equations $e = c/a$ and $b^2 = a^2 - c^2$ to conclude that

$$b^2 = a^2 - c^2 = a^2 - (ea)^2 = a^2(1 - e^2). \qquad \text{Ellipse}$$

Because $e = \frac{2}{3}$, you have

$$b^2 = 9^2\left[1 - \left(\frac{2}{3}\right)^2\right] = 45$$

which implies that $b = \sqrt{45} = 3\sqrt{5}$. Thus, the length of the minor axis is $2b = 6\sqrt{5}$. A similar analysis for hyperbolas yields

$$b^2 = c^2 - a^2 = (ea)^2 - a^2 = a^2(e^2 - 1). \qquad \text{Hyperbola}$$

EXAMPLE 2   ▱   **Analyzing the Graph of a Polar Equation**

Analyze the graph of the polar equation

$$r = \frac{32}{3 + 5 \sin \theta}.$$

Solution

Dividing the numerator and denominator by 3 produces

$$r = \frac{32/3}{1 + (5/3) \sin \theta}.$$

Because $e = 5/3 > 1$, the graph is a hyperbola. The transverse axis of the hyperbola lies on the line $\theta = \pi/2$ and the vertices occur at $(r, \theta) = (4, \pi/2)$ and $(r, \theta) = (-16, 3\pi/2)$. Because the length of the transverse axis is 12, you can see that $a = 6$. To find b, write

$$b^2 = a^2(e^2 - 1) = 6^2\left[\left(\frac{5}{3}\right)^2 - 1\right] = 64.$$

Therefore, $b = 8$. The asymptotes of the hyperbola are $y = 10 \pm \frac{3}{4}x$, as shown in Figure 10.63.   ▱

Figure 10.63

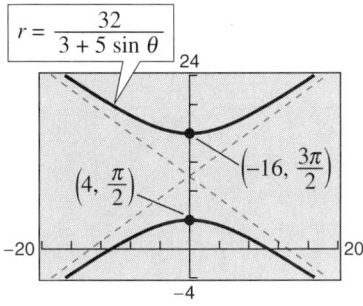

In the next example, you are asked to find a polar equation for a specified conic. To do this, let p be the distance between the pole and the directrix.

1. Horizontal directrix above the pole: $r = \dfrac{ep}{1 + e \sin \theta}$

2. Horizontal directrix below the pole: $r = \dfrac{ep}{1 - e \sin \theta}$

3. Vertical directrix to the right of the pole: $r = \dfrac{ep}{1 + e \cos \theta}$

4. Vertical directrix to the left of the pole: $r = \dfrac{ep}{1 - e \cos \theta}$

▱   **E X P L O R A T I O N**

Try using a graphing utility set in polar mode to verify the four orientations shown above. Remember that e must be positive, but p can be positive or negative.

Figure 10.64

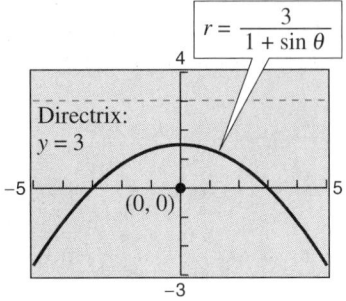

EXAMPLE 3   Finding the Polar Equation of a Conic

Find the polar equation of the parabola whose focus in the pole and whose directrix is the line $y = 3$.

Solution

From Figure 10.64, you can see that the directrix is horizontal. Thus, you can choose an equation of the form

$$r = \frac{ep}{1 + e \sin \theta}.$$

Moreover, because the eccentricity of a parabola is $e = 1$ and the distance between the pole and the directrix is $p = 3$, you have the equation

$$r = \frac{3}{1 + \sin \theta}.$$

Application

Kepler's Laws (listed below), named after the German astronomer Johannes Kepler (1571–1630), can be used to describe the orbits of the planets about the sun.

1. Each planet moves in an elliptical orbit with the sun as a focus.
2. A ray from the sun to the planet sweeps out equal areas of the ellipse in equal times.
3. The square of the period is proportional to the cube of the mean distance between the planet and the sun.

Although Kepler simply stated these laws on the basis of observation, they were later validated by Isaac Newton (1642–1727). In fact, Newton was able to show that each law can be deduced from a set of universal laws of motion and gravitation that govern the movement of all heavenly bodies, including comets and satellites. This is illustrated in the next example, which involves the comet named after the English mathematician and physicist Edmund Halley (1656–1742).

EXPLORATION

Use a graphing utility to compare the graph of the parabola in Example 3 with each of the following polar curves.

a. $r = \dfrac{3}{1 - \sin \theta}$

b. $r = \dfrac{3}{1 + \cos \theta}$

c. $r = \dfrac{3}{1 - \cos \theta}$

Determine the focus and directrix of each parabola.

Note   If you use earth as a reference with a period of 1 year and a distance of 1 astronomical unit, the proportionality constant in Kepler's third law is 1. For example, because Mars has a mean distance to the sun of $d = 1.523$ AU, its period P is given by $d^3 = P^2$. Thus, the period for Mars is $P = 1.88$ years.

Real Life

EXAMPLE 4   **Halley's Comet**

Halley's comet has an elliptical orbit with an eccentricity of $e \approx 0.97$. The length of the major axis of the orbit is approximately 36.18 astronomical units. (An *astronomical unit* is defined as the mean distance between the earth and the sun, 93 million miles.) Find a polar equation for the orbit. How close does Halley's comet come to the sun?

Solution

Using a vertical axis, as shown in Figure 10.65, choose an equation of the form $r = ep/(1 + e \sin \theta)$. Because the vertices of the ellipse occur when $\theta = \pi/2$ and $\theta = 3\pi/2$, you can determine the length of the major axis to be the sum of the r-values of the vertices. That is,

$$2a = \frac{0.97p}{1 + 0.97} + \frac{0.97p}{1 - 0.97} \approx 32.83p \approx 36.18.$$

Thus, $p \approx 1.102$ and $ep \approx (0.97)(1.102) \approx 1.069$. Using this value in the equation, you have

$$r = \frac{1.069}{1 + 0.97 \sin\theta}$$

where r is measured in astronomical units. To find the closest point to the sun (the focus), substitute $\theta = \pi/2$ into this equation to obtain

$$r = \frac{1.069}{1 + 0.97 \sin(\pi/2)}$$

$$\approx 0.54 \text{ AU}$$

$$\approx 50,000,000 \text{ miles.}$$

Figure 10.65

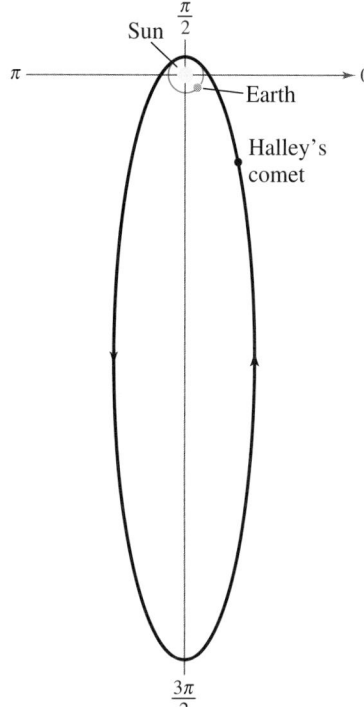

Group Activity   *Comets in Our Solar System*

Halley's comet is not the only spectacular comet that is periodically visible to viewers on earth. Use your school's library to find information about another comet, and write a paragraph describing some of its characteristics.

10.8 /// EXERCISES

Graphical Reasoning In Exercises 1–4, use a graphing utility to graph the polar equation when (a) $e = 1$, (b) $e = 0.5$, and (c) $e = 1.5$.

1. $r = \dfrac{2e}{1 + e \cos \theta}$

2. $r = \dfrac{2e}{1 - e \cos \theta}$

3. $r = \dfrac{2e}{1 - e \sin \theta}$

4. $r = \dfrac{2e}{1 + e \sin \theta}$

In Exercises 5–8, match the polar equation with the correct graph. [The graphs are labeled (a), (b), (c), and (d).]

(a)

(b)

(c)

(d)

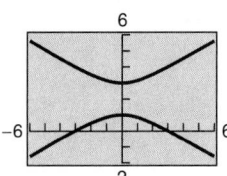

5. $r = \dfrac{4}{1 - \cos \theta}$

6. $r = \dfrac{3}{2 - \cos \theta}$

7. $r = \dfrac{3}{1 + 2 \sin \theta}$

8. $r = \dfrac{4}{1 + \sin \theta}$

In Exercises 9–20, identify and sketch the graph of the polar equation. Use a graphing utility to confirm your sketch.

9. $r = \dfrac{2}{1 - \cos \theta}$

10. $r = \dfrac{3}{1 + \sin \theta}$

11. $r = \dfrac{5}{1 + \sin \theta}$

12. $r = \dfrac{6}{1 + \cos \theta}$

13. $r = \dfrac{2}{2 - \cos \theta}$

14. $r = \dfrac{3}{3 + \sin \theta}$

15. $r = \dfrac{4}{2 + \sin \theta}$

16. $r = \dfrac{6}{3 - 2 \cos \theta}$

17. $r = \dfrac{3}{2 + 4 \sin \theta}$

18. $r = \dfrac{5}{-1 + 2 \cos \theta}$

19. $r = \dfrac{3}{2 - 6 \cos \theta}$

20. $r = \dfrac{3}{2 + 6 \sin \theta}$

In Exercises 21–26, use a graphing utility to graph the polar equation. Identify the graph.

21. $r = \dfrac{-1}{1 - \sin \theta}$

22. $r = \dfrac{-3}{2 + 4 \sin \theta}$

23. $r = \dfrac{3}{-4 + 2 \cos \theta}$

24. $r = \dfrac{4}{1 - 2 \cos \theta}$

25. $r = \dfrac{6}{2 - \cos \theta}$

26. $r = \dfrac{2}{2 + 3 \sin \theta}$

In Exercises 27–30, use a graphing utility to graph the rotated conic.

27. $r = \dfrac{2}{1 - \cos(\theta - \pi/4)}$ (See Exercise 9.)

28. $r = \dfrac{3}{3 + \sin(\theta - \pi/3)}$ (See Exercise 14.)

29. $r = \dfrac{4}{2 + \sin(\theta + \pi/6)}$ (See Exercise 15.)

30. $r = \dfrac{5}{-1 + 2 \cos(\theta + 2\pi/3)}$ (See Exercise 18.)

In Exercises 31–46, find a polar equation of the conic with its focus at the pole.

	Conic	Eccentricity	Directrix
31.	Parabola	$e = 1$	$x = -1$
32.	Parabola	$e = 1$	$y = -2$
33.	Ellipse	$e = \frac{1}{2}$	$y = 1$
34.	Ellipse	$e = \frac{3}{4}$	$y = -2$
35.	Hyperbola	$e = 2$	$x = 1$
36.	Hyperbola	$e = \frac{3}{2}$	$x = -1$

Conic	Vertex or Vertices
37. Parabola	$\left(1, -\dfrac{\pi}{2}\right)$
38. Parabola	$(4, 0)$
39. Parabola	$(5, \pi)$
40. Parabola	$\left(10, \dfrac{\pi}{2}\right)$
41. Ellipse	$(2, 0), (8, \pi)$
42. Ellipse	$\left(2, \dfrac{\pi}{2}\right), \left(4, \dfrac{3\pi}{2}\right)$
43. Ellipse	$(20, 0), (4, \pi)$
44. Hyperbola	$(2, 0), (10, 0)$
45. Hyperbola	$\left(1, \dfrac{3\pi}{2}\right), \left(9, \dfrac{3\pi}{2}\right)$
46. Hyperbola	$\left(4, \dfrac{\pi}{2}\right), \left(-1, \dfrac{3\pi}{2}\right)$

47. Show that the polar equation for the ellipse

$$\frac{x^2}{a^2} + \frac{y^2}{b^2} = 1 \quad \text{is} \quad r^2 = \frac{b^2}{1 - e^2 \cos^2\theta}.$$

48. Show that the polar equation for the hyperbola

$$\frac{x^2}{a^2} - \frac{y^2}{b^2} = 1 \quad \text{is} \quad r^2 = \frac{-b^2}{1 - e^2 \cos^2\theta}.$$

In Exercises 49–54, use the results of Exercises 47 and 48 to write the polar form of the equation of the conic.

49. $\dfrac{x^2}{169} + \dfrac{y^2}{144} = 1$ **50.** $\dfrac{x^2}{25} + \dfrac{y^2}{16} = 1$

51. $\dfrac{x^2}{9} - \dfrac{y^2}{16} = 1$ **52.** $\dfrac{x^2}{36} - \dfrac{y^2}{4} = 1$

53. Hyperbola, One focus: $\left(5, \dfrac{\pi}{2}\right)$; Vertices: $\left(4, \dfrac{\pi}{2}\right), \left(4, -\dfrac{\pi}{2}\right)$

54. Ellipse, One focus: $(4, 0)$; Vertices: $(5, 0), (5, \pi)$

55. *Planetary Motion* The planets travel in elliptical orbits with the sun as a focus. Assume that the focus is at the pole, the major axis lies on the polar axis, and the length of the major axis is $2a$ (see figure).

Show that the polar equation of the orbit is given by

$$r = \frac{(1 - e^2)a}{1 - e\cos\theta}, \text{ where } e \text{ is the eccentricity.}$$

Figure for 55

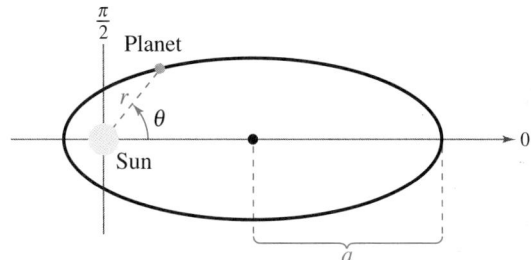

56. *Planetary Motion* Use the result of Exercise 55 to show that the minimum distance (*perihelion distance*) from the sun to the planet is $r = a(1 - e)$ and the maximum distance (*aphelion distance*) is $r = a(1 + e)$.

In Exercises 57 and 58, use the results of Exercises 55 and 56 to find the polar equation of the planet and the perihelion and aphelion distances.

57. Earth $a = 92.957 \times 10^6$ miles
            $e = 0.0167$

58. Pluto $a = 5.900 \times 10^9$ kilometers
            $e = 0.2481$

59. *Explorer 18* On November 26, 1963, the United States launched Explorer 18. Its low and high points over the surface of the earth were 119 miles and 122,000 miles, respectively (see figure). The center of the earth is the focus of the orbit. Find the polar equation for the orbit and find the distance between the surface of the earth and the satellite when $\theta = 60°$.

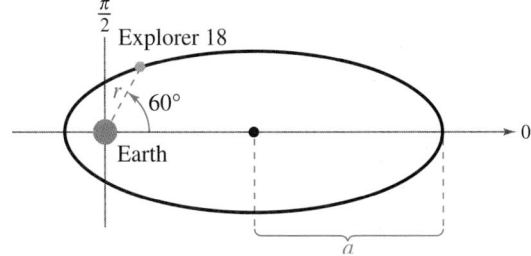

Focus on Concepts

In this chapter, you studied several concepts that are required in the study of algebra. You can use the following questions to check your understanding of several of these basic concepts. The answers to these questions are given in the back of the book.

In Exercises 1 and 2, an equation and four variations are given. In your own words, describe how the graph of each of the variations differs from the graph of the first equation.

1. $y^2 = 8x$

   (a) $(y - 2)^2 = 8x$ (b) $y^2 = 8(x + 1)$

   (c) $y^2 = -8x$ (d) $y^2 = 4x$

2. $\dfrac{x^2}{4} + \dfrac{y^2}{9} = 1$

   (a) $\dfrac{x^2}{9} + \dfrac{y^2}{4} = 1$ (b) $\dfrac{x^2}{4} + \dfrac{y^2}{4} = 1$

   (c) $\dfrac{x^2}{4} + \dfrac{y^2}{25} = 1$ (d) $\dfrac{(x - 3)^2}{4} + \dfrac{y^2}{9} = 1$

3. Explain how the central rectangle of a hyperbola can be used to sketch its asymptotes.

4. Consider an ellipse whose major axis is horizontal and 10 units in length. The number b in the standard form of the ellipse must be less than what real number? Explain the change in the shape of the ellipse as b approaches this number.

5. The curve represented by the parametric equations $x = 2 \sec t$ and $y = 3 \tan t$ is plotted in the figure. Would the curve change for the equations $x = 2 \sec(-t)$ and $y = 3 \tan(-t)$? If so, how would the graph change?

6. The graph of an object moving in an elliptical orbit is modeled by the parametric equations $x = 4 \cos t$ and $y = 3 \sin t$, where t is time (see figure). How would the orbit change for each of the following?

   (a) $x = 4 \cos 2t,$ $y = 3 \sin 2t$

   (b) $x = 5 \cos t,$ $y = 3 \sin t$

7. ***True or False?*** There is only one set of parametric equations that represents the line $y = 3 - 2x$. Explain.

8. ***True or False?*** There is a unique polar coordinate representation of each point in the plane. Explain.

9. Identify the type of symmetry each of the following points has with the point in the figure with polar coordinates $(4, \pi/6)$.

   (a) $\left(-4, \dfrac{\pi}{6}\right)$ (b) $\left(4, -\dfrac{\pi}{6}\right)$ (c) $\left(-4, -\dfrac{\pi}{6}\right)$

10. What is the relationship between the graphs of the rectangular and polar equations?

   (a) $x^2 + y^2 = 25,$ $r = 5$

   (b) $x - y = 0,$ $\theta = \dfrac{\pi}{4}$

2–34 even

10 /// REVIEW EXERCISES

In Exercises 1–4, match the equation with the correct graph. [The graphs are labeled (a), (b), (c), and (d).]

(a)

(b)

(c)

(d)
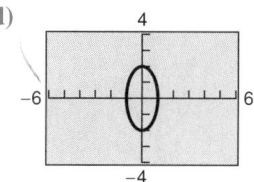

1. $4x^2 + y^2 = 4$ **2.** $4x^2 - y^2 = 4$

3. $y^2 = -4x$ **4.** $y^2 - 4x^2 = 4$

In Exercises 5–16, identify the conic and sketch its graph. Use a graphing utility to verify your sketch.

5. $4x - y^2 = 0$

6. $8y + x^2 = 0$

7. $x^2 - 6x + 2y + 9 = 0$

8. $y^2 - 12y - 8x + 20 = 0$

9. $x^2 + y^2 - 2x - 4y + 5 = 0$

10. $16x^2 + 16y^2 - 16x + 24y - 3 = 0$

11. $4x^2 + y^2 = 16$

12. $2x^2 + 6y^2 = 18$

13. $x^2 + 9y^2 + 10x - 18y + 25 = 0$

14. $4x^2 + y^2 - 16x + 15 = 0$

15. $5y^2 - 4x^2 = 20$

16. $x^2 - 9y^2 + 10x + 18y + 7 = 0$

In Exercises 17–22, solve for y and use a graphing utility to graph the resulting equations. Identify the conic.

17. $3x^2 + 2y^2 - 12x + 12y + 29 = 0$

18. $4x^2 - 4y^2 - 4x + 8y - 11 = 0$

19. $x^2 - 10xy + y^2 + 1 = 0$

20. $40x^2 + 36xy + 25y^2 - 52 = 0$

21. $x^2 + y^2 + 2xy + 2\sqrt{2}x - 2\sqrt{2}y + 2 = 0$

22. $9x^2 + 6y^2 + 4xy - 20 = 0$

In Exercises 23–26, find a rectangular equation of the specified parabola.

23. Vertex: $(4, 2)$ **24.** Vertex: $(2, 0)$
      Focus: $(4, 0)$ Focus: $(0, 0)$

25. Vertex: $(0, 2)$ **26.** Vertex: $(2, 2)$
      Directrix: $x = -3$ Directrix: $y = 0$

In Exercises 27–30, find a rectangular equation of the specified ellipse.

27. Vertices: $(-3, 0), (7, 0)$; Foci: $(0, 0), (4, 0)$

28. Vertices: $(2, 0), (2, 4)$; Foci: $(2, 1), (2, 3)$

29. Vertices: $(0, \pm 6)$; Passes through $(2, 2)$

30. Vertices: $(0, 1), (4, 1)$;
      Minor axis endpoints: $(2, 0), (2, 2)$

In Exercises 31–34, find a rectangular equation of the specified hyperbola.

31. Vertices: $(0, \pm 1)$; Foci: $(0, \pm 3)$

32. Vertices: $(2, 2), (-2, 2)$; Foci: $(4, 2), (-4, 2)$

33. Foci: $(0, 0), (8, 0)$; Asymptotes: $y = \pm 2(x - 4)$

34. Foci: $(3, \pm 2)$; Asymptotes: $y = \pm 2(x - 3)$

In Exercises 35–42, sketch the curve represented by the parametric equations and, where possible, write the corresponding rectangular equation by eliminating the parameter. Verify your result with a graphing utility.

35. $x = 1 + 4t$
   $y = 2 - 3t$

36. $x = t + 4$
   $y = t^2$

37. $x = \dfrac{1}{t}$
   $y = t^2$

38. $x = \dfrac{1}{t}$
   $y = 2t + 3$

39. $x = 6 \cos \theta$
   $y = 6 \sin \theta$

40. $x = 3 + 3 \cos \theta$
   $y = 2 + 5 \sin \theta$

41. $x = \sec \theta$
   $y = \tan \theta$

42. $x = 2\theta - \sin \theta$
   $y = 2 - \cos \theta$

43. *True or False?*   The graph of $(x^2/4) - y^4 = 1$ is a hyperbola. Explain.

44. Find a parametric representation of the ellipse with center at $(-3, 4)$, major axis horizontal and eight units in length, and minor axis six units in length.

45. Find a parametric representation of the hyperbola with vertices $(0, \pm 4)$ and foci $(0, \pm 5)$.

46. *Hypocycloids*   A hypocycloid has the parametric equations

$$x = (a - b) \cos t + b \cos\left(\frac{a - b}{b} t\right)$$

and

$$y = (a - b) \sin t - b \sin\left(\frac{a - b}{b} t\right).$$

Use a graphing utility to obtain a graph of the hypocycloid for the following values of a and b.

   (a) $a = 2, \ b = 1$ (b) $a = 3, \ b = 1$
   (c) $a = 4, \ b = 1$ (d) $a = 10, \ b = 1$
   (e) $a = 3, \ b = 2$ (f) $a = 4, \ b = 3$

In Exercises 47 and 48, convert the rectangular equation to polar form.

47. $(x^2 + y^2)^2 = ax^2y$ **48.** $x^2 + y^2 - 4x = 0$

In Exercises 49–54, convert the polar equation to rectangular form.

49. $r = 3 \cos \theta$

50. $r = 4 \sec\left(\theta - \dfrac{\pi}{3}\right)$

51. $r = \dfrac{2}{1 + \sin \theta}$

52. $r = \dfrac{1}{2 - \cos \theta}$

53. $r^2 = \cos 2\theta$

54. $r = 10$

In Exercises 55–62, identify and sketch the graph of the polar equation. Use a graphing utility to verify your result.

55. $r = 4$

56. $\theta = \dfrac{\pi}{12}$

57. $r = 4 \sin 2\theta$

58. $r = 2\theta$

59. $r = -2(1 + \cos \theta)$

60. $r = 3 - 4 \cos \theta$

61. $r = \dfrac{2}{1 - \sin \theta}$

62. $r = \dfrac{1}{1 + 2 \sin \theta}$

In Exercises 63–66, use a graphing utility to graph the polar equation.

63. $r^2 = 4 \sin^2 2\theta$

64. $r = 3 \csc \theta$

65. $r = \dfrac{3}{\cos(\theta - \pi/4)}$

66. $r = \dfrac{4}{5 - 3 \cos \theta}$

In Exercises 67–72, find a polar equation for the line or conic.

67. Circle   Center: $(5, \pi/2)$
                      Solution point: $(0, 0)$

68. Line   Solution point: $(0, 0)$
                      Slope: $\sqrt{3}$

69. Parabola   Vertex: $(2, \pi)$
                      Focus: $(0, 0)$

70. Parabola   Vertex: $(2, \pi/2)$
                      Focus: $(0, 0)$

71. Ellipse   Vertices: $(5, 0), (1, \pi)$
                      One focus: $(0, 0)$

72. Hyperbola   Vertices: $(1, 0), (7, 0)$
                      One focus: $(0, 0)$

CHAPTER PROJECT   *Polar, Rectangular, and Parametric Forms*

In this project, you will compare the polar, rectangular, and parametric equations for conics.

(a) Consider the polar equation

$$r = \frac{15}{3 - 2\cos\theta}.$$

In Section 10.8, you learned that the graph of this polar equation is an ellipse and that one of the ellipse's foci is at the pole, as shown in the graph at the left. To find the rectangular equation for this ellipse, begin by rewriting the equation as

$$3r - 2r\cos\theta = 15.$$

Then use the substitutions $r = \sqrt{x^2 + y^2}$ and $r\cos\theta = x$ to find the rectangular equation. After you find the rectangular equation, write it in the standard form

$$\frac{(x - h)^2}{a^2} + \frac{(y - k)^2}{b^2} = 1.$$

(b) Use the standard form of the ellipse in part (a) to find the center, foci, and eccentricity of the ellipse. Compare your results with those obtained in Example 1 in Section 10.8. Use the function mode of a graphing utility to graph the ellipse.

(c) Use the standard form of the ellipse and the identity $\sin^2\theta + \cos^2\theta = 1$ to write parametric equations for the ellipse. Then use the parametric mode of a graphing utility to graph the ellipse.

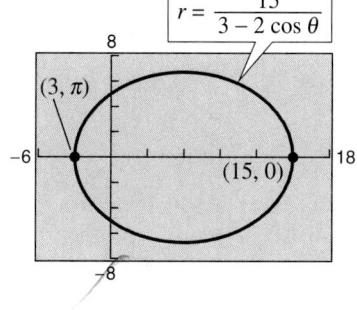

$r = \dfrac{15}{3 - 2\cos\theta}$

$(3, \pi)$

$(15, 0)$

Questions for Further Exploration

1. Sketch the graph of the polar equation

$$r = \frac{32}{3 + 5\sin\theta}.$$

Which type of conic is this?

2. Write the standard form of the rectangular equation of the conic in Question 1. Then sketch the graph using the *function* mode of a graphing utility. Does your graph agree with the graph obtained in Question 1? Which of the two graphs is easier to obtain? Explain your reasoning.

3. Write parametric equations for the conic given in Question 1. Then use the *parametric* mode of a graphing utility to graph the conic. Does your graph agree with the graph obtained in Question 1?

4. Consider the parametric equations

$$x = 4 + 5\cos t \quad \text{and} \quad y = 3\sin t.$$

With appropriate scaling of the coordinate system, could these equations represent the motion of a comet about the sun? Explain your reasoning.

7–10 /// CUMULATIVE TEST

Take this test as you would take a test in class. After you are done, check your work against the answers given in the back of the book.

The *Interactive* CD-ROM provides answers to the Chapter Tests and Cumulative Tests. It also offers Chapter Pre-Tests (that test key skills and concepts covered in previous chapters) and Chapter Post-Tests, both of which have randomly generated exercises with diagnostic capabilities.

In Exercises 1–3, solve the system of equations by the specified method.

1. Substitution

$2x - y^2 = 0$

$x - y = 4$

2. Graphically

$y = \log_3 x$

$y = -\frac{1}{3}x + 2$

3. Elimination

$x - 2y + 3z = 11$

$2x \quad\quad - z = 3$

$3y + z = -8$

4. A merchant plans to sell two models of compact disc players at costs of $275 and $400. The $275 model yields a profit of $55 and the $400 model yields a profit of $75. The merchant estimates that the total monthly demand will not exceed 300 units. No more than $100,000 in inventory should be invested for these products. Find the number of units of each model that should be stocked in order to maximize profit.

5. Find the inverse (if it exists): $\begin{bmatrix} 1 & 2 & -1 \\ 3 & 7 & -10 \\ -5 & -7 & -15 \end{bmatrix}$

6. Use determinants to find the area of the triangle with vertices $(0, 0)$, $(6, 2)$, and $(8, 10)$.

7. Find the sum: $\displaystyle\sum_{i=0}^{\infty} 3\left(\tfrac{1}{2}\right)^i$

8. Use the Binomial Theorem to expand and simplify $(z - 3)^4$.

9. On a game show, the digits 3, 4, and 5 must be arranged in the proper order to form the price of an appliance. If they are arranged correctly, the contestant wins the appliance. What is the probability of winning if the contestant knows that the price is at least $400?

10. Find the mean, median, and standard deviation of the numbers 21, 27, 23, 25, 30, and 19.

11. Sketch the graph of each equation.

(a) $6x - y^2 = 0$ (b) $\dfrac{(x - 2)^2}{4} + \dfrac{(y + 1)^2}{9} = 1$

12. Find an equation of the hyperbola with foci $(0, 0)$ and $(0, 4)$, and asymptotes $y = \pm\frac{1}{2}x + 2$.

13. Find a set of parametric equations of the line passing through the points $(2, -3)$ and $(6, 4)$. (The answer is not unique.)

14. Convert the rectangular equation $x^2 + y^2 - 6y = 0$ to polar form.

15. Graph (a) $r = 4/(1 + \cos \theta)$ and (b) $r = 4/(2 + \cos \theta)$.

Chapter 11

Analytic Geometry in Three Dimensions

Air traffic controlling makes constant use of the motion of objects in three-dimensional space. Here is a typical problem that an air traffic controller faces.

Aircraft #2
20 n mi

Aircraft #1
12 n mi

Airport

Aircraft #1 is flying at a speed of 150 knots, 12 nautical miles from landing and will gradually reduce its speed to 100 knots at touchdown.

Aircraft #2 is flying at a speed of 250 knots, 20 nautical miles behind Aircraft #1, and will gradually reduce its speed to 115 knots at touchdown.

As the controller, you must not let the distance between the two aircraft fall below 3 nautical miles.

Do you need to instruct Aircraft #2 to reduce its speed? If so, at what point does this need to be done? (See Exercises 55 and 56 on page 836.)

Photograph by Randy Mumm

Carol Dayton is shown at her current assignment at the Boise Air Traffic Control Tower. One of the many requirements for the job is the ability to make quick spatial assessments between moving aircraft.

11.1   The Three-Dimensional Coordinate System

The Three-Dimensional Coordinate System   /   The Distance and Midpoint Formulas   /   The Equation of a Sphere

The Three-Dimensional Coordinate System

Recall that the Cartesian plane is determined by two perpendicular number lines called the *x*-axis and the *y*-axis. These axes together with their point of intersection (the origin) allow you to develop a two-dimensional coordinate system for identifying points in a plane. To identify a point in space, you must introduce a third dimension to the model. The geometry of this three-dimensional model is called **solid analytic geometry.**

Figure 11.1

Figure 11.2

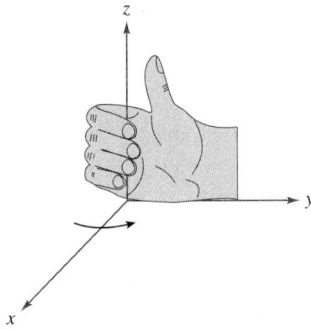

You can construct a **three-dimensional coordinate system** by passing a *z*-axis perpendicular to both the *x*- and *y*-axes at the origin. Figure 11.1 shows the positive portion of each coordinate axis. Taken as pairs, the axes determine three **coordinate planes:** the ***xy*-plane,** the ***xz*-plane,** and the ***yz*-plane.** These three coordinate planes separate the three-dimensional coordinate system into eight **octants.** The first octant is the one for which all three coordinates are positive. In this three-dimensional system, a point *P* in space is determined by an ordered triple (x, y, z), where x, y, and z are as follows.

$x =$ directed distance from *yz*-plane to *P*

$y =$ directed distance from *xz*-plane to *P*

$z =$ directed distance from *xy*-plane to *P*

A three-dimensional coordinate system can have either a **left-handed** or a **right-handed** orientation. In this text, you will work exclusively with right-handed systems, as illustrated in Figure 11.2.

Note   In a right-handed system, Octants II, III, and IV are found by rotating counterclockwise around the positive *z*-axis. Octant V is vertically below Octant I. Octants VI, VII, and VIII are then found by rotating counterclockwise around the negative *z*-axis.

Figure 11.3

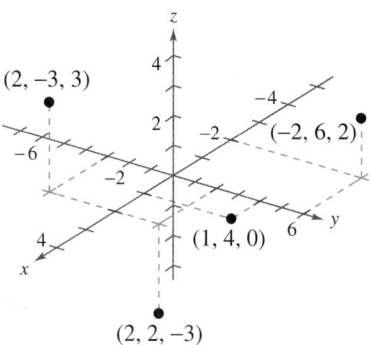

(2, −3, 3)

(−2, 6, 2)

(1, 4, 0)

(2, 2, −3)

EXAMPLE 1   Plotting Points in Space

Plot the following points in space.

a. $(2, -3, 3)$ **b.** $(-2, 6, 2)$ **c.** $(1, 4, 0)$ **d.** $(2, 2, -3)$

Solution

To plot the point $(2, -3, 3)$, notice that $x = 2$, $y = -3$, and $z = 3$. To help visualize the point, locate the point $(2, -3)$ in the xy-plane (denoted by a cross in Figure 11.3). The point $(2, -3, 3)$ lies three units above the cross. The other three points are also shown in Figure 11.3.

The Distance and Midpoint Formulas

Many of the formulas established for the two-dimensional coordinate system can be extended to three dimensions. For example, to find the distance between two points in space, you can use the Pythagorean Theorem twice, as shown in Figure 11.4. By doing this, you will obtain the formula for the distance between two points in space.

Figure 11.4

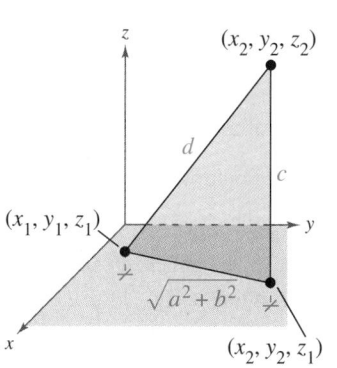

Distance Formula in Space

The distance between the points (x_1, y_1, z_1) and (x_2, y_2, z_2) is

$$d = \sqrt{(x_2 - x_1)^2 + (y_2 - y_1)^2 + (z_2 - z_1)^2}.$$

EXAMPLE 2   Finding the Distance Between Two Points

Find the distance between $(1, 0, 2)$ and $(2, 4, -3)$.

Solution

$$
\begin{aligned}
d &= \sqrt{(x_2 - x_1)^2 + (y_2 - y_1)^2 + (z_2 - z_1)^2} \\
&= \sqrt{(2 - 1)^2 + (4 - 0)^2 + (-3 - 2)^2} \qquad \text{Substitute.} \\
&= \sqrt{1 + 16 + 25} \qquad \text{Simplify.} \\
&= \sqrt{42} \qquad \text{Simplify.}
\end{aligned}
$$

Notice the similarity between the Distance Formulas in the plane and in space. The Midpoint Formulas in the plane and in space are also similar.

The *Interactive* CD-ROM shows every example with its solution; clicking on the *Try It!* button brings up similar problems. Guided Examples and Integrated Examples show step-by-step solutions to additional examples. Integrated Examples are related to several concepts in the section.

Figure 11.5

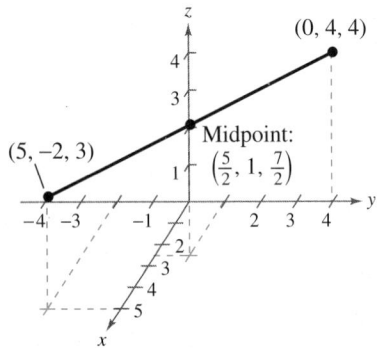

Midpoint Formula in Space

The midpoint of the line segment joining the points (x_1, y_1, z_1) and (x_2, y_2, z_2) is

$$\left(\frac{x_1 + x_2}{2}, \frac{y_1 + y_2}{2}, \frac{z_1 + z_2}{2}\right).$$

EXAMPLE 3   **Using the Midpoint Formula**

Find the midpoint of the line segment joining $(5, -2, 3)$ and $(0, 4, 4)$.

Solution
Using the Midpoint Formula, the midpoint is

$$\left(\frac{5 + 0}{2}, \frac{-2 + 4}{2}, \frac{3 + 4}{2}\right) = \left(\frac{5}{2}, 1, \frac{7}{2}\right)$$

as shown in Figure 11.5.

The Equation of a Sphere

A **sphere** with center at (h, k, j) and radius r is defined as the set of all points (x, y, z) such that the distance between (x, y, z) and (h, k, j) is r, as shown in Figure 11.6. Using the Distance Formula, this condition can be written as

$$\sqrt{(x - h)^2 + (y - k)^2 + (z - j)^2} = r.$$

By squaring both sides of this equation, you obtain the standard equation of a sphere.

Figure 11.6

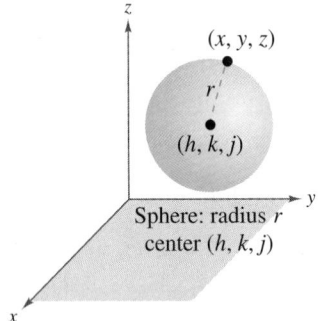

Sphere: radius r
center (h, k, j)

Standard Equation of a Sphere

The **standard equation of a sphere** whose center is (h, k, j) and whose radius is r is

$$(x - h)^2 + (y - k)^2 + (z - j)^2 = r^2.$$

Notice the similarity of this formula to the equation of a circle in the plane.

$(x - h)^2 + (y - k)^2 + (z - j)^2 = r^2$ Equation of sphere in space

$(x - h)^2 + (y - k)^2 = r^2$ Equation of circle in the plane

As is true with the equation of a circle, the equation of a sphere is simplified when the center lies at the origin. Is this case, the equation is

$$x^2 + y^2 + z^2 = r^2.$$ Sphere with center at origin

Figure 11.7

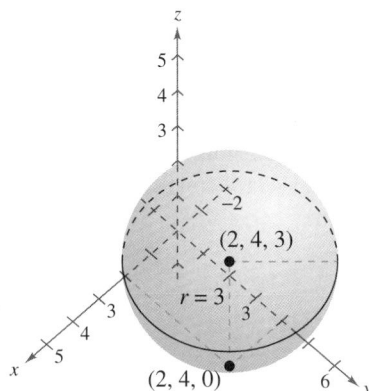

(2, 4, 3)

$r = 3$

(2, 4, 0)

EXAMPLE 4 　　**Finding the Equation of a Sphere**

Find the standard equation for the sphere whose center is (2, 4, 3) and whose radius is 3. Does this sphere intersect the *xy*-plane?

Solution

$$(x - h)^2 + (y - k)^2 + (z - j)^2 = r^2 \qquad \text{Standard equation}$$

$$(x - 2)^2 + (y - 4)^2 + (z - 3)^2 = 3^2 \qquad \text{Substitute.}$$

From the graph shown in Figure 11.7, you can see that the center of the sphere lies three units above the *xy*-plane. Because the sphere has a radius of 3, you can conclude that it does intersect the *xy*-plane—at the point (2, 4, 0). 　■

■　　**EXPLORATION**

Find the equation of the sphere that has the points (3, −2, 6) and (−1, 4, 2) as endpoints of a diameter. Explain how this problem gives you a chance to use all three formulas discussed so far in this section: the Distance Formula in space, the Midpoint Formula in Space, and the Standard Equation of a Sphere.

EXAMPLE 5 　　**Finding the Center and Radius of a Sphere**

Figure 11.8

Find the center and radius of the sphere whose equation is

$$x^2 + y^2 + z^2 - 2x + 4y - 6z + 8 = 0.$$

Sphere:
$(x - 1)^2 + (y + 2)^2 + (z - 3)^2 = 6$

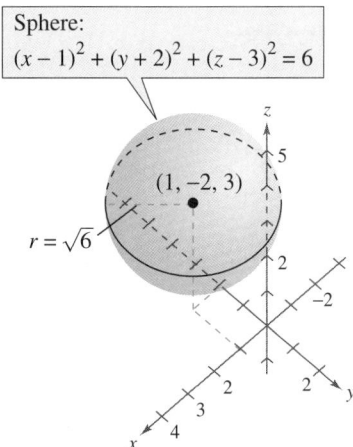

(1, −2, 3)

$r = \sqrt{6}$

Solution

You can obtain the standard equation of this sphere by completing the square, as follows.

$$x^2 + y^2 + z^2 - 2x + 4y - 6z + 8 = 0$$

$$\left(x^2 - 2x + \right) + \left(y^2 + 4y + \right) + \left(z^2 - 6z + \right) = -8$$

$$(x^2 - 2x + 1) + (y^2 + 4y + 4) + (z^2 - 6z + 9) = -8 + 1 + 4 + 9$$

$$(x - 1)^2 + (y + 2)^2 + (z - 3)^2 = 6$$

Therefore, the center of the sphere is (1, −2, 3), and its radius is $\sqrt{6}$, as shown in Figure 11.8. 　■

Note in Example 5 that the points satisfying the equation of the sphere are "surface points," not "interior points." In general, the collection of points satisfying the equation involving x, y, and z is called a **surface in space.**

Finding the intersection of a surface with one of the three coordinate planes (or with a plane parallel to one of the three coordinate planes) helps one visualize the surface. Such an intersection is called a **trace** of the surface. For example, the xy-trace of a surface consists of all points that are common to both the surface *and* the xy-plane. Similarly, the xz-trace of a surface consists of all points that are common to both the surface and xz-plane.

EXAMPLE 6 ▬ **Finding a Trace of a Surface**

Sketch the xy-trace of the sphere whose equation is

$$(x - 3)^2 + (y - 2)^2 + (z + 4)^2 = 5^2.$$

Solution

To find the xy-trace of this surface, use the fact that every point in the xy-plane has a z-coordinate of zero. This means that if you substitute $z = 0$ into the given equation, the resulting equation will represent the intersection of the surface with the xy-plane.

$$(x - 3)^2 + (y - 2)^2 + (0 + 4)^2 = 25$$
$$(x - 3)^2 + (y - 2)^2 + 16 = 25$$
$$(x - 3)^2 + (y - 2)^2 = 9$$
$$(x - 3)^2 + (y - 2)^2 = 3^2$$

From this form, you can see that the xy-trace is a circle of radius 3, as shown in Figure 11.9. ▬

Figure 11.9

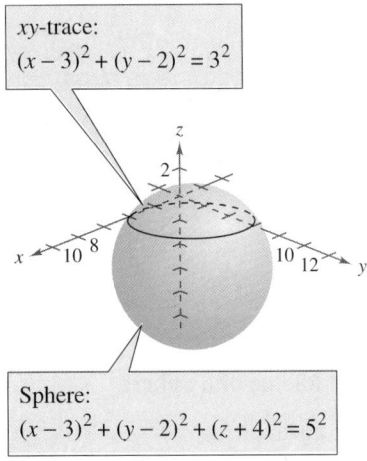

xy-trace:
$(x - 3)^2 + (y - 2)^2 = 3^2$

Sphere:
$(x - 3)^2 + (y - 2)^2 + (z + 4)^2 = 5^2$

Group Activity   *Comparing Two and Three Dimensions*

In this section, you saw similarities between formulas in two-dimensional coordinate geometry and three-dimensional coordinate geometry. In two-dimensional coordinate geometry, the graph of the equation $ax + by + c = 0$ is a line. In three-dimensional coordinate geometry, what is the graph of the equation $ax + by + cz = 0$? Is it a line? Explain your reasoning to another group member.

11.1 /// EXERCISES

In Exercises 1–4, approximate the coordinates of the points.

1.

2.

3.

4.
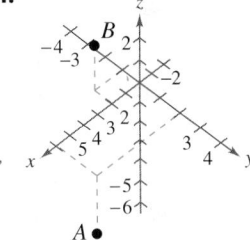

In Exercises 5–8, plot the points on the same three-dimensional coordinate system.

5. (a) $(2, 1, 3)$
   (b) $(-1, 2, 1)$

6. (a) $(3, -2, 5)$
   (b) $\left(\frac{3}{2}, 4, -2\right)$

7. (a) $(5, -2, 2)$
   (b) $(5, -2, -2)$

8. (a) $(0, 4, -5)$
   (b) $(4, 0, 5)$

In Exercises 9–12, find the coordinates of the point.

9. The point is located three units behind the yz-plane, four units to the right of the xz-plane, and five units above the xy-plane.

10. The point is located seven units in front of the yz-plane, two units to the left of the xz-plane, and one unit below the xy-plane.

11. The point is located on the x-axis, 10 units in front of the yz-plane.

12. The point is located in the yz-plane, three units to the right of the xz-plane, and two units above the xy-plane.

13. *Think About It* What is the z-coordinate of any point in the xy-plane? What is the y-coordinate of any point in the xz-plane? What is the x-coordinate of any point in the yz-plane?

14. *Think About It* When plotting points on the three-dimensional coordinate system, is it true that the scales on each of the three axes must be the same? Explain.

In Exercises 15–20, determine the octant(s) in which (x, y, z) is located so that the condition(s) is (are) satisfied.

15. $x > 0, y < 0, z > 0$
16. $x < 0, y > 0, z < 0$
17. $z > 0$
18. $y < 0$
19. $xy < 0$
20. $yz > 0$

In Exercises 21–24, find the lengths of the sides of the right triangle. Show that these lengths satisfy the Pythagorean Theorem.

21.

22.

23.

24. $(2, -1, 2)$

The *Interactive* CD-ROM contains step-by-step solutions to all odd-numbered Section and Review Exercises. It also provides Tutorial Exercises, which link to Guided Examples for additional help.

In Exercises 25–28, find the lengths of the sides of the triangle with the indicated vertices, and determine whether the triangle is a right triangle, an isosceles triangle, or neither.

25. $(0, 0, 0)$, $(2, 2, 1)$, $(2, -4, 4)$

26. $(5, 3, 4)$, $(7, 1, 3)$, $(3, 5, 3)$

27. $(1, -3, -2)$, $(5, -1, 2)$, $(-1, 1, 2)$

28. $(5, 0, 0)$, $(0, 2, 0)$, $(0, 0, -3)$

In Exercises 29–32, find the coordinates of the midpoint of the line segment joining the points.

29. $(3, -6, 10)$, $(-3, 2, 2)$ **30.** $(2, -2, -8)$, $(6, 6, 18)$

31. $(4, -2, 5)$, $(-4, 2, 8)$ **32.** $(-3, 5, 7)$, $(-6, 4, 10)$

33. A line segment has (x_1, y_1, z_1) as one endpoint and (x_m, y_m, z_m) as its midpoint. Find the other endpoint (x_2, y_2, z_2) of the line segment in terms of $x_1, y_1, z_1, x_m, y_m,$ and z_m.

34. Use the result of Exercise 33 to find the coordinates of the endpoint of a line segment if the coordinates of the other endpoint and the midpoint are $(3, 0, 2)$ and $(5, 8, 7)$, respectively.

In Exercises 35–42, find the standard form of the equation of the sphere.

35.   **36.**

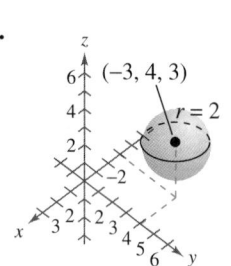

37. Center: $(0, 4, 3)$; Radius: 4

38. Center: $(1, -2, 3)$; Radius: 5

39. Center: $(-3, 7, 5)$; Diameter: 10

40. Center: $(0, 5, -9)$; Diameter: 6

41. Endpoints of a diameter: $(3, 0, 0)$, $(0, 0, 6)$

42. Endpoints of a diameter: $(2, -2, 2)$, $(-1, 4, 6)$

In Exercises 43–48, find the center and radius of the sphere.

43. $x^2 + y^2 + z^2 - 4x + 2y - 6z + 10 = 0$

44. $x^2 + y^2 + z^2 - 6x + 4y + 9 = 0$

45. $x^2 + y^2 + z^2 + 4x - 8z + 19 = 0$

46. $x^2 + y^2 + z^2 - 8y - 6z + 13 = 0$

47. $9x^2 + 9y^2 + 9z^2 - 18x - 6y - 72z + 73 = 0$

48. $2x^2 + 2y^2 + 2z^2 - 2x - 6y - 4z + 5 = 0$

In Exercises 49 and 50, sketch the graph of the equation and sketch the specified traces.

49. $x^2 + y^2 + z^2 = 16$

   (a) xz-trace

   (b) xy-trace

50. $x^2 + y^2 + (z - 3)^2 = 9$

   (a) yz-trace

   (b) xz-trace

In Exercises 51–54, use a graphing utility to graph the sphere. (*Hint:* Solve for z and graph the two resulting expressions in x and y.)

51. $x^2 + y^2 + z^2 - 16 = 0$

52. $x^2 + y^2 + z^2 - 4y - 4 = 0$

53. $(x - 3)^2 + (y - 4)^2 + (z - 5)^2 = 4$

54. $x^2 + y^2 + z^2 + 6y - 8z + 21 = 0$

Chapter Opener In Exercises 55 and 56 the distance from the aircraft to the airport can be modeled by $d_1(t) = 0.07t^2 - 2.43t + 12$ (Aircraft #1) and $d_2(t) = 0.16t^2 - 3.93t + 20$ (Aircraft #2), where d represents the distance in nautical miles and t represents the time in minutes. ($t = 0$ is when the aircraft are 12 and 20 nautical miles, respectively, from the airport.)

55. Find a model that represents the distance between the two aircraft.

56. Will the two aircraft be less than three miles apart before the first aircraft lands? If so, when will that happen?

11.2   Vectors in Space

Vectors in Space / *Parallel Vectors* / *Application*

Vectors in Space

Figure 11.10

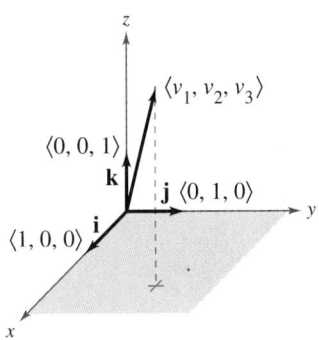

Physical forces and velocities are not confined to the plane, so it is natural to extend the concept of vectors from two-dimensional space to three-dimensional space. In space, vectors are denoted by ordered triples

$$\mathbf{v} = \langle v_1, v_2, v_3 \rangle. \qquad \text{Component form}$$

The **zero vector** is denoted by $\mathbf{0} = \langle 0, 0, 0 \rangle$. Using the unit vectors $\mathbf{i} = \langle 1, 0, 0 \rangle$, $\mathbf{j} = \langle 0, 1, 0 \rangle$, and $\mathbf{k} = \langle 0, 0, 1 \rangle$ in the direction of the positive z-axis, the **standard unit vector notation** for $\mathbf{v}$ is

$$\mathbf{v} = v_1\mathbf{i} + v_2\mathbf{j} + v_3\mathbf{k} \qquad \text{Unit vector form}$$

as shown in Figure 11.10. If $\mathbf{v}$ is represented by the directed line segment from $P(p_1, p_2, p_3)$ to $Q(q_1, q_2, q_3)$, as shown in Figure 11.11, the component form of $\mathbf{v}$ is produced by subtracting the coordinates of the initial point from the coordinates of the terminal point, as follows.

$$\mathbf{v} = \langle v_1, v_2, v_3 \rangle = \langle q_1 - p_1, q_2 - p_2, q_3 - p_3 \rangle$$

Figure 11.11

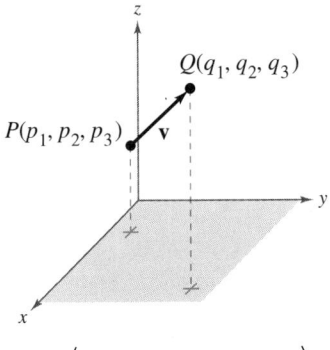

$$\mathbf{v} = \langle q_1 - p_1, q_2 - p_2, q_3 - p_3 \rangle$$

Note Note how similar these definitions are to corresponding definitions of vectors in the plane. The properties of vector operations discussed in Sections 6.3 and 6.4 are also valid for vectors in space.

Vectors in Space

1. Two vectors are equal if and only if their corresponding components are equal.
2. The length of $\mathbf{u} = \langle u_1, u_2, u_3 \rangle$ is

$$\|\mathbf{u}\| = \sqrt{u_1^2 + u_2^2 + u_3^2}. \qquad \text{Vector length}$$

3. A unit vector $\mathbf{u}$ in the direction of $\mathbf{v}$ is given by

$$\mathbf{u} = \frac{\mathbf{v}}{\|\mathbf{v}\|}, \qquad \mathbf{v} \neq \mathbf{0}. \qquad \text{Unit vector}$$

4. The sum of $\mathbf{u} = \langle u_1, u_2, u_3 \rangle$ and $\mathbf{v} = \langle v_1, v_2, v_3 \rangle$ is

$$\mathbf{u} + \mathbf{v} = \langle u_1 + v_1, u_2 + v_2, u_3 + v_3 \rangle. \qquad \text{Vector addition}$$

5. The scalar multiple of the real number c and $\mathbf{u} = \langle u_1, u_2, u_3 \rangle$ is

$$c\mathbf{u} = \langle cu_1, cu_2, cu_3 \rangle. \qquad \text{Scalar multiplication}$$

6. The dot product of $\mathbf{u} = \langle u_1, u_2, u_3 \rangle$ and $\mathbf{v} = \langle v_1, v_2, v_3 \rangle$ is

$$\mathbf{u} \cdot \mathbf{v} = u_1 v_1 + u_2 v_2 + u_3 v_3. \qquad \text{Dot product}$$

Some graphing utilities have the capability to perform vector operations, such as the dot product. For instance, on the *TI-85,* you can use the VECTR menu to verify the dot product in Example 2 as follows.

dot([0, 3, −2], [4, −2, 3])

On the *TI-92,* you can use the MATH/Matrix/Vector ops menu to verify the dot product as follows.

dotP({0, 3, −2}, {4, −2, 3})

The *Interactive* CD-ROM offers graphing utility emulators of the *TI-82* and *TI-83,* which can be used with the Examples, Explorations, Technology notes, and Exercises.

Figure 11.12

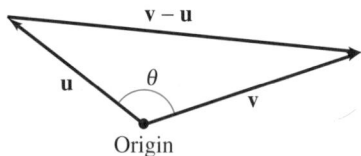

Origin

EXAMPLE 1   Finding the Component Form of a Vector

Find the component form and length of the vector **v** having initial point $(3, 4, 2)$ and terminal point $(3, 6, 4)$. Then find a unit vector in the direction of **v**.

Solution

The component form of **v** is

$$\mathbf{v} = \langle 3 - 3, 6 - 4, 4 - 2 \rangle = \langle 0, 2, 2 \rangle$$

which implies that its length is

$$\|\mathbf{v}\| = \sqrt{0^2 + 2^2 + 2^2} = \sqrt{8} = 2\sqrt{2}.$$

The unit vector in the direction of **v** is

$$\mathbf{u} = \frac{\mathbf{v}}{\|\mathbf{v}\|} = \frac{1}{2\sqrt{2}} \langle 0, 2, 2 \rangle = \left\langle 0, \frac{1}{\sqrt{2}}, \frac{1}{\sqrt{2}} \right\rangle.$$

EXAMPLE 2   Finding the Dot Product of Two Vectors

Find the dot product of $\langle 0, 3, -2 \rangle$ and $\langle 4, -2, 3 \rangle$.

Solution

$$\begin{aligned}
\langle 0, 3, -2 \rangle \cdot \langle 4, -2, 3 \rangle &= 0(4) + 3(-2) + (-2)(3) \\
&= 0 - 6 - 6 \\
&= -12
\end{aligned}$$

Note that the dot product of two vectors is a real number, not a vector.

As was discussed in Section 6.4, the **angle between two nonzero vectors** is the angle θ, $0 \le \theta \le \pi$, between its respective standard position vectors. (See Figure 11.12.) This angle can be found using the dot product. (Note that the angle between the zero vector and another vector is not defined.)

Angle Between Two Vectors
If θ is the angle between two nonzero vectors **u** and **v**, then $$\cos \theta = \frac{\mathbf{u} \cdot \mathbf{v}}{\|\mathbf{u}\| \|\mathbf{v}\|}.$$

If the dot product of two nonzero vectors is zero, the angle between the vectors is 90°. Such vectors are called **orthogonal.** For instance, the standard unit vectors **i**, **j**, and **k** are orthogonal to each other.

Figure 11.13

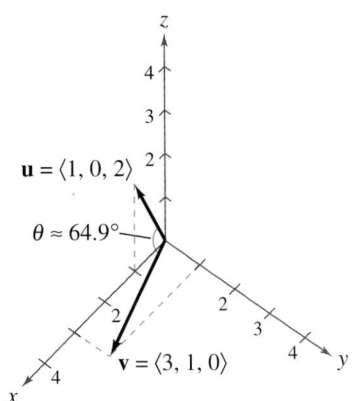

EXAMPLE 3   Finding the Angle Between Two Vectors

Find the angle between $\mathbf{u} = \langle 1, 0, 2 \rangle$ and $\mathbf{v} = \langle 3, 1, 0 \rangle$.

Solution

$$\cos \theta = \frac{\mathbf{u} \cdot \mathbf{v}}{\|\mathbf{u}\| \|\mathbf{v}\|} = \frac{\langle 1, 0, 2 \rangle \cdot \langle 3, 1, 0 \rangle}{\|\langle 1, 0, 2 \rangle\| \|\langle 3, 1, 0 \rangle\|} = \frac{3}{\sqrt{50}}$$

This implies that the angle between the two vectors is

$$\theta = \arccos \frac{3}{\sqrt{50}} \approx 64.9°$$

as shown in Figure 11.13.

Parallel Vectors

Recall from the definition of scalar multiplication that positive scalar multiples of a nonzero vector $\mathbf{v}$ have the same direction as $\mathbf{v}$, whereas negative multiples have the direction opposite that of $\mathbf{v}$. In general, two nonzero vectors $\mathbf{u}$ and $\mathbf{v}$ are **parallel** if there is some scalar c such that $\mathbf{u} = c\mathbf{v}$. For example, in Figure 11.14, the vectors $\mathbf{u}$, $\mathbf{v}$, and $\mathbf{w}$ are parallel because $\mathbf{u} = 2\mathbf{v}$ and $\mathbf{w} = -\mathbf{v}$.

Figure 11.14

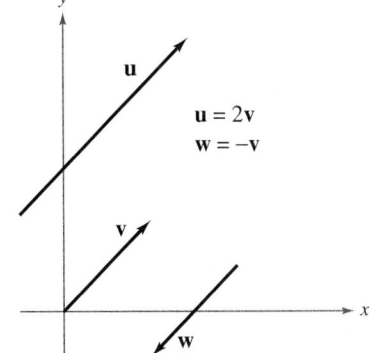

EXAMPLE 4   Parallel Vectors

Vector $\mathbf{w}$ has initial point $(1, -2, 0)$ and terminal point $(3, 2, 1)$. Which of the following vectors is parallel to $\mathbf{w}$?

a. $\mathbf{u} = \langle 4, 8, 2 \rangle$

b. $\mathbf{v} = \langle 4, 8, 4 \rangle$

Solution
Begin by writing $\mathbf{w}$ in component form.

$$\mathbf{w} = \langle 3 - 1, 2 - (-2), 1 - 0 \rangle = \langle 2, 4, 1 \rangle$$

a. The vector $\mathbf{u}$ *is* parallel to $\mathbf{w}$ because

$$\mathbf{u} = \langle 4, 8, 2 \rangle = 2\langle 2, 4, 1 \rangle = 2\mathbf{w}.$$

b. In this case, you need to find a scalar c such that

$$\langle 4, 8, 4 \rangle = c\langle 2, 4, 1 \rangle.$$

However, equating corresponding components produces $c = 2$ for the first two components and $c = 4$ for the third. Hence, the equation has no solution, and the vectors are *not* parallel.

You can use vectors to determine whether three points are collinear (lie on the same line). The points P, Q, and R are collinear if and only if the vectors $\overrightarrow{PQ}$ and $\overrightarrow{PR}$ are parallel.

EXAMPLE 5   **Using Vectors to Determine Collinear Points**

Determine whether the following points lie on the same line.

$$P(2, -1, 4), \quad Q(5, 4, 6), \quad \text{and} \quad R(-4, -11, 0)$$

Figure 11.15

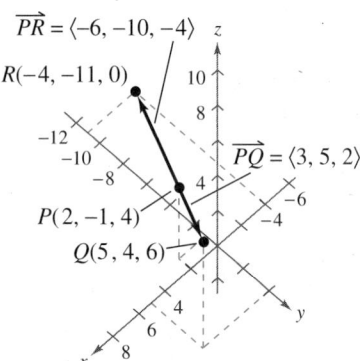

$\overrightarrow{PR} = \langle -6, -10, -4 \rangle$

$R(-4, -11, 0)$

$\overrightarrow{PQ} = \langle 3, 5, 2 \rangle$

$P(2, -1, 4)$

$Q(5, 4, 6)$

Solution

The component forms of $\overrightarrow{PQ}$ and $\overrightarrow{PR}$ are

$$\overrightarrow{PQ} = \langle 5 - 2, 4 + 1, 6 - 4 \rangle$$
$$= \langle 3, 5, 2 \rangle$$

and

$$\overrightarrow{PR} = \langle -4 - 2, -11 + 1, 0 - 4 \rangle$$
$$= \langle -6, -10, -4 \rangle.$$

Because $\overrightarrow{PR} = -2\overrightarrow{PQ}$, you can conclude that they are parallel. Therefore, the points P, Q, and R lie on the same line, as shown in Figure 11.15.

EXAMPLE 6   **Finding the Terminal Point of a Vector**

The initial point of the vector $\mathbf{v} = 4\mathbf{i} + 2\mathbf{j} - \mathbf{k}$ is $P(3, -1, 6)$. What is the terminal point of this vector?

Solution

Using the component form of the vector whose initial point is P and whose terminal point is Q, you can write

$$\overrightarrow{PQ} = \langle q_1 - p_1, q_2 - p_2, q_3 - p_3 \rangle$$
$$= \langle q_1 - 3, q_2 + 1, q_3 - 6 \rangle$$
$$= \langle 4, 2, -1 \rangle.$$

This implies that

$$q_1 - 3 = 4, \quad q_2 + 1 = 2, \quad \text{and} \quad q_3 - 6 = -1.$$

The solutions of these three equations are $q_1 = 7$, $q_2 = 1$, and $q_3 = 5$. Thus, the terminal point is $Q(7, 1, 5)$.

Application

In Section 6.3, you saw how to use vectors to solve an equilibrium problem in a plane. The next example shows how to use vectors to solve an equilibrium problem in space.

Real Life

EXAMPLE 7   **Solving an Equilibrium Problem**

A weight of 480 pounds is supported by three ropes. As shown in Figure 11.16, the weight is located at $S(0, 2, -1)$. The ropes are tied to the points $P(2, 0, 0)$, $Q(0, 4, 0)$, and $R(-2, 0, 0)$. Find the force (or tension) on each rope.

Figure 11.16

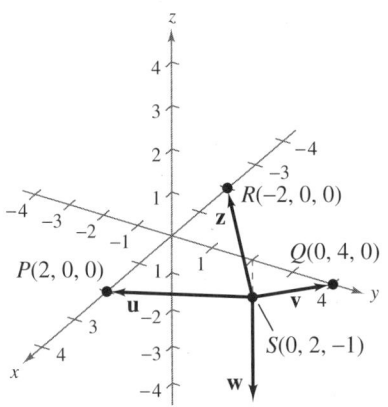

Solution

The (downward) force of the weight is represented by the vector $\mathbf{w} = \langle 0, 0, -480 \rangle$. The force vectors corresponding to the ropes are as follows.

$$\mathbf{u} = \|\mathbf{u}\| \frac{\overrightarrow{SP}}{\|\overrightarrow{SP}\|} = \|\mathbf{u}\| \frac{\langle 2 - 0, 0 - 2, 0 + 1 \rangle}{3} = \|\mathbf{u}\| \left\langle \frac{2}{3}, -\frac{2}{3}, \frac{1}{3} \right\rangle$$

$$\mathbf{v} = \|\mathbf{v}\| \frac{\overrightarrow{SQ}}{\|\overrightarrow{SQ}\|} = \|\mathbf{v}\| \frac{\langle 0 - 0, 4 - 2, 0 + 1 \rangle}{\sqrt{5}} = \|\mathbf{v}\| \left\langle 0, \frac{2}{\sqrt{5}}, \frac{1}{\sqrt{5}} \right\rangle$$

$$\mathbf{z} = \|\mathbf{z}\| \frac{\overrightarrow{SR}}{\|\overrightarrow{SR}\|} = \|\mathbf{z}\| \frac{\langle -2 - 0, 0 - 2, 0 + 1 \rangle}{3} = \|\mathbf{z}\| \left\langle -\frac{2}{3}, -\frac{2}{3}, \frac{1}{3} \right\rangle$$

For the system to be in equilibrium, it must be true that

$$\mathbf{u} + \mathbf{v} + \mathbf{z} + \mathbf{w} = \mathbf{0} \quad \text{or} \quad \mathbf{u} + \mathbf{v} + \mathbf{z} = -\mathbf{w}.$$

This yields the following system of linear equations.

$$\frac{2}{3}\|\mathbf{u}\| \qquad\qquad -\frac{2}{3}\|\mathbf{z}\| = \quad 0$$

$$-\frac{2}{3}\|\mathbf{u}\| + \frac{2}{\sqrt{5}}\|\mathbf{v}\| - \frac{2}{3}\|\mathbf{z}\| = \quad 0$$

$$\frac{1}{3}\|\mathbf{u}\| + \frac{1}{\sqrt{5}}\|\mathbf{v}\| + \frac{1}{3}\|\mathbf{z}\| = 480$$

The solution of this system is

$$\|\mathbf{u}\| = 360.0, \quad \|\mathbf{v}\| \approx 536.7, \quad \text{and} \quad \|\mathbf{z}\| = 360.0.$$

Thus, the rope attached at point P has 360 pounds of tension, the rope attached at point Q has about 536.7 pounds of tension, and the rope attached at point R has 360 pounds of tension.

Group Activity   *Properties of Vectors in Space*

The properties of vectors in the plane that you studied earlier in the text also apply to vectors in space. Choose one of the following properties. Then show how the properties of real numbers can be used to prove the property for vectors in space. Exchange proofs with a partner and test their validity.

1. $\mathbf{u} + \mathbf{v} = \mathbf{v} + \mathbf{u}$

2. $(\mathbf{u} + \mathbf{v}) + \mathbf{w} = \mathbf{u} + (\mathbf{v} + \mathbf{w})$

3. $\mathbf{u} + \mathbf{0} = \mathbf{u}$

4. $\mathbf{u} + (-\mathbf{u}) = \mathbf{0}$

5. $c(d\mathbf{u}) = (cd)\mathbf{u}$

6. $(c + d)\mathbf{u} = c\mathbf{u} + d\mathbf{u}$

7. $c(\mathbf{u} + \mathbf{v}) = c\mathbf{u} + c\mathbf{v}$

8. $1(\mathbf{u}) = \mathbf{u}$

9. $0(\mathbf{u}) = \mathbf{0}$

10. $\|c\mathbf{u}\| = |c|\,\|\mathbf{u}\|$

11.2 /// EXERCISES

In Exercises 1 and 2, (a) find the component form of the vector v and (b) sketch the vector with its initial point at the origin.

1.

2.

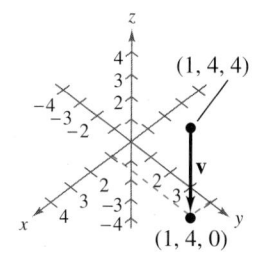

In Exercises 3 and 4, (a) write the component form of v. Then write the component form of (b) a vector parallel to v and (c) a vector in the opposite direction of v. (There are many correct answers.)

3. Initial point: $(-1, -2, 1)$

Terminal point: $(3, 2, 5)$

4. Initial point: $(-4, 5, 5)$

Terminal point: $(4, 0, 0)$

In Exercises 5 and 6, write the component form of v.

5. v lies in the *yz*-plane, has magnitude 4, and makes an angle of 45° with the positive *y*-axis.

6. v lies in the *xz*-plane, has magnitude 10, and makes an angle of 60° with the positive *z*-axis.

In Exercises 7 and 8, sketch each scalar multiple of v.

7. $\mathbf{v} = \langle 1, 2, 2 \rangle$

   (a) $2\mathbf{v}$ (b) $-\mathbf{v}$ (c) $\frac{3}{2}\mathbf{v}$ (d) $0\mathbf{v}$

8. $\mathbf{v} = \langle 2, -2, 1 \rangle$

   (a) $-\mathbf{v}$ (b) $2\mathbf{v}$ (c) $\frac{1}{2}\mathbf{v}$ (d) $\frac{5}{2}\mathbf{v}$

In Exercises 9–12, find the vector z, given $\mathbf{u} = \langle -1, 3, 2 \rangle$, $\mathbf{v} = \langle 1, -2, -2 \rangle$, and $\mathbf{w} = \langle 5, 0, -5 \rangle$. Use a graphing utility to verify your work.

9. $\mathbf{z} = \mathbf{u} - 2\mathbf{v}$

10. $\mathbf{z} = -7\mathbf{u} + \mathbf{v} - \frac{1}{5}\mathbf{w}$

11. $2\mathbf{z} - 4\mathbf{u} = \mathbf{w}$

12. $\mathbf{u} + \mathbf{v} - 2\mathbf{w} + \mathbf{z} = \mathbf{0}$

In Exercises 13–16, find the length of v.

13. $\mathbf{v} = \langle 4, 1, 4 \rangle$

14. $\mathbf{v} = 4\mathbf{i} - 3\mathbf{j} - 7\mathbf{k}$

15. Initial point of $\mathbf{v}$: $(1, -3, 4)$

Terminal point of $\mathbf{v}$: $(1, 0, -1)$

16. Initial point of $\mathbf{v}$: $(0, -1, 0)$

Terminal point of $\mathbf{v}$: $(1, 2, -2)$

In Exercises 17 and 18, find a unit vector (a) in the direction of u and (b) in the direction opposite of u.

17. $\mathbf{u} = 8\mathbf{i} + 3\mathbf{j} - \mathbf{k}$ **18.** $\mathbf{u} = -3\mathbf{i} + 5\mathbf{j} + 10\mathbf{k}$

In Exercises 19–22, use a graphing utility to determine the specified quantity where $\mathbf{u} = \langle -1, 3, 4 \rangle$ and $\mathbf{v} = \langle 5, 4.5, -6 \rangle$.

19. $6\mathbf{u} - 4\mathbf{v}$ **20.** $2\mathbf{u} + \frac{5}{2}\mathbf{v}$

21. $\|\mathbf{u} + \mathbf{v}\|$ **22.** $\dfrac{\mathbf{v}}{\|\mathbf{v}\|}$

In Exercises 23 and 24, find the dot product of u and v. Use a graphing utility to verify your work.

23. $\mathbf{u} = \langle 4, 4, -1 \rangle$ **24.** $\mathbf{u} = 2\mathbf{i} + 5\mathbf{j} - 3\mathbf{k}$

$\mathbf{v} = \langle 2, -5, -8 \rangle$ $\mathbf{v} = 9\mathbf{i} - 3\mathbf{j} + \mathbf{k}$

In Exercises 25 and 26, find the angle θ between the two vectors.

25. $\mathbf{u} = \langle 0, 2, 2 \rangle$ **26.** $\mathbf{u} = 10\mathbf{i} + 40\mathbf{j}$

$\mathbf{v} = \langle 3, 0, -4 \rangle$ $\mathbf{v} = -3\mathbf{j} + 8\mathbf{k}$

In Exercises 27 and 28, determine whether u and v are orthogonal, parallel, or neither. Use a graphing utility to verify your answer.

27. $\mathbf{u} = \langle -12, 6, 15 \rangle$ **28.** $\mathbf{u} = \frac{3}{4}\mathbf{i} - \frac{1}{2}\mathbf{j} + 2\mathbf{k}$

$\mathbf{v} = \langle 8, -4, -10 \rangle$ $\mathbf{v} = 4\mathbf{i} + 10\mathbf{j} + \mathbf{k}$

In Exercises 29–32, use vectors to determine whether the points lie in a straight line.

29. $(5, 4, 1), (7, 3, -1), (4, 5, 3)$

30. $(-2, 7, 4), (-4, 8, 1), (0, 6, 7)$

31. $(1, 3, 2), (-1, 2, 5), (3, 4, -1)$

32. $(0, 4, 4), (-1, 5, 6), (-2, 6, 7)$

In Exercises 33 and 34, use vectors to show that the points form the vertices of a parallelogram.

33. $(2, 9, 1), (3, 11, 4), (0, 10, 2), (1, 12, 5)$

34. $(1, 1, -3), (9, -1, 2), (11, 2, 1), (3, 4, -4)$

In Exercises 35 and 36, the vector v and its initial point are given. Find the terminal point.

35. $\mathbf{v} = \langle 2, -4, 7 \rangle$ **36.** $\mathbf{v} = \langle 4, \frac{3}{2}, -\frac{1}{4} \rangle$

Initial point: $(1, 5, 0)$ Initial point: $\left(2, 1, -\frac{3}{2}\right)$

37. Determine the values of c such that $\|c\mathbf{u}\| = 3$ where $\mathbf{u} = \mathbf{i} + 2\mathbf{j} + 3\mathbf{k}$.

38. *Exploration*   Let $\mathbf{u} = \mathbf{i} + \mathbf{j}$, $\mathbf{v} = \mathbf{j} + \mathbf{k}$, and $\mathbf{w} = a\mathbf{u} + b\mathbf{v}$.

(a) Sketch $\mathbf{u}$ and $\mathbf{v}$.

(b) If $\mathbf{w} = \mathbf{0}$, show that a and b must both be zero.

(c) Find a and b such that $\mathbf{w} = \mathbf{i} + 2\mathbf{j} + \mathbf{k}$.

(d) Show that no choice of a and b yields $\mathbf{w} = \mathbf{i} + 2\mathbf{j} + 3\mathbf{k}$.

39. *Think About It*   The initial and terminal points of the vector $\mathbf{v}$ are (x_1, y_1, z_1) and (x, y, z), respectively. Describe the set of all points (x, y, z) such that $\|\mathbf{v}\| = 4$.

40. *Light Installation*   The lights in an auditorium are 30-pound disks of radius 24 inches. Each disk is supported by three equally spaced 60-inch wires to the ceiling. Find the tension in each wire.

11.3   The Cross Product of Two Vectors

The Cross Product / Geometric Properties of the Cross Product / The Triple Scalar Product

The Cross Product

Many applications in physics, engineering, and geometry involve finding a vector in space that is orthogonal to two given vectors. In this section you will study a product that will yield such a vector. It is called the **cross product,** and it is most conveniently defined and calculated using the standard unit vector form.

Note Be sure you see that this definition applies only to three-dimensional vectors. The cross product is not defined for two-dimensional vectors.

Definition of Cross Product of Two Vectors in Space

Let $\mathbf{u} = u_1\mathbf{i} + u_2\mathbf{j} + u_3\mathbf{k}$ and $\mathbf{v} = v_1\mathbf{i} + v_2\mathbf{j} + v_3\mathbf{k}$ be vectors in space. The **cross product** of $\mathbf{u}$ and $\mathbf{v}$ is the vector

$$\mathbf{u} \times \mathbf{v} = (u_2v_3 - u_3v_2)\mathbf{i} - (u_1v_3 - u_3v_1)\mathbf{j} + (u_1v_2 - u_2v_1)\mathbf{k}.$$

A convenient way to calculate $\mathbf{u} \times \mathbf{v}$ is to use the following *determinant form* with cofactor expansion. (This 3×3 determinant form is used simply to help remember the formula for the cross product—it is technically not a determinant because not all the entries of the corresponding matrix are real numbers.)

$$\mathbf{u} \times \mathbf{v} = \begin{vmatrix} \mathbf{i} & \mathbf{j} & \mathbf{k} \\ u_1 & u_2 & u_3 \\ v_1 & v_2 & v_3 \end{vmatrix}$$

⇐ Put **u** in Row 2.
⇐ Put **v** in Row 3.

$$= \begin{vmatrix} u_2 & u_3 \\ v_2 & v_3 \end{vmatrix}\mathbf{i} - \begin{vmatrix} u_1 & u_3 \\ v_1 & v_3 \end{vmatrix}\mathbf{j} + \begin{vmatrix} u_1 & u_2 \\ v_1 & v_2 \end{vmatrix}\mathbf{k}$$

$$= (u_2v_3 - u_3v_2)\mathbf{i} - (u_1v_3 - u_3v_1)\mathbf{j} + (u_1v_2 - u_2v_1)\mathbf{k}$$

Note the minus sign in front of the **j**-component.

American physicist Josiah Willard Gibbs (1839–1903) developed three-dimensional vector analysis. He introduced the dot product and the cross product.

EXPLORATION

Try finding the following cross products. What can you conclude?

a. $\mathbf{i} \times \mathbf{j}$ **b.** $\mathbf{i} \times \mathbf{k}$ **c.** $\mathbf{j} \times \mathbf{k}$

EXAMPLE 1 ▭ **Finding Cross Products**

Given $\mathbf{u} = \mathbf{i} + 2\mathbf{j} + \mathbf{k}$ and $\mathbf{v} = 3\mathbf{i} + \mathbf{j} + 2\mathbf{k}$, find the following.

a. $\mathbf{u} \times \mathbf{v}$ **b.** $\mathbf{v} \times \mathbf{u}$ **c.** $\mathbf{v} \times \mathbf{v}$

Solution

Some graphing utilities have the capability to perform vector operations, such as the cross product. For instance, on the *TI-85*, you can use the VECTR menu to verify the cross product in Example 1(a) as follows.

cross ([1, 2, 1], [3, 1, 2])

On the *TI-92*, you can use the MATH/Matrix/Vector ops menu to verify the cross product as follows.

crossP({1, 2, 1}, {3, 1, 2})

a. $\mathbf{u} \times \mathbf{v} = \begin{vmatrix} \mathbf{i} & \mathbf{j} & \mathbf{k} \\ 1 & 2 & 1 \\ 3 & 1 & 2 \end{vmatrix} = \begin{vmatrix} 2 & 1 \\ 1 & 2 \end{vmatrix}\mathbf{i} - \begin{vmatrix} 1 & 1 \\ 3 & 2 \end{vmatrix}\mathbf{j} + \begin{vmatrix} 1 & 2 \\ 3 & 1 \end{vmatrix}\mathbf{k}$

$= (4 - 1)\mathbf{i} - (2 - 3)\mathbf{j} + (1 - 6)\mathbf{k}$

$= 3\mathbf{i} + \mathbf{j} - 5\mathbf{k}$

b. $\mathbf{v} \times \mathbf{u} = \begin{vmatrix} \mathbf{i} & \mathbf{j} & \mathbf{k} \\ 3 & 1 & 2 \\ 1 & 2 & 1 \end{vmatrix} = \begin{vmatrix} 1 & 2 \\ 2 & 1 \end{vmatrix}\mathbf{i} - \begin{vmatrix} 3 & 2 \\ 1 & 1 \end{vmatrix}\mathbf{j} + \begin{vmatrix} 3 & 1 \\ 1 & 2 \end{vmatrix}\mathbf{k}$

$= (1 - 4)\mathbf{i} - (3 - 2)\mathbf{j} + (6 - 1)\mathbf{k}$

$= -3\mathbf{i} - \mathbf{j} + 5\mathbf{k}$

Note that this result is the negative of that in part **a.**

c. $\mathbf{v} \times \mathbf{v} = \begin{vmatrix} \mathbf{i} & \mathbf{j} & \mathbf{k} \\ 3 & 1 & 2 \\ 3 & 1 & 2 \end{vmatrix} = \mathbf{0}$ ▭

The results obtained in Example 1 suggest some interesting *algebraic* properties of the cross product. For instance,

$$\mathbf{u} \times \mathbf{v} = -(\mathbf{v} \times \mathbf{u}) \quad \text{and} \quad \mathbf{v} \times \mathbf{v} = \mathbf{0}.$$

These properties, and several others, are summarized in the following list.

Algebraic Properties of the Cross Product

Let $\mathbf{u}$, $\mathbf{v}$, and $\mathbf{w}$ be vectors in space and let c be a scalar.

1. $\mathbf{u} \times \mathbf{v} = -(\mathbf{v} \times \mathbf{u})$
2. $\mathbf{u} \times (\mathbf{v} + \mathbf{w}) = (\mathbf{u} \times \mathbf{v}) + (\mathbf{u} \times \mathbf{w})$
3. $c(\mathbf{u} \times \mathbf{v}) = (c\mathbf{u}) \times \mathbf{v} = \mathbf{u} \times (c\mathbf{v})$
4. $\mathbf{u} \times \mathbf{0} = \mathbf{0} \times \mathbf{u} = \mathbf{0}$
5. $\mathbf{u} \times \mathbf{u} = \mathbf{0}$
6. $\mathbf{u} \cdot (\mathbf{v} \times \mathbf{w}) = (\mathbf{u} \times \mathbf{v}) \cdot \mathbf{w}$

Geometric Properties of the Cross Product

The first property listed on the previous page indicates that the cross product is *not commutative*. In particular, this property indicates that the vectors $\mathbf{u} \times \mathbf{v}$ and $\mathbf{v} \times \mathbf{u}$ have equal lengths but opposite directions. The following list gives some other *geometric* properties of the cross product of two vectors.

Figure 11.17

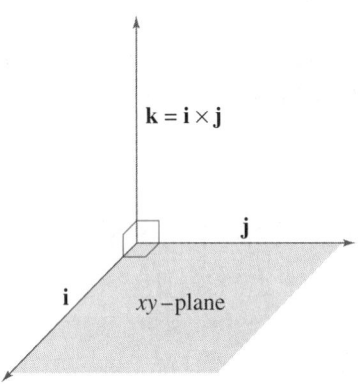

$\mathbf{k} = \mathbf{i} \times \mathbf{j}$

$\mathbf{j}$

$\mathbf{i}$ *xy*-plane

Geometric Properties of the Cross Product

Let $\mathbf{u}$ and $\mathbf{v}$ be nonzero vectors in space, and let θ be the angle between $\mathbf{u}$ and $\mathbf{v}$.

1. $\mathbf{u} \times \mathbf{v}$ is orthogonal to both $\mathbf{u}$ and $\mathbf{v}$.
2. $\|\mathbf{u} \times \mathbf{v}\| = \|\mathbf{u}\| \, \|\mathbf{v}\| \sin \theta$.
3. $\mathbf{u} \times \mathbf{v} = \mathbf{0}$ if and only if $\mathbf{u}$ and $\mathbf{v}$ are scalar multiples.
4. $\|\mathbf{u} \times \mathbf{v}\|$ = area of parallelogram having $\mathbf{u}$ and $\mathbf{v}$ as adjacent sides.

Both $\mathbf{u} \times \mathbf{v}$ and $\mathbf{v} \times \mathbf{u}$ are perpendicular to the plane determined by $\mathbf{u}$ and $\mathbf{v}$. One way to remember the orientation of the vectors $\mathbf{u}$, $\mathbf{v}$, and $\mathbf{u} \times \mathbf{v}$ is to compare them with the unit vectors $\mathbf{i}$, $\mathbf{j}$, and $\mathbf{k} = \mathbf{i} \times \mathbf{j}$, as shown in Figure 11.17. The three vectors $\mathbf{u}$, $\mathbf{v}$, and $\mathbf{u} \times \mathbf{v}$ form a *right-handed system*.

EXAMPLE 2   Using the Cross Product

Find a unit vector that is orthogonal to both

$$\mathbf{u} = 3\mathbf{i} - 4\mathbf{j} + \mathbf{k} \quad \text{and} \quad \mathbf{v} = -3\mathbf{i} + 6\mathbf{j}.$$

Solution

The cross product $\mathbf{u} \times \mathbf{v}$, as shown in Figure 11.18, is orthogonal to both $\mathbf{u}$ and $\mathbf{v}$.

Figure 11.18

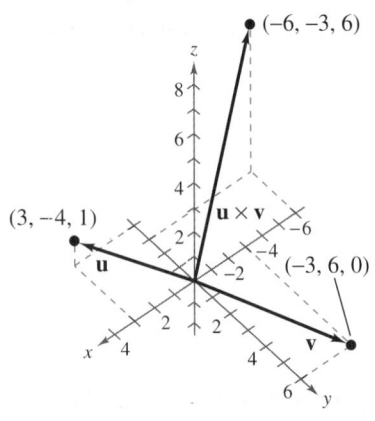

$(-6, -3, 6)$

$(3, -4, 1)$

$\mathbf{u} \times \mathbf{v}$

$(-3, 6, 0)$

$\mathbf{u}$

$\mathbf{v}$

$$\mathbf{u} \times \mathbf{v} = \begin{vmatrix} \mathbf{i} & \mathbf{j} & \mathbf{k} \\ 3 & -4 & 1 \\ -3 & 6 & 0 \end{vmatrix}$$

$$= -6\mathbf{i} - 3\mathbf{j} + 6\mathbf{k}$$

Because

$$\|\mathbf{u} \times \mathbf{v}\| = \sqrt{(-6)^2 + (-3)^2 + 6^2} = \sqrt{81} = 9$$

a unit vector orthogonal to both $\mathbf{u}$ and $\mathbf{v}$ is

$$\frac{\mathbf{u} \times \mathbf{v}}{\|\mathbf{u} \times \mathbf{v}\|} = -\frac{2}{3}\mathbf{i} - \frac{1}{3}\mathbf{j} + \frac{2}{3}\mathbf{k}.$$

In Example 2, note that you could have used the cross product $\mathbf{v} \times \mathbf{u}$ to form a unit vector that is orthogonal to both $\mathbf{u}$ and $\mathbf{v}$. With that choice, you would have obtained the *negative* of the unit vector found in the example.

The fourth geometric property of the cross product states that $\|\mathbf{u} \times \mathbf{v}\|$ is the area of the parallelogram that has $\mathbf{u}$ and $\mathbf{v}$ as adjacent sides. A simple example of this is given by the unit square with adjacent sides of $\mathbf{i}$ and $\mathbf{j}$. Because

$$\mathbf{i} \times \mathbf{j} = \mathbf{k}$$

and $\|\mathbf{k}\| = 1$, it follows that the square has an area of 1. This geometric property of the cross product is illustrated further in the next example.

EXAMPLE 3   **Geometric Application of the Cross Product**

Show that the quadrilateral with vertices at the following points is a parallelogram. Then find the area of the parallelogram.

$$A(5, 2, 0), \quad B(2, 6, 1), \quad C(2, 4, 7), \quad D(5, 0, 6)$$

Solution
From Figure 11.19 you can see that the sides of the quadrilateral correspond to the following four vectors.

$$\overrightarrow{AB} = -3\mathbf{i} + 4\mathbf{j} + \mathbf{k}$$
$$\overrightarrow{CD} = 3\mathbf{i} - 4\mathbf{j} - \mathbf{k} = -\overrightarrow{AB}$$
$$\overrightarrow{AD} = 0\mathbf{i} - 2\mathbf{j} + 6\mathbf{k}$$
$$\overrightarrow{CB} = 0\mathbf{i} + 2\mathbf{j} - 6\mathbf{k} = -\overrightarrow{AD}$$

Because $\overrightarrow{AB}$ is parallel to $\overrightarrow{CD}$ and $\overrightarrow{AD}$ is parallel to $\overrightarrow{CB}$, it follows that the quadrilateral is a parallelogram with $\overrightarrow{AB}$ and $\overrightarrow{AD}$ as adjacent sides. Moreover, because

$$\overrightarrow{AB} \times \overrightarrow{AD} = \begin{vmatrix} \mathbf{i} & \mathbf{j} & \mathbf{k} \\ -3 & 4 & 1 \\ 0 & -2 & 6 \end{vmatrix}$$

$$= 26\mathbf{i} + 18\mathbf{j} + 6\mathbf{k}$$

the area of the parallelogram is

$$\|\overrightarrow{AB} \times \overrightarrow{AD}\| = \sqrt{1036} \approx 32.19.$$

Is the parallelogram a rectangle? You can tell whether it is by finding the angle between the vectors $\overrightarrow{AB}$ and $\overrightarrow{AD}$.

Figure 11.19

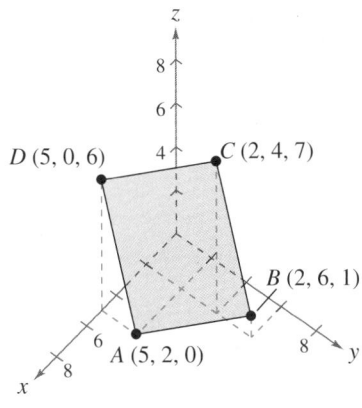

The Triple Scalar Product

For vectors **u**, **v**, and **w** in space, the dot product of **u** and **v** × **w** is called the **triple scalar product** of **u**, **v**, and **w**.

Figure 11.20

Area of base $= \|\mathbf{v} \times \mathbf{w}\|$

Volume of parallelepiped $= |\mathbf{u} \cdot (\mathbf{v} \times \mathbf{w})|$

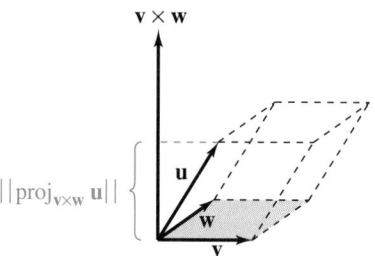

The Triple Scalar Product

The **triple scalar product** of **u**, **v**, and **w** is given by

$$\mathbf{u} \cdot (\mathbf{v} \times \mathbf{w}) = \begin{vmatrix} u_1 & u_2 & u_3 \\ v_1 & v_2 & v_3 \\ w_1 & w_2 & w_3 \end{vmatrix}.$$

If the vectors **u**, **v**, and **w** do not lie in the same plane, the triple scalar product **u** · (**v** × **w**) can be used to determine the volume of the parallelepiped with **u**, **v**, and **w** as adjacent edges, as shown in Figure 11.20.

Geometric Property of Triple Scalar Product

The volume V of a parallelepiped with vectors **u**, **v**, and **w** as adjacent edges is given by

$$V = |\mathbf{u} \cdot (\mathbf{v} \times \mathbf{w})|.$$

EXAMPLE 4   Volume by the Triple Scalar Product

Find the volume of the parallelepiped having

$$\mathbf{u} = 3\mathbf{i} - 5\mathbf{j} + \mathbf{k}, \quad \mathbf{v} = 2\mathbf{j} - 2\mathbf{k}, \quad \text{and} \quad \mathbf{w} = 3\mathbf{i} + \mathbf{j} + \mathbf{k}$$

as adjacent edges, as shown in Figure 11.21.

Figure 11.21

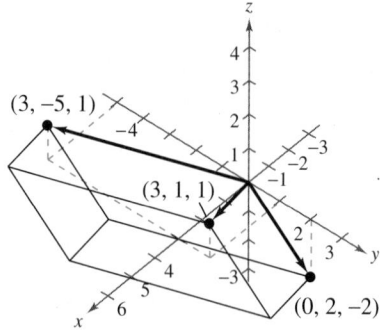

Solution

The value of the triple scalar product is

$$\mathbf{u} \cdot (\mathbf{v} \times \mathbf{w}) = \begin{vmatrix} 3 & -5 & 1 \\ 0 & 2 & -2 \\ 3 & 1 & 1 \end{vmatrix}$$

$$= 3\begin{vmatrix} 2 & -2 \\ 1 & 1 \end{vmatrix} - (-5)\begin{vmatrix} 0 & -2 \\ 3 & 1 \end{vmatrix} + 1\begin{vmatrix} 0 & 2 \\ 3 & 1 \end{vmatrix}$$

$$= 3(4) + 5(6) + 1(-6)$$

$$= 36.$$

Thus, the volume of the parallelepiped is $|\mathbf{u} \cdot (\mathbf{v} \times \mathbf{w})| = |36| = 36$.

Group Activity

Coplanar Vectors

Consider three vectors **u**, **v**, and **w** with the same initial point. It can be shown that they are coplanar if and only if

$$\mathbf{u} \cdot (\mathbf{v} \times \mathbf{w}) = \begin{vmatrix} u_1 & u_2 & u_3 \\ v_1 & v_2 & v_3 \\ w_1 & w_2 & w_3 \end{vmatrix} = 0.$$

Explain why this test for coplanar vectors is valid, and compare your explanation with that of another group member.

11.3 /// EXERCISES

In Exercises 1–4, find the cross product of the unit vectors and sketch the result.

1. $\mathbf{j} \times \mathbf{i}$
2. $\mathbf{j} \times \mathbf{k}$
3. $\mathbf{i} \times \mathbf{k}$
4. $\mathbf{k} \times \mathbf{i}$

In Exercises 5–12, find $\mathbf{u} \times \mathbf{v}$ and show that it is orthogonal to both **u** and **v**.

5. $\mathbf{u} = \langle 1, -4, 0 \rangle$
   $\mathbf{v} = \langle 2, 6, 0 \rangle$
6. $\mathbf{u} = \langle -3, 2, 3 \rangle$
   $\mathbf{v} = \langle 0, 1, 0 \rangle$
7. $\mathbf{u} = \langle 7, -5, 2 \rangle$
   $\mathbf{v} = \langle -1, 4, -1 \rangle$
8. $\mathbf{u} = \langle -5, 5, 11 \rangle$
   $\mathbf{v} = \langle 2, 2, 3 \rangle$
9. $\mathbf{u} = 6\mathbf{i} + 2\mathbf{j} + \mathbf{k}$
   $\mathbf{v} = \mathbf{i} + 3\mathbf{j} - 2\mathbf{k}$
10. $\mathbf{u} = 6\mathbf{k}$
    $\mathbf{v} = -\mathbf{i} + 3\mathbf{j} + \mathbf{k}$
11. $\mathbf{u} = \mathbf{i} + \frac{3}{2}\mathbf{j} - \frac{5}{2}\mathbf{k}$
    $\mathbf{v} = \frac{1}{2}\mathbf{i} - \frac{3}{4}\mathbf{j} + \frac{1}{4}\mathbf{k}$
12. $\mathbf{u} = \frac{2}{3}\mathbf{i}$
    $\mathbf{v} = \frac{1}{3}\mathbf{j} - 3\mathbf{k}$

In Exercises 13–16, use a graphing utility to find $\mathbf{u} \times \mathbf{v}$.

13. $\mathbf{u} = \langle 2, 4, 3 \rangle$
    $\mathbf{v} = \langle 0, -2, 1 \rangle$
14. $\mathbf{u} = \langle 4, -2, 6 \rangle$
    $\mathbf{v} = \langle -1, 5, 7 \rangle$
15. $\mathbf{u} = 6\mathbf{i} - 5\mathbf{j} + \mathbf{k}$
    $\mathbf{v} = \frac{1}{3}\mathbf{i} - \frac{1}{3}\mathbf{j} + \frac{2}{3}\mathbf{k}$
16. $\mathbf{u} = -\mathbf{i} + \mathbf{k}$
    $\mathbf{v} = \mathbf{j} - 2\mathbf{k}$

In Exercises 17–22, find a unit vector orthogonal to u and v.

17. $\mathbf{u} = 3\mathbf{i} + \mathbf{j}$
    $\mathbf{v} = \mathbf{j} + \mathbf{k}$
18. $\mathbf{u} = \mathbf{i} + 2\mathbf{j}$
    $\mathbf{v} = \mathbf{i} - 3\mathbf{j}$
19. $\mathbf{u} = -2\mathbf{i} + \mathbf{j} + 3\mathbf{k}$
    $\mathbf{v} = \mathbf{i} + 4\mathbf{j} + 6\mathbf{k}$
20. $\mathbf{u} = 7\mathbf{i} - 14\mathbf{j} + 5\mathbf{k}$
    $\mathbf{v} = 14\mathbf{i} + 28\mathbf{j} - 15\mathbf{k}$
21. $\mathbf{u} = \mathbf{i} + \mathbf{j} - \mathbf{k}$
    $\mathbf{v} = \mathbf{i} + \mathbf{j} + \mathbf{k}$
22. $\mathbf{u} = \mathbf{i} - 2\mathbf{j} + 2\mathbf{k}$
    $\mathbf{v} = 2\mathbf{i} - \mathbf{j} - 2\mathbf{k}$

In Exercises 23–28, find the area of the parallelogram that has the vectors as adjacent sides.

23. $\mathbf{u} = \mathbf{k}$
    $\mathbf{v} = \mathbf{i} + \mathbf{k}$
24. $\mathbf{u} = \mathbf{i} + 2\mathbf{j} + 2\mathbf{k}$
    $\mathbf{v} = \mathbf{i} + \mathbf{k}$
25. $\mathbf{u} = 3\mathbf{i} + 4\mathbf{j} + 6\mathbf{k}$
    $\mathbf{v} = 2\mathbf{i} - \mathbf{j} + 5\mathbf{k}$
26. $\mathbf{u} = \langle -2, 3, 2 \rangle$
    $\mathbf{v} = \langle 1, 2, 4 \rangle$
27. $\mathbf{u} = \langle 2, 2, -3 \rangle$
    $\mathbf{v} = \langle 0, 2, 3 \rangle$
28. $\mathbf{u} = \langle 4, -3, 2 \rangle$
    $\mathbf{v} = \langle 5, 0, 1 \rangle$

In Exercises 29 and 30, verify that the points are the vertices of a parallelogram and find its area.

29. $A(2, -1, 4)$, $B(3, 1, 2)$, $C(0, 5, 6)$, $D(-1, 3, 8)$
30. $A(3, 5, 0)$, $B(-1, 8, 5)$, $C(1, 3, 11)$, $D(5, 0, 6)$

In Exercises 31–34, find the area of the triangle with the given vertices. (The area of the triangle having u and v as adjacent sides is $\frac{1}{2}\|u \times v\|$.)

31. $(0, 0, 0), (4, -2, 6), (-4, 0, 3)$

32. $(1, -4, 3), (2, 0, 2), (-2, 2, 0)$

33. $(2, 3, -5), (-2, -2, 0), (3, 0, 6)$

34. $(2, 4, 0), (-2, -4, 0), (0, 0, 4)$

In Exercises 35 and 36, find $u \cdot (v \times w)$.

35. $u = \langle 2, 3, 3 \rangle, v = \langle 4, 4, 0 \rangle, w = \langle 0, 0, 4 \rangle$

36. $u = \langle 20, 10, 10 \rangle, v = \langle 1, 4, 4 \rangle, w = \langle 0, 2, 2 \rangle$

In Exercises 37 and 38, use the triple scalar product to find the volume of the parallelepiped having adjacent edges u, v, and w.

37. $u = i + j$
   $v = j + k$
   $w = i + k$

38. $u = \langle 1, 1, 3 \rangle$
   $v = \langle 0, 3, 3 \rangle$
   $w = \langle 3, 0, 3 \rangle$

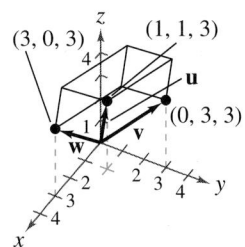

In Exercises 39 and 40, find the volume of the parallelepiped with the given vertices (see figures).

39. $A(0, 0, 0), B(4, 0, 0), C(4, -2, 3), D(0, -2, 3),$
   $E(4, 5, 3), F(0, 5, 3), G(0, 3, 6), H(4, 3, 6)$

40. $A(3, 0, 0), B(4, 1, 2), C(3, -1, 4), D(2, -2, 2),$
   $E(-1, 5, 4), F(0, 6, 6), G(-1, 4, 8), H(-2, 3, 6)$

Figure for 39

Figure for 40

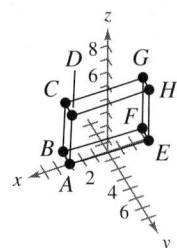

41. *Torque*   A child applies the brakes on a bicycle by applying a downward force of 20 pounds on the pedal when the 6-inch crank makes a 40° angle with the horizontal (see figure). Vectors representing the crank and the force are $V = \frac{1}{2}(\cos 40° j + \sin 40° k)$ and $F = -20k$, respectively. Find the torque on the crank if it is given by $\|V \times F\|$.

Figure for 41

Figure for 42

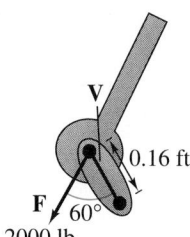

42. *Torque*   Both the magnitude and direction of the force on a crankshaft change as the crankshaft rotates. Use the technique shown in Exercise 41 to find the torque on the crankshaft using the position and data shown in the figure.

In Exercises 43 and 44, prove the property of the cross product where $u = \langle u_1, u_2, u_3 \rangle$ and $v = \langle v_1, v_2, v_3 \rangle$.

43. $u \times u = 0$

44. $u \times v$ is orthogonal to both **u** and **v**.

45. Consider the vectors $u = \langle \cos \alpha, \sin \alpha, 0 \rangle$ and $v = \langle \cos \beta, \sin \beta, 0 \rangle$ where $\alpha > \beta$. Find the cross product of the vectors and use the result to prove the identity $\sin(\alpha - \beta) = \sin \alpha \cos \beta - \cos \alpha \sin \beta$.

11.4 Lines and Planes in Space

Lines in Space / Planes in Space / Sketching Planes in Space / Distance Between a Point and a Plane

Lines in Space

Figure 11.22

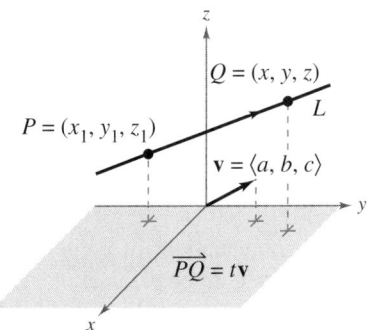

In the plane, *slope* is used to determine an equation of a line. In space, it is more convenient to use *vectors* to determine the equation of a line. In Figure 11.22, consider the line L through the point $P = (x_1, y_1, z_1)$ and parallel to the vector

$$\mathbf{v} = \langle a, b, c \rangle. \qquad \text{Direction vector for } L$$

The vector $\mathbf{v}$ is the **direction vector** for the line L, and a, b, and c are the **direction numbers.** One way of describing the line L is to say that it consists of all points $Q = (x, y, z)$ for which the vector $\overrightarrow{PQ}$ is parallel to $\mathbf{v}$. This means that $\overrightarrow{PQ}$ is a scalar multiple of $\mathbf{v}$, and you can write $\overrightarrow{PQ} = t\mathbf{v}$, where t is a scalar.

$$
\begin{aligned}
\overrightarrow{PQ} &= \langle x - x_1, y - y_1, z - z_1 \rangle \\
&= \langle at, bt, ct \rangle \\
&= t\mathbf{v}
\end{aligned}
$$

By equating corresponding components, you can obtain the **parametric equations** of a line in space.

> ### Parametric Equations of a Line in Space
>
> A line L parallel to the vector $\mathbf{v} = \langle a, b, c \rangle$ and passing through the point $P = (x_1, y_1, z_1)$ is represented by the **parametric equations**
>
> $$x = x_1 + at$$
> $$y = y_1 + bt$$
> $$z = z_1 + ct.$$

If the direction numbers a, b, and c are all nonzero, you can eliminate the parameter t to obtain the **symmetric equations** of a line.

$$\frac{x - x_1}{a} = \frac{y - y_1}{b} = \frac{z - z_1}{c} \qquad \text{Symmetric equations}$$

Figure 11.23

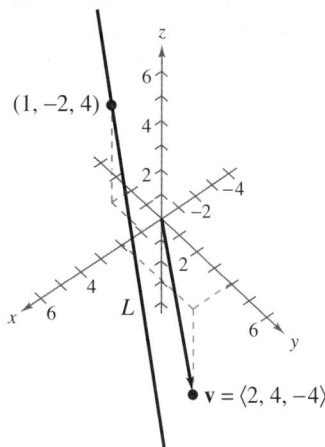

$v = \langle 2, 4, -4 \rangle$

EXAMPLE 1   **Finding Parametric and Symmetric Equations**

Find parametric and symmetric equations of the line L that passes through the point $(1, -2, 4)$ and is parallel to $\mathbf{v} = \langle 2, 4, -4 \rangle$.

Solution

To find a set of parametric equations of the line, use the coordinates $x_1 = 1$, $y_1 = -2$, and $z_1 = 4$ and direction numbers $a = 2$, $b = 4$, and $c = -4$ (see Figure 11.23).

$$x = 1 + 2t, \quad y = -2 + 4t, \quad z = 4 - 4t \qquad \text{Parametric equations}$$

Because a, b, and c are all nonzero, a set of symmetric equations is

$$\frac{x - 1}{2} = \frac{y + 2}{4} = \frac{z - 4}{-4}. \qquad \text{Symmetric equations}$$

Neither the parametric equations nor the symmetric equations of a given line are unique. For instance, in Example 1, by letting $t = 1$ in the parametric equations you would obtain the point $(3, 2, 0)$. Using this point with the direction numbers $a = 2$, $b = 4$, and $c = -4$ produces the parametric equations

$$x = 3 + 2t, \quad y = 2 + 4t, \quad z = -4t.$$

Study Tip

You can check the answer to Example 2 by verifying that the two original points lie on the line. To see this, substitute $t = 0$ and $t = 1$ into the parametric equations.

EXAMPLE 2   **Parametric Equations of a Line Through Two Points**

Find a set of parametric equations of the line that passes through the points $(-2, 1, 0)$ and $(1, 3, 5)$.

Solution

Begin by letting $P = (-2, 1, 0)$ and $Q = (1, 3, 5)$. Then a direction vector for the line passing through P and Q is given by

$$\begin{aligned}
\mathbf{v} &= \overrightarrow{PQ} \\
&= \langle 1 - (-2), 3 - 1, 5 - 0 \rangle \\
&= \langle 3, 2, 5 \rangle \\
&= \langle a, b, c \rangle.
\end{aligned}$$

Using the direction numbers $a = 3$, $b = 2$, and $c = 5$, with the point $P = (-2, 1, 0)$, you can obtain the parametric equations

$$x = -2 + 3t, \quad y = 1 + 2t, \quad z = 5t.$$

Planes in Space

You have seen how an equation of a line in space can be obtained from a point on the line and a vector *parallel* to it. You will now see that an equation of a plane in space can be obtained from a point in the plane and a vector *normal* (perpendicular) to it.

Figure 11.24

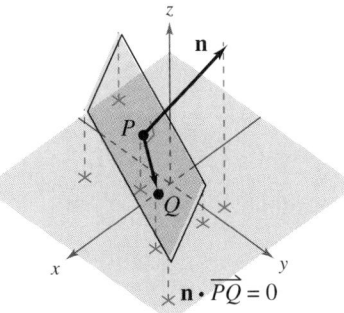

Consider the plane containing the point $P = (x_1, y_1, z_1)$ having a nonzero normal vector $\mathbf{n} = \langle a, b, c \rangle$, as shown on Figure 11.24. This plane consists of all points $Q = (x, y, z)$ for which the vector $\overrightarrow{PQ}$ is orthogonal to $\mathbf{n}$. Using the dot product, you can write the following.

$$\mathbf{n} \cdot \overrightarrow{PQ} = 0$$
$$\langle a, b, c \rangle \cdot \langle x - x_1, y - y_1, z - z_1 \rangle = 0$$
$$a(x - x_1) + b(y - y_1) + c(z - z_1) = 0$$

The latter equation of the plane is said to be in **standard form.**

EXPLORATION

Consider the following four planes.

$$2x + 3y - z = 2$$
$$4x + 6y - 2z = 5$$
$$-2x - 3y + z = -2$$
$$-6x - 9y + 3z = 11$$

What are the normal vectors for each plane? What can you say about the relative positions of these planes in space?

> ### Standard Equation of a Plane in Space
>
> The plane containing the point (x_1, y_1, z_1) and having normal vector $\mathbf{n} = \langle a, b, c \rangle$ can be represented, in **standard form,** by the equation
>
> $$a(x - x_1) + b(y - y_1) + c(z - z_1) = 0.$$

By regrouping terms, you obtain the **general form** of the equation of a plane in space,

$$ax + by + cz + d = 0. \qquad \text{General form of equation of plane}$$

Given the general form of the equation of a plane, it is easy to find a normal vector to the plane. Simply use the coefficients of x, y, and z and write $\mathbf{n} = \langle a, b, c \rangle$.

Figure 11.25

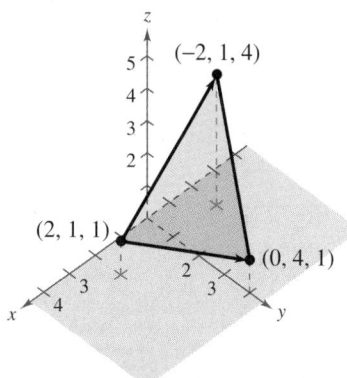

EXAMPLE 3   ▱   **Finding an Equation of a Plane in Three-Space**

Find the general equation of the plane containing the points $(2, 1, 1)$, $(0, 4, 1)$, and $(-2, 1, 4)$.

Solution

To find the equation of the plane, you need a point in the plane and a vector that is normal to the plane. There are three choices for the point, but no normal vector is given. To obtain a normal vector, use the cross product of vectors $\mathbf{u}$ and $\mathbf{v}$ extending from the point $(2, 1, 1)$ to the points $(0, 4, 1)$ and $(-2, 1, 4)$, as shown in Figure 11.25. The component forms of $\mathbf{u}$ and $\mathbf{v}$ are

$$\mathbf{u} = \langle 0 - 2, 4 - 1, 1 - 1 \rangle = \langle -2, 3, 0 \rangle$$
$$\mathbf{v} = \langle -2 - 2, 1 - 1, 4 - 1 \rangle = \langle -4, 0, 3 \rangle$$

and it follows that

$$\mathbf{n} = \mathbf{u} \times \mathbf{v} = \begin{vmatrix} \mathbf{i} & \mathbf{j} & \mathbf{k} \\ -2 & 3 & 0 \\ -4 & 0 & 3 \end{vmatrix}$$

$$= 9\mathbf{i} + 6\mathbf{j} + 12\mathbf{k}$$
$$= \langle a, b, c \rangle$$

Note In Example 3, check to see that each of the three points satisfies the equation $3x + 2y + 4z - 12 = 0$.

is normal to the given plane. Using the direction numbers for $\mathbf{n}$ and the point $(x_1, y_1, z_1) = (2, 1, 1)$, you can determine an equation of the plane to be

$$a(x - x_1) + b(y - y_1) + c(z - z_1) = 0$$
$$9(x - 2) + 6(y - 1) + 12(z - 1) = 0 \qquad \text{Standard form}$$
$$9x + 6y + 12z - 36 = 0$$
$$3x + 2y + 4z - 12 = 0. \qquad \text{General form} \qquad ▱$$

Figure 11.26

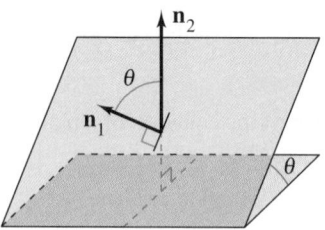

Two distinct planes in three-space either are parallel or intersect in a line. If they intersect, you can determine the angle between them from the angle between their normal vectors, as shown in Figure 11.26. Specifically, if vectors $\mathbf{n}_1$ and $\mathbf{n}_2$ are normal to two intersecting planes, the angle θ between the normal vectors is equal to the angle between the two planes and is given by

$$\cos \theta = \frac{|\mathbf{n}_1 \cdot \mathbf{n}_2|}{\|\mathbf{n}_1\| \, \|\mathbf{n}_2\|}. \qquad \text{Angle between two planes}$$

Consequently, two planes with normal vectors $\mathbf{n}_1$ and $\mathbf{n}_2$ are

1. *perpendicular* if $\mathbf{n}_1 \cdot \mathbf{n}_2 = 0$.
2. *parallel* if $\mathbf{n}_1$ is a scalar multiple of $\mathbf{n}_2$.

EXAMPLE 4   ⬛   **Finding the Line of Intersection of Two Planes**

Find the angle between the two planes given by

$$x - 2y + z = 0 \qquad \text{Equation for plane 1}$$
$$2x + 3y - 2z = 0 \qquad \text{Equation for plane 2}$$

and find parametric equations of their line of intersection (see Figure 11.27).

Solution

The normal vectors for the planes are $\mathbf{n}_1 = \langle 1, -2, 1 \rangle$ and $\mathbf{n}_2 = \langle 2, 3, -2 \rangle$. Consequently, the angle between two planes is determined as follows.

$$\cos \theta = \frac{|\mathbf{n}_1 \cdot \mathbf{n}_2|}{\|\mathbf{n}_1\| \|\mathbf{n}_2\|} = \frac{|-6|}{\sqrt{6}\sqrt{17}} = \frac{6}{\sqrt{102}} \approx 0.59409$$

This implies that the angle between the two planes is $\theta \approx 53.55°$. You can find the line of intersection of the two planes by simultaneously solving the two linear equations representing the planes. One way to do this is to multiply the first equation by -2 and add the result to the second equation.

$$x - 2y + z = 0 \quad \Longrightarrow \quad -2x + 4y - 2z = 0$$
$$2x + 3y - 2z = 0 \qquad\qquad\quad\; \underline{2x + 3y - 2z = 0}$$
$$\qquad\qquad\qquad\qquad\qquad\qquad\quad 7y - 4z = 0 \quad \Longrightarrow \quad y = \frac{4z}{7}$$

Substituting $y = 4z/7$ back into one of the original equations, you can determine that $x = z/7$. Finally, by letting $t = z/7$, you obtain the parametric equations

$$x = x_1 + at = t, \quad y = y_1 + bt = 4t, \quad z = z_1 + ct = 7t$$

which indicates that $a = 1$, $b = 4$, and $c = 7$ are direction numbers for the line of intersection.   ⬛

Note that the direction numbers in Example 4 can be obtained from the cross product of the two normal vectors as follows.

$$\mathbf{n}_1 \times \mathbf{n}_2 = \begin{vmatrix} \mathbf{i} & \mathbf{j} & \mathbf{k} \\ 1 & -2 & 1 \\ 2 & 3 & -2 \end{vmatrix}$$

$$= \begin{vmatrix} -2 & 1 \\ 3 & -2 \end{vmatrix} \mathbf{i} - \begin{vmatrix} 1 & 1 \\ 2 & -2 \end{vmatrix} \mathbf{j} + \begin{vmatrix} 1 & -2 \\ 2 & 3 \end{vmatrix} \mathbf{k}$$

$$= \mathbf{i} + 4\mathbf{j} + 7\mathbf{k}$$

This means that the line of intersection of the two planes is parallel to the cross product of their normal vectors.

Figure 11.27

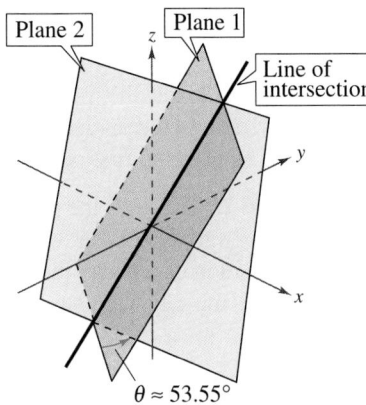

$\theta \approx 53.55°$

Sketching Planes in Space

A graphing utility can be used to graph a plane in space. To graph the plane at the right with the *TI-92*, use the following steps.

1. Set the mode to 3D.
2. Solve for z and enter the equation into $z1$.
3. Use the following viewing cube.

eyeθ° = 20, eyeϕ = 70,
xmin = -10, xmax = 10,
xgrid = 14, ymin = -10,
ymax = 10, ygrid = 14,
zmin = -10, zmax = 20,
zscl = 1

4. Press ◇ F to turn the axes on.
5. Press ◇ GRAPH.

If a plane in space intersects one of the coordinate planes, we call the line of intersection the **trace** of the given plane in the coordinate plane. To sketch a plane in space, it is helpful to find its points of intersection with the coordinate axes and its traces in the coordinate planes. For example, consider the plane given by

$$3x + 2y + 4z = 12. \qquad \text{Equation of plane}$$

You can find the xy-trace by letting $z = 0$ and sketching the line

$$3x + 2y = 12 \qquad \qquad xy\text{-trace}$$

in the xy-plane. This line intersects the x-axis at $(4, 0, 0)$ and the y-axis at $(0, 6, 0)$. In Figure 11.28, this process is continued by finding the yz-trace and the xz-trace, and then shading in the triangular region lying in the first octant.

If the equation of a plane has a missing variable such as $2x + z = 1$, the plane must be *parallel to the axis* represented by the missing variable, as shown in Figure 11.29. If two variables are missing from the equation of a plane, then it is *parallel to the coordinate plane* represented by the missing variables, as shown in Figure 11.30.

Figure 11.28

Figure 11.29

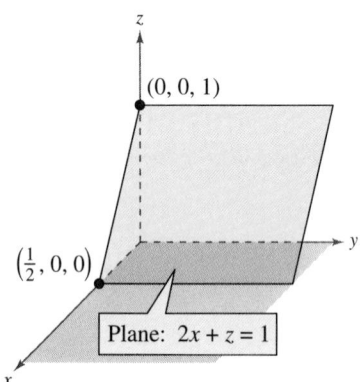

(a) xy-trace($z = 0$): $3x + 2y = 12$

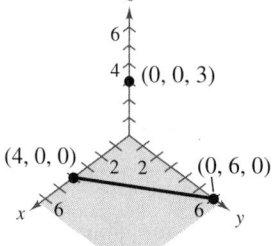

(b) yz-trace($x = 0$): $2y + 4z = 12$

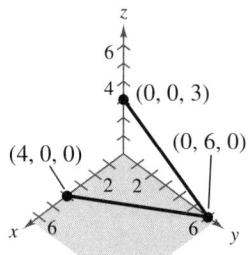

(c) xz-trace($y = 0$): $3x + 4z = 12$

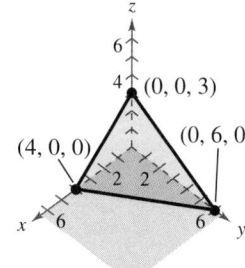

Figure 11.30

(a) Plane $ax + d = 0$ is parallel to yz-plane.

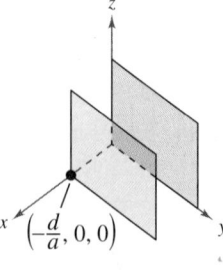

(b) Plane $by + d = 0$ is parallel to xz-plane.

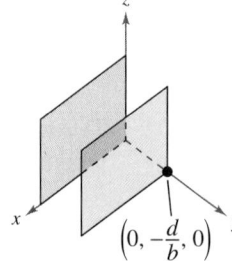

(c) Plane $cz + d = 0$ is parallel to xy-plane.

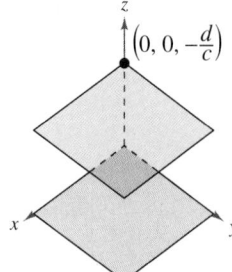

Distance Between a Point and a Plane

Figure 11.31

$$D = \|\operatorname{proj}_{\mathbf{n}} \overrightarrow{PQ}\|$$

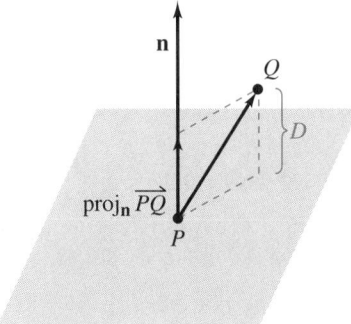

The distance D between a point Q and a plane is the length of the shortest line segment connecting Q to the plane, as shown in Figure 11.31. If P is *any* point in the plane, you can find this distance by projecting the vector $\overrightarrow{PQ}$ onto the normal vector $\mathbf{n}$. The length of this projection is the desired distance.

Distance Between a Point and a Plane

The distance between a plane and a point Q (not in the plane) is

$$D = \|\operatorname{proj}_{\mathbf{n}} \overrightarrow{PQ}\| = \frac{|\overrightarrow{PQ} \cdot \mathbf{n}|}{\|\mathbf{n}\|}$$

where P is a point in the plane and $\mathbf{n}$ is normal to the plane.

To find a point in the plane given by $ax + by + cz + d = 0$, where $a \neq 0$, let $y = 0$ and $z = 0$. Then, from the equation $ax + d = 0$, you can conclude that the point $(-d/a, 0, 0)$ lies in the plane.

EXAMPLE 5   **Finding the Distance Between a Point and a Plane**

Find the distance between the point $Q = (1, 5, -4)$ and the plane given by $3x - y + 2z = 6$.

Solution

You know that $\mathbf{n} = \langle 3, -1, 2 \rangle$ is normal to the given plane. To find a point in the plane, let $y = 0$ and $z = 0$, and obtain the point $P = (2, 0, 0)$. The vector from P to Q is given by

$$\overrightarrow{PQ} = \langle 1 - 2, 5 - 0, -4 - 0 \rangle$$
$$= \langle -1, 5, -4 \rangle.$$

The formula for the distance between a point and a plane produces

Note The choice of the point P in Example 5 is arbitrary. Try choosing a different point to verify that you obtain the same distance.

$$D = \frac{|\overrightarrow{PQ} \cdot \mathbf{n}|}{\|\mathbf{n}\|}$$

$$= \frac{|\langle -1, 5, -4 \rangle \cdot \langle 3, -1, 2 \rangle|}{\sqrt{9 + 1 + 4}}$$

$$= \frac{|-3 - 5 - 8|}{\sqrt{14}}$$

$$= \frac{16}{\sqrt{14}}.$$

Group Activity   *Writing the Equation of a Plane*

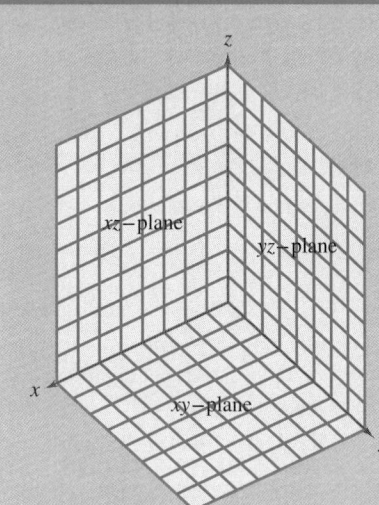

With a partner, fold a piece of graph paper as shown at the left. Label the positive *x*-, *y*-, and *z*-axes. Locate three points, as follows.

• One in the *xy*-plane
• One in the *yz*-plane
• One in the *xz*-plane

a. Find an equation of the plane that contains the three points.

b. Find the points at which the plane intersects the axes. Label these points on your folded graph. Describe any pattern you see in the six points.

11.4 /// EXERCISES

Exploration   In Exercises 1–4, consider the line given by the parametric equations $x = 1 + 3t, y = 2 - t$, and $z = 2 + 5t$, as shown in the figure.

1. Sketch the line represented by the parametric equations. Draw an arrow to indicate its orientation.

2. Find the coordinates of the two points, *P* and *Q*, on the line and determine the vector from *P* to *Q*. What is the relationship of the components of the vector with the coefficients of *t* in the parametric equations?

3. Determine the coordinates of the point where the line intersects the *xz*-plane.

4. What is the relationship between the given line and the line given by the parametric equations $x = -2 + 3s, y = 5 - s$, and $z = 5s$? Explain.

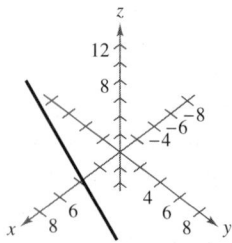

In Exercises 5–10, find a set of (a) parametric equations and (b) symmetric equations for the line through the point and parallel to the specified vector or line. (For each line, express the direction numbers as integers.)

Point	*Parallel to*
5. $(0, 0, 0)$	$\mathbf{v} = \langle -2, 4, 1 \rangle$
6. $(0, 0, 0)$	$\mathbf{v} = \langle 3, -7, -10 \rangle$
7. $(-4, 1, 0)$	$\mathbf{v} = \frac{1}{2}\mathbf{i} + \frac{4}{3}\mathbf{j} - \mathbf{k}$
8. $(5, 0, 10)$	$\mathbf{v} = 4\mathbf{i} + 3\mathbf{k}$
9. $(2, -3, 5)$	$x = 5 + 2t$
	$y = 7 - 3t$
	$z = -2 + t$
10. $(10, -18, 36)$	$\dfrac{x - 2}{3} = \dfrac{y + 2}{-3} = z - 5$

11. Find a set of (a) parametric equations and (b) symmetric equations for the line through the points $(3, -5, -4)$ and $\left(-\frac{3}{2}, \frac{3}{2}, 2\right)$. (Express the direction numbers as integers.)

12. Find a set of parametric equations for the line that passes through the point $(-3, 8, 15)$ and is parallel to the xz-plane and the yz-plane.

13. Determine which of the points lie on the line that passes through the point $(-4, -1, 7)$ and is parallel to the vector $\mathbf{v} = 3\mathbf{i} - \mathbf{j}$.

(a) $(-4, -1, 0)$ (b) $(-1, -2, 7)$

(c) $(-10, 1, 7)$ (d) $(4, 1, -7)$

14. Find the point of intersection of the lines and the angle between the lines.

$x = 6 + t, y = -5 - 2t, z = 1 + 3t$

$x = 7 + 5s, y = 5 + 2s, z = -8 + 3s$

In Exercises 15 and 16, sketch a graph of the line.

15. $x = 2t, y = 2 + t,$ **16.** $x = 5 - 2t, y = 1 + t,$

$z = 1 + \frac{1}{2}t$ $z = 5 - \frac{1}{2}t$

In Exercises 17–22, find an equation of the plane passing through the point and perpendicular to the specified vector or line.

Point	Perpendicular to
17. $(3, 4, -2)$	$\mathbf{n} = \mathbf{j}$
18. $(2, 3, 5)$	$\mathbf{n} = \mathbf{k}$
19. $(5, 6, 3)$	$\mathbf{n} = -2\mathbf{i} + \mathbf{j} - 2\mathbf{k}$
20. $(0, 0, 0)$	$\mathbf{n} = -3\mathbf{j} + 5\mathbf{k}$
21. $(2, 0, 0)$	$x = 3 - t$
	$y = 2 - 2t$
	$z = 4 + t$
22. $(1, 3, 1)$	$\dfrac{x - 2}{3} = y + 1 = \dfrac{z + 2}{-2}$

In Exercises 23 and 24, find an equation of the plane passing through the three points.

23. $(0, 0, 0), (2, 1, 3), (-2, 1, 3)$

24. $(4, -1, 3), (2, 5, 1), (-1, 2, 1)$

In Exercises 25–28, find an equation of the plane.

25. The plane passes through the point $(2, 5, 3)$ and is parallel to the xz-plane.

26. The plane passes through the point $(2, 5, 3)$ and is parallel to the xy-plane.

27. The plane contains the x-axis and makes an angle of $45°$ with the positive y-axis.

28. The plane passes through the points $(4, 0, 0)$ and $(0, 2, 0)$ and is perpendicular to the plane $x + 2y + 2z = 4$.

In Exercises 29–32, determine whether the planes are parallel, orthogonal, or neither. If they are neither parallel nor orthogonal, find the angle of intersection.

29.	$3x + y - 4z = 3$	**30.**	$3x + 2y - z = 7$
	$-9x - 3y + 12z = 4$		$x - 4y + 2z = 0$
31.	$2x \quad\quad - z = 1$	**32.**	$x - 5y - z = 1$
	$4x + y + 8z = 10$		$5x - 25y - 5z = -3$

In Exercises 33–36, mark the intercepts and sketch a graph of the plane.

33. $x + 2y + 3z = 6$ **34.** $2x - y + 4z = 4$

35. $x + z = 3$ **36.** $y + 2z = 4$

In Exercises 37–40, use a graphing utility to graph the plane.

37. $3x + 2y - z = 6$ **38.** $x - 3z = 6$

39. $x + 2y - 6z = 8$ **40.** $3x - 4y - z = -12$

In Exercises 41 and 42, find the distance between the point and the plane.

41. $(0, 0, 0)$ **42.** $(1, 2, 3)$

$3x + 2y + z = 12$ $2x - y + z = 4$

43. *Exploration*

(a) Describe, and find an equation for, the surface generated by all points (x, y, z) that are two units from the point $(4, -1, 1)$.

(b) Describe, and find an equation for, the surface generated by all points (x, y, z) that are two units from the plane $4x - 3y + z = 10$.

44. *True or False?* Every two lines in space are either intersecting or parallel.

Focus on Concepts

In this chapter, you studied analytic geometry in three dimensions. You can use the following questions to check your understanding of several of the basic concepts presented. The answers to these questions are given in the back of the book.

1. In a three-dimensional coordinate system, a point P is determined by an ordered triple (x, y, z). Describe the directed distance measured by each coordinate.

2. Describe the surface consisting of all points (x, y, z) that are a distance r from the point (h, k, j).

3. What is meant by the xz-trace of a surface? Describe how the equation for this trace is found.

4. A sphere intersects the yz-plane. Describe the trace.

5. A plane intersects the xy-plane. Describe the trace.

6. What characterizes a vector in space?

7. Which vectors in the figure appear to be equivalent?

8. Give a geometric description of the scalar multiple $k\mathbf{u}$ of the nonzero vector $\mathbf{u}$.

9. What is known about the nonzero vectors $\mathbf{u}$ and $\mathbf{v}$ if $\mathbf{u} \cdot \mathbf{v} < 0$? Explain.

10. Explain why the cross product is not defined for two-dimensional vectors.

11. *True or False?* If $\mathbf{u}$ and $\mathbf{v}$ are vectors in space that are nonzero and not parallel, then $\mathbf{u} \times \mathbf{v} = \mathbf{v} \times \mathbf{u}$. Explain.

12. Which of the figures shows the cross product $\mathbf{w}$ of the vectors $\mathbf{u}$ and $\mathbf{v}$ in a right-handed system? Explain.

(a)　　　　　　　　　　　(b)

(c)　　　　　　　　　　　(d)

13. The direction numbers of two distinct lines in space are 10, -18, 20, and -15, 27, -30. What is the relationship between the lines? Explain.

14. *True or False?* Two nonparallel lines in space will always intersect. Explain.

15. *True or False?* Two nonparallel planes in space will always intersect. Explain.

16. Consider the two nonzero vectors $\mathbf{u}$ and $\mathbf{v}$. Describe the geometric figure generated by the terminal points of the following vectors, where s and t represent real numbers.

　(a) $t\mathbf{v}$　　(b) $\mathbf{u} + t\mathbf{v}$　　(c) $s\mathbf{u} + t\mathbf{v}$

11 /// REVIEW EXERCISES

In Exercises 1 and 2, approximate the coordinates of the points.

1.

2.

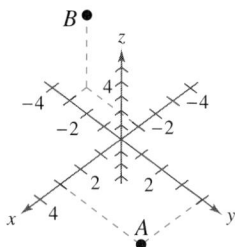

3. Find the coordinates of the point in the *xy*-plane four units to the right of the *xz*-plane and five units behind the *yz*-plane.

4. Find the coordinates of the point located on the *y*-axis and seven units to the left of the *xz*-plane.

In Exercises 5 and 6, determine the octant(s) in which (x, y, z) is located so that the condition(s) is (are) satisfied.

5. $yz > 0$

6. $x < 0, z > 0$

In Exercises 7 and 8, find the lengths of the sides of the right triangle. Show that these lengths satisfy the Pythagorean Theorem.

7.

8.

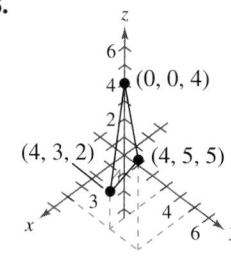

In Exercises 9 and 10, find the coordinates of the midpoint of the line segment joining the points.

9. $(8, -2, 3), (5, 6, 7)$

10. $(6.2, 4.5, -3), (3.6, -3.5, 10)$

In Exercises 11–14, find the standard form of the equation of the sphere.

11.

12.

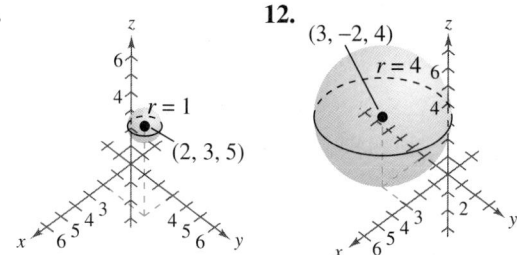

13. Center: $(3, -2, 6)$; Diameter: 15

14. Endpoints of a diameter: $(0, 0, 4), (4, 6, 0)$

In Exercises 15 and 16, find the center and radius of the sphere and sketch its graph.

15. $x^2 + y^2 + z^2 - 4x - 6y + 4 = 0$

16. $x^2 + y^2 + z^2 - 10x + 6y - 4z + 34 = 0$

In Exercises 17–20, find the component form of the vector with initial and terminal points P and Q, respectively.

17.

18.

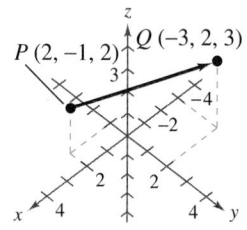

19. $P(7, -4, 3)$
  $Q(-3, 2, 10)$

20. $P(0, 3, -1)$
  $Q(5, -8, 6)$

In Exercises 21 and 22, use a graphing utility to determine whether u and v are orthogonal, parallel, or neither.

21. $\mathbf{u} = \langle 39, -12, 21 \rangle$
  $\mathbf{v} = \langle -26, 8, -14 \rangle$

22. $\mathbf{u} = \langle 8, 5, -8 \rangle$
  $\mathbf{v} = \langle -2, 4, \frac{1}{2} \rangle$

In Exercises 23 and 24, find the angle θ between the vectors u and v.

23. $\mathbf{u} = \langle 2\sqrt{2}, -4, 4 \rangle$
    $\mathbf{v} = \langle -\sqrt{2}, 1, 2 \rangle$

24. $\mathbf{u} = \langle 3, 1, -1 \rangle$
    $\mathbf{v} = \langle 4, 5, 2 \rangle$

In Exercises 25 and 26, find u × v. Use a graphing utility to verify your answer.

25. $\mathbf{u} = \langle -2, 8, 2 \rangle$
    $\mathbf{v} = \langle 1, 1, -1 \rangle$

26. $\mathbf{u} = \langle 10, 15, 5 \rangle$
    $\mathbf{v} = \langle 5, -3, 0 \rangle$

In Exercises 27–36, let $\mathbf{u} = \langle 2, -3, 3 \rangle$, $\mathbf{v} = \langle 1, 4, -2 \rangle$, and $\mathbf{w} = \langle -2, 1, 2 \rangle$.

27. Find $\|\mathbf{u}\|$.

28. Find $2\mathbf{u} + 3\mathbf{v} - 5\mathbf{w}$.

29. Find $\mathbf{u} \cdot \mathbf{v}$.

30. Find the angle between $\mathbf{u}$ and $\mathbf{v}$.

31. Show that $\mathbf{u} \cdot \mathbf{u} = \|\mathbf{u}\|^2$.

32. Find $\mathbf{v} \times \mathbf{w}$.

33. Show that $\mathbf{u} \times \mathbf{v} = -(\mathbf{v} \times \mathbf{u})$.

34. Show that $\mathbf{u} \cdot (\mathbf{v} + \mathbf{w}) = \mathbf{u} \cdot \mathbf{v} + \mathbf{u} \cdot \mathbf{w}$.

35. Show that $\mathbf{u} \times (\mathbf{v} + \mathbf{w}) = (\mathbf{u} \times \mathbf{v}) + (\mathbf{u} \times \mathbf{w})$.

36. Find the volume of the solid with edges $\mathbf{u}$, $\mathbf{v}$, and $\mathbf{w}$.

37. *Volume* Use the triple scalar product to find the volume of the parallelepiped with vertices $(3, 2, -1)$, $(2, 5, 1)$, $(3, 2, 8)$, $(4, -1, 6)$, $(-5, 4, 2)$, $(-6, 7, 4)$, $(-5, 4, 11)$, and $(-4, 1, 9)$.

In Exercises 38 and 39, find the component form of u.

38. $\mathbf{u}$ is perpendicular to the plane $x - 3y + 4z = 0$, and $\|\mathbf{u}\| = 3$.

39. $\mathbf{u}$ is a unit vector perpendicular to each of the lines.
    $x = 4 - t$ $x = -3 + 7s$
    $y = 3 + 3t$ $y = -2 + s$
    $z = 1 + 5t$ $z = 1 + 2s$

In Exercises 40–45, find a set of parametric equations for the specified line.

40. The line passes through the points $(-1, 3, 5)$ and $(3, 6, -1)$.

41. The line passes through the points $(0, -10, 3)$ and $(5, 10, 0)$.

42. The line passes through the point $(2, 1, 4)$ and is perpendicular to the xz-plane.

43. The line passes through the point $(3, 1, 2)$ and is parallel to the line given by $x = y = z$.

44. The line passes through the point $(-5, 6, 7)$ and is perpendicular to the vectors $\mathbf{u} = \langle 0, 2, 2 \rangle$ and $\mathbf{v} = \langle 2, 0, 2 \rangle$.

45. The line is the intersection of the planes given by
    $$2x - y - 2z = 4$$
    $$x - y - z = 8.$$

In Exercises 46–49, find an equation of the plane.

46. The plane passes through the points $(0, 0, 0)$, $(5, 0, 2)$, and $(2, 3, 8)$.

47. The plane passes through the points $(-1, 3, 4)$, $(4, -2, 2)$, and $(2, 8, 6)$.

48. The plane passes through the point $(5, 3, 2)$ and is parallel to the xy-plane.

49. The plane passes through the point $(3, 1, 2)$ and is orthogonal to the line given by $x = y = z$.

In Exercises 50 and 51, find the distance from the point to the plane.

50. $(2, 3, 10)$
    $2x - 20y + 6z = 6$

51. $(0, 0, 0)$
    $x - 10y + 3z = 2$

In Exercises 52 and 53, find the distance between the parallel planes.

52. $5x - 3y + z = 2$
    $5x - 3y + z = -3$

53. $3x - 2y - z = -1$
    $6x - 4y - 2z = 1$

54. Sketch a graph of the plane $x + 2y + 3z = 6$.

CHAPTER PROJECT   *Precalculus with Limits*

In Section 11.4, you learned how to find the distance between a point and a plane. A related problem is to find the distance between a point and a *line*. You will see that the formula resembles that for the distance between a point and a plane, except that you replace the dot product by the cross product and the normal vector **n** by a direction vector **v** for the line.

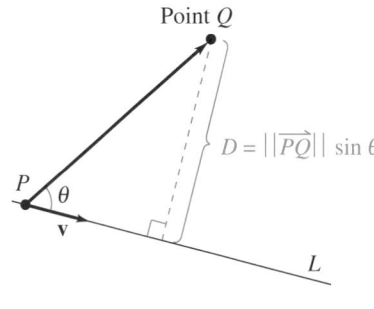

Point Q

$D = \|\overrightarrow{PQ}\| \sin \theta$

P θ

v

L

In order to develop this new formula, suppose you are given a line L and a point Q not on the line. Let D be the distance between the point and the line, as indicated in the figure at the left. Let $\mathbf{v} = \langle a, b, c \rangle$ be the direction vector for L and select any point P on the line. Then $D = \|\overrightarrow{PQ}\| \sin \theta$ where θ is the angle between **v** and $\overrightarrow{PQ}$. Using the second geometric property of the cross product listed on page 846, you know that

$$\|\mathbf{v}\| \|\overrightarrow{PQ}\| \sin \theta = \|\mathbf{v} \times \overrightarrow{PQ}\|.$$

Consequently,

$$D = \|\overrightarrow{PQ}\| \sin \theta = \frac{\|\mathbf{v}\| \|\overrightarrow{PQ}\| \sin \theta}{\|\mathbf{v}\|} = \frac{\|\mathbf{v} \times \overrightarrow{PQ}\|}{\|\mathbf{v}\|}.$$

Questions for Further Exploration

1. Show that the distance between the point $Q = (3, -1, 4)$ and the line given by $x = -2 + 3t$, $y = -2t$, and $z = 1 + 4t$ is $\sqrt{6}$.

2. Find the distance between the point $(10, 3, -2)$ and the line through the points $(-2, 3, 1)$ and $(2, 3, 0)$.

3. Find the distance between the two parallel lines $x = 3t - 1$, $y = -t + 2$, $z = 4t$, and $x = 6t + 2$, $y = -2t$, $z = 8t - 3$.

4. Find the distance between the following two *skew* lines (lines that are neither parallel nor intersecting).

$$x = \frac{y}{2} = \frac{z}{3}$$

$$\frac{x - 1}{-1} = y - 4 = z + 1$$

11 /// CHAPTER TEST

Take this test as you would take a test in class. After you are done, check your work against the answers given in the back of the book.

1. Find the coordinates of the point three units in front of the yz-plane, nine units to the left of the xz-plane, and on the xy-plane.

2. Determine the octants in which the point (x, y, z) is located if $xy = 4$ and $z = 2$.

In Exercises 3–6, use the points $A(8, -2, 5)$, $B(6, 4, -1)$, and $C(-4, 3, 0)$.

3. Find the coordinates of the midpoint of the line segment joining points A and B.

4. Find the standard form of the equation of a sphere of which A and B are the endpoints of a diameter.

5. Consider the triangle with vertices A, B, and C. Is it a right triangle? Explain.

6. Let **u** and **v** be the vectors from A to B and A to C, respectively.

   (a) Find **u** $\cdot$ **v**.

   (b) Find **u** $\times$ **v**.

   (c) Find the angle between **u** and **v**.

   (d) Use **u** and **v** to find the area of the triangle with vertices A, B, and C.

   (e) Find a set of (a) parametric equations and (b) symmetric equations for the line through points A and B.

   (f) Find an equation of the plane containing the points A, B, and C.

7. Find the component form of the vector **v** in the direction of the diagonal of the parallelepiped shown in the figure. Use the vector to find the angle between the diagonal and the edges of the parallelepiped lying on the coordinate axes.

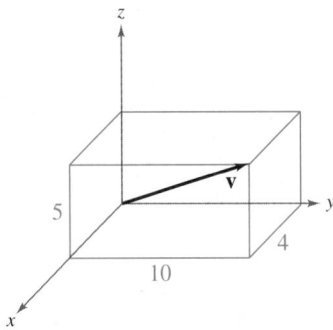

8. Find the distance between the point $(4, 3, 8)$ and the plane $3x - 2y + z = 6$.

9. Use a graphing utility to graph the line represented by the parametric equations $x = 4 - t$, $y = 1 + 2t$, and $z = 2 + t$.

Chapter 12

Limits and an Introduction to Calculus

One of the classic problems in calculus is finding the velocity of an accelerating object.

For instance, consider the velocity of a space shuttle at lift-off. The shuttle begins to rise slowly. As the shuttle moves up, its velocity increases. The height of the shuttle during its first 5 seconds of motion is shown below.

Time (sec)	0	1	2	3	4	5
Height (ft)	0	4	23	60	115	188

During the first 5 seconds, the shuttle has an average velocity of 37.6 feet per second. But what is its actual velocity at the end of 5 seconds? (See Exercises 41 and 42 on page 913.)

Astronaut Mae Jemison is shown floating in the science module of *Spacelab J* aboard the orbiting shuttle *Endeavor*. Of her flight into space, she says, "I always knew I would go—I didn't know how—I just knew I would go."

12.1   Introduction to Limits

The Limit Concept / Definition of Limit / Limits That Fail to Exist /
Properties of Limits

The Limit Concept

Figure 12.1

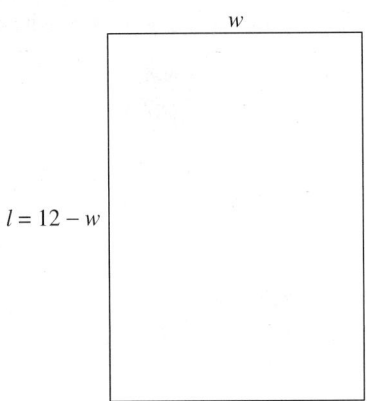

$l = 12 - w$

The notion of a limit is a *fundamental* concept of calculus. In this chapter, you will learn how to evaluate limits and how they are used in the two basic problems of calculus: "the tangent line problem" and "the area problem."

EXAMPLE 1   Finding a Rectangle of Maximum Area

You are given 24 inches of wire and are asked to form a rectangle whose area is as large as possible. What dimensions should the rectangle have?

Solution
Let w represent the width of the rectangle and let l represent the length of the rectangle, as shown in Figure 12.1. Because

$$2w + 2l = 24 \qquad \text{Perimeter is 24}$$

it follows that $l = 12 - w$ and that the area of the rectangle is

$$
\begin{aligned}
A &= wl & &\text{Formula for area}\\
&= w(12 - w) & &\text{Substitute } 12 - w \text{ for } l.\\
&= 12w - w^2. & &\text{Simplify.}
\end{aligned}
$$

Using this model for area, you can experiment with different values of w to see how to obtain the maximum area. After trying several values, it appears that the maximum area occurs when $w = 6$.

Width, w	5.0	5.5	5.9	6.0	6.1	6.5	7.0
Area, A	35.0	35.75	35.99	36.0	35.99	35.75	35.0

In limit terminology, we say that the limit of A as w approaches 6 is 36. This is written as

$$\lim_{w \to 6} A = 36.$$

You can construct a table of values using the table feature of a graphing utility. For instance, to verify the table at the right on a *TI-82* or *TI-83*, let $Y_1 = 12X - X^2$. Then set TblMin = 5 and ΔTbl = .1 in the TblSet menu and select $\boxed{\text{TABLE}}$. You could also choose (Indpnt:Ask) to construct the table.

The *Interactive* CD-ROM shows every example with its solution; clicking on the *Try It!* button brings up similar problems. Guided Examples and Integrated Examples show step-by-step solutions to additional examples. Integrated Examples are related to several concepts in the section.

Definition of Limit

EXAMPLE 2   **Finding a Limit That Can Be Reached**

Use a table to estimate numerically the limit $\lim\limits_{x \to 2} (3x - 2)$.

Solution
Let $f(x) = 3x - 2$. Then construct a table that shows values of $f(x)$ when x is close to 2.

Figure 12.2

$f(x) = 3x - 2$

x	1.9	1.99	1.999	2.0	2.001	2.01	2.1
$f(x)$	3.7	3.97	3.997	?	4.003	4.03	4.3

From the table, it appears that the closer x gets to 2, the closer $f(x)$ gets to 4. Thus, you can estimate the limit to be 4. For this particular function, you can obtain the limit simply by substituting 2 for x to obtain

$$\lim\limits_{x \to 2} (3x - 2) = 3(2) - 2 = 4.$$

Figure 12.2 adds further support to this conclusion.

In Figure 12.2 note that the graph of $f(x) = 3x - 2$ is continuous. That is, it has no gaps or jumps. For graphs that are not continuous, finding a limit can be more difficult.

EXAMPLE 3   **Finding a Limit That Cannot Be Reached**

Use a table to estimate numerically the limit $\lim\limits_{x \to 1} \dfrac{x^3 - x^2 + x - 1}{x - 1}$.

Solution
Let $f(x) = (x^3 - x^2 + x - 1)/(x - 1)$. Then construct a table that shows values of $f(x)$ when x is close to 1.

Figure 12.3

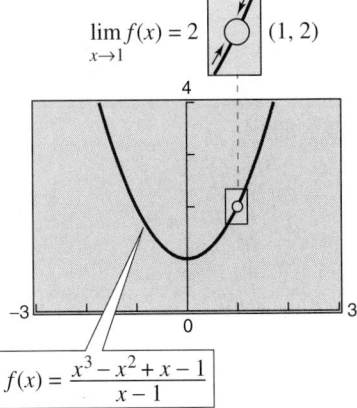

$\lim\limits_{x \to 1} f(x) = 2$ $(1, 2)$

$f(x) = \dfrac{x^3 - x^2 + x - 1}{x - 1}$

x	0.9	0.99	0.999	1.0	1.001	1.01	1.1
$f(x)$	1.81	1.9801	1.998	?	2.002	2.0201	2.21

From the table, it appears that the limit is 2. The graph in Figure 12.3 supports this conclusion. In this case, notice that you cannot obtain the limit simply by evaluating $f(x)$ when $x = 1$.

Definition of Limit

If $f(x)$ becomes arbitrarily close to a unique number L as x approaches c from either side, the **limit** of $f(x)$ as x approaches c is L. This is written as

$$\lim_{x \to c} f(x) = L.$$

EXAMPLE 4 ◢ Using a Graphing Utility to Find a Limit

Use a table to estimate numerically the limit $\lim\limits_{x \to 0} \dfrac{x}{\sqrt{x + 1} - 1}$.

Figure 12.4

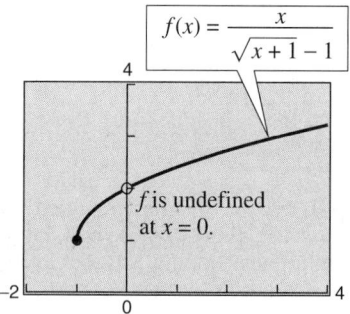

$$f(x) = \frac{x}{\sqrt{x + 1} - 1}$$

f is undefined at $x = 0$.

Solution

Let $f(x) = x/\left(\sqrt{x + 1} - 1\right)$. Then use a graphing utility to construct a table that shows values of $f(x)$ when x is close to 0.

x	-0.01	-0.001	-0.0001	0	0.0001	0.001	0.01
$f(x)$	1.995	1.9995	1.9999	?	2.0001	2.0005	2.005

From the table, it appears that the limit is 2. Try using the zoom and trace features of a graphing utility to verify graphically that the limit is 2 (see Figure 12.4). ◢

In Example 4, note that $f(x)$ has a limit when $x \to 0$ although the function is not defined when $x = 0$. This often happens and it is important to realize that *the existence or nonexistence of $f(x)$ when $x = c$ has no bearing on the existence of the limit of $f(x)$ as x approaches c.*

Figure 12.5

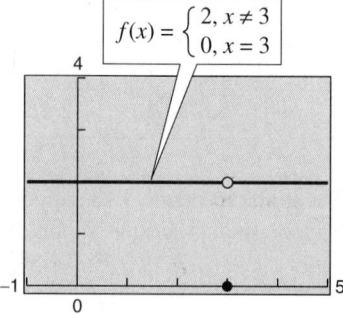

$$f(x) = \begin{cases} 2, & x \neq 3 \\ 0, & x = 3 \end{cases}$$

EXAMPLE 5 ◢ Using a Graph to Find a Limit

Find the limit of $f(x)$ as x approaches 3, where $f(x) = \begin{cases} 2, & x \neq 3 \\ 0, & x = 3. \end{cases}$

Solution

Because $f(x) = 2$ for all x other than $x = 3$ and because the value of $f(3)$ is immaterial, it follows that the limit is 2, as shown in Figure 12.5. Thus, you can write

$$\lim_{x \to 3} f(x) = 2.$$ ◢

Limits That Fail to Exist

In the next three examples, you will examine some functions for which limits do not exist.

Figure 12.6

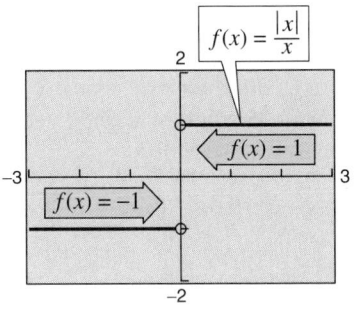

$f(x) = \dfrac{|x|}{x}$

$f(x) = 1$

$f(x) = -1$

EXAMPLE 6   **Comparing Left and Right Behavior**

Show that the following limit does not exist.

$$\lim_{x \to 0} \frac{|x|}{x}$$

Solution
Consider the graph of the function $f(x) = |x|/x$. From Figure 12.6, you can see that for positive x-values

$$\frac{|x|}{x} = 1, \quad x > 0$$

and for negative x-values

$$\frac{|x|}{x} = -1, \quad x < 0.$$

This means that no matter how close x gets to 0, there will be both positive and negative x-values that yield $f(x) = 1$ and $f(x) = -1$. This implies that the limit does not exist.

EXAMPLE 7   **Unbounded Behavior**

Discuss the existence of the limit $\lim\limits_{x \to 0} \dfrac{1}{x^2}$.

Solution
Let $f(x) = 1/x^2$. In Figure 12.7, note that as x approaches 0 from either the right or the left, $f(x)$ increases without bound. This means that by choosing x close enough to 0, you can force $f(x)$ to be as large as you want. For instance, $f(x)$ will be larger than 100 if you choose x that is within $\frac{1}{10}$ of 0. That is,

Figure 12.7

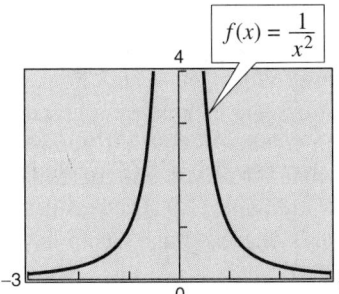

$f(x) = \dfrac{1}{x^2}$

$$0 < |x| < \frac{1}{10} \quad \Longrightarrow \quad f(x) = \frac{1}{x^2} > 100.$$

Similarly, you can force $f(x)$ to be larger than 1,000,000, as follows.

$$0 < |x| < \frac{1}{1000} \quad \Longrightarrow \quad f(x) = \frac{1}{x^2} > 1,000,000$$

Because $f(x)$ is not approaching a real number L as x approaches 0, you can conclude that the limit does not exist.

Figure 12.8

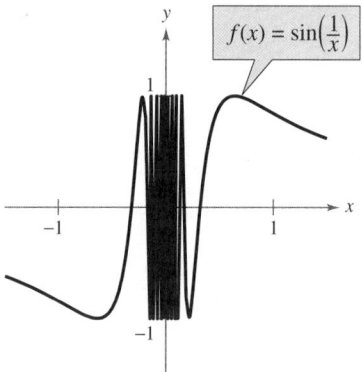

EXAMPLE 8 ▱ **Oscillating Behavior**

Discuss the existence of the limit

$$\lim_{x \to 0} \sin\left(\frac{1}{x}\right).$$

Solution

Let $f(x) = \sin(1/x)$. In Figure 12.8, you can see that as x approaches 0, $f(x)$ oscillates between -1 and 1. Therefore, the limit does not exist because no matter how close you are to 0, it is possible to choose values of x_1 and x_2 such that $\sin(1/x_1) = 1$ and $\sin(1/x_2) = -1$, as indicated in the table.

x	$\dfrac{2}{\pi}$	$\dfrac{2}{3\pi}$	$\dfrac{2}{5\pi}$	$\dfrac{2}{7\pi}$	$\dfrac{2}{9\pi}$	$\dfrac{2}{11\pi}$	$x \to 0$
$\sin\left(\dfrac{1}{x}\right)$	1	-1	1	-1	1	-1	Limit does not exist.

▱

Examples 6, 7, and 8 show three of the most common types of behavior associated with the nonexistence of a limit.

Conditions Under Which Limits Do Not Exist
The limit of $f(x)$ as $x \to c$ does not exist if any of the following conditions is true.

1. $f(x)$ approaches a different number from the right   Example 6
side of c than from the left side of c.
2. $f(x)$ increases or decreases without bound as x Example 7
approaches c.
3. $f(x)$ oscillates between two fixed values as x Example 8
approaches c.

Figure 12.9

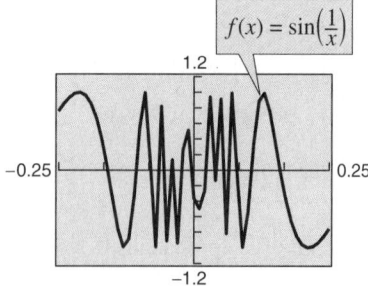

A graphing utility can help you discover the behavior of a function near the x-value at which you are trying to evaluate a limit. When you do this, however, you should realize that you can't always trust the graphs that graphing utilities show. For instance, if you use a graphing utility to sketch the graph of the function in Example 8 over an interval containing 0, you will most likely obtain an incorrect graph—such as that shown in Figure 12.9. The reason that a graphing utility can't show the correct graph is that the graph has infinitely many oscillations over any interval that contains 0.

Properties of Limits

You have seen that sometimes the limit of $f(x)$ as $x \to c$ is simply $f(c)$. In such cases, we say that the limit can be evaluated by *direct substitution*. That is,

$$\lim_{x \to c} f(x) = f(c) \qquad\qquad \text{Substitute } c \text{ for } x.$$

There are many "well-behaved" functions that have this property. Some of the basic ones are included in the following list.

EXPLORATION

Use a graphing utility to graph the tangent function. What are $\lim_{x \to 0} \tan x$ and $\lim_{x \to \pi/4} \tan x$? What can you say about the existence of the limit $\lim_{x \to \pi/2} \tan x$?

Properties of Limits

Let b and c be real numbers and let n be a positive integer.

1. $\lim\limits_{x \to c} b = b$

2. $\lim\limits_{x \to c} x = c$

3. $\lim\limits_{x \to c} x^n = c^n$

4. $\lim\limits_{x \to c} \sqrt[n]{x} = \sqrt[n]{c}$, for n even and $c \geq 0$

Trigonometric functions could also be included in this list. For instance,

$$\lim_{x \to \pi} \sin x = \sin \pi = 0$$

and

$$\lim_{x \to 0} \cos x = \cos 0 = 1.$$

By combining the Properties of Limits with the following operations, you can find limits for a wide variety of functions.

Operations with Limits

Let b and c be real numbers, let n be a positive integer, and let f and g be functions with the following limits.

$$\lim_{x \to c} f(x) = L \qquad \text{and} \qquad \lim_{x \to c} g(x) = K$$

1. Scalar multiple: $\lim\limits_{x \to c} [b f(x)] = bL$

2. Sum or difference: $\lim\limits_{x \to c} [f(x) \pm g(x)] = L \pm K$

3. Product: $\lim\limits_{x \to c} [f(x) g(x)] = LK$

4. Quotient: $\lim\limits_{x \to c} \dfrac{f(x)}{g(x)} = \dfrac{L}{K}$, provided $K \neq 0$

5. Power: $\lim\limits_{x \to c} [f(x)]^n = L^n$

EXAMPLE 9 ▱ **Finding Limits by Direct Substitution**

Find each of the following limits.

a. $\lim_{x \to 4} x^2$

b. $\lim_{x \to 4} 5$

c. $\lim_{x \to 4} (x^2 + 2x - 5)$

d. $\lim_{x \to 4} \dfrac{x^2 + 2x - 5}{x - 1}$

Solution

You can use the Properties of Limits and direct substitution to evaluate each limit.

a. $\lim_{x \to 4} x^2 = (4)^2 = 16$

b. $\lim_{x \to 4} 5 = 5$

c. $\lim_{x \to 4} (x^2 + 2x - 5) = (4)^2 + 2(4) - 5 = 16 + 8 - 5 = 19$

d. $\lim_{x \to 4} \dfrac{x^2 + 2x - 5}{x - 1} = \dfrac{(4)^2 + 2(4) - 5}{4 - 1} = \dfrac{16 + 8 - 5}{3} = \dfrac{19}{3}$ ▱

Group Activity

Graphs with Gaps

Match each graph with one of the following functions.

$$f(x) = x + 1 \qquad g(x) = \frac{x^2 - 1}{x - 1} \qquad h(x) = \frac{x^3 - 2x^2 - x + 2}{x^2 - 3x + 2}$$

Find the limits of each function as x approaches 1 and as x approaches 2. What conclusion can you make?

(a)

(b)

(c)

Graph these functions on a graphing utility to see if it distinguishes among the graphs. Explain your answers to a group member.

12.1 /// EXERCISES

1. *Exploration* Consider an open box that is to be made from a square piece of material, 24 centimeters on a side, by cutting equal squares from the corners and turning up the sides.

(a) Draw a figure that represents the box.

(b) Verify that the volume of the box is $V = 4x(12 - x)^2$.

(c) The box has a maximum volume when $x = 4$. Use a graphing utility to complete the table and observe the behavior of the function as x approaches 4.

x	3	3.5	3.9	4	4.1	4.5	5
V							

Use the table to find $\lim\limits_{x \to 4} V$.

(d) Use a graphing utility to graph the volume function. Verify that the volume is maximum when $x = 4$.

2. *Essay* Write a brief description of the meaning of the notation $\lim\limits_{x \to 5} f(x) = 12$.

In Exercises 3–6, use a graphing utility to complete the table and use the result to estimate the limit. Determine whether or not the limit can be reached.

3. $\lim\limits_{x \to 3} (3 - 2x)$

x	2.9	2.99	2.999	3	3.001	3.01	3.1
$f(x)$				?			

4. $\lim\limits_{x \to 4} \left(\frac{1}{2}x^2 - 2x + 3\right)$

x	3.9	3.99	3.999	4	4.001	4.01	4.1
$f(x)$				?			

The *Interactive* CD-ROM contains step-by-step solutions to all odd-numbered Section and Review Exercises. It also provides Tutorial Exercises, which link to Guided Examples for additional help.

5. $\lim\limits_{x \to 2} \dfrac{x - 2}{x^2 - 4}$

x	1.9	1.99	1.999	2	2.001	2.01	2.1
$f(x)$				?			

6. $\lim\limits_{x \to -1} \dfrac{x + 1}{x^2 - x - 2}$

x	-1.1	-1.01	-1.001	-1	-0.999
$f(x)$				?	

x	-0.99	-0.9
$f(x)$		

7. *Think About It* From Exercises 3 to 6, select a limit that can be reached and one that cannot be reached.

(a) Use a graphing utility to graph the corresponding functions in the standard viewing rectangle. Do the graphs reveal whether or not the limit can be reached? Explain.

(b) Use a graphing utility to graph the corresponding functions using the zoom decimal setting. Do the graphs reveal whether or not the limit can be reached? Explain.

8. *Think About It* Use the results from Exercise 7 to draw a conclusion as to whether or not you can use the graph generated by a graphing utility to determine reliably whether or not a limit can be reached.

In Exercises 9–16, use a graphing utility to complete the table and use the result to estimate the limit. Use the graphing utility to graph the corresponding function to confirm your result.

9. $\lim\limits_{x \to 1} \dfrac{x - 1}{x^2 + 2x - 3}$

x	0.9	0.99	0.999	1	1.001	1.01	1.1
$f(x)$				?			

10. $\lim\limits_{x \to -3} \dfrac{x+3}{x^2+x-6}$

x	-3.1	-3.01	-3.001	-3	-2.999
$f(x)$				?	

x	-2.99	-2.9
$f(x)$		

11. $\lim\limits_{x \to 0} \dfrac{\sqrt{x+3} - \sqrt{3}}{x}$

x	-0.1	-0.01	-0.001	0	0.001
$f(x)$				?	

x	0.01	0.1
$f(x)$		

12. $\lim\limits_{x \to -3} \dfrac{\sqrt{1-x}-2}{x+3}$

x	-3.1	-3.01	-3.001	-3	-2.999
$f(x)$				?	

x	-2.99	-2.9
$f(x)$		

13. $\lim\limits_{x \to -4} \dfrac{\dfrac{x}{x+2}-2}{x+4}$

x	-4.1	-4.01	-4.001	-4	-3.999
$f(x)$				?	

x	-3.99	-3.9
$f(x)$		

14. $\lim\limits_{x \to 2} \dfrac{\dfrac{1}{x+2}-\dfrac{1}{4}}{x-2}$

x	1.9	1.99	1.999	2	2.001	2.01	2.1
$f(x)$				?			

15. $\lim\limits_{x \to 0} \dfrac{\sin x}{x}$

x	-0.1	-0.01	-0.001	0	0.001
$f(x)$				?	

x	0.01	0.1
$f(x)$		

16. $\lim\limits_{x \to 0} \dfrac{\cos x - 1}{x}$

x	-0.1	-0.01	-0.001	0	0.001
$f(x)$				?	

x	0.01	0.1
$f(x)$		

In Exercises 17–24, use the graph to find the limit (if it exists).

17. $\lim\limits_{x \to 2} (3 - x)$

18. $\lim\limits_{x \to -4} (x^2 - 3)$

19. $\lim\limits_{x \to -1} \sin \dfrac{\pi x}{2}$

20. $\lim\limits_{x \to 2} \dfrac{3x^2 - 12}{x - 2}$

21. $\lim\limits_{x \to -2} \dfrac{|x+2|}{x+2}$

22. $\lim\limits_{x \to 1} \dfrac{1}{x-1}$

23. $\lim\limits_{x \to 0} 2 \cos \dfrac{\pi}{x}$

24. $\lim\limits_{x \to \pi/3} \sec x$

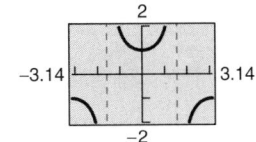

41. $\lim\limits_{x \to \pi} \sin 2x$

42. $\lim\limits_{x \to 1/4} \dfrac{\tan \pi x}{2}$

43. $\lim\limits_{x \to 1/2} \arcsin x$

44. $\lim\limits_{x \to 1} \arccos \dfrac{x}{2}$

In Exercises 45 and 46, find the limit (if it exists) as x approaches 2.

45. $f(x) = \begin{cases} 2x + 1, & x < 2 \\ 2x + 2, & x \geq 2 \end{cases}$

46. $f(x) = \begin{cases} x^2 - 4, & x < 2 \\ x - 2, & x \geq 2 \end{cases}$

In Exercises 25–30, use a graphing utility to graph the function and use the graph to determine whether the specified limit exists.

25. $f(x) = \dfrac{8}{2 + e^{1/x}}$, $\lim\limits_{x \to 0} f(x)$

26. $f(x) = \cos \dfrac{1}{x}$, $\lim\limits_{x \to 0} f(x)$

27. $f(x) = \dfrac{\sqrt{x + 5} - 3}{x - 4}$, $\lim\limits_{x \to 4} f(x)$

28. $f(x) = \dfrac{x - 3}{x^2 - 4x + 3}$, $\lim\limits_{x \to 3} f(x)$

29. $f(x) = \dfrac{5}{x - 3}$, $\lim\limits_{x \to 3} f(x)$

30. $f(x) = \ln(x + 3)$, $\lim\limits_{x \to 4} f(x)$

31. *Think About It* If $f(2) = 4$, can you conclude anything about $\lim\limits_{x \to 2} f(x)$? Give reasons for your answer.

32. *Think About It* If $\lim\limits_{x \to 2} f(x) = 4$, can you conclude anything about $f(2)$? Give reasons for your answer.

47. If $\lim\limits_{x \to c} f(x) = 4$ and $\lim\limits_{x \to c} g(x) = 5$, find

(a) $\lim\limits_{x \to c} [-2g(x)]$.

(b) $\lim\limits_{x \to c} [f(x) + g(x)]$.

(c) $\lim\limits_{x \to c} \dfrac{f(x)}{g(x)}$.

(d) $\lim\limits_{x \to c} \sqrt{f(x)}$.

48. If $\lim\limits_{x \to c} f(x) = \frac{3}{2}$ and $\lim\limits_{x \to c} g(x) = -\frac{1}{2}$, find

(a) $\lim\limits_{x \to c} [f(x) + g(x)]^2$.

(b) $\lim\limits_{x \to c} [6f(x)g(x)]$.

(c) $\lim\limits_{x \to c} \dfrac{5g(x)}{4f(x)}$.

(d) $\lim\limits_{x \to c} \dfrac{1}{\sqrt{f(x)}}$.

In Exercises 33–44, find the limit by direct substitution.

33. $\lim\limits_{x \to 2} (x^2 + 3x - 4)$

34. $\lim\limits_{x \to 2} \sqrt[3]{10x + 7}$

35. $\lim\limits_{x \to 3} \dfrac{12}{x}$

36. $\lim\limits_{x \to -5} \dfrac{4}{x + 2}$

37. $\lim\limits_{x \to -1} \dfrac{x^2 - 1}{x}$

38. $\lim\limits_{x \to 8} \dfrac{\sqrt{x + 1}}{x - 4}$

39. $\lim\limits_{x \to 2} e^x$

40. $\lim\limits_{x \to e} \ln x$

Essay In Exercises 49 and 50, use a graphing utility to graph the function and estimate the limit (if it exists). What is the domain of the function? Can you detect a possible danger in determining the domain of the function solely by analyzing the graph generated by a graphing utility? Write a short paragraph about the importance of examining a function analytically as well as graphically.

49. $f(x) = \dfrac{x - 9}{\sqrt{x} - 3}$

$\lim\limits_{x \to 9} f(x)$

50. $f(x) = \dfrac{x - 3}{x^2 - 9}$

$\lim\limits_{x \to 3} f(x)$

12.2   Techniques for Evaluating Limits

Limits of Polynomial and Rational Functions / Cancellation Technique /
Rationalizing Technique / Using Technology / One-Sided Limits /
A Limit from Calculus

Limits of Polynomial and Rational Functions

In Section 12.1, you saw how direct substitution and operations with limits can be used to evaluate limits of certain well-behaved functions, such as polynomial functions and rational functions with nonzero denominators. This result is summarized as follows.

Limits of Polynomial and Rational Functions

1. If p is a polynomial function and c is a real number, then

$$\lim_{x \to c} p(x) = p(c).$$

2. If r is a rational function given by $r(x) = p(x)/q(x)$, and c is a real number such that $q(c) \neq 0$, then

$$\lim_{x \to c} r(x) = r(c) = \frac{p(c)}{q(c)}, \quad q(c) \neq 0.$$

The *Interactive* CD-ROM offers graphing utility emulators of the *TI-82* and *TI-83*, which can be used with the Examples, Explorations, Technology notes, and Exercises.

EXPLORATION

Use a graphing utility to graph the function $f(x) = \dfrac{x^2 + x - 6}{x + 3}$ in Example 1(b). Use the trace feature to approximate $\lim_{x \to -1} f(x)$. What do you think $\lim_{x \to -3} f(x)$ equals? Is f defined at $x = -3$? Does this affect the existence of the limit as x approaches -3?

EXAMPLE 1   Evaluating Limits by Direct Substitution

Find each of the following limits.

a. $\lim\limits_{x \to -1} (x^2 + x - 6)$

b. $\lim\limits_{x \to -1} \dfrac{x^2 + x - 6}{x + 3}$

Solution

The first function is a polynomial function and the second is a rational function (with a nonzero denominator at $x = -1$). Thus, you can evaluate the limits by direct substitution.

a. $\lim\limits_{x \to -1} (x^2 + x - 6) = (-1)^2 + (-1) - 6 = -6$

b. $\lim\limits_{x \to -1} \dfrac{x^2 + x - 6}{x + 3} = \dfrac{-6}{-1 + 3} = -3$

Cancellation Technique

In Example 1(b), suppose you were asked to find the limit as $x \to -3$.

$$\lim_{x \to -3} \frac{x^2 + x - 6}{x + 3}$$

Direct substitution would fail because -3 is a zero of the denominator. By using a table, however, it appears that the limit of the function as $x \to -3$ is -5.

x	-3.01	-3.001	-3.0001	-3	-2.9999	-2.999	-2.99
$\dfrac{x^2 + x - 6}{x + 3}$	-5.01	-5.001	-5.0001	?	-4.9999	-4.999	-4.99

Another way to find the limit of this function is to factor the numerator and cancel common factors, as shown in Example 2.

EXAMPLE 2   **Cancellation Technique**

Find the limit: $\lim\limits_{x \to -3} \dfrac{x^2 + x - 6}{x + 3}$.

Solution

Begin by factoring the numerator and canceling any common factors.

$$\lim_{x \to -3} \frac{x^2 + x - 6}{x + 3} = \lim_{x \to -3} \frac{(x - 2)(x + 3)}{x + 3} \qquad \text{Factor numerator.}$$

$$= \lim_{x \to -3} \frac{(x - 2)(x + 3)}{x + 3} \qquad \text{Cancel common factor.}$$

$$= \lim_{x \to -3} (x - 2) \qquad \text{Simplify.}$$

$$= -5 \qquad \text{Direct substitution}$$

This procedure for evaluating a limit is called the *cancellation technique*. The validity of the procedure stems from the fact that if two functions agree at all but a single number c, they must have identical limit behavior at $x = c$. In Example 2, the functions

$$f(x) = \frac{x^2 + x - 6}{x + 3} \qquad \text{and} \qquad g(x) = x - 2$$

agree at all values of x other than $x = -3$. Hence, you can use $g(x)$ to find the limit of $f(x)$.

EXPLORATION

Use a graphing utility to compare the graphs of

$$y = \frac{x^2 + x - 6}{x + 3}$$

and

$$y = x - 2.$$

Use the value feature to evaluate the first graph at $x = -3$. Press $\boxed{\triangle}$ to evaluate the second graph at $x = -3$. What can you conclude?

The cancellation technique should be applied only when direct substitution produces 0 in both the numerator *and* the denominator. The resulting fraction, $\frac{0}{0}$, has no meaning as a real number. It is called an **indeterminate form** because you cannot, from the form alone, determine the limit. When you encounter this form by direct substitution into a rational function, you can conclude that the numerator and denominator must have a common factor. After factoring and canceling, you should try direct substitution again.

EXAMPLE 3   Cancellation Technique

Find the following limit.

$$\lim_{x \to 1} \frac{x - 1}{x^3 - x^2 + x - 1}$$

Solution

Begin by substituting $x = 1$ into the numerator and denominator.

$$1 - 1 = 0 \qquad \text{Numerator is 0 when } x = 1.$$
$$(1)^3 - (1)^2 + 1 - 1 = 0 \qquad \text{Denominator is 0 when } x = 1.$$

Because both the numerator and denominator are zero when $x = 1$, direct substitution will not yield the limit. To find the limit, you should factor the numerator and denominator, cancel any common factors, and then try direct substitution again.

Figure 12.10

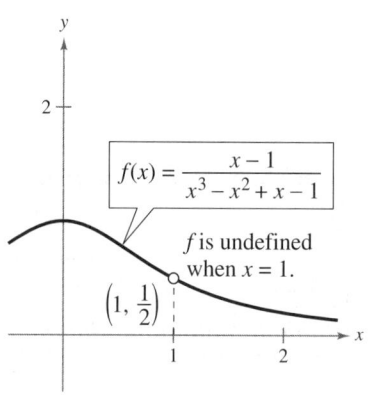

$$f(x) = \frac{x - 1}{x^3 - x^2 + x - 1}$$

f is undefined when $x = 1$.

$\left(1, \frac{1}{2}\right)$

$$\lim_{x \to 1} \frac{x - 1}{x^3 - x^2 + x - 1} = \lim_{x \to 1} \frac{x - 1}{(x - 1)(x^2 + 1)} \qquad \text{Factor denominator.}$$

$$= \lim_{x \to 1} \frac{\cancel{x - 1}}{\cancel{(x - 1)}(x^2 + 1)} \qquad \text{Cancel common factor.}$$

$$= \lim_{x \to 1} \frac{1}{x^2 + 1} \qquad \text{Simplify.}$$

$$= \frac{1}{(1)^2 + 1} \qquad \text{Substitute.}$$

$$= \frac{1}{2} \qquad \text{Simplify.}$$

This result is shown graphically in Figure 12.10.

Note   In Example 3, the factorization of the denominator can be obtained by dividing by $(x - 1)$ or by grouping as follows.

$$x^3 - x^2 + x - 1 = x^2(x - 1) + (x - 1)$$
$$= (x - 1)(x^2 + 1)$$

Rationalizing Technique

EXAMPLE 4   **Rationalizing Technique**

Find the following limit.

$$\lim_{x \to 0} \frac{\sqrt{x + 1} - 1}{x}$$

Solution

By direct substitution, you obtain the indeterminate form $\frac{0}{0}$.

$$\lim_{x \to 0} \frac{\sqrt{x + 1} - 1}{x} = \frac{0}{0} \qquad \text{Indeterminate form}$$

In this case, you can rewrite the fraction by rationalizing the numerator.

$$\frac{\sqrt{x + 1} - 1}{x} = \left(\frac{\sqrt{x + 1} - 1}{x} \right) \left(\frac{\sqrt{x + 1} + 1}{\sqrt{x + 1} + 1} \right)$$

$$= \frac{(x + 1) - 1}{x \left(\sqrt{x + 1} + 1 \right)} \qquad \text{Multiply.}$$

$$= \frac{x}{x \left(\sqrt{x + 1} + 1 \right)} \qquad \text{Simplify.}$$

$$= \frac{\cancel{x}}{\cancel{x} \left(\sqrt{x + 1} + 1 \right)} \qquad \text{Cancel common factor.}$$

$$= \frac{1}{\sqrt{x + 1} + 1}, \quad x \neq 0 \qquad \text{Simplify.}$$

Now you can evaluate the limit by direct substitution.

$$\lim_{x \to 0} \frac{\sqrt{x + 1} - 1}{x} = \lim_{x \to 0} \frac{1}{\sqrt{x + 1} + 1} = \frac{1}{1 + 1} = \frac{1}{2}$$

You can reinforce your conclusion that the limit is $\frac{1}{2}$ by constructing a table, or by sketching a graph, as shown in Figure 12.11.

x	-0.1	-0.01	-0.001	0	0.001	0.01	0.1
$f(x)$	0.5132	0.5013	0.5001	?	0.4999	0.4988	0.4881

Figure 12.11

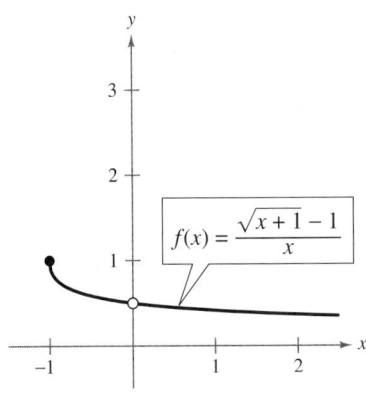

$$f(x) = \frac{\sqrt{x + 1} - 1}{x}$$

Note   The rationalizing technique for evaluating limits is based on multiplication by a convenient form of 1. In Example 4, the convenient form is

$$1 = \frac{\sqrt{x + 1} + 1}{\sqrt{x + 1} + 1}.$$

Using Technology

The cancellation and rationalizing techniques work well for finding the limit of a rational function or for finding the limit of a function involving a radical. To find limits of nonalgebraic functions, you often need to use more sophisticated analytic techniques, as shown in Examples 5 and 6.

EXAMPLE 5   Approximating a Limit

Approximate the limit $\lim\limits_{x \to 0} (1 + x)^{1/x}$.

Figure 12.12

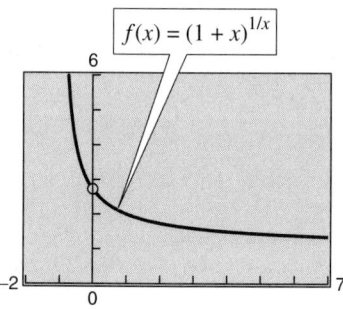

Solution

To approximate the limit graphically, enter the function $f(x) = (1 + x)^{1/x}$ in a graphing utility, as shown in Figure 12.12. Using the zoom and trace features of the graphing utility, choose two points on the graph of f, such as

$$(-0.001645, 2.7205) \quad \text{and} \quad (0.001645, 2.7160).$$

Because the x-coordinates of these two points are equidistant from 0, you can approximate the limit to be the average of the y-coordinates. That is,

$$\lim\limits_{x \to 0} (1 + x)^{1/x} \approx \frac{2.7205 + 2.7160}{2} = 2.71825.$$

Using analytic means, it can be shown that this approximation is accurate to four decimal places. The actual limit is $e \approx 2.71828$.

EXAMPLE 6   Approximating a Limit

Approximate the limit $\lim\limits_{x \to 0} \dfrac{\sin x}{x}$.

Figure 12.13

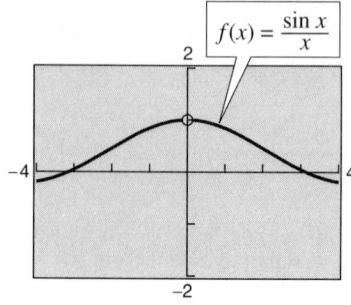

Solution

Notice that direct substitution does not work because it produces the indeterminate form $\frac{0}{0}$. To approximate the limit, use the procedure described in Example 5. Begin by sketching the graph of $f(x) = (\sin x)/x$, as shown in Figure 12.13. Then use the zoom and trace features of the graphing utility to choose a point on each side of 0, such as

$$(-0.001234, 0.9999998) \quad \text{and} \quad (0.001234, 0.9999998).$$

Finally, use interpolation to approximate the limit as the average of the y-coordinates of these two points.

$$\lim\limits_{x \to 0} \frac{\sin x}{x} \approx 0.9999998$$

Using analytic means, it can be shown that this limit is exactly 1.

One-Sided Limits

In Section 12.1, you saw that one way in which a limit can fail to exist is when a function approaches a different value from the left side of c than it approaches from the right side of c. This type of behavior can be described more concisely with the concept of a **one-sided limit.**

$$\lim_{x \to c^-} f(x) = L \qquad\qquad \text{Limit from the left}$$

$$\lim_{x \to c^+} f(x) = L \qquad\qquad \text{Limit from the right}$$

EXAMPLE 7 **Evaluating One-Sided Limits**

Find the limit as $x \to 0$ from the left and the limit as $x \to 0$ from the right for the function given by

$$f(x) = \frac{|2x|}{x}.$$

Figure 12.14

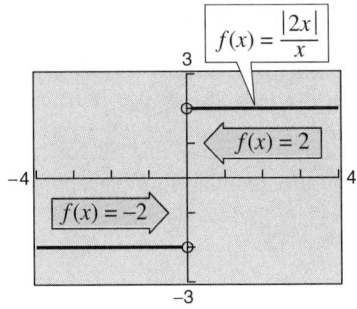

Solution
From the graph of f, shown in Figure 12.14, you can see that $f(x) = -2$ for all $x < 0$. Therefore, the limit from the left is

$$\lim_{x \to 0^-} \frac{|2x|}{x} = -2. \qquad\qquad \text{Limit from the left}$$

Because $f(x) = 2$ for all $x > 0$, the limit from the right is

$$\lim_{x \to 0^+} \frac{|2x|}{x} = 2. \qquad\qquad \text{Limit from the right}$$

In Example 7, note that the function approaches different limits from the left and from the right. In such cases, the limit of $f(x)$ as $x \to c$ does not exist. For the limit of a function to exist as $x \to c$, it must be true that both one-sided limits exist and are equal.

Existence of a Limit

If f is a function and c and L are real numbers, then

$$\lim_{x \to c} f(x) = L$$

if and only if both the left and right limits are equal to L.

EXAMPLE 8 ▭   **Evaluating One-Sided Limits**

Find the limit of $f(x)$ as x approaches 1.

$$f(x) = \begin{cases} 4 - x, & x < 1 \\ 4x - x^2, & x > 1 \end{cases}$$

Solution

Remember that you are concerned about the value of f near $x = 1$ rather than at $x = 1$. Thus, for $x < 1$, $f(x)$ is given by $4 - x$, and you can use direct substitution to obtain

$$\lim_{x \to 1^-} f(x) = \lim_{x \to 1^-} (4 - x) = 4 - 1 = 3.$$

For $x > 1$, $f(x)$ is given by $4x - x^2$, and you can use direct substitution to obtain

$$\lim_{x \to 1^+} f(x) = \lim_{x \to 1^+} (4x - x^2) = 4 - 1 = 3.$$

Because the one-sided limits both exist and are equal to 3, it follows that

$$\lim_{x \to 1} f(x) = 3.$$

The graph in Figure 12.15 confirms this conclusion.   ▭

Figure 12.15

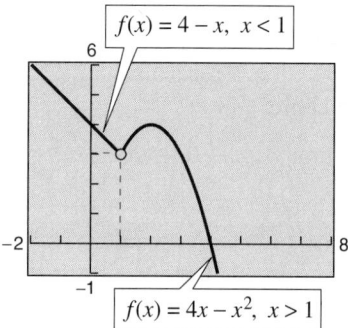

$f(x) = 4 - x, \ x < 1$

$f(x) = 4x - x^2, \ x > 1$

Real Life

EXAMPLE 9 ▭   **Comparing Limits from the Left and Right**

An overnight delivery service charges \$8 for the first pound and \$2 for each additional pound. Let x represent the weight of a parcel and let $f(x)$ represent the shipping cost. Show that the limit of $f(x)$ as $x \to 2$ does not exist.

$$f(x) = \begin{cases} 8, & 0 < x \le 1 \\ 10, & 1 < x \le 2 \\ 12, & 2 < x \le 3 \end{cases}$$

Figure 12.16

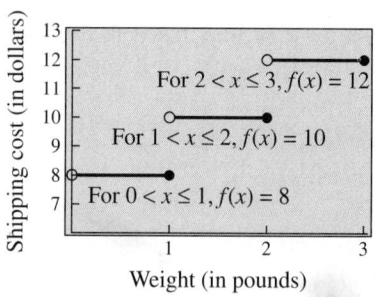

For $2 < x \le 3$, $f(x) = 12$

For $1 < x \le 2$, $f(x) = 10$

For $0 < x \le 1$, $f(x) = 8$

Weight (in pounds)

Solution

The graph of f is shown in Figure 12.16. The limit of $f(x)$ as x approaches 2 from the left is

$$\lim_{x \to 2^-} f(x) = 10$$

whereas the limit of $f(x)$ as x approaches 2 from the right is

$$\lim_{x \to 2^+} f(x) = 12.$$

Because these one-sided limits are not equal, the limit of $f(x)$ as $x \to 2$ does not exist.   ▭

A Limit from Calculus

In the next section, you will study an important type of limit from calculus—the limit of a *difference quotient*.

Note The limit of a difference quotient is an expression of the form

$$\lim_{h \to 0} \frac{f(x + h) - f(x)}{h}.$$

Direct substitution into the difference quotient always produces the indeterminate form $\frac{0}{0}$.

EXAMPLE 10 ▱ **Evaluating a Limit from Calculus**

For the function $f(x) = x^2 - 1$, find $\lim\limits_{h \to 0} \dfrac{f(3 + h) - f(3)}{h}$.

Solution
Direct substitution produces an indeterminant form.

$$\lim_{h \to 0} \frac{f(3 + h) - f(3)}{h} = \lim_{h \to 0} \frac{[(3 + h)^2 - 1] - [(3)^2 - 1]}{h}$$

$$= \lim_{h \to 0} \frac{9 + 6h + h^2 - 1 - 9 + 1}{h}$$

$$= \lim_{h \to 0} \frac{6h + h^2}{h}$$

$$= \frac{0}{0}$$

By factoring and canceling, you can obtain the following.

$$\lim_{h \to 0} \frac{f(3 + h) - f(3)}{h} = \lim_{h \to 0} \frac{h(6 + h)}{h} = \lim_{h \to 0} (6 + h) = 6$$

Thus, the limit is 6. ▱

Group Activity

Limits of Rational Functions

Consider the limit of the rational function $p(x)/q(x)$. What conclusion can you make if direct substitution produces the given expression? Explain your reasoning to a group member.

a. $\lim\limits_{x \to c} \dfrac{p(x)}{q(x)} = \dfrac{0}{1}$ **b.** $\lim\limits_{x \to c} \dfrac{p(x)}{q(x)} = \dfrac{1}{1}$

c. $\lim\limits_{x \to c} \dfrac{p(x)}{q(x)} = \dfrac{1}{0}$ **d.** $\lim\limits_{x \to c} \dfrac{p(x)}{q(x)} = \dfrac{0}{0}$

12.2 /// EXERCISES

In Exercises 1–8, evaluate the limit by direct substitution.

1. $\lim\limits_{x\to5} (12 - x^2)$

2. $\lim\limits_{x\to-2} \left(\frac{1}{4}x^3 - 3x\right)$

3. $\lim\limits_{x\to-3} \dfrac{2x}{x^2 + 1}$

4. $\lim\limits_{x\to4} \dfrac{x - 1}{x^2 + 2x + 3}$

5. $\lim\limits_{x\to-3} \ln e^x$

6. $\lim\limits_{x\to-1} e^{x^2 - 1}$

7. $\lim\limits_{\theta\to1} \sin\dfrac{2\pi\theta}{3}$

8. $\lim\limits_{t\to\pi/2} \cot t$

In Exercises 9–12, use the graph to determine the limit visually (if it exists). Then identify another function that agrees with the given function at all but one point.

9. $g(x) = \dfrac{-2x^2 + x}{x}$

10. $h(x) = \dfrac{x^2 - 3x}{x}$

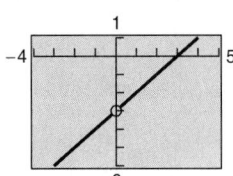

(a) $\lim\limits_{x\to0} g(x)$

(b) $\lim\limits_{x\to-1} g(x)$

(c) $\lim\limits_{x\to-2} g(x)$

(a) $\lim\limits_{x\to-2} h(x)$

(b) $\lim\limits_{x\to0} h(x)$

(c) $\lim\limits_{x\to3} h(x)$

11. $g(x) = \dfrac{x^3 - x}{x - 1}$

12. $f(x) = \dfrac{x^2 - 1}{x + 1}$

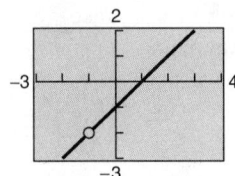

(a) $\lim\limits_{x\to1} g(x)$

(b) $\lim\limits_{x\to-1} g(x)$

(c) $\lim\limits_{x\to0} g(x)$

(a) $\lim\limits_{x\to1} f(x)$

(b) $\lim\limits_{x\to2} f(x)$

(c) $\lim\limits_{x\to-1} f(x)$

In Exercises 13–24, find the limit (if it exists). Use a graphing utility to graph the function to verify your result.

13. $\lim\limits_{x\to7} \dfrac{x - 7}{x^2 - 49}$

14. $\lim\limits_{x\to3} \dfrac{3 - x}{x^2 - 9}$

15. $\lim\limits_{x\to-1} \dfrac{1 - 2x - 3x^2}{1 + x}$

16. $\lim\limits_{t\to-2} \dfrac{t^3 + 8}{t + 2}$

17. $\lim\limits_{y\to0} \dfrac{\sqrt{3 + y} - \sqrt{3}}{y}$

18. $\lim\limits_{z\to0} \dfrac{\sqrt{6 - z} - \sqrt{6}}{z}$

19. $\lim\limits_{x\to-3} \dfrac{\sqrt{x + 7} - 2}{x + 3}$

20. $\lim\limits_{x\to2} \dfrac{4 - \sqrt{18 - x}}{x - 2}$

21. $\lim\limits_{x\to0} \dfrac{\dfrac{1}{x + 1} - 1}{x}$

22. $\lim\limits_{x\to0} \dfrac{\dfrac{1}{5 - x} - \dfrac{1}{5}}{x}$

23. $\lim\limits_{x\to0} \dfrac{\sec x}{\tan x}$

24. $\lim\limits_{x\to\pi/2} \dfrac{1 - \sin x}{\cos x}$

Graphical, Numerical, and Analytical Analysis In Exercises 25–28, use a graphing utility to graph the function and approximate the limit. Construct a table to reinforce your conclusion. Then find the limit by analytic methods.

25. $\lim\limits_{x\to0} \dfrac{\sqrt{1 - x} - 1}{x}$

26. $\lim\limits_{x\to2} \dfrac{\sqrt{x + 14} - 4}{x - 2}$

27. $\lim\limits_{x\to0} \dfrac{\dfrac{1}{3 + x} - \dfrac{1}{3}}{x}$

28. $\lim\limits_{x\to0} \dfrac{\dfrac{1}{2x - 5} + \dfrac{1}{5}}{x}$

In Exercises 29–32, use a graphing utility to graph the function and approximate the limit.

29. $\lim\limits_{x\to0} \dfrac{\sqrt{x + 2} - \sqrt{2}}{x}$

30. $\lim\limits_{x\to1} \dfrac{4 - \sqrt{x}}{x - 16}$

31. $\lim\limits_{x\to2} \dfrac{x^5 - 32}{x - 2}$

32. $\lim\limits_{x\to0} \dfrac{\dfrac{1}{2 + x} - \dfrac{1}{2}}{x}$

In Exercises 33–38, use a graphing utility to graph the function. Determine the limit (if it exists) by evaluating the corresponding one-sided limits.

33. $\lim\limits_{x\to 3} \dfrac{|x-3|}{x-3}$

34. $\lim\limits_{x\to 3} |x-3|$

35. $\lim\limits_{x\to 1} \dfrac{1}{x^2+1}$

36. $\lim\limits_{x\to 1} \dfrac{1}{x^2-1}$

37. $\lim\limits_{x\to 2} f(x)$

$$f(x) = \begin{cases} x-1, & x \le 2 \\ 2x-3, & x > 2 \end{cases}$$

38. $\lim\limits_{x\to 1} f(x)$

$$f(x) = \begin{cases} 4-x^2, & x \le 1 \\ 3-x, & x > 1 \end{cases}$$

In Exercises 39–42, find $\lim\limits_{h\to 0} \dfrac{f(x+h)-f(x)}{h}$.

39. $f(x) = 3x - 1$

40. $f(x) = \sqrt{x}$

41. $f(x) = x^2 - 3x$

42. $f(x) = 4 - 2x - x^2$

In Exercises 43–50, use a graphing utility to graph the function and approximate the limit. Write an approximation that you believe is accurate to three decimal places.

43. $\lim\limits_{x\to 0} (x \ln x)$

44. $\lim\limits_{x\to 0} (x^2 \ln x)$

45. $\lim\limits_{x\to 0} \dfrac{\sin 2x}{x}$

46. $\lim\limits_{x\to 0} \dfrac{\sin 3x}{x}$

47. $\lim\limits_{x\to 0} \dfrac{\tan x}{x}$

48. $\lim\limits_{x\to 0} \dfrac{1-\cos 2x}{x}$

49. $\lim\limits_{x\to 1} \dfrac{1-\sqrt[3]{x}}{1-x}$

50. $\lim\limits_{x\to 0} (1+2x)^{1/x}$

In Exercises 51–56, use a graphing utility to graph the function and the equations $y = x$ and $y = -x$ in the same viewing rectangle. Use the graph to find $\lim\limits_{x\to 0} f(x)$ visually.

51. $f(x) = x \cos x$

52. $f(x) = |x \sin x|$

53. $f(x) = |x| \sin x$

54. $f(x) = |x| \cos x$

55. $f(x) = x \sin \dfrac{1}{x}$

56. $h(x) = x \cos \dfrac{1}{x}$

In Exercises 57 and 58, state which limit can be evaluated using direct substitution. Then evaluate or approximate each limit.

57. (a) $\lim\limits_{x\to 0} x^2 \sin x^2$

(b) $\lim\limits_{x\to 0} \dfrac{\sin x^2}{x^2}$

58. (a) $\lim\limits_{x\to 0} \dfrac{x}{\cos x}$

(b) $\lim\limits_{x\to 0} \dfrac{1-\cos x}{x}$

In Exercises 59 and 60, use the position function $s(t) = -16t^2 + 128$, which gives the height (in feet) of a free-falling object. The velocity at time $t = a$ seconds is given by

$$\lim\limits_{t\to a} \dfrac{s(a)-s(t)}{a-t}.$$

59. Find the velocity when $t = 1$ second.

60. Find the velocity when $t = 2$ seconds.

12.3   The Tangent Line Problem

*Tangent Line to a Graph / Slope of a Graph / Slope and the Limit Process /
The Derivative of a Function*

Tangent Line to a Graph

Calculus is a branch of mathematics that studies rates of change of functions. If you go on to take a course in calculus, you will learn that rates of change have many applications in real life.

Earlier in the text, you learned how the slope of a line indicates the rate at which a line rises or falls. For a line, this rate (or slope) is the same at every point on the line. For graphs other than lines, the rate at which the graph rises or falls changes from point to point. For instance, in Figure 12.17, the parabola is rising more quickly at the point (x_1, y_1) than it is at the point (x_2, y_2). At the vertex (x_3, y_3), the graph levels off, and at the point (x_4, y_4), the graph is falling.

To determine the rate at which a graph rises or falls at a *single point*, you can find the slope of the tangent line at that point. In simple terms, the **tangent line** to a graph of a function f at a point $P(x_1, y_1)$ is the line that best approximates the graph at the point. Figure 12.18 shows other examples of tangent lines.

Figure 12.17

Figure 12.18

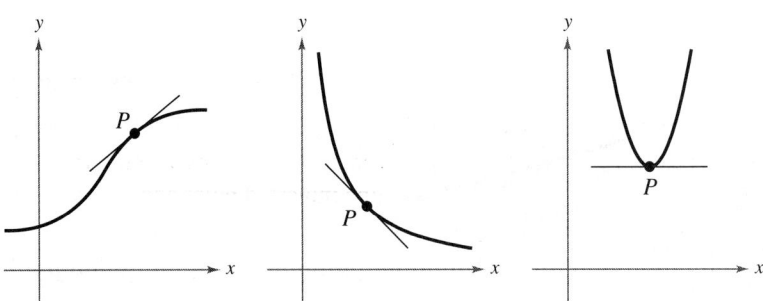

From geometry, you know that a line is tangent to a circle if the line intersects the circle at only one point. Tangent lines to noncircular graphs can intersect the graph at more than one point. For instance, in the first graph in Figure 12.18, if the tangent line were extended, it would intersect the graph at a point other than the point of tangency.

Slope of a Graph

Because a tangent line approximates the graph at a point, the problem of finding the slope of a graph at a point becomes one of finding the slope of the tangent line at the point.

EXAMPLE 1 📦 **Visually Approximating the Slope of a Graph**

Use the graph in Figure 12.19 to approximate the slope of the graph of $f(x) = x^2$ at the point $(1, 1)$.

Figure 12.19

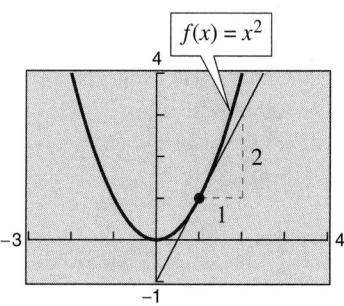

Solution
From the graph of $f(x) = x^2$, you can see that the tangent line at $(1, 1)$ rises approximately two units for each unit change in x. Thus, the slope of the tangent line at $(1, 1)$ is given by

$$\text{Slope} = \frac{\text{change in } y}{\text{change in } x} \approx \frac{2}{1} = 2.$$

Because the tangent line at the point $(1, 1)$ has a slope of about 2, you can conclude that the graph has a slope of about 2 at the point $(1, 1)$. 📦

When you are visually approximating the slope of a graph, remember that the scales on the horizontal and vertical axes may differ. When this happens (as it frequently does in applications), the slope of the tangent line is distorted, and you must be careful to account for the difference in scales.

Real Life

EXAMPLE 2 📦 **Approximating the Slope of a Graph**

Figure 12.20 graphically depicts the average daily temperature (in degrees Fahrenheit) for each month in Dallas, Texas. Estimate the slope of this graph at the indicated point and give a physical interpretation of the result.

Figure 12.20

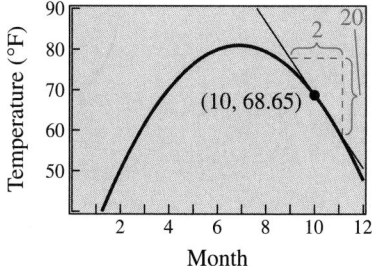

Month

Solution
From the graph, you can see that the tangent line at the given point falls approximately 20 units for each two-unit change in x. Thus, you can estimate the slope at the given point to be

$$\text{Slope} = \frac{\text{change in } y}{\text{change in } x} \approx \frac{-20}{2} = -10 \text{ degrees per month.}$$

This means that you can expect the average daily temperature in November to be about 10 degrees lower than the corresponding temperature in October. 📦

Slope and the Limit Process

Figure 12.21

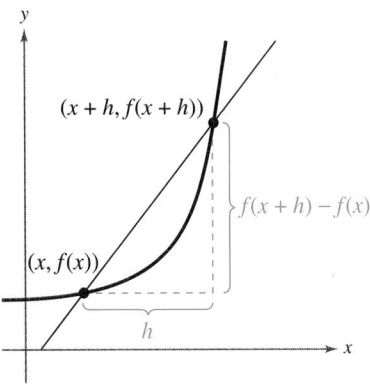

In Examples 1 and 2, you approximated the slope of a graph at a point by making a careful graph and then "eyeballing" the tangent line at the point of tangency. A more precise method of approximating tangent lines makes use of a **secant line** through the point of tangency and a second point on the graph, as shown in Figure 12.21. If $(x, f(x))$ is the point of tangency and $(x + h, f(x + h))$ is a second point on the graph of f, the slope of the secant line through the two points is

$$m_{\text{sec}} = \frac{f(x + h) - f(x)}{h}. \qquad \text{Slope of secant line}$$

The right side of this equation is called the **difference quotient.** The denominator h is the **change in x,** and the numerator is the **change in y.** The beauty of this procedure is that you obtain better and better approximations of the slope of the tangent line by choosing the second point closer and closer to the point of tangency, as shown in Figure 12.22.

 A computer animation of this concept appears in the *Interactive* **CD-ROM.**

Figure 12.22

As h approaches 0, the secant line approaches the tangent line.

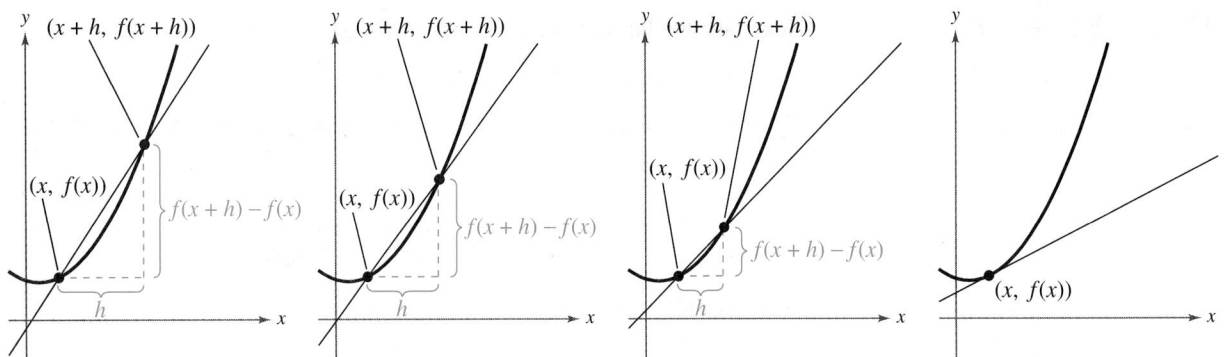

Using the limit process, you can find the *exact* slope of the tangent line at $(x, f(x))$.

The **slope** m of the graph of f at the point $(x, f(x))$ is equal to the slope of its tangent line at $(x, f(x))$, and is given by

$$m = \lim_{h \to 0} m_{\text{sec}} = \lim_{h \to 0} \frac{f(x + h) - f(x)}{h}$$

provided this limit exists.

EXAMPLE 3 ▱ **Finding the Slope of a Graph**

Find the slope of the graph of $f(x) = x^2$ at the point $(-2, 4)$.

Solution

Begin by finding an expression that represents the slope of a secant line at the point $(-2, 4)$.

$$m_{\text{sec}} = \frac{f(-2 + h) - f(-2)}{h} \qquad \text{Set up difference quotient.}$$

$$= \frac{(-2 + h)^2 - (-2)^2}{h} \qquad \text{Use } f(x) = x^2.$$

$$= \frac{4 - 4h + h^2 - 4}{h} \qquad \text{Expand terms.}$$

$$= \frac{-4h + h^2}{h} \qquad \text{Simplify.}$$

$$= \frac{\cancel{h}(-4 + h)}{\cancel{h}} \qquad \text{Factor and cancel.}$$

$$= -4 + h, \quad h \neq 0 \qquad \text{Simplify.}$$

Next, take the limit of m_{sec} as h approaches 0.

$$m = \lim_{h \to 0} m_{\text{sec}} = \lim_{h \to 0} (-4 + h) = -4$$

The graph has a slope of -4 at the point $(-2, 4)$, as shown in Figure 12.23.

▱

Figure 12.23

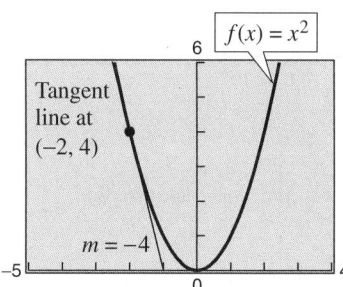

EXAMPLE 4 ▱ **Finding the Slope of a Graph**

Find the slope of $f(x) = -2x + 4$.

Solution

You know from your study of linear functions that the line given by $f(x) = -2x + 4$ has a slope of -2, as shown in Figure 12.24. This conclusion is consistent with that obtained by the limit definition of slope.

$$m = \lim_{h \to 0} \frac{f(x + h) - f(x)}{h}$$

$$= \lim_{h \to 0} \frac{[-2(x + h) + 4] - (-2x + 4)}{h}$$

$$= \lim_{h \to 0} \frac{-2x - 2h + 4 + 2x - 4}{h}$$

$$= \lim_{h \to 0} \frac{-2\cancel{h}}{\cancel{h}}$$

$$= -2, \quad h \neq 0$$

▱

Figure 12.24

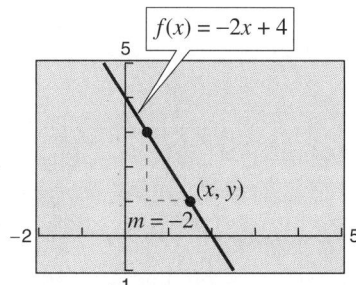

It is important that you see the difference between the ways the difference quotients were set up in Examples 3 and 4. In Example 3, you were finding the slope of a graph at a specific point $(c, f(c))$. To find the slope, you can use the following form of the difference quotient.

$$m = \lim_{h \to 0} \frac{f(c + h) - f(c)}{h} \qquad \text{Slope at specific point}$$

In Example 4, however, you were finding a formula for the slope at *any* point on the graph. In such cases, you should use x, rather than c, in the difference quotient.

$$m = \lim_{h \to 0} \frac{f(x + h) - f(x)}{h} \qquad \text{Formula for slope}$$

Except for linear functions, this form will always produce a function of x, which can then be evaluated to find the slope at any desired point.

Try verifying the result in Example 5 by graphing the three functions

$$y_1 = x^2 + 1$$
$$y_2 = -2x$$
$$y_3 = 4x - 3$$

in the same viewing rectangle. Some graphing utilities even have the capability of automatically graphing the tangent line to a curve at a given point. If you have such a graphing utility, try verifying Example 5.

Figure 12.25

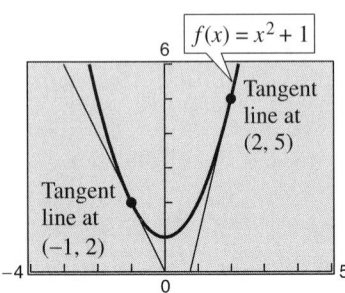

EXAMPLE 5   ◻   **Finding a Formula for the Slope of a Graph**

Find a formula for the slope of the graph of $f(x) = x^2 + 1$. What is the slope at the points $(-1, 2)$ and $(2, 5)$?

Solution

$$m_{\sec} = \frac{f(x + h) - f(x)}{h} \qquad \text{Set up difference quotient.}$$

$$= \frac{[(x + h)^2 + 1] - (x^2 + 1)}{h} \qquad \text{Use } f(x) = x^2 + 1.$$

$$= \frac{x^2 + 2xh + h^2 + 1 - x^2 - 1}{h} \qquad \text{Expand terms.}$$

$$= \frac{2xh + h^2}{h} \qquad \text{Simplify.}$$

$$= \frac{h(2x + h)}{h} \qquad \text{Factor and cancel.}$$

$$= 2x + h, \quad h \neq 0 \qquad \text{Simplify.}$$

Next, take the limit of $m_{\sec}$ as h approaches 0.

$$m = \lim_{h \to 0} m_{\sec}$$
$$= \lim_{h \to 0} (2x + h)$$
$$= 2x$$

Using the formula $m = 2x$ for the slope at $(x, f(x))$, you can find the slope at the specified points. At $(-1, 2)$, the slope is $m = 2(-1) = -2$, and at $(2, 5)$, the slope is $m = 2(2) = 4$. The graph of f is shown in Figure 12.25.   ◻

The Derivative of a Function

In Example 5, you started with the function $f(x) = x^2 + 1$, and used the limit process to derive another function, $m = 2x$, that represents the slope of the graph of f at the point $(x, f(x))$. This derived function is called the **derivative** of f at x. It is denoted by $f'(x)$, which is read as "f prime of x."

Definition of the Derivative

The **derivative** of f at x is given by

$$f'(x) = \lim_{h \to 0} \frac{f(x + h) - f(x)}{h}$$

provided this limit exists.

Remember that the derivative $f'(x)$ is a formula for the slope of the tangent line to the graph of f at the point $(x, f(x))$.

Augustin Louis Cauchy (1789–1857), a leading French mathematician, developed many of the basic concepts of limit and continuity. He defined the derivative.

EXAMPLE 6 ◻ **Finding a Derivative**

Find the derivative of $f(x) = 3x^2 - 2x$.

Solution

$$f'(x) = \lim_{h \to 0} \frac{f(x + h) - f(x)}{h}$$

$$= \lim_{h \to 0} \frac{[3(x + h)^2 - 2(x + h)] - (3x^2 - 2x)}{h}$$

$$= \lim_{h \to 0} \frac{3x^2 + 6xh + 3h^2 - 2x - 2h - 3x^2 + 2x}{h}$$

$$= \lim_{h \to 0} \frac{6xh + 3h^2 - 2h}{h}$$

$$= \lim_{h \to 0} \frac{h(6x + 3h - 2)}{h}$$

$$= \lim_{h \to 0} (6x + 3h - 2)$$

$$= 6x - 2$$

Thus, the derivative of $f(x) = 3x^2 - 2x$ is

$$f'(x) = 6x - 2. \quad ◻$$

EXPLORATION

Use a graphing utility to graph the function $f(x) = 3x^2 - 2x$. Use the trace feature to approximate the coordinates of the vertex of this parabola. Then use the derivative of $f(x) = 3x^2 - 2x$ to find the slope of the tangent line at the vertex. Make a conjecture about the slope of the tangent line at the vertex of an arbitrary parabola.

EXAMPLE 7 ▱ **Using the Derivative**

Find $f'(x)$ for $f(x) = \sqrt{x}$. Then find the slope of the graph of f at the points $(1, 1)$ and $(4, 2)$.

Solution

Use the procedure for rationalizing numerators, as discussed in Section 12.2.

Figure 12.26

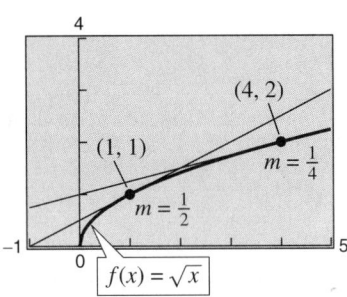

$f(x) = \sqrt{x}$

$$f'(x) = \lim_{h \to 0} \frac{f(x + h) - f(x)}{h}$$

$$= \lim_{h \to 0} \frac{\sqrt{x + h} - \sqrt{x}}{h}$$

$$= \lim_{h \to 0} \left(\frac{\sqrt{x + h} - \sqrt{x}}{h} \right) \left(\frac{\sqrt{x + h} + \sqrt{x}}{\sqrt{x + h} + \sqrt{x}} \right)$$

$$= \lim_{h \to 0} \frac{(x + h) - x}{h\left(\sqrt{x + h} + \sqrt{x}\right)}$$

$$= \lim_{h \to 0} \frac{h}{h\left(\sqrt{x + h} + \sqrt{x}\right)}$$

$$= \lim_{h \to 0} \frac{1}{\sqrt{x + h} + \sqrt{x}}$$

$$= \frac{1}{2\sqrt{x}}$$

At the point $(1, 1)$, the slope is $f'(1) = \frac{1}{2}$. At the point $(4, 2)$, the slope is $f'(4) = \frac{1}{4}$. The graph of f is shown in Figure 12.26. ▱

Group Activity

Using a Derivative to Find Slope

In many applications, it is convenient to use a variable other than x as the independent variable. Complete the following limit process to find the derivative of $f(t) = 3/t$. Then use the result to find the slope of the graph of $f(t) = 3/t$ at the point $(3, 1)$.

$$f'(t) = \lim_{h \to 0} \frac{f(t + h) - f(t)}{h} = \lim_{h \to 0} \frac{\dfrac{3}{t + h} - \dfrac{3}{t}}{h} = \cdots$$

12.3 /// EXERCISES

In Exercises 1–4, estimate the slope of the curve at the point (x, y).

1.

2.

3.

4.

In Exercises 5 and 6, use the graph shown in the figure.

5. Identify or sketch each of the following quantities.

(a) $f(1)$ and $f(3)$

(b) $f(3) - f(1)$

(c) $y = \dfrac{f(3) - f(1)}{3 - 1}(x - 1) + f(1)$

6. Insert the proper inequality symbol ($<$ or $>$) between the expressions.

(a) $\dfrac{f(3) - f(1)}{3 - 1}$ ▮ $\dfrac{f(3) - f(2)}{3 - 2}$

(b) $\dfrac{f(3) - f(1)}{3 - 1}$ ▮ $f'(1)$

Figure for 5 and 6

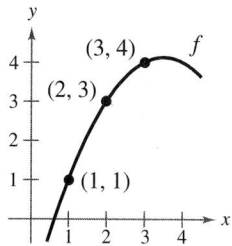

In Exercises 7–10, use a graphing utility to graph the function and the tangent line at the point $(1, f(1))$. Use the graph to estimate the slope of the tangent line.

7. $f(x) = x^2 - 2$

8. $f(x) = x^2 - 2x + 1$

9. $f(x) = \sqrt{2 - x}$

10. $f(x) = \dfrac{3}{2 - x}$

11. *Federal Debt*   The table gives the per capita federal debt for the United States for several years. (Source: U.S. Treasury Department)

Year	1950	1960	1970
Per Capita Debt	$1688	$1572	$1807

Year	1980	1990	1994
Per Capita Debt	$3981	$12,848	$15,750

(a) Use the regression capabilities of a graphing utility to fit a third-degree polynomial to the data. Let t be the time in years, with $t = 0$ corresponding to 1950.

(b) Use a graphing utility to graph the model found in part (a). Estimate the slope of the graph when $t = 30$ and give an interpretation of the result.

(c) Use a graphing utility to graph the tangent line to the model when $t = 30$. Compare the slope given by the graphing utility with the estimate in part (b).

12. *Market Research*   The data in the table gives the number N (in thousands) of books sold when the price is p (in dollars).

p	$10	$15	$20	$25	$30	$35
N	900	630	396	227	102	36

(a) Use the regression capabilities of a graphing utility to fit a second-degree polynomial to the data.

(b) Use a graphing utility to graph the model found in part (a). Estimate the slope of the graph when $p = \$15$ and $p = \$30$.

(c) Use a graphing utility to graph the tangent line to the model when $p = \$15$ and $p = \$30$. Compare the slopes given by the graphing utility with your estimates in part (b).

(d) The slopes of the tangent lines at $p = \$15$ and $p = \$30$ are not the same. Explain what this means to the company selling the books.

In Exercises 13–20, use the limit process to find the slope of the graph of the function at the specified point. Use a graphing utility to confirm your result.

13. $g(x) = 4 - 3x, (1, 1)$ **14.** $h(x) = 2x + 5, (-1, 3)$

15. $g(x) = x^2 - 2x, (3, 3)$

16. $f(x) = 10x - 2x^2, (3, 12)$

17. $g(x) = \dfrac{4}{x}, (2, 2)$ **18.** $g(x) = \dfrac{1}{x - 2}, \left(4, \dfrac{1}{2}\right)$

19. $h(x) = \sqrt{x}, (9, 3)$ **20.** $h(x) = \sqrt{x + 10}, (-1, 3)$

In Exercises 21–24, find a formula for the slope of the graph. Then use it to find the slope at the two points.

21. $g(x) = 4 - x^2$ **22.** $g(x) = x^3$
     (a) $(0, 4)$ (b) $(-1, 3)$ (a) $(1, 1)$ (b) $(-2, -8)$

23. $g(x) = \dfrac{1}{x + 4}$ **24.** $g(x) = \sqrt{x - 1}$
     (a) $\left(0, \dfrac{1}{4}\right)$ (b) $\left(-2, \dfrac{1}{2}\right)$ (a) $(5, 2)$ (b) $(10, 3)$

In Exercises 25–30, find the derivative of the function.

25. $f(x) = 5$ **26.** $f(x) = -4x + 2$

27. $g(x) = 6 - \dfrac{2}{3}x$ **28.** $f(x) = x^2 - 2x + 3$

29. $f(x) = \dfrac{1}{x^2}$ **30.** $h(s) = \dfrac{1}{\sqrt{s + 1}}$

In Exercises 31–34, find the slope of the graph of f at the given point. Use the result to find an equation of the tangent line to the graph at the point. Use a graphing utility to graph the function and the tangent line.

31. $f(x) = x^2 - 1, (2, 3)$ **32.** $f(x) = x^3 - x, (2, 6)$

33. $f(x) = \sqrt{x + 1}, (3, 2)$ **34.** $f(x) = 2x + \dfrac{4}{x}, (2, 6)$

In Exercises 35–38, use a graphing utility to graph f over the interval $[-2, 2]$ and complete the table. Compare the value of the first derivative with a visual estimate of the slope of the graph.

x	-2	-1.5	-1	-0.5	0	0.5	1	1.5	2
$f(x)$									
$f'(x)$									

35. $f(x) = \dfrac{1}{2}x^2$ **36.** $f(x) = \sqrt{x + 3}$

37. $f(x) = \dfrac{1}{4}x^3$ **38.** $f(x) = \dfrac{x^2 - 4}{x + 4}$

In Exercises 39–42, find the derivative of f. Use the derivative to determine any points on the graph of f where the tangent line is horizontal. Use a graphing utility to verify your results.

39. $f(x) = x^2 - 4x + 3$ **40.** $f(x) = x^3 + 3x$

41. $f(x) = 3x^3 - 9x$ **42.** $f(x) = 3x^4 + 4x^3$

In Exercises 43–46, match the function with the graph of its *derivative*. It is not necessary to find the derivative of the function. [The graphs are labeled (a), (b), (c), and (d).]

(a)

(b)

(c)

(d)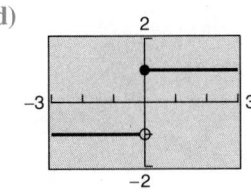

43. $f(x) = \sqrt{x}$ **44.** $f(x) = \dfrac{1}{x}$

45. $f(x) = |x|$ **46.** $f(x) = x^3$

12.4 Limits at Infinity and Limits of Sequences

Limits at Infinity and Horizontal Asymptotes / *Limits of Sequences*

Limits at Infinity and Horizontal Asymptotes

As pointed out at the beginning of this chapter, there are two basic problems in calculus: finding tangent lines and finding the area of a region. In Section 12.3, you saw how limits can be used to solve the tangent line problem. In this section and the next, you will see how a different type of limit, a *limit at infinity*, can be used to solve the area problem. To get an idea of what is meant by a limit at infinity, consider the function

$$f(x) = \frac{x+1}{2x}.$$

Figure 12.27

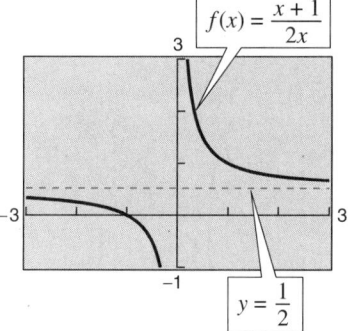

The graph of *f* is shown in Figure 12.27. From earlier work, you know that

$$y = \frac{1}{2} \qquad \text{Horizontal asymptote}$$

is a horizontal asymptote of the graph of this function. Using limit notation, this can be written as follows.

$$\lim_{x \to -\infty} f(x) = \frac{1}{2} \qquad \text{Horizontal asymptote to the left}$$

$$\lim_{x \to \infty} f(x) = \frac{1}{2} \qquad \text{Horizontal asymptote to the right}$$

These limits mean that the value of $f(x)$ gets arbitrarily close to $\frac{1}{2}$ as *x* decreases or increases without bound.

Definition of Limits at Infinity

If *f* is a function and L_1 and L_2 are real numbers, the statements

$$\lim_{x \to -\infty} f(x) = L_1 \qquad \text{Limit as } x \text{ approaches } -\infty$$

and

$$\lim_{x \to \infty} f(x) = L_2 \qquad \text{Limit as } x \text{ approaches } \infty$$

denote the **limits at infinity.** The first is read as *the limit of $f(x)$ as x approaches* $-\infty$ *is* L_1, and the second is read as *the limit of $f(x)$ as x approaches* ∞ *is* L_2.

To help evaluate limits at infinity, you can use the following.

Limits at Infinity

If r is a positive real number, then

$$\lim_{x \to \infty} \frac{1}{x^r} = 0. \qquad \text{Limit toward the right}$$

Furthermore, if x^r is defined when $x < 0$, then

$$\lim_{x \to -\infty} \frac{1}{x^r} = 0. \qquad \text{Limit toward the left}$$

Limits at infinity share many of the properties of limits listed in Section 12.1. Some of these properties are demonstrated in the next example.

EXAMPLE 1 ▱ **Evaluating a Limit at Infinity**

Find the limit.

$$\lim_{x \to \infty} \left(4 - \frac{3}{x^2} \right)$$

Solution

Try to identify the properties of limits that are used in the following steps.

$$\lim_{x \to \infty} \left(4 - \frac{3}{x^2} \right) = \lim_{x \to \infty} 4 - \lim_{x \to \infty} \frac{3}{x^2}$$

$$= \lim_{x \to \infty} 4 - 3 \left(\lim_{x \to \infty} \frac{1}{x^2} \right)$$

$$= 4 - 3(0)$$

$$= 4$$

Thus, the limit of $f(x) = 4 - (3/x^2)$ as x approaches ∞ is 4. This conclusion is confirmed graphically in Figure 12.28. Note that the line $y = 4$ is a horizontal asymptote to the right. ▱

Figure 12.28

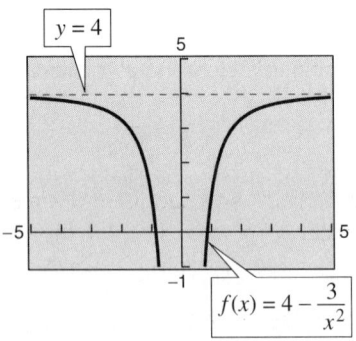

In Figure 12.28, it appears that the line $y = 4$ is also a horizontal asymptote *to the left*. You can verify this by showing that

$$\lim_{x \to -\infty} \left(4 - \frac{3}{x^2} \right) = 4.$$

The graph of a rational function need not have a horizontal asymptote. If it does, however, its left and right asymptotes must be the same.

EXAMPLE 2 ▭ **Comparing Limits at Infinity**

Find the limit as x approaches ∞ for each of the following functions.

a. $f(x) = \dfrac{-2x + 3}{3x^2 + 1}$ **b.** $f(x) = \dfrac{-2x^2 + 3}{3x^2 + 1}$ **c.** $f(x) = \dfrac{-2x^3 + 3}{3x^2 + 1}$

Solution

In each case, begin by dividing both the numerator and denominator by x^2, the highest-powered term in the denominator.

EXPLORATION

Use a graphing utility to graph the rational functions

$$y_1 = \frac{-2x + 3}{3x^2 + 1}$$

$$y_2 = \frac{-2x^2 + 3}{3x^2 + 1}$$

$$y_3 = \frac{-2x^3 + 3}{3x^2 + 1}.$$

In each case, make a conjecture about the limit as x approaches ∞. Compare your answers with the analytic results shown in Example 2.

a. $\displaystyle\lim_{x \to \infty} \frac{-2x + 3}{3x^2 + 1} = \lim_{x \to \infty} \frac{-\dfrac{2}{x} + \dfrac{3}{x^2}}{3 + \dfrac{1}{x^2}}$

$$= \frac{-0 + 0}{3 + 0}$$

$$= 0$$

b. $\displaystyle\lim_{x \to \infty} \frac{-2x^2 + 3}{3x^2 + 1} = \lim_{x \to \infty} \frac{-2 + \dfrac{3}{x^2}}{3 + \dfrac{1}{x^2}}$

$$= \frac{-2 + 0}{3 + 0}$$

$$= -\frac{2}{3}$$

c. $\displaystyle\lim_{x \to \infty} \frac{-2x^3 + 3}{3x^2 + 1} = \lim_{x \to \infty} \frac{-2x + \dfrac{3}{x^2}}{3 + \dfrac{1}{x^2}}$

In this case, you can conclude that the limit does not exist because the numerator increases without bound as the denominator approaches 3. ▭

In Example 2, observe that when the degree of the numerator is less than the degree of the denominator, the limit is 0. When the degrees of the numerator and denominator are equal, the limit is the ratio of the coefficients of the highest-powered terms. When the degree of the numerator is greater than the degree of the denominator, the limit does not exist.

This result seems reasonable when you realize that for large values of x, the highest-powered term of a polynomial is the most "influential" term. That is, a polynomial tends to behave as its highest-powered term behaves as x approaches positive or negative infinity.

Limits at Infinity for Rational Functions

For the rational function $f(x) = p(x)/q(x)$, where

$$p(x) = a_n x^n + \cdots + a_0$$

and

$$q(x) = b_m x^m + \cdots + b_0$$

the limit as x approaches positive or negative infinity is as follows.

$$\lim_{x \to \pm\infty} f(x) = \begin{cases} 0, & n < m \\ \dfrac{a_n}{b_m}, & n = m \end{cases}$$

If $n > m$, the limit does not exist.

Real Life

EXAMPLE 3   Finding the Average Cost

You are manufacturing a product that costs $0.50 per unit to produce. Your initial investment is $5000, which implies that the total cost of producing x units is $C = 0.5x + 5000$. The average cost per unit is given by

$$\overline{C} = \frac{C}{x} = \frac{0.5x + 5000}{x}.$$

Find the average cost per unit when (a) $x = 1000$, (b) $x = 10{,}000$, and (c) $x = 100{,}000$. (d) What is the limit of $\overline{C}$ when x approaches infinity?

Solution

a. When $x = 1000$, the average cost per unit is

$$\overline{C} = \frac{0.5(1000) + 5000}{1000} = \$5.50. \qquad x = 1000$$

b. When $x = 10{,}000$, the average cost per unit is

$$\overline{C} = \frac{0.5(10{,}000) + 5000}{10{,}000} = \$1.00. \qquad x = 10{,}000$$

c. When $x = 100{,}000$, the average cost per unit is

$$\overline{C} = \frac{0.5(100{,}000) + 5000}{100{,}000} = \$0.55. \qquad x = 100{,}000$$

d. As x approaches infinity, the limit of $\overline{C}$ is

$$\lim_{x \to \infty} \frac{0.5x + 5000}{x} = \$0.50. \qquad x \to \infty$$

The graph of $\overline{C}$ is shown in Figure 12.29.

Figure 12.29

As $x \to \infty$ the average cost per unit approaches $0.50.

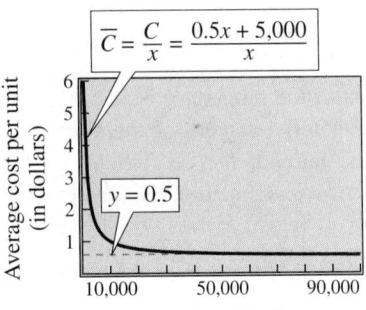

$$\overline{C} = \frac{C}{x} = \frac{0.5x + 5{,}000}{x}$$

Average cost per unit (in dollars)

$y = 0.5$

Number of units

Limits of Sequences

Limits of sequences have many of the same properties as limits of functions. For instance, consider the sequence whose nth term is $a_n = 1/2^n$.

$$\frac{1}{2}, \frac{1}{4}, \frac{1}{8}, \frac{1}{16}, \frac{1}{32}, \cdots$$

As n increases without bound, the terms of this sequence gets closer and closer to 0, and the sequence is said to **converge** to 0. Using limit notation, you can write

$$\lim_{n \to \infty} \frac{1}{2^n} = 0.$$

The following relationship shows how limits of functions of x can be used to evaluate the limit of a sequence.

There are a number of ways to use a graphing utility to generate the terms of a sequence. For instance, you can display the first 10 terms of the sequence

$$\frac{1}{2}, \frac{1}{4}, \frac{1}{8}, \frac{1}{16}, \frac{1}{32}, \cdots$$

on the *TI-82* or *TI-83* using the command seq(1/2^N,N,1,10,1). Or, you can use the table feature with $Y_1 = 1/2^X$, TblMin = 1, and ΔTbl = 1.

Limit of a Sequence

Let f be a function of a real variable, such that

$$\lim_{x \to \infty} f(x) = L.$$

If $\{a_n\}$ is a sequence such that $f(n) = a_n$ for every positive integer n, then

$$\lim_{n \to \infty} a_n = L.$$

A sequence that does not converge is said to **diverge.** For instance, the sequence $1, -1, 1, -1, 1, \ldots$ diverges.

EXAMPLE 4 ▰ **Finding the Limit of a Sequence**

Find the limit of each of the following sequences.

a. $a_n = \dfrac{2n + 1}{n + 4}$ **b.** $b_n = \dfrac{2n + 1}{n^2 + 4}$

Solution

a. $\displaystyle\lim_{n \to \infty} \frac{2n + 1}{n + 4} = 2$ $\dfrac{3}{5}, \dfrac{5}{6}, \dfrac{7}{7}, \dfrac{9}{8}, \dfrac{11}{9}, \dfrac{13}{10}, \cdots \to 2$

b. $\displaystyle\lim_{n \to \infty} \frac{2n + 1}{n^2 + 4} = 0$ $\dfrac{3}{5}, \dfrac{5}{8}, \dfrac{7}{13}, \dfrac{9}{20}, \dfrac{11}{29}, \dfrac{13}{40}, \cdots \to 0$ ▰

In the next section, you will encounter limits of sequences such as that shown in Example 5. A strategy for evaluating such limits is to begin by writing the nth term in standard rational function form. Then you can determine the limit by comparing the degrees of the numerator and denominator.

EXAMPLE 5   Finding the Limit of a Sequence

Find the limit of the sequence whose nth term is

$$a_n = \frac{8}{n^3}\left[\frac{n(n+1)(2n+1)}{6}\right].$$

Solution

Begin by writing the nth term in standard rational function form—as the ratio of two polynomials.

$$a_n = \frac{8}{n^3}\left[\frac{n(n+1)(2n+1)}{6}\right].$$ Original nth term

$$= \frac{8(n)(n+1)(2n+1)}{6n^3}$$ Multiply fractions.

$$= \frac{8n^3 + 12n^2 + 4n}{3n^3}$$ Standard rational form

From this form, you can see that the degree of the numerator is equal to the degree of the denominator. Thus, the limit of the sequence is the ratio of the coefficients of the highest-powered terms.

$$\lim_{n\to\infty} \frac{8n^3 + 12n^2 + 4n}{3n^3} = \frac{8}{3}$$

Group Activity

Comparing Rates of Convergence

In the table in the Study Tip above, the values of a_n are approaching their limit of $\frac{8}{3}$ rather slowly. (The first term to be accurate to three decimal places is $a_{4801} \approx 2.667$.) Each of the following sequences converges to 0. Which converges most rapidly? Which converges most slowly? Why? Discuss your conclusions.

a. $a_n = \dfrac{1}{n}$ b. $b_n = \dfrac{1}{n^2}$ c. $c_n = \dfrac{1}{2^n}$ d. $d_n = \dfrac{1}{n!}$ e. $h_n = \dfrac{2^n}{n!}$

12.4 /// EXERCISES

In Exercises 1–6, match the function with its graph using horizontal asymptotes as aids. [The graphs are labeled (a), (b), (c), (d), (e), and (f).]

(a)

(b)

(c)

(d)

(e)

(f)

1. $f(x) = \dfrac{4}{x}$

2. $f(x) = \dfrac{-2x}{x + 1}$

3. $f(x) = \dfrac{4x^2}{x^2 + 1}$

4. $f(x) = \dfrac{1}{x^2 + 1}$

5. $f(x) = 4 - \dfrac{1}{x^2}$

6. $f(x) = \dfrac{x^2}{x + 1}$

In Exercises 7–20, find the limit (if it exists). Use a graphing utility to verify your result.

7. $\lim\limits_{x \to \infty} \dfrac{2}{x^2}$

8. $\lim\limits_{x \to \infty} \dfrac{4}{2x + 3}$

9. $\lim\limits_{x \to \infty} \dfrac{2 + x}{2 - x}$

10. $\lim\limits_{x \to \infty} \dfrac{1 - 5x}{1 + 2x}$

11. $\lim\limits_{x \to -\infty} \dfrac{4x - 3}{2x + 1}$

12. $\lim\limits_{x \to -\infty} \dfrac{x^2 + 1}{x^2}$

13. $\lim\limits_{t \to \infty} \dfrac{t^2}{t + 2}$

14. $\lim\limits_{y \to \infty} \dfrac{4y^4}{y^2 + 1}$

15. $\lim\limits_{x \to -\infty} \left(-2 + \dfrac{2}{x} \right)$

16. $\lim\limits_{x \to \infty} \dfrac{2x^2 - 6}{(x - 1)^2}$

17. $\lim\limits_{x \to -\infty} \dfrac{x}{(x + 1)^2}$

18. $\lim\limits_{x \to \infty} \left[3 + \dfrac{2x^2}{(x + 5)^2} \right]$

19. $\lim\limits_{t \to \infty} \left(\dfrac{1}{3t^2} - \dfrac{5t}{t + 2} \right)$

20. $\lim\limits_{x \to \infty} \left[\dfrac{x}{2x + 1} + \dfrac{3x^2}{(x - 3)^2} \right]$

In Exercises 21–24, use a graphing utility to graph the function and verify that the horizontal asymptote corresponds to the limit at infinity.

21. $y = \dfrac{2x}{1 - x}$

22. $y = \dfrac{x^2}{x^2 + 9}$

23. $y = \dfrac{2x}{1 - x^2}$

24. $y = 2 - \dfrac{3}{x^2}$

In Exercises 25 and 26, use a graphing utility to graph the function and verify that it has two horizontal asymptotes.

25. $f(x) = \dfrac{4|x|}{x + 1}$

26. $f(x) = \dfrac{2x}{\sqrt{x^2 + 1}}$

Numerical and Graphical Analysis In Exercises 27–30, use a graphing utility to complete the table and estimate the limit as x approaches infinity. Then use the graphing utility to graph the function and use the graph to estimate the limit.

x	10^0	10^1	10^2	10^3	10^4	10^5	10^6
$f(x)$							

27. $f(x) = x - \sqrt{x^2 + 2}$ **28.** $f(x) = 3x - \sqrt{9x^2 + 1}$

29. $f(x) = 3\left(2x - \sqrt{4x^2 + x}\right)$

30. $f(x) = 4\left(4x - \sqrt{16x^2 - x}\right)$

31. *Average Cost* The cost function for a certain product is given by

$$C = 1.35x + 4570$$

where C is measured in dollars and x is the number of units produced.

(a) Find the average cost per unit when $x = 100$ and $x = 1000$.

(b) Determine the limit of the average cost function as x approaches infinity.

32. *Demographics* The table gives the number N (in thousands) of high school graduates at the end of each decade for the years 1900 through 1990. (Source: Department of Education)

Year	1900	1910	1920	1930	1940
N	62	111	231	592	1140

Year	1950	1960	1970	1980	1990
N	1063	1627	2589	2748	2505

A model for this data is given by

$$N = \frac{68{,}436.82 + 4731.82t}{1000 - 23.37t + 0.16t^2}, \quad 0 \le t \le 90$$

where t is the time in years, with $t = 0$ corresponding to 1900.

(a) Use a graphing utility to create a scatter plot of the data and graph the model. How do they compare?

(b) Use the model to estimate the number of high school graduates in 1975.

(c) Why should this model *not* be used to predict the numbers of graduates in future years?

In Exercises 33–44, write the first five terms of the sequence and find the limit of the sequence (if it exists). Assume n begins with 1.

33. $a_n = \dfrac{n + 2}{2n - 1}$

34. $a_n = \dfrac{3n - 1}{n + 2}$

35. $a_n = \dfrac{n + 1}{n^2 + 1}$

36. $a_n = \dfrac{2n^2}{n^4 + n + 1}$

37. $a_n = \dfrac{n^2}{5n + 2}$

38. $a_n = 2n - 5$

39. $a_n = \dfrac{(n + 1)!}{n!}$

40. $a_n = \dfrac{(2n - 1)!}{(2n + 1)!}$

41. $a_n = 2[5 + (n - 1)^3]$

42. $a_n = \dfrac{(-1)^{n+1}}{n^2}$

43. $a_n = (-1)^{n-1}\left(\dfrac{n}{2n + 1}\right)$

44. $a_n = (-1)^{n+1}\left(\dfrac{2n - 1}{n + 4}\right)$

In Exercises 45–50, find the limit of the sequence and use a graphing utility to complete the table.

n	10^0	10^1	10^2	10^3	10^4	10^5	10^6
a_n							

45. $a_n = \dfrac{1}{n}\left(n + \dfrac{1}{n}\left[\dfrac{n(n + 1)}{2}\right]\right)$

46. $a_n = \dfrac{4}{n}\left(n + \dfrac{4}{n}\left[\dfrac{n(n + 1)}{2}\right]\right)$

47. $a_n = 6 - \dfrac{4}{n^2}\left[\dfrac{n(n + 1)}{2}\right]$

48. $a_n = \dfrac{16}{n^4}\left[\dfrac{n^2(n + 1)^2}{4}\right]$

49. $a_n = \dfrac{16}{n^3}\left[\dfrac{n(n + 1)(2n + 1)}{6}\right]$

50. $a_n = \dfrac{n(n + 1)}{n^2} - \dfrac{1}{n^4}\left[\dfrac{n(n + 1)}{2}\right]^2$

Exploration **In Exercises 51–54, use a graphing utility to create a scatter plot of the terms of the sequence. Based on the results, decide whether the sequence converges or diverges. If it converges, estimate its limit.**

51. $a_n = 4\left(\tfrac{2}{3}\right)^n$

52. $a_n = 3\left(\tfrac{3}{2}\right)^n$

53. $a_n = \dfrac{3[1 - (1.5)^n]}{1 - 1.5}$

54. $a_n = \dfrac{3[1 - (0.5)^n]}{1 - 0.5}$

12.5 The Area Problem

Limits of Summations / The Area Problem

Limits of Summations

Earlier in the text, you used the concept of a limit to obtain a formula for the sum S of an infinite geometric sequence

$$S = a_1 + a_1 r + a_1 r^2 + \cdots = \sum_{i=1}^{\infty} a_1 r^{i-1} = \frac{a_1}{1 - r}, \quad |r| < 1.$$

Using limit notation, this sum can be written as

$$S = \lim_{n \to \infty} \sum_{i=1}^{n} a_1 r^{i-1} = \lim_{n \to \infty} \frac{a_1 (1 - r^n)}{1 - r} = \frac{a_1}{1 - r}.$$

The following summation formulas and properties are used to evaluate finite and infinite summations.

Some graphing utilities have the capability of computing summations. For instance, on the *TI-85*, you can use the MATH MISC menu to verify the summation in Example 1 as follows.

sum seq(N, N, 1, 200, 1)

On the *TI-92*, you can use the MATH/List menu to verify the summation as follows.

sum(seq(n, n, 1, 200, 1))

Summation Formulas and Properties

1. $\displaystyle\sum_{i=1}^{n} c = cn$

2. $\displaystyle\sum_{i=1}^{n} i = \frac{n(n + 1)}{2}$

3. $\displaystyle\sum_{i=1}^{n} i^2 = \frac{n(n + 1)(2n + 1)}{6}$

4. $\displaystyle\sum_{i=1}^{n} i^3 = \frac{n^2(n + 1)^2}{4}$

5. $\displaystyle\sum_{i=1}^{n} (a_i \pm b_i) = \sum_{i=1}^{n} a_i \pm \sum_{i=1}^{n} b_i$

6. $\displaystyle\sum_{i=1}^{n} k a_i = k \sum_{i=1}^{n} a_i$

EXAMPLE 1 Evaluating a Summation

Using the second summation formula with $n = 200$, you can write

$$\sum_{i=1}^{200} i = 1 + 2 + 3 + 4 + \cdots + 200$$

$$= \frac{n(n + 1)}{2} = \frac{200(201)}{2} = 20{,}100.$$

EXAMPLE 2   ◼   **Evaluating a Summation**

Evaluate the summation

$$S = \sum_{i=1}^{n} \frac{i+2}{n^2} = \frac{3}{n^2} + \frac{4}{n^2} + \frac{5}{n^2} + \cdots + \frac{n+2}{n^2}$$

for $n = 10, 100, 1000,$ and $10,000$.

Solution

Begin by applying summation formulas and properties to simplify S. In the second line of the solution, note that $1/n^2$ can be factored out of the sum because n is considered to be constant. You could not factor i out of the summation because i is the (variable) index of summation.

$$S = \sum_{i=1}^{n} \frac{i+2}{n^2}$$ Original form of summation

$$= \frac{1}{n^2} \sum_{i=1}^{n}(i+2)$$ Factor constant $1/n^2$ out of sum.

$$= \frac{1}{n^2}\left(\sum_{i=1}^{n}i + \sum_{i=1}^{n}2\right)$$ Write as two sums.

$$= \frac{1}{n^2}\left[\frac{n(n+1)}{2} + 2n\right]$$ Apply Formulas 1 and 2.

$$= \frac{1}{n^2}\left(\frac{n^2+5n}{2}\right)$$ Add fractions.

$$= \frac{n+5}{2n}$$ Simplify.

Now you can evaluate the sum by substituting the appropriate values of n, as shown in the following table.

n	10	100	1000	10,000
$\sum_{i=1}^{n}\dfrac{i+2}{n^2} = \dfrac{n+5}{2n}$	0.75	0.525	0.5025	0.50025

◼

Note   In Example 2, note that the sum appears to approach a limit as n increases. To find the limit of $(n+5)/2n$ as n approaches infinity, you can write

$$\lim_{n \to \infty} \frac{n+5}{2n} = \frac{1}{2}.$$

Be sure you notice the strategy used in Example 2. Rather than separately evaluating the sums

$$\sum_{i=1}^{10} \frac{i+2}{n^2}, \quad \sum_{i=1}^{100} \frac{i+2}{n^2}, \quad \sum_{i=1}^{1000} \frac{i+2}{n^2}, \quad \sum_{i=1}^{10,000} \frac{i+2}{n^2}$$

it was more efficient first to convert to rational form.

$$S = \underbrace{\sum_{i=1}^{n} \frac{i+2}{n^2}}_{\substack{\text{Summation} \\ \text{form}}} = \underbrace{\frac{n+5}{2n}}_{\text{Rational form}}$$

With this rational form, each sum can be evaluated by simply substituting appropriate values of n.

EXAMPLE 3 ◢ **Finding the Limit of a Summation**

Let $S(n)$ be defined by

$$S(n) = \sum_{i=1}^{n} \left(1 + \frac{i}{n}\right)^2 \left(\frac{1}{n}\right)$$

and find the limit of $S(n)$ as $n \to \infty$.

Solution

Begin by rewriting the summation in rational form.

$$S(n) = \sum_{i=1}^{n} \left(1 + \frac{i}{n}\right)^2 \left(\frac{1}{n}\right)$$

$$= \sum_{i=1}^{n} \left(\frac{n^2 + 2ni + i^2}{n^2}\right)\left(\frac{1}{n}\right)$$

$$= \frac{1}{n^3} \sum_{i=1}^{n} (n^2 + 2ni + i^2)$$

$$= \frac{1}{n^3} \left(\sum_{i=1}^{n} n^2 + \sum_{i=1}^{n} 2ni + \sum_{i=1}^{n} i^2\right)$$

$$= \frac{1}{n^3}\left(n^3 + 2n\left[\frac{n(n+1)}{2}\right] + \frac{n(n+1)(2n+1)}{6}\right)$$

$$= \frac{14n^3 + 9n^2 + n}{6n^3}$$

In this rational form, you can now find the limit as $n \to \infty$.

$$\lim_{n\to\infty} S(n) = \lim_{n\to\infty} \frac{14n^3 + 9n^2 + n}{6n^3} = \frac{14}{6} = \frac{7}{3}$$ ◢

The Area Problem

You now have the tools needed to solve the second basic problem of calculus: the area problem. The problem is to find the *area* of the region R bounded by the graph of a nonnegative, continuous function f, the x-axis, and the vertical lines $x = a$ and $x = b$, as shown in Figure 12.30.

If the region R is a square, triangle, trapezoid, or semicircle, you can find its area by using a geometric formula. For more general regions, however, you must use a different approach—one that involves the limit of a summation. The basic strategy is to use a collection of rectangles that approximates the region R, as illustrated in Example 4.

 A computer animation of this concept appears in the *Interactive* CD-ROM.

Figure 12.30

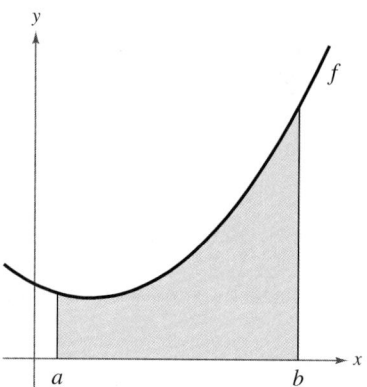

EXAMPLE 4 ▭ **Approximating the Area of a Region**

Use the five rectangles in Figure 12.31 to approximate the area of the region bounded by the graph of $f(x) = 6 - x^2$, the x-axis, and the lines $x = 0$ and $x = 2$.

Figure 12.31

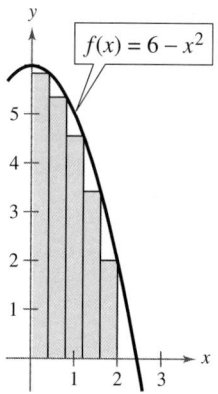

Solution

The right endpoints of the five intervals are $\frac{2}{5}i$, where $i = 1, 2, 3, 4, 5$. The width of each rectangle is $\frac{2}{5}$, and the height of each rectangle can be obtained by evaluating f at the right endpoint of each interval. The five intervals are

$$\left[0, \frac{2}{5}\right], \left[\frac{2}{5}, \frac{4}{5}\right], \left[\frac{4}{5}, \frac{6}{5}\right], \left[\frac{6}{5}, \frac{8}{5}\right], \left[\frac{8}{5}, \frac{10}{5}\right].$$

The sum of the areas of the five rectangles is

$$\sum_{i=1}^{5} f\left(\overbrace{\frac{2i}{5}}^{\text{Height}}\right)\left(\overbrace{\frac{2}{5}}^{\text{Width}}\right) = \sum_{i=1}^{5}\left[6 - \left(\frac{2i}{5}\right)^2\right]\left(\frac{2}{5}\right) = \frac{212}{25} = 8.48.$$

Thus, you can approximate the area of R as 8.48. ▭

By increasing the number of rectangles used in Example 4, you can obtain closer and closer approximations of the area of the region. For instance, using 25 rectangles of width $\frac{2}{25}$ each, you can approximate the area to be $A \approx 9.17$. The following table shows even better approximations.

n	5	25	100	1000	5000
Approximate Area	8.48	9.17	9.29	9.33	9.33

The *exact* area of the region R is given by the limit of the sum of n rectangles as n approaches ∞.

Area of a Plane Region

Let f be continuous and nonnegative on the interval $[a, b]$. The **area** A of the region bounded by the graph of f, the x-axis, and the vertical lines $x = a$ and $x = b$ is

$$A = \lim_{n \to \infty} \sum_{i=1}^{n} f\left(a + \frac{(b-a)i}{n}\right)\left(\frac{b-a}{n}\right).$$

EXAMPLE 5   **Finding the Area of a Region**

Find the area of the region bounded by the graph of $f(x) = x^2$ and the x-axis between $x = 0$ and $x = 1$.

Solution

Begin by finding the dimensions of the rectangles.

$$\text{Width: } \frac{b-a}{n} = \frac{1}{n} \qquad \text{Height: } f\left(a + \frac{(b-a)i}{n}\right) = f\left(\frac{i}{n}\right) = \frac{i^2}{n^2}$$

Next, approximate the area as the sum of the areas of n rectangles.

$$A \approx \sum_{i=1}^{n} f\left(a + \frac{(b-a)i}{n}\right)\left(\frac{b-a}{n}\right)$$

$$= \sum_{i=1}^{n} \left(\frac{i^2}{n^2}\right)\left(\frac{1}{n}\right)$$

$$= \sum_{i=1}^{n} \frac{i^2}{n^3}$$

$$= \frac{1}{n^3} \sum_{i=1}^{n} i^2$$

$$= \frac{1}{n^3} \left[\frac{n(n+1)(2n+1)}{6}\right]$$

$$= \frac{2n^3 + 3n^2 + n}{6n^3}$$

Finally, find the exact area by taking the limit as n approaches ∞.

$$A = \lim_{n \to \infty} \frac{2n^3 + 3n^2 + n}{6n^3} = \frac{1}{3}$$

Figure 12.32 shows the region R.

Figure 12.32

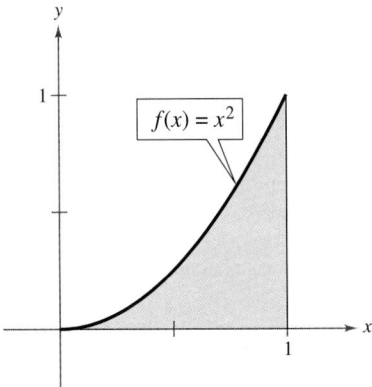

EXAMPLE 6   **Finding the Area of a Region**

Find the area of the region bounded by the graph of $f(x) = 3x - x^2$ and the x-axis between $x = 1$ and $x = 2$.

Solution

Begin by finding the dimensions of the rectangles.

Width: $\dfrac{b - a}{n} = \dfrac{1}{n}$

Height: $f\left(a + \dfrac{(b - a)i}{n}\right) = f\left(1 + \dfrac{i}{n}\right)$

$$= 3\left(1 + \frac{i}{n}\right) - \left(1 + \frac{i}{n}\right)^2$$

$$= 3 + \frac{3i}{n} - \left(1 + \frac{2i}{n} + \frac{i^2}{n^2}\right)$$

$$= 2 + \frac{i}{n} - \frac{i^2}{n^2}$$

Next, approximate the area as the sum of the areas of n rectangles.

$$A \approx \sum_{i=1}^{n} f\left(a + \frac{(b - a)i}{n}\right)\left(\frac{b - a}{n}\right)$$

$$= \sum_{i=1}^{n} \left(2 + \frac{i}{n} - \frac{i^2}{n^2}\right)\left(\frac{1}{n}\right)$$

$$= \frac{1}{n}\sum_{i=1}^{n} 2 + \frac{1}{n^2}\sum_{i=1}^{n} i - \frac{1}{n^3}\sum_{i=1}^{n} i^2$$

$$= \frac{1}{n}(2n) + \frac{1}{n^2}\left[\frac{n(n + 1)}{2}\right] - \frac{1}{n^3}\left[\frac{n(n + 1)(2n + 1)}{6}\right]$$

$$= 2 + \frac{n^2 + n}{2n^2} - \frac{2n^3 + 3n^2 + n}{6n^3}$$

$$= 2 + \frac{1}{2} + \frac{1}{2n} - \frac{1}{3} - \frac{1}{2n} - \frac{1}{6n^2}$$

$$= \frac{13}{6} - \frac{1}{6n^2}$$

Finally, find the exact area by taking the limit as n approaches ∞.

$$A = \lim_{n \to \infty} \left(\frac{13}{6} - \frac{1}{6n^2}\right) = \frac{13}{6}$$

Figure 12.33 shows the region R.

Figure 12.33

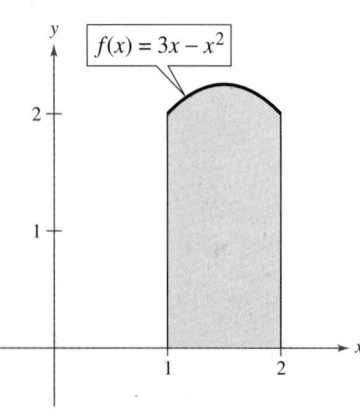

Group Activity   *Comparing Geometry and Calculus*

Show how to use a formula from geometry to find the area of the region bounded by the graph of

$$f(x) = 2 - x$$

and the x-axis between $x = 0$ and $x = 2$. Then show how the same area can be found using the limit procedure described in this section.

12.5 /// EXERCISES

In Exercises 1–6, find the sum using the summation formulas and properties.

1. $\displaystyle\sum_{i=1}^{60} i$

2. $\displaystyle\sum_{i=1}^{30} i^2$

3. $\displaystyle\sum_{k=1}^{20} k^3$

4. $\displaystyle\sum_{k=1}^{50} (2k + 1)$

5. $\displaystyle\sum_{j=1}^{25} (j^2 + j)$

6. $\displaystyle\sum_{j=1}^{10} (j^3 - 3j^2)$

In Exercises 7–16, use the summation formulas and properties to rewrite the sum as a rational function $S(n)$. Use a graphing utility and the function to complete the table. Then find $\lim\limits_{n\to\infty} S(n)$.

n	10^0	10^1	10^2	10^3
$S(n)$				

7. $\displaystyle\sum_{i=1}^{n} \frac{4i^2}{n^3}$

8. $\displaystyle\sum_{i=1}^{n} \frac{i}{n^2}$

9. $\displaystyle\sum_{i=1}^{n} \frac{3}{n^3}(1 + i^2)$

10. $\displaystyle\sum_{i=1}^{n} \frac{i^3}{n^4}$

11. $\displaystyle\sum_{i=1}^{n} \frac{2i + 3}{n^2}$

12. $\displaystyle\sum_{i=1}^{n} \frac{24i^2}{n^3}$

13. $\displaystyle\sum_{i=1}^{n} \left(\frac{i^2}{n^3} + \frac{2}{n}\right)\left(\frac{1}{n}\right)$

14. $\displaystyle\sum_{i=1}^{n} \left[3 - 2\left(\frac{i}{n}\right)\right]\left(\frac{1}{n}\right)$

15. $\displaystyle\sum_{i=1}^{n} \left[1 - \left(\frac{i}{n}\right)^2\right]\left(\frac{1}{n}\right)$

16. $\displaystyle\sum_{i=1}^{n} \left[\frac{4}{n} + \left(\frac{2i}{n^2}\right)\right]\left(\frac{2i}{n}\right)$

In Exercises 17–20, approximate the area of the region using the indicated number of rectangles of equal width.

17. $f(x) = x + 1$

18. $f(x) = x^2 + 1$

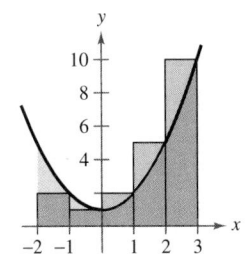

19. $f(x) = \frac{1}{4}x^3$ **20.** $f(x) = \frac{1}{2}(x-2)^3$

     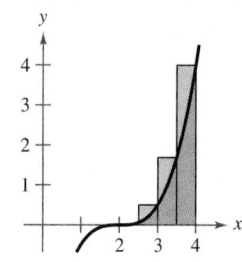

Function	Interval
25. $f(x) = 3x$	$[0, 1]$
26. $f(x) = x + 2$	$[-2, 2]$
27. $f(x) = 3 - x^2$	$[-1, 1]$
28. $f(x) = \frac{1}{4}x^2 + 1$	$[1, 4]$
29. $g(x) = 8 - x^3$	$[1, 2]$
30. $g(x) = 4x - x^3$	$[0, 2]$
31. $f(x) = \frac{1}{4}(x^2 + 4x)$	$[1, 4]$
32. $f(x) = x^2 - x^3$	$[-1, 1]$

In Exercises 21–24, complete the table showing the approximate area of the region in the graph using *n* rectangles of equal width.

n	4	8	20	50
Approximate Area				

21. $f(x) = -\frac{1}{2}x + 4$ **22.** $f(x) = 9 - x^2$

     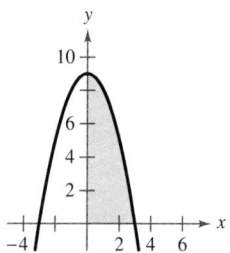

23. $f(x) = \frac{1}{9}x^3$ **24.** $f(x) = 3 - \frac{1}{4}x^3$

     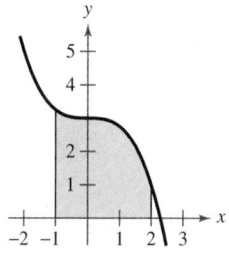

In Exercises 25–32, use the limit process to find the area of the region between the graph of the function and the *x*-axis over the specified interval. Use a graphing utility to shade the region.

Think About It   In Exercises 33 and 34, determine the value that best approximates the area of the region between the *x*-axis and the function over the indicated interval. (Make your selection on the basis of a sketch of the region and *not* by performing any calculations.)

33. $f(x) = 4 - x^2$, $[0, 2]$
    (a) -2 (b) 6   (c) 10   (d) 4   (e) 8

34. $f(x) = 4x - x^2$, $[0, 4]$
    (a) 10   (b) -5 (c) -9 (d) 5   (e) 14

In Exercises 35–38, use a graphing utility to find the area of the region between the graph of the function and the *x*-axis over the specified interval. (*Hint:* Enter the function as Y₁, select [CALC] and select ∫ f (x) dx, enter the endpoints of the interval as the lower and upper bounds, and press [ENTER].)

Function	Interval
35. $y = x - x^2$	$[0, 1]$
36. $y = (3 - x)\sqrt{x}$	$[0, 3]$
37. $y = \cos x$	$\left[0, \dfrac{\pi}{2}\right]$
38. $y = 6e^{-x}$	$[0, 2]$

39. *Real Estate*   The boundaries of a parcel of land are two edges modeled by the coordinate axes and a stream modeled by the equation

$$y = (-3.0 \times 10^{-6})x^3 + 0.002x^2 - 1.05x + 400.$$

Use a graphing utility to graph the equation and find the area of the property. Assume all distances are measured in feet.

Focus on Concepts

In this chapter, you have studied limits and basic concepts in calculus. You can use the following questions to check your understanding of these basic concepts. The answers to these questions are given in the back of the book.

In Exercises 1–4, use the graph to determine if the limit exists as x approaches 2.

1.

2.

3.

4.

5. If $\lim_{x \to 3} f(x) = 5$, can you conclude anything about $f(3)$? Explain.

6. What is meant by an indeterminate form? Give an example.

7. The functions $f(x) = (x^2 - 4)/(x + 2)$ and $g(x) = x - 2$ agree at all real numbers x except one. Identify this value of x.

8. *True or False?* A tangent line to a graph can intersect the graph only at the point of tangency. Explain.

9. What line is used to determine the rate at which a graph rises or falls at a single point?

In Exercises 10 and 11, estimate the slope of the tangent line to the graph at the point.

10.

11.

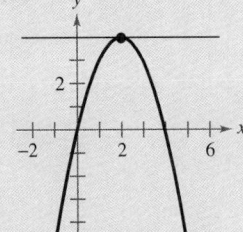

12. *True or False?* Every rational function has a horizontal asymptote. Explain.

13. When the degree of the numerator of a rational function is less than the degree of the denominator, what is the limit at infinity? What is the limit at infinity if the degrees of the numerator and denominator are equal?

14. Describe the process of finding the area of a region bounded by the graph of a nonnegative, continuous function f, the x-axis, and the vertical lines $x = a$ and $x = b$.

15. Determine which value best approximates the area of the region shown in the figure. (Make your selection on the basis of the sketch of the region and *not* by performing any calculations.)

(a) -2 (b) 1   (c) 4   (d) 6   (e) 9

---

12 /// REVIEW EXERCISES

In Exercises 1 and 2, use a graphing utility to complete the table and use the result to estimate the limit.

1. $\lim\limits_{x\to2} \dfrac{x-2}{3x^2-4x-4}$

x	1.9	1.99	1.999	2	2.001	2.01	2.1
$f(x)$				?			

2. $\lim\limits_{x\to3} \dfrac{\dfrac{x}{x+1}-\dfrac{3}{4}}{x-3}$

x	2.9	2.99	2.999	3	3.001	3.01	3.1
$f(x)$				?			

In Exercises 3–18, find the limit (if it exists). Use a graphing utility to verify your answer.

3. $\lim\limits_{x\to3} (5x-4)$

4. $\lim\limits_{x\to-2} (5-2x-x^2)$

5. $\lim\limits_{x\to2} (5x-3)(3x+5)$

6. $\lim\limits_{x\to2} \dfrac{3x+5}{5x-3}$

7. $\lim\limits_{t\to3} \dfrac{t^2+1}{t}$

8. $\lim\limits_{t\to3} \dfrac{t^2-9}{t-3}$

9. $\lim\limits_{t\to-2} \dfrac{t+2}{t^2-4}$

10. $\lim\limits_{x\to5} \dfrac{x-5}{x^2+5x-50}$

11. $\lim\limits_{x\to-2} \dfrac{x^2-4}{x^3+8}$

12. $\lim\limits_{x\to4} \dfrac{x^3-64}{x^2-16}$

13. $\lim\limits_{u\to0} \dfrac{\sqrt{4+u}-2}{u}$

14. $\lim\limits_{v\to0} \dfrac{\sqrt{v+7}-3}{v}$

15. $\lim\limits_{x\to-1} \dfrac{\dfrac{1}{x+2}-1}{x+1}$

16. $\lim\limits_{x\to0} \dfrac{\dfrac{1}{x+1}-1}{x}$

17. $\lim\limits_{x\to6^-} \dfrac{|x-6|}{x-6}$

18. $\lim\limits_{x\to2^+} f(x)$ where
$$f(x)=\begin{cases}5-x, & x\le2\\ x^2-3, & x>2\end{cases}$$

In Exercises 19–22, use a graphing utility to graph the function and use the graph to determine whether the specified limit exists.

19. $f(x)=\begin{cases}3x, & x\le1\\ 2-x, & x>1\end{cases}$, $\lim\limits_{x\to1} f(x)$

20. $g(x)=e^{-2/x}$, $\lim\limits_{x\to0} g(x)$

21. $g(x)=\dfrac{\sin 4x}{2x}$, $\lim\limits_{x\to0} g(x)$

22. $f(t)=\dfrac{t^3+125}{t+5}$, $\lim\limits_{t\to-5} f(t)$

In Exercises 23 and 24, (a) use a graphing utility to complete the table to approximate the limit, and (b) rationalize the numerator to find the exact value of the limit.

x	1.1	1.01	1.001	1.0001
$f(x)$				

23. $\lim\limits_{x\to1^+} \dfrac{\sqrt{2x+1}-\sqrt{3}}{x-1}$

24. $\lim\limits_{x\to1^+} \dfrac{1-\sqrt{x}}{x-1}$

In Exercises 25 and 26, use the given information to evaluate the limits.

25. $\lim\limits_{x\to c} f(x)=4$
$\lim\limits_{x\to c} g(x)=5$
(a) $\lim\limits_{x\to c} [f(x)]^3$
(b) $\lim\limits_{x\to c} [3f(x)-g(x)]$
(c) $\lim\limits_{x\to c} [f(x)g(x)]$
(d) $\lim\limits_{x\to c} \dfrac{f(x)}{g(x)}$

26. $\lim\limits_{x\to c} f(x)=27$
$\lim\limits_{x\to c} g(x)=12$
(a) $\lim\limits_{x\to c} \sqrt[3]{f(x)}$
(b) $\lim\limits_{x\to c} \dfrac{f(x)}{18}$
(c) $\lim\limits_{x\to c} [f(x)g(x)]$
(d) $\lim\limits_{x\to c} [f(x)-2g(x)]$

In Exercises 27 and 28, find $\lim\limits_{h \to 0} \dfrac{f(x+h) - f(x)}{h}$.

27. $f(x) = 3x - x^2$ **28.** $f(x) = x^3 + 5$

In Exercises 29 and 30, use a graphing utility to graph the function and the tangent line at the point $(2, f(2))$. Use the graph to estimate the slope of the tangent line.

29. $f(x) = x^2 - 2x$ **30.** $f(x) = \sqrt{x + 2}$

In Exercises 31–34, find a formula for the slope of the graph. Use the formula to find the slope at the two points.

31. $g(x) = x^2 - 4x$ **32.** $g(x) = \sqrt{x}$
   (a) $(0, 0)$ (b) $(5, 5)$ (a) $(1, 1)$ (b) $(4, 2)$

33. $g(x) = \dfrac{4}{x - 6}$ **34.** $g(x) = \dfrac{1}{4}x^4$
   (a) $(7, 4)$ (b) $(8, 2)$ (a) $(-2, 4)$ (b) $\left(1, \frac{1}{4}\right)$

In Exercises 35–40, find the derivative of the function.

35. $g(x) = -4$ **36.** $f(x) = 3x$
37. $h(x) = 5 - \frac{1}{2}x$ **38.** $f(x) = \frac{1}{2}x^2 + 3$
39. $f(t) = \sqrt{t + 5}$ **40.** $g(s) = \dfrac{4}{s + 5}$

41. *Chapter Opener* Use the regression capability of a graphing utility to fit the quadratic model $y = at^2 + bt + c$ to the data in the Chapter Opener on page 865.

42. *Chapter Opener* The value of the derivative of the model in Exercise 41 at time $t = 5$ is the rate of change of height with respect to time (velocity) at that instant. Find the velocity of the shuttle after 5 seconds.

In Exercises 43–46, use your knowledge of the graph of the function and the geometric interpretation of the derivative to match the function with the graph of its *derivative*. It is not necessary to find the derivative of the function. [The graphs are labeled (a), (b), (c), and (d).]

(a)

(b)

(c)

(d)
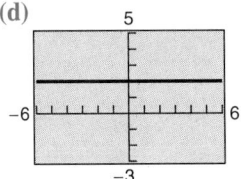

43. $f(x) = 2x$ **44.** $f(x) = \frac{1}{4}x^2$

45. $f(x) = |x - 2|$ **46.** $f(x) = \dfrac{1}{x - 2}$

In Exercises 47–52, find the limit (if it exists). Use a graphing utility to verify your answer.

47. $\lim\limits_{x \to \infty} \dfrac{4x}{2x - 3}$ **48.** $\lim\limits_{x \to \infty} \dfrac{7x}{14x + 2}$

49. $\lim\limits_{x \to \infty} \dfrac{2x}{x^2 - 25}$ **50.** $\lim\limits_{x \to \infty} \dfrac{x^2}{2x + 3}$

51. $\lim\limits_{x \to \infty} \left(4 - \dfrac{7}{x^3}\right)$ **52.** $\lim\limits_{x \to \infty} (x - 2)^{-3}$

In Exercises 53 and 54, find the limit of the sequence.

53. $a_n = \dfrac{1}{2n^2}[3 - 2n(n + 1)]$

54. $a_n = \left(\dfrac{2}{n}\right)\left(n + \dfrac{2}{n}\left[\dfrac{n(n - 1)}{2} - n\right]\right)$

In Exercises 55 and 56, use the summation formulas and properties to rewrite the sum as a rational function $S(n)$. Use a graphing utility and the function to complete the table. Then find $\lim\limits_{n \to \infty} S(n)$.

x	10^0	10^1	10^2	10^3
$S(n)$				

55. $\displaystyle\sum_{i=1}^{n}\left(\frac{4i^2}{n^2} - \frac{i}{n}\right)\left(\frac{1}{n}\right)$

56. $\displaystyle\sum_{i=1}^{n}\left[4 - \left(\frac{3i}{n}\right)^2\right]\left(\frac{3i}{n^2}\right)$

In Exercises 57 and 58, approximate the area of the region using the indicated number of rectangles of equal width.

57. $f(x) = 4 - x$ **58.** $f(x) = 4 - x^2$

   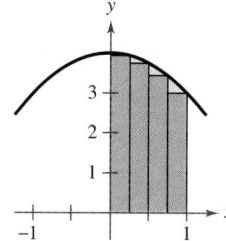

In Exercises 59 and 60, complete the table showing the approximate area of the region in the graph using n rectangles of equal width.

n	4	8	20	50
Approximate Area				

59. $f(x) = \frac{1}{4}x^2$ **60.** $f(x) = 4x - x^2$

   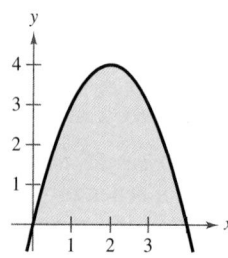

In Exercises 61–66, approximate the area of the region using the indicated number of rectangles of equal width. Then use the limit process to find the exact area.

61. $y = 10 - x$ **62.** $y = 2x - 6$

   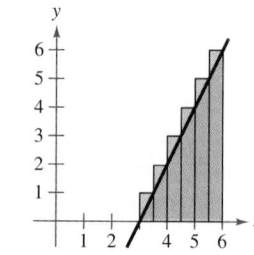

63. $y = x^2 + 4$ **64.** $y = 8(x - x^2)$

   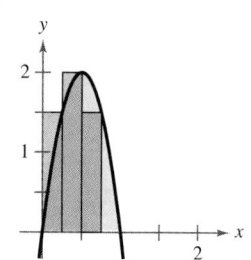

65. $y = 2(x^2 - x^3)$ **66.** $y = 4 - (x - 2)^2$

   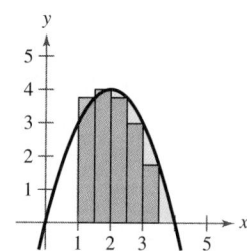

67. *Real Estate* Estimate the number of square feet of land in a parcel of property where x and y are measured in feet, as shown in the figure. The land is bounded by a stream and two straight roads that meet at right angles.

x	y	x	y
0	125	600	95
100	125	700	88
200	120	800	75
300	112	900	35
400	90	1000	0
500	90		

CHAPTER PROJECT   *Tangent Lines to Sine Curves*

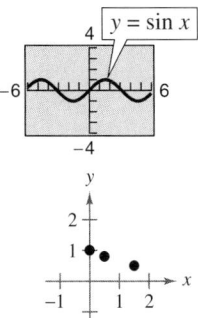

In this project you will develop a formula for the slope of the tangent line to the sine curve at an arbitrary point (x, y). In other words, you will derive a formula for the derivative of the function $f(x) = \sin x$. The first approach will be graphical, the second will be numerical, and the third will be analytical.

(a) *Graphical Approach* Use a graphing utility to graph the function $f(x) = \sin x$ shown at the left. On a separate sheet of paper, plot your estimates of the *slope* of this curve at various x-values. For instance, some values of the slope underneath the sine curve have been plotted at the left. The slope is approximately 1 at $x = 0$, 0.8 at $x = 0.5$, and 0 at $x = 1.5$. After you have plotted 15 or 20 points, connect them with a continuous curve. Do you recognize this curve?

(b) *Numerical Approach* The slope of the tangent line was given by the formula on page 888. The difference quotient $[f(x + h) - f(x)]/h$ will be a good approximation of the slope if h is small. If you select $h = 0.01$, you have the following approximation for the slope of the sine curve at x.

$$m \approx \frac{\sin(x + 0.01) - \sin x}{0.01}$$

Use the table feature of a graphing utility to complete the table at the left. Plot these points and compare your results with part (a).

(c) *Analytical Approach* To calculate the slope of the tangent line to $f(x) = \sin x$ analytically, you need two trigonometric limits. (See Questions 1 and 2 below.)

$$\text{Limit 1: } \lim_{h \to 0} \frac{\sin h}{h} = 1 \qquad \text{Limit 2: } \lim_{h \to 0} \frac{\cos h - 1}{h} = 0$$

Use these limits to find a formula for the slope of the sine curve.

x	m
0	
0.5	
1.0	
1.5	
2.0	
2.5	
3.0	
3.5	
4.0	

Questions for Further Exploration

1. Use a graphing utility to estimate Limit 1 in part (c) above *graphically* and *numerically*.

$$\lim_{h \to 0} \frac{\sin h}{h} = 1$$

2. Use a graphing utility to estimate Limit 2 in part (c) above *graphically* and *numerically*.

$$\lim_{h \to 0} \frac{\cos h - 1}{h} = 0$$

3. Follow the steps used in parts (a), (b), and (c) above to develop a formula for the derivative of the function

$$g(x) = \cos x.$$

Is it true that the sine and cosine functions are each other's derivatives? If not, which is the derivative of the other?

11–12 /// CUMULATIVE TEST

Take this test as you would take a test in class. After you are done, check your work against the answers given in the back of the book.

The *Interactive* CD-ROM provides answers to the Chapter Tests and Cumulative Tests. It also offers Chapter Pre-Tests (that test key skills and concepts covered in previous chapters) and Chapter Post-Tests, both of which have randomly generated exercises with diagnostic capabilities.

1. Find an equation of the sphere for which the endpoints of a diameter are $(0, 0, 0)$ and $(4, 4, 8)$.

2. Find $\mathbf{u} \cdot \mathbf{v}$ and $\mathbf{u} \times \mathbf{v}$ if $\mathbf{u} = \langle -3, 4, 1 \rangle$ and $\mathbf{v} = \langle 5, 0, 2 \rangle$.

3. Find a set of parametric equations of the line passing through the points $(-2, 3, 0)$, and $(5, 8, 25)$.

4. Find an equation of the plane passing through the points $(0, 0, 0)$, $(-2, 3, 0)$, and $(5, 8, 25)$.

5. Use vectors to find the area of the triangle with vertices $(3, 0, 0)$, $(6, 6, 4)$, and $(3, 10, 6)$.

6. Find the distance between the point $(0, 0, 25)$ and the plane $2x - 5y + z = 10$.

In Exercises 7–12, find the limit (if it exists).

7. $\lim\limits_{x \to 4} (5x - x^2)$

8. $\lim\limits_{x \to -2} \dfrac{x + 2}{x^2 + x - 2}$

9. $\lim\limits_{x \to 0} \dfrac{\sqrt{x + 4} - 2}{x}$

10. $\lim\limits_{x \to 4} \dfrac{|x - 4|}{x - 4}$

11. $\lim\limits_{x \to \infty} \dfrac{6x}{3x + 5}$

12. $\lim\limits_{x \to \infty} \dfrac{2x}{x^2 + 3x - 2}$

In Exercises 13 and 14, use the limit process to find the slope of the graph of the function at the specified point. Use a graphing utility to confirm your result.

13. $h(x) = 3 - x^2$, $\quad (1, 2)$

14. $f(x) = \sqrt{x + 3}$, $\quad (-2, 1)$

15. Find the sum using the summation formulas and properties.

$$\sum_{k=1}^{20} (3k^2 - 2k)$$

16. Complete the table showing the approximate area of the region in the figure using n rectangles of equal width.

Figure for 16

n	4	8	20	50
Approximate Area				

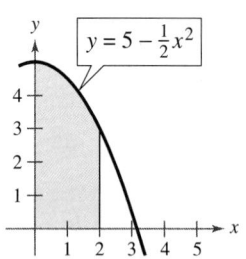

17. Use the limit process to find the area of the region between the graph of the function $f(x) = 1 - x^3$ and the x-axis over the interval $[0, 1]$. Use a graphing utility to verify your result.

Appendix A   Think About the Proof

Section 2.3, Page 185

The Remainder Theorem

If a polynomial $f(x)$ is divided by $x - k$, the remainder is

$$r = f(k).$$

Proof /// From the Division Algorithm, you have

$$f(x) = (x - k)q(x) + r(x)$$

and because either $r(x) = 0$ or the degree of $r(x)$ is less than the degree of $x - k$, you know that $r(x)$ must be a constant. That is, $r(x) = r$. Now, by evaluating $f(x)$ at $x = k$, you have

$$f(k) = (k - k)q(k) + r = (0)q(k) + r = r. \qquad\qquad ///$$

Section 2.3, Page 185

The Factor Theorem

A polynomial $f(x)$ has a factor $(x - k)$ if and only if $f(k) = 0$.

Proof /// Using the Division Algorithm with the factor $(x - k)$, you have

$$f(x) = (x - k)q(x) + r(x).$$

By the Remainder Theorem, $r(x) = r = f(k)$, and you have

$$f(x) = (x - k)q(x) + f(k)$$

where $q(x)$ is a polynomial of lesser degree than $f(x)$. If $f(k) = 0$, then

$$f(x) = (x - k)q(x)$$

and you see that $(x - k)$ is a factor of $f(x)$. Conversely, if $(x - k)$ is a factor of $f(x)$, division of $f(x)$ by $(x - k)$ yields a remainder of 0. Hence, by the Remainder Theorem, you have $f(k) = 0$. ///

A1

Section 2.5, Pages 204 and 205

> ### Linear Factorization Theorem
>
> If $f(x)$ is a polynomial of degree n
>
> $$f(x) = a_n x^n + a_{n-1} x^{n-1} + \cdots + a_1 x + a_0$$
>
> where $n > 0$, f has precisely n linear factors
>
> $$f(x) = a_n(x - c_1)(x - c_2) \cdots (x - c_n)$$
>
> where $c_1, c_2, \ldots, c_n$ are complex numbers and a_n is the leading coefficient of $f(x)$.

Proof **///** Using the Fundamental Theorem, you know that f must have at least one zero, c_1. Consequently, $(x - c_1)$ is a factor of $f(x)$, and you have

$$f(x) = (x - c_1)f_1(x).$$

If the degree of $f_1(x)$ is greater than zero, you again apply the Fundamental Theorem to conclude that f_1 must have a zero c_2, which implies that

$$f(x) = (x - c_1)(x - c_2)f_2(x).$$

It is clear that the degree of $f_1(x)$ is $n - 1$, that the degree of $f_2(x)$ is $n - 2$, and that you can repeatedly apply the Fundamental Theorem n times until you obtain

$$f(x) = a_n(x - c_1)(x - c_2) \cdots (x - c_n)$$

where a_n is the leading coefficient of the polynomial $f(x)$. **///**

Section 2.5, Page 208

Factors of a Polynomial

Every polynomial of degree $n > 0$ with real coefficients can be written as the product of linear and quadratic factors with real coefficients, where the quadratic factors have no real zeros.

Proof *///* To begin, you use the Linear Factorization Theorem to conclude that $f(x)$ can be *completely* factored in the form

$$f(x) = d(x - c_1)(x - c_2)(x - c_3) \cdots (x - c_n).$$

If each c_i is real, there is nothing more to prove. If any c_i is complex ($c_i = a + bi$, $b \neq 0$), then, because the coefficients of $f(x)$ are real, you know that the conjugate $c_j = a - bi$ is also a zero. By multiplying the corresponding factors, you obtain

$$(x - c_i)(x - c_j) = [x - (a + bi)][x - (a - bi)]$$
$$= x^2 - 2ax + (a^2 + b^2)$$

where each coefficient is real. *///*

Section 3.3, Page 264

Properties of Logarithms

Let a be a positive number such that $a \neq 1$, and let n be a real number. If u and v are positive real numbers, the following properties are true.

1. $\log_a(uv) = \log_a u + \log_a v$ 1. $\ln(uv) = \ln u + \ln v$

2. $\log_a \dfrac{u}{v} = \log_a u - \log_a v$ 2. $\ln \dfrac{u}{v} = \ln u - \ln v$

3. $\log_a u^n = n \log_a u$ 3. $\ln u^n = n \ln u$

Proof *///* To prove Property 1, let

$$x = \log_a u \quad \text{and} \quad y = \log_a v.$$

The corresponding exponential forms of these two equations are

$$a^x = u \quad \text{and} \quad a^y = v.$$

Multiplying u and v produces $uv = a^x a^y = a^{x+y}$. The corresponding logarithmic form of $uv = a^{x+y}$ is $\log_a(uv) = x + y$. Hence, $\log_a(uv) = \log_a u + \log_a v$. *///*

Section 6.1, Page 460

Law of Sines

If *ABC* is a triangle with sides *a*, *b*, and *c*, then

$$\frac{a}{\sin A} = \frac{b}{\sin B} = \frac{c}{\sin C}.$$

Oblique Triangles

A is acute.   *A* is obtuse.

Proof ///   Let *h* be the altitude of either triangle found in the figure showing oblique triangles, above. Then, you have

$$\sin A = \frac{h}{b} \qquad \text{or} \qquad h = b \sin A$$

$$\sin B = \frac{h}{a} \qquad \text{or} \qquad h = a \sin B.$$

Equating these two values of *h*, you have

$$a \sin B = b \sin A \qquad \text{or} \qquad \frac{a}{\sin A} = \frac{b}{\sin B}.$$

Note that $\sin A \neq 0$ and $\sin B \neq 0$ because no angle of a triangle can have a measure of 0° or 180°. In a similar manner, by constructing an altitude from vertex *B* to side *AC* (extended), you can show that

$$\frac{a}{\sin A} = \frac{c}{\sin C}.$$

Hence, the Law of Sines is established.   **///**

Section 6.4, Page 491

Properties of the Dot Product

Let $\mathbf{u}$, $\mathbf{v}$, and $\mathbf{w}$ be vectors in the plane or in space and let c be a scalar.

1. $\mathbf{u} \cdot \mathbf{v} = \mathbf{v} \cdot \mathbf{u}$
2. $\mathbf{0} \cdot \mathbf{v} = 0$
3. $\mathbf{u} \cdot (\mathbf{v} + \mathbf{w}) = \mathbf{u} \cdot \mathbf{v} + \mathbf{u} \cdot \mathbf{w}$
4. $\mathbf{v} \cdot \mathbf{v} = \|\mathbf{v}\|^2$
5. $c(\mathbf{u} \cdot \mathbf{v}) = c\mathbf{u} \cdot \mathbf{v} = \mathbf{u} \cdot c\mathbf{v}$

Proof /// To prove the second property, let $\mathbf{0} = \langle 0, 0 \rangle$ and $\mathbf{v} = \langle v_1, v_2 \rangle$. Then,

$$\mathbf{0} \cdot \mathbf{v} = \langle 0, 0 \rangle \cdot \langle v_1, v_2 \rangle$$
$$= 0 \cdot v_1 + 0 \cdot v_2$$
$$= 0 + 0$$
$$= 0.$$

To prove the third property, let $\mathbf{u} = \langle u_1, u_2 \rangle$, $\mathbf{v} = \langle v_1, v_2 \rangle$, and $\mathbf{w} = \langle w_1, w_2 \rangle$. Then,

$$\mathbf{u} \cdot (\mathbf{v} + \mathbf{w}) = \langle u_1, u_2 \rangle \cdot (\langle v_1, v_2 \rangle + \langle w_1, w_2 \rangle)$$
$$= \langle u_1, u_2 \rangle \cdot \langle v_1 + w_1, v_2 + w_2 \rangle$$
$$= u_1(v_1 + w_1) + u_2(v_2 + w_2)$$
$$= u_1 v_1 + u_2 v_2 + u_1 w_1 + u_2 w_2$$
$$= \langle u_1, u_2 \rangle \cdot \langle v_1, v_2 \rangle + \langle u_1, u_2 \rangle \cdot \langle w_1, w_2 \rangle$$
$$= \mathbf{u} \cdot \mathbf{v} + \mathbf{u} \cdot \mathbf{w}.$$

To prove the fifth property, let $\mathbf{u} = \langle u_1, u_2 \rangle$ and $\mathbf{v} = \langle v_1, v_2 \rangle$ and let c be a scalar. Then,

$$c(\mathbf{u} \cdot \mathbf{v}) = c(\langle u_1, u_2 \rangle \cdot \langle v_1, v_2 \rangle)$$
$$= c(u_1 v_1 + u_2 v_2)$$
$$= (cu_1)v_1 + (cu_2)v_2$$
$$= \langle cu_1, cu_2 \rangle \cdot \langle v_1, v_2 \rangle = c\mathbf{u} \cdot \mathbf{v}$$
$$= u_1(cv_1) + u_2(cv_2)$$
$$= \langle u_1, u_2 \rangle \cdot \langle cv_1, cv_2 \rangle = \mathbf{u} \cdot c\mathbf{v}. \qquad \textbf{///}$$

Section 9.1, Page 661

Properties of Sums

1. $\displaystyle\sum_{i=1}^{n} ca_i = c\sum_{i=1}^{n} a_i,$ c is any constant

2. $\displaystyle\sum_{i=1}^{n} (a_i + b_i) = \sum_{i=1}^{n} a_i + \sum_{i=1}^{n} b_i$

3. $\displaystyle\sum_{i=1}^{n} (a_i - b_i) = \sum_{i=1}^{n} a_i - \sum_{i=1}^{n} b_i$

Proof **///**   Each of these properties follows directly from the Associative Property of Addition, the Commutative Property of Addition, and the Distributive Property of multiplication over addition. For example, note the use of the Distributive Property in the proof of Property 1.

$$\sum_{i=1}^{n} ca_i = ca_1 + ca_2 + ca_3 + \cdots + ca_n$$

$$= c(a_1 + a_2 + a_3 + \cdots + a_n) = c\sum_{i=1}^{n} a_i \qquad\qquad \text{///}$$

Section 9.2, Page 670

The Sum of a Finite Arithmetic Sequence

The sum of a finite arithmetic sequence with n terms is given by

$$S_n = \frac{n}{2}(a_1 + a_n).$$

Proof **///** Begin by generating the terms of the arithmetic sequence in two ways. In the first way, repeatedly add d to the first term to obtain

$$S_n = a_1 + a_2 + a_3 + \cdots + a_{n-2} + a_{n-1} + a_n$$
$$= a_1 + [a_1 + d] + [a_1 + 2d] + \cdots + [a_1 + (n-1)d].$$

In the second way, repeatedly subtract d from the nth term to obtain

$$S_n = a_n + a_{n-1} + a_{n-2} + \cdots + a_3 + a_2 + a_1$$
$$= a_n + [a_n - d] + [a_n - 2d] + \cdots + [a_n - (n-1)d].$$

If you add these two versions of S_n, the multiples of d cancel and you obtain

$$2S_n = \overbrace{(a_1 + a_n) + (a_1 + a_n) + (a_1 + a_n) + \cdots + (a_1 + a_n)}^{n \text{ terms}}$$
$$= n(a_1 + a_n).$$

Thus, you have

$$S_n = \frac{n}{2}(a_1 + a_n).$$ **///**

Section 9.3, Page 679

The Sum of a Finite Geometric Sequence

The sum of the geometric sequence

$$a_1, \ a_1r, \ a_1r^2, \ a_1r^3, \ a_1r^4, \ldots, a_1r^{n-1}$$

with common ratio $r \neq 1$ is given by

$$S_n = a_1\left(\frac{1 - r^n}{1 - r}\right).$$

Proof   ///   Begin by writing out the nth partial sum.

$$S_n = a_1 + a_1r + a_1r^2 + \cdots + a_1r^{n-2} + a_1r^{n-1}$$

Multiplication by r yields

$$rS_n = a_1r + a_1r^2 + a_1r^3 + \cdots + a_1r^{n-1} + a_1r^n.$$

Subtracting the second equation from the first yields

$$S_n - rS_n = a_1 - a_1r^n.$$

Therefore, $S_n(1 - r) = a_1(1 - r^n)$, and, because $r \neq 1$, you have

$$S_n = a_1\left(\frac{1 - r^n}{1 - r}\right). \qquad\qquad ///$$

Section 9.5, Page 697

> **The Binomial Theorem**
>
> In the expansion of $(x + y)^n$
>
> $$(x + y)^n = x^n + nx^{n-1}y + \cdots + {}_nC_r\, x^{n-r}y^r + \cdots + nxy^{n-1} + y^n$$
>
> the coefficient of $x^{n-r}y^r$ is given by
>
> $$_nC_r = \frac{n!}{(n-r)!r!}.$$

Proof /// The Binomial Theorem can be proved quite nicely using mathematical induction. The steps are straightforward but look a little messy, so we will present only an outline of the proof.

1. If $n = 1$, you have

$$(x + y)^1 = x^1 + y^1 = {}_1C_0x + {}_1C_1y$$

and the formula is valid.

2. Assuming that the formula is true for $n = k$, the coefficient of $x^{k-r}y^r$ is given by

$$_kC_r = \frac{k!}{(k-r)!r!} = \frac{k(k-1)(k-2)\cdots(k-r+1)}{r!}.$$

To show that the formula is true for $n = k + 1$, look at the coefficient of $x^{k+1-r}y^r$ in the expansion of

$$(x + y)^{k+1} = (x + y)^k(x + y).$$

From the right-hand side, you can determine that the term involving $x^{k+1-r}y^r$ is the sum of two products.

$$({}_kC_r x^{k-r}y^r)(x) + ({}_kC_{r-1}x^{k+1-r}y^{r-1})(y)$$

$$= \left[\frac{k!}{(k-r)!r!} + \frac{k!}{(k-r+1)!(r-1)!}\right]x^{k+1-r}y^r$$

$$= \left[\frac{(k+1-r)k!}{(k+1-r)!r!} + \frac{k!r}{(k+1-r)!r!}\right]x^{k+1-r}y^r$$

$$= \left[\frac{k!(k+1-r+r)}{(k+1-r)!r!}\right]x^{k+1-r}y^r$$

$$= \left[\frac{(k+1)!}{(k+1-r)!r!}\right]x^{k+1-r}y^r$$

$$= {}_{k+1}C_r x^{k+1-r}y^r$$

Thus, by mathematical induction, the Binomial Theorem is valid for all positive integers n. **///**

Section 10.1, Page 755

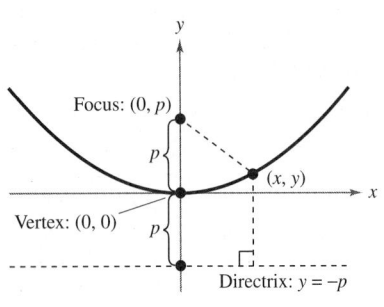

Standard Equation of a Parabola (Vertex at Origin)

The **standard form of the equation of a parabola** with vertex at $(0, 0)$ and directrix $y = -p$ is

$$x^2 = 4py, \qquad p \neq 0. \qquad\qquad \text{Vertical axis}$$

For directrix $x = -p$, the equation is

$$y^2 = 4px, \qquad p \neq 0. \qquad\qquad \text{Horizontal axis}$$

The focus is on the axis p units (directed distance) from the vertex.

Proof /// Because the two cases are similar, a proof will be given for the first case only. Suppose the directrix $(y = -p)$ is parallel to the x-axis. In the figure, you assume that $p > 0$, and because p is the directed distance from the vertex to the focus, the focus must lie above the vertex. Because the point (x, y) is equidistant from $(0, p)$ and $y = -p$, you can apply the Distance Formula to obtain

$$\sqrt{(x - 0)^2 + (y - p)^2} = y + p$$
$$x^2 + (y - p)^2 = (y + p)^2$$
$$x^2 + y^2 - 2py + p^2 = y^2 + 2py + p^2$$
$$x^2 = 4py. \qquad\qquad\qquad ///$$

Appendix B   Programs

Reflections and Shifts Program (Section 1.3)

This program, referenced in the Group Activity on page 115, will sketch a graph of the function $y = R(x + H)^2 + V$, where $R = \pm1$, H is an integer between -6 and 6, and V is an integer between -3 and 3. This program gives you practice working with reflections, horizontal shifts, and vertical shifts.

Note: On the *TI-83* and the *TI-82*, the "int" and "rand" commands may be entered through the "NUM" and "PRB" menus, respectively, accessed by pressing the MATH key. The "=" and "<" symbols may be entered through the "TEST" menu accessed by pressing the TEST key. Other commands, such as "If," "Then," "Else," and "End," may be entered through the "CTL" menu accessed by pressing the PRGM key. The commands "Xmin," "Xmax," "Xscl," "Ymin," "Ymax," and "Yscl" may be entered through the "Window" menu accessed by pressing the VARS key. The commands "DispGraph" and "Pause" may be entered through the "I/O" and "CTL" menus, respectively, accessed by pressing the PRGM key. For additional keystroke instructions, see previous programs in this appendix. Keystroke sequences for similar commands on other calculators will vary. Consult the user manual for your calculator.

TI-80

```
PROGRAM:PARABOL
:-6+INT (12RAND)→H
:-3+INT (6RAND)→V
:RAND→R
:IF R <.5
:THEN
:-1→R
:ELSE
:1→R
:END
:"R(X+H)²+V"→Y1
:-9→XMIN
:9→XMAX
:1→XSCL
:-6→YMIN
:6→YMAX
:1→YSCL
:DISPGRAPH
:PAUSE
:DISP "Y=R(X+H)²+V²"
:DISP "R=",R
:DISP "H=",H
:DISP "V=",V
:PAUSE
```

Press ENTER after viewing the graph to display the values of the integers.

TI-81

```
Prgm2:PARABOLA
:Rand→H
:-6+Int (12H)→H
:Rand→V
:-3+Int (6V)→V
:Rand→R
:If R <.5
:-1→R
:If R >.49
:1→R
:"R(X+H)²+V"→Y1
:-9→Xmin
:9→Xmax
:1→Xscl
:-6→Ymin
:6→Ymax
:1→Yscl
:DispGraph
:Pause
:Disp "Y=R(X+H)²+V"
:Disp "R="
:Disp R
:Disp "H="
:Disp H
:Disp "V="
:Disp V
:End
```

Press ENTER after viewing the graph to display the values of the integers.

A11

TI-83
TI-82
PROGRAM:PARABOLA
:-6+int (12rand)→H
:-3+int (6rand)→V
:rand→R
:If R < .5
:Then
:-1→R
:Else
:1→R
:End
:"R(X+H)^2+V"→Y1
:-9→Xmin
:9→Xmax
:1→Xscl
:-6→Ymin
:6→Ymax
:1→Yscl
:DispGraph
:Pause
:Disp "Y=R(X+H)^2+V"
:Disp "R=",R
:Disp "H=",H
:Disp "V=",V
:Pause

Press ENTER after viewing the graph to display the values of the integers.

TI-85
PROGRAM:PARABOLA
:rand→H
:-6+int (12H)→H
:rand→V
:-3+int (6V)→V
:rand→R
:If R < .5
:-1→R
:If R > .49
:1→R
:y1=R(x+H)^2+V
:-9→xMin
:9→xMax
:1→xScl
:-6→yMin
:6→yMax
:1→yScl
:DispG
:Pause
:Disp "Y=R(X+H)^2+V"
:Disp "R=",R
:Disp "H=",H
:Disp "V=",V
:Pause

Press ENTER after viewing the graph to display the values of the integers.

TI-92

Parabola( )
Prgm
ClrHome
ClrIO
setMode("Split Screen",
  "Left-Right")
setMode("Split 1 App","Home")
setMode("Split 2 App","Graph")
-6+int (12rand( ))→h
-3+int (6rand( ))→v
rand( )→r
If r < .5 Then
  -1→r
    Else
  1→r
EndIf
r*(x+h)^2+v→y1(x)
-9→xmin
9→xmax
1→xscl
-6→ymin
6→ymax
1→yscl
DispG
Disp "y1(x)=r(x+h)^2+v"
Output 20,1, "r=":Output 20,11,r
Output 40,1, "h=":Output 40,11,h
Output 60,1, "v=":Output 60,11,v
Pause
setMode("Split Screen","Full")
EndPrgm

Casio fx-7700G

PARABOLA
-6+INT (12Ran#)→H
-3+INT (6Ran#)→V
-1→R:Ran#<0.5 ⇒1→R
Range -9,9,1,-6,6,1
Graph Y=R(X+H)^2+V◢
"Y=R(X+H)^2+V"
"R=":R◢
"H=":H◢
"V=":V

Press ⬛EXE⬛ after viewing the graph to display the values of the integers.

Casio fx-7700GE
Casio fx-9700GE
Casio CFX-9800G

PARABOLA↵
-6+Int (12Ran#)→H↵
-3+Int (6Ran#)→V↵
Ran#→R↵
R< .5⇒-1→R↵
R≥ .5⇒1→R↵
Range -9,9,1,-6,6,1↵
Graph Y=R(X+H)^2+V◢
"Y=R(X+H)^2+V"↵
"R=":R◢
"H=":H◢
"V=":V

Press ⬛EXE⬛ after viewing the graph to display the values of the integers.

Sharp EL-9200C
Sharp EL-9300C

parabola
——————REAL
h=int (random*12) -6
v=int (random*6) -3
s=(random*2) -1
r=s/abs s
Range -9,9,1,-6,6,1
Graph r(X+h)^2+v
Wait
Print "y=r(X+h)^2+v
Print r
Print h
Print v
End

Press ⬛ENTER⬛ after viewing the graph to display the values of the integers.

HP-38G

PARABOLA

PARABOLA PROGRAM
```
-6+INT(12RANDOM)▶H:
-3+INT(6RANDOM)▶V:
RANDOM ▶R:
IF R>.5
   THEN -1▶R:
   ELSE 1▶R:
END:
'R*(X+H)²+V'▶F1(X):
CHECK 1:
```

PARANS PROGRAM
```
ERASE:
DISP 2;"Y=R(X+H)²+V":
DISP 3;"R=":R:
DISP 4;"H=":H:
DISP 5;"V=":V:
FREEZE:
```

PARABOLA.SV PROGRAM
```
SETVIEWS "RUN
PARABOLA";PARABOLA;1;
"ANSWER";PARANS;1;
"  ";PARABOLA.SV;0:
```

1. Press ⌑LIB⌑. Highlight the Function aplet. Press {{SAVE}}. Enter the name PARABOLA for the new aplet and press {{OK}}.
2. Press ■ [SETUP-PLOT] and set XRNG: from −12 to 12, YRNG: from −6 to 6, and XTICK: and YTICK: to 1.
3. Enter the 3 programs PARABOLA, PARANS, PARABOLA.SV.
4. Run the program PARABOLA.SV.
5. Enter the PARABOLA aplet.
6. Press ■ [VIEWS]. Highlight RUN PARABOLA and press {{OK}}.
7. After viewing the graph press ■ [VIEWS]. Highlight ANSWER and press {{OK}} to see the values of the integers.
8. Press {{OK}} to return to the graph.
9. Repeat steps 6, 7, and 8 for a new parabola.

Graph Reflection Program (Section 1.5)

This program, referenced in the Technology note on page 134, will graph a function *f* and its reflection in the line $y = x$.

Note: On the *TI-83* and the *TI-82*, the "While" command may be entered through the "CTL" menu accessed by pressing the PRGM key. The "Pt-On(" command may be entered through the "POINTS" menu accessed by pressing the DRAW key. For additional keystroke instructions, see previous programs in this appendix. Keystroke sequences required for similar commands on other calculators will vary. Consult the user manual for your calculator.

TI-80

PROGRAM:REFLECT
:47XMIN/63→YMIN
:47XMAX/63→YMAX
:XSCL→YSCL
:"X"→Y2
:DISPGRAPH
:(XMAX−XMIN)/62→I
:XMIN→X
:LBL A
:PT-ON(Y1,X)
:X+I→X
:If X>XMAX
:STOP
:GOTO A

To use this program, enter the function in Y1 and set a viewing rectangle.

TI-81

Prgm3:REFLECT
:2Xmin/3→Ymin
:2Xmax/3→Ymax
:Xscl→Yscl
:"X"→Y2
:DispGraph
:(Xmax−Xmin)/95→I
:Xmin→X
:Lbl 1
:Pt-On(Y1,X)
:X+I→X
:If X>Xmax
:End
:Goto 1

To use this program, enter the function in Y1 and set a viewing rectangle.

TI-83
TI-82
PROGRAM:REFLECT
:63Xmin/95→Ymin
:63Xmax/95→Ymax
:Xscl→Yscl
:"X"→Y_2
:DispGraph
:(Xmax−Xmin)/94→I
:Xmin→X
:While X≤Xmax
:Pt-On(Y_1,X)
:X+I→X
:End

To use this program, enter the function in Y_1 and set a viewing rectangle.

TI-85
PROGRAM:REFLECT
:63*xMin/127→yMin
:63*xMax/127→yMax
:xScl→yScl
:y2=x
:DispG
:(xMax−xMin)/126→I
:xMin→x
:Lbl A
:PtOn(y1,x)
:x+I→x
:If x>xMax
:Stop
:Goto A

To use this program, enter the function in y1 and set a viewing rectangle.

TI-92
Prgm
103xmin/239→ymin
103xmax/239→ymax
xscl→yscl
x→y2(x)
DispG
(xmax−xmin)/238→n
xmin→x
While x<xmax
    PtOn y1(x),x
    x+n→x
EndWhile
EndPrgm

To use this program, enter the function in y1 and set an appropriate viewing window.

Casio fx-7700G
REFLECTION
"GRAPH -A TO A"
"A="?→A
Range -A,A,1,-2A÷3,2A÷3,1
Graph Y=f_1
-A→B
Lbl 1
B→X
Plot f_1,B
B+A÷32→B
B≤A⇒Goto1 :Graph Y=X

To use this program, enter the function in f_1.

Casio fx-7700GE

REFLECTION
"GRAPH -A TO A"↵
"A="?→A↵
Range -A,A,1,-2A÷3,2A÷3,1↵
Graph Y=f₁↵
-A→B↵
Lbl 1↵
B→X↵
Plot f₁,B↵
B+A÷32→B↵
B≤A⇒Goto1:Graph Y=X

To use this program, enter the function in f₁.

Casio fx-9700GE

REFLECTION↵
63Xmin÷127→A↵
63Xmax÷127→B↵
Xscl→C↵
Range , , , A, B, C↵
(Xmax−Xmin)÷126→I↵
Xmax→M↵
Xmin→D↵
Graph Y=f₁↵
Lbl 1↵
D→X↵
Plot f₁,D↵
D+I→D↵
D≤M⇒Goto 1:Graph Y=X

To use this program, enter the function in f₁ and set a viewing rectangle.

Casio CFX-9800G

REFLECTION↵
63Xmin÷95→A↵
63Xmax÷95→B↵
Xscl→C↵
Range , , , A, B, C↵
(Xmax−Xmin)÷94→I↵
Xmax→M↵
Xmin→D↵
Graph Y=f₁↵

Casio CFX-9800G

(Continued)

Lbl 1↵
D→X↵
Plot f₁,D↵
D+I→D↵
D≤M⇒Goto 1:Graph Y=X

To use this program, enter the function in f₁ and set a viewing rectangle.

Sharp EL-9200C
Sharp EL-9300C

reflection
————REAL
Goto top
Label equation
Y=f(X)
Return
Label rng
xmin=-10
xmax=10
xstp=(xmax−xmin)/10
ymin=2xmin/3
ymax=2xmax/3
ystp=xstp
Range xmin,xmax,xstp,ymin,
    ymax,ystp
Return
Label top
Gosub rng
Graph X
step=(xmax−xmin)/(94*2)
X=xmin
Label 1
Gosub equation
Plot X,Y
Plot Y, X
X=X+step
If X<=xmax Goto 1
End

To use this program, replace f(X) with your expression in X.

Angles and Their Measures (Section 4.1)

This program, found in the Group Activity on page 317, An Angle-Drawing Program, can be used to draw several different angles in either radian or degree mode.

TI-80

```
PROGRAM:ANGDRAW
:CLRDRAW
:CLRHOME
:FNOFF
:RADIAN
:PARAM
:DISP "ENTER MODE"
:DISP "0 RADIAN"
:DISP "1 DEGREE"
:INPUT M
:INPUT "ENTER ANGLE",T
:IF M=1
:πT/180→T
:-1.5→XMIN
:1.5→XMAX
:1→XSCL
:-1→YMIN
:1→YMAX
:1→YSCL
:0→TMIN
:ABS T→TMAX
:.15→TSTEP
:COS T→A
:SIN T→B
:1→S
:IF T<0
:-1→S
:"(.25+.04T)COS T"→X1T
:"S(.25+.04T)SIN T"→Y1T
:DISPGRAPH
:LINE(0,0,A,B)
:PAUSE
:FUNC
:STOP
```

TI-81

```
PrgmD:ANGDRAW
:ClrDraw
:ClrHome
:All-Off
:Rad
:Param
:Disp "ENTER MODE"
:Disp "0 RADIAN"
:Disp "1 DEGREE"
:Input M
:Input "ENTER ANGLE"
:Input T
:If M=1
:πT/180→T
:-1.5→Xmin
:1.5→Xmax
:1→Xscl
:-1→Ymin
:1→Ymax
:1→Yscl
:0→Tmin
:abs T→Tmax
:.15→Tstep
:cos T→A
:sin T→B
:1→S
:If T<0
:-1→S
:"(.25+.04T)cos T"→X1T
:"S(.25+.04T)sin T"→Y1T
:DispGraph
:Line(0,0,A,B)
:Pause
:Function
:End
```

TI-83
TI-82

PROGRAM:ANGDRAW
:ClrDraw
:ClrHome
:FnOff
:Radian
:Param
:Disp "ENTER MODE"
:Disp "0 RADIAN"
:Disp "1 DEGREE"
:Input M
:Input "ENTER ANGLE",T
:If M=1
:πT/180$\rightarrow$T
:-1.5$\rightarrow$Xmin
:1.5$\rightarrow$Xmax
:1$\rightarrow$Xscl
:-1$\rightarrow$Ymin
:1$\rightarrow$Ymax
:1$\rightarrow$Yscl
:0$\rightarrow$Tmin
:abs (T)$\rightarrow$Tmax
:.15$\rightarrow$Tstep
:cos (T)$\rightarrow$A
:sin (T)$\rightarrow$B
:1$\rightarrow$S
:If T<0
:-1$\rightarrow$S
:"(.25+.04T)cos T"$\rightarrow$X$_{1T}$
:"S(.25+.04T)sin T"$\rightarrow$Y$_{1T}$
:DispGraph
:Line(0,0,A,B)
:Pause
:Func
:Stop

TI-85

PROGRAM:ANGDRAW
:ClDrw
:ClLCD
:FnOff
:Radian
:Param
:Disp "Enter mode"
:Disp "0 radian"
:Disp "1 degree"
:Input M
:Input "enter angle",T
:If M==1
:πT/180$\rightarrow$T
:-1.5$\rightarrow$xMin
:1.5$\rightarrow$xMax
:1$\rightarrow$xScl
:-1$\rightarrow$yMin
:1$\rightarrow$yMax
:1$\rightarrow$yScl
:0$\rightarrow$tMin
:abs T$\rightarrow$tMax
:.15$\rightarrow$tStep
:cos T$\rightarrow$A
:sin T$\rightarrow$B
:1$\rightarrow$S
:If T<0
:-1$\rightarrow$S
:xt1=(.25+.04T)cos t
:yt1=S(.25+.04T)sin t
:DispG
:Line(0,0,A,B)
:Pause
:Func

TI-92

```
angdraw()
:Prgm
:ClrHome
:ClrGraph
:setMode("Angle","RADIAN")
:setMode("Graph",
    "PARAMETRIC")
:setMode("Split Screen",
    "LEFT-RIGHT")
:setMode("Split 1 App","Home")
:setMode("Split 2 App","Graph")
:Disp "enter mode"
:Disp "0 radian"
:Disp "1 degree"
:Input m
:Input "enter angle",t
:If m=1
:π∗t/180→t
:-1.5→xmin
:1.5→xmax
:1→xscl
:-1→ymin
:1→ymax
:1→yscl
:0→tmin
:abs(t)→tmax
:.15→tstep
:cos (t)→a
:sin (t)→b
:1→s
:If t<0
:-1→s
:Graph(.25+.04∗t)∗cos (t),
    s∗(.25+.04∗t)∗sin (t),t
:Line 0,0,a,b
:Pause
:setMode("Graph","FUNCTION")
:setMode("Split Screen","FULL")
:EndPrgm
```

Casio fx-7700G

```
ANGDRAW
"ENTER MODE"
"0 RADIAN"
"1 DEGREE"
?→M
"ENTER ANGLE"?→T
M=1⇒πT÷180→T
Rad
Cls
Range -1.5,1.5,1,-1,1,1,0,
    Abs T,.15
cos T→A
sin T→B
1→S
T<0⇒-1→S
Graph(X,Y)=((.25+.04T)cos T,
    S(.25+.04T)sin T)
Plot 0,0
Plot A,B
Line
```

Press MODE SHIFT ✕ to change to parametric mode before writing this program.

Casio fx-7700GE
Casio fx-9700GE
Casio CFX-9800G

ANGDRAW.↵
"ENTER MODE"↵
"0 RADIAN"↵
"1 DEGREE"↵
?→M ↵
"ENTER ANGLE"↵
?→T↵
M=1⇒πT÷180→T↵
Rad↵
Cls↵
Range -1.5,1.5,1,-1,1,1,0,
    Abs T,.15↵
cos T→A↵
sin T→B↵
1→S↵
T<0⇒-1→S↵
Graph(X,Y)=((.25+.04T)cos T,
    S(.25+.04T)sin T)↵
Plot 0,0↵
Plot A,B↵
Line

Use SET UP to change the
GRAPH TYPE to parametric
before writing this program.

SHARP EL-9200C
SHARP EL-9300C

angdraw
————————REAL
Print "enter mode"
Print "0 radian"
Print "1 degree"
Input mode
Input angle
If mode=0 Goto 1
angle=angle$*\pi$/180
Label 1
Range -2.25,2.25,1.5,-1.5,1.5,
    1.5
s=angle/abs angle
loops=ipart (s$*$angle/(2π))+1
θ=0
ostep=angle/(30$*$loops)
r=1/loops
rstep=(1−r)/(30$*$loops)
xo=r
yo=0
n=0
Label top
r=r+rstep
θ=θ+ostep
x=r$*$cos (θ)
y=r$*$sin (θ)
Line xo,yo,x,y
xo=x
yo=y
n=n+1
If n<30$*$loops Goto top
Line 0,0,1.5x,1.5y

Use SET UP to set the mode
to radians before running the
program.

HP-38G

ANGDRAW PROGRAM
```
INPUT M;"ENTER MODE";
   "0 OR 1";"0 RADIAN,
   1 DEGREE";1:
INPUT T;"ENTER ANGLE";
   "ANGLE?";" ";1:
IF M==1
   THEN T*π/180▶T:
END:
T/ABS(T)▶S:
0▶K:
INT(ST/(2π)+1▶L:
T/(30L)▶P:
1/L▶R:
(1-R/(30L)▶Q:
R▶A:
0▶B:
0▶N:
ERASE:
LINE -2.6;0;2.6;0:
LINE 0;-1.2;0;1.2:
FOR N=0 TO (29L)
   STEP 1;
   RUN "DRAWCURVE":
END:
LINE 0;0;1.5X;1.5Y:
FREEZE
```

DRAWCURVE PROGRAM
```
R+Q▶R:
K+P▶K:
R*COS(K)▶X:
R*SIN(K)▶Y:
LINE A;B;X;Y:
X▶A:
Y▶B:
```

1. Enter the two programs ANGDRAW and DRAWCURVE.
2. From the HOME screen set the mode to radians. Set the plot range in the Function aplet to $-2.6 \leq x \leq 2.6$ and $-1.2 \leq y \leq 1.2$. Set the angle measure to radians.
3. Run the ANGDRAW program.

Graphing a Sine Function (Section 4.5)

The program, shown in the Group Activity on page 360, will simultaneously draw a unit circle and the corresponding points on the sine curve. After the circle and sine curve are drawn, you can connect the points on the unit circle with their corresponding points on the sine curve by pressing ENTER or EXE .

TI-80

```
PROGRAM:SINESHO
:RADIAN
:CLRDRAW:FNOFF
:PARAM:SIMUL
:-2.25→XMIN
:π/2→XMAX
:3→XSCL
:-1.5→YMIN
:1.5→YMAX
:1→YSCL
:0→TMIN
:6.3→TMAX
:.15→TSTEP
:"-1.25+COS T"→X1T
:"SIN T"→Y1T
:"T/4"→X2T
:"SIN T"→Y2T
:DISPGRAPH
:FOR(N,1,12)
:Nπ/6.5→T
:"-1.25+COS T"→A
:SIN T→B
:T/4→C
:LINE(A,B,C,B)
:PAUSE
:END
:PAUSE:FUNC
:SEQUENTIAL:DISP
```

TI-81

```
PrgmA:SINESHOW
:Rad
:ClrDraw
:Param
:Simul
:-2.25→Xmin
:π/2→Xmax
:3→Xscl
:-1.19→Ymin
:1.19→Ymax
:1→Yscl
:0→Tmin
:6.3→Tmax
:.15→Tstep
:"-1.25+cos T"→X1T
:"sin T"→Y1T
:"T/4"→X2T
:"sin T"→Y2T
:DispGraph
:1→N
:Lbl 1
:IS>(N,12)
:Goto 2
:Pause
:Function
:Sequence
:Disp " "
:End
:Lbl 2
:Nπ/6.5→T
:-1.25+cos T→A
:sin T→B
:T/4→C
:Line(A,B,C,B)
:Pause
:Goto 1
```

TI-83
TI-82

PROGRAM:SINESHOW
:Radian
:ClrDraw:FnOff
:Param:Simul
:-2.25→Xmin
:π/2→Xmax
:3→Xscl
:-1.19→Ymin
:1.19→Ymax
:1→Yscl
:0→Tmin
:6.3→Tmax
:.15→Tstep
:"-1.25+cos (T)"→X1T
:"sin (T)"→Y1T
:"T/4"→X2T
:"sin (T)"→Y2T
:DispGraph
:For(N,1,12)
:Nπ/6.5→T
:-1.25+cos (T)→A
:sin(T)→B
:T/4→C
:Line(A,B,C,B)
:Pause
:End
:Pause :Func
:Sequential:Disp

TI-85

PROGRAM:SINESHOW
:Radian
:ClDrw:FnOff
:Param:SimulG
:-2.25→xMin
:π/2→xMax
:3→xScl
:-1.1→yMin
:1.1→yMax
:1→yScl
:0→tMin
:6.3→tMax
:.15→tStep
:xt1=-1.25+cos t
:yt1=sin t
:xt2=t/4
:yt2=sin t
:For(N,1,12)
:N*π/6.5→t
:-1.25+cos t→A
:sin t→B
:t/4→C
:Line(A,B,C,B)
:Pause
:End
:Pause :Func
:SeqG:Disp

TI-92

sineshow()
Prgm
Disp
ClrDraw:FnOff
setMode("Graph", "Parametric")
setGraph("Graph Order",
    "Simul")
-2.9→xmin
$3\pi/4$→xmax
3→xscl
-1.1→ymin
1.1→ymax
1→yscl
0→tmin
6.3→tmax
.15→tstep
-1.25+cos(t)→xt1(t)
sin(t)→yt1(t)
t/4→xt2(t)
sin(t)→yt2(t)
DispG
For N,1,12
N*π/6.5→t
-1.25+cos(t)→A
sin(t)→B
t/4→C
Line A,B,C,B
Pause
EndFor
Pause
setMode("Graph", "Function")
setGraph("Graph order",
    "Seq")
setMode("Split 1 App",
    "Home")
EndPrgm

Casio fx-7700G

SINESHOW
Rad
Range -2.25,$\pi\div2$,3,-1.19,1.19,
    10,6.3,.15
Graph(X,Y)=(-1.25+cos T,sinT)
Graph(X,Y)=(T$\div$4,sinT)
0→N
Lbl 1
N+1→N
N$\pi\div$6.5→T
-1.25+cos T→A
sin T→B
T$\div$4→C
Plot A,B
Plot C,B
Line
N<12⇒Goto 1

Press $\boxed{\text{MODE}}$ $\boxed{\text{SHIFT}}$ $\boxed{\times}$ to change
to parametric mode when starting
to write this program.

Casio fx-7700GE
Casio fx-9700GE
Casio CFX-9800G

SINESHOW↵
Rad↵
Range -2.25,$\pi\div$2,3,-1.19,1.19,
   1,0,6.3,.15↵
Graph(X,Y)=(-1.25+cos T,sin T)↵
Graph(X,Y)=(T÷4,sin T)↵
0→N↵
Lbl 1↵
N+1→N↵
Nπ÷6.5→T↵
-1.25+cosT→A↵
sin T→B↵
T÷4→C↵
Plot A,B↵
Plot C,B↵
Line
N<12⇒Goto 1↵
Cls

When starting to write this program press SET UP and select PRM or PARM for the GRAPH TYPE to change to parametric mode.

Sharp EL-9200C
Sharp EL-9300C

sineshow
————————REAL
m=sin^{-1} 1/(π/2)
Range -2.25,π/2,3,-1.19,1.19,1
step=π/15
θ=0
xco=-.25
xso=0
yo=0
Label 1
$\theta=\theta$+step
xc=cos(mθ)$-$1.25
xs=θ/4
y=sin (mθ)
Line xco,yo,xc,y
Line xso,yo,xs,y
xco=xc
xso=xs
yo=y
If θ<(2π) Goto 1
step=π/6
θ=0
Label 2
$\theta=\theta$+step
xc=cos (mθ)$-$1.25
xs=θ/4
y=sin (mθ)
Line xc,y,xs,y
Wait
If θ<2π Goto 2
End

HP-38G Programs

SINESHOW PROGRAM
ASIN(1)/(π/2)►M:
0►T:
-.25►A:
0►B:
0►C:
LINE -3;0;π/2;0:
LINE 0;-1.1;0;1.1:
FOR T=0 TO 31π/15
  STEP π/15;
  RUN "DRAW.SINE":
END:
0►T:
FOR T=0 TO 2π
  STEP π/6;
  RUN "DRAW.LINE":
END

DRAW.SINE PROGRAM
COS(MT)$-$1.25►D:
T/4►E:
SIN(MT)►F:
LINE A;C;D;F:
LINE B;C;E;F:
D►A:
E►B:
F►C:

DRAW.LINE PROGRAM
COS(MT)$-$1.25►D:
T/4►E:
SIN(MT)►F:
LINE D;F;E;F:
FREEZE

1. Enter the 3 programs
   SINESHOW, DRAW.SINE, and
   DRAW.LINE.
2. Set the plot range in the
   Function aplet to $-3 \le x \le \pi/2$
   and $-1.1 \le y \le 1.1$. Set the angle
   measure to radians.
3. Run the SINESHOW program.

Finding the Angle between Two Vectors (Section 6.4)

The program, shown in the Technology note on page 492, will sketch two vectors and calculate the measure of the angle between the vectors. Be sure to set an appropriate viewing rectangle.

TI-80

```
:PROGRAM:VECANGL
:CLRHOME
:DEGREE
:DISP "ENTER (A,B)"
:INPUT "ENTER A",A
:INPUT "ENTER B",B
:CLRHOME
:DISP "ENTER (C,D)"
:INPUT "ENTER C",C
:INPUT "ENTER D",D
:LINE(0,0,A,B)
:LINE(0,0,C,D)
:PAUSE
:AC+BD→E
:√ (A²+B²)→U
:√ (C²+D²)→V
:COS⁻¹(E/(UV))→θ
:DISP "θ=", θ
:CLRDRAW
```

TI-81

```
:PrgmB:VECANGL
:ClrHome
:Deg
:Disp "ENTER (A,B)"
:Disp "ENTER A"
:Input A
:Disp "ENTER B"
:Input B
:ClrHome
:Disp "ENTER (C,D)"
:Disp "ENTER C"
:Input C
:Disp "ENTER D"
:Input D
:Line(0,0,A,B)
:Line(0,0,C,D)
:Pause
:AC+BD→E
:√ (A²+B²)→U
:√ (C²+D²)→V
:cos⁻¹(E/(UV))→θ
:Disp "θ="
:Disp θ
:ClrDraw
:End
```

TI-83
TI-82

:PROGRAM:VECANGL
:ClrHome
:Degree
:Disp "ENTER (A,B)"
:Input "ENTER A",A
:Input "ENTER B",B
:ClrHome
:Disp "ENTER (C,D)"
:Input "ENTER C",C
:Input "ENTER D",D
:Line(0,0,A,B)
:Line(0,0,C,D)
:Pause
:AC+BD→E
:$\sqrt{\ }\ (A^2+B^2)$→U
:$\sqrt{\ }\ (C^2+D^2)$→V
:$\cos^{-1}(E/(UV))$→θ
:ClrDraw:ClrHome
:Disp "θ=",θ
:Stop

TI-85

:PROGRAM:VECANGL
:ClLCD
:Radian
:Disp "enter (A,B)"
:Input "enter A",A
:Input "enter B",B
:ClLCD
:Disp "enter (C,D)"
:Input "enter C",C
:Input "enter D",D
:Line(0,0,A,B)
:Line(0,0,C,D)
:Pause
:A*C+B*D→E
:$\sqrt{\ }\ (A^2+B^2)$→U
:$\sqrt{\ }\ (C^2+D^2)$→V
:$\cos^{-1}(E/(U*V))$→T
:T*180/π→T
:Disp "T=",T
:ClDrw

TI-92	**Casio fx-7700G**
vecangl( )	VECANGL
Prgm	Cls
FnOff	Deg
ClrHome:ClrDraw	"ENTER (A,B)"
SetMode("Split Screen",	"A="?→A
"Left-Right")	"B="?→B
SetMode("Split 1 App", "Home")	"ENTER (C,D)"
SetMode("Split 2 App", "Graph")	"C="?→C
SetMode("Exact/Approx",	"D="?→D
"Approximate")	Plot 0,0
ClrIO	Plot A,B
Disp "ENTER (A,B)"	Line
Input "ENTER A", A	Plot 0,0
Input "ENTER B", B	Plot C,D
Line(0,0,A,B)	Line ◢
Pause	AC+BD→E
ClrIO	$\sqrt{\ }\ (A^2+B^2)$→U
Disp "ENTER (C,D)"	$\sqrt{\ }\ (C^2+D^2)$→V
Input "ENTER C",C	$\cos^{-1}(E\div UV)$→θ
Input "ENTER D",D	"θ="
Line(0,0,C,D)	θ
Pause	
ClrIO	
A*C+B*D→E	
$\sqrt{\ }\ (A\char94 2+B\char94 2)$→U	
$\sqrt{\ }\ (C\char94 2+D\char94 2)$→V	
$\cos^{-1}(E/(U*V))$→θ	
Disp "θ=",θ	
Pause	
SetMode("Exact/Approx", "Auto")	
SetMode("Split Screen", "Full")	
SetMode("Split 1 App", "Home")	
Stop	
EndPrgm	

Casio fx-7700GE
Casio fx-9700GE
Casio CFX-9800G

VECANGL↵
Cls↵
Deg↵
"ENTER (A,B)"↵
"A="?→A↵
"B="?→B↵
"ENTER (C,D)"↵
"C="?→C↵
"D="?→D↵
Plot 0,0↵
Plot A,B↵
Line↵
Plot 0,0↵
Plot C,D↵
Line
AC+BD→E↵
$\sqrt{}$ (A^2+B^2)→U↵
$\sqrt{}$ (C^2+D^2)→V↵
cos^{-1}(E÷UV)→θ↵
"θ="↵
θ

Sharp EL-9200C
Sharp EL-9300C

vecangl
───────────REAL
ClrG
ClrT
Print"enter (a,b)"
Input a
Input b
ClrT
Print"enter (c,d)"
Input c
Input d
Line 0,0,a,b
Line 0,0,c,d
Wait
e=a*c+b*d
u=$\sqrt{}$ (a^2+b^2)
v=$\sqrt{}$ (c^2+d^2)
t=cos^{-1}(e/(u*v))
Print t
End

Set the calculator to degree mode before running the program.

HP-38G

```
VECANGL PROGRAM
INPUT A; "ENTER (A,B)";
    "ENTER A";;1:
INPUT B; "ENTER (A,B)";
    "ENTER B";;1:
INPUT C; "ENTER (C,D)";
    "ENTER C";;1:
INPUT D; "ENTER (C,D)";
    "ENTER D";;1:
ERASE:
LINE-10;0;10;0:
LINE 0;-10;0;10:
LINE 0;0;A;B:
LINE 0;0;C;D:
FREEZE:
AC+BD► E
√ (A²+B²)►U:
√ (C²+D²)►V:
ACOS(E/(UV))►T:
ERASE:
DISP 3; "ANGLE= "T:
FREEZE
```

The Function aplet should have a plot range of $-10 \le x \le 10$ and $-10 \le y \le 10$. Set the MODE to degrees before running the program.

Adding Vectors Graphically (Chapter 6 Project)

The program, shown in the Chapter 6 Project on page 517, will sketch two vectors in standard position. Using the parallelogram law for the vector addition, the program also sketches the vector sum. Be sure to set an appropriate viewing rectangle.

TI-80

```
:PROGRAM:ADDVECT
:CLRDRAW
:DISP "ENTER(A,B)"
:INPUT "ENTER A",A
:INPUT "ENTER B",B
:DISP "ENTER (C,D)"
:INPUT "ENTER C",C
:INPUT "ENTER D",D
:LINE(0,0,A,B)
:LINE(0,0,C,D)
:A+C→E
:B+D→F
:LINE(0,0,E,F)
:LINE(A,B,E,F)
:LINE(C,D,E,F)
:PAUSE
```

TI-81

```
:PrgmC:ADDVECT
:ClrDraw
:Disp "ENTER(A,B)"
:Disp "ENTER A"
:Input A
:Disp "ENTER B"
:Input B
:Disp "ENTER (C,D)"
:Disp "ENTER C"
:Input C
:Disp "ENTER D"
:Input D
:Line(0,0,A,B)
:Line(0,0,C,D)
:A+C→E
:B+D→F
:Line(0,0,E,F)
:Line(A,B,E,F)
:Line(C,D,E,F)
:Pause
:End
```

TI-83
TI-82

```
:PROGRAM:ADDVECT
:ClrDraw
:Input "ENTER A",A
:Input "ENTER B",B
:Input "ENTER C",C
:Input "ENTER D",D
:Line(0,0,A,B)
:Line(0,0,C,D)
:A+C→E
:B+D→F
:Line(0,0,E,F)
:Line(A,B,E,F)
:Line(C,D,E,F)
:Pause
:Stop
```

TI-85

```
:PROGRAM:ADDVECT
:ClrDraw
:Input "enter A",A
:Input "enter B",B
:Input "enter C",C
:Input "enter D",D
:Line(0,0,A,B)
:Line(0,0,C,D)
:A+C→E
:B+D→F
:Line(0,0,E,F)
:Line(A,B,E,F)
:Line(C,D,E,F)
:Pause
:Disp
```

TI-92	Casio fx-7700G
addvect( )	ADDVECT
Prgm	Cls
ClrIO	"A="?→A
Input "ENTER a ",a	"B="?→B
Input "ENTER b ",b	"C="?→C
Input "ENTER c ",c	"D="?→D
Input "ENTER d ",d	Plot 0,0
ClrDraw	Plot A,B
Line(0,0,a,b)	Line
Line(0,0,c,d)	Plot 0,0
a+c→e	Plot C,D
b+d→f	Line ◢
Line 0,0,e,f	A+C→E
Line a,b,e,f	B+D→F
Line c,d,e,f	Plot 0,0
Pause	Plot E,F
setMode("Split 1 App","Home")	Line
Stop	Plot A,B
EndPrgm	Plot E,F
	Line
	Plot C,D
	Plot E,F
	Line ◢

Casio fx-7700GE
Casio fx-9700GE
Casio CFX-9800G

```
ADDVECT⏎
Cls⏎
"A="?→A⏎
"B="?→B⏎
"C="?→C⏎
"D="?→D⏎
Plot 0,0⏎
Plot A,B⏎
Line⏎
Plot 0,0⏎
Plot C,D⏎
Line ◢
A+C→E⏎
B+D→F⏎
Plot 0,0⏎
Plot E,F⏎
Line⏎
Plot A,B⏎
Plot E,F⏎
Line⏎
Plot C,D⏎
Plot E,F⏎
Line ◢
```

Sharp EL-9200C
Sharp EL-9300C

```
addvect
────────REAL
ClrG
Input a
Input b
Input c
Input d
Line 0,0,a,b
Line 0,0,c,d
e=a+c
f=b+d
Line 0,0,e,f
Line a,b,e,f
Line c,d,e,f
Wait
End
```

HP-38G PROGRAMS

```
ADDVECT PROGRAM
INPUT A;; "ENTER A";;1:
INPUT B;; "ENTER B";;1:
INPUT C;; "ENTER C";;1:
INPUT D;; "ENTER D";;1:
ERASE:
LINE−10;0;10;0:
LINE 0;−10;0;10:
LINE 0;0;A;B:
LINE 0;0;C;D:
FREEZE:
A+C▶ E
B+D▶ F
LINE 0;0;E;F:
LINE A;B;E;F:
LINE C;D;E;F:
FREEZE
```

The Function aplet should have a plot range of $-10 \le x \le 10$ and $-10 \le y \le 10$.

Systems of Linear Equations (Section 7.2)

This program, shown in the Technology note on page 536, will display the solution of a system of two linear equations in two variables of the form

$$ax + by = c$$
$$dx + ey = f$$

if a unique solution exists.

Note: For help with *TI-83* or *TI-82* keystrokes, see previous programs in this appendix.

TI-80

```
PROGRAM:SOLVE
:DISP "AX+BY=C"
:INPUT "ENTER A",A
:INPUT "ENTER B",B
:INPUT "ENTER C",C
:DISP "DX+EY=F"
:INPUT "ENTER D",D
:INPUT "ENTER E",E
:INPUT "ENTER F",F
:IF AE−DB=0
:THEN
:DISP "NO UNIQUE"
:DISP "SOLUTION"
:ELSE
:(CE−BF)/(AE−DB)→X
:(AF−CD)/(AE−DB)→Y
:DISP X
:DISP Y
:END
```

TI-81

```
Prgm4:SOLVE
:Disp "AX+BY=C"
:Input A
:Input B
:Input C
:Disp "DX+EY=F"
:Input D
:Input E
:Input F
:If AE−DB=0
:Goto 1
:(CE−BF)/(AE−DB)→X
:(AF−CD)/(AE−DB)→Y
:Disp X
:Disp Y
:End
:Lbl 1
:Disp "NO UNIQUE SOLUTION"
:End
```

TI-83
TI-82

```
PROGRAM:SOLVE
:Disp "AX+BY=C"
:Prompt A
:Prompt B
:Prompt C
:Disp "DX+EY=F"
:Prompt D
:Prompt E
:Prompt F
:If AE−DB=0
:Then
:Disp "NO UNIQUE"
:Disp "SOLUTION"
:Else
:(CE−BF)/(AE−DB)→X
:(AF−CD)/(AE−DB)→Y
:Disp X
:Disp Y
:End
```

TI-85

```
PROGRAM:SOLVE
:Disp "AX+BY=C"
:Input "ENTER A",A
:Input "ENTER B",B
:Input "ENTER C",C
:Disp "DX+EY=F"
:Input "ENTER D",D
:Input "ENTER E",E
:Input "ENTER F",F
:If A*E−D*B==0
:Goto A
:(C*E−B*F)/(A*E−D*B)→X
:(A*F−C*D)/(A*E−D*B)→Y
:Disp X
:Disp Y
:Stop
:Lbl A
:Disp "NO UNIQUE SOLUTION"
```

TI-92

```
Solve( )
Prgm
ClrIO
Disp "Ax+By=C"
Input "Enter A.",a
Input "Enter B.",b
Input "Enter C.",c
ClrIO
Disp "Dx+Ey=F"
Input "Enter D.",d
Input "Enter E.",e
Input "Enter F.",f
If a*e−d*b=0 Then
     Disp "No unique solution"
   Else
     (c*e−b*f)/(a*e−d*b)→x
     (a*f−c*d)/(a*e−d*b)→y
     Disp x
     Disp y
EndIf
EndPrgm
```

Casio fx-7700G

```
SOLVE
"AX+BY=C"
"A="?→A
"B="?→B
"C="?→C
"DX+EY=F"
"D="?→D
"E="?→E
"F="?→F
AE−DB=0 ⇒ Goto 1
"X=":(CE−BF)÷(AE−DB) ◢
"Y=":(AF−CD)÷(AE−DB)
Goto 2
Lbl 1
"NO UNIQUE SOLUTION"
Lbl 2
```

Casio fx-7700GE
Casio fx-9700GE
Casio CFX-9800G

SOLVE⏎
"AX+BY=C"⏎
"A":?→A⏎
"B":?→B⏎
"C":?→C⏎
"DX+EY=F"⏎
"D":?→D⏎
"E":?→E⏎
"F":?→F⏎
AE−DB=0⇒Goto 1⏎
"X=":(CE−BF)÷(AE−DB)◢
"Y=":(AF−CD)÷(AE−DB)⏎
Goto 2⏎
Lbl 1⏎
"NO UNIQUE SOLUTION"⏎
Lbl 2

Solutions to systems of linear equations are also available directly from the Casio calculator's EQUATION MENU.

Sharp EL-9200C
Sharp EL-9300C

solve
─────────REAL
Print "AX+BY=C"
Input A
Input B
Input C
Print "DX+EY=F"
Input D
Input E
Input F
If A*E−D*B=0 Goto 1
X=(C*E−B*F)/(A*E−D*B)
Y=(A*F−C*D)/(A*E−D*B)
Print X
Print Y
End
Label 1
Print "no unique solution"
End

Equations must be entered in the form: AX + BY = C; DX + EY = F. Uppercase letters are used so that the values can be accessed in the calculation mode of the calculator.

HP-38G

SOLVE

SOLVE PROGRAM
INPUT A;"AX+BY=C";
    "ENTER A";" ";1:
INPUT B;"AX+BY=C";
    "ENTER B";" ";1:
INPUT C;"AX+BY=C";
    "ENTER C";" ";1:
INPUT D;"DX+EY=F";
    "ENTER D";" ";1:
INPUT E;"DX+EY=F";
    "ENTER E";" ";1:
INPUT F;"DX+EY=F";
    "ENTER F";" ";1:
ERASE:
IF AE−DB==0
THEN DISP 3; "NO UNIQUE
    SOLUTION":
ELSE RUN "SOLVE.SOLN":
END:
FREEZE:

SOLVE.SOLN PROGRAM
(CE−BF)/(AE−DB)▶X:
(AF−CD)/(AE−DB)▶Y:
DISP 3;"X="X:
DISP 5;"Y="Y:

1. Input the 2 programs SOLVE
   and SOLVE.SOLN.
2. Run the SOLVE program.

Visualizing Row Operations (Section 8.1)

This program, referenced in the Technology note on page 597, demonstrates how elementary matrix row operations used in Gauss-Jordan elimination may be interpreted graphically. It asks the user to enter a 2×3 matrix that corresponds to a system of two linear equations. (The matrix entries should not be equivalent to either vertical or horizontal lines. This demonstration is also most effective if the y-intercepts of the lines are between -10 and 10.)

While the demonstration is running, you should notice that each elementary row operation creates an equivalent system. This equivalence is reinforced graphically because while the equations of the lines change with each elementary row operation, the point of intersection remains the same. You may want to run this program a second time to notice the relationship between the row operations and the graphs of the lines of the system.

TI-81

Prgm6:ROWOPS
:Disp "ENTER A"
:Disp "2 BY 3 MATRIX"
:Disp "A B C"
:Disp "D E F"
:Input A
:Input B
:Input C
:Input D
:Input E
:Input F
:A→[A](1,1)
:B→[A](1,2)
:C→[A](1,3)
:D→[A](2,1)
:E→[A](2,2)
:F→[A](2,3)
:ClrHome
:Disp "ORIGINAL MATRIX"
:Disp [A]
:Pause
:"B^{-1}(C−AX)"→Y2
:"E^{-1}(F−DX)"→Y1
:-10→Xmin
:10→Xmax
:1→Xscl
:-10→Ymin
:10→Ymax
:1→Yscl

:DispGraph
:Pause
:ClrHome
:Disp "OBTAIN LEADING"
:Disp "1 IN ROW 1"
:*row(A^{-1},[A],1)→[A]
:Disp [A]
:Pause
:ClrDraw
:"(A/B)(C/A−X)"→Y2
:DispGraph
:Pause
:ClrHome
:Disp "OBTAIN 0 BELOW"
:Disp "LEADING 1 IN"
:Disp "COLUMN 1"
:*row+(-D,[A],1,2)→[A]
:Disp[A]
:Pause
:ClrDraw
:"(E−(BD/A))$^{-1}$(F−(DC/A))"→Y1
:DispGraph
:Pause
:ClrHome
:[A](2,2)→G
:If G=0
:Goto 1
:*row(G^{-1},[A],2)→[A]
:Disp "OBTAIN LEADING"

(Continued on next page)

```
:Disp "1 IN ROW 2"
:Disp [A]
:Pause
:ClrDraw
:DispGraph
:Pause
:ClrHome
:Disp "OBTAIN 0 ABOVE"
:Disp "LEADING 1 IN"
:Disp "COLUMN 2"
:[A](1,2)→H
:*row+(-H,[A],2,1)→[A]
:Disp [A]
:Pause
:ClrDraw
:Y2-Off
:Line([A](1,3),-10,[A](1,3),10)
:DispGraph
:Pause
:ClrHome
:Disp "THE POINT OF"
:Disp "INTERSECTION IS"
:Disp "X="
:Disp [A](1,3)
:Disp "Y="
:Disp [A](2,3)
:End
:Lbl 1
:If [A](2,3)=0
:Disp "INFINITELY MANY"
:Disp "SOLUTIONS"
:If [A](2,3) ≠ 0
:Disp "INCONSISTENT"
:Disp "SYSTEM"
:End
```

To use this program, dimension matrix [A] as a 2×3 matrix. Press ENTER after each screen display to continue the program.

TI-83
TI-82

PROGRAM: ROWOPS
:Disp "ENTER A"
:Disp "2 BY 3 MATRIX:"
:Disp "A B C"
:Disp "D E F"
:Prompt A,B,C
:Prompt D,E,F
:A→[A](1,1):B→[A](1,2)
:C→[A](1,3):D→[A](2,1)
:E→[A](2,2):F→[A](2,3)
:ClrHome
:Disp "ORIGINAL MATRIX:"
:Pause [A]
:"B^{-1}(C−AX)"→Y2
:"E^{-1}(F−DX)"→Y1
:ZStandard:Pause:ClrHome
:Disp "OBTAIN LEADING"
:Disp "1 IN ROW 1"
:*row(A^{-1},[A],1)→[A]
:Pause [A]:ClrDraw
:"(A/B)(C/A−X)"→Y2
:DispGraph:Pause:ClrHome
:Disp "OBTAIN 0 BELOW"
:Disp "LEADING 1 IN"
:Disp "COLUMN 1"
:*row+(-D,[A],1,2)→[A]
:Pause [A]:ClrDraw
:"(E−(BD/A))$^{-1}$(F−(DC/A))"→Y1
:DispGraph:Pause:ClrHome
:[A](2,2)→G
:If G=0
:Goto 1
:*row(G^{-1},[A],2)→[A]
:Disp "OBTAIN LEADING"
:Disp "1 IN ROW 2"
:Pause [A]:ClrDraw
:DispGraph:Pause:ClrHome
:Disp "OBTAIN 0 ABOVE"
:Disp "LEADING 1 IN"
:Disp "COLUMN 2"
:[A](1,2)→H
:*row+(-H,[A],2,1)→[A]
:Pause [A]:ClrDraw:FnOff 2

:Vertical -(B/A)(E−(BD/A))$^{-1}$
  (F−DC/A)+C/A
:DispGraph:Pause:ClrHome
:Disp "THE POINT OF"
:Disp "INTERSECTION IS"
:Disp "X=",[A](1,3),"Y=",[A](2,3)
:Stop
:Lbl 1
If [A](2,3)=0
:Then
:Disp "INFINITELY MANY"
:Disp "SOLUTIONS"
:Else
:Disp "INCONSISTENT"
:Disp "SYSTEM"
:End

To use this program, dimension matrix [A] as a 2×3 matrix. Press ENTER after each screen display to continue the program.

TI-85

```
PROGRAM:ROWOPS
:Disp "enter a"
:Disp "2 by 3 matrix:"
:Disp "A B C"
:Disp "D E F"
:Prompt A,B,C
:Prompt D,E,F
:A→TEMP(1,1):B→TEMP(1,2)
:C→TEMP(1,3):D→TEMP(2,1)
:E→TEMP(2,2):F→TEMP(2,3)
:CILCD
:Disp "original matrix:"
:Disp TEMP
:Pause
:y2=B⁻¹(C−A∗x)
:y1=E⁻¹(F−D∗x)
:ZStd:Pause:CILCD
:Disp "obtain leading"
:Disp "1 in row 1"
:multR(A⁻¹,TEMP,1)→TEMP
:DispTEMP:Pause
:"(A/B)(C/A−X)"→y
:ClDrw:DispG:Pause:CILCD
:Disp "obtain 0 below"
:Disp "leading 1 in"
:Disp "column 1"
:mRAdd(-D,TEMP,1,2)→TEMP
:Disp TEMP:Pause
:If TEMP(2,2)==0
:Goto A
:y1=(E−(B∗D/A))⁻¹(F−(D∗C/A))
:ClDrw:DispG:Pause:CILCD
:TEMP(2,2)→G
:multR(G⁻¹,TEMP,2)→TEMP
:Disp "obtain leading"
:Disp "1 in row 2"
:Disp TEMP:Pause
:ClDrw:DispG:Pause:CILCD
:Disp "obtain 0 above"
:Disp "leading 1 in"
:Disp "column 2"
:TEMP(1,2)→H
:mRAdd(-H,TEMP,2,1)→TEMP
:Disp TEMP
```

```
:Pause:FnOff 2:ClDrw
:Vert -(B/A)(E−(B∗D/A))⁻¹
    (F−D∗C/A)+C/A
:DispG:Pause:CILCD
:Disp "the point of"
:Disp "intersection is"
:Disp "X=",TEMP(1,3),
    "Y=",TEMP(2,3)
:Stop
:Lbl A
:If TEMP(2,3)==0
:Then
:Disp "infinitely many"
:Disp "solutions"
:Else
:Disp "inconsistent"
:Disp "system"
:End
```

To use this program, dimension matrix TEMP as a 2×3 matrix. Press ENTER after each screen display to continue the program.

TI-92

```
rowops( )
Prgm
ClrIO
ClrHome
setMode("Split Screen","Left-
Right")
setMode("Split 1 App","Home")
setMode("Split 2 App","Graph")
Disp "ENTER A"
Disp "2 BY 3 MATRIX:"
Disp "A B C"
Disp "D E F"
Prompt a,b,c
Prompt d,e,f
[[a,b,c][d,e,f]]→mat1
ClrIO
b^(-1)*(c-a*x)→y2(x)
e^(-1)*(f-d*x)→y1(x)
ZoomStd
Disp "ORIGINAL MATRIX:"
Pause mat1
ClrIO
a/b*(c/a-x)→y2(x)
Disp "OBTAIN LEADING"
Disp "1 IN ROW 1"
mRow(a^(-1),mat1,1)→mat1
Pause mat1
ClrIO
(e-b*d/a)^(-1)*(f-d*c/a)→y1(x)
DispG
Disp "OBTAIN 0 BELOW"
Disp "LEADING 1 IN"
Disp "COLUMN 1"
mRowAdd(-d,mat1,1,2)→mat1
Pause mat1
ClrIO
mat1[2,2]→g
If g=0
Goto a1
Disp "OBTAIN LEADING"
Disp "1 IN ROW 2"
mRow(g^(-1),mat1,2)→mat1
Pause mat1
ClrIO
mat1[1,2]→h
```

```
FnOff 2
LineVert -b/a*(e-b*d/a)^(-1)*
    (f-d*c/a)+c/a
Disp "OBTAIN 0 ABOVE"
Disp "LEADING 1 IN"
Disp "COLUMN 2"
mRowAdd(-h,mat1,2,1)→mat1
Pause mat1
ClrIO
Disp "THE POINT OF"
Disp "INTERSECTION IS"
Disp "X=",mat1[1,3],"Y=",
    mat1[2,3]
Goto A2
Lbl a1
If mat1[2,3]=0 Then
Disp "INFINITELY MANY"
Disp "SOLUTIONS"
Else
Disp "INCONSISTENT"
Disp "SYSTEM"
EndIf
Lbl A2
Pause
setMode("Split Screen","Full")
EndPrgm
```

Press ENTER after each screen display to continue the program.

Casio fx-7700GE
Casio fx-9700GE
Casio CFX-9800G

ROWOPS ↵
"ENTER A"↵
"2 BY 3 MATRIX:"↵
"A B C"↵
"D E F"↵
"A="?→A:"B="?→B:
"C="?→C:"D="?→D:
"E="?→E:"F="?→F:↵
[[A,B,C][D,E,F]]→Mat A↵
Cls↵
"ORIGINAL MATRIX:"◢
Mat A ◢
Range -10,10,1,-10,10,1↵
Graph Y=B^{-1}(C−AX)↵
Graph Y=E^{-1}(F−DX) ◢
Cls↵
"OBTAIN LEADING"↵
"1 IN ROW 1"◢
*Row A^{-1},A,1↵
Mat A◢
Graph Y=(A÷B)(C÷A−X)↵
Graph Y=E^{-1}(F−DX) ◢
Cls↵
"OBTAIN 0 BELOW"↵
"LEADING 1 IN"↵
"COLUMN 1"◢
*Row+ -D,A,1,2↵
Mat A ◢
Graph Y=(A÷B)(C÷A−X)↵
GraphY=(E−(BD÷A))$^{-1}$
    (F−(DC÷A)) ◢
Cls↵
Mat A[2,2]→G↵
G=0 ⇒ Goto 1↵
*Row G^{-1},A,2↵
"OBTAIN LEADING"↵
"1 IN ROW 2"◢
Mat A◢
Graph Y=(A÷B)(C÷A−X)↵
Graph Y=(E−(BD÷A))$^{-1}$
    (F−(DC÷A)) ◢

Cls↵
"OBTAIN 0 ABOVE"↵
"LEADING 1 IN"↵
"COLUMN 2"◢
Mat A[1,2]→H↵
*Row+ -H,A,2,1↵
Mat A ◢
Mat A[1,3]→J↵
Mat A[2,3]→K↵
Graph Y=K↵
Plot J,-10:Plot J,10:Line
"THE POINT OF"↵
"INTERSECTION IS"↵
"X=":J ◢
"Y=":K ◢
Goto 3↵
Lbl 1↵
Mat A[2,3]=0 ⇒ Goto 2↵
"INCONSISTENT"↵
"SYSTEM"↵
Goto 3↵
Lbl 2↵
"INFINITELY MANY"↵
"SOLUTIONS"↵
Lbl 3

To use this program, dimension
Mat A as a 2×3 matrix. Press
EXE after each screen display
to continue the program.

Sum Program (Section 9.1, page 663)

TI-80

PROGRAM:SUM
:INPUT "ENTER M", M
:INPUT "ENTER N", N
:0→S
:FOR(X,M,N)
:S+Y1→S
:DISP S
:END

To use this program, first store the nth term of the sequence in Y1 (in terms of X).

TI-81

Prgm5:SUM
:Disp "ENTER M"
:Input M
:Disp "ENTER N"
:Input N
:0→S
:Lbl 1
:M→X
:S+Y1→S
:Disp S
:IS>(M,N)
:Goto 1
:End

To use this program, first store the nth term of the sequence in Y1 (in terms of X).

TI-83
TI-82

PROGRAM:SUM
:Prompt M
:Prompt N
:0→S
:For(X,M,N)
:S+Y1→S
:Disp S
:End

To use this program, first store the nth term of the sequence in Y1 (in terms of X).

TI-85

PROGRAM:SUMS
:Prompt M
:Prompt N
:0→S
:For(x,M,N)
:S+y1→S
:Disp S
:End

To use this program, first store the nth term of the sequence in y1 (in terms of x).

TI-92

Summatn( )
Prgm
Input "Enter lower limit.",m
Input "Enter upper limit.",n
Σ(y1(x),x,m,n)→s
Disp "The partial sum is",s
EndPrgm

Sharp EL-9200C
Sharp EL-9300C

sum

───────REAL

Goto top
Label sumit
s=s+(nth term)
Print s
m=m+1
Goto next
Label top
Input m
Input n
s=0
Label next
If m<=n Goto sumit
End

To use this program, first replace (nth term) with the *n*th term of the sequence in terms of m. For example, s=s+2^m, where 2^n is the *n*th term of the sequence.

Casio fx-7700G

SUM
"M="?→M
"N="?→N
0→S
Lbl 1
M→X
S+f₁→S
"S=":S◢
M+1→M
M≤N⇒1 Goto 1

To use this program, enter the *n*th term of the sequence into f₁ (in terms of X).

Casio fx-7700GE
Casio fx-9700GE
Casio CFX-9800G

SUM ↵
"M="?→M ↵
"N="?→N ↵
0→S ↵
Lbl 1 ↵
M→X ↵
S+f₁→S ↵
"S=":S◢
M+1→M ↵
M≤N⇒1 Goto 1

To use this program, enter the *n*th term of the sequence into f₁ (in terms of X).

HP-38G

SUM
SUM PROGRAM
INPUT M;"LOWER BOUND";
    "ENTER M";"";1:
INPUT N;"UPPER BOUND";
    "ENTER N";"";1:
0▶S:
ERASE:
SELECT "Function":
FOR I=M TO N
STEP 1;
RUN "SUM.STEP":
END:

SUM.STEP PROGRAM
S+F1(I)▶S:
DISP 4;"   "S:
FREEZE:

1. Input the 2 programs SUM and SUM.STEP.
2. Store the *n*th term of the sequence in the F1 function (in terms of *x*) in the Function aplet. Be sure that F1 is checked.
3. Run the SUM program.

Darts Program (Section 9.7, page 723)

TI-80

```
PROGRAM:DARTS
:CLRHOME
:0→K
:0→W
:INPUT "HOW MANY
    THROWS?",N
:LBL 1
:K+1→K
:RAND→X:RAND→Y
:IF X<.25:GOTO 2
:IF X>.75:GOTO 2
:IF Y<.3:GOTO 2
:IF Y>.65:GOTO 2
:W+1→W
:LBL 2
:IF K=N
:THEN
:DISP "WINS=",W
:DISP "LOSSES=",(N−W)
:STOP
:END
:GOTO 1
```

TI-81

```
Prgm6:DARTS
:ClrHome
:0→K
:0→W
:Disp "HOW MANY
    THROWS?"
:Input N
:Lbl 1
:K+1→K
:Rand→X
:Rand→Y
:If X<.25
:Goto 2
:If X>.75
:Goto 2
:If Y<.3
:Goto 2
:If X>.65
:Goto 2
:W+1→W
:Lbl 2
:If K≠N
:Goto 1
:DISP "WINS="
:Disp W
:DISP "LOSSES="
:N−W→L
:Disp L
:End
```

TI-83
TI-82
PROGRAM;DARTS
:ClrHome
:0→K
:0→W
:Input "HOW MANY
    THROWS?",N
:Lbl 1
:K+1→K
:rand→X:rand→Y
:IF X<.25 or X>.75
:Goto 2
:IF Y<.3 or Y>.65
:Goto 2
:W+1→W
:Lbl 2
:If K=N
:Then
:Disp "WINS=",W
:Disp "LOSSES=",(N−W)
:Stop
:End
:Goto 1

TI-85
PROGRAM:DARTS
:CILCD
:0→K
:0→W
:Input "HOW MANY
    THROWS?",N
:Lbl A
:K+1→K
:rand→X:rand→Y
:If X<.25
:Goto B
:If X>.75
:Goto B
:If Y<.3
:Goto B
:If Y>.65
:Goto B
:W+1→W
:Lbl B
:If K≠N
:Goto A
:DISP "WINS=",W
:DISP "LOSSES=",(N−W)
:Stop

TI-92

```
darts( )
Prgm
ClrIO
0→k
0→w
Input "How many throws?",n
Lbl a
k+1→k
rand( )→x
rand( )→y
If x<.25 or x>.75
Goto b
If y<.3 or y>.65
Goto b
w+1→w
Lbl b
If k=n Then
    Disp "Wins=",w
    Disp "Losses=",(n−w)
    Stop
EndIf
Goto a
EndPrgm
```

Casio fx-7700G

```
DARTS
Cls
0→K
0→W
"HOW MANY THROWS?"
?→N
Lbl 1
K+1→K
Ran#→X
Ran#→Y
X<.25⇒Goto 2
X>.75⇒Goto 2
Y<.3⇒Goto 2
Y>.65⇒Goto 2
W+1→W
Lbl 2
K≠N⇒Goto 1
"WINS="
W◢
"LOSSES="
N−W
```

Casio fx-7700GE
Casio fx-9700GE
Casio CFX-9800G

```
DARTS⏎
Cls⏎
0→K⏎
0→W⏎
"HOW MANY THROWS?"⏎
?→N⏎
Lbl 1⏎
K+1→K⏎
Ran#→X⏎
Ran#→Y⏎
X<.25⇒Goto 2⏎
X>.75⇒Goto 2⏎
Y<.3⇒Goto 2⏎
Y>.65⇒Goto 2⏎
W+1→W⏎
Lbl 2⏎
K≠N⇒Goto 1⏎
"WINS="⏎
W◢
"LOSSES="⏎
N−W⏎
```

Sharp EL-9200C
Sharp EL-9300C

darts

————————REAL

ClrT
k=0
w=0
Print "how many throws?"
Input n
Label a
k=k+1
x=random
y=random
If x<.25 Goto b
If x>.75 Goto b
If y<.3 Goto b
If y>.65 Goto b
w=w+1
If k≠n Goto a
Print "wins="
Print w
l=n−w
Print "losses="
Print 1

HP-38G
DARTS

DARTS PROGRAM
0►L:
INPUT N;"";"THROWS?";
"HOW MANY THROWS?";1:
FOR I=1 TO N
    STEP 1;
    RUN "DART.THROW":
END:
ERASE:
DISP 3;"WINS="N−L:
DISP 5;"LOSSES="L:
FREEZE:

DARTS.THROW PROGRAM
RANDOM►X:
RANDOM►Y:
CASE
IF X<.25 OR X>.75
THEN L+1►L:
IF Y<.3 OR Y>.65
THEN L+1►L:
END:

1. Input the 2 programs DARTS
   and DARTS.THROW.
2. Run the DARTS program.

CHAPTER P

Section P.1   *(page 8)*

1.

3.

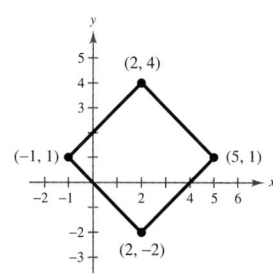

5. A: $(2, 6)$; B: $(-6, -2)$; C: $(4, -4)$; D: $(-3, 2)$

7. A: $(0, 5)$; B: $(-3, -6)$; C: $(1, -4.5)$; D: $(-4, 2)$

9. $(-3, 4)$ **11.** $(-5, -5)$

13. Point on x-axis: $y = 0$; Point on y-axis: $x = 0$

15. Quadrant IV   **17.** Quadrant II

19. Quadrant III or IV   **21.** Quadrant III

23. Quadrants I and III

25.

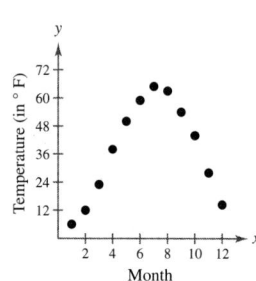

27. $(0, 1)$, $(4, 2)$, $(1, 4)$

29.

x	-2	-1	$-\frac{1}{2}$	0	$\frac{1}{2}$	1	2
y	3	$\frac{5}{2}$	$\frac{9}{4}$	2	$\frac{7}{4}$	$\frac{3}{2}$	1

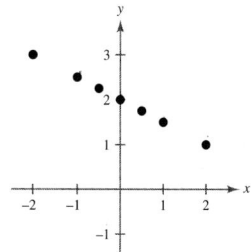

31. 1900%   **33.** 1990: $13.70 per 100 pounds

35. 1970s   **37.** 65   **39.** $4^2 + 3^2 = 5^2$

41. $10^2 + 3^2 = \left(\sqrt{109}\right)^2$ **43.** 8   **45.** 5

47. (a)

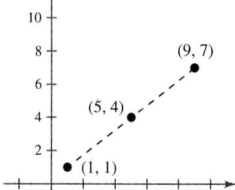

(b) 10

(c) $(5, 4)$

49. (a)

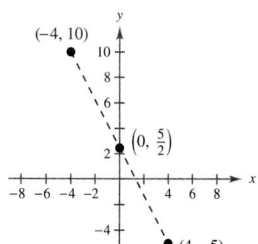

(b) 17

(c) $\left(0, \frac{5}{2}\right)$

51. (a)

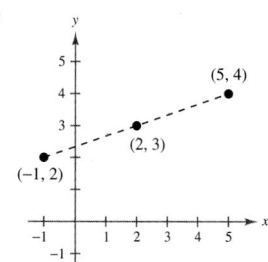

(b) $2\sqrt{10}$

(c) $(2, 3)$

A53

53. (a)

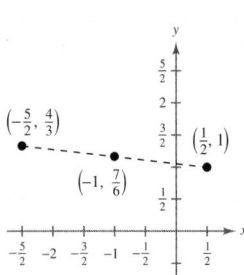

(b) $\dfrac{\sqrt{82}}{3}$

(c) $\left(-1, \dfrac{7}{6}\right)$

55. (a)

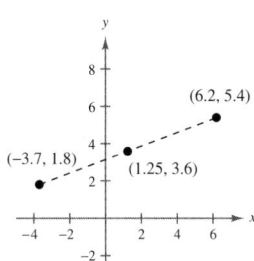

(b) $\sqrt{110.97}$

(c) $(1.25, 3.6)$

57. (a)

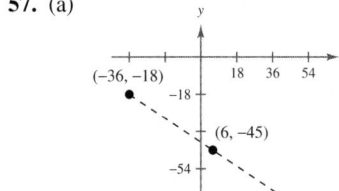

(b) $6\sqrt{277}$

(c) $(6, -45)$

59. $(1993, \$630{,}000)$ **61.** $\left(\sqrt{5}\right)^2 + \left(\sqrt{45}\right)^2 = \left(\sqrt{50}\right)^2$

63. All sides have a length of $\sqrt{5}$.

65. Opposite sides have equal lengths of $2\sqrt{5}$ and $\sqrt{85}$.

67. $x^2 + y^2 = 9$ **69.** $(x - 2)^2 + (y + 1)^2 = 16$

71. $(x + 1)^2 + (y - 2)^2 = 5$

73. $(x - 3)^2 + (y - 4)^2 = 25$

75. Center: $(0, 0)$; Radius $= 2$

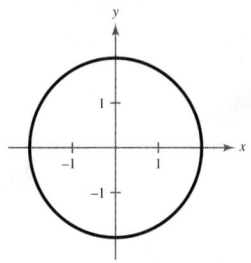

77. Center: $(1, -3)$; Radius $= 2$

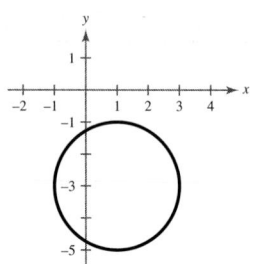

79. Center: $\left(\dfrac{1}{2}, \dfrac{1}{2}\right)$; Radius $= \dfrac{3}{2}$

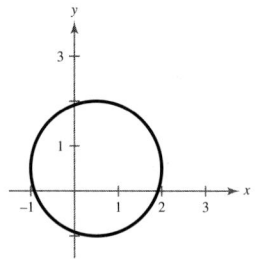

81. $(2x_m - x_1,\ 2y_m - y_1)$

83. $\left(\dfrac{3x_1 + x_2}{4}, \dfrac{3y_1 + y_2}{4}\right), \left(\dfrac{x_1 + x_2}{2}, \dfrac{y_1 + y_2}{2}\right),$

$\left(\dfrac{x_1 + 3x_2}{4}, \dfrac{y_1 + 3y_2}{4}\right)$

85. $5\sqrt{74} \approx 43$ yards

87.

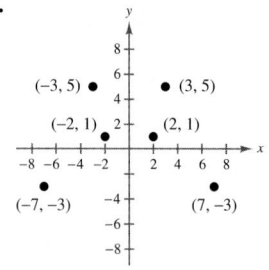

The points are reflected through the y-axis.

89. (a) 7   (b) Answers will vary.

Section P.2   *(page 20)*

1. (a) Yes   (b) Yes   **3.** (a) No   (b) Yes

5. (a) No   (b) Yes   **7.** (a) Yes   (b) Yes

9.

x	-1	0	1	$\frac{3}{2}$	2
y	5	3	1	0	-1

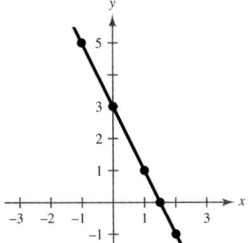

15.

x	1	2	5	10	17
y	0	1	2	3	4

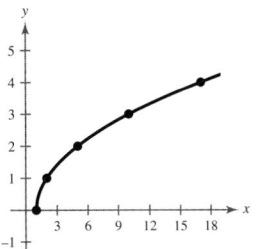

11.

x	-1	0	1	2	3
y	3	0	-1	0	3

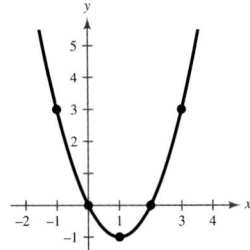

17.

x	-2	-1	0	1	2
y	$-\frac{5}{2}$	$-\frac{11}{4}$	-3	$-\frac{13}{4}$	$-\frac{7}{2}$

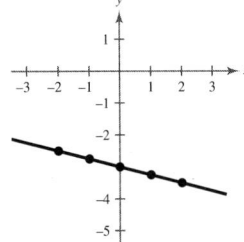

This line slopes downward to the right rather than upward.

13.

x	-2	-1	0	1	2
y	$-\frac{7}{2}$	$-\frac{13}{4}$	-3	$-\frac{11}{4}$	$-\frac{5}{2}$

19.

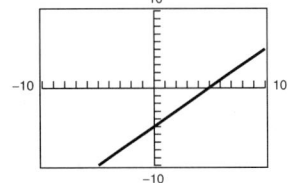

$(5, 0), (0, -5)$

21.

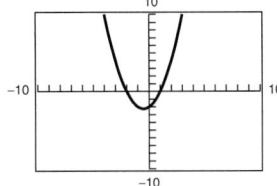

$(1, 0), (-2, 0), (0, -2)$

23.

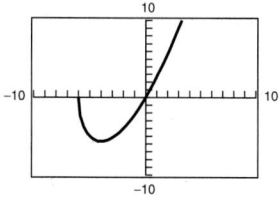

$(0, 0), (-6, 0)$

25.

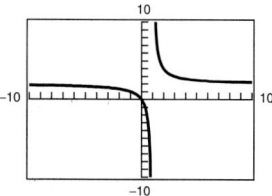

$(0, 0)$

27. (d)   **29.** (f)   **31.** (a)

33.

No symmetry

35.

Symmetry: y-axis

37.

No symmetry

39.

No symmetry

41.

No symmetry

43.

No symmetry

45.

Symmetry: x-axis

47.

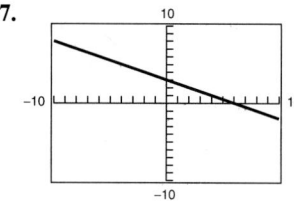

Intercepts: $(6, 0), (0, 3)$

49.

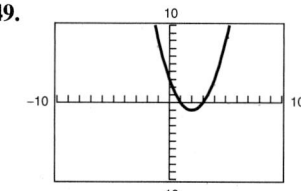

Intercepts: $(3, 0), (1, 0), (0, 3)$

51.

Intercepts: $(0, 0), (2, 0)$

53.

Intercept: $(0, 0)$

55.

 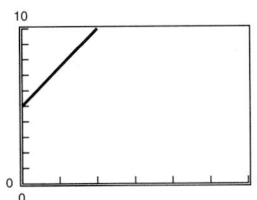

The standard setting gives a more complete graph.

57.

 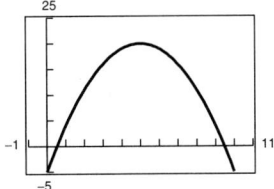

The specified setting gives a more complete graph.

59.
Xmin = -5
Xmax = 5
Xscl = 1
Ymin = -30
Ymax = 10
Yscl = 5

61.
Xmin = -30
Xmax = 30
Xscl = 5
Ymin = -10
Ymax = 50
Yscl = 5

63.

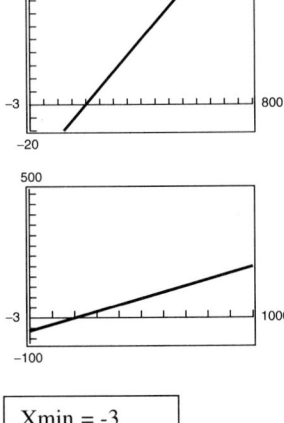

Xmin = -3
Xmax = 800
Xscl = 50
Ymin = -20
Ymax = 100
Yscl = 10

65. $y_1 = \sqrt{64 - x^2}$
$y_2 = -\sqrt{64 - x^2}$

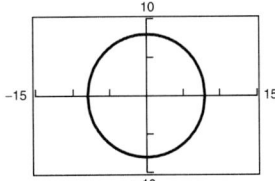

67. $y_1 = \sqrt{49 - x^2}$
$y_2 = -\sqrt{49 - x^2}$

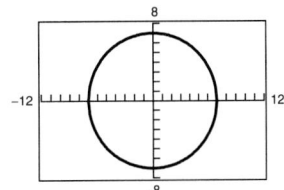

69. The graphs are identical. Distributive Property

71. The graphs are identical. Associative Property

73. (a)
Xmin = -1
Xmax = 9
Xscl = 1
Ymin = 60000
Ymax = 230000
Yscl = 10000

(b)

75. Change the viewing rectangle.

77. (a)   (b) 77.7   (c) 78.0

79.

81.

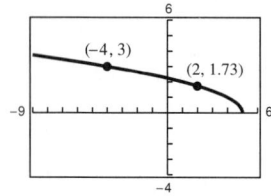

   (a) $(2, 1.73)$

   (b) $(-4, 3)$

83.

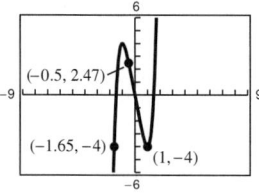

   (a) $(-0.5, 2.47)$

   (b) $(1, -4), (-1.65, -4)$

Section P.3   *(page 34)*

1. (a) L_2 (b) L_3 (c) L_1 **3.** $\frac{8}{5}$ **5.** 0   **7.** -4

9.

11. $m = 2$

13. m is undefined.

15. $m = \frac{4}{3}$

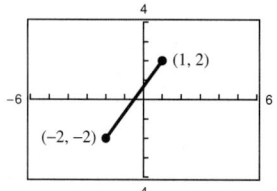

17. $(0, 1), (3, 1), (-1, 1)$ **19.** $(6, -5), (7, -4), (8, -3)$

21. $(-8, 0), (-8, 2), (-8, 3)$ **23.** Perpendicular

25. Parallel   **27.** Yes. The rate of change remains the same on a line.

29. (a) Sales increasing 135 units per year. (b) No change in sales. (c) Sales decreasing 40 units per year.

31. (a)

   (b) Decreased most rapidly: 1989

      Increased most rapidly: 1988

33.

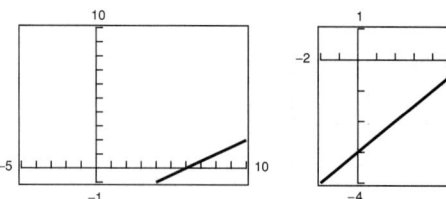

The second setting gives a more complete graph.

35. $m = 5$; Intercept: $(0, 3)$

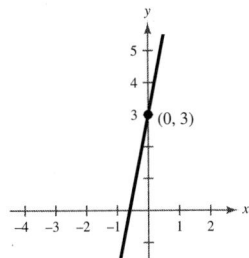

37. m is undefined.
There is no y-intercept.

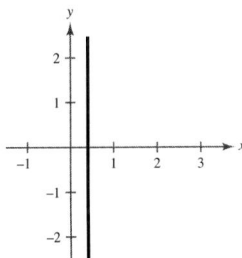

39. $m = -\frac{7}{6}$; Intercept: $(0, 5)$

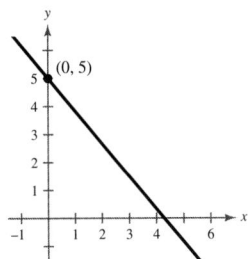

41. $3x + 5y - 10 = 0$

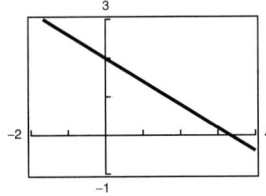

43. $x + 2y - 3 = 0$

45. $x + 8 = 0$

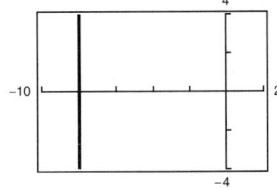

47. $2x - 5y + 1 = 0$

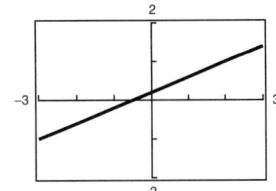

49. $16{,}666\frac{2}{3}$ feet ≈ 3.16 miles

51. $3x - y - 2 = 0$

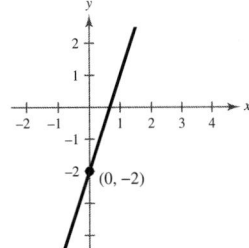

53. $2x + y = 0$

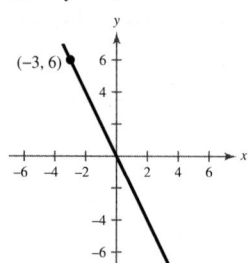

55. $x + 3y - 4 = 0$

57. $x - 6 = 0$

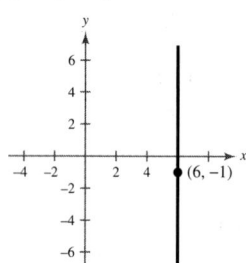

59. $8x - 6y - 17 = 0$

61.

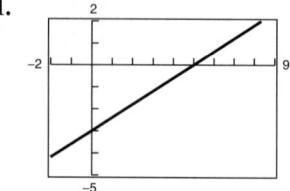

x- and *y*-intercepts

63. $3x + 2y - 6 = 0$

65. (a) $2x - y - 3 = 0$ (b) $x + 2y - 4 = 0$

67. (a) $y = 0$ (b) $x + 1 = 0$

69.

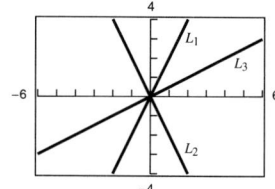

L_2 is perpendicular to L_3.

71.

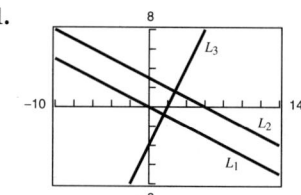

L_1 is parallel to L_2.

L_3 is perpendicular to L_1 and L_2.

73. $V = 1790 + 125t$ **75.** $V = -2000t + 32,400$

77. (b) **79.** (a) **81.** $3x - 2y - 1 = 0$

83. $F = \frac{9}{5}C + 32$ **85.** $\$41,700$

87. (a) $V = -175t + 875$

(b)

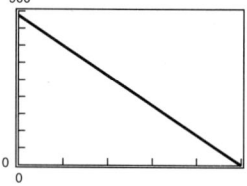

t	0	1	2	3	4	5
V	875	700	525	350	175	0

89. $S = 0.85L$

91. (a) $C = 16.75t + 36,500$ (b) $R = 27t$
    (c) $P = 10.25t - 36,500$ (d) $t \approx 3561$ hours

93. (a) $y = 71.08t - 10.29$
    (b)

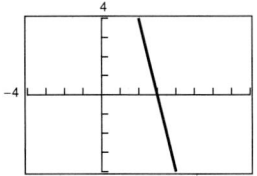

    (c) $\$1,411,000$
    (d) Average increase per year

95. (a) $x = 50 - \dfrac{p - 580}{15}$ (b) 45   (c) 49

Section P.4   (page 51)

1. (a) No   (b) No   (c) Yes   (d) No

3. (a) No   (b) No   (c) No   (d) Yes

5. Identity   **7.** Conditional

9. The equations have the same solutions, and the one is derived from the other by the steps for generating equivalent equations given in this section.
    $2x = 5, \quad 2x + 3 = 8$

11. (a) $x = 5$ (b) $t = 14$ (c) $s = 6$ (d) $u = 14$

13. 3   **15.** 9   **17.** -26 **19.** -4 **21.** $-\frac{6}{5}$

23. 9   **25.** 10   **27.** 4   **29.** 5   **31.** No solution

33. $\frac{11}{6}$ **35.** $\frac{5}{3}$ **37.** No solution   **39.** 0

41. $x = 6$ feet   **43.** $(5, 0), (0, -5)$ **45.** $(-2, 0), (0, 0)$

47. $(-1, 0), (5, 0), (0, -1)$

49.

51.

53.

55.

$(3, 0)$

57.

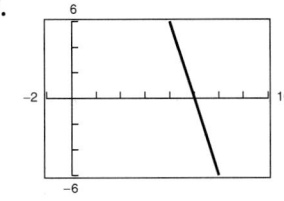

$(10, 0)$

59. 6   **61.** $-\frac{5}{8}$ **63.** $(1, 1)$ **65.** $(6, 4)$ **67.** $(2, 2)$

69. $(4, 1)$ **71.** $(-2, 8), (1.333, 8)$

73. $(0, 0), (-2, 8), (2, 8)$ **75.** $1 \pm \sqrt{2}$ **77.** $6, -12$

79. $0, \pm\dfrac{3\sqrt{2}}{2}$ **81.** $\pm\sqrt{3}, \pm 1$ **83.** $-\dfrac{1}{5}, -\dfrac{1}{3}$

85. $\frac{1}{4}$ **87.** 26   **89.** -16 **91.** 0   **93.** 9

95. $-59, 69$ **97.** $4, -5$ **99.** $\dfrac{-3 \pm \sqrt{21}}{6}$

101. $2, -\frac{3}{2}$ **103.** $1, -3$ **105.** $3, -2$ **107.** $\sqrt{3}, -3$

109.

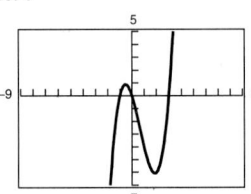

$x = 0, 3, -1$

111.

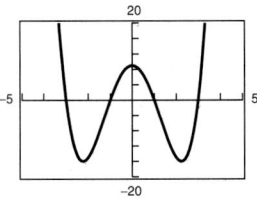

$x = \pm 3, \pm 1$

113.

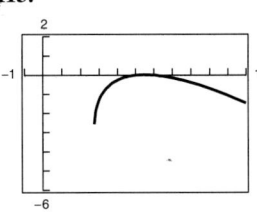

$x = 5, 6$

115.

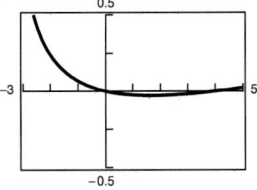

$x = 0, 4$

117.

$x = -1$

119.

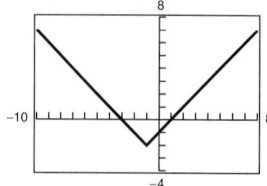

$x = 1, -3$

121. $\dfrac{2A}{b}$ **123.** $\dfrac{2A - ah}{h}$ **125.** $h = \dfrac{1}{\pi r}\sqrt{S^2 - \pi^2 r^4}$

127. (a) $A(x) = \frac{8}{3}x(25 - x)$

(b)

x	y	Area
2	$\frac{92}{3}$	$\frac{368}{3} \approx 123$
4	28	224
6	$\frac{76}{3}$	304
8	$\frac{68}{3}$	$\frac{1088}{3} \approx 363$
10	20	400
12	$\frac{52}{3}$	416
14	$\frac{44}{3}$	$\frac{1232}{3} \approx 411$

Approximate dimensions for maximum area: $24 \times \frac{52}{3}$

(c)

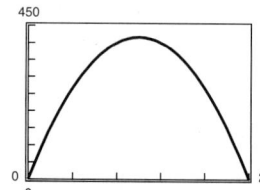

Approximate dimensions for maximum area: $25 \times \frac{50}{3}$

(d) and (e) $15 \times 23\frac{1}{3}$ or 35×10

Section P.5   *(page 64)*

1. (d)   **3.** (c)

5. (a) Yes   (b) No   (c) Yes   (d) No

7. (a) Yes   (b) Yes   (c) Yes   (d) No

9. $x > -4$

11. $x < -\frac{1}{2}$

13. $-1 < x < 3$

15. $-\frac{9}{2} < x < \frac{15}{2}$

17. $-3 < x < 3$

19.

21.

23.

25.

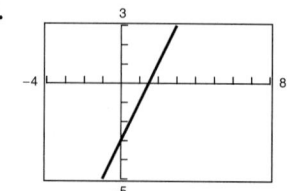

(a) $x \geq 2$

(b) $x \leq \frac{3}{2}$

27.

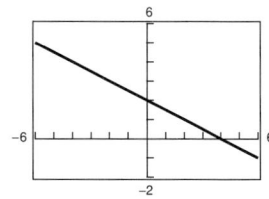

(a) $-2 \le x \le 4$

(b) $x \le 4$

29. $[5, \infty)$

31. $x < -6, \ x > 6$

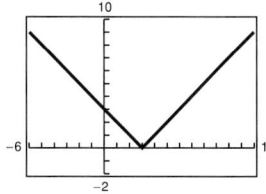

33. $16 \le x \le 24$

35. $x \le 16, \ x \ge 24$

37. $x \le -7, \ x \ge 13$

39. No solution

41.

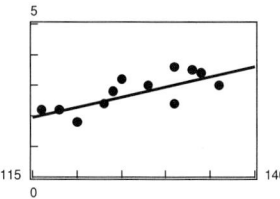

(a) $1 \le x \le 5$

(b) $x \le -1, \ x \ge 7$

43. $|x| \le 3$ **45.** $|x - 7| \ge 3$ **47.** $|x - 12| \le 10$

49. (a)

Linear

(b) $y = 0.067x - 5.638$

(c) $x \ge 129$

(d) IQ scores are not a particularly good predictor of GPA. Other factors, such as study habits, class attendance, and attitude, influence college performance.

51. $(-7, 3)$

53. $(-\infty, -5], [1, \infty)$

55. $(-3, 2)$

57. $[-2, 0], [2, \infty)$

59.

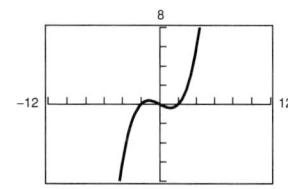

(a) $x \le -1, \ x \ge 3$

(b) $0 \le x \le 2$

61.

(a) $-2 \le x \le 0,$
$2 \le x < \infty$

(b) $x \le 4$

63. $(-\infty, -1), \ (0, 1)$

65. $(-\infty, -1), \ (4, \infty)$

67.

(a) $0 \le x < 2$

(b) $2 < x \le 4$

69.

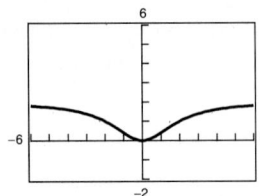

(a) $2 \leq |x|$

(b) $-\infty < x < \infty$

71. $[-2, 2]$

73. (a)

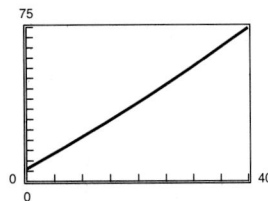

(b) 1962

Section P.6 *(page 73)*

1. (a) $\$1.109$ (b) $[\$0.999, \$1.189]$

3.

15

5.

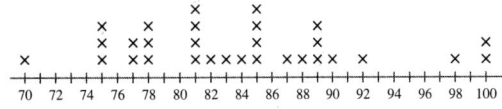

81 and 85

7.

Stems	Leaves
7	0 5 5 5 7 7 8 8 8
8	1 1 1 1 2 3 4 5 5 5 5 7 8 9 9 9
9	0 2 8
10	0 0

9.

Stems	Leaves
6	63 76 92
7	62 64
8	04 18 34 34 41 48 50 63 64 65 85 98
9	03 13 15 16 19 19 32 40 43 64 64 74 74
10	03 24 24 29 45 49 56 61 96 98
11	23 24 31 38 55 97
12	52 65
13	36 51
18	00

11. 500%

13.

15.

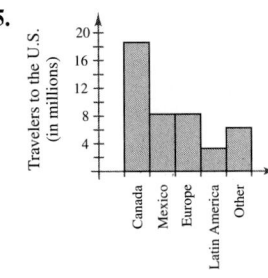

17. (a) 55.6%

(b) No. The trend limits the amount of funds available for capital improvements in industry.

19.

21.

99.

Cha

1.

N

2.

3.

Syn

(0, 4

FOCUS ON CONCEPTS (page 76)

1. (b) **2.** (c) **3.** (d) **4.** (a)

5. An identity is true for all values of the variable, and a conditional equation is true for only some values of the variable.

6. (a) Negative (b) Positive

7. (a) Neither (b) Both (c) Quadratic (d) Neither

8. Dividing by x does not yield an equivalent equation. $x = 0$ is also a solution.

9. (b)

10. (a) $x = a$, $x = b$

(b)

(c) The real zeros of the polynomial.

Review Exercises (page 77)

1.

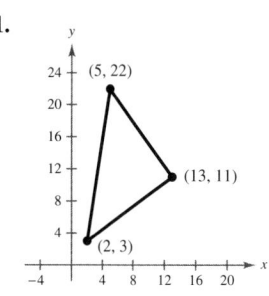

$$2\left(\sqrt{185}\right)^2 = \left(\sqrt{370}\right)^2$$

3. Quadrant IV

5. (a)

(b) 5

(c) $\left(-1, \frac{13}{2}\right)$

7.

9.

11.

13.

15.

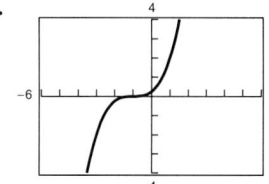

Intercepts: $(-1, 0)$, $\left(0, \frac{1}{4}\right)$

17.

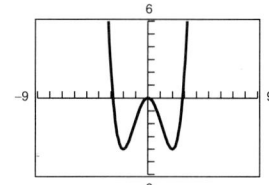

Intercepts: $(0, 0)$, $\left(\pm 2\sqrt{2}, 0\right)$

19.

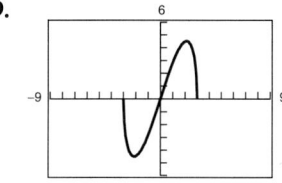

Intercepts: $(0, 0)$, $(\pm 3, 0)$

CHAPTER 1

Section 1.1   *(page 92)*

1. Yes   **3.** No   **5.** Yes   **7.** No

9. (a) Function

   (b) Not a function, because the element 1 in A corresponds to two elements, -2 and 1, in B

   (c) Function

11. Each is a function. To each year there corresponds one and only one circulation.

13. Not a function   **15.** Function   **17.** Function

19. Not a function   **21.** Function

23. (a) $\dfrac{1}{5}$ (b) 1   (c) $\dfrac{1}{4x+1}$ (d) $\dfrac{1}{x+c+1}$

25. (a) -1 (b) -9 (c) $2x-5$

27. (a) 0   (b) -0.75 (c) x^2+2x

29. (a) 1   (b) 2.5   (c) $3-2|x|$

31. (a) $-\dfrac{1}{9}$ (b) Undefined   (c) $\dfrac{1}{y^2+6y}$

33. (a) 1   (b) -1 (c) $\dfrac{|x-1|}{x-1}$

35. (a) -1 (b) 2   (c) 6

37.

x	-2	-1	0	1	2
$f(x)$	1	-2	-3	-2	1

39.

t	-5	-4	-3	-2	-1
$h(t)$	1	$\frac{1}{2}$	0	$\frac{1}{2}$	1

41.

x	-2	-1	0	1	2
$f(x)$	5	$\frac{9}{2}$	4	1	0

43. 5   **45.** ± 3 **47.** 2, -1 **49.** 3, 0

51. All real numbers x **53.** All real numbers except $t=0$

55. $y \ge 10$ **57.** $-1 \le x \le 1$

59. All real numbers except $x = 0, -2$

61. $(-2,4), (-1,1), (0,0), (1,1), (2,4)$

63. $(-2,0), (-1,1), (0,\sqrt{2}), (1,\sqrt{3}), (2,2)$

65. The domain is the set of inputs of the function, and the range is the set of outputs.

67. $g(x) = -2x^2$ **69.** $r(x) = \dfrac{32}{x}$ **71.** $3+h$

73. $3x^2 + 3xc + c^2$ **75.** 3   **77.** $A = \dfrac{C^2}{4\pi}$

79. (a)

Height, x	Length and Width	Volume, V
1	$24 - 2(1)$	$1[24-2(1)]^2 = 484$
2	$24 - 2(2)$	$2[24-2(2)]^2 = 800$
3	$24 - 2(3)$	$3[24-2(3)]^2 = 972$
4	$24 - 2(4)$	$4[24-2(4)]^2 = 1024$
5	$24 - 2(5)$	$5[24-2(5)]^2 = 980$
6	$24 - 2(6)$	$6[24-2(6)]^2 = 864$

Maximum when $x = 4$

(b)

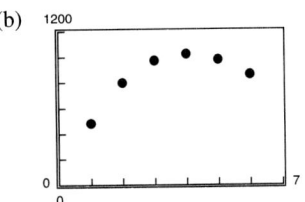

V is a function of x. $V = x(24-2x)^2$, $0 < x < 12$

81. $A = \dfrac{x^2}{2(x-2)}$, $x > 2$

83. $V = x^2y = x^2(108-4x) = 108x^2 - 4x^3$, $0 < x < 27$

85. (a) $C = 12.30x + 98{,}000$ (b) $R = 17.98x$

   (c) $P = 5.68x - 98{,}000$

87. (a) 28

   (b) -17. Average decrease per year in the population.

   (c)

t	1988	1989	1990	1991
N	9.0	19.8	43.9	53.6

t	1992	1993	1994	1995
N	30.9	16.7	10.2	6.9

Section 1.2   *(page 105)*

1. Domain: $(-\infty, -1], [1, \infty)$

   Range: $[0, \infty)$

3. Domain: $[-4, 4]$

   Range: $[0, 4]$

5.

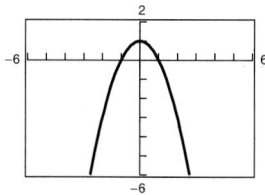

    Domain: All real numbers

    Range: $(-\infty, 1]$

7.

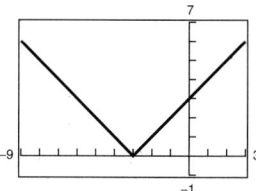

    Domain: All real numbers

    Range: $[0, \infty)$

9. Function. Graph the given function over the window shown in the figure.

11. Not a function. Solve for y and graph the resulting two functions.

13. Function. Solve for y and graph the resulting function.

15. Second setting   **17.** First setting

19. (a) Increasing on $(-\infty, \infty)$ (b) Odd function

21. (a) Increasing on $(-\infty, 0)$, $(2, \infty)$

         Decreasing on $(0, 2)$

    (b) Neither even nor odd

23. Yes. To each value of y there corresponds one and only one value of x.

25. (a)

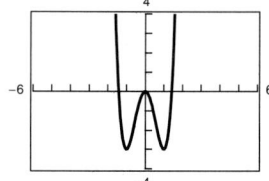

    (b) Increasing on $(-1, 0)$, $(1, \infty)$

         Decreasing on $(-\infty, -1)$, $(0, 1)$

    (c) Even function

27. (a)

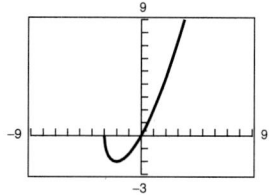

    (b) Increasing on $(-2, \infty)$

         Decreasing on $(-3, -2)$

    (c) Neither even nor odd

29. Even function   **31.** Odd function

33. Neither even nor odd   **35.** (a) $\left(\frac{3}{2}, 4\right)$ (b) $\left(\frac{3}{2}, -4\right)$

37. Even function

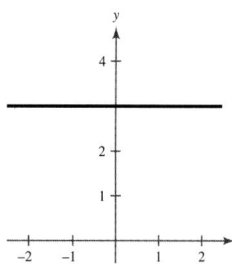

39. Neither even nor odd

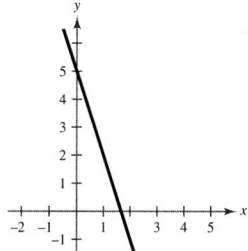

41. Neither even nor odd

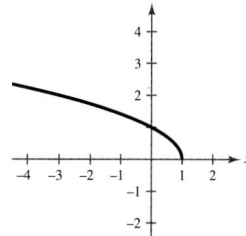

43. Neither even nor odd

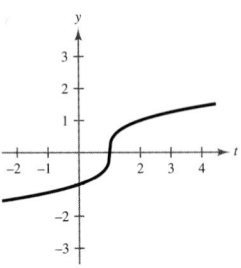

45. Neither even nor odd

47.

49.

51.

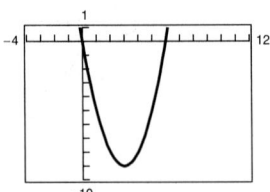

Relative minimum: $(3, -9)$

53.

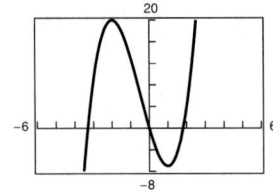

Relative minimum: $(1, -7)$

Relative maximum: $(-2, 20)$

55.

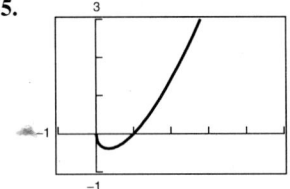

Minimum: $(0.33, -0.38)$

57. (a) Answers will vary.

(b)

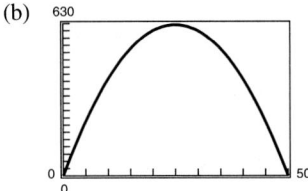

(c) 25×25 meters; 625 square meters

59.

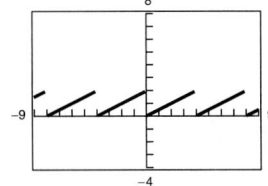

Domain: $(-\infty, \infty)$

Range: $[0, 2)$

Sawtooth pattern

61. (a) C_2 is the appropriate model because the cost does not increase until after the next minute of conversation has started.

(b) $7.85

63. $(-\infty, 4]$

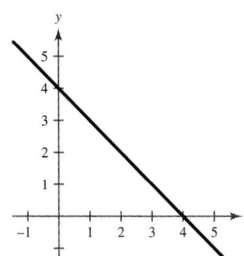

65. $(-\infty, -3], [3, \infty)$

3.

67. $[-1, 1]$

69. $(-\infty, \infty)$

5.

7.

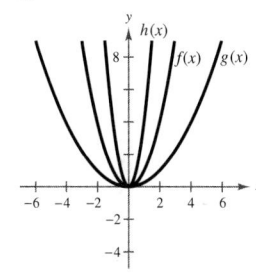

71. $h = -x^2 + 4x - 3$ **73.** $h = 2x - x^2$

75. $L = \frac{1}{2}y^2$

77. (a) $y = -87.49 + 16.28t - 4.82t^2 - 1.17t^3$

    Domain: $-4 \le t \le 3$

(b)

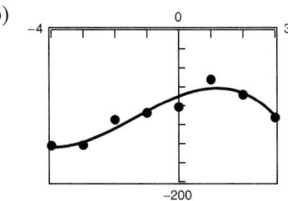

(c) 1992, 1990

(d) Because the balance would continue to decrease

79. Answers will vary.

9.

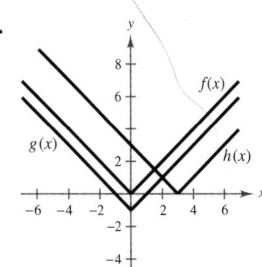

Section 1.3   *(page 116)*

1.

11. (a)

(b)

(c)

(d)

(e)

(f)

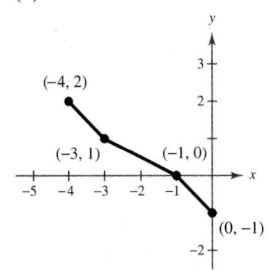

13. Horizontal shift of $y = x^3$
$y = (x - 2)^3$

15. Reflection in the x-axis of $y = x^2$
$y = -x^2$

17. Reflection in the x-axis and a vertical shift of $y = \sqrt{x}$
$y = 1 - \sqrt{x}$

19. Vertical shift of $y = x^2$
$y = x^2 - 1$

21. Reflection in the x-axis of $y = x^3$ followed by a vertical shift
$y = 1 - x^3$

23. Vertical shift of $y = x$
$y = x + 3$

25. Vertical shift   **27.** Horizontal shift

29. Vertical stretch   **31.** Horizontal shift

33. Reflection in the x-axis   **35.** Vertical shrink

37.

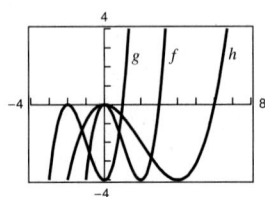

g is a horizontal shift and h is a horizontal stretch.

39.

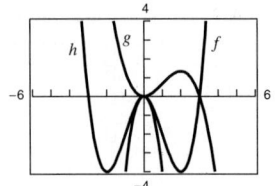

g is a vertical shrink and a reflection in the x-axis and h is a reflection in the y-axis.

41. Reflection in the x-axis followed by a vertical shift

43. A horizontal shift followed by a vertical shrink

45. A horizontal stretch followed by a vertical shift

47. $y = -(x^3 - 3x^2) + 1$

49. (a)

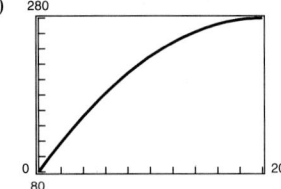

(b) $P(x) = -2420 + 20x - 0.5x^2$; vertical shift

(c) $P(x) = 80 + \dfrac{1}{5}x - \dfrac{x^2}{20{,}000}$; horizontal stretch

51. (a)   (b)

(c)   (d)

(e)

(f)

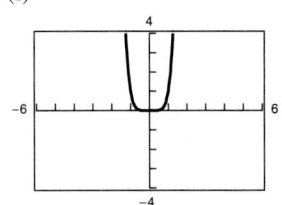

All the graphs pass through the origin. The graphs of the odd powers of x are symmetric to the origin and the graphs of the even powers are symmetric to the y-axis. As the powers increase, the graphs become flatter in the interval $-1 < x < 1$.

53.

55.

57.

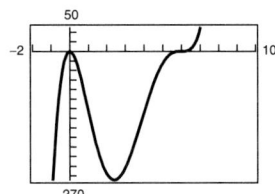

59. (a) To each time t there corresponds one and only one temperature T.

   (b) $60°, 72°$

   (c) All the temperature changes would be 1 hour later.

   (d) The temperature would be decreased by 1 degree.

61.

63.

65.

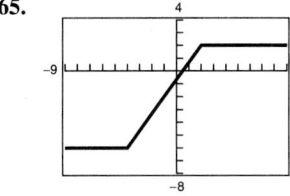

Section 1.4   (page 126)

1.

3.

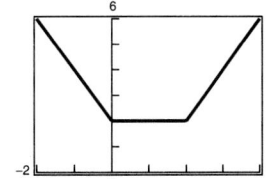

5. (a) $2x$ (b) 2 (c) $x^2 - 1$ (d) $\dfrac{x + 1}{x - 1}$, $x \neq 1$

7. (a) $x^2 - x + 1$ (b) $x^2 + x - 1$ (c) $x^2 - x^3$

   (d) $\dfrac{x^2}{1 - x}$, $x \neq 1$

9. (a) $x^2 + 5 + \sqrt{1 - x}$ (b) $x^2 + 5 - \sqrt{1 - x}$

   (c) $(x^2 + 5)\sqrt{1 - x}$ (d) $\dfrac{x^2 + 5}{\sqrt{1 - x}}$, $x < 1$

11. (a) $\dfrac{x+1}{x^2}$ (b) $\dfrac{x-1}{x^2}$ (c) $\dfrac{1}{x^3}$ (d) $x, \quad x \neq 0$

13. 9   **15.** 5   **17.** $4t^2 - 2t + 5$ **19.** 0   **21.** 26

23. $\frac{3}{5}$

25.

27.

29.

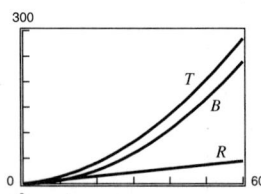

$f(x), g(x)$

31. (a) $T = \frac{3}{4}x + \frac{1}{15}x^2$

(b)

(c) B

33.

35. (a) $(x-1)^2$ (b) $x^2 - 1$ (c) x^4

37. (a) $20 - 3x$ (b) $-3x$ (c) $9x + 20$

39. (a) $(f \circ g)(x) = \sqrt{x^2 + 4}$

$(g \circ f)(x) = x + 4, \quad x \geq -4$

Not equal

(b)

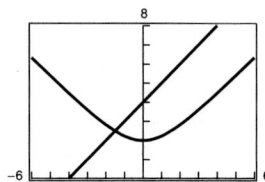

41. (a) $(f \circ g)(x) = x - \frac{8}{3}$

$(g \circ f)(x) = x - 8$

Not equal

(b)

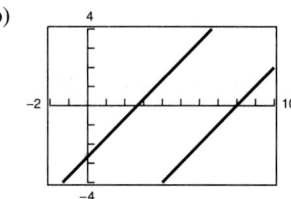

43. (a) $(f \circ g)(x) = x^4$

$(g \circ f)(x) = x^4$

Equal

(b)

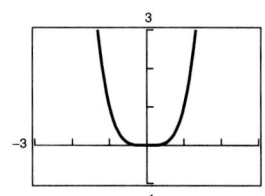

45. (a) 3   (b) 0   **47.** (a) 0   (b) 4

49. $f(x) = x^2, \quad g(x) = 2x + 1$

51. $f(x) = \sqrt[3]{x}, \quad g(x) = x^2 - 4$

53. $f(x) = \dfrac{1}{x}, \quad g(x) = x + 2$

55. $f(x) = x^2 + 2x, \quad g(x) = x + 4$

57. (a) $x \geq 0$ (b) All real numbers   (c) All real numbers

59. (a) All real numbers except $x = \pm 1$

(b) All real numbers

(c) All real numbers except $x = -2, 0$

61. 3   **63.** $\dfrac{-4}{x(x + h)}$

65. $(A \circ r)(t) = 0.36\pi t^2$

$A \circ r$ represents the area of the circle at time t.

67. (a) $(C \circ x)(t) = 3000t + 750$

$C \circ x$ represents the cost after t production hours.

(b)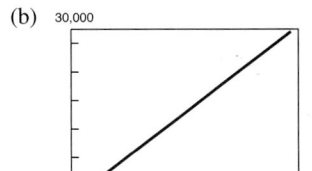

4.75 hours

69. $g(f(x))$ represents 3 percent of an amount over $\$500,000$.

71. Answers will vary.   **73.** Answers will vary.

75. (a) $f(x) = (x^2 + 1) + (-2x)$

(b) $f(x) = \dfrac{-1}{(x + 1)(x - 1)} + \dfrac{x}{(x + 1)(x - 1)}$

77.

79.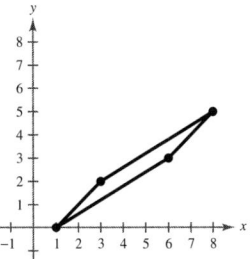

81. Quadrants I and III

11. (a) $f(g(x)) = f\left(\dfrac{x}{2}\right) = 2\left(\dfrac{x}{2}\right) = x$

$g(f(x)) = g(2x) = \dfrac{(2x)}{2} = x$

(b)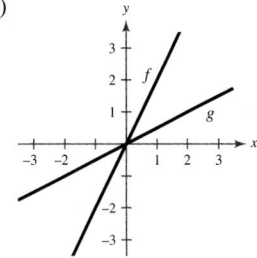

13. (a) $f(g(x)) = f\left(\dfrac{x - 1}{5}\right) = 5\left(\dfrac{x - 1}{5}\right) + 1 = x$

$g(f(x)) = g(5x + 1) = \dfrac{(5x + 1) - 1}{5} = x$

(b)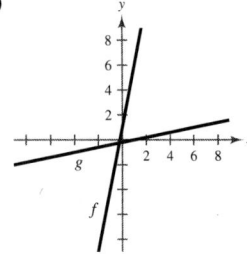

15. $f(g(x)) = f\left(\sqrt[3]{x}\right) = \left(\sqrt[3]{x}\right)^3 = x$

$g(f(x)) = g(x^3) = \sqrt[3]{x^3} = x$

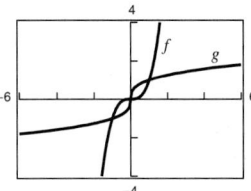

Reflections in the line $y = x$

Section 1.5   (page 139)

1. (c)   **3.** (a)   **5.** $f^{-1}(x) = \frac{1}{8}x$ **7.** $f^{-1}(x) = x - 10$

9. $f^{-1}(x) = x^3$

17. $f(g(x)) = f(x^2 + 4), \quad x \geq 0$
$$= \sqrt{(x^2 + 4) - 4} = x$$
$$g(f(x)) = g(\sqrt{x - 4})$$
$$= (\sqrt{x - 4})^2 + 4 = x$$

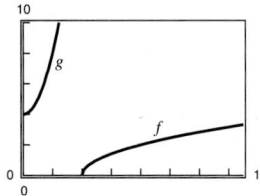

Reflections in the line $y = x$

19. $f(g(x)) = f(\sqrt[3]{1 - x})$
$$= 1 - (\sqrt[3]{1 - x})^3 = x$$
$$g(f(x)) = g(1 - x^3)$$
$$= \sqrt[3]{1 - (1 - x^3)} = x$$

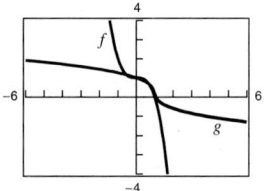

Reflections in the line $y = x$

21.

Yes

23.

No

25.

Yes

27.

No

29.

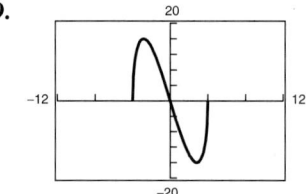

No

31. $f^{-1}(x) = \dfrac{x + 3}{2}$

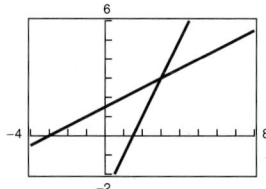

Reflections in the line $y = x$

33. $f^{-1}(x) = \sqrt[5]{x}$

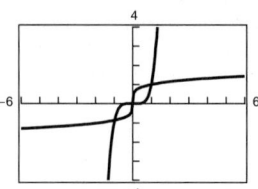

Reflections in the line $y = x$

35. $f^{-1}(x) = x^2, \quad x \geq 0$

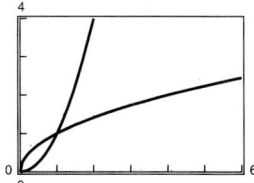

Reflections in the line $y = x$

37. $f^{-1}(x) = \sqrt{4 - x^2}, \quad 0 \leq x \leq 2$

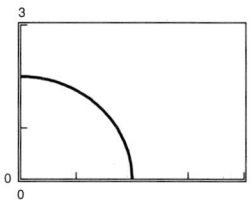

Reflections in the line $y = x$

39. $f^{-1}(x) = x^3 + 1$

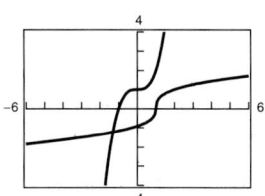

Reflections in the line $y = x$

41. No inverse   **43.** $g^{-1}(x) = 8x$ **45.** No inverse

47. $f^{-1}(x) = \sqrt{x} - 3, \quad x \geq 0$ **49.** No inverse

51. $f^{-1}(x) = \dfrac{x^2 - 3}{2}, \quad x \geq 0$ **53.** No inverse

55. $f^{-1}(x) = -\sqrt{25 - x}, \quad x \leq 25$

57. $y = \sqrt{x} + 2, \quad x \geq 0$ **59.** $y = x - 2, \quad x \geq 0$

61.

x	-4	-2	2	3
$f^{-1}(x)$	-2	-1	1	3

63. (a) and (b)

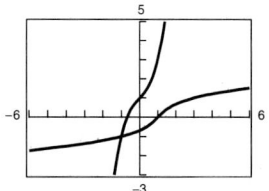

(c) Inverse function because it satisfies the Vertical Line Test

65. (a) and (b)

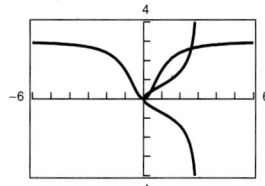

(c) Not an inverse function because it does not satisfy the Vertical Line Test

67. False, $f(x) = x^2$ **69.** True   **71.** 32   **73.** 600

75. $2\sqrt[3]{x + 3}$ **77.** $\dfrac{x + 1}{2}$ **79.** $\dfrac{x + 1}{2}$

81. Answers will vary.

83. (a) $y = \dfrac{x - 8}{0.75}$

y = number of units produced

x = hourly wage

(b)

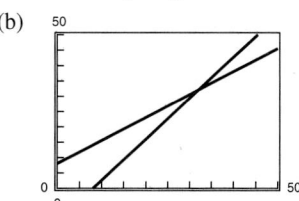

(c) $15.50

(d) 19

85. (a) $y = \sqrt{\dfrac{x - 254.50}{0.03}}$

x = degrees Fahrenheit

y = % load

(b)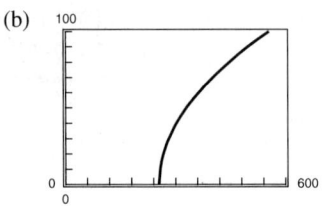

(c) $0 < x < 90.46$

Section 1.6   *(page 148)*

1. Negative correlation   **3.** No correlation

5. (a)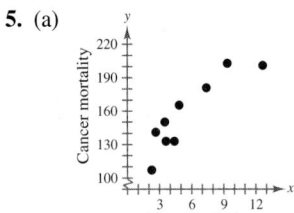

(b) Yes. The cancer mortality increases linearly with increased exposure to the carcinogenic substance.

7. (a) $y = 0.42x + 1.50$

(b)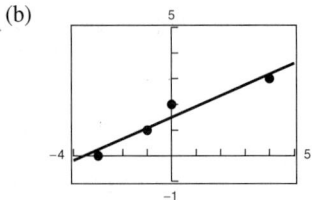

(c) The model appears valid.

9. (a) $y = 0.95x + 0.92$

(b)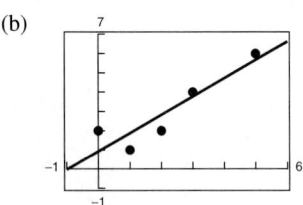

(c) The model appears valid.

11. (a) $d = 0.066F$

(b)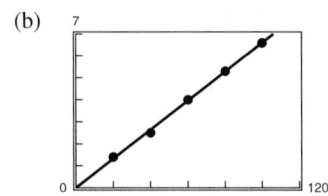

The model fits well.

(c) 3.63

13. (a) $R = 1.21t + 4.73$

(b)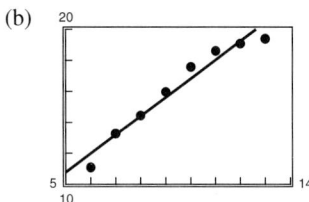

(c) The average annual increase in the monthly basic rate

(d) $\$28.84$

15. (a) $y = 47.77x + 103.77$

(b)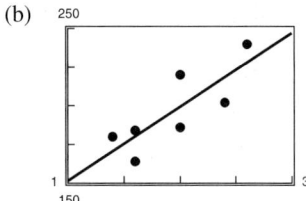

(c) The average increase in sales for increases in advertising expenditures

(d) $\$175{,}000$

17. (a) $y = -0.64x + 1321.56$

(b)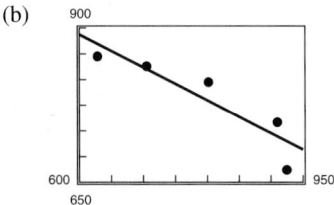

(c) The negative slope indicates that the amount of mortgage debt held by commercial banks is increasing as the amount held by savings institutions decreases.

FOCUS ON CONCEPTS   *(page 151)*

1. No. The element 3 in the domain corresponds to two elements in the range.

2. (a)

| Xmin = -15 |
| Xmax = 6 |
| Xscl = 3 |
| Ymin = -18 |
| Ymax = 6 |
| Yscl = 3 |

(b)

| Xmin = -24 |
| Xmax = 36 |
| Xscl = 6 |
| Ymin = -54 |
| Ymax = 12 |
| Yscl = 6 |

3. (a) Even. The graph is a reflection in the x-axis.

(b) Even. The graph is a reflection in the y-axis.

(c) Even. The graph is a vertical translation of f.

(d) Neither. The graph is a horizontal translation of f.

4. (a) $g(t) = \frac{3}{4} f(t)$

(b) $g(t) = f(t) + 10{,}000$

(c) $g(t) = f(t - 2)$

(d) $g(t) = f\left(\dfrac{t}{2}\right)$

5. False. Implies that the model fits the data well and the slope of the model is negative.

6. True.

Review Exercises   *(page 152)*

1. No. To the input values of 5 and 10 there correspond two output values.

3. No   **5.** Yes   **7.** (a) 5   (b) $t^4 + 1$ (c) $-x^2 - 1$

9. (a) -14 (b) $-5x^2 - 30x - 39$ (c) $-5(2x + t)$

11. (a) 0   (b) 2   (c) 8   **13.** $[-5, 5]$

15. All real numbers except $s = 3$

17. $(-\infty, -2), (-2, 3), (3, \infty)$ **19.** Second setting

21.   **23.**

25.

27.

29.

31.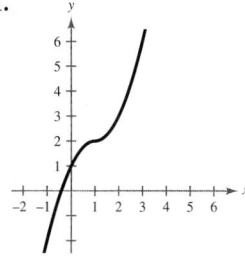

33. Reflection in the x-axis

35. Horizontal shift

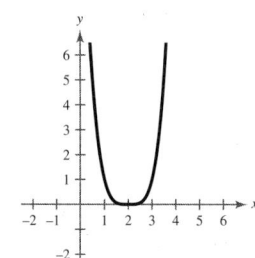

37. Vertical shift
$y = \sqrt{x} - 2$

39. Reflection in the x-axis
$y = -\sqrt{x}$

41.

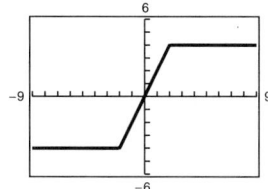

(a) Constant: $(-\infty, -2), (2, \infty)$

    Increasing: $(-2, 2)$

(b) Maximum: 4

    Minimum: -4

(c) Odd

43.

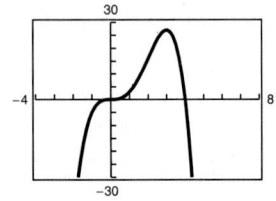

(a) Increasing: $(-\infty, 3)$

   Decreasing: $(3, \infty)$

(b) Relative maximum: $(3, 27)$

(c) Neither

45. (a) $C = 5.35x + 16{,}000$

   (b) $P = 2.85x - 16{,}000$

47. (a) $f^{-1}(x) = 2(x + 3)$

   (b)

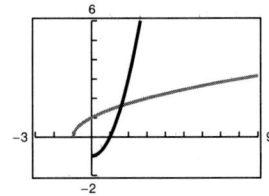

49. (a) $f^{-1}(x) = x^2 - 1, \ x \geq 0$

   (b)

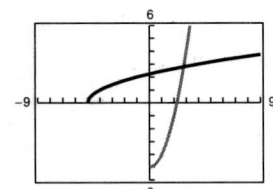

51. (a) $f^{-1}(x) = \sqrt{x + 5}$

   (b)

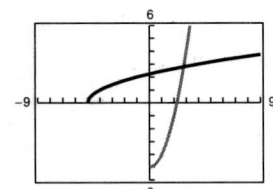

53. $x \geq 4, \ f^{-1}(x) = \sqrt{\dfrac{x}{2}} + 4$

55. $x \geq 2, \ f^{-1}(x) = \sqrt{x^2 + 4}, \ x \geq 0$

57. -7 **59.** 5   **61.** 23   **63.** 9

65. (a)

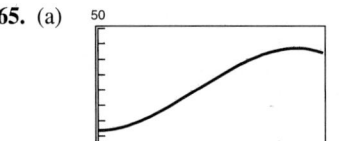

(b) $S = 34.4$

(c) The model will continue to decrease.

67. (c), (b), (d), (a)

Chapter Test   *(page 156)*

1. 7   **2.** $10 - \sqrt{6 - t}$ **3.** $\dfrac{\sqrt{3 - x} - 1}{2 - x}$

4. $(-\infty, 3]$

5. No. To some x there correspond more than one value of y.

6.

    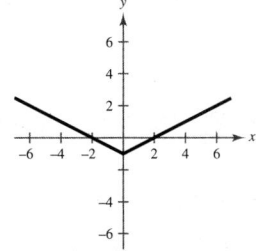

7. $C = 5.60x + 24{,}000$

   $P = 3.60x - 24{,}000$

8.

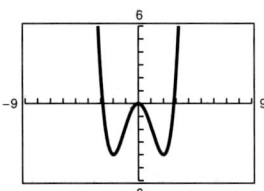

Even

Increasing: $(-2, 0), (2, \infty)$

Decreasing: $(-\infty, -2), (0, 2)$

9.

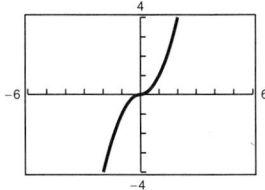

Odd

Increasing: $(-\infty, \infty)$

10. (a) $x^2 - \sqrt{2 - x}$, $(-\infty, 2]$

(b) $\dfrac{x^2}{\sqrt{2 - x}}$, $(-\infty, 2)$

(c) $2 - x$, $(-\infty, \infty)$

(d) $2 - x^2$, $[0, \infty)$

11. Its graph must satisfy the Horizontal Line Test.

12. Reflections in the line $y = x$.

13. $y = 0.0568x - 1.5808$, 3.4

CHAPTER 2

Section 2.1   *(page 165)*

1. (g)   **3.** (b)   **5.** (f)   **7.** (e)

9. (a)

Vertical shrink

(b)

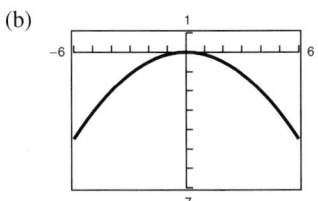

Vertical shrink and reflection in the *x*-axis

(c)

Vertical stretch

(d)

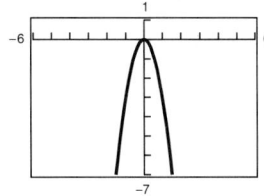

Vertical stretch and reflection in the *x*-axis

11. (a)

Horizontal translation

(b)

Horizontal translation

(c)

Horizontal translation

(d)

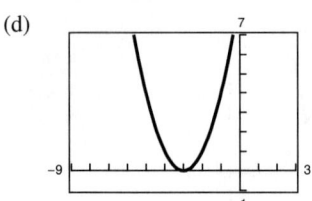

Horizontal translation

13. Vertex: $(0, 16)$

Intercepts: $(\pm 4, 0)$, $(0, 16)$

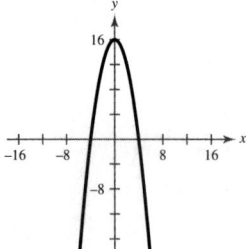

15. Vertex: $(4, 0)$

Intercepts: $(4, 0)$, $(0, 16)$

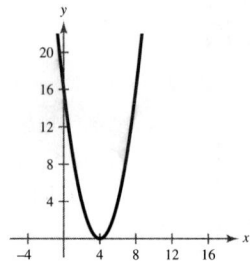

17. Vertex: $\left(\frac{1}{2}, 1\right)$

Intercept: $\left(0, \frac{5}{4}\right)$

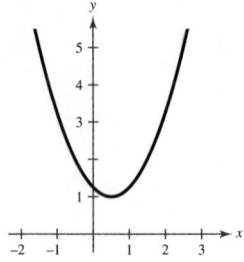

19. Vertex: $(1, 6)$

Intercepts: $\left(1 \pm \sqrt{6}, 0\right)$, $(0, 5)$

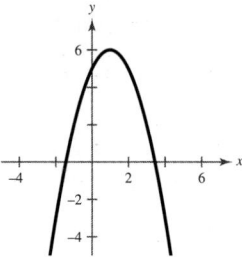

21. Vertex: $\left(\frac{1}{2}, 20\right)$

Intercept: $(0, 21)$

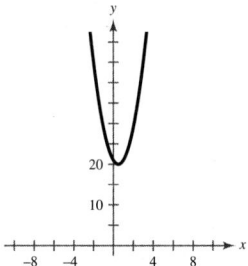

23. Vertex: $(-1, 4)$

Intercepts: $(1, 0)$, $(-3, 0)$, $(0, 3)$

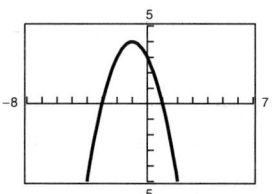

25. Vertex: $(4, -1)$

Intercepts: $\left(4 \pm \frac{1}{2}\sqrt{2}, 0\right)$, $(0, 31)$

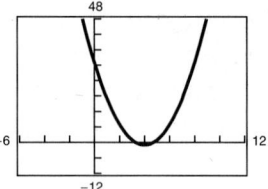

27. $y = (x - 1)^2$ **29**. $y = -2(x + 2)^2 + 2$

31. $f(x) = (x + 2)^2 + 5$ **33**. $f(x) = -\frac{1}{2}(x - 3)^2 + 4$

35. $f(x) = \frac{3}{4}(x - 5)^2 + 12$ **37**. $(\pm 4, 0)$

39. $(5, 0), (-1, 0)$

41.

43.

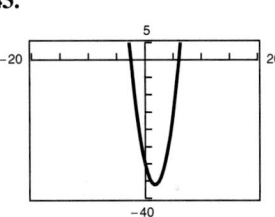

$(0, 0), (4, 0)$ $\left(-\frac{5}{2}, 0\right), (6, 0)$

45. $f(x) = x^2 - 2x - 3$ **47.** $f(x) = 2x^2 + 7x + 3$
$\quad g(x) = -x^2 + 2x + 3$ $\quad g(x) = -2x^2 - 7x - 3$

49. $55, 55$

51. (a) $A = x(50 - x), 0 < x < 50$ **53.** 4500 units

(b)

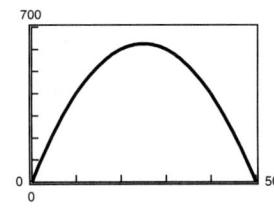

(c) 25 feet $\times$ 25 feet

55. (a)

x	y	Area
2	$\frac{1}{3}[200 - 4(2)]$	$(2)(2)\left(\frac{1}{3}\right)[200 - 4(2)] = 256$
4	$\frac{1}{3}[200 - 4(4)]$	$(2)(4)\left(\frac{1}{3}\right)[200 - 4(4)] \approx 491$
6	$\frac{1}{3}[200 - 4(6)]$	$(2)(6)\left(\frac{1}{3}\right)[200 - 4(6)] = 704$
8	$\frac{1}{3}[200 - 4(8)]$	$(2)(8)\left(\frac{1}{3}\right)[200 - 4(8)] = 896$
10	$\frac{1}{3}[200 - 4(10)]$	$(2)(10)\left(\frac{1}{3}\right)[200 - 4(10)] \approx 1067$
12	$\frac{1}{3}[200 - 4(12)]$	$(2)(12)\left(\frac{1}{3}\right)[200 - 4(12)] = 1216$

(b)

x	y	Area
20	$\frac{1}{3}[200 - 4(20)]$	$(2)(20)\left(\frac{1}{3}\right)[200 - 4(20)] = 1600$
22	$\frac{1}{3}[200 - 4(22)]$	$(2)(22)\left(\frac{1}{3}\right)[200 - 4(22)] \approx 1643$
24	$\frac{1}{3}[200 - 4(24)]$	$(2)(24)\left(\frac{1}{3}\right)[200 - 4(24)] = 1664$
26	$\frac{1}{3}[200 - 4(26)]$	$(2)(26)\left(\frac{1}{3}\right)[200 - 4(26)] = 1664$
28	$\frac{1}{3}[200 - 4(28)]$	$(2)(28)\left(\frac{1}{3}\right)[200 - 4(28)] \approx 1643$
30	$\frac{1}{3}[200 - 4(30)]$	$(2)(30)\left(\frac{1}{3}\right)[200 - 4(30)] = 1600$

(c) $A = \dfrac{8x(50 - x)}{3}$

(d)

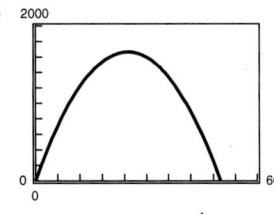

$x = 25$ feet, $y = 33\frac{1}{3}$ feet

(e) $A = -\frac{8}{3}(x - 25)^2 + \frac{5000}{3}$

57. (a)

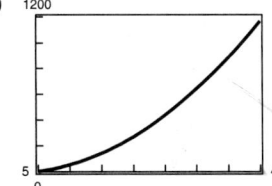

(b) 4 feet   (c) 16 feet   (d) 25.86 feet

59. (a)

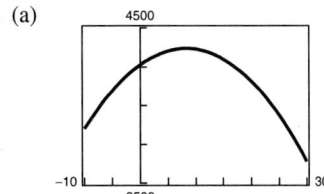

(b) 166.69 board feet

(c) 26.6 inches

61. (a)

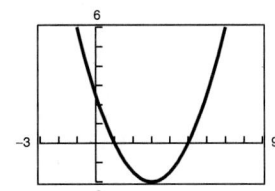

(b) $1968.$ Yes

(c) 4024.5 annually, 11 daily

63. (a) Answers will vary.

(b)

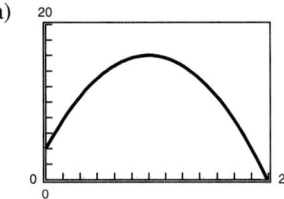

Zeros: $x = 1, 5$

x-coordinate of the vertex: 3

Section 2.2   *(page 177)*

1. (f)   **3.** (c)   **5.** (e)   **7.** (g)

9. (a)   (b)

  (c)   (d)

11. (a)   (b)

  (c)   (d)

13.   **15.**

17. Rises to the left.   **19.** Falls to the left.
Rises to the right.   Falls to the right.

21. Falls to the left.   **23.** Rises to the left.
Rises to the right.   Falls to the right.

25. Falls to the left.   **27.** ± 5 **29.** 3
Falls to the right.

31. $1, -2$ **33.** $2, 0$

35.   **37.**

 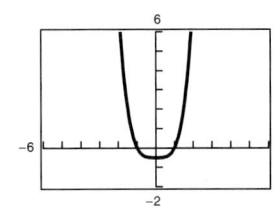

$(0.268, 0), (3.732, 0)$ $(-1, 0), (1, 0)$

39.   **41.**

 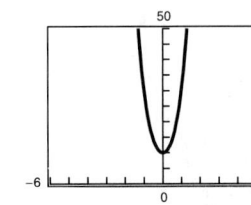

$(-2.236, 0), (2.236, 0)$ No x-intercepts

43.   **45.**

 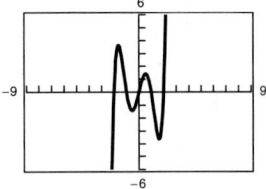

$(0, 0), \left(\frac{5}{2}, 0\right)$ $(0, 0), (\pm 1, 0), (\pm 2, 0)$

47. $f(x) = x^2 - 10x$ **49.** $f(x) = x^2 + 4x - 12$

51. $f(x) = x^3 + 5x^2 + 6x$

53. $f(x) = x^4 - 4x^3 - 9x^2 + 36x$ **55.** $f(x) = x^2 - 2x - 2$

57. (a)

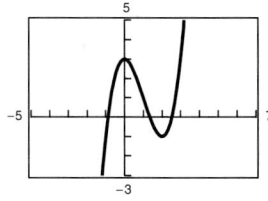

(b) $-0.879, 1.347, 2.532$

$(-1, 0), (1, 2), (2, 3)$

59. (a)

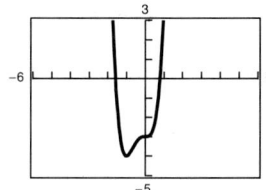

(b) $-1.585, 0.779$

$(-2, -1), (0, 1)$

61.

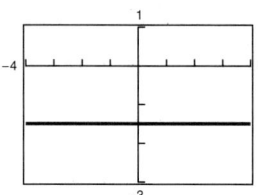

Xmin = -4
Xmax = 4
Xscl = 1
Ymin = -3
Ymax = 1
Yscl = 1

63.

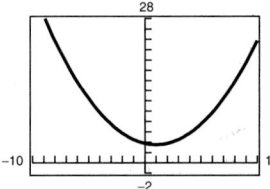

Xmin = -10
Xmax = 10
Xscl = 1
Ymin = -2
Ymax = 28
Yscl = 2

65.

Two x-intercepts

67.

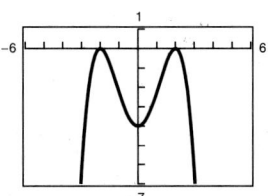

y-axis symmetry
Two x-intercepts

69.

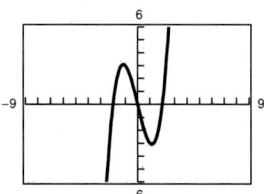

Symmetric to the origin
Three x-intercepts

71.

Three x-intercepts

73.

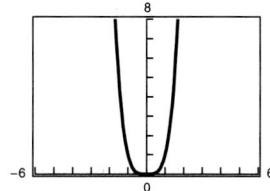

(a) Vertical shift of two units; even
(b) Horizontal shift of two units; neither even nor odd
(c) Reflection in the y-axis; even
(d) Reflection in the x-axis; even
(e) Horizontal stretch; even
(f) Vertical shrink; even
(g) $g(x) = x^3$; odd
(h) $g(x) = x^{16}$; even

75. (a) Answers will vary.   (b) $0 < x < 6$

(c)

$x \approx 2.54$

77. $(200, 160)$

79. (a) $y_1 = -0.158t^3 + 2.850t^2 - 3.814t + 74.703$

(b) $y_2 = -0.007t^3 + 0.196t^2 + 2.533t + 60.844$

(c)

The median price of homes in the South is less than in the Northeast.

Section 2.3   *(page 191)*

1.

3.

5.

7. $2x + 4$ **9.** $x^2 - 3x + 1$ **11.** $x^3 + 3x^2 - 1$

13. $7 - \dfrac{11}{x + 2}$ **15.** $3x + 5 - \dfrac{2x - 3}{2x^2 + 1}$

17. $x^2 + 2x + 4 + \dfrac{2x - 11}{x^2 - 2x + 3}$ **19.** $2x - \dfrac{17x - 5}{x^2 - 2x + 1}$

21. $x^{2n} + 6x^n + 9$ **23.** $4x^2 - 9$

25. $5x^2 + 14x + 56 + \dfrac{232}{x - 4}$ **27.** $x^2 - 8x + 64$

29. $-3x^3 - 6x^2 - 12x - 24 - \dfrac{48}{x - 2}$

31. $4x^2 + 14x - 30$ **33.** The remainder is 0.

35. $f(x) = (x - 4)(x^2 + 3x - 2) + 3/(x - 4),\ f(4) = 3$

37. $f(x) = \left(x - \sqrt{2}\right)\left[x^2 + \left(3 + \sqrt{2}\right)x + 3\sqrt{2}\right] - 8/\left(x - \sqrt{2}\right),\ f\left(\sqrt{2}\right) = -8$

39. (a) 1   (b) 4   (c) 4   (d) 1954

41. $(x - 2)(x + 3)(x - 1)$ **43.** $(2x - 1)(x - 5)(x - 2)$

     Zeros: $2, -3, 1$ Zeros: $\frac{1}{2}, 5, 2$

45. $\left(x + \sqrt{3}\right)\left(x - \sqrt{3}\right)(x + 2)$

     Zeros: $\pm\sqrt{3}, -2$

47. $(x - 1)\left(x - 1 - \sqrt{3}\right)\left(x - 1 + \sqrt{3}\right)$

     Zeros: $1 \pm \sqrt{3}, 1$

49. (a) $R = 16.823 + 1.415t - 0.115t^2 - 0.023t^3$

(b)

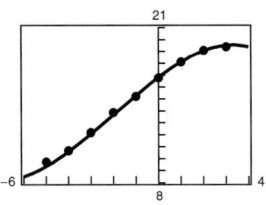

(c)

t	-5	-4	-3	-2	-1
R	9.75	10.80	12.16	13.72	15.32

t	0	1	2	3
R	16.82	18.10	19.01	19.41

(d) 16.205. No; the model will fall to the right and become negative.

51. $\pm1, \pm3$

53. $\pm1, \pm3, \pm5, \pm9, \pm15, \pm45, \pm\frac{1}{2}, \pm\frac{3}{2}, \pm\frac{5}{2}, \pm\frac{9}{2}, \pm\frac{15}{2}, \pm\frac{45}{2}$

55. (a) $\pm1, \pm2, \pm4$

(b)

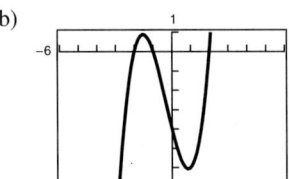

(c) $-2, -1, 2$

57. (a) $\pm1, \pm3, \pm\frac{1}{2}, \pm\frac{3}{2}, \pm\frac{1}{4}, \pm\frac{3}{4}$

(b)

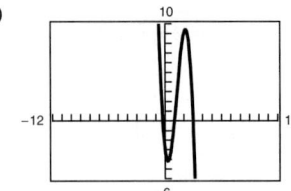

(c) $-\frac{1}{4}, 1, 3$

59. (a) $\pm 1, \pm 2, \pm 4, \pm 8, \pm \frac{1}{2}$

(b)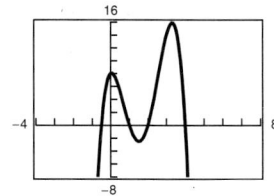

(c) $-\frac{1}{2}, 1, 2, 4$

61. (a) $\pm 1, \pm 3, \pm \frac{1}{2}, \pm \frac{3}{2}, \pm \frac{1}{4}, \pm \frac{3}{4}, \pm \frac{1}{8}, \pm \frac{3}{8}, \pm \frac{1}{16}, \pm \frac{3}{16}, \pm \frac{1}{32}, \pm \frac{3}{32}$

(b)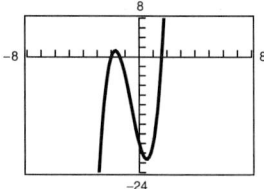

(c) $1, \frac{3}{4}, -\frac{1}{8}$

63. (a) $\pm 1, \pm 2, \pm 3, \pm 6, \pm 9, \pm \frac{1}{2}, \pm \frac{3}{2}, \pm \frac{9}{2}, \pm \frac{1}{4}, \pm \frac{3}{4}, \pm \frac{9}{4}, \pm 18$

(b)

(c) $-2, \dfrac{1}{8} \pm \dfrac{\sqrt{145}}{8}$

65. $-1, 2$ **67.** $0, -1, -3, 4$ **69.** $-2, -\frac{1}{2}, 4$

71. $\pm 1, \pm \sqrt{2}$

$f(x) = (x + 1)(x - 1)(x + \sqrt{2})(x - \sqrt{2})$

73. $x = 0, 3, 4, \pm \sqrt{2}$

$h(x) = x(x - 3)(x - 4)(x + \sqrt{2})(x - \sqrt{2})$

75.–77. Answers will vary.   **79.** $\pm 2, \pm \frac{3}{2}$ **81.** $\pm 1, \frac{1}{4}$

83. (d)   **85.** (b)   **87.** r_1, r_2, r_3

89. $r_1 + 5, r_2 + 5, r_3 + 5$ **91.** Not possible

93. (a)

(b) $V = x(9 - 2x)(15 - 2x)$

Domain: $0 < x < \frac{9}{2}$

(c)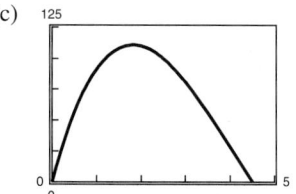

$1.82 \times 5.36 \times 11.36$

(d) $\frac{1}{2}, \frac{7}{2}, 8$

8 is not in the domain of V.

95. (a) 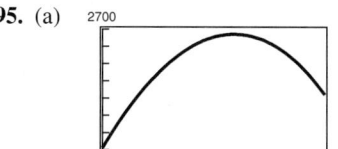   (b) 16.89

(c) 16.89

97. (a)   (b) 1991

(c) Yes

Section 2.4   *(page 202)*

1. $a = -10, \ b = 6$ **3.** $a = 6, \ b = 5$

5. $4 + 3i$ **7.** $2 - 3\sqrt{3}i$ **9.** $5\sqrt{3}i$

11. $-1 - 6i$ **13.** 8   **15.** $0.3i$ **17.** $11 - i$

19. 4   **21.** $3 - 3\sqrt{2}i$ **23.** $-14 + 20i$

25. $\frac{1}{6} + \frac{7}{6}i$ **27.** $-2\sqrt{3}$ **29.** -10 **31.** $5 + i$

33. $12 + 30i$ **35.** 24   **37.** $-9 + 40i$ **39.** -10

41. $\sqrt{-6}\sqrt{-6} = \sqrt{6}i\sqrt{6}i = 6i^2 = -6$

43. 34   **45.** 9   **47.** 400   **49.** 8   **51.** $-6i$

53. $\frac{16}{41} + \frac{20}{41}i$ **55.** $\frac{3}{5} + \frac{4}{5}i$ **57.** $-7 - 6i$

59. $-\frac{9}{1681} + \frac{40}{1681}i$ **61.** $-\frac{1}{2} - \frac{5}{2}i$ **63.** $\frac{62}{949} + \frac{297}{949}i$

65. $i, -1, -i, 1, i, -1, -i, 1, i, -1, -i, 1, i, -1, -i, 1$

67. $-1 + 6i$ **69.** $-5i$ **71.** $-375\sqrt{3}i$

73. i **75.** $4 + 3i$ **77.** $6i$

79.

81.

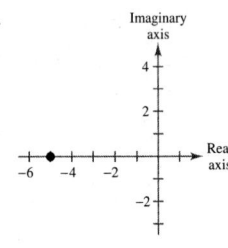

83. $0, 0, 0, 0, 0, 0$

85. $0.5i, -0.25 + 0.5i, -0.1875 + 0.25i, -0.0273 + 0.4063i,$
    $-0.1643 + 0.4778i, -0.2013 + 0.3430i$

87. $1, 2, 5, 26, 677, 458,330$

89. $8, 8, 8$ **91.–93.** Answers will vary.

95. $x^3 + x^2 + 2x - 6$ **97.** $3x^2 + 11\frac{1}{2}x - 2$

99. $x^2 + y^2 + 2xy + 6x + 6y + 9$

101. $r = \sqrt{\dfrac{\alpha m_1 m_2}{F}}$ **103.** 88.9 kilometers per hour

Section 2.5   *(page 210)*

1. $0, 6, 6$ **3.** $3, 2, \pm 3i$ **5.** $(4, 0)$; same

7. No intercepts; same

9. $\pm 5i$ **11.** $2 \pm \sqrt{3}$
    $(x + 5i)(x - 5i)$ $\left(x - 2 - \sqrt{3}\right)\left(x - 2 + \sqrt{3}\right)$

13. $\pm 3, \pm 3i$
    $(x + 3)(x - 3)(x + 3i)(x - 3i)$

15. $1 \pm i$
    $(z - 1 + i)(z - 1 - i)$

17. $-5, 4 \pm 3i$
    $(t + 5)(t - 4 + 3i)(t - 4 - 3i)$

19. $-\frac{3}{4}, 1 \pm \frac{1}{2}i$
    $(4x + 3)\left(x - 1 - \frac{1}{2}i\right)\left(x - 1 + \frac{1}{2}i\right)$

21. $-\frac{1}{5}, 1 \pm \sqrt{5}i$
    $(5x + 1)\left(x - 1 + \sqrt{5}i\right)\left(x - 1 - \sqrt{5}i\right)$

23. $\pm i, \pm 3i$
    $(x + i)(x - i)(x + 3i)(x - 3i)$

25. $2, 2, \pm 2i$
    $(x - 2)^2(x + 2i)(x - 2i)$

27. $-2, -\frac{1}{2}, \pm i$
    $(x + 2)(2x + 1)(x - i)(x + i)$

29. $x^3 - x^2 + 25x - 25$ **31.** $x^3 - 10x^2 + 33x - 34$

33. $x^4 + 37x^2 + 36$ **35.** $x^4 + 8x^3 + 9x^2 - 10x + 100$

37. (a) $(x^2 + 9)(x^2 - 3)$ (b) $(x^2 + 9)\left(x + \sqrt{3}\right)\left(x - \sqrt{3}\right)$
    (c) $(x + 3i)(x - 3i)\left(x + \sqrt{3}\right)\left(x - \sqrt{3}\right)$

39. (a) $(x^2 - 2x - 2)(x^2 - 2x + 3)$
    (b) $\left(x - 1 + \sqrt{3}\right)\left(x - 1 - \sqrt{3}\right)(x^2 - 2x + 3)$
    (c) $\left(x - 1 + \sqrt{3}\right)\left(x - 1 - \sqrt{3}\right) \cdot$
        $\left(x - 1 + \sqrt{2}i\right)\left(x - 1 - \sqrt{2}i\right)$

41. $-\frac{3}{2}, \pm 5i$ **43.** $\pm 2i, 1, -\frac{1}{2}$ **45.** $-3 \pm i, \frac{1}{4}$

47. (a) Answers will vary.   (b) $1, 2, -3 \pm \sqrt{2}i$

49. (a) Answers will vary.   (b) $\dfrac{3}{4}, \dfrac{1}{2} \pm \dfrac{\sqrt{5}}{2}i$

51. f does not have real coefficients   **53.** (a) No (b) No

55. No. Setting $P = 9,000,000$ and solving the resulting equation yields imaginary roots.

57. $x^2 - 2ax + a^2 + b^2$

59.

Section 2.6   *(page 218)*

1. (a)

x	$f(x)$	x	$f(x)$	x	$f(x)$
0.5	-2	1.5	2	5	0.25
0.9	-10	1.1	10	10	$0.\overline{1}$
0.99	-100	1.01	100	100	$0.\overline{01}$
0.999	-1000	1.001	1000	1000	$0.\overline{001}$

(b) Vertical asymptote: $x = 1$
    Horizontal asymptote: $y = 0$
(c) Domain: all $x \neq 1$

3. (a)

x	$f(x)$	x	$f(x)$	x	$f(x)$
0.5	3	1.5	9	5	3.75
0.9	27	1.1	33	10	$3.\overline{3}$
0.99	297	1.01	303	100	$3.\overline{03}$
0.999	2997	1.001	3003	1000	$3.\overline{003}$

(b) Vertical asymptote: $x = 1$
    Horizontal asymptotes: $y = \pm 3$
(c) Domain: all $x \neq 1$

5. (a)

x	$f(x)$	x	$f(x)$	x	$f(x)$
0.5	-1	1.5	5.4	5	3.125
0.9	-12.79	1.1	17.29	10	$3.0\overline{3}$
0.99	-147.8	1.01	152.3	100	3.0003
0.999	-1498	1.001	1502.3	1000	3

  (b) Vertical asymptotes: $x = \pm 1$

  Horizontal asymptote: $y = 3$

  (c) Domain: all $x \neq \pm 1$

7. (a)   **9.** (c)   **11.** (b)

13. Domain: all $x \neq 0$

Vertical asymptote: $x = 0$

Horizontal asymptote: $y = 0$

15. Domain: all $x \neq 2$

Vertical asymptote: $x = 2$

Horizontal asymptote: $y = -1$

17. Domain: all $x \neq \pm 1$

Vertical asymptotes: $x = \pm 1$

19. Domain: all reals

Horizontal asymptote: $y = 3$

21. (a) Domain of f: all $x \neq -2$

  Domain of g: all real numbers

  (b) Vertical asymptote: None

  (c)

x	-4	-3	-2.5	-2	-1.5	-1	0
$f(x)$	-6	-5	-4.5	Undef.	-3.5	-3	-2
$g(x)$	-6	-5	-4.5	-4	-3.5	-3	-2

  (d) Differ only where f is undefined.

23. (a) Domain of f: all $x \neq 0, 3$; Domain of g: all $x \neq 0$

  (b) Vertical asymptote: $x = 0$

  (c)

x	-1	-0.5	0	0.5	2	3	4
$f(x)$	-1	-2	Undef.	2	$\frac{1}{2}$	Undef.	$\frac{1}{4}$
$g(x)$	-1	-2	Undef.	2	$\frac{1}{2}$	$\frac{1}{3}$	$\frac{1}{4}$

  (d) Differ only where f is undefined and g is defined.

25. $f(x) = \dfrac{1}{x^2 + x - 2}$ **27.** $f(x) = \dfrac{2x^2}{1 + x^2}$

29. (a) 4   (b) Less than   (c) Greater than

31. (a) 2   (b) Greater than   (c) Less than

33. ± 2 **35.** 5

37. (a) \$28.33 million   (b) \$170 million

  (c) \$765 million

  (d) No. The function is undefined.

39. (a)

M	200	400	600	800	1000
t	0.472	0.596	0.710	0.817	0.916

M	1200	1400	1600	1800	2000
t	1.009	1.096	1.178	1.255	1.328

The greater the mass, the more time required per oscillation.

  (b) $M \approx 1306$ grams

41. (a) 333 deer, 500 deer, 800 deer   (b) 1500

43. (a)

n	1	2	3	4	5
P	0.50	0.74	0.82	0.86	0.89

n	6	7	8	9	10
P	0.91	0.92	0.93	0.94	0.95

The percentage approaches 1 as n increases.

  (b) 100%

45. $\frac{9}{2}$ **47.** $-\frac{7}{2}, 5$

Section 2.7   *(page 227)*

1.

Vertical shift

3.

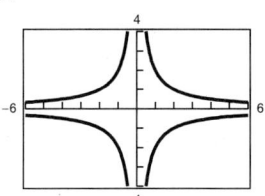

Reflection in the x-axis

5.

Vertical shift

7.

Horizontal shift

9.

Horizontal shift

11.

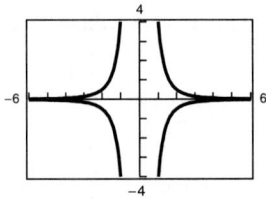

Reflection in the *x*-axis

13.

15.

17.

19.

21.

23.

25.

27.

29.

31.

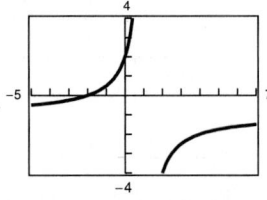

Domain: $(-\infty, 1), (1, \infty)$

$x = 1, y = -1$

33.

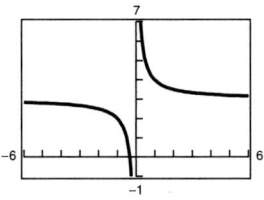

Domain: $(-\infty, 0), (0, \infty)$

$x = 0, y = 3$

35.

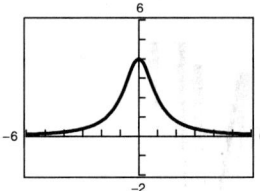

Domain: $(-\infty, \infty)$

$y = 0$

37.

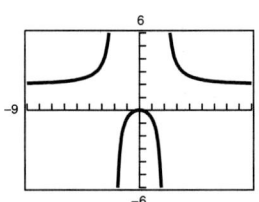

Domain: $(-\infty, -2), (-2, 2),$
       $(2, \infty)$

$x = \pm 2, y = 2$

39.

Domain: $(-\infty, 0), (0, \infty)$

     $x = 0, y = 0$

41.

There are two horizontal
asymptotes.

43.

The graph crosses the horizontal asymptote.

45.

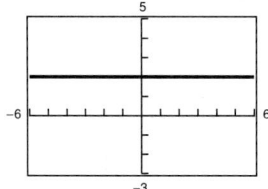

The fraction is not reduced.

47. No. There are two distinct branches of the graph.

49.   **51.**

53.   **55.**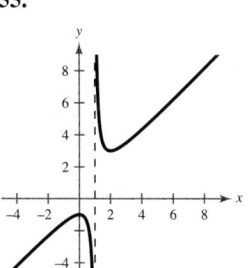

57. $(-1, 0)$ **59.** $(1, 0), (-1, 0)$

61.   **63.**

Domain: $(-\infty, -1), (-1, \infty)$ Domain: $(-\infty, 0), (0, \infty)$

Vertical asymptote: $x = -1$ Vertical asymptote: $x = 0$

$y = 2x - 1$ $y = -x + 3$

65.   **67.**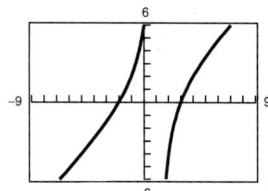

$(-4, 0)$ $(3, 0), (-2, 0)$

69. (a) Answers will vary.   (b) $[0, 950]$

(c)

Increases more slowly; 0.75

71. 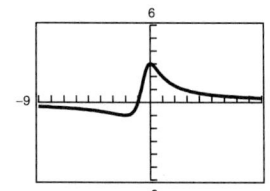   Minimum: $(-2, -1)$

Maximum: $(0, 3)$

73. (a) Answers will vary.   (b) $(2, \infty)$

(c)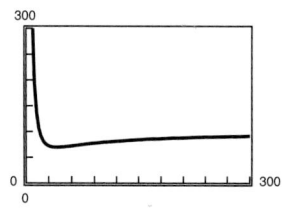

11.75×5.9 inches

75.

$x \approx 40$

77. (a) $C = 0$. The chemical will eventually dissipate.

(b)

$t \approx 4.5$

79. (a)

(b) $y = 2.81t + 68.77$

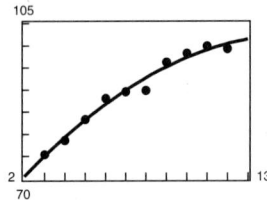

(c) $y = -0.18t^2 + 5.52t + 60.12$

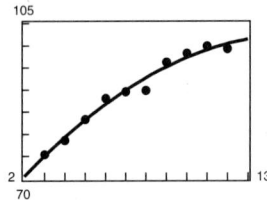

(d) The quadratic and rational models fit the data better than the line. The rational model may be a better predictor because the parabola is near its maximum.

81. $f(x) = \dfrac{x^2 - x - 6}{x - 2}$

FOCUS ON CONCEPTS   *(page 231)*

1. Prefer the conditions (a) and (b) because profits would be increasing.

2. (a) Degree: 3; leading coefficient: positive
(b) Degree: 2; leading coefficient: positive
(c) Degree: 4; leading coefficient: positive
(d) Degree: 5; leading coefficient: positive

3. (a) No
(b) Yes   (c) Yes

(d) Yes

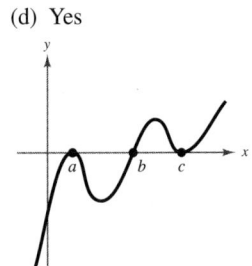

4. $f(x) = \dfrac{2(x - 3)}{x - 2}$ **5.** $f(x) = \dfrac{2}{x^2 - 2}$

6. $f(x) = \dfrac{2}{x - 2}$ **7.** $f(x) = \dfrac{2x}{x - 2}$

Review Exercises   *(page 232)*

1.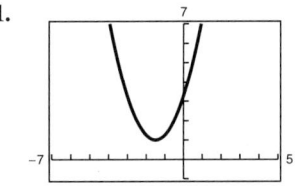

Vertex: $\left(-\frac{3}{2}, 1\right)$

Intercept: $\left(0, \frac{13}{4}\right)$

3.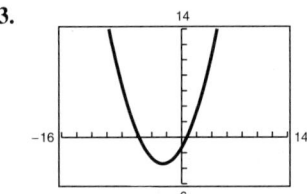

Vertex: $\left(-\frac{5}{2}, -\frac{41}{12}\right)$

Intercepts: $\left(0, -\frac{4}{3}\right), \left(\dfrac{-5 \pm \sqrt{41}}{2}, 0\right)$

5. $f(x) = (x - 1)^2 - 4$

7. (a)

(b)

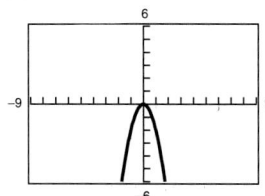

Vertical stretch

Vertical stretch and reflection in the x-axis

(c)

(d)

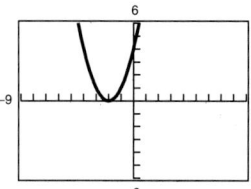

Vertical translation

Horizontal translation

9. Minimum: $(1, -1)$ **11.** Maximum: $(3, 9)$

13. Maximum: $(1, 3)$ **15.** Minimum: $\left(-\frac{5}{2}, -\frac{41}{4}\right)$

17. (a)

x	y	Area
1	$4 - \frac{1}{2}(1)$	$(1)[4 - \frac{1}{2}(1)] = \frac{7}{2}$
2	$4 - \frac{1}{2}(2)$	$(2)[4 - \frac{1}{2}(2)] = 6$
3	$4 - \frac{1}{2}(3)$	$(3)[4 - \frac{1}{2}(3)] = \frac{15}{2}$
4	$4 - \frac{1}{2}(4)$	$(4)[4 - \frac{1}{2}(4)] = 8$
5	$4 - \frac{1}{2}(5)$	$(5)[4 - \frac{1}{2}(5)] = \frac{15}{2}$
6	$4 - \frac{1}{2}(6)$	$(6)[4 - \frac{1}{2}(6)] = 6$

(b) $x = 4, y = 2$ (c) $A = x\left(\dfrac{8 - x}{2}\right), 0 < x < 8$

(d)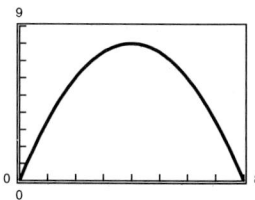

(e) $A = -\frac{1}{2}x^2 + 4x$

$x = 4, y = 2$

19. Falls to the left.   **21.** Rises to the left.
Falls to the right.   Rises to the right.

23.

25.

27.

29.

31.

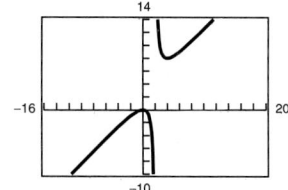

33. $8x + 5 + \dfrac{2}{3x - 2}$ **35.** $x^2 - 2, x \neq \pm 1$

37. $0.25x^2 - 3.5x^2 - 7x - 14 - \dfrac{28}{x - 2}$ **39.** $6x^3 - 27x$

41. $3 + 7i$ **43.** $40 + 65i$ **45.** $1 - 6i$

47. $6x^4 + 13x^3 + 7x^2 - x - 1$ **49.** $1, \frac{3}{4}$ **51.** $\frac{5}{6}, \pm 2i$

53. $-1, \frac{3}{2}, 3, \frac{2}{3}$

55. (a)

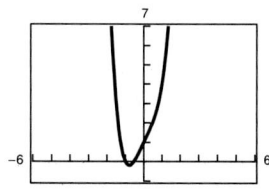

(b) 2
(c) $-1, -0.54$

57. (a)

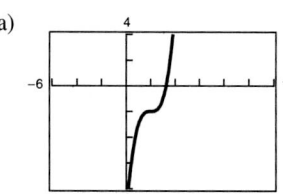

(b) 1
(c) 3.26

59. (a)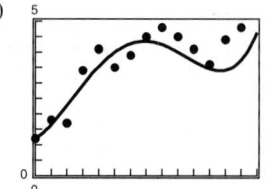

(b) Recession. Yes

(c) 0.3 billion. More

(d) 4.8

(b) Answers will vary.

(c)

x	2.5	3	3.5	4	4.5
A	18.75	13.50	12.25	12	12.15

$x = 4$

(d)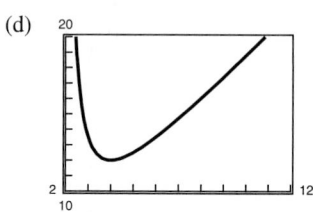

$x = 4$

(e) $y = \frac{3}{2}(x + 2)$. The area increases without bound as x increases.

61.

63.

65.

67.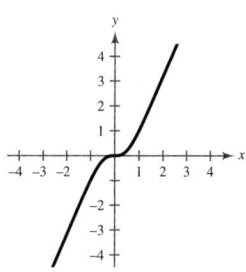

Chapter Test   *(page 236)*

1. (a) Reflection in the x-axis followed by a vertical translation

(b) Horizontal translation

2. Vertex: $(-2, -1)$

Intercepts: $(0, 3)$, $(-3, 0)$, $(-1, 0)$

3. $y = (x - 3)^2 - 6$

4. (a) 50 feet

(b) 5. Changing the constant term results in a vertical translation of the graph and therefore changes the maximum height.

5. $3x + \dfrac{x - 1}{x^2 + 1}$ **6.** $-4 - 46i$

7. $\pm 1, \pm 2, \pm 3, \pm 4, \pm 6, \pm 8,$ **8.** $\pm 1, \pm 2, \pm\frac{1}{3}, \pm\frac{2}{3}$
$\pm 12, \pm 24, \pm\frac{1}{2}, \pm\frac{3}{2}$

69.

71.

 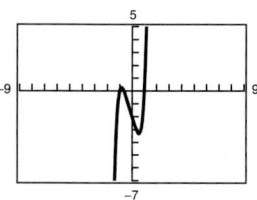

$-2, \frac{3}{2}$ $\pm 1, -\frac{2}{3}$

9. $-0.819, 1.380$ **10.** $-1.414, -0.667, 1.414$

11. $f(x) = x^4 - 9x^3 + 28x^2 - 30x$

12. $f(x) = x^4 - 6x^3 + 16x^2 - 24x + 16$

73. $f(x) = \dfrac{2x^2}{x^2 - x - 12}$

75. (a)

13.

14.

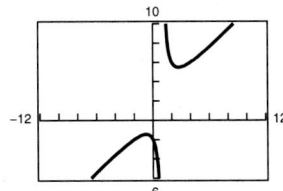

15. $f(x) = \dfrac{4x^2}{x^2 - 9}$

CHAPTER 3

Section 3.1 *(page 247)*

1. 946.852   **3.** 7.352   **5.** 0.006

7. 673.639   **9.** 0.472   **11.** $f(x) = h(x)$

13. $f(x) = g(x) = h(x)$

15.

17.

19.

21.

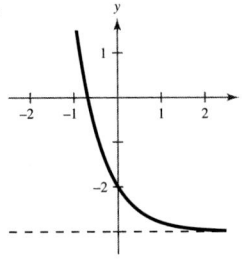

23. (c)   **25.** (e)   **27.** (g)   **29.** (a)

31.

33.

35.

37.

39.

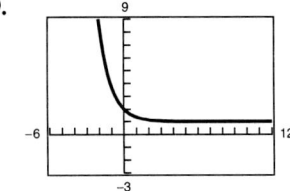

41. (a)

x	-1	-0.5	0	0.5	1
$f(x)$	0.3333	0.5774	1	1.7321	3
$g(x)$	0.25	0.5	1	2	4

$x < 0$

(b)

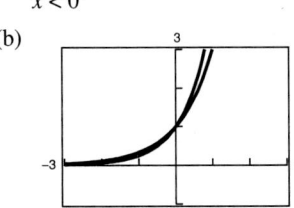

(i) $4^x < 3^x$ when $x < 0$ (ii) $4^x > 3^x$ when $x > 0$

43. (a)

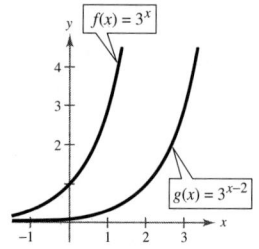

Horizontal shift two units to the right

(b)

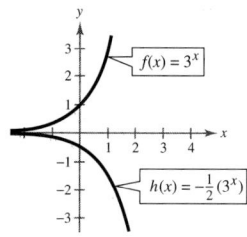

Vertical shrink and a reflection about the x-axis

(c)

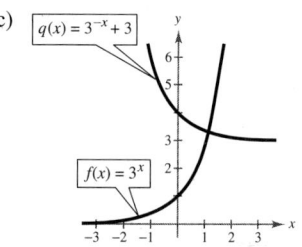

Reflection about the y-axis and a vertical translation

45. (a)

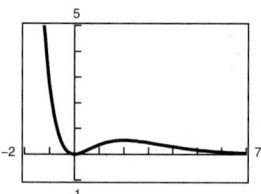

Decreasing: $(-\infty, 0), (2, \infty)$

Increasing: $(0, 2)$

Relative maximum: $(2, 4e^{-2})$

Relative minimum: $(0, 0)$

(b)

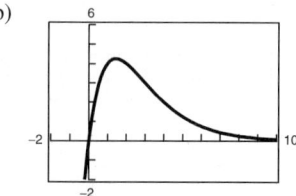

Decreasing: $(1.44, \infty)$

Increasing: $(-\infty, 1.44)$

Relative maximum: $(1.44, 4.25)$

47. The exponential function increases at a faster rate.

49.

$f(x)$ approaches $g(x) = 1.6487$

51. \$212,605.41

53.

n	1	2	4
A	\$7764.62	\$8017.84	\$8155.09

n	12	365	Continuous
A	\$8250.97	\$8298.66	\$8300.29

55.

n	1	2	4
A	\$24,115.73	\$25,714.29	\$26,602.23

n	12	365	Continuous
A	\$27,231.38	\$27,547.07	\$27,557.94

57.

t	1	10	20
P	\$91,393.12	\$40,656.97	\$16,529.89

t	30	40	50
P	\$6720.55	\$2732.37	\$1110.90

59.

t	1	10	20
P	\$90,521.24	\$36,940.70	\$13,646.15

t	30	40	50
P	\$5040.98	\$1862.17	\$687.90

61. (a) 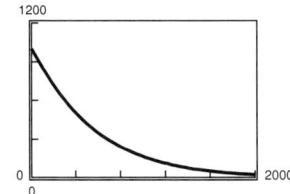   (b) \$421.12
(c) \$350.13

63. (a) 100   (b) 300   (c) 900
65. (a) 25 units
(b) 16.30 units
(c)

67. (a) and (b)

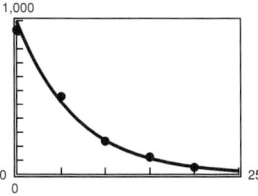

(c)

h	0	5	10	15	20
P	10,958	5176	2445	1155	546

(d) 3300 kilograms per square meter
(e) 11.3 kilometers

69. (a) $T = -1.239t + 73.021$

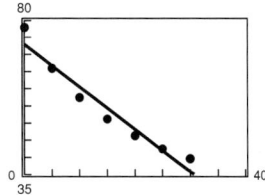

The temperature decreases at a slower rate as it approaches the room temperature.
(b) $T = 0.034t^2 - 2.264t + 77.295$

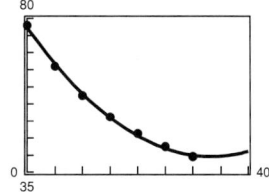

The parabola is increasing when $t = 60$.
(c) $T = 54.438(0.964)^t + 21$

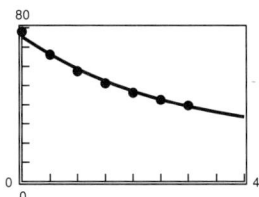

(d) The horizontal asymptote of the exponential model is $T = 0$.

71.

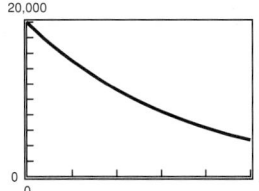

\$11,250
73. $1 < \sqrt{2} < 2$
$2^1 < 2^{\sqrt{2}} < 2^2$

75.

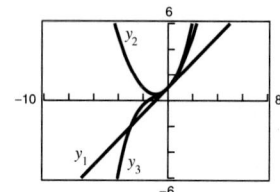

77. (a) $f(u + v) = a^{u+v} = a^u \cdot a^v = f(u) \cdot f(v)$
(b) $f(2x) = a^{2x} = (a^x)^2 = [f(x)]^2$

Section 3.2   *(page 259)*

1. $4^3 = 64$ **3.** $7^{-2} = \frac{1}{49}$ **5.** $32^{2/5} = 4$ **7.** $e^0 = 1$

9. $\log_5 125 = 3$ **11.** $\log_{81} 3 = \frac{1}{4}$

13. $\log_6 \frac{1}{36} = -2$ **15.** $\ln 20.0855 = 3$

17. $\ln 4 = x$ **19.** 4   **21.** $\frac{1}{2}$ **23.** 0

25. -2 **27.** 3   **29.** 2   **31.** 2.538   **33.** 2.161

35. 2.913   **37.** 1.005   **39.** -1.139

41.

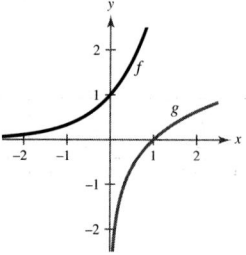

Reflections in the line $y = x$
$g = f^{-1}$

43.

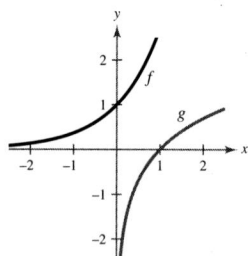

Reflections in the line $y = x$
$g = f^{-1}$

45. (c)   **47.** (d)   **49.** (b)

51. Domain: $(0, \infty)$
Vertical asymptote: $x = 0$
Intercept: $(1, 0)$

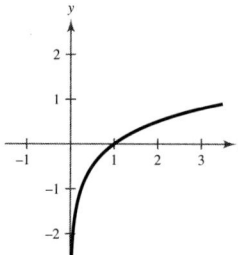

53. Domain: $(3, \infty)$
Vertical asymptote: $x = 3$
Intercept: $(4, 0)$

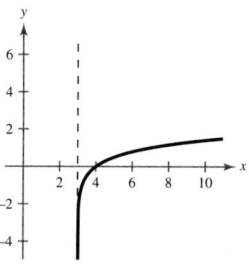

55. Domain: $(0, \infty)$
Vertical asymptote: $x = 0$
Intercept: $(9, 0)$

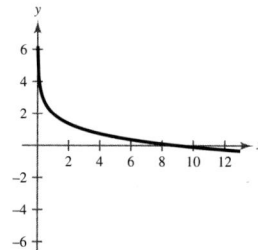

57. Domain: $(0, \infty)$

Vertical asymptote: $x = 0$

Intercept: $(5, 0)$

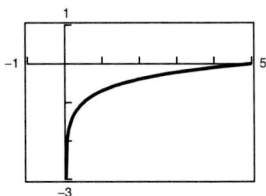

59. Domain: $(2, \infty)$

Vertical asymptote: $x = 2$

Intercept: $(3, 0)$

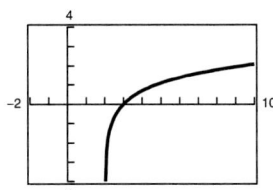

61. Domain: $(-\infty, 0)$

Vertical asymptote: $x = 0$

Intercept: $(-1, 0)$

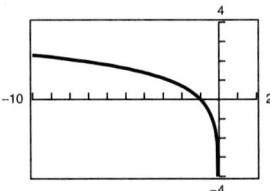

63. Domain: $(0, \infty)$

Decreasing: $(0, 2)$

Increasing: $(2, \infty)$

Relative minimum: $(2, 1.693)$

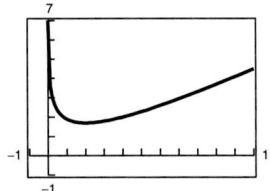

65. Domain: $(0, \infty)$

Decreasing: $(0, 0.37)$

Increasing: $(0.37, \infty)$

Relative minimum: $(0.37, -1.47)$

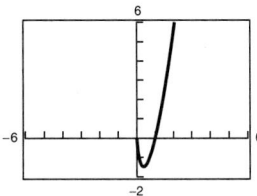

67. 23.68 years

69. (a) False　(b) True　(c) True　(d) False

71. $y = (x - 1) - \frac{1}{2}(x - 1)^2 + \frac{1}{3}(x - 1)^3 - \frac{1}{4}(x - 1)^4$

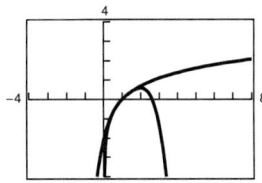

73. (a)

K	1	2	4	6	8	10	12
t	0	7.3	14.6	18.9	21.9	24.2	26.2

(b)

75. (a)

(b) $\ln P = -0.1499h + 9.3018$

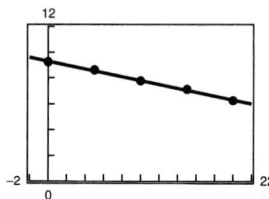

(c) $P = 10{,}957.7e^{-0.1499h}$

77. (a)

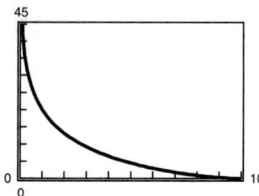

Domain: $0 < x \le 10$

(b) $x = 0$

(c) 22.9

(d) $p = 10 \ln\left(\dfrac{\sqrt{100 - x^2} + 10}{x}\right)$

(e)

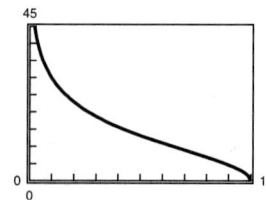

Beginning of the process

79.

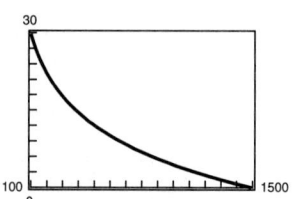

17.66 cubic feet per minute

81. 21,357 foot-pounds   **83.** 30 years

85. Total amount: $473,886

Interest: $323,886

87. (a) $(0, \infty)$ (b) $f^{-1} = 10^x$ (c) $3 < x < 4$

(d) $0 < x < 1$ (e) 10 (f) $10^{2n} : 1$

Section 3.3   *(page 267)*

1.

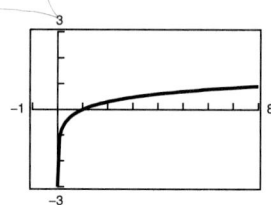

$f = g$

3. $\dfrac{\log_{10} 5}{\log_{10} 3}$ **5.** $\dfrac{\log_{10} x}{\log_{10} 2}$ **7.** $\dfrac{\ln 5}{\ln 3}$ **9.** $\dfrac{\ln x}{\ln 2}$

11. 1.771 **13.** -2.000 **15.** -0.417 **17.** 2.633

19. $\log_{10} 5 + \log_{10} x$ **21.** $\log_{10} 5 - \log_{10} x$

23. $4\log_8 x$ **25.** $\frac{1}{2}\ln z$ **27.** $\ln x + \ln y + \ln z$

29. $\frac{1}{2}\ln(a - 1)$ **31.** $\ln z + 2\ln(z - 1)$

33. $\frac{1}{3}\ln x - \frac{1}{3}\ln y$ **35.** $4\ln x + \frac{1}{2}\ln y - 5\ln z$

37. $2\log_b x - 2\log_b y - 3\log_b z$

39.

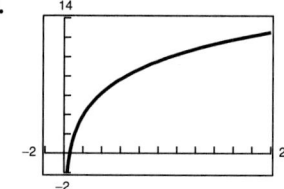

$y_1 = y_2$

41. $\ln 2x$ **43.** $\log_4 \dfrac{z}{y}$ **45.** $\log_2(x + 4)^2$

47. $\log_3 \sqrt[3]{5x}$ **49.** $\ln \dfrac{x}{(x + 1)^3}$ **51.** $\ln \dfrac{x - 2}{x + 2}$

53. $\ln \dfrac{x}{(x^2 - 4)^2}$ **55.** $\ln \sqrt[3]{\dfrac{x(x + 3)^2}{x^2 - 1}}$

57. $\ln \dfrac{\sqrt[3]{y(y + 4)^2}}{y - 1}$ **59.** $\ln \dfrac{9}{\sqrt{x^2 + 1}}$

61.

$y_1 = y_2$

63.

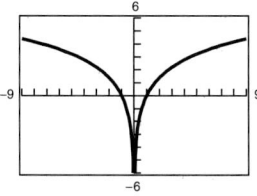

No. The domains differ.

65.

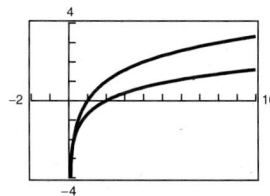

$f(x) = h(x)$

67. 2   **69.** 2.4

71. -9 is not in the domain of $\log_3 x$.

73. 2   **75.** -3

77. 0 is not in the domain of $\log_{10} x$.

79. 4.5   **81.** $\frac{3}{2}$ **83.** $\frac{1}{2}(1 + \log_7 10)$

85. $-3 - \log_5 2$ **87.** $6 + \ln 5$

89. $\beta = 10(\log_{10} I + 16)$, 60 decibels

91. (a)

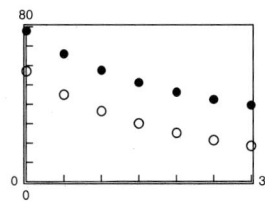

(b) $T - 21 = 54.4380(0.9635^t)$

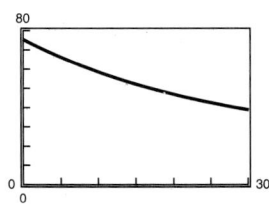

(c) $\ln(T - 21) = -0.0372t + 3.9971$

(d) $T = \dfrac{4960}{6t + 80} + 21$

93.   **95.**

        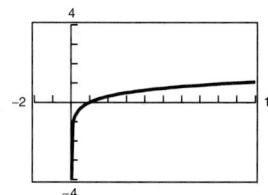

97. False. $\ln 1 = 0$ **99.** False. $f(x) - f(2) = \ln \dfrac{x}{2}$

101. False. $u = v^2$ **103.** Answers will vary.

105. $\dfrac{3x^4}{2y^3}$ **107.** 1

Section 3.4   *(page 277)*

1. (a) Yes   (b) No   **3.** (a) No   (b) Yes   (c) Yes

5. (a) No   (b) No   (c) Yes

7. (3, 8)   **9.** (9, 2)   **11.** 2   **13.** -2

15. 3   **17.** 64   **19.** $\frac{1}{10}$ **21.** x^2 **23.** $5x + 2$

25. x^2 **27.** $\ln 10 \approx 2.303$ **29.** 0

31. $\ln \frac{5}{3} \approx 0.511$ **33.** $\log_{10} 42 \approx 1.623$

35.

x	0.6	0.7	0.8	0.9	1.0
$f(x)$	6.05	8.17	11.02	14.88	20.09

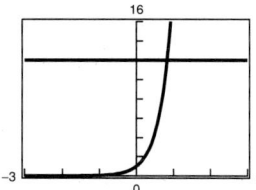

0.828

37.

x	5	6	7	8	9
$f(x)$	1756	1598	1338	908	200

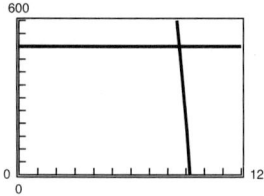

8.635

39. $\ln 5 \approx 1.609$ **41.** $\dfrac{\ln 80}{2 \ln 3} \approx 1.994$ **43.** 2

45. $\dfrac{\ln 8 - \ln 565}{\ln 2} \approx -6.142$ **47.** 0.059 **49.** 6.960

51.

-0.427

53.

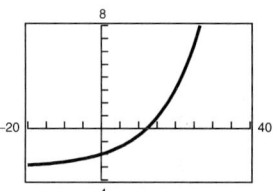

12.207

55. $e^{-3} \approx 0.050$ **57.** $e^2 - 2 \approx 5.389$

59. $\dfrac{-1 + \sqrt{17}}{2} \approx 1.562$

61.

x	2	3	4	5	6
$f(x)$	1.39	1.79	2.08	2.30	2.49

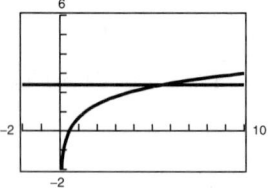

5.512

63.

x	12	13	14	15	16
$f(x)$	9.79	10.22	10.63	11.00	11.36

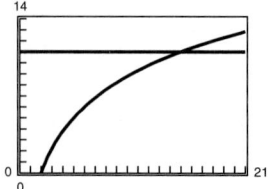

14.988

65. 103 **67.** 33.115 **69.** $1 + \sqrt{1 + e} \approx 2.928$
71. 4 **73.** No solution **75.** 14.369
77. **79.**

  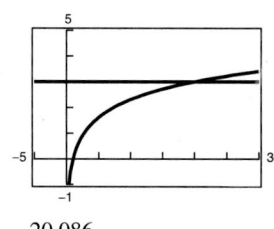

2.807 20.086

81. 8.2 years

83. Yes. Doubling time: $t = \dfrac{\ln 2}{r}$

Quadrupling time: $t = \dfrac{\ln 4}{r} = 2\left(\dfrac{\ln 2}{r}\right)$

85. 12.9 years

87. (a)

(b) $y = 100$ and $y = 0$
(c) Males: 69.71 inches
   Females: 64.51 inches

89. (a) 1426 units (b) 1498 units

91. (a)

(b) $y = 6.7$. Yield will approach 6.7 million cubic feet per acre.

(c) 29.3 years

93. (a)

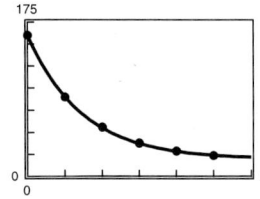

(b) $y = 20$. Room temperature

(c) 0.81 hour

95. (b)   **97.** (f)   **99.** (d)

Section 3.5   (page 288)

1. (b)   **3.** (e)   **5.** (f)

Initial Investment	Annual % Rate	Time to Double	Amount After 10 Years
7. $1000	12%	5.78 yr	$3320.12
9. $750	8.94%	7.75 yr	$1833.67
11. $500	9.5%	7.30 yr	$1292.85
13. $6376.28	4.5%	15.4 yr	$10,000.00

15. $112,087.09

17. (a) 6.642 years   (b) 6.330 years
(c) 6.302 years   (d) 6.301 years

19.

r	2%	4%	6%	8%	10%	12%
t	54.93	27.47	18.31	13.73	10.99	9.16

21.

r	2%	4%	6%	8%	10%	12%
t	55.47	28.01	18.85	14.27	11.53	9.69

23.

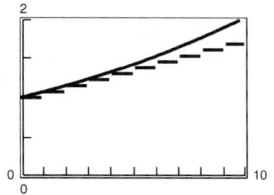

Continuous compounding

Isotope	Half-Life (years)	Initial Quantity	Amount After 1000 Years
25. Ra^{226}	1620	10 g	6.52 g
27. C^{14}	5730	3 g	2.66 g

29. $y = e^{0.7675x}$ **31.** $y = e^{-0.4621x}$ **33.** 2013

35. $k = 0.0137$; 3288   **37.** $y = 4.22e^{0.0430t}$; 9.97 million

39. $y = 3e^{-0.0091t}$; 2.50 million

41. The greater the rate of growth, the greater the value of b.

43. 3.15 hours   **45.** 95.8%   **47.** $2423

49. (a) $S(t) = 100(1 - e^{-0.1625t})$

(b)

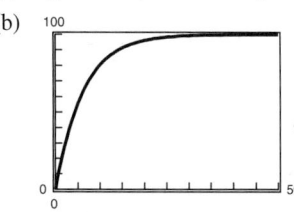

(c) 55,625

51. (a) $S = 10(1 - e^{-0.0575x})$ (b) 3314

53. (a) $N = 30(1 - e^{-0.050t})$

(b) 36 days

(c) No. It is not a linear function.

55. (a) 7.91   (b) 7.68

57. (a) 20   (b) 70   (c) 95   (d) 120

59. 95%   **61.** 4.64   **63.** 1.58×10^{-6} moles per liter

65. 10,000,000 times

67. (a)

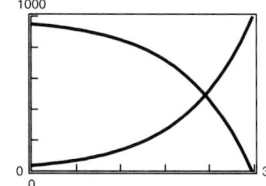

(b) Interest, $t \approx 28$ years

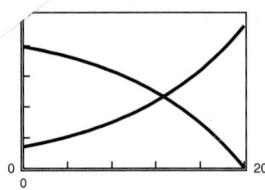

Interest, $t \approx 12.7$ years

69. (a) $t_3 = 0.2729s - 6.0143$

$t_4 = 1.5385e^{0.0291s}$

(b)

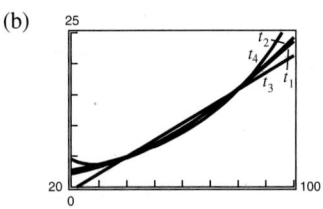

(c)

s	30	40	50	60	70	80	90
t_1	3.6	4.7	6.7	9.4	12.5	15.9	19.6
t_2	3.3	4.9	7.0	9.5	12.5	15.9	19.9
t_3	2.2	4.9	7.6	10.4	13.1	15.8	18.5
t_4	3.7	4.9	6.6	8.8	11.8	15.8	21.1

(d) Model: t_1; Sum ≈ 1.9

Model: t_2; Sum ≈ 1.2

Model: t_3; Sum ≈ 5.6

Model: t_4; Sum ≈ 2.6

Quadratic model fits best.

71. Answers will vary.   **73.** 7:30 A.M.

75. $8x^2 - 24x + 18$

77. $x^3 - 5x^2 + 25x - 128 + \dfrac{641}{x + 5}$

Section 3.6   *(page 298)*

1. Logarithmic model   **3.** Gaussian model

5. Exponential model   **7.** Gaussian model

9.

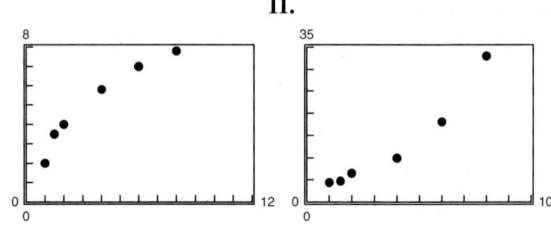

Logarithmic model

11.

Exponential model

13.

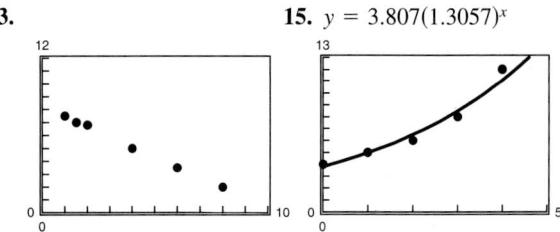

Linear model

15. $y = 3.807(1.3057)^x$

17. $y = 8.463(0.7775)^x$

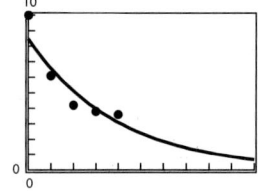

19. $y = 2.083 + 1.257 \ln x$

21. $y = 9.826 - 4.097 \ln x$

23. $y = 1.985x^{0.760}$

25. $y = 16.103x^{-3.174}$

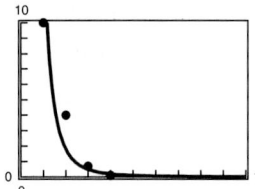

27. (a) $y = 38.233d^{1.955}$

(b)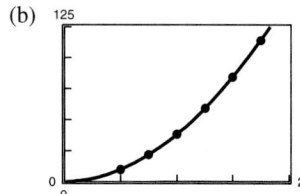

(c) 148.2

29. (a) $y = 0.088x + 4.413$ (b) $y = 4.456(1.017)^x$

(c) Linear

(d) Linear: 6.26

    Exponential: 6.35

31. (a) $y = 5.088x^{0.645}$

(b)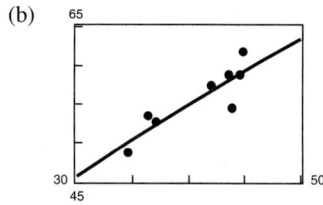

(c) 63.4 million board feet

33. (a) $f = 396.48T^{0.5055}$ (b) 487   (c) $x^{1/2}$

35. (a) $y_1 = 7.01x + 41.14$

    $y_2 = 5.31 + 45.83 \ln x$

    $y_3 = 51.26(1.08)^x$

    $y_4 = 34.00x^{0.513}$

(b)

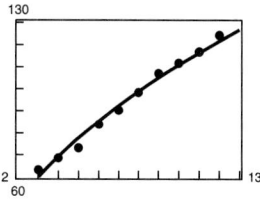

Linear model

(c)

x	y	$y - y_1$	$(y - y_1)^2$	$y - y_2$	$(y - y_2)^2$
3	63.2	1.03	1.06	7.54	56.86
4	68.6	−0.58	0.34	−0.24	0.06
5	73.2	−2.99	8.94	−5.87	34.46
6	83.9	0.70	0.49	−3.53	12.44
7	90.3	0.09	0.01	−4.19	17.57
8	98.4	1.18	1.39	−2.21	4.89
9	107.0	2.77	7.67	0.99	0.98
10	111.7	0.46	0.21	0.86	0.74
11	116.8	−1.45	2.10	1.59	2.54
12	124.3	−0.96	0.92	5.11	26.08

x	$y - y_3$	$(y - y_3)^2$	$y - y_4$	$(y - y_4)^2$
3	−1.37	1.88	3.46	11.99
4	−1.14	1.30	−0.64	0.41
5	−2.12	4.48	−4.43	19.66
6	2.56	6.54	−1.35	1.81
7	2.45	6.00	−1.96	3.84
8	3.52	12.40	−0.40	0.16
9	4.53	20.53	2.04	4.18
10	1.03	1.07	0.92	0.84
11	−2.72	7.40	0.46	0.22
12	−4.78	22.86	2.65	7.04

(d) Linear model

  y_1: 23.14; y_2: 156.62; y_3: 84.46; y_4: 50.15

(e) Sum of the squares of the errors

FOCUS ON CONCEPTS   *(page 302)*

1. $b < d < a < c$

  b and d are negative.

2. (a) True. $\log_b uv = \log_b u + \log_b v$

  (b) False. $2.04 \approx \log_{10}(10 + 100) \neq (\log_{10} 10)(\log_{10} 100)$
    $= 2$

  (c) False. $1.95 \approx \log_{10}(100 - 10) \neq \log_{10} 100 - \log_{10} 10$
    $= 1$

  (d) True. $\log_b \dfrac{u}{v} = \log_b u - \log_b v$

3. Double the interest rate or time as it doubles the exponent in the exponential function.

4. (a) Logarithmic   (b) Logistic   (c) Exponential
  (d) Linear   (e) None of the above   (f) Exponential

Review Exercises   *(page 303)*

1. (e)   3. (b)   5. (a)

7.

9.

11.

13.

15.
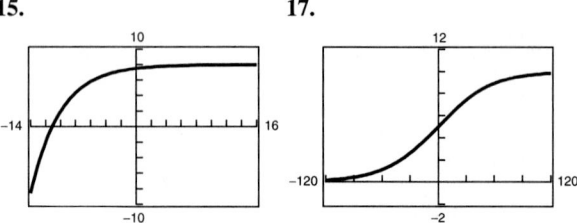

$y = 8$

17.

$y = 0, y = 10$

19.

n	1	2	4
A	$9499.28	$9738.91	$9867.22

n	12	365	Continuous
A	$9956.20	$10,000.27	$10,001.78

21.

t	1	10	20
P	$184,623.27	$89,865.79	$40,379.30

t	30	40	50
P	$18,143.59	$8152.44	$3663.13

23. (a)
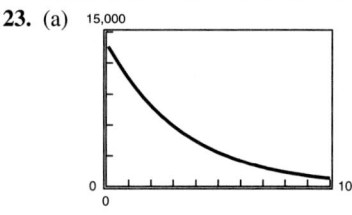

  (b) $7875

  (c) At the beginning. Yes

25.
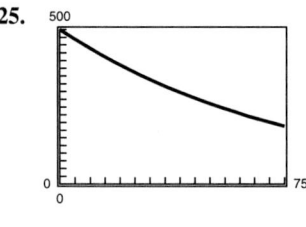

229.2

27.

Speed	50	55	60	65	70
Miles Per Gallon	28	26.4	24.8	23.4	22.0

29.   **31.**

101. $y = 234.684(0.8746)^x$

33.   **35.**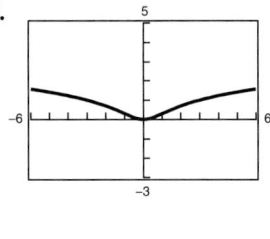

103. (a) $y = 12.907x^{0.5102}$

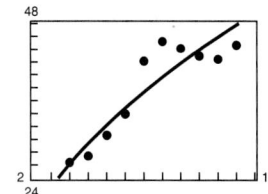

(b) $61.0 billion

37. $\log_4 64 = 3$ **39.** 3   **41.** -2 **43.** 7   **45.** 0

47. 1.585   **49.** 2.132   **51.** $1 + 2\log_5 x$

53. $\log_{10} 5 + \frac{1}{2}\log_{10} y - 2\log_{10} x$

55. $\ln(x^2 + 1) + \ln(x - 1)$ **57.** $\log_2 5x$

59. $\ln\dfrac{\sqrt{|2x - 1|}}{(x + 1)^2}$ **61.** $\ln\dfrac{3\sqrt[3]{4 - x^2}}{x}$ **63.** True

65. False   **67.** False   **69.** 1.6542   **71.** 0.2823

73. (a) $0 \le h < 18{,}000$

(b)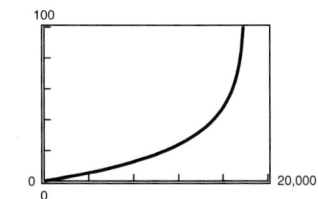

(c) Climbing at a slower rate

(d) 5.46 minutes

75. $\ln 12 \approx 2.485$ **77.** $-\dfrac{\ln 44}{5} \approx -0.757$

79. No solution

81. $\frac{1}{3}e^{8.2} \approx 1213.650$ **83.** $3e^2 \approx 22.167$

85. No solution   **87.** 0.39, 7.48   **89.** 1.64

91. $y = 2e^{0.1014x}$ **93.** $y = 4e^{-0.4159x}$

95. (a) 1151 units   (b) 1325 units

97. (a) 8.94%   (b) $1834.37   (c) 9.36%   **99.** $10^{-3.5}$

Cumulative Test for Chapters P–3   *(page 308)*

1.   **2.**

3.   **4.**

5.   **6.**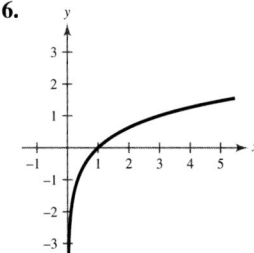

7. $2x - y + 2 = 0$

8. To some values of x there correspond more than one value of y.

9. (a) Vertical shrink   (b) Vertical shift
   (c) Horizontal shift

10. $h^{-1}(x) = \frac{1}{5}(x + 2)$ **11.** 7   **12.** 5

13. $\dfrac{-3 \pm \sqrt{3}}{3}$ **14.** 6   **15.** $\dfrac{1}{2} \ln 12 \approx 1.242$

16. $\frac{64}{5}$ **17.** \$2000   **18.** $-2, \pm 2i$ **19.** 1.20

20. $\ln \dfrac{x^2}{\sqrt{x + 5}}$

21. $y = 5.280(1.4455)^x$

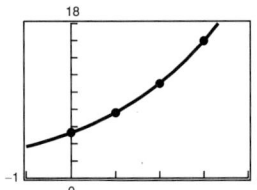

CHAPTER 4

Section 4.1   *(page 317)*

1. 2   **3.** -3 **5.** (a) Quadrant I   (b) Quadrant III

7. (a) Quadrant IV   (b) Quadrant II

9. (a) Quadrant III   (b) Quadrant II

11.
(a)   (b)

13.
(a)   (b)

    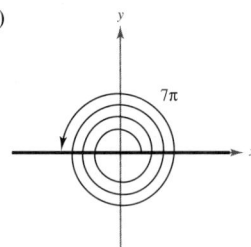

15. (a) $\dfrac{25\pi}{12}, -\dfrac{23\pi}{12}$ (b) $\dfrac{8\pi}{3}, -\dfrac{4\pi}{3}$

17. (a) $\dfrac{7\pi}{4}, -\dfrac{\pi}{4}$ (b) $\dfrac{28\pi}{15}, -\dfrac{32\pi}{15}$ **19.** $210°$

21. $-45°$ **23.** (a) Quadrant II   (b) Quadrant IV

25. (a) Quadrant III   (b) Quadrant I

27.
(a)   (b)

29.
(a)   (b)

    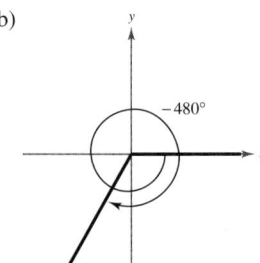

31. (a) $405°, -315°$ (b) $324°, -396°$

33. (a) $660°, -60°$ (b) $20°, -340°$

35. (a) $54.75°$ (b) $-128.5°$

37. (a) $85.308°$ (b) $330.007°$

39. (a) $240°36'$ (b) $-145°48'$

41. (a) $143°14'22''$ (b) $-205°7'8''$

43. (a) Complement: $72°$; Supplement: $162°$
   (b) Complement: none; Supplement: $65°$

45. (a) Complement: $\dfrac{\pi}{6}$; Supplement: $\dfrac{2\pi}{3}$

(b) Complement: none; Supplement: $\dfrac{\pi}{4}$

47. (a) $\dfrac{\pi}{6}$ (b) $\dfrac{5\pi}{6}$ **49.** (a) $-\dfrac{\pi}{9}$ (b) $-\dfrac{4\pi}{3}$

51. (a) $270°$ (b) $210°$ **53.** (a) $420°$ (b) $-66°$

55. 2.007 **57.** -3.776 **59.** 9.285 **61.** -0.014

63. $25.714°$ **65.** $337.5°$ **67.** $-756°$

69. $-114.592°$ **71.** $\frac{6}{5}$ rad **73.** $4\frac{4}{7}$ rad **75.** $\frac{4}{15}$ rad

77. 1.724 rad **79.** 15π inches ≈ 47.12 inches

81. 12 meters **83.** 591.72 miles **85.** 1141.02 miles

87. 0.094 rad $\approx 5.39°$ **89.** $\frac{5}{12}$ rad

91. (a) 560.2 revolutions per minute

(b) 3520 rad per minute

93. Radian. 1 rad $\approx 57.3°$

95. (a) 80π rad per second (b) 78.54 feet per second

97. (a) $\dfrac{14\pi}{3}$ feet per second (b) ≈ 10 miles per hour

99. $\dfrac{50\pi}{3}$ square meters

101. Two angles in standard position are coterminal angles if they have the same initial and terminal sides.

103. 0.0003 rad or $0.0172°$

Section 4.2 (page 328)

1. $\sin t = \frac{4}{5}$
$\cos t = -\frac{3}{5}$
$\tan t = -\frac{4}{3}$
$\csc t = \frac{5}{4}$
$\sec t = -\frac{5}{3}$
$\cot t = -\frac{3}{4}$

3. $\sin t = -\frac{15}{17}$
$\cos t = \frac{8}{17}$
$\tan t = -\frac{15}{8}$
$\csc t = -\frac{17}{15}$
$\sec t = \frac{17}{8}$
$\cot t = -\frac{8}{15}$

5. $\sin t = -\frac{1}{2}$
$\cos t = -\dfrac{\sqrt{3}}{2}$
$\tan t = \dfrac{\sqrt{3}}{3}$
$\csc t = -2$
$\sec t = -\dfrac{2\sqrt{3}}{3}$
$\cot t = \sqrt{3}$

7. $\left(\dfrac{\sqrt{2}}{2}, \dfrac{\sqrt{2}}{2}\right)$ **9.** $\left(-\dfrac{\sqrt{3}}{2}, \dfrac{1}{2}\right)$ **11.** $\left(-\dfrac{1}{2}, -\dfrac{\sqrt{3}}{2}\right)$

13. $(0, -1)$

15. $\sin \dfrac{\pi}{4} = \dfrac{\sqrt{2}}{2}$

$\cos \dfrac{\pi}{4} = \dfrac{\sqrt{2}}{2}$

$\tan \dfrac{\pi}{4} = 1$

17. $\sin\left(-\dfrac{11\pi}{6}\right) = -\dfrac{1}{2}$

$\cos\left(-\dfrac{11\pi}{6}\right) = \dfrac{\sqrt{3}}{2}$

$\tan\left(-\dfrac{11\pi}{6}\right) = -\dfrac{\sqrt{3}}{3}$

19. $\sin\left(-\dfrac{5\pi}{4}\right) = \dfrac{\sqrt{2}}{2}$

$\cos\left(-\dfrac{5\pi}{4}\right) = -\dfrac{\sqrt{2}}{2}$

$\tan\left(-\dfrac{5\pi}{4}\right) = -1$

21. $\sin \dfrac{11\pi}{6} = -\dfrac{1}{2}$

$\cos \dfrac{11\pi}{6} = \dfrac{\sqrt{3}}{2}$

$\tan \dfrac{11\pi}{6} = -\dfrac{\sqrt{3}}{3}$

23. $\sin \dfrac{4\pi}{3} = -\dfrac{\sqrt{3}}{2}$

$\cos \dfrac{4\pi}{3} = -\dfrac{1}{2}$

$\tan \dfrac{4\pi}{3} = \sqrt{3}$

25. $\sin\left(-\dfrac{3\pi}{2}\right) = 1$

$\cos\left(-\dfrac{3\pi}{2}\right) = 0$

$\tan\left(-\dfrac{3\pi}{2}\right)$ is undefined.

27. $\sin \dfrac{3\pi}{4} = \dfrac{\sqrt{2}}{2}$

$\cos \dfrac{3\pi}{4} = -\dfrac{\sqrt{2}}{2}$

$\tan \dfrac{3\pi}{4} = -1$

$\csc \dfrac{3\pi}{4} = \sqrt{2}$

$\sec \dfrac{3\pi}{4} = -\sqrt{2}$

$\cot \dfrac{3\pi}{4} = -1$

29. $\sin \dfrac{\pi}{2} = 1$

$\cos \dfrac{\pi}{2} = 0$

$\tan \dfrac{\pi}{2}$ is undefined.

$\csc \dfrac{\pi}{2} = 1$

$\sec \dfrac{\pi}{2}$ is undefined.

$\cot \dfrac{\pi}{2} = 0$

31. $\sin\left(-\dfrac{4\pi}{3}\right) = \dfrac{\sqrt{3}}{2}$

$\cos\left(-\dfrac{4\pi}{3}\right) = -\dfrac{1}{2}$

$\tan\left(-\dfrac{4\pi}{3}\right) = -\sqrt{3}$

$\csc\left(-\dfrac{4\pi}{3}\right) = \dfrac{2\sqrt{3}}{3}$

$\sec\left(-\dfrac{4\pi}{3}\right) = -2$

$\cot\left(-\dfrac{4\pi}{3}\right) = -\dfrac{\sqrt{3}}{3}$

33. 0   **35.** $-\dfrac{1}{2}$ **37.** $-\dfrac{\sqrt{3}}{2}$ **39.** $-\dfrac{\sqrt{2}}{2}$

41. (a) $-\dfrac{1}{3}$ (b) -3 **43.** (a) $-\dfrac{7}{8}$ (b) $-\dfrac{8}{7}$

45. (a) $\dfrac{4}{5}$ (b) $-\dfrac{4}{5}$ **47.** 0.7071   **49.** -0.9900

51. -0.1288 **53.** 1.3940   **55.** -1.4486

57. (a) -1 (b) -0.4 **59.** (a) 0.25, 2.89   (b) 1.82, 4.46

61. If $t = \dfrac{\pi}{4}$, then $0 = \cos 2t \neq 2\cos t = \sqrt{2}$.

63. (a) y-axis   (b) $\sin t_1 = \sin(\pi - t_1)$
(c) $\cos(\pi - t_1) = -\cos t_1$

65. (a) 0.2500 foot   (b) 0.0177 foot   (c) -0.2475 foot

67. 0.794   **69.** Odd

71. $f^{-1}(x) = \dfrac{2}{3}(x + 1)$

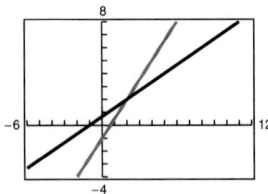

73. $f^{-1}(x) = \sqrt{x^2 + 4},\quad x \geq 0$

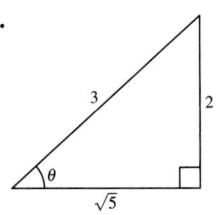

Section 4.3   *(page 338)*

1. $\sin\theta = \dfrac{1}{2}$

$\cos\theta = \dfrac{\sqrt{3}}{2}$

$\tan\theta = \dfrac{\sqrt{3}}{3}$

$\csc\theta = 2$

$\sec\theta = \dfrac{2\sqrt{3}}{3}$

$\cot\theta = \sqrt{3}$

3. $\sin\theta = \dfrac{8}{17}$

$\cos\theta = \dfrac{15}{17}$

$\tan\theta = \dfrac{8}{15}$

$\csc\theta = \dfrac{17}{8}$

$\sec\theta = \dfrac{17}{15}$

$\cot\theta = \dfrac{15}{8}$

5. $\sin\theta = \dfrac{1}{3}$

$\cos\theta = \dfrac{2\sqrt{2}}{3}$

$\tan\theta = \dfrac{\sqrt{2}}{4}$

$\csc\theta = 3$

$\sec\theta = \dfrac{3\sqrt{2}}{4}$

$\cot\theta = 2\sqrt{2}$

7. $\sin\theta = \dfrac{3}{5}$

$\cos\theta = \dfrac{4}{5}$

$\tan\theta = \dfrac{3}{4}$

$\csc\theta = \dfrac{5}{3}$

$\sec\theta = \dfrac{5}{4}$

$\cot\theta = \dfrac{4}{3}$

The triangles are similar and corresponding sides are proportional.

The triangles are similar and corresponding sides are proportional.

9.

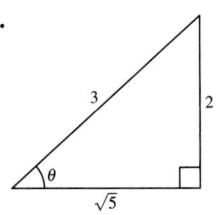

$\cos\theta = \dfrac{\sqrt{5}}{3}$

$\tan\theta = \dfrac{2\sqrt{5}}{5}$

$\csc\theta = \dfrac{3}{2}$

$\sec\theta = \dfrac{3\sqrt{5}}{5}$

$\cot\theta = \dfrac{\sqrt{5}}{2}$

11.

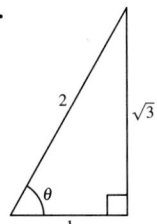

$\sin\theta = \dfrac{\sqrt{3}}{2}$

$\cos\theta = \dfrac{1}{2}$

$\tan\theta = \sqrt{3}$

$\csc\theta = \dfrac{2\sqrt{3}}{3}$

$\cot\theta = \dfrac{\sqrt{3}}{3}$

13.

$$\sin \theta = \frac{3\sqrt{10}}{10}$$

$$\cos \theta = \frac{\sqrt{10}}{10}$$

$$\csc \theta = \frac{\sqrt{10}}{3}$$

$$\sec \theta = \sqrt{10}$$

$$\cot \theta = \frac{1}{3}$$

15.

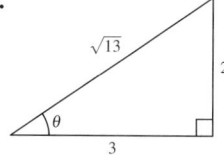

$$\sin \theta = \frac{2\sqrt{13}}{13}$$

$$\cos \theta = \frac{3\sqrt{13}}{13}$$

$$\tan \theta = \frac{2}{3}$$

$$\csc \theta = \frac{\sqrt{13}}{2}$$

$$\sec \theta = \frac{\sqrt{13}}{3}$$

17. (a) $\sqrt{3}$ (b) $\frac{1}{2}$ (c) $\frac{\sqrt{3}}{2}$ (d) $\frac{\sqrt{3}}{3}$

19. (a) $\frac{1}{3}$ (b) $\frac{2\sqrt{2}}{3}$ (c) $\frac{\sqrt{2}}{4}$ (d) 3

21. (a) 4 (b) $\pm\frac{\sqrt{15}}{4}$ (c) $\pm\frac{\sqrt{15}}{15}$ (d) $\frac{1}{4}$

23.–31. Answers will vary.

33. (a) $\frac{1}{2}$ (b) $\frac{\sqrt{3}}{3}$ **35.** (a) 1 (b) $\frac{\sqrt{2}}{2}$

37. (a) 0.1736 (b) 0.1736 **39.** (a) 0.2815 (b) 3.5523

41. (a) 1.3499 (b) 1.3432 **43.** (a) 5.0273 (b) 0.1989

45. (a) 1.1884 (b) 0.5463

47. (a) $30° = \frac{\pi}{6}$ (b) $30° = \frac{\pi}{6}$

49. (a) $60° = \frac{\pi}{3}$ (b) $45° = \frac{\pi}{4}$

51. (a) $60° = \frac{\pi}{3}$ (b) $45° = \frac{\pi}{4}$

53. (a) $55° \approx 0.960$ (b) $89° \approx 1.553$

55. (a) $50° \approx 0.873$ (b) $25° \approx 0.436$

57. $25\sqrt{3}$ **59.** $\frac{32\sqrt{3}}{3}$ **61.** 23.3 **63.** 6.1

65. $17\frac{1}{4}$ feet

67. (a)

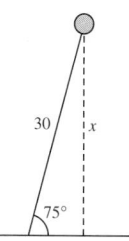

(b) $\sin 75° = \dfrac{x}{30}$

(c) 28.98 meters

69. 1144.87 feet

71. $(x_1, y_1) = \left(28\sqrt{3}, 28\right)$
$(x_2, y_2) = \left(28, 28\sqrt{3}\right)$

73. $\sin 20° \approx 0.34$
$\cos 20° \approx 0.94$
$\tan 20° \approx 0.36$
$\csc 20° \approx 2.92$
$\sec 20° \approx 1.06$
$\cot 20° \approx 2.75$

75. (a)

θ	0	0.1	0.2	0.3	0.4	0.5
$\sin \theta$	0	0.0998	0.1987	0.2955	0.3894	0.4794

(b) θ

(c) $\sin \theta$ approaches θ as θ approaches 0.

77. (a)

θ	0°	20°	40°	60°	80°
$\sin \theta$	0	0.3420	0.6428	0.8660	0.9848
$\cos \theta$	1	0.9397	0.7660	0.5000	0.1736
$\tan \theta$	0	0.3640	0.8391	1.7321	5.6713

(b) Sine: increasing
Cosine: decreasing
Tangent: increasing

79. (a)

θ	20°	30°	40°	50°	60°	70°
R	170.3	229.5	261.0	261.0	229.5	170.3

(b) The range increases as θ increases from 20° through 40° and decreases as θ increases from 50° through 70°.

(c) 45°

(d)

θ	43°	44°	45°	46°	47°
R	264.4	264.8	265.0	264.8	264.4

81. True, $\csc x = \dfrac{1}{\sin x}$ **83.** False, $\dfrac{\sqrt{2}}{2} + \dfrac{\sqrt{2}}{2} \neq 1$

85. False, $1.7321 \neq 0.0349$

Section 4.4   *(page 349)*

1. (a) $\sin\ \theta = \frac{3}{5}$

$\cos\ \theta = \frac{4}{5}$

$\tan\ \theta = \frac{3}{4}$

$\csc\ \theta = \frac{5}{3}$

$\sec\ \theta = \frac{5}{4}$

$\cot\ \theta = \frac{4}{3}$

(b) $\sin\ \theta = -\frac{15}{17}$

$\cos\ \theta = -\frac{8}{17}$

$\tan\ \theta = \frac{15}{8}$

$\csc\ \theta = -\frac{17}{15}$

$\sec\ \theta = -\frac{17}{8}$

$\cot\ \theta = \frac{8}{15}$

3. (a) $\sin\ \theta = -\dfrac{1}{2}$

$\cos\ \theta = -\dfrac{\sqrt{3}}{2}$

$\tan\ \theta = \dfrac{\sqrt{3}}{3}$

$\csc\ \theta = -2$

$\sec\ \theta = -\dfrac{2\sqrt{3}}{3}$

$\cot\ \theta = \sqrt{3}$

(b) $\sin\ \theta = \dfrac{\sqrt{2}}{2}$

$\cos\ \theta = -\dfrac{\sqrt{2}}{2}$

$\tan\ \theta = -1$

$\csc\ \theta = \sqrt{2}$

$\sec\ \theta = -\sqrt{2}$

$\cot\ \theta = -1$

5. (a) $\sin\ \theta = \frac{24}{25}$

$\cos\ \theta = \frac{7}{25}$

$\tan\ \theta = \frac{24}{7}$

$\csc\ \theta = \frac{25}{24}$

$\sec\ \theta = \frac{25}{7}$

$\cot\ \theta = \frac{7}{24}$

(b) $\sin\ \theta = -\frac{24}{25}$

$\cos\ \theta = \frac{7}{25}$

$\tan\ \theta = -\frac{24}{7}$

$\csc\ \theta = -\frac{25}{24}$

$\sec\ \theta = \frac{25}{7}$

$\cot\ \theta = -\frac{7}{24}$

7. (a) $\sin\ \theta = \dfrac{5\sqrt{29}}{29}$

$\cos\ \theta = -\dfrac{2\sqrt{29}}{29}$

$\tan\ \theta = -\dfrac{5}{2}$

$\csc\ \theta = \dfrac{\sqrt{29}}{5}$

$\sec\ \theta = -\dfrac{\sqrt{29}}{2}$

$\cot\ \theta = -\dfrac{2}{5}$

(b) $\sin\ \theta = -\dfrac{5\sqrt{34}}{34}$

$\cos\ \theta = \dfrac{3\sqrt{34}}{34}$

$\tan\ \theta = -\dfrac{5}{3}$

$\csc\ \theta = -\dfrac{\sqrt{34}}{5}$

$\sec\ \theta = \dfrac{\sqrt{34}}{3}$

$\cot\ \theta = -\dfrac{3}{5}$

9. (a) Quadrant III   **(b)** Quadrant II

11. (a) Quadrant II   **(b)** Quadrant IV

13. $\sin\ \theta = \frac{3}{5}$

$\cos\ \theta = -\frac{4}{5}$

$\tan\ \theta = -\frac{3}{4}$

$\csc\ \theta = \frac{5}{3}$

$\sec\ \theta = -\frac{5}{4}$

$\cot\ \theta = -\frac{4}{3}$

15. $\sin\ \theta = -\frac{15}{17}$

$\cos\ \theta = \frac{8}{17}$

$\tan\ \theta = -\frac{15}{8}$

$\csc\ \theta = -\frac{17}{15}$

$\sec\ \theta = \frac{17}{8}$

$\cot\ \theta = -\frac{8}{15}$

17. $\sin \theta = -\dfrac{\sqrt{10}}{10}$

$\cos \theta = \dfrac{3\sqrt{10}}{10}$

$\tan \theta = -\dfrac{1}{3}$

$\csc \theta = -\sqrt{10}$

$\sec \theta = \dfrac{\sqrt{10}}{3}$

$\cot \theta = -3$

19. $\sin \theta = \dfrac{\sqrt{3}}{2}$

$\cos \theta = -\dfrac{1}{2}$

$\tan \theta = -\sqrt{3}$

$\csc \theta = \dfrac{2\sqrt{3}}{3}$

$\sec \theta = -2$

$\cot \theta = -\dfrac{\sqrt{3}}{3}$

21. $\sin \theta = 0$

$\cos \theta = -1$

$\tan \theta = 0$

$\csc \theta$ is undefined.

$\sec \theta = -1$

$\cot \theta$ is undefined.

23. $\sin \theta = \dfrac{\sqrt{2}}{2}$

$\cos \theta = -\dfrac{\sqrt{2}}{2}$

$\tan \theta = -1$

$\csc \theta = \sqrt{2}$

$\sec \theta = -\sqrt{2}$

$\cot \theta = -1$

25. $\sin \theta = -\dfrac{2\sqrt{5}}{5}$

$\cos \theta = -\dfrac{\sqrt{5}}{5}$

$\tan \theta = 2$

$\csc \theta = \dfrac{-\sqrt{5}}{2}$

$\sec \theta = -\sqrt{5}$

$\cot \theta = \dfrac{1}{2}$

27. -1 **29.** -1 **31.** Undefined   **33.** 0

35. (a) $\theta' = 23°$ (b) $\theta' = 53°$

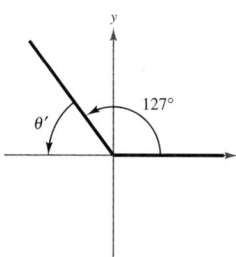

37. (a) $\theta' = 65°$ (b) $\theta' = 72°$

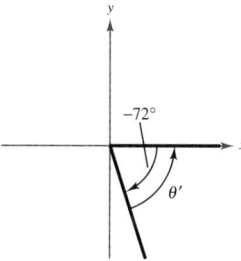

39. (a) $\theta' = \dfrac{\pi}{3}$ (b) $\theta' = \dfrac{\pi}{6}$

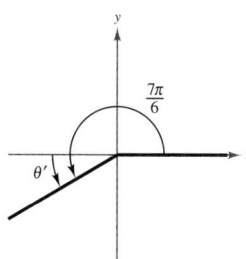

41. (a) $\theta' = 3.5 - \pi$ (b) $\theta' = 2\pi - 5.8$

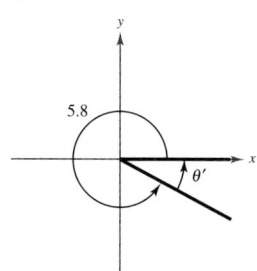

43. (a) $\sin 225° = -\dfrac{\sqrt{2}}{2}$ (b) $\sin(-225°) = \dfrac{\sqrt{2}}{2}$

$\cos 225° = -\dfrac{\sqrt{2}}{2}$ $\cos(-225°) = -\dfrac{\sqrt{2}}{2}$

$\tan 225° = 1$ $\tan(-225°) = -1$

45. (a) $\sin 750° = \dfrac{1}{2}$ (b) $\sin 510° = \dfrac{1}{2}$

$\cos 750° = \dfrac{\sqrt{3}}{2}$ $\cos 510° = -\dfrac{\sqrt{3}}{2}$

$\tan 750° = \dfrac{\sqrt{3}}{3}$ $\tan 510° = -\dfrac{\sqrt{3}}{3}$

47. (a) $\sin \dfrac{4\pi}{3} = -\dfrac{\sqrt{3}}{2}$ (b) $\sin \dfrac{2\pi}{3} = \dfrac{\sqrt{3}}{2}$

$\cos \dfrac{4\pi}{3} = -\dfrac{1}{2}$ $\cos \dfrac{2\pi}{3} = -\dfrac{1}{2}$

$\tan \dfrac{4\pi}{3} = \sqrt{3}$ $\tan \dfrac{2\pi}{3} = -\sqrt{3}$

49. (a) $\sin\left(-\dfrac{\pi}{6}\right) = -\dfrac{1}{2}$ (b) $\sin\left(\dfrac{5\pi}{6}\right) = \dfrac{1}{2}$

$\cos\left(-\dfrac{\pi}{6}\right) = \dfrac{\sqrt{3}}{2}$ $\cos\left(\dfrac{5\pi}{6}\right) = -\dfrac{\sqrt{3}}{2}$

$\tan\left(-\dfrac{\pi}{6}\right) = -\dfrac{\sqrt{3}}{3}$ $\tan\left(\dfrac{5\pi}{6}\right) = -\dfrac{\sqrt{3}}{3}$

51. (a) $\sin \dfrac{11\pi}{4} = \dfrac{\sqrt{2}}{2}$ (b) $\sin\left(-\dfrac{13\pi}{6}\right) = -\dfrac{1}{2}$

$\cos \dfrac{11\pi}{4} = -\dfrac{\sqrt{2}}{2}$ $\cos\left(-\dfrac{13\pi}{6}\right) = \dfrac{\sqrt{3}}{2}$

$\tan \dfrac{11\pi}{4} = -1$ $\tan\left(-\dfrac{13\pi}{6}\right) = -\dfrac{\sqrt{3}}{3}$

53. (a) 0.1736   (b) 5.7588
55. (a) −0.3420   (b) −0.3420
57. (a) 1.7321   (b) 1.7321
59. (a) 0.3640   (b) 0.3640
61. (a) 0.6052   (b) 0.6077
63. (a) $30° = \dfrac{\pi}{6}$, $150° = \dfrac{5\pi}{6}$

(b) $210° = \dfrac{7\pi}{6}$, $330° = \dfrac{11\pi}{6}$

65. (a) $60° = \dfrac{\pi}{3}$, $120° = \dfrac{2\pi}{3}$

   (b) $135° = \dfrac{3\pi}{4}$, $315° = \dfrac{7\pi}{4}$

67. (a) $45° = \dfrac{\pi}{4}$, $225° = \dfrac{5\pi}{4}$

   (b) $150° = \dfrac{5\pi}{6}$, $330° = \dfrac{11\pi}{6}$

69. (a) 54.99°, 125.01°   **71.** (a) 0.175, 6.109
   (b) 195.00°, 345.00°   (b) 2.201, 4.083

73. (a) 0.873, 4.014
   (b) 1.693, 4.835

75. $\dfrac{4}{5}$ **77.** $-\dfrac{\sqrt{13}}{2}$ **79.** $\dfrac{8}{5}$

81. (a) 25.2° F   (b) 65.1° F   (c) 50.8° F
83. (a) 12 miles   (b) 6 miles   (c) 6.9 miles
85. -0.766

87.   **89.**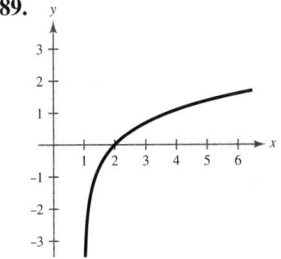

Section 4.5   *(page 361)*

1. Period: π **3.** Period: 4π
   Amplitude: 3   Amplitude: $\frac{5}{2}$

Xmin = -2π	Xmin = -4π
Xmax = 2π	Xmax = 4π
Xscl = $\pi/2$	Xscl = π
Ymin = -4	Ymin = -3
Ymax = 4	Ymax = 3
Yscl = 1	Yscl = 1

5. Period: 2   **7.** Period: 2π
   Amplitude: $\frac{2}{3}$ Amplitude: 2

Xmin = $-\pi$
Xmax = π
Xscl = $\pi/2$
Ymin = -1
Ymax = 1
Yscl = .5

9. Period: $\dfrac{\pi}{5}$ **11.** Period: 3π
   Amplitude: 3   Amplitude: $\frac{1}{2}$

13. Period: $\frac{1}{2}$
   Amplitude: 3

15. g is a shift of f π units to the right.
17. g is a reflection of f about the x-axis.
19. g is a reflection in the x-axis and three times the amplitude of f.
21. Shift the graph of f two units up to obtain the graph of g.
23. The graph of g has twice the amplitude as the graph of f.
25. The graph of g is a horizontal shift of the graph of f π units to the right.

27.

Amplitude changes

29.

Period changes

31.

33.

47.

49.

35.

37.

51.

53.

39.

41.

55.

57.

59.

43.

45.

61.

63.

65.

83. (a)

(b)

67.

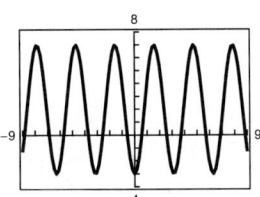

69. $y = 4 - 4 \cos x$

(c) $-\dfrac{x^7}{7!}, \ -\dfrac{x^6}{6!}$

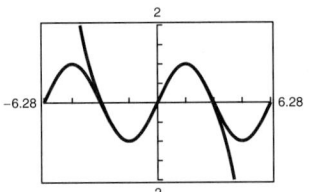

71. $y = -3 \sin(2x)$ **73.** $y = \sin\left(x - \dfrac{\pi}{4}\right)$

75.

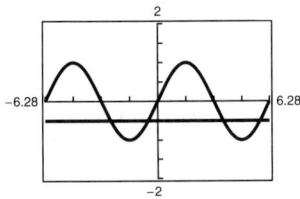

$$x = -\frac{\pi}{6}, \ -\frac{5\pi}{6}, \ \frac{7\pi}{6}, \ \frac{11\pi}{6}$$

(d)

The accuracy increased.

77.

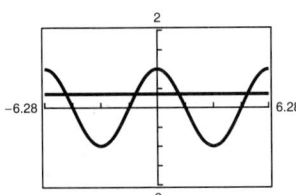

$$x = \pm\frac{\pi}{4}, \ \pm\frac{7\pi}{4}$$

79. (a) Even
　　(b) Even
　　(c) Odd

85.

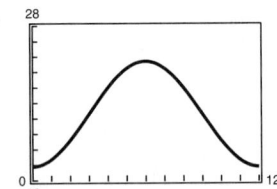

Maximum sales: June
Minimum sales: December

81. (a)

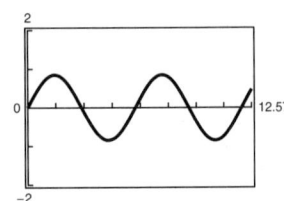

(b) 6 seconds
(c) 10 cycles per minute

87. (a) 365. Yes, one year (365 days)
　　(b) The constant 30.3 gallons

(c)

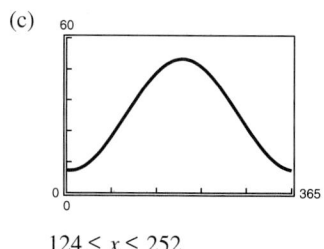

$124 \leq x \leq 252$

89. (a) $C(t) = 56.35 + 27.35 \sin\left(\dfrac{\pi t}{6} + 4.19\right)$

(b)

(c)

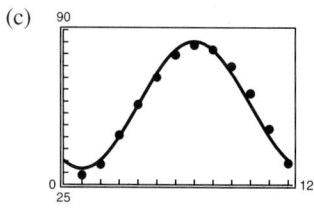

(d) Honolulu: 84.40; Chicago: 56.35;
  vertical translation *(d)*

(e) 12. Yes, one full period is one year.

(f) Chicago, amplitude

Section 4.6 *(page 372)*

1. (g) 4π **3.** (f) $\dfrac{\pi}{2}$ **5.** (b) 2   **7.** (e) 2π

9.

11.

13.

15.

17.

19.

21.

23.

25.

27.

29.

31.

33.

35.

37.

39.

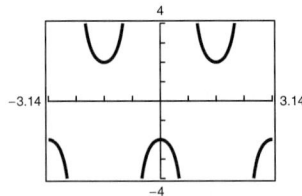

41. $-\dfrac{7\pi}{4}, -\dfrac{3\pi}{4}, \dfrac{\pi}{4}, \dfrac{5\pi}{4}$ **43.** $-\dfrac{4\pi}{3}, -\dfrac{2\pi}{3}, \dfrac{2\pi}{3}, \dfrac{4\pi}{3}$

45. Even

47. As x approaches $\pi/2$ from the left, f approaches ∞. As x approaches $\pi/2$ from the right, f approaches $-\infty$.

49. (a)

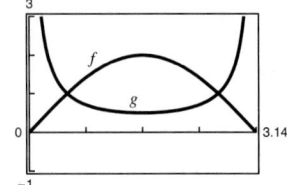

(b) $\dfrac{\pi}{6} < x < \dfrac{5\pi}{6}$

(c) Sine approaches 0 and cosecant approaches ∞ because the cosecant is the reciprocal of the sine.

51.

Not equivalent

53.

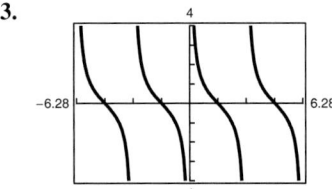

Equivalent

55. (d); as x approaches 0, $f(x)$ approaches 0.

57. (b); as x approaches 0, $g(x)$ approaches 0.

59.

Equal

61.

Equal

63.

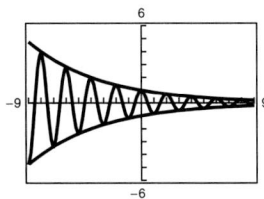

$f \to 0$ as $x \to \infty$

65.

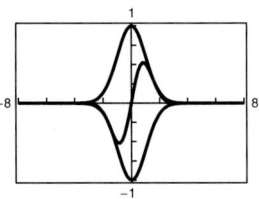

$g \to 0$ as $x \to \infty$

67. $d = 5 \cot x$

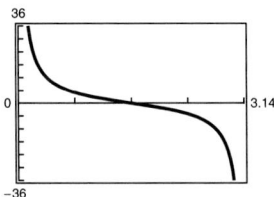

69. As the predator population increases, the number of prey decrease. When the number of prey is small, the number of predators decreases.

71. (a)

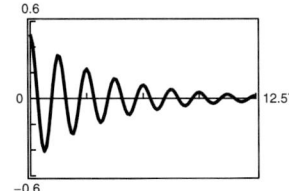

(b) Periodic but damped; goes to 0 as t increases.

73.

75. (a)

 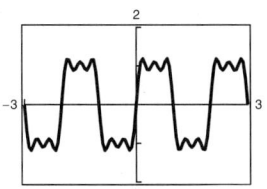

(b) $y_3 = \dfrac{4}{\pi}\left[\sin(\pi x) + \dfrac{1}{3}\sin(3\pi x)\right.$

$\left. + \dfrac{1}{5}\sin(5\pi x) + \dfrac{1}{7}\sin(7\pi x)\right]$

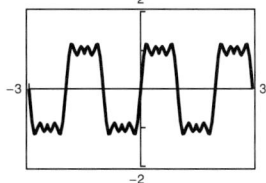

(c) $y_4 = \dfrac{4}{\pi}\left[\sin(\pi x) + \dfrac{1}{3}\sin(3\pi x) + \dfrac{1}{5}\sin(5\pi x)\right.$

$\left. + \dfrac{1}{7}\sin(7\pi x) + \dfrac{1}{9}\sin(9\pi x)\right]$

77.

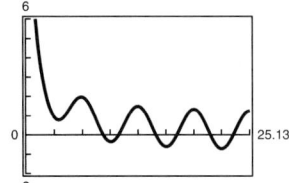

Approaches $\pm\infty$

79. (a) Yes. For each t there corresponds one and only one value of y.

(b) 1.27 oscillations per second

(c) $y = 12(0.2231)^t \cos 8t$

(d) $y = 12e^{-1.5t} \cos 8t$

(e)

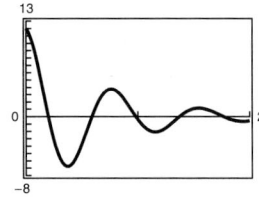

81. (a) 850 revolutions per minute

(b) The direction of the saw is reversed.

(c) $L = 60\left[\left(\dfrac{\pi}{2} + \phi\right) + \cot\phi\right],\quad 0 < \phi < \dfrac{\pi}{2}$

(d)

ϕ	0.3	0.6	0.9	1.2	1.5
L	306.2	217.9	195.9	189.6	188.5

(e) faster

(f)

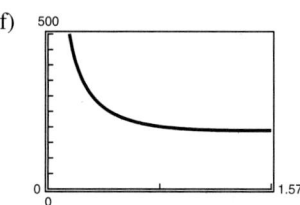

83. -0.693 **85.** 2

Section 4.7   *(page 383)*

1. (a)

x	-1	-0.8	-0.6	-0.4	-0.2
y	-1.5708	-0.9273	-0.6435	-0.4115	-0.2014

x	0	0.2	0.4	0.6	0.8	1
y	0	0.2014	0.4115	0.6435	0.9273	1.5708

(b)

(c)

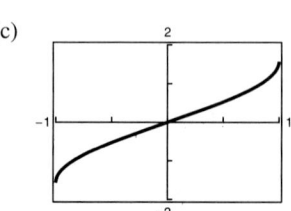

(d) $(0, 0)$. Symmetric to the origin

3. (a)

x	-10	-8	-6	-4	-2
y	-1.4711	-1.4464	-1.4056	-1.3258	-1.1071

x	0	2	4	6	8	10
y	0	1.1071	1.3258	1.4056	1.4464	1.4711

(b)

(c)

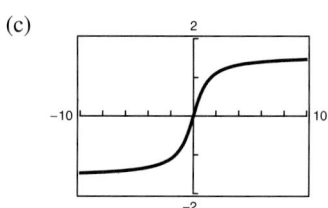

(d) $y = \pm\dfrac{\pi}{2}$

5. $-\dfrac{\pi}{3},\ \dfrac{-\sqrt{3}}{3},\ 1$ **7.** (a) $\dfrac{\pi}{6}$ (b) 0

9. (a) $\dfrac{\pi}{6}$ (b) $-\dfrac{\pi}{4}$ **11.** (a) $-\dfrac{\pi}{3}$ (b) $\dfrac{\pi}{3}$

13. (a) $\dfrac{\pi}{3}$ (b) $-\dfrac{\pi}{6}$ **15.** (a) 1.29   (b) 0.47

17. (a) -1.25 (b) 1.50   **19.** (a) 1.99   (b) -0.13

21.

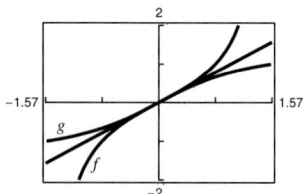

23. $\theta = \arctan\dfrac{x}{4}$ **25.** $\theta = \arcsin\dfrac{x+2}{5}$ **27.** 0.3

29. -0.1 **31.** 0   **33.** $\dfrac{3}{5}$ **35.** $\dfrac{\sqrt{5}}{5}$ **37.** $\dfrac{12}{13}$

39. $\dfrac{\sqrt{34}}{5}$ **41.** $\dfrac{\sqrt{5}}{3}$ **43.** $\dfrac{1}{x}$ **45.** $\sqrt{1-4x^2}$

47. $\sqrt{1-x^2}$ **49.** $\dfrac{\sqrt{9-x^2}}{x}$ **51.** $\dfrac{\sqrt{x^2+2}}{x}$

53.

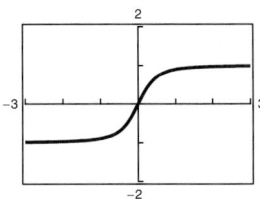

$y = \pm 1$

55. $\dfrac{9}{\sqrt{x^2+81}}, x>0; \dfrac{-9}{\sqrt{x^2+81}}, x<0$ **57.** $\dfrac{|x-1|}{\sqrt{x^2-2x+10}}$

59.

61.

63.

65.

67. $3\sqrt{2}\,\sin\!\left(2t+\dfrac{\pi}{4}\right)$

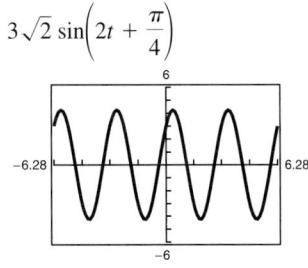

69. (a) $f \circ f^{-1}$ $f^{-1} \circ f$

  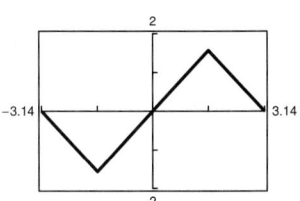

(b) Differ because of the domain and range of f and f^{-1}.

71. (a) $\theta = \arcsin\dfrac{10}{s}$ (b) 0.21 rad, 0.43 rad

73. (a)

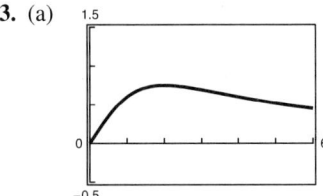

(b) 2 feet   (c) $\beta = 0$

75. (a) $\theta = \arctan\dfrac{5}{x}$ (b) 0.46 rad, 1.03 rad

77.   **79.**

  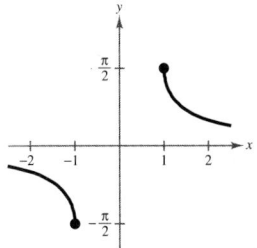

81.–85. Answers will vary.

87. Buy now   **89.** 8

Section 4.8   *(page 394)*

1. $a \approx 3.64$ **3.** $a \approx 8.26$ **5.** $c \approx 11.66$
    $c \approx 10.64$ $c \approx 25.38$ $A \approx 30.96°$
    $B = 70°$ $A = 19°$ $B \approx 59.04°$

7. $a \approx 49.48$ **9.** $a \approx 91.34$
    $A \approx 72.08°$ $b \approx 420.70$
    $B \approx 17.92°$ $B = 77°45'$

11. 2.56 inches

13. (a) $L = 60 \cot \theta$

(b)

θ	10°	20°	30°	40°	50°
L	340	165	104	72	50

(c) No. Cotangent is not a linear function.

15. (a) $h = 20 \sin \theta$

(b)

θ	60°	65°	70°	75°	80°
h	17.3	18.1	18.8	19.3	19.7

17. (a)

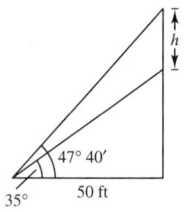

(b) $h = 50(\tan 47°40' - \tan 35°)$

(c) 19.9 feet

19. 2236.8 feet   **21.** 56.3°   **23.** 15.5°

25. 5099 feet   **27.** 0.73 miles

29. 508 miles north; 650 miles east

31. (a) N 58° E   (b) 68.82 meters   **33.** N 56.3° W

35. 1933.3 feet   **37.** 17,054 feet $\approx$ 3.23 miles

39. 78.7°   **41.** 35.3°   **43.** $y = \sqrt{3}\,r$

45. 29.4 inches   **47.** 7, 12.2

49. (a) 4   (b) 4   (c) $\frac{1}{16}$ **51.** (a) $\frac{1}{16}$ (b) 60   (c) $\frac{1}{120}$

53. $y = 4 \sin(\pi t)$ **55.** $y = 3 \cos\left(\dfrac{4\pi t}{3}\right)$

57. $\omega = 528\pi$

59. (a)

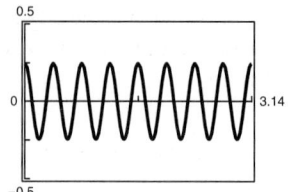

(b) $\dfrac{\pi}{8}$ seconds   (c) $\dfrac{\pi}{32}$ seconds

61. (a)

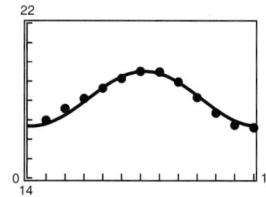

(b) 12 months. Yes, 1 period is 1 year.

(c) 1.41 hours. 1.41 represents the maximum change in time from the average time ($d = 18.09$) of sunset.

63. (a)

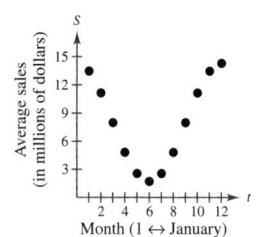

(b) $S = 8 + 6.3 \cos\left(\dfrac{\pi t}{6}\right)$

(c) 12 months. Sales of outerwear is seasonal.

(d) Maximum displacement of 6.3 from the average sales of 8.

FOCUS ON CONCEPTS   *(page 400)*

1. (a) The vertex is at the origin and the initial side is on the positive x-axis.

(b) Clockwise rotation of the terminal side

(c) Two angles in standard position where the terminal sides coincide

5. Determine the trigonometric function of the reference angle and, depending on the quadrant in which the obtuse angle lies, prefix the appropriate sign.

6. (d); the period is 2π and the amplitude is 3.

7. (a); the period is 2π and, because $a < 0$, the graph is reflected about the x-axis.

8. (b); the period is 2 and the amplitude is 2.

9. (c); the period is 4π and the amplitude is 2.

10. (a) Equal; two-period shift

    (b) Not equal; $f\left(t + \frac{1}{2}c\right)$ is a horizontal translation and $f\left(\frac{1}{2}t\right)$ is a period change.

    (c) Equal; the period change is the same in each.

11. Their range is $(-\infty, \infty)$.

12. (a) The displacement is increased.

    (b) The friction damps the oscillations more quickly.

    (c) The frequency of the oscillations increases.

13. False. $3\pi/4$ is not in the range of the arctangent function.

Review Exercises   *(page 401)*

1.

$\dfrac{3\pi}{4}, -\dfrac{5\pi}{4}$

3.

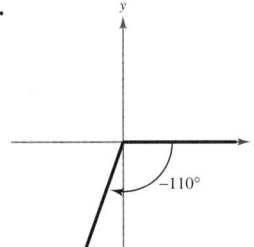

$250°, -470°$

5. $135.28°$ **7.** $5.38°$ **9.** $135°16'12''$ **11.** $-85°9'$

13. $128.57°$ **15.** $-200.54°$ **17.** 8.3776

19. -0.5890 **21.** $72°$ **23.** $\dfrac{\pi}{5}$ **25.** $\dfrac{25}{12}$ **27.** 48.20

29. $\sin\theta = \dfrac{5\sqrt{61}}{61}$

$\cos\theta = \dfrac{6\sqrt{61}}{61}$

$\tan\theta = \dfrac{5}{6}$

$\csc\theta = \dfrac{\sqrt{61}}{5}$

$\sec\theta = \dfrac{\sqrt{61}}{6}$

$\cot\theta = \dfrac{6}{5}$

31. $\sin\theta = \dfrac{\sqrt{5}}{3}$

$\cos\theta = \dfrac{2}{3}$

$\tan\theta = \dfrac{\sqrt{5}}{2}$

$\csc\theta = \dfrac{3\sqrt{5}}{5}$

$\sec\theta = \dfrac{3}{2}$

$\cot\theta = \dfrac{2\sqrt{5}}{5}$

33. $\sin\theta = \dfrac{4}{5}$

$\cos\theta = \dfrac{3}{5}$

$\tan\theta = \dfrac{4}{3}$

$\csc\theta = \dfrac{5}{4}$

$\sec\theta = \dfrac{5}{3}$

$\cot\theta = \dfrac{3}{4}$

35. $\sin\theta = \dfrac{2\sqrt{53}}{53}$

$\cos\theta = -\dfrac{7\sqrt{53}}{53}$

$\tan\theta = -\dfrac{2}{7}$

$\csc\theta = \dfrac{\sqrt{53}}{2}$

$\sec\theta = -\dfrac{\sqrt{53}}{7}$

$\cot\theta = -\dfrac{7}{2}$

37. $\sin\theta = -\dfrac{3\sqrt{13}}{13}$

$\cos\theta = -\dfrac{2\sqrt{13}}{13}$

$\tan\theta = \dfrac{3}{2}$

$\csc\theta = -\dfrac{\sqrt{13}}{3}$

$\sec\theta = -\dfrac{\sqrt{13}}{2}$

$\cot\theta = \dfrac{2}{3}$

39. $\sin\theta = -\dfrac{\sqrt{11}}{6}$

$\cos\theta = \dfrac{5}{6}$

$\tan\theta = -\dfrac{\sqrt{11}}{5}$

$\csc\theta = -\dfrac{6\sqrt{11}}{11}$

$\cot\theta = -\dfrac{5\sqrt{11}}{11}$

41. $\cos \theta = -\dfrac{\sqrt{55}}{8}$

$\tan \theta = -\dfrac{3\sqrt{55}}{55}$

$\csc \theta = \dfrac{8}{3}$

$\sec \theta = -\dfrac{8\sqrt{55}}{55}$

$\cot \theta = -\dfrac{\sqrt{55}}{3}$

43. $\sqrt{3}$ **45.** $-\dfrac{\sqrt{3}}{2}$ **47.** $-\dfrac{\sqrt{2}}{2}$ **49.** 0.65

51. 3.24   **53.** $135° = \dfrac{3\pi}{4}$, $225° = \dfrac{5\pi}{4}$

55. $210° = \dfrac{7\pi}{6}$, $330° = \dfrac{11\pi}{6}$

57. $57°$; ≈ 0.9949; $123°$; ≈ 2.1467

59. $165°$; ≈ 2.8797; $195°$; ≈ 3.4035

61. Period: 2; Amplitude: 5

63. Period: π; Amplitude: 3.4

65.

67.

69.

71.

73.

75.

77.

79.

Not periodic

81.

Not periodic

83.

Not periodic

85.

Not periodic

87.

Periodic: 6

89.

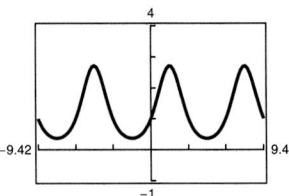

Periodic: $\left(\dfrac{\pi}{2}, e\right)$, $\left(\dfrac{3\pi}{2}, e^{-1}\right)$

91.

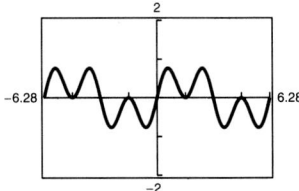

Periodic: $(0.61, 0.77)$, $(2.53, 0.77)$, $\left(\dfrac{3\pi}{2}, 0\right)$, $\left(\dfrac{\pi}{2}, 0\right)$,

$(3.76, -0.77)$, $(5.67, -0.77)$

93. $f(x) = -2\cos\left(x - \dfrac{\pi}{4}\right)$ **95.** $f(x) = -4\cos\left(2x - \dfrac{\pi}{2}\right)$

97. $f(x) = \dfrac{1}{2}\tan\dfrac{x}{2}$ **99.** $\dfrac{\sqrt{-x^2 + 2x}}{-x^2 + 2x}$

101. $\dfrac{2\sqrt{4 - 2x^2}}{4 - x^2}$ **103.** 9.2 meters **105.** 1.2 miles

107. 0.071 kilometer or 71 meters

109. (a) $A = 72(\tan\theta - \theta)$

(b)

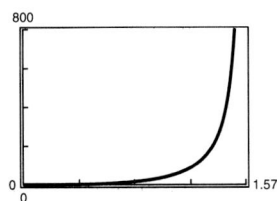

Area increases without bound as θ approaches $\pi/2$.

Chapter Test *(page 406)*

1. (a)

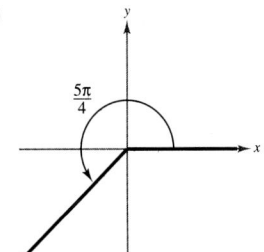

(b) $\dfrac{13\pi}{4}$, $-\dfrac{3\pi}{4}$

(c) $225°$

2. 3000 rad per minute

3. $\sin\theta = \dfrac{4\sqrt{17}}{17}$

$\cos\theta = -\dfrac{\sqrt{17}}{17}$

$\tan\theta = -4$

$\csc\theta = \dfrac{\sqrt{17}}{4}$

$\sec\theta = -\sqrt{17}$

$\cot\theta = -\dfrac{1}{4}$

4. $\sin\theta = \pm\dfrac{3\sqrt{13}}{13}$

$\cos\theta = \pm\dfrac{2\sqrt{13}}{13}$

$\csc\theta = \pm\dfrac{\sqrt{13}}{3}$

$\sec\theta = \pm\dfrac{\sqrt{13}}{2}$

$\cot\theta = \dfrac{2}{3}$

5. $70°$

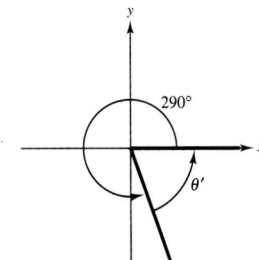

6. III   **7.** 150°, 210°   **8.** 1.33, 1.81

9.

10.

11.

12.

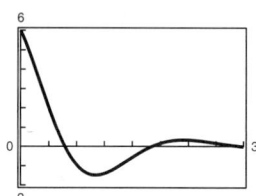

Period: 2

13. $y = -2 \sin\left(\dfrac{x}{2} - \dfrac{\pi}{4}\right)$ **14.** $\dfrac{\sqrt{5}}{2}$

15.

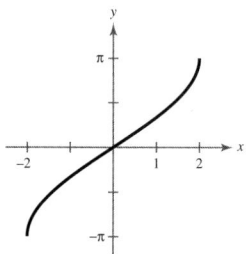

16. $(-14.88, 51.91)$

CHAPTER 5

Section 5.1   *(page 413)*

1. $\sin x = \dfrac{1}{2}$

$\cos x = \dfrac{\sqrt{3}}{2}$

$\tan x = \dfrac{\sqrt{3}}{3}$

$\csc x = 2$

$\sec x = \dfrac{2\sqrt{3}}{3}$

$\cot x = \sqrt{3}$

3. $\sin \theta = -\dfrac{\sqrt{2}}{2}$

$\cos \theta = \dfrac{\sqrt{2}}{2}$

$\tan \theta = -1$

$\csc \theta = -\sqrt{2}$

$\sec \theta = \sqrt{2}$

$\cot \theta = -1$

---

5. $\sin x = -\dfrac{5}{13}$

$\cos x = -\dfrac{12}{13}$

$\tan x = \dfrac{5}{12}$

$\csc x = -\dfrac{13}{5}$

$\sec x = -\dfrac{13}{12}$

$\cot x = \dfrac{12}{5}$

7. $\sin \phi = 0$

$\cos \phi = -1$

$\tan \phi = 0$

$\csc \phi$ is undefined.

$\sec \phi = -1$

$\cot \phi$ is undefined.

9. $\sin x = \dfrac{2}{3}$

$\cos x = -\dfrac{\sqrt{5}}{3}$

$\tan x = -\dfrac{2\sqrt{5}}{5}$

$\csc x = \dfrac{3}{2}$

$\sec x = -\dfrac{3\sqrt{5}}{5}$

$\cot x = -\dfrac{\sqrt{5}}{2}$

11. $\sin \theta = -\dfrac{2\sqrt{5}}{5}$

$\cos \theta = -\dfrac{\sqrt{5}}{5}$

$\tan \theta = 2$

$\csc \theta = -\dfrac{\sqrt{5}}{2}$

$\sec \theta = -\sqrt{5}$

$\cot \theta = \dfrac{1}{2}$

13. $\sin \theta = -1$

$\cos \theta = 0$

$\tan \theta$ is undefined.

$\csc \theta = -1$

$\sec \theta$ is undefined.

$\cot \theta = 0$

15. 1, 1   **17.** $\infty, 0$ **19.** (d)   **21.** (a)   **23.** (e)

25. (b)   **27.** (f)   **29.** (e)   **31.** $\sec \phi$ **33.** $\sin \beta$

35. $\cos x$ **37.** 1   **39.** $-\tan x$ **41.** $\tan x$

43. $1 + \sin y$ **45.** $\sin^2 x$ **47.** $\sin^2 x \tan^2 x$

49. $\sec^4 x$ **51.** $\sin^2 x - \cos^2 x$ **53.** $1 + 2 \sin x \cos x$

55. $\tan^2 x$ **57.** $2 \csc^2 x$ **59.** $2 \sec x$ **61.** $1 + \cos y$

63. $3(\sec x + \tan x)$

65.

x	0.2	0.4	0.6	0.8	1.0
y_1	0.1987	0.3894	0.5646	0.7174	0.8415
y_2	0.1987	0.3894	0.5646	0.7174	0.8415

x	1.2	1.4
y_1	0.9320	0.9854
y_2	0.9320	0.9854

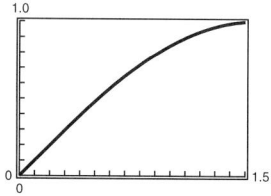

$y_1 = y_2$

67.

x	0.2	0.4	0.6	0.8	1.0
y_1	1.2230	1.5085	1.8958	2.4650	3.4082
y_2	1.2230	1.5085	1.8958	2.4650	3.4082

x	1.2	1.4
y_1	5.3319	11.6814
y_2	5.3319	11.6814

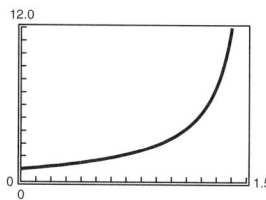

$y_1 = y_2$

69. csc x **71.** 5 cos θ **73.** 3 tan θ **75.** 5 sec θ

77. $0 \le \theta \le \pi$ **79.** $0 \le \theta < \dfrac{\pi}{2}, \ \dfrac{3\pi}{2} < \theta < 2\pi$

81. ln $|\cot \theta|$ **83.** Not an identity because $\dfrac{\sin k\theta}{\cos k\theta} = \tan k\theta$

85. Identity because $\sin \theta \cdot \dfrac{1}{\sin \theta} = 1$

87. (a) $\csc^2 132° - \cot^2 132° \approx 1.8107 - 0.8107 = 1$

(b) $\csc^2 \dfrac{2\pi}{7} - \cot^2 \dfrac{2\pi}{7} \approx 1.6360 - 0.6360 = 1$

89. (a) $\cos(90° - 80°) = \sin 80° \approx 0.9848$

(b) $\cos\left(\dfrac{\pi}{2} - 0.8\right) = \sin 0.8 \approx 0.7174$

91. $\cos \theta = \pm\sqrt{1 - \sin^2\theta}$

$\tan \theta = \pm\dfrac{\sin \theta}{\sqrt{1 - \sin^2\theta}}$

$\csc \theta = \dfrac{1}{\sin \theta}$

$\sec \theta = \pm\dfrac{1}{\sqrt{1 - \sin^2\theta}}$

$\cot \theta = \pm\dfrac{\sqrt{1 - \sin^2\theta}}{\sin \theta}$

The sign depends on the choice of θ.

93. Seward; 6.4 and 1.9

Section 5.2   *(page 421)*

1.–9. Answers will vary.

11.

x	0.2	0.4	0.6	0.8	1.0
y_1	4.835	2.1785	1.2064	0.6767	0.3469
y_2	4.835	2.1785	1.2064	0.6767	0.3469

x	1.2	1.4
y_1	0.1409	0.0293
y_2	0.1409	0.0293

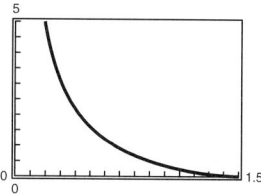

13.

x	0.2	0.4	0.6	0.8	1.0
y_1	4.835	2.1785	1.2064	0.6767	0.3469
y_2	4.835	2.1785	1.2064	0.6767	0.3469

x	1.2	1.4
y_1	0.1409	0.0293
y_2	0.1409	0.0293

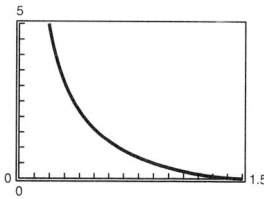

15.

x	0.2	0.4	0.6	0.8	1.0
y_1	1.0203	1.0857	1.2116	1.4353	1.8508
y_2	1.0203	1.0857	1.2116	1.4353	1.8508

x	1.2	1.4
y_1	2.7597	5.8835
y_2	2.7597	5.8835

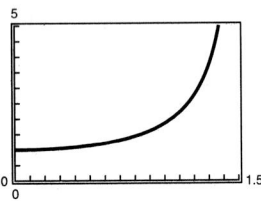

17.

x	0.2	0.4	0.6	0.8	1.0
y_1	5.1359	2.7880	2.1458	2.0009	2.1995
y_2	5.1359	2.7880	2.1458	2.0009	2.1995

x	1.2	1.4
y_1	2.9609	5.9704
y_2	2.9609	5.9704

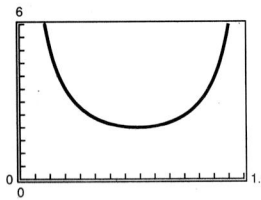

19.– 47. Answers will vary.

49.

1

51.

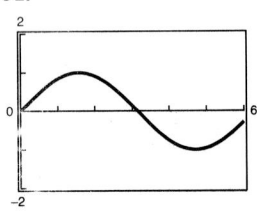

$\sin x$

53.–55. Answers will vary.

57. $\sqrt{\tan^2 x} = |\tan x|$; $\dfrac{3\pi}{4}$

59. $|\tan \theta| = \sqrt{\sec^2 \theta - 1}$; $\dfrac{3\pi}{4}$ **61.** 1   **63.** 2

65. Answers will vary.   **67.** $\mu = \tan \theta$

Section 5.3   *(page 431)*

1. $x = -1,\ 3$ **3.** $x = \pm 2$ **5.–9.** Answers will vary.

11. $\dfrac{2\pi}{3} + 2n\pi,\ \dfrac{4\pi}{3} + 2n\pi$ **13.** $\dfrac{\pi}{3} + 2n\pi,\ \dfrac{2\pi}{3} + 2n\pi$

15. $\dfrac{\pi}{6} + n\pi,\ \dfrac{5\pi}{6} + n\pi$

17. $\dfrac{\pi}{8} + n\pi,\ \dfrac{3\pi}{8} + n\pi,\ \dfrac{5\pi}{8} + n\pi,\ \dfrac{7\pi}{8} + n\pi$

19. $\dfrac{\pi}{3} + n\pi,\ \dfrac{2\pi}{3} + n\pi$ **21.** $\dfrac{\pi}{3} + n\pi,\ \dfrac{2\pi}{3} + n\pi$

23. $\dfrac{\pi}{6} + n\pi,\ \dfrac{5\pi}{6} + n\pi,\ \dfrac{\pi}{3} + n\pi,\ \dfrac{2\pi}{3} + n\pi$

25. $0,\ \dfrac{\pi}{2},\ \pi,\ \dfrac{3\pi}{2}$ **27.** $0,\ \pi,\ \dfrac{\pi}{6},\ \dfrac{5\pi}{6},\ \dfrac{7\pi}{6},\ \dfrac{11\pi}{6}$

29. $\dfrac{\pi}{3},\ \dfrac{5\pi}{3},\ \pi$ **31.** No solution   **33.** $\dfrac{\pi}{2}$ **35.** $\dfrac{\pi}{2}$

37. π **39.** $\dfrac{\pi}{6},\ \dfrac{5\pi}{6},\ \dfrac{7\pi}{6},\ \dfrac{11\pi}{6}$ **41.** 0.5236, 2.6180

43. 0.7854, 2.3562, 3.9270, 5.4978, 3.6652, 5.7596

45. 0, 1.8955   **47.** 0, 2.6779, 3.1416, 5.8195

49. 0.3398, 0.8481, 2.2935, 2.8018

51. (a)

x	0	1	2	3	4
$f(x)$	Undef.	0.83	-1.36	-2.93	-4.46

x	5	6
$f(x)$	-6.34	-13.02

The zero is in the interval $(1, 2)$ because f changes signs in the interval.

(b)

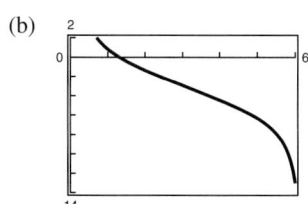

The interval is the same as in part (a).

(c) 1.3065

53. (a)

x	0	1	2	3	4
$f(x)$	-1	1.39	1.65	-0.70	-1.94

x	5	6
$f(x)$	-2.00	-1.48

The zeros are in the intervals $(0, 1)$ and $(2, 3)$ because f changes signs in these intervals.

(b)

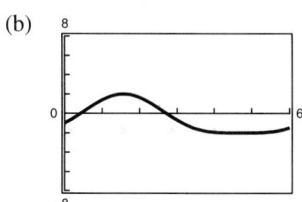

The intervals are the same as in part (a).

(c) 0.4271, 2.7145

55.

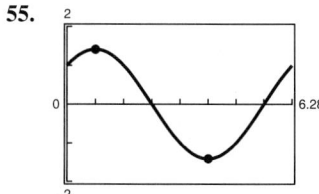

Maximum: $\left(\dfrac{\pi}{4}, \sqrt{2}\right)$

Minimum: $\left(\dfrac{5\pi}{4}, -\sqrt{2}\right)$

57. 1

59. (a) All real numbers except $x = 0$

(b) y-axis symmetry; horizontal asymptote: $y = 1$

(c) Oscillates

(d) Infinite solutions

(e) Yes, 0.6366

61. 0.04, 0.43, 0.83   **63.** 37°, 53°

65. (a)

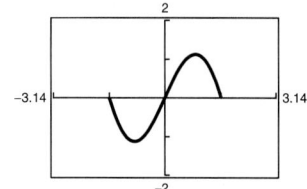

$x \approx 0.86, A \approx 1.12$

(b) $0.6 < x < 1.1$

67. (a)

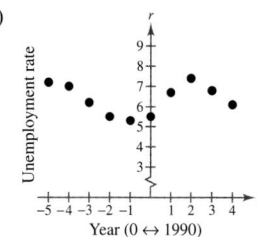

(b) (iii)

(c) Constant: 6.20%

(d) 7 years

(e) 2000

Section 5.4   *(page 440)*

1. (a) $\dfrac{\sqrt{2} - \sqrt{6}}{4}$ (b) $\dfrac{\sqrt{2} + 1}{2}$

3. (a) $\dfrac{1}{2}$ (b) $\dfrac{-\sqrt{3} - 1}{2}$

5. False. Parts (a) and (b) are unequal in Exercises 1–4.

7. $\sin 75° = \dfrac{\sqrt{2}}{4}\left(1 + \sqrt{3}\right)$

$\cos 75° = \dfrac{\sqrt{2}}{4}\left(\sqrt{3} - 1\right)$

$\tan 75° = \sqrt{3} + 2$

9. $\sin 105° = \dfrac{\sqrt{2}}{4}\left(\sqrt{3} + 1\right)$

$\cos 105° = \dfrac{\sqrt{2}}{4}\left(1 - \sqrt{3}\right)$

$\tan 105° = -2 - \sqrt{3}$

11. $\sin 195° = \dfrac{\sqrt{2}}{4}\left(1 - \sqrt{3}\right)$

$\cos 195° = -\dfrac{\sqrt{2}}{4}\left(\sqrt{3} + 1\right)$

$\tan 195° = 2 - \sqrt{3}$

13. $\sin \dfrac{11\pi}{12} = \dfrac{\sqrt{2}}{4}\left(\sqrt{3} - 1\right)$

$\cos \dfrac{11\pi}{12} = -\dfrac{\sqrt{2}}{4}\left(\sqrt{3} + 1\right)$

$\tan \dfrac{11\pi}{12} = -2 + \sqrt{3}$

15. $\sin \dfrac{17\pi}{12} = -\dfrac{\sqrt{2}}{4}\left(\sqrt{3} + 1\right)$

$\cos \dfrac{17\pi}{12} = \dfrac{\sqrt{2}}{4}\left(1 - \sqrt{3}\right)$

$\tan \dfrac{17\pi}{12} = 2 + \sqrt{3}$

17. $\cos 40°$ **19.** $\sin 200°$ **21.** $\tan 239°$ **23.** $\sin 1.8$

25. $\tan 3x$

27.

x	0.2	0.4	0.6	0.8	1.0
y_1	0.9801	0.9211	0.8253	0.6967	0.5403
y_2	0.9801	0.9211	0.8253	0.6967	0.5403

x	1.2	1.4
y_1	0.3624	0.1700
y_2	0.3624	0.1700

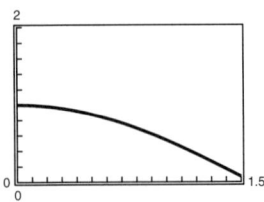

29.

x	0.2	0.4	0.6	0.8	1.0
y_1	0.6621	0.7978	0.9017	0.9696	0.9989
y_2	0.6621	0.7978	0.9017	0.9696	0.9989

x	1.2	1.4
y_1	0.9883	0.9384
y_2	0.9883	0.9384

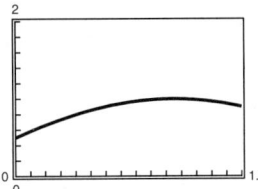

31.

x	0.2	0.4	0.6	0.8	1.0
y_1	0.9605	0.8484	0.6812	0.4854	0.2919
y_2	0.9605	0.8484	0.6812	0.4854	0.2919

x	1.2	1.4
y_1	0.1313	0.0289
y_2	0.1313	0.0289

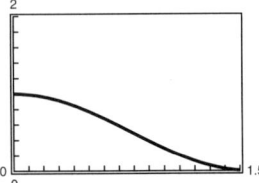

33. $\dfrac{33}{65}$ **35.** $-\dfrac{56}{65}$ **37.** $-\dfrac{3}{5}$ **39.** $\dfrac{44}{125}$

41.–47. Answers will vary.

49.

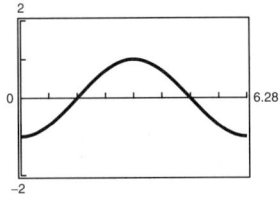

$g(x) = -\cos x$

51. (a) $\sqrt{2}\sin\left(\theta + \dfrac{\pi}{4}\right)$ (b) $\sqrt{2}\cos\left(\theta - \dfrac{\pi}{4}\right)$

53. (a) $13\sin(3\theta + 0.3948)$ (b) $13\cos(3\theta - 1.1760)$

55. $2\cos\theta$ **57.** 1 **59.** $\dfrac{\pi}{2}$

61. $\dfrac{5\pi}{4}, \dfrac{7\pi}{4}$ **63.** $\dfrac{\pi}{4}, \dfrac{7\pi}{4}$ **65.** Answers will vary.

67. (a)

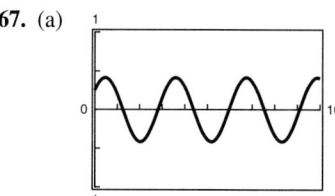

(b) $y = \dfrac{5}{12}\sin(2t + 0.6435)$ (c) $\dfrac{5}{12}$ (d) $\dfrac{1}{\pi}$

69. (a) $(-\infty, 0),\ (0, \infty)$

(b)

h	0.01	0.02	0.05	0.1
$f(h)$	-0.5043	-0.5086	-0.5214	-0.5424
$g(h)$	-0.5043	-0.5086	-0.5214	-0.5424

h	0.2	0.5
$f(h)$	-0.5830	-0.6915
$g(h)$	-0.5830	-0.6915

(c)

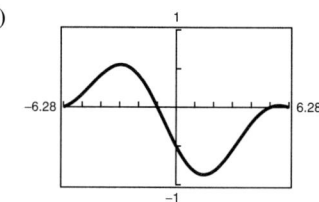

(d) $-\dfrac{1}{2}$

Section 5.5   *(page 451)*

1. $\dfrac{3}{5}$ **3.** $\dfrac{7}{25}$ **5.** $\dfrac{24}{7}$ **7.** $\dfrac{25}{24}$ **9.** $0, \dfrac{\pi}{3}, \pi, \dfrac{5\pi}{3}$

11. $\dfrac{\pi}{12}, \dfrac{5\pi}{12}, \dfrac{13\pi}{12}, \dfrac{17\pi}{12}$ **13.** $0, \dfrac{2\pi}{3}, \dfrac{4\pi}{3}$

15. $\dfrac{\pi}{2}, \dfrac{\pi}{6}, \dfrac{5\pi}{6}, \dfrac{7\pi}{6}, \dfrac{3\pi}{2}, \dfrac{11\pi}{6}$ **17.** $0, \dfrac{\pi}{2}, \pi, \dfrac{3\pi}{2}$

19. $f(x) = 3\sin 2x$ **21.** $g(x) = 4\cos 2x$

23. $\sin 2u = \dfrac{24}{25}$ **25.** $\sin 2u = \dfrac{4}{5}$

$\cos 2u = \dfrac{7}{25}$ $\cos 2u = \dfrac{3}{5}$

$\tan 2u = \dfrac{24}{7}$ $\tan 2u = \dfrac{4}{3}$

27. $\frac{1}{8}(3 + 4\cos 2x + \cos 4x)$ **29.** $\frac{1}{8}(1 - \cos 4x)$

31. $\frac{1}{32}(2 + \cos 2x - 2\cos 4x - \cos 6x)$

33. $\dfrac{5}{\sqrt{26}}$ **35.** $\dfrac{1}{5}$ **37.** $\sqrt{26}$

39. $\sin 105° = \frac{1}{2}\sqrt{2 + \sqrt{3}}$

$\cos 105° = -\frac{1}{2}\sqrt{2 - \sqrt{3}}$

$\tan 105° = -2 - \sqrt{3}$

41. $\sin 112°\,30' = \frac{1}{2}\sqrt{2 + \sqrt{2}}$

$\cos 112°\,30' = -\frac{1}{2}\sqrt{2 - \sqrt{2}}$

$\tan 112°\,30' = -1 - \sqrt{2}$

43. $\sin\dfrac{\pi}{8} = \dfrac{1}{2}\sqrt{2 - \sqrt{2}}$ **45.** $\sin\dfrac{u}{2} = \dfrac{5\sqrt{26}}{26}$

$\cos\dfrac{\pi}{8} = \dfrac{1}{2}\sqrt{2 + \sqrt{2}}$ $\cos\dfrac{u}{2} = \dfrac{\sqrt{26}}{26}$

$\tan\dfrac{\pi}{8} = \sqrt{2} - 1$ $\tan\dfrac{u}{2} = 5$

47. $\sin\dfrac{u}{2} = \sqrt{\dfrac{89 - 8\sqrt{89}}{178}}$

$\cos\dfrac{u}{2} = -\sqrt{\dfrac{89 + 8\sqrt{89}}{178}}$

$\tan\dfrac{u}{2} = \dfrac{8 - \sqrt{89}}{5}$

49. $|\sin 3x|$ **51.** $-|\tan 4x|$ **53.** π **55.** $\dfrac{\pi}{3}, \pi, \dfrac{5\pi}{3}$

57. $3\left(\sin\dfrac{\pi}{2} + \sin 0\right)$ **59.** $\dfrac{1}{2}(\sin 8\theta + \sin 2\theta)$

61. $\frac{5}{2}(\cos 8\beta + \cos 2\beta)$ **63.** $2\sin 45°\cos 15°$

65. $-2\sin\dfrac{\pi}{2}\sin\dfrac{\pi}{4}$ **67.** $2\cos 4x\cos 2x$

69. $2\cos\alpha\sin\beta$ **71.** $2\cos(\phi + \pi)\cos\pi$

73. $0, \dfrac{\pi}{4}, \dfrac{\pi}{2}, \dfrac{3\pi}{4}, \pi, \dfrac{5\pi}{4}, \dfrac{3\pi}{2}, \dfrac{7\pi}{4}$ **75.** $\dfrac{\pi}{6}, \dfrac{5\pi}{6}$

77. $\dfrac{25}{169}$ **79.** $\dfrac{4}{13}$

81.–93. Answers will vary.

95.

97.

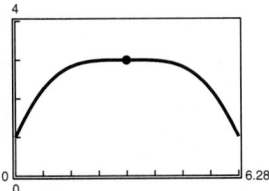

Maximum: $(\pi, 3)$

99. (a)

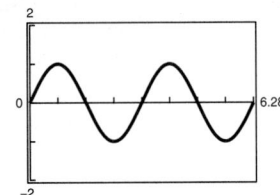

(b) $g(x) = \sin 2x$

101. (a) $f(x) = \frac{1}{4}(3 + \cos 4x)$

(b) $f(x) = 2\cos^4 x - 2\cos^2 x + 1$

(c) $f(x) = 1 - 2\sin^2 x \cos^2 x$

(d) $f(x) = 1 - \frac{1}{2}\sin^2 2x$

(e) There is often more than one way to rewrite a trigonometric expression.

103. $2x\sqrt{1 - x^2}$

FOCUS ON CONCEPTS   *(page 454)*

1. An identity is true for all values of the variable and a conditional equation is true for some values of the variable.

2. When proving an identity, use the fundamental identities and rules of algebra to transform one expression into another. To solve a trigonometric equation, use standard algebraic techniques and identities to isolate a trigonometric function involved in the equation. Find the value of the variable by using the inverse of the trigonometric function.

3. Reciprocal identities:
$$\csc \theta = \frac{1}{\sin \theta}, \quad \sec \theta = \frac{1}{\cos \theta}, \quad \cot \theta = \frac{1}{\tan \theta}$$

Quotient identities: $\tan \theta = \dfrac{\sin \theta}{\cos \theta}, \quad \cot \theta = \dfrac{\cos \theta}{\sin \theta}$

Pythagorean identities:
$$\sin^2 \theta + \cos^2 \theta = 1, \quad \tan^2 \theta + 1 = \sec^2 \theta,$$
$$1 + \cot^2 \theta = \csc^2 \theta$$

4. No. $\cos \theta = \pm\sqrt{1 - \sin^2 \theta}$

5. False. The order in which algebraic operations and fundamental identities are used to verify an identity may vary.

6. (a) True. The period of the tangent is π.

(b) False. The period of the cosine is 2π.

(c) False. $\sec \theta \cos \theta = 1$

(d) True

(e) True. $\sin(-\alpha) = -\sin \alpha$

7. $y_1 = y_2 + 1$ **8.** $y_1 = 1 - y_2$ **9.** 1   **10.** 3

11. 5   **12.** 4   **13.** 2π

Review Exercises   *(page 455)*

1. $\sin^2 x$ **3.** $1 + \cot \alpha$ **5.** 1   **7.** $\tan(2x + 2)$

9.–25. Answers will vary.

27.

29.

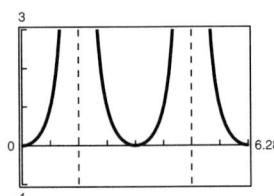

31. $\dfrac{\sqrt{2}}{4}(\sqrt{3} + 1)$ **33.** $-\dfrac{1}{2}\sqrt{2 + \sqrt{2}}$

35. $-\dfrac{3}{52}(5 + 4\sqrt{7})$ **37.** $\dfrac{1}{52}(36 + 5\sqrt{7})$

39. $\frac{1}{4}\sqrt{2\left(4 - \sqrt{7}\right)}$

41. False. If $\frac{\pi}{2} < \theta < \pi$, then $\cos\frac{\theta}{2} > 0$.

43. True   **45.** $0,\ \pi$ **47.** $0,\ \dfrac{3\pi}{4},\ \pi,\ \dfrac{5\pi}{4}$

49. $0,\ \dfrac{\pi}{2},\ \pi$ **51.** $\dfrac{\pi}{3},\ \dfrac{5\pi}{3}$ **53.** $\dfrac{\pi}{4},\ \dfrac{5\pi}{4}$

55. False. $\sin\theta = \frac{1}{2}$ has an infinite number of solutions but is not an identity.

57. $2\cos\dfrac{5\theta}{2}\cos\dfrac{\theta}{2}$ **59.** $\dfrac{1}{2}(\cos\alpha - \cos 5\alpha)$

61. $8x^2 - 1$ **63.** Answers will vary.

65. (a) $y = \frac{1}{2}\sqrt{10}\sin\left(8t - \arctan\frac{1}{3}\right)$

(b)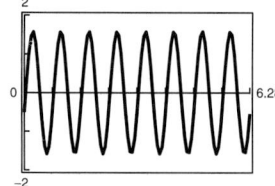

(c) $\dfrac{1}{2}\sqrt{10}$ (d) $\dfrac{4}{\pi}$

Chapter Test   *(page 458)*

1. $\sin\theta = -\dfrac{3}{\sqrt{13}}$ **2.** 1   **3.** 1   **4.** $\dfrac{1}{\sin\theta\cos\theta}$

$\cos\theta = -\dfrac{2}{\sqrt{13}}$

$\csc\theta = -\dfrac{\sqrt{13}}{3}$

$\sec\theta = -\dfrac{\sqrt{13}}{2}$

$\cot\theta = \dfrac{2}{3}$

5. $\dfrac{\pi}{2} < \theta \le \pi$

$\dfrac{3\pi}{2} < \theta < 2\pi$

6.

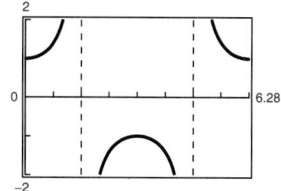

$y_1 = y_2$

7.–12. Answers will vary.

13. $0,\ \dfrac{3\pi}{4},\ \pi,\ \dfrac{7\pi}{4}$ **14.** $\dfrac{\pi}{6},\ \dfrac{\pi}{2},\ \dfrac{5\pi}{6},\ \dfrac{3\pi}{2}$

15. $\dfrac{\pi}{6},\ \dfrac{5\pi}{6},\ \dfrac{7\pi}{6},\ \dfrac{11\pi}{6}$ **16.** $\dfrac{\pi}{6},\ \dfrac{5\pi}{6},\ \dfrac{3\pi}{2}$

17. $-2.938,\ -2.663,\ 1.170$

18. $\left|\cos^2 x + \cos x\right| \le 2$ for all x **19.** $\dfrac{\sqrt{2} - \sqrt{6}}{4}$

20. $\sin 2u = \frac{4}{5},\ \tan 2u = -\frac{4}{3}$

CHAPTER 6

Section 6.1   *(page 466)*

1. $C = 105°,\ b \approx 28.28,\ c \approx 38.64$

3. $C = 110°,\ b \approx 37.40,\ c \approx 40.59$

5. $B \approx 21.55°,\ C \approx 122.45°,\ c \approx 11.49$

7. $B = 10°,\ b \approx 69.46,\ c \approx 136.81$

9. $B = 42°4',\ a \approx 22.05,\ b \approx 14.88$

11. $A \approx 10°11',\ C \approx 154°19',\ c \approx 11.03$

13. $B \approx 18°13',\ C \approx 51°32',\ c \approx 40.05$

15. No solution

17. Two solutions

$B \approx 70.4°,\ C \approx 51.6°,\ c \approx 4.16$

$B \approx 109.6°,\ C \approx 12.4°,\ c \approx 1.14$

19. (a) $b \le 5,\ b = \dfrac{5}{\sin 36°}$ (b) $5 < b < \dfrac{5}{\sin 36°}$

(c) $b > \dfrac{5}{\sin 36°}$

21. (a)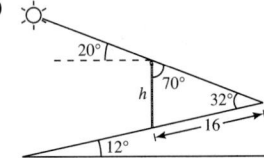

(b) $\dfrac{16}{\sin 70°} = \dfrac{h}{\sin 32°}$ (c) 9 meters

23. 16.1°   **25.** 77 meters

27. (a)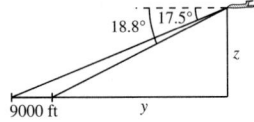

(b) 22.6 miles   (c) 21.4 miles   (d) 38,443 feet

29. 25.8 kilometers, 42.3 kilometers   **31.** 4.55 miles

33. (a) $\alpha = \arcsin(0.5 \sin \beta)$

(b)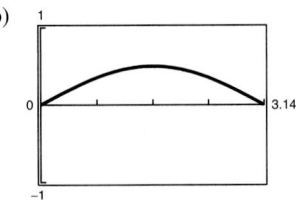

Domain: $0 \le \beta \le \pi$

Range: $0 \le \alpha \le \dfrac{\pi}{6}$

(c) $c = \dfrac{18 \sin[\pi - \beta - \arcsin(0.5 \sin \beta)]}{\sin \beta}$

(d)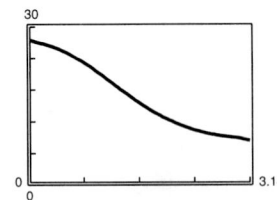

Domain: $0 < \beta < \pi$

Range: $9 < c < 27$

(e)

β	0	0.4	0.8	1.2	1.6
α	0	0.1960	0.3669	0.4848	0.5234
c	Undef.	25.95	23.07	19.19	15.33

β	2.0	2.4	2.8
α	0.4720	0.3445	0.1683
c	12.29	10.31	9.27

35. 10.4   **37.** 1675.2   **39.** 474.9

41. (a) $20\left[15 \sin \dfrac{3\theta}{2} - 4 \sin \dfrac{\theta}{2} - 6 \sin \theta\right]$

(b)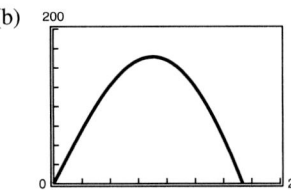

(c) Domain: $0 \le \theta \le 1.6690$

The domain would increase in length and the area would increase.

Section 6.2   (page 474)

1. $A \approx 26.4°$, $B \approx 36.3°$, $C \approx 117.3°$

3. $B \approx 23.8°$, $C \approx 126.2°$, $a \approx 18.6$

5. $A \approx 36.9°$, $B \approx 53.1°$, $C \approx 90°$

7. $A \approx 92.9°$, $B \approx 43.53°$, $C \approx 43.53°$

9. $A \approx 158°37'$, $C \approx 12°38'$, $b \approx 10.4$

	a	b	c	d	θ	ϕ
11.	4	6	9.67	3.23	30°	150°
13.	10	14	20	13.86	68.2°	111.8°
15.	10	11.58	18	12	67.1°	112.9°

17.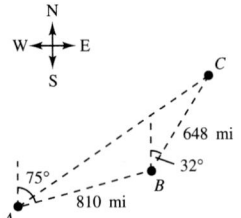

1357.8 miles, N 56° E

19. 422.5 meters   **21.** 43.3 miles

23. (a) N 58.4° W   (b) S 81.5° W

25. 63.7 feet   **27.** $\overline{PQ} \approx 9.4$, $\overline{QS} \approx 5.0$, $\overline{RS} \approx 12.8$

29. (a) $49 = 2.25 + x^2 - 3x \cos \theta$

(b) $x = \frac{1}{2}\left(3 \cos \theta + \sqrt{9 \cos^2 \theta + 187}\right)$

(c)

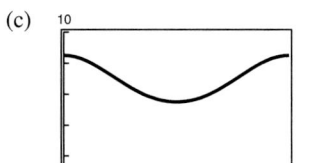

(d) 6 inches

31. 2.76 feet   **33.** (a) 570.60   (b) 5910.68   (c) 177.09

35. 16.25   **37.** 54   **39.** 96.82

41. Answers will vary.

Section 6.3   *(page 486)*

1. $\langle 4, 3 \rangle$, $\|\mathbf{v}\| = 5$ **3.** $\langle -3, 2 \rangle$, $\|\mathbf{v}\| = \sqrt{13}$

5. $\langle 0, 5 \rangle$, $\|\mathbf{v}\| = 5$ **7.** $\mathbf{v} = \langle -\frac{9}{2}, -\frac{5}{2} \rangle$, $\|\mathbf{v}\| = \frac{1}{2}\sqrt{106}$

9. $\mathbf{v} = \langle 8, -6 \rangle$, $\|\mathbf{v}\| = 10$

11.

13.

15.

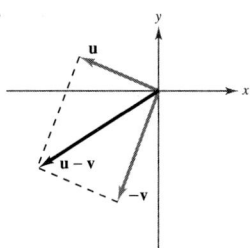

17. (a) $\langle 4, 3 \rangle$ (b) $\langle -2, 1 \rangle$ (c) $\langle -7, 1 \rangle$

19. (a) $3\mathbf{i} - 2\mathbf{j}$ (b) $-\mathbf{i} + 4\mathbf{j}$ (c) $-4\mathbf{i} + 11\mathbf{j}$

21. $\langle 1, 0 \rangle$ **23.** $\left\langle -\frac{1}{\sqrt{2}}, \frac{1}{\sqrt{2}} \right\rangle$ **25.** $\frac{4}{5}\mathbf{i} - \frac{3}{5}\mathbf{j}$

27. $\mathbf{j}$ **29.** $\left\langle \frac{5}{\sqrt{2}}, \frac{5}{\sqrt{2}} \right\rangle$ **31.** $\left\langle -\frac{21}{5}, \frac{28}{5} \right\rangle$

33. $\mathbf{v} = \left\langle 3, -\frac{3}{2} \right\rangle$ **35.** $\mathbf{v} = \langle 4, 3 \rangle$

    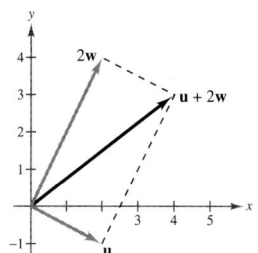

37. $\mathbf{v} = \left\langle \frac{7}{2}, -\frac{1}{2} \right\rangle$

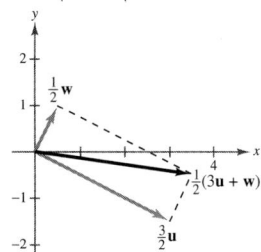

39. $\|\mathbf{v}\| = 5$, $\theta = 30°$ **41.** $\|\mathbf{v}\| = 6\sqrt{2}$, $\theta = 315°$

43. $\mathbf{v} = \langle 3, 0 \rangle$

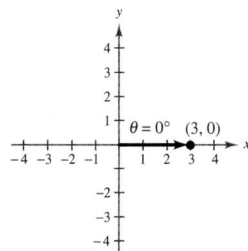

45. $\mathbf{v} = \left\langle -\frac{3\sqrt{6}}{2}, \frac{3\sqrt{2}}{2} \right\rangle$

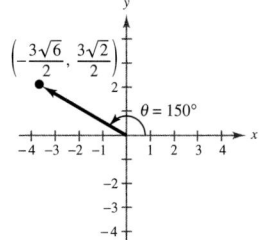

47. $\mathbf{v} = \left\langle \dfrac{\sqrt{10}}{5}, \dfrac{3\sqrt{10}}{5} \right\rangle$

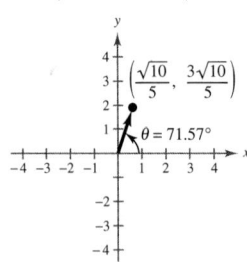

49. $\langle 5, 5 \rangle$　　**51.** $\langle 10\sqrt{2} - 50, \ 10\sqrt{2} \rangle$　　**53.** $90°$

55. $63.4°$　　**57.** $62.7°$

59. (a) $0°$　(b) $180°$　(c) No. Equal to the sum when the angle between the vectors is $0°$.

61. (a) $12.1°$, 357.85 newtons

(b) $M = 10\sqrt{660 \cos \theta + 709}$,

$\alpha = \arctan \dfrac{15 \sin \theta}{15 \cos \theta + 22}$

(c)

θ	$0°$	$30°$	$60°$	$90°$	$120°$
M	370.0	357.9	322.3	266.3	194.7
α	$0°$	$12.1°$	$23.8°$	$34.3°$	$41.9°$

θ	$150°$	$180°$
M	117.2	70.0
α	$39.8°$	$0°$

(d)

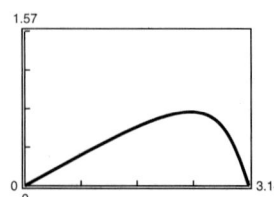

(e) For increasing θ, the two vectors tend to work against each other, resulting in a decrease in the magnitude of the resultant.

63. $71.3°$, 228.5 pounds

65.

67. Horizontal component: $80 \cos 40° \approx 61.28$ feet per second

Vertical component: $80 \sin 40° \approx 51.42$ feet per second

69. $T_{AC} \approx 1758.8$ pounds

$T_{BC} \approx 1305.4$ pounds

71. (a) 3192.5 pounds

(b) $T = 3000 \sec \theta$

(c)

θ	$10°$	$20°$	$30°$	$40°$	$50°$	$60°$
T	3046.3	3192.5	3464.1	3916.2	4667.2	6000.0

(d)

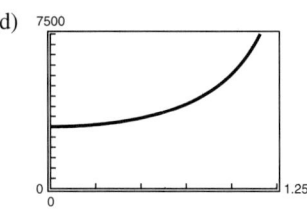

(e) The component in the direction of the motion of the barge decreases.

73. N $21.4°$ E, 138.7 kilometers per hour

75. 850 foot-pounds　　**77.** True　　**79.** False. $a = b = 0$

81. Answers will vary.　　**83.** $\langle 1, 3 \rangle$ or $\langle -1, -3 \rangle$

Section 6.4　(page 498)

1. -6　　**3.** 6　　**5.** 8, scalar　　**7.** $\langle -6, 8 \rangle$, vector

9. 13　　**11.** $5\sqrt{41}$　　**13.** \$37,289, total revenue

15. $90°$　　**17.** $143.13°$　　**19.** $\dfrac{5\pi}{12}$

21.

$91.33°$

23.

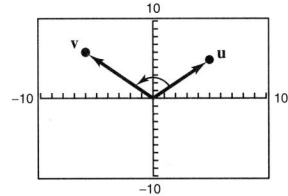

90°

25. 26.6°, 63.4°, 90° **27.** −20 **29.** Parallel

31. Neither **33.** Orthogonal **35.** $\frac{16}{17}\langle 4, 1\rangle$, $\frac{13}{17}\langle -1, 4\rangle$

37. $\frac{45}{229}\langle 2, 15\rangle$, $\frac{6}{229}\langle -15, 2\rangle$ **39.** u **41.** 0

43. $\langle -5, 3\rangle$, $\langle 5, -3\rangle$ **45.** $\frac{2}{3}\mathbf{i} + \frac{1}{2}\mathbf{j}$, $-\frac{2}{3}\mathbf{i} - \frac{1}{2}\mathbf{j}$

47. (a) 6251.3 pounds (b) 35,453.1 pounds

49. (a) $\theta = \frac{\pi}{2}$ (b) $0 \le \theta < \frac{\pi}{2}$ (c) $\frac{\pi}{2} < \theta \le \pi$

51. 735 newton-meters **53.** 779.4 foot-pounds **55.** 32

57.– 61. Anwers will vary. **63.** (a) $\dfrac{x^2 - 1}{x^2}$ (b) $\sin^2 x$

65. (a) $\dfrac{y^2 + (z + 1)^2}{y(z + 1)}$ (b) $2 \csc x$

Section 6.5 *(page 509)*

1. 5

3. $4\sqrt{2}$

5. $\sqrt{85}$

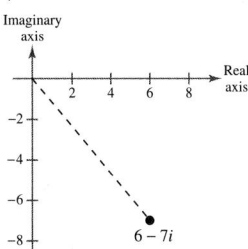

7. $3\left(\cos\dfrac{\pi}{2} + i\sin\dfrac{\pi}{2}\right)$ **9.** $2\sqrt{2}\left(\cos\dfrac{5\pi}{4} + i\sin\dfrac{5\pi}{4}\right)$

11. $3\sqrt{2}\left(\cos\dfrac{7\pi}{4} + i\sin\dfrac{7\pi}{4}\right)$

13. $2\left(\cos\dfrac{\pi}{6} + i\sin\dfrac{\pi}{6}\right)$

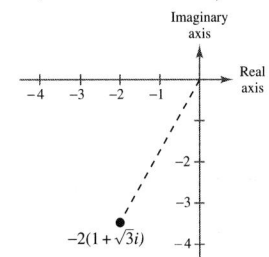

15. $4\left(\cos\dfrac{4\pi}{3} + i\sin\dfrac{4\pi}{3}\right)$

17. $6\left(\cos\dfrac{\pi}{2} + i\sin\dfrac{\pi}{2}\right)$

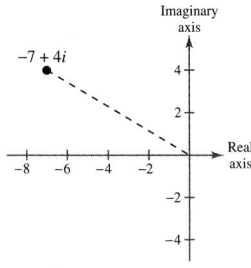

19. $\sqrt{65}\,(\cos 2.62 + i\sin 2.62)$

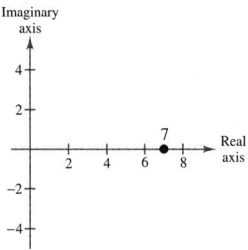

21. $7\,(\cos 0 + i\sin 0)$

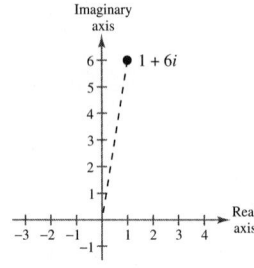

23. $\sqrt{37}\,(\cos 1.41 + i\sin 1.41)$

25. $\sqrt{10}\,(\cos 3.46 + i\sin 3.46)$

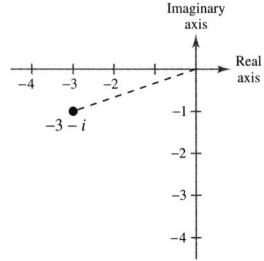

27. $5.39\,(\cos 0.38 + i\sin 0.38)$

29. $8.19\,(\cos 5.26 + i\sin 5.26)$

31. $-\sqrt{3} + i$

33. $\dfrac{3}{4} - \dfrac{3\sqrt{3}}{4}\,i$

35. $\dfrac{-15\sqrt{2}}{8} + \dfrac{15\sqrt{2}}{8}\,i$

37. $-4i$

39. $2.8408 + 0.9643i$

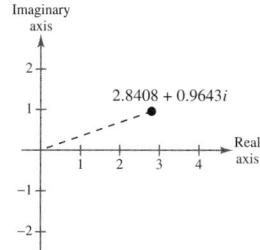

41. $4.70 + 1.71i$ **43.** $-3.71 + 11.41i$

45.

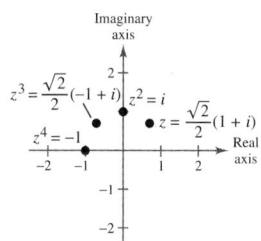

The absolute value of each is 1.

47. $12\left(\cos\dfrac{\pi}{2} + i\sin\dfrac{\pi}{2}\right)$ **49.** $\dfrac{10}{9}(\cos 200° + i\sin 200°)$

51. $0.27\,(\cos 150° + i\sin 150°)$ **53.** $\cos 30° + i\sin 30°$

55. $\frac{1}{2}\,(\cos 80° + i\sin 80°)$

57. $4[\cos(-58°) + i\sin(-58°)]$

59. (a) $2\sqrt{2}\,(\cos 45° + i\sin 45°)$;

$\sqrt{2}\,[\cos(-45°) + i\sin(-45°)]$

(b) $4(\cos 0° + i\sin 0°) = 4$

(c) 4

61. (a) $2[\cos(-90°) + i\sin(-90°)]$;

$\sqrt{2}\,(\cos 45° + i\sin 45°)$

(b) $2\sqrt{2}\,[\cos(-45°) + i\sin(-45°)] = 2 - 2i$

(c) $-2i - 2i^2 = -2i + 2 = 2 - 2i$

63. (a) $5(\cos 0° + i\sin 0°)$;

$\sqrt{13}\,(\cos 56.31° + i\sin 56.31°)$

(b) $\dfrac{5}{\sqrt{13}}[\cos(-56.31°) + i\sin(-56.31°)] \approx$

$0.7692 - 1.154i$

(c) $\frac{10}{13} - \frac{15}{13}i \approx 0.7692 - 1.154i$

65.–67. Answers will vary.

69.

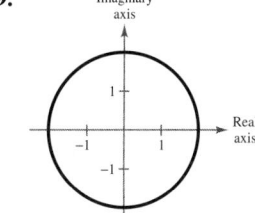

71. $-4 - 4i$ **73.** $-32i$ **75.** $-128\sqrt{3} - 128i$

77. $\dfrac{125}{2} + \dfrac{125\sqrt{3}}{2}i$ **79.** i **81.** $608.02 + 144.69i$

83. $-597 - 122i$ **85.** $\dfrac{81}{2} + \dfrac{81\sqrt{3}}{2}i$

87. Answers will vary.

89. (a) $2(\cos 30° + i\sin 30°)$ (b) $8i$

$2(\cos 150° + i\sin 150°)$

$2(\cos 270° + i\sin 270°)$

91. (a) $\sqrt{5}(\cos 60° + i\sin 60°)$

$\sqrt{5}(\cos 240° + i\sin 240°)$

(b)

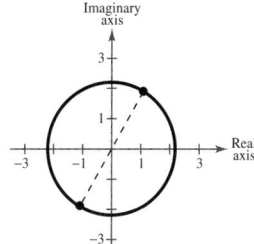

(c) $\dfrac{\sqrt{5}}{2} + \dfrac{\sqrt{15}}{2}i,\quad -\dfrac{\sqrt{5}}{2} - \dfrac{\sqrt{15}}{2}i$

93. (a) $2\left(\cos\dfrac{\pi}{3} + i\sin\dfrac{\pi}{3}\right)$

$2\left(\cos\dfrac{5\pi}{6} + i\sin\dfrac{5\pi}{6}\right)$

$2\left(\cos\dfrac{4\pi}{3} + i\sin\dfrac{4\pi}{3}\right)$

$2\left(\cos\dfrac{11\pi}{6} + i\sin\dfrac{11\pi}{6}\right)$

(b)

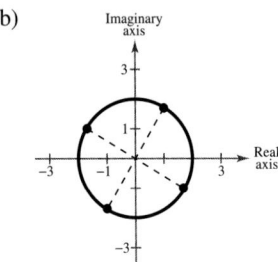

(c) $1 + \sqrt{3}\,i,\ -\sqrt{3} + i,\ -1 - \sqrt{3}\,i,\ \sqrt{3} - i$

95. (a) $5\left(\cos\dfrac{3\pi}{4} + i\sin\dfrac{3\pi}{4}\right)$

$5\left(\cos\dfrac{7\pi}{4} + i\sin\dfrac{7\pi}{4}\right)$

(b)

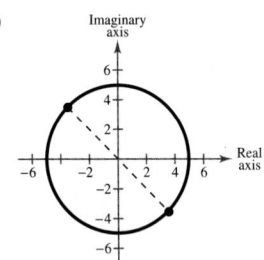

(c) $-\dfrac{5\sqrt{2}}{2} + \dfrac{5\sqrt{2}}{2}\,i,\ \dfrac{5\sqrt{2}}{2} - \dfrac{5\sqrt{2}}{2}\,i$

97. (a) $5\left(\cos\dfrac{4\pi}{9} + i\sin\dfrac{4\pi}{9}\right)$

$5\left(\cos\dfrac{10\pi}{9} + i\sin\dfrac{10\pi}{9}\right)$

$5\left(\cos\dfrac{16\pi}{9} + i\sin\dfrac{16\pi}{9}\right)$

(b)

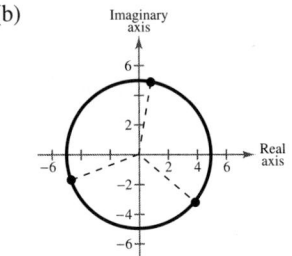

(c) $0.8682 + 4.924i,\ -4.698 - 1.710i,\ 3.830 - 3.214i$

99. (a) $2(\cos 0 + i\sin 0)$

$2\left(\cos\dfrac{2\pi}{3} + i\sin\dfrac{2\pi}{3}\right)$

$2\left(\cos\dfrac{4\pi}{3} + i\sin\dfrac{4\pi}{3}\right)$

(b)

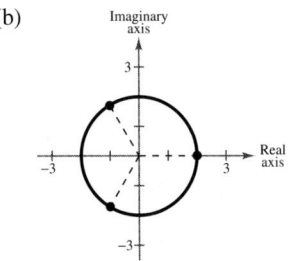

(c) $2,\ -1 + \sqrt{3}\,i,\ -1 - \sqrt{3}\,i$

101. (a) $\cos 0 + i\sin 0$

$\cos\dfrac{2\pi}{5} + i\sin\dfrac{2\pi}{5}$

$\cos\dfrac{4\pi}{5} + i\sin\dfrac{4\pi}{5}$

$\cos\dfrac{6\pi}{5} + i\sin\dfrac{6\pi}{5}$

$\cos\dfrac{8\pi}{5} + i\sin\dfrac{8\pi}{5}$

(b)

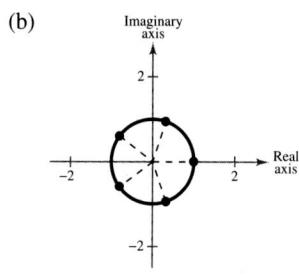

(c) 1, 0.3090 + 0.9511i, −0.8090 + 0.5878i,
−0.8090 − 0.5878i, 0.3090 − 0.9511i

103. (a) 5(cos 60° + i sin 60°)
5(cos 180° + i sin 180°)
5(cos 300° + i sin 300°)

(b)

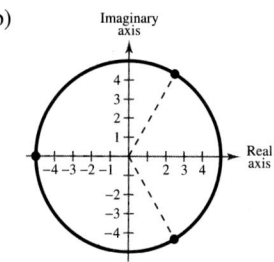

(c) −5, $\dfrac{5}{2} \pm \dfrac{5\sqrt{3}}{2}i$

105. (a) $2\sqrt[5]{4\sqrt{2}}$ (cos 27° + i sin 27°)
$2\sqrt[5]{4\sqrt{2}}$ (cos 99° + i sin 99°)
$2\sqrt[5]{4\sqrt{2}}$ (cos 171° + i sin 171°)
$2\sqrt[5]{4\sqrt{2}}$ (cos 243° + i sin 243°)
$2\sqrt[5]{4\sqrt{2}}$ (cos 315° + i sin 315°)

(b)

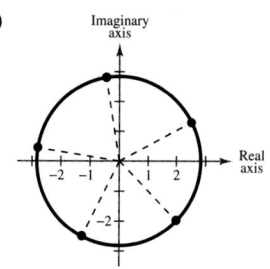

(c) 2.52 + 1.28i, −0.44 + 2.79i, −2.79 + 0.44i,
−1.28 − 2.52i, 2 − 2i

107. $\cos \dfrac{\pi}{8} + i \sin \dfrac{\pi}{8}$

$\cos \dfrac{5\pi}{8} + i \sin \dfrac{5\pi}{8}$

$\cos \dfrac{9\pi}{8} + i \sin \dfrac{9\pi}{8}$

$\cos \dfrac{13\pi}{8} + i \sin \dfrac{13\pi}{8}$

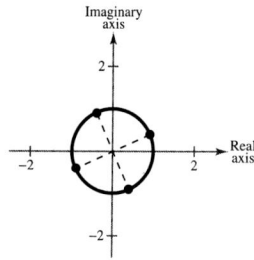

109. $3\left(\cos \dfrac{\pi}{5} + i \sin \dfrac{\pi}{5}\right)$

$3\left(\cos \dfrac{3\pi}{5} + i \sin \dfrac{3\pi}{5}\right)$

$3(\cos \pi + i \sin \pi)$

$3\left(\cos \dfrac{7\pi}{5} + i \sin \dfrac{7\pi}{5}\right)$

$3\left(\cos \dfrac{9\pi}{5} + i \sin \dfrac{9\pi}{5}\right)$

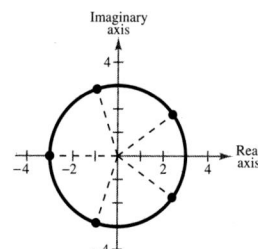

111. $4\left(\cos\dfrac{\pi}{2} + i\sin\dfrac{\pi}{2}\right)$

$4\left(\cos\dfrac{7\pi}{6} + i\sin\dfrac{7\pi}{6}\right)$

$4\left(\cos\dfrac{11\pi}{6} + i\sin\dfrac{11\pi}{6}\right)$

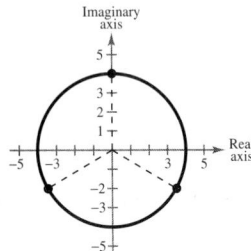

113. $\sqrt[6]{2}\,(\cos 105° + i\sin 105°)$

$\sqrt[6]{2}\,(\cos 225° + i\sin 225°)$

$\sqrt[6]{2}\,(\cos 345° + i\sin 345°)$

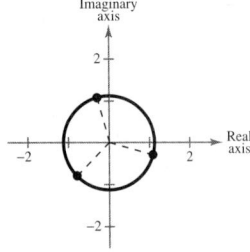

115. 8.48

FOCUS ON CONCEPTS   *(page 512)*

1. $\dfrac{a}{\sin A} = \dfrac{b}{\sin B} = \dfrac{c}{\sin C}$

2. $a^2 = b^2 + c^2 - 2bc\cos A$, $b^2 = a^2 + c^2 - 2ac\cos B$,

$c^2 = a^2 + b^2 - 2ab\cos C$

3. True   **4.** Pythagorean Theorem

5. False. There may be no solution, one solution, or two solutions.

6. Direction and magnitude   **7.** A, C

8. (a) The angle between the vectors is acute.

9. If $k > 0$, the direction is the same and the magnitude is k times as great.

If $k < 0$, the result is a vector in the opposite direction and the magnitude is k times as great.

10. The diagonal of the parallelogram with **u** and **v** as its adjacent sides

11. (b) Visualize the sum of **u** and $-\mathbf{v}$.

12. $z_1 z_2 = -4$

$\dfrac{z_1}{z_2} = -\dfrac{1}{4}z_1{}^2$

13. (a) 3

(b) The modulus of each is 2, and the arguments are 120°, 210°, and 300°.

Review Exercises   *(page 513)*

1. $A \approx 29.7°$, $B \approx 52.4°$, $C \approx 97.9°$

3. $C = 110°$, $b \approx 20.4$, $c \approx 22.6$

5. $A = 35°$, $C = 35°$, $b \approx 6.6$

7. No solution   **9.** $A \approx 25.9°$, $C \approx 39.1°$, $c \approx 10.1$

11. $B \approx 31.2°$, $C \approx 133.8°$, $c \approx 13.9$

$B \approx 148.8°$, $C \approx 16.2°$, $c \approx 5.39$

13. $A \approx 9.9°$, $C \approx 20.1°$, $b \approx 29.1$

15. $A \approx 40.9°$, $C \approx 114.1°$, $c \approx 8.6$

$A \approx 139.1°$, $C \approx 15.9°$, $c \approx 2.6$

17. 9.798   **19.** 9.08   **21.** 31.1 meters

23. 31.0 feet   **25.** 1135 miles   **27.** 14.7 meters

29. $\langle 7, -5\rangle$ **31.** $\langle 7, -7\rangle$ **33.** $\langle -4, 4\sqrt{3}\rangle$

35. $10\sqrt{2}(\mathbf{i}\sin 135° + \mathbf{j}\cos 135°)$

37. $\left\langle \dfrac{6}{\sqrt{61}}, -\dfrac{5}{\sqrt{61}}\right\rangle$

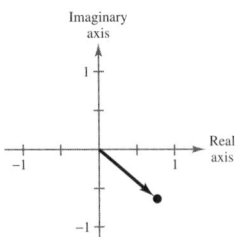

39. $\langle -26, -35 \rangle$

41.

Magnitude: 32.62

Direction: 44.72°

43. 92.2 pounds, 79.9°   **45.** 180 pounds

47. 104 pounds   **49.** 740.5 kilometers per hour, N 32.1° E

51. $\dfrac{1}{\sqrt{34}} \langle -5, 3 \rangle$ **53.** Parallel   **55.** $\dfrac{11\pi}{12}$

57. 160.5°

59.

90°

61.

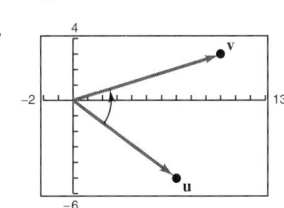

52.2°

63. $-\frac{13}{17}\langle 4, 1 \rangle$ **65.** $-\frac{5}{2}\langle 1, -1 \rangle$

67. $3(\cos 180° + i \sin 180°)$

69. $\sqrt{29}(\cos 338.2° + i \sin 338.2°)$

71. $5\sqrt{2}(\cos 315° + i \sin 315°)$

73. $13(\cos 67.38° + i \sin 67.38°)$

75. $-50 - 50\sqrt{3}i$ **77.** 13

79. (a) $z_1 = 4(\cos 330° + i \sin 330°)$

$z_2 = 10(\cos 270° + i \sin 270°)$

(b) $z_1 z_2 = 40(\cos 240° + i \sin 240°)$

$\dfrac{z_1}{z_2} = \dfrac{2}{5}(\cos 60° + i \sin 60°)$

81. $\dfrac{625}{2} + \dfrac{625\sqrt{3}}{2}i$ **83.** $2035 - 828i$

85. (a) $4(\cos 60° + i \sin 60°)$

$4(\cos 180° + i \sin 180°)$

$4(\cos 300° + i \sin 300°)$

(b) -64

87. (a) 2

$2(\cos 72° + i \sin 72°)$

$2(\cos 144° + i \sin 144°)$

$2(\cos 216° + i \sin 216°)$

$2(\cos 288° + i \sin 288°)$

(b) 32

89. $3\left(\cos \dfrac{\pi}{4} + i \sin \dfrac{\pi}{4}\right)$

$3\left(\cos \dfrac{7\pi}{12} + i \sin \dfrac{7\pi}{12}\right)$

$3\left(\cos \dfrac{11\pi}{12} + i \sin \dfrac{11\pi}{12}\right)$

$3\left(\cos \dfrac{5\pi}{4} + i \sin \dfrac{5\pi}{4}\right)$

$3\left(\cos \dfrac{19\pi}{12} + i \sin \dfrac{19\pi}{12}\right)$

$3\left(\cos \dfrac{23\pi}{12} + i \sin \dfrac{23\pi}{12}\right)$

91. $3\left(\cos\dfrac{\pi}{4} + i\sin\dfrac{\pi}{4}\right) = \dfrac{3\sqrt{2}}{2} + \dfrac{3\sqrt{2}}{2}i$

$3\left(\cos\dfrac{3\pi}{4} + i\sin\dfrac{3\pi}{4}\right) = -\dfrac{3\sqrt{2}}{2} + \dfrac{3\sqrt{2}}{2}i$

$3\left(\cos\dfrac{5\pi}{4} + i\sin\dfrac{5\pi}{4}\right) = -\dfrac{3\sqrt{2}}{2} - \dfrac{3\sqrt{2}}{2}i$

$3\left(\cos\dfrac{7\pi}{4} + i\sin\dfrac{7\pi}{4}\right) = \dfrac{3\sqrt{2}}{2} - \dfrac{3\sqrt{2}}{2}i$

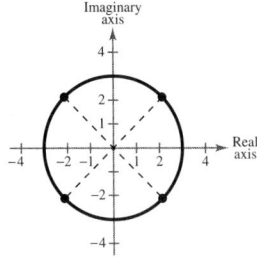

93. $2\left(\cos\dfrac{\pi}{2} + i\sin\dfrac{\pi}{2}\right) = 2i$

$2\left(\cos\dfrac{7\pi}{6} + i\sin\dfrac{7\pi}{6}\right) = -\sqrt{3} - i$

$2\left(\cos\dfrac{11\pi}{6} + i\sin\dfrac{11\pi}{6}\right) = \sqrt{3} - i$

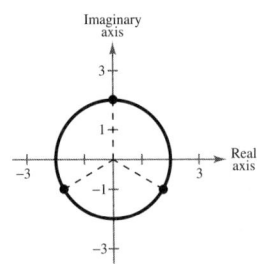

Cumulative Test for Chapters 4–6 *(page 518)*

1. (a)

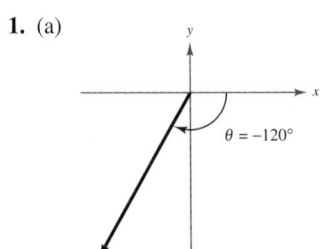

(b) $240°$

(c) $-\dfrac{2\pi}{3}$

(d) $60°$

(e) $\sin(-120°) = -\dfrac{\sqrt{3}}{2}$

$\cos(-120°) = -\dfrac{1}{2}$

$\tan(-120°) = \sqrt{3}$

$\csc(-120°) = -\dfrac{2\sqrt{3}}{3}$

$\sec(-120°) = -2$

$\cot(-120°) = \dfrac{\sqrt{3}}{3}$

2. $134.6°$ **3.** $\dfrac{3}{5}$

4. (a)

(b)

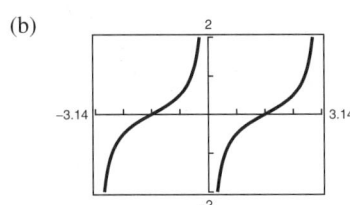

5. $a = -3, b = \pi, c = 0$ **6.** $\sqrt{1 - 4x^2}$

7. $2\tan\theta$ **8.** (a) and (b) Answers will vary.

9. (a) $\dfrac{\pi}{3}, \dfrac{\pi}{2}, \dfrac{3\pi}{2}, \dfrac{5\pi}{3}$ (b) $\dfrac{\pi}{6}, \dfrac{5\pi}{6}, \dfrac{7\pi}{6}, \dfrac{11\pi}{6}$

10. (a) $B \approx 26.4°,\ C \approx 123.6°,\ c \approx 15.0$

(b) $a \approx 5.0,\ B \approx 52.5°,\ C \approx 97.5°$

11. $2\sqrt{2}(\cos 135° + i\sin 135°)$

12. $-12\sqrt{3} + 12i$

13. 1

$\cos 120° + i\sin 120° = -\dfrac{1}{2} + \dfrac{\sqrt{3}}{2}i$

$\cos 240° + i\sin 240° = -\dfrac{1}{2} - \dfrac{\sqrt{3}}{2}i$

14. 5 feet   **15.** N 32.6° E, 543.9 kilometers per hour

CHAPTER 7

Section 7.1 *(page 527)*

1. $(2, 2)$ **3.** $(2, 6), (-1, 3)$ **5.** $(3, -4), (0, -5)$

7. $(0, 0), (2, -4)$ **9.** $(-2, 1), (16, 4)$ **11.** $(5, 5)$

13. $\left(\frac{1}{2}, 3\right)$ **15.** $(1, 1)$ **17.** $\left(\frac{20}{3}, \frac{40}{3}\right)$ **19.** No solution

21. $(0, 0)$ **23.** $(4, 3)$ **25.** $\left(\frac{5}{2}, \frac{3}{2}\right)$ **27.** $(2, 2), (4, 0)$

29.

$(4, -0.5)$

31.

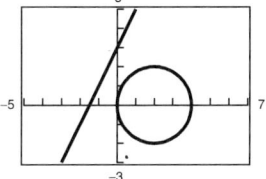

No points of intersection

33.

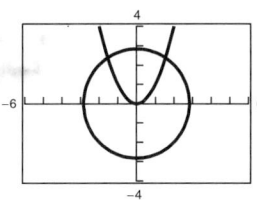

$(\pm 1.540, \ 2.372)$

35.

$(0, 1)$

37.

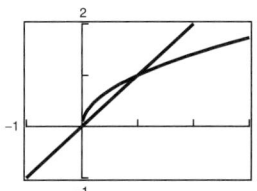

$(0, 0), (1, 1)$

39.

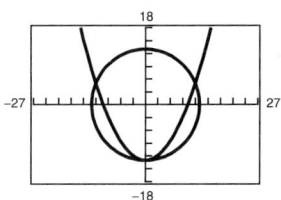

$(0, -13), (\pm 12, 5)$

41. $(1, 2)$ **43.** $(-2, 0), \left(\frac{29}{10}, \frac{21}{10}\right)$ **45.** No solution

47. $(0.287, 1.75)$ **49.** $(-1, 0), (0, 1), (1, 0)$

51. $\left(\frac{1}{2}, 2\right), \left(-4, -\frac{1}{4}\right)$

53. For a linear system the result will be a contradictory equation such as $0 = N$, where N is a nonzero real number. For nonlinear systems there may be an equation with imaginary roots.

55.

192 units

57.

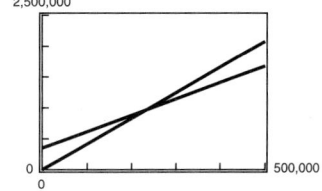

233,333 units

59. (a) $C = 3.45x + 16,000$
  $R = 5.95x$

(b)

6400 units

61. $\$5000$ **63.** $\$8333.33$

65.

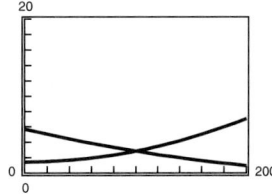

$x = 100$

67. 6×9 meters **69.** 9×12 inches **71.** 8×12 miles

73. (a) $y = 2x$ (b) $y = 0$ (c) $y = x - 2$

75. (a) $f(t) = 0.18t + 6.63$
  $g(t) = -0.05t^2 + 0.33t + 6.58$

(b)

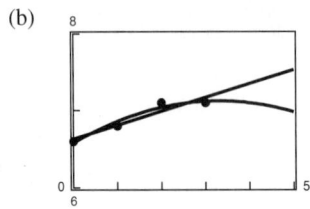

(c) (0.382, 6.70), (2.618, 7.10)

(d) $f(4) = 7.35$

$g(4) = 7.1$

$g(x)$ gives a more accurate prediction.

77. $2x + 7y - 45 = 0$ **79.** $y = 3$

81. $30x - 17y - 18 = 0$

Section 7.2   *(page 539)*

1. $(2, 1)$ **3.** $(1, -1)$ **5.** Inconsistent

7. Infinitely many solutions   **9.** $\left(\frac{1}{3}, -\frac{2}{3}\right)$ **11.** $\left(\frac{5}{2}, \frac{3}{4}\right)$

13. $(3, 4)$ **15.** $(4, -1)$ **17.** $(40, 40)$

19. $\left(\frac{1}{2}, 0\right)$ **21.** $\left(\frac{18}{5}, \frac{3}{5}\right)$ **23.** $(5, -2)$

25. Infinitely many solutions   **27.** $\left(\frac{90}{31}, -\frac{67}{31}\right)$

29. $\left(-\frac{6}{35}, \frac{43}{35}\right)$

31.

Inconsistent

33.

Consistent, one solution

35.

(3, 2)

37.

(3, -2)

39. $(4, 1)$ **41.** $(2, -1)$ **43.** $2x + 2y = 11$

$x - 4y = -7$

45. (39,600, 398). It is necessary to change the scale on the axes to see the point of intersection.

47. No. Two lines will intersect only once or coincide and the system will have infinitely many solutions.

49. $k = -4$ **51.** 550 miles per hour, 50 miles per hour

53. (a) $x + \quad y = 10$

$0.2x + 0.5y = \quad 3$

(b)

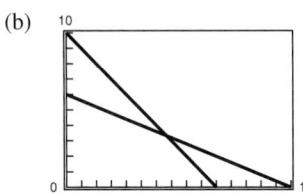

Decreases.

(c) 20% solution: $6\frac{2}{3}$ liters

50% solution: $3\frac{1}{3}$ liters

55. $\$6000$ **57.** 375 adults, 125 children

59. 225 kilometers and 75 kilometers

61. $y = 0.97x + 2.10$ **63.** $y = 0.318x + 4.061$

65. $y = \frac{3}{4}x + \frac{4}{3}$ **67.** $y = -2x + 4$

69. $y = 14x + 19$, 41.4   **71.** $(80, 10)$

73. $(2,000,000, 100)$ **75.** $u = 1$, $v = -\tan x$

77. $u = -x \ln x$, $v = \ln x$ **79.** All real numbers

81. $-5 \le x \le 5$

Section 7.3   *(page 550)*

1. $(1, -2, 4)$ **3.** $(1, 2, -2)$ **5.** $\left(\frac{1}{2}, -2, 2\right)$

7. $x - 2y + 3z = 5$

$y - 2z = 9$

$2x \qquad - 3z = 0$

First step in putting the system in row-echelon form

9. $(1, 2, 3)$ **11.** $(-4, 8, 5)$ **13.** $(5, -2, 0)$

15. Inconsistent   **17.** $\left(1, -\frac{3}{2}, \frac{1}{2}\right)$

19. $(-3a + 10, 5a - 7, a)$ **21.** $(-a + 3, a + 1, a)$

23. $(2a, 21a - 1, 8a)$ **25.** $\left(\frac{1}{2} - \frac{3}{2}a, 1 - \frac{2}{3}a, a\right)$

27. $(1, 1, 1, 1)$ **29.** Inconsistent   **31.** $(0, 0, 0)$

33. $(9a, -35a, 67a)$

35. No. There are two arithmetic errors. They are the constant in the second equation and the coefficient of z in the third equation.

37. $3x + \quad y - \quad z = 9$

$x + 2y - \quad z = 0$

$-x + \quad y + 3z = 1$

39. $y = \frac{1}{2}x^2 - 2x$

41. $y = x^2 - 6x + 8$

43. $x^2 + y^2 - 4x = 0$

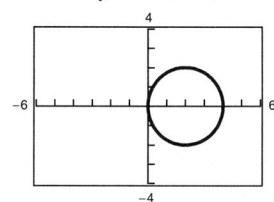

45. $x^2 + y^2 + 6x - 8y = 0$

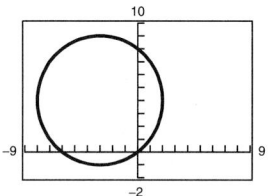

47. $s = -16t^2 + 144$ **49.** $s = 24t^2 - 264t + 692$

51. $4000 at 5%   **53.** $366,666.67 at 8%

$5000 at 6% $316,666.67 at 9%

$7000 at 7% $91,666.67 at 10%

55. $250,000 - \frac{1}{2}s$ in certificates of deposit

$125,000 + \frac{1}{2}s$ in municipal bonds

$125,000 - s$ in blue-chip stocks

s in growth stocks

57. 20 liters of spray X

18 liters of spray Y

16 liters of spray Z

59. Use four medium trucks or two large trucks.
One medium truck and two small trucks

61. $t_1 = 96$ pounds

$t_2 = 48$ pounds

$a = -16$ feet per second squared

63.

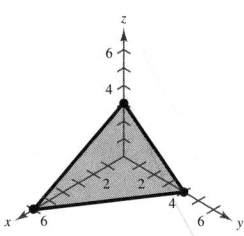

$(6, 0, 0)$, $(0, 4, 0)$, $(0, 0, 3)$, $(4, 0, 1)$

65.

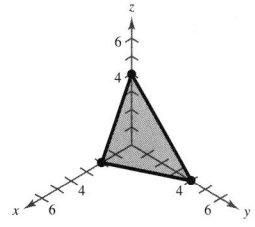

$(2, 0, 0)$, $(0, 4, 0)$, $(0, 0, 4)$, $(0, 2, 2)$

67. $y = -\frac{5}{24}x^2 - \frac{3}{10}x + \frac{41}{6}$ **69.** $y = x^2 - x$

71. (a) $y = 0.14x^2 - 4.43x + 58.40$

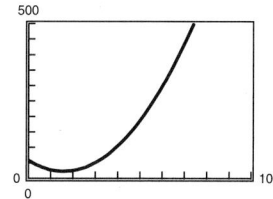

(b) 434.3

73. $x = 5$ **75.** $x = \sqrt{2}/2$ $x = -\sqrt{2}/2$ $x = 0$

$y = 5$ $y = \frac{1}{2}$ $y = \frac{1}{2}$ $y = 0$

$\lambda = -5$ $\lambda = 1$ $\lambda = 1$ $\lambda = 0$

77. The average increase in sales per year

79. 6.375   **81.** 80,000

Section 7.4   *(page 560)*

1. $\dfrac{A}{x} + \dfrac{B}{x - 14}$ **3.** $\dfrac{A}{x} + \dfrac{B}{x^2} + \dfrac{C}{x - 10}$

5. $\dfrac{A}{x} + \dfrac{Bx + C}{x^2 + 10}$ **7.** $\dfrac{1}{2}\left(\dfrac{1}{x - 1} - \dfrac{1}{x + 1}\right)$

9. $\dfrac{1}{x} - \dfrac{1}{x + 1}$ **11.** $\dfrac{1}{x} - \dfrac{2}{2x + 1}$ **13.** $\dfrac{1}{x - 1} - \dfrac{1}{x + 2}$

15. $-\dfrac{3}{x} - \dfrac{1}{x + 2} + \dfrac{5}{x - 2}$ **17.** $\dfrac{3}{x} - \dfrac{1}{x^2} + \dfrac{1}{x + 1}$

19. $\dfrac{3}{x - 3} + \dfrac{9}{(x - 3)^2}$ **21.** $-\dfrac{1}{x} + \dfrac{2x}{x^2 + 1}$

23. $\dfrac{1}{3(x^2 + 2)} - \dfrac{1}{6(x + 2)} + \dfrac{1}{6(x - 2)}$

25. $\dfrac{1}{8(2x + 1)} + \dfrac{1}{8(2x - 1)} - \dfrac{x}{2(4x^2 + 1)}$

27. $\dfrac{1}{x+1} + \dfrac{2}{x^2 - 2x + 3}$

29. $x + 3 + \dfrac{6}{x-1} + \dfrac{4}{(x-1)^2} + \dfrac{1}{(x-1)^3}$

31. $\dfrac{3}{2x-1} - \dfrac{2}{x+1}$ **33.** $\dfrac{2}{x} - \dfrac{1}{x^2} - \dfrac{2}{x+1}$

35. $\dfrac{1}{x^2+2} + \dfrac{x}{(x^2+2)^2}$ **37.** $2x + \dfrac{1}{2}\left(\dfrac{3}{x-4} - \dfrac{1}{x+2}\right)$

39. $\dfrac{1}{2a}\left(\dfrac{1}{a+x} + \dfrac{1}{a-x}\right)$ **41.** $\dfrac{1}{a}\left(\dfrac{1}{y} + \dfrac{1}{a-y}\right)$

43. $\dfrac{3}{x} - \dfrac{2}{x-4}$

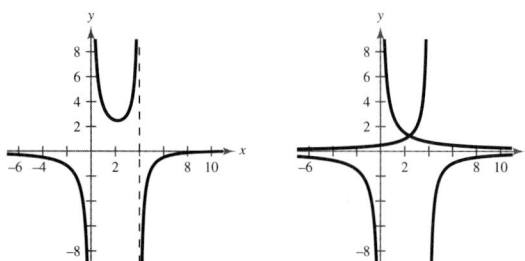

Vertical asymptotes are the same.

45. $\dfrac{3}{x-3} + \dfrac{5}{x+3}$

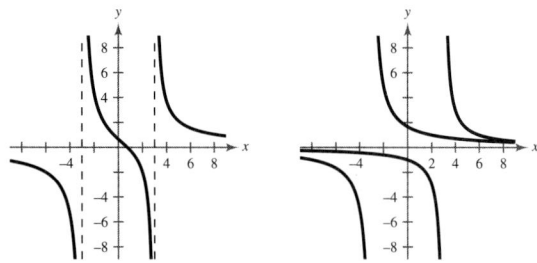

Vertical asymptotes are the same.

47. (a) $\dfrac{2000}{7-4x} - \dfrac{2000}{11-7x}, \quad 0 \le x \le 1$

(b)
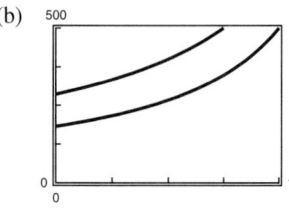

Section 7.5   *(page 569)*

1. (g) **3.** (a) **5.** (e) **7.** (f)

9.

11.

13. **15.**

17.

19. **21.**

23.
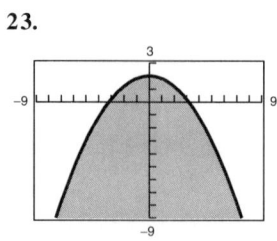

25. $y \le \frac{1}{2}x + 2$ **27.** $\frac{x}{3} + \frac{y}{2} \ge 1$ **29.** $y < 4 - |x|$

31.   **33.**

35.   **37.**

39.   **41.**

43.   **45.**

47.   **49.**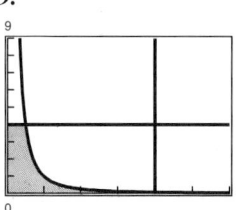

51. $\frac{1}{4}x + \frac{1}{4}y \le 1$ **53.** $y \ge 4 - x$
   $x \ge 0$ $y \ge 2 - \frac{1}{4}x$
   $y \ge 0$ $x \ge 0, \quad y \ge 0$

55. $x^2 + y^2 \le 16$ **57.** $2 \le x \le 5$ **59.** $y \le \frac{3}{2}x$
   $x \ge 0, \quad y \ge 0$ $1 \le y \le 7$ $y \le -x + 5$
                                      $y \ge 0$

61. (a) $x + y \le 20{,}000$
   $y \ge \quad 2x$
   $x \quad \ge \quad 5000$
   $y \ge \quad 5000$

(b)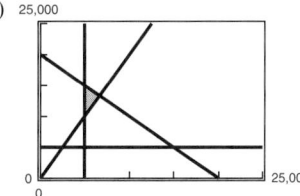

63. (a) $x + \frac{3}{2}y \le 12$
   $\frac{4}{3}x + \frac{3}{2}y \le 15$
   $x \quad \ge \quad 0$
   $y \ge \quad 0$

(b)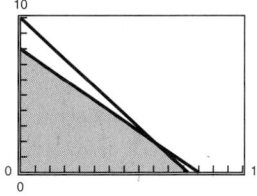

65. (a) $20x + 10y \geq 280$

$\qquad 15x + 10y \geq 160$

$\qquad 10x + 20y \geq 180$

$\qquad x \qquad\qquad \geq \quad 0$

$\qquad\qquad y \geq \quad 0$

(b)

67. (a) $\qquad xy \geq 500$

$\qquad 2x + \pi y \geq 125$

$\qquad x \qquad\qquad \geq \quad 0$

$\qquad\qquad y \geq \quad 0$

(b)

69.

Consumer surplus: 1600
Producer surplus: 400

71.

Consumer surplus: 5000
Producer surplus: 5000

73. Test a point on either side.

Section 7.6   *(page 579)*

1. Minimum at $(0, 0)$: 0

Maximum at $(0, 6)$: 30

3. Minimum at $(0, 0)$: 0

Maximum at $(6, 0)$: 60

5. Minimum at $(0, 0)$: 0

Maximum at $(3, 4)$: 17

7. Minimum at $(0, 0)$: 0

Maximum at $(4, 0)$: 20

9. Minimum at $(0, 0)$: 0

Maximum at $(60, 20)$: 740

11. Minimum at $(0, 0)$: 0

Maximum at any point on the line segment connecting $(60, 20)$ and $(30, 45)$: 2100

13.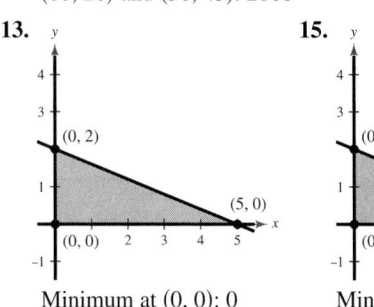

Minimum at $(0, 0)$: 0

Maximum at $(5, 0)$: 30

15.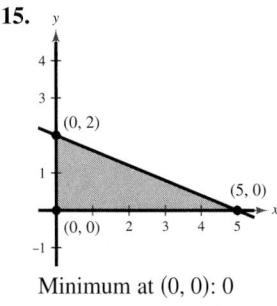

Minimum at $(0, 0)$: 0

Maximum at $(0, 2)$: 48

17.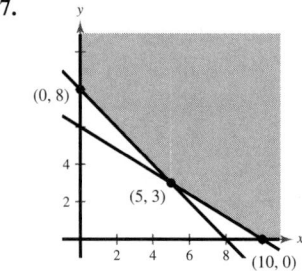

Minimum at $(5, 3)$: 35

No maximum

19.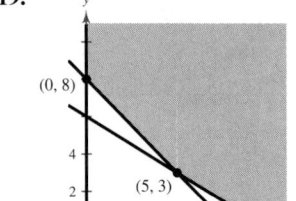

Minimum at $(10, 0)$: 20

No maximum

21.

23.

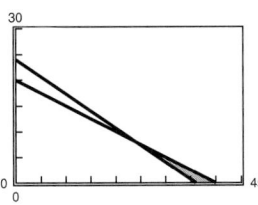

Minimum at $(24, 8)$: 104

Maximum at $(40, 0)$: 160

Minimum at $(36, 0)$: 36

Maximum at $(24, 8)$: 56

25.

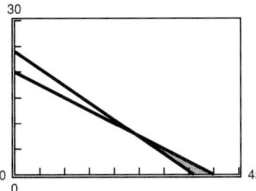

Minimum at any point on the line segment connecting $(24, 8)$ and $(36, 0)$: 72

Maximum at $(40, 0)$: 80

27. (a) and (b)

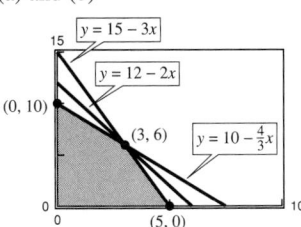

(c) $(3, 6)$

29. (a) and (b)

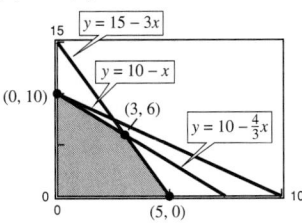

(c) $(0, 10)$

31. (a) and (b)

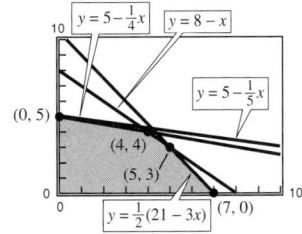

(c) $(0, 5)$

33. (a) and (b)

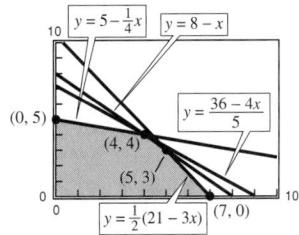

(c) $(4, 4)$

35. $z = x + 5y$ **37.** $z = 4x + y$

39. 200 units of the $250 model

50 units of the $400 model

Maximum profit: $11,500

41. A: $\frac{1}{6}$ gallon

B: $\frac{5}{6}$ gallon

Minimum cost: $1.255 per gallon

43. No audits

40 tax returns

Maximum revenue: $12,000

45. 1000 units of Model A

500 units of Model B

Maximum profit: $76,000

47.

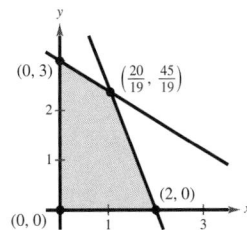

z is maximum at any point on the line segment connecting $(2, 0)$ and $\left(\frac{20}{19}, \frac{45}{19}\right)$.

49.

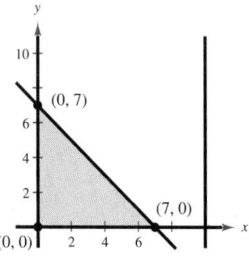

The constraint $x \leq 10$ is extraneous. Maximum 14 at (0, 7)

51.

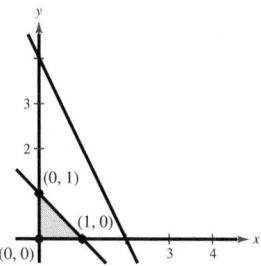

The constraint $2x + y \leq 4$ is extraneous.

Maximum at (0, 1): 4

53. (a) $t > 9$ (b) $\frac{3}{4} < t < 9$ **55.** $\dfrac{9}{2(x + 3)}$

57. $\dfrac{x^2 + 2x - 13}{x(x - 2)}$

FOCUS ON CONCEPTS   *(page 583)*

1. A solution of a system is an ordered pair that satisfies each equation in the system.

2. For a linear system the result will be a contradictory equation such as $0 = N$, where N is a nonzero real number. For nonlinear systems there may be an equation with imaginary roots.

3. There will be a contradictory equation of the form $0 = N$, where N is a nonzero real number.

4. The algebraic methods yield exact solutions.

5. (a) One   (b) Two   (c) Four

6. The system has no solution.

7. The lines are distinct and parallel.

$$x + 2y = 3$$
$$2x + 4y = 9$$

8. (a) Interchange any two equations.

(b) Multiply an equation by a nonzero constant.

(c) Add a multiple of one equation to any other equation in the system.

9. No. When -2 times Equation 1 is added to Equation 2, the constant is -11.

10. The first system is inconsistent because 4 times Equation 1 added to Equation 2 yields $0 = 20$.

11. (d)   **12.** (b)   **13.** (c)   **14.** (a)

15. (a) The boundary would be included in the solution.

(b) The solution would be the half-plane on the opposite side of the boundary.

Review Exercises   *(page 584)*

1. (1, 1)   **3.** (5, 4)   **5.** (0, 0), (2, 8), (−2, 8)

7. (0, 0), (−3, 3)   **9.** (4, 4)   **11.** $\left(\frac{5}{2}, 3\right)$

13. (−0.5, 0.8)   **15.** (0, 0)   **17.** $\left(\frac{14}{5} + \frac{8}{5}a, a\right)$

19. $3x + y = 7$ **21.** 4762 units
$-6x + 3y = 1$

23. 218.75 miles per hour, 193.75 miles per hour

25. $\left(\dfrac{500{,}000}{7}, \dfrac{159}{7}\right)$ **27.** $\left(\dfrac{38}{17}, \dfrac{40}{17}, -\dfrac{63}{17}\right)$

29. $(3a + 4, 2a + 5, a)$ **31.** $2x + y - 2z = 1$
$x + y - z = 0$
$2x - 3y - 2z = 5$

33. $y = 2x^2 + x - 5$

35. $\$8000$ at 7%
$\$5000$ at 9%
$\$7000$ at 11%

37. $\dfrac{3}{x + 2} - \dfrac{4}{x + 4}$ **39.** $1 - \dfrac{25}{8(x + 5)} + \dfrac{9}{8(x - 3)}$

41. $\dfrac{1}{2}\left(\dfrac{3}{x - 1} - \dfrac{x - 3}{x^2 + 1}\right)$

43.   **45.**

47.

49.

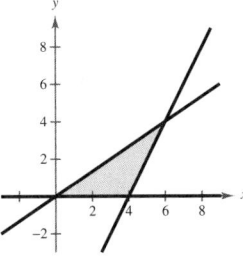

51. $-x + y \leq 4$
$2x + y \leq 22$
$-x + y \geq -2$
$2x + y \geq 7$

53. $x + y \leq 1500$
$x \quad \geq 400$
$y \geq 600$

55.

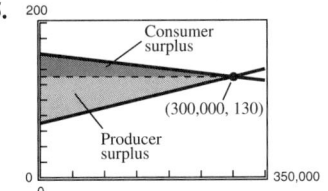

Consumer surplus: 4,500,000
Producer surplus: 9,000,000

57.

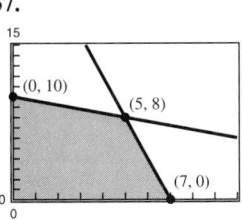

Maximum at $(5, 8)$: 47

59.

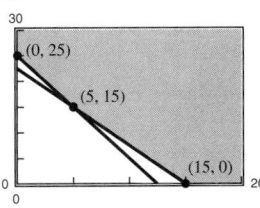

Minimum at $(15, 0)$: 26.25

61. 20 perms

63. Three bags of Brand X
Two bags of Brand Y
Minimum cost per bag: $21

65. $k_1 = -\frac{10}{3}$
$k_2 = \frac{16}{3}$

Chapter Test *(page 588)*

1. $(2, -2)$ **2.** $(0, -1), (1, 0), (2, 1)$

3. $(8, 4), (2, -2)$ **4.** $(3, 2)$ **5.** $(-3, 0), (2, 5)$

6. $(3, 1)$ **7.** $(1, 5)$ **8.** $(2, -3, 1)$

9. $3x - y = 9$ **10.** $y = -\frac{1}{2}x^2 + x + 6$
$6x + y = 3$

11. **12.**

 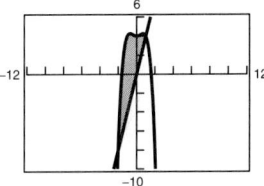

13. $x + 3y \leq 45$ **14.** $(12, 0); z = 240$
$7x + 3y \leq 99$
$x \quad \leq 12$
$x \quad \geq 0$
$y \geq 0$

15. \$275 model: 160 units
\$400 model: 140 units

CHAPTER 8

Section 8.1 *(page 600)*

1. 3×2 **3.** 3×1 **5.** 2×2

7. $\begin{bmatrix} 4 & -3 & \vdots & -5 \\ -1 & 3 & \vdots & 12 \end{bmatrix}$ **9.** $\begin{bmatrix} 1 & 10 & -2 & \vdots & 2 \\ 5 & -3 & 4 & \vdots & 0 \\ 2 & 1 & 0 & \vdots & 6 \end{bmatrix}$

11. $x + 2y = 7$ **13.** $2x \qquad + 5z = -12$
$2x - 3y = 4$ $\qquad y - 2z = 7$
$\qquad 6x + 3y \qquad = 2$

15. Reduced row-echelon form

17. Not in row-echelon form **19.** $\begin{bmatrix} 1 & 4 & 3 \\ 0 & 2 & -1 \end{bmatrix}$

21. $\begin{bmatrix} 1 & 1 & 4 & -1 \\ 0 & 5 & -2 & 6 \\ 0 & 3 & 20 & 4 \end{bmatrix}, \begin{bmatrix} 1 & 1 & 4 & -1 \\ 0 & 1 & -\frac{2}{5} & \frac{6}{5} \\ 0 & 3 & 20 & 4 \end{bmatrix}$

23. (a) $\begin{bmatrix} 1 & 2 & 3 \\ 0 & -5 & -10 \\ 3 & 1 & -1 \end{bmatrix}$ (b) $\begin{bmatrix} 1 & 2 & 3 \\ 0 & -5 & -10 \\ 0 & -5 & -10 \end{bmatrix}$ (c) $\begin{bmatrix} 1 & 2 & 3 \\ 0 & -5 & -10 \\ 0 & 0 & 0 \end{bmatrix}$

(d) $\begin{bmatrix} 1 & 2 & 3 \\ 0 & 1 & 2 \\ 0 & 0 & 0 \end{bmatrix}$ (e) $\begin{bmatrix} 1 & 0 & -1 \\ 0 & 1 & 2 \\ 0 & 0 & 0 \end{bmatrix}$

25. $\begin{bmatrix} 1 & 1 & 0 & 5 \\ 0 & 1 & 2 & 0 \\ 0 & 0 & 1 & -1 \end{bmatrix}$ **27.** $\begin{bmatrix} 1 & -1 & -1 & 1 \\ 0 & 1 & 6 & 3 \\ 0 & 0 & 0 & 0 \end{bmatrix}$

29. $\begin{bmatrix} 1 & 0 & 0 \\ 0 & 1 & 0 \\ 0 & 0 & 1 \end{bmatrix}$ **31.** $\begin{bmatrix} 1 & 2 & 0 & 0 \\ 0 & 0 & 1 & 0 \\ 0 & 0 & 0 & 1 \\ 0 & 0 & 0 & 0 \end{bmatrix}$

33. $\begin{array}{rcl} x - 2y &=& 4 \\ y &=& -3 \end{array}$ **35.** $\begin{array}{rcl} x - y + 2z &=& 4 \\ y - z &=& 2 \\ z &=& -2 \end{array}$

$(-2, -3)$

$(8, 0, -2)$

37. $(7, -5)$ **39.** $(-4, -8, 2)$ **41.** $(3, 2)$

43. $(4, -2)$ **45.** $\left(\frac{1}{2}, -\frac{3}{4}\right)$ **47.** Inconsistent

49. $(4, -3, 2)$ **51.** $(2a + 1, 3a + 2, a)$

53. $(5a + 4, -3a + 2, a)$ **55.** $(0, 0)$

57. $(0, 2 - 4a, a)$ **59.** $(1, 0, 4, -2)$ **61.** $(-2a, a, a)$

63. $\begin{array}{rcl} x + y + 7z &=& -1 \\ x + 2y + 11z &=& 0 \\ 2x + y + 10z &=& -3 \end{array}$ **65.** $800,000 at 8%
$500,000 at 9%
$200,000 at 12%

67. $\dfrac{4x^2}{(x + 1)^2(x - 1)} = \dfrac{1}{x - 1} + \dfrac{3}{x + 1} - \dfrac{2}{(x + 1)^2}$

69. $y = x^2 + 2x + 5$

71. $y = \frac{1}{4}x^3 + x^2 - x - 2$ **73.** $y = -x^4 + 4x^2$

75. (a) $y = 170.5t^2 - 560.5t + 670$

(b) 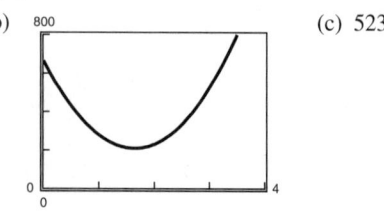   (c) 523

77. (a) $x_1 = s, x_2 = t, x_3 = 600 - s,$
$x_4 = s - t, x_5 = 500 - t, x_6 = s, x_7 = t$
(b) $x_1 = 0, x_2 = 0, x_3 = 600, x_4 = 0, x_5 = 500,$
$x_6 = 0, x_7 = 0$
(c) $x_1 = 0, x_2 = -500, x_3 = 600, x_4 = 500,$
$x_5 = 1000, x_6 = 0, x_7 = -500$

79. (a) $x_1 = 100 + t, x_2 = -100 + t, x_3 = 200 + t, x_4 = t$
(b) $x_1 = 100, x_2 = -100, x_3 = 200, x_4 = 0$
(c) $x_1 = 200, x_2 = 0, x_3 = 300, x_4 = 100$

81. Men: 1.171 minutes
Women: 1.239 minutes

83.   **85.**

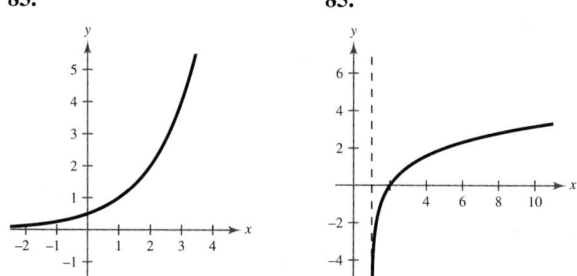

Section 8.2   *(page 615)*

1. $x = -4, y = 22$ **3.** $x = 2, y = 3$

5. (a) $\begin{bmatrix} 3 & -2 \\ 1 & 7 \end{bmatrix}$ (b) $\begin{bmatrix} -1 & 0 \\ 3 & -9 \end{bmatrix}$

(c) $\begin{bmatrix} 3 & -3 \\ 6 & -3 \end{bmatrix}$ (d) $\begin{bmatrix} -1 & -1 \\ 8 & -19 \end{bmatrix}$

7. (a) $\begin{bmatrix} 7 & 3 \\ 1 & 9 \\ -2 & 15 \end{bmatrix}$ (b) $\begin{bmatrix} 5 & -5 \\ 3 & -1 \\ -4 & -5 \end{bmatrix}$

(c) $\begin{bmatrix} 18 & -3 \\ 6 & 12 \\ -9 & 15 \end{bmatrix}$ (d) $\begin{bmatrix} 16 & -11 \\ 8 & 2 \\ -11 & -5 \end{bmatrix}$

9. (a) $\begin{bmatrix} 3 & 3 & -2 & 1 & 1 \\ -2 & 5 & 7 & -6 & -8 \end{bmatrix}$

(b) $\begin{bmatrix} 1 & 1 & 0 & -1 & 1 \\ 4 & -3 & -11 & 6 & 6 \end{bmatrix}$

(c) $\begin{bmatrix} 6 & 6 & -3 & 0 & 3 \\ 3 & 3 & -6 & 0 & -3 \end{bmatrix}$

(d) $\begin{bmatrix} 4 & 4 & -1 & -2 & 3 \\ 9 & -5 & -24 & 12 & 11 \end{bmatrix}$

11. $\begin{bmatrix} -6 & -9 \\ -1 & 0 \\ 17 & -10 \end{bmatrix}$ **13.** $\begin{bmatrix} 3 & 3 \\ -\frac{1}{2} & 0 \\ -\frac{13}{2} & \frac{11}{2} \end{bmatrix}$

15. (a) $\begin{bmatrix} 0 & 15 \\ 6 & 12 \end{bmatrix}$ (b) $\begin{bmatrix} -2 & 2 \\ 31 & 14 \end{bmatrix}$ (c) $\begin{bmatrix} 9 & 6 \\ 12 & 12 \end{bmatrix}$

17. (a) $\begin{bmatrix} 0 & -10 \\ 10 & 0 \end{bmatrix}$ (b) $\begin{bmatrix} 0 & -10 \\ 10 & 0 \end{bmatrix}$ (c) $\begin{bmatrix} 8 & -6 \\ 6 & 8 \end{bmatrix}$

19. (a) $\begin{bmatrix} 6 & -21 & 15 \\ 8 & -23 & 19 \\ 4 & 7 & 5 \end{bmatrix}$ (b) $\begin{bmatrix} 9 & 0 & 13 \\ 7 & -2 & 21 \\ 1 & 4 & -19 \end{bmatrix}$

(c) $\begin{bmatrix} 20 & 7 & -8 \\ 24 & 7 & -2 \\ 2 & -5 & 30 \end{bmatrix}$

21. Not possible **23.** $\begin{bmatrix} -1 & 19 \\ 4 & -27 \\ 0 & 14 \end{bmatrix}$ **25.** $\begin{bmatrix} 1 & 0 & 0 \\ 0 & 1 & 0 \\ 0 & 0 & \frac{7}{2} \end{bmatrix}$

27. $\begin{bmatrix} 0 & 0 & 0 \\ 0 & 0 & 0 \\ 0 & 0 & 0 \end{bmatrix}$ **29.** $\begin{bmatrix} 41 & 7 & 7 \\ 42 & 5 & 25 \\ -10 & -25 & 45 \end{bmatrix}$

31. $\begin{bmatrix} 151 & 25 & 48 \\ 516 & 279 & 387 \\ 47 & -20 & 87 \end{bmatrix}$ **33.** Not possible

35. $A = \begin{bmatrix} -1 & 1 \\ -2 & 1 \end{bmatrix}$ **37.** $A = \begin{bmatrix} 2 & 3 \\ 1 & 4 \end{bmatrix}$

$X = \begin{bmatrix} x \\ y \end{bmatrix}$ $X = \begin{bmatrix} x \\ y \end{bmatrix}$

$B = \begin{bmatrix} 4 \\ 0 \end{bmatrix}$ $B = \begin{bmatrix} 5 \\ 10 \end{bmatrix}$

$x = 4, y = 8$ $x = -2, y = 3$

39. $\begin{bmatrix} -4 & 0 \\ 8 & 2 \end{bmatrix}$ **41.** $\begin{bmatrix} 0 & 0 & 0 \\ 0 & 0 & 0 \\ 0 & 0 & 0 \end{bmatrix}$

43. $AC = BC = \begin{bmatrix} 2 & 3 \\ 2 & 3 \end{bmatrix}$ **45.** Not possible

47. Not possible **49.** 2×2 **51.** Not possible

53. 2×3 **55.** $\begin{bmatrix} 72 & 48 & 24 \\ 36 & 108 & 72 \end{bmatrix}$

57. $BA = [\$1250 \quad \$1331.25 \quad \$981.25]$

The entries represent the profits from the two products at each of the three outlets.

59. $A^2 = \begin{bmatrix} -1 & 0 \\ 0 & -1 \end{bmatrix}$

$A^3 = \begin{bmatrix} -i & 0 \\ 0 & -i \end{bmatrix}$

$A^4 = \begin{bmatrix} 1 & 0 \\ 0 & 1 \end{bmatrix}$

61. $\begin{bmatrix} \$15,770 & \$18,300 \\ \$26,500 & \$29,250 \\ \$21,260 & \$24,150 \end{bmatrix}$

The entries are the wholesale and retail prices of the inventory at each outlet.

63. $\begin{bmatrix} 0.40 & 0.15 & 0.15 \\ 0.28 & 0.53 & 0.17 \\ 0.32 & 0.32 & 0.68 \end{bmatrix}$

65. Diagonal matrix whose entries are the products of the corresponding entries of A and B

67. $\frac{5}{4}$ **69.** $2 + \log_2 3 - \log_2 5$

Section 8.3 *(page 625)*

1.–7. Answers will vary.

9. $\begin{bmatrix} \frac{1}{2} & 0 \\ 0 & \frac{1}{3} \end{bmatrix}$ **11.** $\begin{bmatrix} -3 & 2 \\ -2 & 1 \end{bmatrix}$ **13.** $\begin{bmatrix} 1 & -1 \\ 2 & -1 \end{bmatrix}$

15. Does not exist **17.** Does not exist

19. $\begin{bmatrix} 1 & 1 & -1 \\ -3 & 2 & -1 \\ 3 & -3 & 2 \end{bmatrix}$ **21.** $\begin{bmatrix} 1 & 0 & 0 \\ -0.75 & 0.25 & 0 \\ 0.35 & -0.25 & 0.2 \end{bmatrix}$

23. $\begin{bmatrix} -\frac{1}{8} & 0 & 0 & 0 \\ 0 & 1 & 0 & 0 \\ 0 & 0 & \frac{1}{4} & 0 \\ 0 & 0 & 0 & -\frac{1}{5} \end{bmatrix}$ **25.** $\begin{bmatrix} -175 & 37 & -13 \\ 95 & -20 & 7 \\ 14 & -3 & 1 \end{bmatrix}$

27. $\frac{1}{2}\begin{bmatrix} -3 & 3 & 2 \\ 9 & -7 & -6 \\ -2 & 2 & 2 \end{bmatrix}$ **29.** $\frac{5}{11}\begin{bmatrix} 0 & -4 & 2 \\ -22 & 11 & 11 \\ 22 & -6 & -8 \end{bmatrix}$

31. Does not exist **33.** $\begin{bmatrix} -24 & 7 & 1 & -2 \\ -10 & 3 & 0 & -1 \\ -29 & 7 & 3 & -2 \\ 12 & -3 & -1 & 1 \end{bmatrix}$

35. Answers will vary. **37.** $(5, 0)$ **39.** $(-8, -6)$

41. $(3, 8, -11)$ **43.** $(2, 1, 0, 0)$ **45.** $(2, -2)$

47. No solution **49.** $\left(3, -\frac{1}{2}\right)$ **51.** $(-1, 3, 2)$

53. No solution **55.** $(5, 0, -2, 3)$

57. $10,000 in AAA-rated bonds

$5000 in A-rated bonds

$10,000 in B-rated bonds

59. $9000 in AAA-rated bonds

$1000 in A-rated bonds

$2000 in B-rated bonds

61. Answers will vary.

63. $I_1 = -3$ amperes

$I_2 = 8$ amperes

$I_3 = 5$ amperes

65. $A^{-1} = \begin{bmatrix} \frac{1}{a_{11}} & 0 & 0 & 0 & \cdots & 0 \\ 0 & \frac{1}{a_{22}} & 0 & 0 & \cdots & 0 \\ 0 & 0 & \frac{1}{a_{33}} & 0 & \cdots & 0 \\ \vdots & \vdots & \vdots & \vdots & \cdots & \vdots \\ 0 & 0 & 0 & 0 & \cdots & \frac{1}{a_{nn}} \end{bmatrix}$

Section 8.4   (page 634)

1. 5   **3.** 5   **5.** 27   **7.** -24 **9.** 6   **11.** 0

13. 0   **15.** -9 **17.** -0.002 **19.** 0

21. (a) $M_{11} = -5, M_{12} = 2, M_{21} = 4, M_{22} = 3$

    (b) $C_{11} = -5, C_{12} = -2, C_{21} = -4, C_{22} = 3$

23. (a) $M_{11} = 30, M_{12} = 12, M_{13} = 11, M_{21} = -36,$

       $M_{22} = 26, M_{23} = 7, M_{31} = -4, M_{32} = -42, M_{33} = 12$

    (b) $C_{11} = 30, C_{12} = -12, C_{13} = 11, C_{21} = 36, C_{22} = 26,$

       $C_{23} = -7, C_{31} = -4, C_{32} = 42, C_{33} = 12$

25. -75 **27.** 96   **29.** 170   **31.** -58

33. -30 **35.** -168 **37.** 0   **39.** 412

41. -126 **43.** 0   **45.** -336 **47.** 410

49.–53. Answers will vary.   **55.** $-1, 4$

57. $8uv - 1$ **59.** e^{5x} **61.** $1 - \ln x$

63. (a) -3 (b) -2

    (c) $\begin{bmatrix} -2 & 0 \\ 0 & -3 \end{bmatrix}$ (d) 6

65. (a) 2   (b) -6

    (c) $\begin{bmatrix} 1 & 4 & 3 \\ -1 & 0 & 3 \\ 0 & 2 & 0 \end{bmatrix}$ (d) -12

67. $A = \begin{bmatrix} 1 & 3 \\ -2 & 4 \end{bmatrix}, B = \begin{bmatrix} -4 & 0 \\ 3 & 5 \end{bmatrix}$

    $|A + B| = -30, |A| + |B| = -10$

69. A square matrix is a square array of numbers. A determinant of a square matrix is a real number.

71. (a) Columns 2 and 3 are interchanged.

    (b) Rows 1 and 3 are interchanged.

73. (a) 5 is factored from the first row of the matrix.

    (b) 4 and 3 are factored from the second and third columns.

Section 8.5   (page 646)

1. 14   **3.** 7   **5.** $\frac{33}{8}$ **7.** 10   **9.** 28

11. $\frac{16}{5}$ or 0   **13.** $(2, -2)$ **15.** $(-1, 3, 2)$

17. $\left(0, -\frac{1}{2}, \frac{1}{2}\right)$ **19.** 250 square miles   **21.** Collinear

23. Not collinear   **25.** Collinear   **27.** $3x - 5y = 0$

29. $x + 3y - 5 = 0$ **31.** $2x + 3y - 8 = 0$

33. $x = -3$

35. Uncoded: $[20, 18, 15], [21, 2, 12], [5, 0, 9], [14, 0, 18],$

             $[9, 22, 5], [18, 0, 3], [9, 20, 25]$

    Encoded: $[-52, 10, 27], [-49, 3, 34], [-49, 13, 27],$

            $[-94, 22, 54], [1, 1, -7], [0, -12, 9],$

            $[-121, 41, 55]$

37. 1 -25 -65 17 15 -9 -12 -62 -119 27 51

    48 43 67 48 57 111 117

39. -5 -41 -87 91 207 257 11 -5 -41 40 80

    84 76 177 227

41. HAPPY NEW YEAR   **43.** SEND PLANES

45. MEET ME TONIGHT RON

FOCUS ON CONCEPTS   (page 648)

1. Interchange two rows.

Multiply a row by a nonzero constant.

Add a multiple of a row to another row.

2. They are the same.

3. A matrix in row-echelon form is in reduced row-echelon form if every column that has a leading 1 has zeros in every position above and below its leading 1.

4. Consistent—an infinite number of solutions

5. Inconsistent   **6.** Consistent—a unique solution

7. Consistent—an infinite number of solutions

8. (a) The operation can be performed.

    (b) The operation can be performed.

9. (a) The operation can be performed.

    (b) The operation cannot be performed. The number of rows in B must be the same as the number of columns in A.

10. (a) The operation cannot be performed. The order of A and B must be the same to perform the operation.

(b) The operation can be performed.

11. The matrix must be square and its determinant nonzero.

12. A square matrix is a square array of numbers and a determinant is a real number associated with a square matrix.

13. No. The matrix must be square.

14. If A is a square matrix, the cofactor C_{ij} of the entry a_{ij} is $(-1)^{i+j} M_{ij}$, where M_{ij} is the determinant obtained by deleting the ith row and jth column of A. The determinant of A is the sum of the entries of any row or column of A multiplied by their respective cofactors.

15. No. The first two yield a unique solution to the system and the third yields an infinite number of solutions.

Review Exercises   *(page 649)*

1. $\begin{bmatrix} 3 & -10 & \vdots & 15 \\ 5 & 4 & \vdots & 22 \end{bmatrix}$
3. $\begin{aligned} 5x + y + 7z &= -9 \\ 4x + 2y &= 10 \\ 9x + 4y + 2z &= 3 \end{aligned}$

5. $\begin{bmatrix} 1 & 0 & 0 \\ 0 & 1 & 0 \\ 0 & 0 & 1 \end{bmatrix}$
7. $\begin{bmatrix} 1 & 0 & 3 & -2 \\ 0 & 1 & 4 & -3 \end{bmatrix}$

9. $\begin{bmatrix} 1 & 0 & 1 \\ 0 & 1 & 1 \\ 0 & 0 & 0 \end{bmatrix}$
11. $(10, -12)$ **13.** $(-0.2, 0.7)$

15. $\left(-2a + \frac{3}{2}, 2a + 1, a\right)$ **17.** $\left(\frac{31}{42}, \frac{5}{14}, \frac{13}{84}\right)$

19. $(2, -3, 3)$ **21.** Inconsistent

23. The part of the matrix corresponding to the coefficients of the system reduces to a matrix in which the number of rows with nonzero entries is the same as the number of variables.

25. $(1, 1)$ **27.** $(3, 0, -4)$ **29.** $\begin{bmatrix} -13 & -8 & 18 \\ 0 & 11 & -19 \end{bmatrix}$

31. $\begin{bmatrix} 14 & -2 & 8 \\ 14 & -10 & 40 \\ 36 & -12 & 48 \end{bmatrix}$
33. $\begin{bmatrix} 44 & 4 \\ 20 & 8 \end{bmatrix}$

35. $\begin{bmatrix} 4 & 6 & 3 \\ 0 & 6 & -10 \\ 0 & 0 & 6 \end{bmatrix}$
37. $\begin{bmatrix} 48 & -18 & -3 \\ 15 & 51 & 33 \end{bmatrix}$

39. $\begin{bmatrix} 14 & -22 & 22 \\ 19 & -41 & 80 \\ 42 & -66 & 66 \end{bmatrix}$
41. $\begin{bmatrix} -14 & -4 \\ 7 & -17 \\ -17 & -2 \end{bmatrix}$

43. $\frac{1}{3}\begin{bmatrix} 9 & 2 \\ -4 & 11 \\ 10 & 0 \end{bmatrix}$
45. $\begin{aligned} 5x + 4y &= 2 \\ -x + y &= -22 \end{aligned}$

47. $\begin{bmatrix} \frac{1}{5} & \frac{1}{5} \\ \frac{1}{10} & -\frac{1}{15} \end{bmatrix}$
49. $\begin{bmatrix} \frac{1}{2} & -1 & -\frac{1}{2} \\ \frac{1}{2} & -\frac{2}{3} & -\frac{5}{6} \\ 0 & \frac{2}{3} & \frac{1}{3} \end{bmatrix}$
51. 550

53. 8   **55.** -3 **57.** 279   **59.** $(-3, 1)$

61. $(1, 1, -2)$ **63.** $(2. -4, 6)$ **65.** Inconsistent

67. $(1, 2)$ **69.** $\left(\frac{3}{4}, -\frac{1}{2}\right)$ **71.** $\left(\frac{2}{3}, \frac{1}{2}\right)$

73. Cramer's Rule does not apply.

75. Eight carnations, four roses   **77.** $y = x^2 + 2x + 3$

79. (a) $a \approx 59.9, b = 3.6$; $y = 59.9 + 3.6t$

(b)
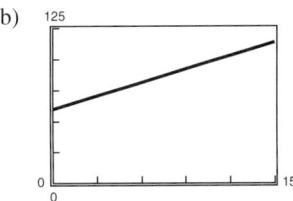

(c) The median price was increasing by an average of \$3600 per year.   (d) \$113,900

81. 16   **83.** 7   **85.** $x - 2y + 4 = 0$

87. $2x + 6y - 13 = 0$ **89.** Answers will vary.

91. Because each of the three rows is multiplied by 4, $|4A| = 4^3|A| = 128$.

Chapter Test   *(page 654)*

1. $\begin{bmatrix} 1 & 0 & 0 \\ 0 & 1 & 0 \\ 0 & 0 & 1 \end{bmatrix}$
2. $\begin{bmatrix} 1 & 0 & -1 & 2 \\ 0 & 1 & 0 & -1 \\ 0 & 0 & 0 & 0 \\ 0 & 0 & 0 & 0 \end{bmatrix}$

3. $\left(1, 3, -\frac{1}{2}\right)$ **4.** $y = -\frac{1}{2}x^2 + x + 2$

5. (a) $\begin{bmatrix} 1 & 5 & -2 \\ 0 & -4 & 3 \end{bmatrix}$ (b) $\begin{bmatrix} 15 & 12 & 12 \\ -12 & -12 & 0 \end{bmatrix}$

(c) $\begin{bmatrix} 7 & 14 & 0 \\ -4 & -12 & 6 \end{bmatrix}$

6. $\begin{bmatrix} 8 & -8 \\ 16 & -4 \\ 6 & 12 \end{bmatrix}$
7. $\begin{bmatrix} \frac{1}{2} & \frac{2}{5} \\ 1 & \frac{3}{5} \end{bmatrix}$
8. $(13, 22)$

9. -2 **10.** 7

CHAPTER 9

Section 9.1   (page 664)

1. 3, 5, 7, 9, 11   **3.** 2, 4, 8, 16, 32

5. $-2, 4, -8, 16, -32$ **7.** $2, \frac{3}{2}, \frac{4}{3}, \frac{5}{4}, \frac{6}{5}$

9. $3, \frac{12}{11}, \frac{9}{13}, \frac{24}{47}, \frac{15}{37}$ **11.** $0, 1, 0, \frac{1}{2}, 0$

13. $\frac{5}{2}, \frac{11}{4}, \frac{23}{8}, \frac{47}{16}, \frac{95}{32}$

15. $1, \dfrac{1}{2^{3/2}}, \dfrac{1}{3^{3/2}}, \dfrac{1}{4^{3/2}}, \dfrac{1}{5^{3/2}}$ **17.** $3, \dfrac{9}{2}, \dfrac{9}{2}, \dfrac{27}{8}, \dfrac{81}{40}$

19. $-1, \frac{1}{4}, -\frac{1}{9}, \frac{1}{16}, -\frac{1}{25}$ **21.** $\frac{2}{3}, \frac{2}{3}, \frac{2}{3}, \frac{2}{3}, \frac{2}{3}$

23. -73 **25.** 28, 24, 20, 16, 12   **27.** 3, 4, 6, 10, 18

29.

31.

33.

35. (c)   **37.** (d)   **39.** $\frac{1}{30}$ **41.** 90   **43.** $n + 1$

45. $\dfrac{1}{2n(2n + 1)}$ **47.** $a_n = 3n - 2$

49. $a_n = n^2 - 1$ **51.** $a_n = \dfrac{n + 1}{n + 2}$

53. $a_n = \dfrac{(-1)^{n+1}}{2^n}$ **55.** $a_n = 1 + \dfrac{1}{n}$ **57.** $a_n = \dfrac{1}{n!}$

59. $a_n = (-1)^{n+1}$

61. 6, 8, 10, 12, 14
$a_n = 2n + 4$

63. 81, 27, 9, 3, 1
$a_n = \dfrac{243}{3^n}$

65. 35   **67.** 40   **69.** 30   **71.** $\frac{9}{5}$ **73.** 238

75. 30   **77.** 81   **79.** $\frac{47}{60}$ **81.** $\displaystyle\sum_{i=1}^{9} \frac{1}{3i}$

83. $\displaystyle\sum_{i=1}^{8} \left[2\left(\frac{i}{8}\right) + 3 \right]$ **85.** $\displaystyle\sum_{i=1}^{6} (-1)^{i+1} 3i$

87. $\displaystyle\sum_{i=1}^{20} \frac{(-1)^{i+1}}{i^2}$ **89.** $\displaystyle\sum_{i=1}^{5} \frac{2^i - 1}{2^{i+1}}$

91. (a) $A_1 = \$5100.00$, $A_2 = \$5202.00$, $A_3 = \$5306.04$,
   $A_4 = \$5412.16$, $A_5 = \$5520.40$, $A_6 = \$5630.81$,
   $A_7 = \$5743.43$, $A_8 = \$5858.30$
(b) $\$11,040.20$

93. $A_1 = 529.73$, $A_2 = 555.79$, $A_3 = 588.31$,
   $A_4 = 627.29$, $A_5 = 672.73$, $A_6 = 724.63$,
   $A_7 = 782.99$, $A_8 = 847.81$, $A_9 = 919.09$,
   $A_{10} = 996.83$, $A_{11} = 1081.03$

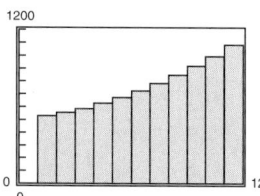

95. $\$11,131.5$ million

97. 1, 1, 2, 3, 5, 8, 13, 21, 34, 55, 89, 144
$1, 2, \frac{3}{2}, \frac{5}{3}, \frac{8}{5}, \frac{13}{8}, \frac{21}{13}, \frac{34}{21}, \frac{55}{34}, \frac{89}{55}$

Section 9.2   (page 673)

1. Arithmetic sequence, $d = -2$

3. Not an arithmetic sequence

5. Arithmetic sequence, $d = -\frac{1}{4}$

7. Arithmetic sequence, $d = 4$

9. Arithmetic sequence, $d = 0.4$

11. 8, 11, 14, 17, 20
Arithmetic sequence, $d = 3$

13. $\frac{1}{2}, \frac{1}{3}, \frac{1}{4}, \frac{1}{5}, \frac{1}{6}$
Not an arithmetic sequence

15. 97, 94, 91, 88, 85
Arithmetic sequence, $d = -3$

17. $1, 4, \frac{7}{3}, \frac{7}{2}, \frac{13}{5}$
Not an arithmetic sequence

19. 15, 19, 23, 27, 31
$a_n = 11 + 4n$

21. 200, 190, 180, 170, 160
$a_n = 210 - 10n$

23. $\frac{3}{2}, \frac{5}{4}, 1, \frac{3}{4}, \frac{1}{2}$
$a_n = \frac{7}{4} - \frac{1}{4}n$

25. 5, 11, 17, 23, 29   **27.** $-2.6, -3.0, -3.4, -3.8, -4.2$

29. 2, 6, 10, 14, 18   **31.** $-2, 2, 6, 10, 14$

33. $a_n = 1 + (n - 1)3$ **35.** $a_n = 100 + (n - 1)(-8)$

37. $a_n = x + (n - 1)(2x)$ **39.** $a_n = 4 + (n - 1)\left(-\frac{5}{2}\right)$

41. $a_n = 5 + (n - 1)\left(\frac{10}{3}\right)$ **43.** $a_n = 100 + (n - 1)(-3)$

45. (b)   **47.** (c)

49.   **51.**

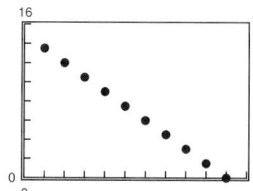

53. Linear   **55.** 620   **57.** 4600   **59.** 265

61. 4000   **63.** 1275   **65.** 25,250   **67.** 355

69. 126,750   **71.** 520   **73.** 44,625   **75.** 10,120

77. 10,000   **79.** (a) $40,000   (b) $217,500

81. 2340   **83.** 405 bricks   **85.** 585 seats

87. 156 times

89. (a) 4, 9, 16, 25, 36   (b) n^2

(c) $\frac{n}{2}[1 + (2n - 1)] = n^2$

91. 4

Section 9.3   *(page 682)*

1. Geometric sequence, $r = 3$

3. Not a geometric sequence

5. Geometric sequence, $r = -\frac{1}{2}$

7. Not a geometric sequence   **9.** Not a geometric sequence

11. 2, 6, 18, 54, 162   **13.** $1, \frac{1}{2}, \frac{1}{4}, \frac{1}{8}, \frac{1}{16}$

15. $5, -\frac{1}{2}, \frac{1}{20}, -\frac{1}{200}, \frac{1}{2000}$ **17.** $1, e, e^2, e^3, e^4$

19. $3, \frac{3x}{2}, \frac{3x^2}{4}, \frac{3x^3}{8}, \frac{3x^4}{16}$

21. 64, 32, 16, 8, 4   **23.** 4, 12, 36, 108, 324
$a_n = 128\left(\frac{1}{2}\right)^n$ $a_n = \frac{4}{3}(3)^n$

25. $6, -9, \frac{27}{2}, -\frac{81}{4}, \frac{243}{8}$
$a_n = 6\left(-\frac{3}{2}\right)^{n-1}$

27. $\left(\frac{1}{2}\right)^7$ **29.** $-\frac{2}{3^{10}}$ **31.** $100e^{8x}$ **33.** $500(1.02)^{39}$

35. 9   **37.** $-\frac{2}{9}$ **39.** (a)   **41.** (b)

43.   **45.**

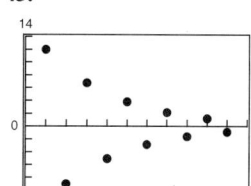

47. Increased powers of real numbers between -1 and 1 approach zero.

49. (a) $2593.74   (b) $2653.30   (c) $2685.06
(d) $2707.04   (e) $2717.91

51. $22,689.45   **53.** 8, 4, 6, 5   **55.** 511

57. 43   **59.** 29,921.31   **61.** 6.4   **63.** 2092.60

65. $\sum_{n=1}^{7} 5(3)^{n-1}$ **67.** $7808.24   **69.** Answers will vary.

71. (a) $26,198.27   (b) $26,263.88

73. (a) $637,678.02   (b) $645,861.43

75. Answers will vary.   **77.** 126 square inches

79. (a) $5,368,709.11   (b) $10,737,418.23
(c) $21,474,836.47

81. 2   **83.** $\frac{2}{3}$ **85.** $\frac{16}{3}$ **87.** 32   **89.** $\frac{4}{11}$ **91.** $\frac{7}{22}$

93.

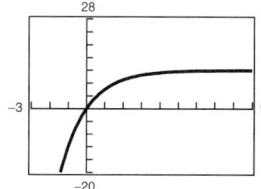

Horizontal asymptote: $y = 12$
Corresponds to the sum of the series

95. (a) 152.42 feet   (b) 19 seconds

97. 45.65 miler per hour   **99.** 2.4 hours

Section 9.4   *(page 695)*

1. $\dfrac{5}{(k + 1)(k + 2)}$ **3.** $\dfrac{(k + 1)^2(k + 2)^2}{4}$

5.–17. Answers will vary.   **19.** 210   **21.** 91

23. 979   **25.** 70   **27.** -3402

29. $\displaystyle\sum_{n=0}^{\infty} (1 + 4n)$ **31.** $\displaystyle\sum_{n=1}^{\infty} \left(\tfrac{9}{10}\right)^{n-1}$

33. $\displaystyle\sum_{n=2}^{\infty} \frac{1}{2n(n-1)}$ **35.–47.** Answers will vary.

49. See the domino illustration and Figure 7.4.

51. 1, 3, 5, 7, 9   **53.** 4, 2, -2, -4, -2

55. 0, 3, 6, 9, 12

First differences: 3, 3, 3, 3

Second differences: 0, 0, 0

Linear

57. 3, 1, -2, -6, -11

First differences: $-2, -3, -4, -5$

Second differences: $-1, -1, -1$

Quadratic

59. 0, 1, 3, 6, 10

First differences: 1, 2, 3, 4

Second differences: 1, 1, 1

Quadratic

61. 2, 4, 6, 8, 10

First differences: 2, 2, 2, 2

Second differences: 0, 0, 0

Linear

63. 1, 2, 6, 15, 31

First differences: 1, 4, 9, 16

Second differences: 3, 5, 7

Neither

65. $a_n = n^2 - n + 3$ **67.** $a_n = \tfrac{1}{2}n^2 + n - 3$

69. $(2, 4), \left(-\tfrac{1}{2}, \tfrac{1}{4}\right)$ **71.** $(1, 2, 1)$

Section 9.5   *(page 702)*

1. 10   **3.** 1   **5.** 15,504   **7.** 4950   **9.** 4950

11. The first and last numbers in each row are 1. Every other number in each row is formed by adding the two numbers immediately above the number.

13. 35   **15.** 56   **17.** $x^4 + 4x^3 + 6x^2 + 4x + 1$

19. $a^3 + 6a^2 + 12a + 8$

21. $y^4 - 8y^3 + 24y^2 - 32y + 16$

23. $x^5 + 5x^4y + 10x^3y^2 + 10x^2y^3 + 5xy^4 + y^5$

25. $r^6 + 18r^5s + 135r^4s^2 + 540r^3s^3 + 1215r^2s^4$

$+ 1458rs^5 + 729s^6$

27. $x^5 - 5x^4y + 10x^3y^2 - 10x^2y^3 + 5xy^4 - y^5$

29. $1 - 6x + 12x^2 - 8x^3$

31. $x^8 + 20x^6 + 150x^4 + 500x^2 + 625$

33. $\dfrac{1}{x^5} + \dfrac{5y}{x^4} + \dfrac{10y^2}{x^3} + \dfrac{10y^3}{x^2} + \dfrac{5y^4}{x} + y^5$

35. $2x^4 - 24x^3 + 113x^2 - 246x + 207$

37. $32t^5 - 80t^4s + 80t^3s^2 - 40t^2s^3 + 10ts^4 - s^5$

39. $81 - 216z + 216z^2 - 96z^3 + 16z^4$

41. 1,732,104   **43.** 180   **45.** $-326{,}592$ **47.** 210

49. $n + 1$ terms

51. $x^2 + 12x^{3/2} + 54x + 108x^{1/2} + 81$

53. $x^2 - 3x^{4/3}y^{1/3} + 3x^{2/3}y^{2/3} - y$

55. $3x^2 + 3xh + h^2$ **57.** $\dfrac{\sqrt{x+h} - \sqrt{x}}{h}$ **59.** -4

61. $2035 + 828i$ **63.** 1   **65.** 0.273   **67.** 0.171

69. 1.172   **71.** 510,568.785

73. $g(x) = x^3 + 12x^2 + 44x + 48$

Shifted four units to the left

75. $g(x) = -x^2 + 7x - 8$

Shifted two units to the right

77. Answers will vary.   **79.** Answers will vary.

81. (a) 792   (b) 36   (c) 792   (d) 12

83.

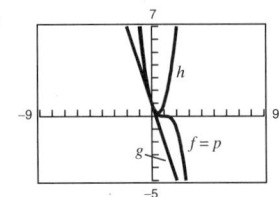

$p(x)$ is the expansion of $f(x)$.

85. (a) $g(t) = 0.1506t^2 + 3.7481t + 43.5584$

(b)

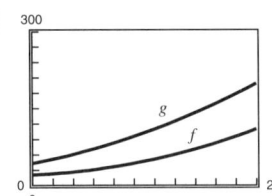

87. $g(x)$ is shifted eight units up from $f(x)$.

89. $g(x)$ is the reflection of $f(x)$ in the y-axis.

Section 9.6 *(page 712)*

1. 6 **3.** 5 **5.** 3 **7.** 7 **9.** 48 **11.** 12

13. 64 **15.** 12 **17.** 6,760,000

19. (a) 900 (b) 648 (c) 180 (d) 600

21. 64,000 **23.** (a) 720 (b) 48

25. 24 **27.** 336 **29.** 120

31. $n = 5$ or $n = 6$ **33.** 1,860,480

35. 970,200 **37.** 15,504

39. For some calculators the number is too great.

41. ABCD, ABDC, ACBD, ACDB, ADBC, ADCB, BACD, BADC, CABD, CADB, DABC, DACB

BCAD, BDAC, CBAD, CDAB, DBAC, DCAB, BCDA, BDCA, CBDA, CDBA, DBCA, DCBA

43. 120 **45.** 11,880 **47.** 420 **49.** 2520

51. AB, AC, AD, AE, AF, BC, BD, BE, BF, CD, CE, CF, DE, DF, EF

53. 4845 **55.** 3,838,380 **57.** 3,921,225 **59.** 21

61. (a) 70 (b) 30 **63.** (a) 70 (b) 54 (c) 16

65. 5 **67.** 20 **69.** Answers will vary.

71. Answers will vary. **73.** 8.30 **75.** 35

Section 9.7 *(page 724)*

1. $\{(H, 1), (H, 2), (H, 3), (H, 4), (H, 5), (H, 6),$
$(T, 1), (T, 2), (T, 3), (T, 4), (T, 5), (T, 6)\}$

3. $\{ABC, ACB, BAC, BCA, CAB, CBA\}$

5. $\{(A, B), (A, C), (A, D), (A, E), (B, C), (B, D),$
$(B, E), (C, D), (C, E), (D, E)\}$

7. $\frac{3}{8}$ **9.** $\frac{7}{8}$ **11.** $\frac{3}{13}$ **13.** $\frac{3}{26}$ **15.** $\frac{1}{12}$ **17.** $\frac{11}{12}$

19. $\frac{1}{3}$ **21.** $\frac{1}{5}$ **23.** $\frac{2}{5}$ **25.** 0.3 **27.** 0.85

29. (a) 925,000 (b) 0.18

31. (a) 0.58 (b) 0.956 (c) 0.004

33. (a) $\frac{672}{1254}$ (b) $\frac{582}{1254}$ (c) $\frac{548}{1254}$

35. $P(\{\text{Taylor wins}\}) = \frac{1}{2}$
$P(\{\text{Moore wins}\}) = P(\{\text{Jenkins wins}\}) = \frac{1}{4}$

37. (a) $\frac{21}{1292} \approx 0.016$ (b) $\frac{225}{646} \approx 0.348$ (c) $\frac{49}{323} \approx 0.152$

39. (a) $\frac{1}{3}$ (b) $\frac{5}{8}$ **41.** (a) $\frac{1}{120}$ (b) $\frac{1}{24}$

43. (a) $\frac{1}{169}$ (b) $\frac{1}{221}$ **45.** (a) $\frac{14}{55}$ (b) $\frac{12}{55}$ (c) $\frac{54}{55}$

47. (a) $\frac{1}{4}$ (b) $\frac{1}{2}$ (c) $\frac{9}{100}$ (d) $\frac{1}{30}$

49. (a) 0.9702 (b) 0.9998 (c) 0.0002

51. (a) $\frac{1}{1024}$ (b) $\frac{243}{1024}$ (c) $\frac{781}{1024}$

53. 0.1024 **55.** $\frac{7}{16}$

57. (a) As you consider successive people with distinct birthdays, the probabilities must decrease to take into account the birth dates already used. Because the birth dates of people are independent events, multiply the respective probabilities of distinct birthdays.

(b) $\dfrac{365}{365} \cdot \dfrac{364}{365} \cdot \dfrac{363}{365} \cdot \dfrac{362}{365}$

(c) Answers will vary.

(d) Q_n is the probability that the birthdays are *not* distinct, which is equivalent to at least two people having the same birthday.

(e)

n	10	15	20	23	30	40	50
P_n	0.88	0.75	0.59	0.49	0.29	0.11	0.03
Q_n	0.12	0.25	0.41	0.51	0.71	0.89	0.97

(f) 23

59. 0.454

Section 9.8 *(page 734)*

1. Mean: 8.86; Median: 8; Mode: 7

3. Mean: 10.29; Median: 8; Mode: 7

5. Mean: 9; Median: 8; Mode: 7

7. The mean, because the magnitude of the extreme value is entered into the algebraic formula used to compute the mean.

9. Mean: 26.2; Median: 25.5; Mode: 23

11. Mean: 30.36; Median: 30; Mode: none

13. Mean: $77.21; Median: $75.35

15. (a) 200 (b) Mean: 3.07; Median: 3; Mode: 3

17. Mode

19. (a) (b) 36,540 miles

21. (a)

(b)

Width of the intervals is changed.

(c)

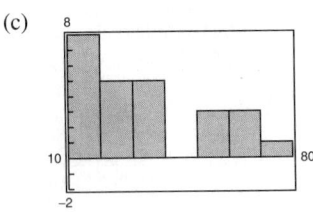

Endpoints of the intervals are changed.

(d) 27   (e) 34.12

Section 9.9   *(page 743)*

1. $\bar{x} = 6$; $v = 10$; $\sigma = 3.16$ **3.** $\bar{x} = 2$; $v = \frac{4}{3}$; $\sigma = 1.15$

5. $\bar{x} = 4$; $v = 4$; $\sigma = 2$ **7.** 3.42   **9.** 5.75

11. $\bar{x} = 47$; $v = 226$; $\sigma = 15.03$

13. $\bar{x} = 1.1$; $v = 0.38$; $\sigma = 0.62$

15. $\bar{x} = 5.8$; $v = 2.71$; $\sigma = 1.65$

17. (a) $\bar{x} = 12$; $\sigma = 2.83$ (b) $\bar{x} = 20$; $\sigma = 2.83$
   (c) $\bar{x} = 12$; $\sigma = 1.41$ (d) $\bar{x} = 9$; $\sigma = 1.41$

19. The mean will increase by 5. The dispersion of the scores will remain the same.

21. C, D, A, B

23. (a) At least $\frac{3}{4}$ of the scores: $[179, 291]$
      At least $\frac{8}{9}$ of the scores: $[151, 319]$
   (b) At least $\frac{3}{4}$ of the scores: $[203, 267]$
      At least $\frac{8}{9}$ of the scores: $[187, 283]$
   (c) At least $\frac{3}{4}$ of the scores: $[169, 281]$
      At least $\frac{8}{9}$ of the scores: $[141, 309]$

25. (a) $\bar{x} = 197.32$; $\sigma = 14.81$ (b) 96%

27. (a) $\bar{x} = 2.63$; $\sigma = 0.12$ (b) Hawaii, Utah

29.

31.

33.

35.

37.

From the plots, you can see that the lifetimes of the units in the sample made by the new design are greater than the lifetimes of the units in the sample made by the original design. (The median increased by more than 1 year.)

FOCUS ON CONCEPTS   *(page 746)*

1. Natural numbers

2. (a) Odd-numbered terms are negative.
   (b) Even-numbered terms are negative.

3. True   **4.** True   **5.** True   **6.** True

7. (a) Each term is obtained by adding the same constant (common difference) to the previous term.
   (b) Each term is obtained by multiplying the same constant (common ratio) to the previous term.

8. (a) Arithmetic. There is a constant difference between consecutive terms.
   (b) Geometric. Each term is a constant multiple of the previous term. In this case the common ratio is greater than 1.

9. Each term of the sequence is defined in terms of the previous term.

10. Increased powers of real numbers between 0 and 1 approach zero.

11. (d)   **12.** (a)   **13.** (b)   **14.** (c)

15. The signs of the terms alternate in the expansion $(x - y)^n$.

16. Same

17. $_{10}P_6 > {_{10}C_6}$. Changing the order of any of the six elements selected results in a different permutation but the same combination.

18. Meteorological records indicate that over an extended period of time with similar weather conditions it will rain 60% of the time.

19. Mean   **20.** The numbers in the set are identical.

Review Exercises   *(page 747)*

1. 8, 5, 4, $\frac{7}{2}$, $\frac{16}{5}$ **3.** 72, 36, 12, 3, $\frac{3}{5}$

5.   **7.**

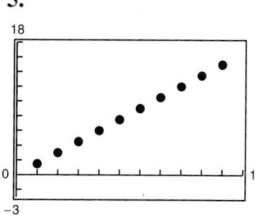

9. $\displaystyle\sum_{k=1}^{20} \frac{1}{2k}$ **11.** $\displaystyle\sum_{k=1}^{9} \frac{k}{k+1}$ **13.** 30   **15.** $\dfrac{205}{24}$

17. 6050   **19.** 418   **21.** 3, 7, 11, 15, 19

23. 1, 4, 7, 10, 13

25. 35, 32, 29, 26, 23   **27.** 9, 16, 23, 30, 37

   $a_n = 38 - 3n$ $a_n = 2 + 7n$

29. $a_n = 103 - 3n$ **31.** 80   **33.** 88   **35.** 25,250

   1430

37. (a) $43,000   (b) $192,500   **39.** 4, -1, $\frac{1}{4}$, $-\frac{1}{16}$, $\frac{1}{64}$

41. 9, 6, 4, $\frac{8}{3}$, $\frac{16}{9}$ or 9, -6, 4, $-\frac{8}{3}$, $\frac{16}{9}$

43. 120, 40, $\frac{40}{3}$, $\frac{40}{9}$, $\frac{40}{27}$ **45.** 25, -15, 9, $-\frac{27}{5}$, $\frac{81}{25}$

   $a_n = 120\left(\frac{1}{3}\right)^{n-1}$ $a_n = 25\left(-\frac{3}{5}\right)^{n-1}$

47. $a_n = 16\left(-\frac{1}{2}\right)^{n-1}$, 10.67   **49.** 127   **51.** 8

53. 12   **55.** 24.849

57. (a) $a_t = 120,000(0.7)^t$ (b) $20,168.40

59. $5111.82   **61.** Answers will vary.

63. Answers will vary.   **65.** 15   **67.** 6720

69. $\dfrac{x^4}{16} + \dfrac{x^3 y}{2} + \dfrac{3x^2 y^2}{2} + 2xy^3 + y^4$

71. $\dfrac{64}{x^6} - \dfrac{576}{x^4} + \dfrac{2160}{x^2} - 4320 + 4860x^2 - 2916x^4 + 729x^6$

73. $41 + 840i$ **75.** 118,813,760

77.

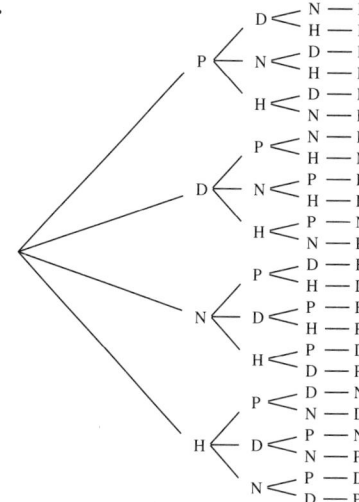

79. $\frac{1}{9}$ **81.** No. Rolling a 3 with one die

83. $\frac{31}{32}$ **85.** 0.0475

87. (a) $\bar{x} = 4$; $\sigma = 1.79$ (b) $\bar{x} = 10$; $\sigma = 1.79$

   (c) $\bar{x} = 8$; $\sigma = 0.89$ (d) $\bar{x} = 20$; $\sigma = 0.89$

89. Mean: 10.7; Median: 11.5; Standard deviation: 4.65

91. Mean: 166.28; Median: 174.75; Standard deviation: 34.16

93. Mean: 50.36; Median: 51; Standard deviation: 7.39; 1

95. (a) Mean: 30.47; Median: 30.5; Modes: 30, 32

   (b) Standard deviation: 2.79

   (c) $33\frac{1}{3}\%$

Chapter Test   *(page 752)*

1. 1, $-\frac{2}{3}$, $\frac{4}{9}$, $-\frac{8}{27}$, $\frac{16}{81}$ **2.** 12, 16, 20, 24, 28

3. $a_n = 5000 - 100(n-1)$ **4.** $a_n = 4\left(\frac{1}{2}\right)^{n-1}$

5. $\displaystyle\sum_{n=1}^{12} \frac{2}{3n+1}$ **6.** 3825   **7.** $-\dfrac{3}{2}$ **8.** 765

9. $47,868.33   **10.** 1140   **11.** 56   **12.** 26,000

13. 12,650   **14.** $\frac{1}{6}$ **15.** $\bar{x} = 10$; $\sigma = 1.34$

16.

24 26 28   43 46

CHAPTER 10

Section 10.1 *(page 759)*

1. (e)   **3.** (d)   **5.** (a)

7. Vertex: $(0, 0)$
Focus: $\left(0, \frac{1}{2}\right)$
Directrix: $y = -\frac{1}{2}$

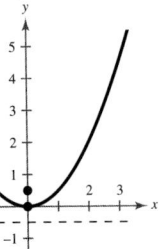

9. Vertex: $(0, 0)$
Focus: $\left(-\frac{3}{2}, 0\right)$
Directrix: $x = \frac{3}{2}$

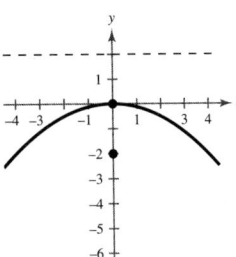

11. Vertex: $(0, 0)$
Focus: $(0, -2)$
Directrix: $y = 2$

13. Vertex: $(1, -2)$
Focus: $(1, -4)$
Directrix: $y = 0$

15. Vertex: $\left(5, -\frac{1}{2}\right)$
Focus: $\left(\frac{11}{2}, -\frac{1}{2}\right)$
Directrix: $x = \frac{9}{2}$

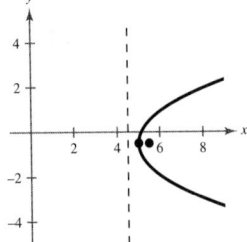

17. Vertex: $(1, 1)$
Focus: $(1, 2)$
Directrix: $y = 0$

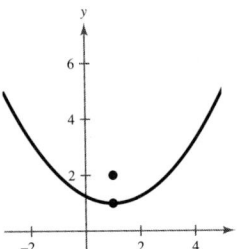

19. Vertex: $(-2, -3)$
Focus: $(-4, -3)$
Directrix: $x = 0$

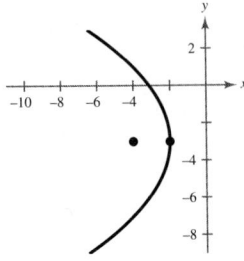

21. Vertex: $(-2, 1)$
Focus: $\left(-2, -\frac{1}{2}\right)$
Directrix: $y = \frac{5}{2}$

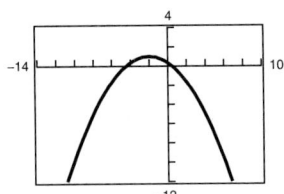

23. Vertex: $\left(\frac{1}{4}, -\frac{1}{2}\right)$
Focus: $\left(0, -\frac{1}{2}\right)$
Directrix: $x = \frac{1}{2}$

25.

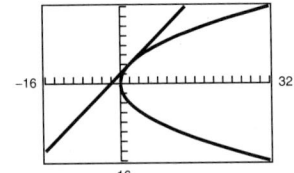

$(2, 4)$

27. $y = -\sqrt{-6x}$

29. (a)

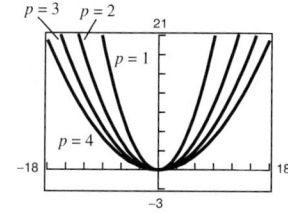

As p increases the parabola becomes wider.

(b) $(0, 1), (0, 2), (0, 3), (0, 4)$

(c) $4, 8, 12, 16; 4p$

(d) Easy way to determine two additional points on the graph

31. $y = \frac{2}{3}x^2$ **33.** $x^2 = -6y$ **35.** $y^2 = -8x$

37. $x^2 = 4y$ **39.** $x^2 = -8y$ **41.** $y^2 = 9x$

43. $(x - 3)^2 = -(y - 1)$ **45.** $y^2 = 2(x + 2)$

47. $(y - 2)^2 = -8(x - 3)$ **49.** $x^2 = 8(y - 4)$

51. $(y - 2)^2 = 8x$ **53.** $y = \sqrt{6(x + 1)} + 3$

55. $y = \frac{1}{14}x^2$ **57.** (a) $y = \frac{3x^2}{640,000}$ (b) 462 centimeters

59. $y^2 = 640x$

61. 43.3 seconds prior to being over the target

63. $\frac{x_1}{2p}$ **65.** $4x - y - 8 = 0; (2, 0)$

67. $4x - y + 2 = 0; \left(-\frac{1}{2}, 0\right)$

69. (a) $y = -\frac{1}{64}x^2 + 75$ (b) 69.3 feet

71. $y = x^3 - 7x^2 + 17x - 15$ **73.** $\pm 2, \frac{1}{2}, -\frac{5}{3}$

Section 10.2   *(page 769)*

1. (b)   **3.** (d)   **5.** (a)

7. Center: $(0, 0)$ **9.** Center: $(0, 0)$

Vertices: $(\pm 5, 0)$ Vertices: $(0, \pm 5)$

Foci: $(\pm 3, 0)$ Foci: $(0, \pm 3)$

Eccentricity: $\frac{3}{5}$ Eccentricity: $\frac{3}{5}$

   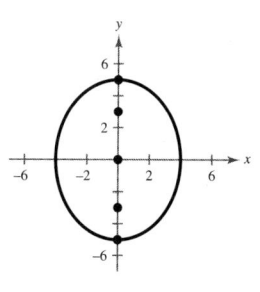

11. Center: $(0, 0)$

Vertices: $(\pm 3, 0)$

Foci: $(\pm 2, 0)$

Eccentricity: $\frac{2}{3}$

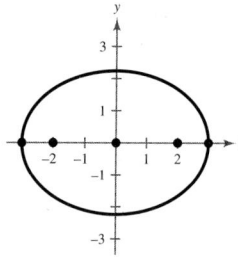

13. Center: $(1, 5)$

Vertices: $(1, 10), (1, 0)$

Foci: $(1, 9), (1, 1)$

Eccentricity: $\frac{4}{5}$

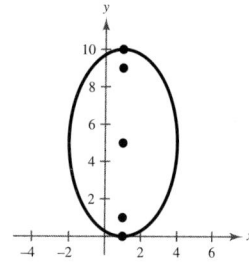

15. Center: $(-2, 3)$

Vertices: $(-2, 6), (-2, 0)$

Foci: $\left(-2, 3 \pm \sqrt{5}\right)$

Eccentricity: $\frac{\sqrt{5}}{3}$

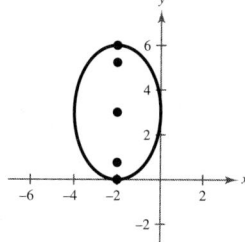

17. Center: $(1, -1)$

Vertices: $\left(\frac{9}{4}, -1\right), \left(-\frac{1}{4}, -1\right)$

Foci: $\left(\frac{7}{4}, -1\right), \left(\frac{1}{4}, -1\right)$

Eccentricity: $\frac{3}{5}$

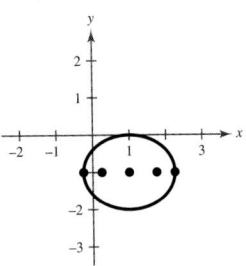

19. Center: $(0, 0)$

Vertices: $\left(0, \pm\sqrt{5}\right)$

Foci: $\left(0, \pm\sqrt{2}\right)$

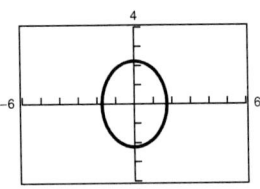

21. Center: $\left(\frac{1}{2}, -1\right)$

Vertices: $\left(\frac{1}{2} \pm \sqrt{5}, -1\right)$

Foci: $\left(\frac{1}{2} \pm \sqrt{2}, -1\right)$

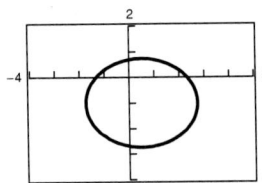

23. $x = \frac{3}{2}\left(2 + \sqrt{4 - y^2}\right)$ **25.** $\dfrac{x^2}{25} + \dfrac{y^2}{21} = 1$

27. $\dfrac{x^2}{1} + \dfrac{y^2}{4} = 1$ **29.** $\dfrac{x^2}{36} + \dfrac{y^2}{11} = 1$

31. $\dfrac{21x^2}{400} + \dfrac{y^2}{25} = 1$ **33.** $\dfrac{(x-2)^2}{1} + \dfrac{(y-3)^2}{9} = 1$

35. $\dfrac{(x-2)^2}{9} + \dfrac{(y-2)^2}{4} = 1$

37. $\dfrac{(x-2)^2}{4} + \dfrac{(y-2)^2}{1} = 1$

39. $\dfrac{x^2}{48} + \dfrac{(y-4)^2}{64} = 1$

41. $\dfrac{(x-3)^2}{9} + \dfrac{(y-5)^2}{16} = 1$

43. $\dfrac{x^2}{16} + \dfrac{(y-4)^2}{12} = 1$

45. (a) $2a$

(b) The sum of the distances from the two fixed points is constant.

47. (a) $A = \pi a(20 - a)$ (b) $\dfrac{x^2}{196} + \dfrac{y^2}{36} = 1$

(c)

a	8	9	10	11	12	13
A	301.6	311.0	314.2	311.0	301.6	285.9

$a = 10$. Circle

(d)

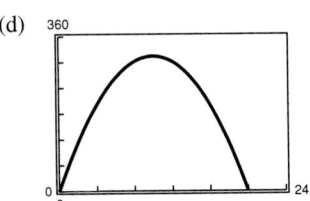

$a = 10$

49. $\dfrac{x^2}{25} + \dfrac{y^2}{16} = 1$ **51.** $\dfrac{x^2}{328.15} + \dfrac{y^2}{19.39} = 1$

53. (a) Answers will vary.

(b)

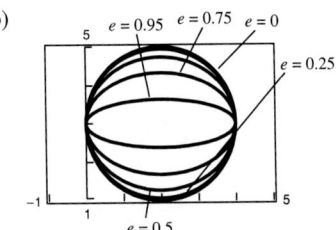

(c) Ellipse becomes more circular.

55.

57.

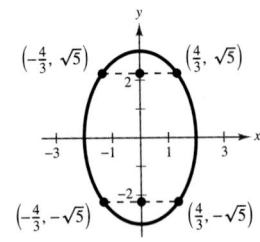

59. False. The equation of an ellipse is second degree in x and y.

Section 10.3   *(page 779)*

1. (a)   **3.** (d)

5. Center: $(0, 0)$
  Vertices: $(\pm 1, 0)$
  Foci: $\left(\pm \sqrt{2}, 0\right)$
  Asymptotes: $y = \pm x$

7. Center: $(0, 0)$
  Vertices: $(0, \pm 1)$
  Foci: $\left(0, \pm \sqrt{5}\right)$
  Asymptotes: $y = \pm \frac{1}{2}x$

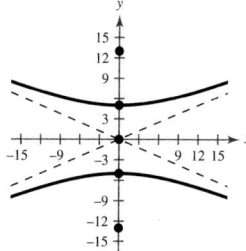

9. Center: $(0, 0)$
  Vertices: $(0, \pm 5)$
  Foci: $(0, \pm 13)$
  Asymptotes: $y = \pm \frac{5}{12}x$

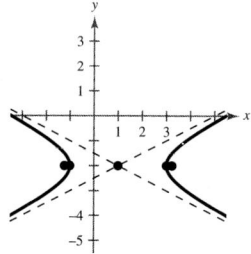

11. Center: $(1, -2)$
  Vertices: $(3, -2), (-1, -2)$
  Foci: $\left(1 \pm \sqrt{5}, -2\right)$
  Asymptotes: $y = -2 \pm \frac{1}{2}(x - 1)$

13. Center: $(2, -6)$
  Vertices: $(2, -5), (2, -7)$
  Foci: $\left(2, -6 \pm \sqrt{2}\right)$
  Asymptotes: $y = -6 \pm (x - 2)$

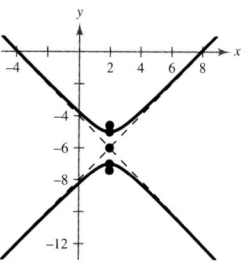

15. Center: $(2, -3)$
  Vertices: $(3, -3), (1, -3)$
  Foci: $\left(2 \pm \sqrt{10}, -3\right)$
  Asymptotes: $y = -3 \pm 3(x - 2)$

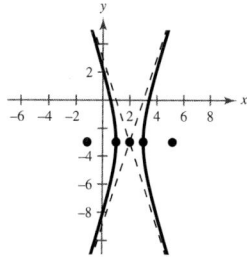

17. The graph of this equation is two lines intersecting at $(-1, -3)$.

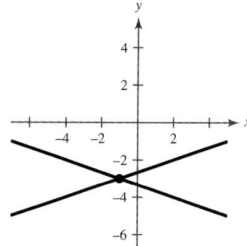

19. Center: $(0, 0)$
Vertices: $\left(\pm\sqrt{3}, 0\right)$
Foci: $\left(\pm\sqrt{5}, 0\right)$

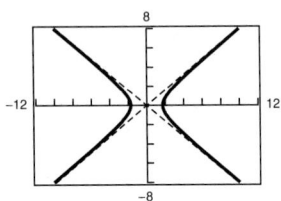

21. Center: $(1, -3)$
Vertices: $\left(1, -3 \pm \sqrt{2}\right)$
Foci: $\left(1, -3 \pm 2\sqrt{5}\right)$

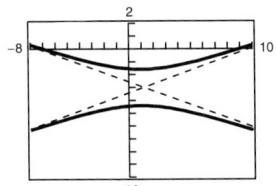

23. $\dfrac{y^2}{4} - \dfrac{x^2}{12} = 1$ **25.** $\dfrac{x^2}{1} - \dfrac{y^2}{9} = 1$

27. $\dfrac{17y^2}{1024} - \dfrac{17x^2}{64} = 1$ **29.** $\dfrac{(x-4)^2}{4} - \dfrac{y^2}{12} = 1$

31. $\dfrac{(y-5)^2}{16} - \dfrac{(x-4)^2}{9} = 1$ **33.** $\dfrac{y^2}{9} - \dfrac{4(x-2)^2}{9} = 1$

35. $\dfrac{(x-3)^2}{9} - \dfrac{(y-2)^2}{4} = 1$ **37.** Left half

39. $x \approx 110.3$ miles   **41.** $\left(12(\sqrt{5}-1), 0\right) \approx (14.83, 0)$

43. Circle   **45.** Hyperbola   **47.** Ellipse   **49.** Parabola

Section 10.4   *(page 789)*

1. $(-3, 0)$ **3.** $\left(\frac{1}{2}(\sqrt{3}-4), \frac{1}{2}(1+4\sqrt{3})\right)$

5. $\dfrac{(y')^2}{2} - \dfrac{(x')^2}{2} = 1$ **7.** $\dfrac{(x')^2}{1/4} - \dfrac{(y')^2}{1/6} = 1$

9. $\dfrac{\left(x' - 3\sqrt{2}\right)^2}{16} - \dfrac{\left(y' - \sqrt{2}\right)^2}{16} = 1$

11. $\dfrac{(x')^2}{3} + \dfrac{(y')^2}{2} = 1$ **13.** $x' = -(y')^2$

    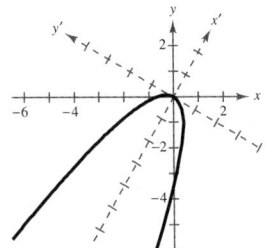

15. $y' = \frac{1}{6}(x')^2 - \frac{1}{3}x'$ **17.** $\theta = 45°$

    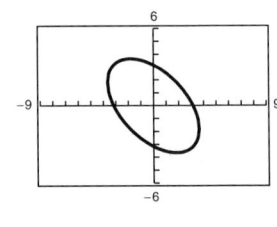

19. $\theta = 26.57°$ **21.** $\theta = 31.72°$

    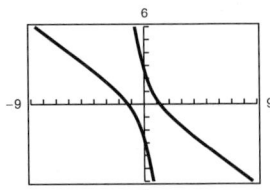

23. e   **25.** b   **27.** d   **29.** Parabola   **31.** Ellipse
33. Hyperbola   **35.** Parabola

37.

39.

41.

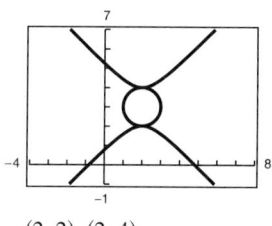

$(2, 2), (2, 4)$

43.

$(-10, 12)$

45.

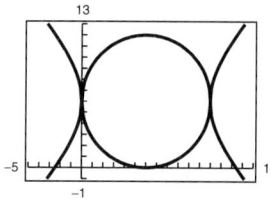

$(0, 6), (12, 6)$

47.

$(0, 4)$

49.

$(4, \pm 3)$

51.

No solution

53.

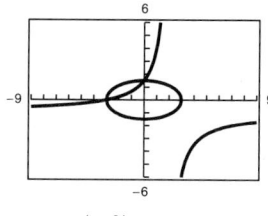

$(-3, 0), \left(0, \frac{3}{2}\right)$

55. Answers will vary.

57.

59.

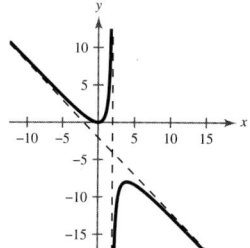

Section 10.5   *(page 797)*

1. (a)

t	0	1	2	3	4
x	0	1	$\sqrt{2}$	$\sqrt{3}$	2
y	1	0	-1	-2	-3

(b)

(c) $y = 1 - x^2$

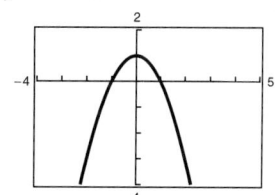

Entire parabola rather than just the right half

3.

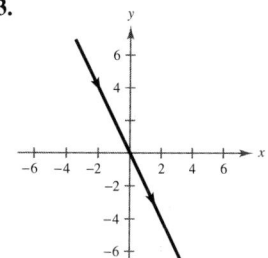

$2x + y = 0$

5.

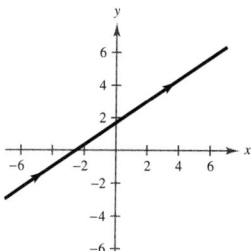

$2x - 3y + 5 = 0$

7.

$y = 16x^2$

9.

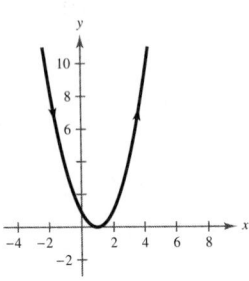

$y = (x - 1)^2$

11.

$y = \frac{1}{2}\left(\sqrt[3]{x}\right)^2$

13.

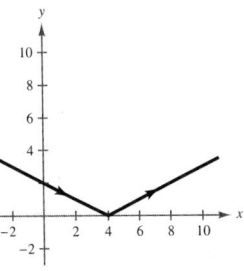

$y = \frac{1}{2}|x - 4|$

15.

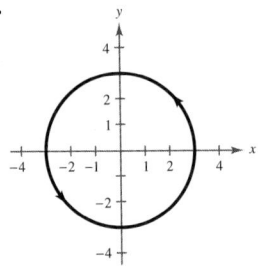

$x^2 + y^2 = 9$

17.

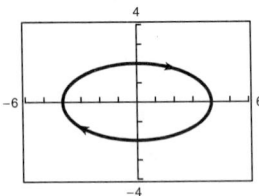

$\dfrac{x^2}{16} + \dfrac{y^2}{4} = 1$

19.

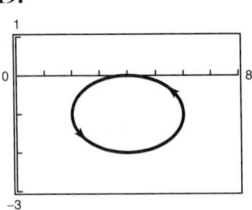

$\dfrac{(x - 4)^2}{4} + (y + 1)^2 = 1$

21.

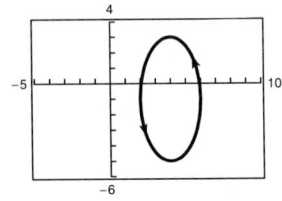

$\dfrac{(x - 4)^2}{4} + \dfrac{(y + 1)^2}{16} = 1$

23.

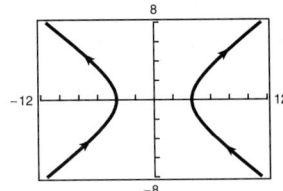

$\dfrac{x^2}{16} - \dfrac{y^2}{9} = 1$

25.

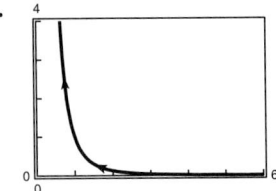

$y = \dfrac{1}{x^3}, \quad x > 0, y > 0$

27.

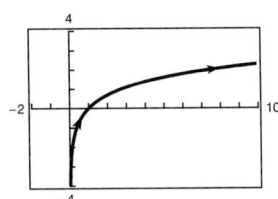

$y = \ln x$

29. Each curve represents a portion of the line $y = 2x + 1$.

Domain	Orientation
(a) $(-\infty, \infty)$	Left to right
(b) $[-1, 1]$	Depends on θ
(c) $(0, \infty)$	Right to left
(d) $(0, \infty)$	Left to right

31. $y - y_1 = \dfrac{y_2 - y_1}{x_2 - x_1}(x - x_1)$

33. $\dfrac{(x - h)^2}{a^2} + \dfrac{(y - k)^2}{b^2} = 1$

35. $x = 5t$

$\quad\ y = -2t$

37. $x = 2 + 4\cos\theta$

$\quad\ y = 1 + 4\sin\theta$

39. $x = 5\cos\theta$

$\quad\ y = 3\sin\theta$

41. $x = t,\ y = 3t - 2$

$\quad\ x = 2t,\ y = 6t - 2$

43.

45.

47.

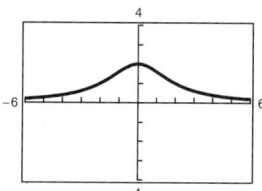

49. (b)   **51.** (d)

53. (a) $x = (146.67\cos\theta)t$

$\quad\quad y = 3 + (146.67\sin\theta)t - 16t^2$

(b) $x = 141.7t$

$\quad\ y = 3 + 38.0t - 16t^2$

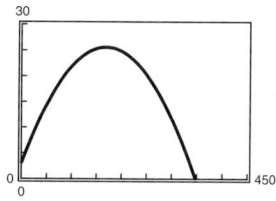

No

(c) $x = 135.0t$

$\quad\ y = 3 + 57.3t - 16t^2$

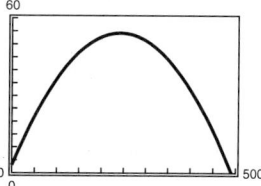

Yes

(d) About $19.4°$

55. Answers will vary.

57. False. x and y cannot have negative values.

Section 10.6   *(page 805)*

1. $(0, 4)$ **3.** $\left(\dfrac{\sqrt{2}}{2}, \dfrac{\sqrt{2}}{2}\right)$

5.

$\left(2, -2\sqrt{3}\right)$

7.

$(0, 0)$

9.

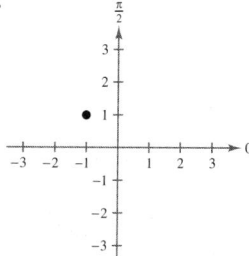

$(-1.004, 0.996)$

11. $\left(-\sqrt{2}, \sqrt{2}\right)$ **13.** $(-1.204, -4.336)$

15.

17.

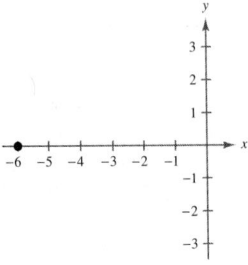

$\left(\sqrt{2}, \dfrac{\pi}{4}\right), \left(-\sqrt{2}, \dfrac{5\pi}{4}\right)$ $(6, \pi), (-6, 0)$

19.

21.

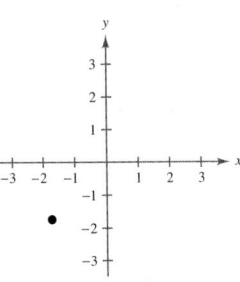

$(5, 2.214), (-5, 5.356)$ $\left(\sqrt{6}, \dfrac{5\pi}{4}\right), \left(-\sqrt{6}, \dfrac{\pi}{4}\right)$

23.

$\left(2\sqrt{13}, 0.983\right), \left(-2\sqrt{13}, 4.124\right)$

25. $\left(\sqrt{13}, -0.588\right)$ **27.** $\left(\sqrt{7}, 0.857\right)$ **29.** $\left(\frac{17}{6}, 0.490\right)$

31. True   **33.** $r = 3$ **35.** $r = 2a \cos \theta$

37. $r = 4 \csc \theta$ **39.** $r = 10 \sec \theta$

41. $r = \dfrac{-2}{3 \cos \theta - \sin \theta}$

43. $r^2 = 4 \sec \theta \csc \theta = 8 \csc 2\theta$ **45.** $r^2 = 9 \cos 2\theta$

47. $x^2 + y^2 - 4y = 0$ **49.** $\sqrt{3}x - 3y = 0$

51. $y = 2$ **53.** $(x^2 + y^2)^2 = 6x^2y - 2y^3$

55. $4x^2 - 5y^2 - 36y - 36 = 0$

57. $x^2 + y^2 = 9$

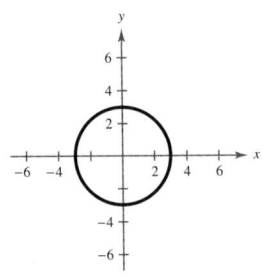

59. $x - y = 0$ **61.** $x - 3 = 0$

    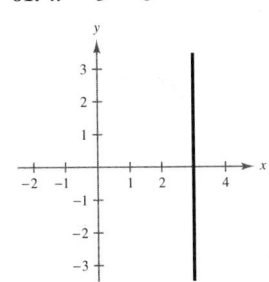

63. $(x - h)^2 + (y - k)^2 = h^2 + k^2$; $\sqrt{h^2 + k^2}$; (h, k)

65. (a) Answers will vary.

(b) The points lie on a line.
$$d = \sqrt{r_1^2 + r_2^2 - 2r_1 r_2} = |r_1 - r_2|$$

(c) $d = \sqrt{r_1^2 + r_2^2}$ (Pythagorean Theorem)

(d) Answers will vary. The distance formulas should give the same results.

67. $(2, 3)$ **69.** $\left(\frac{8}{7}, \frac{88}{35}, \frac{8}{5}\right)$ **71.** $(1, -4, 1, 2)$

Section 10.7   *(page 814)*

1. Rose curve   **3.** Limaçon   **5.** Rose curve

7. Polar axis   **9.** $\theta = \dfrac{\pi}{2}$

11. $\theta = \dfrac{\pi}{2}$, polar axis, pole

13. Maximum: $|r| = 20$ when $\theta = \dfrac{3\pi}{2}$

Zero: $r = 0$ when $\theta = \dfrac{\pi}{2}$

15. Maximum: $|r| = 4$ when $\theta = 0, \dfrac{\pi}{3}, \dfrac{2\pi}{3}$

Zero: $r = 0$ when $\theta = \dfrac{\pi}{6}, \dfrac{\pi}{2}, \dfrac{5\pi}{6}$

17.

19.

21.

23.

25.

27.

29.

31.

33.

35.

37.

39.

41.

43.

45.

47.

49.

51.

53.

55.

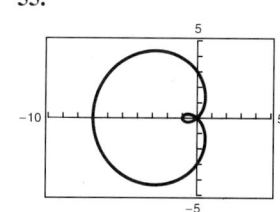

$0 \le \theta < 2\pi$

57.

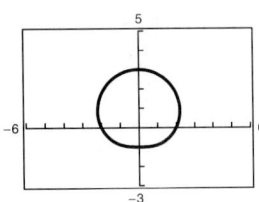

$$0 \le \theta < 2\pi$$

59.

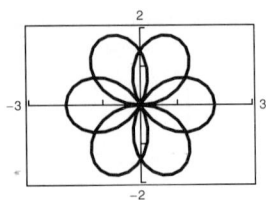

$$0 \le \theta < 4\pi$$

61.

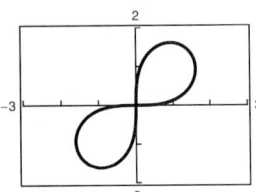

$$0 \le \theta < \frac{\pi}{2}$$

63.

65.

67. (a)

Semicircle

(b)

Semicircle

(c)

Circle

(d)

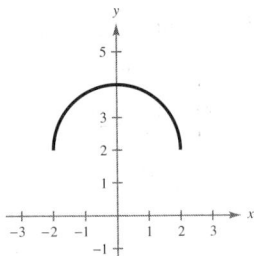

Semicircle

69. Answers will vary.

71. (a) $r = 2 - \sin\left(\theta - \dfrac{\pi}{4}\right)$

(b) $r = 2 + \cos\theta$

(c) $r = 2 + \sin\theta$

(d) $r = 2 - \cos\theta$

73. (a)

(b)

75.

$k = 0$: Circle

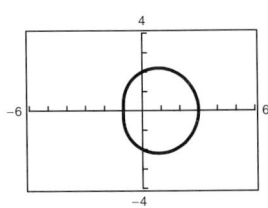

$k = 1$: Convex limaçon

$k = 2$: cardioid

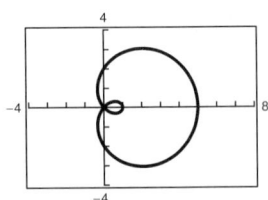

$k = 3$: limaçon with inner loop

Section 10.8   *(page 822)*

1.

3.

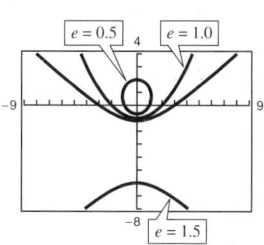

5. (b)   **7.** (d)

9. Parabola

11. Parabola

13. Ellipse

15. Ellipse

17. Hyperbola

19. Hyperbola

21.

Parabola

23.

Ellipse

25.

Ellipse

27.

29.

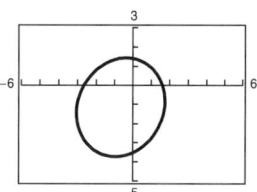

31. $r = \dfrac{1}{1 - \cos \theta}$ **33.** $r = \dfrac{1}{2 + \sin \theta}$

35. $r = \dfrac{2}{1 + 2 \cos \theta}$ **37.** $r = \dfrac{2}{1 - \sin \theta}$

39. $r = \dfrac{10}{1 - \cos \theta}$ **41.** $r = \dfrac{16}{5 + 3 \cos \theta}$

43. $r = \dfrac{20}{3 - 2 \cos \theta}$ **45.** $r = \dfrac{9}{4 - 5 \sin \theta}$

47. Answers will vary.

49. $r^2 = \dfrac{24{,}336}{169 - 25 \cos^2 \theta}$ **51.** $r^2 = \dfrac{144}{25 \cos^2 \theta - 9}$

53. $r^2 = \dfrac{144}{25 \sin^2 \theta - 16}$ **55.** Answers will vary.

57. $r = \dfrac{9.2931 \times 10^7}{1 - 0.0167 \cos \theta}$

9.1405×10^7 miles

9.4509×10^7 miles

59. $r = \dfrac{7977.22}{1 - 0.93669 \cos \theta}$

$11{,}004.5$ miles

FOCUS ON CONCEPTS   *(page 824)*

1. (a) Vertical translation   (b) Horizontal translation

(c) Reflection in the y-axis

(d) The parabola is narrower.

2. (a) Major axis horizontal

(b) Circle   (c) Ellipse is flatter.

(d) Horizontal translation

3. The extended diagonals of the central rectangle are asymptotes of the hyperbola.

4. 5. The ellipse becomes more circular and approaches a circle of radius 5.

5. The orientation would be reversed.

6. (a) The speed would double.

(b) The elliptical orbit would be flatter. The length of the major axis is greater.

7. False. The following are two sets of parametric equations for the line.

$x = t, \quad y = 3 - 2t$

$x = 3t, \quad y = 3 - 6t$

8. False. $(2, \pi/4)$, $(-2, 5\pi/4)$, and $(2, 9\pi/4)$ all represent the same point.

9. (a) Symmetric to the pole

(b) Symmetric to the polar axis

(c) Symmetric to $\pi/2$

10. (a) The graphs are the same.

(b) The graphs are the same.

Review Exercises   *(page 825)*

1. (d)   **3.** (a)

5. Parabola

7. Parabola

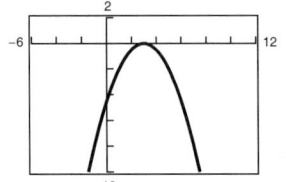

9. Degenerate Circle (a point)

11. Ellipse

13. Ellipse

15. Hyperbola

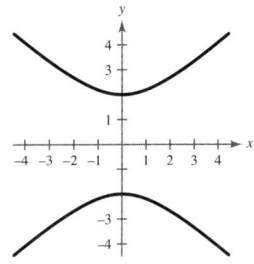

17. $y = -3 \pm \dfrac{\sqrt{2}}{2}\sqrt{-3x^2 + 12x - 11}$

Ellipse

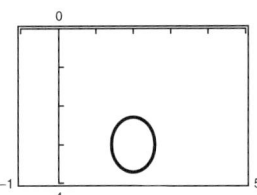

19. $y = 5x \pm \sqrt{24x^2 - 1}$

Hyperbola

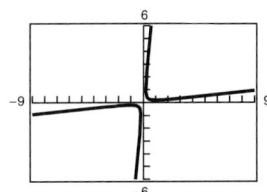

21. $y = \left(\sqrt{2} - x\right) \pm 2\sqrt[4]{2}\sqrt{-x}$

Parabola

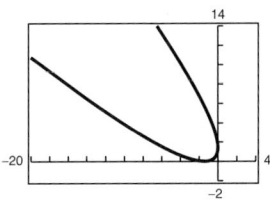

23. $(x - 4)^2 = -8(y - 2)$　　**25.** $(y - 2)^2 = 12x$

27. $\dfrac{(x - 2)^2}{25} + \dfrac{y^2}{21} = 1$　　**29.** $\dfrac{2x^2}{9} + \dfrac{y^2}{36} = 1$

31. $y^2 - \dfrac{x^2}{8} = 1$　　**33.** $\dfrac{5(x - 4)^2}{16} - \dfrac{5y^2}{64} = 1$

35.

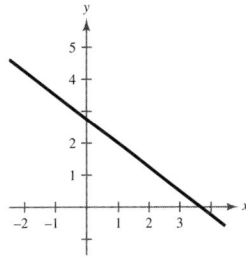

$3x + 4y - 11 = 0$

37.

$y = \dfrac{1}{x^2}$

39.

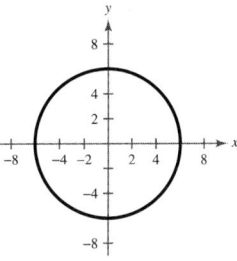

$x^2 + y^2 = 36$

41.

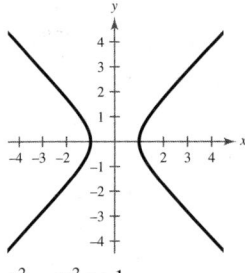

$x^2 - y^2 = 1$

43. False. The equation of a hyperbola is second degree in x and y.

45. $x = 3 \tan \theta, \ y = 4 \sec \theta$

47. $r = a \cos^2\theta \sin \theta$　　**49.** $x^2 - 3x + y^2 = 0$

51. $x^2 + 4y - 4 = 0$　　**53.** $(x^2 + y^2)^2 - x^2 + y^2 = 0$

55. Circle

57. Rose curve

59. Cardioid   **61.** Parabola

63.   **65.**

                    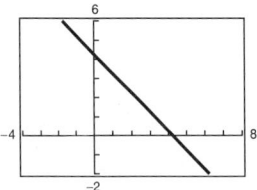

67. $r = 10 \sin \theta$ **69.** $r = \dfrac{4}{1 - \cos \theta}$

71. $r = \dfrac{5}{3 - 2 \cos \theta}$

Cumulative Test for Chapters 7–10   *(page 828)*

1. $(2, -2), (8, 4)$ **2.** $(3, 1)$ **3.** $(2, -3, 1)$

4. $275 model: 160 units; $400 model: 140 units

5. $\begin{bmatrix} -175 & 37 & -13 \\ 95 & -20 & 7 \\ 14 & -3 & 1 \end{bmatrix}$ **6.** 22   **7.** 6

8. $z^4 - 12z^3 + 54z^2 - 108z + 81$ **9.** $\frac{1}{4}$

10. Mean: 24.17; Median: 24; Standard deviation: 3.67

11. (a)   (b)

                    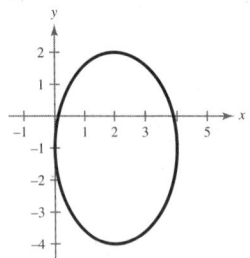

12. $\dfrac{(y - 2)^2}{4/5} - \dfrac{x^2}{16/5} = 1$

13. $x = 2 + 4t$ **14.** $r = 6 \sin \theta$
       $y = -3 + 7t$

15. (a)   (b)

                    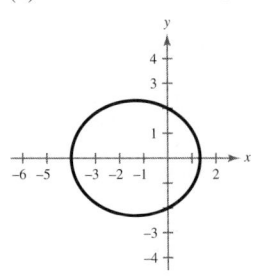

CHAPTER 11

Section 11.1   *(page 835)*

1. $A(2, 1, 3), B(1, -2, 2)$

3. $A(-1, -3, 2), B(3, -2, -4)$

5.   **7.**

                    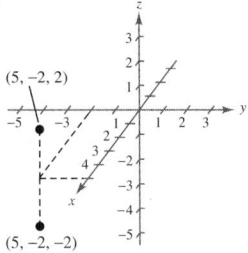

9. $(-3, 4, 5)$ **11.** $(10, 0, 0)$

13. z-coordinate: 0; y-coordinate: 0; x-coordinate: 0

15. Octant IV

17. Octants I, II, III, and IV   **19.** Octants II, IV, VI, and VIII

21. $\left(3\sqrt{2}\right)^2 + 2^2 = \left(\sqrt{22}\right)^2$ **23.** $\left(2\sqrt{5}\right)^2 + 3^2 = \left(\sqrt{29}\right)^2$

25. $3^2 + 6^2 = \left(3\sqrt{5}\right)^2$ **27.** $6, 6, 2\sqrt{10}$
       Right triangle   Isosceles triangle

29. $(0, -2, 6)$ **31.** $\left(0, 0, \frac{13}{2}\right)$

33. $(2x_m - x_1, 2y_m - y_1, 2z_m - z_1)$

35. $(x - 3)^2 + (y - 2)^2 + (z - 4)^2 = 16$

37. $x^2 + (y - 4)^2 + (z - 3)^2 = 16$

39. $(x + 3)^2 + (y - 7)^2 + (z - 5)^2 = 25$

41. $\left(x - \frac{3}{2}\right)^2 + y^2 + (z - 3)^2 = \frac{45}{4}$

43. Center: $(2, -1, 3)$; Radius: 2

45. Center: $(-2, 0, 4)$; Radius: 1

47. Center: $(1, \frac{1}{3}, 4)$; Radius: 3

49. (a)   (b)

51.   **53.**

 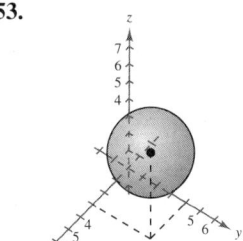

55. $d = 0.09t^2 - 1.50t + 8$

Section 11.2   *(page 842)*

1. (a) $\langle -2, 3, 1 \rangle$

(b)

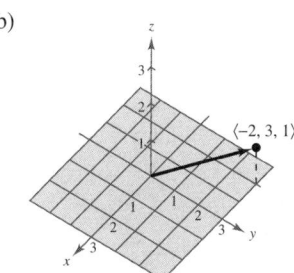

3. (a) $\langle 4, 4, 4 \rangle$ (b) $\langle 1, 1, 1 \rangle$ (c) $\langle -1, -1, -1 \rangle$

5. $\langle 0, 2\sqrt{2}, 2\sqrt{2} \rangle$

7. (a)   (b)

(c)   (d)

 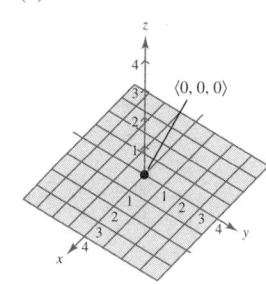

9. $\mathbf{z} = \langle -3, 7, 6 \rangle$ **11.** $\mathbf{z} = \langle \frac{1}{2}, 6, \frac{3}{2} \rangle$ **13.** $\sqrt{33}$

15. $\sqrt{34}$ **17.** (a) $\dfrac{1}{\sqrt{74}}(8\mathbf{i} + 3\mathbf{j} - \mathbf{k})$

            (b) $-\dfrac{1}{\sqrt{74}}(8\mathbf{i} + 3\mathbf{j} - \mathbf{k})$

19. $\langle -26, 0, 48 \rangle$ **21.** $\frac{1}{2}\sqrt{305}$ **23.** -4

25. $\approx 124.45°$ **27.** Parallel   **29.** Not collinear

31. Collinear   **33.** Answers will vary.   **35.** $(3, 1, 7)$

37. $\dfrac{3\sqrt{14}}{14}$ **39.** Sphere of radius 4 centered at (x_1, y_1, z_1)

Section 11.3   *(page 849)*

1. $-\mathbf{k}$ **3.** $-\mathbf{j}$ **5.** $\langle 0, 0, 14 \rangle$ **7.** $\langle -3, 5, 23 \rangle$

9. $-7\mathbf{i} + 13\mathbf{j} + 16\mathbf{k}$ **11.** $-\frac{3}{2}\mathbf{i} - \frac{3}{2}\mathbf{j} - \frac{3}{2}\mathbf{k}$

13. $\langle 10, -2, -4 \rangle$ **15.** $-3\mathbf{i} - \frac{11}{3}\mathbf{j} - \frac{1}{3}\mathbf{k}$

17. $\dfrac{1}{\sqrt{19}}(\mathbf{i} - 3\mathbf{j} + 3\mathbf{k})$ **19.** $\dfrac{1}{\sqrt{342}}(-6\mathbf{i} + 15\mathbf{j} - 9\mathbf{k})$

21. $\dfrac{1}{2\sqrt{2}}(2\mathbf{i} - 2\mathbf{j})$ **23.** 1   **25.** $\sqrt{806}$ **27.** 14

29. $\overrightarrow{AB} = \langle 1, 2, -2 \rangle$ is parallel to $\overrightarrow{DC} = \langle 1, 2, -2 \rangle$.

     $\overrightarrow{AD} = \langle -3, 4, 4 \rangle$ is parallel to $\overrightarrow{BC} = \langle -3, 4, 4 \rangle$.

     Area is $\| \overrightarrow{AB} \times \overrightarrow{AD} \| = 6\sqrt{10}$.

31. $\sqrt{349}$ **33.** $\frac{1}{2}\sqrt{4290}$ **35.** -16 **37.** 2

39. 84 **41.** $10\cos 40° \approx 7.66$ foot-pounds

43. Answers will vary.

45. $\mathbf{v} \times \mathbf{u} = (\sin\alpha\cos\beta - \cos\alpha\sin\beta)\mathbf{k}$; Answers will vary.

Section 11.4 *(page 858)*

1.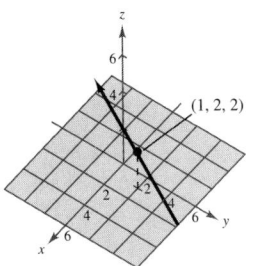

3. $(7, 0, 12)$

5. (a) $x = -2t$, $y = 4t$, $z = t$

(b) $-\dfrac{x}{2} = \dfrac{y}{4} = z$

7. (a) $x = -4 + 3t$ (b) $\dfrac{x+4}{3} = \dfrac{y-1}{8} = -\dfrac{z}{6}$
$y = 1 + 8t$
$z = -6t$

9. (a) $x = 2 + 2t$ (b) $\dfrac{x-2}{2} = -\dfrac{y+3}{3} = z - 5$
$y = -3 - 3t$
$z = 5 + t$

11. (a) $x = 3 + 9t$ (b) $\dfrac{x-3}{9} = -\dfrac{y+5}{13} = -\dfrac{z+4}{12}$
$y = -5 - 13t$
$z = -4 - 12t$

13. (b), (c)

15.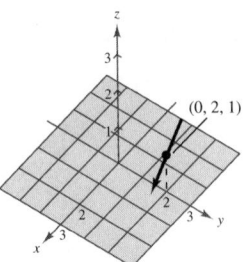

17. $y - 4 = 0$

19. $-2x + y - 2z + 10 = 0$

21. $-x - 2y + z + 2 = 0$ **23.** $-3y + z = 0$

25. $y - 5 = 0$ **27.** $y - z = 0$ **29.** Parallel

31. Orthogonal

33.

35.

37.

39.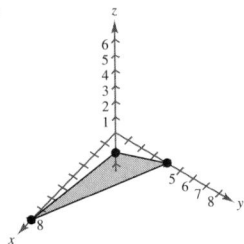

41. $\dfrac{12}{\sqrt{14}}$

43. (a) Sphere: $(x - 4)^2 + (y + 1)^2 + (z - 1)^2 = 4$

(b) Two planes: $4x - 3y + z = 10 \pm 2\sqrt{26}$

FOCUS ON CONCEPTS *(page 860)*

1. $x =$ directed distance from yz-plane to P
$y =$ directed distance from xz-plane to P
$z =$ directed distance from xy-plane to P

2 Sphere: $(x - h)^2 + (y - k)^2 + (z - j)^2 = r^2$

3. The xz-trace of a surface consists of all points that are common to both the surface and the xz-plane. To find the equation for this trace, substitute $y = 0$ into the equation for the surface.

4. Circle **5.** Line **6.** Direction and magnitude

7. B and **D**

8. The vector has the same direction or opposite direction depending on whether k is positive or negative, respectively. The magnitude is changed by the factor $|k|$.

9. The angle θ between the vectors is in the interval $90° < \theta < 180°$.

10. The vectors must be in three space because the cross product is orthogonal to the two given vectors.

11. False. $\mathbf{u} \times \mathbf{v} = -(\mathbf{v} \times \mathbf{u})$ **12.** (c)

13. Parallel because the direction numbers are proportional

14. False   **15.** True   **16.** (a) Line   (b) Line   (c) Plane

Review Exercises   *(page 861)*

1. $A(2, -3, -2)$, $B(-2, 4, 4)$ **3.** $(-5, 4, 0)$

5. Octants I, II, VII, and VIII

7. $\left(\sqrt{38}\right)^2 + \left(\sqrt{29}\right)^2 = \left(\sqrt{67}\right)^2$ **9.** $\left(\frac{13}{2}, 2, 5\right)$

11. $(x - 2)^2 + (y - 3)^2 + (z - 5)^2 = 1$

13. $(x - 3)^2 + (y + 2)^2 + (z - 6)^2 = \frac{225}{4}$

15. Center: $(2, 3, 0)$; Radius: 3

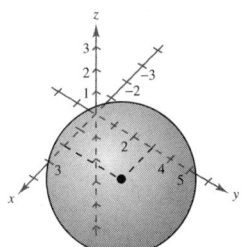

17. $\langle 1, 4, -4 \rangle$ **19.** $\langle -10, 6, 7 \rangle$ **21.** Parallel

23. $90°$ **25.** $\langle -10, 0, -10 \rangle$ **27.** $\sqrt{22}$ **29.** -16

31. Answers will vary.   **33–35.** Answers will vary.

37. 198   **39.** $\dfrac{1}{\sqrt{1854}}\langle -1, 37, -15 \rangle$

41. $x = 5t$
    $y = -10 + 20t$
    $z = 3 - 3t$

43. $x = 3 + t$
    $y = 1 + t$
    $z = 2 + t$

45. $x = -4 + t$
    $y = -12$
    $z = t$

47. $-2y + 5z - 14 = 0$ **49.** $x + y + z - 6 = 0$

51. $\dfrac{2}{\sqrt{110}}$ **53.** $\dfrac{3}{2\sqrt{14}}$

Chapter Test   *(page 864)*

1. $(3, -9, 0)$ **2.** Octants I and III   **3.** $(7, 1, 2)$

4. $(x - 7)^2 + (y - 1)^2 + (z - 2)^2 = 19$

5. No. $\left(\sqrt{76}\right)^2 + \left(\sqrt{102}\right)^2 \ne \left(\sqrt{194}\right)^2$

6. (a) 84   (b) $\langle 0, 62, 62 \rangle$ (c) $46.23°$ (d) $31\sqrt{2}$

    (e) $x = 8 - 2t, y = -2 + 6t, z = 5 - 6t$

    (f) $y + z - 3 = 0$

7. $70.3°, 32.6°, 65.1°$ **8.** $\dfrac{4\sqrt{14}}{7}$

9.

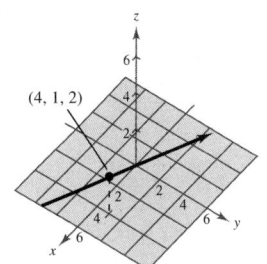

CHAPTER 12

Section 12.1   *(page 873)*

1. (a)

(b) Answers will vary.

(c)

x	3	3.5	3.9	4	4.1	4.5	5
V	972.0	1011.5	1023.5	1024.0	1023.5	1012.5	980.0

$\lim\limits_{x \to 4} V = 1024$

(d)

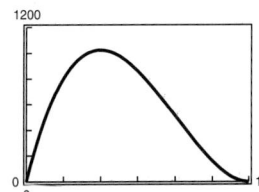

3.

x	2.9	2.99	2.999	3	3.001	3.01	3.1
$f(x)$	−2.8	−2.98	−2.998	−3	−3.002	−3.02	−3.2

$\lim\limits_{x \to 3} (3 - 2x) = -3$. The limit is reached.

5.

x	1.9	1.99	1.999	2
$f(x)$	0.2564	0.2506	0.2501	?

x	2.001	2.01	2.1
$f(x)$	0.2499	0.2494	0.2439

$\lim\limits_{x \to 2} \dfrac{x - 2}{x^2 - 4} = \dfrac{1}{4}$. The limit is not reached.

7. (a)

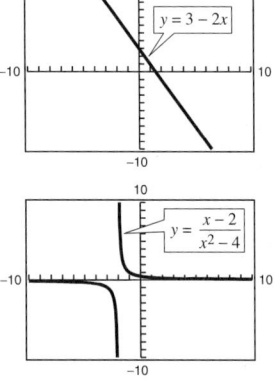

No. In each case, x is approaching an integer and the utility may not be evaluating the function at integer values.

(b)

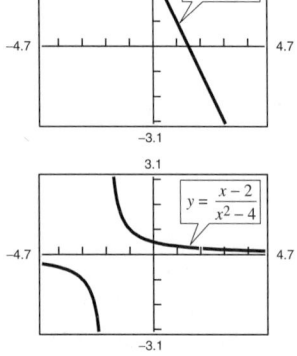

Yes. In each case, x is approaching an integer and the utility is attempting to evaluate the function at integer values.

9.

x	0.9	0.99	0.999	1
$f(x)$	0.2564	0.2506	0.2501	?

x	1.001	1.01	1.1
$f(x)$	0.2499	0.2494	0.2439

$\lim\limits_{x \to 1} \dfrac{x - 1}{x^2 + 2x - 3} = \dfrac{1}{4}$

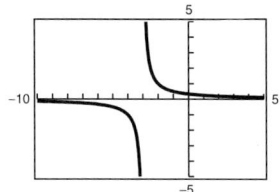

11.

x	−0.1	−0.01	−0.001	0
$f(x)$	0.2911	0.2889	0.2887	?

x	0.001	0.01	0.1
$f(x)$	0.2887	0.2884	0.2863

$\lim\limits_{x \to 0} \dfrac{\sqrt{x + 3} - \sqrt{3}}{x} \approx 0.2887$. [Actual limit is $1/(2\sqrt{3})$.]

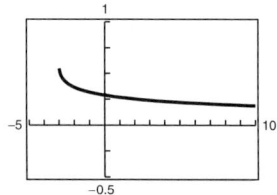

13.

x	-4.1	-4.01	-4.001	-4
$f(x)$	0.4762	0.4975	0.4998	?

x	-3.999	-3.99	-3.9
$f(x)$	0.5003	0.5025	0.5263

$$\lim_{x \to -4} \frac{[x/(x + 2)] - 2}{x + 4} = \frac{1}{2}$$

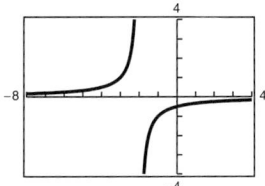

15.

x	-0.1	-0.01	-0.001	0
$f(x)$	0.9983	0.99998	0.9999998	?

x	0.001	0.01	0.1
$f(x)$	0.9999998	0.99998	0.9983

$$\lim_{x \to 0} \frac{\sin x}{x} = 1$$

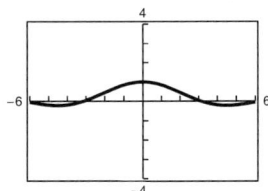

17. 1   **19.** -1 **21.** The limit does not exist.

23. The limit does not exist.

25.

The limit does not exist.

27.

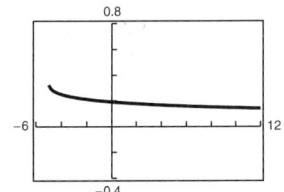

$$\lim_{x \to 4} f(x) = \frac{1}{6}$$

29.

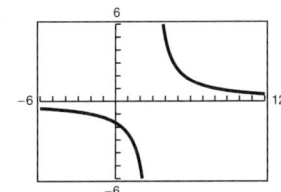

The limit does not exist.

31. No. The limit may or may not exist.   **33.** 6   **35.** 4

37. 0   **39.** $e^2 \approx 7.389$ **41.** 0   **43.** $\arcsin \frac{1}{2} \approx 0.524$

45. The limit does not exist.

47. (a) -10 (b) 9   (c) $\frac{4}{5}$ (d) 2

49.

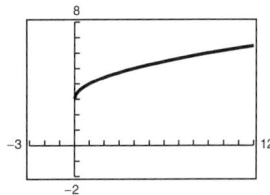

$$\lim_{x \to 9} f(x) = 6$$

Domain: $[0, 9)$, $(9, \infty)$

It is difficult to determine the domain solely by the graph because it is not obvious that the function is undefined at $x = 9$.

Section 12.2   *(page 884)*

1. -13 **3.** $-\frac{3}{5}$ **5.** -3 **7.** $\dfrac{\sqrt{3}}{2}$

9. (a) 1   **11.** (a) 2

(b) 3   (b) 0

(c) 5   (c) 0

$g_2(x) = -2x + 1$ $g_2(x) = x(x + 1)$

13. $\dfrac{1}{14}$ **15.** 4   **17.** $\dfrac{1}{2\sqrt{3}}$ **19.** $\dfrac{1}{4}$ **21.** -1

23. Does not exist

25.

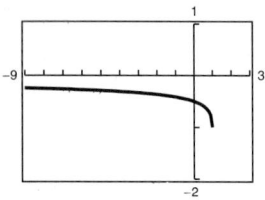

x	-0.1	-0.01	-0.001	0
$f(x)$	-0.4881	-0.4988	-0.5000	?

x	0.001	0.01	0.1
$f(x)$	-0.5001	-0.5013	-0.5132

$$\lim_{x \to 0} \frac{\sqrt{1-x}-1}{x} = -0.500$$

27.

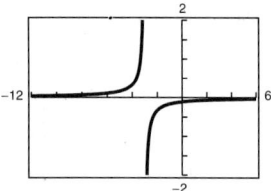

x	-0.1	-0.01	-0.001	0
$f(x)$	-0.1149	-0.1115	-0.1111	?

x	0.001	0.01	0.1
$f(x)$	-0.1111	-0.1107	-0.1075

$$\lim_{x \to 0} \frac{1/(3+x)-(1/3)}{x} = -0.1111$$

29.

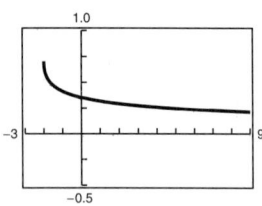

$$\lim_{x \to 0} \frac{\sqrt{x+2}-\sqrt{2}}{x} \approx 0.3536$$

31.

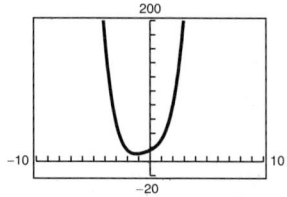

$$\lim_{x \to 2} \frac{x^5 - 32}{x - 2} = 80$$

33.

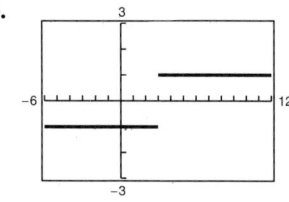

The limit does not exist.

35.

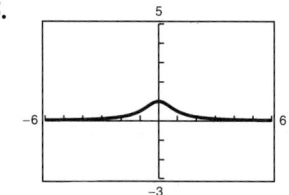

$$\lim_{x \to 1} f(x) = \tfrac{1}{2}$$

37.

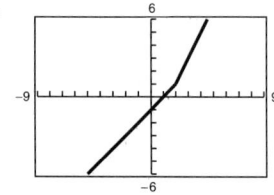

$$\lim_{x \to 2} f(x) = 1$$

39. 3   **41.** $2x - 3$

43.

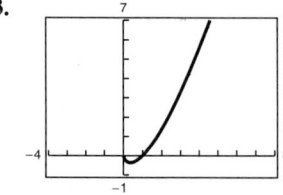

$$\lim_{x \to 0} f(x) = 0$$

45.

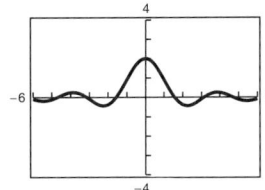

$$\lim_{x \to 0} f(x) = 2$$

47.

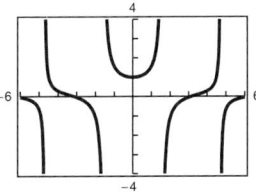

$$\lim_{x \to 0} f(x) = 1$$

49.

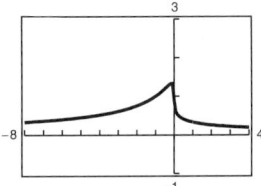

$$\lim_{x \to 1} f(x) \approx 0.333$$

51.

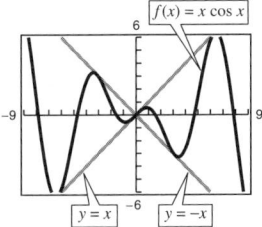

$$\lim_{x \to 0} f(x) = 0$$

53.

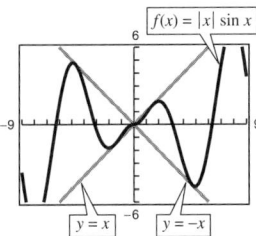

$$\lim_{x \to 0} f(x) = 0$$

55.

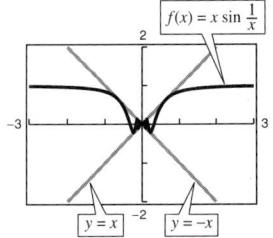

$$\lim_{x \to 0} f(x) = 0$$

57. (a) 0. Direct substitution   (b) 1

59. -32 feet per second

Section 12.3   *(page 893)*

1. 0   **3.** $\frac{1}{2}$

5. (a)

(b)

(c)

7.

9.

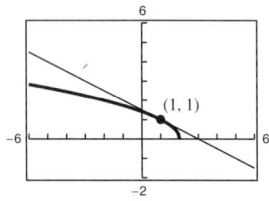

11. (a) $y = 0.301t^3 - 5.380t^2 - 14.495t + 1767.199$

(b) 475. When $t = 30$ the per capita debt is increasing at the rate of \$475 per year.

(c)

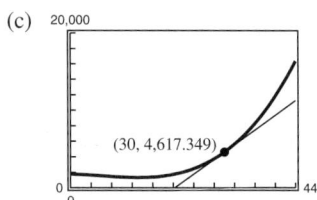

13. -3 **15.** 4   **17.** -1 **19.** $\frac{1}{6}$

21. $-2x$;   (a) 0   (b) 2

23. $-\dfrac{1}{(x+4)^2}$;   (a) $-\frac{1}{16}$ (b) $-\frac{1}{4}$

25. 0   **27.** $-\dfrac{2}{3}$ **29.** $-\dfrac{2}{x^3}$

31. $4; y = 4x - 5$

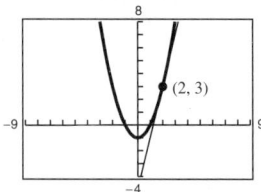

33. $\frac{1}{4}; y = \frac{1}{4}x + \frac{5}{4}$

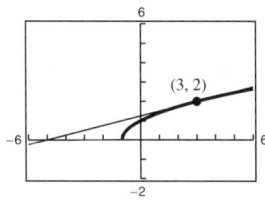

35.

x	-2	-1.5	-1	-0.5	0	0.5	1	1.5	2
$f(x)$	2	1.125	0.5	0.125	0	0.125	0.5	1.125	2
$f'(x)$	-2	-1.5	-1	-0.5	0	0.5	1	1.5	2

37.

x	-2	-1.5	-1	-0.5	0	0.5
$f(x)$	-2	-0.844	-0.25	-0.031	0	0.031
$f'(x)$	3	1.688	0.75	0.188	0	0.188

x	1	1.5	2
$f(x)$	0.25	0.844	2
$f'(x)$	0.75	1.688	3

39. $f'(x) = 2x - 4$; Horizontal tangent at $(2, -1)$

41. $f'(x) = 9x^2 - 9$

Horizontal tangents at $(-1, 6)$ and $(1, -6)$

43. (b)   **45.** (d)

Section 12.4   *(page 901)*

1. (b)   **3.** (e)   **5.** (f)   **7.** 0   **9.** -1 **11.** 2

13. Does not exist   **15.** -2 **17.** 0   **19.** -5

21.

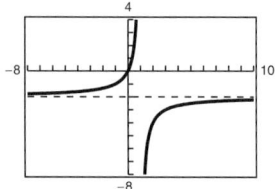

Horizontal asymptote: $y = -2$

23.

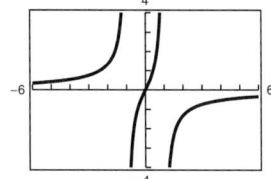

Horizontal asymptote: $y = 0$

25.

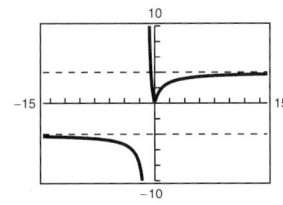

Horizontal asymptotes: $y = \pm 4$

27.

x	10^0	10^1	10^2	10^3
$f(x)$	-0.7321	-0.0995	-0.0100	-0.0010

x	10^4	10^5	10^6
$f(x)$	-1.0×10^{-4}	-1.0×10^{-5}	-1.0×10^{-6}

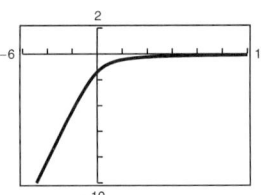

$\lim\limits_{x \to \infty} f(x) = 0$

29.

x	10^0	10^1	10^2	10^3
$f(x)$	-0.7082	-0.7454	-0.7495	-0.74995

x	10^4	10^5	10^6
$f(x)$	-0.749995	-0.7499995	-0.75

$\lim\limits_{x \to \infty} f(x) = -\frac{3}{4}$

31. (a) \$47.05, \$5.92 (b) \$1.35

33. $3, \frac{4}{3}, 1, \frac{6}{7}, \frac{7}{9}$
Limit: $\frac{1}{2}$

35. $1, \frac{3}{5}, \frac{2}{5}, \frac{5}{17}, \frac{3}{13}$
Limit: 0

37. $\frac{1}{7}, \frac{1}{3}, \frac{9}{17}, \frac{8}{11}, \frac{25}{27}$
Limit does not exist.

39. $2, 3, 4, 5, 6$
Limit does not exist.

41. $10, 12, 26, 64, 138$
Limit does not exist.

43. $\frac{1}{3}, -\frac{2}{5}, \frac{3}{7}, -\frac{4}{9}, \frac{5}{11}$
Limit does not exist.

45.

n	10^0	10^1	10^2	10^3	10^4
a_n	2	1.55	1.505	1.5005	1.50005

n	10^5	10^6
a_n	1.500005	1.5000005

$\lim\limits_{n \to \infty} a_n = \frac{3}{2}$

47.

n	10^0	10^1	10^2	10^3	10^4
a_n	2	3.8	3.98	3.998	3.9998

n	10^5	10^6
a_n	3.99998	3.999998

$\lim\limits_{n \to \infty} a_n = 4$

49.

n	10^0	10^1	10^2	10^3	10^4
a_n	16	6.16	5.4136	5.3413	5.3341

n	10^5	10^6
a_n	5.33341	5.333341

$\lim\limits_{n \to \infty} a_n = \frac{16}{3}$

51.

Converges to 0

53.

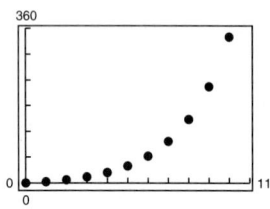

Diverges

Section 12.5 *(page 909)*

1. 1830   **3.** 44,100   **5.** 5850

7.

n	10^0	10^1	10^2	10^3
$S(n)$	4	1.54	1.353	1.335

Limit: $\frac{4}{3}$

9.

n	10^0	10^1	10^2	10^3
$S(n)$	6	1.185	1.0154	1.0015

Limit: 1

11.

n	10^0	10^1	10^2	10^3
$S(n)$	5	1.4	1.04	1.004

Limit: 1

13.

n	10^0	10^1	10^2	10^3
$S(n)$	3	0.2385	0.02338	0.00233

Limit: 0

15.

n	10^0	10^1	10^2	10^3
$S(n)$	0	0.615	0.66165	0.66617

Limit: $\frac{2}{3}$

17. 4.5   **19.** 1.257

21.

n	4	8	20	50
Approximate Area	12	14	15.2	15.68

23.

n	4	8	20	50
Approximate Area	3.52	2.85	2.48	2.34

25. $\frac{3}{2}$

27. $\frac{16}{3}$

29. $\frac{17}{4}$

31. $\frac{51}{4}$

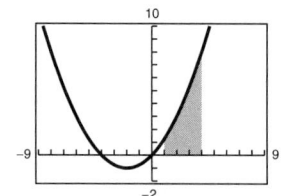

33. (b)   **35.** $\frac{1}{6}$ **37.** 1

39. 1.052×10^5 square feet ≈ 2.415 acres

FOCUS ON CONCEPTS *(page 911)*

1. The limit exists.   **2.** The limit does not exist.

3. The limit does not exist.   **4.** The limit exists.

5. No. $f(3)$ may or may not exist because the limit only explains the behavior of the function as x approaches 3.

6. When direct substitution in a fractional function produces 0 in both the numerator and denominator, the resulting fraction $\frac{0}{0}$ is called an indeterminate form. (There are other indeterminate forms studied in calculus.)

7. $x = -2$ **8.** False. Graph $y = x^3$ and $y = 3x - 2$.

9. Tangent line   **10.** 2   **11.** 0

12. False. Graph $y = \dfrac{x^2}{x + 1}$.

13. 0. Ratio of the coefficients of the highest-powered terms.

14. Divide the interval from a to b into n subintervals of equal width. Form a collection of rectangles over the subintervals whose heights are given by the function. The exact area is the limit of the sum of the areas of the n rectangles as n approaches ∞.

15. (c)

Review Exercises   *(page 912)*

1.

x	1.9	1.99	1.999	2
$f(x)$	0.1299	0.1255	0.1250	?

x	2.001	2.01	2.1
$f(x)$	0.1250	0.1245	0.1205

$$\lim_{x \to 2} \frac{x - 2}{3x^2 - 4x - 4} = \frac{1}{8}$$

3. 11   **5.** 77   **7.** $\frac{10}{3}$ **9.** $-\frac{1}{4}$ **11.** $-\frac{1}{3}$ **13.** $\frac{1}{4}$

15. -1 **17.** -1 **19.** Does not exist   **21.** 2

23.

x	1.1	1.01	1.001	1.0001
$f(x)$	0.5680	0.5764	0.5772	0.5773

$$\lim_{x \to 1^+} \frac{\sqrt{2x + 1} - \sqrt{3}}{x - 1} \approx 0.577. \left(\text{Actual limit is } \sqrt{3}/3.\right)$$

25. (a) 64   (b) 7   (c) 20   (d) $\frac{4}{5}$ **27.** $3 - 2x$

29.

31. $2x - 4$;   (a) -4 (b) 6

33. $-\dfrac{4}{(x - 6)^2}$;   (a) -4 (b) -1

35. $g'(x) = 0$ **37.** $h'(x) = -\dfrac{1}{2}$ **39.** $f'(t) = \dfrac{1}{2\sqrt{t + 5}}$

41. $y = 8.73t^2 - 6.23t + 0.54$ **43.** (d)   **45.** (b)

47. 2   **49.** 0   **51.** 4   **53.** -1

55. $\dfrac{(n + 1)(5n + 4)}{6n^2}$

n	10^0	10^1	10^2	10^3
$S(n)$	3	0.99	0.8484	0.8348

Limit: $\frac{5}{6}$

57. $\frac{27}{4} = 6.75$

59.

n	4	8	20	50
Approximate Area	6.7266	5.9707	5.5341	5.3629

61. Approximation: 40; Exact: 50

63. Approximation: 15.875; Exact: 15

65. Approximation: 0.5; Exact: $\frac{4}{3}$ **67.** 83,000 square feet

Cumulative Test for Chapters 11–12   *(page 916)*

1. $(x - 2)^2 + (y - 2)^2 + (z - 4)^2 = 24$

2. -13; $\langle 8, 11, -20 \rangle$

3. $x = -2 + 7t$ **4.** $75x + 50y - 31z = 0$
$y = 3 + 5t$
$z = 25t$

5. $\sqrt{310} \approx 17.61$ **6.** $\dfrac{\sqrt{30}}{2} \approx 2.74$ **7.** 4   **8.** $-\dfrac{1}{3}$

9. $\frac{1}{4}$ **10.** Does not exist   **11.** 2   **12.** 0   **13.** -2

14. $\frac{1}{2}$ **15.** 8190

16.

n	4	8	20	50
Approximate Area	8.1250	8.4063	8.5650	8.6264

17. $\frac{3}{4}$

2

Index of Applications

Index